D0163017

# Beginning Algebra

**Andrea Hendricks**
*Georgia Perimeter College*

**Oiyin Pauline Chow**
*Harrisburg Area Community College*

McGraw Hill

Connect
Learn
Succeed™

BEGINNING ALGEBRA

Published by McGraw-Hill, a business unit of The McGraw-Hill Companies, Inc., 1221 Avenue of the Americas, New York, NY 10020. Copyright © 2013 by The McGraw-Hill Companies, Inc. All rights reserved. Printed in the United States of America. No part of this publication may be reproduced or distributed in any form or by any means, or stored in a database or retrieval system, without the prior written consent of The McGraw-Hill Companies, Inc., including, but not limited to, in any network or other electronic storage or transmission, or broadcast for distance learning.

Some ancillaries, including electronic and print components, may not be available to customers outside the United States.

This book is printed on acid-free paper.

1 2 3 4 5 6 7 8 9 0 DOW/DOW 1 0 9 8 7 6 5 4 3 2

ISBN 978-0-07-338427-6
MHID 0-07-338427-5

ISBN 978-0-07-336662-3 (Annotated Instructor's Edition)
MHID 0-07-336662-5

Vice President, Editor-in-Chief: *Marty Lange*
Vice President, EDP: *Kimberly Meriwether David*
Senior Director of Development: *Kristine Tibbetts*
Editorial Director: *Stewart K. Mattson*
Executive Editor: *Dawn R. Bercier*
Sponsoring Editor: *Mary Ellen Rahn*
Developmental Editor: *Emily Williams*
Marketing Manager: *Peter A. Vanaria*
Senior Project Manager: *Vicki Krug*
Senior Buyer: *Sherry L. Kane*
Senior Media Project Manager: *Sandra M. Schnee*
Senior Designer: *Laurie B. Janssen*
Cover Illustration: *Imagineering Media Services Inc.*
Senior Photo Research Coordinator: *Lori Hancock*
Photo Research: *Danny Meldung/Photo Affairs, Inc*
Compositor: *Cenveo Publisher Services*
Typeface: *10.5 Times LT Std*
Printer: *R. R. Donnelley*

All credits appearing on page or at the end of the book are considered to be an extension of the copyright page.

**Library of Congress Cataloging-in-Publication Data**

Hendricks, Andrea.
    Beginning algebra / Andrea Hendricks, Oiyin Pauline Chow.—1st ed.
        p. cm.
    Includes index.
    ISBN 978-0-07-338427-6—ISBN 0-07-338427-5 (hard copy : alk. paper) 1. Algebra. I. Chow, Oiyin Pauline. II. Title.
    QA152.3.H335 2013
    512—dc23
                                                                                2011030527

www.mhhe.com

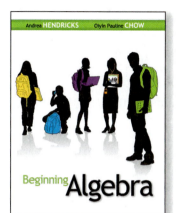

**In developmental math,** the focus needs to be about the "whole student" and providing students with "more than just the math." As Andrea and Pauline say, we want students to know that we care about their success. Therefore, we chose to include students on the covers to show how interested we are in their pursuit to succeed and persist through their math courses. We also wanted to visually represent today's students in their current environments. Our authors have made this their focus too by providing an entire chapter of robust resources for teaching and learning success strategies beyond just the math. And, no matter the course format, Andrea and Pauline have provided purposeful examples and exercises, current and relevant applications, and critical thinking exercises to reach today's whole student. Our hope is that students will want to envision themselves as the successful and confident math students on these covers.

## Dedications

This text is dedicated to my wonderful family. Thank you, Todd, for your support and encouragement through this process and my wonderful boys, Andy, Charlie, and Cory, for understanding all of those times that Mommy had to work on the computer.

　　—*Andrea Hendricks*

Chow Cheung—my father, who recognized the importance of education and sent all seven of his children abroad to pursue a college education in the 70s. Wong Shing—my mother, who constantly provides us with love and nurture. Michael, Amy, and Andrew—my family, pride and joy.

　　—*Oiyin Pauline Chow*

# Hendricks & Chow

## Developmental Math Hardcover Series

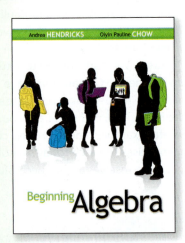

### Beginning Algebra

**Andrea Hendricks and Oiyin Pauline Chow, ©2013**

ISBN: 978-0-07-338427-6
MHID: 0-07-338427-5

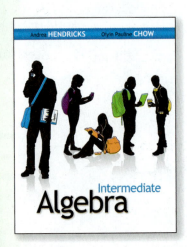

### Intermediate Algebra

**Andrea Hendricks and Oiyin Pauline Chow, ©2013**

ISBN: 978-0-07-338426-9
MHID: 0-07-338426-7

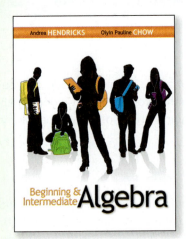

### Beginning & Intermediate Algebra

**Andrea Hendricks and Oiyin Pauline Chow, ©2013**

ISBN: 978-0-07-338453-5
MHID: 0-07-338453-4

# Brief Contents

**Interested in customizing your own Table of Contents?**
Contact your McGraw-Hill representative at **www.mhhe.com**
to learn more about our custom print and digital solutions.

# About the Authors

**Andrea Hendricks** I am an Associate Professor of Mathematics at Georgia Perimeter College (GPC), Online Campus. I have been teaching at the college level since 1992 and have taught the full range of math courses. In 2008, I joined the online campus of GPC and teach exclusively online. Prior to joining the online campus, I taught traditional face-to-face classes, hybrid classes, and online classes. During my tenure at GPC, I have served as Assistant Department Chair for one of the ground campuses and am currently Assistant Department Chair for the online math department, which gives me responsibility for the part-time faculty members and for managing and overseeing the developmental math courses. In addition, I have chaired and served on various curriculum committees, the Faculty Senate, Peer Review Committees, Promotion and Tenure Panels, and the Math Conference Committee. I have received the NISOD Teaching Excellence Award, the Faculty Teaching and Service Award, and the GPC Collegiality Award. I particularly enjoy teaching developmental math classes so that I can help students overcome their fear and anxiety toward learning math. I look forward to sharing my strategies and teaching moments on a larger scale within this Developmental Math series. My husband, Todd, is also a professor of mathematics at GPC, and we have three growing boys under the age of 13: Andy, Charlie, and Cory. In additon to teaching, authoring, and being a mom, I enjoy golf, playing the piano, and am involved with my church through teaching a 4-year-old Sunday school class and also AWANA classes.

**Oiyin Pauline Chow** As a Senior Professor and Chair of the Mathematics and Computer Science Department at Central Pennsylvania's Community College (HACC), I have approximately 30 years of teaching experience, and I have taught a full curriculum—from Developmental Math to Calculus, Linear Algebra, and Differential Equations in both traditional and online formats. Currently, I serve as Mathematics Department Chair at HACC's five regional campuses and their virtual campus. I also participate in many active roles on campus to support the Faculty Council and various developmental education and department committees. My other interests include serving on various state and national math organizations: the executive boards of AMATYC (secretary) and three Pennsylvania math organizations, PCTM (president), PADE (member at large), and PSMATYC (president). I have received the NISOD Teaching Excellent Award, The PADE Exemplary Teaching in Developmental Education award, and the PXTM Outstanding Contribution to Mathematics Education Award. The greatest joy in this profession is being able to work with students, empower them to take ownership in learning, and help them see their own accomplishments. In particular, at the Developmental Math level, I take pride if I can change students' outlook on mathematics, help them achieve an appreciation of math that they never had, and gain a better understanding of math. My husband Mike and I are experiencing an official empty nest, but we enjoy visiting and keeping tabs on our son, Andrew, at his software developer job in Palo Alto California and our daughter, Amy, in New York City at a digital marketing firm.

> **"Students want to know that someone is interested in their success."**
>
> —Andrea Hendricks and Pauline Chow

# Our Development Story

From our years of teaching developmental math students, we have observed that more and more students need a review of study skills and how these can be applied to equal success in college. So, when we teach, we teach more than just the math—we teach life skills. Accordingly, more and more schools are integrating study skills into their developmental math classes to help retain students and improve their success rates. Students want to know that someone is interested in their success and can make the math attainable without sacrificing the integrity of the course. Therefore, we felt that it is our role to provide a new, purposeful developmental math text series. We have thoroughly enjoyed partnering in this quest of authoring and have formed a great team with Andrea creating the organization and framework for the text and writing the narrative and explanations/examples while Pauline manages the exercise sets and oversees the digital content. Together, we have focused on the following three key course needs to provide an accessible, relevant, and motivating series that will help drive students to succeed.

## Success Strategies For Today's Students

Since student success is critical, we incorporated a separate chapter on student success strategies including topics like time management, note taking, and preparing for tests. We integrated these topics throughout the text by highlighting each chapter opener with a characteristic of a successful student and a motivational quote. Therefore, students will find new resources needed to be successful at their fingertips while studying math.

## Student-Centered Examples and Exercise Sets

Today, we must address and serve a diverse population of developmental math students with various learning styles. When students realize the relevance of math in their lives, they are motivated to learn the material and are prepared to move into various academic disciplines. We have worked diligently to ensure that our clear, concise explanations, exercises, and applications mean something to today's students. Our numerous worked examples are organized by objectives with detailed, step-by-step procedures wherever possible. We also have worked to provide multiple ways for students to learn—troubleshooting their own mistakes, writing about the math, advancing through various levels of problems, identifying other's errors, and critically taking ownership of the material that they are learning.

## Digital Solutions For Every Teaching Environment

Because of the growth of online learning, hybrid classes, and course redesigns, it has become even more important to create materials that help students grow into successful, independent learners. This book is designed not only to motivate and inspire students but students can also learn from it in any type of classroom environment. From our experience teaching online, it is clear to us that students depend on the textbook more so than in a traditional classroom. We also worked to take an active role in the creation of the digital content, and while developing the organization and progression of our exercise sets, we considered how a student works through online homework for additional practice. Each and every exercise has a purpose both in the text and online to support the objectives. We were involved in the process of ensuring that the guided solutions in the online homework represent the same author voice and narrative as the printed book to eliminate any inconsistencies.

*Andrea M. Hendricks* and *Pauline Chow*

# Contents

# Success Strategies for Today's Students

Today's students need more guidance than ever before on how to succeed in both their current math course and future courses.

▶ Do your students come to class prepared to study, take good notes, make time for homework, and study for tests?

▶ How do you cover all of the math curriculum and teach these success strategies?

The Hendricks/Chow series provides materials dedicated to student success strategies that are integrated within both the text and the instructor and student supplements.

## ▶ How do you incorporate study skills or success strategies into your course?

An entire chapter, entitled **Strategies to Succeed in Math**, is dedicated to study skills, such as time management, test taking, note taking, and tips to succeed in online/hybrid courses.

A **Learning Style Inventory Quiz** will help students identify if they are a visual, auditory, or kinesthetic learner and then provides specific math strategies to help support the various learning styles.

**ACTIVITY 1** Complete the following survey to determine your math learning style.

**Math Learning Styles Survey**

Answer each question with the number "3" if you agree most of the time, "2" if you agree sometimes, and "1" if you agree rarely or do not agree.

_____ 1. When there is talking or noise in class, I get easily distracted.

_____ 2. If a problem is written on the board, I have difficulty following the steps unless the teacher verbally explains the steps.

_____ 3. I find it easier to have someone explain something to me than to read it in my math book.

_____ 4. If I know how the math is used in real life, it is easier for me to learn.

_____ 5. To remember formulas and definitions, I need to write them down.

_____ 6. I prefer listening to the lecture rather than taking notes.

_____ 7. Using manipulatives, hands-on activities, or games helps me learn math concepts.

_____ 8. When solving a problem, I try to picture working it out in my mind first.

**SECTION S.3** | Study Skills

**▶ OBJECTIVES**

As a result of completing this section, you will be able to

1. Identify habits that good math students employ.
2. Read the textbook more effectively and efficiently.
3. Take quality notes.
4. Organize materials in a notebook or portfolio.
5. Complete homework in ways that maximize retention.
6. Establish an action plan.
7. Troubleshoot common errors.

**Study skills** are the skills students need to improve their learning capacity and to acquire new knowledge. No two students are going to employ the same set of study techniques, though there are some that every student should utilize when studying math. This section will address some of the skills that will enhance the success of all students.

**Habits of Successful Math Students**

Good math students study with purpose. When you underline something, it should be because it is important. If you write out a note card, it should be because it is something you want to remember long term. When you work a homework problem, it should lead to a better understanding of the process used to solve that problem. If you can't answer "Why am I doing this?" then what you are doing may not be a productive use of your time. You should always have goals to achieve when you sit down to study and everything you do during your study time should work toward those goals. The following suggestions should help you approach your math class with purpose.

**Objective 1 ▶**
Identify habits that good math students employ.

1. *Successful students are responsible.*
   ▶ Attend every class meeting. Try to be a few minutes early so that your materials are out and you are ready for class to begin.
   ▶ Find an accountability partner in class. Encourage one another and use each other as a resource if one of you is absent from class.
   ▶ Adhere to all deadlines for assignments.

2. *Successful students make the most of their time in class.*
   ▶ When you are in class, "be in class." Try not to worry about other things going on in your life. Focus your attention on the topics being discussed. (No texting, surfing, or doing other homework.)
   ▶ Either take notes in class or record the lecture so that you can refer to it later. Some instructors even post their class notes on their website.
   ▶ Actively participate in class. Follow along with the instructor. Ask questions

Instructors and students will benefit from several helpful strategies to use while in these courses.

---

**Strategies to Succeed in Math**

CHAPTER

**S**

### Success

It is with great delight that we welcome you to this course and the materials we have provided you. You are embarking on perhaps the most rewarding experience of your life. College is a series of challenges that will prepare you for a lifetime of learning and accomplishments. At the end of your college education, you will be awarded a degree—a degree that establishes your ability to learn and to persevere. It is a statement that you can work to accomplish a goal, no matter what the obstacles.

At this point, you must evaluate the reason you are here. Are you here to fulfill your dreams or the dreams of someone else? To successfully complete this journey, you must be here to fulfill your own desires, not those of a friend or family member. It will be most difficult to withstand the trials of college if you do not have a personal desire to see it through.

A semester or quarter can be very overwhelming. Focus on one day at a time and not on everything that you must learn throughout the entire course. Before you know it, the course will be over. We know that it is very easy to get distracted from your goals. Stay committed and motivated by remembering your ultimate reason for attending school.

We wish you success in this course and in your future educational endeavors. It is our hope that you are successful, not only this semester or quarter, but every term until your ultimate goal is achieved. This course will pave the way for that success. It is the door to achieving your dreams.

*Mrs. Andrea Hendricks and Mrs. Pauline Chow*

*Question for Thought:* What do you dream about doing? How does attending college make that dream possible? How does this course help you meet your goals? What is your biggest obstacle in being successful in this course? What can you do to overcome that obstacle? What is your plan for succeeding in this course?

**Chapter S** will introduce some important strategies that can enhance your performance in this course.

### Chapter Outline

Section S.1   Time Management and Goal Setting

Section S.2   Learning Styles

Section S.3   Study Skills

# ▶ Are your students coming to your class prepared to practice appropriate study skills?

> Each chapter also contains additional materials on success strategies, such as motivational quotes and chapter openers dedicated to a characteristic of a successful student.

## Real Numbers and Algebraic Expressions

CHAPTER 1

### Time Management

As you begin this chapter, we want you to focus on how you are managing your time. Your success in this course depends on your ability to devote the appropriate time to learning and practicing the material.

- The general rule of thumb is that you should spend 2 hours studying for each hour you are in class.
- Prepare a calendar of your week and schedule an appointment with yourself for study time.
- Review how you are using your time through the week and make an honest assessment of whether you have the time to make this class a priority.
- See the Preface for additional resources concerning time management.

> "Imagination is more important than knowledge. For while knowledge defines all we currently know and understand, imagination points to all we might yet discover and create."
>
> — Albert Einstein (Mathematician and scientist)

Section 1.2  Fractions Review  13
Section 1.3  The Order of Operations, Algebraic Expressions, and Equations  27

> "The key is not to prioritize what's on your schedule, but to schedule your priorities."
>
> —Stephen Covey

### Coming Up...

In Section 1.3, we will learn that the number of viewers (in millions) for the season premiere of Fox Network's *American Idol* for Seasons 1 to 8 can be approximated by the expression $0.1x^3 + 18.4x - 2.54x^2 - 4.6$, where $x$ is the number of seasons aired. We will learn how to use this expression to estimate the number of viewers for future seasons.

1

## Resources Available in Success Strategies Manual

- Overcoming Math Anxiety Strategies
- Note Taking and Homework Strategies
- Test Preparation Worksheets
- Time Management Activities
- Goal Management
- How to Navigate a Math Textbook
- Organizing a Portfolio
- Printable Math Study Sheets
- Error Analysis Activities
- Math Term Glossaries

> In addition to *Sucess Strategies* and the chapter opener materials, more resources can be found in our robust *Success Strategies Manual,* with both Student and Instructor versions available. The manual will provide additional materials geared toward success strategies, including worksheets, tips, handouts and templates.

> "This is a GREAT chapter. All math classes should start with this information!"
>
> —Elise Price, *Tarrant County College*

# Relevant Examples and Exercises for Today's Students

Every example and exercise has a purpose! The Hendricks/Chow series offers unique types of examples and exercises to capture students' interest and help them build different skill sets as they move throughout the chapter.

## ▶ How Relevant are the applications in your text?

---

### SECTION 1.2 | Fractions Review

#### ▶ OBJECTIVES

As a result of completing this section, you will be able to

1. Write the prime factorization of a number.
2. Define and write fractions.
3. Simplify fractions.
4. Multiply fractions.
5. Divide fractions.
6. Add or subtract fractions with common denominators.
7. Add or subtract fractions with unlike denominators.
8. Troubleshoot common errors.

#### ▶ Objective 1 ▶

Write the prime factorization of a number.

As of 2009, there were approximately 309 million people living in the United States and approximately 116 million Facebook users in the United States. Write a fraction that represents the part of the United States population that were Facebook users in 2009. (Source: http://flavorwire .com/82308/awesome-infographic-facebook-vs-the-united-states)

In this section, we will learn how to use a fraction to represent such information. We will also learn how to simplify fractions and perform operations with fractions.

#### Factoring Numbers

To express a number in *factored form* is to write the number as a product of two or more numbers. For example, in the statement $4 \cdot 5 = 20$, the numbers 4 and 5 are called **factors** and 20 is the *product*. Factors divide a number evenly without any remainder.

The numbers 2, 3, 5, 7, 11, 13, 17, and 19 have something in common. The factors of these numbers are only 1 and the number itself. These types of numbers are called *prime numbers*. The following prime numbers are written in factored form.

$$2 = 1 \cdot 2, 3 = 1 \cdot 3, 5 = 1 \cdot 5$$

> Each section opens with a relevant application that is then revisited in the worked examples.

---

**29.** The top 10 U.S. Internet search providers processed a total of approximately 9,200,000,000 search requests, during August 2010. Google processed about 6,000,000,000 and Yahoo! processed about 1,210,000,000 search requests. Write fractions that represent the portion of Google search requests

**84.** The total number of individual songs purchased digitally in 2008 and 2009 was 2.231 billion. The total number of individual songs purchased digitally in 2008 was 0.089 billion less than the total number in 2009. Let $x$ = total number of individual songs purchased digitally in 2008 (in billions) and let $y$ = total number in 2009 (in billions). (Source: http://www.ritholtz.com)

**a.** Write an equation using $x$ and $y$ that represents the combined total number of individual songs purchased digitally in 2008 and 2009.

**b.** Write an equation that relates the total number of songs purchased digitally in 2008 to the total number of songs purchased digitally in 2009.

> How often do you find the applications in your current text irrelevant and stale? In order to promote active learning, the authors have worked to provide current and relevant applications that are interesting to students and contain subjects that are not always in math books, such as social media, smart phones, consumer topics, and current events.

**93.** The table shows the U.S. unemployment rates between 2001 and 2010. (Source: Bureau of Labor Statistics)

| Years After 2001 | 0 | 1 | 2 | 3 | 4 | 5 | 6 | 7 | 8 | 9 |
|---|---|---|---|---|---|---|---|---|---|---|
| Unemployment Rate (percent) | 4.7 | 5.8 | 6.0 | 5.5 | 5.1 | 4.6 | 4.6 | 5.8 | 9.3 | 9.6 |

**a.** Write the ordered pairs $(x, y)$ that correspond to the data in the table, where $x$ is the years after 2001 and $y$ is the unemployment rate.

**b.** Interpret the meaning of the first and last ordered pairs in the context of the problem.

**c.** In what year was the unemployment rate the highest? The lowest?

... a scatter plot of the data.

... lists the approximate salary of Peyton ... quarterback for the Indianapolis Colts, for ... 9. (Source: http://content.usatoday.com/sportsdata/ ...salaries/player/Peyton-Manning)

| 03 | 0 | 1 | 2 | 3 | 4 | 5 | 6 |
|---|---|---|---|---|---|---|---|
| ns) | \$11.3 | \$35 | \$0.7 | \$10 | \$11 | \$11.5 | \$14 |

... he ... d pairs that ... ond to the ... the table, ... is years ... 03 and $y$ is Manning's salary (in millions of dollars).

**b.** Interpret the meaning of first and last ordered pairs in the context of the problem.

**c.** In what year was his salary the highest? The lowest?

**d.** Make a scatter plot of the data.

---

> " I like the application problems included here—very appropriate for this level student: connected to their real-world future occupations. Kudos to you! "
>
> —Vicki Schell, *Pensacola State College*

# ▶ How do your current resources help your students progress to become critical thinkers?

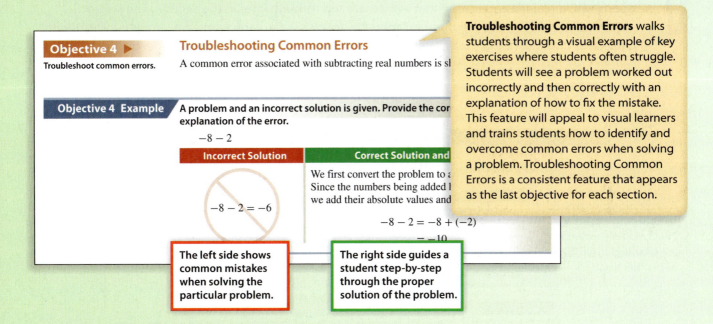

**Troubleshooting Common Errors** walks students through a visual example of key exercises where students often struggle. Students will see a problem worked out incorrectly and then correctly with an explanation of how to fix the mistake. This feature will appeal to visual learners and trains students how to identify and overcome common errors when solving a problem. Troubleshooting Common Errors is a consistent feature that appears as the last objective for each section.

The left side shows common mistakes when solving the particular problem.

The right side guides a student step-by-step through the proper solution of the problem.

**You Be the Teacher!** exercises reinforce and build critical thinking skills by teaching students to constantly reevaluate their work. In these exercises, students are asked to check a student's work and make necessary corrections to ensure the problem is answered correctly. Like Troubleshooting Common Errors, this feature emphasizes Error Analysis and trains students to take ownership and identify their own mistakes.

### You Be the Teacher!

Correct the student's errors, if any.

103. Determine if the following relations are functions.
    a. $\{(3, 1), (7, 5), (1, 2), (5, 2), (0, -2)\}$
    b. $\{(1, 3), (5, 7), (2, 1), (2, 5), (-2, 0)\}$

    Vivian's work: **a.** a function and **b.** not a function

104. Find $f(-3)$ if $f(x) = 12x - 5$.

    William's work:
    $f(-3) = -3(12x - 5) = -36x + 15$

> "The Troubleshooting Common Errors examples make this text stand out among other books. I use this concept in my classes now and would continue to reinforce the material with this feature. Critical thinking is so important for students and this idea helps students to think more about process rather than obtaining the 'right' answer."
>
> —Carol Ann Poore, *Hinds Community College, Ramond Campus*

# Chapter Walkthrough & Organization

When developing the organization and framework for this series, Andrea considered her teaching methodologies. **She organized each section with a five-step process, just as she presents her lectures: (1) lead-in, (2) objectives, (3) lesson, (4) check for understanding, and (5) summary.** This five-step process has served as the framework of each section of the textbook. Most every section opens with a real-life application that will be worked through later in the section. The section's objectives, stated in clear, measurable language, follows with many worked examples and student checks for understanding. Finally each section closes with a summary of key concepts.

Each section is organized by **Learning Objectives.** Then, multiple examples are provided to illustrate each objective.

Each objective is followed by thorough and **multiple examples,** numbered exactly as the objective, to reinforce the concepts and show the stepped out solutions.

**Procedure boxes** are included whenever possible to help verbalize the math steps and processes.

The examples provide a high level of **step-by-step** detail so that students do not lose track of the various steps.

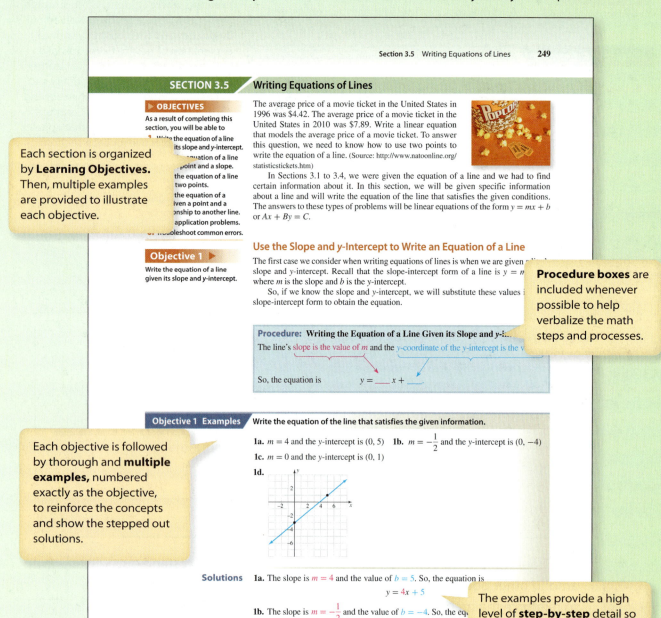

Section 3.5   Writing Equations of Lines        249

### SECTION 3.5   Writing Equations of Lines

#### OBJECTIVES

As a result of completing this section, you will be able to

1. Write the equation of a line given its slope and y-intercept.
2. [...] equation of a line [...] point and a slope.
3. [...] the equation of a line [...] two points.
4. [...] the equation of a [...] given a point and a [...] relationship to another line.
5. [...] application problems.
6. Troubleshoot common errors.

The average price of a movie ticket in the United States in 1996 was \$4.42. The average price of a movie ticket in the United States in 2010 was \$7.89. Write a linear equation that models the average price of a movie ticket. To answer this question, we need to know how to use two points to write the equation of a line. (Source: http://www.natoonline.org/statisticstickets.htm)

In Sections 3.1 to 3.4, we were given the equation of a line and we had to find certain information about it. In this section, we will be given specific information about a line and will write the equation of the line that satisfies the given conditions. The answers to these types of problems will be linear equations of the form $y = mx + b$ or $Ax + By = C$.

#### Use the Slope and y-Intercept to Write an Equation of a Line

The first case we consider when writing equations of lines is when we are given [the] slope and y-intercept. Recall that the slope-intercept form of a line is $y = m[x + b]$ where $m$ is the slope and $b$ is the y-intercept.

So, if we know the slope and y-intercept, we will substitute these values i[nto] slope-intercept form to obtain the equation.

**Objective 1**

Write the equation of a line given its slope and y-intercept.

**Procedure:  Writing the Equation of a Line Given its Slope and y-i[ntercept]**

The line's slope is the value of $m$ and the y-coordinate of the y-intercept is the v[alue]

So, the equation is         $y = \_\_\_ x + \_\_\_.$

**Objective 1  Examples**    Write the equation of the line that satisfies the given information.

1a.  $m = 4$ and the y-intercept is $(0, 5)$    1b.  $m = -\dfrac{1}{2}$ and the y-intercept is $(0, -4)$

1c.  $m = 0$ and the y-intercept is $(0, 1)$

1d.

**Solutions**    1a. The slope is $m = 4$ and the value of $b = 5$. So, the equation is
$$y = 4x + 5$$

1b. The slope is $m = -\dfrac{1}{2}$ and the value of $b = -4$. So, the eq[uation is]
$$y = -\dfrac{1}{2}x - 4$$

After each set of examples for an objective, there is a **Student Check** feature that provides additional exercises for students to try on their own, similar to the examples shown.

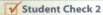

**✓ Student Check 2**  Write the equation of the line that has the given slope and passes through the given point. Write the answer in both slope-intercept form and standard form.

**a.** $m = -5$; $(3, 1)$    **b.** $m = -\frac{2}{7}$; $(4, 5)$

**c.** $m = 0$; $(-6, 9)$    **d.** undefined slope; $(8, 1)$

**Note:** Using method 1 or 2 produces the same linear equation. When using the slope-intercept form, we must substitute the values of m and b into $y = mx + b$ to obtain the equation. When using the point-slope form, the equation is obtained through the process.

**Notes** from the author appear throughout the text to give students valuable tips and information.

The end of each section contains the **Answers to the Student Checks**, along with a **Summary of Key Concepts** and most include **Graphing Calculator Skills** for those instructors who incorporate calculator use.

## ANSWERS TO STUDENT CHECKS

**Student Check 1**  **a.** $y = -2x + 9$  **b.** $y = \frac{7}{3}x - 1$
**c.** $y = \frac{2}{3}$  **d.** $y = \frac{3}{2}x + 1$
**Student Check 2**  **a.** $y = -5x + 16$  **b.** $y = -\frac{2}{7}x + \frac{43}{7}$
**c.** $y = 9$  **d.** $x = 8$
**Student Check 3**  **a.** $y = -5x + 8$  **b.** $y = -\frac{4}{3}x + \frac{2}{3}$
**c.** $y = -1$  **d.** $x = -8$

**Student Check 4**  **a.** $y = \frac{1}{2}x - 4$  **b.** $y = -2x + 1$
**c.** $x = 2$  **d.** $y = -3$
**Student Check 5**  **a.** **i.** $y = -3000x + 30,000$
**ii.** \$18,000  **iii.** 7 yr
**b.** **i.** $y = 0.54x + 15.3$  **ii.** 21.78 million  **iii.** 2024

## SUMMARY OF KEY CONCEPTS

1. If the slope and *y*-intercept of an equation are known, writing the equation that satisfies this information is immediate. The value of *m* and *b* are substituted into the slope-intercept form $y = mx + b$.

2. There are three other situations that provide enough information for an equation of a line to be written. They are:
   - a point and a slope
   - two points
   - a point and a line parallel or perpendicular

   In each of these situations, the slope must be determined. If it is not given, use the slope formula or the relationship to a given line to find it. After the slope is found, use it

   with one of the points in either the point-slope form or the slope-intercept form to write the equation of the line.

3. If a line is described as vertical or with undefined slope, the equation of the line will be of the form $x = h$, where *h* is the *x*-coordinate of the given point.

4. If a line is described as horizontal or with zero slope, the equation of the line will be of the form $y = k$, where *k* is the *y*-coordinate of the given point.

5. In application problems, we will either know the slope (how the values change) and *y*-intercept (initial value) from the problem or we will be given two points that enable us to write the equation.

## GRAPHING CALCULATOR SKILLS

The graphing calculator has the ability to calculate the equation of the line if provided enough information. At this point, it is more beneficial to use the calculator to check our work instead of allowing it to do the work for us.

**Example:** Use the calculator to verify that $y = \frac{3}{2}x + 6$ is the equation of the line that goes through the points $(-4, 0)$ and $(4, 12)$.

> "I think the Hendricks/Chow series will improve the way I teach. The authors offer techniques I've thought about but wasn't sure how to implement. The resources and tips are very helpful and will enhance my teaching!"
>
> —Edward Ennels, *Baltimore City Community College*

# Exercise Sets for Today's Whole Student

The Exercise Sets contain a wide quantity and variety of exercise types that purposefully help students persist from basic skills through differentiation of topics, to practicing critical thinking and evaluating.

**Write About It!** exercises help students practice their vocabulary and verbal skills.

---

SECTION 3.5 / **EXERCISE SET**

 **Write About It!**

Use complete sentences in your answer to each question.

1. How can you determine the equation of a line if you know its slope and $y$-intercept?
2. How can you determine the equation of a line if you know two points on the line?
3. Which method do you prefer to find the equation of a line—using the slope-intercept form or using the point-slope form? Why?

**Practice Makes Perfect!**

Write the equation of the line that satisfies the given information. (*See Objective 1.*)

9. $m = 4, b = 5$
10. $m = -2, b = 1$
11. $m = 3$, passes through $(0, -4)$
12. $m = \frac{1}{2}$, passes through $(0, 7)$

---

**Mix 'Em Up!** is another layer of practice that mirrors what students might see on a test and also mixes together exercises from various objectives when possible.

 **Mix 'Em Up!**

Determine if the ordered pair is a solution of the given inequality.

43. $x + 3y < 6$; $(0, 0)$
44. $x - 7y < 1$; $(10, 3)$
45. $y > \frac{1}{2}x + 8$; $(8, -5)$
46. $y < -\frac{1}{3}x - 11$; $(-9, 0)$
47. $x > 12y$; $(1, 0)$
48. $y > 6x$; $(0, -1)$
49. $y - 11 < 4$; $(-3, 25)$
50. $x + 5 \geq 1$; $(-2, 4)$
51. $6x - y \leq 1$; $(0.2, -0.5)$
52. $x + 8y > 3$; $(-1.4, 0.5)$

Graph the solut... variables.

53. $x + 3y <$
55. $5x - 2y >$
57. $x + 3 \geq 0$

at least $350 per week.

64. If Brasil wants to make at least $450 per week, (a) write a linear inequality in two variables that represents this situation, (b) graph the linear inequality, and (c) give three possibl... of hours that Brasil could work at... at least $450 per week.

 **You Be the Teacher**

Correct each student's error, if any.

65. Graph $4x + y < 8$.
Jamie's work: The boundary line is $4x + y = 8$ and my test point is $(0, 0)$.
$4(0) + 0 < 8$
$0 < 8$
True

**You Be the Teacher!** exercises have students correct another students' work and help students analyze common errors.

---

 **Calculate It!**

109. Each table shows the points on the graph of a given line. Determine which graph contains the points $(0, 2)$ and $(-1, -3)$.

a.
| X | Y₁ |
|---|---|
| | 12 |
| | 7 |

X = -4

b.
| X | Y₁ |
|---|---|
| | -18.5 |
| | -8.5 |
| | -5 |

---

**Think About It!** exercises asks students to think critically about conceptual problems.

 **Think About It!**

73. Give an example of a linear inequality for which the point $(0, 0)$ cannot be used as the test point to determine which half-plane to shade.
74. Is it possible for a linear inequality in two variables to have one point as a solution? Explain.
75. Is it possible for a linear inequality in two variables to have only its boundary line as a solution? Explain.
76. Graph the inequalities in parts a–d and use your results to answer parts e–g.

g. How does the graph of $y \geq mx + b$ differ from the graph of $y > mx + b$?

77. Graph the inequalities in parts a–d and use your results to answer parts e–g.
a. $y < -2x + 4$      b. $y < \frac{1}{3}x - 1$
c. $y < 4x$            d. $y < 1$
e. Which half-plane was shaded in each of these inequalities?
f. What can you conclude about the graph of the solution of $y < mx + b$?
g. How does the graph of $y \leq mx + b$ differ from the graph of $y < mx + b$?

---

**PIECE IT TOGETHER** / **SECTIONS 3.1–3.3 Review**

Determine algebraically if the ordered pair is a solution of the equation. (*Section 3.1, Objective 1.*)

1. $(6, 2)$; $y = |x - 4|$          2. $(7, -4)$; $8x - 2y = 62$

Identify the quadrant or axis where the point is located. (*Section 3.1, Objective 2.*)

3. $(-8.9, -1)$          4. $\left(\frac{7}{8}, -\frac{5}{6}\right)$
5. $(-5, -1)$            6. $(9, 4)$

Use the given graph of an equation to determine if each ordered pair is a solution of the equation. (*Section 3.1, Objective 4.*)

7. $(0, 3)$
8. ...

Graph each equation. (*Section 3.2, Objectives 2–4.*)

9. $y = 4x + 2$  10. $y = -5x + 4$  11. $4x - y = 0$
12. $y = -7x$    13. $3x + 8 = 0$    14. $2y - 11 = 0$

Use the slope formula to determine the slope of the line between each pair of points. (*Section 3.3, Objective 1.*)

15. $(-2, 3)$ and $(4, 7)$          16. $(-2, 8)$ and $(5, 8)$

Write the equation in slope-intercept form, if necessary. State the slope and $y$-intercept of the line from its equation. Write the $y$-intercept as an ordered pair. Explain how the $x$- and $y$-values change with respect to one another. (*Section 3.3, Objective 2.*)

17. $3x - 2y = 6$

At an appropriate mid-point in the chapter, a review, **Piece It Together**, allows students to review and show mastery of the concepts presented thus far in the chapter.

---

**Group Activity** features provide multistep projects for groups of students to complete. This feature appears at the end of each chapter.

 **GROUP ACTIVITY** / **Use a Linear Equation to Model Population Growth**

1. Go to http://www.google.com/publicdata?ds=uspopulation and select a state or expand the state to select a specific county.
2. After you make your selection, a graph will appear in the main window. Use your cursor to trace the graph to see the population for each year. Write two ordered pairs of the form (year, population) for the state or county that you have selected.
3. Use the two ordered pairs from step 2 to write a linear equation that models the population of the selected state/county.
4. What is the slope of the model? What does it mean in the context of the problem?
5. What is the $y$-intercept of the model? What does it mean in the context of the problem?
6. Use the model to predict the population of your selected state/county in the year 2020.

# Review Material

The Review Material has been designed purposefully for student interaction. Students are provided a list of key terms, formulas, and properties from the chapter, as well as a Chapter Summary that students can quickly complete to review the main concepts that were presented. This section ends with Chapter Review Exercises, a Chapter Test, and Cumulative Review Exercises.

**What's the big idea?** helps students understand the "why" behind what they learned in that particular chapter.

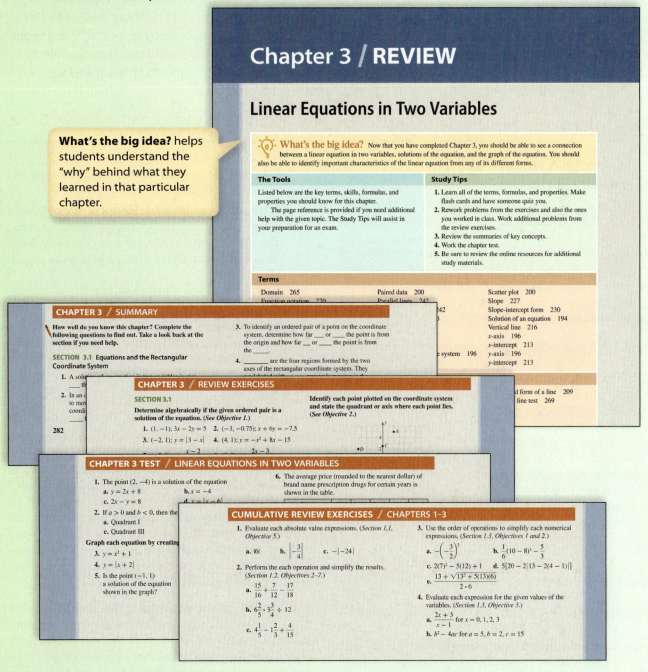

## Chapter 3 / REVIEW

### Linear Equations in Two Variables

**What's the big idea?** Now that you have completed Chapter 3, you should be able to see a connection between a linear equation in two variables, solutions of the equation, and the graph of the equation. You should also be able to identify important characteristics of the linear equation from any of its different forms.

**The Tools**

Listed below are the key terms, skills, formulas, and properties you should know for this chapter.

The page reference is provided if you need additional help with the given topic. The Study Tips will assist in your preparation for an exam.

**Study Tips**

1. Learn all of the terms, formulas, and properties. Make flash cards and have someone quiz you.
2. Rework problems from the exercises and also the ones you worked in class. Work additional problems from the review exercises.
3. Review the summaries of key concepts.
4. Work the chapter test.
5. Be sure to review the online resources for additional study materials.

**Terms**

| | | |
|---|---|---|
| Domain 265 | Paired data 200 | Scatter plot 200 |
| Function notation 270 | Parallel lines 242 | Slope 227 |
| | | Slope-intercept form 230 |
| | | Solution of an equation 194 |
| | | Vertical line 216 |
| | coordinate system 196 | x-axis 196 |
| | | x-intercept 213 |
| | | y-axis 196 |
| | | y-intercept 213 |

### CHAPTER 3 / SUMMARY

How well do you know this chapter? Complete the following questions to find out. Take a look back at the section if you need help.

**SECTION 3.1 Equations and the Rectangular Coordinate System**

1. A solution of an equation ...
   ___ th...
2. In an ...
   to mov...
   coordi...

3. To identify an ordered pair of a point on the coordinate system, determine how far ___ or ___ the point is from the origin and how far ___ or ___ the point is from the _____.

4. _____ are the four regions formed by the two axes of the rectangular coordinate system. They ...

282

### CHAPTER 3 / REVIEW EXERCISES

**SECTION 3.1**

Determine algebraically if the given ordered pair is a solution of the equation. (See Objective 1.)

1. $(1, -1)$; $3x - 2y = 5$  2. $(-3, -0.75)$; $x + 6y = -7.5$
3. $(-2, 1)$; $y = |3 - x|$  4. $(4, 1)$; $y = -x^2 + 8x - 15$

Identify each point plotted on the coordinate system and state the quadrant or axis where each point lies. (See Objective 2.)

### CHAPTER 3 TEST / LINEAR EQUATIONS IN TWO VARIABLES

1. The point $(2, -4)$ is a solution of the equation
   a. $y = 2x + 8$  b. $x = -4$
   c. $2x - y = 8$  d. $y = |x - 6|$

2. If $a > 0$ and $b < 0$, then the ...
   a. Quadrant I
   c. Quadrant III

Graph each equation by creating ...
3. $y = x^2 + 1$
4. $y = |x + 2|$
5. Is the point $(-1, 1)$ a solution of the equation shown in the graph?

6. The average price (rounded to the nearest dollar) of brand name prescription drugs for certain years is shown in the table.

### CUMULATIVE REVIEW EXERCISES / CHAPTERS 1–3

1. Evaluate each absolute value expressions. (Section 1.1, Objective 5.)
   a. $|6|$  b. $\left|-\dfrac{3}{4}\right|$  c. $-|-24|$

2. Perform the each operation and simplify the results. (Section 1.2, Objectives 2–7.)
   a. $\dfrac{15}{16} + \dfrac{7}{12} - \dfrac{17}{18}$
   b. $6\dfrac{2}{5} \cdot 3\dfrac{3}{4} \div 12$
   c. $4\dfrac{1}{5} - 1\dfrac{2}{3} + \dfrac{4}{15}$

3. Use the order of operations to simplify each numerical expressions. (Section 1.3, Objectives 1 and 2.)
   a. $-\left(-\dfrac{3}{2}\right)^3$  b. $\dfrac{1}{6}(10 - 8)^3 - \dfrac{5}{3}$
   c. $2(7)^2 - 5(12) + 1$  d. $5[20 - 2|13 - 2(4 - 1)|]$
   e. $\dfrac{13 + \sqrt{13^2 + 5(13)(6)}}{2 \cdot 6}$

4. Evaluate each expression for the given values of the variables. (Section 1.3, Objective 3.)
   a. $\dfrac{2x + 3}{x - 1}$ for $x = 0, 1, 2, 3$
   b. $b^2 - 4ac$ for $a = 5, b = 2, c = 15$

---

"This is a very good book and I would like to use it to teach my courses as soon as possible!"
—Zakia Ibaroudene, *Northeast Lakeview College*

# Supplements

## Comprehensive Resources for Every Teaching and Learning Environment

Teaching to today's whole student, beyond just the math, forces instructors to find more resources and tools to implement both inside and outside of class. As virtual classrooms continue to evolve, these types of resources become even more necessary. The Hendricks/Chow series has a number of quality, relevant supplements designed to save instructors preparation time and motivate students to learn and succeed.

## Supplements for the Student

### McGraw-Hill Connect Math
### Hosted by ALEKS Corp.

Connect Math Hosted by ALEKS Corp. is an exciting, new assignment and assessment ehomework platform. Starting with an easily viewable, intuitive interface, students will be able to access key information, complete homework assignments, and utilize an integrated, media-rich eBook.

**ALEKS** is a unique, online program that dramatically raises student proficiency and success rates in mathematics, while reducing faculty workload and office-hour lines. ALEKS uses artificial intelligence and adaptive questioning to assess precisely a student's knowledge, and deliver individualized learning tailored to the student's needs. ALEKS offers instructors robust course management tools, including automated reports and automatically-graded assignments. With a comprehensive course library that includes the developmental math sequence, ALEKS provides a dynamic assessment and learning system that can be used for a variety of instructional purposes.

- **Artificial Intelligence** and adaptive questioning determine precisely what each student knows, doesn't know, and is most ready to learn. ALEKS can then successfully target knowledge gaps and guide student learning.
- **Individualized Assessment and Learning** ensure student mastery of course material. ALEKS delivers highly individualized instruction on the exact topics each student is most **ready to learn**. The student is periodically reassessed to fill knowledge gaps and ensure long-term retention.
- **Adaptive, Open-Response Environment** avoids multiple-choice questions and includes comprehensive practice problems, explanations, and immediate feedback.

- **Dynamic, Automated Reports** track detailed student and class progress toward course mastery. With these reports, instructors can effectively direct instruction by identifying what students know and are ready to learn.
- **Robust Course Management Tools** include textbook integration, automatically-graded assignments, a customizable gradebook, and more. These tools allow instructors to spend less time on administrative tasks and more time directing student learning.

### ALEKS Prep

**ALEKS Prep for Beginning Algebra** and **Prep for Intermediate Algebra** focus on prerequisite and introductory material, and can be used during the first six weeks of the term to ensure student success in Beginning and Intermediate Algebra courses. ALEKS Prep quickly fills gaps in prerequisite knowledge by assessing precisely each student's preparedness and delivering individualized instruction on the exact topics students are most **ready to learn**. As a result, instructors can focus on core course concepts and see improved student performance with fewer drops.

**Student Solution Manual** The student solution manual provides comprehensive, worked-out solutions to the odd-numbered exercises in the section exercises, review exercises, piece-it-together exercises, chapter tests, and the cumulative review. The steps shown in the solutions match the style of solved examples in the textbook.

**Lecture and Exercise Videos** Online lecture and exercise videos will feature Andrea Hendricks and other instructors guiding students through the learning objectives, examples, and also exercises using the same methodology

from the text. Additionally, Andrea Hendricks developed a subset of **Troubleshooting Common Errors videos** that show students how to learn from and overcome common mistakes. These videos support the last objective of each section. All videos are available online as part of Connect Math Hosted by ALEKS Corp. or within ALEKS 360. Other supplemental videos include eProfessor videos, which are animations based on examples in the book. The videos are closed-captioned for the hearing impaired, and meet the Americans with Disabilities Act Standards for Accessible Design.

### Student Success Strategies Manual

To support the Chapter S materials in the text, Kelly Jackson from Camden County College, has developed a practical manual of activities, worksheets, tips, and strategies to support Chapter S (*Success Strategies*). This manual is available online as a resource for Connect Math Hosted by ALEKS Corp.

### Guided Student Workbook

Developmental Math students often struggle with taking quality notes while in class or listening to a lecture. To support the Hendricks & Chow text, this guided workbook provides a template of a lecture for the students to fill-in the important topics, terms, and procedures so that they spend less time creating notes and more time engaged in class. Also, in addition to the notes sections, students can then practice with additional student check exercises, additional problems similar to those in the text, and extra You Be the Teacher! problems. The Guided Student Workbook is an excellent companion that instructors and students can use to support the main text. This workbook is fully editable and available online as part of Connect Math Hosted by ALEKS Corp. or also available for custom packages. Students can download and print as needed.

## Supplements for the Instructor

**McGraw-Hill Connect® Math**
**Hosted by ALEKS Corp.**

Connect Math Hosted by ALEKS Corp. is an exciting, new assignment and assessment ehomework platform. Instructors can assign an AI-driven ALEKS Assessment to identify the strengths and weaknesses of each student at the beginning of the term rather than after the first exam. Assignment creation and navigation is efficient and intuitive. The grade, based on instructor feedback, has a straightforward design and allows flexibility to import and export additional grades.

**Instructor's Resource Manual (IRM)** The IRM is an excellent resource of additional materials for full- and part-time instructors to incorporate into their courses. This robust manual includes teaching strategies, historical notes, extra group and classroom activities, concept reviews, and activities that support the materials in Chapter S to help teach study skills including note taking, test preparation, and studying, etc. This manual is downloadable at Connect Math Hosted by ALEKS Corp.

**Annotated Instructor's Edition** In the Annotated Instructor's Edition (AIE), answers to exercises, review, and tests appear adjacent to each exercise set, in a color used only for annotations. Instructors will also find

helpful hints and notes within the margins to consider while teaching.

**Instructor's Solution Manual** The instructor's solution manual provides comprehensive, worked-out solutions to all exercises in the section exercises, review exercises, piece-it-together exercises, chapter tests, and the cumulative review. The steps shown in the solutions match the style and methodology of solved examples in the textbook.

**Powerpoints** These powerpoints will present key concepts and definitions with fully editable slides that follow the textbook. Project in class or post to a website in an online course.

**Instructor's Testing and Resource Online** Among the supplements is a computerized test bank utilizing Brownstone Diploma algorithm-based testing software to create customized exams quickly. This user-friendly program enables instructors to search for questions by topic, format, or difficulty level; to edit existing questions, or to add new ones; and to scramble questions and answer keys for multiple versions of a single test. Hundreds of text-specific, open-ended, and multiple-choice questions are included in the question bank. Sample chapter tests are also provided. CDs are available upon request.

# ALEKS 360: A Total Course Solution

**A cost-effective total course solution: fully integrated, interactive eBook combined with ALEKS individualized assessment and learning.**

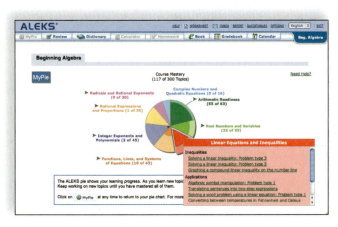

## Individualized Learning

- The ALEKS Pie summarizes a student's current knowledge and provides individualized learning on the exact topics the student is **ready to learn**

- Artificial intelligence successfully targets gaps by assessing precisely a student's knowledge and periodically reassessing for long-term retention

- Adaptive, open-response environment avoids multiple-choice and includes problems, explanations, and realistic answer input tools

## Interactive eBook

- eBook access provides worked examples, videos, and additional support

- Robust virtual features include highlighting, bookmarking, and note-taking capabilities

- Students can easily access the eBook, multimedia resources, and their notes from within their ALEKS Student Accounts

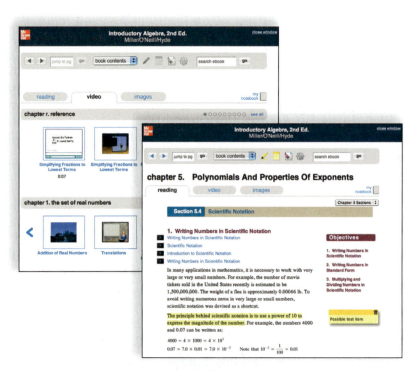

**Learn More: www.aleks.com/highered/math/aleks360**

# ALEKS Course Management Tools

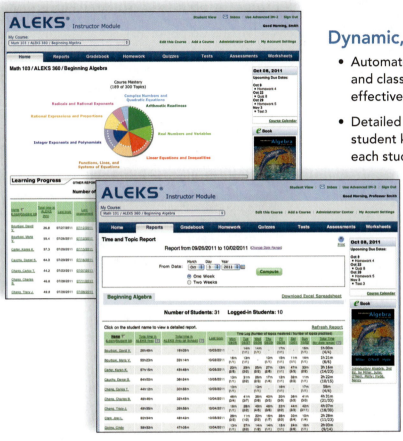

## Dynamic, Automated Reporting

- Automated reports dynamically track student and class learning progress so instructors can effectively direct classroom instruction

- Detailed reports identify precisely what each student knows, and more importantly, what each student is ready to learn next

- Time and Topic Report offers up-to-the-minute daily progress, including time logged, topics attempted, and topics mastered

## Course Control and Customization

- Align ALEKS topics with a textbook or course syllabus

- Create and customize course objectives and modules

- Set due dates for course objectives to pace student progress

- Assign automatically-graded homework, quizzes, and tests

- Seamlessly track and adjust student scores with the customizable gradebook

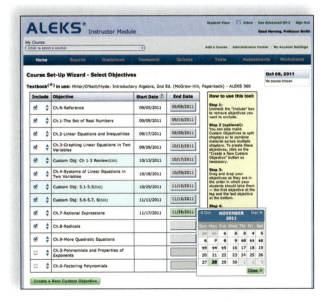

**For more information, contact your McGraw-Hill sales representative or visit www.aleks.com**

# McGraw Hill | connect®

## |MATH

### Hosted by **ALEKS Corp.**

**Connect Math Hosted by ALEKS Corporation** is an exciting, new ehomework platform combining the strengths of McGraw-Hill Higher Education and ALEKS Corporation. Connect Math Hosted by ALEKS Corporation is the first platform on the market to combine an artificially-intelligent, diagnostic assessment with an intuitive ehomework platform designed to meet your needs.

Connect Math Hosted by ALEKS Corporation is the culmination of a one-of-a-kind market development process involving full-time and adjunct Math faculty at every step of the process. This process enables us to provide you with a solution that best meets your needs.

Connect Math Hosted by ALEKS Corporation is built by Math educators for Math educators!

**1** *Your students want a well-organized homepage where key information is easily viewable.*

### Modern Student Homepage

▶ This homepage provides a dashboard for students to immediately view their assignments, grades, and announcements for their course. (Assignments include HW, quizzes, and tests.)

▶ Students can access their assignments through the course Calendar to stay up-to-date and organized for their class.

*Modern, intuitive, and simple interface.*

**2** *You want a way to identify the strengths and weaknesses of your class at the beginning of the term rather than after the first exam.*

### Integrated ALEKS® Assessment

▶ This artificially-intelligent (AI), diagnostic assessment identifies precisely what a student knows and is ready to learn next.

▶ Detailed assessment reports provide instructors with specific information about where students are struggling most.

▶ This AI-driven assessment is the only one of its kind in an online homework platform.

*Recommended to be used as the first assignment in any course.*

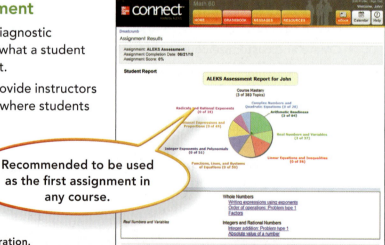

ALEKS is a registered trademark of ALEKS Corporation.

# Resources for Online Homework

**3** *Your students want an assignment page that is easy to use and includes lots of extra help resources.*

## Efficient Assignment Navigation

▶ Students have access to immediate feedback and help while working through assignments.

▶ Students have direct access to a media-rich eBook for easy referencing.

▶ Students can view detailed, step-by-step solutions written by instructors who teach the course, providing a unique solution to each and every exercise.

Students can easily monitor and track their progress on a given assignment.

**4** *You want a more intuitive and efficient assignment creation process because of your busy schedule.*

## Assignment Creation Process

▶ Instructors can select textbook-specific questions organized by chapter, section, and objective.

▶ Drag-and-drop functionality makes creating an assignment quick and easy.

▶ Instructors can preview their assignments for efficient editing.

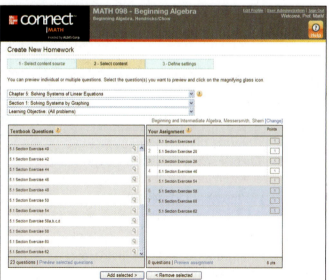

Connect
Learn
Succeed™

**connect** ®

**|MATH**

Hosted by **ALEKS Corp.**

---

**5** *Your students want an interactive eBook with rich functionality integrated into the product.*

 **connect** plus+

**|MATH**

Hosted by **ALEKS Corp.**

### Integrated Media-Rich eBook

▶ A Web-optimized eBook is seamlessly integrated within ConnectPlus Math Hosted by ALEKS Corp. for ease of use.

▶ Students can access videos, images, and other media in context within each chapter or subject area to enhance their learning experience.

▶ Students can highlight, take notes, or even access shared instructor highlights/notes to learn the course material.

▶ The integrated eBook provides students with a cost-saving alternative to traditional textbooks.

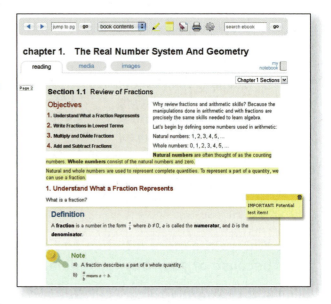

---

**6** *You want a flexible gradebook that is easy to use.*

### Flexible Instructor Gradebook

▶ Based on instructor feedback, Connect Math Hosted by ALEKS Corp.'s straightforward design creates an intuitive, visually pleasing grade management environment.

▶ Assignment types are color-coded for easy viewing.

▶ The gradebook allows instructors the flexibility to import and export additional grades.

Instructors have the ability to drop grades as well as assign extra credit.

**7** *You want algorithmic content that was developed by math faculty to ensure the content is pedagogically sound and accurate.*

## Digital Content Development Story

As the usage of online homework progresses and evolves, McGraw-Hill understands the need to have author involvement and author approval of the digital content to ensure that what students see in the online homework system is consistent with what they see in their textbooks. For this new developmental math series, co-author Pauline Chow has not only been closely involved with writing exercises for the text but also has overseen and led the creation of the digital content to ensure a seamless transition from print to digital offerings.

The development of McGraw-Hill's Connect Math Hosted by ALEKS Corporation content involved collaboration between McGraw-Hill, our authors, experienced instructors, and ALEKS Corporation, a company known for its high-quality digital content. The result of this process, outlined below, is accurate content created with your students in mind. It is available in a simple-to-use interface with all the functionality tools needed to manage your course.

1. McGraw-Hill partnered with author Pauline Chow to lead and oversee the digital content development.
2. Pauline Chow selected the textbook exercises to be included in the algorithmic content to ensure appropriate coverage of the textbook content.
3. McGraw-Hill auditioned and selected experienced instructors to work as digital contributors and represent the author's voice.
4. These digital contributors created detailed solutions for use in the Guided Solution and Solve It features, matching the voice of authors Andrea Hendricks and Pauline Chow.
5. Pauline and the digital contributors provided detailed instructions for authoring the algorithm specific to each exercise to maintain the original intent and integrity of each unique exercise.
6. Each algorithm was reviewed by Pauline and the contributors, then went through a detailed quality control process by ALEKS Corporation before being copyedited and posted live.

*Solutions in Connect Math Hosted by ALEKS Corp. match the procedure and language of the text.*

| Previous | 1 | 2 | 3 | 4 |

Question 6 of 7 (1 point)

**Factor completely.**

$$100u^2 - 80u + 16$$

Step 1:
Factor out the GCF of 4.

Step 2:
$$100u^2 - 80u + 16 = 4(25u^2 - 20u + 4)$$

Step 3:
Can we factor again? Yes.

Step 4:
Notice that the first and last terms in the trinomial are perfect squares.
$$25u^2 - 20u + 4 = (5u)^2 - 20u + (-2)^2$$
The middle term is $2 \cdot 5u \cdot (-2) = -20u$.

Print whole assignment | Print this page only     Tutorial help

Try Another
Solve It
Guided Solution
Show Example
Ask My Instructor
Link to Textbook

## RESULT = Truly Vetted, Consistent Digital Content That Is Approved by the Authors and Supported by ALEKS Corporation.

**Author and Lead Digital Contributor, O. Pauline Chow,**
*Harrisburg Area Community College*

**Lead Digital Contributor,** Amy Naughten

**Digital Contributors**
Dihema Ferguson, *Georgia Perimeter College*
Marianne Rosato, *Massasoit Community College*
Chris Yarrish, *Harrisburg Area Community College*
Allison Williams, *Georgia Perimeter College*
Katy Cryer

**McGraw Hill**

*Connect Learn Succeed*™

# Market Development

## Our Commitment to Market Development and Accuracy

McGraw-Hill's Development Process is an ongoing, never-ending, market-oriented approach to building accurate and innovative print and digital products. We begin developing a series by partnering with authors that desire to make an impact within their discipline to help students succeed. Next, we share these ideas and manuscript with instructors for review for feedback and to ensure that the authors' ideas represent the needs within that discipline. Throughout multiple drafts, we help our authors adapt to incorporate ideas and suggestions from reviewers to ensure that the series carries the same pulse as today's classrooms. With any new series, we commit to accuracy across the series and its supplements. In addition to involving instructors as we develop our content, we also utilize accuracy checks through our various stages of development and production. The following is a summary of our commitment to market development and accuracy:

1. 3 drafts of author manuscript
2. 5 rounds of manuscript review
3. 2 focus groups
4. 1 consultative, expert review
5. 3 accuracy checks
6. 3 rounds of proofreading and copyediting
7. Towards the final stages of production, we are able to incorporate additional rounds of quality assurance from instructors as they help contribute towards our digital content and print supplements

This process then will start again immediately upon publication in anticipation of the next edition. With our commitment to this process, we are confident that our series has the most developed content the industry has to offer, thus pushing our desire for quality and accurate content that meets the needs of today's students and instructors.

## Acknowledgements

Paramount to the development of *Beginning Algebra* was the invaluable feedback provided by the instructors from around the country that reviewed the manuscript or attended a market development event over the course of the several years the text was in development.

### A Special Thanks To All of The Event Attendees Who Helped Shape Beginning Algebra.

Focus groups and symposia were conducted with instructors from around the country to provide feedback to editors and the authors and ensure the direction of the text was meeting the needs of students and instructors.

Mihaela Blanariu, *Columbia College Chicago*
Eddie Ennels, *Baltimore City Community College*
Dihema Ferguson, *Georgia Perimeter College–*
    *Decatur Campus*
Stephanie Fernandez, *Lewis and Clark College*
Cathy Hoffmaster, *Thomas Nelson Community College*
Joe Howe, *St. Charles Community College*
Kelly Jackson, *Camden County College*
Jason King, *Moraine Valley Community College*
Rob King, *Harrisburg Area Community College*
Michael Kirby, *Tidewater Community College*
Viktoriya Lanier, *Middle Georgia College*
Cindy Light, *Indiana University–Southeast*

Catherine Moushon, *Elgin Community College*
Sandi Nieto, *Santa Rosa Junior College*
Toni Parise, *Southern Maine Community College*
Mari Peddycoart, *Lone Star College–Kingwood*
David Price, *Tarrant County College*
Elise Price, *Tarrant County College*

Amber Rust, *University of Maryland*
Mark Schwartz, *Southern Maine Community College*
Andrew Stephan, *St. Charles Community College*
Brad Stetson, *Schoolcraft College*

Richard Watkins, *Tidewater Community College*
Karen Watson, *Cypress College*
Carol White, *Tarrant County College–Southeast Campus*
Joanna Wilson, *Georgia Perimeter College*

## Manuscript Review Panels

Over 200 instructors reviewed the various drafts of manuscript to give feedback on content, design, pedagogy, and organization. Their reviews were used to guide the direction of the text.

Ricki Alexander, *Harrisburg Area Community College*
Marie Aratari, *Oakland Community College*
Dr. Eric Aurand, *Mohave Community College–Lake Havasu Campus*
Chris Barker, *San Joaquin Delta College*
Scott Barnett, *Henry Ford Community College*
Disa Beaty, *Rose State College*
David Behrman, *Somerset Community College*
Sandra Belcher, *Midwestern State University*
Monika Bender, *Central Texas College*
Teresa Betkowski, *Gordon College*
Katrina Bishop, *Craven Community College*
Bret Black, *Oxnard College*
Gregory Bloxom, *Pensacola State College*
Stacey Boggs, *Allegany College of Maryland*
Karen Bond, *Pearl River Community College*
Anthony Bottone, *Arizona Western College*
Cynthia Box, *Georgia Perimeter College*
Susan P. Bradley, *Angelina College*
Chandra Breaux, *Georgia Perimeter College*
Kirby Bunas, *Santa Rosa Junior College*
Julia Burch, *Central Michigan University*
Rebecca Burkala, *Rose State College*
David Busekist, *Southeastern Louisiana University*
Susan Byars, *Gordon College*
Nick Bykov, *Delta College*
Lynn Cade, *Pensacola State College*
Yungchen Cheng, *Missouri State University*
Kim Clark, *Wayne Community College*
Adam Cloutier, *Henry Ford Community College*
Delaine Cochran, *Indiana University Southeast*
David Cooper, *Wake Technical Community College*
Wendy Davidson, *Georgia Perimeter College*
Carlos de la Lama, *San Diego City College*
Marlene Dean, *Oxnard College*
Robert Diaz, *Fullerton College*
Paul Diehl, *Indiana University Southeast*
David Dillard, *Patrick Henry Community College*
Michael Dubrowsky, *Wayne Community College*
Scott M. Dunn, *Central Michigan University*
John Edwards, *University of Oklahoma*
Cheryl Eichenseer, *St. Charles Community College*
Mark Ellis, *Central Piedmont Community College*

Marcos Enriquez, *Moorpark College*
Paul Farnham, *Fullerton College*
Dale Felkins, *Arkansas Tech University*
Dihema Ferguson, *Georgia Perimeter College*
Jacqui Fields, *Wake Technical Community College*
Rhoderick Fleming, *Wake Technical Community College*
Cynthia Fletcher, *Pulaski Technical College*
Donna Flint, *South Dakota State University*
Dorothy French, *Community College of Philadelphia*
John Fulk, *Georgia Perimeter College*
Jenine Galka, *Moraine Valley Community College*
Angela Gallant, *Inver Hills Community College*
Sunshine Gibbons, *Southeast Missouri State University*
Sharon L. Giles, *Grossmont Community College*
Suzette Goss, *Lone Star College–Kingwood*
Kathleen Grigsby, *Moraine Valley Community College*
Kathryn Gunderson, *Three Rivers Community College*
Jin Ha, *Northeast Lakeview College*
Shawna Haider, *Salt Lake Community College*
Mark Harbison, *Sacramento City College*
Jennifer Hastings, *Northeast Mississippi Community College*
Dr. Annette Hawkins, *Wayne Community College*
Alan Hayashi, *Oxnard College*
Kristy Hill, *Hinds Community College–Rankin Campus*
Irene Hollman, *Southwestern College*
Teresa Houston, *East Mississippi Community College*
Heidi Howard, *Florida State College at Jacksonville*
Steven Howard, *Rose State College*
Joe Howe, *St. Charles Community College*
Susan Howell, *University of Southern Mississippi*
Denise Hum, *Canada College*
Zakia Ibaroudene, *Northeast Lakeview College*
Sally Jackman, *Madisonville Community College*
Tina Johnson, *Midwestern State University*
Nancy Johnson, *State College of Florida–Manatee, Sarasota*
Linda Jones, *Vincennes University*
Dynechia Jones, *Baton Rouge Community College*
Paul Jones, *University of Cincinnati*
Diane Joyner, *Wayne Community College*
Laura Kalbaugh, *Wake Technical Community College*
Edward Kavanaugh, *Schoolcraft College*
Pallavi Ketkar, *Arkansas Tech University*

Rob King, *Harrisburg Area Community College*

Jason King, *Moraine Valley Community College*

Jeff Koleno, *Lorain County Community College*

Jacek Kostyrko, *San Joaquin Delta College*

Eugene Kramer, *University of Cincinnati; Raymond Walters College*

Jason Lachowicz, *Patrick Henry Community College*

Marsha Lake, *Brevard Community College*

Debra Landre, *San Joaquin Delta College*

Carol Lanfear, *Central Michigan University*

Betty J. Larson, *South Dakota State University*

Lonnie Larson, *Sacramento City College*

Sungwook Lee, *University of Southern Mississippi*

Lisa Lindloff, *McLennan Community College*

Barbara Little, *Central Texas College*

Wanda J. Long, *St. Charles Community College*

Francine Long, *Edgecombe Community College*

Mike Long, *Shippensburg University of Pennsylvania*

Yixia Lu, *South Suburban College*

Amy Marolt, *Northeast Mississippi Community College*

Dorothy S. Marshall, *Edison College*

Abbas Masum, *Houston Community College & Alvin Community College*

Barabara Maurice, *Three Rivers Community College*

Julie Mays, *Angelina College*

Toni McCall, *Angelina College*

Roger McCoach, *County College of Morris*

Michael McComas, *Marshall Community & Technical College*

Mikal McDowell, *Cedar Valley College*

Bridget Middleton, *Santa Fe Community College*

Edward Migliore, *University of California–Santa Cruz*

Shahnaz Milani, *Blinn College*

Bronte Miller, *Patrick Henry Community College*

Phillip Miller, *Indiana University Southeast*

Jon David Miller, *Lone Star College–CyFair*

Dennis Monbrod, *South Suburban College*

Roya Namavar, *Rogers State University*

Martha Nega, *Georgia Perimeter College*

Cao Nguyen, *Central Piedmont Community College*

Kevin Olwell, *San Joaquin Delta College*

Priti Patel, *Tarrant County College*

Curtis Paul, *Moorpark College*

Mari Peddycoart, *Lone Star College, Kingwood*

Karen Pender, *Chaffey College*

Vic Perera, *Kent State University–Trumbull*

Michele Poast, *Dixie State College of Utah*

Carol Ann Poore, *Hinds Community College–Rankin Campus*

David Price, *Tarrant County College*

Elise Price, *Tarrant County College*

Cynthia Reed, *Moorpark College*

Pamelyn Reed, *Lone Star College, CyFair*

Lynn Rickabaugh, *Aiken Technical College*

Dianne Robinson, *Ivy Tech Community College*

Cosmin Roman, *The Ohio State University*

Jody Rooney, *Jackson Community College*

Elaine Russel, *Angelina College*

Amber Rust, *University of Maryland*

Kristina Sampson, *Lone Star College–CyFair*

Vicki Schell, *Pensacola State College*

Laura Schoppmann, *Seton Hall University*

Mark Schwartz, *Southern Maine Community College*

Daniel Seaton, *University of Maryland Eastern Shore*

Jerry Shawyer, *Florida State College at Jacksonville*

Jenny Shotwell, *Central Texas College*

Jean Shutters, *Harrisburg Area Community College*

Craig Slocum, *Moraine Valley Community College*

Jennifer Smeal, *Wake Technical Community College*

Brad Stetson, *Schoolcraft College*

Mark Stigge, *Baton Rouge Community College*

Daniela Stoevska-Kojouharov, *Tarrant County College*

Panyada Sullivan, *Yakima Valley Community College*

Sharon L. Sweet, *Brevard Community College*

Marcia Swope, *Santa Fe Community College*

Nader Taha, *Kent State University*

M. Kaye Tanner, *Linn Benton Community College*

Linda Tansil, *Southeast Missouri State University*

Carolyn Thomas, *San Diego City College*

Lee Topham, *Lone Star College–Kingwood*

Scott Travis, *Lone Star College–Tomball*

Barbara Jo Tucker, *Tarrant County College–SE Campus*

Laura Tucker, *Central Piedmont Community College*

Chris Turner, *Pensacola State College*

Jewell Valrie, *Wake Technical Community College*

Terry R. Varvil Jr., *Hillsborough Community College–Ybor Campus*

Mansoor Vejdani, *University of Cincinnati*

Mildred Vernia, *Indiana University Southeast*

Carol Walker, *Hinds Community College*

Jimmy Walker, *Hill College*

Jane Wampler, *Housatonic Community College*

Michelle Watts, *Lone Star College, Tomball*

Gail Whitaker, *Wharton County Junior College*

Robert White, *Allan Hancock College*

Suzanne Williams, *Central Piedmont Community College*

Olga Cynthia Wilson Harrison, *Baton Rouge Community College*

Jackie Wing, *Angelina College*

Rick Woodmansee, *Sacramento City College*

Grethe Wygant, *Moorpark College*

Tzu-Yi Alan Yang, *Columbus State Community College*

Mina Yavari, *Allan Hancock College*

Loris Zucca, *Lone Star College–Kingwood*

# Acknowledgments

This first edition would not have been possible without the encouragement and support of our families. There really are not words that convey our thanks and appreciation for their understanding as we worked seemingly night and day on the manuscript these last few years. We love you Todd, Andy, Charlie, and Cory Hendricks and Michael, Amy, and Andrew Ko.

Just as it takes a village to raise children, it takes a village to write a book. There are some important members of this village we would like to personally thank. We first extend our thanks to our dear friend and first sponsoring editor at McGraw-Hill, David Millage, who had the foresight to bring us together for this exciting journey. We also thank our Sponsoring Editor, Mary Ellen Rahn, for her support in this project. The resources, guidance, and insight you have provided through this project have been invaluable. We extend our gratitude to our Developmental Editors, Adam Fischer and most recently Emily Williams, for keeping us on task and coordinating all of the helpful feedback we have received through this process. We also truly appreciate the many hours that the production team has spent with us to make this series come together. We thank Vicki Krug for teaching us the world of production and we thank Laurie Janssen for the beautiful, current and creative design. Many thanks also goes to Emilie Berglund, Director of Digital Content, for overseeing the process of making the book come alive through Connect Math Hosted by ALEKS. Thanks for your countless hours and dedication to this book. We would like to thank every other member of the McGraw-Hill team for their excitement and enthusiasm in this project.

Some of the key contributors that we would like to thank are Calandra Davis, Kelly Jackson, Andrew Stephan, Joe Howe, Lisa Collette, Pat Steele, and Bea Sussman. Thanks, Calandra, for being there and offering your support to get this project going. Thanks, Kelly, for your wonderful insight and enhancements for Chapter S. You really helped our vision for this chapter become a reality. Thanks, Andy and Joe, for your contribution of Chapter 12 for Intermediate Algebra. Your work provided a great addition to the text. Thanks, Lisa, for your priceless input and suggestions for the final manuscript of this text. Your detailed comments were very helpful. Thanks, Pat and Bea, for your keen eyes and thorough review of the final manuscript. Your suggestions have provided the polishing touch to the book for which we are most appreciative.

We also thank all of the supplements and digital content contributors that helped complete this series. Thanks, Emily Whaley, for your dedicated and detailed efforts on the Solutions Manuals. Also, we must thank all of the several instructors that have reviewed this series throughout various stages of manuscript. We appreciate your guidance, feedback and support to help this series come to life. We look forward to you seeing the completed project to see how your suggestions have shaped this series.

Lastly, we extend our thanks to all of the students we have taught in our collective 45 years of teaching for making us the teacher we are today.

# Application Index

# Strategies to Succeed in Math

## Success

It is with great delight that we welcome you to this course and the materials we have provided you. You are embarking on perhaps the most rewarding experience of your life. College is a series of challenges that will prepare you for a lifetime of learning and accomplishments. At the end of your college education, you will be awarded a degree—a degree that establishes your ability to learn and to persevere. It is a statement that you can work to accomplish a goal, no matter what the obstacles.

At this point, you must evaluate the reason you are here. Are you here to fulfill your dreams or the dreams of someone else? To successfully complete this journey, you must be here to fulfill your own desires, not those of a friend or family member. It will be very difficult to withstand the trials of college if you do not have a personal desire to see it through.

A semester or quarter can be very overwhelming. Focus on one day at a time and not on everything that you must learn throughout the entire course. Before you know it, the course will be over. We know that it is very easy to get distracted from your goals. Stay committed and motivated by remembering your ultimate reason for attending school.

We wish you success in this course and in your future educational endeavors. It is our hope that you are successful, not only this semester or quarter, but every term until your ultimate goal is achieved. This course will pave the way for that success. It is the door to achieving your dreams.

*Mrs. Andrea Hendricks   and   Mrs. Pauline Chow*

**?** *Question for Thought:* What do you dream about doing? How does attending college make that dream possible? How does this course help you meet your goals? What is your biggest obstacle in being successful in this course? What can you do to overcome that obstacle? What is your plan for succeeding in this course?

**Chapter S** will introduce some important strategies that can enhance your performance in this course.

## Chapter Outline

"Continuous effort—not strength or intelligence—is the key to unlocking our potential."

Winston Churchill

## Time Management and Goal Setting

> **OBJECTIVES**
>
> As a result of completing this section, you will be able to
>
> **1.** Set realistic grade and attendance goals.
> **2.** Establish a schedule for studying math.
> **3.** Establish an action plan.
> **4.** Troubleshoot common errors.

A student might have great study skills but without good utilization of time, those skills are a wasted treasure. Time is like money; we should know how every moment is spent so that we do not waste it, and we should spend it wisely. The great news is that everyone has exactly the same amount of time. The bad news is that once the time has passed, it can never be recaptured. In this section, we will learn some helpful ways to manage our time and to set appropriate goals that make the best use of our time.

### Setting Realistic Goals

> **Objective 1** ▶
>
> Set realistic grade and attendance goals.

As we begin a new course, we have a "clean slate." No grades are in the grade book, no absences have been recorded, and no assignments are late or outstanding. We can control our behavior in such a way that we maximize our chance of success in this course. The beginning of a semester is an optimal time to set realistic goals that we can achieve by the end of the semester. Setting goals provides us with a sure focus and clarifies the direction in which we are headed. Activity 1 will provide us a chance to think about the goals we have for this course and for college in general.

> **INSTRUCTOR NOTE:**
> Tell students the importance of writing down their goals. Writing down goals makes them more tangible to us and it can also help motivate us.

### ACTIVITY 1    Answer each question.

1. What was the last math course you took? When did you take it? Was it a successful experience? Why or why not? _____
_____

2. What grade would you like to earn in this course? How many hours a week do you think you will need to devote outside of class to meet this goal?
_____

3. How many absences do you think could put your goal at risk? What other activities could put your goal at risk?_____
_____

4. What is your short-term goal? What do you hope to accomplish this semester?
_____

5. What is your long-term goal? Why are you in college?_____
_____

6. What are some things you can do that will enable you to reach your goals?
_____

7. Are there any other activities that you need to avoid or limit that would prevent you from reaching your goals?_____
_____

**Note:** *Keep in mind that being late or leaving class early can be just as harmful as missing class completely. Attending class regularly, punctually, and for the entire time is a goal that will lead us to success in math.*

## Plan Study Time

Before we can establish study time, we need to know how we currently use and manage our time. One way to do this is by recording our activities in a calendar or planner. The key is to record all of our activities, including time for driving, sleeping, and eating, as shown in Figure S.1. When we do not take the time to record our activities, we are more apt to waste time and to spend time on things that are not aligned with our goals. Once we understand and realize how we use our time, we can be more deliberate in planning a schedule that is beneficial to our goals.

**Figure S.1**

| Time | Monday, 9/22 |
|------|--------------|
| 7:00 | Shower/get dressed |
| 7:30 | Breakfast |
| 8:00 | Leave for school |
| 8:30 | |
| 9:00 | Math 1001 (9–10:15) |
| 9:30 | |
| 10:00 | |
| 10:30 | Engl 1101 (10:30–11:30) |
| 11:00 | |
| 11:30 | |
| 12:00 | Lunch |
| 12:30 | |
| 1:00 | Drive to work |
| 1:30 | Work (1:30–6) |
| 2:00 | |
| 2:30 | |
| 3:00 | |
| 3:30 | |
| 4:00 | |
| 4:30 | |
| 5:00 | |
| 5:30 | |
| 6:00 | Drive home |
| 6:30 | Dinner |
| 7:00 | Watch TV |
| 7:30 | |
| 8:00 | Study |
| 8:30 | |
| 9:00 | Watch TV |
| 9:30 | |
| 10:00 | Exercise |
| 10:30 | |
| 11:00 | Go to bed |

Following are some suggested guidelines for planning study time.

1. Make a daily appointment to study for this class.
   a. A general guideline is to spend 2 hr studying for each hour in class. For example, a 3-credit-hour class will require an average of 6 hr of study time each week.

         **b.** Devote some time each day to studying math. Do not cram the suggested hours into one day.

         **c.** The best time to study is the hour immediately following class time. If this is not doable, study time should be as soon after class as possible.

    **2.** Multitask when feasible.

         **a.** When waiting in line or waiting for an appointment, review your class notes.

         **b.** When riding a bus, listen to a lecture or review note cards.

    **3.** Prioritize tasks. Differentiate between the things that must be done and the things that you just want to do.

    **4.** Maintain a balanced schedule by not overcommitting your time. Say no when you don't have the time to devote to a task.

    **5.** Make a habit of writing things down in a calendar or a to-do list.

**Note:** *Doing something right the first time takes less time in the long run than doing it poorly and having to redo it later.*

## ACTIVITY 2    Answer each question.

**1.** Without using a calendar or planner, estimate how much time you spend sleeping, eating, watching TV, playing on the computer, working, socializing with friends, sitting in classes, studying, taking care of children, running errands, traveling from one place to another, exercising, and so on.

    **a.** What did you discover about yourself and how you spend your time? _____

_____

    **b.** Is there anything that you want to spend more time doing? _____

_____

    **c.** Is there anything you want to spend less time doing?_____

_____

**2.** Create a planned time chart for next week using one similar to Figure S.1.

    **a.** First record large blocks of time commitments (class time, work time, driving time, church activities, and so on)._____

_____

    **b.** Schedule necessary activities like sleeping and eating._____

_____

    **c.** Schedule regular activities such as grocery shopping, going to the gym, cleaning house, putting kids to bed, and the like._____

_____

    **d.** Based on suggested guidelines, you should study 2 hr for each hour you are in class. If you are enrolled in 12-credit hours, you should have 24 hr of study time. Distribute this time over the week in reasonable blocks of time. Do not set yourself up for cram sessions. _____

_____

    **e.** Schedule yourself some fun time and time to decompress._____

_____

    **f.** Allow flexibility in your schedule. Things will come up, so you shouldn't have every moment scheduled._____

_____

**3.** Are there things in your life that need to be removed to improve your chances of success?_____

_____

## Action Plan

**1.** Utilize the print and digital resources provided to you that accompany your textbook. Work with your instructor to access the available Success Strategies Manual. You will find things such as
- a time tracker.
- a tip sheet for getting to know your teacher.
- a shopping list for math success that includes a list of supplies you might need.
- positive affirmations that you can recite to help motivate you.
- other resources to get organized and to help you have a productive semester.

**2.** Set some specific goals for this course.
- I will get a grade of _____ in this course.
- Three things I can do to ensure that I meet this goal are
  1. _____
  2. _____
  3. _____

- I plan to spend _____ hours per week outside of class for this course.
- Three things I can do to be sure that I have enough time to devote to math are
  1. _____
  2. _____
  3. _____
- At most I will miss _____classes this semester.
- Three things I can do to be sure I attend each class meeting are
  1. _____
  2. _____
  3. _____
- What could interfere with my ability to meet these goals? _____
- What can I do to overcome or prevent this challenge?_____

## Troubleshooting Common Errors

Some of the common errors associated with time management are shown in the following table along with a more appropriate behavior.

| Poor Time Management Behaviors | Better Time Management Behaviors |
| --- | --- |
| ▶ Student waits until Sunday afternoon to do his homework for the week. | ▶ Student sets aside 1 hr per night for math and gets a little bit done each day. |
| ▶ Student arrives late to class each day. | ▶ Student arrives on time and is ready for class to start. |
| ▶ Student has no idea when the next test is scheduled. | ▶ Student uses a planner to record due dates and exam dates. |
| ▶ Student plans a vacation starting a week before the semester ends. | ▶ Student checks the Academic Calendar for the college before making travel plans. |

## SECTION S.2   Learning Styles

### OBJECTIVES

As a result of completing this section, you will be able to

1. Identify your dominant learning style(s).
2. Implement strategies to maximize success in math based on your learning style(s).
3. Establish an action plan.
4. Troubleshoot common errors.

People learn in many different ways. The way a person best gathers, processes, organizes, and remembers information refers to their **learning style.** There are three basic learning styles—auditory, visual, and kinesthetic. If you do not know how you best learn, it will be helpful for you to complete a learning styles inventory. One is included in this section, but there are many inventories available online.

### Objective 1 ▶

Identify your dominant learning style(s).

### Learning Styles

The following survey has been created to identify a student's learning style as it pertains to mathematics. Following the survey is a tally sheet for your responses. The category with the highest percentage represents your dominant math learning style.

## ACTIVITY 1   Complete the survey.

### Math Learning Styles Survey

Answer each question with the number "3" if you agree most of the time, "2" if you agree sometimes, and "1" if you agree rarely or do not agree.

_____ 1. When there is talking or noise in class, I get easily distracted.

_____ 2. If a problem is written on the board, I have difficulty following the steps unless the teacher verbally explains the steps.

_____ 3. I find it easier to have someone explain something to me than to read it in my math book.

_____ 4. If I know how the math is used in real life, it is easier for me to learn.

_____ 5. To remember formulas and definitions, I need to write them down.

_____ 6. I prefer listening to the lecture rather than taking notes.

_____ 7. Using manipulatives, hands-on activities, or games helps me learn math concepts.

_____ 8. When solving a problem, I try to picture working it out in my mind first.

_____ 9. Quiet places are the best places to study math.

_____ 10. Talking myself through a math problem helps me solve it.

_____ 11. When I write things down I remember them better.

_____ 12. I have to do a math problem myself to learn it. I can't really know it by watching someone else do it.

_____  **13.** If someone lectures about math without writing things down, I find it difficult to follow.

_____  **14.** When I take a math test, I recall more of what was said to me than what I read in my notes or in my math book.

_____  **15.** When I take a math test I can picture my notes or problems on the board in my head.

_____  **16.** I can do math problems sometimes, but can't verbally explain what I did.

_____  **17.** Reading math makes my eyes feel tired or strained.

_____  **18.** When I study math I need to take a lot of breaks.

_____  **19.** I am good at using my intuition to know how to solve a math problem.

_____  **20.** I can learn math fast when someone explains it to me.

_____  **21.** Puzzles and games are a good way to practice math.

_____  **22.** When I work a math problem I say the numbers I am working with to myself.

_____  **23.** I try to write down everything and take a lot of notes in math class.

_____  **24.** I do best with math if I just roll up my sleeves and work on problems.

*Developed by Kelly Jackson, Camden County College*

## Math Learning Styles Tally Sheet

Carefully copy your responses (1, 2, or 3) to each question on the survey into the spaces provided.

**Auditory:**  ___ + ___ + ___ + ___ + ___ + ___ + ___ + ___ = _____
          2     3     6    10   14   17   20   22   A-Total

**Visual:**  ___ + ___ + ___ + ___ + ___ + ___ + ___ + ___ = _____
          1     5     8     9   11   13   15   23   V-Total

**Kinesthetic:**  ___ + ___ + ___ + ___ + ___ + ___ + ___ + ___ = _____
          4     7   12   16   18   19   21   24   K-Total

Overall Total = A-Total + V-Total + K-Total = _____

Auditory Percentage = (A-Total ÷ Overall Total) × 100 _____

Visual Percentage = (V-Total ÷ Overall Total) × 100 _____

Kinesthetic Percentage = (K-Total ÷ Overall Total) × 100 _____

**Note:** *Very few people learn math only one way. We hear things, see things, and do things that help us understand math better. One style may dominate the others though.*

## Strategies to Enhance Your Learning

**Objective 2** ▶

Implement strategies to maximize success in math that are specific to your learning style(s).

Now that you have identified your math learning style, some suggestions for things you can do to increase your success in math class based on which style(s) suits you best are provided. A summary of each type of learner and strategies to apply are shown next. General success strategies that apply to all students will be presented in a later section of this chapter.

> **Definition:** An **auditory learner** learns best through listening to lectures, having discussions, talking things through, and listening to what others have to say. These learners often benefit from reading the text aloud and using a recording device.

### Strategies to Assist Auditory Learners

✔ Sit near the front of the class so you can hear your instructor.

✔ Sit away from auditory distractions (air conditioner, door, window, and so on).

✔ Participate in class discussions. Ask and answer questions.

✔ Listen carefully to what the teacher says, their tone of voice, inflection, and volume will give cues about what is important.

✔ Record lectures to aid in taking notes.

✔ Read the text out loud.

✔ Use computer tutorials, online sites, or videos with an audio track.

✔ Use songs, rhymes, and other auditory memory devices.

✔ Discuss your ideas verbally.

✔ Dictate to someone while they write down your thoughts.

✔ Study with a classmate, which allows you to talk about and hear the information.

✔ Recite out loud the information you want to remember several times.

✔ Make recordings of important points you want to remember and listen to them repeatedly.

✔ Ask your teacher to repeat or restate something, as needed.

✔ Verbalize your goals for completing your assignments and say your goals out loud each time you begin work on a particular assignment.

> **Definition:** A **visual learner** learns best through seeing. These learners prefer to sit at the front of the classroom to avoid visual obstructions and do best when the teacher provides detailed notes. They generally think in terms of pictures and learn best from visual displays, such as diagrams, illustrated textbooks, videos, PowerPoint slides, and handouts. During a lecture or classroom discussion, visual learners often prefer to take notes to help them grasp the material.

### Strategies to Assist Visual Learners

✔ Use visual materials such as pictures, charts, graphs, and the like.

✔ Have a clear view of your teacher when she is speaking so you can see body language and facial expressions.

✔ Use colored markers, Post-it notes, or highlighters to identify important points in the text.

✔ Illustrate your ideas as a picture or brainstorming bubble before writing them down.

✔ Use multimedia (e.g., computers and videos) learning tools.

✔ Use different colors and pictures in your notes, exercise books, and the like.

✔ Study in a quiet place away from auditory disturbances.

✔ Visualize information as a picture to aid memorization.

✔ Write down things that you want to remember.

✔ Take many notes and write down lots of details.

✔ Learn new material by writing out notes, cover your notes then rewrite them.

✔ Write your goals down and read them as you complete your assignments.

✔ Take advantage of study guides, handouts, and workbooks that include worked out solutions.

✔ Choose a seat away from visual distractions and close to the front of the class.

✔ Prepare flashcards and review them often.

**Definition:** A **kinesthetic learner** learns best through moving, doing, and touching. These students learn best through a hands-on approach. It is difficult for these students to sit still for long periods.

## Strategies to Assist Tactile/Kinesthetic Learners

✔ Practice, practice, practice. Because you prefer a hands-on approach, you need to work as many problems as possible.

✔ Take a 3- to 5-min study break every 15–25 min.

✔ Use manipulatives that can help you investigate the topic you are studying (algebra tiles, base 10 blocks, and the like).

✔ Work in a standing position.

✔ Chew gum while studying.

✔ Use bright colors to highlight what you are reading.

✔ If you wish, listen to music while you study (be sure it is not distracting though).

✔ Make or use a model.

✔ Pace or walk around while reciting to yourself or using flashcards or notes.

✔ If the opportunity arises, volunteer to go to the board to work problems.

✔ Study while sitting in a comfortable lounge chair or on cushions or a beanbag.

✔ Cover your desk with your favorite colored construction paper or even decorate your area to help you focus.

✔ Memorize information by closing your eyes and writing the information in the air, try to picture and hear the words in your head as you are doing this.

✔ Make flashcards, card games, floor games, and the like to help you process information.

✔ When working with someone, after they show you a problem, ask if you can try one.

✔ Make a graphic organizer showing the connections between topics.

✔ Get a study group together in a classroom where you can write on the board, walk around, talk about the math, but do problems.

✔ Use applets or computer tutorials with guided solutions in which you have to enter information throughout.

**Note:** There is no right or wrong learning style. Determine what you have a tendency toward and use strategies to enhance your performance in class. You cannot control the way your instructor delivers information, but you can control what you do to best receive the information.

## Action Plan

**1.** My dominant learning style(s) is(are) dominant _____

_____.

**2.** Five strategies that I will implement immediately to improve my chances of success in math based on my learning style(s) are

    **a.** _____.

    **b.** _____.

    **c.** _____.

    **d.** _____.

    **e.** _____.

**3.** Does it seem like my teacher's style of delivery matches my style or is it a mismatch? (Example: you prefer to learn visually but your teacher does not write on the board a lot)_____

_____.

## Troubleshooting Common Errors

One common error is how students deal with the situation in which their teacher's teaching style does not match their learning style. Students can control their own behavior but not those of others. Following are some suggestions on how to deal with this situation.

| Learning Style/Teaching Style Mismatch | What Can You Do? |
|---|---|
| Visual learner with a teacher who talks a lot but doesn't write down a lot. | **1.** Record the lesson. Later go back and listen to the lesson and fill in your notes with anything you missed the first time. <br> **2.** Read the section in the book prior to class so you know the topic that will be discussed and write down some of the vocabulary and steps. <br> **3.** Visit your instructor or a tutor in a setting where you can ask questions and have time to write things down. |
| Kinesthetic learner with a teacher who does not use activities, does not have you work problems in class, and does not use manipulatives. | **1.** Rework the problems that the teacher did in class, check your work with your instructors, and then try some similar problems on your own. <br> **2.** Use the Explain or Show me features in your online homework system that will ask you to enter each step. <br> **3.** Meet with your instructor or a tutor. Each time they show you a problem ask, "Can I try one now?" |
| Auditory learner in a blended or online class, with limited access to "lectures." | **1.** Take advantage of lecture and exercise videos that accompany your textbook or from popular Internet sites that have great educational videos. <br> **2.** Make audio notes for yourself, reading important rules, definitions, and processes into a recorder. Listen to your recordings as often as possible. <br> **3.** Have a study group or tutoring session where you can discuss the math concepts with someone. |

| SECTION S.3 | Study Skills |
| --- | --- |

### OBJECTIVES

As a result of completing this section, you will be able to

1. Identify habits that good math students employ.
2. Read the textbook more effectively and efficiently.
3. Take quality notes.
4. Organize materials in a notebook or portfolio.
5. Complete homework in ways that maximize retention.
6. Establish an action plan.
7. Troubleshoot common errors.

**Study skills** are the skills students need to improve their learning capacity and to acquire new knowledge. No two students are going to employ the same set of study techniques, though there are some that every student should utilize when studying math. This section will address some of the skills that will enhance the success of all students.

## Habits of Successful Math Students

Good math students study with purpose. When you underline something, it should be because it is important. If you write out a note card, it should be because it is something you want to remember long term. When you work a homework problem, it should lead to a better understanding of the process used to solve that problem. If you can't answer "Why am I doing this?" then what you are doing may not be a productive use of your time. You should always have goals to achieve when you sit down to study and everything you do during your study time should work toward those goals. The following suggestions will help you approach your math class with purpose.

### Objective 1

Identify habits that good math students employ.

1. *Successful students are responsible.*
   ▶ Attend every class meeting. Try to be a few minutes early so that your materials are out and you are ready for class to begin.
   ▶ Find an accountability partner in class. Encourage one another and use each other as a resource if one of you is absent from class.
   ▶ Adhere to all deadlines for assignments.

2. *Successful students make the most of their time in class.*
   ▶ When you are in class, "be in class." Try not to worry about other things going on in your life. Focus your attention on the topics being discussed. (No texting, surfing, or doing other homework.)
   ▶ Either take notes in class or record the lecture so that you can refer to it later. Some instructors even post their class notes on their website.
   ▶ Actively participate in class. Follow along with the instructor. Ask questions when you don't understand and be prepared to answer questions that you know.

3. *Successful students dedicate an appropriate amount of time to math outside of class.*
   ▶ Immediately after class, take a few minutes to write down a summary of the key concepts that you remember. (It has been shown that students who write down what they know retain 1½ times as much as those who don't, 6 weeks later.)
   ▶ Review your class notes as soon after class as possible.
   ▶ Begin your homework assignment as soon as you can.
   ▶ Review notes from previous classes on a regular basis so that you do not forget material covered earlier in the course.
   ▶ Devote some time every day to your math class.

**4.** *Successful students aren't afraid to ask for help.*

▶ Math does not need to be a solo activity. Ask your instructor, a tutor, or a classmate about problems you do not understand. Form a study group.

▶ Take advantage of your college's tutoring center and your instructor's office hours.

▶ Use self-help books, other texts, online websites, computer programs, DVDs, or any other outside sources you can find to supplement your course materials.

**5.** *Successful students are persistent.*

▶ If you get "stuck" on a problem, refer to your notes, the book, or other help resources. Rework the problem until you can do it without referring to these things.

▶ Be patient with yourself in learning the material. Don't get frustrated that other students may seem to learn more quickly than you. Every student comes to class with a different mathematical background. Some classmates just graduated from high school, while others may be returning after several years.

▶ Be willing to try new approaches to solving a problem. If your first attempt fails, continue trying other possible strategies. Often there is more than one way to get to the solution.

Study skills consist of the things you do in class as well as the things you do outside of class. We will take a closer look at four of these topics: reading a math textbook, taking notes, organizing your notebook/portfolio, and completing homework.

## Reading a Math Textbook

**Objective 2** ▶

**Read the textbook more effectively and efficiently.**

The textbook is a great resource to aid in the mastery of material presented in class. Many students pay a lot of money for a textbook but do not take the time to read it and often have never been shown how to read a math textbook. Before you attempt homework problems, it is important that you carefully read the relevant sections of your math textbook. The examples should be studied and definitions, properties, and formulas should be learned.

Reading a math textbook is very different than reading a psychology or history text and especially different from reading a novel or a newspaper. Math textbooks generally do not have a whole lot of prose. Textbooks are typically organized around new definitions, new procedures, and worked examples that illustrate these concepts. Graphs, tables, diagrams, equations, formulas, and notations are used to visualize some of the concepts, but these are no more and no less important that the worded passages themselves. Together, they explain the complete math concept you are reading about. When reading a page in your math textbook, you will be reading from left to right, right to left, up and down (think tables), diagonally (think graphs), and from the inside out (think equation solutions).

### Some guidelines for reading a math textbook are

1. Skim a new section briefly to identify new vocabulary or processes. Identify the objectives to be presented in the section, which are located in the headings and subheadings.
2. Read the section a second time, using more time and concentration. Try to connect your prior knowledge with the new material you are reading. You will need to read slowly, reread sections, and constantly ask yourself if you understand.
3. When you get to the worked examples, use your own paper to work the problems in detail, making sure that you understand how each step follows from the previous one.
4. Make a list of concepts, formulas, and vocabulary you do *not* understand and seek help.
5. Make a list of formulas and vocabulary that you *do* know and understand.

6. Make sure you understand and use the mathematically correct definition for each vocabulary word. Often words are "borrowed" from everyday language and the meaning can be the same or different in mathematics.

7. You must be able to decode and understand the math notation along with the vocabulary.

8. Learn the formal definitions and properties. It is a good idea to be able to paraphrase into your own words but you don't want to create "new" math rules that may not always work.

9. Get help if you do not understand the reading material.
   - Go back and review the previous section to see if it might be related.
   - Review your instructor's notes on the material.
   - Use the resources that came with your book (videos, CDs, animations, etc.).
   - Ask a classmate, a tutor, or your instructor. Be specific with what you do not understand.

## ACTIVITY 1    Get to know the features of this book.

1. Where are the answers located and which answers are available?_____
   _____

2. Where is the chapter review? What is included in the review?_____
   _____

3. What do the different icons used in the book mean?_____
   _____

4. What colors are used to set off definitions, formulas, tips, and other important information?_____
   _____

## Taking Notes

**Objective 3** ▶

**Take quality notes.**

Effective note taking begins when you enter the classroom. After you get to your seat, prepare to take notes by taking out your paper, pencil, and book. Taking notes enables you to record how your instructor explains processes, to record examples that are worked, and to identify important class information, such as homework assignments and test dates. Note taking also helps you listen more attentively and to be actively engaged in the learning process. Studies show that people may forget 50% of a lecture within 24 hr, 80% in 2 weeks, and 95% within 1 month if they do not take notes.

Some guidelines for taking notes are as follows.

1. To be an effective note taker, it is important that you attend class.
   **Don't be late.**
   **Don't leave early.**

2. To be an effective note taker, you must be a good listener.
   a. Be actively involved in the lecture so that you stay focused on what is being presented.
   b. Sit near the front of the class so that distractions are minimized.
   c. Try to relate the new topics to the material that you have already learned.
   d. Ask your instructor for clarification, if you do not understand. For example, "I don't understand how you got from step 2 to step 3…", "I don't know what the symbol _____ means.", and "What is the difference between _____ and _____?"
   e. Listen for words that signal important information.
   f. Notice how your teacher uses formal math vocabulary when speaking.

3. To be an effective note taker, you must actually take notes.
4. Bring pencils and paper to class and take math notes in pencil, not pen.
   a. Each day start a new set of notes on a new page. Date and label them with the chapter (and section) that is being discussed.
   b. Copy down *everything* that the instructor writes on the board. If the instructor takes the time to write something, it is important.
   c. Take notes, even though your understanding may not be complete.
   d. Leave a space where you may have missed steps or have questions so that you can get these filled in after class. Then actually get them filled in by reading your book, going to a tutor, visiting your instructor's office, or asking a study buddy.
   e. Develop a good note-taking system. Ideas for note-taking systems can be found on the Web or also available with the optional Success Strategies Manual.
5. To be an effective note taker, you should review your notes.
   a. Review and reorganize your notes as soon as possible after class. Fill in any steps you missed.
   b. Write clearly and legibly so you can understand what you have written later.
   c. Rewrite ideas in your own words.
   d. Highlight important ideas, examples and issues with colored pens, pencils, or highlighters.
   e. Review your class notes before the next class period.
   f. Ask questions during office hours or the next class period if there are items that are unclear.
   g. Review all of your notes at least once each week to get a perspective on the course.

The following is an example of a page of effective notes.

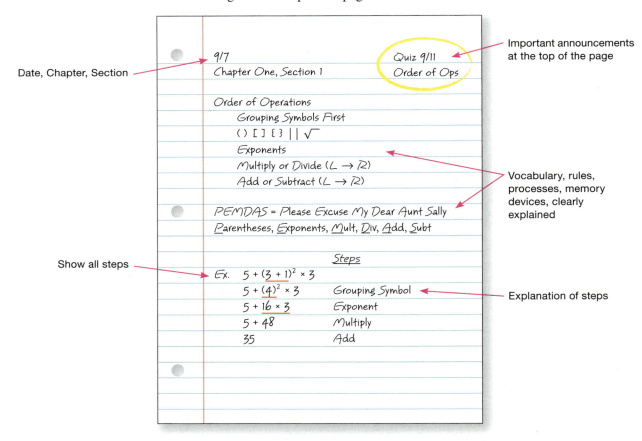

## Organizing a Notebook

You attend class, take notes, get handouts, complete homework assignments, and so on. All of this information can be overwhelming if it is not kept in an orderly fashion. The purpose of organizing your course materials is so that you can use them as tools to prepare for tests and other assignments. There are many different ways that you can organize your materials. The main thing is that you keep them all together and in a logical manner.

Some organization suggestions are presented next. Use what seems useful to your situation and adapt them as necessary. The key is to have a system that works for you.

1. Keep materials in a three-ring binder with pockets.
2. At the front of the binder, have a calendar and a copy of your syllabus, a list of homework assignments for the term, and other handouts your instructor gives you on the first day of class.
3. In the front pocket, keep a to-do list of what needs to be done in class.
4. Use dividers with tabs to divide the materials into different sections. Have a section for class notes and handouts, homework, study guides and test reviews, and tests and quizzes.

   - Class notes should be ordered by date. If your instructor provides a handout with additional notes or instructions for a section, keep this with your class notes for the day. (Use a three-hole punch on handouts so they fit in the binder.)
   - Homework should be written out neatly and brought to class each day.
   - If your instructor provides a study guide or test reviews, keep these together.
   - Finally, be sure to keep all of your tests and any other graded assignments to assist you in preparing for the final exam.

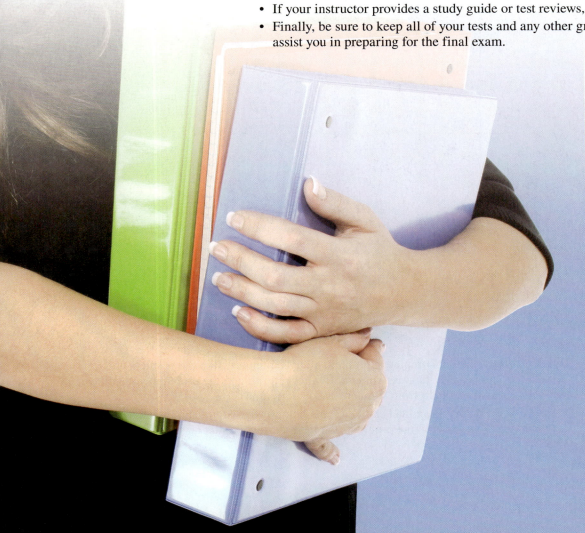

## Completing Homework

**Objective 5** ▶

**Complete homework in ways that maximize retention.**

Have you heard the saying, "Math is not a spectator sport?" What this means is that you cannot expect to watch someone else do it and master the material. A colleague used the following example with her class. One summer, she attended 37 Atlanta Braves baseball games. By the end of the summer, her ability to watch and enjoy baseball improved greatly. However, her ability to perform baseball did not improve at all, because she did not play baseball; she only observed it.

While math is not a baseball game, the same general rule applies. Attending class every day is definitely helpful (and fun!), but it is not the only thing needed to successfully master the subject. You must practice. This is where homework comes in. Homework is not just a necessary "evil." This is your opportunity to apply the new definitions, formulas, and procedures to gain deeper understanding.

Some guidelines for completing your homework are as follows.

1. Start your homework as soon after class as possible. The longer the time between class and homework, the more difficult it will be to remember everything you learned in class.

2. Complete your homework neatly and in order. It might look like this.

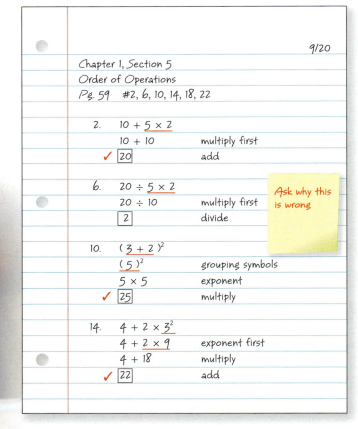

9/20

Chapter 1, Section 5
Order of Operations
Pg. 59   #2, 6, 10, 14, 18, 22

2.    $10 + 5 \times 2$
      $10 + 10$          multiply first
✓     $\boxed{20}$       add

6.    $20 \div 5 \times 2$
      $20 \div 10$       multiply first     *Ask why this is wrong*
      $\boxed{2}$        divide

10.   $(3 + 2)^2$
      $(5)^2$            grouping symbols
      $5 \times 5$       exponent
✓     $\boxed{25}$       multiply

14.   $4 + 2 \times 3^2$
      $4 + 2 \times 9$   exponent first
      $4 + 18$           multiply
✓     $\boxed{22}$       add

3. If you cannot complete a problem after 10 minutes, skip the problem and continue with your assignment. If, after returning to the problem, you still cannot work it, get help from someone.

4. If you cannot even get started on the homework assignment, review your class notes and rework the problems the instructor worked in class.

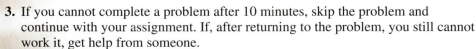

5. At the top of the first page of your homework exercises, list the problems that you have questions on. At the next class meeting or during your instructor's office hours, ask for help with these exercises.

6. After completing your homework assignment, compare your answers with those in the back of the book. Most books contain answers to the odd-numbered problems. Rework any problems that you missed.

7. After completing your homework, answer the following questions.
   a. What do I understand clearly?
   b. What do I not understand?
   c. What new definitions, rules, or formulas did I apply?
   d. What types of mistakes did I make on problems I worked incorrectly?
      i. Secretarial: miscopied, misread handwriting, misaligned
      ii. Computational: arithmetic mistake (like $8 \times 7 = 54$)
      iii. Procedural: missed a step, steps out of order, stopped too soon
      iv. Conceptual: no idea how to start, wrong method or formula used

8. Review your homework each day until class meets again so you do not forget your newly acquired skills.

**Note:** If your instructor requires you to complete online homework, print a copy of the homework and work it offline. Most systems will allow you to log back in and enter your answers. There really is no substitute for working problems by hand. As your brain directs the motor skills involved in writing the exercises out, a connection is made that impacts your ability to remember the academic concept.

## Action Plan

**Objective 6** ▶

**Establish an action plan.**

Check out additional homework strategies in the supplemental Success Strategies Manual for examples of a "Homework Cover Sheet," a "Chapter Preview" note-taking tool for use when you read your textbook, a tip sheet for "Making Note Cards," "Creating Memory Devices and Mnemonics," "Getting the Most Out of Tutoring," and several other study aids.

List one new strategy under each heading that you will implement immediately to improve your chances of success in your math course:

1. I will make it a habit to _____
   _____

2. From now on when I read my book I will _____
   _____

3. My notes would be better if I _____
   _____

4. I will make my notebook more organized by _____
   _____

5. When I do my homework I will _____
   _____

## Troubleshooting Common Errors

Think about the behaviors on the left compared with those on the right. Which habits seem more likely to lead to success in your course?

| Poor Habits | Better Habits |
| --- | --- |
| A student shows up 10 minutes late for class with a latte in hand but no pen or paper. | A student shows up 10 minutes early for class and has his supplies out and has some questions ready for the instructor. |
| A student texts friends during class and surfs the net. | A student records the lesson, noting in her notebook the counter number that goes with each problem she is working on. |
| A student emails her instructor with a message that looks like a text… "i missed ur class pls send HW" | Good Morning Dr. Jones,<br>I missed the 10:00 Algebra class this morning. I got the homework and notes from a classmate. Would it be possible for you to email me a copy of the handout that you distributed in class? |
| A student comes to class and asks, "I wasn't here for the last class; what did I miss?" | A student misses class and contacts a classmate prior to the next class meeting to get the notes, HW assignment, and reads the book to try to learn what she missed. |
| A student starts his homework and feels stuck. He closes the book and decides to ask his teacher about it next class. | A student starts his homework and feels stuck. He looks to his notebook for a similar problem; he checks the textbook examples to see if he can find a similar problem. He checks the tutoring schedule to see when he can get in for some help. |

## SECTION S.4 | Test Taking

### OBJECTIVES

As a result of completing this section, you will be able to

1. Prepare for tests and quizzes.
2. Maximize success during the test-taking process.
3. Analyze mistakes on a test.
4. Establish an action plan.
5. Troubleshoot common errors.

Tests are used to demonstrate mastery or knowledge. Tests can show that we can apply what we have learned. With adequate preparation, tests should simply be an extension of homework. Test taking is only successful if you have employed successful study strategies prior to the test. As the saying goes, "practice makes perfect."

### Before the Test

1. Complete your homework assignments regularly.
2. Review completed sections on a regular basis. Use the weekend to review the week's sections.
3. Create a practice test from the problems your teacher worked in class. Work the problems without referring to your notes or books so that you simulate a real test environment. Include problems at all levels of difficulty on the practice test. Check the answers by reviewing your notes or have a classmate check your work.
4. Know which formulas will be provided and which ones must be memorized.
5. Understand all formulas and definitions you need to know for the test. Flashcards are very helpful with this task. Use the front of an index card for the name of the formula or definition and write the formula or definition on the back of the card. Be sure you know what each symbol in a formula represents.
6. Begin preparing a week before an exam. Use your weekly planner to assist with study time. Record the date of the exam on your calendar and then schedule test preparation time beginning a week earlier.
7. Find out as much information about the test as you can:
   a. How many questions will be on the test?
   b. What types of questions (multiple choice, free response) will be included?
   c. How long will you be given to complete the test?
   d. What materials are you allowed to use?
   e. Will there be bonus or extra credit problems?
   f. What chapter/sections are covered?

#### Objective 1 ▶

Prepare for tests and quizzes.

### During the Test

#### Objective 2 ▶

Maximize success during the test-taking process.

1. Arrive at class early with all the necessary supplies and any aids you are permitted to use, and be ready to begin.
2. When you get your exam, write down all of the formulas on the top or sides of the first page. If your test is computerized, use scratch paper to write down the formulas.
3. Read the instructions carefully.
4. Review the test and complete the problems you know how to do first. This will build your confidence and bring to mind other things you have learned.
5. Keep an eye on the time. Do not spend too much time on one problem (especially if it is a low-points question).

6. Check your answers to make sure they are reasonable. For example, if you are solving for a length, a negative answer would not make sense.

7. Show all of your work neatly and clearly.

8. If you have time, review your work. Don't change any answers unless you have good reason. Often times, first instincts are correct. Double check your signs and arithmetic.

9. Use all of the allotted time to take your test. There is no prize for finishing first nor is there a penalty for being the last one to turn in the exam.

10. Mark up the test paper, if you are allowed, with information that will help you save time.

   a. Circle what you are looking for.

   b. With multiple-choice questions, if you know an answer is impossible, cross it out.

   c. If you have checked an answer, put a check mark next to the problem.

   d. If you skipped a problem but want to come back to it, put a plus sign next to it.

   e. If you see a problem that you don't recognize at all, put a minus sign next to it and come back to it last.

   f. If you have eliminated some options in a multiple-choice question, put down how many options are left. When running out of time, go back to the ones with the fewest options first.

Here is what your test might look like after the first time through.

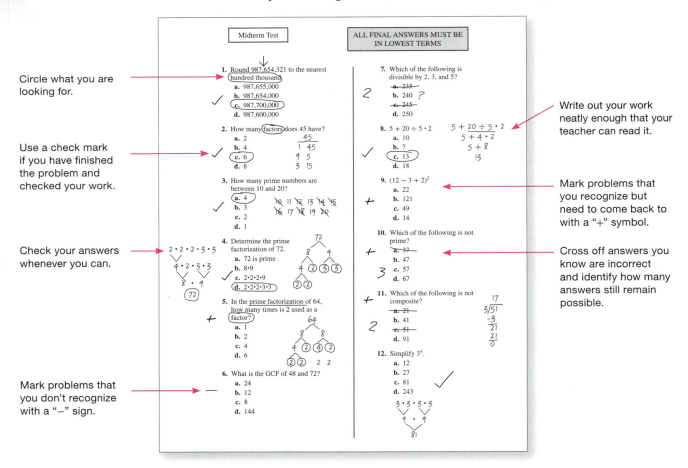

## After the Test

**Objective 3** ▶

Analyze mistakes on a test.

1. Look through the test for the problems you missed. Rework these problems until you get them correct.
2. Keep your test corrections with the exam (if it is returned to you) for future studying.
3. Make an appointment to meet with your instructor to review any questions you cannot figure out.
4. Most likely, the material on your test will be covered on your final, so it is important that you keep the exam if your instructor allows you to do so. File it in your notebook so you can review it later in the course.
5. Here is a strategy for test corrections:
   a. Divide a piece of paper in thirds.
   b. On the left side, write out the original problem with your original work.
   c. In the middle, write out the correct answer to the problem.
   d. On the right side, work a similar problem to ensure that you can do this type of problem correctly.

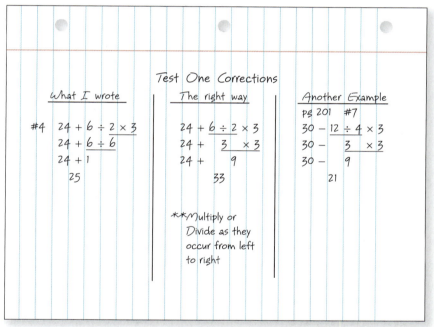

## Action Plan

**Objective 4** ▶

Establish an action plan.

Check out additional homework strategies in the supplemental Success Strategies Manual for a "Test Preparation Worksheet," a "Post-Test Debriefing" form, suggestions for "How to Reduce Test Anxiety," and tips for "Problem-Solving Strategies."

1. Two things that I will commit to doing before a test to improve my chances of success:

   _____

   _____

2. Two strategies that I will employ during a test to keep my anxiety down:

   _____

   _____

3. After a test I will do the following to learn from my mistakes:

   _____

   _____

## Troubleshooting Common Errors

After you check your homework or take a quiz or test, you should review it to determine the types of mistakes you made and think about how you can fix them. It is important to understand whether the errors you make are conceptual, procedural, computational, or secretarial.

A **conceptual error** shows little or no understanding of the underlying concept or procedure.

In a **procedural error**, the correct process is used but there is a mistake made with the steps.

In a **computational error,** an arithmetic mistake is made.

In a **secretarial error**, a miscopy of some kind is made.

| Evaluate | |
|---|---|
| $24 + 6 \div 2 \times 3$ | |
| $\underline{24 + 6} \div 2 \times 3$ | Add |
| $\underline{30 \div 2} \times 3$ | Divide |
| $15 \times 3$ | Multiply |
| 45 | |

The student shows no understanding of the concept of "order of operations." He just works straight through from left to right. This is an example of a conceptual error.

| Evaluate | |
|---|---|
| $24 + 6 \div 2 \times 3$ | *PEMDAS* |
| $24 + 6 \div \underline{2 \times 3}$ | Multiply |
| $24 + \underline{6 \div 6}$ | Divide |
| $24 + 1$ | Add |
| 25 | |

The student clearly identifies that the order of operations is being used as she uses the acronym PEMDAS. She knows the right process. However, she thinks that she must multiply then divide, rather than the correct rule: Multiplication and division are done as they occur, from left to right. This is an example of a procedural error.

| Evaluate | |
|---|---|
| $24 + 6 \div 2 \times 3$ | *PEMDAS* |
| $24 + \underline{6 \div 2} \times 3$ | Divide |
| $24 + \underline{3 \times 3}$ | Multiply |
| $24 + 9$ | Add |
| 32 | |

The student clearly identifies that "order of operations" is being used as he uses the acronym PEMDAS. He does all of the right steps, but then adds incorrectly. This is an example of a computational error.

A fourth type of mistake is *simply secretarial*. This type of mistake includes miscopying any part of the problem, misreading your handwriting, misaligning the problem, or any other non-math-related issue with how you write out your work.

## How do I correct these errors?

A conceptual error needs to be corrected before the test begins. This mistake is typically about lack of preparation for the test.

> **Note:** *Remember P⁵*
>
> *Proper Preparation Prevents Poor Performance*

▶ If you have not been doing any homework, do some. If you have been doing some, do more. If you have been doing all of the homework and are still struggling with the concepts, you may need some tutoring.

▶ Sometimes the issue is problem recognition. Perhaps you can do all of the problems when they are organized into separate sections in the book, but not when they are jumbled together on a test.

  • A great way to mimic this situation is to mix up your notecards.

  • Don't study everything in the same order each time.

  • Mix problems from different sections and chapters together.

  • Work any cumulative reviews in the text.

  • Review old homework each time you complete a new homework assignment to compare how the new assignment is the same or different from the old topics.

A procedural mistake usually requires only a minor fix. You understand the concept; you know what the steps are, but have confused them in some way. This mistake is also about preparation; the steps should be natural by the time you get to a test. If you have repeated a process enough times, you will not even think about the steps any more, you will just know what to do. Whether it is playing music, throwing a baseball, making a pie, or doing a math problem, practice makes perfect!

Computational and secretarial mistakes both have similar fixes. Check your work and slow down. These mistakes are typically made on problems that you are very confident with and are working through quickly. You do not want quickly to mean *carelessly*!

▶ For many problem types, you can check your answers. When this is possible you should always do so. Check the solution of a problem with the original problem. In this way even a miscopy can be found.

▶ Estimate your answer before you begin your work. If your answer is not near the estimate, investigate why.

▶ Think about the reasonableness of your answer. If you made a secretarial error with a decimal point and have an answer where a car is driving 500 mph, it should be clear to you that this type of answer is not reasonable.

▶ If you are permitted to use a calculator, check hand calculations with the calculator. Also, check the calculator output with your reasonableness criteria. If you press the wrong buttons, a calculator can still give you a wrong answer.

▶ Use grid paper instead of lined paper to help with misalignment issues. You can also turn lined paper horizontally to create columns.

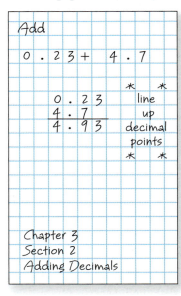

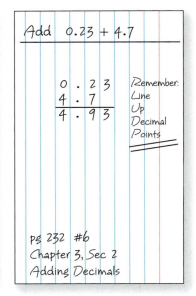

 **Note:** *If you can't read it, you can't review or study it. If your teacher can't read it, you may get it marked wrong.*

**Be Neat!**

---

**Procedure:  Determining the Type of Mistake on a Test or Quiz**

**Step 1:** Conceptual: Left it blank, wrong process, mis-memorized rule
**Step 2:** Procedural: Performed steps out of order, missed a step, or couldn't finish all steps
**Step 3:** Computational: Added, subtracted, multiplied, or divided wrong
**Step 4:** Secretarial: Miscopy, omission, misalignment, misread handwriting

---

After a test, review each problem that you got wrong and evaluate what went wrong. What types of mistakes did you make? Is there a trend in which type of mistakes you make? What will you do to eliminate these mistakes in the future? Also, reflect about whether the test is helping you meet your grade goal or harming you in making your grade goal. After each test, you should evaluate your progress toward meeting your grade goal.

## SECTION S.5 / Blended and Online Classes

### OBJECTIVES

**As a result of completing this section, you will be able to:**

1. Differentiate between blended, hybrid, and online classes.
2. Use techniques to maximize your success in blended and online settings.
3. Establish an action plan.
4. Troubleshoot common errors.

The desire for flexibility and convenience in scheduling classes has led to a continued increase in blended and online classes.

### Objective 1 ▶

Differentiate between blended, hybrid, and online classes.

## Identifying Different Delivery Methods

**Blended** classes, or **hybrid** classes, incorporate traditional classroom experiences with online learning activities. Students in these types of classes are required to attend a specific number of on-campus class meetings. The remaining portion of the course is delivered online, thereby reducing the amount of time a student spends in class. In summary, blended classes strive to combine the best elements of traditional face-to-face instruction with the best aspects of learning at a distance.

**Online** classes, unlike blended classes, deliver entire courses at a distance. Some may have required meetings for exams, but typically all aspects of the course are provided online.

With a portion of a course or an entire course delivered at a distance, the roles of the teacher and student change. The teacher becomes more of a facilitator and the student becomes more like a teacher in these environments. While in a traditional classroom, the student's learning is his or her responsibility, this responsibility becomes even more important in blended and online courses. The information provided in the previous sections certainly applies to students in any type of class, but there are some additional tips for students in blended and online classes.

## Tips for Blended and Online Classes

### Objective 2 ▶

Use techniques to maximize your success in blended and online settings.

1. The first day of class is extremely important. If you are in a blended class, be sure to attend the first on-campus meeting. If you are in an online class, be sure to log into the course the first day the course is available.
2. Review the syllabus and course policies to determine what is expected of you.
   a. How many required on-campus meetings are there?
   b. Are there proctored exams?
   c. Upon what activities is your grade based (e.g., online homework, discussion board postings, online quizzes and/or tests, group projects, and so on)?
3. Familiarize yourself with the course resources.
   a. Does your instructor post notes or videos, solutions for quizzes/tests, or other material?
   b. Does your instructor hold online office hours through chat, email, or a live classroom? If so, when?
   c. What learning materials (videos, practice problems, ebook, and the like) are available with your textbook?

4. Make a calendar for the semester (one may be available by your instructor) and write down due dates for homework, quizzes, tests, discussion posts, and other assignments.

5. Determine a time each week to have class with yourself.
   a. If the face-to-face version of the class meets, for example, 2 days a week for 1 hr and 45 min, then you should set aside this same amount of time to "have class" on your own.
   b. Class time is time for you to watch videos, to read the textbook, to do what is necessary to learn the material, and to complete quizzes, tests, and discussions. As you watch videos or read the textbook, take notes as you would if you were attending a traditional class.

6. Determine a time each week to have study time. In addition to "class," you need study time.
   a. For each credit or contact hour for the course, you should have three hours of class plus study time (3 credit class = 9 hr, 3 hr for "class" and 6 hr of study time).
   b. Study time is time for you to complete homework. So that you do not get dependent on the help resources online, we suggest that you print the homework assignment, work it offline, and then log back in to enter your answers. For the problems that you answered incorrectly, use the help features to assist you.

7. Remember that you are not alone in the online classroom. Interact with your instructor and other classmates through discussion boards, email, and other features that are available in your class.

8. Ask your instructor for help when you do not understand the material or the technology.

9. Maintain your notes, homework, printed copies of assignments, and any other printables in your notebook.

10. Check in frequently, daily if possible, for announcements, changes to the calendar or assignments, and for any updates you need to know.

## Action Plan

**Objective 3** ▶

**Establish an action plan.**

Set some goals for how you will be successful in a class with a nontraditional delivery method.

1. I will "have class with myself" _____ times a week for _____ minutes each time.

2. If I find I can't do it alone, I will get help by _____
   _____.

3. Five things that I am committed to do to ensure that I maximize my success are
   _____.
   _____
   _____
   _____
   _____

## Troubleshooting Common Errors

**Objective 4** ▶

Troubleshoot common errors.

Blended and online classes present some different challenges than face-to-face classes. Here are some suggestions to overcome those challenges from students who have taken blended and/or online classes.

| Study Skills | Suggestions |
| --- | --- |
| Homework | "Plan ahead and don't procrastinate in completing your homework. And try to do the take home assignments without looking at notes . . . it will help a lot at test time!"<br><br>"Follow the study guide, do all the online homework and take-home work. Try the practice tests to help with confidence if you get nervous taking tests." |
| Textbook | "Read the book first, then do book questions, then do all online homework until a score of 100 percent is reached. Review all materials again before an exam. You have to be disciplined in getting into a routine just like a classroom based course." |
| Communication | "I would recommend keeping in constant contact with the instructor. Monitor your progress and make sure you are constantly evaluating your study habits. Reviewing feedback is another great way to improve your understanding." |
| Time Management | "You will need a lot of time to succeed in this course. Just because it's an online class doesn't mean you can sign on for 5 minutes do some work and fly by. Time is essential to succeeding in this course."<br><br>"This course requires a good bit of time, and dedication. Had I not spend much time on the course at the beginning, I would have not been able to perform well for the remainder of the course."<br><br>"It is very flexible but students have to be prepared to discipline themselves enough to do the work on time and not wait until the last minute. Don't try to do all the problems at one time, it is very overwhelming. It's not always possible but try to treat it like a normal class and set aside an hour a day or every other day, to work on it. Sometimes you have a busy week and that's what's nice about having an online class, the flexibility to work on the material when you have time."<br><br>"In order to succeed you have to put the time into this class as if it were on campus. I feel that if individuals apply themselves to this course they will succeed!!!!" |

# Strategies to Succeed in Math

 **What's the big idea?** Chapter S provides many strategies for maximizing your success in math class. No one will implement all of these ideas, but using even some will improve your chances of success. The more you implement, the better your chance of success.

## The Tools

Additional student resources can be found in the supplemental Success Strategies Manual which is available in a workbook or also through your online homework system.

*You can learn math!*

## Summary of Success Strategies

▶ Set goals.
▶ Manage your time.
▶ Adopt habits of successful math students.
▶ Read your book with purpose.
▶ Take meaningful notes.
▶ Keep an organized notebook/portfolio.
▶ Do homework to ensure deep understanding.
▶ Adopt good test-taking skills before, during, and after an exam or quiz.
▶ If you are in a blended or online class, adapt your strategies.

## What's Next?

Throughout this text we will continue to support your efforts to succeed. With each chapter we will revisit one of the characteristics of success discussed in this opening chapter. Some chapter openers will focus on skills and others attitudes. When you need a little extra motivation, scan through this chapter or through the chapter openers to find suggestions.

The topics for each chapter opener are as follows.

**Chapter 1** Time Management

**Chapter 2** Organization

**Chapter 3** Commitment and Perseverance

**Chapter 4** Study Skills

**Chapter 5** Goal Setting

**Chapter 6** Motivation

**Chapter 7** Learning Strategies

**Chapter 8** Refocus

**Chapter 9** Reflection

# Real Numbers and Algebraic Expressions

## Time Management

As you begin this chapter, we want you to focus on how you are managing your time. Your success in this course depends on your ability to devote the appropriate time to learning and practicing the material.

- The general rule of thumb is that you should spend 2 hours studying for each hour you are in class.

- Prepare a calendar of your week and schedule an appointment with yourself for study time.

- Review how you are using your time through the week and make an honest assessment of whether you have the time to make this class a priority.

- See the Preface for additional resources concerning time management.

**Question For Thought:** Of the activities you participate in, which are essential? Can you give up any of these activities to provide more time for your studies?

## Chapter Outline

## Coming Up...

In Section 1.3, we will learn that the number of viewers (in millions) for the season premiere of Fox Network's *American Idol* for Seasons 1 to 8 can be approximated by the expression $0.1x^3 + 18.4x - 2.54x^2 - 4.6$, where $x$ is the number of seasons aired. We will learn how to use this expression to estimate the number of viewers for future seasons.

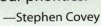
"The key is not to prioritize what's on your schedule, but to schedule your priorities."

—Stephen Covey

## SECTION 1.1 | The Set of Real Numbers

**Chapter 1** lays the foundation of algebra. In this chapter, we will explore the set of real numbers, operations with real numbers, and numerical and algebraic expressions. The skills learned in this chapter will be applied in every chapter of this text. These are the skills that will enable us to solve equations, simplify expressions, and solve application problems.

**Objective 1** ▶

Classify a number as a natural number, whole number, integer, rational number, or irrational number.

How would you classify the following numbers?

12 hr: the number of semester hours to be a full-time student

$-70°F$: the coldest recorded temperature in the continental United States (1971)

$7\frac{1}{2}$ ft: the height of one of the tallest NBA players, Yao Ming

$\pi$: the distance around a circle divided by its diameter

In this section, we will learn the different types of real numbers and will classify numbers such as the ones shown.

### The Set of Real Numbers

Numbers are used to represent so many things in our lives—the amount of money in a bank account, hours in a day, temperature, weight, height, number of credit hours, GPA, salaries, measurements, and so on. In addition, computers store all of their data as numbers. The world as we know it would basically cease to exist without numbers. So, it seems logical to begin with a review of the different types of numbers.

We first need to define a few terms related to sets. A **set** is a collection of objects. Each object in a set is called a **member** or an **element**. A set is written in braces, { }, and is usually denoted with a capital letter. Sets can be either **finite** or **infinite**. A finite set has a specific number of elements. An example of a finite set is $A = \{1, 2, 3, 4, 5\}$. An infinite set has infinitely many elements. An example of an infinite set is $B = \{1, 3, 5, 7, 9, \ldots\}$. The three dots at the end are called an **ellipsis** and indicate that the set continues indefinitely.

In sets $A$ and $B$, the elements of the sets are listed explicitly. This method of listing each element of the set is called the **roster method**. Another method, the **set-builder notation**, states the conditions the elements must satisfy to be included in the set. Set-builder notation is written in the form:

$$\{x \mid \text{condition } x \text{ must satisfy}\}$$

This is read as "the set of $x$ such that _____." In the blank, we insert the condition that $x$ must satisfy. The letter $x$ is a **variable** and represents some unknown number.

An example of set-builder notation is $C = \{x \mid x \text{ is a positive odd number}\}$. The given condition tells us that $x$ can be 1, 3, 5, 7, 9, . . . . Note that set $C$ is the same as set $B$.

To denote that a number is or is not a member of a set, we use the notation shown in the following table.

| Symbol | Meaning | Example | Verbal Statement |
|--------|---------|---------|------------------|
| $\in$ | Element of | $4 \in A$ | "4 is an element of $A$." or "4 is a member of $A$." |
| $\notin$ | Is not an element of | $6 \notin A$ | "6 is not an element of $A$." or "6 is not a member of $A$." |

Numbers belong to certain sets. The next table defines these sets.

| Natural numbers (or counting numbers) | $\mathbb{N} = \{1, 2, 3, 4, 5, \dots\}$ |
|---|---|
| Whole numbers | $\mathbb{W} = \{0, 1, 2, 3, 4, 5, \dots\}$ |
| Integers | $\mathbb{Z} = \{\dots, -5, -4, -3, -2, -1, 0, 1, 2, 3, 4, 5, \dots\}$ |
| Rational numbers | $\mathbb{Q} = \left\{ \dfrac{p}{q} \middle\vert p, q \in \mathbb{Z} \text{ with } q \neq 0 \right\}$ |
| Irrational numbers | $\mathbb{I} = \{\text{numbers that are not rational}\}$ |
| Real numbers | $\mathbb{R} = \{\text{rational or irrational numbers}\}$ |

The set of **rational numbers** is given in set-builder notation. It is read as "the set of numbers $p$ over $q$, such that $p$ and $q$ are **integers** with $q$ not equal to 0." That is, the values of $p$ and $q$ can be replaced with any integer except that $q$ cannot be replaced with zero. So, a rational number is the quotient, or ratio, of two integers whose denominator is not zero. It is impossible to list all of the rational numbers, since there are infinitely many of them. Some examples of rational numbers are

$$\frac{10}{1} = 10, \quad \frac{37}{7} = 5\frac{2}{7}, \quad -\frac{3}{5} = -0.6, \quad \frac{1}{3} = 0.3333\dots$$

Rational numbers include **integers, mixed numbers, decimals that terminate,** and **repeating decimals**. Note all integers are rational numbers since each integer can be written as its value divided by 1, as illustrated by $10 = \dfrac{10}{1}$.

The set of **irrational numbers** arose out of the study of geometry, specifically triangles and circles. A simple definition of an irrational number is a number that is not rational. This means that the number cannot be written as a quotient of two integers.

**Note:** *Irrational numbers are numbers whose values are nonterminating and nonrepeating decimals.*

The number $\pi$ (pi) may be the most commonly known irrational number. We often use the value 3.14 for $\pi$ but this is only a decimal approximation. The exact value of $\pi$ is $3.1415926535\dots$. This value continues indefinitely with no repeating pattern. Another irrational number used in advanced math classes is the number $e$, Euler's number. It is used in many important formulas, such as those involving population growth. Its value is approximately 2.72.

Other examples of irrational numbers involve square roots. The **square root** of a number is the number that must be multiplied by itself to get the original number. For instance, $\sqrt{9} = 3$ since $3 \cdot 3 = 9$. The square root of 9 is a rational number since its value is equivalent to 3, which can be expressed as a quotient of two integers, $\dfrac{3}{1}$. However, not all square roots are rational numbers.

We can use the calculator to approximate square roots to determine if the number is rational or irrational. If the value on the calculator is a decimal that repeats or terminates, then the number is rational. If the value on the calculator is a decimal that is nonterminating and nonrepeating, then the number is irrational. Some irrational numbers are

$$\sqrt{3} = 1.7320508075\ldots$$
$$\sqrt{8} = 2.8284271247\ldots$$

A real number is either rational or irrational. It cannot be both. The next diagram illustrates how the sets of numbers relate to one another.

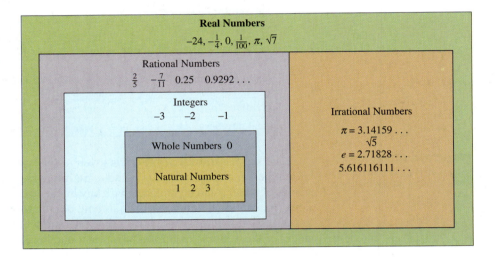

This diagram shows us the following facts.

- All natural numbers are whole numbers, integers, and rational numbers.
- All whole numbers are integers and rational numbers.
- All integers are rational numbers.
- Natural numbers, whole numbers, integers, and rational numbers are *not* irrational numbers.
- Natural numbers, whole numbers, integers, rational numbers, and irrational numbers are all real numbers.

In Chapter 9, another set of numbers, *complex numbers*, is introduced. This set contains all of the real numbers.

---

**Procedure: Classifying a Real Number**

**Step 1:** If the number is 1, 2, 3, . . . , then the number is a natural number.

**Step 2:** If the number is 0, 1, 2, 3, . . . , then the number is a whole number.

**Step 3:** If the number is . . . , $-3, -2, -1, 0, 1, 2, 3, \ldots$ , then the number is an integer.

**Step 4:** Determine if the number is rational or irrational.
   **a.** If the number is equivalent to a decimal that terminates or repeats, then the number is rational.
   **b.** If the number is equivalent to a decimal that does not terminate or repeat, then the number is irrational.

**Step 5:** If the number is rational or irrational, it is a real number.

**Objective 1 Examples**

Classify each number as a natural number, whole number, integer, rational number, irrational number, and/or real number. If the number is a rational number, write it in fractional form. If the number is irrational, approximate its value to two decimal places.

**1a.** 0.75　　**1b.** $-3\frac{1}{2}$　　**1c.** $\sqrt{11}$　　**1d.** 0

**1e.** $\sqrt{25}$　　**1f.** $\frac{11}{16}$　　**1g.** $-7$

**Solutions**

| | Number | Natural | Whole | Integer | Rational | Irrational | Real |
|---|---|---|---|---|---|---|---|
| **1a.** | $0.75 = \frac{75}{100} = \frac{3}{4}$ | | | | X | | X |
| **1b.** | $-3\frac{1}{2} = -\frac{7}{2}$ | | | | X | | X |
| **1c.** | $\sqrt{11} = 3.316624\ldots$ $\approx 3.32$ | | | | | X | X |
| **1d.** | $0 = \frac{0}{1}$ | | X | X | X | | X |
| **1e.** | $\sqrt{25} = 5 = \frac{5}{1}$ | X | X | X | X | | X |
| **1f.** | $\frac{11}{16}$ | | | | X | | X |
| **1g.** | $-7 = -\frac{7}{1}$ | | | X | X | | X |

**✓ Student Check 1**

Classify each number as a natural number, whole number, integer, rational number, irrational number, and/or real number. If the number is a rational number, write it in fractional form. If the number is irrational, approximate its value to two decimal places.

**a.** 0.4　　**b.** $-6\frac{4}{7}$　　**c.** $\sqrt{36}$　　**d.** 10

**e.** $\sqrt{21}$　　**f.** $\frac{5}{6}$　　**g.** $-5$　　**h.** 1.232232223 . . .

**Objective 2** ▶

Graph real numbers on a real number line.

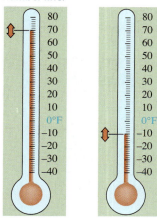

**Figure 1.1**　　**Figure 1.2**

## The Real Number Line

Graphing or plotting real numbers on a number line is a skill that will be used in later sections when we solve inequalities and graph equations. The act of reading a thermometer can assist us in plotting real numbers. Figure 1.1 illustrates a thermometer reading of 70°F and Figure 1.2 illustrates a thermometer reading of −10°F.

If we rotate the thermometer, we get a number line. If we put a dot at 70°F, we have graphed the real number 70. If we put a dot at −10°F, we have graphed the real number −10, as shown.

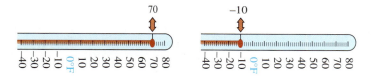

The **real number line** is similar to this rotated thermometer. It is a horizontal line drawn with arrows on both ends to indicate that the real numbers are infinite. Tick marks are used to divide the number line into equal segments. Positive numbers are located to the right of 0 and negative numbers are located to the left of 0. 0 is neither positive nor negative.

**Procedure: Graphing or Plotting Real Numbers on a Number Line**

**Step 1:** If the number is not an integer, approximate its value to two decimal places.
**Step 2:** Place a dot on its position on the number line. Make the dot easy to distinguish from the number line itself.

**Objective 2 Examples**    Graph each number on a real number line.

**2a.** 5     **2b.** $-3$     **2c.** $\dfrac{2}{3}$     **2d.** $-4\dfrac{1}{2}$     **2e.** 6.1     **2f.** $\sqrt{2}$

**Solutions**    **2a.** 5                          Graph at the appropriate tick mark.
**2b.** $-3$                       Graph at the appropriate tick mark.
**2c.** $\dfrac{2}{3} \approx 0.67$          Graph this value between 0 and 1 but closer to 1.
**2d.** $-4\dfrac{1}{2} = -4.50$       Graph this value exactly halfway between $-4$ and $-5$.
**2e.** 6.1                        Graph this value between 6 and 7 but closer to 6.
**2f.** $\sqrt{2} \approx 1.41$         Graph this value close to halfway between 1 and 2.

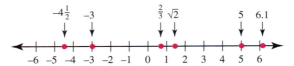

**✓ Student Check 2**    Graph each number on a real number line.

**a.** $-3$     **b.** $-1$     **c.** $-\dfrac{5}{2}$     **d.** $1\dfrac{1}{4}$     **e.** $-5.2$     **f.** $\sqrt{24}$

## Ordering Real Numbers

**Objective 3** ▶

Compare the value of two real numbers.

We can express the relationship between two real numbers $a$ and $b$ using symbols of equality and inequality. The equality symbol is the equals sign, $=$. The inequality symbols are $<$, $>$, $\leq$, $\geq$, or $\neq$. The equality and inequality symbols can be used to write mathematical statements. The statement is either true or false. Some examples of true statements are shown.

**INSTRUCTOR NOTE:**
Point out that in a true statement, the inequality symbol points to the smaller number.

| Verbal Statement | Mathematical Statement | Example | Read as |
|---|---|---|---|
| $a$ is equal to $b$ | $a = b$ | $3 = 3$ | "3 is equal to 3" |
| $a$ is less than $b$ | $a < b$ | $3 < 4$ | "3 is less than 4" |
| $a$ is greater than $b$ | $a > b$ | $3 > 2$ | "3 is greater than 2" |
| $a$ is less than or equal to $b$ | $a \leq b$ | $3 \leq 4$ | "3 is less than or equal to 4" |
| $a$ is greater than or equal to $b$ | $a \geq b$ | $3 \geq 2$ | "3 is greater than or equal to 2" |
| $a$ is not equal to $b$ | $a \neq b$ | $1 \neq 3$ | "1 is not equal to 3" |

**INSTRUCTOR NOTE:**
The arrows at the ends of the number line can serve as a memory aid. As we move to the right, the numbers are larger. As we move to the left, the numbers are smaller.

To determine how the numbers $-10$ and $-20$, for example, relate to one another, we will use a thermometer. As we move up a thermometer, the temperatures get larger. As we move down a thermometer, the temperatures get smaller. Since $-20$ lies below $-10$ on the thermometer, $-20$ is less than $-10$. That is, $-20 < -10$.

When the thermometer, or number line, is rotated horizontally, we see that $-20$ is to the left of $-10$.

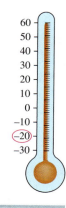

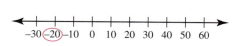

---

**Property: Order of Real Numbers**

If a number $a$ lies to the **left** of a number $b$ on a real number line, then $a < b$ or $a \le b$.

If a number $a$ lies to the **right** of a number $b$ on a real number line, then $a > b$ or $a \ge b$.

If a number $a$ lies in the **same** location as a number $b$ on a real number line, then $a = b$.

---

**Objective 3 Examples**    **Use the number line to compare the numbers. Use a <, >, or = symbol to make a true statement.**

| Problems | Solutions |
|---|---|
| **3a.** $-3$ ___ $-4$ | (number line from −5 to 1) <br> $-3$ lies to the **right** of $-4 \Rightarrow -3 > -4$. |
| **3b.** $-2$ ___ $0$ | (number line from −5 to 1) <br> $-2$ lies to the **left** of $0 \Rightarrow -2 < 0$. |
| **3c.** $\pi$ ___ $\sqrt{5}$ | Note that $\pi \approx 3.14$ and $\sqrt{5} \approx 2.24$. <br> (number line from 0 to 6 with $\sqrt{5}$ and $\pi$ marked) <br> $\pi$ lies to the **right** of $\sqrt{5} \Rightarrow \pi > \sqrt{5}$. |
| **3d.** $3.5$ ___ $\dfrac{7}{2}$ | Note that $\dfrac{7}{2} = 3.5$. <br> The two numbers are **equal**, so $3.5 = \dfrac{7}{2}$. |

✓ **Student Check 3**    Use the number line to compare the numbers. Use a <, >, or = symbol to make a true statement.

    **a.** $-5$ ___ $-4$     **b.** $\sqrt{2}$ ___ $\dfrac{1}{2}$     **c.** $1.25$ ___ $\dfrac{5}{4}$

---

**Note:** *The inequality statements in Example 3 can also be written by reversing the inequality symbol. In either case, the arrow points to the smaller number.*

*The statement $-3 > -4$ can also be written as $-4 < -3$.*

*The statement $-2 < 0$ can also be written as $0 > -2$.*

*The statement $\pi > \sqrt{5}$ can also be written as $\sqrt{5} < \pi$.*

## Opposites of Real Numbers

**Objective 4** ▶

Find the opposite of a real number.

A number line not only provides us a way to order real numbers, it also helps us visualize the distance between numbers. Two numbers that lie equal distances from zero and lie on opposite sides of zero are *opposites* or *additive inverses* of one another.

> **Definition: Opposite of a Real Number**
>
> The **opposite** or **additive inverse** of a real number $a$ is the number that has the same distance from 0 but lies on the opposite side of 0. The opposite of a real number $a$ is denoted as $-a$.

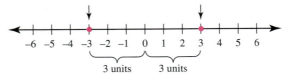

This number line illustrates that $-3$ and $3$ are opposites of one another since they are the same distance from zero but on opposite sides of zero. We state these facts as shown in the table.

| Verbal Statement | Mathematical Statement |
|---|---|
| The *opposite of* negative three is three. | $-(-3) = 3$ |
| The *opposite of* three is negative three. | $-(3) = -3$ |

**Facts:**
- The opposite of a positive real number is a negative number.
- The opposite of a negative real number is a positive number.

**Objective 4  Examples**   **Find the opposite of each number. Write a mathematical statement to represent the answer.**

| | Solution | |
|---|---|---|
| **Problem** | **Opposite** | **Mathematical Statement** |
| **4a.** $-2$ | The *opposite* of $-2$ is $2$. | $-(-2) = 2$ |
| **4b.** $16$ | The *opposite* of $16$ is $-16$. | $-(16) = -16$ |
| **4c.** $\dfrac{3}{4}$ | The *opposite* of $\dfrac{3}{4}$ is $-\dfrac{3}{4}$. | $-\left(\dfrac{3}{4}\right) = -\dfrac{3}{4}$ |
| **4d.** $-\dfrac{9}{5}$ | The *opposite* of $-\dfrac{9}{5}$ is $\dfrac{9}{5}$. | $-\left(-\dfrac{9}{5}\right) = \dfrac{9}{5}$ |
| **4e.** $\pi$ | The *opposite* of $\pi$ is $-\pi$. | $-(\pi) = -\pi$ |
| **4f.** $-0.23$ | The *opposite* of $-0.23$ is $0.23$. | $-(-0.23) = 0.23$ |

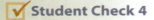 **Student Check 4**   Find the opposite of each number. Write a mathematical statement to represent the answer.

**a.** $-7$   **b.** $4$   **c.** $\dfrac{1}{2}$   **d.** $-\dfrac{25}{6}$   **e.** $\sqrt{6}$   **f.** $8.2$

**Objective 5** ▶

Find the absolute value of a real number.

## Absolute Value

We just defined numbers that are opposites as numbers that have the same *distance* from zero on a number line. A number's distance from zero is called its *absolute value*.

> **Definition:** The **absolute value** of a real number $a$ is the distance between 0 and $a$ on the real number line. The absolute value of $a$ is denoted $|a|$.

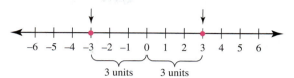

3 units    3 units

From the graph, we see that the distance between 3 and 0 is 3 units and the distance between $-3$ and 0 is 3 units. We can state these facts as shown in the table.

| Verbal Statement | Mathematical Statement |
|---|---|
| The *absolute value* of three is three. | $\lvert 3 \rvert = 3$ |
| The *absolute value* of negative three is three. | $\lvert -3 \rvert = 3$ |

**INSTRUCTOR NOTE:**
Use the example of distance from home to school to help students understand this concept. If the distance from home to school is 10 miles, what is the distance from school back home?

Because absolute value refers to distance, the absolute value of a number is always greater than or equal to zero. The following property summarizes this fact.

> **Property:  Absolute Value**
> If $a \geq 0$, then $|a| = a$.
> If $a < 0$, then $|a| = -a$.

This property states that if a number is greater than or equal to zero, its absolute value is equal to that number. For instance, $5 \geq 0$ and 5 is 5 units from zero, so $|5| = 5$. If a number is less than zero, its absolute value is equal to the opposite of the number. For instance, $-2 < 0$ and $-2$ is 2 units from zero, so $|-2| = 2$. Note that $2 = -(-2)$.

> **Procedure:  Finding the Absolute Value of a Number**
> **Step 1:** Find the number's distance from zero.
> **Step 2:** The absolute value of a number is always greater than or equal to 0.

**Objective 5  Examples**    Simplify each absolute value expression.

| Problem | Solution |
|---|---|
| **5a.** $\lvert 6 \rvert$ | $\lvert 6 \rvert = 6$, since 6 is 6 units from 0. |
| **5b.** $\lvert -8 \rvert$ | $\lvert -8 \rvert = 8$, since $-8$ is 8 units from 0. |
| **5c.** $\lvert 0 \rvert$ | $\lvert 0 \rvert = 0$, since 0 is 0 units from 0. |
| **5d.** $\left\lvert -\dfrac{1}{2} \right\rvert$ | $\left\lvert -\dfrac{1}{2} \right\rvert = \dfrac{1}{2}$, since $-\dfrac{1}{2}$ is $\dfrac{1}{2}$ unit from 0. |
| **5e.** $-\lvert 2 \rvert$ | First, find the absolute value of 2, which is 2. Then find the opposite of 2, which is $-2$. $$-\lvert 2 \rvert = -(2) = -2$$ |
| **5f.** $-\lvert -3 \rvert$ | First, find the absolute value of $-3$, which is 3. Then find the opposite of 3, which is $-3$. $$-\lvert -3 \rvert = -(3) = -3$$ |

✓ **Student Check 5**    Simplify each absolute value expression.

 **a.** $\lvert 10 \rvert$    **b.** $\lvert -6 \rvert$    **c.** $\lvert \pi \rvert$    **d.** $\left\lvert -\dfrac{2}{5} \right\rvert$    **e.** $-\lvert 4 \rvert$    **f.** $-\lvert -7 \rvert$

**Objective 6** ▶

Troubleshoot common errors.

## Troubleshooting Common Errors

Some common errors associated with the objectives in this section are shown next.

**Objective 6  Examples**   **A problem and an incorrect solution are given. Provide the correct solution and an explanation of the error.**

**6a.** Classify the number $\sqrt{49}$.

| Incorrect Solution | Correct Solution and Explanation |
|---|---|
| 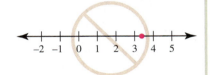 $\sqrt{49}$ is an irrational number since it is a square root. | $\sqrt{49} = 7 = \dfrac{7}{1}$, so this number is a rational number, an integer, a whole number, and a natural number. The square root of a number is irrational only if the number has a decimal value that is nonrepeating and nonterminating. |

**6b.** Plot the number $\dfrac{3}{4}$ on a real number line.

| Incorrect Solution | Correct Solution and Explanation |
|---|---|
|  | $\dfrac{3}{4} = 0.75$ So, the plot of the point is between 0 and 1. |

**6c.** Simplify $-\lvert -10 \rvert$.

| Incorrect Solution | Correct Solution and Explanation |
|---|---|
| 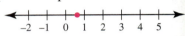 $-\lvert -10 \rvert = 10$ | This problem means to find the opposite of the absolute value of $-10$. So, we first find the absolute value of $-10$ and then take its opposite. $$-\lvert -10 \rvert = -(10) = -10$$ |

## ANSWERS TO STUDENT CHECKS

**Student Check 1**   **a.** $0.4 = \dfrac{4}{10}$ rational, real

**b.** $-6\dfrac{4}{7} = -\dfrac{46}{7}$ rational, real

**c.** $\sqrt{36} = 6 = \dfrac{6}{1}$ rational, integer, whole, natural, real

**d.** $10 = \dfrac{10}{1}$ rational, integer, whole, natural, real

**e.** $\sqrt{21} \approx 4.58$ irrational, real

**f.** $\dfrac{5}{6}$ rational, real

**g.** $-5 = \dfrac{-5}{1}$ integer, rational, real

**h.** $1.232232223\ldots$ irrational, real

**Student Check 2**

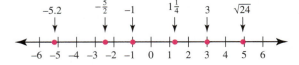

**Student Check 3**   **a.** $-5 < -4$   **b.** $\sqrt{2} > \dfrac{1}{2}$

**c.** $1.25 = \dfrac{5}{4}$

**Student Check 4**   **a.** $-(-7) = 7$   **b.** $-(4) = -4$

**c.** $-\left(\dfrac{1}{2}\right) = -\dfrac{1}{2}$   **d.** $-\left(-\dfrac{25}{6}\right) = \dfrac{25}{6}$

**e.** $-\left(\sqrt{6}\right) = -\sqrt{6}$   **f.** $-(8.2) = -8.2$

**Student Check 5**   **a.** 10   **b.** 6   **c.** $\pi$

**d.** $\dfrac{2}{5}$   **e.** $-4$   **f.** $-7$

## SUMMARY OF KEY CONCEPTS

1. Real numbers consist of rational numbers and irrational numbers. Rational numbers are made up of numbers that can be written as the ratio of two integers. All of the integers are elements of the set of rational numbers. The integers are positive and negative whole numbers. Whole numbers are the natural numbers combined with zero. The natural numbers are counting numbers.

2. Every real number can be graphed on a real number line.

3. For any two numbers $a$ and $b$, $a = b$, $a < b$, or $a > b$. To determine the order of numbers, place them on the number line. The number to the right is the larger number.

4. The numbers $a$ and $-a$ are opposites of one another. These numbers have the same distance from zero and are on opposite sides of zero.

5. The absolute value of a number measures the number's distance from zero on the real number line. The absolute value of a number is always zero or positive. Vertical bars, $|\ |$, denote absolute value.

## GRAPHING CALCULATOR SKILLS

There are three basic skills that the calculator can be applied to in this section: approximating an irrational number, finding the opposite of a number, and finding the absolute value of a number. The keystrokes correspond to a TI-83 Plus or TI-84 Plus. Similar keystrokes can be used with other calculators as well.

1. Approximate the values of $\sqrt{3}$ and $\pi$.

**Solution:** Enter the expressions on the calculator.

So, $\sqrt{3} \approx 1.73$ and $\pi \approx 3.14$.

2. Find the opposite of $-6$.

**Solution:** The symbol $(-)$ denotes the opposite of a number and is also used to enter a negative number. So, we must use it twice to find the opposite of $-6$.

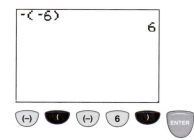

So, $-(-6) = 6$.

3. Determine the values of $|-5|$ and $-|-7|$.

**Solution:** The absolute value operation is found under the MATH menu. Press the MATH menu and then arrow right to go to the NUM menu. The first option is abs, which denotes absolute value.

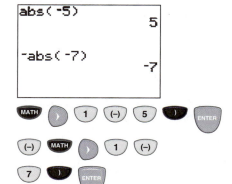

So, $|-5| = 5$ and $-|-7| = -7$.

---

## SECTION 1.1 / EXERCISE SET

### Write About It!

Use complete sentences to explain the meaning of each term or phrase.

1. Rational number  A rational number is a number that can be written as a ratio of two integers, $p$ and $q$, with $q$ not equal to zero.
2. Irrational number  An irrational number is a number that cannot be written as a ratio of two integers.
3. The opposite of a number  The opposite of a number is the number that has the same distance from 0 but lies on the opposite side of 0.

Additional answers can be found in the Instructor Answer Appendix.

4. The absolute value of a number  The absolute value of a number is the distance between 0 and the number on the number line.

**Determine if the statement is true or false. If a statement is false, provide an example that contradicts the statement.**

5. Every rational number is also an integer.  False, for example, $\frac{5}{2}$ is a rational number but is not an integer.
6. All real numbers can be classified as either a rational number or an irrational number.  True

**7.** All whole numbers are natural numbers.
False, for example, 0 is a whole number but is not a natural number.
**8.** All irrational numbers are real numbers.   True

**9.** A decimal that doesn't terminate is an irrational number.

**10.** The sum of a number and its additive inverse is 1.
False, for example, 3 + (−3) = 0.

## Practice Makes Perfect!

**Classify each number as a natural number, whole number, integer, rational number, irrational number, and/or real number. If the number is a rational number, write it in fractional form. If the number is irrational, approximate its value to two decimal places. (*See Objective 1.*)**

**11.** −7   integer, $-\dfrac{7}{1}$ rational, real
**12.** −15   integer, $-\dfrac{15}{1}$ rational, real

**13.** 3   natural, whole, integer, rational, real
**14.** 10

**15.** 2.5
**16.** 7.3   $\dfrac{73}{10}$ rational, real

**17.** $1\dfrac{7}{8}$   $\dfrac{15}{8}$ rational, real
**18.** $-4\dfrac{1}{7}$   $-\dfrac{29}{7}$ rational, real

**19.** 0.1   $\dfrac{1}{10}$ rational, real
**20.** 2.1   $\dfrac{21}{10}$ rational, real

**21.** $\sqrt{81}$
**22.** $\sqrt{49}$   natural, whole, integer, $\dfrac{7}{1}$ rational, real

**23.** $\sqrt{15}$   irrational; 3.87, real
**24.** $\sqrt{21}$   irrational; 4.58, real

**25.** $-\pi$   irrational; −3.14, real
**26.** $-\sqrt{7}$   irrational; −2.65, real

**27.** 365 days   natural, whole, integer, rational, real

**28.** 52 weeks   natural, whole, integer, rational, real

**29.** −80°F: the coldest recorded temperature in Prospect Creek Camp, Alaska, on January 23, 1971
integer, rational, real

**30.** −28°F: the coldest recorded temperature in Caesars Head, South Carolina, on January 21, 1985
integer, rational, real

**31.** 109°F: the hottest recorded temperature in Monticello, Florida, on June 29, 1931   natural, whole, integer, rational, real

**32.** 134°F: the hottest recorded temperature in Greenland Ranch, California, on July 10, 1913
natural, whole, integer, rational, real

**Give an example of a real number that satisfies each condition. (*See Objective 1.*)**

**33.** A rational number that is not an integer   Answers vary.

**34.** A rational number that is an integer   Answers vary.

**35.** An irrational number between 2 and 3   Answers vary.

**36.** An irrational number between −2 and −1   Answers vary.

**37.** A rational number between −3 and −4   Answers vary.

**38.** A rational number between 4 and 5   Answers vary.

**39.** An integer between $\sqrt{3}$ and $\sqrt{6}$   2

**40.** An integer between $\sqrt{5}$ and $\sqrt{10}$   3

**Graph each number on a real number line. (*See Objective 2.*)**

**41.** −4      **42.** −1      **43.** $\dfrac{1}{2}$      **44.** $\dfrac{8}{3}$

**45.** 3.5      **46.** 1.5      **47.** −4.5      **48.** −2.5

**49.** $1.3\overline{3}$   **50.** $0.29\overline{29}$   **51.** $\dfrac{\pi}{2}$   **52.** $2\pi$

**53.** $\sqrt{8}$      **54.** $\sqrt{14}$      **55.** $\sqrt{16}$      **56.** $-\sqrt{25}$

**Compare the values of each pair of numbers. Use a <, >, or = symbol to make the statement true. (*See Objective 3.*)**

**57.** −7 ___ −8   >
**58.** −3 ___ −2   <

**59.** $\dfrac{2}{5}$ ___ 0.4   =
**60.** $-\dfrac{7}{4}$ ___ −1.75   =

**61.** $\sqrt{3}$ ___ 1.75   <
**62.** $\sqrt{11}$ ___ 3   >

**63.** −(−2) ___ 2   =
**64.** −(4) ___ 4   <

**65.** 0 ___ −2   >
**66.** −(−6) ___ 0   >

**67.** $\pi$ ___ 3.1   >
**68.** 1.4 ___ $\sqrt{2}$   <

**Find the opposite of each real number. (*See Objective 4.*)**

**69.** 6   −6
**70.** 14   −14
**71.** −1   1

**72.** −5   5
**73.** $\dfrac{1}{4}$   $-\dfrac{1}{4}$
**74.** $-\dfrac{10}{3}$   $\dfrac{10}{3}$

**75.** −5.1   5.1
**76.** 10.5   −10.5
**77.** $-\pi$   $\pi$

**78.** $\sqrt{13}$   $-\sqrt{13}$
**79.** $-1.\overline{33}$   $1.\overline{33}$
**80.** $2.2\overline{323}$   $-2.2\overline{323}$

**Simplify each absolute value expression. (*See Objective 5.*)**

**81.** $|5|$   5
**82.** $|-3.8|$   3.8
**83.** $|-7.5|$   7.5

**84.** $|6.7|$   6.7
**85.** $\left|\dfrac{1}{3}\right|$   $\dfrac{1}{3}$
**86.** $-\left|\dfrac{5}{4}\right|$   $-\dfrac{5}{4}$

**87.** $-\left|\dfrac{9}{11}\right|$   $-\dfrac{9}{11}$
**88.** $\left|-\dfrac{4}{7}\right|$   $\dfrac{4}{7}$
**89.** $-|7|$   −7

**90.** $-|-14|$   −14

##  Mix 'Em Up!

**Classify each real number as a natural number, whole number, integer, rational number, irrational number, and/or real number. If the number is a rational number, write it in fractional form. If the number is irrational, approximate its value to two decimal places. (*See Objective 1.*)**

**91.** 12.5   rational, real; $\dfrac{25}{2}$
**92.** −17.5   rational, real; $-\dfrac{35}{2}$

**93.** $-\dfrac{11}{3}$   rational, real
**94.** $\dfrac{11}{5}$   rational, real

**95.** $96 billion: the cost of the damage from hurricane Katrina in Louisiana in 2005

**96.** 4160 mi—the length of the Great Wall of China, the world's longest man-made structure

**97.** 6.3   rational, real; $\dfrac{63}{10}$
**98.** −3.47   rational, real; $-\dfrac{347}{100}$

**99.** $\sqrt{12}$   irrational, real; 3.46
**100.** $-\sqrt{36}$   rational, integer, real; $-\dfrac{6}{1}$

**101.** 1.37%: interest rate for a 12-month certificate of deposit at Ally bank   rational, real; $\dfrac{137}{10000}$

**102.** 5.75%: a 30-yr mortgage rate   rational, real; $\dfrac{23}{400}$

**103.** $-\pi$   irrational, real; −3.14
**104.** $\sqrt{31}$   irrational, real; 5.57

**105.** −1293 ft: elevation of the Dead Sea

**106.** −86 m: elevation of Death Valley   integer, rational, real; $-\dfrac{86}{1}$

**Graph each number on a real number line. (*See Objective 2.*)**

**107.** $\left\{-5.2, -2\dfrac{1}{2}, 1, \sqrt{6}\right\}$
**108.** $\left\{-\sqrt{18}, -3\dfrac{1}{3}, 1.5, 2\pi\right\}$

**109.** $\left\{ |-5|, -1.5, \dfrac{4}{3}, \sqrt{9} \right\}$ **110.** $\left\{ -|-8|, -\dfrac{\pi}{2}, \sqrt{4}, 5\dfrac{1}{2} \right\}$

**Compare the value of each pair of numbers. Use a <, >, or = symbol to make the statement true. (*See Objective 3.*)**

**111.** $\dfrac{99}{70}$ ___ $\sqrt{2}$  $>$  **112.** $\dfrac{21}{7}$ ___ $\pi$  $<$

**113.** $-|-9|$ ___ $\sqrt{81}$  $<$  **114.** $\sqrt{16}$ ___ $-|-4|$  $>$

**115.** $-\left|-\dfrac{10}{3}\right|$ ___ $-\dfrac{\pi}{2}$  $<$  **116.** $\dfrac{12}{6}$ ___ $\sqrt{4}$  $=$

**Find the opposite of each real number. (*See Objective 4.*)**

**117.** $\dfrac{14}{3}$  $-\dfrac{14}{3}$  **118.** $-\dfrac{3}{11}$  $\dfrac{3}{11}$

**119.** $4.29\overline{29}$  $-4.29\overline{29}$  **120.** $-1.5\overline{5}$  $1.5\overline{5}$

**Simplify each absolute value expression. (*See Objective 5.*)**

**121.** $-|6|$  $-6$  **122.** $|16|$  $16$

**123.** $|-9.5|$  $9.5$  **124.** $-\left|-\dfrac{9}{8}\right|$  $-\dfrac{9}{8}$

 **You Be the Teacher!**

**Correct each student's errors, if any.**

**125.** Is the opposite of a number always negative?

Betty's work: Yes, if $x$ is a number, the opposite is $-x$.

No. The opposite of a positive number is negative but the opposite of a negative number is positive.

**126.** Graph the number on the real number line: $-1.99$.

Gloria's work:

**127.** Fill in the blank with the appropriate symbol to make the statement true: $\dfrac{5}{7}$ ___ $\dfrac{7}{10}$.

Enefiok's work: $\dfrac{5}{7} < \dfrac{7}{10}$  Since $\dfrac{5}{7} \approx 0.71$ and $\dfrac{7}{10} = 0.7$, $\dfrac{5}{7} > \dfrac{7}{10}$.

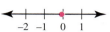 **Calculate It!**

**Approximate the value of each real number to the nearest hundredth. Specify if the number is rational or irrational.**

**128.** $\dfrac{\pi}{4}$  $0.79$, irrational  **129.** $\dfrac{\pi}{6}$  $0.52$, irrational

**130.** $1 + \sqrt{5}$  $3.24$, irrational  **131.** $2 + \sqrt{3}$  $3.73$, irrational

**132.** $2 + \dfrac{5}{6}$  $2.83$, rational  **133.** $4 + \dfrac{2}{9}$  $4.22$, rational

**134.** $6 + \sqrt{81}$  $15$, rational  **135.** $9 - \sqrt{4}$  $7$, rational

**136.** $|-142|$  $142$, rational  **137.** $-|23|$  $-23$, rational

---

**SECTION 1.2**  **Fractions Review**

**▶ OBJECTIVES**

**As a result of completing this section, you will be able to**

**1.** Write the prime factorization of a number.

**2.** Define and write fractions.

**3.** Simplify fractions.

**4.** Multiply fractions.

**5.** Divide fractions.

**6.** Add or subtract fractions with common denominators.

**7.** Add or subtract fractions with unlike denominators.

**8.** Troubleshoot common errors.

As of 2009, there were approximately 309 million people living in the United States and approximately 116 million Facebook users in the United States. Write a fraction that represents the part of the United States population that were Facebook users in 2009. (Source: http://flavorwire .com/82308/awesome-infographic-facebook-vs-the-united-states)

In this section, we will learn how to use a fraction to represent such information. We will also learn how to simplify fractions and perform operations with fractions.

**Factoring Numbers**

To express a number in *factored form* is to write the number as a product of two or more numbers. For example, in the statement $4 \cdot 5 = 20$, the numbers 4 and 5 are called **factors** and 20 is the *product*. Factors divide a number evenly without any remainder.

The numbers 2, 3, 5, 7, 11, 13, 17, and 19 have something in common. The factors of these numbers are only 1 and the number itself. These types of numbers are called *prime numbers*. The following prime numbers are written in factored form.

$$2 = 1 \cdot 2, \; 3 = 1 \cdot 3, \; 5 = 1 \cdot 5$$

**Objective 1 ▶**

Write the prime factorization of a number.

> **Definition:** A number $p$ is a **prime number** if its whole number factors are only 1 and the number $p$ itself.

A number that is not prime is called **composite**. This means that the number has factors other than 1 and itself. Examples of composite numbers are 4, 6, and 16, and their factored forms are shown.

| Number | Factored Forms |
|--------|----------------|
| 4 | 1 · 4 or 2 · 2 |
| 6 | 1 · 6 or 2 · 3 |
| 16 | 1 · 16 or 2 · 8 or 4 · 4 |

The number 1 is neither prime nor composite. Every number, other than 1, can be expressed as a product of prime factors. This factorization is called the **prime factorization** of the number. We can find the prime factorization of 60, for example, using a **factor tree**. We begin with any two factors of 60, such as 6 and 10 or 2 and 30. We find factors of any number that is not prime and continue the process until all factors are prime numbers. When we find a prime factor, we circle it for easy reference, as shown.

So, the prime factorization of 60 is 2 · 3 · 2 · 5 or $2^2$ · 3 · 5 because the factors 2, 3, and 5 are prime numbers. Note that it doesn't matter which two factors we choose to begin the factor tree, because we will always obtain the same factorization except for the order of the factors.

> **Procedure:  Writing the Prime Factorization of a Number**
>
> **Step 1:** Write the number as a product of any two of its factors.
> **Step 2:** If both factors are prime, then the factorization is complete.
> **Step 3:** If either of the factors is not prime, rewrite that factor as a product. If these factors are prime, then the factorization is complete.
> **Step 4:** Continue this process until all factors are prime.

**Objective 1 Examples**    Write the prime factorization of each number.

**1a.** 42                     **1b.**  120

**Solutions**    **1a.**

          42

Write 42 as a product of any two of its factors. Circle the prime factor, 7.

Rewrite 6 as 2 · 3.

The prime factorization of 42 is 2 · 3 · 7.

**1b.**

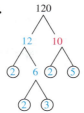

120

12      10

②  6  ②  ⑤

②  ③

Write 120 as a product of any two of its factors.

Rewrite 12 as 2 · 6 and 10 as 2 · 5. Circle the prime factors 2, 2, and 5.

Rewrite 6 as 2 · 3.

The prime factorization of 120 is 2 · 2 · 3 · 2 · 5 or $2^3 \cdot 3 \cdot 5$.

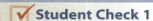

**✓ Student Check 1**      Write the prime factorization of each number.
  **a.** 24                            **b.** 128

## Fractions

**Objective 2** ▶

Define and write fractions.

In Section 1.1, we defined a rational number as a number that can be written as the quotient of two integers. A *fraction* is a quotient of two real numbers.

> **Definition:** A **fraction** is a number of the form $\frac{a}{b}$, where $a$ and $b$ are real numbers with $b \neq 0$. The number $a$ is called the **numerator** and $b$ is called the **denominator**.

A fraction represents a division of a whole into parts. The denominator of a fraction represents the "total number of parts in the whole" while the numerator represents the "number of parts chosen." For example, the fraction $\frac{3}{4}$ represents 3 parts out of 4. We can visualize this as shown in Figure 1.1.

**Figure 1.1**

It is important to note that a fraction may or may not be a rational number. For example,

| $\frac{3}{4}$ | This fraction is a rational number since it is a quotient of two integers. |
|---|---|
| $\frac{\pi}{2}$ | This fraction is not a rational number since $\pi$ is not an integer. |

If the numerator of a fraction is less than the denominator, the fraction is called a **proper fraction**. Some examples are $\frac{1}{2}$ and $\frac{4}{15}$. If the numerator is greater than the denominator, the fraction is called an **improper fraction**. Some examples are $\frac{5}{3}$ and $\frac{17}{2}$. An improper fraction can also be written as a **mixed number**. For instance,

$$\frac{5}{3} = 1\frac{2}{3} \quad \text{and} \quad \frac{17}{2} = 8\frac{1}{2}$$

**Objective 2 Examples**    **Write a fraction that represents each quantity.**

**2a.** Prior to the 2010 gubernatorial elections, 23 Republicans, 1 Independent, and 26 Democrats held the office of governor in the United States. Write fractions that represent the portion of the U.S. governors who were Republican, Independent, and Democratic.

**Solution**    **2a.** The whole is the total number of governors in the United States, 50. The part is the number of governors from each party. So, the fraction of United States governors for each party is

| Republican | $\dfrac{23}{50}$ |
|---|---|
| Independent | $\dfrac{1}{50}$ |
| Democrat | $\dfrac{26}{50}$ |

**2b.** As of 2009, there were approximately 309 million people living in the United States and approximately 116 million Facebook users in the United States. Write a fraction that represents the portion of the U.S. population who were Facebook users in 2009. (Source: http://flavorwire.com/82308/awesome-infographic-facebook-vs-the-united-states)

**Solution**    **2b.** The whole is the population of the United States, 309 million. The part is the number of Facebook users, 116 million. So, the fraction of the U.S. population who were Facebook users is

$$\frac{116,000,000}{309,000,000}$$

**INSTRUCTOR NOTE:**
Simplifying fractions is covered in Objective 2.

✔ **Student Check 2**    Write a fraction that represents each quantity.

**a.** Thirty-nine states elected a governor during the 2010 gubernatorial elections. Write a fraction that represents the portion of states that went through a gubernatorial election. (Source: http://statehouserock.com/)

**b.** The National Center for Education Statistics published the number of bachelor's degrees conferred by major for the 2007–2008 academic year. The pie graph to the right shows this information. Write a fraction that represents the portion of Education degrees that was awarded.

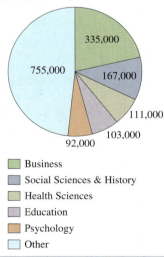

- ▇ Business
- ▇ Social Sciences & History
- ▇ Health Sciences
- ▇ Education
- ▇ Psychology
- ▇ Other

## Writing Fractions in Simplified Form

**Objective 3** ▶

Simplify fractions.

A fraction is **simplified** or in **lowest terms** if the numerator and denominator share no common factors other than 1. The following property provides us a method to write a fraction in lowest terms.

> **Property: The Fundamental Property of Fractions**
> If $\dfrac{a}{b}$ is a fraction, where $b \neq 0$ and $c$ is a nonzero real number, then
> $$\frac{a \cdot c}{b \cdot c} = \frac{a}{b}$$

The reason the fundamental property works is that $\dfrac{c}{c} = 1$.

$$\frac{a \cdot c}{b \cdot c} = \frac{a}{b} \cdot \frac{c}{c}$$

$$= \frac{a}{b} \cdot 1$$

$$= \frac{a}{b}$$

**Procedure:  Simplifying Fractions**

**Step 1:** Write the numerator and denominator with their prime factorizations.
**Step 2:** Divide the numerator and denominator by their common factors.
**Step 3:** Multiply the remaining factors.

**Objective 3 Examples**   Simplify each fraction.

**3a.** $\dfrac{28}{30}$          **3b.** $\dfrac{144}{99}$          **3c.** $\dfrac{116,000,000}{309,000,000}$

**Solutions**   **3a.**

$$\frac{28}{30} = \frac{2 \cdot 2 \cdot 7}{2 \cdot 3 \cdot 5}$$          Write the prime factorization of 28 and 30.

$$= \frac{\cancel{2} \cdot 2 \cdot 7}{\cancel{2} \cdot 3 \cdot 5}$$          Divide out the common factor of 2.

$$= \frac{14}{15}$$          Multiply the remaining factors.

**Note:** *It is not necessary to write the prime factorization of the numerator and denominator of a fraction to simplify it. For instance, observe that both 28 and 30 are divisible by 2. So, we can divide them each by 2 to obtain the simplified form of the fraction.*

$$\frac{\overset{14}{\cancel{28}}}{\underset{15}{\cancel{30}}} = \frac{14}{15}$$

**3b.**

$$\frac{144}{99} = \frac{2 \cdot 2 \cdot 2 \cdot 2 \cdot 3 \cdot 3}{3 \cdot 3 \cdot 11}$$          Write the prime factorization of 144 and 99.

$$= \frac{2 \cdot 2 \cdot 2 \cdot 2 \cdot \cancel{3} \cdot \cancel{3}}{\cancel{3} \cdot \cancel{3} \cdot 11}$$          Divide out the common factors.

$$= \frac{16}{11}$$          Multiply the remaining factors.

**3c.** $\dfrac{116,000,000}{309,000,000} = \dfrac{116,000,000 \div 1,000,000}{309,000,000 \div 1,000,000}$          Divide the numerator and denominator by their common factor of 1,000,000.

$$= \frac{116}{309}$$          Simplify.

✔ **Student Check 3**   Simplify each fraction.

**a.** $\dfrac{20}{32}$          **b.** $\dfrac{180}{200}$          **c.** $\dfrac{104,000}{624,000}$

## Multiplying Fractions

**Objective 4** ▶

Multiply fractions.

We can multiply fractions by multiplying the numerators of the fractions and by multiplying the denominators of the fractions. We always want to express the answers in lowest terms.

> **Property: Multiplying Fractions**
>
> For $b, d \neq 0$,
>
> $$\frac{a}{b} \cdot \frac{c}{d} = \frac{a \cdot c}{b \cdot d}$$

**Objective 4  Examples**   Multiply the fractions and write each answer in lowest terms.

**4a.** $\dfrac{3}{4} \cdot \dfrac{5}{2}$     **4b.** $\dfrac{10}{3} \cdot \dfrac{18}{25}$     **4c.** $2\dfrac{2}{3} \cdot 5\dfrac{3}{4}$

**Solutions**   **4a.**   $\dfrac{3}{4} \cdot \dfrac{5}{2} = \dfrac{3 \cdot 5}{4 \cdot 2}$     Multiply the numerators and denominators.

$= \dfrac{15}{8}$     Simplify the products.

**4b.** $\dfrac{10}{3} \cdot \dfrac{18}{25} = \dfrac{10 \cdot 18}{3 \cdot 25}$     Multiply the numerators and denominators.

$= \dfrac{2 \cdot 5 \cdot 2 \cdot 3 \cdot 3}{3 \cdot 5 \cdot 5}$     Write the prime factorization of each number.
$10 = 2 \cdot 5, \ 18 = 2 \cdot 3 \cdot 3, \ 25 = 5 \cdot 5$

$= \dfrac{2 \cdot 2 \cdot 3}{5}$     Divide out the common factors of 5 and 3.

$= \dfrac{12}{5}$     Multiply the remaining factors.

**4c.** $2\dfrac{2}{3} \cdot 5\dfrac{3}{4} = \dfrac{8}{3} \cdot \dfrac{23}{4}$     Convert the mixed fractions to improper fractions.
$2\dfrac{2}{3} = \dfrac{8}{3} \quad \text{and} \quad 5\dfrac{3}{4} = \dfrac{23}{4}$

$= \dfrac{\overset{2}{\cancel{8}} \cdot 23}{3 \cdot \underset{1}{\cancel{4}}}$     Multiply the numerators and denominators. Divide out the common factor of 4 from the numerator and denominator.

$= \dfrac{46}{3}$     Multiply the remaining factors.

✔ **Student Check 4**   Multiply the fractions and write each answer in lowest terms.

**a.** $\dfrac{5}{8} \cdot \dfrac{3}{7}$     **b.** $\dfrac{4}{9} \cdot \dfrac{15}{14}$     **c.** $1\dfrac{3}{7} \cdot 2\dfrac{1}{10}$

## Dividing Fractions

**Objective 5** ▶

Divide fractions.

Before we can divide fractions, we must discuss the concept of reciprocals. Two numbers are *reciprocals* if their product is 1. Some examples are shown.

| Number | Reciprocal | Their Product |
|--------|-----------|---------------|
| 6 | $\dfrac{1}{6}$ | $6 \cdot \dfrac{1}{6} = \dfrac{6}{1} \cdot \dfrac{1}{6} = 1$ |
| $\dfrac{4}{5}$ | $\dfrac{5}{4}$ | $\dfrac{4}{5} \cdot \dfrac{5}{4} = 1$ |

INSTRUCTOR NOTE:
Remind students that reciprocals are not the same as opposites.

**Definition:** Two numbers are **reciprocals** if their product is 1. The reciprocal of $\dfrac{c}{d}$ is $\dfrac{d}{c}$ since

$$\frac{c}{d} \cdot \frac{d}{c} = \frac{c \cdot d}{d \cdot c} = 1$$

Dividing fractions is related to multiplying fractions. For example, we know

$$30 \div 6 = 5, \text{ but we also know } 30 \cdot \frac{1}{6} = \frac{30}{1} \cdot \frac{1}{6} = \frac{30}{6} = 5$$

Therefore,

$$30 \div 6 = 30 \cdot \frac{1}{6}$$

This illustrates that dividing by a number is the same as multiplying by the reciprocal of the number.

**Property: Dividing Fractions**

For $b, c, d \neq 0$,

$$\frac{a}{b} \div \frac{c}{d} = \frac{a}{b} \cdot \frac{d}{c}$$

**Objective 5 Examples**   Divide the fractions and write each answer in lowest terms.

**5a.** $\dfrac{8}{7} \div \dfrac{12}{14}$        **5b.** $\dfrac{3}{5} \div 9$        **5c.** $3\dfrac{3}{7} \div \dfrac{7}{15}$

**Solutions**   **5a.**

$$\frac{8}{7} \div \frac{12}{14} = \frac{8}{7} \cdot \frac{14}{12}$$    Multiply by the reciprocal of $\dfrac{12}{14}$, which is $\dfrac{14}{12}$.

$$= \frac{8 \cdot 14}{7 \cdot 12}$$    Multiply the numerators and denominators.

$$= \frac{2 \cdot 2 \cdot 2 \cdot 2 \cdot 7}{7 \cdot 2 \cdot 2 \cdot 3}$$    Rewrite each number as a product of prime factors. $8 = 2 \cdot 2 \cdot 2$, $14 = 2 \cdot 7$, $12 = 2 \cdot 2 \cdot 3$

$$= \frac{2 \cdot 2}{3}$$    Divide out the common factors of 2, 2, and 7.

$$= \frac{4}{3}$$    Multiply the remaining factors.

**5b.**

$$\frac{3}{5} \div 9 = \frac{3}{5} \cdot \frac{1}{9}$$    Multiply by the reciprocal of 9, which is $\dfrac{1}{9}$.

$$= \frac{3 \cdot 1}{5 \cdot 9}$$    Multiply the numerators and denominators.

$$= \frac{3 \cdot 1}{5 \cdot 3 \cdot 3}$$    $9 = 3 \cdot 3$. Rewrite each number as a product of prime factors.

$$= \frac{1}{5 \cdot 3}$$    Divide out the common factor of 3.

$$= \frac{1}{15}$$    Multiply the remaining factors.

**5c.** $3\dfrac{3}{7} \div \dfrac{7}{15} = \dfrac{24}{7} \cdot \dfrac{15}{7}$    Write $3\dfrac{3}{7}$ as $\dfrac{24}{7}$ and multiply by the reciprocal of $\dfrac{7}{15}$, which is $\dfrac{15}{7}$.

$\qquad\qquad = \dfrac{24 \cdot 15}{7 \cdot 7}$    Multiply the numerators and denominators.

$\qquad\qquad = \dfrac{360}{49}$ or $7\dfrac{17}{49}$    Simplify the products.

The answer can be written as either an improper fraction or as a mixed number.

---

✔ **Student Check 5**    Divide the fractions and write each answer in lowest terms.

**a.** $\dfrac{12}{15} \div \dfrac{30}{24}$    **b.** $\dfrac{6}{7} \div 12$    **c.** $5\dfrac{2}{3} \div 4\dfrac{1}{9}$

---

## Adding and Subtracting Fractions with Like Denominators

**Objective 6** ▶

Add or subtract fractions with like denominators.

One common method for adding or subtracting fractions requires that the denominators of the fractions be the same. If the denominators are the same, we simply add or subtract the numerators and place this number over their common denominator.

For example,

$$\frac{2}{7} + \frac{1}{7} = \frac{2+1}{7} = \frac{3}{7}$$

> **Property:  Adding or Subtracting Fractions with Common Denominators**
>
> $$\frac{a}{b} + \frac{c}{b} = \frac{a+c}{b}, \quad \text{for } b \neq 0$$
>
> $$\frac{a}{b} - \frac{c}{b} = \frac{a-c}{b}, \quad \text{for } b \neq 0$$

---

**Objective 6  Examples**    Add or subtract the fractions and simplify the result.

**6a.** $\dfrac{5}{8} + \dfrac{2}{8}$    **6b.** $1\dfrac{1}{3} + 6\dfrac{2}{3}$    **6c.** $\dfrac{8}{9} - \dfrac{2}{9}$    **6d.** $5\dfrac{1}{7} - 4\dfrac{3}{7}$

**Solutions**    **6a.**    $\dfrac{5}{8} + \dfrac{2}{8} = \dfrac{5+2}{8}$    Add the numerators and place over the common denominator.

$\qquad\qquad\quad = \dfrac{7}{8}$    Simplify the numerator.

**6b.** $1\dfrac{1}{3} + 6\dfrac{2}{3} = \dfrac{4}{3} + \dfrac{20}{3}$    Write each mixed number as an improper fraction.

$\qquad\qquad\quad = \dfrac{4+20}{3}$    Add the numerators and place over the common denominator.

$\qquad\qquad\quad = \dfrac{24}{3}$    Simplify the numerator.

$\qquad\qquad\quad = 8$    Simplify the fraction.

**6c.**    $\dfrac{8}{9} - \dfrac{2}{9} = \dfrac{8-2}{9}$    Subtract the numerators and place over the common denominator.

$\qquad\qquad\quad = \dfrac{6}{9}$    Simplify the numerator.

$\qquad\qquad\quad = \dfrac{2 \cdot 3}{3 \cdot 3}$    Factor the numerator and denominator.

$\qquad\qquad\quad = \dfrac{2}{3}$    Divide out the common factor of 3.

**6d.** $5\frac{1}{7} - 4\frac{3}{7} = \frac{36}{7} - \frac{31}{7}$    Write each mixed number as an improper fraction.

$= \frac{36 - 31}{7}$    Subtract the numerators and place over the common denominator.

$= \frac{5}{7}$    Simplify the numerator.

☑ **Student Check 6**    Add or subtract the fractions and simplify the result.

**a.** $\frac{4}{5} + \frac{2}{5}$      **b.** $\frac{3}{8} + \frac{1}{8}$      **c.** $\frac{4}{5} - \frac{1}{5}$      **d.** $9\frac{2}{9} - 7\frac{4}{9}$

## Adding and Subtracting Fractions with Different Denominators

**Objective 7** ▶

Add or subtract fractions with unlike denominators.

If the fractions we need to add do not have the same denominator, we must first rewrite each fraction as an *equivalent fraction* with the same denominator, or a *common denominator*.

**Equivalent fractions** are fractions that represent the same quantity. For example, $\frac{3}{4}$ and $\frac{6}{8}$ are equivalent fractions as shown.

We can obtain equivalent fractions by multiplying a given fraction by 1. Recall multiplying by 1 does not change the value of the fraction. When we multiply a fraction by 1, we are applying the Fundamental Property of Fractions. This property not only enables us to simplify fractions, but also enables us to multiply the numerator and denominator by the same nonzero number—that is, by a form of 1. For example, to rewrite $\frac{3}{4}$ as a fraction with a denominator of 8, we can multiply the numerator and denominator by 2 as shown.

$$\frac{3}{4} = \frac{3}{4} \cdot 1 = \frac{3}{4} \cdot \frac{2}{2} = \frac{3 \cdot 2}{4 \cdot 2} = \frac{6}{8}$$

If we want to add $\frac{4}{9}$ and $\frac{3}{2}$, for example, we must first convert these fractions to equivalent fractions with the same, or common denominator. If we find the *least common denominator*, that will make the process somewhat easier.

**INSTRUCTOR NOTE:**
Point out that the least common denominator is the same as the least common multiple.

The **least common denominator (LCD)** is the smallest number that all denominators divide into evenly. A *common denominator* is a number that all the denominators divide evenly but may not be the smallest such number. So, if the denominators are 9 and 2, the LCD is the smallest number that both 9 and 2 divide into—that is, the number 18.

If the LCD is not easily identified, we can use the prime factorizations of the denominators to determine it. For example, the least common denominator of 6 and 20 is 60. We know this by their prime factorizations as shown.

$$6 = 2 \cdot ③$$
$$20 = ②\cdot②\cdot⑤ \quad \longrightarrow \quad LCD = 2 \cdot 2 \cdot 3 \cdot 5 = 60$$

Notice that 6 has a prime factor of 2 and 20 has two prime factors of 2. Since 2 appears twice in the prime factorization of 20, we use two factors of 2 in the LCD. Since 3 and 5 appear only once in the prime factorizations of 6 and 20, respectively, we include one factor of each of them in the product of the LCD.

**Procedure: Finding the LCD**

**Step 1:** Write the prime factorization of each denominator.
**Step 2:** For each of the different factors, circle the largest number of occurrences of this factor found in the factorizations.
**Step 3:** The product of the circled factors is the LCD.

**Fact:** *The LCD must include all of the different factors found in the denominators. The factor must repeat the largest number of times it appears in any of the denominators.*

**Procedure: Adding or Subtracting Fractions with Unlike Denominators**

**Step 1:** Determine the LCD.
**Step 2:** Convert each fraction to an equivalent fraction with the LCD as its denominator. Multiply the numerator and denominator by the number that is needed to make the denominator the LCD.
**Step 3:** Add or subtract the fractions.
**Step 4:** Simplify the result, if necessary.

**Objective 7 Examples**    **Add or subtract the fractions and simplify the result.**

**7a.** $\dfrac{2}{3} + \dfrac{1}{4}$     **7b.** $\dfrac{5}{12} - \dfrac{3}{8}$     **7c.** $2 + \dfrac{5}{6}$     **7d.** $4\dfrac{1}{12} - 1\dfrac{9}{14}$

**Solutions**    **7a.**   $\dfrac{2}{3} + \dfrac{1}{4} = \dfrac{2}{3} \cdot \dfrac{4}{4} + \dfrac{1}{4} \cdot \dfrac{3}{3}$    Multiply each fraction by a form of 1 to obtain the LCD, 12.

$$= \dfrac{8}{12} + \dfrac{3}{12}$$    Simplify each product.

$$= \dfrac{8 + 3}{12}$$    Add the numerators and place over the common denominator.

$$= \dfrac{11}{12}$$    Simplify the numerator.

**7b.** The LCD of 12 and 8 can be found as follows.

$$12 = 2 \cdot 2 \cdot ③$$
$$8 = ②\cdot 2 \cdot 2$$
$$\longrightarrow \qquad \text{LCD} = 2 \cdot 2 \cdot 2 \cdot 3 = 24$$

$$\dfrac{5}{12} - \dfrac{3}{8} = \dfrac{5}{12} \cdot \dfrac{2}{2} - \dfrac{3}{8} \cdot \dfrac{3}{3}$$    Multiply each fraction by a form of 1 to obtain the LCD, 24.

$$= \dfrac{10}{24} - \dfrac{9}{24}$$    Simplify each product.

$$= \dfrac{10 - 9}{24}$$    Subtract the numerators and place over the common denominator.

$$= \dfrac{1}{24}$$    Simplify the numerator.

**7c.**   $2 + \dfrac{5}{6} = \dfrac{2}{1} + \dfrac{5}{6}$    Write 2 as $\dfrac{2}{1}$.

$$= \dfrac{2}{1} \cdot \dfrac{6}{6} + \dfrac{5}{6}$$    Multiply the first fraction by a form of 1 to obtain the LCD, 6.

$$= \frac{12}{6} + \frac{5}{6} \qquad \text{Simplify the product.}$$

$$= \frac{12 + 5}{6} \qquad \text{Add the numerators and place over the common denominator.}$$

$$= \frac{17}{6} \qquad \text{Simplify the numerator.}$$

**7d.** We first convert each mixed number to an improper fraction, $4\frac{1}{12} = \frac{49}{12}$ and $1\frac{9}{14} = \frac{23}{14}$. We find the LCD, as follows.

$$14 = 2 \cdot \boxed{7}$$
$$12 = \boxed{2 \cdot 2} \cdot \boxed{3} \qquad\longrightarrow\qquad \text{LCD} = 2 \cdot 2 \cdot 3 \cdot 7 = 84$$

$$4\frac{1}{12} - 1\frac{9}{14} = \frac{49}{12} - \frac{23}{14} \qquad \text{Write each fraction as an improper fraction.}$$

$$= \frac{49}{12} \cdot \frac{7}{7} - \frac{23}{14} \cdot \frac{6}{6} \qquad \text{Multiply each fraction by a form of 1 to obtain the LCD, 84.}$$

$$= \frac{343}{84} - \frac{138}{84} \qquad \text{Simplify each product.}$$

$$= \frac{343 - 138}{84} \qquad \text{Subtract the numerators and place over the common denominator.}$$

$$= \frac{205}{84} \qquad \text{Simplify the numerator.}$$

$$= 2\frac{37}{84} \qquad \text{Convert to a mixed number.}$$

✓ **Student Check 7**   Add or subtract the fractions and simplify the result.

**a.** $\dfrac{2}{7} + \dfrac{1}{3}$    **b.** $\dfrac{5}{24} - \dfrac{1}{30}$    **c.** $7 + \dfrac{1}{3}$    **d.** $6\dfrac{3}{16} - 4\dfrac{9}{20}$

---

**Objective 8** ▶

Troubleshoot common errors.

## Troubleshooting Common Errors

Some common errors associated with fractions are shown next.

**Objective 8  Examples**   A problem and an incorrect solution are given. Provide the correct solution and an explanation of the error.

**8a.** Simplify $\dfrac{2}{3} \cdot \dfrac{1}{5}$.

| Incorrect Solution | Correct Solution and Explanation |
|---|---|
| $\dfrac{2}{3} \cdot \dfrac{1}{5} = \dfrac{2}{3} \cdot \dfrac{5}{5} \times \dfrac{1}{5} \cdot \dfrac{3}{3}$ $= \dfrac{10}{15} \times \dfrac{3}{15}$ $= \dfrac{30}{225}$ $= \dfrac{2}{15}$ | We do not need to find the LCD when multiplying fractions, we can simply multiply the numerators and denominators. $$\dfrac{2}{3} \cdot \dfrac{1}{5} = \dfrac{2 \cdot 1}{3 \cdot 5} = \dfrac{2}{15}$$ |

**8b.** Simplify $\frac{4}{9} + \frac{1}{9}$.

| Incorrect Solution | Correct Solution and Explanation |
|---|---|
| $\frac{4}{9} + \frac{1}{9} = \frac{4+1}{9+9} = \frac{5}{18}$ | To add fractions with the same denominator, we add their numerators and place this sum over the common denominator. $$\frac{4}{9} + \frac{1}{9} = \frac{4+1}{9} = \frac{5}{9}$$ |

## ANSWERS TO STUDENT CHECKS

**Student Check 1** **a.** $24 = 2 \cdot 2 \cdot 2 \cdot 3$
**b.** $128 = 2 \cdot 2 \cdot 2 \cdot 2 \cdot 2 \cdot 2 \cdot 2$

**Student Check 2** **a.** $\frac{39}{50}$ **b.** $\frac{103,000}{1,563,000}$

**Student Check 3** **a.** $\frac{5}{8}$ **b.** $\frac{9}{10}$ **c.** $\frac{1}{6}$

**Student Check 4** **a.** $\frac{15}{56}$ **b.** $\frac{10}{21}$ **c.** $3$

**Student Check 5** **a.** $\frac{16}{25}$ **b.** $\frac{1}{14}$ **c.** $\frac{51}{37}$

**Student Check 6** **a.** $\frac{6}{5}$ **b.** $\frac{1}{2}$ **c.** $\frac{3}{5}$ **d.** $\frac{16}{9}$ or $1\frac{7}{9}$

**Student Check 7** **a.** $\frac{13}{21}$ **b.** $\frac{7}{40}$ **c.** $\frac{22}{3}$ **d.** $\frac{139}{80}$ or $1\frac{59}{80}$

## SUMMARY OF KEY CONCEPTS

1. The prime factorization of a number is a way to write a number as a product of prime numbers. Prime numbers are 2, 3, 5, 7, 11, 13, and so on.
2. A fraction represents a part of a whole and is written as $\frac{a}{b}$.
3. Fractions can be simplified by dividing the numerator and denominator by their common factor.
4. Multiply fractions by multiplying their numerators and their denominators together.
5. Divide fractions by multiplying by the reciprocal. The reciprocal of the fraction $\frac{a}{b}$ is $\frac{b}{a}$ for $a$ and $b$ not zero.
6. We can only add and subtract fractions with a common denominator.
7. If the given fractions do not have the same denominator, convert each fraction to an equivalent fraction with the LCD as its denominator.

## GRAPHING CALCULATOR SKILLS

Practice operations with fractions by hand before relying on the calculator to perform the work for you. At this point, use the calculator only as a check of your work. The calculator will display all answers as decimals. To convert an answer to a fraction, use the [MATH] menu.

**1.** Simplify $\frac{12}{18}$.

```
12/18
         .6666666667
Ans▶Frac
               2/3
```

**2.** $\frac{1}{4} \cdot \frac{12}{15}$

```
1/4*12/15
              .2
Ans▶Frac
            1/5
```

**3.** $\dfrac{8}{7} \div \dfrac{12}{14}$

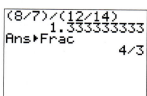

**4.** $\dfrac{5}{12} - \dfrac{3}{8}$

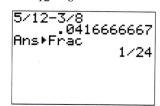

## SECTION 1.2  /  EXERCISE SET

### ✍ Write About It!

Use complete sentences to explain the meaning of the given term or process.

1. Fraction   *Answers vary.*
2. Simplifying a fraction to lowest terms   *Answers vary.*
3. Prime number   *Answers vary.*
4. Composite number   *Answers vary.*
5. Prime factorization of a number   *Answers vary.*
6. (a) Equivalent fractions   *Answers vary.*
   (b) Converting a fraction to an equivalent fraction   *Answers vary.*
7. (a) Least common denominator   *Answers vary.*
   (b) Finding the LCD between two fractions   *Answers vary.*
8. Multiplying fractions   *Answers vary.*
9. Dividing fractions   *Answers vary.*
10. (a) Adding fractions   *Answers vary.*
    (b) Subtracting fractions   *Answers vary.*

Determine if the statement is true or false. If a statement is false, explain why.

11. The fraction $\dfrac{2+3}{5+3}$ simplifies to $\dfrac{2}{5}$.   *False. It simplifies to $\dfrac{5}{8}$.*
12. All odd numbers are prime numbers.   *False. 9 is an odd number but is not prime.*
13. The only even prime number is the number 2.   *True.*
14. The LCD of $\dfrac{3}{4}$, $\dfrac{5}{6}$, and $\dfrac{1}{12}$ is 24.   *False. The LCD of 4, 6, and 12 is 12.*
15. The number 0 has a reciprocal.   *False. The reciprocal of 0 would be $\dfrac{1}{0}$ which is undefined.*
16. $4 \div \dfrac{1}{8} = \dfrac{1}{4} \times \dfrac{1}{8}$   *False. It becomes $4 \times \dfrac{8}{1}$ or $4 \times 8 = 32$.*
17. $\dfrac{1}{2} + \dfrac{1}{5} = \dfrac{2}{7}$   *False. The LCD of 2 and 5 is 10. We don't add the denominators.*
18. $\dfrac{3}{7} - \dfrac{1}{3} = \dfrac{2}{4} = \dfrac{1}{2}$   *False. The LCD of 7 and 3 is 21. We don't subtract the denominators.*

*Additional answers can be found in the Instructor Answer Appendix.*

### 🎹 Practice Makes Perfect!

Write the prime factorization of each number. (*See Objective 1.*)

19. 12   $2 \cdot 2 \cdot 3$
20. 27   $3 \cdot 3 \cdot 3$
21. 80   $2 \cdot 2 \cdot 2 \cdot 2 \cdot 5$
22. 84   $2 \cdot 2 \cdot 3 \cdot 7$
23. 144   $2 \cdot 2 \cdot 2 \cdot 2 \cdot 3 \cdot 3$
24. 210   $2 \cdot 3 \cdot 5 \cdot 7$

Write a fraction that represents the given quantity. (*See Objective 2.*)

25. As of June 30, 2010, there were approximately 3835 million people living in Asia and approximately 825 million Internet users in Asia. Write a fraction that represents the portion of Internet users in Asia. (Source: http://internetworldstats.com/america.htm)   $\dfrac{165}{767}$

26. As of June 30, 2010, there were approximately 344 million people living in North America and approximately 226 million Internet users. Write a fraction that represents the portion of Internet users in North America. (Source: http://internetworldstats.com/america.htm)   $\dfrac{113}{172}$

27. California had a total of 53 members of the U.S. House of Representatives serving in the 111th Congress and 34 representatives were Democrats. Write a fraction that represents the portion of Democrats who served as representatives for California. (http://www.house.gov)   $\dfrac{34}{53}$

28. Georgia had a total of 13 members of the U.S. House of Representatives serving in the 111th Congress and 7 representatives were Republicans. Write a fraction that represents the portion of Republicans who served as representatives for Georgia. (http://www.house.gov)   $\dfrac{7}{13}$

29. The top 10 U.S. Internet search providers processed a total of approximately 9,200,000,000 search requests, during August 2010. Google processed about 6,000,000,000 and Yahoo! processed about 1,210,000,000 search requests. Write fractions that represent the portion of Google search requests

and Yahoo! search requests among the top 10 U.S. providers. (http://en-us.nielsen.com/content/nielsen/en_us/insights/rankings/internet.html) Google: $\frac{15}{23}$; Yahoo! $\frac{121}{920}$

**30.** Senators in Class III took office in 2005 and their terms expired in 2011. With a total of 34 senators in Class III, 16 are Democrats and 18 are Republicans. Write fractions that represent the portion of Democrats and the portion of Republicans in Class III. (http://www.senate.gov)

**Simplify each fraction.** (*See Objective 3.*)

**31.** $\frac{8}{24}$ $\frac{1}{3}$    **32.** $\frac{12}{60}$ $\frac{1}{5}$    **33.** $\frac{64}{8}$ 8

**34.** $\frac{70}{35}$ 2    **35.** $\frac{192,000}{256,000}$ $\frac{3}{4}$  **36.** $\frac{126,000}{432,000}$ $\frac{7}{24}$

**Multiply the fractions and simplify each result.**
(*See Objective 4.*)

**37.** $\frac{1}{2} \cdot \frac{3}{5}$ $\frac{3}{10}$    **38.** $\frac{4}{7} \cdot \frac{2}{11}$ $\frac{8}{77}$    **39.** $\frac{2}{3} \cdot \frac{9}{28}$ $\frac{3}{14}$

**40.** $\frac{4}{15} \cdot \frac{5}{24}$ $\frac{1}{18}$    **41.** $\frac{3}{5} \cdot \frac{5}{3}$ 1    **42.** $\frac{11}{10} \cdot \frac{10}{11}$ 1

**43.** $\frac{5}{4} \cdot 20$ 25    **44.** $12 \cdot \frac{3}{2}$ 18    **45.** $2\frac{2}{3} \cdot 3\frac{3}{4}$ 10

**46.** $3\frac{1}{2} \cdot 1\frac{1}{7}$ 4

**Divide the fractions and simplify each result.**
(*See Objective 5.*)

**47.** $\frac{1}{2} \div \frac{3}{5}$ $\frac{5}{6}$    **48.** $\frac{4}{7} \div \frac{2}{11}$ $\frac{22}{7}$

**49.** $\frac{2}{3} \div \frac{9}{28}$ $\frac{56}{27}$    **50.** $\frac{7}{15} \div \frac{14}{36}$ $\frac{6}{5}$

**51.** $\frac{1}{3} \div 3$ $\frac{1}{9}$    **52.** $7 \div \frac{1}{7}$ 49

**53.** $20 \div \frac{5}{4}$ 16    **54.** $12 \div \frac{3}{5}$ 20

**55.** $6\frac{2}{3} \div 1\frac{1}{9}$ 6    **56.** $7\frac{1}{2} \div 2\frac{1}{2}$ 3

**Add or subtract the fractions and simplify each result.**
(*See Objective 6.*)

**57.** $\frac{1}{2} + \frac{7}{2}$ 4    **58.** $\frac{2}{3} + \frac{10}{3}$ 4    **59.** $\frac{3}{7} + \frac{1}{7}$ $\frac{4}{7}$

**60.** $\frac{5}{10} + \frac{1}{10}$ $\frac{3}{5}$    **61.** $\frac{9}{10} - \frac{1}{10}$ $\frac{4}{5}$    **62.** $\frac{5}{12} - \frac{3}{12}$ $\frac{1}{6}$

**63.** $7\frac{3}{4} + 1\frac{3}{4}$ $9\frac{1}{2}$    **64.** $6\frac{1}{5} - 3\frac{4}{5}$ $2\frac{2}{5}$

**Find the least common denominator (LCD) of the fractions.** (*See Objective 7.*)

**65.** $\frac{2}{3}, \frac{3}{4}$ 12    **66.** $\frac{1}{5}, \frac{1}{2}$ 10    **67.** $\frac{5}{8}, \frac{1}{2}$ 8

**68.** $\frac{3}{5}, \frac{7}{20}$ 20    **69.** $\frac{3}{18}, \frac{4}{27}$ 54    **70.** $\frac{5}{6}, \frac{2}{21}$ 42

**71.** $\frac{1}{2}, 8$ 2    **72.** $\frac{4}{7}, 5$ 7

**Add or subtract the fractions and simplify each result.**
(*See Objective 7.*)

**73.** $\frac{1}{2} + \frac{1}{7}$ $\frac{9}{14}$    **74.** $\frac{1}{4} + \frac{1}{5}$ $\frac{9}{20}$    **75.** $\frac{3}{8} + 4$ $\frac{35}{8}$

**76.** $\frac{6}{11} + 2$ $\frac{28}{11}$    **77.** $\frac{7}{2} - \frac{7}{3}$ $\frac{7}{6}$    **78.** $\frac{6}{4} - \frac{6}{5}$ $\frac{3}{10}$

**79.** $5 - \frac{1}{2}$ $\frac{9}{2}$    **80.** $4 - \frac{2}{3}$ $\frac{10}{3}$    **81.** $\frac{3}{14} - \frac{1}{12}$ $\frac{11}{84}$

**82.** $\frac{4}{10} - \frac{2}{15}$ $\frac{4}{15}$    **83.** $2\frac{1}{6} + 3\frac{2}{3}$ $5\frac{5}{6}$   **84.** $6\frac{1}{4} - 3\frac{1}{2}$ $2\frac{3}{4}$

 **Mix 'Em Up!**

**Simplify each fraction.**

**85.** $\frac{243}{405}$ $\frac{3}{5}$    **86.** $\frac{11400}{60800}$ $\frac{3}{16}$

**87.** $\frac{336}{384}$ $\frac{7}{8}$    **88.** $\frac{144}{464}$ $\frac{9}{29}$

**Perform the operation and simplify each result.**

**89.** $\frac{4}{15} + \frac{13}{30}$ $\frac{7}{10}$  **90.** $\frac{1}{3} + \frac{5}{12}$ $\frac{3}{4}$   **91.** $4\frac{1}{5} \cdot 1\frac{5}{7}$ $\frac{36}{5}$

**92.** $3\frac{3}{10} \cdot 1\frac{2}{3}$ $\frac{11}{2}$  **93.** $\frac{5}{3} - \frac{1}{5}$ $\frac{22}{15}$  **94.** $\frac{7}{2} - \frac{1}{3}$ $\frac{19}{6}$

**95.** $6\frac{7}{8} \div 2\frac{3}{4}$ $\frac{5}{2}$  **96.** $1\frac{2}{7} \div 6\frac{3}{7}$ $\frac{1}{5}$  **97.** $9\frac{1}{4} - 6\frac{2}{3}$ $2\frac{7}{12}$

**98.** $7\frac{2}{5} + 8\frac{4}{5}$ $16\frac{1}{5}$  **99.** $\frac{5}{18} \cdot \frac{2}{15}$ $\frac{1}{27}$  **100.** $\frac{2}{3} \cdot \frac{21}{16}$ $\frac{7}{8}$

**101.** $10\frac{2}{5} - 8\frac{4}{5}$ $\frac{8}{5}$  **102.** $2\frac{5}{6} - 1\frac{7}{8}$ $\frac{23}{24}$

 **You Be the Teacher!**

**Correct each student's errors, if any.**

**103.** Find the numerator that makes the fractions equivalent: $\frac{3}{8} = \frac{}{16}$.

Mary's work:

$\frac{3}{8} = \frac{3+8}{8+8} = \frac{11}{16}$. So, $\frac{3}{8} = \frac{11}{16}$.   $\frac{3}{8} = \frac{3 \cdot 2}{8 \cdot 2} = \frac{6}{16}$

**104.** Multiply and simplify the result: $\frac{2}{5} \times \frac{3}{10}$.

Jennifer's work:

$\frac{2}{5} \times \frac{3}{10} = \frac{2 \times 2}{5 \times 2} \cdot \frac{3}{10} = \frac{4}{10} \cdot \frac{3}{10} = \frac{12}{10} = \frac{2 \times 6}{2 \times 5} = \frac{6}{5}$

**105.** Multiply and simplify the result: $2\frac{1}{4} \times 6\frac{3}{5}$.

Scott's work:

$$2\frac{1}{3} \times 3\frac{3}{4} = 6\frac{3}{12} = 6\frac{1}{4} = \frac{25}{4}$$

**106.** Subtract and simplify the result: $\frac{3}{8} - \frac{1}{12}$.

Pauline's work: $8 \times 12 = 96$

$$\frac{3}{8} - \frac{1}{12} = \frac{3 \times 12}{8 \times 12} - \frac{1 \times 8}{12 \times 8} = \frac{36}{96} - \frac{8}{96}$$

$$= \frac{28}{96} = \frac{4 \times 7}{4 \times 24} = \frac{7}{24}$$

 **Calculate It!**

Use a calculator to perform the operation. Verify the answer by performing the operation by hand.

**107.** $\frac{3}{7} + \frac{12}{11}$   $\frac{117}{77}$   **108.** $\frac{8}{13} \cdot 6$   $\frac{48}{13}$   **109.** $3 - \frac{4}{3}$   $\frac{5}{3}$

**110.** $\frac{12}{35} \cdot \frac{14}{3}$   $\frac{8}{5}$   **111.** $\frac{3}{14} \cdot 3$   $\frac{9}{14}$   **112.** $5 + \frac{6}{5}$   $\frac{31}{5}$

---

| SECTION  1.3 | The Order of Operations, Algebraic Expressions, and Equations |

▶ **OBJECTIVES**

As a result of completing this section, you will be able to

**1.** Evaluate exponential expressions.

**2.** Use the order of operations to simplify numerical expressions.

**3.** Evaluate algebraic expressions.

**4.** Determine if a value is a solution of an equation.

**5.** Express relationships mathematically.

**6.** Solve application problems.

**7.** Troubleshoot common errors.

While we may not simplify numerical expressions on a daily basis, many of the things we encounter in life are based on this skill. One example involves saving money. There is an equation that can be used to calculate how much money we will accumulate if we invest a specific amount of money in a savings account. For example, if we invest $5000 in a savings account that earns 6% annual interest for 3 yr, the total amount saved is given by the equation

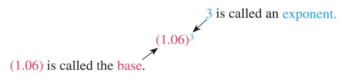

$$\text{Amount saved} = 5000(1.06)^3$$

To find the value of this expression, we must know the order in which to perform the operations. In this section, we will learn how to simplify numerical expressions using the accepted order of operations. We will also learn how to apply this skill to evaluate algebraic expressions and equations.

## Exponential Expressions

The equation shown in the section opener contains an expression of the form $(1.06)^3$. This is an example of an *exponential expression*.

3 is called an exponent.

$$(1.06)^3$$

$(1.06)$ is called the base.

**Objective 1** ▶

Evaluate exponential expressions.

The exponent implies that we use the base as a repeated factor 3 times. That is,

$$(1.06)^3 = (1.06)(1.06)(1.06) \approx 1.19$$

**Definition:  Exponential Notation**

For $b$ a real number and $n$ a natural number,

$$b^n = \underbrace{b \cdot b \cdot b \cdots b}_{n \text{ times}}$$

The number $b$ is called the **base** of the **exponential expression** and the number $n$ is called the **exponent**.

INSTRUCTOR NOTE:
Illustrate the difference between $3^2$ and $3 \cdot 2$.

When the number $n$ is a natural number, it indicates the number of times $b$ is multiplied by itself or used as a factor. Some examples are illustrated next.

| Phrase | Mathematical Expression |
|---|---|
| 3 squared | $3^2 = 3 \cdot 3 = 9$ |
| 5 cubed | $5^3 = 5 \cdot 5 \cdot 5 = 125$ |
| 6 to the 4th | $6^4 = 6 \cdot 6 \cdot 6 \cdot 6 = 1296$ |

**Procedure: Evaluating an Exponential Expression**

**Step 1:** Identify the base and the exponent.
**Step 2:** Rewrite the expression as repeated multiplication as indicated by the exponent.
**Step 3:** Simplify the result.

**Objective 1  Examples**  Complete the chart by identifying the base, the exponent, and then evaluate the expression.

| Problems | Base | Exponent | Evaluate |
|---|---|---|---|
| **1a.** $2^5$ | 2 | 5 | $2^5 = 2 \cdot 2 \cdot 2 \cdot 2 \cdot 2 = 32$ |
| **1b.** $4.2^3$ | 4.2 | 3 | $4.2^3 = (4.2)(4.2)(4.2) = 74.088$ |
| **1c.** $\left(\dfrac{2}{3}\right)^4$ | $\dfrac{2}{3}$ | 4 | $\left(\dfrac{2}{3}\right)^4 = \dfrac{2}{3} \cdot \dfrac{2}{3} \cdot \dfrac{2}{3} \cdot \dfrac{2}{3} = \dfrac{16}{81}$ |
| **1d.** $-6^2$ | 6 | 2 | We must find the *opposite* of 6 squared. $-6^2 = -(6)(6) = -36$ |

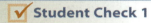 **Student Check 1**  Identify the base, the exponent, and then evaluate the expression.

    **a.** $10^3$         **b.** $1.5^2$         **c.** $\left(\dfrac{6}{7}\right)^3$         **d.** $-8^4$

## The Order of Operations

**Objective 2 ▶**

Use the order of operations to simplify numerical expressions.

The expression $5000(1.06)^3$, found in the section opener, is an example of an expression involving a combination of operations. This expression involves both multiplication and exponents. To simplify this expression, we need to know the order in which to perform the operations. Without an order to perform the operations, we would get different values for the same expression.

If we multiply 5000 and 1.06 first in the expression $5000(1.06)^3$, we get

$$[5000(1.06)]^3 = (5300)^3 = 148,877,000,000$$

If we evaluate the exponential expression $(1.06)^3$ first, we get

$$5000(1.06)^3 = 5000(1.191016) = 5955.08$$

Sometimes expressions include *grouping symbols* such as parentheses ( ), brackets [ ], braces { }, absolute value symbols | |, square roots $\sqrt{\phantom{x}}$, or the fraction bar. Depending on where the grouping symbols are placed in an expression, different results may be obtained.

To avoid confusion and errors, mathematicians developed an order for which we perform operations when simplifying expressions. This is called the *order of operations* and it is stated next.

**Procedure: Order of Operations**

When simplifying a numerical expression, perform the operations in the following order:

**Step 1:** Simplify expressions inside grouping symbols first, starting with the innermost set. If there is a fraction bar, simplify the numerator and denominator separately.

**Step 2:** Simplify any exponential expressions.

**Step 3:** Perform multiplication or division in order from left to right.

**Step 4:** Perform addition or subtraction in order from left to right.

---

**Objective 2 Examples**   **Use the order of operations to simplify each expression.**

**2a.** $9 + 5(6)$

**2b.** $4 \cdot 5 \div 2 + 3 \cdot 8$

**2c.** $\dfrac{2}{3}(6-2)^2 - \dfrac{1}{3}$

**2d.** $\dfrac{6 - \sqrt{6^2 - 4(5)(1)}}{2(1)}$

**2e.** $4[7 - |9 - 4(5-3)|]$

**Solutions**

**2a.** $9 + 5(6)$

$\quad 9 + 30$      Multiply 5 and 6.

$\quad 39$      Add.

**2b.** $4 \cdot 5 \div 2 + 3 \cdot 8$

$\quad 20 \div 2 + 3 \cdot 8$      Multiply 4 and 5.

$\quad 10 + 3 \cdot 8$      Divide 20 by 2.

$\quad 10 + 24$      Multiply 3 and 8.

$\quad 34$      Add the numbers.

**2c.** $\dfrac{2}{3}(6-2)^2 - \dfrac{1}{3}$

$\quad \dfrac{2}{3}(4)^2 - \dfrac{1}{3}$      Simplify inside parentheses.

$\quad \dfrac{2}{3}(16) - \dfrac{1}{3}$      Simplify $4^2$.

$\quad \dfrac{32}{3} - \dfrac{1}{3}$      Multiply: $\dfrac{2}{3}\left(\dfrac{16}{1}\right) = \dfrac{32}{3}$.

$\quad \dfrac{31}{3}$      Subtract the resulting fractions.

**2d.** $\dfrac{6 - \sqrt{6^2 - 4(5)(1)}}{2(1)}$

$\quad \dfrac{6 - \sqrt{36 - 4(5)(1)}}{2}$      Simplify $6^2$. Multiply the numbers in the denominator.

$\quad \dfrac{6 - \sqrt{36 - 20}}{2}$      Multiply: $4(5)(1) = 20(1) = 20$.

$\quad \dfrac{6 - \sqrt{16}}{2}$      Subtract the numbers in the square root.

$\quad \dfrac{6 - 4}{2}$      Simplify $\sqrt{16}$.

$\quad \dfrac{2}{2}$      Subtract the numbers in the numerator.

$\quad 1$      Divide the numbers.

**2e.** $4[7 - |9 - 4(5 - 3)|]$

| | |
|---|---|
| $4[7 - |9 - 4(2)|]$ | Simplify inside parentheses. |
| $4[7 - |9 - 8|]$ | Multiply 4 and 2 in the absolute value. |
| $4[7 - |1|]$ | Subtract the numbers in the absolute value. |
| $4[7 - 1]$ | Simplify the absolute value of 1. |
| $4(6)$ | Subtract the numbers in the brackets. |
| $24$ | Multiply. |

✓ **Student Check 2**    Use the order of operations to simplify each expression.

**a.** $3 + 2(9)$     **b.** $8 \cdot 3 \div 6 - 2 \cdot 2$     **c.** $\frac{1}{5}(8 - 6)^3 - \frac{3}{5}$

**d.** $\dfrac{4 + \sqrt{8^2 - 4(7)(1)}}{2(7)}$     **e.** $2[11 - 4|8 - 3(5 - 3)|]$

## Evaluating Algebraic Expressions

**Objective 3** ▶

Evaluate algebraic expressions.

We will now apply the order of operations to evaluating algebraic expressions. An **algebraic expression** is an expression that involves variables and/or numbers joined by arithmetic operations. Recall a *variable* is a letter or symbol that represents some unknown number. Examples of algebraic expressions are

$$2x \qquad 4y - 5 \qquad 3x^2 - 5x + 7$$

When we evaluate an algebraic expression, we assign a specific value for each of the variables and determine the value of the resulting expression.

> **Procedure: Evaluating an Algebraic Expression**
>
> **Step 1:** Replace the variable(s) with the given number(s).
> **Step 2:** Use the order of operations to simplify the resulting expression.

**Objective 3 Examples**    **Evaluate each expression for the given values.**

**3a.** Find the value of $2x + 3$ when $x = 0, \frac{1}{2}, 1$, and 2.

**3b.** Find the value of $\dfrac{x - 1}{x}$ for $x = 1, 1.5, 2$, and 3.

**3c.** Find the value of $b^2 - 4ac$ when $a = 3, b = 5$, and $c = 2$.

**Solutions**    **3a.** Since we are evaluating the same expression for more than one value, we can organize the information in a chart.

**INSTRUCTOR NOTE:**
To prepare students for graphing equations in two variables, use a table and evaluate the expression for more than one value of the variable. This will enable them to transition more easily to finding solutions of equations of the form $y = 2x + 3$.

| $x$ | Value of $2x + 3$ | |
|---|---|---|
| 0 | $2(0) + 3 = 0 + 3 = 3$ | Replace $x$ with 0. |
| $\frac{1}{2}$ | $2\left(\frac{1}{2}\right) + 3 = 1 + 3 = 4$ | Replace $x$ with $\frac{1}{2}$. |
| 1 | $2(1) + 3 = 2 + 3 = 5$ | Replace $x$ with 1. |
| 2 | $2(2) + 3 = 4 + 3 = 7$ | Replace $x$ with 2. |

**3b.**

| $x$ | Value of $\dfrac{x-1}{x}$ | |
|---|---|---|
| 1 | $\dfrac{1-1}{1} = \dfrac{0}{1} = 0$ | Replace $x$ with 1. |
| 1.5 | $\dfrac{1.5-1}{1.5} = \dfrac{0.5}{1.5} = \dfrac{1}{3}$ | Replace $x$ with 1.5. |
| 2 | $\dfrac{2-1}{2} = \dfrac{1}{2}$ | Replace $x$ with 2. |
| 3 | $\dfrac{3-1}{3} = \dfrac{2}{3}$ | Replace $x$ with 3. |

**3c.**

$$b^2 - 4ac = (5)^2 - 4(3)(2) \qquad \text{Replace } b \text{ with 5, } a \text{ with 3, and } c \text{ with 2.}$$
$$= 25 - 4(3)(2) \qquad \text{Simplify the exponential expression.}$$
$$= 25 - 12(2) \qquad \text{Multiply 4 and 3.}$$
$$= 25 - 24 \qquad \text{Multiply 12 and 2.}$$
$$= 1 \qquad \text{Subtract 25 and 24.}$$

✓ **Student Check 3**   Evaluate each expression for the given values.
  **a.** Find the value of $3x + 5$ for $x = 0$, 1, and 2.
  **b.** Find the value of $\dfrac{x-4}{x+1}$ for $x = 4$, 5, 5.2, and 6.
  **c.** Find the value of $b^2 - 4ac$ when $a = 6$, $b = 8$, and $c = 2$.

## Solutions of Equations

**Objective 4** ▶

Determine if a value is a solution of an equation.

A statement that shows two expressions are equal is an **equation**. The equals sign "=" is used to denote this equality. We will study many different kinds of equations throughout this text. Some examples of equations are

$$4x + 5 = 9, \qquad x^2 - x = 6, \qquad \sqrt{y+4} = 2y, \qquad P = 2l + 2w$$

One of the goals in algebra is to **solve an equation**. To solve an equation means to find the value(s) of the variable(s) that satisfies the equation, that is, makes the equation true. Each such value is called a **solution of an equation**. For instance,

| $x = 1$ is a solution of $4x + 5 = 9$: | $x = 2$ is not a solution of $4x + 5 = 9$: |
|---|---|
| $4(1) + 5 = 9$ | $4(2) + 5 = 9$ |
| $4 + 5 = 9$ | $8 + 5 = 9$ |
| $9 = 9$ | $13 = 9$ |
| True | False |

> **Procedure:  Determining if a Value is a Solution of an Equation**
>
> **Step 1:** Replace the variable(s) with the given values.
> **Step 2:** Simplify each side of the equation.
> **Step 3:** If the resulting equation is true, then the value is a solution. If the resulting equation is false, then the value is *not* a solution of the equation.

**Objective 4 Examples** / Determine if the given value is a solution of the equation.

**4a.** $4y + 5 = y^2$; $y = 5$      **4b.** $3x - 2(x + 1) = x + 5$; $x = 4$

**Solutions**    **4a.**

$$4y + 5 = y^2$$
$$4(5) + 5 = (5)^2 \qquad \text{Replace } y \text{ with 5.}$$
$$20 + 5 = 25 \qquad \text{Multiply 4 and 5 and simplify } 5^2.$$
$$25 = 25 \qquad \text{Add.}$$

Since $y = 5$ makes the equation true, it is a solution of $4y + 5 = y^2$.

**4b.**

$$3x - 2(x + 1) = x + 5$$
$$3(4) - 2(4 + 1) = 4 + 5 \qquad \text{Replace } x \text{ with 4.}$$
$$12 - 2(5) = 9 \qquad \text{Multiply 3 and 4. Add in parentheses and on the right side..}$$
$$12 - 10 = 9 \qquad \text{Multiply 2 and 5.}$$
$$2 = 9 \qquad \text{Simplify.}$$

Since $x = 4$ makes the equation false, it is *not* a solution of $3x - 2(x + 1) = x + 5$.

✓ **Student Check 4**    Determine if the given value is a solution of the equation.

**a.** $a^2 - a + 6 = a - 2$; $a = 4$      **b.** $7b - 5(b + 2) = b + 3$; $b = 13$

## Translate Expressions into Symbols

**Objective 5** ▶

Express relationships mathematically.

Now that we know how to evaluate an algebraic expression, we will turn our focus on how to express or write mathematical relationships given certain phrases or sentences. The following table shows some common phrases for the basic mathematical expressions.

**INSTRUCTOR NOTE:**
Point out that order matters when subtracting or dividing, so students should be careful when translating with these operations. Show the difference between "4 less than a number" and "a number less than 4." Point out that sometimes we must decide whether a word translates to an operation or not. Discuss the word "of" in the following examples: "$\frac{1}{2}$ of 24" and "the difference of 10 and 5."

| | | |
|---|---|---|
| **Addition** | $a + b$ | sum of $a$ and $b$<br>$a$ increased by $b$<br>$b$ more than $a$<br>$a$ added to $b$<br>$a$ plus $b$ |
| **Subtraction** | $a - b$ | difference of $a$ and $b$<br>$b$ subtracted from $a$<br>$b$ less than $a$<br>$a$ minus $b$<br>$a$ decreased by $b$<br>from $a$, subtract $b$ |
| **Multiplication** | $ab$ | product of $a$ and $b$<br>$a$ times $b$<br>$a$ multiplied by $b$<br>$a$ of $b$ |
| **Division** | $\dfrac{a}{b}$ | $a$ divided by $b$<br>quotient of $a$ and $b$<br>ratio of $a$ to $b$ |
| **Equation** | $a = b$ | $a$ is $b$<br>$a$ is equal to $b$<br>$a$ yields $b$<br>The result of $a$ is the same as $b$ |

Many of the problems that we encounter in math require us to have the ability to express mathematical relationships correctly. Example 5 practices this skill. When one of the numbers is not known, a variable is used to represent this unknown number.

**Objective 5  Examples**  Translate each phrase into an algebraic expression, equation, or inequality. Use the variable $x$ to represent the unknown number.

| Problems | Solutions |
|---|---|
| **5a.** A number increased by 7 | Since "increase" means add, the expression is<br>$$x + 7$$ |
| **5b.** Five less than a number | Since "less than" means subtract, the expression is<br>$$x - 5$$ |
| **5c.** The difference of a number and 2 | Since "difference" means subtract, the expression is<br>$$x - 2$$ |
| **5d.** The product of 5 and a number | Since "product" means multiply, the expression is<br>$$5 \cdot x \text{ or } 5x$$ |
| **5e.** The quotient of a number and 3 | Since "quotient" means divide, the expression is<br>$$\frac{x}{3}$$ |
| **5f.** The sum of twice a number and 4 | Since "twice a number" means multiply and "sum" means add, the expression is<br>$$2x + 4$$ |
| **5g.** The difference of three times a number and 1 | Since "three times" means multiply and "difference" means subtract, the expression is<br>$$3x - 1$$ |
| **5h.** Twice the sum of a number and 6 | Since "sum" means add and "twice" means multiply, we have<br>$$2(x + 6)$$ |
| **5i.** Three less than one-half of a number | Since "less than" means subtract and "$\frac{1}{2}$ of" means multiply we have<br>$$\frac{1}{2}x - 3$$ |
| **5j.** A number less than 8 is the same as three times the number. | Since "is the same as" means equal to, the translation is an equation of the form<br>$$8 - x = 3x$$ |
| **5k.** A number is less than 3. | Since "is less than" denotes inequality, we have<br>$$x < 3$$ |

✔ **Student Check 5**  Translate each phrase or sentence into an algebraic expression, equation, or inequality. Use the variable $x$ to represent the unknown number.

**a.** A number decreased by 2

**b.** Three more than a number

**c.** The sum of three times a number and 5

**d.** The quotient of a number and 2

**e.** The difference of a number and 4

**f.** The product of a number and 6

**g.** The difference of twice a number and 9

**h.** Three times the sum of a number and 1

**i.** Five less than one-third of a number

**j.** One more than four times a number is 15.

**k.** Twice a number is less than 4.

**Objective 6** ▶

Solve application problems.

## Applications

The skills of simplifying numerical expressions, evaluating algebraic expressions or equations, and expressing relationships mathematically will be used in the context of real-life situations.

One of the specific applications deals with calculating the *perimeter* of polygons. Polygons are two-dimensional closed shapes made up of straight lines, such as triangles, squares, rectangles, trapezoids, and the like.

> **Definition: Perimeter of a Polygon**
>
> The **perimeter** of a polygon is the total distance around the outside of the polygon.

For example, the perimeter of a rectangular yard with width 100 ft and length 250 ft is

$$P = 100 + 250 + 100 + 250$$

$$P = 700$$

So, the perimeter is 700 ft.

100 ft — 250 ft — 250 ft — 100 ft

**Objective 6  Examples**  Solve each problem.

**6a.** If $5000 is invested in a savings account that earns 6% annual interest for 3 yr, the total amount saved is given by the equation.

$$\text{Amount saved} = 5000(1.06)^3$$

Find the amount that would be saved.

**Solution**    **6a.** Simplifying the expression on the right side of the equation gives us that

$$\text{Amount saved} = 5955.08$$

So, in 3 yr, a total of $5955.08 will be in the account.

**6b.** After renovations in 2010, the Michigan Stadium at the University of Michigan in Ann Arbor, Michigan, became the largest football stadium in the United States with a seating capacity of 108,000. The football field is in the shape of a rectangle. The perimeter $P$ of a rectangle is the sum of 2 times its length and 2 times its width.

  **i.** Write an equation that denotes the perimeter of a rectangle, where $l$ is the length and $w$ is the width.

  **ii.** If the football field has a length of 360 ft and a width of 160 ft, what is the field's perimeter?

**Solution**    **6b.**    **i.** The equation that represents the perimeter is $P = 2l + 2w$.

    **ii.** $P = 2l + 2w$      Use the relationship defined in part (i).

         $P = 2(360) + 2(160)$      Replace *l* with 360 and *w* with 160.

         $P = 720 + 320$      Simplify each product.

         $P = 1040$      Add.

The perimeter of the football field is 1040 ft.

**6c.** The number of viewers (in millions) for the season premiere of Fox Network's *American Idol* for Seasons 1 to 8 is shown in the table.

| Season | Viewers (in millions) |
|--------|-----------------------|
| 1 (2002) | 9.9 |
| 2 (2003) | 26.5 |
| 3 (2004) | 28.56 |
| 4 (2005) | 33.58 |
| 5 (2006) | 35.53 |
| 6 (2007) | 37.7 |
| 7 (2008) | 33.4 |
| 8 (2009) | 30.4 |

   **i.** The number of viewers can be approximated by the expression $0.1x^3 + 18.4x - 2.54x^2 - 4.6$, where $x$ is the number of seasons aired. Write an equation that represents the approximate number of viewers, where $v$ is the number of viewers in millions.

   **ii.** Use the equation to estimate the number of viewers for Season 9 (Jan. 2010).

**Solution**   **6c.**   **i.** The equation that represents the number of viewers, in millions, is
$$v = 0.1x^3 + 18.4x - 2.54x^2 - 4.6.$$

   **ii.**
$$v = 0.1x^3 + 18.4x - 2.54x^2 - 4.6$$
$$v = 0.1(9)^3 + 18.4(9) - 2.54(9)^2 - 4.6 \qquad \text{Replace the variable } x \text{ with 9.}$$
$$v = 0.1(729) + 18.4(9) - 2.54(81) - 4.6 \qquad \text{Evaluate the exponential expressions.}$$
$$v = 72.9 + 165.6 - 205.74 - 4.6 \qquad \text{Add or subtract from left to right.}$$
$$v = 28.16$$

The expression estimates that there were approximately 28.16 million, or 28,160,000, viewers for Season 9.

**6d.** The basal metabolic rate (BMR) is the number of calories a person burns to sustain life. It is the number of calories burned even if a person stayed in bed all day. It is based on a person's height, weight, and age. The BMR for women is 655 plus 4.35 times a person's weight in pounds plus 4.7 times a person's height in inches minus 4.7 times a person's age.

   **i.** Write an equation that represents a woman's BMR, where $w$ is the woman's weight (in pounds), $h$ is the woman's height (in inches), and $a$ is the woman's age.

   **ii.** If Susan is a 35-yr-old who weighs 145 lb and is 5 ft 2 in. tall, what is her BMR? Round to the nearest integer.

**Solution**   **6d.**   **i.** The equation that represents a woman's basal metabolic rate is given by

BMR is  655  plus 4.35 times plus 4.7 times minus 4.7 times
                        weight          height         age
$$\text{BMR} = 655 + 4.35w + 4.7h - 4.7a$$

   **ii.** To find Susan's BMR, we must make appropriate substitutions for the variables. We are given that $w = 145$, $h = 62$, and $a = 35$. So, her BMR is

$$\text{BMR} = 655 + 4.35w + 4.7h - 4.7a$$
$$\text{BMR} = 655 + 4.35(145) + 4.7(62) - 4.7(35)$$
$$\text{BMR} = 655 + 630.75 + 291.4 - 164.5$$
$$\text{BMR} = 1412.65$$

Susan's BMR is approximately 1413 calories.

✓ **Student Check 6**    Solve each problem.

a. Suppose $2000 is invested in a certificate of deposit, CD, account that earns 5% annual interest for 4 yr. The amount of money in the account after 4 yr is given by Amount $= 2000(1.05)^4$. Simplify this expression to determine the amount of money in the account in 4 yr.

b. The Singapore Flyer is the world's largest Ferris wheel. The Ferris wheel is in the shape of a circle. The distance around a circle is called its *circumference*. The circumference of a circle is twice the product of $\pi$ and the radius of the circle, where $r$ is the radius of the circle.

   i. Write an equation that represents the circumference of the circle.

   ii. If the radius of the Singapore Flyer is 75 m, find its circumference. Use 3.14 for $\pi$.

c. The number of bachelor's degrees conferred in the Computer and Information Sciences (CIS) majors from 1970 to 2005 are shown in the table. The number of degrees conferred in this field can be approximated by the expression $0.3x^4 - 18x^3 + 308.4x^2 + 273.3x + 1096.6$, where $x$ is the number of years after 1970.
(Source: National Center for Education Statistics)

| Years after 1970 | Degrees Conferred |
|---|---|
| 0 | 2388 |
| 5 | 5652 |
| 10 | 15121 |
| 15 | 42337 |
| 20 | 25159 |
| 25 | 24506 |
| 30 | 44142 |
| 35 | 47480 |

   i. Write an equation that can be used to approximate the number of CIS degrees conferred, where $d$ is the number of degrees.

   ii. Use the equation to estimate the number of degrees conferred in 2010. Round the answer to the nearest integer.

d. The basal metabolic rate (BMR) for men is 66 plus 6.23 times a person's weight in pounds $+12.7$ times a person's height in inches minus 6.8 times a person's age in years.

   i. Write an equation that represents a man's BMR, where $w$ is the man's weight (in pounds), $h$ is the man's height (in inches), and $a$ is the man's age.

   ii. If Todd is a 43-yr-old who weighs 250 lb and is 6 ft 4 in. tall, what is his BMR? Round to the nearest integer.

**Objective 7** ▶
Troubleshoot common errors.

## Troubleshooting Common Errors

Some of the common errors associated with numerical expressions and algebraic expressions are shown next.

**Objective 7  Examples**    A problem and an incorrect solution are given. Provide the correct solution and an explanation of the error.

a. Simplify $3(4)^2$.

| Incorrect Solution | Correct Solution and Explanation |
|---|---|
| $3(4)^2$ <br> $12^2$ <br> 144 | Based on the order of operations, we must evaluate the exponent first and then multiply. <br><br> $3(4)^2$ <br> $3(16)$ <br> 48 |

**b.** Translate the phrase: 6 less than a number.

| Incorrect Solution | Correct Solution and Explanation |
|---|---|
| The translation is $6 - x$. | The phrase "6 less than a number" is translated as $x - 6$. The order of this translation is very important. Think of how we find 6 less than 9—that is, $9 - 6$ or 3. Note that the expression $6 - x$ is the translation of "6 less a number" or "the difference of 6 and a number." |

## ANSWERS TO STUDENT CHECKS

**Student Check 1**   **a.** 1000   **b.** 2.25   **c.** $\frac{216}{343}$   **d.** $-4096$

**e.** $x - 4$   **f.** $6x$   **g.** $2x - 9$   **h.** $3(x + 1)$
**i.** $\frac{1}{3}x - 5$   **j.** $4x + 1 = 15$   **k.** $2x < 4$

**Student Check 2**   **a.** 21   **b.** 0   **c.** 1   **d.** $\frac{5}{7}$   **e.** 6

**Student Check 6**   **a.** \$2431.01   **b.** $C = 2\pi r$; 471 m

**Student Check 3**   **a.** 5, 8, 11   **b.** $0, \frac{1}{6}, \frac{6}{31}, \frac{2}{7}$   **c.** 16

**c.** $d = 0.3x^4 - 18x^3 + 308.4x^2 + 273.3x + 1096.6$; 121,469 degrees conferred

**Student Check 4**   **a.** no   **b.** yes

**d.** BMR $= 66 + 6.23w + 12.7h - 6.8a$; 2296 cal

**Student Check 5**   **a.** $x - 2$   **b.** $x + 3$   **c.** $3x + 5$   **d.** $\frac{x}{2}$

## SUMMARY OF KEY CONCEPTS

**1.** An exponent is a way of writing repeated multiplication. The exponent indicates the number of times to repeat the base as a factor.

**2.** The order of operations is used to simplify numerical expressions. The order is: parentheses (or any grouping), exponents, multiplication/division in order from left to right, and addition/subtraction in order from left to right.

**3.** To evaluate an algebraic expression, replace the variable with the given value and use the order of operations to simplify the resulting expression. It is helpful to put parentheses around the number being substituted into the expression.

**4.** An equation is a statement that two expressions are equal. To determine if a number is a solution of an equation, substitute it into the equation. If the resulting statement is true, the number is a solution. If it is not true, the number is not a solution.

**5.** Be familiar with the common phrases for addition, subtraction, multiplication, division, and equality. Use the phrases to express relationships mathematically.

**6.** When solving application problems, we must be able to write an appropriate equation that represents the problem and then use the equation to determine values when given specific information.

## GRAPHING CALCULATOR SKILLS

We can use the calculator for simplifying exponential expressions, simplifying numerical expressions, and evaluating algebraic expressions. The most important part of entering a numerical expression is "telling" the calculator where the grouping symbols occur in the problem. We use parentheses to indicate grouping symbols on the calculator. If parentheses are not used correctly, an incorrect order of operations is applied.

**Example:** Use the calculator to find the value of the numerical expressions.

**1.** $\left(\dfrac{2}{3}\right)^4$

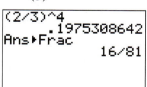

**2.** $4 \cdot 5 \div 2 + 3 \cdot 8$

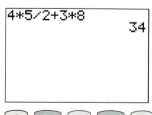

**3.** $\frac{2}{3}(6-2)^2 - \frac{1}{3}$

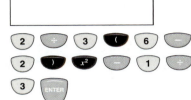

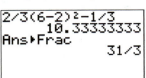

**4.** $\frac{6 - \sqrt{6^2 - 4(1)(5)}}{2(1)}$

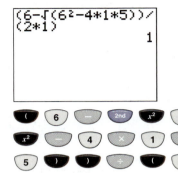

**5.** $|9 - 4(5-3)|$

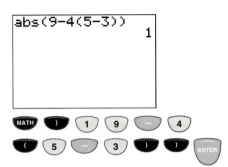

---

## SECTION 1.3 / EXERCISE SET

### Write About It!

Use complete sentences in your answers to the following exercises.

1. Explain how to apply the order of operations to simplify numerical expressions.  Answers vary.

2. How does a numerical expression differ from an algebraic expression?  Answers vary.

3. Explain how to apply the order of operations to simplify the expressions $(4+3)^2$ and $4^2 + 3^2$.  Answers vary.

4. Explain how to apply the order of operations to simplify the expressions $\frac{6+4}{3+2}$ and $\frac{6}{3} + \frac{4}{2}$.  Answers vary.

5. Explain what it means to evaluate an algebraic expression.  Answers vary.

6. What is the first step in translating phrases into algebraic expressions?  Answers vary.

7. Explain how to translate an English phrase involving "more than" into an algebraic expression.  Answers vary.

8. Explain how to translate an English phrase involving "subtracted from" into an algebraic expression.  Answers vary.

**Determine if the statement is true or false. If false, explain the error.**

9. The phrase "four less than three times a number" can be translated into $4 - 3x$.  False, the translation is $3x - 4$.

10. When the expression $x^2 - 2x$ is evaluated for $x = -4$, the result is $-8$.  False, the result is 24 because $(-4)$ squared is 16 not $-16$.

### Practice Makes Perfect!

**Identify the base, the exponent, and then evaluate the expression. (See Objective 1.)**

11. $4^4$  256

12. $5^4$  625

13. $-3^3$  $-27$

Additional answers can be found in the Instructor Answer Appendix.

14. $-7^3$  $-343$

15. $2.5^3$  15.625

16. $1.2^3$  1.728

17. $-0.6^3$  $-0.216$

18. $-0.5^3$  $-0.125$

19. $-\left(\frac{3}{5}\right)^2$  $-\frac{9}{25}$

20. $-\left(\frac{4}{7}\right)^2$  $-\frac{16}{49}$

21. $\left(\frac{2}{3}\right)^5$  $\frac{32}{243}$

22. $\left(\frac{1}{4}\right)^3$  $\frac{1}{64}$

**Use the order of operations to simplify each expression. (See Objective 2.)**

23. $5 + 4(3)$  17

24. $6 + 2(10)$  26

25. $4(7) - (3)(2)$  22

26. $11(4) - (7)(5)$  9

27. $12 \cdot 4 \div 8 - 2 \cdot 2$  2

28. $21 \cdot 4 \div 6 - 5 \cdot 2$  4

29. $\frac{1}{7}(6-4)^3 - \frac{2}{5}$  $\frac{26}{35}$

30. $\frac{1}{4}(8-5)^3 - \frac{5}{2}$  $\frac{17}{4}$

31. $\frac{16 - 11}{5 - 1}$  $\frac{5}{4}$

32. $\frac{5 + 2}{4 - 1}$  $\frac{7}{3}$

33. $2(4)^2 - 5(4) + 7$  19

34. $5(2)^2 + 4(2) - 1$  27

35. $4 \cdot 3^2 - 5 \cdot 3 + 1$  22

36. $6 \cdot 5^2 - 7 \cdot 5 + 2$  117

37. $(4 \cdot 2)^2 - 8 \div 2 \cdot 7$  36

38. $(2 \cdot 3)^2 - 4 \cdot 2 \div 8 + 3$  38

39. $(5)^2 - 4(2)(3)$  1

40. $3 \cdot 7^2 - 4(5)(6)$  27

41. $2[44 - 2|35 - 5(6-2)|]$  28

42. $5[18 - 6|14 - 2(9-3)|]$  30

43. $\frac{13 - \sqrt{13^2 - 4(4)(9)}}{2 \cdot 9}$  $\frac{4}{9}$

44. $\frac{10 - \sqrt{10^2 - 4(2)(8)}}{2 \cdot 8}$  $\frac{1}{4}$

**Evaluate each expression for the specified values of $x$. Organize the information in a chart. (See Objective 3.)**

45. $3x + 2$ for $x = 0, \frac{1}{3}, 2,$ and 4

46. $4x + 1$ for $x = \frac{1}{4}, 1, 2,$ and 4

47. $\frac{3x - 2}{x}$ for $x = 1, 2, 3,$ and 4

**48.** $\dfrac{5x - 2}{x}$ for $x = 1, 2, 3,$ and $4$

**49.** $\dfrac{x + 3}{x - 1}$ for $x = 2, 3, 4,$ and $5$

**50.** $\dfrac{x + 1}{x - 2}$ for $x = 3, 4, 5,$ and $6$

**Evaluate each expression for the given values of the variables. (See Objective 3.)**

**51.** $2x + 3y$ for $x = 0$ and $y = 2$   6

**52.** $4x + y$ for $x = 2$ and $y = 0$   8

**53.** $\dfrac{y_2 - y_1}{x_2 - x_1}$ for $x_1 = 2, x_2 = 5,$
$y_1 = 4,$ and $y_2 = 6$   $\dfrac{2}{3}$

**54.** $\dfrac{y_2 - y_1}{x_2 - x_1}$ for $x_1 = 0, x_2 = 3,$
$y_1 = 2,$ and $y_2 = 7$   $\dfrac{5}{3}$

**55.** $\dfrac{1}{2}bh$ for $b = 4$ and $h = 5$   10

**56.** $\dfrac{1}{2}bh$ for $b = 3$ and $h = 6$   9

**57.** $2l + 2w$ for $l = 8$ and $w = 10$   36

**58.** $2l + 2w$ for $l = 4$ and $w = 12$   32

**59.** $\sqrt{(x_1 - x_2)^2 + (y_1 - y_2)^2}$ for $x_1 = 1,$
$x_2 = 4, y_1 = 12, y_2 = 16$   5

**60.** $\sqrt{(x_1 - x_2)^2 + (y_1 - y_2)^2}$ for $x_1 = 0,$
$x_2 = 6, y_1 = 3, y_2 = 11$   10

**61.** $b^2 - 4ac$ for $a = 2, b = 5, c = 3$   1

**62.** $b^2 - 4ac$ for $a = 1, b = 6, c = 2$   28

**Determine if the given value is a solution of the equation. (See Objective 4.)**

**63.** $2x^2 - 3x - 1 = x + 5; x = 2$   $x = 2$ is not a solution.

**64.** $y^2 + 3y - 6 = y + 9; y = 3$   $y = 3$ is a solution.

**65.** $a^2 - 2a + 3 = 2a + 8; a = 5$   $a = 5$ is a solution.

**66.** $3b^2 - 6b - 10 = 5b - 6; b = 4$   $b = 4$ is a solution.

**67.** $9x - 6(x - 1) = 4x + 2; x = 3$   $x = 3$ is not a solution.

**68.** $4z - 2(z + 1) = z + 2; z = 5$   $z = 5$ is not a solution.

**69.** $6y - 2(y - 3) = y + 9; y = 1$   $y = 1$ is a solution.

**70.** $8z - 4(z + 1) = 2z + 8; z = 6$   $z = 6$ is a solution.

**Translate each phrase into an algebraic expression. Let $x$ represent the unknown number. (See Objective 5.)**

**71.** The sum of a number and 4   $x + 4$

**72.** The sum of a number and 12   $x + 12$

**73.** 2 more than three times a number   $3x + 2$

**74.** 6 more than four times a number   $4x + 6$

**75.** The difference of a number and 9   $x - 9$

**76.** The difference of a number and 15   $x - 15$

**77.** 3 less than four times a number   $4x - 3$

**78.** 3 decreased by twice a number   $3 - 2x$

**79.** The product of a number and 10   $10x$

**80.** The product of a number and 7   $7x$

**81.** The product of 3 and the sum of a number and 15   $3(x + 15)$

**82.** The product of 4 and a number increased by 6   $4(x + 6)$

**83.** The sum of four times a number and 3   $4x + 3$

**84.** The sum of twice a number and 4   $2x + 4$

**85.** Twice the difference of a number and 7   $2(x - 7)$

**86.** Twice the difference of a number and 3   $2(x - 3)$

**87.** The quotient of three times a number and 4   $\dfrac{3x}{4}$

**88.** The quotient of four times a number and 5   $\dfrac{4x}{5}$

**Write an algebraic expression that represents each unknown quantity. (See Objective 6.)**

**89.** Shanika invested money in two different accounts. She invested a total of $5000. If $x$ represents the amount she invested in the first account, write an expression that represents the amount she invested in the second account.   $5000 - x$

**90.** David invested money in two different accounts. He invested a total of $2000. If $x$ represents the amount he invested in the first account, write an expression that represents the amount he invested in the second account.   $2000 - x$

**91.** In a rectangle, the length is three less than twice the width. If $w$ represents the width, write an expression that represents the length of the rectangle.   $2w - 3$

**92.** In a rectangle, the width is four more than the length. If $l$ represents the length, write an expression that represents the width of the rectangle.   $l + 4$

**93.** In 2011, Oprah Winfrey and Tiger Woods were both among the 100 highest paid celebrities. Oprah Winfrey was paid $215 million more than Tiger Woods. If $x$ represents the amount Tiger Woods was paid in millions, write an algebraic expression that represents how much Oprah Winfrey was paid. (Source: http://www.forbes.com)   $x + 215$

**94.** In 2011, pop stars Lady Gaga and Justin Bieber were among the 100 highest paid celebrities. Justin Bieber was paid $37 million less than Lady Gaga. If $x$ represents the amount Lady Gaga was paid in millions, write an algebraic expression that represents how much Justin Bieber was paid. (Source: http://www.forbes.com)   $x - 37$

**Solve each problem. (See Objective 6.)**

**95.** Find the perimeter of the Lincoln Memorial Reflecting Pool in Washington, D.C. The reflecting pool is approximately 2029 ft long and 167 ft wide.   4392 ft

**96.** The perimeter of a pentagon is five times the length of a side of the pentagon. If each outside wall of the Pentagon is 921 ft long, what is the perimeter of the Pentagon?   4605 ft

**97.** The expression $P(1 + r)^t$ represents the amount of money in an account when $P$ dollars is invested at an annual interest rate of $r$ (in decimal form) for $t$ yr. Find the amount of money that will be in an account at the end of 5 yr if $1000 is invested at 5% annual interest. (Hint: Convert 5% to a decimal value before substituting this value in the expression.)   $1276.28

**98.** The expression $P(1 + r)^t$ represents the amount of money in an account when $P$ dollars is invested at an annual interest rate of $r$ (in decimal form) for $t$ yr. Find the amount of money that will be in an account at the end of 3 yr if $5000 is invested at 3% annual interest. (Hint: Convert 3% to a decimal value before substituting this value in the expression.)   $5463.64

**99.** In 2011, the *U.S. News & World Report Guide to America's Best Colleges* ranked Princeton University as one of the top national universities. Yearly tuition and fees at Princeton University is represented by the expression $1100c + 9200$, where $c$ is the number of credit hours taken in a year. What is the yearly tuition and fees if 30 credit hours are taken in a year?   $42,200

**100.** In 2011, the *U.S. News & World Report Guide to America's Best Colleges* ranked the University of Pennsylvania as the top Business School in the United States. Yearly tuition and fees at University of Pennsylvania is represented by the expression $3105c + 9804$, where $c$ is the number of credit units taken each year. What is the yearly tuition and fees if 11 credit units are taken in 1 yr?   $43,959

**101.** According to the Bureau of Labor Statistics, the average hourly rate of registered nurses in the state of Illinois is $25. The expression $25x + 37.5y$ represents the average weekly earnings of a registered nurse in Illinois, where $x$ is the number of regular hours worked in a week and $y$ is the number of overtime hours worked in a week. Find the weekly salary of a registered nurse if a nurse works 36 regular hours and 10 overtime hours.   $1275

**102.** According to the Bureau of Labor Statistics, airplane pilots and navigators earned an average of $95.80 per hour. The expression $95.80x$ represents the average weekly earnings of an airplane pilot and navigator, where $x$ is the number of hours worked in a week. What are the weekly earnings for an airplane pilot who works 30 hr per week?   $2874

**103.** The height of a baseball hit upward with an initial velocity of 100 ft/sec from an initial height of 2.5 ft is represented by the expression $-16t^2 + 100t + 2.5$, where $t$ is the number of seconds after the ball has been hit. What is the height of the ball after 4 sec?   146.5 ft

**104.** The USS *Constitution* is the oldest commissioned warship afloat in the world. The height of a cannonball fired from the USS *Constitution* with an initial velocity

of 388 ft/sec from a height of 24 ft is represented by the expression $-16t^2 + 388t + 24$, where $t$ is the number of seconds after the cannonball has been fired. What is the height of the cannonball after 12 sec?   2376 ft

## Mix 'Em Up!

**Use the order of operations to simplify each expression.**

**105.** $6^3$   216

**106.** $-3^2$   $-9$

**107.** $|-12(7) + 8(4)|$   52

**108.** $|-6(-13) + 8(-7)|$   22

**109.** $1.4^3$   2.744

**110.** $-\left(\dfrac{3}{4}\right)^3$   $-\dfrac{27}{64}$

**111.** $5[20 - |3 \cdot 8 - 2(11 - 7)|]$   20

**112.** $6[18 - 7|4 \cdot 3 - 2(6 - 1)|]$   24

**113.** $\dfrac{25 + \sqrt{25^2 - 4(16)(9)}}{2 \cdot 9}$   $\dfrac{16}{9}$

**114.** $\dfrac{25 - \sqrt{25^2 - 4(8)(18)}}{2 \cdot 18}$   $\dfrac{1}{2}$

**115.** $\dfrac{1}{15}(45) - 1\dfrac{7}{15}$   $1\dfrac{8}{15}$

**116.** $\dfrac{1}{19}(76) - 2\dfrac{17}{19}$   $1\dfrac{2}{19}$

**117.** $\dfrac{21 - 13}{3 \cdot 4 - 7}$   $\dfrac{8}{5}$

**118.** $\dfrac{18 \div 3 + 9}{27 + 13}$   $\dfrac{3}{8}$

**119.** $10 \cdot 3^2 - 28 \div 4 \cdot 7 + 1$   42

**120.** $3 \cdot 5^2 - 32 \div 4 \cdot 8 + 6$   17

**121.** $\dfrac{1}{6}(14 - 11)^3 - 12 \div 3$   $\dfrac{1}{2}$

**122.** $(1 \cdot 3)^2 + 6 \cdot 4 \div 2 + 10$   31

**Evaluate each expression for the given values of the variables. When multiple values are given, organize the information in a chart.**

**123.** $5x - 3$ for $x = 1, 2, 3,$ and $4$

**124.** $\dfrac{3x + 2}{x + 1}$ for $x = 0, 1, 2,$ and $3$

**125.** $7x - 6y$ for $x = 3$ and $y = 2$   9

**126.** $\dfrac{y_2 - y_1}{x_2 - x_1}$ for $x_1 = 2, x_2 = 14, y_1 = 13,$ and $y_2 = 22$   $\dfrac{3}{4}$

**127.** $\dfrac{1}{2}(b_1 + b_2)h$ for $b_1 = 10, b_2 = 6, h = 3$   24

**128.** $2l + 2w$ for $l = 9$ and $w = 17$   52

**129.** $\sqrt{(x_1 - x_2)^2 + (y_1 - y_2)^2}$ for $x_1 = 2,$ $x_2 = 14, y_1 = 17, y_2 = 22$   13

**130.** $b^2 - 4ac$ for $a = 2, b = 10, c = 5$   60

**Determine if the given value is a solution of the equation.**

**131.** $x^2 + 3x - 6 = x + 9; x = -3$   $x = 3$ is not a solution.

**132.** $6y + 2(5y - 3) = y + 9; y = 1$   $y = 1$ is a solution.

**Translate each phrase into an algebraic expression. Let $x$ represent the unknown number.**

**133.** Five times the sum of a number and 11   $5(x + 11)$

**134.** Three times the sum of a number and 6   $3(x + 6)$

**135.** The difference of six times a number and 18   $6x - 18$

**136.** 6 less than four times the sum of a number and 1
$4(x + 1) - 6$

**137.** Twice the product of a number and 15   $2(x)(15)$

**138.** The product of 18 and the sum of a number and 4   $18(x + 4)$

**139.** The product of 14 and a number increased by 9   $14(x + 9)$

**140.** The quotient of four times a number and 6   $\dfrac{4x}{6}$

**Write an algebraic expression that represents each unknown quantity.**

**141.** Norma mixes a 5% salt solution with a 40% solution to get a 25% solution. Let $x$ oz represent the volume of the 5% salt solution. If Norma wants 50 oz of 25% salt solution, write an expression that represents the volume of the 40% salt solution.   $50 - x$ oz

**142.** David mixes a 10% alcohol solution with a 50% solution to get a 20% solution. Let $x$ mL be the volume of the 50% alcohol solution. If David wants to get 30 mL of 20% solution, write an expression that represents the volume of the 10% solution.   $30 - x$ mL

**143.** In a rectangle, the length is two more than three times the width. If $w$ represents the width, write an expression that represents the length of the rectangle.   $3w + 2$

**144.** In a rectangle, the width is four less than twice the length. If $l$ represents the length, write an expression that represents the width of the rectangle.   $2l - 4$

**Solve each problem.**

**145.** The expression $P(1 + r)^t$ represents the amount of money in an account when $P$ dollars is invested at an annual interest rate of $r$ (in decimal form) for $t$ yr. Find the amount of money that will be in an account at the end of 5 yr if $1500 is invested at 2% annual interest. (Hint: Convert 2% to a decimal value before substituting this value in the expression.)   $1656.12

**146.** The expression $P(1 + r)^t$ represents the amount of money in an account when $P$ dollars is invested at an annual interest rate of $r$ (in decimal form) for $t$ yr. Find the amount of money that will be in an account at the end of 2 yr if $3200 is invested at 4% annual interest. (Hint: Convert 4% to a decimal value before substituting this value in the expression.)   $3461.12

**147.** The height of a baseball hit upward with an initial velocity of 120 ft/sec from an initial height of 6 ft is represented by the expression $-16t^2 + 120t + 6$, where $t$ is the number of seconds after the ball has been hit. What is the height of the ball after 5 sec?   206 ft

**148.** The height of a baseball hit upward with an initial velocity of 95 ft/sec from an initial height of 4.5 ft is represented by the expression $-16t^2 + 95t + 4.5$, where $t$ is the number of seconds after the ball has been hit. What is the height of the ball after 4 sec?
128.5 ft

**149.** At Central Pennsylvania Community College, in-state-resident tuition and fees for the academic year 2010–2011 were calculated by the expression $211c$, where $c$ is the number of credit units taken per semester. If Angela took 13 credits in Spring 2011, what was her tuition and fees?   $2743

**150.** At Los Angeles Valley College, in-state-resident tuition and fees for academic year 2010–2011 were calculated by the expression $26c + 42$, where $c$ is the number of units taken per semester. If Goran took 15 units in Spring 2011, what was his tuition and fees?   $432

 **You Be the Teacher!**

**Correct each student's errors, if any.**

**151.** Evaluate the expression $3x^2 - 2x + 1$ for $x = 2$.

Keith's work:

$3 \cdot 2^2 - 2 \cdot 2 + 1 = 6^2 - 4 + 1 = 36 - 4 + 1 = 33$
$3 \cdot 2^2 - 2 \cdot 2 + 1 = 3 \cdot 4 - 4 + 1 = 12 - 4 + 1 = 9$

**152.** Evaluate the expression $-\dfrac{2}{3}x + 5$ for $x = 3$.

Frankie's work:

$$-\frac{2}{3} \cdot 3 + 5 = -\frac{2}{3} \cdot 8 = -\frac{2}{3} \cdot \frac{24}{3} = -\frac{48}{9}$$

**153.** Evaluate the expression $30 - 2x^2$ for $x = 3$.

Lestine's work:

$30 - 2(3)^2 = 28(3^2) = 28(9) = 252$
$30 - 2(3)^2 = 30 - 2(9) = 30 - 18 = 12$

**154.** Evaluate the expression $|15 - 4x|$ for $x = 3$.

Ian's work:

$|15 - 4(3)| = |11(3)| = |33| = 33$
$|15 - 4(3)| = |15 - 12| = |3| = 3$

**155.** Simplify the expression $-3 \cdot 2^2 + 28 \div 7 \cdot 4 + 36$.

Basil's work:

$$-3 \cdot 2^2 + 28 \div 7 \cdot 4 + 36 = -6^2 + 28 \div 28 + 36$$
$$= -36 + 1 + 36$$
$$= 1$$

**156.** Simplify the expression $38 - 2^2 - 45 \div 5 \cdot 3$.

Abul's work:

$$38 - 2^2 - 45 \div 5 \cdot 3 = 38 - 4 - 45 \div 15$$
$$= 38 - 4 - 3$$
$$= 31$$

 **Calculate It!**

**Use a calculator to evaluate each algebraic expression.**

**157.** $x^3 - 2x + 1$ for $x = 0, 2.5,$ and $3.2$   1, 11.625, 27.368

**158.** $\dfrac{5x + 1}{x - 1}$ for $x = 1.5, 1.8,$ and $4$   17, 12.5, 7

**159.** $\sqrt{x^2 - 9}$ for $x = 3, 5,$ and $9$   0, 4, 8.49

**160.** $|500 - 3x^2|$ for $x = 8.5, 10,$ and $12$   283.25, 200, 68

## SECTION 1.4     Addition of Real Numbers

▶ **OBJECTIVES**

As a result of completing this
section, you will be able to

**1.** Add numbers with the same
sign.

**2.** Add numbers with different
signs.

**3.** Solve real-life problems.

**4.** Troubleshoot common
errors.

**Objective 1** ▶

Add numbers with the same
sign.

**INSTRUCTOR NOTE:**
A thermometer can also be used to
demonstrate the concept of adding
signed numbers.

Have students come up with other
examples of adding two positive
numbers and adding two negative
numbers.

The operating income for Delta Airlines for the years
2002 to 2009 is shown in the table. Find the total
operating income for the years 2002–2009. (Source:
http://www.airlinefinancials.com/uploads/2002–2009_Delta
_mainline.pdf)

To answer this question, we must know how to
perform operations with real numbers. In this section,
we will examine addition of real numbers.

| Year | Operating Income (in millions) |
|------|-------------------------------|
| 2002 | −$1309 |
| 2003 | −$ 786 |
| 2004 | −$5985 |
| 2005 | −$1605 |
| 2006 | −$1139 |
| 2007 | $   78 |
| 2008 | −$  58 |
| 2009 | $  605 |

### Adding Numbers with the Same Signs

The real number line can help us visualize the process of adding real numbers. Adding two positive
numbers is a skill that we encounter on a daily basis. For example, if a student enrolls
in 6 hr of classes and then adds 4 more hours during the drop/add period, the student
enrolls in a total of $6 + 4 = 10$ hr of classes.

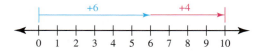

Adding two negative numbers is not something that we encounter as often as adding
two positive numbers. Suppose we owe a friend $5 and then we borrow $2 more. At
this point, we owe our friend a total of $7. Mathematically, the total we owe can be
represented as

$$-5 + (-2) = -7$$

On the number line, we can visualize this situation as beginning at 0, moving to the left
5 units, and moving left again 2 units until we arrive at −7.

**INSTRUCTOR NOTE:**
Give the students a helpful hint for
adding numbers with the same sign.
SSS = **S**ame sign, **S**um the values,
keep the **S**ame sign.

While number lines provide us the ability to visualize the operations on the numbers, using them every time we add two numbers can be very time consuming.
From the previous examples, we can conclude that adding two positive numbers
results in a positive number and adding two negative numbers results in a negative
number.

> **Procedure:  Adding Numbers with the Same Sign**
>
> **Step 1:** Add the absolute values of the numbers.
> **Step 2:** The sign of the result is the same as the sign of the numbers in the
> original expression.

**Objective 1  Examples**     Add. For parts (a) and (b), use a number line to add the numbers.

**1a.** $7 + 2$     **1b.** $(-1) + (-4)$     **1c.** $(-2) + (-7)$     **1d.** $(-1.5) + (-3)$

**1e.** $\dfrac{1}{5} + \dfrac{3}{10}$     **1f.** $(-7) + \left(-\dfrac{2}{3}\right)$     **1g.** $(-8) + (-2) + (-4)$

**Solutions**   **1a.** We begin at 0, move right 7 units, and then right 2 more units.

So, $7 + 2 = 9$.

**1b.** We begin at 0, move left 1 unit, and then left 4 more units.

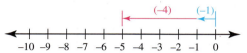

So, $(-1) + (-4) = -5$.

**1c.**      $(-2) + (-7) = -9$

Add the absolute values.
$|-2| + |-7| = 2 + 7 = 9$
Use the sign of the original numbers.

**1d.**      $(-1.5) + (-3) = -4.5$

Add the absolute values.
$|-1.5| + |-3| = 1.5 + 3 = 4.5$
Use the sign of the original numbers.

**1e.**
$$\frac{1}{5} + \frac{3}{10} = \frac{2}{10} + \frac{3}{10}$$
$$= \frac{5}{10}$$
$$= \frac{1}{2}$$

Convert $\frac{1}{5}$ to a fraction with a denominator of 10, $\frac{1}{5} = \frac{2}{10}$.

Add the absolute values.

$\left|\frac{2}{10}\right| + \left|\frac{3}{10}\right| = \frac{2}{10} + \frac{3}{10} = \frac{5}{10}$

Keep the sign of the original numbers.
Simplify the result.

**1f.**      $(-7) + \left(-\frac{2}{3}\right) = \left(-\frac{7}{1}\right) + \left(-\frac{2}{3}\right)$
$$= \left(-\frac{21}{3}\right) + \left(-\frac{2}{3}\right)$$
$$= -\frac{23}{3}$$

Rewrite $-7$ as a fraction, $-\frac{7}{1}$, and convert it to a fraction with a denominator of 3, $-\frac{7}{1} = -\frac{21}{3}$.

Add the absolute values of the two numbers.

$\left|\frac{-21}{3}\right| + \left|\frac{-2}{3}\right| = \frac{21}{3} + \frac{2}{3} = \frac{23}{3}$

Use the sign of the original numbers.

**1g.** $(-8) + (-2) + (-4) = -14$

Add the absolute values.
$|-8| + |-2| + |-4| = 8 + 2 + 4 = 14$
Use the sign of the original numbers.

---

✔ **Student Check 1**   Add. For parts (a) and (b), use a number line to add the numbers.

**a.** $3 + 5$      **b.** $(-4) + (-3)$      **c.** $(-2.6) + (-1.3)$      **d.** $\frac{2}{3} + \frac{1}{6}$

**e.** $\left(-\frac{4}{5}\right) + \left(-\frac{7}{15}\right)$      **f.** $(-5) + \left(-\frac{1}{2}\right)$      **g.** $(-5) + (-1) + (-6)$

---

## Adding Numbers with Different Signs

**Objective 2** ▶

Add numbers with different signs.

Adding real numbers with different signs occurs in many different situations. An everyday occurrence deals with temperature changes. Suppose the temperature at 7:00 A.M. in Duluth, Minnesota, one winter morning was 2°F below zero and the temperature rose 5°F after 3 hr. The temperature at 10 A.M. can be found by simplifying the expression $-2°F + 5°F$. To perform this operation, we will use a thermometer. We begin at 0°F and go down to $-2°F$ and then go up by 5°F. This takes us to 3°F. So, we have

$$-2 + 5 = 3$$

We can also obtain the same result by using a number line as well.

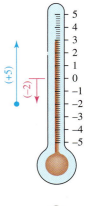

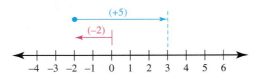

Suppose the temperature is 5°F at 12:00 A.M. and the temperature fell 8°F in 2 hr. To determine the temperature at 2:00 A.M., we need to find the result of $5°F + (-8)°F$. The thermometer can assist us again. In this case, we begin at 0°F, and go up by 5°F and then go down 8°F. This takes us to $-3°F$. So, we have

$$5 + (-8) = -3$$

Again, we can obtain the same result by using a number line.

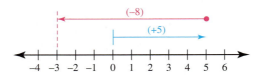

From these examples, we conclude that adding numbers with opposite signs can result in either a positive or negative number. The final sign is actually determined by the sign of the number with the larger absolute value. In the first example, 5 is the number with the larger absolute value and the sign of the $-2 + 5$ is positive. In the second example, $-8$ is the number with the larger absolute value and the sign of $5 + (-8)$ is negative.

> **Procedure:  Adding Numbers with Different Signs**
>
> **Step 1:** Subtract the absolute values of the numbers.
> **Step 2:** The sign of the result is the same as the sign of the number with the larger absolute value.

**Objective 2  Examples** | **Add. For parts (a) and (b), use a number line to add the numbers.**

**2a.** $3 + (-2)$ **2b.** $(-7) + 4$ **2c.** $20 + (-35)$ **2d.** $-14 + 15.5$

**2e.** $-7 + 7$ **2f.** $\left(-\dfrac{1}{4}\right) + \dfrac{2}{3}$ **2g.** $(-9) + 5 + (-2)$

**Solutions** | **2a.** We begin at 0 and move right 3 units and then left 2 units.

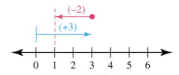

So, $3 + (-2) = 1$.

**2b.** We begin at 0 and move left 7 units and then right 4 units.

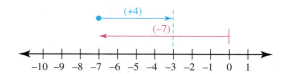

So, $(-7) + 4 = -3$.

**2c.** $20 + (-35) = -15$    Subtract the absolute values of the numbers.
$|20| = 20, |-35| = 35, 35 - 20 = 15$
Since $|-35| > |20|$, the sign of the answer is negative.

**2d.** $-14 + 15.5 = 1.5$    Subtract the absolute values of the numbers.
$|-14| = 14, |15.5| = 15.5, 15.5 - 14 = 1.5$
Since $|15.5| > |-14|$, the sign of the answer is positive.

**2e.** $-7 + 7 = 0$    Subtract the absolute values of the numbers.
$|-7| = 7, |7| = 7, 7 - 7 = 0$
Since the difference of the absolute values is zero, the sign of the answer is neither positive nor negative.

**2f.** $\left(-\dfrac{1}{4}\right) + \dfrac{2}{3} = -\dfrac{3}{12} + \dfrac{8}{12}$    Rewrite each fraction with a denominator of 12.
$-\dfrac{1}{4} = -\dfrac{3}{12}$ and $\dfrac{2}{3} = \dfrac{8}{12}$
$\qquad = \dfrac{5}{12}$    Subtract the absolute values of the numbers.
$\left|-\dfrac{3}{12}\right| = \dfrac{3}{12}, \left|\dfrac{8}{12}\right| = \dfrac{8}{12}, \dfrac{8}{12} - \dfrac{3}{12} = \dfrac{5}{12}$
Since $\left|\dfrac{8}{12}\right| > \left|-\dfrac{3}{12}\right|$, the sign of the answer is positive.

**2g.** $(-9) + 5 + (-2)$    Add the first two numbers. Their signs are different so
$\qquad = -4 + (-2)$    we subtract their absolute values and keep the sign of
$\qquad = -6$    the number with the larger absolute value.
$|-9| = 9, |5| = 5, 9 - 5 = 4$
$|-9| > |5| \rightarrow (-9) + 5 = -4$
Add the result and the last number. Their signs are the same, so we add their absolute values and keep the sign of the original numbers.
$|-4| = 4, |-2| = 2, 4 + 2 = 6$
Since the numbers are negative, the result is negative.

✓ **Student Check 2**    Add the numbers.

**a.** $2 + (-10)$    **b.** $-4 + 8$    **c.** $(-16) + 4$    **d.** $18.4 + (-6.2)$

**e.** $3 + (-3)$    **f.** $\dfrac{3}{8} + \left(-\dfrac{5}{4}\right)$    **g.** $12 + (-2) + (-6)$

## Applications

**Objective 3** ▶

Solve real-life problems.

Addition of real numbers occurs in many different aspects of real life. We find adding real numbers when balancing checkbooks, determining net worth, determining values of stocks, calculating golf scores, calculating total gain/loss in yardage for football plays, changes in temperature, changes in elevation, and so on.

> **Procedure: Solving Applications of Adding Signed Numbers**
> **Step 1:** Express the problem mathematically.
> **Step 2:** Apply the rules for adding signed numbers.

**Objective 3 Examples**

**Write a mathematical equation that models the given situation and then find the result.**

The operating income for Delta Airlines for the years 2002 to 2009 is shown in the table. A negative number represents a loss in operating income. (Source: http://www.airlinefinancials.com/uploads/2002–2009_Delta_mainline.pdf)

**3a.** Find the total operating income (TOI) for the years 2002–2005.

**3b.** Find the average operating income (AOI) for the years 2007–2009.

| Year | Operating Income (in millions) |
|------|-------------------------------|
| 2002 | –$1309 |
| 2003 | –$ 786 |
| 2004 | –$5985 |
| 2005 | –$1605 |
| 2006 | –$1139 |
| 2007 | $  78 |
| 2008 | –$  58 |
| 2009 | $ 605 |

**Solutions**

**3a.** We add the values for the 4 yr. Since the numbers all have the same sign, we add their absolute values and keep the sign.

$$\text{TOI} = (-1309) + (-786) + (-5985) + (-1605)$$
$$\text{TOI} = -9685$$

So, Delta Airlines had a total loss of $9685 million during the years 2002 to 2005.

**3b.** We add the values for the given years and divide by the number of years, 3.

$$\text{AOI} = \frac{78 + (-58) + 605}{3}$$
$$\text{AOI} = \frac{625}{3}$$
$$\text{AOI} = 208.33$$

So, Delta Airlines had an average operating income of $208.33 million during the years 2007 to 2009.

**✓ Student Check 3**

Write a mathematical equation that models the given situation and then find the result.

The table shows the average monthly low temperature for one of the northernmost points in the United States: Barrow, Alaska. Find the average yearly low temperature.

| Month | °F | Month | °F |
|-------|-----|-------|-----|
| Jan. | −20 | July | 34 |
| Feb. | −22 | Aug. | 34 |
| Mar. | −22 | Sept. | 28 |
| Apr. | −7 | Oct. | 10 |
| May | 15 | Nov. | −6 |
| June | 30 | Dec. | −16 |

**Objective 4** ▶

Troubleshoot common errors.

## Troubleshooting Common Errors

Some common errors associated with adding real numbers are shown next.

**Objective 4  Examples**  A problem and an incorrect solution are given. Provide the correct solution and an explanation of the error.

**4a.** $7 + (-11)$

| Incorrect Solution | Correct Solution and Explanation |
|---|---|
| $7 + (-11) = 4$ | Since the signs are different, we should subtract the absolute values of the numbers and take the sign of the number with the larger absolute value. The difference of the absolute values is 4 but $-11$ has the larger absolute value, so the sign of the answer should be negative. $$7 + (-11) = -4$$ |

**4b.** $-5 + (-4)$

| Incorrect Solution | Correct Solution and Explanation |
|---|---|
| $-5 + (-4) = -1$ | Since the signs are the same, we should add the absolute values of the numbers and keep the sign the same. So, the correct solution is $$-5 + (-4) = -9$$ |

## ANSWERS TO STUDENT CHECKS

**Student Check 1**   **a.** 8   **b.** $-7$   **c.** $-3.9$   **d.** $\dfrac{5}{6}$

**e.** $-\dfrac{11}{15}$   **f.** $-\dfrac{11}{2}$   **g.** $-12$

**Student Check 2**   **a.** $-8$   **b.** 4   **c.** $-12$   **d.** 12.2

**e.** 0   **f.** $-\dfrac{7}{8}$   **g.** 4

**Student Check 3**   4.83°F

## SUMMARY OF KEY CONCEPTS

1. To add numbers with the same sign, add their absolute values and keep the sign.
2. To add numbers with different signs, subtract the absolute values of the numbers. The sign of the answer is the sign of the number with the larger absolute value.
3. When solving real-life applications, look for the key phrases to determine how to solve the problem.

## GRAPHING CALCULATOR SKILLS

In this section, the calculator will help you confirm your work. You should be able to enter operations with signed numbers.

**Example:** Simplify $(-4) + (-7)$.

```
( -4 )+( -7 )
            -11
```

Very important: Please note the difference between a negative sign and the subtraction sign. Use the negative sign $(-)$ to denote the sign of a number and the subtraction sign to perform an operation with numbers.

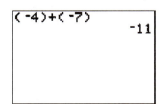

## SECTION 1.4 / EXERCISE SET

 **Write About It!**

Use complete sentences to explain how to perform the given operation.

1. Add two negative numbers   *Answers vary.*

2. Add a positive number and a negative number   *Answers vary.*

 **Practice Makes Perfect!**

Perform the indicated operation and simplify.
(*See Objective 1.*)

3. $1.95 + 10.26$   $12.21$

4. $1.37 + 2.56$   $3.93$

5. $-5 + (-4)$   $-9$

6. $-7 + (-9)$   $-16$

7. $-1.68 + (-6.77)$   $-8.45$

8. $-4.33 + (-11.86)$   $-16.19$

9. $\frac{9}{10} + \frac{3}{5}$   $\frac{3}{2}$

10. $\frac{13}{14} + \frac{3}{4}$   $\frac{47}{28}$

11. $-2 + \left(-\frac{3}{5}\right)$   $-\frac{13}{5}$

12. $-1 + \left(-\frac{5}{7}\right)$   $-\frac{12}{7}$

13. $\left(-\frac{1}{10}\right) + \left(-\frac{3}{4}\right)$   $-\frac{17}{20}$

14. $\left(-\frac{1}{3}\right) + \left(-\frac{3}{8}\right)$   $-\frac{17}{24}$

15. $\left(-3\frac{1}{4}\right) + \left(-1\frac{1}{4}\right)$   $-4\frac{1}{2}$

16. $\left(-2\frac{1}{5}\right) + \left(-1\frac{3}{5}\right)$   $-3\frac{4}{5}$

17. $-4 + (-12) + (-9)$   $-25$

18. $-13 + (-8) + (-14)$   $-35$

19. $-5.5 + (-2.4) + (-3.1)$   $-11$

20. $-6.2 + (-4.2) + (-5.8)$   $-16.2$

Perform the indicated operation and simplify.
(*See Objective 2.*)

21. $1 + (-3)$   $-2$

22. $4 + (-7)$   $-3$

23. $-6 + 10$   $4$

24. $-30 + 15$   $-15$

25. $6 + (-10)$   $-4$

26. $30 + (-18)$   $12$

27. $(-1.5) + 3.6$   $2.1$

28. $(-4.1) + 0.8$   $-3.3$

29. $1.6 + (-4.2)$   $-2.6$

30. $3.7 + (-1.9)$   $1.8$

31. $\frac{2}{3} + \left(-\frac{4}{5}\right)$   $-\frac{2}{15}$

32. $\frac{4}{7} + \left(-\frac{1}{14}\right)$   $\frac{1}{2}$

33. $1\frac{1}{3} + \left(-2\frac{2}{3}\right)$   $-1\frac{1}{3}$

34. $3\frac{3}{8} + \left(-4\frac{5}{8}\right)$   $-1\frac{1}{4}$

35. $15 + (-9) + (-2)$   $4$

36. $18 + (-8) + (-5)$   $5$

37. $-21 + 9 + (-3)$   $-15$

38. $-19 + 11 + (-6)$   $-14$

Solve each problem by first writing a mathematical expression that represents the situation. (*See Objective 3.*)

39. Daimler reported a net income of $-332$ million euros in 2008, $-4765$ million euros in 2009 and 5399 million euros in 2010. What was Daimler's total net income for 2008 to 2010? (Source: www.daimler.com)   *302 million euros*

40. United Airlines reported a total operating income of $1037 million in 2007, $-$4438 million in 2008, $-$161 million in 2009, and $976 million in 2010. What was United's total operating income for 2007 to 2010? (Source: www.united.com)   $-$2586 million

 **Mix 'Em Up!**

Perform the indicated operation and simplify.
(*See Objectives 1 and 2, 4 and 5.*)

41. $-1.5 + (-3.6)$   $-5.1$

42. $(-4.1) + (-0.8)$   $-4.9$

43. $1.27 + (-3.73)$   $-2.46$

44. $1.98 + (-3.19)$   $-1.21$

45. $-\frac{10}{3} + \frac{1}{6}$   $-\frac{19}{6}$

46. $\left(-\frac{3}{2}\right) + \left(-\frac{10}{3}\right)$   $-\frac{29}{6}$

47. $2 + \left(-\frac{3}{7}\right)$   $\frac{11}{7}$

48. $\left(\frac{1}{2}\right) + \left(-\frac{5}{6}\right)$   $-\frac{1}{3}$

49. $\left(-3\frac{1}{2}\right) + 1\frac{1}{4}$   $-2\frac{1}{4}$

50. $\left(-1\frac{2}{5}\right) + 2\frac{4}{5}$   $1\frac{2}{5}$

51. $(-6) + 5$   $-1$

52. $(-9) + 4$   $-5$

53. $5.6 + (-6.2)$   $-0.6$

54. $1\frac{5}{8} + \left(-2\frac{3}{8}\right)$   $-\frac{3}{4}$

Translate each phrase into a mathematical expression and then simplify the result.

55. 2 more than $-5$
$-5 + 2 = -3$

56. 6 more than $-1$
$-1 + 6 = 5$

57. The sum of $-3$ and $-4$
$-3 + (-4) = -7$

58. The sum of 7 and $-10$
$7 + (-10) = -3$

59. 12 increased by $-4$
$12 + (-4) = 8$

60. $-7$ increased by 3
$-7 + 3 = -4$

61. 9 added to $-2$
$-2 + 9 = 7$

62. $-4$ added to $-1$
$-1 + (-4) = -5$

Write the expression needed to solve each problem and then answer the question. (*See Objective 3.*)

63. According to an article in the *Washington Post* in April 2005, Apple Computer Inc. had a profit of $46 million in 2004. In 2005, the profits of Apple were $244 million more than the profit in 2004. This boost in Apple's profits was due to the increased popularity of the iPod. What was the profit of Apple Computer in 2005?   $290 million

64. On August 1, 2006, the average retail price of regular gas in Atlanta, Georgia, was $2.93 per gal. Over the next 2 months, gas prices fell by $0.93 per gal. Gas prices then rose by $0.21 per gal during the following 2 months. By January 2007, gas prices fell $0.25 per gal. What was the average retail price of regular gas in Atlanta, Georgia, in January 2007?   $1.96

65. On August 1, 2006, the average retail price of regular gas in San Francisco, California, was $3.15 per gal. Over the next 2 months, gas prices fell by $0.58 per gal. Prices fell by $0.15 per gal the following months. By January 2007, gas prices rose by $0.25 per gal. What was the average retail price of regular gas in San Francisco, California, in January 2007?   $2.67

**66.** Rhonda has $152.35 in her checking account. She deposits her paycheck of $423.45. Rhonda writes a check for daycare for $135, for groceries for $93.58, and then for rent for $350. What is Rhonda's checking account balance?   −$2.78

**67.** Mohammad has $52.75 in his checking account. He writes a check for groceries for $48.21. He writes a check for books for $475 and then deposits his student loan check for $750. What is Mohammad's checking account balance?   $279.54

**68.** American Airlines reported an operating income of $965 million in 2007, −$1889 million in 2008, −$1004 million in 2009, and $308 million in 2010. What was American's total operating income for 2007 to 2010? (Source: www.aa.com)   −$1620 million

**69.** Ford Motor Company reported a net income of −$14,766 million in 2008, $2717 million in 2009, and $6561 million in 2010. What was Ford's total net income for 2008 to 2010? (Source: http://corporate.ford.com/investors)   −$5488 million

 ## You Be the Teacher!

**Answer each student's question.**

**70.** Benjamin: *I am trying to solve the following problem:*

**Simplify the following expression:** $-\frac{3}{4} + \left(-\frac{1}{2}\right).$

*Please check my work and tell me if I did it correctly.*

Since $-\frac{3}{4}$ and $-\frac{1}{2}$ are the same sign, I just need to add their absolute values together. I know that $\left|-\frac{3}{4}\right| = \frac{3}{4}$ and $\left|-\frac{1}{2}\right| = \frac{1}{2}.$

So, $-\frac{3}{4} + \left(-\frac{1}{2}\right) = \frac{3}{4} + \frac{1}{2} = \frac{3}{4} + \frac{1 \times 2}{2 \times 2} = \frac{3}{4} + \frac{2}{4} = \frac{5}{4}.$   Answers vary.

**71.** Renae: *I am trying to solve the following problem:*

**Simplify the following expression:** $\frac{1}{4} + \left(-\frac{1}{2}\right).$

*I'm having trouble visualizing what this means in terms of the number line. Can you explain this problem to me using a number line?*   Answers vary.

 ## Calculate It!

**Use a calculator to simplify each expression. Express each answer in fractional form, when applicable.**

**72.** $4.56 + (-8.27)$   −3.71

**73.** $-3.12 + (-10.89)$   −14.01

**74.** $-\frac{1}{3} + 12$   $11\frac{2}{3}$

**75.** $-4\frac{1}{2} + \frac{3}{5}$   $-3\frac{9}{10}$

**76.** $5\frac{1}{7} + \left(-2\frac{3}{7}\right)$   $2\frac{5}{7}$

**77.** $\left(-4\frac{1}{4}\right) + \left(-6\frac{1}{5}\right)$   $-10\frac{9}{20}$

---

## PIECE IT TOGETHER   **SECTIONS 1.1–1.4 Review**

**Graph each number on a real number line and classify the number as natural, whole, integer, rational, irrational, and/or real.** (*Section 1.1, Objectives 1 and 2*)

**1.** $\left\{-\frac{15}{2}, -\sqrt{3}, 2.5, 6\frac{1}{2}\right\}$   $-\frac{15}{2}, 2.5, 6\frac{1}{2}$: rational and real; $-\sqrt{3}$: irrational and real

**Compare the values of each pair of numbers. Use a <, >, or, = symbol to make the statement true.** (*Section 1.1, Objective 3*)

**2.** $-5$ _____ $-(-5)$   <

**3.** $7$ _____ $-(-7)$   =

**Perform the operation and simplify each result.** (*Section 1.2, Objectives 4–6*)

**4.** $\frac{2}{3} \cdot \frac{6}{10}$   $\frac{2}{5}$

**5.** $\frac{7}{12} \cdot \frac{2}{7}$   $\frac{1}{6}$

**6.** $\frac{1}{3} \div 3$   $\frac{1}{9}$

**7.** $7 \div \frac{1}{7}$   49

**8.** $\frac{9}{10} - \frac{1}{10}$   $\frac{4}{5}$

**9.** $\frac{5}{18} - \frac{3}{12}$   $\frac{1}{36}$

**10.** $\frac{7}{12} - \frac{1}{6}$   $\frac{5}{12}$

**11.** $5 - \frac{1}{2}$   $\frac{9}{2}$

**Identify the base, the exponent, and then simplify the expression.** (*Section 1.3, Objective 1*)

**12.** $2.1^2$   4.41

**13.** $-3^4$   −81

**Use the order of operations to simplify each expression.** (*Section 1.3, Objective 2*)

**14.** $7[40 - 2|5 + 3(15 - 11)|]$   42

**15.** $\dfrac{3 + \sqrt{3^2 - 4(2)(1)}}{2 \cdot 2}$   1

**16.** $2(16) - 7(2)$   18

**Evaluate each expression for the specified values of x. When multiple values are given, organize the information in a chart.** (*Section 1.3, Objective 3*)

**17.** $\frac{2}{5}x + 3$ for $x = 0, 1, 5,$ and 10

**18.** $2x - y$ for $x = 3$ and $y = 1$   5

**Translate each phrase into an algebraic expression. Let x represent the unknown number.** (*Section 1.3, Objective 5*)

**19.** The quotient of a number and 6   $x \div 6$

**20.** Twice the sum of a number and 4   $2(x + 4)$

**21.** 3 more than one-fifth of a number   $\frac{1}{5}x + 3$

**22.** 8 less than twice a number   $2x - 8$

**Perform the indicated operation and simplify.** (*Section 1.4, Objectives 1 and 2*)

**23.** $(-8) + (-2)$   −10

**24.** $-19 + 11 + (-6)$   −14

**25.** $\left(-\frac{3}{5}\right) + \frac{19}{15}$   $\frac{2}{3}$

Additional answers can be found in the Instructor Answer Appendix.

| SECTION 1.5 | **Subtraction of Real Numbers** |

### ▶ OBJECTIVES

As a result of completing this section, you will be able to

1. Subtract real numbers.
2. Apply the order of operations.
3. Solve real-life problems.
4. Troubleshoot common errors.

The world's coldest recorded temperature is $-129°F$ in Vostok, Antarctica. The world's hottest recorded temperature is $134°F$ in Death Valley, California. What is the difference between the hottest and coldest temperatures?

To answer this question, we must know how to subtract real numbers.

## Subtracting Real Numbers

**Objective 1** ▶

Subtract two real numbers.

To understand how to subtract two real numbers, consider the following expressions.

$$7 - 2 = 5 \qquad 7 + (-2) = 5$$

These two expressions are equivalent, which indicates that subtraction is really a form of addition. Subtraction is equivalent to adding the opposite of a number.

---

**Property: Subtraction of Real Numbers**

For all real numbers $a$ and $b$,

$$a - b = a + (-b)$$

---

**Procedure: Subtracting Real Numbers**

**Step 1:** Rewrite the expression by adding the opposite of the second number.
**Step 2:** Use the rules for addition to simplify the expression.

---

**Objective 1 Examples** | Perform the indicated operation.

**1a.** $-6 - 2$      **1b.** $3.5 - (-2.5)$      **1c.** $5 - 12$

**1d.** $-100 - (-25)$      **1e.** $-\dfrac{1}{2} - \dfrac{3}{5}$

**Solutions**

**1a.**
$$-6 - 2 = -6 + (-2)$$
            Rewrite as adding the opposite of 2, which is $-2$.
$$= -8$$
            Add the numbers.

**1b.**
$$3.5 - (-2.5) = 3.5 + (2.5)$$
            Rewrite as adding the opposite of $-2.5$, which is 2.5.
$$= 6$$
            Add the numbers.

**1c.**
$$5 - 12 = 5 + (-12)$$
            Rewrite as adding the opposite of 12, which is $-12$.
$$= -7$$
            Add the numbers.

**1d.**
$$-100 - (-25) = -100 + (25)$$
            Rewrite as adding the opposite of $-25$, which is 25.
$$= -75$$
            Add the numbers.

**1e.**  $-\dfrac{1}{2} - \dfrac{3}{5} = -\dfrac{1}{2} + \left(-\dfrac{3}{5}\right)$    Rewrite as adding the opposite of $\dfrac{3}{5}$, which is $-\dfrac{3}{5}$.

$= -\dfrac{5}{10} + \left(-\dfrac{6}{10}\right)$    Convert the fractions to equivalent fractions with their common denominator of 10.

$= -\dfrac{11}{10}$    Add the numbers.

---

✓ **Student Check 1**    Perform the indicated operation.

**a.** $1 - 4$    **b.** $11.3 - (-4.3)$    **c.** $-75 - 15$

**d.** $-32 - (-8)$    **e.** $-\dfrac{2}{3} - \left(-\dfrac{1}{4}\right)$

---

**Note:** *Instead of rewriting a subtraction problem as adding the opposite, we observe that, after any double negative signs are eliminated, the operation connecting the numbers is equivalent to the sign of the number being added. This enables us to use the rules for addition to simplify the expression.*

*For instance,*

$-6 - 2 \rightarrow$ *Both numbers are negative, so we get* $-8$.

$6 - 10 \rightarrow$ *These numbers have opposite signs, so we get* $-4$.

$-1 - (-5) = -1 + 5 \rightarrow$ *These numbers have opposite signs, so we get* 4.

*Also, note that* $-(-a) = a$.

## More about the Order of Operations

**Objective 2** ▶

Apply the order of operations.

In Section 1.3, we learned the order of operations. We will use that again to simplify expressions, but this time, we will apply the rules for adding and subtracting signed numbers.

**Procedure:  Order of Operations**

When simplifying a numerical expression, perform the operations in the following order.

**Step 1:**  Simplify expressions inside grouping symbols first, starting with the innermost set. If there is a fraction bar, simplify the numerator and denominator separately.

**Step 2:**  Evaluate any exponential expressions.

**Step 3:**  Perform multiplication or division in order from left to right.

**Step 4:**  Perform addition or subtraction in order from left to right.

**Objective 2  Examples**    **Simplify each expression.**

**2a.** $-4 - (-3) + 5 - 7$    **2b.** $-4^2 + 6 - |-10 + 3|$

**2c.** $5 - \{4 - [6 - (1 - \sqrt{2 - (-2)})]\}$

---

**Solutions**    **2a.** $-4 - (-3) + 5 - 7 = -4 + 3 + 5 + (-7)$    Rewrite.

$= -1 + 5 + (-7)$    Add the first two numbers, $-4 + 3 = -1$.

$= 4 + (-7)$    Add the first two numbers, $-1 + 5 = 4$.

$= -3$    Add the final numbers.

**2b.** $-4^2 + 6 - |-10 + 3|$

$\qquad = -4^2 + 6 - |-7|$          Add inside the absolute value.

$\qquad = -16 + 6 - (7)$          Simplify $4^2$ and $|-7|$.

$\qquad = -16 + 6 - 7$

$\qquad = -10 - 7$          Add from left to right.

$\qquad = -17$

**2c.** $5 - \{4 - [6 - (1 - \sqrt{2 - (-2)})]\}$      Rewrite the subtraction inside the square root symbol as addition.

$\qquad = 5 - \{4 - [6 - (1 - \sqrt{2 + 2})]\}$      Add inside the square root.

$\qquad = 5 - \{4 - [6 - (1 - \sqrt{4})]\}$      Simplify the square root.

$\qquad = 5 - \{4 - [6 - (1 - 2)]\}$      Subtract the numbers in parentheses.

$\qquad = 5 - \{4 - [6 - (-1)]\}$      Rewrite the expression inside the brackets as addition.

$\qquad = 5 - \{4 - [6 + 1]\}$      Add the numbers inside the brackets.

$\qquad = 5 - \{4 - [7]\}$      Subtract the numbers inside the braces.

$\qquad = 5 - \{-3\}$      Rewrite the subtraction as addition.

$\qquad = 5 + 3$      Add.

$\qquad = 8$

---

✓ **Student Check 2**     Simplify each expression.

    **a.** $-2 - 1 - (-3) + 5$          **b.** $-3^3 + 15 - |4 - (-2)|$

    **c.** $-8 - \{2 - [7 - (3 - \sqrt{25 - 16})]\}$

---

## Applications

**Objective 3** ▶

**Solve real-life problems.**

Subtracting real numbers occurs in real life when we need to find changes in temperatures, elevation, the stock market, bank accounts, and so on. It is important to be able to express the given change mathematically. We will use the skills from Section 1.3 to express mathematical relationships together with the skills for adding and subtracting real numbers.

    Other types of applications come from the study of geometry. There are some special angle relationships that are helpful to know.

---

**Definition:** The sum of the measure of the angles in a *triangle* is 180°.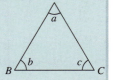

$$a + b + c = 180°$$

---

**Definition: Complementary Angles**

Complementary angles are two angles whose sum is 90°. Complementary angles form a **right angle** when they are adjacent to one another. A right angle is an angle whose measure is 90°. If $a$ and $b$ are complementary angles, then the sum of their measures is 90°.

$$a + b = 90°$$

Some examples of complementary angles are shown in the following table.

| $a$ | $b$ | $a + b$ |
|-----|-----|---------|
| 20° | 70° | 20° + 70° = 90° |
| 30° | 60° | 30° + 60° = 90° |
| 45° | 45° | 45° + 45° = 90° |
| 50° | 40° | 50° + 40° = 90° |
| $x°$ | $(90 - x)°$ | $x° + (90 - x)° = 90°$ |

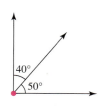

### Definition:  Supplementary Angles

Supplementary angles are two angles whose sum is 180°. Supplementary angles form a **straight angle** when they are adjacent to each other. A straight angle is an angle whose measure is 180°. If $a$ and $b$ are supplementary angles, then the sum of their measures is 180°.

$$a + b = 180°$$

Some examples of supplementary angles are shown in the following table.

| $a$ | $b$ | $a + b$ |
|-----|-----|---------|
| 20° | 160° | 20° + 160° = 180° |
| 30° | 150° | 30° + 150° = 180° |
| 45° | 135° | 45° + 135° = 180° |
| 50° | 130° | 50° + 130° = 180° |
| $x°$ | $(180 - x)°$ | $x° + (180 - x)° = 180°$ |

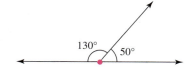

**Objective 3  Examples**

Vostok, Antarctica

**Write a mathematical expression that represents the given situation and then find the result.**

**3a.** The world's coldest recorded temperature is $-129°F$ in Vostok, Antarctica. The world's hottest recorded temperature is 134°F in Death Valley, California. What is the difference between the hottest and coldest temperatures?

**3b.** Find the measure of the unknown angle of the given triangle.

**3c.** Find the complement and supplement of an angle whose measure is 25°.

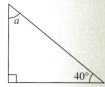

**Solutions**

**3a.** We must find the difference between the hottest and coldest temperatures. This can be translated as "hottest − coldest."

Since the hottest temperature is 134°F and the coldest is −129°F, we need to find

$$134 - (-129) = 134 + 129 = 263°F$$

Therefore, the difference between the hottest and coldest temperatures is 263°F.

**3b.** The two known angles have measures of 90° and 40°. So, the sum of the known angles is 90° + 40° = 130°.

Since the total of the three angles must equal 180°, the remaining angle is the difference of 180° and the sum of the known angles. So,

$$a = 180° - 130° = 50°$$

The measure of the third angle is 50°.

**3c.** The complement of a 25° angle is
$90° - 25° = 65°$

The supplement of a 25° angle is
$180° - 25° = 155°$

✓ **Student Check 3**    Write a mathematical expression that represents the given situation and then find the result.

**a.** The highest point on Earth is the peak of Mount Everest in the Himalaya Mountains. The peak is 29,028 ft above sea level. The lowest point on land is the Dead Sea at 1312 ft below sea level. What is the difference in these altitudes?

**b.** Find the measure of the unknown angle of the given triangle.

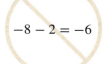

**c.** Find the complement and the supplement of an angle whose measure is 85°.

**Objective 4** ▶

Troubleshoot common errors.

## Troubleshooting Common Errors

A common error associated with subtracting real numbers is shown.

**Objective 4  Example**    **A problem and an incorrect solution is given. Provide the correct solution and an explanation of the error.**

$-8 - 2$

| Incorrect Solution | Correct Solution and Explanation |
|---|---|
| 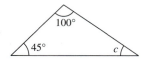 | We first convert the problem to an addition problem. Since the numbers being added have the same signs, we add their absolute values and keep the sign. $$-8 - 2 = -8 + (-2)$$ $$= -10$$ |

## ANSWERS TO STUDENT CHECKS

**Student Check 1    a.** $-3$   **b.** $15.6$   **c.** $-90$
   **d.** $-24$   **e.** $-\dfrac{5}{12}$

**Student Check 2    a.** $5$   **b.** $-18$   **c.** $-3$
**Student Check 3    a.** $30{,}340$   **b.** $35°$   **c.** $5°, 95°$

## SUMMARY OF KEY CONCEPTS

**1.** Subtracting real numbers is equivalent to adding the opposite of a number.

**2.** We can apply the order of operations to simplify expressions that involve adding and subtracting real numbers.

**3.** We should be able to solve real-life applications by first expressing the situation mathematically and then simplifying the expression. Some applications involve triangles and angles. The sum of the measure of the three angles in a triangle is 180°. Complementary angles have a sum of 90°; supplementary angles have a sum of 180°.

## GRAPHING CALCULATOR SKILLS

When using the graphing calculator for subtraction, it is important to distinguish between the negative sign, $(-)$, and the subtraction sign, $-$. Use the negative sign to denote the sign of a number and the subtraction sign to perform an operation with numbers.

**Example:** $-4 - (-7)$

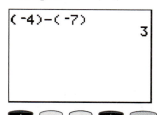

## SECTION 1.5 / EXERCISE SET

 ### Write About It!

Use complete sentences in your answer to each exercise.

**1.** Explain how to subtract two real numbers.   Answers vary.

**2.** Explain the difference between complementary angles and supplementary angles.   Answers vary.

**Determine if the statement is true or false. If a statement is false, provide an example with real numbers that contradicts the statement.**

**3.** $a - (-a) = 0$ for all real numbers $a$.
   False, $3 - (-3) = 3 + 3 = 6$
**4.** $a - a = 0$ for all real numbers $a$.   True

 ### Practice Makes Perfect!

**Perform the indicated operation and simplify. (See Objective 1.)**

**5.** $2 - 9$   $-7$
**6.** $5 - 11$   $-6$
**7.** $2.1 - (-4.7)$   6.8
**8.** $3.7 - (-9.4)$   13.1
**9.** $-2 - 6$   $-8$
**10.** $-3 - 9$   $-12$
**11.** $-43 - 35$   $-78$
**12.** $-15 - 19$   $-34$
**13.** $2 - (-6)$   8
**14.** $3 - (-9)$   12
**15.** $-2 - (-6)$   4
**16.** $-3 - (-9)$   6
**17.** $\frac{2}{3} - \frac{7}{3}$   $-\frac{5}{3}$
**18.** $\frac{3}{5} - \frac{3}{10}$   $\frac{3}{10}$
**19.** $-\frac{2}{3} - \frac{7}{3}$   $-3$
**20.** $-\frac{3}{5} - \frac{3}{10}$   $-\frac{9}{10}$

**21.** $\frac{2}{3} - \left(-\frac{7}{3}\right)$   3
**22.** $\frac{3}{5} - \left(-\frac{3}{10}\right)$   $\frac{9}{10}$
**23.** $-\frac{2}{3} - \left(-\frac{7}{3}\right)$   $\frac{5}{3}$
**24.** $-\frac{3}{5} - \left(-\frac{3}{10}\right)$   $-\frac{3}{10}$
**25.** $1\frac{1}{5} - 6\frac{4}{5}$   $-5\frac{3}{5}$
**26.** $3\frac{1}{7} - 4\frac{6}{7}$   $-1\frac{5}{7}$
**27.** $-1\frac{1}{5} - 6\frac{4}{5}$   $-8$
**28.** $-3\frac{1}{7} - 4\frac{6}{7}$   $-8$
**29.** $1\frac{1}{5} - \left(-6\frac{4}{5}\right)$   8
**30.** $3\frac{1}{7} - \left(-4\frac{6}{7}\right)$   8
**31.** $-1\frac{1}{5} - \left(-6\frac{4}{5}\right)$   $5\frac{3}{5}$
**32.** $-3\frac{1}{7} - \left(-4\frac{6}{7}\right)$   $1\frac{5}{7}$
**33.** $-2.1 - (-4.7)$   2.6
**34.** $-3.7 - (-9.4)$   5.7

**Simplify each expression. (See Objective 2.)**

**35.** $1 - (-5) + (-3) - 7$   $-4$
**36.** $-2 + 4 - 5 - (-6)$   3
**37.** $-8 - 9 + 4 - 5$   $-18$
**38.** $3 - (-7) - 6 - 4$   0
**39.** $5 - 15 - (-12) + 25$   27
**40.** $-8 - (-12) - (-6)$   10
**41.** $-1.9 - (-0.6) - 2.8$   $-4.1$
**42.** $5.8 - (-2.2) + (-1.3)$   6.7
**43.** $-\frac{1}{2} + \frac{3}{4} - \left(-\frac{3}{8}\right)$   $\frac{5}{8}$
**44.** $\frac{3}{5} + \left(-\frac{1}{5}\right) - \left(-\frac{7}{10}\right)$   $\frac{11}{10}$
**45.** $-4^2 + 13 - |7 - (-2)|$   $-12$
**46.** $-7^2 + 11 - |8 - (-5)|$   $-51$
**47.** $(-3)^2 - (-6) - |2 - 11|$   6
**48.** $(-6)^2 - (-10) - |3 - 15|$   34

Additional answers can be found in the Instructor Answer Appendix.

**49.** $-12 - \{4 - [1 - (6 - \sqrt{19 - 10})]\}$ $-18$

**50.** $-10 - \{7 - [2 - (3 - \sqrt{29 - 13})]\}$ $-14$

**Solve the problem by first writing a mathematical expression that represents the situation.**
(*See Objective 3.*)

**51.** The lowest recorded temperature in Vermont was $-46°C$ at the Bloomfield weather station on December 30, 1933. The highest recorded temperature was $41°C$ at the Vernon weather station on July 4, 1911. What is the difference between the highest and lowest temperatures? 87°C

**52.** The lowest recorded temperature in Alabama was $-33°C$ at the New Market weather station on January 30, 1966. The highest recorded temperature was $44°C$ at the Centerville weather station on Sept. 5, 1925. What is the difference between the highest and lowest temperatures? 77°C

**53.** The highest point on the North American continent is the peak of Mount McKinley. The peak is 20,320 ft above sea level. The lowest point on land in Syria is near Lake Tiberias with an elevation of 656.2 ft below sea level. What is the difference between these altitudes? 20,976.2 ft

**54.** The highest point on the Antarctica continent is the peak of Vinson Massif. The peak is 4897 meters above sea level. The lowest point on land in Russia is the Caspian Sea shoreline 28 m below sea level. What is the difference between these altitudes? 4925 m

**55.** The table shows the annual net income, in thousands, for AirTran Airways for 2005–2009. Find the change in net income for each year. (Source: http://investor.airtran.com, 2009 Annual Report)

| Year | Net Income | Change in Net Income |
|------|-----------|---------------------|
| 2005 | $ 9,320 | n/a |
| 2006 | $ 14,494 | $5174 |
| 2007 | $ 50,545 | $36,051 |
| 2008 | −$266,334 | −$316,879 |
| 2009 | $134,662 | $400,996 |

**56.** The operating profit of General Motors (GM) Corporation was $10,705 million in December 2005. By March 2006, the operating profit decreased by $5904 million. In June 2006, the operating profit decreased another $4990 million. By September 2006, the operating profit increased by $5583 million. What was the operating profit of GM in September 2006? (Source: CNNMoney.com) $5394 million

**Find the measure of the unknown angle of the given triangle.**

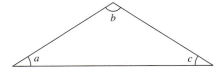

**57.** $a = 74°$, $b = 38°$ $c = 68°$    **58.** $b = 114°$, $c = 21°$ $a = 45°$

**59.** $a = 23°$, $b = 97°$ $c = 60°$    **60.** $b = 61°$, $c = 53°$ $a = 66°$

**61.** Find the complement and the supplement of an angle whose measure is $52°$. 38° and 128°

**62.** Find the complement and the supplement of an angle whose measure is $26°$. 64° and 154°

**63.** Find the complement and the supplement of an angle whose measure is $73°$. 17° and 107°

**64.** Find the complement and the supplement of an angle whose measure is $34°$. 56° and 146°

 **Mix 'Em Up!**

**Perform the indicated operation and simplify.**

**65.** $(-3.6) - 5.1$ 8.7    **66.** $(-5.78) - 7.15$ $-12.93$

**67.** $-\dfrac{5}{13} - \dfrac{1}{2}$ $-\dfrac{23}{26}$    **68.** $-\dfrac{9}{14} + \dfrac{1}{2}$ $-\dfrac{1}{7}$

**69.** $14 - 8 + (-5) - (-7)$ 8    **70.** $22 + (-5) - (-10)$ 26 −1

**71.** $-4 - \left(-\dfrac{5}{4}\right)$ $-\dfrac{11}{4}$    **72.** $3\dfrac{2}{3} - \left(-6\dfrac{1}{2}\right)$ $10\dfrac{1}{6}$

**73.** $-2^2 + 11 - |-8 - (-4)|$ 3

**74.** $-3^2 - 16 - |-8 - (-2)|$ $-31$

**75.** $4.5 - (-1.8) + 3.2 - 0.1$ 9.4

**76.** $2.9 + (-1.1) - 3.6 - 0.5$ $-2.3$

**77.** $18 - \{10 - [4 - (7 - \sqrt{30 - 5})]\}$ 10

**78.** $20 - \{6 - [5 - (1 - \sqrt{20 - 4})]\}$ 22

**79.** $-2.6 - (-1.2) - 2.7$ $-4.1$    **80.** $9.5 - 2.9 + (-0.1)$ 6.5

**81.** $-\dfrac{5}{6} + \dfrac{1}{2} - \left(-\dfrac{2}{3}\right)$ $\dfrac{1}{3}$    **82.** $\dfrac{5}{7} + \left(-\dfrac{3}{4}\right) - \left(-\dfrac{9}{14}\right)$ $\dfrac{17}{28}$

**83.** $(-1.5)^2 - (-4.5) - |3.2 - 7.2|$ 2.75

**84.** $(-2.5)^2 + (-6.4) - |5.7 - 11.7|$ $-6.15$

**85.** $1 - 18 - (-10) - 34$ $-41$    **86.** $-6 - (-17) - 1$ 10

**Write an expression needed to solve each problem and then answer the question.**

**87.** The northernmost city in the United States is Barrow, Alaska. The southernmost city in the Continental United States is Key West, Florida. The coldest recorded temperature in Barrow, Alaska, is $-47°F$.

The coldest recorded temperature in Key West, Florida, is 41°F. What is the difference between the coldest recorded temperature in Key West and the coldest recorded temperature in Barrow?  88°F

**88.** The record low temperature in New York City is 1 degree Fahrenheit below zero. The record low temperature in Palm Springs, California, is 23°F. What is the difference between the record low temperatures of New York City and Palm Springs?  24°F

**Find the measure of the unknown angle of the given triangle.**

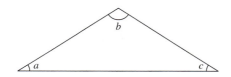

**89.** $a = 41°$ $b = 99°$  $c = 40°$

**90.** $b = 82°$, $c = 27°$  $a = 71°$

**91.** Find the complement and the supplement of an angle whose measure is 69°.  21° and 111°

**92.** Find the complement and the supplement of an angle whose measure is 18°.  72° and 162°

### You Be the Teacher!

**Correct each student's errors, if any.**

**93.** Simplify the following expression: $\frac{1}{5} - \left(-1\frac{3}{4}\right)$.

Elvernice's work:

$$\frac{1}{5} - \left(-1\frac{3}{4}\right) = \frac{1}{5} - \left(-\frac{7}{4}\right) = \frac{1}{5} - \frac{7}{4} = \frac{1 \times 4}{5 \times 4} - \frac{7 \times 5}{4 \times 5}$$

$$= \frac{4}{20} - \frac{35}{20} = -\frac{31}{20}.$$

**94.** Simplify the following expression: $-\frac{1}{4} - (-3)$.

Sherry's work:

$$-\frac{1}{4} \times 3 = -\frac{1}{4} \times \frac{3}{1} = -\frac{3}{4}.$$

### Calculate It!

**Use a calculator to simplify each expression. Express answers in fractional form, when applicable.**

**95.** $25.19 - 17.26$  7.93

**96.** $-2.45 - 5.68$  −8.13

**97.** $-\frac{5}{6} - 5$  $-5\frac{5}{6}$

**98.** $3\frac{2}{5} - \frac{3}{5}$  $2\frac{4}{5}$

---

| **SECTION 1.6** | **Multiplication and Division of Real Numbers** |

▶ **OBJECTIVES**

As a result of completing this section, you will be able to

**1.** Multiply real numbers.

**2.** Simplify exponential expressions.

**3.** Divide real numbers.

**4.** Evaluate algebraic expressions.

**5.** Solve real-life problems.

**6.** Troubleshoot common errors.

Ryan decides to play some poker. To play a hand, he must bet $5. He plays seven hands and does not win any of them. How much money did Ryan lose? To solve this problem, we need to know how to multiply positive and negative numbers together. If Ryan lost $5 for each of seven hands, he lost a total of $35. This can be determined mathematically by finding the product of 7 and −5.

In this section, we will discuss multiplying and dividing real numbers along with real-life illustrations.

### Multiplying Real Numbers

**Objective 1** ▶

Multiply real numbers.

**INSTRUCTOR NOTE:**
Remind students that in the expression 2(3), 2 and 3 are called factors and 6 is their product.

Multiplication of real numbers is just repeated addition. For example, 2(3) means that 3 is added two times or $3 + 3 = 6$. Similarly, 3(2) means that 2 is added three times or $2 + 2 + 2 = 6$. Using this concept, we can find the product 2(−3). This expression means that −3 is added two times or

$$2(-3) = (-3) + (-3) = -6$$

Now we need to find out what happens when two negative numbers are multiplied. Consider the product $(-2)(-3)$.

$$(-2)(-3) = -(2)(-3)$$
$$= -[2(-3)]$$
$$= -(-6)$$
$$= 6$$

Rewrite negative 2 as the opposite of 2.
Rewrite as the opposite of 2 times −3.
The product of 2 and −3 is −6.
The opposite of −6 is 6.

> **Procedure: Multiplying Real Numbers**
> **Step 1:** Determine the sign of the product.
>    **a.** The product of two real numbers with the *same* sign is positive.
>    **b.** The product of two real numbers with *different* signs is negative.
> **Step 2:** Multiply the numbers.

**Objective 1 Examples**   Multiply the real numbers.

**1a.** $7(-5)$      **1b.** $(-4)(-6)$      **1c.** $\left(-\dfrac{3}{5}\right)\left(\dfrac{5}{3}\right)$      **1d.** $(-4)\left(-\dfrac{1}{8}\right)$

**1e.** $(-2.35)(10)$    **1f.** $(-4)(-8)(-3)$   **1g.** $(-5)(-4)(-2)(-1)$

**Solutions**   **1a.**         $7(-5) = -35$         The product of two numbers with different signs is negative.

**1b.**         $(-4)(-6) = 24$         The product of two numbers with the same sign is positive.

**1c.**         $\left(-\dfrac{3}{5}\right)\left(\dfrac{5}{3}\right) = -\dfrac{3(5)}{5(3)}$         The product of numbers with different signs is negative.

$= -\dfrac{15}{15}$         Multiply the numerators and denominators.

$= -1$         Simplify.

**1d.**         $(-4)\left(-\dfrac{1}{8}\right) = \left(-\dfrac{4}{1}\right)\left(-\dfrac{1}{8}\right)$         Rewrite $-4$ as $-\dfrac{4}{1}$.

$= \dfrac{4(1)}{1(8)}$         The product of numbers with the same sign is positive.

$= \dfrac{4}{8}$         Simplify.

$= \dfrac{1}{2}$

**1e.**         $(-2.35)(10) = -23.5$         The product of numbers with different signs is negative.

**1f.**         $(-4)(-8)(-3) = 32(-3)$         Multiply from left to right. The product of the first two numbers is positive.

$= -96$         The final product is negative.

**1g.**   $(-5)(-4)(-2)(-1) = (20)(-2)(-1)$         Multiply from left to right. The product of the first two numbers is positive.

$= (-40)(-1)$         The product of the numbers is negative.

$= 40$         The final product is positive.

☑ **Student Check 1**   Multiply the real numbers.

     **a.** $(-5)(3)$      **b.** $(-12)(-2)$      **c.** $\dfrac{2}{3}\left(-\dfrac{3}{2}\right)$      **d.** $(-6)\left(-\dfrac{1}{18}\right)$

     **e.** $-5.71(100)$      **f.** $(-7)(-3)(-4)$      **g.** $(-8)(-9)(-2)\left(-\dfrac{1}{2}\right)$

> **Note:** *Example 1(f) contains a product of three negative numbers and their product is negative.*
> *Example 1(g) contains a product of four negative numbers and their product is positive.*
>
> *The product of an even number of negative numbers is positive.*
> *The product of an odd number of negative numbers is negative.*

## Simplifying Exponential Expressions

**Objective 2** ▶

Simplify exponential expressions.

In Section 1.3, we defined an exponent. Recall the following definition of an exponent.

> **Definition: Exponents**
>
> For $b$ a real number and $n$ a natural number,
>
> $$b^n = \underbrace{b \cdot b \cdot b \cdots b}_{n \text{ times}}$$
>
> where $b$ is the base and $n$ is the exponent.

We will now examine what happens when a negative base is raised to an even or odd exponent.

| Negative Base Raised to an Even Exponent (2, 4, 6, . . .) | Negative Base Raised to an Odd Exponent (1, 3, 5, . . .) |
|---|---|
| $(-5)^2 = (-5)(-5)$ | $(-4)^3 = (-4)(-4)(-4)$ |
| $\quad\quad\;\; = 25$ | $\quad\quad\;\; = 16(-4)$ |
| | $\quad\quad\;\; = -64$ |
| $(-2)^4 = (-2)(-2)(-2)(-2)$ | $(-3)^5 = (-3)(-3)(-3)(-3)(-3)$ |
| $\quad\quad\;\; = 4(-2)(-2)$ | $\quad\quad\;\; = 9(-3)(-3)(-3)$ |
| $\quad\quad\;\; = -8(-2)$ | $\quad\quad\;\; = -27(-3)(-3)$ |
| $\quad\quad\;\; = 16$ | $\quad\quad\;\; = (81)(-3)$ |
| | $\quad\quad\;\; = -243$ |
| A negative base raised to an even exponent is *positive*. | A negative base raised to an odd exponent is *negative*. |

Identifying the base of an exponent incorrectly can cause the result to be incorrect. The error most often occurs with expressions of the form $-3^2$ and $(-3)^2$. While these expressions look very similar, they represent two very different things.

| Expression | Verbal Representation | Mathematical Representation | Base |
|---|---|---|---|
| $-3^2$ | Opposite of 3 squared | $-3^2 = -(3 \cdot 3)$ <br> $= -(9)$ or $-9$ | 3 |
| $(-3)^2$ | Negative 3 squared | $(-3)^2 = (-3)(-3)$ <br> $= 9$ | $-3$ |

The examples in the table illustrate the following facts.

1. If a negative sign is in front of an exponential expression, the negative sign indicates the opposite of the value of the exponential expression. In other words, the negative sign is not part of the base of the exponent. That is,

$$-b^n = -\underbrace{(b \cdot b \cdot b \cdots b)}_{n \text{ times}}$$

The base $b$ is repeated as a factor $n$ times.

2. If there are parentheses around a negative number raised to an exponent, then the exponent is applied to the negative number. The negative number is repeated as a factor. That is,

$$(-b)^n = \underbrace{(-b)(-b)(-b) \cdots (-b)}_{n \text{ times}}$$

The base $-b$ is repeated as a factor $n$ times.

---

**Procedure: Simplifying an Exponential Expression**

**Step 1:** Identify the base and exponent.
**Step 2:** Rewrite the expression as repeated multiplication.
**Step 3:** Simplify the result.

---

**Objective 2 Examples**  Identify the base, determine if the negative sign repeats in the product, and simplify each exponential expression.

| Problems | Base | Does negative sign repeat? | Evaluate |
|---|---|---|---|
| 2a. $(-4)^3$ | $-4$ | yes | $(-4)^3 = (-4)(-4)(-4)$ <br> $= -64$ |
| 2b. $-5^4$ | 5 | no | $-5^4 = -(5 \cdot 5 \cdot 5 \cdot 5)$ <br> $= -625$ |
| 2c. $\left(-\dfrac{1}{2}\right)^2$ | $-\dfrac{1}{2}$ | yes | $\left(-\dfrac{1}{2}\right)^2 = \left(-\dfrac{1}{2}\right)\left(-\dfrac{1}{2}\right)$ <br> $= \dfrac{1}{4}$ |
| 2d. $-(-3)^5$ | $-3$ | yes | We must find the *opposite* of $-3$ raised to the fifth power. <br> $-(-3)^5 = -[(-3)(-3)(-3)(-3)(-3)]$ <br> $= -(-243)$ <br> $= 243$ |

☑ **Student Check 2** Identify the base, determine if the negative sign repeats in the product, and simplify each exponential expression.

**a.** $(-8)^2$ **b.** $-8^2$ **c.** $\left(-\frac{2}{3}\right)^5$ **d.** $-(-0.5)^3$

### Dividing Real Numbers

**Objective 3** ▶

Divide real numbers.

As discussed earlier in this chapter, division by a number is equivalent to multiplying by the *reciprocal* of the number. For example,

$$4 \div 2 = 2 \text{ is equivalent to } 4 \cdot \frac{1}{2} = 2$$

**Definition: Division of Real Numbers**

For real numbers $a$ and $b$, with $b \neq 0$,

$$a \div b = \frac{a}{b} = a \cdot \frac{1}{b}$$

The value $a$ is called the **dividend** and the value $b$ is called the **divisor**.

Since the quotient $a \div b$ can be written as the product $a \cdot \frac{1}{b}$, the sign rules for multiplication also apply to division.

**Procedure: Dividing Real Numbers**

**Step 1:** Determine the sign of the quotient.
 **a.** The quotient of two numbers with the *same* sign is positive.
 **b.** The quotient of two numbers with *different* signs is negative.
**Step 2:** Simplify the quotient.

Recall that division can always be checked by multiplication. For example,

$$\frac{-6}{3} = -2 \text{ because } 3(-2) = -6$$

Recall also that $\frac{0}{6} = 0$ because $0 \cdot 6 = 0$. This result is stated below.

**Property: Zero as Dividend**

The quotient of 0 and any nonzero real number, $b$, is 0.

$$\frac{0}{b} = 0, b \neq 0$$

If we try to find $\frac{6}{0}$, we would have to find a number that, when multiplied by 0, is 6. There is no such number since 0 times any real number is 0. Therefore, we say that this quotient is undefined.

**Property: Zero as Divisor**

The quotient of any nonzero real number, $b$, and 0 is *undefined*.

$$\frac{b}{0} \text{ is undefined}$$

One last fact about negative signs and fractions is important to note. Consider the following four quantities.

$$-\frac{4}{2} = -2 \qquad \frac{-4}{2} = -2 \qquad \frac{4}{-2} = -2 \qquad \frac{-4}{-2} = 2$$

So,

$$-\frac{4}{2} = \frac{-4}{2} = \frac{4}{-2} \quad \text{but} \quad -\frac{4}{2} \neq \frac{-4}{-2}$$

The placement of the negative sign in a fraction can be done in three ways. The negative sign can be placed in the numerator of the fraction, *or* the denominator of the fraction, but *not* both, *or* in front of the fraction.

---

**Property: Negative Signs in Fractions**

For real numbers $a$ and $b$ with $b \neq 0$,

$$-\frac{a}{b} = \frac{-a}{b} = \frac{a}{-b}$$

---

**Objective 3 Examples**    Perform the indicated operation.

| Problems | Solutions | |
|---|---|---|
| **3a.** $\dfrac{24}{-6}$ | $\dfrac{24}{-6} = -4$ | |
| **3b.** $\dfrac{-40}{0}$ | $\dfrac{-40}{0}$ is undefined | |
| **3c.** $\dfrac{-35}{-7}$ | $\dfrac{-35}{-7} = 5$ | |
| **3d.** $\dfrac{-6}{-7}$ | $\dfrac{-6}{-7} = \dfrac{6}{7}$ | |
| **3e.** $\dfrac{0}{-3}$ | $\dfrac{0}{-3} = 0$ | |
| **3f.** $20 \div \left(-\dfrac{2}{5}\right)$ | $20 \div \left(-\dfrac{2}{5}\right) = 20 \cdot \left(-\dfrac{5}{2}\right)$ | Rewrite as multiplying by the reciprocal. |
| | $= \dfrac{20}{1} \cdot -\dfrac{5}{2}$ | Rewrite 20 as a fraction. |
| | $= -\dfrac{100}{2}$ | Multiply the numerators and denominators. |
| | $= -50$ | Simplify. |
| **3g.** $-\dfrac{3}{4} \div \dfrac{5}{8}$ | $-\dfrac{3}{4} \div \dfrac{5}{8} = -\dfrac{3}{4} \cdot \dfrac{8}{5}$ | Rewrite as multiplying by the reciprocal. |
| | $= -\dfrac{24}{20}$ | Multiply the numerators and denominators. |
| | $= -\dfrac{4 \cdot 6}{4 \cdot 5}$ | Factor the numerator and denominator. |
| | $= -\dfrac{6}{5}$ | Divide out the common factor of 4. |

---

✓ **Student Check 3**    Perform the indicated operation.

a. $\dfrac{30}{-5}$    b. $\dfrac{0}{-4}$    c. $\dfrac{-60}{-12}$    d. $\dfrac{3}{5} \div (-9)$    e. $\dfrac{-11}{-15}$    f. $\left(-\dfrac{1}{2}\right) \div \left(-\dfrac{9}{10}\right)$

### Algebraic Expressions

Now that we know how to perform operations with signed numbers, we will revisit evaluating algebraic expressions. Recall that we evaluate an algebraic expression by replacing the variable(s) with the given values. Then we use the order of operations to simplify the resulting expression.

---

**Objective 4  Examples**   **Evaluate each expression at the given values.**

**4a.** $\dfrac{-b + \sqrt{b^2 - 4ac}}{2a}$, for $a = 3$, $b = -11$, $c = -20$

**4b.** $3x^2 - x + 5$, for $x = -2, -1, 0, 1, 2$

---

**Solutions**   **4a.** $\dfrac{-b + \sqrt{b^2 - 4ac}}{2a}$

$= \dfrac{-(-11) + \sqrt{(-11)^2 - 4(3)(-20)}}{2(3)}$    Let $a = 3$, $b = -11$, $c = -20$.

$= \dfrac{11 + \sqrt{121 - 4(3)(-20)}}{6}$    Simplify the opposite of $-11$ and the exponential expression. Simplify the denominator.

$= \dfrac{11 + \sqrt{121 - 12(-20)}}{6}$    Multiply from left to right inside the square root.

$= \dfrac{11 + \sqrt{121 + 240}}{6}$    Multiply $-12$ and $-20$.

$= \dfrac{11 + \sqrt{361}}{6}$    Add the numbers inside the square root.

$= \dfrac{11 + 19}{6}$    Simplify the square root.

$= \dfrac{30}{6}$    Add the numerators.

$= 5$    Divide the numbers.

**4b.** Make a chart to organize the information. Make the appropriate substitution for the variable. Simplify the expression.

| $x$ | $3x^2 - x + 5$ | |
|---|---|---|
| $-2$ | $3(-2)^2 - (-2) + 5 = 3(4) - (-2) + 5$ <br> $= 12 + 2 + 5$ <br> $= 19$ | Replace $x$ with $-2$. Simplify the exponent. <br> Multiply. <br> Add or subtract from left to right. |
| $-1$ | $3(-1)^2 - (-1) + 5 = 3(1) - (-1) + 5$ <br> $= 3 + 1 + 5$ <br> $= 9$ | Replace $x$ with $-1$. Simplify the exponent. <br> Multiply. <br> Add or subtract from left to right. |
| $0$ | $3(0)^2 - (0) + 5 = 3(0) - (0) + 5$ <br> $= 0 - 0 + 5$ <br> $= 5$ | Replace $x$ with $0$. Simplify the exponent. <br> Multiply. <br> Add or subtract from left to right. |
| $1$ | $3(1)^2 - (1) + 5 = 3(1) - (1) + 5$ <br> $= 3 - 1 + 5$ <br> $= 7$ | Replace $x$ with $1$. Simplify the exponent. <br> Multiply. <br> Add or subtract from left to right. |
| $2$ | $3(2)^2 - (2) + 5 = 3(4) - (2) + 5$ <br> $= 12 - 2 + 5$ <br> $= 15$ | Replace $x$ with $2$. Simplify the exponent. <br> Multiply. <br> Add or subtract from left to right. |

✓ **Student Check 4** Evaluate each expression at the given values.

a. $\sqrt{(x_1 - x_2)^2 + (y_1 - y_2)^2}$, for $x_1 = -3$, $x_2 = 2$, $y_1 = 8$, $y_2 = -4$

b. $\dfrac{|x - 2|}{x + 2}$, for $x = -2, -1, 0, 1, 2$

## Applications

**Objective 5** ▶

Solve real-life problems.

Multiplication and division of real numbers occur in real life through various types of problems. Multiplication is needed to calculate the *area* and *volume* of geometric shapes. Division is needed to calculate the *average* of a group of numbers.

> **Definition:** The **average** of a group of two or more values is the sum of the values divided by the number of values.

For example, the average of 50 and 100 is $\dfrac{50 + 100}{2} = \dfrac{150}{2} = 75$.

Division is also needed to find the percent change in a value.

> **Property: Percent Change**
> If an original value has taken on a new value, then the percent change is calculated as
> $$\frac{\text{new value} - \text{original value}}{\text{original value}} \quad \text{or} \quad \frac{\text{change in value}}{\text{original value}}$$

For example, if the price of a $30 item is on sale for $24, then the percent change in the price is

$$\frac{\text{new} - \text{original}}{\text{original}} = \frac{24 - 30}{30} = \frac{-6}{30} = -0.20 \quad \text{or} \quad \text{a decrease of } 20\%$$

**Objective 5 Examples** | Solve each problem.

**5a.** The area of a rectangle is length times width. Write an equation that represents the area of a rectangle, where $l$ is the length and $w$ is the width of the rectangle. Use the equation to find the area of an iPad which has a length of 9.56 in. and a width of 7.47 in. Round to two decimal places.

**Solution** | **5a.** The area of a rectangle can be written as $A = (l)(w)$ or $A = lw$.
To find the area of an iPad, replace $l$ with 9.56 and $w$ with 7.47.

$$A = lw$$
$$A = (9.56)(7.47)$$
$$A = 71.41$$

So, the area of an iPad is 71.41 in.$^2$

**5b.** The following chart shows the closing price of Apple's stock for August 23 to August 27, 2010. (Source: http://www.google.com/finance/historical?q=NASDAQ:AAPL)

| Date | Closing Price | Daily Change in Price | Percent Change in Price |
|------|---------------|----------------------|------------------------|
| 8/27/2010 | $241.62 | | |
| 8/26/2010 | $240.28 | | |
| 8/25/2010 | $242.89 | | |
| 8/24/2010 | $239.93 | | |
| 8/23/2010 | $245.80 | n/a | n/a |

   **i.** Complete the chart by showing how the price of the stock changed from the previous day to the next and by calculating the percent change in the price for each day.

   **ii.** After completing the chart, find the average daily change in price.

**Solution**    **5b.**    **i.** The daily change in price is the difference in one day's closing price and the previous day's closing price. The percent change is the daily change divided by the previous day's closing price.

| Date | Closing Price | Daily Change in Price | Percent Change in Price |
|------|---------------|----------------------|------------------------|
| 8/27 | $241.62 | $241.62 - 240.28 = \$1.34$ | $\dfrac{1.34}{240.28} = 0.0056$ or $0.56\%$ |
| 8/26 | $240.28 | $240.28 - 242.89 = -\$2.61$ | $\dfrac{-2.61}{242.89} = -0.0107$ or $-1.07\%$ |
| 8/25 | $242.89 | $242.89 - 239.93 = \$2.96$ | $\dfrac{2.96}{239.93} = 0.0123$ or $1.23\%$ |
| 8/24 | $239.93 | $239.93 - 245.80 = -\$5.87$ | $\dfrac{-5.87}{245.80} = -0.0239$ or $-2.39\%$ |
| 8/23 | $245.80 | n/a | n/a |

   **ii.** The average daily change in the closing price is the sum of the daily changes divided by the number of days.

$$\frac{-5.87 + 2.96 + (-2.61) + 1.34}{4} = \frac{-4.18}{4} = -1.045$$

On average between August 23 and August 27, the closing price decreased by approximately $1.05 each day.

✓ **Student Check 5**    Solve each problem.

  **a.** The volume of a right circular cylinder is the square of the radius, times the product of $\pi$ and the height of the cylinder. Write an equation that represents the volume of a right circular cylinder, where $r$ is the radius and $h$ is the height. Then use the equation to find the volume of a fuel oil storage tank that is a right circular cylinder with radius 24 in. and height 72 in. If there are 231 in.$^3$ in a gallon, how many gallons will this tank hold?

  **b.** The following chart shows the closing price of Callaway Golf Company's stock for August 9 to August 13, 2010. (Source: http://www.google.com/finance/historical?q=NYSE:ELY)

| Date | Closing Price | Daily Change in Price | Percent Change in Price |
|------|---------------|----------------------|------------------------|
| 8/13/2010 | $6.41 | | |
| 8/12/2010 | $6.64 | | |
| 8/11/2010 | $6.68 | | |
| 8/10/2010 | $7.07 | | |
| 8/9/2010 | $7.25 | n/a | n/a |

i. Complete the chart by showing how the price of the stock changed from the previous day to the next and by calculating the percent change in the price for each day.

ii. After completing the chart, find the average daily change in price.

## Troubleshooting Common Errors

**Objective 6** ▸

Troubleshoot common errors.

Some common errors associated with multiplying and dividing real numbers are shown next. One of the main errors arises when negative signs are involved in exponential expressions. Other errors occur when division with zero is involved.

**Objective 6 Examples**    A problem and an incorrect solution are given. Provide the correct solution and an explanation of the error.

**6a.** Simplify $-7^2$.

| Incorrect Solution | Correct Solution and Explanation |
|---|---|
| $-7^2 = 49$ | The base of the exponential expression is 7. The negative sign indicates that we find the opposite of squared.  $$-7^2 = -(7 \cdot 7) = -49$$ |

**6b.** Simplify $\dfrac{-2}{0}$.

| Incorrect Solution | Correct Solution and Explanation |
|---|---|
| $\dfrac{-2}{0} = 0$ | When we divide by zero, the expression is undefined. So, $\dfrac{-2}{0}$ is undefined. |

**6c.** Simplify $\dfrac{0}{-2}$.

| Incorrect Solution | Correct Solution and Explanation |
|---|---|
| $\dfrac{0}{-2}$ is undefined | When zero is divided by any nonzero number, the result is zero. So, $$\dfrac{-2}{0} = 0$$ |

## ANSWERS TO STUDENT CHECKS

**Student Check 1**   **a.** $-15$   **b.** 24   **c.** $-1$   **d.** $\dfrac{1}{3}$   **e.** $-571$   **f.** $-84$   **g.** 72

**Student Check 2**   **a.** 64   **b.** $-64$   **c.** $-\dfrac{32}{243}$   **d.** 0.125

**Student Check 3**   **a.** $-6$   **b.** 0   **c.** 5   **d.** $-\dfrac{1}{15}$   **e.** $\dfrac{11}{15}$   **f.** $\dfrac{5}{9}$

**Student Check 4**   **a.** 13   **b.** undefined, 3, 1, $\dfrac{1}{3}$, 0

**Student Check 5** **a.** $V = \pi r^2 h$; $V = 130{,}288$ in.³ or approximately 564 gal

**b. i.** in table below **ii.** The average daily change in price is −$0.21.

| Date | Closing Price | Daily Change in Price | Percent Change in Price |
|------|--------------|----------------------|------------------------|
| 8/13/2010 | $6.41 | −$0.23 | −0.0346 or −3.46% |
| 8/12/2010 | $6.64 | −$0.04 | −0.0060 or −0.60% |
| 8/11/2010 | $6.68 | −$0.39 | −0.0552 or −5.52% |
| 8/10/2010 | $7.07 | −$0.18 | −0.0248 or −2.48% |
| 8/9/2010 | $7.25 | n/a | n/a |

## SUMMARY OF KEY CONCEPTS

1. The product or quotient of two real numbers with the same sign is positive. The product or quotient of two real numbers with different signs is negative.

2. The product of an even number of negative numbers is positive. The product of an odd number of negative numbers is negative.

3. An exponent indicates how many times a number is multiplied by itself or used as a factor.

   a. A negative number raised to an odd exponent is negative.

   b. A negative number raised to an even exponent is positive.

   c. When a negative sign is in the base of an exponential expression, it should be repeated in the multiplication.

   d. When a negative sign is not in parentheses of an exponential expression, we find the opposite of the value of the exponential expression. The negative sign is not repeated.

4. Division by zero is undefined but 0 divided by a nonzero number is 0.

5. Evaluate an algebraic expression by replacing the variable with the given number. Simplify using the order of operations.

6. To solve real-life problems, express the problem mathematically and then simplify.

## GRAPHING CALCULATOR SKILLS

The calculator can assist us with multiplying and dividing real numbers as well as evaluating exponential expressions. The caret symbol  is used with exponents. To enter the power of 2, you may either press the $x^2$ key or you may use ^ and 2. If parentheses are given in the problem, you must enter them on the calculator.

**Examples:**

1. $(-4)(-5)$

2. $-\dfrac{60}{15}$

3. $(-6)^2$

4. $-6^2$

5. $(-4)^3$

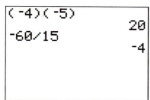

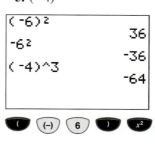

## SECTION 1.6 / EXERCISE SET

 **Write About It!**

**Use complete sentences to explain the process or problem.**

1. Multiply signed numbers   Answers vary.

2. Simplify an exponential expression   Answers vary.

3. Divide signed numbers (where the denominator is nonzero)   Answers vary.

4. Explain why division by zero is undefined.   Answers vary.

**Determine if the statement is true or false. If a statement is false, provide an example with real numbers that contradicts the statement.**

5. The expression $(-b)^n$, where $n$ is a natural number, is always positive.   Answers vary.

6. The expression $-b^n$ is equivalent to multiplying $-b$ by itself $n$ times.   Answers vary.

7. The product of an odd number and a negative number is negative.   Answers vary.

8. A negative base raised to an odd exponent is always negative.   True

9. The product of a nonzero real number and its additive inverse is 1.   Answers vary.

10. Zero divided by a nonzero real number is always 0.   True

**Practice Makes Perfect!**

**Perform the indicated operation and simplify. (See Objective 1.)**

11. $8(-6)$   $-48$

12. $(-3)(5)$   $-15$

13. $\dfrac{1}{3}(-9)$   $-3$

14. $\left(-\dfrac{1}{4}\right)(20)$   $-5$

15. $\left(-\dfrac{1}{3}\right)(-9)$   $3$

16. $\left(-\dfrac{1}{4}\right)(-20)$   $5$

17. $\left(\dfrac{2}{5}\right)\left(-\dfrac{4}{7}\right)$   $-\dfrac{8}{35}$

18. $\left(-\dfrac{3}{8}\right)\left(\dfrac{4}{9}\right)$   $-\dfrac{1}{6}$

19. $\left(-\dfrac{2}{5}\right)\left(-\dfrac{4}{7}\right)$   $\dfrac{8}{35}$

20. $\left(-\dfrac{3}{8}\right)\left(-\dfrac{4}{9}\right)$   $\dfrac{1}{6}$

21. $5(-2.3)$   $-11.5$

22. $(-6)(1.4)$   $-8.4$

23. $-3.24(1000)$   $-3240$

24. $4.98(-100)$   $-498$

25. $-5\left(-\dfrac{1}{5}\right)$   $1$

26. $\left(-\dfrac{4}{5}\right)\left(-\dfrac{5}{4}\right)$   $1$

27. $(-1)(-2)(3)$   $6$

28. $(-4)(5)(-6)$   $120$

29. $-7(-4)(-9)\left(\dfrac{2}{3}\right)$   $-168$

30. $-3(-4)(-7)\left(\dfrac{1}{14}\right)$   $-6$

31. $(-9)(-8)(0)(-1)$   $0$

32. $(-4)(0)(-5)(3)$   $0$

33. $(-8.2)(10)$   $-82$

34. $(-7.25)(100)$   $-725$

Additional answers can be found in the Instructor Answer Appendix.

**Simplify each exponential expression. (See Objective 2.)**

35. $8^2$   $64$

36. $12^2$   $144$

37. $5^3$   $125$

38. $7^3$   $343$

39. $(-9)^2$   $81$

40. $(-11)^2$   $121$

41. $(-4)^3$   $-64$

42. $(-5)^3$   $-125$

43. $(-2)^4$   $16$

44. $(-1)^4$   $1$

45. $-13^2$   $-169$

46. $-2^3$   $-8$

47. $-(-5)^4$   $-625$

48. $-(-10)^4$   $-10,000$

49. $\left(\dfrac{1}{4}\right)^2$   $\dfrac{1}{16}$

50. $\left(-\dfrac{3}{8}\right)^2$   $\dfrac{9}{64}$

51. $-\left(-\dfrac{3}{2}\right)^3$   $\dfrac{27}{8}$

52. $-\left(-\dfrac{5}{4}\right)^3$   $\dfrac{125}{64}$

53. $(0.8)^2$   $0.64$

54. $(0.5)^2$   $0.25$

55. $-(-1.2)^3$   $1.728$

56. $-(-1.5)^3$   $3.375$

**Perform the indicated operation and simplify. (See Objective 3.)**

57. $\dfrac{-12}{3}$   $-4$

58. $\dfrac{-40}{10}$   $-4$

59. $\dfrac{14}{-2}$   $-7$

60. $\dfrac{18}{-9}$   $-2$

61. $\dfrac{-16}{-2}$   $8$

62. $\dfrac{-1}{-7}$   $\dfrac{1}{7}$

63. $-5 \div \dfrac{1}{5}$   $-25$

64. $-7 \div \dfrac{1}{7}$   $-49$

65. $\dfrac{4}{7} \div \left(-\dfrac{1}{21}\right)$   $-12$

66. $\dfrac{5}{6} \div \left(-\dfrac{1}{30}\right)$   $-25$

67. $\left(-\dfrac{2}{3}\right) \div \left(-\dfrac{2}{3}\right)$   $1$

68. $\left(-\dfrac{3}{5}\right) \div \left(-\dfrac{3}{5}\right)$   $1$

69. $\dfrac{0}{-7}$   $0$

70. $\dfrac{0}{-3}$   $0$

71. $\dfrac{-5}{0}$   undefined

72. $\dfrac{-9}{0}$   undefined

**Evaluate each expression for the given values. (See Objective 4.)**

73. $\dfrac{-b + \sqrt{b^2 - 4ac}}{2a}$ for $a = 2, b = -9, c = 10$   $\dfrac{5}{2}$

74. $\dfrac{-b + \sqrt{b^2 - 4ac}}{2a}$ for $a = 3, b = 14, c = 8$   $-\dfrac{2}{3}$

75. $\dfrac{-b - \sqrt{b^2 - 4ac}}{2a}$ for $a = 2, b = -1, c = -28$   $-\dfrac{7}{2}$

76. $\dfrac{-b - \sqrt{b^2 - 4ac}}{2a}$ for $a = -6, b = -5, c = 4$   $\dfrac{1}{2}$

77. $2x^2 - 5x + 1$ for $x = -2, 0, \dfrac{1}{2}, 2$

78. $3x^2 + x - 2$ for $x = -2, 0, \dfrac{1}{2}, 2$

**79.** $-2x^2 + 4x + 3$ for $x = -2, -1, \frac{1}{2}, 2$

**80.** $-x^2 - 3x + 4$ for $x = -2, -1, \frac{1}{2}, 2$

**81.** $\sqrt{(x_1 - x_2)^2 + (y_1 - y_2)^2}$ for $x_1 = -1$, $x_2 = -4$, $y_1 = -3$, $y_2 = 1$   5

**82.** $\sqrt{(x_1 - x_2)^2 + (y_1 - y_2)^2}$ for $x_1 = -7$, $x_2 = 13$, $y_1 = 9$, $y_2 = -6$   25

**83.** $\dfrac{|x - 2|}{x + 1}$ for $x = -2, -1, 0, 1, 2$

**84.** $\dfrac{|x - 2|}{x + 2}$ for $x = -2, -1, 0, 1, 2$

**85.** $\dfrac{y_2 - y_1}{x_2 - x_1}$ for $x_1 = -2$, $x_2 = 8$, $y_1 = -4$, and $y_2 = -6$   $-\dfrac{1}{5}$

**86.** $\dfrac{y_2 - y_1}{x_2 - x_1}$ for $x_1 = 5$, $x_2 = -3$, $y_1 = -7$, and $y_2 = 2$   $-\dfrac{9}{8}$

**87.** $b^2 - 4ac$ for $a = 3, b = -5, c = -6$   97

**88.** $b^2 - 4ac$ for $a = -1, b = 6, c = -3$   24

**89.** The volume of an inverted cone is the square of the radius, times the product of $\pi$ and $\frac{1}{3}$ and the height of the cone. Write an equation that represents the volume of an inverted cone, where $r$ is the radius and $h$ is the height. Then use the equation to find the volume of a liquid storage tank that is an inverted cone with radius 30 in. and height 48 in. Round to the nearest integer.
$V = \frac{1}{3}\pi r^2 h$; 45,239 in.$^3$

**90.** The volume of a rectangular box is the product of the length, width, and height. Write an equation that represents the volume of a rectangular box, where $l$ is the length, $w$ is the width, and $h$ is the height. Then use the equation to find the volume of a liquid storage tank that is a rectangular box with the length 24.5 ft, width 36.5 ft, and height 56 ft. Round to the nearest integer.   $V = lwh$; 50,078 ft$^3$

**91.** The chart shows the closing daily price of Hershey Company's stock for October 4 to October 8, 2010. (Source: http://www.google.com/finance/historical?q=NYSE:ELY)

| Date | Closing Price | Daily Change in Price | Percent Change in Price |
|------|---------------|----------------------|-------------------------|
| 10/8/2010 | $48.75 | | |
| 10/7/2010 | $47.88 | | |
| 10/6/2010 | $48.17 | | |
| 10/5/2010 | $48.10 | | |
| 10/4/2010 | $47.63 | n/a | n/a |

   **a.** Complete the chart by showing how the price of the stock changed from the previous day to the next and by calculating the percent change in the price for each day.

   **b.** After completing the chart, find the average daily change in price.   $0.28

**92.** The following chart shows the closing weekly price of Walmart Stores' stock for September 10 to October 8, 2010. (Source: http://www.google.com/finance/historical?q=NYSE:ELY)

| Date | Closing Price | Weekly Change in Price | Percent Change in Price |
|------|---------------|------------------------|-------------------------|
| 10/8/2010 | $54.41 | | |
| 10/1/2010 | $53.36 | | |
| 9/24/2010 | $54.08 | | |
| 9/17/2010 | $53.01 | | |
| 9/10/2010 | $51.97 | | |

   **a.** Complete the chart by showing how the price of the stock changed from the previous week to the next and by calculating the percent change in the price for each week.

   **b.** After completing the chart, find the average weekly change in price.   $0.61

 **Mix 'Em Up!**

**Perform the indicated operation and simplify.**

**93.** $(-7)(5)$   $-35$

**94.** $(-13)(-11)$   143

**95.** $(-3.03)(10)$   $-30.3$

**96.** $(-7.36)(-1000)$   7360

**97.** $(-55)\left(-\dfrac{7}{11}\right)$   35

**98.** $-4\left(\dfrac{1}{20}\right)$   $-\dfrac{1}{5}$

**99.** $3^4$   81

**100.** $6^4$   1296

**101.** $\left(\dfrac{4}{7}\right)\left(-\dfrac{3}{16}\right)$   $-\dfrac{3}{28}$

**102.** $\left(-\dfrac{10}{7}\right)\left(-\dfrac{14}{25}\right)$   $\dfrac{4}{5}$

**103.** $\dfrac{-63}{-7}$   9

**104.** $\dfrac{-44}{11}$   $-4$

**105.** $-5^4$   $-625$

**106.** $10^3$   1000

**107.** $\left(\dfrac{2}{5}\right)^3$   $\dfrac{8}{125}$

**108.** $\left(\dfrac{3}{7}\right)^3$   $\dfrac{27}{343}$

**109.** $(-4)^4$   256

**110.** $-(-3)^5$   243

**111.** $-0.2^4$   $-0.0016$

**112.** $-(-0.5)^3$   0.125

**113.** $(4.2)(-3.5)$   $-14.7$

**114.** $(-1.2)(-5.2)$   6.24

**115.** $\left(-\dfrac{2}{3}\right)\left(-\dfrac{3}{2}\right)$   1

**116.** $\left(-\dfrac{7}{5}\right)\left(-\dfrac{5}{7}\right)$   1

**117.** $-14^2$   $-196$

**118.** $-6^3$   $-216$

**119.** $\dfrac{0}{-2}$   0

**120.** $\dfrac{-2}{0}$   undefined

**121.** $2 \div 0$   undefined

**122.** $0 \div (-9)$   0

**123.** $\dfrac{10}{-9} \div (-25)$   $\dfrac{2}{45}$

**124.** $\dfrac{-12}{5} \div (-10)$   $\dfrac{6}{25}$

**125.** $\left(-\dfrac{15}{19}\right) \div \left(-\dfrac{6}{19}\right)$   $\dfrac{5}{2}$

**126.** $\left(-\dfrac{8}{21}\right) \div \left(\dfrac{12}{21}\right)$   $-\dfrac{2}{3}$

**127.** $\left(-\dfrac{5}{6}\right)^3$   $-\dfrac{125}{216}$

**128.** $-\left(-\dfrac{8}{11}\right)^2$   $-\dfrac{64}{121}$

**129.** $6(10)(-1)$   $-60$

**130.** $10(-6)(-9)$   $540$

**Evaluate each expression for the given values.**

**131.** $\sqrt{(x_1 - x_2)^2 + (y_1 - y_2)^2}$ for $x_1 = 17$, $x_2 = -3, y_1 = -11, y_2 = 4$   $25$

**132.** $\sqrt{(x_1 - x_2)^2 + (y_1 - y_2)^2}$ for $x_1 = -10$, $x_2 = 2, y_1 = 3, y_2 = -2$   $13$

**133.** $2l + 2w$ for $l = 11$ and $w = 15$   $52$

**134.** $2l + 2w$ for $l = 3$ and $w = 8$   $22$

**135.** $b^2 - 4ac$ for $a = -4, b = -7, c = -9$   $-95$

**136.** $b^2 - 4ac$ for $a = -7, b = -8, c = 2$   $120$

**137.** $\dfrac{1}{2}bh$ for $b = 4$ and $h = 5$   $10$

**138.** $\dfrac{1}{2}bh$ for $b = 3$ and $h = 6$   $9$

**139.** $\dfrac{-b - \sqrt{b^2 - 4ac}}{2a}$ for $a = 5, b = -7, c = -6$   $-\dfrac{3}{5}$

**140.** $\dfrac{-b + \sqrt{b^2 - 4ac}}{2a}$ for $a = 5, b = -7, c = -6$   $2$

**141.** $\dfrac{y_2 - y_1}{x_2 - x_1}$ for $x_1 = -4, x_2 = 13, y_1 = -1$, and $y_2 = -9$   $-\dfrac{8}{17}$

**142.** $\dfrac{y_2 - y_1}{x_2 - x_1}$ for $x_1 = 2, x_2 = -3, y_1 = -6$, and $y_2 = -11$   $1$

**143.** $\dfrac{|x - 2|}{x + 3}$ for $x = -3, -2, 0, 1, 2$

**144.** $\dfrac{|2x - 3|}{x}$ for $x = -2, -1, 0, 1, 2$

**Solve each problem.**

**145.** The volume of a rectangular box is the product of the length, width, and height. Write an equation that represents the volume of a rectangular box, where $l$ is the length, $w$ is the width, and $h$ is the height. Then use the equation to find the volume of a liquid storage tank that is a rectangular box with the length 32.4 in., width 26.6 in., and height 64.5 in. If there are 231 in.$^3$ in a gallon, how many gallons will this tank hold? Round to the nearest integer.   $V = lwh$; 55,589 in.$^3$ 241 gal

**146.** The following chart shows the closing weekly price of Dow Jones Industrial Average prices for September 10 to October 8, 2010. (Source: http://www.google.com/finance/historical?q=NYSE:ELY)

| Date | Closing Price | Weekly Change in Price | Percent Change in Price |
|---|---|---|---|
| 10/8/2010 | $11,006.48 | | |
| 10/1/2010 | $10,829.68 | | |
| 9/24/2010 | $10,860.26 | | |
| 9/17/2010 | $10,607.85 | | |
| 9/10/2010 | $10,462.77 | n/a | n/a |

**a.** Complete the chart by showing how the price of the stock changed from the previous week to the next and by calculating the percent change in the price for each week.

**b.** After completing the chart, find the average weekly change in price.   $135.93

 **You Be the Teacher!**

**Correct each student's errors, if any.**

**147.** Determine if $\dfrac{0}{a}$ or $\dfrac{a}{0}$ is undefined.

Marcia's work:

$\dfrac{0}{a}$ = undefined and $\dfrac{a}{0}$ = 0   $\dfrac{0}{a} = 0$ and $\dfrac{a}{0}$ = undefined

**148.** Simplify $-(-12)^2$.

Basil's work:

$-(-12)^2 = 12^2 = 144$   $-(-12)^2 = -(-12)(-12) = -144$

**149.** Simplify $\dfrac{12}{15} \div \left(-\dfrac{10}{9}\right)$.

Holly's work:

$\dfrac{12}{15} \div \left(-\dfrac{10}{9}\right) = -\dfrac{12}{15} \cdot \dfrac{10}{9} = -\dfrac{\overset{4}{\cancel{12}}}{\underset{5}{\cancel{15}}} \cdot \dfrac{\overset{2}{\cancel{10}}}{9} = -\dfrac{8}{9}$

$\dfrac{12}{15} \div \left(-\dfrac{10}{9}\right) = -\dfrac{12}{15} \cdot \dfrac{9}{10} = -\dfrac{\overset{6}{\cancel{12}}}{\underset{5}{\cancel{15}}} \cdot \dfrac{\overset{3}{\cancel{9}}}{\underset{5}{\cancel{10}}} = -\dfrac{18}{25}$

 **Calculate It!**

**Use a calculator to simplify each expression. Express each answer in fractional form, when applicable.**

**150.** $(-0.4)^5$   $-0.01024$

**151.** $(-1.2)^2$   $1.44$

**152.** $-\left(-\dfrac{2}{3}\right)^5$   $\dfrac{32}{243}$

**153.** $\left(-\dfrac{3}{4}\right)^2$   $\dfrac{9}{16}$

**154.** $\dfrac{7}{30} \div \left(-\dfrac{1}{21}\right)$   $-\dfrac{49}{10}$

**155.** $-\dfrac{15}{8} \div \left(-\dfrac{20}{16}\right) \cdot 12$   $18$

| SECTION 1.7 | **Properties of Real Numbers** |

▶ **OBJECTIVES**

As a result of completing this section, you will be able to

1. Apply the identity and inverse properties.
2. Apply the commutative and associative properties.
3. Apply the distributive property.

The set of real numbers contains some interesting properties. We use some of the properties without even thinking about them. These properties are very important to the study of algebra, so we will state them explicitly in this section.

## The Identity and Inverse Properties

The first property that we will discuss is the *identity property*. An **identity element** is a number which leaves another number unchanged when an operation is performed on it.

- When we add numbers, the only number that can be added to another number without changing its value is *zero*.
- When we multiply numbers, the only number that can be multiplied to another number without changing its value is *one*.

**Objective 1** ▶

Apply the identity and inverse properties.

|  | **Identity Property** | **Identity Element** | **Example ($a = 6$)** |
|---|---|---|---|
| **Addition** | For all real numbers $a$, $a + 0 = 0 + a = a$ | Zero is the **additive identity**. | $6 + 0 = 0 + 6 = 6$ |
| **Multiplication** | For all real numbers $a$, $a \cdot 1 = 1 \cdot a = a$ | One is the **multiplicative identity**. | $6 \cdot 1 = 1 \cdot 6 = 6$ |

A related property is the *inverse property*. An **inverse** is a number which produces the identity element when an operation is performed on it.

- The *additive inverse* of a number is its opposite since adding a number and its opposite results in 0.
- The *multiplicative inverse* of a number is its reciprocal since multiplying a number and its reciprocal results in 1.

**INSTRUCTOR NOTE:**
Help students understand that when we apply the inverse, we always get the identity element.

|  | **Inverse Property** | **Inverse Element** | **Example ($a = 5$)** |
|---|---|---|---|
| **Addition** | For all real numbers $a$, $a + (-a) = (-a) + a$ $= 0$ | $-a$ is the **additive inverse** (or opposite) of $a$. | The opposite of 5 is $-5$. $5 + (-5) = (-5) + 5$ $= 0$ |
| **Multiplication** | For all real numbers $a \neq 0$, $a \cdot \dfrac{1}{a} = \dfrac{1}{a} \cdot a$ $= 1$ | $\dfrac{1}{a}$ is the **multiplicative inverse** (or reciprocal) of $a$, $a \neq 0$. | The reciprocal of 5 is $\dfrac{1}{5}$. $5\left(\dfrac{1}{5}\right) = \left(\dfrac{1}{5}\right)(5)$ $= 1$ |

**Objective 1 Examples** **Find both the additive and multiplicative inverses of each number. Assume any variables are nonzero.**

| Problems | **Solutions** | |
|---|---|---|
|  | **Additive Inverse, $-a$** | **Multiplicative Inverse, $\dfrac{1}{a}$** |
| **1a.** $-6$ | $-(-6) = 6$ | $\dfrac{1}{-6} = -\dfrac{1}{6}$ |
| **1b.** $\dfrac{3}{4}$ | $-\left(\dfrac{3}{4}\right) = -\dfrac{3}{4}$ | $\dfrac{1}{\frac{3}{4}} = 1 \cdot \dfrac{4}{3} = \dfrac{4}{3}$ |
| **1c.** $2x$ | $-(2x) = -2x$ | $\dfrac{1}{2x} = \dfrac{1}{2x}$ |
| **1d.** $-3y$ | $-(-3y) = 3y$ | $\dfrac{1}{-3y} = -\dfrac{1}{3y}$ |
| **1e.** $\dfrac{x}{7}$ | $-\left(\dfrac{x}{7}\right) = -\dfrac{x}{7}$ | $\dfrac{1}{\frac{x}{7}} = 1 \cdot \dfrac{7}{x} = \dfrac{7}{x}$ |

☑ **Student Check 1**    Find both the additive and multiplicative inverses of each number. Assume any variables are nonzero.

   **a.** $-10$    **b.** $\dfrac{7}{8}$    **c.** $4y$    **d.** $-9b$    **e.** $\dfrac{a}{3}$

## The Commutative and Associative Properties

**Objective 2** ▶

Apply the commutative and associative properties.

Additional properties of the real numbers relate to how we add and multiply the numbers. These properties form the foundation of how we work with algebraic expressions.

The *commutative property* of the real numbers states that the order in which we add real numbers or multiply real numbers doesn't change the result.

**INSTRUCTOR NOTE:**
Use the concept of commuting to work and back home as an aid to help students remember the commutative property. We go from home to work and then from work to home. The order changed but the distance traveled was the same.

| | Commutative Property | Examples ($a = 2, b = 3$) |
|---|---|---|
| **Addition** | For all real numbers $a$ and $b$, $a + b = b + a$ | $2 + 3 = 3 + 2 = 5$ |
| **Multiplication** | For all real numbers $a$ and $b$, $a \cdot b = b \cdot a$ | $2 \cdot 3 = 3 \cdot 2 = 6$ |

The *associative property* of the real numbers states that the way numbers are grouped when they are added or multiplied doesn't change the outcome.

**INSTRUCTOR NOTE:**
Introduce this concept by asking students "what is an association?" Since it deals with a group of people, help them see the connection with the associative property and changing the grouping.

| | Associative Property | Examples ($a = 2, b = 3, c = 4$) |
|---|---|---|
| **Addition** | For all real numbers $a$, $b$, and $c$, $a + (b + c) = (a + b) + c$ | $2 + (3 + 4) = (2 + 3) + 4$<br>$2 + (7) = (5) + 4$<br>$9 = 9$ |
| **Multiplication** | For all real numbers $a$, $b$, and $c$, $(a \cdot b) \cdot c = a \cdot (b \cdot c)$ | $(2 \cdot 3) \cdot 4 = 2 \cdot (3 \cdot 4)$<br>$(6) \cdot 4 = 2 \cdot (12)$<br>$24 = 24$ |

**Objective 2 Examples**    **Apply the commutative and associative properties to rewrite each expression and simplify the result.**

   **2a.** $2 + y + 7$    **2b.** $2(y)(7)$    **2c.** $(x + 6) + 4$    **2d.** $4(6x)$

**Solutions**    **2a.**    $2 + y + 7 = y + 2 + 7$       Apply the commutative property of addition.

$= y + 9$       Add the numbers.

   **2b.**    $2(y)(7) = 2(7)y$       Apply the commutative property of multiplication.

$= 14y$       Multiply the numbers.

   **2c.** $(x + 6) + 4 = x + (6 + 4)$       Apply the associative property of addition.

$= x + 10$       Add the numbers.

   **2d.**    $4(6x) = (4 \cdot 6)x$       Apply the associative property of multiplication.

$= 24x$       Multiply the numbers.

☑ **Student Check 2**    Apply the commutative or associative properties to rewrite each expression and simplify the result.

   **a.** $3 + x + 5$    **b.** $3(x)(5)$    **c.** $(b + 2) + 9$    **d.** $2(9b)$

> **Note:** *Subtraction and division are* not *commutative. We cannot change the order of the numbers being subtracted or divided and obtain the same result.*
>
> *For instance,* $5 - 3 = 2$ *but* $3 - 5 = 3 + (-5) = -2$.
>
> *Also,* $\dfrac{6}{3} = 2$ *but* $\dfrac{3}{6} = \dfrac{1}{2}$.

### The Distributive Property

**Objective 3** ▶

**Apply the distributive property.**

The distributive property is a property that is used extensively in algebra. It provides a way for us to multiply a group of numbers by a number—that is, it enables us to rewrite a product as a sum or a sum as a product.

To illustrate, we will simplify the two expressions below using the order of operations.

$$3(4 + 5) = 3(9) \qquad\qquad 3(4) + 3(5) = 12 + 15$$
$$= 27 \qquad\qquad\qquad\qquad\quad = 27$$

These two expressions are equivalent—that is,

$$3(4 + 5) = 3(4) + 3(5)$$

The preceding example illustrates that when a number is multiplied by a sum, it is equivalent to the sum of the products.

Also, consider a difference multiplied by a number.

$$3(4 - 5) = 3(-1) \qquad\qquad 3(4) - 3(5) = 12 - 15$$
$$= -3 \qquad\qquad\qquad\qquad\quad = -3$$

These two expressions are equivalent, so

$$3(4 - 5) = 3(4) - 3(5)$$

The preceding example illustrates that when a number is multiplied by a difference, it is equivalent to the difference of the products.

These properties make up the distributive property over addition and over subtraction.

> **Property: Distributive Property over Addition**
>
> For all real numbers $a$, $b$, and $c$,
>
> $$a(b + c) = ab + ac$$

> **Property: Distributive Property over Subtraction**
>
> $$a(b - c) = ab - ac$$

When asked to apply the distributive property, our goal is to multiply the factor outside of the parentheses by each of the terms inside the parentheses.

Because multiplication is commutative, we can also distribute from the left and write the distributive property as follows.

> **Property: Alternate Form of the Distributive Property**
>
> $$(b + c)a = ba + ca$$

An illustration of this alternate form is $(2 + 6)4 = 2(4) + 6(4) = 8 + 24 = 32$.

The distributive property can be applied when there are more than two terms inside parentheses. We just multiply the factor by each term in parentheses.

| **Objective 3 Examples** | **Apply the distributive property to rewrite each expression. Simplify the result.** |

**3a.** $2(x + 4)$    **3b.** $3(x - 5)$    **3c.** $4(2x + 7)$    **3d.** $-5(x + 2)$

**3e.** $-3(x - 7)$    **3f.** $8(2x + 3y - 5)$    **3g.** $-(6x - 9)$    **3h.** $10\left(\dfrac{1}{5}x + \dfrac{1}{2}\right)$

**Solutions**

**3a.**
$$2(x + 4) = 2(x) + 2(4)$$ — Apply the distributive property.
$$= 2x + 8$$ — Simplify.

**3b.**
$$3(x - 5) = 3(x) - 3(5)$$ — Apply the distributive property.
$$= 3x - 15$$ — Simplify.

**3c.**
$$4(2x + 7) = 4(2x) + 4(7)$$ — Apply the distributive property.
$$= 8x + 28$$ — Simplify.

**3d.**
$$-5(x + 2) = -5(x) + (-5)(2)$$ — Apply the distributive property.
$$= -5x - 10$$ — Simplify.

**3e.**
$$-3(x - 7) = -3(x) - (-3)(7)$$ — Apply the distributive property.
$$= -3x + 21$$ — Simplify.

**3f.** $8(2x + 3y - 5) = 8(2x) + 8(3y) - 8(5)$ — Apply the distributive property.
$$= 16x + 24y - 40$$ — Simplify.

**INSTRUCTOR NOTE:**
Example 3g can also be thought of as finding the opposite of the expression, $6x - 9$. Point out that finding the opposite is equivalent to multiplying the expression by $-1$.

**3g.**
$$-1(6x - 9) = -1(6x) - (-1)(9)$$ — Distribute $-1$ to each term.
$$= -6x + 9$$ — Simplify.

 **Note:** Multiplying by $-1$ changes the signs of the original terms in parentheses.

**3h.**
$$10\left(\dfrac{1}{5}x + \dfrac{1}{2}\right) = 10\left(\dfrac{1}{5}x\right) + 10\left(\dfrac{1}{2}\right)$$ — Apply the distributive property.
$$= 2x + 5$$ — Simplify.

| ✔ **Student Check 3** | **Apply the distributive property to rewrite each expression. Simplify the result.** |

**a.** $5(x + 6)$    **b.** $7(x - 3)$    **c.** $2(3x + 1)$    **d.** $-3(x + 2)$

**e.** $-6(x - 4)$    **f.** $9(4x + 5y - 3)$    **g.** $-(8x + 2)$    **h.** $15\left(-\dfrac{1}{3}x + \dfrac{2}{5}\right)$

**Objective 4**

Troubleshoot common errors.

## Troubleshooting Common Errors

Some common errors associated with the properties of real numbers are shown next.

**Objective 4  Examples**   **A problem and an incorrect solution are given. Provide the correct solution and an explanation of the error.**

**a.** Simplify $(2x + 6) + 3$.

| Incorrect Solution | Correct Solution and Explanation |
|---|---|
| $(2x + 6) + 3$ $6x + 18$ | To simplify this expression, we must apply the associative property of addition. The distributive property does not apply since 3 is connected by addition, not multiplication. $$(2x + 6) + 3$$ $$2x + (6 + 3)$$ $$2x + 9$$ |

**b.** Simplify $4(6x + 7)$.

| Incorrect Solution | Correct Solution and Explanation |
|---|---|
| $4(6x + 7)$ $24x + 7$ | We must distribute 4 to each term in parentheses, not just the first term. $$4(6x + 7)$$ $$24x + 28$$ |

## ANSWERS TO STUDENT CHECKS

**Student Check 1**   **a.** $10, -\dfrac{1}{10}$   **b.** $-\dfrac{7}{8}, \dfrac{8}{7}$   **c.** $-4y, \dfrac{1}{4y}$
  **d.** $9b, -\dfrac{1}{9b}$   **e.** $-\dfrac{a}{3}, \dfrac{3}{a}$

**Student Check 2**   **a.** $x + 8$   **b.** $15x$   **c.** $b + 11$   **d.** $18b$
**Student Check 3**   **a.** $5x + 30$   **b.** $7x - 21$   **c.** $6x + 2$
  **d.** $-3x - 6$   **e.** $-6x + 24$   **f.** $36x + 45y - 27$
  **g.** $-8x - 2$   **h.** $-5x + 6$

## SUMMARY OF KEY CONCEPTS

1. The identity element is the number that leaves another number unchanged when an operation is performed on it.
   a. For instance, 0 is the identity element for addition since any number added to zero is the original number.
   b. Also, 1 is the identity element for multiplication since any number multiplied by 1 is the original number.
2. An inverse is a number which produces the identity element when an operation is performed on it.
   a. The additive inverse is the opposite of a number.

   b. The multiplicative inverse is the reciprocal of a number.
3. Addition and multiplication are both commutative and associative. We can change the order and the grouping of the numbers being added or multiplied and obtain the same result.
4. The distributive property enables us to multiply a group of numbers by another number. The outer number distributes to all terms inside parentheses.

## SECTION 1.7   EXERCISE SET

 **Write About It!**

**Use complete sentences to explain the term or the process.**

1. Identity element for addition   Answers vary.
2. Identity element for multiplication   Answers vary.
3. Additive inverse   Answers vary.

4. Multiplicative inverse   Answers vary.
5. Commutative property for addition   Answers vary.
6. Commutative property for multiplication   Answers vary.
7. Associative property for addition   Answers vary.
8. Associative property for multiplication   Answers vary.

Additional answers can be found in the Instructor Answer Appendix.

**9.** Distributive property  *Answers vary.*

**10.** How to apply distributive property to simplify an expression  *Answers vary.*

## Practice Makes Perfect!

Find the additive inverse and the multiplicative inverse of each number. Assume all variables are nonzero. (*See Objective 1.*)

**11.** $-7$  $7, -\dfrac{1}{7}$  **12.** $8$  $-8, \dfrac{1}{8}$  **13.** $25$  $-25, \dfrac{1}{25}$

**14.** $-12$  $12, -\dfrac{1}{12}$  **15.** $\dfrac{3}{5}$  $-\dfrac{3}{5}, \dfrac{5}{3}$  **16.** $\dfrac{7}{4}$  $-\dfrac{7}{4}, \dfrac{4}{7}$

**17.** $-\dfrac{4}{9}$  $\dfrac{4}{9}, -\dfrac{9}{4}$  **18.** $-\dfrac{3}{10}$  $\dfrac{3}{10}, -\dfrac{10}{3}$  **19.** $3x$  $-3x, \dfrac{1}{3x}$

**20.** $12y$  $-12y, \dfrac{1}{12y}$  **21.** $-6a$  $6a, -\dfrac{1}{6a}$  **22.** $-5b$  $5b, -\dfrac{1}{5b}$

**23.** $\dfrac{x}{6}$  $-\dfrac{x}{6}, \dfrac{6}{x}$  **24.** $\dfrac{2y}{5}$  $-\dfrac{2y}{5}, \dfrac{5}{2y}$

**25.** $-\dfrac{7a}{2}$  $\dfrac{7a}{2}, -\dfrac{2}{7a}$  **26.** $-\dfrac{4b}{3}$  $\dfrac{4b}{3}, -\dfrac{3}{4b}$

Apply the commutative and associative properties to rewrite each expression and simplify the result. (*See Objective 2.*)

**27.** $4 + x + 7$  $x + 11$  **28.** $9 + y + 5$  $y + 14$

**29.** $-12 + a - 1$  $a - 13$  **30.** $-15 + b - 7$  $b - 22$

**31.** $-6 + x + (-10)$  $x - 16$  **32.** $-18 + y + (-12)$  $y - 30$

**33.** $3(x)(9)$  $27x$  **34.** $5(y)(15)$  $75y$

**35.** $(-1)(a)(-16)$  $16a$  **36.** $(-1)(b)(-24)$  $24b$

**37.** $5(x)(-14)$  $-70x$  **38.** $-9(y)(4)$  $-36y$

**39.** $(c - 6) + (-3)$  $c - 9$  **40.** $(d - 12) + (-9)$  $d - 21$

**41.** $[x - (-2)] - 12$  $x - 10$  **42.** $[y - (-28)] - (-15)$  $y + 43$

**43.** $-2(3x)$  $-6x$  **44.** $-8(2y)$  $-16y$

**45.** $-7(-3a)$  $21a$  **46.** $-4(-5b)$  $20b$

**47.** $\left(-\dfrac{1}{2}\right) + x + \dfrac{3}{4}$  $x + \dfrac{1}{4}$  **48.** $\dfrac{2}{3} + y - \dfrac{5}{12}$  $y + \dfrac{1}{4}$

**49.** $\left(a + \dfrac{5}{7}\right) + 2$  $a + \dfrac{19}{7}$  **50.** $\left(b - \dfrac{3}{5}\right) + 2$  $b + \dfrac{7}{5}$

**51.** $12(x)\left(\dfrac{1}{6}\right)$  $2x$  **52.** $18(y)\left(\dfrac{5}{6}\right)$  $15y$

**53.** $-6\left(\dfrac{1}{4}x\right)$  $-\dfrac{3x}{2}$  **54.** $-5\left(\dfrac{7}{10}y\right)$  $-\dfrac{7y}{2}$

Apply the distributive property to rewrite each expression. Simplify the result. (*See Objective 3.*)

**55.** $4(x - 7)$  $4x - 28$  **56.** $5(x - 9)$  $5x - 45$

**57.** $3(x + 5)$  $3x + 15$  **58.** $6(x + 2)$  $6x + 12$

**59.** $11(3x - 4)$  $33x - 44$  **60.** $12(5x - 3)$  $60x - 36$

**61.** $6(2x + 3y - 5)$  $12x + 18y - 30$  **62.** $3(5x - 2y + 10)$  $15x - 6y + 30$

**63.** $-(-5x + 2y - 9)$  $5x - 2y + 9$  **64.** $-11(3x + 4y - 6)$  $-33x - 44y + 66$

**65.** $-(16x - 3)$  $-16x + 3$  **66.** $-(23x - 12)$  $-23x + 12$

**67.** $12\left(-\dfrac{1}{2}x + \dfrac{1}{3}\right)$  $-6x + 4$  **68.** $10\left(-\dfrac{1}{5}x + \dfrac{1}{2}\right)$  $-2x + 5$

**69.** $-12\left(\dfrac{3}{2}x - \dfrac{1}{4}\right)$  $-18x + 3$  **70.** $-24\left(\dfrac{1}{6}x - \dfrac{3}{8}\right)$  $-4x + 9$

## Mix 'Em Up!

Find the additive inverse and the multiplicative inverse of each number. Assume all variables are nonzero.

**71.** $15x$  $-15x, \dfrac{1}{15x}$  **72.** $-2a$  $2a, -\dfrac{1}{2a}$  **73.** $\dfrac{6x}{7}$  $-\dfrac{6x}{7}, \dfrac{7}{6x}$

**74.** $\dfrac{3c}{5}$  $-\dfrac{3c}{5}, \dfrac{5}{3c}$  **75.** $-11$  $11, -\dfrac{1}{11}$  **76.** $21$  $-21, \dfrac{1}{21}$

**77.** $\dfrac{9}{5}$  $-\dfrac{9}{5}, \dfrac{5}{9}$  **78.** $-\dfrac{6}{5}$  $\dfrac{6}{5}, -\dfrac{5}{6}$

Apply the commutative, associative, and/or distributive properties to rewrite each expression and simplify the result.

**79.** $8 + x + 12$  $x + 20$  **80.** $-12 + b - 4$  $b - 16$

**81.** $2(x)(14)$  $28x$  **82.** $3(y)(19)$  $57y$

**83.** $-8 + x + (-28)$  $x - 36$  **84.** $-4 + y + (-22)$  $y - 26$

**85.** $(-1)(a)(-18)$  $18a$  **86.** $-7(y)(14)$  $-98y$

**87.** $[x - (-5)] - 19$  $x - 14$  **88.** $[y - (-17)] - (-1)$  $y + 18$

**89.** $\left(-\dfrac{3}{5}\right) + a + \dfrac{1}{2}$  $a - \dfrac{1}{10}$  **90.** $-\dfrac{1}{3} + b - \dfrac{5}{6}$  $b - \dfrac{7}{6}$

**91.** $4(2.1a)$  $8.4a$  **92.** $3(1.2b)$  $3.6b$

**93.** $\left(a - \dfrac{7}{8}\right) + 1$  $a + \dfrac{1}{8}$  **94.** $\left(b - \dfrac{7}{10}\right) + 1$  $b + \dfrac{3}{10}$

**95.** $-(-x - 12y + 5)$  $x + 12y - 5$  **96.** $8(2x - 3y + 1)$  $16x - 24y + 8$

**97.** $-2(7x)$  $-14x$  **98.** $-9(-5b)$  $45b$

**99.** $5(-0.3x + 0.2)$  $-1.5x + 1$  **100.** $2(-0.9x + 0.1)$  $-1.8x + 0.2$

**101.** $-5(2x + 3)$  $-10x - 15$  **102.** $-(12x + 5)$  $-12x - 5$

**103.** $12(x)\left(-\dfrac{1}{4}\right)$  $-3x$  **104.** $-5\left(-\dfrac{3}{10}y\right)$  $\dfrac{3y}{2}$

**105.** $6\left(-\dfrac{1}{6}x + \dfrac{1}{3}\right)$  $-x + 2$  **106.** $0.4(1.5x - 0.3)$  $0.6x - 0.12$

**107.** $0.8(0.7x - 1.2)$  $0.56x - 0.96$  **108.** $5\left(-\dfrac{1}{10}x + \dfrac{2}{5}\right)$  $-\dfrac{1}{2}x + 2$

## You Be the Teacher!

Correct each student's errors, if any.

**109.** Simplify $3(2a)$.

Josh's work:

$3(2a) = 3(2)(3a) = 6(3a) = 18a$  $3(2a) = 6a$

**110.** Simplify $3(6x + 2y - 4)$.

Grace's work:

$3(6x + 2y - 4) = 18x + 2y - 4$  $3(6x + 2y - 4) = 18x + 6y - 12$

**111.** Simplify $-(2 - x + 3y)$.

Charlotte's work:

$-(2 - x + 3y) = -2 - x + 3y$  $-(2 - x + 3y) = -2 + x - 3y$

**112.** Simplify $-6(-4b)$.

Mary's work:

$-6(-4b) = -6(-4)(-6b) = 24(-6b) = -144b$  $-6(-4b) = 24b$

## SECTION 1.8 / Algebraic Expressions

### OBJECTIVES

As a result of completing this section, you will be able to

1. Identify terms and coefficients of algebraic expressions.
2. Identify like terms.
3. Combine like terms.
4. Simplify algebraic expressions.
5. Troubleshoot common errors.

### Objective 1 ▶

Identify terms and coefficients of algebraic expressions.

**INSTRUCTOR NOTE:**
When we discuss polynomials, these concepts will be discussed again, along with the idea of degree.

In the previous sections, we have evaluated algebraic expressions and also expressed relationships mathematically as algebraic expressions. We will now discuss these types of expressions in more detail and learn some terminology that we use to describe them.

### Terminology Dealing with Algebraic Expressions

A **term** is a number or a product of a number and variables raised to exponents. Terms are separated by addition or subtraction signs. Algebraic expressions consist of **variable terms** and **constant terms**. Variable terms are terms that contain a variable. Constant terms do not contain variables. Each variable term has a numerical **coefficient**. The coefficient is the number that is multiplied by the variable.

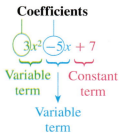

**Procedure:** **Identifying Terms and Coefficients of an Algebraic Expression**

**Step 1:** Determine how many expressions are connected by an addition or subtraction sign. This gives us how many terms the expression contains.
  **a.** The terms with variables are the variable terms.
  **b.** The terms with numbers only are the constant terms.

**Step 2:** The coefficients of the variable terms are the numbers multiplied by the variable. Be careful to identify the sign of the coefficient. If the variable term is connected by a subtraction sign, then the coefficient of that term is a negative value. Remember that subtraction is adding the opposite. Also, remember the Identity Property of Multiplication: $x = 1 \cdot x$.

### Objective 1 Examples

Complete the table to identify the number of terms, the variable terms, the constant terms, and the coefficients of the variable terms of each expression.

| Expression | Total Number of Terms | Variable Term(s) | Constant Term(s) | Coefficients of Each Variable Term |
|---|---|---|---|---|
| **1a.** $2x$ | 1 | $2x$ | None | 2 |
| **1b.** $-x$ | 1 | $-x = -1x$ | None | $-1$ |
| **1c.** $x + 2$ | 2 | $x = 1x$ | 2 | 1 |
| **1d.** $\dfrac{y}{2} - 5$ | 2 | $\dfrac{y}{2} = \dfrac{1y}{2} = \dfrac{1}{2}y$ | $-5$ | $\dfrac{1}{2}$ |
| **1e.** $3x^2 - 5x + 7$ | 3 | $3x^2, -5x$ | 7 | $3, -5$ |
| **1f.** $-2y^3 + 3y^2 - y - 6$ | 4 | $-2y^3, 3y^2, -y$ | $-6$ | $-2, 3, -1$ |

### ✔ Student Check 1

Complete the table to identify the number of terms, the variable terms, the constant terms, and the coefficients of the variable terms of each expression.

| Expression | Total Number of Terms | Variable Term(s) | Constant Term(s) | Coefficients of Each Variable Term |
|---|---|---|---|---|
| **a.** $7a$ | | | | |
| **b.** $-y$ | | | | |
| **c.** $x - 7$ | | | | |
| **d.** $\dfrac{x}{3} - 4$ | | | | |
| **e.** $6y^2 - y + 3$ | | | | |
| **f.** $4y^3 - y^2 + 2y - 1$ | | | | |

## Like Terms

**Objective 2** ▶

Identify like terms.

Terms that have the same variables with the same exponents are **like terms**.
Terms that do not have the same variables with the same exponents are **unlike terms**.

| Like Terms | Unlike Terms |
|---|---|
| $2x, -5x$ | $2x, -5$ |
| $y^2, 4y^2$ | $y^2, 4y$ |
| $-3xy, 7yx$ | $-3xy, 7x^2y$ |

> **Procedure:  Identifying Like Terms**
>
> **Step:  a.** If the variables and exponents are the same, the terms are *like*.
> **b.** If the variables or exponents are different, the terms are *unlike*.

**Objective 2  Examples**    Determine if the terms are like or unlike and explain why.

| Problems | Like Terms | Unlike Terms | Why? |
|---|---|---|---|
| **2a.**  $-9y$ and $-9$ | | X | One term contains the variable $y$ and the other term does not contain a variable. |
| **2b.**  $x$ and $-8x$ | X | | Each term contains the variable $x$. |
| **2c.**  $2ba, 3ab, ab$ | X | | Each term contains the variables $ab$. Recall $ba = ab$ by the commutative property of multiplication. |
| **2d.**  $6xy^2$ and $2x^2y$ | | X | These terms are unlike terms because the variables have different exponents. |

✓ **Student Check 2**    Determine if the terms are like or unlike and explain why.
**a.** $4h$ and $4$      **b.** $2y$ and $y$      **c.** $6rh$ and $-6hr$      **d.** $x^2y$ and $xy^2$

## Combining Like Terms

**Objective 3** ▶

Combine like terms.

Algebraic expressions that contain like terms can be simplified by combining their like terms. There are two ways to think about combining like terms. Consider the algebraic expression $4x + 2x$. The terms $4x$ and $2x$ are like terms.

To simplify this expression, we can think of the problem in this way.

$$4x + 2x = x + x + x + x + x + x = 6x$$

We can also use the distributive property to combine like terms.

$$4x + 2x = (4 + 2)x = 6x$$

In either method, the result is the same, $4x + 2x = 6x$.

**INSTRUCTOR NOTE:**
Remind students that multiplication by an integer represents repeated addition.
So,
$4x = 4 \cdot x = x + x + x + x$ and
$2x = 2 \cdot x = x + x$.

> **Procedure:  Combining Like Terms**
>
> **Step 1:** Apply the distributive property to determine the coefficient of the like terms.
> **Step 2:** Simplify. Keep the variable component the same.

**Objective 3  Examples**    Simplify each expression by combining like terms, if possible.

**3a.** $2x + x$          **3b.** $8x - x$          **3c.** $4y^2 + 3y^2$

**3d.** $3 + 5x$          **3e.** $\dfrac{x}{2} + 5x$          **3f.** $-2a^2 - 4a^2 + 5a - 3a$

**Solutions**   **3a.**

$$2x + x = 2x + 1x \qquad \text{Recall } x = 1x.$$
$$= (2 + 1)x \qquad \text{Apply the distributive property.}$$
$$= 3x \qquad \text{Simplify.}$$

It is not necessary to show the distributive property step. We can simply add the coefficients of the like terms mentally to get

$$2x + x = 2x + 1x$$
$$= 3x$$

**3b.**

$$8x - x = 8x - 1x \qquad\qquad 8x - x = 8x - 1x$$
$$= (8 - 1)x \qquad\qquad\qquad\qquad = 7x$$
$$= 7x$$

**3c.**

$$4y^2 + 3y^2 = (4 + 3)y^2 \qquad 4y^2 + 3y^2 = 7y^2$$
$$= 7y^2$$

**3d.** The terms are unlike and therefore cannot be combined.

**3e.**

$$\frac{x}{2} + 5x = \frac{1}{2}x + 5x \qquad \text{Recall that } \frac{x}{2} = \frac{1x}{2} = \frac{1}{2}x$$

$$= \left(\frac{1}{2} + 5\right)x \qquad \text{Apply the distributive property.}$$

$$= \left(\frac{1}{2} + \frac{10}{2}\right)x \qquad \text{Write 5 as } \frac{10}{2}$$

$$= \frac{11}{2}x \qquad \text{Add.}$$

**3f.** There are two sets of like terms. Apply the distributive property for each set of like terms to get

$$-2a^2 - 4a^2 + 5a - 3a = (-2 - 4)a^2 + (5 - 3)a$$
$$= -6a^2 + 2a$$

Adding the coefficients of the like terms and keeping the variables of the like terms the same gives us

$$-2a^2 - 4a^2 + 5a - 3a = -6a^2 + 2a$$

✓ **Student Check 3**   Simplify each expression by combining like terms, if possible.

**a.** $11y + y$        **b.** $10b - b$        **c.** $2a^2 + 9a^2$

**d.** $6 - 2x$        **e.** $\dfrac{a}{3} + 4a$        **f.** $-3x^2 - 4x^2 + 7x - 6x$

## Simplifying Algebraic Expressions

**Objective 4** ▶

Simplify algebraic expressions.

Simplifying algebraic expressions is a skill that we use in most aspects of algebra. Simplifying algebraic expressions involves clearing parentheses and combining any like terms.

> **Procedure: Simplifying Algebraic Expressions**
>
> **Step 1:** Clear any parentheses by applying the distributive property.
> **Step 2:** Combine like terms.

**Objective 4 Examples**    **Simplify each algebraic expression.**

**4a.** $3y - 5 + 8y - 1$        **4b.** $3(4x + 2) - 8$

**4c.** $2(4x - 5) + 3(2x - 1)$        **4d.** $7 - 2(8x + 6)$

**Solutions**    **4a.**

$$3y - 5 + 8y - 1 = 3y + 8y - 5 - 1$$    Apply the commutative property.

$$= 11y - 6$$    Combine like terms.

**4b.**

$$3(4x + 2) - 8 = 12x + 6 - 8$$    Apply the distributive property.

$$= 12x - 2$$    Combine like terms.

**4c.**

$$2(4x - 5) + 3(2x - 1) = 8x - 10 + 6x - 3$$    Apply the distributive property.

$$= 8x + 6x - 10 - 3$$    Apply the commutative property.

$$= 14x - 13$$    Combine like terms.

**4d.**

$$7 - 2(8x + 6) = 7 - 16x - 12$$    Apply the distributive property.

$$= -16x + 7 - 12$$    Apply the commutative property.

$$= -16x - 5$$    Combine like terms.

**✓ Student Check 4**    Simplify each algebraic expression.

**a.** $4y - 8 - 2y$        **b.** $5(6x + 3) - 4$

**c.** $6(3x + 1) + 4(5x - 3)$        **d.** $5 - 3(2x + 4)$

---

**Objective 5 ▶**

Troubleshoot common errors.

## Troubleshooting Common Errors

Some common errors associated with simplifying algebraic expressions are shown next.

**Objective 5 Examples**    **A problem and an incorrect solution are given. Provide the correct solution and an explanation of the error.**

**5a.** Combine like terms: $5 + 2x$

| Incorrect Solution | Correct Solution and Explanation |
|---|---|
| $5 + 2x$ <br> $7x$ | The terms 5 and $2x$ are not like terms. Therefore, they cannot be combined. The problem would have to be $5x + 2x$ to get an answer of $7x$. |

**5b.** Combine like terms: $x^2 + x^2$

| Incorrect Solution | Correct Solution and Explanation |
|---|---|
| $x^2 + x^2$ <br> $x^4$ | The terms are like. To add them, we must add their coefficients, not their exponents. Recall $x^2 = 1x^2$. <br><br> $$x^2 + x^2$$ $$1x^2 + 1x^2$$ $$2x^2$$ |

**5c.** Simplify $4 - 3(x - 2)$.

| Incorrect Solution | Correct Solution and Explanation |
|---|---|
| $4 - 3(x - 2)$ <br> $1(x - 2)$ <br> $x - 2$ | The distributive property must be applied first because, according to the order of operations, multiplication comes before addition or subtraction. After distributing, we can combine like terms. <br><br> $4 - 3(x - 2)$ <br> $4 - 3x + 6$ <br> $-3x + 4 + 6$ <br> $-3x + 10$ |

## ANSWERS TO STUDENT CHECKS

|  | Expression | Total Number of Terms | Variable Term(s) | Constant Term(s) | Coefficients of Each Variable Term |
|---|---|---|---|---|---|
| **Student Check 1** | **a.** $7a$ | 1 | $7a$ | None | 7 |
|  | **b.** $-y$ | 1 | $-y$ | None | $-1$ |
|  | **c.** $x - 7$ | 2 | $x$ | $-7$ | 1 |
|  | **d.** $\dfrac{x}{3} - 4$ | 2 | $\dfrac{1}{3}x$ | $-4$ | $\dfrac{1}{3}$ |
|  | **e.** $6y^2 - y + 3$ | 3 | $6y^2, -y$ | 3 | $6, -1$ |
|  | **f.** $4y^3 - y^2 + 2y - 1$ | 4 | $4y^3, -y^2, 2y$ | $-1$ | $4, -1, 2$ |

**Student Check 2**   **a.** unlike   **b.** like   **c.** like   **d.** unlike

**Student Check 3**   **a.** $12y$   **b.** $9b$   **c.** $11a^2$

    **d.** can't be combined   **e.** $\dfrac{13}{3}a$   **f.** $-7x^2 + x$

**Student Check 4**   **a.** $2y - 8$   **b.** $30x + 11$

    **c.** $38x - 6$   **d.** $-6x - 7$

## SUMMARY OF KEY CONCEPTS

1. A term is a number or a product of a number and a variable raised to exponents. The terms in an algebraic expression are separated by addition or subtraction. The coefficient of a variable term is the number multiplied by the variable. Remember that a variable by itself is understood to have a coefficient of 1 since $1 \cdot x = x$.

2. Like terms are terms with the same variables and same exponents.

3. To combine like terms, combine the coefficients of the like terms and keep the variable the same.

4. To simplify an algebraic expression, clear any parentheses by applying the distributive property and then combine like terms.

## SECTION 1.8 / EXERCISE SET

 **Write About It!**

**Use complete sentences to explain each term or process.**

1. Difference of a variable term and a constant term   *Answers vary.*
2. Numerical coefficient   *Answers vary.*
3. Terms and coefficients of an algebraic expression   *Answers vary.*
4. Like terms   *Answers vary.*

*Additional answers can be found in the Instructor Answer Appendix.*

5. How to apply the distributive property to combine like terms   *Answers vary.*
6. How to simplify an algebraic expression   *Answers vary.*
7. Determine if the following statement is true or false: An algebraic expression must have a variable term.   *Answers vary.*
8. Determine if the following statement is true or false; if false, explain the error: $-2(3x - 4) = -6x - 8$   *Answers vary.*

## Practice Makes Perfect!

For each algebraic expression, complete the chart by identifying the total number of terms, the variable terms, the constant terms, and the coefficients of the variable terms. (*See Objective 1.*)

| Expression | Total Number of Terms | Variable Term(s) | Constant Term(s) | Coefficient of Each Variable Term |
|---|---|---|---|---|
| 9. $-5x$ | 1 | $-5x$ | None | $-5$ |
| 10. $6x$ | 1 | $6x$ | None | $6$ |
| 11. $2y - 5$ | 2 | $2y$ | $-5$ | $2$ |
| 12. $-y + 4$ | 2 | $-1y$ | $4$ | $-1$ |
| 13. $x^2 - 5x + 3$ | 3 | $x^2, -5x$ | $3$ | $1, -5$ |
| 14. $-x^2 + 6x - 7$ | 3 | $-x^2, 6x$ | $-7$ | $-1, 6$ |
| 15. $-3x^4 - x^2 + 4x - 1$ | 4 | $-3x^4, -1x^2, 4x$ | $-1$ | $-3, -1, 4$ |
| 16. $6x^3 - 2x^2 + 4x - 5$ | 4 | $6x^3, -2x^2, 4x$ | $-5$ | $6, -2, 4$ |
| 17. $\dfrac{x}{2}$ | 1 | $\dfrac{1}{2}x$ | None | $\dfrac{1}{2}$ |
| 18. $\dfrac{x}{4} - 2$ | 2 | $\dfrac{1}{4}x$ | $-2$ | $\dfrac{1}{4}$ |

**Determine if the terms are like or unlike.**
(*See Objective 2.*)

19. $5k$ and $k$    like terms
20. $h$ and $-3h$    like terms
21. $3r^2h$ and $r^2h$    like terms
22. $2\pi rh$ and $\pi rh$    like terms
23. $5xy$ and $-6x^2y$    unlike terms
24. $3ab^2$ and $a^2b$    unlike terms
25. $\dfrac{1}{2}c^3$ and $-c^3d$    unlike terms
26. $\dfrac{4}{3}\pi r^3$ and $4\pi r^2$    unlike terms

**Simplify each expression by combining like terms, if possible. (*See Objective 3.*)**

27. $12x + x$    $13x$
28. $20y + y$    $21y$
29. $12a - a$    $11a$
30. $23b - b$    $22b$
31. $14x - 26x$    $-12x$
32. $-9y - 12y$    $-21y$
33. $4x^2 + 15x^2$    $19x^2$
34. $5y^2 + 8y^2$    $13y^2$
35. $5x^2 - 23x^2$    $-18x^2$
36. $-11y^2 + 18y^2$    $7y^2$
37. $2x - 3y$    not like terms
38. $4a + 6b$    not like terms
39. $9xy - 3y$    not like terms
40. $4ab + 5b$    not like terms
41. $\dfrac{x}{4} + 2x$    $\dfrac{9x}{4}$
42. $\dfrac{2y}{5} + 3y$    $\dfrac{17y}{5}$

43. $2a - \dfrac{3a}{5}$    $\dfrac{7a}{5}$
44. $3b - \dfrac{5b}{7}$    $\dfrac{16b}{7}$
45. $-4x^2 + 5x + 2x^2 - 8x$    $-2x^2 - 3x$
46. $6y^2 - 9y - 8y^2 + 7y$    $-2y^2 - 2y$
47. $6x^2 - 21x - 18x^2 - 4x$    $-12x^2 - 25x$
48. $-9y^2 + 12y - 7y^2 - 2y$    $-16y^2 + 10y$

**Simplify each expression. (*See Objective 4.*)**

49. $8x - 7 - 6x$    $2x - 7$
50. $12x - 3 - x$    $11x - 3$
51. $5(3x + 4) - 10$    $15x + 10$
52. $6(10x + 5) - 12$    $60x + 18$
53. $-5(2x + 4) + 3(4x - 1)$    $2x - 23$
54. $-6(7x + 1) + 2(3x - 1)$    $-36x - 8$
55. $4(x - 3) + 2(-3x - 5)$    $-2x - 22$
56. $-3(4x + 1) - 7(5x + 6)$    $-47x - 45$
57. $8\left(\dfrac{1}{4}x - 2\right) - 3x$    $-x - 16$
58. $-9\left(\dfrac{1}{3}x + 2\right) + 5x$    $2x - 18$
59. $18 - 3(-5x + 4)$    $15x + 6$
60. $7 - 2(3x + 6)$    $-6x - 5$
61. $-12\left(\dfrac{5}{6}y - \dfrac{1}{12}\right) - 10\left(\dfrac{1}{5}y + \dfrac{1}{2}\right)$    $-12y - 4$
62. $-10\left(\dfrac{1}{10}x - \dfrac{1}{10}\right) + 6\left(\dfrac{5}{6}x - \dfrac{2}{3}\right)$    $4x - 3$

## Mix 'Em Up!

For each algebraic expression, complete the chart by identifying the total number of terms, the variable terms, the constant terms, and the coefficients of the variable terms.

| Expression | Total Number of Terms | Variable Term(s) | Constant Term(s) | Coefficient of Each Variable Term |
|---|---|---|---|---|
| 63. $-x + 12$ | 2 | $-1x$ | $12$ | $-1$ |
| 64. $6x - 13$ | 2 | $6x$ | $-13$ | $6$ |
| 65. $-3x^2 + 7x - 2$ | 3 | $-3x^2, 7x$ | $-2$ | $-3, 7$ |

| Expression | Total Number of Terms | Variable Term(s) | Constant Term(s) | Coefficient of Each Variable Term |
|---|---|---|---|---|
| **66.** $4x^2 - x + 8$ | 3 | $4x^2, -x$ | 8 | $4, -1$ |
| **67.** $-x^4 + 3x^2 - 5x + 6$ | 4 | $-x^4, 3x^2, -5x$ | 6 | $-1, 3, -5$ |
| **68.** $-6x^3 - 7x^2 + x - 1$ | 4 | $-6x^3, -7x^2, x$ | $-1$ | $-6, -7, 1$ |
| **69.** $\dfrac{3x}{2} - \dfrac{5}{4}$ | 2 | $\dfrac{3}{2}x$ | $-\dfrac{5}{4}$ | $\dfrac{3}{2}$ |
| **70.** $\dfrac{7x}{2} + \dfrac{1}{3}$ | 2 | $\dfrac{7}{2}x$ | $\dfrac{1}{3}$ | $\dfrac{7}{2}$ |
| **71.** $0.5x^2 - 1.8x - 2.1$ | 3 | $0.5x^2, -1.8x$ | $-2.1$ | $0.5, -1.8$ |
| **72.** $-0.1x^2 + 2.5x + 3.6$ | 3 | $-0.1x^2, 2.5x$ | $3.6$ | $-0.1, 2.5$ |

**Simplify each algebraic expression.**

**73.** $-5(x + 6) + 12$   $-5x - 18$

**74.** $-2(b + 7) + 9$   $-2b - 5$

**75.** $-7(x - 1) + 3(2x - 5)$   $-x - 8$

**76.** $-2(6x + 5) - 3(4x - 1)$   $-24x - 7$

**77.** $1.8 - 3(-0.2x + 1.6)$   $0.6x - 3$

**78.** $2.7 - 0.5(1.3x + 6)$   $-0.65x - 0.3$

**79.** $14\left(\dfrac{1}{2}x - \dfrac{3}{7}\right) + 8$   $7x + 2$

**80.** $20\left(-\dfrac{2}{5}x + \dfrac{3}{4}\right) - 24$   $-8x - 9$

**81.** $-18\left(\dfrac{2}{9}y - \dfrac{1}{2}\right) + 6\left(-\dfrac{2}{3}y + \dfrac{1}{2}\right)$   $-8y + 12$

**82.** $-21\left(\dfrac{1}{3}x + \dfrac{1}{7}\right) + 4\left(\dfrac{1}{2}x - \dfrac{3}{4}\right)$   $-5x - 6$

**83.** $0.5(2x - 6) + 0.4(x + 4) - 1.5$   $1.4x - 2.9$

**84.** $-0.6(1.5x - 2) + 0.7(3x + 6) + 2.9$   $1.2x + 8.3$

**85.** $10(0.5x - 2) + 10(0.4)$   $5x - 16$

**86.** $100(0.6x + 0.25) + 100(0.35x)$   $95x + 25$

 **You Be the Teacher!**

**Correct each student's errors, if any.**

**87.** Simplify $5 - 3(2 + 4)$.

Brian's work:

$5 - 3(2 + 4) = 2(2 + 4) = 2(6) = 12$
$5 - 3(2 + 4) = 5 - 3(6) = 5 - 18 = -13$

**88.** Simplify $3a - a$.

Shirl's work:

$3a - a = 3a$    $3a - a = 3a - 1a = 2a$

**89.** Simplify $x^2 + x^2$.

Molly's work:

$x^2 + x^2 = x^4$    $x^2 + x^2 = 1x^2 + 1x^2 = 2x^2$

**90.** Simplify $3(6x + 2) - 4$.

Isaac's work:

$3(6x + 2) - 4 = 18x + 2 - 4 = 18x - 2$
$3(6x + 2) - 4 = 18x + 6 - 4 = 18x + 2$

**91.** Simplify $4 + 2(5 + x)$.

Warren's work:

$4 + 2(5 + x) = 6(5 + x) = 30 + x$
$4 + 2(5 + x) = 4 + 10 + 2x = 14 + 2x$

**92.** Simplify $6 - (2 + x)$.

Ashley's work:

$6 - (2 + x) = 4 - 2 + x = 2 + x$
$6 - (2 + x) = 6 - 2 - x = 4 - x$

**93.** Simplify $-6b - 4b$.

Sue's work:

$-6b - 4b = 2b^2$    $-6b - 4b = -10b$

**94.** Simplify $9a - 2(a + 5) + 7$.

Rob's work:

$9a - 2(a + 5) + 7 = -19a - 10 + 7 = -19a - 3$
$9a - 2(a + 5) + 7 = 9a - 2a - 10 + 7 = 7a - 3$

 **GROUP ACTIVITY**   **Mathematics of Being Fit**

**1.** One method of determining if you are at a healthy weight for your height is by finding your body mass index (BMI). The BMI is 703 times the quotient of a person's weight (in pounds) and the square of a person's height (in inches). Write an equation that represents BMI. Let $w$ represent weight and $h$ represent height.   Answers vary.

**2.** Use the equation from Step 1 to determine your BMI. Use the chart to determine your weight status. (You do not have to share this information with your group members.)   Answers vary.

**INSTRUCTOR NOTE:**
Inform groups that they can choose a random height and weight to use for these problems.

| BMI | Weight Status |
|---|---|
| Below 18.5 | Underweight |
| 18.5–24.9 | Normal |
| 25.0–29.9 | Overweight |
| 30.0 and above | Obese |

**3.** In Section 1.3, the basal metabolic rate was presented. (See pages 35 and 36.) Recall this calculates the number of calories that your body needs each day to perform essential functions. Calculate your BMR now, if you have not already done so. Based on the level of activity you perform each day, your body needs additional calories to perform its basic functions and the activities in which you participate.    Answers vary.

| Activity Level | Calories Needed |
|---|---|
| Sedentary (little or no exercise, desk job) | 1.2 times BMR |
| Lightly active (light exercise/sports 1–3 days/week) | 1.375 times BMR |
| Moderately active (moderate exercise/sports 3–5 days/week) | 1.55 times BMR |
| Very active (hard exercise/sports 6–7 days/week) | 1.725 times BMR |
| Extra active (hard daily exercise/ sports and physical job or training for a marathon) | 1.9 times BMR |

What activity level describes your lifestyle? Calculate the number of calories you need to consume for your body to sustain this activity level and to maintain your current weight.

4. If you want to lose weight, you must take in fewer calories than what you found in Step 3. One pound of body fat is equal to 3500 cal. So, if you want to lose 1 lb in a week, you will need to decrease your intake by 3500 cal or by an average of 500 cal per day. If you want to lose 2 lb in a week, you will need to decrease your intake by 1000 cal per day. Assume you want to lose 1 lb a week, how many calories should you consume each day?    Answers vary.

5. In addition to decreasing caloric intake, you can also increase your physical activity to burn off additional calories. List a number of physical activities you do during a week, such as walking, jogging, or climbing stairs.    Answers vary.

6. Research the Internet to find estimates for the number of calories you burn per minute for each of these activities.    Answers vary.

7. Develop an expression for the estimated number of calories you burn doing each of these activities in a week, using a variable for the number of minutes you do each activity.    Answers vary.

8. Evaluate the expression you found in Step 3 to determine the amount of calories burned by engaging in these activities for 30 min, 1 hr, 2 hr, 3 hr, 4 hr, 5 hr, and 6 hr per week.    Answers vary.

9. How long would you have to engage in these activities to burn 3500 cal in a week?    Answers vary.

# Chapter 1 / REVIEW

# Real Numbers and Algebraic Expressions

**What's the big idea?** Chapter 1 provides us with the skills to simplify numerical expressions by applying the rules for signed numbers, the properties of real numbers, and the order of operations. These rules and properties also provide the framework for us to simplify and evaluate algebraic expressions. We also learned how to express relationships in mathematical form. These skills form the foundation for algebra and will be used in every section of this text. Success in this course depends on mastering these skills.

## The Tools

Listed below are the key terms, skills, formulas, and properties you should know for this chapter.

The page reference is provided if you need additional help with the given topic. The Study Tips will assist in your preparation for an exam.

## Study Tips

1. Learn all of the terms, formulas, and properties. Make flash cards and have someone quiz you.
2. Rework problems from the exercises and also the ones you worked in class. Work additional problems from the review exercises.
3. Review the summaries of key concepts.
4. Work the chapter test.
5. Be sure to review the online resources for additional study materials.

## Terms

| | | | |
|---|---|---|---|
| Absolute value   9 | Factor   13 | Multiplicative identity   71 | Right angle   52 |
| Additive identity   71 | Factor tree   14 | Multiplicative inverse   71 | Roster method   2 |
| Additive inverse   71 | Finite   2 | Natural number   3 | Set   2 |
| Algebraic expression   30 | Fraction   15 | Numerator and denominator   15 | Set-builder notation   2 |
| Average   64 | Identity element   71 | | Simplified   16 |
| Base   27 | Improper fraction   15 | | Solve an equation   31 |
| Coefficient   77 | Infinite   2 | Opposite   8 | Solution of an equation   31 |
| Composite   14 | Integer   3 | Perimeter   34 | Square root   3 |
| Constant terms   77 | Inverse   71 | Prime factorization   14 | Straight angle   53 |
| Dividend   61 | Irrational number   3 | Prime number   14 | Term   77 |
| Divisor   61 | Least common denominator (LCD)   21 | Proper fraction   15 | Unlike terms   78 |
| Equation   31 | Like terms   78 | Rational number   3 | Variable   2 |
| Ellipsis   2 | Lowest terms   16 | Real number   3 | Variable term   77 |
| Equivalent fractions   21 | Member or element   2 | Real number line   5 | Whole number   3 |
| Exponential expression   27 | Mixed number   15 | Reciprocal   19 | |

## Formulas and Properties

- Associative property of addition   72
- Associative property of multiplication   72
- Average   64
- Commutative property of addition   72
- Commutative property of multiplication   72
- Complementary angles   52
- Distributive property   73
- Fundamental property of fractions   16

- Identity property of addition   71
- Identity property of multiplication   71
- Inverse property of addition   71
- Inverse property of multiplication   71
- Order of operations   29, 51
- Percent change   64
- Supplementary angles   53

## CHAPTER 1 / SUMMARY

*How well do you know this chapter? Complete the following questions to find out. Take a look back at the section if you need help.*

### SECTION 1.1  The Set of Real Numbers

1. A(n) <u>set</u> is a collection of objects.

2. The set of <u>natural (or counting)</u> numbers is $\{1, 2, 3, \ldots\}$.

3. The set of <u>whole</u> numbers is $\{0, 1, 2, 3, \ldots\}$.

4. The set of <u>integers</u> is the set of positive and negative whole numbers.

5. A(n) <u>rational</u> number is a number that can be written as the quotient of integers.

6. A(n) <u>irrational</u> number is a number whose decimal form continues indefinitely without a repeating pattern.

7. The statement $a < b$ is read as $a$ <u>is less than</u> $b$. The statement $a > b$ is read as $a$ <u>is greater than</u> $b$.

8. The <u>opposite</u> of a number is a number with the same distance from zero but lies on the other side of zero on the number line.

9. The <u>absolute value</u> of a number is the distance the number is from zero on the real number line.

### SECTION 1.2  Fractions Review

10. A number with factors of 1 and the number itself is a(n) <u>prime</u> number.

11. A fraction is a(n) <u>quotient</u> of <u>real</u> numbers. The top is called the <u>numerator</u> and the bottom is called the <u>denominator</u>.

12. A fraction is in <u>simplest form</u> if the top and bottom of the fraction do not have any common factors.

13. The <u>fundamental</u> <u>property</u> of <u>fractions</u> enables us to divide the top and bottom of the fraction by their common factor. $\dfrac{a \cdot c}{b \cdot c} =$ <u>Answers vary</u>.

14. The numbers $a$ and $\dfrac{1}{a}$ are <u>reciprocals</u>.

15. To divide fractions, <u>multiply</u> by the <u>reciprocal</u> of the second fraction.

16. We can add or subtract fractions with <u>common</u> denominators.

17. The <u>least</u> <u>common</u> <u>denominator</u> is the number that all denominators divide into evenly.

### SECTION 1.3  The Order of Operations, Algebraic Expressions, and Equations

18. A(n) <u>exponent</u> indicates repeated multiplication.

19. In the expression $b^n$, $b$ is called the <u>base</u> and $n$ is called the <u>exponent</u> or <u>power</u>.

20. The <u>order</u> of <u>operations</u> provides us a way to simplify numerical expressions. First, simplify what is in <u>grouping</u> symbols, then evaluate <u>exponents</u>, <u>multiply</u> or <u>divide</u> from left to right, and finally <u>add</u> or <u>subtract</u> from left to right.

21. A(n) <u>variable</u> represents an unknown number and is represented by a <u>letter</u>.

22. A(n) <u>algebraic</u> <u>expression</u> is an expression that involves variables and/or numbers.

23. To evaluate an algebraic expression, replace the <u>variable</u> with the <u>given</u> <u>value</u> and simplify.

24. A statement that two expressions are equal is a(n) <u>equation</u>.

25. A(n) <u>solution</u> of an equation is a value that makes the equation true.

### SECTION 1.4  Addition of Real Numbers

26. To add numbers with the same sign, add their <u>absolute</u> <u>values</u> and keep the <u>sign</u> the same.

27. To add numbers with different signs, subtract their <u>absolute</u> <u>values</u>. The sign of the answer has the same sign as the number with the <u>larger</u> <u>absolute</u> <u>value</u>.

### SECTION 1.5  Subtraction of Real Numbers

28. Subtracting real numbers is the same as <u>adding</u> the <u>opposite</u> of a number. $a - b =$ <u>$a + (-b)$</u>.

### SECTION 1.6  Multiplication and Division of Real Numbers

29. The product of real numbers with the same signs is <u>positive</u>. The product of real numbers with opposite signs is <u>negative</u>.

30. A negative base raised to an even power is <u>positive</u>.

31. A negative base raised to an odd power is <u>negative</u>.

32. To divide real numbers is the same as <u>multiplying</u> by the <u>reciprocal</u>. $a \div b = $ <u>$a \cdot \dfrac{1}{b}$</u>.

33. Zero divided by a nonzero number is <u>zero</u>. A number divided by zero is <u>undefined</u>.

### SECTION 1.7  Properties of Real Numbers

34. The <u>commutative</u> property of addition states that changing the order of the things being added doesn't change the result. $a + b =$ <u>$b + a$</u>.

35. The <u>associative</u> property of addition states that the grouping of the things being added doesn't change the result. $a + (b + c) =$ <u>$(a + b) + c$</u>.

36. <u>zero</u> is the identity element of addition.

37. The <u>identity</u> property of addition states that adding <u>zero</u> to a number doesn't change its value. $a +$ <u>$0$</u> $= a$.

38. The <u>inverse</u> property of addition states that adding a number and its <u>opposite</u> is zero. $a +$ <u>$-a$</u> $= 0$.

39. The <u>commutative</u> property of multiplication states that changing the order of the things being multiplied doesn't change the result. $ab =$ <u>$ba$</u>.

**40.** The _associative_ property of multiplication states that the grouping of the things being multiplied doesn't change the result. $a(bc) = $ _(ab)c_ .

**41.** _One_ is the identity element of multiplication.

**42.** The _identity_ property of multiplication states that multiplying a number by _one_ doesn't change its value. $a \cdot$ _1_ $= a$.

**43.** The _inverse_ property of multiplication states that multiplying a number and its _reciprocal_ is one. $a \cdot \frac{1}{a} = 1$.

**44.** The _distributive_ property enables us to multiply a number by a sum or difference. $a(b + c) = $ _ab + ac_ .

**45.** When a(n) _negative_ number is distributed to a sum, the signs of each of the terms _change_ .

## SECTION **1.8** Algebraic Expressions

**46.** A(n) _term_ is a number or a product of a number and a variable raised to powers. The _terms_ in an algebraic expression are separated by addition or subtraction.

**47.** A(n) _variable_ _term_ is a term that contains a variable.

**48.** A(n) _constant_ _term_ is a term that does not contain a variable.

**49.** The _coefficient_ of a variable term is the number multiplied by the variable.

**50.** Terms that have the same variables with the same powers are _like_ terms. Terms that do not have the same variables or same powers are _unlike_ terms.

**51.** To combine like terms, add their _coefficients_ and keep the _variable_ the same.

**52.** To simplify algebraic expressions, clear any _parentheses_ and then combine _like_ _terms_ .

## CHAPTER 1 / REVIEW EXERCISES

### SECTION 1.1

**Classify each number as a natural number, whole number, integer, rational number, irrational number, and/or real number. If the number is an irrational number, approximate its value to the nearest hundredth. (*See Objective 1.*)**

**1.** $5\frac{1}{3}$   rational, real

**2.** $7.1\overline{9}$   rational, real

**3.** $10\pi$   irrational, real; 31.42

**4.** $\sqrt{51}$   irrational, real; 7.14

**Give an example of a real number that satisfies each condition. (*See Objective 1.*)**

**5.** An integer that is not a whole number.   Answers vary, 0

**6.** An irrational number that is between 2 and 3.
Answers vary, $\sqrt{5}$

**7.** A rational number that is between $\frac{3}{2}$ and 2.
Answers vary, 1.8

**8.** An integer that is not a natural number. Answers vary, $-1$

**Graph each number on a real number line. (*See Objective 2.*)**

**9.** $\left\{ -2.3\overline{3}, 2\frac{1}{4}, 5 \right\}$

**10.** $\left\{ -\pi, \sqrt{7}, \frac{20}{3} \right\}$

**Compare the values of each pair of numbers. Use a $<$, $>$, or $=$ symbol to make the statement true. (*See Objectives 3 and 5.*)**

**11.** $\frac{22}{7}$ _____ $\pi$   $>$

**12.** $\sqrt{49}$ _____ $-|-7|$   $>$

**Find the opposite of each real number. (*See Objective 4.*)**

**13.** $1.12\overline{12}$   $-1.12\overline{12}$

**14.** $5.2$   $-5.2$

**15.** $-3\frac{1}{5}$   $3\frac{1}{5}$

**16.** $\frac{7}{9}$   $-\frac{7}{9}$

**Simplify each absolute value expression. (*See Objective 5.*)**

**17.** $|-5.8|$   5.8

**18.** $-|1.9|$   $-1.9$

**19.** $-\left| -\frac{6}{13} \right|$   $-\frac{6}{13}$

**20.** $-\left| \frac{5}{4} \right|$   $-\frac{5}{4}$

### SECTION 1.2

**Write the prime factorization of each number. (*See Objective 1.*)**

**21.** $90$   $2 \cdot 3 \cdot 3 \cdot 5$

**22.** $560$   $2 \cdot 2 \cdot 2 \cdot 2 \cdot 5 \cdot 7$

**23.** $945$   $3 \cdot 3 \cdot 3 \cdot 5 \cdot 7$

**24.** $200$   $2 \cdot 2 \cdot 2 \cdot 5 \cdot 5$

**Simplify each fraction to lowest terms. (*See Objective 3.*)**

**25.** $\frac{30}{45}$   $\frac{2}{3}$

**26.** $\frac{168}{288}$   $\frac{7}{12}$

**27.** $\frac{336}{378}$   $\frac{8}{9}$

**28.** $\frac{126}{360}$   $\frac{7}{20}$

**Perform the indicated operation. Express answers in lowest terms, when applicable. (*See Objectives 4 and 5.*)**

**29.** $\frac{7}{12} + \frac{11}{28}$   $\frac{41}{42}$

**30.** $\frac{1}{2} - \frac{3}{10}$   $\frac{1}{5}$

**31.** $3\frac{1}{5} \cdot 1\frac{1}{4}$   $4$

**32.** $2\frac{1}{7} \div 6\frac{3}{7}$   $\frac{1}{3}$

**33.** $\frac{7}{12} \cdot \frac{9}{14}$   $\frac{3}{8}$

**34.** $6\frac{1}{5} - 7\frac{3}{5}$   $-\frac{7}{5}$

### SECTION 1.3

**Use the order of operations to simplify each expression. (*See Objectives 1 and 2.*)**

**35.** $4[(-11) - 9](-2)$   160

**36.** $2.15(-1)^2 + (-3.16)(0)$   2.15

**37.** $[-5 - (-5)][-6 + (-2)]$   0

**38.** $-\frac{3}{14}(-12) + \frac{10}{7}$   4

**39.** $\frac{12 + (-4)}{-21 + (-9)}$   $-\frac{4}{15}$

**40.** $-1.5(-4)^2 + 3.2(2) + 6.5$   $-11.1$

**41.** $(-1 \cdot 6)^2 - 5 \cdot 12 \div 6 - (-3.6)$   29.6

**42.** $(-6)^2 - (-15) \div 5 \cdot 3 - 7$   38

**43.** $|-8(-12) + 2(-13)|$   70

**44.** $\sqrt{(14 - 10)^2 + (-18 + 15)^2}$   5

**Translate each phrase into a mathematical expression and then simplify the resulting expression, if applicable. Use the variable $x$ for any unknown number. (See Objective 5.)**

**45.** 7 less than $-18$   $-18 - 7 = -25$

**46.** The difference of 22 and $-10$   $22 - (-10) = 32$

**47.** 6 subtracted from $-11$   $-11 - 6 = -17$   **48.** $-10$ decreased by 4   $-10 - 4 = -14$

**49.** 9 less than the sum of $-13$ and 7   $(-13 + 7) - 9 = -15$

**50.** 21 more than the difference of 18 and $-4$   $21 + [18 - (-4)] = 43$

**51.** 5 more than $-2$   $5 + (-2) = 3$   **52.** The sum of 12 and $-25$   $12 + (-25) = -13$

**53.** 15 increased by $-3$   $15 + (-3) = 12$   **54.** $-19$ added to 16   $16 + (-19) = -3$

**55.** 7 less than $-18$   $-18 - 7 = -25$

**56.** The product of $-1$, $-5$, and 2   $(-1)(-5)(2) = 10$

**57.** Twice $-35$   $2(-35) = -70$   **58.** Three times 25   $3(25) = 75$

**59.** 30% of $-12$   $0.3(12) = -3.6$   **60.** One-half of 15   $\frac{1}{2}(15) = 7.5$

**61.** The ratio of 28 and 42   **62.** 54 divided by $-2$   $\frac{54}{-2} = -27$

**63.** The quotient of $-32$ and $-48$   $\frac{-32}{-48} = \frac{2}{3}$

**64.** Four times the sum of a number and 34   $4(x + 34)$

**65.** The difference of seven times a number and 12   $7x - 12$

**66.** 2 less than five times the sum of a number and 6   $5(x + 6) - 2$

**67.** Twice the product of a number and 45   $2(x)(45)$

**68.** The product of 12 and the sum of a number and $-7$   $12[x + (-7)]$

**69.** Six times the quotient of a number and $-9$   $6\left(\frac{x}{-9}\right)$

**Solve each problem. (See Objective 6.)**

**70.** The expression $P(1 + r)^t$ represents the amount of money in an account when $P$ dollars is invested at an annual interest rate of $r$ (in decimal form) for $t$ years. Find the amount of money in the account at the end of 5 yr if $1250 is invested at 1.8% annual interest.   $1366.62

**71.** The height of a baseball hit upward with an initial velocity of 128 ft/s from an initial height of 16 ft is represented by the expression $-16t^2 + 128t + 16$, where $t$ is the number of seconds after the ball has been hit. What is the height of the ball after 3 sec?   256 ft

**Write an algebraic expression that represents the unknown quantity. (See Objective 6.)**

**72.** Norma mixes a 6% salt solution with a 35% solution to get a 22% solution. Let $x$ oz represent the volume of the 6% salt solution. If Norma wants 60 oz of 22% salt solution, write an expression that represents the volume of the 35% salt solution.   $60 - x$ oz

**73.** In a rectangle, the length is two more than four times the width. If $w$ represents the width, write an expression that represents the length of the rectangle.   $4w + 2$

## SECTION 1.4

**Perform the indicated operation and simplify. (See Objectives 1 and 2.)**

**74.** $-5.6 + (-2.1)$   $-7.7$   **75.** $3.12 + (-2.85)$   $0.27$

**76.** $-\frac{11}{12} + \frac{17}{18}$   $\frac{1}{36}$   **77.** $\left(-\frac{7}{3}\right) + \left(-\frac{13}{6}\right)$   $-\frac{9}{2}$

**78.** $\left(\frac{3}{4}\right) + \left(-\frac{2}{5}\right)$   $\frac{7}{20}$   **79.** $\left(-5\frac{5}{6}\right) + 3\frac{1}{4}$   $-2\frac{7}{12}$

## SECTION 1.5

**Perform the indicated operation. (See Objectives 1 and 2.)**

**80.** $(-13.8) - 9.4$   $-23.2$   **81.** $2.62 - (-1.79)$   $4.41$

**82.** $3 - \frac{8}{5}$   $1\frac{2}{5}$   **83.** $1\frac{3}{5} - \left(-3\frac{4}{5}\right)$   $5\frac{2}{5}$

**84.** $19 - 3 + (-7) - (-11)$   $20$

**85.** $13 + (-10) - (-9) - 5$   $7$

**86.** $-4 + (-14) - (-6)$   $-12$

**87.** $-9 + (-16) - (-1)$   $-24$

**88.** $3.2 - (-2.7) + 1.9 - 0.6$   $7.2$

**89.** $5.1 + (-3.4) - 4.2 - 0.1$   $-2.6$

**Write the mathematical expression needed to solve each problem and then answer the question asked. (See Objective 3.)**

**90.** The highest point on the South American continent is the peak of Mount Aconcagua. The peak is 22,834 ft above sea level. The lowest point on land in the United States is in Death Valley at 282.2 ft below sea level. What is the difference between these elevations?   23,116.2 ft

**91.** The Hershey Company reported an annual net sales of $5671 million, total cost and expenses of $4766 million, interest expense of $96 million, and income taxes of $299 million in 2010. What was Hershey Company's total net income in 2010? (Source: http://www.thehersheycompany.com/investors/financialreports.aspx)   $510 million

**92.** The lowest recorded temperature in Illinois was $-36°$F at the Congerville weather station on January 5, 1999. The highest recorded temperature was $117°$F at the East St. Louis weather station on July 14, 1954. What is the difference between the highest and lowest temperatures?   153°F

**93.** Wayne has $163.23 in his checking account. He deposits his paycheck of $545.85. Wayne writes a check for phone bill for $85.55, for groceries for $113.43, and then for rent for $455. What is Wayne's checking account balance?   $55.10

## SECTION 1.6

**Perform the indicated operation and simplify. (See Objectives 1, 2, and 3.)**

**94.** $(-12)(-20)$   $240$   **95.** $(-2.5)(0.8)$   $-2$

**96.** $(-6)\left(\frac{7}{20}\right)$   $-\frac{21}{10}$   **97.** $\left(-\frac{5}{12}\right)\left(-\frac{16}{25}\right)$   $\frac{4}{15}$

**98.** $\frac{35}{-14}$   $-\frac{5}{2}$   **99.** $-6^4$   $-1296$

**100.** $-(-2)^4$   $-16$   **101.** $-\left(-\frac{2}{3}\right)^3$   $\frac{8}{27}$

**102.** $\frac{0}{5}$   $0$   **103.** $6 \div 0$   undefined

**104.** $\frac{28}{-6} \div (-4)$   $\frac{7}{6}$   **105.** $\left(-\frac{42}{18}\right) \div \left(\frac{8}{14}\right)$   $-\frac{49}{12}$

**106.** $9(-12)(-2)$   $216$   **107.** $-6(-3)(-5)$   $-90$

**Evaluate each expression for the specified values of x. Organize the information in a chart.** (*See Objective 4.*)

**108.** $-2x + 7$ for $x = -4, -3, -2, -1,$ and $0$

**109.** $\dfrac{2x + 1}{x - 3}$ for $x = -2, -1, 0, 1,$ and $2$

**110.** $-x^2 + 5$ for $x = -2, -1, 0, 1,$ and $2$

**111.** $x^2 - 2$ for $x = -2, -1, 0, 1,$ and $2$

**112.** $x^2 - x + 1,$ for $x = -2, -1, 0, 1$ and $2$

**113.** $|5x - 1|$ for $x = -4, -2, 0, 2,$ and $4$

**Evaluate each expression for the given values of the variables.** (*See Objective 4.*)

**114.** $2x - 5y$ for $x = -1$ and $y = 2$   $-12$

**115.** $\dfrac{y_2 - y_1}{x_2 - x_1}$ for $x_1 = -1, x_2 = -9, y_1 = 1, y_2 = -15$   $2$

**116.** $2l + 2w$ for $l = 23$ and $w = 12$   $70$

**117.** $\dfrac{1}{2}(b_1 + b_2)h$ for $b_1 = 6, b_2 = 4, h = 7$   $35$

**118.** $\sqrt{(x_1 - x_2)^2 + (y_1 - y_2)^2}$ for $x_1 = -6,$ $x_2 = 6, y_1 = -5, y_2 = 0$   $13$

**119.** $b^2 - 4ac$ for $a = -4, b = 8, c = -3$   $16$

## SECTION 1.7

**Identify the property of addition or multiplication applied in each problem.** (*See Objectives 1 and 2.*)

**120.** $0 + (-8) = -8$   identity property of addition

**121.** $\dfrac{1}{6} + \left(-\dfrac{3}{10}\right) = \left(-\dfrac{3}{10}\right) + \dfrac{1}{6}$   commutative property of addition

**122.** $\left[2 + \left(-\dfrac{1}{3}\right)\right] + \dfrac{4}{5} = 2 + \left(-\dfrac{1}{3} + \dfrac{4}{5}\right)$
associative property of addition

**123.** $12 + (-7 + (-5)) = 0$   inverse property of addition

**124.** $\left(-\dfrac{4}{3}\right)\left(-\dfrac{3}{4}\right) = 1$   inverse property of multiplication

**125.** $-24(6 \cdot 7) = (-24 \cdot 6)(7)$   associative property of multiplication

**126.** $5(-1 + 1) = 0$   zero property of multiplication

**127.** $-6\left(5 \cdot \dfrac{1}{5}\right) = -6$   identity property of multiplication

**128.** $(-1.7)(0.8) = (0.8)(-1.7)$   commutative property of multiplication

**129.** $(-4 \cdot 5)(9) = -4(5 \cdot 9)$   associative property of multiplication

**Use the distributive property to rewrite each expression and simplify the result.** (*See Objective 3.*)

**130.** $-7(x - 12)$
$-7x + 84$

**131.** $-5(3x - 4y + 2)$
$-15x + 20y - 10$

**132.** $0.5(1.6x - 0.4)$
$0.8x - 0.2$

**133.** $9\left(-\dfrac{1}{3}x + \dfrac{1}{6}\right)$   $-3x + \dfrac{3}{2}$

## SECTION 1.8

**For each algebraic expression, complete the chart by identifying the total number of terms, the variable terms, the constant terms, and the coefficients of the variable terms.** (*See Objective 1.*)

| Expression | Total Number of Terms | Variable Term(s) | Constant Term(s) | Coefficient of Each Variable Term |
|---|---|---|---|---|
| **134.** $-5x + 11$ | 2 | $-5x$ | 11 | $-5$ |
| **135.** $-20x^4 + 13x^2 + 5$ | 3 | $-20x^4, 13x^2$ | 5 | $-20, 13$ |
| **136.** $\dfrac{5x}{2} - \dfrac{3}{2}$ | 2 | $\dfrac{5}{2}x$ | $-\dfrac{3}{2}$ | $\dfrac{5}{2}, -\dfrac{3}{2}$ |
| **137.** $-0.2x^2 + 3.2x + 3.6$ | 3 | $-0.2x^2, 3.2x$ | 3.6 | $-0.2, 3.2$ |

**Determine if the terms are like or unlike.** (*See Objective 2.*)

**138.** $1.9h, 2.4h,$ and $-h$
like terms

**139.** $5x^3y$ and $-10x^2y$
unlike terms

**148.** $4\left(\dfrac{1}{8}x - \dfrac{3}{4}\right) + 5$   $\dfrac{1}{2}x + 2$

**140.** $\dfrac{1}{4}cd$ and $-\dfrac{1}{8}c^2d$
unlike terms

**141.** $\pi r^2 h$ and $3\pi r^2$
unlike terms

**149.** $-15\left(\dfrac{2}{3}y - \dfrac{1}{5}\right) + 8\left(\dfrac{1}{4}y - \dfrac{1}{2}\right)$   $-8y - 1$

**Simplify each algebraic expression.** (*See Objectives 3 and 4.*)

**142.** $-(1 - 5x) + (6x + 4)$
$11x + 3$

**143.** $-4(b)(7)$   $-28b$

**150.** $5(2x^2 - 1) - 3(x^2 + 4)$   **151.** $7x^2 + 2(x + 3x^2) - 2x$
$7x^2 - 17$                                             $13x^2$

**144.** $10(3x - 7) - 26x$
$4x - 70$

**145.** $4x - 3(1 - 4x)$   $16x - 3$

**152.** $0.3(3x - 1) + 0.2(x + 5) - 2.6$   $1.1x - 1.9$

**146.** $-3(x - 5) - 2(4x + 5)$
$-11x + 5$

**147.** $2.6 - 0.2(-0.3x + 1.5)$
$0.06x - 2.3$

**153.** $2.8 + 0.3(4x - 1)$   **154.** $3.2 - 0.4(2x + 1.5)$
$1.2x + 2.5$                              $-0.8x - 2.6$

**155.** $3(0.7x - 6) + 20(0.3)$   $2.1x - 12$

## CHAPTER 1 TEST / REAL NUMBERS AND ALGEBRAIC EXPRESSIONS

**1.** The number that is not a natural number is

 **a.** $(-2)^2$   **b.** $\sqrt{25}$   **c.** $\dfrac{9 - 3}{4 - 2}$   **d.** $3 - 4(2)$

**2.** Provide an example of a rational number that is also an integer.   Answers vary, $-5$

**3.** The number that is an irrational number is

 **a.** $\sqrt{6^2 + 8^2}$   **b.** $4\pi$
 **c.** $\dfrac{4 + 5(-2)}{3^2}$   **d.** $\dfrac{0}{-4}$

**4.** Place an appropriate symbol that makes the statement true.

   **a.** $-3 \ge -5$        **b.** $\sqrt{2} \le \pi$

**5.** Simplify each expression.

   **a.** $-\left(-\dfrac{7}{2}\right)$   $\dfrac{7}{2}$       **b.** $-(2-5)$   $3$

   **c.** $|4 - 5(3^2)|$   $41$       **d.** $|(-4)^3|$   $64$

**6.** Find the prime factorization of 180.   $2 \cdot 2 \cdot 3 \cdot 3 \cdot 5$

**7.** Use the fundamental property of fractions to write each fraction in lowest terms.

   **a.** $\dfrac{124}{200}$   $\dfrac{31}{50}$       **b.** $\dfrac{-56}{-14}$   $4$

**8.** Perform each operation. Write answers in lowest terms, if appropriate.

   **a.** $\dfrac{12}{25} \cdot -\dfrac{5}{3}$   $-\dfrac{4}{5}$       **b.** $32 \div \dfrac{1}{4}$   $128$

**10.** Complete the table for the given algebraic expressions.

| Expression | Total Number of Terms | Variable Term(s) | Constant Term(s) | Coefficients of Each Variable Term |
|---|---|---|---|---|
| **a.** $-3p$ | 1 | $-3p$ | None | $-3$ |
| **b.** $-a$ | 1 | $-a$ | None | $-1$ |
| **c.** $y - 1$ | 2 | $y$ | $-1$ | $1$ |
| **d.** $\dfrac{x}{2} + 3$ | 2 | $\dfrac{x}{2}$ | $3$ | $\dfrac{1}{2}$ |
| **e.** $-2y^2 + y - 7$ | 3 | $-2y^2, y$ | $-7$ | $-2, 1$ |
| **f.** $9y^3 - y^2 + 8y - 4$ | 4 | $9y^3, -y^2, 8y$ | $-4$ | $9, -1, 8$ |

**11.** Simplify each expression.

   **a.** $4(3x + 7)$   $12x + 28$       **b.** $8(6x - 5)$   $48x - 40$

   **c.** $-6\left(\dfrac{1}{3}x + \dfrac{1}{2}\right)$   $-2x - 3$    **d.** $x + 2 + 5x - 4$   $6x - 2$

   **e.** $2(5x - 3) - (7x - 9)$   $3x + 3$

   **f.** $2x^2 - 8x + 1 + x^2 - x - 3$   $3x^2 - 9x - 2$

   **g.** $x + \dfrac{3}{2}x$   $\dfrac{5}{2}x$       **h.** $5 + 4(3x + 2)$   $12x + 13$

**12.** Evaluate the expression $4x^2 - 2x + 1$ for $x = -3$.   $43$

**13.** Translate each phrase into a mathematical expression.

   **a.** Two more than a number   $x + 2$

   **b.** The difference of a number and 5   $x - 5$

   **c.** The sum of three times a number and 4   $3x + 4$

   **d.** Twice the sum of a number and 7   $2(x + 7)$

   **e.** The ratio of a 4 times a number and 3   $\dfrac{4x}{3}$

   **f.** Six less than one-fourth of a number   $\dfrac{1}{4}x - 3$

**Solve each problem.**

**14.** The country with the highest rate of decrease in the natural birth rate is the Ukraine. Its population is currently 46.8 million. It is expected to be 33.4 million in 2050. What is the difference between the expected population and the current population? (Source: http://geography.about.com/od/populationgeography/a/zero.htm)   $-13.4$ million

**9.** Identify the property that is illustrated.

   **c.** $-\dfrac{5}{9} + \left(-\dfrac{4}{9}\right)$   $-1$    **d.** $\dfrac{1}{4} - \dfrac{2}{7}$   $-\dfrac{1}{28}$

   **e.** $\dfrac{6}{11} + \left(-\dfrac{1}{2}\right)\left(\dfrac{2}{5}\right) \div 3\dfrac{79}{165}$   **f.** $(-4) + (-5)(2)$   $-14$

   **g.** $(9 - 12) - (7 - 9)$   $-1$    **h.** $(-2)^3 - 4 \div 2 \cdot 6$   $-20$

   **i.** $\sqrt{(1 - (-4))^2 + (-4 - 8)^2}$   $13$

   **j.** $\dfrac{6 - 3(-2)}{4 - 2^2}$   undefined

**9.** Identify the property that is illustrated.

   **a.** $5(6 + 2) = (6 + 2)5$   commutative property of multiplication    **b.** $4 \cdot \dfrac{1}{4} = 1$   inverse property of multiplication

   **c.** $3 + (4 + 7) = (3 + 4) + 7$   associative property of addition

   **d.** $\dfrac{1}{5} + \left(-\dfrac{1}{5}\right) = 0$   inverse property of addition    **e.** $3 \cdot 1 = 3$

   **f.** $-9 + 0 = -9$   identity property of addition    **g.** $4 + (-3) = (-3) + 4$   commutative property of addition

**15.** The following chart shows the closing price of Google's stock for October 26 to November 2, 2009. (Source: http://finance.yahoo.com/q/hp?s=GOOG&a=09&b=19&c=2009&d=10&e=3&f=2009&g=d)

| Date | Closing Price of Google Stock | Daily Change in Price |
|---|---|---|
| 11/2/2009 | $533.99 | -2.13 |
| 10/30/2009 | $536.12 | -14.93 |
| 10/29/2009 | $551.05 | 10.75 |
| 10/28/2009 | $540.30 | -7.99 |
| 10/27/2009 | $548.29 | -5.92 |
| 10/26/2009 | $554.21 | n/a |

   **a.** Complete the chart by showing how the price of the stock changed from the previous day to the next.

   **b.** After completing the chart, find the average daily change in price.   $-\$4.044$

**16.** If $5000 is invested in a savings account that earns 4.25% annual interest for 3 yr, the amount in the account at the end of the 3 yr is given by $5000(1.0425)^3$. Simplify this expression to determine how much is in the account after 3 yr.   $\$5664.98$

**17.** The volume of a sphere is given by $\dfrac{4}{3}\pi r^3$, where $r$ is the radius of the sphere. Find the volume of a basketball that has a radius of 4.7 in. Round the answer to the nearest hundredth.   $434.89$ in.$^3$

# Linear Equations and Inequalities in One Variable

## Organization

Organization is critical for your success in college. Good organizational skills will assist you in

- meeting assignment deadlines.
- keeping you focused on your tasks.
- memory skills.
- note taking.
- test preparation and test taking.
- writing.

Organization is needed for your personal study space and your course information. Your personal study space should be organized so that it is

- free from clutter.
- free from distractions.
- easy to access class materials.

Organize the course information with a

- binder with dividers for notes, exams, homework, handouts, and course information.
- different color binder for each course.
- course calendar placed at the front of each binder.

There are many websites that contain helpful information on how to get organized. Take a few moments to review these if you need assistance.

*Question For Thought:* Do you consider yourself organized? If not, what is one area that could use improvement?

## Chapter Outline

## Coming Up...

In Section 2.7, we will learn how to use linear inequalities in one variable to determine the number of attendees a couple can have at their wedding reception.

"In reading the lives of great men, I found that the first victory they won was over themselves . . . self-discipline with all of them came first."

—Harry S. Truman (U.S. President)

---

**Equations and Their Solutions**

In this chapter, we will learn how to solve equations and inequalities using two important properties—the addition property of equality and the multiplication property of equality. We will also learn real-life applications of these equations and inequalities. We will rely heavily on the skills we obtained from Chapter 1.

▶ **OBJECTIVES**

As a result of completing this section, you will be able to

1. Identify an equation and expression.
2. Determine if a number is a solution of an equation.
3. Express statements as mathematical equations.
4. Set up equations for application problems.
5. Troubleshoot common errors.

From June 2009 to June 2010, Oprah Winfrey and James Cameron, director of the movies *Avatar* and *Titanic*, were the two highest paid celebrities. They earned a combined total of $525 million. Winfrey earned $105 million more than Cameron. How much did each of them earn? (Source: http://www.forbes.com/lists/)

In this section, we will learn how to express this type of information as a mathematical equation. We will learn how to solve these types of equations in Section 2.2.

## Equations and Expressions

In Chapter 1, we worked with numerical expressions and algebraic expressions. In this section, we turn our attention to equations. An *equation* is a mathematical statement that two expressions are equal. Equations can be numerical or algebraic. Some examples of each are shown in the following table.

| Numerical Equations | Algebraic Equations |
|---|---|
| $2 + 3 = 5$ | $x + 4 = 2$ |
| $-7 + 6 = 1$ | $x^2 = x$ |

*Numerical equations* are either true or false. In the above example, $2 + 3$ equals 5, so this equation is true. The next example is false since $-7 + 6$ does not equal 1. *Algebraic equations* are neither true nor false until we assign a value to the variable. The value(s) of the variable that makes the equation true is called a **solution of the equation**.

It is helpful to understand the difference between an equation and an expression. Equations are solved but expressions are simplified.

**Objective 1** ▶

Identify an equation and expression.

**INSTRUCTOR NOTE:**
Make sure that students know that expressions are simplified and equations are solved.

> **Definition: Equations and Expressions**
>
> 1. An **expression** consists of terms that are combinations of letters and numbers. Recall from Section 1.8 that terms are connected by addition or subtraction signs.
> 2. If two expressions are connected by an equals sign, then it is an **equation**.

**Objective 1 Examples** — Determine if each problem is an expression or equation.

| Problems | Solutions |
|---|---|
| **1a.** $2x + 4 - 6x + 3$ | $2x + 4 - 6x + 3$ is an **expression** because it does not contain an equals sign. |
| **1b.** $2x + 4 = 6x + 3$ | $2x + 4 = 6x + 3$ is an **equation** because it contains an equals sign. |
| **1c.** $4y^2 + 2y = 1$ | $4y^2 + 2y = 1$ is an **equation** because it contains an equals sign. |
| **1d.** $a^2 + (a + 1)^2 - 6$ | $a^2 + (a + 1)^2 - 6$ is an **expression** because it does not contain an equals sign. |

✓ **Student Check 1**     Determine if each problem is an expression or equation.

   **a.** $-4y + 2 = 3y + 1$     **b.** $-4y + 2 - 3y + 1$
   **c.** $-2x^2 + 4x = 3x$     **d.** $5r^2 - r - 4$

## Solutions of Equations

**Objective 2** ▶

Determine if a number is a solution of an equation.

As we have learned, a *solution*, or *root*, of an equation is a value of the variable that makes the equation true. The process of finding the value(s) of the variable that makes the equation true is called *solving* an equation for the given variable.

   The process of solving equations is one of the most important skills in algebra. Two fundamental properties of equation solving will be presented in Section 2.2. Before we actually solve equations, we will review how to check if a number is a solution of an equation. We did this in Section 1.3 but it is worth repeating.

> **Procedure: Determining if a Number Is a Solution of an Equation**
>
> **Step 1:** Evaluate each side of the equation for the given value.
> **Step 2:** Determine if the two values are equal.
>    **a.** If the two values are equal, then the given number makes the equation true and is a solution of the equation.
>    **b.** If the two values are not equal, then the given number makes the equation false and is *not* a solution of the equation.

**Objective 2 Examples**     **Determine if the given numbers are solutions of the equation.**

   **2a.** $-4x + 2 = 2x - 1$; $x = -2, 4$, or $\dfrac{1}{2}$     **2b.** $x^2 = x$; $x = -0.5, 0, 1$

**Solutions**   **2a.**

**INSTRUCTOR NOTE:**
Point out that when we need to evaluate an expression for multiple values, a table is helpful in organizing the information.

| $x$ | $-4x + 2 = 2x - 1$ | |
|-----|--------------------|--|
| $-2$ | $-4(-2) + 2 \overset{?}{=} 2(-2) - 1$ <br> $8 + 2 \overset{?}{=} -4 - 1$ <br> $10 \neq -5$ | Since $10 \neq -5$, $x = -2$ is *not* a solution. |
| $4$ | $-4(4) + 2 \overset{?}{=} 2(4) - 1$ <br> $-16 + 2 \overset{?}{=} 8 - 1$ <br> $-14 \neq 7$ | Since $-14 \neq 7$, $x = 4$ is *not* a solution. |
| $\dfrac{1}{2}$ | $-4\left(\dfrac{1}{2}\right) + 2 \overset{?}{=} 2\left(\dfrac{1}{2}\right) - 1$ <br> $-2 + 2 \overset{?}{=} 1 - 1$ <br> $0 = 0$ | Since $0 = 0$, $x = \dfrac{1}{2}$ is a solution. |

Of the given values, $x = \dfrac{1}{2}$ is a solution of the equation $-4x + 2 = 2x - 1$.

**2b.**

| $x$ | $x^2 = x$ | |
|-----|-----------|--|
| $-0.5$ | $(-0.5)^2 \overset{?}{=} -0.5$ <br> $0.25 \neq -0.5$ | Since $0.25 \neq -0.5$, $x = -0.5$ is *not* a solution. |
| $0$ | $(0)^2 \overset{?}{=} 0$ <br> $0 = 0$ | Since $0 = 0$, $x = 0$ is a solution. |
| $1$ | $(1)^2 \overset{?}{=} 1$ <br> $1 = 1$ | Since $1 = 1$, $x = 1$ is a solution. |

Of the given values, $x = 0$ and $x = 1$ are solutions of the equation $x^2 = x$.

✓ **Student Check 2** Determine if the given numbers are solutions of the equation.

**a.** $1 - 2x = -5x - 8$; $x = -3$, $\frac{1}{4}$, or 2    **b.** $x^2 = 2x$; $x = \frac{1}{2}$, 0, or 2

## Translating Phrases into Mathematical Equations

**Objective 3** ▶

Express statements as mathematical equations.

In Chapter 1, we learned how to translate phrases into mathematical expressions and equations. The phrases for equality are shown in the following table.

| Phrases for " = " |
| --- |
| is equal to |
| the result is |
| is the same as |
| equals |
| is |

In many sections of this book, we will be required to solve application problems. The most difficult part of solving an application problem is setting up the equation correctly. Example 3 illustrates the process of setting up an equation to solve an application problem. We will only set up the equation. Solving the equations will come later.

> **Procedure: Expressing Statements as Mathematical Equations**
>
> **Step 1:** Read the problem carefully.
> **Step 2:** Determine the unknown and assign a variable to it. (If there is more than one unknown, the other unknowns will be represented in terms of the same variable initially chosen.)
> **Step 3:** From the given statement, determine the phrase that represents the equals sign. The expression before this phrase is the left side of the equation. The expression after the phrase is the right side of the equation. Translate both expressions and set them equal to obtain the equation.

**Objective 3 Examples** **For each problem, define a variable and write an equation that can be used to solve the problem.**

**3a.** Twice the sum of a number and 3 is the same as 4 less than the number.

**Solution** **3a.** What is unknown? The number is unknown. Let $n$ represent the number. What is known? Twice the sum of a number and 3 is the same as 4 less than the number.

The phrase "is the same as" represents the equals sign.
"Twice the sum of a number and 3" represents the left side of the equation.
"Four less than the number" represents the right side of the equation.

| Twice the sum of a number and 3 | is the same as | 4 less than the number |
| --- | --- | --- |
| $2(n + 3)$ | $=$ | $n - 4$ |

So, the equation used to solve this problem is $2(n + 3) = n - 4$.

**3b.** The difference of three times a number and 6 is the same as the quotient of the number and 9.

**Solution** **3b.** What is unknown? The number is unknown. Let $n$ represent the number. What is known? The difference of three times a number and 6 is the same as the quotient of the number and 9.

The phrase "is the same as" represents the equals sign.

"The difference of three times a number and 6" represents the left side of the equation.

"The quotient of the number and 9" represents the right side of the equation.

| The difference of three times a number and 6 | is the same as | the quotient of the number and 9 |
|:---:|:---:|:---:|
| $3n - 6$ | $=$ | $\dfrac{n}{9}$ |

So, the equation used to solve this problem is $3n - 6 = \dfrac{n}{9}$.

---

✔ **Student Check 3** For each problem, define a variable and write an equation that can be used to solve the problem.

   **a.** Three times the sum of a number and five is the same as four times the number.

   **b.** The sum of twice a number and one is equal to two more than the number.

---

## Representing Applications Algebraically

**Objective 4** ▶

Set up equations for application problems.

Setting up equations for application problems is very similar to the previous objective. The main difference is that we may have to rely on key words in the problem that define the relationships implicitly. For instance, we might be told the perimeter of an object or that two angles are complementary. In these cases, we have to know how to calculate the perimeter of the object or what it means for two angles to be complementary to set up the equation.

**Objective 4 Examples** **For each problem, define a variable and write an equation that can be used to solve the problem.**

   **4a.** One angle is 30° more than another angle. The angles are complementary. Find the measure of each angle.

**Solution** **4a.** What is unknown? The measure of the two angles is unknown.

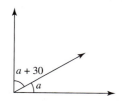

     Let $a$ represent the measure of one of the angles.

     The other angle is 30° more, so $a + 30$ represents the measure of the other angle.

What is known? The angles are complementary, so their sum is 90°.

The phrase "is" represents the equals sign.

"Their sum" represents the left side of the equation.

"90°" represents the right side of the equation.

| Their sum | is | 90° |
|:---:|:---:|:---:|
| $a + a + 30$ | $=$ | 90° |

So, the equation used to solve this problem is $a + a + 30 = 90°$.

   **4b.** From June 2009 to June 2010, Oprah Winfrey and James Cameron, director of the movies *Avatar* and *Titanic*, were the two highest paid celebrities. They earned a combined total of $525 million. Winfrey earned $105 million more than Cameron. How much did each of them earn (in millions)? (Source: http://www.forbes.com/lists/)

**Solution**     **4b.** What is unknown? The earnings of each celebrity are unknown.

Let $j$ represent James Cameron's earnings. Winfrey's earnings are $105 million more than Cameron's, so $j + 105$ represents Winfrey's earnings.

What is known? They earned a combined total of $525 million.

The phrase "of" represents the equals sign.

"They earned a combined total" represents the left side of the equation.

"525" represents the right side of the equation.

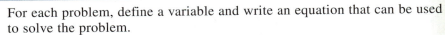

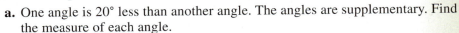

So, the equation used to solve this problem is $j + j + 105 = 525$.

---

**✓ Student Check 4**

For each problem, define a variable and write an equation that can be used to solve the problem.

**a.** One angle is 20° less than another angle. The angles are supplementary. Find the measure of each angle.

**b.** The total number of U.S. deaths from World War I and World War II was approximately 522 thousand. The number of deaths from World War I was 288 thousand less than the number of deaths from World War II. How many deaths occurred in each world war? (Source: http://www.fas.org/sgp/crs/natsec/RL32492.pdf )

---

**Objective 5** ▶

Troubleshoot common errors.

## Troubleshooting Common Errors

Some common errors associated with determining if a value is a solution of an equation and translating expressions are shown next.

---

**Objective 5 Examples**

**A problem and an incorrect solution are given. Provide the correct solution and an explanation of the error.**

**5a.** Is $4x - 2 - 6x + 4$ an equation or expression?

| Incorrect Solution | Correct Solution and Explanation |
|---|---|
| $4x - 2 - 6x + 4$ is an equation. | $4x - 2 - 6x + 4$ does not contain an equals sign, so this is an expression, *not* an equation. |

**5b.** Determine if $x = -5$ is a solution of $3 - 2(x + 4) = 5$.

| Incorrect Solution | Correct Solution and Explanation |
|---|---|
| $3 - 2(x + 4) = 5$ $3 - 2(-5 + 4) = 5$ $3 - 2(-1) = 5$ $1(-1) = 5$ $-1 = 5$ Since this is a false statement, $x = -5$ is not a solution. | The error was made in evaluating the left side of the equation. After simplifying what is in parentheses, we should multiply before adding. $3 - 2(x + 4) = 5$ $3 - 2(-5 + 4) = 5$ $3 - 2(-1) = 5$ $3 + 2 = 5$ $5 = 5$ Since $x = -5$ makes the equation true, it is a solution. |

**c.** Three less than a number is twice the number. Find the number. Write an equation to solve this problem.

| Incorrect Solution | Correct Solution and Explanation |
|---|---|
| Let $x$ be the number. $3 - x = x^2$ | Three less than a number is expressed as $x - 3$ and twice a number is $2x$. So, the equation is $x - 3 = 2x$. |

## ANSWERS TO STUDENT CHECKS

**Student Check 1**   **a.** equation   **b.** expression   **c.** equation
    **d.** expression

**Student Check 2**   **a.** $x = -3$ is a solution.   **b.** $x = 0$ and
    $x = 2$ are solutions.

**Student Check 3**   **a.** $3(n + 5) = 4n$   **b.** $2n + 1 = n + 2$
**Student Check 4**   **a.** $a + a - 20 = 180$
    **b.** $x + x - 288 = 522$

## SUMMARY OF KEY CONCEPTS

**1.** An equation is a statement that two expressions are equal. Equations can be solved, whereas expressions can be simplified or evaluated.

**2.** The number(s) that make an equation a true numerical statement are called solutions of the equation.

**3.** Expressing statements as mathematical equations takes a lot of practice. It is very important to recognize the key phrases for the different operations and for the equals sign.

**4.** Writing equations that can be used to solve application problems requires us to know some essential information about the situation. This essential information can be gathered from key words in the problem—perimeter, complementary, supplementary, and the like.

## GRAPHING CALCULATOR SKILLS

The graphing calculator can be used to determine whether a number is a solution of an equation. There are several ways to use the calculator to perform this skill.

**Example:** Determine if $x = -2$ is a solution of $-4x + 2 = 2x - 1$.

**Method 1:** Use the calculator to evaluate the left and right sides of the equation.

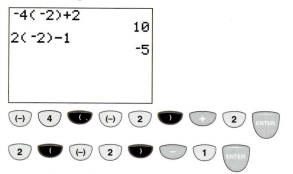

Since the left side, 10, does not equal the right side, $-5$, the value $x = -2$ is not a solution of the equation.

**Method 2:** Enter the left side and right side of the equation in the equation editor and use the table to determine the value of each side.

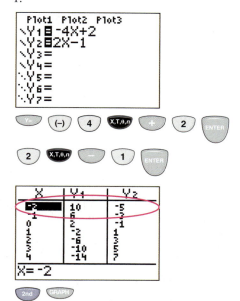

When $x = -2$, the values in the columns for $Y_1$ and $Y_2$ are not equal. This shows that $x = -2$ is not a solution.

## SECTION 2.1   /   EXERCISE SET

 **Write About It!**

Use complete sentences in your answer to the following exercises. Provide a specific example of each term or process.

1. Explain the meaning of an equation.   Answers vary.

2. Explain what it means for a number to be the solution of an equation.   Answers vary.

3. Explain the process of verifying if a number is a solution of an equation.   Answers vary.

4. Explain the difference between an expression and an equation.   Answers vary.

5. Explain the difference between a numerical equation and an algebraic equation.   Answers vary.

6. Explain the difference between the two statements: "4 less than a number" and "4 minus a number".   Answers vary.

 **Practice Makes Perfect!**

**Determine whether each of the following is an expression or an equation. (See Objective 1.)**

7. $3x + 2$   expression

8. $4y + 6 = 3$   equation

9. $2x - 1 = -5$   equation

10. $2x - 1$   expression

11. $-7x$   expression

12. $7x = 0$   equation

13. $x^2 - x - 6 = 0$   equation

14. $x^2 - x - 6$   expression

15. $3(x - 8) = 4$   equation

16. $3(x - 8)$   expression

**Determine if the given numbers are solutions of the equation. (See Objective 2.)**

17. Is $x = -2$, $x = -1$, $x = 0$, $x = 1$, or $x = 2$ a solution of $x - 1 = -3$?

18. Is $x = -2$, $x = -1$, $x = 0$, $x = 1$, or $x = 2$ a solution of $x + 1 = 0$?

19. Is $x = -2$, $x = -1$, $x = 0$, $x = 1$, or $x = 2$ a solution of $3x = 0$?

20. Is $x = -2$, $x = -1$, $x = 0$, $x = 1$, or $x = 2$ a solution of $-2x = -4$?

21. Is $x = -2$, $x = -1$, $x = 0$, $x = 1$, or $x = 2$ a solution of $x^2 - x - 2 = 0$?

22. Is $x = -2$, $x = -1$, $x = 0$, $x = 1$, or $x = 2$ a solution of $x^2 + x - 2 = 0$?

23. Is $x = -2$, $x = -1$, $x = 0$, $x = 1$, or $x = 2$ a solution of $x^2 = 1$?

24. Is $x = -2$, $x = -1$, $x = 0$, $x = 1$, or $x = 2$ a solution of $x^2 = 2x$?

25. Is $x = -2$, $x = -1$, $x = 0$, $x = 1$, or $x = 2$ a solution of $x - 2 = x$?

26. Is $x = -2$, $x = -1$, $x = 0$, $x = 1$, or $x = 2$ a solution of $3x - 8 - x = 2(x - 4)$?

Additional answers can be found in the Instructor Answer Appendix.

**For each exercise, define a variable and write an equation that can be used to solve the problem. (See Objective 3.)**

27. The sum of a number and 4 is $-3$.   $x + 4 = -3$

28. The sum of a number and $-6$ is 8.   $x + (-6) = 8$

29. The difference of a number and 7 is 4.   $x - 7 = 4$

30. The difference of a number and 10 is $-2$.   $x - 10 = -2$

31. Twice a number is 18.   $2x = 18$

32. Three times a number is $-12$.   $3x = -12$

33. The quotient of a number and 4 is 8.   $\frac{x}{4} = 8$

34. The quotient of a number and 2 is $-5$.   $\frac{x}{2} = -5$

35. The sum of twice a number and 7 is the same as one less than the number.   $2x + 7 = x - 1$

36. Two more than three times a number equals the number decreased by 9.   $3x + 2 = x - 9$

37. Three divided by a number gives the same result as one-third of the number.   $\frac{3}{x} = \frac{1}{3}x$

38. The ratio of a number to 5 equals twice the number.   $\frac{x}{5} = 2x$

39. Four less than three times a number is six more than the number.   $3x - 4 = 6 + x$

40. Six less than five times a number is two more than the number.   $5x - 6 = 2 + x$

41. If three times a number is added to four times the same number, the result is the same as seven less than five times the number.   $3x + 4x = 5x - 7$

42. If twice a number is added to three times the same number, the result is the same as seven more than four times the number.   $2x + 3x = 4x + 7$

43. Four times the sum of a number and $-2$ is the same as twice the number.   $4[x - (-2)] = 2x$

44. Six times the difference of a number and $-10$ is the same as four times the number.   $6[x - (-10)] = 4x$

45. The difference of four times a number and $-12$ is equal to 50 more than the number.   $4x - (-12) = 50 + x$

46. The difference of twice a number and $-7$ is equal to 18 more than the number.   $2x - (-7) = 18 + x$

47. The sum of twice a number and 18 is equal to 27 less than the number.   $2x + 18 = x - 27$

48. The sum of eight times a number and 24 is equal to 32 less than the number.   $8x + 24 = x - 32$

**For each problem, define a variable and write an equation that can be used to solve the problem. (See Objective 4.)**

49. One angle is $13°$ less than another angle. Their sum is $90°$. Find the measure of each angle.   $x + x - 13° = 90°$

50. One angle is $16°$ more than another angle. Their sum is $180°$. Find the measure of each angle.   $x + x + 16° = 180°$

**51.** From June 2008 to June 2009, Carl Icahn and Larry Page were the two of the 31 richest Americans. Their combined net worth was $25.8 billion. Carl Icahn earned $4.8 billion less than Larry Page. How much did each of them earn (in billions of dollars)? (Source: http://www.forbes.com/lists/) $x + x - 4.8 = 25.8$

**52.** From June 2008 to June 2009, William Gates III and Jacqueline Mars were the two of the 31 richest Americans. Their combined net worth was $61 billion. William Gates III earned $39 billion more than Jacqueline Mars. How much did each of them earn (in billions of dollars)? (Source: http://www.forbes.com/lists/) $x + x + 39 = 61$

**53.** The perimeter of a rectangular garden is 32 ft. The width of the garden is 4 ft less than the length. Let $l$ represent the length of the garden and let $w$ represent the width of the garden.

**a.** Write an equation in terms of $l$ and $w$ that represents the perimeter of the garden. $2l + 2w = 32$

**b.** Write an equation that relates the length of the garden to the width of the garden. $l = w + 4$ or $w = l - 4$

**54.** The perimeter of a regulation size basketball court is 288 ft. The length of the court is 44 ft more than the width. Let $l$ represent the length of the court and let $w$ represent the width of the court.

**a.** Write an equation in terms of $l$ and $w$ that represents the perimeter of the court. $2l + 2w = 288$

**b.** Write an equation that relates the length of the court to the width of the court. $l = w + 44$ or $w = l - 44$

**55.** The Dixie Chicks reached number 1 on the Billboard 200 charts for their albums *Taking the Long Way* (2006) and *Home* (2002). The two albums sold a total of 1.31 million albums. The *Taking the Long Way* album sold 0.254 million albums less than the *Home* album. Let $l$ represent the number of *Taking the Long Way* albums sold (in millions) and let $h$ represent the number of *Home* albums sold (in millions). (Source: www.billboard.com)

**a.** Write an equation in terms of $l$ and $h$ that represents the combined sales of the two albums. $l + h = 1.31$

**b.** Write an equation that relates the number of *Taking the Long Way* albums sold to the number of *Home* albums sold. $l = h - 0.254$

**56.** In 2008 and 2009, a total of 802.2 million albums were sold in the United States. The number of albums sold in 2009 was 54.4 million less than the number of albums sold in 2008. Let $x$ represent the number of albums sold in 2008 (in millions) and let $y$ represent the number of albums sold in 2009 (in millions). (Source: http://washingtonpost.com)

**a.** Write an equation using $x$ and $y$ that represents the total number of albums sold in 2008 and 2009. $x + y = 802.2$

**b.** Write an equation that relates the number of albums sold in 2009 to the number of albums sold in 2008. $y = x - 54.4$

**57.** There are about 3.1 million elementary and middle school teachers in the United States. Elementary and middle

school teachers make up about half of all teachers in the United States. Let $e$ represent the number of elementary teachers (in millions) and let $m$ represent the number of middle school teachers (in millions).

**a.** Write an equation using $e$ and $m$ that represents the number of elementary and middle school teachers in the United States. $e + m = 3.1$

**b.** If $t$ represents the total number of teachers in the United States, write an equation that relates the total number of teachers in the United States to the number of elementary and middle school teachers in the United States. (Source: U.S. Census) $e + m = \frac{1}{2}t$

**58.** Financial planners often recommend that a monthly house payment be no more than about $\frac{1}{3}$ of your monthly income. If $x$ represents a person's monthly income and $y$ represents the maximum recommended monthly house payment, write an equation that relates $x$ and $y$. $y = \frac{1}{3}x$

 ## Mix 'Em Up!

**Determine whether each of the following is an expression or an equation.**

**59.** $6z - 8 = 8z - 6$    equation

**60.** $-2r^2 + 12 - 13r - 46$    expression

**61.** $3x + 10 - 8y + 5$    expression

**62.** $5y^2 - 20 = 18y - 35$    equation

**63.** $-6r^2 + 5 = 11r + 4$    equation

**64.** $-7y + 16 + 11y + 14$    expression

**Determine if the given numbers are solutions of the equation.**

**65.** Is $x = -4$, $x = -1$, $x = 0$, $x = 4$, or $x = \frac{5}{14}$ a solution of $8x - 8 = -6x - 3$?

**66.** Is $x = -7$, $x = -2$, $x = 0$, $x = 1$, or $x = -\frac{11}{4}$ a solution of $-5x + 10 = -9x - 1$?

**67.** Is $x = -4$, $x = -2$, $x = 0$, $x = 5$, or $x = \frac{1}{2}$ a solution of $x^2 = x + 20$?

**68.** Is $x = -2$, $x = -1$, $x = 0$, $x = 2$, or $x = \frac{1}{3}$ a solution of $x^2 = -3x - 2$?

**For each exercise, define a variable and write an equation that can be used to solve the problem.**

**69.** Twice the sum of a number and 11 is the same as five times the number. $2(x + 11) = 5x$

**70.** Five times the difference of a number and 20 is the same as three times the number. $5(x - 20) = 3x$

**71.** The sum of four times a number and $-10$ is equal to 20 more than the number. $4x + (-10) = 20 + x$

**72.** The difference of twice a number and −4 is equal to 14 less that the number.   $2x - (-4) = x - 14$

**73.** Four more than twice a number equals three times the number decreased by 5.   $2x + 4 = 3x - 5$

**74.** The ratio of a number to 4 gives the same result as one-half of the number plus 3.   $\frac{x}{4} = \frac{1}{2}x + 3$

**75.** If five times a number is subtracted from four, the result is the same as six less than twice the number.   $4 - 5x = 2x - 6$

**76.** If three times a number is added to ten, the result is the same as six more than five times the number.   $3x + 10 = 5x + 6$

**77.** One angle is 25° more than another angle. Their sum is 180°. Find the measure of each angle.   $x + x + 25° = 180°$

**78.** One angle is 40° less than another angle. Their sum is 90°. Find the measure of each angle.   $x + x - 40° = 90°$

**79.** One angle is 19° less than another angle. Their sum is 90°. Find the measure of each angle.   $x + x - 19° = 90°$

**80.** One angle is 37° more than another angle. Their sum is 180°. Find the measure of each angle.   $x + x + 37° = 180°$

**81.** From June 2008 to June 2009, Alice Walton and Lawrence Ellison were two of the 31 richest Americans. Their combined net worth was $46.3 billion. Alice Walton earned $7.7 billion less than Lawrence Ellison. How much did each of them earn (in billions of dollars)? (Source: http://www.forbes.com/lists/)   $x + x - 7.7 = 46.3$

**82.** From June 2008 to June 2009, Christy Walton and Forrest Edward Mars were two of the 31 richest Americans. Their combined net worth was $32.5 billion. Christy Walton earned $10.5 billion more than Forrest Edward Mars. How much did each of them earn (in billions of dollars)? (Source: http://www.forbes.com/lists/)   $x + x + 10.5 = 32.5$

**83.** The combined total sales of a corporation in 2009 and 2010 was $69.1 billion. The total sales in 2009 was $1.5 billion less than the total sales in 2010. Let $x$ = total sales in 2009 (in billions of dollars) and let $y$ = total sales in 2010 (in billions of dollars).

   **a.** Write an equation using $x$ and $y$ that represents the combined total sales in 2009 and 2010.   $x + y = 69.1$

   **b.** Write an equation that relates the total sales in 2010 to the total sales in 2009.   $x = y - 1.5$

**84.** The total number of individual songs purchased digitally in 2008 and 2009 was 2.231 billion. The total number of individual songs purchased digitally in 2008 was 0.089 billion less than the total number in 2009. Let $x$ = total number of individual songs purchased digitally in 2008 (in billions) and let $y$ = total number in 2009 (in billions). (Source: http://www.ritholtz.com)

   **a.** Write an equation using $x$ and $y$ that represents the combined total number of individual songs purchased digitally in 2008 and 2009.   $x + y = 2.231$

   **b.** Write an equation that relates the total number of songs purchased digitally in 2008 to the total number of songs purchased digitally in 2009.   $x = y - 0.089$

 **You Be the Teacher!**

**Anita and Maria are studying together. They are solving problems requiring them to translate phrases into mathematical statements. However, the wording of the problem is ambiguous so that both of their answers are correct. Change the wording of each statement so that only Anita's answer is correct.**

**85.** The square of a number subtracted by 3 is 10.

   Anita: $x^2 - 3 = 10$

   Maria: $(x - 3)^2 = 10$   Answers vary. Three less than the square of a number is 10.

**86.** The absolute value of a number added to 7 is 14.

   Anita: $|x| + 7 = 14$

   Maria: $|x + 7| = 14$   Answers vary. Seven more than the absolute value of a number is 14.

**87.** Half of a number plus 2 is 6.

   Anita: $\frac{1}{2}x + 2 = 6$

   Maria: $\frac{1}{2}(x + 2) = 6$   Answers vary. Two more than half a number is 6.

**88.** Twice a number subtracted from 12 is 23.

   Anita: $2(12 - x) = 23$

   Maria: $12 - 2x = 23$   Answers vary. Twice the difference of 12 and a number is 23.

 **Calculate It!**

**Use a graphing calculator to determine if the given numbers are solutions of the equation. Specify the equation(s) that were entered into the calculator.**

**89.** $3x^2 - x = 2x; x = -1, 0, 1$

**90.** $|x - 3| = 5; x = -2, 0, 2$

**91.** $4x = x + \frac{3}{2}; x = 0, \frac{1}{2}, 1$

**92.** $x + \frac{1}{2} = x^2 - \frac{1}{4}; x = -1, -\frac{1}{2}, \frac{1}{2}$

 **Think About It!**

**93.** Write a statement that corresponds to the equation $2(x - 5) = x + 6$.   Twice the difference of a number and 5 is 6 more than the number.

**94.** Write a statement that corresponds to the equation $2x - 5 = 6x$.   Five less than twice a number is the same as 6 times the number.

**95.** Write an equation that has $x = 3$ as its solution.   Answers vary. $2x - 1 = 8 - x$

**96.** Write an equation that has $x = -1$ as its solution.   Answers vary. $-4(x + 1) = x + 1$

**SECTION 2.2**   **The Addition Property of Equality**

▶ **OBJECTIVES**

As a result of completing this section, you will be able to

1. Define and recognize a linear equation.
2. Use the addition property of equality to solve a linear equation.
3. Solve equations that require more than one step.
4. Solve applications of linear equations.
5. Troubleshoot common errors.

The selling price of a new car is $24,891. The selling price is the base price plus fees and options. The base price is $22,985. What is the total of the fees and options?

In this section, we use the skills from Section 2.1 to write an equation that represents the situation and to solve the equation using the addition property of equality.

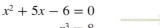

### Linear Equations

In Section 2.1, we defined a mathematical equation. In the remaining sections of this chapter, we focus on one type of equation, a *linear equation*.

**Objective 1** ▶

Define and recognize a linear equation.

> **Definition:** A **linear equation in one variable** is an equation that can be written in the form $ax + b = c$, where $a$, $b$, and $c$ are real numbers and $a \neq 0$.

The main characteristic of a linear equation in one variable is that the largest exponent of the variable is one. Recall that $x = x^1$. The exponent of one is called the *degree* of the equation. Linear equations are also called *first-degree equations*. As we will learn in later chapters, the degree denotes the maximum number of solutions of an equation. Therefore, a linear equation will have one solution except for two special cases, which we will solve in Section 2.4. Some examples of linear equations in one variable are

$$x + 3 = -5 \qquad 2x - 5 = 4x + 1 \qquad 3(x + 4) = 8$$

The last two of these linear equations are not in the form $ax + b = c$. However, we will learn methods that enable us to write these equations in this form.

Examples of equations that are not linear are

$$x^2 + 5x - 6 = 0$$
$$x^3 = 8$$

The largest exponent of the variable is 2, not 1.
The largest exponent of the variable is 3, not 1.

> **Procedure: Recognizing a Linear Equation**
>
> **Step 1:** Identify the largest exponent on the variable.
> **Step 2: a.** If this exponent is one, then the equation is a linear equation.
> **b.** If this exponent is not one, then the equation is not linear.

**Objective 1 Examples**   Determine if the equation is a linear equation in one variable.

| Problems | Solutions |
|---|---|
| **1a.** $x = 5$ | The equation $x^1 = 5$ is a linear equation since the largest exponent of the variable is 1. |
| **1b.** $\dfrac{2}{5} - x = \dfrac{3}{4}(x + 2)$ | The equation $\dfrac{2}{5} - x^1 = \dfrac{3}{4}(x^1 + 2)$ is a linear equation since both occurrences of the variable $x$ have an exponent of 1. |
| **1c.** $3y^2 - 4y = 1$ | The equation $3y^2 - 4y = 1$ is not a linear equation because the largest exponent of the variable $y$ is 2. |

✓ **Student Check 1**    Determine if the equation is a linear equation in one variable.

**a.** $2y - 3 = 7$          **b.** $a - \dfrac{1}{2} = \dfrac{3}{2}$          **c.** $y^4 + 8y^2 = 9$

---

### The Addition Property of Equality

We have learned that a solution of an equation is the value of the variable that makes a true statement. It would be too tedious to use evaluating an equation as a method for solving it. The goal of this section is to determine the solution of a linear equation by performing a series of steps that yield simpler equations that are *equivalent* to the original equation.

**Equivalent equations** are equations that have the same solution set. The process of producing equivalent equations will enable us to isolate the variable to one side of the equation. These steps will ultimately produce an equation of the form

$$x = \text{some number} \qquad \text{or} \qquad \text{some number} = x$$

Since two sides of an equation are equal, or equivalent, to each other, we must perform the same operations to each side of the equation to keep the equation balanced.

- It is helpful to think of a balance scale to see this relationship.
- If each side of the equation represents some weight, these two weights must be the same since they are set equal to one another.
- If we add something to one side of the equation or "scale," we must add it to the other side to keep the scales balanced.
- If we subtract something from one side of the equation or "scale," we must subtract it from the other side to keep the scales balanced.

Consider the equation, $x + 2 = 5$. The following diagram shows each side of the equation on the balance scale. Note the scales are balanced since the two sides are equal.

To isolate the variable $x$ on the left side of the scale, we must take 2 units away from the left side of the scale. To maintain balance, we must also take 2 units away from the right side of the scale.

This leaves us with $x$ on the left side of the scale and 3 on the right side of the scale. This action leads us to the solution of the equation, $x = 3$.

The equations $x + 2 = 5$ and $x = 3$ are called equivalent equations since they have the same solution set. We obtained the second equation by subtracting two from each side of the original equation, that is, $x + 2 - 2 = 5 - 2$ is equivalent to $x = 3$. This leads to the addition property of equality.

---

**Property:  Addition Property of Equality**

If $a = c$, then

$$a + b = c + b \qquad \text{and} \qquad a - b = c - b.$$

This property tells us that if two expressions are equal, we can

- add the same number to both expressions and the expressions will remain equal.
- subtract the same number from both expressions and the expressions will remain equal.

Example 2 illustrates the most basic use of the addition property of equality. We will use this property to solve for the variable in one step.

---

**Procedure: Using the Addition Property of Equality to Solve an Equation**

**Step 1:** Determine the operation that will isolate the variable on one side of the equation. Perform this operation on each side of the equation. Remember the inverse property for addition: $a + (-a) = -a + a = a - a = 0$.

**Step 2:** Simplify each side of the equation, as necessary. The result should be of the form

$$x = \text{some number} \qquad \text{or} \qquad \text{some number} = x$$

**Step 3:** Check the solution by substituting the value into the original equation.

**Step 4:** Write the solution in set notation.

---

**Objective 2 Examples** | Solve each equation using the addition property of equality. Check each answer.

**2a.** $x + 12 = 10$  **2b.** $y - 7 = -13$

**2c.** $b + \dfrac{1}{2} = -\dfrac{3}{2}$  **2d.** $1.8 = 2.1 + r$

**Solutions**

**2a.**
$$x + 12 = 10$$
$$x + 12 - 12 = 10 - 12 \qquad \text{Subtract 12 from each side.}$$
$$x = -2 \qquad \text{Simplify.}$$

**Check:**
$$x + 12 = 10 \qquad \text{Original equation}$$
$$-2 + 12 = 10 \qquad \text{Replace } x \text{ with } -2.$$
$$10 = 10 \qquad \text{Simplify.}$$

Since $x = -2$ makes the equation true, the solution set is $\{-2\}$.

**2b.**
$$y - 7 = -13$$
$$y - 7 + 7 = -13 + 7 \qquad \text{Add 7 to each side.}$$
$$y = -6 \qquad \text{Simplify.}$$

**Check:**
$$y - 7 = -13 \qquad \text{Original equation}$$
$$-6 - 7 = -13 \qquad \text{Replace } y \text{ with } -6.$$
$$-13 = -13 \qquad \text{Simplify.}$$

Since $y = -6$ makes the equation true, the solution set is $\{-6\}$.

**2c.**
$$b + \frac{1}{2} = -\frac{3}{2}$$
$$b + \frac{1}{2} - \frac{1}{2} = -\frac{3}{2} - \frac{1}{2} \qquad \text{Subtract } \frac{1}{2} \text{ from each side of the equation.}$$
$$b = -\frac{4}{2} \qquad \text{Simplify each side.}$$
$$b = -2 \qquad \text{Simplify the fraction.}$$

**Check:**    $b + \dfrac{1}{2} = -\dfrac{3}{2}$    Original equation

$-2 + \dfrac{1}{2} = -\dfrac{3}{2}$    Replace $b$ with $-2$.

$-\dfrac{4}{2} + \dfrac{1}{2} = -\dfrac{3}{2}$    Write $-2$ with the LCD of 2 and simplify.

$-\dfrac{3}{2} = -\dfrac{3}{2}$    Simplify.

Since $b = -2$ makes the equation true, the solution set is $\{-2\}$.

**2d.** Since the variable $r$ is on the right side, we will isolate the variable on the right side by subtracting 2.1 from each side.

$$1.8 = 2.1 + r$$
$$1.8 - 2.1 = 2.1 + r - 2.1 \qquad \text{Subtract 2.1 from each side.}$$
$$-0.3 = r \qquad \text{Simplify each side of the equation.}$$
$$r = -0.3$$

While the equation is solved when $r$ is isolated on the right side, note that we can write the final answer with the variable on the left since $a = b$ is equivalent to $b = a$.

**Check:**    $1.8 = 2.1 + r$    Original equation

$1.8 = 2.1 + (-0.3)$    Replace $r$ with $-0.3$.

$1.8 = 1.8$    Simplify.

Since $r = -0.3$ makes the equation true, the solution set is $\{-0.3\}$.

> ✔ **Student Check 2**    Solve each equation using the addition property of equality. Check each answer.
>
> **a.** $y - 5 = -2$          **b.** $y + 10 = -9$
>
> **c.** $x - \dfrac{1}{3} = \dfrac{5}{3}$          **d.** $4 = 2.4 + x$

## Multistep Equations

**Objective 3** ▶

Solve equations that require more than one step.

The equations we solved in Example 2 were set up to immediately apply the addition property of equality since each side of the equation was as simplified as possible and because each equation contained a single variable term. This is not the case for most equations. Most equations require us to do some work to get them in the form that enables us to apply the addition property of equality.

> **Procedure: Using the Addition Property of Equality**
>
> **Step 1:** Simplify each side as much as possible by combining any like terms, removing parentheses, and the like. Remember that each side must be treated as a separate expression.
> **Step 2:** If there are variable terms on both sides of the equation, use the addition property of equality to remove the variable term from one side of the equation.
> **Step 3:** If there is a term still attached to the variable, use the addition property of equality to isolate the variable.
> **Step 4:** Simplify each side of the equation. The equation should be in the form "$x = $ some number" or "some number $= x$."
> **Step 5:** Check the solution by substituting the value into the original equation.
> **Step 6:** Write the solution set in set notation.

| **Objective 3 Examples** | **Solve each equation.** |
|---|---|

**3a.** $4 + x - 3 = -2$          **3b.** $1.5y - 3 - 4 - 0.5y = 0$

**3c.** $6a - 3 = 5a + 1$          **3d.** $3(x + 2) - (2x - 5) = 0$

**3e.** $3(x + 6) = 4(x - 1)$

**Solutions**

**3a.**

$$4 + x - 3 = -2$$
$$x + 1 = -2 \qquad \text{Combine like terms on the left.}$$
$$x + 1 - 1 = -2 - 1 \qquad \text{Subtract 1 from each side of the equation.}$$
$$x = -3 \qquad \text{Simplify.}$$

**Check:**
$$4 + x - 3 = -2 \qquad \text{Original equation}$$
$$4 + (-3) - 3 = -2 \qquad \text{Replace } x \text{ with } -3.$$
$$-2 = -2 \qquad \text{Simplify.}$$

Since $x = -3$ makes the equation true, the solution set is $\{-3\}$.

**3b.**

$$1.5y - 3 - 4 - 0.5y = 0 \qquad \text{Combine like terms on the left side. Note that}$$
$$y - 7 = 0 \qquad 1.5y - 0.5y = 1y = y.$$
$$y - 7 + 7 = 0 + 7 \qquad \text{Add 7 to each side.}$$
$$y = 7 \qquad \text{Simplify.}$$

**Check:**
$$1.5y - 3 - 4 - 0.5y = 0 \qquad \text{Original equation}$$
$$1.5(7) - 3 - 4 - 0.5(7) = 0 \qquad \text{Replace } y \text{ with } 7.$$
$$10.5 - 3 - 4 - 3.5 = 0 \qquad \text{Simplify each product.}$$
$$0 = 0 \qquad \text{Simplify.}$$

Since $y = 7$ makes the equation true, the solution set is $\{7\}$.

**3c.**

$$6a - 3 = 5a + 1$$
$$6a - 3 - 5a = 5a + 1 - 5a \qquad \text{Subtract } 5a \text{ from each side.}$$
$$a - 3 = 1 \qquad \text{Simplify each side.}$$
$$a - 3 + 3 = 1 + 3 \qquad \text{Add 3 to each side.}$$
$$a = 4 \qquad \text{Simplify.}$$

**Check:**
$$6a - 3 = 5a + 1 \qquad \text{Original equation}$$
$$6(4) - 3 = 5(4) + 1 \qquad \text{Replace } a \text{ with } 4.$$
$$24 - 3 = 20 + 1 \qquad \text{Find each product.}$$
$$21 = 21 \qquad \text{Simplify each side.}$$

Since $a = 4$ makes the equation true, the solution set is $\{4\}$.

**3d.**

$$3(x + 2) - 1(2x - 5) = 0 \qquad \text{Recall } -(2x - 5) = -1(2x - 5).$$
$$3x + 6 - 2x + 5 = 0 \qquad \text{Apply the distributive property.}$$
$$x + 11 = 0 \qquad \text{Combine like terms.}$$
$$x + 11 - 11 = 0 - 11 \qquad \text{Subtract 11 from each side.}$$
$$x = -11 \qquad \text{Simplify.}$$

**Check:**    $3(x + 2) - (2x - 5) = 0$      Original equation

$3(-11 + 2) - [2(-11) - 5] = 0$      Replace $x$ with $-11$.

$3(-9) - (-22 - 5) = 0$      Simplify the left side using the

$-27 - (-27) = 0$      order of operations.

$-27 + 27 = 0$      Add.

$0 = 0$      Simplify.

Since $x = -11$ makes the equation true, the solution set is $\{-11\}$.

**INSTRUCTOR NOTE:**
Point out that we subtract $3x$ from each side to avoid a negative coefficient on the variable.

**3e.**    $3(x + 6) = 4(x - 1)$

$3x + 18 = 4x - 4$      Apply the distributive property.

$3x + 18 - 3x = 4x - 4 - 3x$      Subtract $3x$ from each side.

$18 = x - 4$      Simplify.

$18 + 4 = x - 4 + 4$      Add 4 to each side.

$22 = x$      Simplify.

**Check:**    $3(x + 6) = 4(x - 1)$      Original equation

$3(22 + 6) = 4(22 - 1)$      Replace $x$ with 22.

$3(28) = 4(21)$      Simplify each side.

$84 = 84$      Multiply.

Since $x = 22$ makes the equation true, the solution set is $\{22\}$.

✔ **Student Check 3**    Solve each equation.

**a.** $-3t + 5 + 4t = 0$            **b.** $0.2y + 6 - 3 + 0.8y = 0$

**c.** $5t - 1 = 4t + 4$            **d.** $2(x - 5) - (x + 3) = 5$

**e.** $5(x + 2) = 4(x - 1)$

## Applications

**Objective 4** ▶

Solve applications of linear equations.

The key to solving applications is setting up the equation correctly. In Section 2.1, we practiced setting up the equation but not solving them. Feel free to go back and review this material as needed. The only difference in what we are doing in this section is that we will solve the equation we set up.

**Objective 4 Examples**    **For each problem: (1) assign a variable to the unknown, (2) write an equation that represents the situation, (3) solve the equation, and (4) answer the question using complete sentences.**

**4a.** Four less than twice a number has the same result as eight more than a number. Find the number.

**Solution**    **4a.** What is the unknown? The number is unknown. Let $n$ be a number.

What is known? Four less than twice a number has the same result as eight more than a number.

From the given information, we can write the equation.

Four less        has the        eight
than twice   same result   more than
a number         as         a number.

$$2n - 4 = n + 8$$
$$2n - 4 - n = n + 8 - n \qquad \text{Subtract } n \text{ from each side.}$$
$$n - 4 = 8 \qquad \text{Simplify.}$$
$$n - 4 + 4 = 8 + 4 \qquad \text{Add 4 to each side.}$$
$$n = 12 \qquad \text{Simplify.}$$

**Check:**
$$2n - 4 = n + 8 \qquad \text{Original equation}$$
$$2(12) - 4 = 12 + 8 \qquad \text{Replace } n \text{ with 12.}$$
$$24 - 4 = 20 \qquad \text{Simplify each side.}$$
$$20 = 20 \qquad \text{Simplify.}$$

The solution, $n = 12$, checks in the original equation. So, the number that satisfies the given relationship is 12.

**4b.** The selling price of a new car is $24,891. The selling price is the base price plus fees and options. The base price is $22,985. What is the total of the fees and options?

**Solution**   **4b.** What is unknown? The total fees and options are unknown. Let $x$ represent the fees and options.

What is known?

• The selling price is the base price plus fees and options.
• The selling price is $24,891.
• The base price is $22,985.

From the given information, we can write the equation.

base price plus
Selling price   is   fees and options.

$$24,891 = 22,985 + x$$
$$24,891 - 22,985 = 22,985 + x - 22,985 \qquad \text{Subtract 22,985 from each side.}$$
$$1906 = x \qquad \text{Simplify.}$$

The total of the fees and options is $1906. We can check by replacing $x$ with 1906. With application problems, we also want to check to make sure the answer is reasonable. It wouldn't make sense to have a negative value for this answer nor would it make sense for the fees and options to be more than the selling price.

✓ **Student Check 4**

For each problem: (1) assign a variable to the unknown, (2) write an equation that represents the situation, (3) solve the equation, and (4) answer the question using complete sentences.

a. Twice the sum of a number and 3 is the same as 4 less than the number. Find the number.

b. The tuition for 15 credit hours for a new in-state student at Georgia Perimeter College, a community college in Atlanta, Georgia, is $1235 per semester for students entering in Fall 2012. Tuition plus mandatory fees is $1710 per semester. How much are the fees the student pays each semester? (Source: http://www.usg.edu/fiscal_affairs/documents/tuition_and_fees/FY2012_Tuition_Rates.pdf)

**Objective 5** ►

Troubleshoot common errors.

## Troubleshooting Common Errors

Some common errors associated with the addition property of equality are shown next.

**Objective 5 Examples**   A problem and an incorrect solution are given. Provide the correct solution and an explanation of the error.

**5a.** Solve $4 - x = 6$.

| Incorrect Solution | Correct Solution and Explanation |
|---|---|
| $4 - x = 6$ <br> $4 - x - 4 = 6 - 4$ <br> $x = 2$ <br> The solution set is $\{2\}$. | The first step is correct in that we should subtract 4 from both sides. The resulting equation is incorrect. It should be <br><br> $4 - x = 6$ <br> $4 - x - 4 = 6 - 4$ <br> $-x = 2$ <br><br> This tells us that the opposite of $x$ is 2. Therefore, $x$ must be the opposite of 2, which is $-2$. The solution set is $\{-2\}$. <br><br> The equation $-x = 2$ could also be solved by adding $x$ to both sides and subtracting 2 from each side. <br><br> $-x = 2$ <br> $-x + x = 2 + x$ <br> $0 = x + 2$ <br> $0 - 2 = x + 2 - 2$ <br> $-2 = x$ |

**5b.** Solve $3(x - 2) + 4 = 2x - 5$.

| Incorrect Solution | Correct Solution and Explanation |
|---|---|
| $3(x - 2) + 4 = 2x - 5$ <br> $3x - 2 + 4 = 2x - 5$ <br> $3x + 2 = 2x - 5$ <br> $3x + 2 - 2x = 2x - 5 - 2x$ <br> $x + 2 = -5$ <br> $x + 2 - 2 = -5 - 2$ <br> $x = -7$ <br> The solution set is $\{-7\}$. | Three was not distributed to the second term in the parentheses. <br><br> $3(x - 2) + 4 = 2x - 5$ <br> $3x - 6 + 4 = 2x - 5$ <br> $3x - 2 = 2x - 5$ <br> $3x - 2 - 2x = 2x - 5 - 2x$ <br> $x - 2 = -5$ <br> $x - 2 + 2 = -5 + 2$ <br> $x = -3$ <br><br> The solution set is $\{-3\}$. |

## ANSWERS TO STUDENT CHECKS

**Student Check 1**   a. linear equation   b. linear equation
c. not linear

**Student Check 2**   a. $\{3\}$   b. $\{-19\}$   c. $\{2\}$
d. $\{1.6\}$

**Student Check 3**   a. $\{-5\}$   b. $\{-3\}$   c. $\{5\}$   d. $\{18\}$
e. $\{-14\}$

**Student Check 4**   a. The number is $-10$.   b. The student fees are $475.

## SUMMARY OF KEY CONCEPTS

1. A linear equation is an equation that can be written in the form $ax + b = c$, where $a$, $b$, and $c$ are real numbers and $a \neq 0$. The largest exponent of the variable term is one.

2. The addition property of equality enables us to add or subtract the same number from both sides of an equation. The key is to isolate the variable to one side of the equation.

3. To solve an equation, first simplify each side of the equation as much as possible. Then get variables on one side of the equation and constants on the other using the addition property of equality.

4. The key to solving applications is in setting up the equation that represents the given situation. Review Section 2.1 to assist you with this process.

## GRAPHING CALCULATOR SKILLS

Solving equations is a critical skill that we need in algebra. At this level, the calculator should be only used to check our work. Two methods were shown in Section 2.1 as to how the calculator can be used to verify solutions. Another method is shown next.

**Example:** Verify that $x = -14$ is a solution of $x + 4 = -10$.

**Solution:** Store the value of $x$ into the calculator. Enter the equation and press enter. If "1" is displayed, the statement is true for the stored value of $x$ and this value is a solution of the equation. If "0" is displayed, the statement is false for the stored value of $x$ and this value is *not* a solution of the equation.

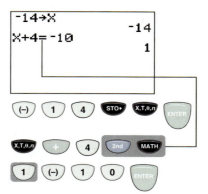

Because a 1 is displayed, $x = -14$ is a solution of $x + 4 = -10$.

---

## SECTION 2.2   EXERCISE SET

 **Write About It!**

**Use complete sentences to explain the meaning of the given term or process.**

1. Linear equation in one variable   Answers vary.

2. The addition property of equality   Answers vary.

3. Solving a linear equation using the addition property of equality   Answers vary.

4. The solution of an equation   Answers vary.

**Determine if each statement is true or false. If a statement is false, explain why.**

5. To isolate the variable term in the equation $7 + x = -4$, add 4 to both sides of the equation.
   False, we must subtract 7 from both sides.

6. To isolate the variable term in the equation $-5x = 2$, add 5 to both sides of the equation.
   False, $-5x + 5 \neq x$.

7. In the equation $-5x = -6x + 4$, the variable term on the right side of the equation can be eliminated by adding $6x$ to both sides.   True

8. In the equation $7x = 9x - 14$, the variable term on the right side of the equation can be eliminated by adding $9x$ to both sides.   False, we must subtract $9x$ from both sides.

 **Practice Makes Perfect!**

**Decide if each equation is linear. If the equation is not linear, explain why. (*See Objective 1.*)**

9. $2x - 5 = 0$
   linear equation

10. $3(4x + 5) - 7 = 2(x + 1)$
    linear equation

11. $2x^2 - x + 1 = 3$
    not a linear equation because the largest exponent is 2

12. $-4x^2 + 3x + 5 = 0$
    not a linear equation because the largest exponent is 2

Additional answers can be found in the Instructor Answer Appendix.

13. $\frac{1}{5}x - 3 + \frac{4}{5}x = 7$
    linear equation

14. $\frac{1}{2}x - 5 = -\frac{1}{2}x + 2$
    linear equation

**Use the addition property of equality to solve each equation. Remember to check each solution. (*See Objective 2.*)**

15. $x + 3 = -5$   {−8}

16. $y + 8 = -1$   {−9}

17. $b - 7 = -3$   {4}

18. $a - 1 = -10$   {−9}

19. $x - 3 = 3$   {6}

20. $x + 12 = 12$   {0}

21. $-3 = x - 4$   {1}

22. $10 = 5 + x$   {5}

23. $x + \frac{6}{7} = -\frac{1}{7}$   {−1}

24. $x - \frac{3}{4} = \frac{5}{4}$   {2}

25. $y + 2.5 = .5$   {−2}

26. $b - 5.4 = 5$   {10.4}

27. $6.5 + c = 2.4$   {−4.1}

28. $-4.2 + a = -7.3$   {−3.1}

**Solve each equation. (*See Objective 3.*)**

29. $5 - x + 2x - 4 = 7$   {6}

30. $-3y + 7 + 4y - 10 = -3$   {0}

31. $-9 - 6y - 7 + 7y = 10 - 13$   {13}

32. $8b + 11 - 7b - 12 = -7 - 5$   {−11}

33. $3x + 6 = 2x - 5$   {−11}

34. $5x - 7 = 4x - 6$   {1}

35. $-2x - 3 = -3x + 4$   {7}

36. $-5y + 1 = -6y + 2$   {1}

37. $\frac{4}{3}x - 5 = \frac{1}{3}x + 2$   {7}

38. $\frac{7}{6}x - 7 = \frac{1}{6}x + 2$   {9}

39. $1.5a + 4 = 0.5a + 2$   {−2}

40. $2.4y + 3 = 1.4y - 7$   {−10}

41. $4(2x + 6) = 7x - 4$   {−28}

42. $3(4x - 5) = 11x + 20$   {35}

43. $3(2x - 1) - 5(x + 3) = 0$   {18}

44. $7(4x + 2) - 9(3x - 1) = 0$   {−23}

45. $\frac{6}{5}(x - 10) = \frac{1}{5}(x + 5)$   {13}

46. $\frac{5}{4}(y - 8) = \frac{1}{4}(y + 8)$   {12}

**For each problem: (1) assign a variable to the unknown, (2) write an equation that represents the situation, (3) solve the equation, and (4) answer the question using complete sentences. (See Objective 4.)**

47. The sum of twice a number and 4 is equal to 19 more than the number. Find the number.    The number is 15.

48. Five times the sum of a number and 6 is the same as six times the number.    The number is 30.

49. The difference of twice a number and −7 is equal to 38 less than the number.    The number is −45.

50. Five times the difference of a number and −9 is the same as six times the number.    The number is 45.

51. Many jewelers mark up the price of engagement rings by as much as 300%! One jeweler buys a ring for $1250 but sells it for $5000. (The selling price is the cost to the jeweler plus the markup.) What is the markup that this jeweler adds to their cost of the ring?

The markup is $3750.

52. Many college bookstores mark up the price of a textbook by 30%. If a math textbook costs the bookstore $80, but the bookstore sells it for $104, what is the markup amount?    The markup is $24.

53. The price of a roundtrip airline ticket from Atlanta, Georgia, to San Francisco, California, is $576.10. This price includes the cost of the ticket plus taxes and fees. What are the taxes and fees if the cost of the ticket is $497.21?    The taxes and fees are $78.89.

54. Many mortgage companies set up escrow accounts to pay a homeowner's property taxes and insurance. With such a setup, a person's mortgage payment includes the home loan payment plus the escrow payment. If Bailey's monthly mortgage payment is $1210.80 and $199.61 of it is his escrow payment, what is Bailey's monthly home loan payment?
The home loan payment is $1011.19.

55. A department store sells a stand mixer for $299.99. This price includes a $50 discount. What is the original price of the mixer?
The original price of the mixer is $349.99.

56. A department store sells a gourmet single cup home brewing system for $105.96. This price includes a $6 sales tax. What is the price of the brewing system before taxes?
The price of the brewing system before taxes is $99.96.

57. The 2010 Forbes 400 Richest Americans list included Facebook founder, Mark Zuckerberg, and Facebook co-founder, Dustin Moskovitz. Their combined net worth was $8.3 billion. Zuckerberg's net worth was $5.5 billion more than Moskovitz's. What was the net worth of each person? (Source: http://www.forbes.com/lists/)

58. The 2010 Forbes 400 Richest Americans list included Apple founder, Steve Jobs, and Dell founder, Michael Dell. Their combined net worth was $20.1 billion. Jobs's net worth was $7.9 billion less than Dell's. What was the net worth of each person? (Source: http://www.forbes.com/lists/)
Dell's net worth was $14 billion and Jobs's net worth was $6.1 billion.

**Mix 'Em Up!**

**Solve each equations.**

59. $3x + 1 = 2x + 1$    {0}

60. $7x + 7 = 6x − 12$    {−19}

61. $3 − x = 11 − 2x$    {8}

62. $10 − 9x = 14 − 10x$    {4}

63. $\frac{3}{4}x + 1 = 6 − \frac{1}{4}x$    {5}

64. $\frac{5}{8}x − 8 = 3 − \frac{3}{8}x$    {11}

65. $\frac{2}{7}x − 1 = 5 + \frac{9}{7}x$    {−6}

66. $−\frac{7}{12}x + 2 = \frac{5}{12}x − 6$    {8}

67. $3(x − 1) + 5 = 4(x − 2) − 3$    {13}

68. $10(2x + 3) − 16 = 19x − 15$    {−29}

69. $5.68x + 1.02 = 4.68x + 3.88$    {2.86}

70. $4.19x + 3.64 = 3.19x − 1.45$    {−5.09}

71. $5.1x − 1.8 + 2.3x = 6.4x − 2.7$    {−0.9}

72. $4.1x − 2.1 + 0.4x = 3.5x − 1$    {1.1}

73. $14 − 11x = 14 − 12x$    {0}

74. $1.2 − 2.5x = 1.2 − 1.5x$    {0}

**For each problem: (1) assign a variable to the unknown, (2) write an equation that represents the problem situation, (3) solve the equation, and (4) answer the question asked using complete sentences.**

75. The difference of a number and 3 is −10. Find the number.
The number is −7.

76. The difference of a number and 10 is −55. Find the number.    The number is −45.

77. Four less than a number is −7. Find the number.
The number is −3.

78. Eight less than a number is −19. Find the number.
The number is −11.

79. Twice the sum of a number and 4 is equal to one more than the number. Find the number.    The number is −7.

80. Three times the difference of a number and 8 equals twice the number. Find the number.    The number is 24.

81. One angle is 12° less than another angle. The angles are supplementary. Find the measure of each angle.
The measures of the angles are 96° and 84°.

82. One angle is 36° more than another angle. The angles are complementary. Find the measure of each angle.
The measures of the angles are 27° and 63°.

83. In 2010, Forbes Celebrity 100 list included the two highest paid female athletes, Serena Williams and Maria Sharapova. Their combined income was $45 million. Williams earned $5 million less than Sharapova. How much did each of them earn? (Source: http://www.forbes.com/lists/)
Williams earned $20 million and Sharapova earned $25 million.

84. In 2010, Forbes Celebrity 100 list included the two of the highest paid teen stars, Taylor Swift and Miley Cyrus. Their combined income was $93 million. Cyrus earned $3 million more than Swift. How much did each of them earn? (Source: http://www.forbes.com/lists/)
Swift earned $45 million and Cyrus earned $48 million.

## You Be the Teacher!

**Correct each student's errors, if any.**

**85.** Solve: $2(x - 1) + 3 = x - 4$

Blanche's work:

$$2(x - 1) + 3 = x - 4$$
$$2x - 1 + 3 = x - 4$$
$$2x + 2 = x - 4$$
$$\underline{\phantom{2x} -2 \qquad -2}$$
$$2x \phantom{+2} = x - 6$$
$$\underline{-x \qquad\quad -x}$$
$$x = -6$$

$$2(x - 1) + 3 = x - 4$$
$$2x - 2 + 3 = x - 4$$
$$2x + 1 = x - 4$$
$$x + 1 = -4$$
$$x = -5$$

**86.** Solve: $5(x - 13) + 3 = 4x - 14$

Darcy's work:

$$5(x - 13) + 3 = 4x - 14$$
$$5x - 65 + 3 = 4x - 14$$
$$5x - 62 = 4x - 14$$
$$\underline{\phantom{5x} -62 \qquad -62}$$
$$5x \phantom{+2} = 4x - 76$$
$$\underline{-4x \qquad\quad -4x}$$
$$x = -76$$

$$5(x - 13) + 3 = 4x - 14$$
$$5x - 65 + 3 = 4x - 14$$
$$5x - 62 = 4x - 14$$
$$x - 62 = -14$$
$$x = 48$$

**87.** Solve:

$$\frac{4}{5}x + 5 + \frac{1}{5}x = 12$$

Charlie's work:

$$\frac{4}{5}x + 5 + \frac{1}{5}x = 12$$

$$\frac{4}{5}x + \frac{1}{5}x + 5 = 12$$

$$\frac{5}{5}x + 5 = 12$$

$$x + 5 = 12$$

$$\underline{\phantom{x} -5 \quad -5}$$

$$x = 7 \quad \text{correct}$$

**88.** Solve:

$$\frac{1}{6}x - 4 = 10 - \frac{5}{6}x$$

Wanda's work:

$$\frac{1}{6}x - 4 = 10 - \frac{5}{6}x$$

$$\underline{\phantom{xx} +4 \quad +4}$$

$$\frac{1}{6}x \phantom{-4} = 14 - \frac{5}{6}x$$

$$\underline{+\frac{5}{6}x \qquad\quad +\frac{5}{6}x}$$

$$\frac{6}{6}x = 14$$

$$x = 14 \quad \text{correct}$$

## Calculate It!

**Solve each equation. Then use a graphing calculator to check the solution.**

**89.** $6(x - 1) + 3 = 5x - 9$  $\{-6\}$

**90.** $3 + 2(x - 1) = x - 3$  $\{-4\}$

**91.** $\dfrac{3}{4}x + 2 = 5 - \dfrac{1}{4}x$  $\{3\}$

**92.** $\dfrac{1}{2}x + 4 + x = \dfrac{1}{2}x - 7$  $\{-11\}$

**93.** $1.2(x - 2.5) - 2.2(x + 10.8) = 4.6$  $\{-31.36\}$

**94.** $2.3(x + 3.1) - 1.3(x - 2.4) = 6.8$  $\{-3.45\}$

## Think About It!

**95.** According to a national retail organization, the average person spent $116.21 on traditional Valentine's Day merchandise in 2011 compared to an average of $103.00 in 2010. Use this information to write a word problem.   Answers vary.

**96.** The following pie chart shows the total number of retailers by type in the state of Georgia in 2010. Use the information to write a word problem that relates two of the quantities. (Source: http://www.nrf.com)   Answers vary.

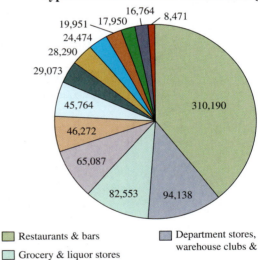

**Types of Retailers in the State of Georgia**

- Restaurants & bars
- Grocery & liquor stores
- Clothing & clothing accessories stores
- Health & personal care stores
- Miscellaneous store retailers
- Sporting goods, hobby, book & music stores
- Catalog, Internet & mail-order retailers
- Department stores, warehouse clubs & super stores
- Motor vehicle & parts dealers
- Building material & garden supply stores
- Gasoline stations
- Furniture & home furnishings stores
- Electronics & appliance stores

## SECTION 2.3 / The Multiplication Property of Equality

### OBJECTIVES

As a result of completing this section, you will be able to

1. Use the multiplication property of equality to solve linear equations.
2. Apply both the addition and multiplication properties of equality to solve linear equations.
3. Solve application problems.
4. Solve consecutive integer problems.
5. Troubleshoot common errors.

To rent a moving truck for one day, a company charges \$44.95 plus \$0.45 per mile. If a family budgets \$300 for a moving truck, how many miles will they be able to drive the truck? To answer this question, we have to solve the equation

$$44.95 + 0.45x = 300,$$

where $x$ is the number of miles driven.

In this section, we will learn how to find the solution of this type of equation where the variable term has a coefficient other than one.

### The Multiplication Property of Equality

The addition property of equality, presented in Section 2.2, only enables us to solve for the variable when its coefficient is 1. We need another property to solve equations which have coefficients other than 1 on the variable. Some examples of equations of this form are

$$3x - 4 = 5 \qquad \frac{2}{3}x = 6$$

Consider how we can solve the equation $2x = 6$ using our reasoning skills.

$2x = 6 \rightarrow$ This equation means 2 times some number is 6. We know that 2 times 3 is 6, so the solution of this equation is $x = 3$.

The equations $2x = 6$ and $x = 3$ are equivalent equations. To derive the first equation from the second equation, we must divide each side of the equation by 2.

$$2x = 6$$
$$\frac{2x}{2} = \frac{6}{2}$$
$$x = 3$$

Note dividing by 2 is the same as multiplying by $\frac{1}{2}$, so we could have also multiplied each side by $\frac{1}{2}$ to obtain the same result. That is,

$$2x = 6$$
$$\frac{1}{2}(2x) = \frac{1}{2}(6)$$
$$x = 3$$

This leads us to another property of solving equations.

---

**Property: Multiplication Property of Equality**

If $a = c$, and $b \neq 0$, then

$$ab = cb$$

and

$$\frac{a}{b} = \frac{c}{b}$$

---

This property tells us that if two expressions are equal or equivalent, we can

- multiply both expressions by the same nonzero number and the expressions will remain equal.
- divide both expressions by the same nonzero number and the expressions will remain equal.

### Objective 1 ▶

Use the multiplication property of equality to solve linear equations.

Example 1 illustrates how we solve an equation that requires us to use only the multiplication property of equality. In the next objective and in Section 2.4, we will solve equations that require both the addition and multiplication properties of equality.

---

**Procedure:  Using the Multiplication Property of Equality to Solve Equations of the Form $ax = b$**

**Step 1:** Determine what operation will isolate the variable on one side of the equation. Remember the inverse property for multiplication:
$$a \cdot \frac{1}{a} = \frac{a}{a} = 1,$$ that is, the product of reciprocals is 1.

**Step 2:** Perform this operation on each side of the equation.

**Step 3:** Simplify each side of the equation, as necessary. The result should be of the form
$$x = \text{some number} \quad \text{or} \quad \text{some number} = x$$

**Step 4:**   Check the solution by substituting the value into the original equation.

**Step 5:**   Write the solution in set notation.

---

**Objective 1  Examples**   **Use the multiplication property of equality to solve each equation.**

**1a.** $3x = -12$          **1b.** $0.05y = 40$          **1c.** $-x = 6$

**1d.** $\dfrac{3}{2}a = 12$          **1e.** $\dfrac{x}{4} = -8$

**Solutions**   **1a.**

$$3x = -12$$

$$\frac{3x}{3} = \frac{-12}{3} \qquad \text{Divide each side by 3.}$$

$$x = -4 \qquad \text{Simplify. Recall } \frac{3x}{3} = \frac{3}{3}x = 1x = x.$$

**Check:**
$$3x = -12 \qquad \text{Original equation}$$
$$3(-4) = -12 \qquad \text{Replace } x \text{ with } -4.$$
$$-12 = -12 \qquad \text{Simplify.}$$

Since $x = -4$ makes the equation true, the solution set is $\{-4\}$.

**1b.**
$$0.05y = 4$$

$$\frac{0.05y}{0.05} = \frac{4}{0.05} \qquad \text{Divide each side by 0.05.}$$

$$y = 80 \qquad \text{Simplify. Recall } \frac{0.05y}{0.05} = \frac{0.05}{0.05}y = 1y = y.$$

**Check:**
$$0.05y = 4 \qquad \text{Original equation}$$
$$0.05(80) = 4 \qquad \text{Replace } y \text{ with } 80.$$
$$4 = 4 \qquad \text{Simplify.}$$

Since $y = 80$ makes the equation true, the solution set is $\{80\}$.

**1c.**
$$-x = 6$$

$$-1x = 6 \qquad \text{Rewrite } -x \text{ as } -1x.$$

$$\frac{-1x}{-1} = \frac{6}{-1} \qquad \text{Divide each side by } -1.$$

$$x = -6 \qquad \text{Simplify.}$$

**Check:**
$$-x = 6 \qquad \text{Original equation}$$
$$-(-6) = 6 \qquad \text{Replace } x \text{ with } -6.$$
$$6 = 6 \qquad \text{Simplify.}$$

Since $x = -6$ makes the equation true, the solution set is $\{-6\}$.

**1d.**      $\dfrac{3}{2}a = 12$

$\dfrac{2}{3}\left(\dfrac{3}{2}a\right) = \dfrac{2}{3}(12)$     Multiply each side by the reciprocal of $\dfrac{3}{2}$, $\dfrac{2}{3}$.

$1a = \dfrac{24}{3}$     Simplify each side.

$a = 8$     Simplify.

**Check:**  $\dfrac{3}{2}a = 12$     Original equation

$\dfrac{3}{2}(8) = 12$     Replace $a$ with 8.

$\dfrac{24}{2} = 12$     Simplify each side.

$12 = 12$     Simplify.

Since $a = 8$ makes the equation true, the solution set is $\{8\}$.

**1e.**      $\dfrac{x}{4} = -8$

$\dfrac{1}{4}x = -8$     Rewrite as $\dfrac{x}{4} - \dfrac{1}{4}x$.

$4\left(\dfrac{1}{4}x\right) = 4(-8)$     Multiply each side by the reciprocal of $\dfrac{1}{4}$, 4.

$x = -32$     Simplify each side.

**Check:**     $\dfrac{x}{4} = -8$     Original equation

$\dfrac{-32}{4} = -8$     Replace $x$ with $-32$.

$-8 = -8$     Simplify.

Since $x = -32$ makes the equation true, the solution set is $\{-32\}$.

✓ **Student Check 1**    Use the multiplication property of equality to solve each equation.

**a.** $2x = -14$    **b.** $0.4x = 24$    **c.** $-x = -8$    **d.** $\dfrac{5}{4}a = -20$    **e.** $\dfrac{y}{3} = -15$

## Solving Linear Equations with the Addition and Multiplication Properties of Equality

**Objective 2** ▶

Apply both the addition and multiplication properties of equality to solve linear equations.

Most linear equations cannot be solved using only the addition property of equality or only the multiplication property of equality; they require the use of both properties. The addition property of equality is used to isolate the variable on one side of an equation and the multiplication property of equality is used to get a coefficient of one on the variable.

Example 2 illustrates the use of both the addition property and multiplication property of equality to solve linear equations.

**Objective 2 Examples**    Solve each equation.

**2a.** $2x + 1 = -5$          **2b.** $4y + 3 = -2y - 9$

**2c.** $5x - 3 - 2x = 8x - 1$          **2d.** $8(2x - 1) = 10x + 7$

**Solutions**   **2a.** We need to isolate the variable term, $2x$, on one side of the equation. One way to do this is to subtract 1 from both sides.

$$2x + 1 = -5$$
$$2x + 1 - 1 = -5 - 1 \qquad \text{Subtract 1 from each side.}$$
$$2x = -6 \qquad \text{Simplify.}$$
$$\frac{2x}{2} = \frac{-6}{2} \qquad \text{Divide each side by 2.}$$
$$x = -3 \qquad \text{Simplify.}$$

**Check:**  $2x + 1 = -5$      Original equation
$$2(-3) + 1 = -5 \qquad \text{Replace } x \text{ with } -3.$$
$$-6 + 1 = -5 \qquad \text{Simplify.}$$
$$-5 = -5 \qquad \text{Simplify.}$$

Since $x = -3$ makes the equation is true, the solution set is $\{-3\}$.

**2b.** We need to remove a variable term from either the left side or the right side of the equation. One way to do this is to add $2y$ to both sides.

$$4y + 3 = -2y - 9$$
$$4y + 3 + 2y = -2y - 9 + 2y \qquad \text{Add } 2y \text{ to each side.}$$
$$6y + 3 = -9 \qquad \text{Simplify.}$$
$$6y + 3 - 3 = -9 - 3 \qquad \text{Subtract 3 from each side.}$$
$$6y = -12 \qquad \text{Simplify.}$$
$$\frac{6y}{6} = \frac{-12}{6} \qquad \text{Divide each side by 6.}$$
$$y = -2 \qquad \text{Simplify.}$$

**Check:**  $4y + 3 = -2y - 9$      Original equation
$$4(-2) + 3 = -2(-2) - 9 \qquad \text{Replace } y \text{ with } -2.$$
$$-8 + 3 = 4 - 9 \qquad \text{Simplify.}$$
$$-5 = -5 \qquad \text{Simplify.}$$

Since $y = -2$ makes the equation true, the solution set is $\{-2\}$.

> **Note:** *We would have obtained the same solution of this equation if we had subtracted 4y from both sides.*
>
> $$4y + 3 = -2y - 9$$
> $$4y + 3 - 4y = -2y - 9 - 4y \qquad \text{Subtract } 4y \text{ from each side.}$$
> $$3 = -6y - 9 \qquad \text{Simplify.}$$
> $$3 + 9 = -6y - 9 + 9 \qquad \text{Add 9 to each side.}$$
> $$12 = -6y \qquad \text{Simplify.}$$
> $$\frac{12}{-6} = \frac{-6y}{-6} \qquad \text{Divide each side by } -6.$$
> $$-2 = y \qquad \text{Simplify.}$$

**2c.** $5x - 3 - 2x = 8x - 1$
$$3x - 3 = 8x - 1 \qquad \text{Combine like terms on the left.}$$
$$3x - 3 - 3x = 8x - 1 - 3x \qquad \text{Subtract } 3x \text{ from each side.}$$
$$-3 = 5x - 1 \qquad \text{Simplify.}$$
$$-3 + 1 = 5x - 1 + 1 \qquad \text{Add 1 to each side.}$$
$$-2 = 5x \qquad \text{Simplify.}$$
$$\frac{-2}{5} = \frac{5x}{5} \qquad \text{Divide each side by 5.}$$
$$-\frac{2}{5} = x \qquad \text{Simplify.}$$

**Check:**    $5x - 3 - 2x = 8x - 1$    Original equation

$$5\left(-\frac{2}{5}\right) - 3 - 2\left(-\frac{2}{5}\right) = 8\left(-\frac{2}{5}\right) - 1$$    Replace $x$ with $-\frac{2}{5}$.

$$-\frac{10}{5} - 3 + \frac{4}{5} = -\frac{16}{5} - 1$$    Multiply.

$$-\frac{10}{5} - \frac{15}{5} + \frac{4}{5} = -\frac{16}{5} - \frac{5}{5}$$    Convert each number to a fraction with the common denominator of 5.

$$-\frac{21}{5} = -\frac{21}{5}$$    Add.

Since $x = -\frac{2}{5}$ makes the equation true, the solution set is $\left\{-\frac{2}{5}\right\}$.

**2d.**

$$8(2x - 1) = 10x + 7$$

$$16x - 8 = 10x + 7$$    Apply the distributive property.

$$16x - 8 - 10x = 10x + 7 - 10x$$    Subtract 10x from each side.

$$6x - 8 = 7$$    Simplify.

$$6x - 8 + 8 = 7 + 8$$    Add 8 to each side.

$$6x = 15$$    Simplify.

$$\frac{6x}{6} = \frac{15}{6}$$    Divide each side by 6.

$$x = \frac{5}{2}$$    Simplify. Recall $\frac{15}{6} = \frac{5}{2}$.

**Check:**    $8(2x - 1) = 10x + 7$    Original equation

$$8\left[2\left(\frac{5}{2}\right) - 1\right] = 10\left(\frac{5}{2}\right) + 7$$    Replace $x$ with $\frac{5}{2}$.

$$8(5 - 1) = 25 + 7$$    Multiply.

$$8(4) = 32$$    Subtract.

$$32 = 32$$    Simplify.

Since $x = \frac{5}{2}$ makes the equation true, the solution set is $\left\{\frac{5}{2}\right\}$.

---

✔ **Student Check 2**    Solve each equation.

**a.** $-3x + 2 = -7$    **b.** $2y - 6 = -9y + 1$

**c.** $7x + 4 + 5x = 14x + 3$    **d.** $6(x - 2) = 9x + 8$

---

## Applications

**Objective 3** ▶

Solve application problems.

As we have already noted, expressing relationships mathematically is a key component to successfully solving word problems. While there may be some problems that we can solve by reasoning, it is important that we be able to justify how we came up with our answer and why we know it is valid. The ability to speak, think, and write mathematically equips us with the tools needed to provide such justification. The more we practice the skills at this level, the easier it is for us to extend them to more difficult concepts.

**Procedure:  Solving Word Problems**

**Step 1:**  Read the problem and determine what is unknown. Assign a variable for the unknown quantity.

**Step 2:** Read the problem and determine what is given. Use this information to write an equation that models the situation.

**Step 3:** Solve the equation using the addition and multiplication properties of equality.

**Step 4:** Verify that the solution is reasonable.

**Step 5:** Write the answer to the problem and attach appropriate units, if needed.

**Objective 3 Examples** | **Write an equation that represents each situation. Solve the equation and explain the answer using a complete sentence.**

**3a.** One less than three times a number is $-13$. Find the number.

Solution **3a.** What is known? The number is unknown. Let $n$ represent the number.

What is known? One less than three times a number is $-13$.
So, the equation is

$$3n - 1 = -13$$

We solve the equation to find the number.

$$3n - 1 = -13$$
$$3n - 1 + 1 = -13 + 1 \qquad \text{Add 1 to each side.}$$
$$3n = -12 \qquad \text{Simplify.}$$
$$\frac{3n}{3} = \frac{-12}{3} \qquad \text{Divide each side by 3.}$$
$$n = -4 \qquad \text{Simplify.}$$

The number is $-4$.

**3b.** How many quarters does it take to make $20.00?

Solution **3b.** What is unknown? The number of quarters is unknown. Let $q$ represent the number of quarters.

**INSTRUCTOR NOTE:**
For the word problems, be sure to remind students to check to see that their answer is reasonable and makes sense in the context of the problem.

What is known? The total value of the quarters is $20.00. To find the value of an unknown number of quarters, we use our knowledge of how we calculate the value of a specific number of quarters.

| Number of Quarters | Total Value |
|--------------------|-------------|
| 1 | $(0.25)(1) = \$0.25$ |
| 2 | $(0.25)(2) = \$0.50$ |
| 3 | $(0.25)(3) = \$0.75$ |
| $q$ | $(0.25)(q) = 0.25q$ |

The total value of a collection of quarters is the value of a single quarter times the number of quarters.

So, to determine how many quarters we need to make $20.00, we can use the equation

$$0.25q = 20$$

We solve the equation to find the number of quarters.

$$0.25q = 20$$
$$\frac{0.25q}{0.25} = \frac{20}{0.25} \qquad \text{Divide each side by 0.25.}$$
$$q = 80 \qquad \text{Simplify.}$$

So, it takes 80 quarters to make $20.00.

**3c.** To rent a moving truck for one day, a company charges $44.95 plus $0.45 per mile. If a family budgets $300 for a moving truck, how many miles will they be able to drive the truck?

**Solution**

**3c.** What is unknown? The number of miles the truck is driven is unknown. Let $m$ represent the miles driven.

What is known? The total cost of renting the truck is $44.95 plus $0.45 per mile and the total budgeted is $300.

To find the cost of renting the truck, we calculate the cost of renting the truck for a specific number of miles and use this to determine an expression for the cost of driving the truck $m$ miles.

| Number of Miles | Cost of Renting |
|:---:|:---|
| 100 | $44.95 + 0.45(100) = \$\ 89.95$ |
| 250 | $44.95 + 0.45(250) =\ \ 157.45$ |
| 400 | $44.95 + 0.45(400) =\ \ 224.95$ |
| $m$ | $44.95 + 0.45m$ |

The cost of renting the truck is the flat fee of $44.95 plus a mileage fee. We multiply the miles driven by $0.45 and add this to the flat fee.

The equation comes from setting the cost of renting equal to the amount budgeted, $300.

$$44.95 + 0.45x = 300$$

Solving the equation enables us to determine how many miles the truck can be driven.

$$44.95 + 0.45x = 300$$
$$44.95 + 0.45x - 44.95 = 300 - 44.95 \qquad \text{Subtract 44.95 from each side.}$$
$$0.45x = 255.05 \qquad \text{Simplify.}$$
$$\frac{0.45x}{0.45} = \frac{255.05}{0.45} \qquad \text{Divide each side by 0.45.}$$
$$x = 566.78 \qquad \text{Simplify.}$$

So, the family can drive approximately 567 miles for a cost of $300.

✓ **Student Check 3**   Write an equation that represents each situation. Solve the equation and explain the answer using a complete sentence.

   **a.** Four more than three times a number is −5. Find the number.

   **b.** How many nickels does it take to make $30?

   **c.** The cost to rent a car for one day is $13.99 plus $0.20 per mile. How many miles can be driven for $50? Round to the nearest mile.

## Consecutive Integer Problems

**Objective 4** ▶

Solve consecutive integer problems.

Consecutive means "successive" or "following one after another without interruption." So, **consecutive integers** are integers that follow one another. Some examples are listed in the table.

| | Specific Example | Variable Representation |
|:---|:---:|:---:|
| **Consecutive integers** | 4　　5　　6　(+1)　(+1) | $x, x + 1, x + 2, \ldots$ |
| **Consecutive even integers** | 10　　12　　14　(+2)　(+2) | $x, x + 2, x + 4, \ldots$ |
| **Consecutive odd integers** | 11　　13　　15　(+2)　(+2) | $x, x + 2, x + 4, \ldots$ |

Note that **consecutive odd** and **consecutive even integers** are represented in the same way since the difference between these types of integers is two units. The value of $x$ determines the numbers that will be generated from the expressions $x + 2$ and $x + 4$.

If $x = 15$, then $x + 2 = 15 + 2 = 17$ and $x + 4 = 15 + 4 = 19$. So, the numbers generated are 15, 17, and 19.

If $x = 22$, then $x + 2 = 22 + 2 = 24$ and $x + 4 = 22 + 4 = 26$. So, the numbers generated are 22, 24, and 26.

---

**Objective 4 Examples**   Write an equation that represents each problem. Solve the equation and explain the answer a complete sentence.

**4a.** The sum of two consecutive integers is $-21$. Find the integers.

**4b.** Three times the smallest of three consecutive odd integers is 17 more than the sum of the other two integers. Find the integers.

**Solutions**   **4a.** What is unknown? Two consecutive integers are unknown. Let $x$ and $x + 1$ represent the two integers.

What is known? The sum of the two integers is $-21$.

We use this statement to write the equation.

The sum of the
two integers is $-21$.

| | |
|---|---|
| $x + (x + 1) = -21$ | Express the relationship. |
| $2x + 1 = -21$ | Combine like terms. |
| $2x + 1 - 1 = -21 - 1$ | Subtract 1 from each side. |
| $2x = -22$ | Simplify. |
| $\dfrac{2x}{2} = \dfrac{-22}{2}$ | Divide each side by 2. |
| $x = -11$ | Simplify. |

Since $x = -11$, the integers are $-11$ and $-11 + 1 = -10$.

**4b.** What is unknown? Three consecutive odd integers are unknown. Let $x$, $x + 2$, and $x + 4$ represent the three odd integers.

What is known? Three times the smallest of three consecutive odd integers is 17 more than the sum of the other two integers.

We use this statement to obtain the equation.

Three times the
smallest of three      17 more than the
consecutive odd        sum of the other
integers is            two integers

| | |
|---|---|
| $3(x) = (x + 2) + (x + 4) + 17$ | Express the relationship. |
| $3x = 2x + 23$ | Combine like terms. |
| $3x - 2x = 2x + 23 - 2x$ | Subtract $2x$ from each side. |
| $x = 23$ | Simplify. |

Since $x = 23$, the integers are 23, $23 + 2 = 25$, and $23 + 4 = 27$.

---

**✓ Student Check 4**   Write an equation that represents each problem. Solve the equation and explain the answer using a complete sentence.

**a.** The sum of two consecutive integers is $-137$. Find the integers.

**b.** Five times the smallest of three consecutive even integers is two more than twice the sum of the other two integers. Find the integers.

### Troubleshooting Common Errors

**Objective 5** ▶

Troubleshoot common errors.

Some common errors associated with solving linear equations are shown next. It is very important to be able to distinguish when the addition property of equality applies and when the multiplication property of equality applies. We use the addition property of equality to remove terms from one side of the equation. The multiplication property of equality is used to obtain a coefficient of one on the variable.

**Objective 5 Examples**    A problem and an incorrect solution are given. Provide the correct solution and an explanation of the error.

**5a.** Solve $-2x = 30$.

| Incorrect Solution | Correct Solution and Explanation |
|---|---|
| $-2x = 30$ <br> $-2x + 2 = 30 + 2$ <br> $x = 32$ <br> The solution set is $\{32\}$. | The expression $-2x + 2$ cannot be combined since they are not like terms. So, adding 2 to both sides will not isolate $x$. Note that $-2$ is the coefficient of $x$. So, we divide both sides by $-2$ to obtain a coefficient of 1 on the variable. $$\frac{-2x}{-2} = \frac{30}{-2}$$ $$x = -15$$ The solution set is $\{-15\}$. |

**5b.** Solve $2x + 8 = 4x + 10$.

| Incorrect Solution | Correct Solution and Explanation |
|---|---|
| $2x + 8 = 4x + 10$ <br> $4x - 2x + 8 = 4x - 4x + 10$ <br> $2x + 8 = 10$ <br> $2x + 8 - 8 = 10 - 8$ <br> $2x = 2$ <br> $x = 1$ <br> The solution set is $\{1\}$. | The first step of subtracting $4x$ from both sides was applied incorrectly on the left side of the equation. We should subtract $4x$ at the end of the terms on the left, not insert it at the beginning. $$2x + 8 = 4x + 10$$ $$2x + 8 - 4x = 4x - 4x + 10$$ $$-2x + 8 = 10$$ $$-2x + 8 - 8 = 10 - 8$$ $$-2x = 2$$ $$\frac{-2x}{-2} = \frac{2}{-2}$$ $$x = -1$$ The solution set is $\{-1\}$. |

## ANSWERS TO STUDENT CHECKS

**Student Check 1**   **a.** $\{-7\}$   **b.** $\{60\}$   **c.** $\{8\}$   **d.** $\{-16\}$
   **e.** $\{-45\}$

**Student Check 2**   **a.** $\{3\}$   **b.** $\left\{\dfrac{7}{11}\right\}$   **c.** $\left\{\dfrac{1}{2}\right\}$

   **d.** $\left\{-\dfrac{20}{3}\right\}$

**Student Check 3**   **a.** The number is $-3$.
   **b.** It takes 600 nickels to make $30.
   **c.** The car can be driven for 180 miles for a cost of $50.

**Student Check 4**   **a.** The integers are $-69$ and $-68$.
   **b.** The integers are 14, 16, and 18.

## SUMMARY OF KEY CONCEPTS

1. The multiplication property of equality enables us to multiply or divide both sides of an equation by the same nonzero number.

2. Most equations are going to require use of both the addition and multiplication properties of equality to solve them. The addition property of equality is used to eliminate terms from one side of the equation. The multiplication property of equality is used to make the coefficient of the variable one.

3. Word problems are solved by expressing the given relationships as a mathematical equation.

4. Consecutive integers are $x, x + 1, x + 2, \ldots$ Consecutive even and odd integers are $x, x + 2, x + 4, \ldots$ The equation used to solve these problems will be a result of direct translation. Refer to the key phrases learned in Chapter 1 and this chapter for assistance.

## GRAPHING CALCULATOR SKILLS

The calculator should only be used to check that an equation was solved correctly at this point.

**Example:** Verify that $x = -\dfrac{2}{5}$ is a solution of $5x - 3 - 2x = 8x - 1$.

**Method 1:** Evaluate each side of the equation at the proposed solution. If each side produces the same number, then the solution is correct.

$$5\left(-\frac{2}{5}\right) - 3 - 2\left(-\frac{2}{5}\right) \stackrel{?}{=} 8\left(-\frac{2}{5}\right) - 1$$

```
5(-2/5)-3-2(-2/5
)
                -4.2
8(-2/5)-1
                -4.2
```

Since each side of the equation produces the same result when evaluated at $x = -\dfrac{2}{5}$, the solution is correct.

**Method 2:** Store the value of $x$ into the calculator. Enter the equation and press enter. If a 1 is displayed, the statement is true for the stored value of $x$. If a 0 is displayed, the statement is false for the stored value of $x$.

```
-2/5→X
                -.4
5X-3-2X=8X-1
                1
```

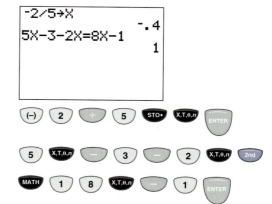

Since 1 is displayed, the solution is correct.

---

## SECTION 2.3 / EXERCISE SET

### Write About It!

Use complete sentences in your answers to the following exercises.

1. Explain the multiplication property of equality.
Answers vary.

2. Give an example of an equation that requires use of the addition property of equality to solve it, but not the multiplication property of equality. Explain your answer. Answers vary.

3. Give an example of an equation that requires use of the multiplication property of equality to solve it, but not the addition property of equality. Explain your answer. Answers vary.

Additional answers can be found in the Instructor Answer Appendix.

4. Give an example of an equation that requires use of both the additional and multiplication properties of equality to solve it. Explain your answer. Answers vary.

**Determine if each statement is true or false. If a statement is false, explain why.**

5. The equation, $-p = 5$, is solved for $p$.
False, we must divide both sides by $-1$ to solve for $p$.

6. To solve the equation $-3x = 6$, add three to both sides. False, we must divide both sides by $-3$ to solve it.

7. The solution of $\dfrac{1}{4}x = 3$ is $x = 12$.    True

8. To make the coefficient of $-\dfrac{3}{5}x$ equal 1, we must

multiply the expression by $\dfrac{3}{5}$. False, we must multiply by the reciprocal of $-\dfrac{3}{5}$, which is $-\dfrac{5}{3}$.

9. To represent consecutive odd integers, use $x$, $x + 1$, $x + 3$, $x + 5$, etc. False, we use $x$, $x + 2$, $x + 4$, etc.

10. To represent consecutive even integers, use $x$, $2x$, $4x$, $6x$, etc. False, we use $x$, $x + 2$, $x + 4$, etc.

## Practice Makes Perfect!

**Solve each equation using the multiplication property of equality. (*See Objective 1.*)**

11. $2x = 10$  {5}
12. $4x = 48$  {12}
13. $5x = -15$  {-3}
14. $8x = -16$  {-2}
15. $-0.3x = 12$  {-40}
16. $-0.7x = 49$  {-70}
17. $-x = 11$  {-11}
18. $-x = 10$  {-10}
19. $-4x = -24$  {6}
20. $-6x = -30$  {5}
21. $\dfrac{x}{5} = 2$  {10}
22. $\dfrac{a}{9} = -5$  {-45}
23. $\dfrac{2}{3}x = 6$  {9}
24. $\dfrac{7}{2}y = 14$  {4}
25. $-\dfrac{5}{6}a = -10$  {12}
26. $-\dfrac{3}{8}b = -27$  {72}
27. $0.25y = 8$  {32}
28. $3.5a = -7$  {-2}

**Solve each equation. (*See Objective 2.*)**

29. $6x + 2 = 14$  {2}
30. $4x + 5 = 21$  {4}
31. $5x - 10 = 20$  {6}
32. $7y - 6 = 36$  {6}
33. $-2a + 4 = -20$  {12}
34. $-9y + 6 = 6$  {0}
35. $-1.4x + 3.7 = 5.38$  {-1.2}
36. $0.2b - 2.8 = -1.18$ {8.1}
37. $3x + 7 = x - 9$  {-8}
38. $11y - 8 = 8y - 5$  {1}
39. $-5a - 6 = -2a + 3$  {-3}
40. $-4x - 3 = 2x + 9$  {-2}
41. $2x + 3 - 4x = 7x - 8$ $\left\{\dfrac{11}{9}\right\}$
42. $8a + 1 - 4a = 13 + 7a$ {-4}
43. $3(2z + 6) = 4(z + 5)$  {1}
44. $5(4x - 1) = 2(3x + 5)$ $\left\{\dfrac{15}{14}\right\}$
45. $-4.5x + 10.5 = -3.1x + 11.06$  {-0.4}
46. $-0.1b + 7 = 3.3b - 1.84$  {2.6}
47. $0.3(1.2z + 5.2) = -0.4(2.5z - 9.0)$  {1.5}
48. $-0.6(1.5x - 4.2) = 0.5(-x + 6.96)$  {-2.4}

**Write an equation that represents each situation. Solve the equation and explain the answer using a complete sentence. (*See Objective 3.*)**

49. The cost to rent a luxury car for one day is \$75 plus \$0.25 per mile. This daily cost is represented by the expression $75 + 0.25x$, where $x$ is the number of miles driven. How many miles has the car been driven if the cost of the rental car is \$112.50? The car was driven for 150 mi.

50. The cost to rent a banquet facility for a wedding reception is \$1500 plus \$35 per person for the food. This is represented by the expression $1500 + 35x$, where $x$ is the number of attendees. If a couple budgets \$5000 for their reception, how many people can attend the reception? 100 can attend.
51. How many nickels does it take to make \$8.75? 175 nickels
52. How many nickels does it take to make \$6.15? 123 nickels
53. How many quarters are needed to make \$19.50? 78 quarters
54. How many quarters are needed to make \$12.25? 49 quarters

**Write an equation that represents each situation. Solve the equation and explain the answer using a complete sentence. (*See Objective 4.*)**

55. The sum of two consecutive even integers is $-78$. Find the integers.  $-40$ and $-38$
56. The sum of three consecutive odd integers is $-309$. Find the integers.  $-105, -103, -101$
57. Four times the largest of three consecutive integers is 47 more than the sum of the other two integers. Find the integers.  20, 21, 22
58. The sum of the two smallest of three consecutive even integers is the same as three times the largest integer. Find the integers.  $-10, -8, -6$
59. The sum of the two largest of three consecutive odd integers is the same as 5 more than the smallest integer. Find the integers.  $-1, 1, 3$
60. The lengths of the sides of a triangle are consecutive integers. If the perimeter of the triangle is 12 ft, what are the lengths of the three sides of the triangle? 3 ft, 4 ft, 5 ft

## Mix 'Em Up!

**Solve each equation.**

61. $5x = 10$  {2}
62. $2x + 5 = 11$  {3}
63. $4y - 1 = 15$  {4}
64. $12y + 3 = 51$  {4}
65. $-4x + 16 = 20$  {-1}
66. $-3x + 15 = 21$  {-2}
67. $\dfrac{a}{4} = -2$  {-8}
68. $-\dfrac{3a}{2} = \dfrac{1}{2}$ $\left\{-\dfrac{1}{3}\right\}$
69. $10z + 3 - 5z = 12 - 4z$  {1}
70. $-3z + 1 - 2z = 10 + 3z + 7$  {-2}
71. $-7x + 2 - 5x = 12 + 4x - 6$ $\left\{-\dfrac{1}{4}\right\}$
72. $4y - 9 - y = 14 - 7y - 8$ $\left\{\dfrac{3}{2}\right\}$
73. $-\dfrac{1}{3}(x + 3) + 2 = x + 1$ {0}
74. $\dfrac{1}{4}(x - 8) = -2x + 3$
75. $-1.2b + 4 = 22$  {-15}
76. $-2.1b + 1.1 = 7.4$ {-3}
77. $-0.7y + 13.5 = 7y + 31.98$  {-2.4}
78. $0.2a - 8.4 = 5.9a + 22.38$  {-5.4}
79. $3.9(1.6x - 0.6) = -0.4(4.2x - 2.07)$  {0.4}
80. $-2(-0.7x + 3) = -0.5(0.2x + 16.8)$  {-1.6}
81. $8(x - 3) = 2x - 42$  {-3}
82. $-5(x + 11) = 3x + 1$ {-7}
83. $-6(2x - 5) = 4x - 2$  {2}
84. $7(4x - 9) = 6x + 3$ {3}

**Write an equation that represents each situation. Solve the equation and explain the answer in a complete sentence.**

**85.** Twice a number is equal to $-18$. Find the number.   $-9$

**86.** One-half of a number is equal to $-10$. Find the number.   $-20$

**87.** Three-fifths of a number is 15. Find the number.   $25$

**88.** Two-sevenths of a number is $-14$. Find the number.   $-49$

**89.** The quotient of a number and 7 is $-21$. Find the number.   $-147$

**90.** The quotient of a number and 3 is 12. Find the number.   $36$

**91.** 30% of a number is 9. Find the number.   $30$

**92.** 15% of a number is 1.5. Find the number.   $10$

**93.** Two less than five times a number is $-7$. Find the number.   $-1$

**94.** Four more than twice a number is $-8$. Find the number.   $-6$

**95.** How many dimes will it take to make \$15?   $150$

**96.** How many half-dollars are needed to make \$75?   $150$

**97.** The cost to rent a midsize car for one day is \$63 plus \$0.28 per mile. This daily cost is represented by the expression $63 + 0.28x$, where $x$ is the number of miles driven. How many miles has the car been driven if the cost of the rental car is \$121.80?   The car was driven 210 mi.

**98.** The cost to rent a midsize car for one day is \$56 plus \$0.32 per mile. This daily cost is represented by the expression $56 + 0.32x$, where $x$ is the number of miles driven. How many miles has the car been driven if the cost of the rental car is \$134.40?   The car was driven 245 mi.

**99.** The sum of four consecutive integers is $-154$. Find the integers.   $-40, -39, -38, -37$

**100.** The sum of three consecutive odd integers is 51. Find the integers.   $15, 17, 19$

**101.** Three times the largest of three consecutive even integers is 2 more than the sum of the other two integers. Find the integers.   $-8, -6, -4$

**102.** Three times the largest of three consecutive odd integers is 13 more than the sum of the other two integers. Find the integers.   $3, 5, 7$

 **You Be the Teacher!**

**Correct each student's errors, if any.**

**103.** Solve: $\dfrac{x}{2} + 3 = 7$

Michael's work:

$$\frac{x}{2} + 3 - 3 = 7 - 3$$

$$\frac{x}{2} = 4$$

$$\frac{x}{2} - 2 = 4 - 2$$

$$x = 2$$

$\dfrac{x}{2} + 3 - 3 = 7 - 3$

$2 \cdot \dfrac{x}{2} = 2 \cdot 4$

$x = 8$

**104.** Solve: $-\dfrac{2}{3}x + 3 = x - 2$

Gabriel's work:

$$-\frac{2}{3}x + 3 - 3 = x - 2 - 3$$

$$-\frac{2}{3}x = x - 5$$

$$-\frac{2}{3}x - x = x - 5 - x$$

$$-\frac{2}{3}x - \frac{3}{3}x = -5$$

$$-\frac{5}{3}x = -5$$

$$-\frac{5}{3}x + \frac{5}{3} = -5 + \frac{5}{3}$$

$$x = -\frac{10}{3}$$

$-\dfrac{2}{3}x + 3 = x - 2$

$-\dfrac{2}{3}x + 3 - 3 = x - 2 - 3$

$-\dfrac{2}{3}x = x - 5$

$-\dfrac{2}{3}x - x = x - x - 5$

$-\dfrac{5}{3}x = -5$

$-\dfrac{3}{5} \cdot -\dfrac{5}{3}x = -5 \cdot -\dfrac{3}{5}$

$x = 3$

**105.** Translate: Five less than twice a number equals four times the number plus ten.

Kristen's answer:

Let $x$ be the number. $5 - 2x = 4x + 10$   $2x - 5 = 4x + 10$

**106.** Translate: Half the sum of 4 and a number equals the number subtracted from 6.

Lauren's answer:

Let $x$ be the number. $\dfrac{1}{2}(4 + x) = x - 6$   $\dfrac{1}{2}(4 + x) = 6 - x$

**107.** Three times the largest of three consecutive odd integers is 87 more than the sum of the other two integers. Find the integers.

Cynthia's work: Let $x, x + 1, x + 2$ be the three consecutive odd integers.

$3(x + 2) = 87 + x + (x + 1)$

$3x + 6 = 88 + 2x$

$x = 82$

Let $x, x + 2$, and $x + 4$ be the three consecutive odd integers.

$3(x + 4) = 87 + x + (x + 2)$

$3x + 12 = 89 + 2x$

$x = 77$

The three odd integers are 77, 79, and 81.

**108.** The sum of the two largest of three consecutive even integers is 30 less than four times the smallest integer. Find the integers.

Ron's work: Let $x, x + 2$, and $x + 4$ be the three consecutive even integers.

$(x + 2) + (x + 4) = 30 - 4x$

$2x + 6 = 30 - 6x$

$8x = 24$

$x = 3$

Let $x, x + 2$, and $x + 4$ be the three consecutive odd integers.

$(x + 2) + (x + 4) = 4x - 30$

$2x + 6 = 4x - 30$

$2x = 36$

$x = 18$

The three even integers are 18, 20, and 22.

 **Calculate It!**

**Solve each equation. Then use a graphing calculator to check the solution.**

**109.** $8x + 20 = 2x - 15 + x$   $\{-7\}$

**110.** $2(x + 10) - 5 = 4x + 18$   $\left\{-\dfrac{3}{2}\right\}$

**111.** $\dfrac{5}{6}x - x = 8 - \dfrac{3}{2}x$   $\{6\}$

**112.** $-3.5(-1.6x + 2.4) = 0.8(-2.1x - 41.44)$   $\{-3.4\}$

## SECTION 2.4     More on Solving Linear Equations

Sections 2.2 and 2.3 laid the groundwork for us to solve linear equations. In this section, we will solve linear equations that require a few more steps than what we have encountered so far and we will also explore equations that have special solution sets.

### Linear Equations Containing Fractions

We have solved equations with fractions in earlier sections in which we simply worked with the fractions to obtain the solution. Working with fractions can often be cumbersome, so we can use the multiplication property of equality to first clear fractions from the equation and then solve it. Consider the expression $\frac{3}{4}x - \frac{1}{6}$. The least common denominator (LCD) of the fractions in this expression is 12. If we multiply the expression by 12, we get

**Objective 1** ▶

Solve linear equations with fractions.

$$12\left(\frac{3}{4}x - \frac{1}{6}\right) = \frac{12}{1}\left(\frac{3}{4}x\right) - \frac{12}{1}\left(\frac{1}{6}\right)$$

$$= \frac{36}{4}x - \frac{12}{6}$$

$$= 9x - 2$$

When the expression is multiplied by its LCD, the fractions are cleared and the expression completely changes. But if we perform this operation to both sides of an equation with fractions, the new equation is equivalent to the original equation.

So, if we solve an equation with fractions, we can clear fractions before we apply the steps to solve the equation. These steps are presented next in a general strategy for solving linear equations in one variable. These steps can be applied to solving any linear equation in one variable.

---

**Procedure:  Solving Linear Equations in One Variable—A General Strategy**

**Step 1:** Write an equivalent equation that doesn't contain fractions or decimals, if necessary.
  **a.** Multiply both sides of the equation by the LCD to clear fractions in the equation if they occur.
  **b.** If decimal numbers appear in the equation, multiply both sides of the equation by the power of 10 that eliminates the number with the most decimal places.

**Step 2:** Apply the distributive property to remove any parentheses if they occur.

**Step 3:** Use the addition property of equality to get all the variable terms on one side of the equation and all constants on the other side of the equation.

**Step 4:** Isolate the variable by applying the multiplication property of equality.

**Step 5:** Check the proposed solution by substituting it into the *original* equation.

---

**Objective 1  Examples**     Solve each equation by first clearing fractions. Check each answer.

**1a.** $\dfrac{3x}{4} - 5 = \dfrac{x}{4}$

**1b.** $\dfrac{a}{6} - \dfrac{1}{8} = \dfrac{a}{12}$

**1c.** $-\dfrac{2}{3}(x - 6) = \dfrac{4}{5}(x + 10)$

**1d.** $\dfrac{y - 5}{4} - \dfrac{2y}{9} = \dfrac{1}{6}$

**Solutions**    **1a.**

$$\frac{3x}{4} - 5 = \frac{x}{4}$$

$$4\left(\frac{3x}{4} - 5\right) = 4\left(\frac{x}{4}\right)$$    Multiply each side by the LCD, 4.

$$4\left(\frac{3x}{4}\right) - 4(5) = x$$    Apply the distributive property. Simplify.

$$3x - 20 = x$$    Simplify.

$$3x - 20 - x = x - x$$    Subtract $x$ from each side.

$$2x - 20 = 0$$    Simplify.

$$2x - 20 + 20 = 0 + 20$$    Add 20 to each side.

$$2x = 20$$    Simplify.

$$\frac{2x}{2} = \frac{20}{2}$$    Divide each side by 2.

$$x = 10$$    Simplify.

**Check:**    $$\frac{3x}{4} - 5 = \frac{x}{4}$$    Original equation

$$\frac{3(10)}{4} - 5 = \frac{10}{4}$$    Replace $x$ with 10.

$$\frac{30}{4} - 5 = \frac{10}{4}$$    Simplify.

$$\frac{30}{4} - \frac{20}{4} = \frac{10}{4}$$    Subtract the numbers on the left by converting 5 to $\frac{20}{4}$.

$$\frac{10}{4} = \frac{10}{4}$$    Simplify.

Since $x = 10$ makes the equation a true statement, the solution set is $\{10\}$.

**1b.**

$$\frac{a}{6} - \frac{1}{8} = \frac{a}{12}$$

$$24\left(\frac{a}{6} - \frac{1}{8}\right) = 24\left(\frac{a}{12}\right)$$    Multiply each side by the LCD, 24.

$$24\left(\frac{a}{6}\right) - 24\left(\frac{1}{8}\right) = 2a$$    Apply the distributive property. Simplify.

$$4a - 3 = 2a$$    Simplify.

$$4a - 3 - 2a = 2a - 2a$$    Subtract $2a$ from each side.

$$2a - 3 = 0$$    Simplify.

$$2a - 3 + 3 = 0 + 3$$    Add 3 to each side.

$$2a = 3$$    Simplify.

$$\frac{2a}{2} = \frac{3}{2}$$    Divide each side by 2.

$$a = \frac{3}{2}$$    Simplify.

**Check:**  $\dfrac{a}{6} - \dfrac{1}{8} = \dfrac{a}{12}$    Original equation

$\dfrac{\frac{3}{2}}{6} - \dfrac{1}{8} = \dfrac{\frac{3}{2}}{12}$    Replace $a$ with $\dfrac{3}{2}$.

$\dfrac{3}{12} - \dfrac{1}{8} = \dfrac{3}{24}$    Simplify. Recall: $\dfrac{3}{2} \div 6 = \dfrac{3}{2} \cdot \dfrac{1}{6} = \dfrac{3}{12}$,

$\dfrac{3}{2} \div 12 = \dfrac{3}{2} \cdot \dfrac{1}{12} = \dfrac{3}{24}$

$\dfrac{6}{24} - \dfrac{3}{24} = \dfrac{3}{24}$    Write equivalent fractions with LCD, 24.

$\dfrac{3}{24} = \dfrac{3}{24}$    Simplify.

Since $a = \dfrac{3}{2}$ makes the equation true, the solution set is $\left\{\dfrac{3}{2}\right\}$.

**1c.**    $-\dfrac{2}{3}(x - 6) = \dfrac{4}{5}(x + 10)$

$15\left[-\dfrac{2}{3}(x - 6)\right] = 15\left[\dfrac{4}{5}(x + 10)\right]$    Multiply each side by the LCD, 15.

$\left[15\left(-\dfrac{2}{3}\right)\right](x - 6) = \left[15\left(\dfrac{4}{5}\right)\right](x + 10)$    Apply the associative property.

$-10(x - 6) = 12(x + 10)$    Multiply the coefficients.

$-10x + 60 = 12x + 120$    Apply the distributive property.

$-10x + 60 - 12x = 12x + 120 - 12x$    Subtract $12x$ from each side.

$-22x + 60 = 120$    Simplify.

$-22x + 60 - 60 = 120 - 60$    Subtract 60 from each side.

$-22x = 60$    Simplify.

$\dfrac{-22x}{-22} = \dfrac{60}{-22}$    Divide each side by $-22$.

$x = -\dfrac{30}{11}$    Simplify.

**Check:**    $-\dfrac{2}{3}(x - 6) = \dfrac{4}{5}(x + 10)$    Original equation

$-\dfrac{2}{3}\left(-\dfrac{30}{11} - 6\right) = \dfrac{4}{5}\left(-\dfrac{30}{11} + 10\right)$    Replace $x$ with $-\dfrac{30}{11}$.

$-\dfrac{2}{3}\left(-\dfrac{30}{11}\right) - \left(-\dfrac{2}{3}\right)(6) = \dfrac{4}{5}\left(-\dfrac{30}{11}\right) + \dfrac{4}{5}(10)$    Distribute.

$\dfrac{20}{11} + 4 = -\dfrac{24}{11} + 8$    Simplify.

$\dfrac{20}{11} + \dfrac{44}{11} = -\dfrac{24}{11} + \dfrac{88}{11}$    Add the fractions.

$\dfrac{64}{11} = \dfrac{64}{11}$    Simplify.

Since $x = -\dfrac{30}{11}$ makes the equation true, the solution set is $\left\{-\dfrac{30}{11}\right\}$.

**1d.**
$$\frac{y-5}{4} - \frac{2y}{9} = \frac{1}{6}$$

$$36\left(\frac{y-5}{4} - \frac{2y}{9}\right) = 36\left(\frac{1}{6}\right)$$   Multiply each side by the LCD, 36.

$$36\left(\frac{y-5}{4}\right) - 36\left(\frac{2y}{9}\right) = 6$$   Apply the distributive property. Simplify.

$$9(y-5) - 4(2y) = 6$$   Simplify each product.

$$9y - 45 - 8y = 6$$   Multiply the remaining factors.

$$y - 45 = 6$$   Simplify each side.

$$y - 45 + 45 = 6 + 45$$   Add 45 to each side.

$$y = 51$$   Simplify.

**Check:**
$$\frac{y-5}{4} - \frac{2y}{9} = \frac{1}{6}$$   Original equation

$$\frac{51-5}{4} - \frac{2(51)}{9} = \frac{1}{6}$$   Replace $y$ with 51.

$$\frac{46}{4} - \frac{102}{9} = \frac{1}{6}$$   Simplify the numerators.

$$\frac{414}{36} - \frac{408}{36} = \frac{6}{36}$$   Write equivalent fractions with LCD, 36.

$$\frac{6}{36} = \frac{6}{36}$$   Simplify.

Since $y = 51$ makes the equation true, the solution set is $\{51\}$.

✔ **Student Check 1**   Solve each equation by first clearing fractions. Check each answer.

**a.** $\dfrac{7x}{6} - 4 = \dfrac{x}{6}$

**b.** $\dfrac{y}{4} - \dfrac{1}{12} = \dfrac{y}{8}$

**c.** $\dfrac{3}{4}(x-8) = \dfrac{2}{7}(x+14)$

**d.** $\dfrac{a+7}{3} - 4 = \dfrac{a}{6}$

**Note:** When we multiply an equation by the LCD of all the fractions in the equation, the resulting equation should not have any fractions. If the resulting equation still contains fractions, then we either simplified incorrectly or calculated the LCD incorrectly.

## Linear Equations Containing Decimals

**Objective 2** ▶

Solve linear equations with decimals.

As with equations that contain fractions, equations containing decimals can also be tedious to solve by hand. So, we will clear the decimals from the equation.

Consider the following products. Recall that multiplying a number by a power of 10 moves the decimal point to the right.

$$10(3.2\,y) = (10)(3.2)y = 32y$$

One decimal place

Multiplying an expression with one decimal place by $10^1$ or 10 clears the decimal from the expression.

$$100\underbrace{(0.05x)}_{\text{Two decimal places}} = (100)(0.05)x = 5x$$

> Multiplying an expression with two decimal places by $10^2$ or 100 clears the decimal from the expression.

$$1000\underbrace{(2.34a + 12.567)}$$
$$= 1000(2.34a) + 1000(12.567)$$
$$= 2340a + 12{,}567$$
Two and three decimal places

> Multiplying an expression with three decimal places by $10^3$ or 1000 clears the decimal from the expression.

Note that when we multiply the preceding expressions by an appropriate power of 10, the decimals are cleared from the expression. We will apply this same concept to remove decimals from an equation; that is, we will multiply each side of the equation by a power of 10 (10, 100, 1000, and so on) that will eliminate the decimal from the number with the most decimal places.

To solve equations with decimals, we apply the general strategy that was stated at the beginning of the section.

---

**Objective 2  Examples**   **Solve each equation by clearing the decimals. Check each answer.**

**2a.** $x + 0.15x = 36.80$          **2b.** $0.04x + 0.05(500 - x) = 23$

**Solutions**   **2a.**

**INSTRUCTOR NOTE:**
Point out that it is not necessary to clear decimals from the equation before we solve it.

| | |
|---|---|
| $x + 0.15x = 36.80$ | |
| $100(x + 0.15x) = 100(36.80)$ | Multiply each side by 100. |
| $100(x) + 100(0.15x) = 3680$ | Distribute and simplify. |
| $100x + 15x = 3680$ | Simplify. |
| $115x = 3680$ | Combine like terms. |
| $\dfrac{115x}{115} = \dfrac{3680}{115}$ | Divide each side by 115. |
| $x = 32$ | Simplify. |

**Check:**

| | |
|---|---|
| $x + 0.15x = 36.80$ | Original equation |
| $32 + 0.15(32) = 36.80$ | Replace $x$ with 32. |
| $32 + 4.8 = 36.80$ | Simplify. |
| $36.8 = 36.80$ | Add. |

Since $x = 32$ makes the equation true, the solution set is $\{32\}$.

> **Note:** We could also solve the equation without removing decimals by combining like terms on the left.
>
> | | |
> |---|---|
> | $1x + 0.15x = 36.80$ | Recall that $x = 1x$. |
> | $1.15x = 36.80$ | Combine like terms. |
> | $\dfrac{1.15x}{1.15} = \dfrac{36.80}{1.15}$ | Divide each side by 1.15. |
> | $x = 32$ | Simplify. |

**2b.**
$$0.04x + 0.05(500 - x) = 23$$

$$100[0.04x + 0.05(500 - x)] = 100(23)$$   Multiply each side by 100.

$$100(0.04x) + 100(0.05)(500 - x) = 2300$$   Distribute and simplify.

$$4x + 5(500 - x) = 2300$$   Simplify each product.

$$4x + 2500 - 5x = 2300$$   Apply the distributive property.

$$-x + 2500 = 2300$$   Combine like terms.

$$-x + 2500 - 2500 = 2300 - 2500$$   Subtract 2500 from each side.

$$-x = -200$$   Simplify.

$$\frac{-x}{-1} = \frac{-200}{-1}$$   Divide each side by −1.

$$x = 200$$   Simplify.

**Check:**
$$0.04x + 0.05(500 - x) = 23$$   Original equation

$$0.04(200) + 0.05(500 - 200) = 23$$   Replace x with 200.

$$8 + 0.05(300) = 23$$   Simplify.

$$8 + 15 = 23$$   Multiply.

$$23 = 23$$   Add.

Since $x = 200$ makes the equation true, the solution set is $\{200\}$.

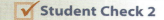 **Student Check 2**   Solve each equation by first clearing the decimals. Check each answer.
**a.** $x + 0.25x = 90$       **b.** $0.05x + 0.07(800 - x) = 54$

## Linear Equations with No Solution

**Objective 3** ▶

Solve linear equations with no solution.

All of the linear equations we have encountered so far have had one solution. When solving these equations, we were able to isolate the variable on one side of the equation and the constant on the other side. These types of equations are called *conditional equations*.

> **Definition:** A **conditional equation** is an equation that is true for some values of the variable and not true for other values of the variable.

We will now turn our attention to a special type of linear equation, one that has no solution. Consider the equation $x = x + 1$. The left and right sides of this equation will never be equal since the right side of the equation is always one more than the left side of the equation as shown in the table.

| $x$ | $x = x + 1$ | |
|---|---|---|
| $-2$ | $-2 = -2 + 1$ <br> $-2 = -1$ | False |
| $-1$ | $-1 = -1 + 1$ <br> $-1 = 0$ | False |
| $0$ | $0 = 0 + 1$ <br> $0 = 1$ | False |
| $1$ | $1 = 1 + 1$ <br> $1 = 2$ | False |
| $2$ | $2 = 2 + 1$ <br> $2 = 3$ | False |

So, $x = x + 1$ is an example of an equation with no solution. It is called a *contradiction*.

> **Definition:** An equation that is not true for any value of the variable is called a **contradiction**. A contradiction has no solution. The solution set is the empty set, or null set, ∅.

Contradictions do not initially look any different from conditional equations. The difference becomes apparent in the equation-solving process. As we solve the equation, the variable terms are eliminated from both sides of the equation and a false statement (e.g., $5 = 7$) remains.

 If we solve an equation and obtain a false statement, then the equation is a contradiction and has no solution. The solution set is the empty set, ∅.

**Objective 3 Examples**  **Solve each equation.**

**3a.** $x + 3 = x + 2$    **3b.** $4x + 5 - x = 3(x + 2)$

**Solutions**    **3a.**
$$x + 3 = x + 2$$
$$x + 3 - x = x + 2 - x \qquad \text{Subtract } x \text{ from each side.}$$
$$3 = 2 \qquad \text{Simplify.}$$

The resulting equation, $3 = 2$, is a contradiction, or false statement. So, the solution set is the empty set, ∅.

**3b.**  $4x + 5 - x = 3(x + 2)$    Apply the distributive property on the right side and
$$3x + 5 = 3x + 6 \qquad \text{combine like terms on the left side.}$$
$$3x + 5 - 3x = 3x + 6 - 3x \qquad \text{Subtract } 3x \text{ from each side.}$$
$$5 = 6 \qquad \text{Simplify.}$$

The resulting equation, $5 = 6$, is a contradiction, so the solution set is the empty set, ∅.

**✓ Student Check 3**    Solve each equation.

 **a.** $2x - 4 = 2x + 1$    **b.** $5(2x - 4) = 3(3x + 1) + x$

> **Note:** Once an equation is simplified on each side, we know we have a contradiction when the variable term on each side of the equation is the same but the constant terms are different.

## Linear Equations with Infinitely Many Solutions

**Objective 4 ▶**

**Solve linear equations with infinitely many solutions.**

We will now turn our attention to solving linear equations that have infinitely many solutions. Consider the equation $x + 1 = x + 1$. The left and right sides of this equation will always be equal since the right side of the equation is the exact same as the left side of the equation as shown in the following table.

| $x$ | $x + 1 = x + 1$ | |
|---|---|---|
| $-2$ | $-2 + 1 = -2 + 1$<br>$-1 = -1$ | True |
| $-1$ | $-1 + 1 = -1 + 1$<br>$0 = 0$ | True |
| $0$ | $0 + 1 = 0 + 1$<br>$1 = 1$ | True |
| $1$ | $1 + 1 = 1 + 1$<br>$2 = 2$ | True |
| $2$ | $2 + 1 = 2 + 1$<br>$3 = 3$ | True |

So, $x + 1 = x + 1$ is an example of an equation with infinitely many solutions. It is true for every value of the variable. This type of equation is called an *identity*.

> **Definition:** An **identity** is an equation that is true for all values of the variable. An identity has infinitely many solutions. The solution set is all real numbers, denoted $\mathbb{R}$.

Again, this type of equation looks no different from a conditional equation. Through the equation-solving process, we obtain a true statement (e.g., $5 = 5$ or $x = x$).

If we solve an equation and a true statement remains, then the equation is an identity and has infinitely many solutions of the equation. The solution set is the set of all real numbers, denoted by $\mathbb{R}$.

| **Objective 4 Examples** | Solve each equation. |
|---|---|

**4a.** $2x + 3 = 2(x - 2) + 7$        **4b.** $4(3x + 5) - (x - 3) = 5(3x + 5) - (4x + 2)$

**Solutions**   **4a.**

$$2x + 3 = 2(x - 2) + 7$$

$$2x + 3 = 2x - 4 + 7 \qquad \text{Apply the distributive property.}$$

$$2x + 3 = 2x + 3 \qquad \text{Combine like terms.}$$

$$2x + 3 - 2x = 2x + 3 - 2x \qquad \text{Subtract } 2x \text{ from each side.}$$

$$3 = 3 \qquad \text{Simplify.}$$

The resulting equation, $3 = 3$, is a true statement, so this equation is an identity. The solution set is all real numbers, or $\mathbb{R}$.

**4b.** $4(3x + 5) - (x - 3) = 5(3x + 5) - (4x + 2)$

$$12x + 20 - x + 3 = 15x + 25 - 4x - 2 \qquad \text{Apply the distributive property.}$$

$$11x + 23 = 11x + 23 \qquad \text{Combine like terms.}$$

$$11x + 23 - 23 = 11x + 23 - 23 \qquad \text{Subtract 23 from each side.}$$

$$11x = 11x \qquad \text{Simplify.}$$

This resulting equation, $11x = 11x$, is always true, so this equation is an identity. The solution set is all real numbers, or $\mathbb{R}$.

| ✓ **Student Check 4** | Solve each equation. |
|---|---|

**a.** $2(6x + 4) = 4(3x + 2)$        **b.** $7x + 3(2x - 1) = 4(3x - 1) + x + 1$

> **Note:** *Once an equation is simplified on each side, we know we have an identity when the variable terms on each side of the equation are the same and the constant terms on each side of the equation are the same.*

| **Objective 5** ▶ | ## Troubleshooting Common Errors |
|---|---|
| Troubleshoot common errors. | Some common errors associated with solving linear equations are shown next. |

**Objective 5 Examples**    A problem and an incorrect solution are given. Provide the correct solution and an explanation of the error.

**5a.** Solve $\dfrac{3x}{2} + 5 = \dfrac{x}{2}$.

| Incorrect Solution | Correct Solution and Explanation |
|---|---|
| $\dfrac{3x}{2} + 5 = \dfrac{x}{2}$ $2\left(\dfrac{3x}{2} + 5\right) = 2\left(\dfrac{x}{2}\right)$ $3x + 5 = x$ $2x = -5$ $x = -\dfrac{5}{2}$ | The error was made in not applying the distributive property to each term on the left side of the equation. $\dfrac{3x}{2} + 5 = \dfrac{x}{2}$ $2\left(\dfrac{3x}{2} + 5\right) = 2\left(\dfrac{x}{2}\right)$ $3x + 10 = x$ $2x = -10$ $x = -\dfrac{10}{2}$ $x = -5$ The solution set is $\{-5\}$. |

**5b.** Solve $2x - 4(x - 1) = -2(x + 2)$.

| Incorrect Solution | Correct Solution and Explanation |
|---|---|
| $2x - 4(x - 1) = -2(x + 2)$ $2x - 4x + 4 = -2x - 4$ $-2x + 4 = -2x - 4$ $\underline{+2x - 4 \quad\quad +2x - 4}$ $0x = -8$ The solution set is $\{-8\}$. | The error was made in not recognizing that $0x = 0$. The resulting statement should be $0 = -8$, which is a contradiction. So, the solution set is the empty set, $\varnothing$. |

## ANSWERS TO STUDENT CHECKS

**Student Check 1**   a. $\{4\}$   b. $\left\{\dfrac{2}{3}\right\}$   c. $\left\{\dfrac{280}{13}\right\}$   d. $\{10\}$   **Student Check 3**   a. $\varnothing$   b. $\varnothing$

**Student Check 2**   a. $\{72\}$   b. $\{100\}$   **Student Check 4**   a. $\mathbb{R}$   b. $\mathbb{R}$

## SUMMARY OF KEY CONCEPTS

1. To solve an equation with fractions, multiply both sides by the LCD of all fractions in the equation. This will create an equation that does not have any fractions. If the resulting equation has fractions, then the LCD was not correct or the equation was simplified incorrectly.

2. To solve an equation with decimals, multiply both sides by a power of 10 that will eliminate decimals in the number with the largest number of decimal places.

3. Equations that produce a false statement have no solution. These are called contradictions and are denoted by the null set, $\varnothing$.

4. Equations that produce a true statement have infinitely many solutions and are called identities. The solution set is the set of all real numbers, $\mathbb{R}$.

## GRAPHING CALCULATOR SKILLS

In previous sections, we learned how to check solutions of equations on the calculator by using the main screen, the store feature, and the table feature. We will now use the table feature to verify solutions when the proposed solution is large or a fraction. We can change the table settings in one of two ways as shown next.

**Example:** Determine if $y = 51$ is a solution of $\dfrac{y - 5}{4} - \dfrac{2y}{9} = \dfrac{1}{6}$.

**Solution:**

**Method 1:** Enter each side of the equation in the equation editor and view the table. Since the calculator accepts only the variable $x$, we will use $x$ in place of $y$ in the equation.

Instead of scrolling to $x = 51$ in the table, we can change the starting value of the table.

When we view the table we see the $x$-value of 51 and find that the corresponding $y$-values are equal, which confirms $x = 51$ is a solution of the equation.

**Method 2:** We can also change the table settings to enable us to enter the value of $x$ that we want to check. Go to the Table Setup and change the Indpnt setting from Auto to Ask. Move the cursor to Ask in the Indpnt row and press enter.

Now go to the table and enter the value for $x$ and press enter.

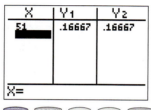

Since the $y_1$ and $y_2$ values are equal for $x = 51$, the solution is correct.

> **Note:** If the equation is an identity, the $Y_1$ and $Y_2$ values are equal for every value of $x$. If the equation is a contradiction, the $Y_1$ and $Y_2$ values are not equal for any value of $x$.

# SECTION 2.4 / EXERCISE SET

## Write About It!

**Use complete sentences to explain the meaning of the given term.**

1. Conditional equation   2. Contradiction   3. Identity
   Answers vary.            Answers vary.        Answers vary.

**Use complete sentences to answer each question.**

4. How can you eliminate fractions from an equation?
   Answers vary.
5. How can you eliminate decimals from an equation?
   Answers vary.
6. When solving an equation, how will you know if the solution is the empty set?   Answers vary.
7. When solving an equation, how will you know if the solution is all real numbers?   Answers vary.

**Determine if each statement is true or false. If a statement is false, explain why.**

8. The solution of the equation $2x = 5x$ is the empty set.
   False, the solution is $x = 0$.
   Additional answers can be found in the Instructor Answer Appendix.

9. If the solution of the equation $x + a = x + b$ is the empty set, then $a \neq b$.   True

10. If the solution of the equation $x + a = x + b$ is all real numbers, then $a = b$.   True

11. An equivalent form of the equation $\frac{y}{4} + 3 = \frac{y}{8}$ is $2y + 3 = y$.   False, we must multiply 3 by 8 to get $2y + 24 = y$.

12. An equivalent form of the equation $-5(2x - 3) = 2x - 9$ is $-10x - 3 = 2x - 9$.
    False, we must distribute $-5$ to $-3$ to get $-10x + 15 = 2x - 9$.

## Practice Makes Perfect!

**Solve each equation by first clearing fractions. (See Objective 1.)**

13. $\frac{x}{2} + \frac{3}{2} = \frac{1}{2}$   $\{-2\}$

14. $\frac{y}{3} - \frac{4}{3} = -\frac{10}{3}$   $\{-6\}$

15. $\frac{a}{6} + 5 = \frac{a}{3}$   $\{30\}$

16. $\frac{x}{5} - 2 = \frac{x}{10}$   $\{20\}$

**17.** $\dfrac{3x}{5} + \dfrac{1}{3} = \dfrac{2x}{15}$ $\left\{-\dfrac{5}{7}\right\}$ **18.** $\dfrac{4y}{7} + \dfrac{1}{2} = \dfrac{5y}{14}$ $\left\{-\dfrac{7}{3}\right\}$

**19.** $2 - \dfrac{x}{6} = \dfrac{x}{3}$ $\{4\}$ **20.** $4 - \dfrac{y}{2} = \dfrac{y}{8}$ $\left\{\dfrac{32}{5}\right\}$

**21.** $\dfrac{2}{9}a - \dfrac{3}{18} = \dfrac{5}{18}a + \dfrac{4}{9}$ $\{-11\}$ **22.** $\dfrac{7}{3}y + \dfrac{5}{6} = \dfrac{1}{6}y - \dfrac{2}{3}$ $\left\{-\dfrac{9}{13}\right\}$

**23.** $-3 = \dfrac{9}{2} + b$ $\left\{-\dfrac{15}{2}\right\}$ **24.** $5 = -\dfrac{4}{7} + b$ $\left\{\dfrac{39}{7}\right\}$

**25.** $\dfrac{3}{2}(x + 4) = \dfrac{1}{3}(x + 9)$ **26.** $-\dfrac{6}{7}(x - 14) = \dfrac{2}{3}(x + 6)$ $\left\{\dfrac{21}{4}\right\}$

**27.** $\dfrac{x + 4}{2} - 5 = \dfrac{x - 1}{6}$ $\left\{\dfrac{17}{2}\right\}$ **28.** $\dfrac{x - 2}{3} + 1 = \dfrac{x + 4}{8}$ $\left\{\dfrac{4}{5}\right\}$

**Solve each equation by first clearing decimals.**
**(See Objective 2.)**

**29.** $x + 0.3x = 65$ $\{50\}$ **30.** $y + 0.1y = 11.88$ $\{10.8\}$

**31.** $a - 0.25a = 48.75$ $\{65\}$ **32.** $x - 0.95x = 6.25$ $\{125\}$

**33.** $0.06x + 0.05(4000 - x) = 235$ $\{3500\}$

**34.** $0.03x + 0.045(10{,}000 - x) = 390$ $\{4000\}$

**35.** $0.6x + 0.25(10) = 0.5(x + 10)$ $\{25\}$

**36.** $0.15x + 0.75(40) = 0.3(x + 40)$ $\{120\}$

**Solve each equation. If the equation is a contradiction, write the solution as ∅. If the equation is an identity, write the solution as ℝ. (See Objectives 3 and 4.)**

**37.** $x - 7 = x + 4$ ∅ **38.** $y + 2 = y - 1$ ∅

**39.** $2x + 3 = 2x - 3$ ∅ **40.** $8a + 2 = 8a - 2$ ∅

**41.** $3(2a + 5) = 2(3a - 4)$ ∅ **42.** $5(x + 2) = 3x + 10 + 2x$ ℝ

**43.** $4x + 1 = 4(x - 2) + 9$ ℝ

**44.** $2 - 5x = 7(1 - x) + 2x - 5$ ℝ

**45.** $-2(y - 7) - (y + 3) = -3(y - 4) - 1$ ℝ

**46.** $-7(2 - a) - (a + 3) = -3(5 - 2a) - 4$ ∅

**47.** $3(2a - 1) - (a + 4) = 5(a - 6) - 8$ ∅

**48.** $12(y - 4) + 3 - y = 5(2y - 9) + y$ ℝ

 **Mix 'Em Up!**

**Solve each equation. If the equation is a contradiction, write the solution as ∅. If the equation is an identity, write the solution as ℝ.**

**49.** $2x + 3 = 3$ $\{0\}$ **50.** $6x - 7 = -7$ $\{0\}$

**51.** $3(y + 1) - (y - 4) = 5(y + 2)$ $\{-1\}$

**52.** $4(x + 6) - (3x - 2) = 2(x + 3)$ $\{20\}$

**53.** $\dfrac{4x}{5} - 3 = \dfrac{x}{10}$ $\left\{\dfrac{30}{7}\right\}$ **54.** $\dfrac{7}{3}y + 1 = \dfrac{5}{9}y$ $\left\{-\dfrac{9}{16}\right\}$

**55.** $x + 0.8x = 900$ $\{500\}$ **56.** $x - 0.30x = 332.5$ $\{475\}$

**57.** $\dfrac{3}{4}b + \dfrac{7}{10} = \dfrac{2}{5}b - \dfrac{1}{4}$ $\left\{-\dfrac{19}{7}\right\}$ **58.** $\dfrac{5}{6}c - \dfrac{2}{3} = \dfrac{1}{2}c - \dfrac{5}{12}$ $\left\{\dfrac{3}{4}\right\}$

**59.** $0.021x + 0.045(2400 - x) = 91.2$ $\{700\}$

**60.** $0.016x + 0.02(8800 - x) = 158.4$ $\{4400\}$

**61.** $\dfrac{y - 2}{7} - 3 = \dfrac{y - 11}{4}$ $\{-5\}$

**62.** $\dfrac{y + 5}{6} - 2 = \dfrac{y - 3}{10}$ $\{13\}$

**63.** $0.046x + 0.05(6900 - x) = 339.4$ $\{1400\}$

**64.** $0.066x + 0.08(9300 - x) = 727.2$ $\{1200\}$

**65.** $\dfrac{2}{3}(x - 1) = \dfrac{1}{6}(4x - 4)$ ℝ

**66.** $\dfrac{1}{4}(2x + 3) = \dfrac{1}{2}(x - 3)$ ∅

**67.** $0.81x + 2.5(-1.06) = 0.2(x + 2)$ $\{5\}$

**68.** $0.31x + 0.5(-6.54) = 0.3(x - 9)$ $\{57\}$

**69.** $3(a + 4) + 2a = 5(a + 2)$ ∅

**70.** $7(2x - 3) - 5x = 3(3x - 7)$ ℝ

**71.** $\dfrac{3}{2}(x - 4) = \dfrac{1}{4}(5x + 1)$ $\{25\}$

**72.** $\dfrac{7}{12}(3x + 5) = \dfrac{5}{6}(x - 2)$ $\{-5\}$

**73.** $0.26x + 0.8(73.5) = 0.75x$ $\{120\}$

**74.** $0.66x + 2.5(-6.448) = 0.2(x - 7)$ $\{32\}$

**75.** $\dfrac{3x - 8}{6} - 1 = \dfrac{x + 3}{4}$ $\left\{\dfrac{37}{3}\right\}$

**76.** $\dfrac{2x - 7}{10} + 1 = \dfrac{x + 3}{2}$ $\{-4\}$

 **You Be the Teacher!**

**Correct each student's errors, if any.**

**77.** Solve: $\dfrac{x}{5} = \dfrac{1}{2} - \dfrac{x}{3}$

Susan's work:

$\dfrac{x}{5} = \dfrac{1}{2} - \dfrac{x}{3}.$ LCD = 30

$30\left(\dfrac{x}{5}\right) = 30\left(\dfrac{1}{2} - \dfrac{x}{3}\right)$

$\dfrac{30x}{5} = \dfrac{30}{2} - \dfrac{x}{3}$

$6x = 15 - \dfrac{x}{3}$

$3(6x) = 3\left(15 - \dfrac{x}{3}\right)$

$18x = 45 - \dfrac{3x}{3}$

$18x + x = 45 - x + x$

$19x = 45$

$\dfrac{19x}{19} = \dfrac{45}{19}$

$x = \dfrac{45}{19}$

$\dfrac{x}{5} = \dfrac{1}{2} - \dfrac{x}{3}.$ LCD = 30

$\dfrac{30x}{5} = \dfrac{30}{2} - \dfrac{30x}{3}$

$6x = 15 - 10x$

$16x = 15$

$x = \dfrac{15}{16}$

**78.** Solve: $\dfrac{5}{6}(x + 12) = \dfrac{1}{8}(2x - 9)$

Weston's work:

$$\dfrac{5}{6}(x + 12) = \dfrac{1}{8}(2x - 9). \text{ LCD} = 24$$

$$24\left(\dfrac{5}{6}(x + 12)\right) = 24\left(\dfrac{1}{8}(2x - 9)\right)$$

$$\left(24 \cdot \dfrac{5}{6}\right)(x + 12) = \left(24 \cdot \dfrac{1}{8}\right)(2x - 9)$$

$$20(x + 12) = 3(2x - 9)$$

$$20x + 12 - 6x = 6x - 27 - 6x$$

$$14x + 12 - 12 = -27 - 12$$

$$14x = -39$$

$$x = -\dfrac{39}{14}$$

$\dfrac{5}{6}(x + 12) = \dfrac{1}{8}(2x - 9).$

LCD $= 24$

$20(x + 12) = 3(2x - 9)$

$20x + 240 = 6x - 27$

$14x + 240 = -27$

$14x = -267$

$x = -\dfrac{267}{14}$

**79.** Solve: $3x - 5 = 3(x - 5)$

Katie's work:

$$3x - 5 = 3(x - 5)$$

$$3x - 5 = 3x - 5$$

$$3x - 5 - 3x = 3x - 5 - 3x$$

$$-5 = -5$$

Since this is true, the solution is $\mathbb{R}$.

$3x - 5 = 3(x - 5)$

$3x - 5 = 3x - 15$

$3x - 5 - 3x = 3x - 15 - 3x$

$-5 \neq -15$

Therefore, the solution is $\varnothing$.

**80.** Solve: $-2(x - 1) + x - 3 = x - 1$

Mitchell's work:

$$-2(x - 1) + x - 3 = x - 1 \qquad \text{The solution is } \{0\}.$$

$$-2x + 2 + x - 3 = x - 1$$

$$-x - 1 = x - 1$$

$$-x - 1 - x = x - 1 - x$$

$$-2x - 1 = -1$$

$$-2x - 1 + 1 = -1 + 1$$

$$-2x = 0$$

$$\dfrac{-2x}{-2} = \dfrac{0}{-2}$$

$$x = 0$$

Therefore, the solution is $\varnothing$.

 **Calculate It!**

Solve each equation. Then use a graphing calculator to check the solution.

**81.** $5x - 1 - (x + 3) = 4x - 4$   $\mathbb{R}$ **82.** $3x - 4 = 2(x + 1) + x$   $\varnothing$

**83.** $\dfrac{2}{3}(x - 6) = x + 3$   $\{-21\}$ **84.** $\dfrac{x}{2} + \dfrac{3}{2} = x - \dfrac{x}{2} + \dfrac{3}{2}$   $\mathbb{R}$

**85.** $0.86x + 1.25(3.344) = 0.7(x + 11)$   $\{22\}$

**86.** $0.56x + 0.4(-31.3) = 0.25(x + 2)$   $\{42\}$

 **Think About It!**

**87.** Write an equation that has no solution.   Answers vary.

**88.** Write an equation that has infinitely many solutions.   Answers vary.

**89.** Write an equation that has 0 as a solution.   Answers vary.

---

## PIECE IT TOGETHER   SECTIONS 2.1–2.4 Review

**Determine whether the following is an expression or an equation.** (*Section 2.1, Objective 1*)

**1.** $2x - 16 = 0$   equation     **2.** $3x + 19$   expression

**Decide if the equation is linear. If the equation is not linear, explain why.** (*Section 2.2, Objective 1*)

**3.** $-x^2 + 4x - 12 = 0$   not a linear equation because the largest exponent is 2     **4.** $\dfrac{3}{2}x - 4 + \dfrac{5}{6}x = 11$   linear equation

**Solve each equation. If the equation is a contradiction, write the solution as $\varnothing$. If the equation is an identity, write the solution as $\mathbb{R}$.** (*Sections 2.2–2.4*)

**5.** $y - 5 = 5$   $\{10\}$

**6.** $8.6 + c = 1.4$   $\{-7.2\}$

**7.** $2(x - 3) = 3x$   $\{-6\}$

**8.** $9(a - 7) = 4(2a + 3)$   $\{75\}$

**9.** $\dfrac{x}{3} = 9$   $\{27\}$

**10.** $-3x + 7 = -14$   $\{7\}$

**11.** $-5.2x + 13.5 = -3.6x + 14.46$   $\{-0.6\}$

**12.** $3(2z - 6) = 4(z + 5)$   $\{19\}$

**13.** $2(x - 3) = 3x$   $\{-6\}$

**14.** $5(3a + 1) = 7(2a - 3)$   $\{-26\}$

**15.** $\dfrac{2}{3}(x + 6) = \dfrac{1}{2}(x - 4)$   $\{-36\}$

**16.** $\dfrac{x + 7}{5} - 2 = \dfrac{x - 2}{2}$   $\left\{\dfrac{4}{3}\right\}$

**17.** $0.6x + 4.2 = 0.3(x + 24)$   $\{10\}$

**18.** $x - 2 + 3x = 2(2x - 1)$   $\mathbb{R}$

**Write an equation that represents each situation. Solve the equation and explain the answer using a complete sentence.** (*Section 2.3, Objectives 3 and 4*)

**19.** The sum of three times a number and 13 is the same as two less than the number. Find the number.   $-\dfrac{15}{2}$

**20.** The sum of the two smaller of three consecutive odd integers is the same as 82 less than six times the largest integer. Find the integers.   15, 17, 19

## SECTION 2.5 / Formulas and Applications from Geometry

As a result of completing this section, you will be able to

**1.** Evaluate formulas.

**2.** Use perimeter, area, and circumference formulas.

**3.** Solve formulas for a specified variable.

**4.** Solve problems involving complementary and supplementary angles.

**5.** Solve problems involving straight and vertical angles.

**6.** Solve problems involving triangles.

**7.** Troubleshoot common errors.

**Objective 1** ▶
Evaluate formulas.

The Columbus Circle in New York City, built in 1905, was the first traffic circle in the United States. It was renovated a century later in 2005. The distance around the inner circle measures approximately 673 ft. The distance around the outer circle is approximately 1364 ft. Knowing these values, we can find the radius of each circle and also the area of each circle. The formulas that enable us to solve this problem are provided in this section. (Source: http://en.wikipedia.org/wiki/Columbus_Circle)

## Formulas

A mathematical **formula** is an equation that expresses the relationship between two or more quantities. These quantities are represented by variables. Mathematical formulas enable us to find valuable information if we know some specific values of some of the variables in the formula. In Section 1.3, we evaluated algebraic expressions; some of these were mathematical formulas.

Some of the commonly used formulas are for calculating the area and perimeter of geometric shapes, converting degrees Fahrenheit to degrees Celsius, and calculating the amount of money in a savings account. Some of these formulas are listed within this section. Evaluating formulas is exactly like evaluating algebraic expressions.

> **Procedure: Evaluating Formulas**
>
> **Step 1:** Substitute known values in place of the appropriate variables.
> **Step 2:** Simplify the resulting expression or solve the resulting equation.

**Objective 1 Examples** / Use the formula and the given values to solve for the missing variable.

**1a.** The formula $A = P(1 + r)^t$ is used to calculate the amount of money in an account at the end of $t$ years if $P$ dollars are initially invested at an annual interest rate, $r$. Find the total value, $A$, if \$10,000 is invested for 5 yr at 4.5% annual interest.

**Solution** 

**1a.** $A = P(1 + r)^t$      State the formula.

$A = 10,000(1 + 0.045)^5$      Let $P = 10,000$, $r = 4.5\% = 0.045$, and $t = 5$.

$A = 10,000(1.045)^5$      Add the expression in parentheses.

$A = 10,000(1.246182)$      Evaluate the exponential expression.

$A = \$12,461.82$      Multiply.

In 5 yr, the total value in the account will be \$12,461.82.

**1b.** The formula $F = \dfrac{9}{5}C + 32$ is used to convert degrees Celsius ($C$) to degrees Fahrenheit ($F$). Find the value of $C$ if $F = 63°$.

**Solution** 

**1b.**

$F = \dfrac{9}{5}C + 32$      State the formula.

$63 = \dfrac{9}{5}C + 32$      Replace $F$ with 63.

$5(63) = 5\left(\dfrac{9}{5}C + 32\right)$      Multiply each side by the LCD, 5.

$315 = 5\left(\dfrac{9}{5}C\right) + 5(32)$      Simplify and apply the distributive property.

$315 = 9C + 160$      Simplify.

$$315 - 160 = 9C + 160 - 160 \quad \text{Subtract 160 from each side.}$$
$$155 = 9C \quad \text{Simplify.}$$
$$\frac{155}{9} = \frac{9C}{9} \quad \text{Divide each side by 9.}$$
$$17.2 = C \quad \text{Simplify.}$$

So, 63°F is equivalent to 17.2°C.

**1c.** The formula $d = rt$ calculates the distance, $d$, traveled by an object going at an average rate $r$ for time $t$. A cruise ship travels from Fort Lauderdale, Florida, to San Juan, Puerto Rico, in 45.5 hr at an average speed of 24 knots (about 27.6 mph). What is the distance the ship traveled?

**Solution    1c.**
$$d = rt$$
$$d = (27.6)(45.5) \quad \text{Replace } r \text{ with 27.6 and } t \text{ with 45.5.}$$
$$d = 1255.8 \quad \text{Multiply.}$$

The cruise ship traveled 1255.8 mi.

✓ **Student Check 1**   Use the formula and the given values to solve for the missing variable.

**a.** The formula $I = Prt$ is used to calculate simple interest for an investment of $P$ dollars at an annual interest rate, $r$, for $t$ years. Find the value of $I$ if $P = \$1000$, $r = 5\%$, and $t = 3$.

**b.** The formula $C = \dfrac{5}{9}(F - 32)$ is used to convert degrees Fahrenheit ($F$) to degrees Celsius ($C$). Find the value of $F$ if $C = 50°$.

**c.** The world's fastest passenger train, the Harmony Express in China, opened in December 2009. It travels at an average speed of 217 mph from Wuhan, China, to Guangzhou, China. It takes 3 hr to complete the trip. Use the formula $d = rt$ to calculate the distance between the cities of Wuhan and Guangzhou. (Source: http://www.telegraph.co.uk/news/worldnews/asia/china/7230137/China-steams-ahead-with-worlds-fastest-train.html)

## Perimeter, Area, and Circumference

**Objective 2** ▶

Use perimeter, area, and circumference formulas.

As presented in Chapter 1, the **perimeter** of a polygon is the distance around the outside of the figure. In other words, it is the sum of the lengths of the sides of the polygon. The distance around a circle is called the **circumference** of the circle.

The **area** of a figure is the number of square units it takes to cover the inside of the figure. (A *square unit* is a square whose length and width are both 1 unit long.)

The formulas for some common shapes are provided.

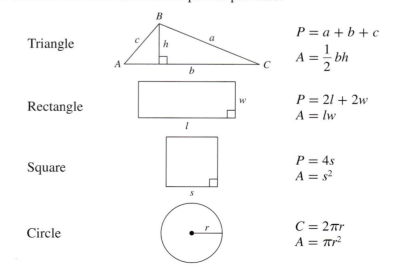

Triangle
$$P = a + b + c$$
$$A = \frac{1}{2}bh$$

Rectangle
$$P = 2l + 2w$$
$$A = lw$$

Square
$$P = 4s$$
$$A = s^2$$

Circle
$$C = 2\pi r$$
$$A = \pi r^2$$

**Objective 2 Examples**    **Use the perimeter, area, or circumference formulas to solve each problem.**

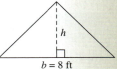

**2a.** Assume that the perimeter of the *Mona Lisa* by Leonardo da Vinci is 260 cm. If the length of the painting is 77 cm, find its width. Then find the area of the painting.

**Solution**    **2a.**

| | |
|---|---|
| $P = 2l + 2w$ | State the perimeter formula. |
| $260 = 2(77) + 2w$ | Replace $P$ with 260 and $l$ with 77. |
| $260 = 154 + 2w$ | Simplify. |
| $260 - 154 = 154 + 2w - 154$ | Subtract 154 from each side. |
| $106 = 2w$ | Simplify. |
| $\dfrac{106}{2} = \dfrac{2w}{2}$ | Divide each side by 2. |
| $53 = w$ | Simplify. |

The width of the painting is 53 cm. Since we know both the length and width of the painting, we can calculate the area using the formula $A = lw$.

| | |
|---|---|
| $A = lw$ | State the area formula. |
| $A = (77)(53)$ | Replace $l$ with 77 and $w$ with 53. |
| $A = 4081$ | Multiply. |

So, the area of the painting is 4081 cm².

**2b.** What is the height of a triangle whose area is 16 ft² and whose base is 8 ft?

**Solution**    **2b.**

| | |
|---|---|
| $A = \dfrac{1}{2}bh$ | State the area formula. |
| $16 = \dfrac{1}{2}(8)h$ | Replace $A$ with 16 and $b$ with 8. |
| $16 = 4h$ | Simplify. |
| $\dfrac{16}{4} = \dfrac{4h}{4}$ | Divide each side by 4. |
| $4 = h$ | Simplify. |

So, the height of the triangle is 4 ft.

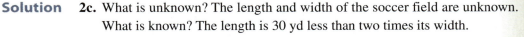

**2c.** Suppose the length of a soccer field is 30 yd less than two times its width. If its perimeter is 390 yd, find the dimensions of the soccer field.

**Solution**    **2c.** What is unknown? The length and width of the soccer field are unknown.
What is known? The length is 30 yd less than two times its width.

Let $w$ = width of the field.
Then $2w - 30$ = length of the field.

We also know the perimeter is 390 yd. So, we use the perimeter formula of a rectangle to write the equation.

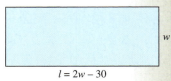

| | |
|---|---|
| $P = 2l + 2w$ | State the perimeter formula. |
| $390 = 2(2w - 30) + 2w$ | Replace $l$ with $2w - 30$. |
| $390 = 4w - 60 + 2w$ | Apply the distributive property. |
| $390 = 6w - 60$ | Combine like terms. |
| $390 + 60 = 6w - 60 + 60$ | Add 60 to each side. |

$$450 = 6w \qquad \text{Simplify.}$$

$$\frac{450}{6} = \frac{6w}{6} \qquad \text{Divide each side by 6.}$$

$$75 = w \qquad \text{Simplify.}$$

So, the width of the soccer field is 75 yd. The length is
$2(75) - 30 = 150 - 30 = 120$ yd.

**2d.** The circumference of the inner circle of the Columbus Circle in New York City measures approximately 673 ft, and the circumference of the outer circle is approximately 1364 ft. Find the radius of the inner and outer circles. Find the area of the inner circle. Round all values to the nearest hundredth.

**Solution**    **2d.** Since we know the circumference of each of the inner and outer circles, we use the circumference formula to find the radii.

Circumference of inner circle             Circumference of outer circle

$$C = 2\pi r \qquad \text{State the circumference formula.} \qquad C = 2\pi r$$

$$673 = 2\pi r \qquad \begin{array}{l}\text{Replace the value of } C\\ \text{with the given circumference.}\end{array} \qquad 1364 = 2\pi r$$

$$\frac{673}{2\pi} = \frac{2\pi r}{2\pi} \qquad \text{Divide each side by } 2\pi. \qquad \frac{1364}{2\pi} = \frac{2\pi r}{2\pi}$$

$$\frac{673}{2\pi} = r \qquad\qquad\qquad\qquad\qquad\qquad \frac{1364}{2\pi} = r$$

$$107.11 \approx r \qquad \text{Simplify.} \qquad\qquad 217.09 \approx r$$

The radius of the inner circle             The radius of the outer circle
is approximately 107.11 ft.             is approximately 217.09 ft.

Now that we know the radius of the inner circle, we can find its area.

$$A = \pi r^2 \qquad \text{State the area formula.}$$

$$A = \pi(107.11)^2 \qquad \text{Replace } r \text{ with 107.11.}$$

$$A = \pi(11{,}472.5521) \qquad \text{Simplify } (107.11)^2.$$

$$A = 36{,}042.09 \qquad \text{Multiply by } \pi.$$

So, the area of the inner circle is approximately 36,042.09 ft².

✔️ **Student Check 2**    Use the perimeter, area, or circumference formulas to solve each problem.

**a.** The largest painting at the Louvre Museum is the *Wedding at Cana* by Veronese. Its perimeter is 1304 in. and its width is 262 in. Find its length and then the area of the painting.

**b.** Find the length of the base of a triangle whose area is 10 ft² and whose height is 4 ft.

**c.** The length of the White House is 3 ft less than twice its width. If the perimeter of the White House is 507 ft, find its length and width.

**d.** Find the area and circumference of the clock face on top of Big Ben in London if the diameter of the clock face is 23 ft.

## Rewriting Formulas

**Objective 3** ▶

Solve formulas for a specified variable.

The Fahrenheit-Celsius formula illustrated in Example 1b can be used to find Fahrenheit or Celsius values. When the Celsius temperature was given, we evaluated the formula to determine the corresponding Fahrenheit value because the formula was already solved for $F$. But when the Fahrenheit temperature was given, we had to solve the equation for $C$ to find the equivalent Celsius value. An alternate way of dealing with this is to first rewrite the Fahrenheit formula so that the variable $C$ is isolated on one side of the equation. Then we can substitute the known Fahrenheit value.

When working with formulas and algebraic equations containing more than one variable, it is sometimes necessary for us to rewrite the formula or equation so that a different variable is isolated.

---

**Procedure: Solving a Formula for a Specified Variable**

**Step 1:** Multiply both sides of the equation by the LCD to clear fractions from the equation, if necessary.

**Step 2:** Apply the distributive property to clear parentheses as needed.

**Step 3:** Apply the addition property of equality to move all terms containing the specified variable to one side of the equation and all other terms to the other side of the equation.

**Step 4:** Apply the multiplication property of equality to obtain a coefficient of 1 on the specified variable.

---

**Objective 3  Examples**    Solve each formula for the specified variable.

**3a.** The formula $A = lw$ is used to find the area of a rectangle. Solve for $w$.

**3b.** The formula $P = 2l + 2w$ gives the perimeter of a rectangle. Solve for $l$.

**3c.** The formula $F = \dfrac{9}{5}C + 32$ converts degrees Celsius to degrees Fahrenheit. Solve for $C$.

**3d.** Solve $3x - y = 6$ for $y$.

**3e.** Solve $2x + 3y = 12$ for $y$.

**Solutions**    **3a.**

$$A = lw$$    Highlight the variable to isolate.

$$\frac{A}{l} = \frac{lw}{l}$$    Divide each side by $l$.

$$\frac{A}{l} = w$$    Simplify.

So, $w = \dfrac{A}{l}$.

**3b.**

$$P = 2l + 2w$$    Highlight the variable to isolate.

$$P - 2w = 2l + 2w - 2w$$    Subtract $2w$ from each side.

$$P - 2w = 2l$$    Simplify.

$$\frac{P - 2w}{2} = \frac{2l}{2}$$    Divide each side by 2.

$$\frac{P - 2w}{2} = l$$    Simplify.

So, $l = \dfrac{P - 2w}{2}$.

**Note:** $\dfrac{P - 2w}{2} = \dfrac{P}{2} - \dfrac{2w}{2} = \dfrac{P}{2} - w$. *It is not equal to* $P - w$.

**3c.**
$$F = \frac{9}{5}C + 32 \qquad \text{Highlight the variable to isolate.}$$

$$5(F) = 5\left(\frac{9}{5}C + 32\right) \qquad \text{Multiply each side by the LCD, 5.}$$

$$5F = 5\left(\frac{9}{5}C\right) + 5(32) \qquad \text{Apply the distributive property.}$$

$$5F = 9C + 160 \qquad \text{Simplify.}$$

$$5F - 160 = 9C + 160 - 160 \qquad \text{Subtract 160 from each side.}$$

$$5F - 160 = 9C \qquad \text{Simplify.}$$

$$\frac{5F - 160}{9} = \frac{9C}{9} \qquad \text{Divide each side by 9.}$$

$$\frac{5F - 160}{9} = C \qquad \text{Simplify.}$$

So, $C = \dfrac{5F - 160}{9}$.

**INSTRUCTOR NOTE:**
Examples 3d and 3e will help students rewrite a linear equation in two variables in the slope-intercept form.

**3d.**
$$3x - y = 6 \qquad \text{Highlight the variable to isolate.}$$

$$3x - y - 3x = 6 - 3x \qquad \text{Subtract } 3x \text{ from each side.}$$

$$-y = -3x + 6 \qquad \text{Note the term containing } y \text{ has a coefficient}$$

$$-1(-y) = -1(-3x + 6) \qquad \text{of } -1, \text{ so multiply each side by } -1.$$

$$y = 3x - 6 \qquad \text{Apply the distributive property.}$$

So, $y = 3x - 6$.

**3e.**
$$2x + 3y = 12 \qquad \text{Highlight the variable to isolate.}$$

$$2x + 3y - 2x = 12 - 2x \qquad \text{Subtract } 2x \text{ from each side.}$$

$$3y = -2x + 12 \qquad \text{Simplify.}$$

$$\frac{3y}{3} = \frac{-2x + 12}{3} \qquad \text{Divide each side by 3.}$$

$$y = \frac{-2x}{3} + \frac{12}{3} \qquad \text{Divide each term by 3.}$$

$$y = -\frac{2}{3}x + 4 \qquad \text{Simplify.}$$

So, $y = -\dfrac{2}{3}x + 4$.

 **Note:** *Parts (d) and (e) will be presented more fully in Chapter 3 when we discuss linear equations in two variables.*

✓ **Student Check 3** Solve each formula for the specified variable.

   **a.** The formula $d = rt$ calculates distance for a given rate and time. Solve for $t$.

   **b.** The formula $V = \dfrac{1}{3}bh$ calculates the volume of a regular pyramid, given the length of its base and height. Solve for $h$.

   **c.** The formula $A = \dfrac{h}{2}(b_1 + b_2)$ calculates the area of a trapezoid. Solve for $b_1$.

   **d.** Solve $4x - y = 8$ for $y$.

   **e.** Solve $4x + 5y = -20$ for $y$.

## Complementary and Supplementary Angles

Recall from Chapter 1 that **complementary angles** are angles whose sum is 90° and **supplementary angles** are angles whose sum is 180°.

A 40° angle and 50° degree angle are complementary because their sum is 90°. The two angles together form a *right angle*. A **right angle** is an angle whose measure is 90°.

A 40° angle and 140° degree angle are supplementary because their sum is 180°. The two angles together form a *straight angle*. A **straight angle** is an angle whose measure is 180°.

Some other examples of complementary and supplementary angles are listed in the table.

| Complementary Angles | Supplementary Angles |
|---|---|
| 20°, 90° − 20° = 70° | 20°, 180° − 20° = 160° |
| 34°, 90° − 34° = 56° | 34°, 180° − 34° = 146° |
| 45°, 90° − 45° = 45° | 45°, 180° − 45° = 135° |
| 73°, 90° − 73° = 17° | 73°, 180° − 73° = 107° |
| $x°$, $(90 − x)°$ | $y°$, $(180 − y)°$ |

In Section 2.1, we practiced setting up equations given information about complementary and supplementary angles. In this section, we will go through the entire process: setting up the equation and solving it.

**Find the measure of each unknown angle. Write an equation that represents the situation and solve it.**

**4a.** Find the measure of an angle whose complement is 15° less than twice the measure of the angle.

**4b.** Find the measure of an angle whose supplement is 40° more than the measure of the angle.

**4c.** The supplement of an angle is 10° more than twice its complement. Find the measure of the angle.

**Solutions**    **4a.** What is unknown? The measure of the angle and its complement are unknown.

$$\text{Let } a = \text{the measure of the angle.}$$
$$\text{Then } 90 − a = \text{measure of the complement.}$$

What is known? The complement is 15° less than twice the measure of the angle. We use this statement to write the equation that we will use to solve the problem.

Complement   is   15° less than twice the measure of the angle.

$$90 − a = 2a − 15 \qquad \text{Express the relationship.}$$
$$90 − a + a = 2a − 15 + a \qquad \text{Add } a \text{ to each side.}$$
$$90 = 3a − 15 \qquad \text{Simplify.}$$
$$90 + 15 = 3a − 15 + 15 \qquad \text{Add 15 to each side.}$$
$$105 = 3a \qquad \text{Simplify.}$$
$$\frac{105}{3} = \frac{3a}{3} \qquad \text{Divide each side by 3.}$$
$$35 = a \qquad \text{Simplify.}$$

So, the measure of the angle is 35°.

**4b.** What is unknown? The measure of the angle and its supplement are unknown.

Let $a =$ the measure of the angle.

Then $180 - a =$ the measure of the supplement.

What is known? The supplement is 40° more than the measure of the angle. We use this statement to write the equation that we will use to solve the problem.

Supplement is $\frac{40° \text{ more than the}}{\text{measure of the angle.}}$

| | |
|---|---|
| $180 - a = a + 40$ | Express the relationship. |
| $180 - a + a = a + 40 + a$ | Add $a$ to each side. |
| $180 = 2a + 40$ | Simplify. |
| $180 - 40 = 2a + 40 - 40$ | Subtract 40 from each side. |
| $140 = 2a$ | Simplify. |
| $\dfrac{140}{2} = \dfrac{2a}{2}$ | Divide each side by 2. |
| $70 = a$ | Simplify. |

So, the measure of the angle is 70°.

**4c.** What is unknown? The measure of the angle, its complement and supplement are unknown.

Let $a =$ the measure of the angle.

Then $90 - a =$ the measure of the complement

and $180 - a =$ the measure of the supplement.

What is known? The supplement of an angle is 10° more than twice its complement. We use this relationship to write the equation to solve the problem.

Supplement is $\frac{10° \text{ more than twice}}{\text{its complement.}}$

| | |
|---|---|
| $180 - a = 2(90 - a) + 10$ | Express the relationship. |
| $180 - a = 180 - 2a + 10$ | Apply the distributive property. |
| $180 - a = 190 - 2a$ | Combine like terms on the right. |
| $180 - a + 2a = 190 - 2a + 2a$ | Add $2a$ to each side. |
| $180 + a = 190$ | Simplify. |
| $180 + a - 180 = 190 - 180$ | Subtract 180 from each side. |
| $a = 10$ | Simplify. |

So, the measure of the angle is 10°.

✓ **Student Check 4** Find the measure of each unknown angle. Write an equation that represents the situation and solve it.

**a.** Find the measure of an angle whose complement is 10° more than three times the angle.

**b.** Find the measure of an angle whose supplement is 60° more than the angle.

**c.** The supplement of an angle is 30° more than twice its complement. Find the measure of the angle.

## Straight and Vertical Angles

**Objective 5** ▶

Solve problems involving straight and vertical angles.

As mentioned in Objective 4, straight angles are angles whose measure is 180°. Straight angles form a straight line.

When two lines intersect one another, four angles are created. Angles opposite from one another are **vertical** or **opposite angles**. These angles have equal measure.

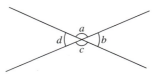

In this figure, angles $a$ and $c$ are vertical angles and angles $b$ and $d$ are vertical angles.

$$a = c \qquad \text{and} \qquad b = d$$

**Note:**

Angles $a$ and $b$ form a straight angle. So, $a + b = 180°$.
Angles $b$ and $c$ form a straight angle. So, $b + c = 180°$.
Angles $c$ and $d$ form a straight angle. So, $c + d = 180°$.
Angles $d$ and $a$ form a straight angle. So, $d + a = 180°$.

**Objective 5 Examples**  **Find the measure of each unknown angle.**

**5a.**

**5b.**

**Solutions**  **5a.** The two angles form a straight angle and, therefore, have a sum of 180°.

| | |
|---|---|
| $5a + a = 180$ | Express the relationship. |
| $6a = 180$ | Combine like terms. |
| $\dfrac{6a}{6} = \dfrac{180}{6}$ | Divide each side by 6. |
| $a = 30$ | Simplify. |

Since $a$ is 30°, the other angle is 5(30) or 150°.

**5b.** The two angles are vertical angles and, therefore, have equal measure.

| | |
|---|---|
| $9x - 6 = 7x + 2$ | Express the relationship. |
| $9x - 6 - 7x = 7x + 2 - 7x$ | Subtract $7x$ from each side. |
| $2x - 6 = 2$ | Simplify. |
| $2x - 6 + 6 = 2 + 6$ | Add 6 to each side. |
| $2x = 8$ | Simplify. |
| $\dfrac{2x}{2} = \dfrac{8}{2}$ | Divide each side by 2. |
| $x = 4$ | Simplify. |

Since $x = 4$, the measures of the angles are $9(4) - 6 = 36 - 6 = 30°$ and $7(4) + 2 = 28 + 2 = 30°$.

**✓ Student Check 5**  Find the measure of each unknown angle.

**a.**

**b.**

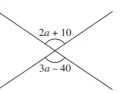

## Triangles

A **triangle** is one of the basic geometric shapes. It has three sides and the sum of its angles is 180°.

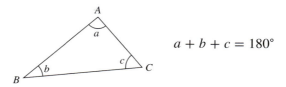

$$a + b + c = 180°$$

We will use this relationship to find the measures of the angles in a triangle.

**Objective 6 Examples**   **Find the measure of each angle in the triangle.**

**6a.**

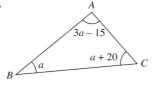

**Solution**   **6a.**

| | |
|---|---|
| $a + a + 20 + 3a - 15 = 180$ | Express the relationship. |
| $5a + 5 = 180$ | Combine like terms. |
| $5a + 5 - 5 = 180 - 5$ | Subtract 5 from each side. |
| $5a = 175$ | Simplify. |
| $\dfrac{5a}{5} = \dfrac{175}{5}$ | Divide each side by 5. |
| $a = 35$ | Simplify. |

Since $a = 35$, the measures of the angles are 35°, 35° + 20° = 55°, and 3(35°) − 15° = 90°. Note that the sum of these measures is 35° + 55° + 90° = 180°.

**6b.** In a triangle $ABC$, angles $B$ and $C$ have the same measure. The measure of angle $A$ is four times the measure of either of the other angles. Find the measure of each angle in the triangle.

**Solution**   **6b.** Draw a diagram to represent the three angles of the triangle. Let $a$ be the measure of angle $B$. Since angle $B$ and angle $C$ have the same measure, $a$ is also the measure of angle $C$. Then the measure of angle $A$ is $4a$.

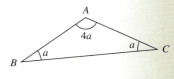

The sum of the measures of the angles in a triangle is 180°, so we have the following equation.

| | |
|---|---|
| $a + a + 4a = 180$ | Express the relationship. |
| $6a = 180$ | Combine like terms. |
| $\dfrac{6a}{6} = \dfrac{180}{6}$ | Divide each side by 6. |
| $a = 30$ | Simplify. |

The measure of angles $B$ and $C$ is 30°. The measure of angle $A$ is 4(30) = 120°.

✓ **Student Check 6**   Find the measure of each angle in the triangle.

**a.**

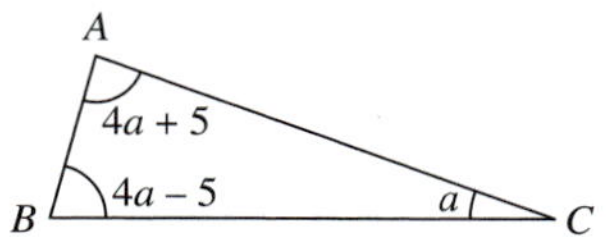

**b.** In a triangle $ABC$, the measure of angle $A$ is twice the measure of angle $C$. The measure of angle $B$ is 20° less than the measure of angle $C$. Find the measure of each angle.

**Objective 7** ▶
Troubleshoot common errors.

## Troubleshooting Common Errors

Some common errors associated with formulas and geometry applications are illustrated next.

**Objective 7 Examples**   A problem and an incorrect solution are given. Provide the correct solution and an explanation of the error.

**7a.** Solve $d = rt$ for $t$.

| Incorrect Solution | Correct Solution and Explanation |
|---|---|
| $d = rt$ <br><br> $d - r = t$ | The variable $r$ is connected to $t$ by multiplication. Therefore, we must divide both sides by $r$ to isolate $t$. <br><br> $d = rt$ <br><br> $\dfrac{d}{r} = \dfrac{rt}{r}$ <br><br> $\dfrac{d}{r} = t$ |

**7b.** The complement of an angle is six degrees less than seven times the measure of the angle. Find the measure of the angle.

| Incorrect Solution | Correct Solution and Explanation |
|---|---|
| Let $a$ be the measure of the angle. Then $a - 90$ is the measure of the complement. <br><br> $a - 90 = 7a - 6$ <br> $-84 = 6a$ <br> $\dfrac{-84}{6} = \dfrac{6a}{6}$ <br> $-14 = a$ <br><br> The angle is 14°. | The error was made in representing the complement of the angle. If $a$ is the measure of the angle, then $90 - a$ is the measure of the complement. Also, it is unreasonable to have a negative value for the measure of an angle. <br><br> $90 - a = 7a - 6$ <br> $96 = 8a$ <br> $\dfrac{96}{8} = \dfrac{8a}{8}$ <br> $12 = a$ <br><br> The angle is 12°. |

## ANSWERS TO STUDENT CHECKS

**Student Check 1   a.** $150   **b.** $F = 122°$   **c.** 651 mi

**Student Check 2   a.** The length of the painting is 390 in. Its area is 102,180 in.² **b.** The base of the triangle is 5 ft.

**c.** The dimensions of the White House are 168 ft by 85.5 ft. **d.** The circumference of the clock is 72.3 ft. The area is 415.5 ft².

**Student Check 3**    **a.** $t = \dfrac{d}{r}$   **b.** $h = \dfrac{3V}{b}$   **c.** $b_1 = \dfrac{2A - b_2 h}{h}$

           **d.** $y = 4x - 8$   **e.** $y = -\dfrac{4}{5}x - 4$

**Student Check 4**    **a.** The angle measures 20°.   **b.** The angle measures 60°.   **c.** The angle measures 30°.

**Student Check 5**    **a.** The measures of the angles are 56° and 124°.   **b.** The angle measure is 110°.

**Student Check 6**    **a.** The angles in the triangle have measures of 85°, 75°, and 20°.   **b.** The angles in the triangle have measures of 100°, 30°, and 50°.

## SUMMARY OF KEY CONCEPTS

1. Evaluating formulas is similar to evaluating algebraic expressions. Substitute the given values for the appropriate variables and simplify. Use the order of operations to simplify the resulting expression. The goal in this objective is to find the value of the missing variable. To find this value, we will either simplify an expression or solve an equation.

2. Perimeter, area, and circumference formulas should be memorized for the basic shapes (triangle, square, rectangle, and circle). If the dimensions of the figure are given, we can find the perimeter, area, or circumference by substituting the given dimensions. We can also find the dimensions of the figure if the perimeter, area, or circumference are provided as well.

3. To solve a formula for a specified variable is to rewrite the equation so that a different variable is isolated to one side. No substitutions are made in this process. We simply apply the addition and multiplication properties of equality to solve for a different variable.

4. Complementary angles and supplementary angles are two angles whose sum is 90° and 180°, respectively. The most common error is misrepresenting the unknown form of these angles. If $a$ is the measure of the angle, then $90 - a$ is the measure of the complement and $180 - a$ is the measure of the supplement.

5. Straight angles measure 180°. Vertical angles are formed when two lines intersect. Vertical angles are equal in measure.

6. The sum of the measures of the angles in a triangle is 180°.

## GRAPHING CALCULATOR SKILLS

We can use the calculator to approximate expressions involving $\pi$.

1. $\dfrac{673}{2\pi}$

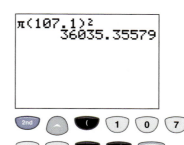

So, $\dfrac{673}{2\pi}$ is approximately equal to 107.11.

2. $\pi(107.1)^2$

So, $\pi(107.1)^2$ is approximately equal to 36,035.36.

## SECTION 2.5 / EXERCISE SET

 **Write About It!**

**Use complete sentences to explain the meaning of each term.**

1. Formula   Answers vary.
2. Perimeter   Answers vary.
3. Area   Answers vary.
4. Circumference   Answers vary.
5. Complementary angles   Answers vary.
6. Supplementary angles   Answers vary.
7. Straight angle   Answers vary.
8. Vertical angles   Answers vary.

Additional answers can be found in the Instructor Answer Appendix.

**Determine if each statement is true or false. If a statement is false, explain why.**

9. When $8x - y = 16$ is solved for $y$, we get $y = -8x$. False, we can't combine $16 - 8x$. We should get $y = 8x - 16$.

10. If the measure of an angle is $a$, then the measure of the angle's supplement is $a - 180$.   False, $180 - a$

11. If the radius of a circle is 10 ft, the area of the circle is $100\pi$ ft².   True

12. If angles with measures $(4a - 5)°$ and $(2a + 20)°$ are vertical angles, then $4a - 5 + 2a + 20 = 180$. False, it means that $4a - 5 = 2a + 20$.

## Practice Makes Perfect!

**The formula $A = P(1 + r)^t$ is used to calculate the amount of money in an account at the end of $t$ years if $P$ dollars are invested at an annual interest rate, $r$. (*See Objective 1.*)**

13. Find the value of $A$ if $8000 is invested for 4 yr at 2.5% annual interest. $8830.50

14. Find the value of $A$ if $6000 is invested for 3 yr at 1.3% annual interest. $6237.06

15. Find the value of $A$ if $12,000 is invested for 2 yr at 2% annual interest. $12,484.80

16. Find the value of $A$ if $15,000 is invested for 2 yr at 1.8% annual interest. $15,544.86

**The formula $d = rt$ calculates the distance, $d$, traveled by an object traveling at an average rate $r$ for time $t$. (*See Objective 1.*)**

17. Find $d$ if $r = 60$ mph and $t = 5$ hr. 300 mi

18. Find $d$ if $r = 75$ mph and $t = 2$ hr. 150 mi

19. Find $r$ if $d = 440$ mi and $t = 8$ hr. 55 mph

20. Find $r$ if $d = 243$ mi and $t = 4.5$ hr. 54 mph

21. Find $t$ if $d = 80$ mi and $r = 40$ mph. 2 hr

22. Find $t$ if $d = 156$ mi and $r = 65$ mph. 2.4 hr

**The formula $V = lwh$ calculates the volume of a box with length $l$, width $w$, and height $h$. (*See Objective 1.*)**

23. Find $V$ if $l = 10$ in., $w = 5$ in., and $h = 2$ in. 100 in.³

24. Find $V$ if $l = 3$ ft, $w = 4$ ft, and $h = 6$ ft. 72 ft³

25. Find $w$ if $V = 28$ yd³, $l = 7$ yd, and $h = 2$ yd. 2 yd

26. Find $h$ if $V = 960$ m³, $l = 12$ m, and $w = 10$ m. 8 m

**Use the perimeter, area, or circumference formula to find the indicated quantity. (*See Objective 2.*)**

27. Find the area of a triangle whose base is 10 in. and whose height is 5 in. 25 in.²

28. Find the area of a triangle whose base is 9 ft and whose height is 14 ft. 63 ft²

29. Find the height of a triangle whose area is 20 in.² and whose base is 5 in. 8 in.

30. Find the base of a triangle whose area is 35 cm² and whose height is 7 cm. 10 cm

31. Find the perimeter of a rectangle whose length is 6 ft and whose width is 8 ft. 28 ft

32. Find the perimeter of a rectangle whose length is 12 in. and whose width is 20 in. 64 in.

33. Find the length of a rectangle if the perimeter is 24 m and the width is 7 m. 5 m

34. Find the width of a rectangle if the perimeter is 180 ft and the length is 40 ft. 50 ft

35. If the area of a rectangle is 32 ft² and the width is 8 ft, find the perimeter of the rectangle. 24 ft

36. If the area of a rectangle is 50 in.² and the length is 10 in., find the perimeter of the rectangle. 30 in.

37. Find the circumference and area of a circle with radius 8 cm. Use 3.14 for the value of $\pi$ to approximate answers to two decimal places. 50.24 cm; 200.96 cm²

38. Find the circumference and area of a circle with radius 12 in. Use 3.14 for the value of $\pi$ to approximate answers to two decimal places. 75.36 in.; 452.16 in.²

39. If the circumference of a circle is 157 ft, find the radius of the circle and the area of the circle. Use 3.14 for the value of $\pi$ to approximate answers to two decimal places. 25 ft; 1962.5 ft²

40. If the circumference of a circle is 200 in., find the radius of the circle and the area of the circle. Use 3.14 for the value of $\pi$ to approximate answers to two decimal places. 31.85 in.; 3185.29 in.²

**Solve each formula for the specified variable. (*See Objective 3.*)**

41. $A = lw$ for $l$ $\quad l = \dfrac{A}{w}$

42. $d = rt$ for $t$ $\quad t = \dfrac{d}{r}$

43. $P = a + b + c$ for $c$ $\quad c = P - a - b$

44. $P = 2l + 2w$ for $w$

45. $x + 2y = 4$ for $x$ $\quad x = -2y + 4$

46. $x + 2y = 4$ for $y$

47. $4x - y = 8$ for $x$ $\quad x = \dfrac{1}{4}y + 2$

48. $4x - y = 8$ for $y$ $\quad y = 4x - 8$

49. $y = \dfrac{1}{4}x + 5$ for $x$ $\quad x = 4y - 20$

50. $y = \dfrac{1}{2}x - 3$ for $x$ $\quad x = 2y + 6$

51. $7x - 3y = -21$ for $x$

52. $3x - 5y = -15$ for $y$

53. $A = P + Prt$ for $r$ $\quad r = \dfrac{A - P}{Pt}$

54. $A = P(1 + rt)$ for $P$

**Find the measure of each unknown angle. (*See Objective 4.*)**

55. Find the measure of an angle whose complement is twice the measure of the angle. 30°

56. Find the measure of an angle whose complement is eight times the measure of the angle. 10°

57. Find the measure of an angle whose complement is 10° more than three times the measure of the angle. 20°

58. Find the measure of an angle whose complement is 15° more than four times the measure of the angle. 15°

59. Find the measure of an angle whose supplement is eight times the measure of the angle. 20°

60. Find the measure of an angle whose supplement is five times the measure of the angle. 30°

61. Find the measure of an angle whose supplement is 16° less than six times the measure of the angle. 28°

62. Find the measure of an angle whose supplement is 12° less than twice the measure of the angle. 64°

63. The supplement of an angle is 20° less than three times its complement. Find the measure of the angle. 35°

64. The supplement of an angle is 18° less than four times its complement. Find the measure of the angle. 54°

**Find the measure of each angle labeled in the figure. (*See Objective 5.*)**

65.                                                                 53°, 127°

$(2a + 21)°$   $a°$

**66.** 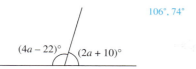   106°, 74°

$(4a - 22)°$   $(2a + 10)°$

**67.** 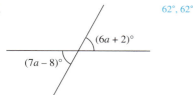   62°, 62°

$(6a + 2)°$

$(7a - 8)°$

**68.** 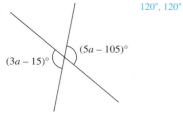   120°, 120°

$(5a - 105)°$

$(3a - 15)°$

**Find the measure of each angle in the triangle that is illustrated or described.** (*See Objective 6.*)

**69.** 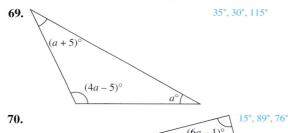   35°, 30°, 115°

$(a + 5)°$

$(4a - 5)°$

$a°$

**70.**   15°, 89°, 76°

$(6a - 1)°$

$(5a + 1)°$

$a°$

**71.** In triangle $ABC$, the measure of angle $B$ is 8° more than the measure of angle $A$. The measure of angle $C$ is 37° more than the measure of angle $A$. Find the measure of each angle in the triangle.   45°, 53°, 82°

**72.** In triangle $ABC$, the measure of angles $A$ and $B$ are the same. The measure of angle $C$ is 105° larger than the each of angle $A$ and $B$. Find the measure of each angle in the triangle.   25°, 25°, 130°

**Mix 'Em Up!**

**Find the requested information.**

**73.** Use the formula $A = P(1 + r)^t$ to find the value of $A$ if $5000 is invested for 2 yr at 4% annual interest.   $5408

**74.** Use the formula $A = P(1 + r)^t$ to find the value of $A$ if $13,500 is invested for 5 yr at 5.5% annual interest.   $17,643.96

**75.** If the area of a rectangle is 1290 m² and the width is 43 m, find the perimeter of the rectangle.   146 m

**76.** If the area of a rectangle is 240 mm² and the length is 40 mm, find the perimeter of the rectangle.   92 mm

**77.** The revenue $R$, unit price $p$, and sale level $x$ of a product are related by the formula $R = xp$. Use this formula to find $R$ if $p = $26 and $x = 120$.   $3120

**78.** The revenue $R$, unit price $p$, and sale level $x$ of a product are related by the formula $R = xp$. Use this formula to find $R$ if $p = $15 and $x = 65$.   $975

**79.** The formula to find the volume of a cylinder with radius $r$ and height $h$ is $V = \pi r^2 h$. Use $\pi \approx 3.14$ to find $V$ if $r = 3$ in. and $h = 2$ in.   56.52 in.³

**80.** The formula to find the volume of a cylinder with radius $r$ and height $h$ is $V = \pi r^2 h$. Use $\pi \approx 3.14$ to find $V$ if $r = 2$ ft and $h = 7$ ft.   87.92 ft³

**81.** Use the formula $d = rt$ to find $d$ if $r = 45$ mph and $t = 3.2$ hr.   144 mi

**82.** Use the formula $d = rt$ to find $r$ if $d = 244.8$ mi and $t = 5.1$ hr.   48 mph

**83.** Solve $y = -\dfrac{3}{2}x + 3$ for $x$.   $x = -\dfrac{2}{3}y + 2$

**84.** Solve $y = \dfrac{7}{2}x - 21$ for $x$.   $x = \dfrac{2}{7}y + 6$

**85.** The formula to find the volume of a cylinder with radius $r$ and height $h$ is $V = \pi r^2 h$. Use $\pi \approx 3.14$ to find $h$ if $V = 1384.74$ m³ and $r = 4.2$ m.   25 m

**86.** The formula to find the volume of a cylinder with radius $r$ and height $h$ is $V = \pi r^2 h$. Use $\pi \approx 3.14$ to find $h$ if $V = 615.44$ yd³ and $r = 3.5$ yd.   16 yd

**87.** Use the formula $A = P(1 + r)^t$ to find the value of $P$ if $23,500 is in an account at the end of 8 yr at 5% annual interest.   $15,905.73

**88.** Use the formula $A = P(1 + r)^t$ to find the value of $P$ if $16,500 is in an account at the end of 4 yr at 5.9% annual interest.   $13,118.98

**89.** Solve $R = xp$ for $x$.   $x = \dfrac{R}{p}$

**90.** Solve $V = IR$ for $I$.   $I = \dfrac{V}{R}$

**91.** Solve $0.4x - y = 100$ for $x$.   $x = 2.5y + 250$

**92.** Solve $x + 0.2y = 12$ for $y$.   $y = -5x + 60$

**93.** Find the measure of an angle whose complement is 12° less than the measure of the angle.   51°

**94.** Find the measure of an angle whose supplement is 40° more than three times the measure of the angle.   35°

**95.** Solve $C = 6l + 4w$ for $w$.   $w = \dfrac{C - 6l}{4}$

**96.** Solve $P = 2l + 2w$ for $l$.   $l = \dfrac{P - 2w}{2}$

**97.** Find the measure of each angle labeled in the figure.

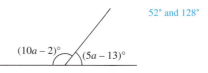   52° and 128°

$(10a - 2)°$   $(5a - 13)°$

**98.** Find the measure of each angle labeled in the figure.

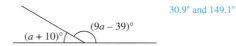   30.9° and 149.1°

$(9a - 39)°$

$(a + 10)°$

**99.** In triangle $ABC$, the measure of angles $A$ and $B$ are the same. The measure of angle $C$ is 27° larger than the measure of angle $A$. Find the measure of each angle in the triangle.   51°, 51°, and 78°

**100.** In triangle $ABC$, the measure of angles $A$ and $B$ are the same. The measure of angle $C$ is 6° smaller than the measure of angle $A$. Find the measure of each angle in the triangle.   62°, 62°, and 56°

**101.** Find the height of a triangle whose area is 537.5 m² and whose base is 25 m.   43 m

**102.** Find the height of a triangle whose area is 693 cm² and whose base is 33 cm.   42 cm

**103.** Solve $0.02x + 0.05y = 80$ for $y$.   $y = -0.4x + 1600$

**104.** Solve $0.05x + 0.01y = -50$ for $x$.   $x = -0.2y - 1000$

**105.** Find the radius and area of a circle with circumference 119.32 cm. Use 3.14 for the value of $\pi$ to approximate answers to two decimal places.   19 cm; 1133.54 cm²

**106.** Find the radius and area of a circle with circumference 175.84 in. Use 3.14 for the value of $\pi$ to approximate answers to two decimal places.   28 in.; 2461.76 in.²

**107.** Find the measure of each angle labeled in the figure.

29° and 29°

**108.** Find the measure of each angle labeled in the figure.

51.5° and 51.5°

**109.** Solve $5x - y = -10$ for $x$.   $x = \frac{1}{5}y - 2$

**110.** Solve $3x + 2y = 12$ for $x$.   $x = -\frac{2}{3}y + 4$

**111.** The supplement of an angle is 24° more than four times its complement. Find the measure of the angle.   68°

**112.** The supplement of an angle is 19° less than five times its complement. Find the measure of the angle.   62.75°

**113.** Find the measure of each angle labeled in the figure.

107°, 50°, and 23°

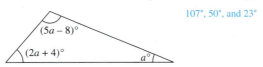

**114.** Find the measure of each angle labeled in the figure.

94.5°, 67.2°, and 18.3°

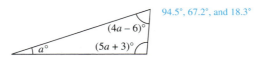

**115.** In triangle $ABC$, the measure of angle $B$ is 9° less than the measure of angle $A$. The measure of angle $C$ is 27° less than the measure of angle $A$. Find the measure of each angle in the triangle.   72°, 63°, and 45°

**116.** In triangle $ABC$, the measure of angle $B$ is 16° more than the measure of angle $A$. The measure of angle $C$ is 2° more than the measure of angle $A$. Find the measure of each angle in the triangle.   54°, 70°, and 56°

**117.** The circumference of the inner circle of a washer measures 50.3 in., and the circumference of the outer circle measures 370.7 in. Find the radius of the inner and outer circles. Find the area of the inner circle. Use 3.14 for the value of $\pi$. Round the values to two decimal places.   8.01 in.; 59.03 in.; 201.46 in.²

**118.** The circumference of the inner circle of a washer measures 88.0 mm, and the circumference of the outer circle measures 157.1 mm. Find the radius of the inner and outer circles. Find the area of the inner circle. Use 3.14 for the value of $\pi$. Round the values to two decimal places.   14.01 mm; 25.02 mm; 616.32 mm²

## You Be the Teacher!

**Correct each student's errors, if any.**

**119.** Find the measure of each angle labeled in the figure.

Emma's work:

$(7a - 20) + (5a + 8) = 180$      $7a - 20 = 5a + 8$

$\qquad\qquad 12a - 12 = 180$       $2a - 20 = 8$

$\qquad\qquad\quad 12a = 192$        $2a = 28$

$\qquad\qquad\qquad a = 16$          $a = 14$

$7a - 20 = 7(16) - 20 = 92$     $7a - 20 = 7(14) - 20 = 78$

$5a + 8 = 5(16) + 8 = 88$       $5a + 8 = 5(14) + 8 = 78$

$\qquad\qquad\qquad\qquad\qquad$ The two angles are both 78°.

**120.** Solve: $F = \frac{9}{5}C + 32$ for $C$.

Cynthia's work:

$$F = \frac{9}{5}C + 32 \qquad\qquad F = \frac{9}{5}C + 32$$

$$5 \cdot F = 5 \cdot \frac{9}{5}C + 32 \qquad 5 \cdot F = 5 \cdot \frac{9}{5}C + 5 \cdot 32$$

$$5F = 9C + 32 \qquad\qquad 5F = 9C + 160$$

$$5F - 32 = 9C \qquad\qquad 5F - 160 = 9C$$

$$C = \frac{5F - 32}{9} \qquad\qquad C = \frac{5F - 160}{9}$$

**121.** Solve: $A = 2\pi r^2 + 2\pi rh$ for $h$.

Kimberly's work:

$$A = 2\pi r^2 + 2\pi rh \qquad\qquad A = 2\pi r^2 + 2\pi rh$$

$$A - 2\pi r^2 = 2\pi rh \qquad\qquad A - 2\pi r^2 = 2\pi rh$$

$$\frac{A - 2\pi r^2}{2\pi rh} = \frac{2\pi rh}{2\pi rh} \qquad \frac{A - 2\pi r^2}{2\pi r} = \frac{2\pi rh}{2\pi r}$$

$$\frac{A - 2\pi r^2}{2\pi rh} = h \qquad\qquad h = \frac{A - 2\pi r^2}{2\pi r}$$

**122.** The supplement of an angle is 14° less than six times its complement. Calculate the measure of the angle.

Francis' work:

$180 - x = 14 - 6(x - 90)$       $180 - x = 6(90 - x) - 14$

$180 - x = 14 - 6x + 540$       $180 - x = 540 - 6x - 14$

$5x = 374$                    $180 + 5x = 526$

$x = 74.8$                    $5x = 346$

                              $x = 69.2$

**Calculate It!**

**Solve each equation. Then use a graphing calculator to check the answer.**

**123.** $0.03(x + 5.4) = 0.02(x - 1.8) + x$   {0.2}

**124.** $180 - x = 5(90 - x) - 24.8$   {61.3}

**125.** $5x - 32.6 + 3(x + 4.2) = 180$   {25}

**126.** The formula to find the volume of a cylinder with radius $r$ and height $h$ is $V = \pi r^2 h$. Use $\pi = 3.14$ to find $V$ if $r = 15.6$ ft and $h = 6.5$ ft.   4966.98 ft³

---

| SECTION 2.6 | **Percent, Rate, and Mixture Problems** |
|---|---|

▶ **OBJECTIVES**

**As a result of completing this section, you will be able to**

**1.** Solve percent applications.

**2.** Solve simple interest applications.

**3.** Solve mixture applications.

**4.** Solve distance, rate, and time applications.

**5.** Troubleshoot common errors.

According to Oprah's Debt Diet, how a family should spend their monthly income is based on this pie chart. Because the pie chart illustrates how a family should allocate all of their monthly income, the percents add to 100%. If a family following this "diet" spends $1400 on housing each month, what is their monthly income? (Source: www.oprah.com)

   We will learn how to solve this problem using a linear equation.

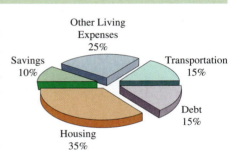

**Percent Applications**

**Objective 1** ▶

Solve percent applications.

Many real-life situations involve percents—paying taxes, sale discounts, investments, commissions, and so on. In this section, we will explore applications of linear equations involving percents. Recall that percent means per hundred. For example, $5\% = \dfrac{5}{100} = 0.05$. Percents can be converted to decimals by dropping the percent sign and moving the decimal two places to the left.

   Following are four specific types of applications of percents.

---

**Type 1: General Percent Problem**

An example of a general percent problem is to answer the question "30 is 15% of what number?" To answer this question, we let $n$ represent the unknown number. Recall that 15% of a number is equivalent to multiplying the number by 0.15. So, we have that

$$30 = 0.15n$$

$$\frac{30}{0.15} = \frac{0.15n}{0.15}$$

$$200 = n$$

The number is 200.

**Type 2: Discount Problem**

Suppose a sweater that is regularly priced at $36 is on sale for 25% off. To find the sale price, we must first find the amount to be taken off the original price, which is the discount amount.

$$\text{Discount} = 25\% \text{ of } \$36$$
$$= 0.25(36)$$
$$= 9$$

$$\text{Sale price} = \text{original price} - \text{discount}$$
$$= \$36 - \$9$$
$$= \$27$$

**Type 3: Markup Problem**

Suppose a bookstore pays $80 for a textbook. Then the bookstore has a markup of 40% to sell to the customer. What is the selling price of the book?

To find the price of the book, we must find the amount of the markup.

$$\text{Markup} = 40\% \text{ of } \$80$$
$$= 0.40(80)$$
$$= 32$$

$$\text{Selling price} = \text{cost} + \text{markup}$$
$$= \$80 + \$32$$
$$= \$112$$

**Type 4: Percent Increase or Decrease**

If a salary increased from $30,000 to $35,000, what percent increase does this represent? To find the percent increase, we must find the change in the values and divide this by the original value.

$$\frac{\text{New amount} - \text{original amount}}{\text{original amount}} = \frac{35,000 - 30,000}{30,000}$$
$$= \frac{5000}{30,000}$$
$$= 0.1667$$

So, the salary increased by 16.67%.

---

**Objective 1 Examples** | **Write an equation that represents each situation. Solve the equation and answer the question in a complete sentence.**

**1a.** According to Oprah's Debt Diet, how a family should spend their monthly income is based on this pie chart. If a family following this "diet" spends $1400 on housing each month, what is their monthly income? (Source: www.oprah.com)

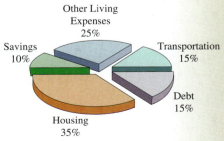

Other Living Expenses 25%
Savings 10%
Transportation 15%
Debt 15%
Housing 35%

**Solution** | **1a.** What is unknown? The monthly income is unknown.

Let $x$ = monthly income.

What is known? 35% of the monthly income should be spent on housing. The amount spent on housing is $1400.

35% of monthly     amount spent on
income   is   housing

$$0.35x = 1400$$          Express the relationship.

$$\frac{0.35x}{0.35} = \frac{1400}{0.35}$$          Divide each side by 0.35.

$$x = 4000$$          Simplify.

So, the monthly income of this family is $4000.

**1b.** An electronics store has a sale price of $1559.99 on a 50-in. plasma HDTV. This price is 40% off the original selling price. What is the original selling price of the television?

**Solution**   **1b.**   What is unknown? The original selling price of the TV is unknown.

Let $x$ = the original selling price of the TV.

What is known? The sale price is $1599.99.

The sale price is 40% off the original selling price.

To write the equation, we use the following relationship.

Sale price = original price − discount amount

$$1559.99 = x - 0.40x$$          Express the relationship. The discount amount is 0.40x.

$$1559.99 = 1x - 0.40x$$          Write x as 1x.

$$1559.99 = 0.60x$$          Combine like terms: 1x − 0.40x = (1 − 0.40)x = 0.60x

$$\frac{1559.99}{0.60} = \frac{0.60x}{0.60}$$          Divide each side by 0.60.

$$2599.98 = x$$          Simplify.

So, the original selling price of the plasma HDTV was $2599.98.

**1c.** Nanette receives a 5% raise at work. She now makes $26,250 per year. What was her annual salary before the raise?

**Solution**   **1c.**   What is unknown? The salary before the raise is unknown.

Let $x$ = Nanette's original salary.

What is known? She received a 5% raise. So, her raise is 0.05x.

This problem is similar to the markup illustration. So, we use the following relationship to write the equation.

Original salary + raise = new salary

$$x + 0.05x = 26{,}250$$          Express the relationship.

$$1x + 0.05x = 26{,}250$$          Write x as 1x.

$$1.05x = 26{,}250$$          Combine like terms.

$$\frac{1.05x}{1.05} = \frac{26{,}250}{1.05}$$          Divide each side by 1.05.

$$x = 25{,}000$$          Simplify.

So, Nanette's salary before her raise was $25,000.

**1d.** In fall 2000, enrollment in public preK–12th grade nationwide was 47.2 million. In fall 2010, the enrollment in preK–12th grade was 50 million. What was the percent increase in enrollment from 2000 to 2010?

**Solution**

**1d.** What is unknown? The percent increase in enrollment between 2000 and 2010 is unknown.

Let $x$ = percent increase in enrollment in schools.

What is known? The original enrollment (in 2000) was 47.2 million. The new enrollment (in 2010) was 50.0 million.

To solve this problem, we use the following relationship.

$$\text{Percent increase or decrease} = \frac{\text{new amount} - \text{original amount}}{\text{original amount}}$$

$$x = \frac{50 - 47.2}{47.2}$$ 　Replace the new amount with 50 and the original amount with 47.2.

$$x = \frac{2.8}{47.2}$$ 　Simplify the numerator.

$$x \approx 0.059$$ 　Divide.

So, the enrollment increased by 5.9% from fall 2000 to fall 2010.

---

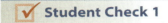

**Student Check 1** Write an equation that represents each situation. Solve the equation and answer the question in a complete sentence.

**a.** Using the percentages from Oprah's Debt Diet, what is a family's monthly income if they spend $450 on transportation each month?

**b.** A shirt is on sale for $21. This is 30% off the original price. What is the original selling price of the shirt?

**c.** Aisha's salary after her raise is $55,640. Prior to the raise, her salary was $52,000. What percent raise did she receive?

**d.** The national debt was $10.6 trillion on the day President Obama took office. The Obama Administration's 4-year estimate shows that by the end of September 2012, the debt will reach $16.2 trillion. What is the projected percent increase in the national debt from 2009 to 2012? (Source: http://www.cbsnews.com/blogs/2009/03/17/politics/politicalhotsheet/entry4872310.shtml)

---

## Simple Interest Problems

**Objective 2** ▶

Solve simple interest applications.

**Interest** is the amount of money collected for investing or borrowing money. Banks pay its customers interest when they invest money into their bank. Customers pay banks interest when they borrow money from the bank.

One type of interest is simple interest. Simple interest is interest earned on the amount that was initially invested or borrowed. Simple interest is computed as follows.

**Property: Simple Interest Formula**

$$\text{Interest} = \text{principal} \times \text{rate} \times \text{time} \qquad \text{or} \qquad I = prt$$

Principal, $p$, is the amount initially invested or borrowed.

Rate, $r$, is the annual interest rate (must be converted to a decimal).

Time, $t$, is the years that the money was invested or borrowed.

| Objective 2 Examples | Use the simple interest formula to solve each problem. |
|---|---|

**2a.** Marcus invested $10,000 in two accounts. He invested $2000 in a savings account that pays 3.5% annual interest and the remaining amount in a CD that pays 6% annual interest. How much interest did he earn from the two accounts in one year?

**Solution**   **2a.** What is unknown? The total interest earned from the two accounts is unknown. What is known? A total of $10,000 was invested, $2000 invested at 3.5% and $8000 invested at 6%.

| Account | Principal | × | Rate | × Time = | Interest |
|---|---|---|---|---|---|
| Savings | 2000 | | 3.5% = 0.035 | 1 yr | (2000) (0.035) (1) = 70 |
| CD | 10,000 − 2000 = 8000 | | 6% = 0.06 | 1 yr | (8000) (0.06) (1) = 480 |

The total interest earned from the two accounts is given by

$$\text{Total interest} = (2000)(0.035)(1) + (8000)(0.06)(1)$$
$$= 70 + 480$$
$$= 550$$

So, Marcus earned a total of $550 in interest from the two accounts.

**2b.** Rosa has $5000 to invest. She invests part of her money in a savings account that earns 3% annual interest and the rest in a CD that pays 5% annual interest. If the total interest earned in 1 yr is $230, how much did she invest in each account?

**Solution**   **2b.** What is unknown? The amount invested in each of the accounts is unknown.
Let $x$ = amount invested in the savings account.
Then $5000 - x$ = amount invested in the CD.

What is known? The total invested is $5000 and the total interest earned from the two accounts is $230.

| Account | Principal | × | Rate | × Time = | Interest |
|---|---|---|---|---|---|
| Savings | $x$ | | 3% = 0.03 | 1 yr | $(x)(0.03)(1)$   or   $0.03x$ |
| CD | $5000 - x$ | | 5% = 0.05 | 1 yr | $(5000 - x)(0.05)(1)$   or   $0.05(5000 - x)$ |
| Total | $5000 | | | | $230 |

The sum of the interest from the savings account and the interest from the CD account is the total interest earned.

| | |
|---|---|
| $0.03x + 0.05(5000 - x) = 230$ | Express the relationship. |
| $0.03x + 250 - 0.05x = 230$ | Apply the distributive property. |
| $-0.02x + 250 = 230$ | Combine like terms. |
| $-0.02x + 250 - 250 = 230 - 250$ | Subtract 250 from each side. |
| $-0.02x = -20$ | Simplify. |
| $\dfrac{-0.02x}{-0.02} = \dfrac{-20}{-0.02}$ | Divide each side by −0.02. |
| $x = 1000$ | Simplify. |

So, Rosa invested $1000 in the 3% account and $5000 − $1000 = $4000 in the 5% account.

✓ **Student Check 2**   Use the simple interest formula to solve each problem.

**a.** Allison invested a total of $2000 in two accounts. She invested $1500 in a retirement account that pays 5.5% annual interest and the rest in a savings account that pays 4% annual interest. How much interest did she earn in 1 yr from these two accounts?

**b.** Huynh has $8000 to invest. She invests part of the money in a money market account that pays 6.5% annual interest and the rest of the money in a CD that pays 5% annual interest. If she earns $490 interest in 1 yr, how much did she invest in each account?

## Mixture Problems

**Objective 3** ▶

Solve mixture applications.

Mixture applications involve combining two substances together to obtain a third substance. For example, combining different types of coins or dollar bills to get a monetary total or mixing two different types of liquid solutions together to make a third type of liquid solution are all mixture problems.

The main point to remember when solving mixture problems is that we cannot add unlike quantities together. We must find a way to convert the different quantities to a like quantity.

When working with monetary mixture problems, we convert the given number of coins or bills to its total value. We discussed this in Section 2.3 but it is worth repeating here.

| Coins/Bills | Value of Coins/Bills |
|---|---|
| 5 quarters | $5(0.25) = \$1.25$ |
| 8 dimes | $8(0.10) = \$0.80$ |
| 4 $10 bills | $4(10) = \$40$ |
| $x$ nickels | $x(0.05) = \$0.05x$ |

To find the value of a collection of coins, multiply the number of coins by the monetary worth of the coin.

When working with liquid mixture problems, we must determine the amount of "pure" substance in each solution. To illustrate this concept, consider the percent alcohol in beer, wine, and liquor. Most law enforcement agencies consider each of the following equivalent to one drink for the purposes of calculating blood alcohol concentration. (Source: http://www.bloodalcoholcontent.org/alcoholinformation.html)

| Solution | Amount of Alcohol |
|---|---|
| 1.5 oz of 80 proof liquor (40% alcohol) | $1.5(0.40) = 0.6$ oz of alcohol |
| 12 oz of regular beer (5% alcohol) | $12(0.05) = 0.6$ oz of alcohol |
| 5 oz of table wine (12% alcohol) | $5(0.12) = 0.6$ oz of alcohol |

Though these three alcoholic beverages vary in quantity and percent alcohol, they each contain the same amount of alcohol. If we combine the beverages, the total amount of alcohol is the sum of the alcohol found in each beverage. This is true not only when dealing with alcoholic beverages, but for any combination of solutions with different concentrations.

**Note:** *To find the amount of substance in a solution, multiply the given amount by the strength of the solution.*

**Objective 3 Examples**   **Write an equation that can be used to solve each problem. Solve the equation and answer the question using a complete sentence.**

**3a.** Teresa has a collection of dimes and quarters. She has 40 less dimes than quarters. The value of her collection is $20.50. Find the number of dimes and quarters she has in her collection.

**Solution**   **3a.** What is unknown? The number of dimes and quarters is unknown.

Let $q$ = number of quarters.

Then $q - 40$ = number of dimes.

What is known? The total value of the collection is $20.50.

| Type of Coin | Number of coins | × Value of coin | = Total value of the coins |
|---|---|---|---|
| Quarters | $q$ | $0.25 | $0.25q$ |
| Dimes | $q - 40$ | $0.10 | $0.10(q - 40)$ |

The total value of the collection is the value of the quarters plus the value of the dimes.

| | |
|---|---|
| $0.25q + 0.10(q - 40) = 20.50$ | Express relationship. |
| $0.25q + 0.10q - 4 = 20.50$ | Apply the distributive property. |
| $0.35q - 4 = 20.50$ | Combine like terms. |
| $0.35q - 4 + 4 = 20.50 + 4$ | Add 4 to each side. |
| $0.35q = 24.50$ | Simplify. |
| $\dfrac{0.35q}{0.35} = \dfrac{24.50}{0.35}$ | Divide each side by 0.35. |
| $q = 70$ | Simplify. |

So, Teresa has 70 quarters and $70 - 40 = 30$ dimes in her collection.

**3b.** Louisa is a chemist and she needs to conduct an experiment with a solution that is 30% copper sulfate. She does not have a solution with this concentration but she does have 40 mL of a 25% copper sulfate solution and she also has a 60% copper sulfate solution. How many milliliters of the 60% copper sulfate solution does she need to mix with the 40 mL of the 25% solution to obtain a 30% copper sulfate solution?

**Solution**   **3b.** What is unknown? The amount of the 60% copper sulfate solution is unknown.

Let $x$ = the amount of the 60% solution.

What is known? She needs a solution that is 30% copper sulfate. She has 40 mL of a 25% copper sulfate solution.

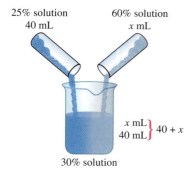

25% solution
40 mL

60% solution
$x$ mL

$x$ mL
40 mL } $40 + x$

30% solution

| Type of Solution | Number of mL | × Solution strength | = Total mL of copper sulfate |
|---|---|---|---|
| 25% | 40 mL | 25% or 0.25 | 0.25(40) |
| 60% | $x$ | 60% or 0.60 | $0.60x$ |
| 30% | $40 + x$ | 30% or 0.30 | $0.30(40 + x)$ |

The amount of copper sulfate in the 25% solution and the 60% solution equals the amount of copper sulfate in the 30% solution. This gives us the following equation.

$$0.25(40) + 0.60(x) = 0.30(40 + x) \qquad \text{Express the relationship.}$$

$$10 + 0.60x = 12 + 0.30x \qquad \text{Simplify.}$$

$$10 + 0.60x - 0.30x = 12 + 0.30x - 0.30x \qquad \text{Subtract } 0.30x \text{ from each side.}$$

$$10 + 0.30x = 12 \qquad \text{Simplify.}$$

$$10 + 0.30x - 10 = 12 - 10 \qquad \text{Subtract 10 from each side.}$$

$$0.30x = 2 \qquad \text{Simplify.}$$

$$\frac{0.30x}{0.30} = \frac{2}{0.30} \qquad \text{Divide each side by 0.30.}$$

$$x = \frac{20}{3} \text{ or } 6\frac{2}{3} \qquad \text{Simplify.}$$

Louisa needs to add $6\frac{2}{3}$ mL of the 60% copper sulfate solution to the 40 mL of 25% copper sulfate solution to obtain a 30% copper sulfate solution.

**3c.** A specialty coffee shop wants to make an After Dinner Blend from two of their coffees, French Roast and Java. The French Roast sells for $14 per pound and Java sells for $8 per pound. How many pounds of French Roast should be mixed with 10 lb of Java to produce an After Dinner Blend that sells for $12 per pound?

**Solution**  **3c.** What is unknown? The pounds of French Roast are unknown.

Let $x =$ pounds of French Roast.

What is known? French Roast sells for $14 per pound, Java sells for $8 per pound, and the blend sells for $12 per pound. There are 10 lb of Java.

| Type of Coffee | Pounds of coffee × | Price per pound = | Total price of coffee |
|---|---|---|---|
| French Roast | $x$ | $14 | $14x$ |
| Java | 10 | $ 8 | 8(10) |
| After Dinner Blend | $x + 10$ | $12 | $12(x + 10)$ |

The combined price of the French Roast and Java equals the total price of the After Dinner Blend. This gives us the following equation.

$$14x + 8(10) = 12(x + 10) \qquad \text{Express the relationship.}$$

$$14x + 80 = 12x + 120 \qquad \text{Simplify and distribute.}$$

$$14x + 80 - 12x = 12x + 120 - 12x \qquad \text{Subtract } 12x \text{ from each side.}$$

$$2x + 80 = 120 \qquad \text{Simplify.}$$

$$2x + 80 - 80 = 120 - 80 \qquad \text{Subtract 80 from each side.}$$

$$2x = 40 \qquad \text{Simplify.}$$

$$\frac{2x}{2} = \frac{40}{2} \qquad \text{Divide each side by 2.}$$

$$x = 20 \qquad \text{Simplify.}$$

To make the blend, 20 lb of French Roast should be mixed with 10 lb of Java.

**✓ Student Check 3**  Write an equation that can be used to solve each problem. Solve the equation and answer the question using a complete sentence.

**a.** Damien has a collection of nickels and dimes. He has twice as many dimes as nickels. The value of his collection is $5.25. Find the number of nickels and dimes in his collection.

**b.** Todd needs to make 4 gal of a 40% antifreeze solution for his truck. How much pure antifreeze and how much water should he mix together to get 4 gal of a 40% antifreeze solution?

**c.** The Nut Company mixes pecans that sell for $8.25 per pound and walnuts that sell for $6.45 per pound to make 50 lb of a mixture that sells for $7.15 per pound. How much of each kind of nut should be put in the mixture?

## Distance Problems

**Objective 4** ▶

Solve distance, rate, and time applications.

As illustrated in Section 2.5, Example 1c, distance traveled depends on one's speed and the time traveled.

> **Property: Distance Formula**
>
> $$\text{Distance} = \text{rate} \times \text{time} \qquad \text{or} \qquad d = rt$$
>
> Distance is measured in miles, feet, kilometers, and the like.
> Rate is measured in miles per hour, feet per second, kilometers per hour, and the like.
> Time is measured in hours, minutes, seconds, and the like.

We can use this formula to determine how far a car travels if it is driven at a speed of 65 mph for 2 hr.

$$d = (65 \text{ mph})(2 \text{ hr}) = 130 \text{ mi}$$

For these word problems, there will be two people/things that are traveling. So, there will be two distances, rates, and times. One of these quantities will be given for each people/things. The other quantity will be unknown. The remaining quantity will be obtained using the distance formula. It is helpful to organize the information in a chart.

**Objective 4  Examples**  Use the distance formula to write an equation that will solve each problem. Solve the equation and answer the question using a complete sentence.

**4a.** Tatiana and Blair live 390 mi apart. They leave their homes at the same time and drive towards one another until they meet. Tatiana is traveling at 60 mph and Blair is traveling at 70 mph. How long will they drive until they meet one another?

**Solution**    **4a.** What is unknown? The time that Tatiana and Blair travel is unknown. Since they leave at the same time and travel until they meet, their times are the same.

Let $t$ = the time Tatiana and Blair travel.

What is known? They live 390 mi apart. Tatiana travels at a rate of 60 mph. Blair travels at a rate of 70 mph.

|         | Rate | ×  Time | = Distance |
|---------|------|---------|------------|
| Tatiana | 60   | $t$     | $60t$      |
| Blair   | 70   | $t$     | $70t$      |

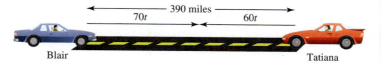

Since Blair and Tatiana are traveling toward each other, their combined distance traveled is 390 mi. We use this to write the equation.

Tatiana's distance + Blair's distance = 390

$$60t + 70t = 390$$    Express the relationship.

$$130t = 390$$    Combine like terms.

$$\frac{130t}{130} = \frac{390}{130}$$    Divide each side by 130.

$$t = 3$$    Simplify.

So, they both traveled for 3 hr until they met. Note that Tatiana drove $60(3) = 180$ miles in 3 hr and Blair drove $70(3) = 210$ miles in 3 hr.

**4b.** Highway inspectors are examining a highway. Two inspectors start from the same point but one travels north and the other travels south. The inspector heading north leaves at 8:00 A.M. and travels at an average speed of 40 mph. The inspector heading south leaves at 10:00 A.M. the same day and travels at an average speed of 50 mph. When will the inspectors finish examining a 260-mi portion of the highway?

**Solution**    **4b.** What is unknown? The time that each inspector travels is unknown. The inspector heading south leaves 2 hr after the inspector heading north.

Let $t$ = time for the inspector heading north and
$t - 2$ = time for the inspector heading south.

What is known? The total distance covered is 260 mi. The north inspector has a rate of 40 mph and the south inspector has a rate of 50 mph.

|  | Rate × | Time = | Distance |
|---|---|---|---|
| North | 40 | $t$ | $40t$ |
| South | 50 | $t - 2$ | $50(t - 2)$ |

Since their total distance is 260 mi, the equation is as follows.

$$40t + 50(t - 2) = 260$$    Express the relationship.

$$40t + 50t - 100 = 260$$    Apply the distributive property.

$$90t - 100 = 260$$    Combine like terms.

$$90t - 100 + 100 = 260 + 100$$    Add 100 to each side.

$$90t = 360$$    Simplify.

$$\frac{90t}{90} = \frac{360}{90}$$    Divide each side by 90.

$$t = 4$$    Simplify.

The inspector heading north will be done in 4 hr and the inspector heading south will be done in $4 - 2 = 2$ hr. So, the job will be complete at 12:00 P.M.

✓ **Student Check 4**    Use the distance formula to write an equation that will solve each problem. Solve the equation and answer the question using a complete sentence.

**a.** Two cars leave the same town heading in opposite directions. One car travels at 50 mph and the other at 70 mph. How long will it take for the cars to be 480 mi apart?

**b.** Tom and Barbara live 500 mi apart. They leave their homes and travel toward one another. Tom travels at an average speed of 65 mph and Barbara travels at an average speed of 60 mph. Tom leaves 1 hr before Barbara. How long will they each travel before they meet?

**Objective 5** ▶

Troubleshoot common errors.

## Troubleshooting Common Errors

The most common errors in solving word problems occur in setting up the equation incorrectly. Example 5 illustrates some of these errors.

**Objective 5   Examples**   **A problem and an incorrect solution are given. Provide the correct solution and an explanation of the error.**

**5a.** A golf club is originally priced at $224.99 and is on sale for $149.99. What is the discount rate?

| Incorrect Solution | Correct Solution and Explanation |
|---|---|
| The original − discount = sale price. $$224.99 - x = 149.99$$ $$224.99 - x - 224.99 = 149.99 - 224.99$$ $$-x = -75$$ $$x = 75$$ So, the discount rate is 75%. | The discount amount is the discount rate times the original amount. So, the equation is $$224.99 - 224.99x = 149.99$$ $$-224.99x = -75$$ $$\frac{-224.99x}{-224.99} = \frac{-75}{-224.99}$$ $$x \approx 0.3333$$ So, the discount rate is 33.3%. |

**5b.** Jackson has $20,000 in a retirement account that he wants to invest in two different ways. He invests some of the money in a money market account that earns 5% annual interest and the other in a high risk stock fund that yields 10% annual interest. If he earned $1600 interest in a year, how much did he invest in each account? Write the equation that can be used to solve the problem.

| Incorrect Solution | Correct Solution and Explanation |
|---|---|
| Let $x$ = amount invested in 5% account. Then $x - 20,000$ = amount invested in 10% account. $$0.05x + 0.10(x - 20,000) = 1600$$ | The error is the amount remaining. If $x$ = amount invested in the 5% account, then $20,000 - x$ is the amount invested in the 10% account. $$0.05x + 0.10(20,000 - x) = 1600$$ |

## ANSWERS TO STUDENT CHECKS

**Student Check 1   a.** $3000   **b.** $30   **c.** 7%   **d.** 52.8%

**Student Check 2   a.** $102.50   **b.** $6000 in the 6.5% account and $2000 in the 5% account

**Student Check 3   a.** 21 nickels and 42 dimes   **b.** 1.6 gal of pure antifreeze and 2.4 gal of water   **c.** 19.44 lb of pecans and 30.56 lb of walnuts

**Student Check 4   a.** 4 hr   **b.** Barbara travels for 3.5 hr and Tom travels for 4.5 hr.

## SUMMARY OF KEY CONCEPTS

**1.** Percent applications arise in many situations. Convert percents to decimals before using them in an equation.

 • Discount problems: Original − discount = sale price

   The discount amount is a percentage of the original amount.

 • Markup: Original + markup = new price

 • Percent increase or decrease: $\dfrac{\text{new amount} - \text{original amount}}{\text{original amount}}$

 • Recall "of" means multiplication.

**2.** Simple interest is given by $I = Prt$. When a total amount of money is given that is split between two accounts,

let $x$ represent one amount and the remaining amount is represented as the "total amount − $x$."

**3.** Mixture problems are set up by adding the values/amounts from collections/solutions to equal a total value/amount. Remember that the quantities being added must be equivalent.

**4.** Distance = rate × time. Organize the information in a chart by using variables for one of the columns, given amounts for another column, and then the formula to derive the third column. The expressions obtained in the last column will be used in the equation.

## GRAPHING CALCULATOR SKILLS

There are no new calculator skills needed for this section.

## SECTION 2.6 / EXERCISE SET

### Write About It!

**Use complete sentences to explain the meaning of each term.**

**1.** Percent   **2.** Simple interest   **3.** Mixture applications
Answers vary.   Answers vary.            Answers vary.

**Determine if each statement is true or false. If a statement is false, explain why.**

**4.** The price of an item increased from $50 to $75. To find the percent increase, we must compute $\dfrac{75 - 50}{75}$.

**5.** If $6000 is split between two different investments, we can represent the two investment amounts as $x$ and $x - 6000$.
False, the amounts are $x$ and $6000 - x$.

**6.** James and Terry live 300 mi apart. They leave their homes at the same time traveling toward one another. James travels at 60 mph, and Terry travels at 72 mph. To find the time when James and Terry meet, we solve the equation $\dfrac{300}{60} = \dfrac{x}{72}$ for $x$.
False, we must solve the equation $60x + 72x = 300$.

### Practice Makes Perfect!

**Solve each problem involving percents. Round answers to two decimal places, if needed. (*See Objective 1.*)**

**7.** A computer store has a deal on a notebook computer. The computer is on sale for $699.99. This is 20% off the original price. What is the original price of the computer?   $874.99

**8.** A suit is on clearance at a department store for $60. If this price is 75% off the original price, what is the original price of the suit?   $240

**9.** A digital camera with a printer dock is regularly priced at $329.99. The camera is on sale for 15% off the regular price. What is the sale price of the camera?   $280.49

**10.** A leather sofa is regularly priced at $999.99. The furniture store is running a sale in which all furniture is 20% off. What is the sale price of the sofa?   $799.99

**11.** A campus bookstore sells a graphing calculator for $104. This is a 30% markup on the cost the bookstore pays for the calculator. What does the bookstore pay for the calculator?   $80

**12.** The price of an MP3 player is $200. This is a markup of approximately 122% from the actual cost to make the player. What is the cost to make the MP3 player?   $90.09

**13.** The price of a luxury car plus 7% sales tax is $45,582. What is the price of the car without tax?   $42,600

Additional answers can be found in the Instructor Answer Appendix.

**14.** Real estate commission is a percentage of the selling price of a house that is paid to realtors for services rendered. Michael is a real estate agent assisting a family in buying a new home. If Michael's commission rate is 7% of the selling price of the home and his total commission is $13,020, what is the selling price of the home?   $186,000

**15.** Selena receives a 5% raise at work, making her salary $31,500. What was her salary before the raise?   $30,000

**16.** Raj receives a 4% raise at work. He now makes $15.60 per hour. What was his hourly wage before the raise?   $15

**The pie chart shows a suggested breakdown for one's monthly income. Use the pie chart to answer each question.**

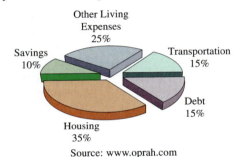

Source: www.oprah.com

**17.** The Donnellys' monthly income is $5000. According to the pie chart, what should they spend on their housing each month?   $1750

**18.** The Andersons' monthly income is $4200. According to the pie chart, what should they spend on their transportation each month?   $630

**19.** Sonja is a single mom and is trying to save money. If she saves $250 each month, what should her monthly income be according to the pie chart?   $2500

**20.** If Brandon spends $550 each month on other living expenses, what should his monthly income be according to the pie chart?   $2200

**According to a national education organization, the average time it takes for students to complete their bachelor's degrees varies based on the type and number of colleges they attend, as shown in the following table. Use this information for Exercises 21 and 22.**

| Students attending only one institution | 51 months |
|---|---|
| Students attending two institutions | 59 months |
| Students attending three or more institutions | 67 months |
| Students beginning at public 2-yr colleges | 71 months |
| Students attending only one private institution | 51 months |
| Students attending only public 4-yr colleges | 55 months |

21. What is the percentage increase in the time it takes to complete a bachelor's degree for students attending one public 4-yr college and students attending three or more institutions? Round to the nearest hundredth of a percent. 21.82%

22. What is the percentage increase in the time it takes to complete a bachelor's degree for students attending only one private institution and students beginning at public 2-yr colleges? Round to the nearest hundredth of a percent.   39.22%

23. New York City is the most populated city in the United States. The 2000 census reported that NYC had a population of 8,008,278. In 2009, the population was estimated to be 8,391,881. By what percent did the population increase from 2000 to 2009? Round to the nearest hundredth of a percent. (Source: www.census.gov) 4.79%

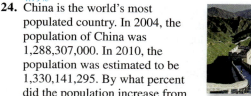

24. China is the world's most populated country. In 2004, the population of China was 1,288,307,000. In 2010, the population was estimated to be 1,330,141,295. By what percent did the population increase from 2004 to 2010? Round to the nearest hundredth of a percent. (Source: http://internetworldstats.com)   3.25%

**For each simple interest problem: (1) use a chart to organize the information, (2) write an appropriate equation, (3) solve the equation, and (4) write the answer in complete sentences. (*See Objective 2.*)**

25. Rob earns a $20,000 bonus for the year. He invests this money in two accounts. One account is highly risky but earns 11% annual interest. The other account is less risky and earns 5% annual interest. If Rob earns a total of $1300 in interest for the year, how much did he invest in each account? $5000 in the high risk account and $15,000 in the less risk account

26. Marissa has saved $4000 over the last several years. She invests part of her money in a CD that earns 6% annual interest and the rest in a savings account that pays 4% annual interest. If Marissa earns $210 in interest for the year, how much did she invest in each account?   $2500 in the CD and $1500 in the savings account

27. Isaiah invests some money into a money market account that earns 10% annual interest. He also invests $3000 less than the amount invested in the money market account in a CD that earns 5.5% annual interest. If he earns $1385 in interest for the year, how much did he invest in each account? $10,000 in the money market account and $7000 in a CD

28. Georgia receives an inheritance of $100,000. She invests this money in two different accounts. She invests part of the money in the stock market and earns an average of 12% annual interest. She invests the rest of the money in a money market account that earns 7% annual interest. If Georgia earns $11,000 in interest for the year, how much did she invest in each account?
$80,000 in the stock market and $20,000 in the money market account

**For each mixture problem: (1) use a chart to organize the information, (2) write an appropriate equation, (3) solve the equation, and (4) write the answer in complete sentences. (*See Objective 3.*)**

29. Thomas has a collection of nickels and quarters. He has three times as many quarters as nickels. If the collection is worth $24, how many nickels and quarters does he have? 30 nickels and 90 quarters

30. Samantha has a total of 300 coins, which consist of dimes and quarters. If her collection is worth $46.50, how many of each coin does she have? 190 dimes and 110 quarters

31. A bank teller has a collection of $10 bills and $20 bills. He has fifteen more $20 bills than $10 bills. The value of his collection is $900. How many of each bill does he have?   20 $10 bills and 35 $20 bills

32. A department store clerk has only $5 bills and $10 bills in her cash register. She has a total of 30 bills and the value of the collection is $215. How many of each bill does she have?   17 $5 bills and 13 $10 bills

33. A lab technician needs a 40% alcohol solution. He has 10 gal of a 35% alcohol solution. How many gallons of a 50% alcohol solution need to be mixed with the 10 gal of the 35% solution to obtain a 40% alcohol solution?   5 gal of 50% alcohol solution

34. A chemist has a 50% acid solution. She needs 30 L of a 15% acid solution. How many liters of water and how many liters of the 50% acid solution should be mixed together to obtain 30 L of a 15% acid solution? 21 L of water and 9 L of 50% acid solution

35. A dairy farmer has two types of milk. How many gallons of a 3% milk fat solution and how many gallons of a 4.5% milk fat solution should be mixed to obtain 750 gal of a 4% milk fat solution? 250 gal of 3% milk fat solution and 500 gal of 4.5% milk fat solution

36. How many liters of a 47% hydrochloric acid solution should be mixed with 4 L of a 30% hydrochloric acid solution to obtain a 35% hydrochloric acid solution? $1\frac{2}{3}$ L of 47% hydrochloric acid solution

**For each problem involving distance, rate, and time: (1) use a chart to organize the information, (2) write an appropriate equation, (3) solve the equation, and (4) write the answer in complete sentences. (*See Objective 4.*)**

37. Andy and Brittany live in towns 420 mi apart. They leave their homes at the same time and begin traveling toward one another. Andy travels at 65 mph and Brittany travels at 55 mph. How long will it take before they meet?   3.5 hr

38. Two planes leave Chicago's Midway airport. One plane travels east at 450 mph and the other plane travels west at 600 mph. How long will it take for the planes to be 2310 mi apart?   2.2 hr

39. Charlie leaves his home for a meeting traveling at an average rate of 50 mph. After his business meeting is over, he travels back home at an average rate of 70 mph. If his total driving time is 9 hr, how long does he drive going to and from his meeting? What is the total distance Charlie travels?
5.25 hr to the meeting; 3.75 hr to home; 525 mi total distance

**40.** Mohammad goes on a leisurely drive on a beautiful day. He leaves his home traveling at an average rate of 55 mph. He travels until he is tired and stops for lunch. After lunch, he decides to drive back home. He travels back home at an average speed of 65 mph. If his total driving time is 4 hr, how long did he drive each way? What is the total distance he traveled?

**41.** Sheree leaves her job and heads east at 55 mph. One hr later, Janis leaves the same workplace and heads east at 75 mph. How long will it take Janis to catch Sheree? 2.75 hr

**42.** A plane leaves Atlanta heading to San Francisco at 9 A.M. traveling at 400 mph. Another plane leaves Atlanta heading to San Francisco at 10 A.M. the same day traveling at 550 mph. How long will it take for the second plane to catch up with the first? 2 hr 40 min

 **Mix 'Em Up!**

**For each problem: (1) write an appropriate equation, (2) solve the equation, and (3) write the answer in complete sentences.**

**43.** A 46-in. LED-LCD HDTV is on sale for $1220.99. This is 45% off the original price. What is the original selling price of the TV? $2219.98

**44.** A 40-in. LCD-HDTV is on sale for $1614.99. This is 15% off the original price. What is the original selling price of the TV? $1899.99

**45.** Lynette's salary after her raise is $62,829. Prior to the raise, her salary was $58,300. What percent raise did she receive? 7.8%

**46.** Tyrone's salary after his raise is $51,450. Prior to the raise, his salary was $50,000. What percent raise did he receive? 2.9%

**47.** A bank teller has a collection of $1 bills and $5 bills. He has seven more $1 bills than $5 bills. The value of his collection is $229. How many of each bill does he have? 44 $1 bills and 37 $5 bills

**48.** A department store clerk has only $5 bills and $10 bills in her cash register. She has a total of 200 bills and the value of the collection is $1,565. How many of each bill does she have? 87 $5 bills and 113 $10 bills

**Use the percentages from Oprah's Debt Diet, to answer Exercises 49 and 50.**

| Oprah's Debt Diet | Percentages |
|---|---|
| Housing | 35% |
| Debt | 15% |
| Saving | 10% |
| Transportation | 15% |
| Other Living Expenses | 25% |

Source: www.oprah.com

**49.** The Basils' monthly income is $3000. According to the table, what should they spend on their debt each month? $450

**50.** The Morans' monthly income is $1714.29. According to the table, what should they spend on their housing each month? $600

**51.** Jim invests $10,200 in two accounts. He invests $4850 in a mutual fund that pays 2.2% annual interest and the remaining amount in a money market fund that pays 3.3% annual interest. How much interest does he earn from the two accounts in 1 yr? $283.25

**52.** Adrianne invests $4250 in two accounts. She invests $1650 in a mutual fund that pays 1.1% annual interest and the remaining amount in a money market fund that pays 5.3% annual interest. How much interest does she earn from the two accounts in 1 yr? $155.95

**53.** Two planes leave Wittman Regional Airport in Oshkosh. One plane travels west at 355 mph and the other plane travels east at 340 mph. How long will it take for the planes to be 3683.5 mi apart? 5.3 hr

**54.** Two cars leave the same town heading in opposite directions. One car travels at 69 mph and the other at 62 mph. How long will it take for the cars to be 275.1 mi apart? 2.1 hr

**55.** A dairy farmer has two types of milk. How many gallons of a 2% milk fat solution and how many gallons of a 4.5% milk fat solution should be mixed to obtain 500 gal of a 2.9% milk fat solution? 320 gal of 2% milk fat solution and 180 gal of 4.5% milk fat solution

**56.** Michael needs to make 243 L of a 58% alcohol solution. How much pure alcohol and how much 37% alcohol solution should be mixed together to obtain 243 L of a 58% alcohol solution? 81 L of pure alcohol solution and 162 L of 37% alcohol solution

**57.** Brandon and Kimberly live 320 mi apart. They leave their homes at the same time and drive toward one another until they meet. Brandon travels at 55 mph and Kimberly travels at 73 mph. How long will it take before they meet? 2.5 hr

**58.** Karen and Evan live 416 mi apart. They leave their homes at the same time and drive towards one another until they meet. Karen travels at 68 mph and Evan travels at 62 mph. How long will it take before they meet? 3.2 hr

**59.** John earns a $9100 bonus for the year. He invests part of this money in two accounts. One account is highly risky but earns 5.7% annual interest. The other account is less risky and earns 2.1% annual interest. If John earns a total of $367.50 in interest for the year, how much did he invest in each account? $4900 in the high risk account and $4200 in the less risky account

**60.** Maria has $11,750 to invest. She invests part of her money in a stock that earns 2.9% annual dividends and the rest in a mutual fund account that pays 5.5% annual dividends. If Maria earns $553.95 in dividends for the year, how much did she invest in each account? $3550 in the stock and $8200 in mutual fund account

**61.** Kevin has a collection of dimes and quarters. He has 10 more dimes than quarters. If the collection is worth $98.30, how many dimes and quarters does he have? 288 dimes and 278 quarters

**62.** Joan has a collection of dimes and nickels. She has 38 fewer dimes than nickels. If the collection is worth $7, how many dimes and nickels does she have? 34 dimes and 72 nickels

**63.** A sweater is on sale for $20.35. If this price is 45% off the original price, what is the original price of the sweater? $37

**64.** A jacket is on sale for $39.60. If this price is 55% off the original price, what is the original price of the jacket? $88

**65.** Alison needs to make 68 L of a 56% hydrochloric acid solution. How much 8% hydrochloric acid solution and how much 72% hydrochloric acid solution should she mix to get 68 L of a 56% hydrochloric acid solution? 17 L of 8% hydrochloric and 51 L of 72% hydrochloric acid solution

**66.** Yarrish has a 78% acid solution. He needs 143 L of a 24% acid solution. How many liters of water and how many liters of the 78% acid solution should he mix together to obtain 143 L of a 24% acid solution? 99 L of water and 44 L of 78% acid solution

**67.** The passenger volume at a major international airport was 41.89 million in 2005 and 47.81 million in 2010. What is the percent increase in the passenger volume from 2005 to 2010? Round to one hundredth of a percent. 14.13% increase

**68.** The passenger volume at an airport in Michigan was 36.39 million in 2005 and 35.14 million in 2011. What is the percent decrease in the passenger volume from 2005 to 2011? Round to one hundredth of a percent. 3.44% decrease

**69.** A campus bookstore sells an elementary algebra text for $93.60. This is a 30% markup on the cost the bookstore pays for the text. What does the bookstore pay for the text? $72

**70.** A campus bookstore sells a chemistry text for $125.35. This is a 15% markup on the cost the bookstore pays for the chemistry text. What does the bookstore pay for the text? $109

**71.** Jeanette leaves her job and heads west at 48 mph. Thirty minutes later, Warren leaves the same workplace and heads west at 68 mph. How long will it take Warren to catch up with Jeanette? 1.2 hr

**72.** A plane leaves Newark heading to Las Vegas at 7 A.M. traveling at 320 mph. A second plane leaves Newark heading to Las Vegas at 8:18 A.M. the same day traveling at 580 mph. How long will it take for the second plane to catch up with the first plane? 1.6 hr

 **You Be the Teacher!**

**Correct each student's errors, if any.**

**73.** The price of a 2010 SUV plus 8% sales tax is $52,629.48. Find the price of the car without tax.

Emma's work:
$$\text{sales tax} = 52,629.48(8\%)$$
$$= 52,629.48(0.08)$$
$$= 4210.3584$$
$$\approx 4210.36$$

$$\text{Price before tax} = 52,629.48 - 4210.36$$
$$= 48,419.12$$

**74.** A 32-in. LCD-HDTV is on sale for $209.99. This is 40% off the original price. Find the original selling price of the TV.

Cynthia's work:

$$\text{Original selling price} = \frac{209.99}{40\%}$$
$$= \frac{209.99}{0.4}$$
$$= 524.975$$
$$\approx 524.98$$

**75.** The passenger volume an international airport in Florida was 30.17 million in 2004 and 34.06 million in 2011. Find the percent increase in the passenger volume from 2004 to 2011.

Ralpha's work:

$$\text{Averge percent} = \frac{34.06 - 30.17}{34.06}$$
$$= \frac{3.89}{34.06}$$
$$= 0.114210$$
$$\approx 11.4\%$$

**76.** Haazim has $13,850 to invest. He invests part of the money in a stock that pays 4.5% annual dividends and the rest of the money in a mutual fund that pays 3.8% annual dividends. If he earns $581.95 in dividends in 1 yr, find the amount invested in each account.

Haazim's work:

Let $x$ be the amount invested in the stock and $x - 13,850$ in the mutual fund account.

$$4.5x + 3.8(x - 13850) = 581.95$$
$$4.5x + 3.8x - 52630 = 581.95$$
$$8.3x = 53211.95$$
$$x \approx 6411.08$$

**77.** Shan needs to make 210 L of a 60% acid solution. Calculate the number of liters of pure acid and the number of liters of 40% solution Shan needs to mix together to get 210 L of a 60% acid solution.

Shan's work:

Let $x$ be the number of liters of pure acid and $x - 210$ be the number of liters of 40%.

$$0x + 0.40(x - 210) = 0.60(210)$$
$$0.40x - 84 = 126$$
$$0.40x = 210$$
$$x = 210/0.40$$
$$x = 525$$

/ **Linear Inequalities in One Variable**

▶ **OBJECTIVES**

As a result of completing this section, you will be able to

1. Graph the solution set of an inequality.
2. Write the solution set of an inequality in interval notation.
3. Write the solution set of an inequality in set-builder notation.
4. Solve linear inequalities.
5. Solve applications of linear inequalities.
6. Troubleshoot common errors.

As math classes come to an end, a prevailing thought of many students is, "What grade do I need to make on the final to pass this class, or to keep my A?" In this section, we will look at one example of how to determine the grade needed on a final exam to achieve a certain grade in a class. This is an application of linear inequalities.

## Graphs of Inequalities

Until now, we have solved only linear equations in one variable. We now turn our focus to solving linear inequalities in one variable. Linear equations and linear inequalities are similar except that a linear inequality has an inequality symbol instead of an equals sign. In Chapter 1, inequality symbols were introduced.

> **Definition:**  A **linear inequality in one variable** is an inequality of the form $ax + b < c$, where $a$, $b$, and $c$ are real numbers with $a \neq 0$. Note that a linear inequality can have any of the inequality symbols: $<$, $\leq$, $>$, or $\geq$.

Some examples of linear inequalities in one variable are

$$2x + 3 < -7 \qquad 7x - 2 \geq 6x + 4 \qquad 4(3y - 1) < -2(y - 9)$$

Like linear equations, linear inequalities are neither true nor false until a value is assigned to the variable. Our job is to determine the values of the variable that make the inequality true. Each such value is a **solution of a linear inequality**.

Linear inequalities generally have infinitely many values that make the statement true. Because we will not be able to list all of these values, we graphically represent the solution set of all the solutions on a number line. So, before we actually solve a linear inequality, we will examine the graphs of possible solution sets.

### **Objective 1**  ▶

Graph the solution set of an inequality.

> **Procedure:  Graphing the Solution Set of a Linear Inequality**
>
> **Step 1:** Draw a number line.
> **Step 2:** Shade the portion of the number line that satisfies the inequality.
> **Step 3:** Put the appropriate symbol on the endpoint.
>   **a.** If the endpoint satisfies the inequality, then it is included in the solution set as denoted by a bracket. A closed circle can also be used in place of the bracket.
>   **b.** If the endpoint does not satisfy the inequality, then it is not included in the solution set as denoted by a parenthesis. An open circle can also be used in place of the parentheses.

**Objective 1  Examples**   **Graph the solution set of each inequality on a number line.**

   **1a.** $x > 3$         **1b.** $x \leq 4$         **1c.** $\dfrac{1}{2} < x$         **1d.** $-2 < x \leq 5$

   **Solutions**   **1a.** To graph the solutions of $x > 3$, we find the values of $x$ that are larger than 3. We know the integers 4, 5, 6, and so on are larger than 3, but we must not forget that there are numbers between 3 and 4 that are also larger than 3 (e.g., 3.9, 3.54, 3.0002, etc.). The number 3 itself is not a solution of this inequality since 3 is not greater than itself. So, we shade the portion of the number line that is to the right of 3. To indicate that 3 is not included in the solution set, put a parenthesis on it.

**1b.** To graph the solution of $x \leq 4$, we shade the portion of the number line that corresponds to values less than 4 or equal to 4. Numbers less than 4 are numbers to the left of 4. A bracket is used to show that 4 is included in the solution set since 4 equals 4.

**1c.** The inequality $\frac{1}{2} < x$ means that $\frac{1}{2}$ is smaller than all of the solutions, or that all of the solutions are larger than $\frac{1}{2}$. So, $\frac{1}{2} < x$ is equivalent to $x > \frac{1}{2}$. Therefore, we shade the portion of the number line to the right of $\frac{1}{2}$ and place a parenthesis on $\frac{1}{2}$ since it is not included in the solution set.

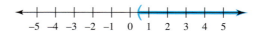

 **Note:** *The inequality $c < x$ is equivalent to $x > c$.*

**1d.** To graph the solutions of $-2 < x \leq 5$, we find the values that are between $-2$ and 5, including 5. So, we shade the portion of the number line that lies between these endpoints. The number $-2$ is not included since $-2$ is not less than $-2$.

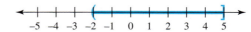

✓ **Student Check 1**   Graph the solution set of each inequality on a number line.

   **a.** $x > -2$     **b.** $x \leq -4.2$     **c.** $\frac{5}{3} > x$     **d.** $-4 \leq x < 2$

Examples 1a–1c illustrate a **simple inequality**. The solutions of simple inequalities have to satisfy one only inequality.

Example 1d illustrates a **compound inequality**. The solutions of compound inequalities must satisfy two inequalities, not just one. For instance,

$$-2 < x \leq 5 \text{ means that "}-2 < x \text{ and } x \leq 5 \text{"} \qquad \text{or} \qquad \text{"} x > -2 \text{ and } x \leq 5 \text{"}$$

### Interval Notation

**Objective 2** ▶

Write the solution set of an inequality in interval notation.

Solutions of inequalities can also be represented in a concise way that represents the smallest and largest values in the solution set. This notation is called **interval notation**.

When the graph of an inequality extends to the right indefinitely, we say that the numbers in the set approach $\infty$ (infinity). This means that the numbers continue getting larger without bound. When the graph of an inequality extends to the left indefinitely, we say that the numbers in the set approach $-\infty$ (negative infinity). This means that the numbers continue getting smaller without bound.

Interval notation makes use of parentheses and brackets like graphing solution sets of inequalities. If the endpoint is included in the solution set of the inequality, then a bracket is used with the endpoint. If the endpoint is not included, then a parenthesis is used.

Positive and negative infinity are always written with parentheses because they don't represent a specific number. Two examples are shown next.

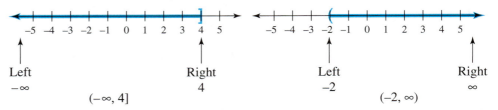

| Left | Right |
| :--- | :--- |
| $-\infty$ | 4 |
| | $(-\infty, 4]$ |

| Left | Right |
| :--- | :--- |
| $-2$ | $\infty$ |
| | $(-2, \infty)$ |

**Objective 2 Examples**

Complete the table by graphing each inequality and writing the interval notation that corresponds to the solution set.

| Inequality | Graph | Interval Notation |
| :--- | :--- | :--- |
| **2a.** $x > 3$ | A number line from 0 to 10 with an open parenthesis at 3 and shading to the right. | $(3, \infty)$ |
| **2b.** $x \leq 1.5$ | A number line from $-5$ to $5$ with a bracket at 1.5 and shading to the left. | $(-\infty, 1.5]$ |
| **2c.** $\dfrac{1}{2} < x$ | A number line from $-5$ to $5$ with an open parenthesis at $\frac{1}{2}$ and shading to the right. | $\left(\dfrac{1}{2}, \infty\right)$ |
| **2d.** $-2 < x \leq 5$ | A number line from $-5$ to $5$ with an open parenthesis at $-2$, a bracket at 5, and shading between. | $(-2, 5]$ |

**✓ Student Check 2**

Graph each inequality and express the solution set in interval notation.

**a.** $x > -2$    **b.** $x \leq -4.2$    **c.** $\dfrac{5}{3} > x$    **d.** $-4 \leq x < 2$

## Set-Builder Notation

**Objective 3** ▶

Write the solution set of an inequality in set-builder notation.

Set-builder notation is another way to represent the solution set of an inequality. Set-builder notation was briefly discussed in Chapter 1. **Set-builder notation** is used to state conditions that the solutions must satisfy. An example is shown.

| Set-builder Notation | Verbal Statement |
| :--- | :--- |
| $\{x \mid x \leq 4\}$ | The set of all $x$ such that $x$ is less than or equal to 4. |

**Objective 3 Examples**

Graph each inequality and express the solution set in set-builder notation.

| Inequality | Graph | Set-Builder Notation |
| :--- | :--- | :--- |
| **3a.** $x > 3$ | A number line from 0 to 10 with an open parenthesis at 3 and shading to the right. | $\{x \mid x > 3\}$ |
| **3b.** $x \leq 1.5$ | A number line from $-5$ to $5$ with a bracket at 1.5 and shading to the left. | $\{x \mid x \leq 1.5\}$ |
| **3c.** $\dfrac{1}{2} < x$ | A number line from $-5$ to $5$ with an open parenthesis at $\frac{1}{2}$ and shading to the right. | $\left\{x \mid x > \dfrac{1}{2}\right\}$ |
| **3d.** $-2 < x \leq 5$ | A number line from $-5$ to $5$ with an open parenthesis at $-2$, a bracket at 5, and shading between. | $\{x \mid -2 < x \leq 5\}$ |

✓ **Student Check 3**   Graph each inequality and express the solution set in set-builder notation.

**a.** $x > -2$      **b.** $x \leq -4.2$      **c.** $\dfrac{5}{3} > x$      **d.** $-4 \leq x < 2$

## Linear Inequalities

**Objective 4** ▶

Solve linear inequalities.

Solving linear inequalities is very similar to solving linear equations. The goal is the same: to isolate the variable to one side of the inequality. The properties that enable us to do this are the addition and multiplication properties of inequality. While only one inequality symbol is used in the statement of the following properties, the properties work with all inequality symbols.

> **Property:  Addition Property of Inequality**
>
> If $a < b$ and $c$ is a real number, then
>
> $$a + c < b + c \qquad \text{and} \qquad a - c < b - c.$$

The property illustrates that adding or subtracting a number from both sides of an inequality produces an equivalent inequality. For an illustration of this property, let $a = -5$, $b = 2$, and $c = 4$.

| $a < b$ | $a + c < b + c$ | $a - c < b - c$ |
|---------|-----------------|-----------------|
| $-5 < 2$ | $-5 + 4 < 2 + 4$ | $-5 - 4 < 2 - 4$ |
| True | $-1 < 6$ | $-9 < -2$ |
|  | True | True |

Adding or subtracting a number from both sides of an inequality maintains the inequality relationship.

> **Property:  Multiplication Property of Inequality**
>
> **1.** If $a < b$ and $c > 0$, then $ac < bc$ and $\dfrac{a}{c} < \dfrac{b}{c}$.
>
> **2.** If $a < b$ and $c < 0$, then $ac > bc$ and $\dfrac{a}{c} > \dfrac{b}{c}$.

This property states the following:

1. Multiplying or dividing by a positive number produces an equivalent inequality.
2. Multiplying or dividing by a negative number produces an equivalent inequality only if the inequality symbol is reversed.

For an illustration of the property, let $a = -6$ and $b = 12$.

| Let $c = 3$. | $a < b$ | $ac < bc$ | $\dfrac{a}{c} < \dfrac{b}{c}$ |
|--------------|---------|-----------|-------------------------------|
|  | $-6 < 12$ | $-6(3) < 12(3)$ | $\dfrac{-6}{3} < \dfrac{12}{3}$ |
|  | True | $-18 < 36$ | $-2 < 4$ |
|  |  | True | True |

Multiplying both sides of an inequality by a positive number maintains the inequality relationship.

Let $c = -1$.      $a < b$         $ac > bc$         $\dfrac{a}{c} > \dfrac{b}{c}$

$-6 < 12$      $-6(-1) > 12(-1)$      $\dfrac{-6}{-1} > \dfrac{12}{-1}$

True          $6 > -12$         $6 > -12$

True          True

Multiplying both sides of an inequality by a negative number requires us to *reverse* the inequality symbol to maintain the inequality relationship.

---

**Procedure: Solving a Linear Inequality**

**Step 1:** Clear any parentheses from the equation by applying the distributive property.

**Step 2:** Remove any fractions by multiplying by the LCD.

**Step 3:** Use the addition property of inequality to collect all variable terms on one side and all constant terms on the other side.

**Step 4:** Use the multiplication property of inequality to get a coefficient of 1 on the variable. Remember that multiplying or dividing by a negative number, reverses the inequality symbol.

**Step 5:** If the inequality is a compound inequality of the form $a < x < b$, then the variable must be eliminated in the middle. Any operation that is required to isolate the variable must be done to all three parts of the inequality.

**Step 6:** Graph the solution set.

**Step 7:** Write the solution set in interval notation or set-builder notation.

---

**Objective 4 Examples**    **Solve each inequality. Graph the solution set and write the solution set in interval and set-builder notation.**

**4a.** $x + 4 < -1$          **4b.** $-4x \leq 4$

**4c.** $3y - 7 > 2$          **4d.** $3(2a + 1) - a \geq 2a - 7$

**4e.** $\dfrac{1}{2}b - \left(b + \dfrac{3}{4}\right) > \dfrac{1}{4}b + \dfrac{1}{2}$      **4f.** $2(4x - 3) > 2x + 3(2x + 1)$

**4g.** $x + 5(x - 2) < 3(2x + 1) + 7$      **4h.** $-3 < 2x + 1 < 7$

**Solutions**    **4a.**      $x + 4 < -1$

$x + 4 - 4 < -1 - 4$          Subtract 4 from each side.

$x < -5$          Simplify.

The graph of the solution set consists of all numbers less than $-5$; that is, the numbers to the left of $-5$ but not including $-5$.

Interval notation: $(-\infty, -5)$

Set-builder notation: $\{x \mid x < -5\}$

**4b.**
$$-4x \leq 4$$

$$\frac{-4x}{-4} \geq \frac{4}{-4} \qquad \text{Divide each side by } -4 \text{ and reverse the inequality symbol.}$$

$$x \geq -1 \qquad \text{Simplify.}$$

The graph consists of all values greater than or equal to $-1$; that is, numbers to the right of $-1$ and including $-1$.

Interval notation: $[-1, -\infty)$

Set-builder notation: $\{x \mid x \geq -1\}$

**4c.**
$$3y - 7 > 2$$

$$3y - 7 + 7 > 2 + 7 \qquad \text{Add 7 to each side.}$$

$$3y > 9 \qquad \text{Simplify.}$$

$$\frac{3y}{3} > \frac{9}{3} \qquad \text{Divide each side by 3.}$$

$$y > 3 \qquad \text{Simplify.}$$

The graph consists of all values greater than 3, or to the right of 3, but not including 3.

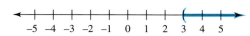

Interval notation: $(3, \infty)$

Set-builder notation: $\{y \mid y > 3\}$

**4d.** $3(2a + 1) - a \geq 2a - 7$

$$6a + 3 - a \geq 2a - 7 \qquad \text{Apply the distributive property.}$$

$$5a + 3 \geq 2a - 7 \qquad \text{Combine like terms.}$$

$$5a + 3 - 2a \geq 2a - 7 - 2a \qquad \text{Subtract } 2a \text{ from each side.}$$

$$3a + 3 \geq -7 \qquad \text{Simplify.}$$

$$3a + 3 - 3 \geq -7 - 3 \qquad \text{Subtract 3 from each side.}$$

$$3a \geq -10 \qquad \text{Simplify.}$$

$$\frac{3a}{3} \geq \frac{-10}{3} \qquad \text{Divide each side by 3.}$$

$$a \geq -\frac{10}{3} \qquad \text{Simplify.}$$

The graph consists of all numbers to the right of $-\dfrac{10}{3}$, including $-\dfrac{10}{3}$.

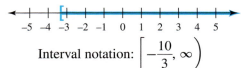

Interval notation: $\left[-\dfrac{10}{3}, \infty\right)$

Set-builder notation: $\left\{a \mid a \geq -\dfrac{10}{3}\right\}$

**4e.** $\dfrac{1}{2}b - 1\left(b + \dfrac{3}{4}\right) > \dfrac{1}{4}b + \dfrac{1}{2}$

$\quad\quad \dfrac{1}{2}b - b - \dfrac{3}{4} > \dfrac{1}{4}b + \dfrac{1}{2}$     Distribute $-1$ to $\left(b + \dfrac{3}{4}\right)$.

$\quad\quad 4\left(\dfrac{1}{2}b - b - \dfrac{3}{4}\right) > 4\left(\dfrac{1}{4}b + \dfrac{1}{2}\right)$     Multiply each side by the LCD, 4.

$\quad\quad\quad 2b - 4b - 3 > b + 2$     Simplify.

$\quad\quad\quad\quad -2b - 3 > b + 2$     Combine like terms.

$\quad\quad -2b - 3 - b > b + 2 - b$     Subtract $b$ from each side.

$\quad\quad\quad\quad\quad -3b - 3 > 2$     Simplify.

$\quad\quad -3b - 3 + 3 > 2 + 3$     Add 3 to each side.

$\quad\quad\quad\quad\quad\quad -3b > 5$     Simplify.

$\quad\quad\quad\quad \dfrac{-3b}{-3} < \dfrac{5}{-3}$     Divide each side by $-3$. Remember to reverse the inequality symbol.

$\quad\quad\quad\quad\quad\quad b < -\dfrac{5}{3}$     Simplify.

The graph consists of all numbers to the left of $-\dfrac{5}{3}$ but not including $-\dfrac{5}{3}$.

Interval notation: $\left(-\infty, -\dfrac{5}{3}\right)$

Set-builder notation: $\left\{b \middle| b < -\dfrac{5}{3}\right\}$

**4f.**     $2(4x - 3) > 2x + 3(2x + 1)$

$\quad\quad 8x - 6 > 2x + 6x + 3$     Apply the distributive property.

$\quad\quad\quad 8x - 6 > 8x + 3$     Combine like terms.

$\quad 8x - 6 - 8x > 8x + 3 - 8x$     Subtract $8x$ from each side.

$\quad\quad\quad\quad\quad -6 > 3$     False.

The resulting inequality, $-6 > 3$, is always false. Therefore, there is no solution of this inequality. The solution set is the empty set, or $\varnothing$. The graph is a blank number line as shown.

**4g.**    $x + 5(x - 2) < 3(2x + 1) + 7$

$\quad\quad x + 5x - 10 < 6x + 3 + 7$     Apply the distributive property.

$\quad\quad\quad 6x - 10 < 6x + 10$     Combine like terms.

$\quad 6x - 10 - 6x < 6x + 10 - 6x$     Subtract $6x$ from each side.

$\quad\quad\quad\quad -10 < 10$     True.

The resulting inequality, $-10 < 10$, is always true. Therefore, any real number is a solution of this inequality. So, the solution set is all real numbers, or $\mathbb{R}$.

Interval notation: $(-\infty, \infty)$

Set-builder notation: $\{x \mid x \text{ is a real number}\}$

**4h.** This is a compound inequality. We must isolate the variable in the middle.

$$3 < \quad 2x + 1 \quad < 7$$

$$-3 - 1 < 2x + 1 - 1 < 7 - 1 \qquad \text{Subtract 1 from each part.}$$

$$-4 < \quad 2x \quad < 6 \qquad \text{Simplify.}$$

$$\frac{-4}{2} < \quad \frac{2x}{2} \quad < \frac{6}{2} \qquad \text{Divide each part by 2.}$$

$$\qquad\qquad\qquad\qquad\qquad \text{Simplify.}$$

$$-2 < \quad x \quad < 3$$

Since $-2 < x < 3$ is equivalent to $x > -2$ and $x < 3$, the graph consists of all the points greater than $-2$, but not including $-2$, and less than 3, but not including 3. This is equivalent to all points between, but not including, $-2$ and 3.

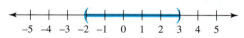

Interval notation: $(-2, 3)$

Set-builder notation: $\{x \mid -2 < x < 3\}$

✓ **Student Check 4** Solve each inequality. Graph the solution set and write the solution set in interval and set-builder notation.

**a.** $y + 3 < -2$          **b.** $-x \le 2$

**c.** $7y - 1 > 6$         **d.** $3(a + 2) - 7 \ge -4a + 10$

**e.** $\frac{1}{3}y - 2\left(y + \frac{1}{2}\right) > \frac{2}{3}y + \frac{1}{6}$      **f.** $4(x - 3) + 1 \ge 5(x + 2) - x$

**g.** $7x - 2(4x + 3) < 3(x + 5) - 4x$     **h.** $7 \le 6x - 3 \le 15$

## Applications

**Objective 5** ▶

Solve applications of linear inequalities.

**INSTRUCTOR NOTE:**
To help students learn the inequality symbol that corresponds to a statement, use the concept of money. Say, "What if I have at most $10 in my pocket, how much money could I have?" or "If I have at least $50, how much could I have?"

There are a few key phrases that we need before we can solve applications relating to inequalities.

| Phrase | Mathematical Statement |
|---|---|
| $a$ is less than $b$ | $a < b$ |
| $a$ is less than or equal to $b$ <br> $a$ is no more than $b$ <br> $a$ is at most $b$ | $a \le b$ |
| $a$ is greater than $b$ | $a > b$ |
| $a$ is greater than or equal to $b$ <br> $a$ is no less than $b$ <br> $a$ is at least $b$ | $a \ge b$ |

These phrases together with the ones presented in Chapter 1 enable us to write an inequality to solve application problems.

**Procedure: Solving Applications of Linear Inequalities**

**Step 1:** Read the problem and determine the unknown. Assign a variable to the unknown value.

**Step 2:** Read the problem and determine the given information.

**Step 3:** Find the statement in the problem that states the inequality relationship, looking for key phrases that were listed in the chart above.

**Step 4:** Use the statement in Step 3 to write the inequality.

**Step 5:** Apply the addition and multiplication properties of inequalities to solve the inequality.

**Step 6:** Answer the question with a complete sentence.

| Objective 5 Examples | **Write an inequality that models each situation. Solve the inequality and answer the question in a complete sentence.** |

**5a.** Juanita has math test scores of 73, 85, and 80. What score does she need on her fourth math test to have a test average of at least 80?

**Solution** **5a.** What is unknown? The score on the fourth test is unknown. Let $x$ represent the score on the fourth math test.

What is known? The first three test scores are 73, 85, and 80.

To obtain the inequality, we use the following statement.

The average of the is at
four test scores least 80.

$$\frac{73 + 85 + 80 + x}{4} \geq 80 \qquad \text{Recall the average is } \frac{\text{the sum of all items}}{\text{the number of items}}.$$

$$\frac{238 + x}{4} \geq 80 \qquad \text{Simplify the numerator of the fraction.}$$

$$4\left(\frac{238 + x}{4}\right) \geq 4(80) \qquad \text{Multiply each side by 4.}$$

$$238 + x \geq 320 \qquad \text{Simplify.}$$

$$238 + x - 238 \geq 320 - 238 \qquad \text{Subtract 238 from each side.}$$

$$x \geq 82 \qquad \text{Simplify.}$$

Juanita needs at least an 82 on her fourth test to have at least an 80 test average.

**5b.** Jamaal's final grade in his math class is based on a weighted average. The weights are as follows:

Homework average 10%      Quiz average 15%

Test average 50%          Final exam 25%

If Jamaal's homework average is 85, his quiz average is 80, and his test average is 75, what grade does he need on his final exam to have a final grade of at least 70 in the class? final grade of at least 80 in the class?

**Solution** **5b.** What is unknown? The final exam grade is unknown. Let $x$ = final exam grade.

What is known? The final grade is calculated by

10% (home work average) + 15% (quiz average) + 50% (test average) + 25% (final exam) = 0.10 (home work average) + 0.15 (quiz average) + 0.50 (test average) + 0.25 (final exam) = 0.10(85) + 0.15(80) + 0.50(75) + 0.25x

To obtain the inequalities, we use the following statements.

| Final grade is at least 70. | Final grade is at least 80. |
|---|---|
| $0.10(85) + 0.15(80)$ | $0.10(85) + 0.15(80)$ |
| $+ 0.50(75) + 0.25x \geq 70$ | $+ 0.50(75) + 0.25x \geq 80$ |
| $8.5 + 12 + 37.5 + 0.25x \geq 70$ | $8.5 + 12 + 37.5 + 0.25x \geq 80$ |
| $58 + 0.25x \geq 70$ | $58 + 0.25x \geq 80$ |
| $58 + 0.25x - 58 \geq 70 - 58$ | $58 + 0.25x - 58 \geq 80 - 58$ |
| $0.25x \geq 12$ | $0.25x \geq 22$ |
| $\dfrac{0.25x}{0.25} \geq \dfrac{12}{0.25}$ | $\dfrac{0.25x}{0.25} \geq \dfrac{22}{0.25}$ |
| $x \geq 48$ | $x \geq 88$ |

Jamaal needs at least a 48 on the final exam to have a grade of at least 70 in the class. | Jamaal needs at least an 88 on the final exam to have a grade of at least 80 in the class.

**5c.** The cost to rent a banquet facility for a wedding reception is $1500 plus $35 per person for the food. What is the number of attendees the couple can have at the reception if they have at most $5000 budgeted for the reception?

**Solution**   **5c.** What is unknown? The number of attendees is unknown. Let $x$ represent the number of people attending the reception.

What is known? The cost of the reception is $1500 plus $35 per person or $1500 + 35x$.

The cost is at most $5000.

$$1500 + 35x \leq 5000 \qquad \text{Express statement mathematically.}$$
$$\text{"at most" is } \leq.$$
$$1500 + 35x - 1500 \leq 5000 - 1500 \qquad \text{Subtract 1500 from each side.}$$
$$35x \leq 3500 \qquad \text{Simplify.}$$
$$\frac{35x}{35} \leq \frac{3500}{35} \qquad \text{Divide each side by 35.}$$
$$x \leq 100 \qquad \text{Simplify.}$$

The couple can have at most 100 people attending their reception to have a cost of no more than $5000.

✓ **Student Check 5**   Write an inequality that models each situation. Solve the inequality and answer the question in a complete sentence.

**a.** Darryl has scores of 90, 85, 83, and 92 on his first four tests in his math class. What score does he need on the fifth test to have a test average of at least a 90?

**b.** Rosita's final grade in her math class is based on a weighted average. The weights are as follows:

Homework average 10%      Test average 45%
Quiz average 15%          Final exam 30%

If Rosita's homework average is 100, her quiz average is 75, and her test average is 80, what does she need on her final exam to have a final grade of at least 70 in the class? a final grade of at least 80 in the class?

**c.** The cost of the Verizon Wireless Nationwide Select 450 is $59.99 per month. The plan includes 450 anytime minutes and unlimited text, picture, video and instant messaging to anyone on any network in the United States. Additional anytime minutes cost $0.45 each. How many additional minutes can Lisa talk so that her total monthly cell phone bill is at most $75? Round to the nearest minute. (Source: www.myrateplan.com)

**Objective 6** ▶
Troubleshoot common errors.

## Troubleshooting Common Errors

Some common errors associated with linear inequalities are shown next.

**Objective 6 Examples**   **A problem and an incorrect solution are given. Provide the correct solution and an explanation of the error.**

**6a.** Write the interval for the graph.

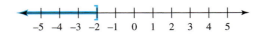

| Incorrect Solution | Correct Solution and Explanation |
|---|---|
| The interval is $[-2, -\infty)$. | The interval must be written in order from smallest to largest or from left to right. Since the graph continues indefinitely to the left, the interval begins at $-\infty$ and ends at the right with $-2$. So, the interval is $(-\infty, -2]$. |

**6b.** Solve $-6x + 2 > 8$.

| Incorrect Solution | Correct Solution and Explanation |
|---|---|
| $$-6x + 2 > 8$$ $$-6x + 2 - 2 > 8 - 2$$ $$-6x > 6$$ $$\frac{-6x}{-6} > \frac{6}{-6}$$ $$x > -1$$ The solution is $(-1, \infty)$. | The only error made was that the inequality symbol was not reversed when both sides were divided by a negative number. $$-6x + 2 > 8$$ $$-6x + 2 - 2 > 8 - 2$$ $$-6x > 6$$ $$\frac{-6x}{-6} < \frac{6}{-6}$$ $$x < -1$$ The solution is $(-\infty, -1)$. |

## ANSWERS TO STUDENT CHECKS

| | | Graph | Interval Notation | Set-Builder Notation |
|---|---|---|---|---|
| Student Check 1<br>Student Check 2<br>Student Check 3 | a.<br>a.<br>a. | | $(-2, \infty)$ | $\{x \mid x > -2\}$ |
| Student Check 1<br>Student Check 2<br>Student Check 3 | b.<br>b.<br>b. | | $(-\infty, -4.2)$ | $\{x \mid x \le -4.2\}$ |
| Student Check 1<br>Student Check 2<br>Student Check 3 | c.<br>c.<br>c. | | $\left(-\infty, \dfrac{5}{3}\right)$ | $\left\{x \mid x < \dfrac{5}{3}\right\}$ |
| Student Check 1<br>Student Check 2<br>Student Check 3 | d.<br>d.<br>d. | | $[-4, 2)$ | $\{x \mid -4 \le x < 2\}$ |
| Student Check 4 | a. | | $(-\infty, -5)$ | $\{y \mid y < -5\}$ |
| | b. | | $[-2, \infty)$ | $\{x \mid x \ge -2\}$ |
| | c. | | $[1, \infty)$ | $\{y \mid y > 1\}$ |
| | d. | | $\left[\dfrac{11}{7}, \infty\right)$ | $\left\{a \mid a \ge \dfrac{11}{7}\right\}$ |
| | e. | | $\left(-\infty, -\dfrac{1}{2}\right)$ | $\left\{y \mid y < -\dfrac{1}{2}\right\}$ |
| | f. | | $\varnothing$ | $\varnothing$ |
| | g. | | $(-\infty, \infty)$ | $\{x \mid x \text{ is a real number}\}$ |
| | h. | | $\left[\dfrac{5}{3}, 3\right]$ | $\left\{x \mid \dfrac{5}{3} \le x \le 3\right\}$ |

**Student Check 5**  **a.** Darryl must make a 100 or better on his fifth test to have a test average of at least a 90.

**b.** Rosita needs at least a 42.5 on her final exam to have a final grade of at least 70. She needs at least an 75.8 on her final exam to have a final grade of at least 80.

**c.** Lisa can talk no more than 33 additional minutes to have a monthly bill less than or equal to $75.

## SUMMARY OF KEY CONCEPTS

1. The graph of the solution set of an inequality is a picture of all real numbers that make the inequality a true statement. A parenthesis (used with < or >) on a number indicates the number is not included in the solution set but every number very close to it is included. A bracket (used with ≤ or ≥) on a number indicates the number is included in the solution set.

2. Interval notation is a concise way to represent the solution set of an inequality. It represents the interval of real numbers in which solutions lie. The interval notation always begins with the left bound of the solution set and ends with the right bound of the solution set. When a set continues indefinitely to the right, the right bound is represented by $\infty$. When a set continues indefinitely to the left, the left bound is represented by $-\infty$. A parenthesis is always used with $\infty$ or $-\infty$.

3. Set-builder notation is another way to represent the solution set of an inequality. It is written using { }. We write {variable|final inequality}. Example: $\{y|y < 5\}$.

4. Linear inequalities are solved using the addition and multiplication properties of inequality. The most important thing to remember is that when you multiply or divide by a negative number, you must also reverse the inequality symbol.

5. Applications are solved by translating the given statements into appropriate inequalities. Key phrases are "is at least" and "is at most." A good way to remember the translations is to think of money. For example, if you have at least $10, you would have $10 or more; if you have at most $10, you would have $10 or less.

## GRAPHING CALCULATOR SKILLS

The calculator can be used as a reference to check the work we have done by hand.

**Example:** $x - 5 > 2$

By adding 5 to both sides, we find the solution of $x - 5 > 2$ to be $x > 7$. The graph is

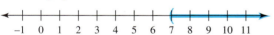

To check our work on the calculator, we should determine if the inequality is true for a number larger than 7, if it is false for a number less than 7, and what happens at 7.

We will use the store feature to test values. We first store in a value for $x$ using the STO> command. Then we enter the original inequality in the calculator and press Enter. The result will either be 1 or 0. If the result is 1, then the stored value satisfies the inequality and is a solution; if the result is 0, then the stored value does not satisfy the inequality and is not a solution.

Check: $x = 8$ (a value larger than 7).

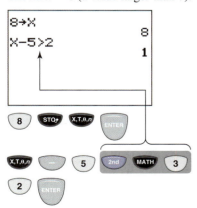

Check: $x = 6$ (a value less than 7).

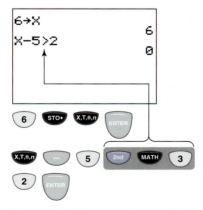

Check: $x = 7$.

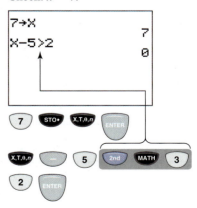

The result of 1 confirms that $x = 8$ is a solution of the inequality. The shaded portion of the graph should contain 8 (to the right of 7).

The result of 0 means that $x = 6$ is not a solution of the inequality. The shaded portion of the graph should not include the side with 6.

The result of 0 means that $x = 7$ is not a solution of the inequality. The graph should not include 7, so we put a parenthesis on 7.

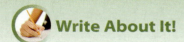

## SECTION 2.7 / EXERCISE SET

### Write About It!

Use complete sentences in your answer to the following exercises.

1. Define an inequality in one variable.   *Answers vary.*

2. When do you use brackets and when do you use parentheses when graphing solutions of inequalities?   *Answers vary.*

3. What operation do you perform to both sides of an inequality that requires you to reverse the inequality symbol?   *Answers vary.*

4. Explain how to write the interval notation for the solution set of an inequality. Use an example.   *Answers vary.*

5. Explain how to write the interval notation for the solution set of a compound inequality, $a \leq x \leq b$. Use an example.   *Answers vary.*

6. Explain how to write the interval notation for the solution set of a compound inequality, $a < x \leq b$. Use an example.   *Answers vary.*

**Determine if each statement is true or false. If the statement is false, explain why.**

7. The value $x = 2$ is a solution of the inequality $x - 3 > -1$.   *False, $2 - 3 = -1$ which is not greater than $-1$.*

8. The solution set of $-2x \geq 4$ is $[-2, \infty)$.   *False, we must reverse the inequality symbol. The solution set is $(-\infty, -2]$.*

9. The inequality $5 > x$ is equivalent to $x < 5$.   *True*

10. The inequality $-3 \leq x \leq 8$ is equivalent to $8 \geq x \geq -3$.   *True*

11. The inequality $-6 \leq -3x \leq 12$ is equivalent to $2 \leq x \leq -4$.   *False, reverse the inequality symbols.*

12. Solving a linear inequality is no different from solving a linear equation.   *False, the difference is that when we multiply or divide by a negative number, we must reverse the inequality symbol.*

### Practice Makes Perfect!

**Fill in the missing information. (*See Objectives 1–3.*)**

| Inequality | Graph | Interval Notation | Set-Builder Notation |
|---|---|---|---|
| 13. $x > -2$ | number line: open circle at $-2$, shaded right; marks $-7$ to $3$ | $(-2, \infty)$ | $\{x \mid x > -2\}$ |
| 14. $y \geq -5$ | number line: bracket at $-5$, shaded right; marks $-10$ to $0$ | $[-5, \infty)$ | $\{y \mid y \geq -5\}$ |
| 15. $a < 1$ | number line: open circle at $1$, shaded left; marks $-4$ to $6$ | $(-\infty, 1)$ | $\{a \mid a < 1\}$ |
| 16. $x \leq 8$ | number line: bracket at $8$, shaded left; marks $3$ to $13$ | $(-\infty, 8)$ | $\{x \mid x \leq 8\}$ |
| 17. $x < -6$ | number line: open circle at $-6$, shaded left; marks $-11$ to $4$ | $(-\infty, -6)$ | $\{x \mid x < -6\}$ |
| 18. $x \leq -\dfrac{1}{2}$ | number line: bracket at $-\frac{1}{2}$, shaded left; marks $-5$ to $5$ | $\left(-\infty, -\dfrac{1}{2}\right]$ | $\left\{x \,\middle|\, x \leq -\dfrac{1}{2}\right\}$ |
| 19. $x > 7$ | number line: open circle at $7$, shaded right; marks $1$ to $11$ | $(7, \infty)$ | $\{x \mid x > 7\}$ |
| 20. $x \geq 4$ | number line: bracket at $4$, shaded right; marks $-1$ to $9$ | $[4, \infty)$ | $\{x \mid x \geq 4\}$ |
| 21. $x \leq -3$ | number line: bracket at $-3$, shaded left; marks $-8$ to $2$ | $(-\infty, -3]$ | $\{x \mid x \leq -3\}$ |
| 22. $x > 2$ | number line: open circle at $2$, shaded right; marks $-3$ to $7$ | $(2, \infty)$ | $\{x \mid x > 2\}$ |
| 23. $x \leq \dfrac{1}{4}$ | number line: bracket at $\frac{1}{4}$, shaded left; marks $-5$ to $5$ | $\left(-\infty, \dfrac{1}{4}\right]$ | $\left\{x \,\middle|\, x \leq \dfrac{1}{4}\right\}$ |

Additional answers can be found in the Instructor Answer Appendix.

| Inequality | Graph | Interval Notation | Set-Builder Notation |
|---|---|---|---|
| **24.** $x \geq \dfrac{3}{5}$ | number line with closed bracket at $\frac{3}{5}$, shaded right | $\left[\dfrac{3}{5}, \infty\right)$ | $\left\{x \mid x \geq \dfrac{3}{5}\right\}$ |
| **25.** $x < 10$ | number line with open parenthesis at 10, shaded left | $(-\infty, 10)$ | $\{x \mid x < 10\}$ |
| **26.** $y > 1$ | number line with open parenthesis at 1, shaded right | $(1, \infty)$ | $\{y \mid y > 1\}$ |
| **27.** $x \leq -\dfrac{5}{4}$ | number line with closed bracket at $-\frac{5}{4}$, shaded left | $\left(-\infty, -\dfrac{5}{4}\right]$ | $\left\{x \mid x \leq -\dfrac{5}{4}\right\}$ |
| **28.** $y \geq -1$ | number line with closed bracket at $-1$, shaded right | $[-1, \infty)$ | $\{y \mid y \geq -1\}$ |
| **29.** $-2 \leq x \leq 5$ | number line with closed brackets at $-2$ and $5$, shaded between | $[-2, 5]$ | $\{x \mid -2 \leq x \leq 5\}$ |
| **30.** $-6 < y \leq 8$ | number line with open paren at $-6$, closed bracket at $8$ | $(-6, 8]$ | $\{y \mid -6 < y \leq 8\}$ |
| **31.** $3 \leq x < 10$ | number line with closed bracket at $3$, open paren at $10$ | $[3, 10)$ | $\{x \mid 3 \leq x < 10\}$ |
| **32.** $0 < a < 12$ | number line with open parens at $0$ and $12$ | $(0, 12)$ | $\{a \mid 0 < a < 12\}$ |
| **33.** $-7 \leq x < 3$ | number line with closed bracket at $-7$, open paren at $3$ | $[-7, 3)$ | $\{x \mid -7 \leq x < 3\}$ |
| **34.** $1 < x \leq 9$ | number line with open paren at $1$, closed bracket at $9$ | $(1, 9]$ | $\{x \mid 1 < x \leq 9\}$ |
| **35.** $-\dfrac{7}{3} \leq x \leq 8$ | number line with closed brackets at $-\frac{7}{3}$ and $8$ | $\left[-\dfrac{7}{3}, 8\right]$ | $\left\{x \mid -\dfrac{7}{3} \leq x \leq 8\right\}$ |
| **36.** $-\dfrac{5}{2} < y < \dfrac{1}{2}$ | number line with open parens at $-\frac{5}{2}$ and $\frac{1}{2}$ | $\left(-\dfrac{5}{2}, \dfrac{1}{2}\right)$ | $\left\{y \mid -\dfrac{5}{2} < y < \dfrac{1}{2}\right\}$ |

**Solve each inequality. Graph the solution set and write the solution set in interval notation and in set-builder notation. (*See Objective 4.*)**

**37.** $x - 3 > -7$

**38.** $y - 6 \geq -15$

**39.** $x + 1 \geq 8$

**40.** $a + 9 \geq -2$

**41.** $y - 4 < -4$

**42.** $a + 5 > 5$

**43.** $x + 10 \leq 6$

**44.** $y - 2 \leq -3$

**45.** $2 < y + 3$

**46.** $-4 \geq a - 7$

**47.** $-3x < 9$

**48.** $7a \geq -14$

**49.** $\dfrac{y}{4} > 1$

**50.** $\dfrac{x}{5} \geq -2$

**51.** $-a > 8$

**52.** $-y \leq -4$

**53.** $-6a \geq -12$

**54.** $-9a < 27$

**55.** $-\dfrac{2}{3}x > 18$

**56.** $-\dfrac{4}{5}a \leq -16$

**57.** $2y - 4 < -8$

**58.** $6a + 1 > -5$

**59.** $-5a + 3 \geq -7$

**60.** $4 - 7x \leq -10$

**61.** $6x - 5 \leq 2x + 8$

**62.** $-9x + 3 > 8x - 1$

**63.** $4 - 3(x - 5) > -2x + 1$

**64.** $8 - (7x + 9) \geq 10x + 4$

**65.** $\dfrac{1}{2}y - 3 \leq \dfrac{1}{6}$

**66.** $\dfrac{3}{4}a + 2 < \dfrac{5}{8}$

**67.** $75 + 0.25x < 100$

**68.** $30 + 0.15a > 51$

**69.** $-9 < a + 4 < 12$

**70.** $-3 \leq b - 7 \leq 6$

**71.** $-13 \leq 3x - 7 \leq 8$

**72.** $-18 < 5y - 8 < 12$

**73.** $18 \leq 6 - 4x < 34$

**74.** $-11 < 3 - 2y \leq 25$

**75.** $5 \leq -6a - 1 \leq 17$

**76.** $-13 < -9b + 5 < 32$

**77.** $\dfrac{1}{8} < \dfrac{3}{4}a + 1 < \dfrac{5}{8}$

**78.** $-\dfrac{5}{12} \leq \dfrac{2}{3}b - 1 \leq \dfrac{7}{6}$

**79.** $-5.2 < 1.2x - 2.8 < 5.6$  **80.** $1.5 \leq 1.8 - 0.3x \leq 7.5$

**81.** $4(x - 2) - 9 \geq -7x + 16$

**82.** $2(x + 5) - 14 \leq -9x + 18$

**83.** $3(x - 1) - 5 < 8(x + 2) - x$

**84.** $4(x - 1) + 10 > 9(x + 3) - 2x$

**85.** $\frac{2}{5}y - 2\left(y - \frac{3}{10}\right) < \frac{1}{2}y + \frac{7}{10}$

**86.** $\frac{3}{4}y - 2\left(y + \frac{1}{3}\right) > \frac{1}{2}y - \frac{5}{6}$

**87.** $12x - 5(2x + 3) > 6(x + 7) - x$

**88.** $16y - 7(y + 4) > 4(y - 2) + y$

**Write an inequality that models each situation. Solve the inequality and answer the question in complete sentences.** (*See Objective 5.*)

**89.** Mykel scored 85, 78, 72, and 81 on four math tests. What score does she need on the fifth test to have at least an 80 average?   at least 84 on the fifth test

**90.** Lucas scored 77, 78, and 75 on three math tests. What score does he need on the fourth test to have at least an 80 average?   at least 90 on the fourth test

**91.** Shanika's final grade in her math class is based on a weighted average. Homework counts as 5%, quizzes count as 10%, projects count as 10%, tests count as 50%, and the final exam counts as 25%. If she has a homework average of 85, a quiz average of 78, a project average of 90, and a test average of 73, what score does she need on her final exam to have at least an 80 for her final grade?   89.8 on the final exam

**92.** Hassan's final grade in his math class is based on a weighted average. Homework counts as 10%, quizzes count as 20%, participation counts as 5%, tests count as 40%, and the final exam counts as 25%. If he has a homework average of 88, a quiz average of 82, a participation grade of 100, and a test average of 80, what score does he need on his final exam to have at least an 80 for his final grade?   71.2 on the final exam

**93.** Skype is a software program that allows users to call any number in the world with their computer. Skype offers 12 months of unlimited phone calls in the United States for $29.95 per year. Sergeant Walker is stationed in Iraq for one year. Skype charges 37.2¢ per minute plus 3.9¢ per international call. If Sergeant Walker's family calls him once a week for the year, the cost of the phone service is given by $C = 29.95 + 0.372m + 0.039(52) = 31.978 + 0.372m$ where $m$ is the number of minutes. How many minutes can the family talk in a year so that their cost in phone calls for the year is at most $1000? (Source: http://www.skype.com)   at most 2602 minutes

**94.** The cost to rent a car for a week from a luxury rental agency is $350 plus 44 cents per mile. How many miles can be driven if you have at most $500 to rent the car for a week? Round to the nearest whole number.   at most 340 mi

 **Mix 'Em Up!**

**Solve each inequality. Graph the solution set and write the solution set in interval notation and in set-builder notation.**

**95.** $-x + 12 < -12$

**96.** $-y + 16 \geq 14$

**97.** $9x - 20 > -2$

**98.** $-10y + 19 \leq 39$

**99.** $0.2a - 7 \leq -1.5$

**100.** $-0.32b + 4 < 10.4$

**101.** $\frac{y}{8} - 2 \leq 6$

**102.** $-\frac{1}{3}x + 5 > \frac{4}{3}$

**103.** $-\frac{5}{8}x - 3 \leq \frac{3}{8}$

**104.** $\frac{1}{4}y + 3 \geq \frac{11}{4}$

**105.** $-11 < 2a - 1 < 17$

**106.** $15 < 6b - 3 \leq 21$

**107.** $-5 \leq 4x + 3 \leq 19$

**108.** $2 \leq 5 - 3y < 26$

**109.** $-9 < -2a + 5 < 5$

**110.** $-6 \leq 4b + 10 < 2$

**111.** $12 - 2(x - 3) > -2x + 1$

**112.** $6x - (3x - 2) > 3x + 1$

**113.** $7x + 2(x + 5) < 9x + 8$

**114.** $x - (4x - 2) > 5 - 3x$

**115.** $\frac{5}{2}x + 3\left(x + \frac{3}{2}\right) \geq \frac{9}{2}x - \frac{7}{2}$

**116.** $\frac{8}{3}y - 2\left(y - \frac{2}{11}\right) \geq \frac{5}{3}y - \frac{2}{11}$

**117.** $-7 \leq -3(x - 3) + 2 \leq -1$

**118.** $-3 < -2(y + 4) + 7 < 5$

**119.** $-\frac{7}{2} < \frac{1}{3}x - 1 < \frac{13}{6}$

**120.** $-\frac{4}{5} \leq \frac{1}{2}y + 2 \leq \frac{11}{5}$

**121.** $-3.8 < 0.25x - 2.8 < 13.2$

**122.** $2.5 \leq 0.12y + 3.1 < 12.7$

**Write an inequality that models each situation. Solve the inequality and answer the question in complete sentences.**

**123.** Azziz scored 57, 97, and 85 on three math tests. What score does he need on the fourth test to have at least an 80 average?   at least 81 on the fourth test

**124.** Elaine scored 84, 60, and 63 on three chemistry tests. What score does she need on the fourth test to have at least an 70 average?   at least 73 on the fourth test

**125.** Derek's final grade in his math class is based on a weighted average. Homework counts as 10%, quizzes count as 20%, tests count as 50%, and the final exam counts as 20%. If he has a homework average of 67, a quiz average of 65, and a test average of 89, what score does he need on his final exam to have at least an 80 for his final grade? *79 on the final exam*

126. Sue's final grade in her math class is based on a weighted average. Homework counts as 15%, quizzes count as 10%, tests count as 55%, and the final exam counts as 20%. If she has a homework average of 91, a quiz average of 42, and a test average of 82, what score does she need on her final exam to have at least an 80 for her final grade? *85.25 on the final exam*

127. The cost to rent a banquet facility for a wedding reception is $1050 plus $36 per person for the food. What is the number of attendees the couple can have at the reception if they have at most $9400 budgeted for the reception? *231 attendees*

128. The cost to rent a banquet facility for a wedding reception is $1030 plus $25 per person for the food. What is the number of attendees the couple can have at the reception if they have at most $4900 budgeted for the reception? *154 attendees*

129. The cost to rent a car for a week from Convenient Rental is $304 plus 30 cents per mile. How many miles can be driven if you have at most $960 to rent the car for a week? Round to the nearest whole number, if necessary. *at most 2187 mi*

130. The cost to rent a car for a week from Economy Rental is $257 plus 30 cents per mile. How many miles can be driven if you have at most $410 to rent the car for a week? Round to the nearest whole number, if necessary. *at most 510 mi*

## You Be the Teacher!

**Correct each student's errors, if any.**

131. Solve the inequality: $-2x + 5 < 11$.

Mike's work:

$$-2x + 5 < 11$$
$$\underline{\quad -5 \quad -5 \quad}$$
$$\frac{-2x}{\quad} < 6$$
$$\frac{-2x}{-2} < \frac{6}{-2}$$
$$x < -3$$

$-2x + 5 - 5 < 11 - 5$
$-2x < 6$
$\frac{-2x}{-2} > \frac{6}{-2}$
$x > -3$
The solution is $(-3, \infty)$

132. Solve the inequality $-9 < 3 - 2x < 13$.

Edna's work:

$$-9 < 3 - 2x < 13$$
$$-12 < -2x < 10$$
$$6 < x < -5$$

$-9 < 3 - 2x < 13$
$-12 < -2x < 10$
$6 > x > -5$
The solution is $(-5, 6)$.

133. Solve the inequality: $3(2x - 5) + 9 > 2(3x + 1)$.

Gabriel's work:

$$3(2x - 5) + 9 > 2(3x + 1)$$
$$6x - 5 + 9 > 6x + 1$$
$$6x + 4 > 6x + 1$$
$$4 > 1$$

$3(2x - 5) + 9 > 2(3x + 1)$
$6x - 15 + 9 > 6x + 2$
$6x - 6 > 6x + 2$
$-6 > 2$
The solution is $\varnothing$.

The solution is $\mathbb{R}$.

134. Solve the in equality: $5(x - 1) + 2x < 7(x + 3) - 10$.

Kristen's work:

$$5(x - 1) + 2x < 7(x + 3) - 10$$
$$5x - 1 + 2x < 7x + 3 - 10$$
$$7x - 1 < 7x - 7$$
$$-1 < -7$$

$5(x - 1) + 2x < 7(x + 3) - 10$
$5x - 5 + 2x < 7x + 21 - 10$
$7x - 5 < 7x + 11$
$-5 < 11$
The solution is $\mathbb{R}$.

The solution is $\varnothing$.

## Calculate It!

**Solve each inequality. Then use a graphing calculator to check the solution.**

135. $8x + 20 < 2x - 15 + x$  *$x < -7$*

136. $2(x + 10) - 5 \geq 4x + 18$  *$x \leq -\frac{3}{2}$*

137. $\frac{5}{6}x - x < 8 - \frac{3}{2}x$  *$x < 6$*

138. $-3.5(-1.6x + 2.4) \geq 0.8(-2.1x - 41.44)$  *$x \geq -3.4$*

## GROUP ACTIVITY    The Growth of Cell Phones and Cell Sites

**Group Project for Chapter 2**

1. The number of U.S. wireless subscriber connections, $w$, in millions is approximately 19.35 more than 17.83 times the number of years after 1995. If $x$ is the number of years after 1995, write an equation that approximates the number of wireless subscriber connections. *$w = 17.83x + 19.35$*

2. Use the equation in Step 1 to determine the number of wireless subscriber connections for the given years. Round to the nearest tenth.

| Year | Years After 1995, $x$ | Number of Connections, $w$ |
|------|------|------|
| 1995 | 0 | 19.4 |
| 2000 | 5 | 108.5 |
| 2005 | 10 | 197.7 |
| 2010 | 15 | 286.8 |

3. In 2010, the number of wireless subscriber connections was 93% of the population of the United States. Use the previous information to approximate the U.S. population. Show the equation that was used to solve the problem.

4. Research the Internet to determine the current U.S. population. Round the population to the nearest million. Use the equation in step 1 to determine when the number of wireless subscriber connections will reach the current U.S. population.

5. In 2010, 24.5% of U.S. households were wireless-only households. Use the previous information to determine the number of wireless only households in the United States. Approximately 75.6 million households are wireless-only.

6. The number of cell sites can be given by $c = 15,552.28x + 19,662.9$, where $x$ is the years after 1995. Complete the table to determine the number of cell sites for the given years.

| Year | Years After 1995, $x$ | Number of Cell Sites, $c$ |
|------|------------------------|----------------------------|
| 1995 | 0 | 19,663 |
| 2000 | 5 | 97,424 |
| 2005 | 10 | 175,186 |
| 2010 | 15 | 252,947 |

7. Determine the percent increase in wireless subscriber connections and cell sites for each 5-yr period. Does the growth of cell sites meet the growth of subscribers?

| 5-yr Period | Percent Increase in Wireless Subscriber Connections | Percent Increase in Cell Sites |
|-------------|------------------------------------------------------|--------------------------------|
| 1995–2000 | 459.3% | 395.5% |
| 2000–2005 | 82.2% | 79.8% |
| 2005–2010 | 45.1% | 44.4% |

Source: www.ctia.org

The growth in cell sites has lagged behind the growth in subscribers but is improving to meet demand.

# Linear Equations and Inequalities in One Variable

> **What's the big idea?** Chapter 2 provides us with the skills to solve equations and inequalities in one variable. Knowing how to solve these equations and inequalities gives us the foundation we need to solve some word problems. The properties that enable us to solve these types of problems also provide the framework for us to solve more difficult equations, which will be encountered in later chapters.

## The Tools

Listed below are the key terms, skills, formulas, and properties you should know for this chapter.

The page reference is provided if you need additional help with the given topic. The Study Tips will assist in your preparation for an exam.

## Study Tips

1. Learn all of the terms, formulas, and properties. Make flash cards and have someone quiz you.
2. Rework problems from the exercises and also the ones you worked in class. Work additional problems from the review exercises.
3. Review the summaries of key concepts.
4. Work the chapter test.
5. Be sure to review the online resources for additional study materials.

## Terms

Area   137
Circumference   137
Complementary angles   142
Compound inequality   167
Conditional equation   129
Consecutive even integers   119
Consecutive integers   118
Consecutive odd integers   119
Contradictions   130
Equation   92

Equivalent equations   102
Expression   92
Formula   136
Identity   131
Interest   154
Interval notation   167
Linear equation in one
   variable   101
Linear inequality in one variable   166
Opposite angles   144

Perimeter   137
Right angle   142
Set-builder notation   168
Simple inequality   167
Solution of a linear inequality   166
Solution of the equation   92
Straight angle   142
Supplementary angles   142
Triangle   145
Vertical angles   144

## Formulas and Properties

- Addition property of equality   102
- Addition property of inequality   169
- Discount problem   152

- Distance formula   159
- General percent problem   151
- Markup problem   152
- Multiplication property of equality   112

- Multiplication property of inequality   169
- Percent increase or decrease   152
- Simple interest formula   154

## CHAPTER 2 / SUMMARY

How well do you know this chapter? Complete the following questions to find out. Take a look back at the section if you need help.

### SECTION 2.1 Equations and Their Solutions

1. A(n) _equation_ is a statement in which two expressions are equal.

2. A(n) _solution_ is a value of the variable that makes the equation true.

3. When translating phrases, the expressions _is equal to_, _the result is_, _is the same as_, _equals_, and _is_ represent the equals sign.

## SECTION 2.2 The Addition Property of Equality

**4.** A linear equation in one variable is an equation that can be written in the form $Ax + B = C$. An example of a linear equation in one variable is $2x + 3 = 7$ (Answers vary.).

**5.** To solve a linear equation in one variable, the goal is to produce an equation of the form $x = c$ or $c = x$.

**6.** Equivalent equations are equations with the same solution set.

**7.** The addition property of equality states that we can add or subtract the same number from both sides of an equation and not change the solution set.

**8.** To solve an equation, we must simplify each side as much as possible. Then we must isolate the variable to one side and the constants to the other.

## SECTION 2.3 The Multiplication Property of Equality

**9.** The multiplication property of equality enables us to multiply or divide both sides of an equation by the same number.

**10.** If the coefficient of the variable is a fraction, we can multiply both sides of the equation by its reciprocal to obtain a coefficient of 1 on the variable.

**11.** Consecutive integers are integers that follow one another. An example of consecutive integers is 1, 2, 3, . . . , an example of consecutive odd integers is 1, 3, 5, . . . , and an example of consecutive even integers is 2, 4, 6, . . . . If $x$ is an integer, then $x + 1$ and $x + 2$ represents the next two consecutive integers.

## SECTION 2.4 More on Solving Linear Equations

**12.** To solve equations with fractions, we can multiply both sides of the equation by the least common denominator to obtain an equivalent equation without fractions.

**13.** To solve equations with decimals, we can multiply both sides of the equation by the power of 10 that eliminates the decimals.

**14.** A(n) conditional equation is an equation that is true for some values of the variable but not true for others.

**15.** A(n) contradiction is an equation that is not true for any values of the variable. The solution set for these equations is the empty set or ∅.

**16.** A(n) identity is an equation that is true for all values of the variable. The solution set for these equations is all real numbers or ℝ.

## SECTION 2.5 Formulas and Applications from Geometry

**17.** A(n) formula is an equation that expresses the relationship between two or more variables.

**18.** To evaluate a formula, we substitute the given values in place of the variable and simplify.

**19.** The perimeter of a polygon is the distance around the figure.

**20.** The distance around a circle is its circumference.

**21.** The area of a figure is the number of square units it takes to cover the inside of the figure.

**22.** The sum of the measures of complementary angles is 90°. The sum of the measures of supplementary angles is 180°.

**23.** A right angle has measure 90° and a straight angle has measure of 180°.

**24.** If $x$ represents the measure of an angle, $90 - x$ represents the measure of its complement and $180 - x$ represents the measure of its supplement.

**25.** Vertical angles, or opposite angles, are formed by intersecting lines. These angles are equal in measure.

**26.** The sum of the measures of the angles in a triangle is 180°.

## SECTION 2.6 Percent, Rate, and Mixture Problems

**27.** Percent means per hundred. So, $30\% = \dfrac{30}{100} = 0.30$.

**28.** Interest is the amount of money collected for using or borrowing money. The simple interest formula is $I = Prt$.

**29.** When working with mixture problems, we must find the amount of pure substance in each solution. We do this by multiplying the strength of the solution by the percentage (or concentration) of the solution.

**30.** To find the value of a collection of coins, multiply the number of coins by the value of each coin.

**31.** Distance traveled is the rate multiplied by the time traveled. In symbols, $d = rt$.

## SECTION 2.7 Linear Inequalities in One Variable

**32.** A linear inequality in one variable is an inequality of the form $ax + b < c$.

**33.** A(n) solution of a linear inequality is a value that makes the inequality true.

**34.** The picture of the solution set of an inequality is the graph of the solution set.

**35.** If an endpoint of a solution set of an inequality is included in the solution, a bracket is used. If an endpoint of a solution set is not included in the solution set, a parenthesis is used.

**36.** A concise way to express the solution set of an inequality is interval notation.

**37.** The symbol ∞ indicates that the solution of an inequality continues to the right indefinitely. The symbol −∞ indicates that the solution of an inequality continues indefinitely to the left. A parenthesis is always used with these notations.

**38.** The set-builder notation is also used to represent the solution set of an inequality. This notation states the conditions the solution must satisfy.

**39.** When solving linear inequalities, we can add or subtract the same number from both sides of an inequality and not change the relationship between the two expressions.

**40.** When solving linear inequalities, we can multiply or divide both sides by a _positive_ number and not change the relationship between the two expressions.

**41.** When solving linear inequalities, we can multiply or divide both sides by a _negative_ number, but we must

also _reverse_ the inequality symbol to maintain the relationship between the two expressions.

**42.** The phrase "is at most" can be translated by the symbol $\leq$. The phrase "is at least" can be translated by the symbol $\geq$.

## CHAPTER 2 / REVIEW EXERCISES

### SECTION 2.1

**Determine whether each of the following is an expression or an equation.** (*See Objective 1.*)

**1.** $7x - 9 = 2x - 12$
equation
**2.** $-5r^2 + 11 - 12r - 24$
expression
**3.** $5x - 11 - 2x + 3$
expression
**4.** $6y^2 - 10 = 12y - 5$
equation

**Determine if the given numbers are solutions of the equation. Organize the information in tables.** (*See Objective 2.*)

**5.** Is $x = -4$, $x = -1$, $x = 0$, $x = 4$, or
$x = -\dfrac{4}{5}$ a solution of $3x - 1 = -2x - 5$?

**6.** Is $x = -6$, $x = -2$, $x = 0$, $x = 3$, or
$x = \dfrac{1}{2}$ a solution of $x^2 + 3x = 18$?

**Assign a variable to the unknown quantity and then write an equation that represents each statement.** (*See Objective 3.*)

**7.** Four times the sum of a number and $-17$ is the same as twice the number. Find the number.  $4[x + (-17)] = 2x$

**8.** The sum of a number and 18 is the same as seven times the number. Find the number.  $x + 18 = 7x$

**9.** The difference of four times a number and $-3$ equals 45 less than the number. Find the number.  $4x - (-3) = x - 45$

**10.** One angle is $6°$ more than another angle. Their sum is $90°$. Find the measure of each angle.  $x + x + 6° = 90°$

**11.** One angle is $17°$ less than another angle. Their sum is $180°$. Find the measure of each angle.  $x + x - 17° = 180°$

**12.** From June 2008 to June 2009, John Mars and Larry Page were the two of the 31 richest Americans. Their combined net worth was $26.3 billion. John Mars's net worth is $4.3 billion less than Larry Page's. Find the net worth of each person. (Source: http://www.forbes.com/lists/)  $x + x - 4.3 = 26.3$

### SECTION 2.2

**Solve each equation.** (*See Objectives 2 and 3.*)

**13.** $5x + 6 = 2x + 18$  {4}
**14.** $3 - x = 11 - 5x$  {2}

**15.** $\dfrac{3}{5}x + 2 = -6 - \dfrac{2}{5}x$  {-8}
**16.** $\dfrac{4}{7} - x = 5x - \dfrac{3}{7}$  $\left\{\dfrac{1}{6}\right\}$

**17.** $5(x - 2) + 13 = 3(x - 2) - 9$  {-9}

**18.** $1.24x + 3.25 = 2.94x - 1$  {2.5}

**19.** $3.5x - 2.6 + 1.9x = 7.3x - 13.24$  {5.6}

**20.** $32 - 10x = 32 - 15x$  {0}

**For each problem: (1) assign a variable to the unknown, (2) write an equation that represents the situation, (3) solve the equation, and (4) answer the question using complete sentences.** (*See Objective 4.*)

**21.** The sum of a number and $-18$ is the same as five times the number. Find the number.  The number is $-\dfrac{9}{2}$.

**22.** Three times the sum of a number and 4 is the same as twice the number. Find the number.  The number is $-12$.

**23.** Twice the difference of a number and $-5$ is the same as four times the number. Find the number.  The number is 5.

**24.** Six more than a number is $-1$. Find the number.  The number is $-7$.

**25.** The difference of twice a number and $-19$ is equal to 47 more than the number. Find the number.  The number is 28.

**26.** One angle is $24°$ more than another angle. Their sum is $90°$. Find the measure of each angle.  The measures of the angles are $33°$ and $57°$.

**27.** From June 2008 to June 2009, Christy Walton and Philip Knight were the two of the 31 richest Americans. Their combined net worth was $31 billion. Christy Walton's net worth is $12 billion more than Philip Knight's. Find the net worth of each person. (Source: http://www.forbes.com/lists/)  The net worth of Christy Walton was $21.5 billion and the net worth of Philip Knight was $9.5 billion.

**28.** From June 2008 to June 2009, Dr. Phil McGraw and Oprah Winfrey were the two of the top 100 celebrities. Their combined earnings were $355 million. Dr. Phil McGraw earned $195 million less than Oprah Winfrey. How much did each of them earn? (Source: http://www.forbes.com/lists/)  Dr. Phil McGraw earned $80 million and Oprah Winfrey earned $275 million.

### SECTION 2.3

**Solve each equation.** (*See Objectives 1 and 2.*)

**29.** $5y - 7 = 23$  {6}
**30.** $-2x + 30 = 4x$  {5}

**31.** $\dfrac{a}{2} = -8$  {-16}
**32.** $8z + 5 - 2z = 15 - 3z$

**33.** $-10x + 9 - 3x = 18 + 5x - 16$  $\left\{\dfrac{7}{18}\right\}$  $\left\{\dfrac{10}{9}\right\}$

**34.** $-\dfrac{1}{2}(x + 6) + 8 = x + 5$  {0}

**35.** $-1.8b + 6 = 42$  {-20}

**36.** $-0.6y + 15.2 = 2.7y + 24.44$  {-2.8}

**37.** $1.2(5.2x - 3.2) = -0.6(2.5x - 0.05)$  {0.5}

**38.** $12(x - 2) = 2(6x - 1)$  ⌀

**For each problem: (1) assign a variable to the unknown, (2) write an equation that represents the situation, (3) solve the equation, and (4) answer the question using complete sentences.** (*See Objectives 3 and 4.*)

**39.** Three times a number is equal to $-54$. Find the number.  $-18$

**40.** Two-fifths of a number is 12. Find the number.  30

**41.** The quotient of a number and 3 is $-12$. Find the number. $-36$

**42.** 60% of a number is 9. Find the number.   15

**43.** The sum of a number and 15 is the same as three more than twice a number. Find the number.   12

**44.** How many nickels will it take to make $8.50?   {170}

**45.** The sum of two consecutive integers is $-97$. Find the integers.   The integers are $-49$ and $-48$.

**46.** The cost to rent a midsize car for one day is $55 plus $0.32 per mile. The daily cost is represented by the expression $55 + 0.32x$, where $x$ is the number of miles driven. How many miles has the car been driven if the cost of the rental car is $167?   The car has been driven 350 mi.

## SECTION 2.4

**Solve each equation. If the equation is a contradiction, write the solution as ∅. If the equation is an identity, write the solution as ℝ. (See Objectives 1–4.)**

**47.** $2x - 5 = 5$   {5}

**48.** $4(y + 1) - (y - 5) = 3(y + 1)$   ∅

**49.** $\frac{2}{5}x - 4 = \frac{3}{10}x$   {40}

**50.** $1.2x + 0.8x = 900$   {450}

**51.** $\frac{x - 8}{14} + 1 = \frac{x + 5}{7}$   {$-4$}

**52.** $0.032x + 0.013(540 - x) = 13.67$   {350}

**53.** $\frac{5}{6}(x - 2) = \frac{1}{6}(5x - 8) - \frac{1}{3}$   ℝ

**54.** $\frac{3}{4}a - \frac{5}{8} = \frac{1}{2}a + \frac{1}{4}$   {$\frac{7}{2}$}

**55.** $0.3(a - 1) + 0.5a = 0.4(2a + 9)$   ∅

**56.** $0.64x + 1.6(-4.5) = 0.4(x - 17.1)$   {1.5}

## SECTION 2.5

**The formula $A = P(1 + r)^t$ is used to calculate the amount of money in an account at the end of $t$ years if $P$ dollars are invested at an annual interest rate, $r$. (See Objective 1.)**

**57.** Find the value of $A$ if $1200 is invested for 2 yr at 1.5% annual interest.   $1236.27

**58.** Find the value of $P$ if $2521.50 is expected to be accumulated at the end of 2 yr at 2.5% annual interest.   $2400

**The distance $d$, rate $r$, and time $t$ of an object in motion can be related by the formula $d = rt$. (See Objective 1.)**

**59.** Find $d$ if $r = 62$ mph and $t = 2.5$ hr.   155 mi

**60.** Find $t$ if $d = 306$ mi and $r = 68$ mph.   4.5 hr

**The revenue $R$, unit price $p$, and sale level $x$ of a product are related by the formula $R = xp$. (See Objective 1.)**

**61.** Find $R$ if $p = 125$ and $x = 56$.   $7000

**62.** Find $p$ if $R = 1392$ and $x = 96$.   $14.50

**The formula to find the volume of a cylinder with radius $r$ and height $h$ is $V = \pi r^2 h$. Use $\pi = 3.14$. (See Objective 1.)**

**63.** Find $V$ if $r = 6$ in. and $h = 15$ in.   1695.6 in.³

**64.** Find $h$ if $V = 157$ m³ and $r = 2.5$ m.   8 m

**Use the perimeter, area, or circumference formula to find the indicated quantity. (See Objective 2.)**

**65.** Find the height of a triangle whose area is 84 m² and whose base is 12 m.   14 m

**66.** Find the perimeter of the rectangle whose area is 432 in.² and whose length is 27 in.   86 in.

**67.** Find the area of a rectangle if the perimeter is 74 m and whose width is 20 m.   340 m²

**68.** Find the radius and area of a circle with circumference 150.72 cm. Use 3.14 for the value of $\pi$ to approximate answers to two decimal places.   24 cm; 1808.64 cm²

**Solve each formula for the specified variable. (See Objective 3.)**

**69.** $R = xp$ for $p$   $p = \dfrac{R}{x}$

**70.** $C = 3l + 4w$ for $l$   $l = \dfrac{C - 4w}{3}$

**71.** $7x + 3y = 21$ for $y$   $y = -\dfrac{7}{3}x + 7$

**72.** $y = \dfrac{1}{4}x - 1$ for $x$   $x = 4y + 4$

**73.** $0.05x + 0.02y = 35$ for $x$   $x = -0.4y + 700$

**74.** $0.36x - 0.2y = 216$ for $y$   $y = 1.8x - 1080$

**Find the measure of each unknown angle. (See Objective 4.)**

**75.** Find the measure of an angle whose complement is 25° less than the measure of the angle.   57.5°

**76.** Find the measure of an angle whose supplement is 54° more than the measure of the angle.   63°

**77.** Find the measure of an angle whose supplement is 20° less than three times the measure of the angle.   50°

**78.** The supplement of an angle is 30° more than three times its complement. Find the measure of the angle.   60°

**Find the measure of each angle labeled in the figure. (See Objectives 5 and 6.)**

**79.**   49° and 131°

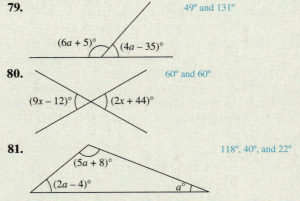

**80.**   60° and 60°

**81.**   118°, 40°, and 22°

**82.** In triangle $ABC$, the measure of angle $B$ is 18° less than the measure of angle $A$. The measure of angle $C$ is 27° more than the measure of angle $A$. Find the measure of each angle in the triangle.   57°, 39°, and 84°

## SECTION 2.6

**Write an equation that represents each situation, solve the equation, and write the answer in complete sentences. (*See Objectives 1–5.*)**

**83.** A 46-in. LED–LCD HDTV is on sale for $1170. This is 35% off the original price. What is the original selling price of the TV?   $1800

**84.** A sweater is on sale for $18.90. If this price is 65% off the original price, what is the original price of the sweater?   $54

**85.** A campus bookstore sells an intermediate algebra text for $101.25. This is a 25% markup on the cost the bookstore pays for the text. What does the bookstore pay for the text?   $81

**86.** Bev's salary after her raise is $44,285. Prior to the raise, her salary was $42,500. What percent raise did she receive?   4.2%

**87.** The passenger volume at an international airport in Florida was 34.13 million in 2005 and 35.66 million in 2011. What is the percent increase in the passenger volume from 2005 to 2011? Round to one-tenth of a percent.   4.5% increase

**88.** Kevin invests $8600 in two accounts. He invests $5350 in a mutual fund that pays 3.4% annual interest and the remaining amount in a money market fund that pays 2.6% annual interest. How much interest did he earn from the two accounts in 1 yr?   $266.40

**89.** Stephanie has a collection of half-dollars and quarters. She has 40 more half-dollars than quarters. If the collection is worth $50, how many half-dollars and quarters does she have?   80 half-dollars and 40 quarters

**90.** Joan has a collection of $20 bills and $1 bills. She has 29 more $20 bills than $1 bills. If the value of her collection is $1756, how many of each bill does she have?   85 $20 bills and 56 $1 bills

**91.** Rachael needs to make 72 L of a 33.5% hydrochloric acid solution. How much 79% hydrochloric acid solution and how much 27% hydrochloric acid solution should she mix to get 72 L of a 33.5% hydrochloric acid solution?   9 L of 79% hydrochloric acid and 63 L of 27% hydrochloric acid solution

**92.** A dairy farmer has two types of milk. How many gallons of a 2.5% milk fat solution and how many gallons of a 4.4% milk fat solution should be mixed to obtain 380 gal of a 2.9% milk fat solution?   300 gal of 2.5% milk fat solution and 80 gal of 4.4% milk fat solution

**93.** Kyle and Bevin live 1026 mi apart. They leave their homes at the same time and drive toward one another until they meet. Kyle travels at 63 mph and Bevin travels at 72 mph. How long will it take before they meet?   7.6 hr

**94.** Two planes leave an international airport in Tennessee at the same time. One plane travels west at 460 mph and the other plane travels east at 375 mph. How long will it take for the planes to be 4258.5 mi apart?   5.1 hr

**95.** The sum of the two smaller of three consecutive even integers is the same as 300 less than nine times the largest integer. Find the integers.   38, 40, and 42

**96.** A plane leaves Atlanta heading to Seattle at 6 A.M. traveling at 340 mph. A second plane leaves Atlanta heading to Seattle at 8 A.M. the same day traveling at 510 mph. How long will it take for the second plane to catch up to the first plane?   4 hr

## SECTION 2.7

**Solve each inequality. Graph the solution set and write the solution set in interval notation and in set-builder notation. (*See Objectives 1–4.*)**

**97.** $a + 5 \geq -9$

**98.** $4x - 18 > -6$

**99.** $0.3a - 7.2 \leq -2.4$

**100.** $\dfrac{2}{7}x + 1 \geq \dfrac{5}{7}$

**101.** $-1 < 3a + 2 < 23$

**102.** $-2 \leq 4x + 10 \leq 26$

**103.** $17 - 2(x - 5) < -6x + 3$

**104.** $6x + 2(x + 12) < 9x + 15$

**105.** $\dfrac{5}{3}y + 3\left(y + \dfrac{1}{6}\right) \geq \dfrac{2}{3}y - \dfrac{1}{2}$

**106.** $-19 \leq -2(x + 1) + 5 \leq -1$

**107.** $-3.5 < 0.5y + 1.4 < 1.2$

**108.** $-1.6 < 0.3x - 2.8 < 12.2$

**Write an inequality that will solve each problem. Solve the inequality and answer the question in complete sentences. (*See Objective 6.*)**

**109.** Baziz scored 67, 85, and 80 on three math tests. What score does he need on the fourth test to have at least an 80 average?   at least 88 on the fourth test

**110.** Shirl's final grade in her math class is based on a weighted average. Homework counts as 10%, quizzes count as 20%, tests count as 50%, and the final exam counts as 20%. If she has a homework average of 72, a quiz average of 70, and a test average of 84, what score does she need to make on her final exam to have at least a 80 for her final grade?   84 on the final exam

**111.** The cost to rent a banquet facility for a wedding reception is $1250 plus $45 per person for the food. What is the number of attendees the couple can have at the reception if they have at most $9600 budgeted for the reception?   185 attendees

**112.** The cost to rent a car for a week is $280 plus $0.32 per mile. How many miles can be driven if you have at most $842 to rent the car for a week? Round to the nearest whole number.   at most 1756 mi

---

## CHAPTER 2 TEST / LINEAR EQUATIONS AND INEQUALITIES IN ONE VARIABLE

**1.** An example of a linear equation in one variable is
   **a.** $3x + 5 - 7x$
   **(b.)** $3x + 5 = 7x$

**2.** A solution of $x^2 = 4x$ is
   **a.** $x = -4$
   **b.** $x = -2$
   **c.** $x = 2$
   **(d.)** $x = 4$

3. The statement "The sum of three times a number and 6 is the same as 4 less than the number." can be written as

   **a.** $3(n + 6) = n - 4$     **b.** $3(n + 6) = 4 - n$

   **c.** $3n + 6 = n - 4$     **d.** $3n + 6 = 4 - n$

4. The equation that can be solved by subtracting 3 from each side is

   **a.** $3x = 21$     **b.** $4 = 3 + x$

   **c.** $x - 3 = 5$     **d.** $\frac{1}{3}x = 4$

5. In the Korean and Vietnam Wars, the United States lost a total of approximately 112,499 people. There were about 4007 fewer deaths in the Korean War than in the Vietnam War. How many U.S. deaths were there in each of these wars? If $x$ represents the number of deaths in the Vietnam War,

   **a.** write an equation that can be used to solve this problem.

   **b.** solve the equation and explain the answer.

**Solve each equation. Write each solution in a solution set.**

6. $7x - 3(2x + 1) = 0$   {3}    7. $4(x + 3) = 5(2 - x)$   $\left\{-\frac{2}{9}\right\}$

8. $3 - x = -7$   {10}    9. $\frac{3}{2}b = 9$   {6}

10. $4 + 3(2x - 5) = 7x + 6 - 2x$   {17}

11. $\frac{y}{6} + \frac{1}{2} = \frac{7}{6}$   {4}    12. $-\frac{1}{5}(x - 5) = \frac{2}{5}(x + 10)$   {−5}

13. $0.05x + 0.03(10,000 - x) = 440$   {7000}

14. $9y - 3(4y + 2) = 2y - 4 - 5y - 2$   $\mathbb{R}$

15. $-2(b - 5) = 4(b + 1) - 6b$   $\varnothing$

16. Solve the equation $4x + 3y = -12$ for $y$.   $y = -\frac{4}{3}x - 4$

17. Evaluate the formula $C = \frac{5}{9}(F - 32)$ if $F = 59°$.   15°C

18. Explain how solving linear equations and solving linear inequalities are the same. Explain how they are different.

19. The graph of $-4 \geq x$ can be represented as

   **a.**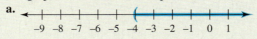

   **b.**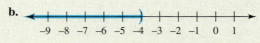

   **c.**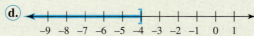

   **d.**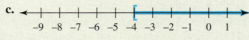

20. The interval notation for the following graph is

   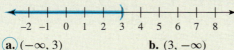

   **a.** $(-\infty, 3)$     **b.** $(3, -\infty)$

   **c.** $(\infty, 3)$     **d.** $(3, \infty)$

**Solve each inequality. Graph the solution set. Write the solution in interval notation and set-builder notation.**

21. $4x - 5(x - 2) \geq -3$   $(-\infty, 13]$

22. $\frac{2}{3}(x - 1) < x + \frac{4}{3}$   $(-6, \infty)$

**Define a variable to represent the unknown quantity. Write an equation (or inequality) that represents each situation. Solve the equation and answer the question in a complete sentence.**

23. A taxicab authority in Nevada has set standard fares and fees. To pick up a passenger, the cab company charges a fee of $3.30. For each mile driven, there is a charge of $2.40.

   **a.** If you have at most $75, how far could you travel in the cab? Round your answer to the nearest tenth.   I could travel at most 29.9 mi.

   **b.** The distance from your hotel to the Hoover Dam is 40.8 mi. Do you have enough money to travel there by taxicab?   No

24. The measure of the complement of an angle is 6° less than three times the measure of the angle. Find the measure of the angle.   The angle has a measure of 24°.

25. The final grade in Joann's math class is based upon a weighted average. Tests count as 50%, quizzes count as 15%, homework counts as 10%, and the final exam counts as 25%. If she has a test average of 84, a quiz average of 86, and a homework average of 100, what does she need on the final exam to have at least an 80 for her final grade?   Joann needs a 60.4 or higher on the final to have at least an 80 in the class.

26. Find the value of $a$ and the measures of the vertical angles shown.

   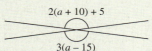

   The measure of angle $a$ is 70° and the measure of each angle is 165°.

27. Two of the angles in a triangle have equal measure. The third angle is 20 less than six times the measure of the two equal angles. Find the measure of each angle in the triangle.   Measures of the angles are 25°, 25°, and 130°.

28. An architect designs a circular stadium. If the footprint of the stadium has a perimeter of $180\pi$ ft, what is the area of the footprint of the stadium? Use 3.14 for the value of $\pi$. Round your answer to the nearest integer. The footprint of the stadium has an area of 25,434 ft².

29. David has $3000 to invest. He invests part of the money in a savings account that earns 4% annual interest and the remaining amount in a CD that earns 6% annual interest. If he earns $164 in interest in 1 yr, how much money did he invest in each account? David invested $800 in the 4% savings account and $2200 in the CD.

30. A department store advertises that they donate 5% of their weekly profit to various charities. Their donation is approximately $3 million each week. What is the department store's weekly profit? The department store's weekly profit is $60 million.

31. Dr. Cleves is a pharmacist who needs a 15% glycerol solution. He has 10 mL of a 50% glycerol solution. How much water does he need to mix with the 10 mL of the 50% solution to obtain a solution that is 15% glycerol? Round to the nearest tenth. Dr. Cleves needs to mix 23.3 mL of water with the 10 mL of 50% glycerol solution to obtain a 15% glycerol solution.

32. The sum of the two smallest of three consecutive odd integers is the same as one less than three times the

third consecutive odd integer. What are the three odd integers? The integers are $-9$, $-7$, and $-5$.

**33.** Gloria and Richard go out for an afternoon jog. They begin running at an average speed of 9 mph. After

some time, they turn around and walk back home at an average speed of 4 mph. If they were gone for a total of 2 hr, how long did they jog before turning around? They jog for approximately 0.62 hr or 37 min before turning around.

## CUMULATIVE REVIEW EXERCISES / CHAPTERS 1 AND 2

**1.** Classify each number as a natural number, whole number, integer, rational number, irrational number, and/or a real number. If the number is a rational number, write it in fractional form. If the number is an irrational number, approximate its value to two decimal places. (*Section 1.1, Objective 1*)

   **a.** $\dfrac{9}{11}$   rational number, real number

   **b.** $\sqrt{13}$   irrational number, real number, 3.61

   **c.** $-47°F$: the coldest recorded temperature in Elkader, Iowa on February 31, 1996   integer, rational number, real number; $-\dfrac{47}{1}$

**2.** Give an example of a real number that satisfies the given conditions. (*Section 1.1, Objective 1*)

   **a.** An integer that is not a whole number.   $-5$; answers vary

   **b.** An irrational number that is between 3 and 4.   $\sqrt{10}$; answers vary

   **c.** A rational number that is between 5/2 and 3.   2.6; answers vary

   **d.** An integer that is not a natural number.   $-10$; answers vary

**3.** Graph each number on a real number line. (*Section 1.1, Objective 2*)

   **a.** $\left\{-4.1\overline{1}, -1, \sqrt{7}, 5\dfrac{1}{2}\right\}$   **b.** $\left\{-2\pi, -3, \sqrt{11}, \dfrac{40}{3}\right\}$

**4.** Compare the values of each pair of numbers. Use a $<$, $>$, or $=$ symbol to make the statement true. (*Section 1.1, Objective 3*)

   **a.** $\pi$ ____ 3.3   $\pi < 3.3$   **b.** $\sqrt{16}$ ____ $-|-4|$   $\sqrt{16} > -|-4|$

**5.** Find the opposite of each real number. (*Section 1.1, Objective 4*)

   **a.** $1.3\overline{3}$   $-1.33$   **b.** 7.5   $-7.5$   **c.** $-1\dfrac{2}{3}$   $1\dfrac{2}{3}$

**6.** Simplify each absolute value expression. (*Section 1.1, Objective 5*)

   **a.** $|5|$   5   **b.** $\left|-\dfrac{4}{7}\right|$   $\dfrac{4}{7}$   **c.** $-|-14|$   $-14$

**7.** Write the prime factorization of each number. (*Section 1.2, Objective 1*)

   **a.** 180   $2^2 \cdot 3^2 \cdot 5$   **b.** 98   $2 \cdot 7^2$   **c.** 500   $2^2 \cdot 5^3$

**8.** For the academic year 2009–2010, it was reported that there were 6896 higher education institutions in the United States and other jurisdictions; 2853 were 4-yr, 2259 were 2-yr, and 1784 were less-than-2-yr institutions. Write a fraction that represents the portion of less-than-2-yr institutions. (*Section 1.2, Objective 2*)   $\dfrac{223}{862}$

**9.** Simplify each fraction. (*Section 1.2, Objective 2*)

   **a.** $\dfrac{36}{90}$   $\dfrac{2}{5}$   **b.** $\dfrac{250}{516}$   $\dfrac{125}{258}$

**10.** Perform each operation and simplify the result. (*Section 1.2, Objectives 2–7*)

   **a.** $\dfrac{13}{18} + \dfrac{25}{42}$   $\dfrac{83}{63}$   **b.** $\dfrac{7}{12} - \dfrac{3}{10}$   $\dfrac{17}{60}$   **c.** $4\dfrac{1}{5} \cdot 3\dfrac{1}{3}$   14

   **d.** $2\dfrac{5}{6} \div 2\dfrac{1}{6}$   $\dfrac{17}{13}$   **e.** $5\dfrac{1}{5} - 4\dfrac{3}{5}$   $\dfrac{3}{5}$

**11.** Use the order of operations to simplify each numerical expression. (*Section 1.3, Objectives 1 and 2*)

   **a.** $3.5^2$   12.25   **b.** $-\left(-\dfrac{2}{7}\right)^3$   $\dfrac{8}{343}$

   **c.** $\dfrac{1}{6}(9-7)^4 - \dfrac{5}{12}$   $\dfrac{9}{4}$   **d.** $2(6)^2 - 5(3) + 7$   64

   **e.** $7[28 - 3|15 - 2(7 - 3)|]$   49

   **f.** $\dfrac{5 + \sqrt{5^2 + 4(3)(8)}}{2 \cdot 3}$   $\dfrac{8}{3}$

**12.** Evaluate each expression for the given values of the variables. (*Section 1.3, Objective 3*)

   **a.** $\dfrac{3x - 5}{x + 1}$ for $x = 0, 1, 2, 3$   $-5, -1, \dfrac{1}{3}, 1$

   **b.** $4x + 6y$ for $x = 2$ and $y = 3$   26

   **c.** $b^2 - 4ac$ for $a = 2, b = 6, c = 3$   12

**13.** Determine if the given value is a solution of the equation. (*Section 1.3, Objective 4*)

   **a.** $x^2 - 5x - 12 = 2x + 6$; $x = 9$   $x = 9$ is a solution.

   **b.** $5y - (y - 3) = y + 9$; $y = 4$   $y = 4$ is not a solution.

**14.** Translate each phrase into an algebraic expression. Let $x$ represent the unknown number. (*Section 1.3, Objective 5*)

   **a.** 16 more than three times a number   $16 + 3x$

   **b.** The difference of a number and 12   $x - 12$

   **c.** 13 less than five times a number   $5x - 13$

   **d.** The sum of four times a number and 8   $4x + 8$

**15.** The height of a baseball hit upward with an initial velocity of 96 ft/sec from an initial height of 6 ft is represented by the expression $-16t^2 + 96t + 6$, where $t$ is the number of seconds after the ball has been hit. What is the height of the ball after 3 sec? (*Section 1.3, Objective 6*)   150 ft

**16** The expression $P(1 + r)^t$ represents the amount of money in an account when $P$ dollars is invested at an annual interest rate of $r$ (in decimal form) for $t$ yr. Find the amount of money that will be in an account at the end of 4 yr if $2500 is invested at 1.2% annual interest. (*Section 1.3, Objective 6*)   $2622.18 is in the account.

**17.** At a community college, tuition and fees are calculated by the expression $135c + 86.75$, where $c$ is the number of credit hours taken per semester. If Bianca takes 10 credit hours next semester, what are her tuition and fees? (*Section 1.3, Objective 6*) $1436.75

**18.** Perform the indicated operation and simplify. (*Sections 1.4 and 1.5, Objectives 1 and 2*)

**a.** $5.16 + (-3.97)$ 1.19   **b.** $\left(-\dfrac{5}{4}\right) + \left(-\dfrac{3}{2}\right)$ $-\dfrac{11}{4}$

**c.** $5 - 9 + (-10) - (-15)$ 1

**d.** $5.4 - (-3.2) + (-2.1) + 2.8$ 9.3

**e.** $\dfrac{3}{4} - \dfrac{5}{8} - \left(-\dfrac{1}{2}\right)$ $\dfrac{5}{8}$

**f.** $28 - \{9 - [5 - (1 - \sqrt{30 - 14})]\}$ 27

**g.** $-6^2 - (-15) - |2 - 10|$ $-29$

**For Exercises 19–22, write a mathematical expression needed to solve each problem and then answer the question. (*Sections 1.4 and 1.5, Objective 3*)**

**19.** The highest point on the Australian continent is the peak of Mount Kosciuszko. The peak is 7310 ft above sea level. The lowest point on land in China is the Turpan Pendi, which is 505.2 ft below sea level. What is the difference between these altitudes? The difference is 7815.2 ft.

**20.** If the lowest recorded temperature in New Mexico is $-50°$F and the highest recorded temperature is $122°$F, what is the difference between the highest and lowest temperatures? The difference between the highest and lowest temperatures is 172°F.

**21.** Dave has $245.75 in his checking account. He deposits his paycheck for $635.25. Dave writes a check for his phone bill for $65.43, for groceries for $125.78, and then for rent for $525. What is Dave's checking account balance? Dave's checking account balance is $164.79.

**22.** Find the measure of the unknown angle of the given triangle.

**a.** $a = 29°, b = 87°$   $c = 64°$

**b.** $b = 103°, c = 43°$   $a = 34°$

**23.** Find the complement and the supplement of an angle whose measure is $78°$. 12°; 102°

**24.** Perform the indicated operation and simplify. (*Section 1.6, Objectives 1–3*)

**a.** $(-3.2)(1.5)$  $-4.8$   **b.** $(-18)\left(\dfrac{-7}{24}\right)$ $\dfrac{21}{4}$

**c.** $-(-3)^4$  $-81$   **d.** $\dfrac{0}{6}$  0

**e.** $10 \div 0$  undefined   **f.** $8(-25)(-3)$  600

**g.** $\left(-\dfrac{5}{12}\right) \div \left(-\dfrac{10}{9}\right)$ $\dfrac{3}{8}$   **h.** $-8 \div \dfrac{20}{7}$ $-\dfrac{14}{5}$

**25.** Evaluate each expression for the given values of the variables. (*Section 1.6, Objective 4*)

**a.** $\sqrt{(x_1 - x_2)^2 + (y_1 - y_2)^2}$ for $x_1 = -2, x_2 = 1, y_1 = 7, y_2 = 3$  5

**b.** $\dfrac{|2x + 3|}{x - 2}$ for $x = -2, -1, 0, 1, 2$   $-\dfrac{1}{4}, -\dfrac{1}{3}, -\dfrac{3}{2}, -5,$ undefined

**26.** Find the additive and multiplicative inverses of each number. Assume all variables are nonzero. (*Section 1.7, Objective 1*)

**a.** $-16x$  $16x, -\dfrac{1}{16x}$   **b.** $\dfrac{5}{8}$  $-\dfrac{5}{8}, \dfrac{8}{5}$   **c.** $6a$  $-6a, \dfrac{1}{6a}$

**27.** Apply the commutative, associative, and/or distributive properties to rewrite each expression and simplify the result. (*Section 1.7, Objectives 2 and 3*)

**a.** $-18 + c - (-15)$  $c - 3$   **b.** $-2(x - 10y - 7)$  $-2x + 20y + 14$

**c.** $\left(b - \dfrac{3}{10}\right) + 2$  $b + \dfrac{17}{10}$   **d.** $10\left(-\dfrac{2}{5}x + \dfrac{3}{2}\right)$  $-4x + 15$

**28.** Determine if the terms are like or unlike. (*Section 1.8, Objective 2*)

**a.** $8ab^2$ and $-34ab^2$ like terms   **b.** $\dfrac{1}{3}cd^2$ and $-\dfrac{5}{7}c^2d$ unlike terms

**c.** $\dfrac{4}{3}\pi r^3$ and $\pi r^2 h$ unlike terms

**29.** Simplify each algebraic expression. (*Section 1.8, Objectives 3 and 4*)

**a.** $-5(x)(14)$  $-70x$   **b.** $15(2x - 1) - 36x$  $-6x - 15$

**c.** $-(x - 12) - 2(5x + 7)$  $-11x - 2$   **d.** $3.7 - 0.8(-0.1x + 2.5)$  $0.08x + 1.7$

**e.** $-12\left(\dfrac{1}{3}a - \dfrac{3}{4}\right) + 14\left(\dfrac{3}{7}a - \dfrac{1}{2}\right)$  $2a + 2$

**f.** $2x^2 - 4(x - 3x^2) - x$  $14x^2 - 5x$

**30.** Determine whether the following is an expression or an equation. (*Section 2.1, Objective 1*)

**a.** $6x - 19 = 4x - 11$  equation   **b.** $7x - 15 - 3x + 4$  expression

**31.** Determine if each number is a solution of the equation. (*Section 2.1, Objective 2*)

**a.** $x = -2, x = 0,$ or $x = 2$; $5x - 2 = -2x - 16$  $x = -2$ is a solution; $x = 0$ is not a solution; $x = 2$ is not a solution

**b.** $x = -2, x = 0,$ or $x = 2$; $8x - 12 = -3x + 10$  $x = -2$ is not a solution; $x = 0$ is not a solution; $x = 2$ is a solution

**32.** For each problem, define the variable, write an equation, and solve the problem. (*Section 2.1, Objectives 3 and 4*)

**a.** Five times the sum of a number and $-26$ is the same as three times the number. Find the number.  $5[x + (-26)] = 3x; x = 65$

**b.** The sum of a number and 28 is the same as five times the number. Find the number.  $x + 28 = 5x; x = 7$

**c.** The difference of twice a number and $-6$ equals 32 less than the number. Find the number.  $2x - (-6) = x - 32; x = -38$

**d.** One angle is $16°$ more than another angle. Their sum is $180°$. Find the measure of each angle.  82° and 98°

**e.** One angle is $12°$ less than another angle. Their sum is $90°$. Find the measure of each angle.  51° and 39°

**f.** One angle is $6°$ more than another angle. Their sum is $90°$. Find the measure of each angle.  42° and 48°

**33.** Solve each equation. (*Section 2.2, Objectives 2 and 3; Sections 2.3 and 2.4, Objectives 1 and 2*)

**a.** $7(x - 3) + 18 = 2(x - 6) - 26$  $\{-7\}$

**b.** $2.46x + 12.25 = 5.28x + 8.02$  $\{1.5\}$

**c.** $4.5x - 3.6 + 2.9x = 8.2x - 5.04$  {1.8}

**d.** $10(3 - x) - 5x = 20 + 5(2 - 3x)$  $\mathbb{R}$

**e.** $4z + 32 - z = 5 - 6z$  {−3}

**f.** $-\dfrac{2}{3}(6x + 12) + 10 = x + 5$  $\left\{-\dfrac{3}{5}\right\}$

**g.** $-0.8y + 18.4 = 3.7y + 5.8$  {2.8}

**h.** $4(3x - 1) = 4(x - 1) + 8x$  $\mathbb{R}$

**For each problem: (1) assign a variable to the unknown, (2) write an equation that represents the situation, (3) solve the equation, and (4) answer the question using complete sentences. (*Section 2.2, Objective 4; Section 2.3, Objectives 3 and 4*)**

**34.** Twice the sum of a number and 6 is the same as three times the number. Find the number.  12

**35.** The cost to rent a midsize car for one day is $58 plus $0.35 per mile. The daily cost is represented by the expression $58 + 0.35x$, where $x$ is the number of miles driven. How many miles has the car been driven if the cost of the rental car is $250.50?  550 mi

**36.** The sum of three consecutive even integers is 102. Find the integers.  32, 34, and 36

**37.** The sum of the two smaller of three consecutive even integers is the same as 40 less than three times the largest integer. Find the integers.  30, 32, and 34.

**38.** Solve each equation. If the equation is a contradiction, write the solution as ∅. If the equation is an identity, write the solution as $\mathbb{R}$. (*Section 2.4, Objectives 1–4*)

**a.** $3(2y + 3) - 2(y + 2) = 4(y + 1) + 1$  $\mathbb{R}$

**b.** $\dfrac{3}{5}x - 4 = \dfrac{7}{10}x$  {−40}

**c.** $0.6x + 1.8x = 960$  {400}

**d.** $0.021x + 0.012(840 - x) = 15.12$  {560}

**e.** $\dfrac{5}{6}(x - 2) = \dfrac{1}{6}(5x - 8) + \dfrac{1}{3}$  ∅

**39.** The formula $A = P(1 + r)^t$ calculates the amount of money in an account at the end of $t$ years if $P$ dollars invested at an annual interest rate, $r$. (*Section 2.5, Objective 1*)

**a.** Find the value of $A$ if $4800 is invested for 2 yr at 1.5% annual interest.  $4945.08

**b.** Find the value of $P$ if $5181.62 is expected to be accumulated at the end of 2 yr at 1.8% annual interest.  $5000.00

**40.** The distance $d$, rate $r$, and time $t$ of an object in motion can be related by the formula $d = rt$. (*Section 2.5, Objective 1*)

**a.** Find $d$ if $r = 65$ mph and $t = 3.2$ hr.  208 mi

**b.** Find $t$ if $d = 203$ mi and $r = 58$ mph.  3.5 hr

**41.** The revenue $R$, unit price $p$, and sale level $x$ of a product are related by the formula $R = xp$. (*Section 2.5, Objective 1*)

**a.** Find $R$ if $p = \$112.50$ and $x = 60$.  $6750

**b.** Find $p$ if $R = \$2249.10$ and $x = 90$.  $24.99

**42.** Use the perimeter, area, or circumference formula to find the indicated quantity. (*Section 2.5, Objective 2*)

**a.** Find the perimeter of a rectangle if its area is 400 in.$^2$ and its length is 16 in.  82 in.

**b.** Find the area of a rectangle if its perimeter is 92 m and its width is 18 m.  504 m²

**43.** Solve each formula for the specified variable. (*Section 2.5, Objective 3*)

**a.** $A = lw$ for $l$  $l = \dfrac{A}{w}$   **b.** $C = 5l + 6w$ for $w$  $w = \dfrac{C - 5l}{6}$

**c.** $3x - 8y = 24$ for $y$   **d.** $0.02x + 0.05y = 4.5$ for $x$

**44.** Find the measure of each unknown angle. (*Section 2.5, Objective 5*)

**a.** Find the measure of an angle whose complement is 20° less than the measure of the angle.  55°

**b.** Find the measure of an angle whose supplement is 64° more than the measure of the angle.  58°

**c.** The supplement of an angle is 42° more than three times its complement. Find the measure of the angle.  66°

**45.** Find the measure of each angle labeled in the figure. (*Section 2.5, Objectives 4 and 5*)

**a.**

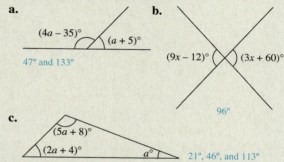

$(4a - 35)°$  $(a + 5)°$  47° and 133°

**b.**

$(9x - 12)°$  $(3x + 60)°$  96°

**c.**

$(5a + 8)°$  $(2a + 4)°$  $a°$  21°, 46°, and 113°

**46.** In triangle $ABC$, the measure of angle $B$ is 18° less than the measure of angle $A$. The measure of angle $C$ is 45° more than the measure of angle $A$. Find the measure of each angle in the triangle. (*Section 2.5, Objective 5*)  51°, 33°, 96°

**Solve each problem. Write an appropriate equation, solve the equation, and write the answer in complete sentences. (*Section 2.6, Objectives 1–5*)**

**47.** A 46-in. LED-LCD HDTV is on sale for $975. This is 35% off the original price. What is the original selling price of the TV?  $1500

**48.** A sweater is on sale for $31.50. If this price is 65% off the original price, what is the original price of the sweater?  $90

**49.** Kye invested $2500 in two accounts. He invested $1500 in a mutual fund that pays 2.4% annual interest and the remaining amount in a money market fund that pays 1.6% annual interest. How much interest did he earn from the two accounts in 1 yr?  $52

**50.** Richard needs to make 72 L of a 25% hydrochloric acid solution. How much 40% hydrochloric acid solution and how much 10% hydrochloric acid solution should be mixed to get 72 L of a 25% hydrochloric acid solution?  36 L of 40% hydrochloric acid solution and 36 L of 10% hydrochloric acid

**51.** Solve each inequality. Graph the solution set and write the solution set in interval notation and in set-builder notation. (*Section 2.7, Objectives 1–4*)

   **a.** $0.4a - 7.7 \le -2.1$

   **b.** $-6 < 4a + 6 < 26$

   **c.** $8x + 3(x + 12) < 9x + 16$

   **d.** $-3.5 \le 0.5y - 1.4 \le 1.2$

**Write an inequality that can be used to solve each problem. Solve the inequality and answer the question in complete sentences. (*Section 2.7, Objective 5*)**

**52.** Myra's final grade in her math class is based on a weighted average. Homework counts as 10%, quizzes count as 20%, tests count as 50%, and the final exam counts as 20%. If she has a homework average of 72, a quiz average of 68, and a test average of 82, what score does she need on her final exam to have at least an 80 for her final grade?   at least 91 on the final

**53.** The cost to rent a car for a week is $320 plus $0.35 per mile. How many miles can be driven if you have at most $740 to rent the car for a week? Round to the nearest whole number, if necessary.   at most 1200 mi

# Linear Equations in Two Variables

### CHAPTER 3

## Commitment and Perseverance

Commitment and perseverance are necessary to be successful in any college classroom. Some courses may require more commitment and perseverance than others, especially if you find the material somewhat difficult to conquer. Your willingness to stick to your goals is far more important than your mathematical abilities. We truly believe that, given the right resources and the right amount of time, you can pass this class. You cannot compare yourself to other students in the classroom since college brings together students of all backgrounds—some may have seen the material last year, some 5 years ago, and some may never have seen the material. So do not get frustrated when others seem to learn the concepts faster than you. This does not mean that you will not learn them; it just means you need to have a little more perseverance until you do.

When we believe we can do something and believe that what we are doing is important, we will be committed to doing it. Know that what you are doing in the classroom is important. It is one step in achieving your college education. Try to remember the big picture as you take these small steps to reach your goal. Find a person that is supportive of your goals and allow them to hold you to the commitment of being successful in this class.

*Question For Thought:* How would you evaluate your ability to see things through to the end? Do you typically give up on your commitments or do you see them through to the end?

## Chapter Outline

## Coming Up...

In Section 3.5, we are going to learn how to write a linear equation that models the average price of a movie ticket given that the average price of a movie ticket in the United States in 1996 was $4.42 and the average price of a movie ticket in the United States in 2010 was $7.89. (Source: http://www.natoonline.org/statisticstickets.htm)

"The will to win, the desire to succeed, the urge to reach your full potential . . . these are the keys that will unlock the door to personal excellence."

—Eddie Robinson (College Football Coach)

## SECTION 3.1 | Equations and the Rectangular Coordinate System

**In Chapter 2,** we studied linear equations in one variable. In this chapter, we will focus on linear equations in two variables. We will also study real-life applications of these equations and introduce the concept of a function.

### ▶ OBJECTIVES

As a result of completing this section, you will be able to

1. Determine algebraically if an ordered pair is a solution of an equation.
2. Plot ordered pairs and identify quadrants.
3. Graph the solutions of an equation.
4. Determine graphically if an ordered pair is a solution of an equation.
5. Solve application problems.
6. Troubleshoot common errors.

The following line graph shows the average cost of a 30-sec Super Bowl advertisement for 2002 to 2010. This is an example of a graph that we will learn how to read to obtain specific information.

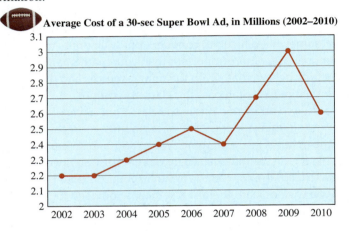

Average Cost of a 30-sec Super Bowl Ad, in Millions (2002–2010)

### Verifying Solutions of Equations Algebraically

**Objective 1** ▶

Determine algebraically if an ordered pair is a solution of an equation.

In Chapter 2, we studied linear equations with one variable and learned how to find solutions of these equations. We found that a linear equation in one variable has one solution with the exception of identities and contradictions. Identities are satisfied by all real numbers and contradictions do not have a solution.

In this chapter, our attention turns to equations with two variables. Some examples of equations in two variables are

$$x + y = 3 \qquad y = 4x - 1 \qquad 2x - 3y = 6 \qquad y = x^2 + 2x - 3$$

Equations containing two variables have pairs of numbers that satisfy the equation. This pair of numbers is called an **ordered pair** and is denoted by $(x, y)$, where $x$ and $y$ are the variables in the equation. A **solution of an equation** containing two variables is an ordered pair $(x, y)$ that satisfies the equation.

Consider the equation $x + y = 3$. Solutions of this equation must satisfy the fact that the sum of the $x$ and $y$ values is 3. Some solutions are $(1, 2)$, $(3, 0)$, $(-1, 4)$, and $(5, -2)$. We know these ordered pairs are solutions because they make the equation true when we replace the variables with their given values as shown in the following table.

| $(x, y)$ | $x + y = 3$ | Solution? |
|---|---|---|
| $(1, 2)$ | $1 + 2 = 3$ <br> $3 = 3$ | Yes |
| $(3, 0)$ | $3 + 0 = 3$ <br> $3 = 3$ | Yes |
| $(-1, 4)$ | $-1 + 4 = 3$ <br> $3 = 3$ | Yes |
| $(5, -2)$ | $5 + (-2) = 3$ <br> $3 = 3$ | Yes |

There are, in fact, infinitely many solutions of an equation in two variables. Just as there are infinitely many solutions of this type of equation, there are infinitely many ordered pairs that do not satisfy the equation $x + y = 3$. Any ordered pair whose values do not add to 3 is not a solution of $x + y = 3$. Some ordered pairs that are not solutions are $(1, 5)$, $(3, -3)$, and $(-1, -2)$.

| $(x, y)$ | $x + y = 3$ | Solution? |
|---|---|---|
| $(1, 5)$ | $1 + 5 = 3$ <br> $6 \neq 3$ | No |
| $(3, -3)$ | $3 + (-3) = 3$ <br> $0 \neq 3$ | No |
| $(-1, -2)$ | $-1 + (-2) = 3$ <br> $-3 \neq 3$ | No |

**Procedure: Determining if an Ordered Pair Is a Solution of an Equation in Two Variables**

**Step 1:** Replace the values of $x$ and $y$ with the numbers given in the ordered pair.
**Step 2:** Simplify each side of the equation.
**Step 3:** If the resulting equation is true, then the ordered pair is a solution of the equation. If it is not true, then the ordered pair is not a solution.

**Objective 1 Examples** Determine if the ordered pair is a solution of the equation.

**1a.** $(3, -1)$; $2x - y = 7$    **1b.** $\left(\dfrac{3}{2}, 0\right)$; $y = 2x + 3$    **1c.** $(0, 5)$; $y = x^2 - 3x + 5$

**Solutions** **1a.**

$$2x - y = 7$$
$$2(3) - (-1) = 7 \qquad \text{Let } x = 3 \text{ and } y = -1.$$
$$6 + 1 = 7 \qquad \text{Simplify.}$$
$$7 = 7 \qquad \text{True}$$

Since $(3, -1)$ makes the equation a true statement, it is a solution of $2x - y = 7$.

**1b.**

$$y = 2x + 3$$
$$0 = 2\left(\dfrac{3}{2}\right) + 3 \qquad \text{Let } x = \dfrac{3}{2} \text{ and } y = 0.$$
$$0 = 3 + 3 \qquad \text{Simplify.}$$
$$0 = 6 \qquad \text{False}$$

Since $\left(\dfrac{3}{2}, 0\right)$ makes the equation a false statement, it is *not* a solution of $y = 2x + 3$.

**1c.**

$$y = x^2 - 3x + 5$$
$$5 = (0)^2 - 3(0) + 5 \qquad \text{Let } x = 0 \text{ and } y = 5.$$
$$5 = 0 + 5 \qquad \text{Simplify.}$$
$$5 = 5 \qquad \text{True}$$

Since $(0, 5)$ makes the equation a true statement, it is a solution of $y = x^2 - 3x + 5$.

**✓ Student Check 1** Determine if the ordered pair is a solution of the equation.

**a.** $(-2, 6)$; $3x - y = 0$    **b.** $\left(\dfrac{1}{4}, 0\right)$; $y = 4x - 1$    **c.** $(-3, 5)$; $y = |x - 2|$

## The Rectangular Coordinate System

As mentioned, there are infinitely many solutions of equations in two variables. Our ultimate goal is to visualize the solutions of these types of equations. Recall that we use a real number line to visualize solutions of equations and inequalities in one variable. To graph solutions of equations in two variables, we need a special system, called the *rectangular coordinate system* or the *Cartesian coordinate system*.

The **rectangular coordinate system** consists of two real number lines intersecting at right angles. The horizontal number line is referred to as the **x-axis** and the vertical number line is referred to as the **y-axis**. The point where the two number lines intersect is called the **origin**.

Every point on the coordinate system has an "address." This address is denoted by an ordered pair $(x, y)$. Each point on the plane is determined by knowing how far left or right the point is from the origin and how far up or down the point is located from the x-axis.

In an ordered pair $(x, y)$, the value $x$ is called the *first coordinate* or *x-coordinate*. The x-coordinate tells us how far left (if $x$ is negative) or right (if $x$ is positive) to move from the origin. The value $y$ is called the *second coordinate* or *y-coordinate*. The y-coordinate tells us how far up (if $y$ is positive) or down (if $y$ is negative) to move from the x-axis.

For example, in the ordered pair $(4, 5)$,

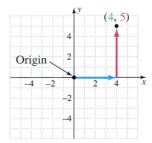

4 is the x-coordinate → move from the origin 4 units right

5 is the y-coordinate → move from the position on the x-axis 5 units up

Note that the ordered pair $(5, 4)$ is different than $(4, 5)$. For the point $(5, 4)$, we move 5 units to the right and 4 units up.

---

**Procedure: Plotting an Ordered Pair**

**Step 1:** From the origin, move left or right to the given x-value in the ordered pair.

**Step 2:** From this x-value, move up or down to the given y-value in the ordered pair.

**Step 3:** The point is located at the result of the two movements.

---

The two numbers lines that form the Cartesian coordinate system divide the plane into four regions called **quadrants**. We label these four quadrants with Roman numerals I, II, III, and IV beginning in the upper right quadrant and rotating counterclockwise.

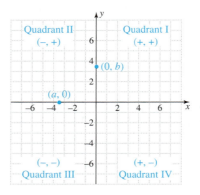

The signs of a point's coordinates determine where on the plane the point is located.

- If both the *x*- and *y*-coordinates are positive, the point lies in Quadrant I.
- If the *x*-coordinate is negative and the *y*-coordinate is positive, the point lies in Quadrant II.
- If both the *x*- and *y*-coordinates are negative, the point lies in Quadrant III.
- If the *x*-coordinate is positive and the *y*-coordinate is negative, the point lies in Quadrant IV.
- If both the *x*- and *y*-coordinates are 0, the point lies on both the *x*- and *y*-axes (or at the origin).
- If the *x*-coordinate is 0 and the *y*-coordinate is nonzero, the point lies on the *y*-axis.
- If the *x*-coordinate is nonzero and the *y*-coordinate is 0, the point lies on the *x*-axis.

**Objective 2   Examples**   Plot each ordered pair on the Cartesian coordinate system and state the quadrant or axis where each point is located.

**2a.** $(4, 3)$   **2b.** $(-2, 5)$   **2c.** $(5, -2)$   **2d.** $(0, 4)$   **2e.** $(-3, 0)$   **2f.** $\left(-\dfrac{3}{2}, -\dfrac{8}{2}\right)$

**Solutions**

**2a.** To plot $(4, 3)$, move 4 units right from the origin and 3 units up. $(4, 3)$ is located in Quadrant I.

**2b.** To plot $(-2, 5)$, move 2 units left from the origin and 5 units up. $(-2, 5)$ is located in Quadrant II.

**2c.** To plot $(5, -2)$, move 5 units right from the origin and 2 units down. $(5, -2)$ is located in Quadrant IV.

**2d.** To plot $(0, 4)$, do not move left or right, move 4 units up. $(0, 4)$ is located on the *y*-axis.

**2e.** To plot $(-3, 0)$, move 3 units left from the origin and do not move up or down. $(-3, 0)$ is located on the *x*-axis.

**2f.** The point $\left(-\dfrac{3}{2}, -\dfrac{8}{3}\right) = (-1.5, -2.7)$. Move 1.5 units left from the origin and 2.7 units down. $\left(-\dfrac{3}{2}, -\dfrac{8}{3}\right)$ is located in Quadrant III.

 **Note:** *The points $(-2, 5)$ and $(5, -2)$ illustrate the fact that the order is important. Changing the order of the numbers in the ordered pairs changes the location of the point.*

**✓ Student Check 2**   Plot each ordered pair on the Cartesian coordinate system and state the quadrant or axis where each point is located.

**a.** $(-5, -1)$   **b.** $(3, -4)$   **c.** $(-4, 3)$   **d.** $(0, -2)$   **e.** $(1, 0)$   **f.** $\left(\dfrac{1}{2}, \dfrac{5}{4}\right)$

## Graphing Equations

**Objective 3** ▶

Graph the solutions of an equation.

The graph of an equation is a picture of all the ordered pairs that satisfy the equation. When an equation contains two variables, there are infinitely many ordered pairs that satisfy the equation. It is impossible to find every solution of an equation in two variables. So, we plot several ordered pairs (solutions) to get an idea of the graph's shape. Recall in Chapter 1 that we evaluated algebraic expressions and recorded our results in a table. We use the table again to record the values for *x* and *y*.

> **Procedure: Graphing Solutions of an Equation**
>
> **Step 1:** Determine at least three solutions of the equation. Use a table to show the solutions.
> **a.** Substitute any value for $x$.
> **b.** Solve the resulting equation or simplify the resulting expression to find the value of $y$.
> **c.** The ordered pair $(x, y)$ is a solution of the equation.
> **Step 2:** Plot the solutions on a coordinate system.
> **Step 3:** Connect the points and use their pattern to sketch the graph.

> **Note:** We will study graphing special types of equations in later sections and chapters. This method is a basic introduction to graphing.

| **Objective 3 Examples** | Graph each equation by completing a table of solutions. |

**3a.** $y = x + 3$      **3b.** $y = x^2 - 1$      **3c.** $y = |x + 2|$

**Solutions**    **3a.**

| $x$ | $y = x + 3$ | $(x, y)$ |
|---|---|---|
| $-2$ | $y = -2 + 3 = 1$ | $(-2, 1)$ |
| $-1$ | $y = -1 + 3 = 2$ | $(-1, 2)$ |
| $0$ | $y = 0 + 3 = 3$ | $(0, 3)$ |
| $1$ | $y = 1 + 3 = 4$ | $(1, 4)$ |
| $2$ | $y = 2 + 3 = 5$ | $(2, 5)$ |

Now plot the solutions and connect them to form the graph.

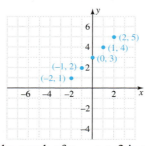

Notice the graph of $y = x + 3$ is a line.

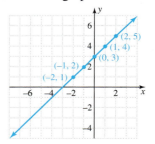

**INSTRUCTOR NOTE:**
This example is to illustrate that not all graphs are lines.

**3b.**

| $x$ | $y = x^2 - 1$ | $(x, y)$ |
|---|---|---|
| $-2$ | $y = (-2)^2 - 1 = 4 - 1 = 3$ | $(-2, 3)$ |
| $-1$ | $y = (-1)^2 - 1 = 1 - 1 = 0$ | $(-1, 0)$ |
| $0$ | $y = (0)^2 - 1 = 0 - 1 = -1$ | $(0, -1)$ |
| $1$ | $y = (1)^2 - 1 = 1 - 1 = 0$ | $(1, 0)$ |
| $2$ | $y = (2)^2 - 1 = 4 - 1 = 3$ | $(2, 3)$ |

Now plot the solutions and connect them to form the graph.

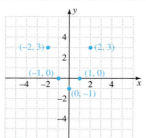

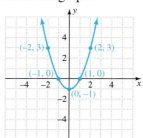

Notice the graph of $y = x^2 - 1$ has a "U-shape" pattern.

3c.

| $x$ | $y = \lvert x + 2 \rvert$ | $(x, y)$ |
|---|---|---|
| $-4$ | $y = \lvert -4 + 2 \rvert = \lvert -2 \rvert = 2$ | $(-4, 2)$ |
| $-3$ | $y = \lvert -3 + 2 \rvert = \lvert -1 \rvert = 1$ | $(-3, 1)$ |
| $-2$ | $y = \lvert -2 + 2 \rvert = \lvert 0 \rvert = 0$ | $(-2, 0)$ |
| $-1$ | $y = \lvert -1 + 2 \rvert = \lvert 1 \rvert = 1$ | $(-1, 1)$ |
| $0$ | $y = \lvert 0 + 2 \rvert = \lvert 2 \rvert = 2$ | $(0, 2)$ |

Now plot the solutions and connect them to form the graph.

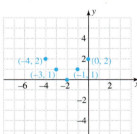 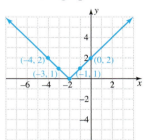

Notice the graph of $y = \lvert x + 2 \rvert$ has a "V-shape" pattern.

✓ **Student Check 3**  Graph each equation by completing a table of solutions.

**a.** $y = x - 1$   **b.** $y = x^2$   **c.** $y = \lvert x - 1 \rvert$

## Verifying Solutions Graphically

**Objective 4** ▶

Determine graphically if an ordered pair is a solution of an equation.

In Objective 1, we determined if an ordered pair is a solution of an equation by substituting the values of $x$ and $y$ in the equation. We can also examine the graph of an equation to determine if an ordered pair is a solution of an equation. If a point lies on the graph, then it is a solution of the equation. If a point does not lie on the graph, then it is not a solution of the equation.

**Objective 4 Examples**  **The graph of the equation $y = x^2 - 4$ is provided. Use the graph to determine if each ordered pair is a solution of the equation.**

**4a.** $(-4, 0)$   **4b.** $(0, -4)$

**4c.** $(2, 0)$   **4d.** $(0, 2)$

**4e.** $(-3, 5)$

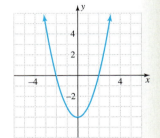

**Solutions**  After plotting the points, we find that $(0, -4)$, $(2, 0)$, and $(-3, 5)$ lie on the graph; thus, they are solutions of the equation. The points $(-4, 0)$ and $(0, 2)$ are not solutions of the equation because they do not lie on the graph.

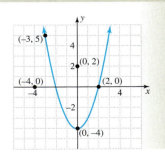

✔ **Student Check 4**    The graph of the equation $y = |x| - 3$ is provided. Use the graph to determine if each ordered pair is a solution of the equation.

a. $(3, 0)$    b. $(0, 3)$

c. $(0, -3)$    d. $(4, -1)$

e. $(-4, 1)$

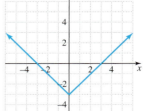

## Applications

**Objective 5** ▶

Solve application problems.

The applications in this section require us to read graphs to obtain information. Data that can be represented as an ordered pair are called **paired data**. For example, a person's weight for a given year can be written as an ordered pair in the form (year, weight).

When we are given real-life data in tables, bar graphs, or line graphs, we can convert the data to ordered pairs and use them to plot the data. The plot of this data is called a **scatter plot** or *scatter diagram*. Scatter plots can be used to look for patterns or trends that occur in paired data.

**Objective 5 Examples**    Use the table or graph to answer each question.

**5a.** The following table provides the number of cell phone subscribers in the United States for the given year. Write an ordered pair for each year, where $x$ is the number of years after 2000 and $y$ is the number of cell phone subscribers (in millions).

| Years After 2000 | 0 | 1 | 2 | 3 | 4 | 5 | 6 | 8 | 9 |
|---|---|---|---|---|---|---|---|---|---|
| Subscribers (in millions) | 109 | 128 | 141 | 159 | 182 | 208 | 233 | 263 | 286 |

(Sources: www.infoplease.com and http://www.ctia.org/media/industry_info/index.cfm/AID/10323)

   **i.** Write the ordered pairs that correspond to the data in the table.

   **ii.** Interpret the meaning of the first and last ordered pairs.

   **iii.** Plot the ordered pairs to form a scatter plot of the given data.

**Solution    5a.    i.** The ordered pairs that correspond to the data in the table are (0, 109), (1, 128), (2, 141), (3, 159), (4, 182), (5, 208), (6, 233), (8, 263), and (9, 286).

   **ii.** The first ordered pair (0, 109) means that 0 yr after 2000, or in 2000, there were approximately 109 million cell phone subscribers in the United States. The last ordered pair (9, 286) means that 9 yr after 2000, or in 2009, there were approximately 286 million cell phone subscribers in the United States.

   **iii.** The scatter plot of the ordered pairs is

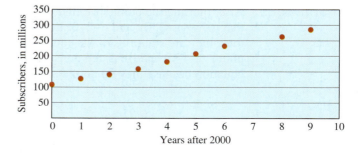

**5b.** The following bar graph shows the amount of money spent, in millions, on Google ads for the top 10 companies in June 2010. (Source: www.adage.com)

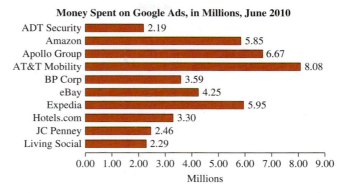

**Money Spent on Google Ads, in Millions, June 2010**

**i.** Write ordered pairs for this bar graph, where $x$ is the company and $y$ is the money spent on advertising, in millions.

**ii.** Which company spent the most money on Google ads in June 2010? How much was spent?

**iii.** Which company spent the least money in Google ads in June 2010? How much was spent?

**iv.** What was the total amount of money spent on Google ads for these companies in June 2010?

**Solution**  **5b.**  **i.** The ordered pairs corresponding to the given data are (ADT Security, 2.19), (Amazon, 5.85), (Apollo Group, 6.67), (AT&T Mobility, 8.08), (BP Corp, 3.59), (eBay, 4.25), (Expedia, 5.95), (Hotels.com, 3.30), (JC Penney, 2.46), and (Living Social, 2.29).

**ii.** The company that spent the most on Google ads was AT&T Mobility, since the horizontal bar is the longest for this company. AT&T Mobility spent $8.08 million.

**iii.** The company that spent the least on Google ads was ADT Security since this horizontal bar is the shortest. ADT Security spent $2.19 million.

**iv.** The total spent on Google ads for these 10 companies in June 2010, was $2.19 + 5.85 + 6.67 + 8.08 + 3.59 + 4.25 + 5.95 + 3.30 + 2.46 + 2.29 = \$44.63$ million.

**5c.** The line graph shows the average cost, in millions, of a 30-sec Super Bowl ad for the years 2002–2010. (Sources: http://www.huffingtonpost.com/2010/01/11/super-bowl-commercial-pri_n_418245.html and www.marketingcharts.com)

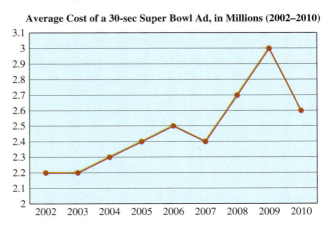

**Average Cost of a 30-sec Super Bowl Ad, in Millions (2002–2010)**

**i.** Write an ordered pair for the points on the graph.

**ii.** What year had the greatest average ad cost? What was it?

**iii.** In what year(s) was the average cost of an ad $2.4 million?

**Solution**    **5c.**    **i.** The ordered pairs on the graph correspond to the points (2002, 2.2), (2003, 2.2), (2004, 2.3), (2005, 2.4), (2006, 2.5), (2007, 2.4), (2008, 2.7), (2009, 3), and (2010, 2.6).

**ii.** The average cost of a Super Bowl ad was greatest in the year 2009. The cost was $3 million.

**iii.** The average cost of an ad was $2.4 million in the years 2005 and 2007.

---

✓ **Student Check 5**    Use the table or graph to answer each question.

**a.** The table provides information on the number of murder victims by juvenile gang killings for the given year. (Source: http://www.fbi.gov/ucr)

| Years After 2000 | 0 | 1 | 2 | 3 | 4 |
|---|---|---|---|---|---|
| Number of Victims | 653 | 862 | 911 | 819 | 804 |

**i.** Write an ordered pair for each year, where $x$ is the number of years after 2000 and $y$ is the number of victims.

**ii.** Interpret the first and last ordered pairs.

**iii.** Plot the ordered pairs to form a scatter plot of the data.

**b.** The bar graph shows the number of unique visitors, in millions, for the given websites for August 2010. (Source: http://www.compete.com)

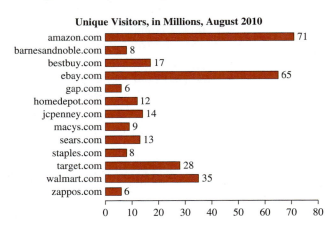

Unique Visitors, in Millions, August 2010

**i.** Write ordered pairs for the bar graph, where $x$ is the website and $y$ is the number of unique visitors, in millions.

**ii.** Which website had the largest number of visitors in August 2010? How many did it have?

**iii.** Which website had the least number of visitors in August 2010? How many did it have?

---

**Objective 6** ▶

Troubleshoot common errors.

## Troubleshooting Common Errors

Some common errors associated with solutions of equations and the rectangular coordinate system are shown next.

**Objective 6 Examples**    **A problem and an incorrect solution are given. Provide the correct solution and an explanation of the error.**

**6a.** Is the ordered pair $(4, -2)$ a solution of $y = 2x - 10$?

| Incorrect Solution | Correct Solution and Explanation |
|---|---|
| $y = 2x - 10$ <br> $4 = 2(-2) - 10$ <br> $4 = -4 - 10$ <br> $4 = -14$ <br><br> Since the point makes the equation false, it is not a solution. | In the ordered pair $(4, -2)$, the $x$-value is 4 and the $y$-value is $-2$. <br><br> $y = 2x - 10$ <br> $-2 = 2(4) - 10$ <br> $-2 = 8 - 10$ <br> $-2 = -2$ <br><br> Since the point makes the equation true, it is a solution of the equation. |

**6b.** Plot the points $(0, 3)$ and $(-2, 0)$.

| Incorrect Solution | Correct Solution and Explanation |
|---|---|
| The points are shown in the graph. <br><br> 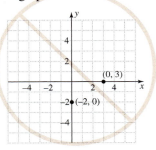 | When the first coordinate is 0, there is no movement left or right from the origin. When the second coordinate is 0, there is no movement up or down from the $x$-axis. The correct plot is shown here. <br><br> 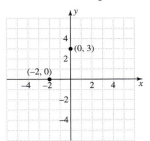 |

**6c.** Graph $y = |x + 3|$.

| Incorrect Solution | Correct Solution and Explanation |
|---|---|
| The table that corresponds to this graph is | While the points in the table are solutions of the equation, we cannot assume that the graph continues in this manner. It is always a good idea to input values for $x$ that make the expression inside the absolute value negative. So, we find a couple more points to get |

| $x$ | $y = |x + 3|$ | $(x, y)$ |
|---|---|---|
| $-2$ | $y = |-2 + 3| = 1$ | $(-2, 1)$ |
| $-1$ | $y = |-1 + 3| = 2$ | $(-1, 2)$ |
| $0$ | $y = |0 + 3| = 3$ | $(0, 3)$ |
| $1$ | $y = |1 + 3| = 4$ | $(1, 4)$ |
| $2$ | $y = |2 + 3| = 5$ | $(2, 5)$ |

| $x$ | $y = |x + 3|$ | $(x, y)$ |
|---|---|---|
| $-4$ | $y = |-4 + 3| = |-1| = 1$ | $(-4, 1)$ |
| $-3$ | $y = |-3 + 3| = 0$ | $(-3, 0)$ |

So, the graph is

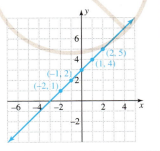

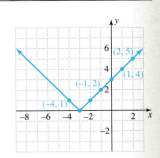

## ANSWERS TO STUDENT CHECKS

**Student Check 1**    **a.** no    **b.** yes    **c.** yes

**Student Check 2**    **a.** Quadrant III

   **b.** Quadrant IV

   **c.** Quadrant II

   **d.** $y$-axis

   **e.** $x$-axis

   **f.** Quadrant I

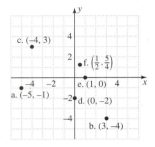

**Student Check 3**    **a.**    **b.**

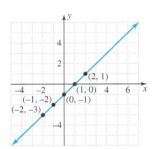    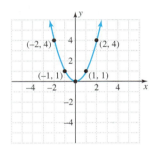

   **c.**

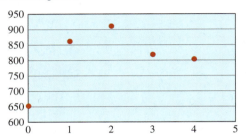

**Student Check 4**    **a.** yes    **b.** no    **c.** yes    **d.** no    **e.** yes

**Student Check 5**    **a. i.**  (0, 653), (1, 862), (2, 911), (3, 819), (4, 804)

   **ii.** The first point means that in 0 yr after 2000, in the year 2000, there were 653 murder victims by juvenile gang killings. The last point means that in 4 yr after 2000, in the year 2004, there were 804 murder victims by juvenile gang killings.

   **iii.** The scatter plot is

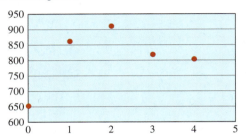

  **b. i.**  (amazon.com, 71), (barnesandnoble.com, 8), (bestbuy.com, 17), (ebay.com, 65), (gap.com, 6), (homedepot.com, 12), (jcpenney.com, 14), (macys.com, 9), (sears.com, 13), (staples.com, 8), (target.com, 28), (walmart.com, 35), (zappos.com, 6)

   **ii.** The website with the largest number of visitors in August 2010 was amazon.com. The site had 71 million visitors.

   **iii.** The websites with the least number of visitors in August 2010 were gap.com and zappos.com, each with 6 million visitors.

## SUMMARY OF KEY CONCEPTS

1. A solution of an equation in two variables is an ordered pair that makes the equation a true statement.

2. Ordered pairs can be plotted on a Cartesian coordinate system. The $x$-value determines the movement from the origin on the $x$-axis. The $y$-value determines how far up or down the point is located from the $x$-axis. Quadrants are the four regions formed by the axes. Quadrant I is in the upper right, Quadrant II is in the upper left, Quadrant III is in the lower left, and Quadrant IV is in the lower right. Points will lie in either one of the four quadrants or on one of the axes.

3. To graph solutions of an equation, find several solutions of the equation by completing a table. Plot the solutions and then connect them to form the graph. There are infinitely many solutions of an equation in two variables. The graph is a picture of these solutions.

4. We can determine if an ordered pair is a solution of an equation if it lies on the graph of the equation.

5. For application problems, it is important to understand what the $x$- and $y$-coordinates represent. If the data are given in a table, bar graph, or line graph, the set of ordered pairs is called paired data. Graphing paired data gives us a scatter plot of the data.

## GRAPHING CALCULATOR SKILLS

The calculator can be used to make a scatter plot and to graph equations. These skills will come in handy in later math courses. At this point, know the features are available, but do not rely on the calculator to plot points or draw graphs.

**Example 1:** Plot the points $(3, -2)$, $(0, 4)$, and $(-2, 0)$.

To plot points on the calculator, the $x$- and $y$-values must be entered into lists. Press STAT, 1 to access the list feature. Then enter the values in the appropriate columns, using L1 as the $x$-values and L2 as the $y$-values.

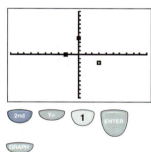

Once the points are entered, turn the STAT PLOT feature ON and graph.

**Example 2:** Graph $y = x + 3$.

To graph an equation, enter the equation into the equation editor by pressing Y = and then press GRAPH to view the graph.

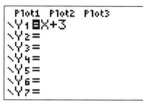

To view a table of specific solutions, access the TABLE feature by pressing 2nd GRAPH.

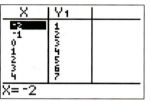

More detail will be provided in Section 3.2 on graphing with the calculator.

---

## SECTION 3.1 / EXERCISE SET

### Write About It!

Use complete sentences to explain the meaning of the given term or process.

1. Solution of an equation in two variables   Answers vary.
2. Plotting a point on a rectangular coordinate system   Answers vary.
3. Identifying a quadrant or axis where a point is located   Answers vary.
4. Graphing the solutions of an equation in two variables   Answers vary.

**Determine if each statement is true or false. If the statement is false, provide an explanation.**

5. The ordered pair $(0, -4)$ is a solution of the equation $x - y = -4$.   False, the ordered pair does not satisfy the equation.
6. The point $(3, 5)$ does not lie on the graph of the equation $y = 2x + 1$.   True
7. Any point located on the $x$-axis has an $x$-value of 0.   False, the y-value is zero.
8. Any point located on the $y$-axis has a $y$-value of 0.   False, the x-value is zero.

Additional answers can be found in the Instructor Answer Appendix.

9. If $a > 0$ and $b < 0$, the point $(a, b)$ is located in Quadrant II.   False, the point is in Quadrant IV.
10. If $a > 0$ and $b < 0$, the point $(-a, -b)$ is located in Quadrant III.   False, the point is in Quadrant II.

### Practice Makes Perfect!

**Determine algebraically if the ordered pair is a solution of the equation. (See Objective 1.)**

11. $(2, 0)$; $3x + y = 6$   yes
12. $(-8, 0)$; $2x - 3y = 16$   no
13. $(0, -3)$; $4x - 2y = -6$   no
14. $(0, 5)$; $x - 2y = -10$   yes
15. $(5, -1)$; $7x + y = 36$   no
16. $(-4, -2)$; $5x - 6y = -8$   yes
17. $\left(\dfrac{1}{2}, -\dfrac{3}{2}\right)$; $2x - 4y = 7$   yes
18. $\left(-\dfrac{3}{5}, \dfrac{1}{4}\right)$; $10x + 4y = 5$   no
19. $\left(-\dfrac{2}{5}, 0\right)$; $y = 5x + 2$   yes
20. $\left(\dfrac{4}{7}, 0\right)$; $y = 7x + 4$   no
21. $(0, 2)$; $y = x^2 - 5x + 2$   yes
22. $(3, 4)$; $y = x^2 - 5x + 2$   no

**23.** $(-3, 8); y = 2x^2 + 3x - 1$   yes   **24.** $(0, 1); y = 2x^2 + 3x - 1$   no

**25.** $(0, -4); y = |x - 4|$   no   **26.** $\left(-\dfrac{1}{2}, 3\right); y = |4x - 1|$   yes

**27.** $(4, 0); y = \dfrac{x - 4}{x + 2}$   yes   **28.** $(-2, 0); y = \dfrac{x - 4}{x + 2}$   no

**Plot each ordered pair on the Cartesian coordinate system and state the quadrant or axis where each point is located. (*See Objective 2.*)**

**29.** $(1, 5)$   I     **30.** $(5, 1)$   I     **31.** $(-6, 2)$   II

**32.** $(2, -6)$   IV     **33.** $(3, -4)$   IV     **34.** $(-4, 3)$   II

**35.** $\left(\dfrac{7}{2}, -\dfrac{1}{2}\right)$   IV     **36.** $\left(\dfrac{1}{2}, \dfrac{7}{2}\right)$   I     **37.** $(-6.2, -2.4)$   III

**38.** $(-3.5, 7.2)$   II     **39.** $(0, -5)$   y-axis   **40.** $(-3, 0)$   x-axis

**41.** $(2, 0)$   x-axis     **42.** $(0, -4)$   y-axis   **43.** $(\pi, -6)$   IV

**44.** $(4, \pi)$   I

**Identify each point plotted on the coordinate system and state the quadrant or axis where each point is located. (*See Objective 2.*)**

**45.** $A$   $(4, -5)$, IV

**46.** $B$   $(6, 0)$, x-axis

**47.** $C$   $(1, 3)$, I

**48.** $D$   $(0, 5)$, y-axis

**49.** $E$   $(-7, 0)$, x-axis

**50.** $F$   $(-4, 4)$, II

**51.** $G$   $(-3, -5)$, III

**52.** $H$   $(0, -4)$, y-axis

**53.** $I$   $(-4, -1)$, III

**54.** $J$   $(2, -2)$, IV

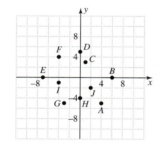

**Identify the quadrant or axis where each point is located. (*See Objective 2.*)**

**55.** $(5, -3)$   IV     **56.** $(-2, -6)$   III     **57.** $(4, 10)$   I

**58.** $(-9, 15)$   II     **59.** $(0, -10)$   y-axis   **60.** $(15, 0)$   x-axis

**61.** $(1.2, 3.7)$   I     **62.** $\left(-\dfrac{1}{5}, 80\right)$   II     **63.** $\left(-\dfrac{1}{4}, 0\right)$   x-axis

**64.** $(0, 20)$   y-axis

**Graph each equation by completing a table of solutions. (*See Objective 3.*)**

**65.** $y = x + 2$     **66.** $y = x - 3$     **67.** $y = -2x + 4$

**68.** $y = -3x - 6$     **69.** $y = x^2 + 2$     **70.** $y = x^2 - 3$

**71.** $y = -x^2 + 4$     **72.** $y = -x^2 - 1$     **73.** $y = |x - 3|$

**74.** $y = |x + 2|$

**Use the given graph of an equation to determine if each ordered pair is a solution of the equation. (*See Objective 4.*)**

**75.** $(-2, -3)$   no

**76.** $(0, -1)$   yes

**77.** $(3, 5)$   yes

**78.** $(-2, 2)$   no

**79.** $\left(\dfrac{1}{2}, 0\right)$   yes

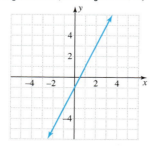

**80.** $(0, 2)$   no

**81.** $(-2, 0)$   yes

**82.** $(3, 1)$   yes

**83.** $(-4, -2)$   no

**84.** $(-6, 4)$   yes

**85.** $(-5, 3)$   yes

**86.** $(3, 5)$   no

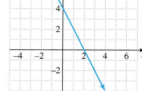

**Use the graph to complete the ordered pair that is a solution. (*See Objective 4.*)**

**87.** $(0, \_\_\_\_)$   $(0, 4)$

**88.** $(-1, \_\_\_\_)$   $(-1, 6)$

**89.** $(\_\_\_\_, 0)$   $(2, 0)$

**90.** $(\_\_\_\_, 2)$   $(1, 2)$

**91.** $(3, \_\_\_\_)$   $(3, -2)$

**92.** $(\_\_\_\_, -6)$   $(5, -6)$

**Use the table or graph to answer each question. (*See Objective 5.*)**

**93.** The table shows the U.S. unemployment rates between 2001 and 2010. (Source: Bureau of Labor Statistics)

| Years After 2001 | 0 | 1 | 2 | 3 | 4 | 5 | 6 | 7 | 8 | 9 |
|---|---|---|---|---|---|---|---|---|---|---|
| Unemployment Rate (percent) | 4.7 | 5.8 | 6.0 | 5.5 | 5.1 | 4.6 | 4.6 | 5.8 | 9.3 | 9.6 |

  **a.** Write the ordered pairs $(x, y)$ that correspond to the data in the table, where $x$ is the years after 2001 and $y$ is the unemployment rate.

  **b.** Interpret the meaning of the first and last ordered pairs in the context of the problem.

  **c.** In what year was the unemployment rate the highest? The lowest?

  **d.** Make a scatter plot of the data.

**94.** The table lists the approximate salary of Peyton Manning, quarterback for the Indianapolis Colts, for 2003–2009. (Source: http://content.usatoday.com/sportsdata/football/nfl/salaries/player/Peyton-Manning)

| Years After 2003 | 0 | 1 | 2 | 3 | 4 | 5 | 6 |
|---|---|---|---|---|---|---|---|
| Salary (in millions) | $11.3 | $35 | $0.7 | $10 | $11 | $11.5 | $14 |

  **a.** Write the ordered pairs that correspond to the data in the table, where $x$ is years after 2003 and $y$ is Manning's salary (in millions of dollars).

  **b.** Interpret the meaning of first and last ordered pairs in the context of the problem.

  **c.** In what year was his salary the highest? The lowest?

  **d.** Make a scatter plot of the data.

**95.** The graph shows the national average hourly salary for registered nurses between 1998 and 2005. (Source: Bureau of Labor Statistics)

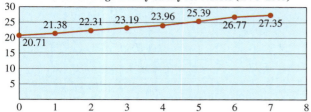

**National Average Hourly Salary for Nurses (1998–2005)**

a. Write an ordered pair for each point labeled on the graph.

b. What was the national average hourly salary for nurses in 1998?

c. What was the national average hourly salary for nurses in 2002?

d. How much did the average hourly salary increase from 1998 to 2005?

e. Based on the graph, how would you describe the average hourly salaries of nurses?

**96.** The bar graph shows the number of unique visitors, in millions, for the given websites for August 2010. (Source: http://www.compete.com)

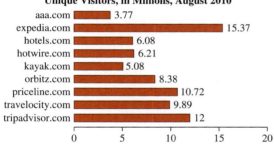

**Unique Visitors, in Millions, August 2010**

a. Write ordered pairs for this bar graph, where $x$ is the website and $y$ is the number of unique visitors, in millions.

b. Which website had the largest number of visitors in August 2010? How many did it have?

c. Which website had the least number of visitors in August 2010? How many did it have?

 **Mix 'Em Up!**

**Determine algebraically if the given ordered pair is a solution of the equation.**

**97.** $(1, -3); 9x - y = 12$   yes

**98.** $(5, 4); 3x + 2y = 25$   no

**99.** $(-0.75, -3); 6x + y = -7.5$   yes

**100.** $(0.7, 0); -5x + 10y = 6.5$   no

**101.** $(-2, 5); y = 2x^2 + 9x + 16$   no

**102.** $(10, 4); y = -x^2 + x + 94$   yes

**103.** $\left(\frac{1}{2}, -1\right); y = -2x^2 + 9x - 5$   yes

**104.** $\left(\frac{2}{3}, -2\right); y = -3x^2 - 7x + 4$   yes

**105.** $(-0.8, 7.4); y = |3x - 5|$   yes   **106.** $(0.6, 3.4); y = |4x + 2|$   no

**107.** $(3, 0); y = \dfrac{10x - 9}{x - 3}$   no   **108.** $(1, 3); y = \dfrac{5x + 4}{-6x + 9}$   yes

**Plot each ordered pair on the Cartesian coordinate system and state the quadrant or axis where each point is located.**

**109.** $\left(-\frac{3}{2}, -5\right)$   III   **110.** $\left(0, -\frac{1}{2}\right)$   y-axis   **111.** $(6.5, 0)$   x-axis

**112.** $\left(3, -\frac{5}{2}\right)$   IV   **113.** $(-4, 5)$   II   **114.** $(0, 3)$   y-axis

**115.** $(2.5, 4.5)$   I   **116.** $(-2, 1.5)$   II

**Identify each point plotted on the coordinate system and state the quadrant or axis where each point is located.**

**117.** $A$   (5, 4), I

**118.** $B$   (−6, −4), III

**119.** $C$   (3, 0), x-axis

**120.** $D$   (0, −2), y-axis

**121.** $E$   (−3, 3), II

**122.** $F$   (0, 6), y-axis

**123.** $G$   (3, −6), IV

**124.** $H$   (−5, 0), x-axis

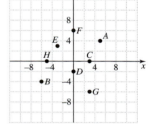

**Graph each equation by completing a table of solutions.**

**125.** $y = 3 - x$   **126.** $y = 1 - 2x$   **127.** $y = |2 - x|$

**128.** $y = 4 - x^2$   **129.** $y = 1 - |x|$   **130.** $y = 1 - x^2$

**Use the given graph of an equation to determine if each ordered pair is a solution of the equation.**

**131.** $(3, 0)$   yes

**132.** $(0, -3)$   yes

**133.** $(4, 6)$   no

**134.** $(-2, 4)$   no

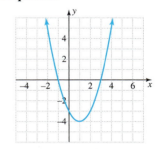

**Use the given graph to complete each ordered pair that is a solution.**

**135.** (____, −3)   (0, −3) and (4, −3)

**136.** $\left(-\frac{1}{2}, \text{____}\right)$   $\left(-\frac{1}{2}, -5\right)$

**137.** (____, 1)   (2, 1)

**138.** (____, −8)   (−1, −8) and (5, −8)

**139.** (1, ____)   (1, 0)

**140.** (3, ____)   (3, 0)

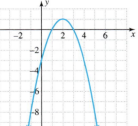

**Use the graphs to answer each question.**

**141.** The graph shows the total expenditures for airline transportation in the United States between 1998 and 2007. (Source: http://www.bts.gov/publications/national_transportation_statistics/)

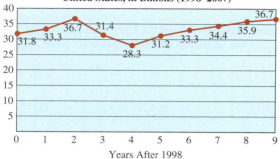

**Yearly Expenditures for Airline Transportation in the United States, in Billions (1998–2007)**

Years After 1998

a. Write an ordered pair for each point labeled on the graph.

b. In what year were the expenditures the greatest? The least?

c. Between what consecutive years did the expenditures decrease? By how much?

d. Based on the graph, how would you describe expenditures for U.S. airline transportation?

**142.** The graph shows the gross public debt amount in the United States between 2000 and 2010. (Source: http://www.treasurydirect.gov/NP/BPDLogin?application=np)

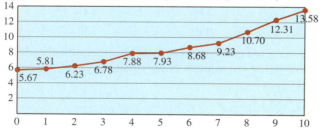

**Gross Public Debt Amount, in Trillions (2000–2010)**

a. Write an ordered pair for each point labeled on the graph.

b. In what year was the public debt amount the greatest? The least?

c. Between what consecutive years did the debt increase most? By how much?

d. Based on the graph, how would you describe public debt amount for the United States?

 **You Be the Teacher!**

**Answer each student's question or correct the errors, if any.**

**143.** Andrea: I always mix up the location of the four quadrants. Can you give me a way to remember which quadrant is which?

**144.** Graph the equation $y = |x - 4|$.

Bernadette's work:

| $x$ | $y = |x - 4|$ | $(x, y)$ |
|---|---|---|
| $-1$ | $-5$ | $(-1, -5)$ |
| $0$ | $-4$ | $(0, -4)$ |
| $1$ | $-3$ | $(1, -3)$ |
| $2$ | $-2$ | $(2, -2)$ |
| $3$ | $-1$ | $(3, -1)$ |

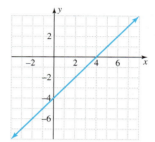

**145.** Graph the equation $y = -x^2 + 1$.

Valerie's work:

| $x$ | $y = -x^2 + 1$ | $(x, y)$ |
|---|---|---|
| $-2$ | $5$ | $(-2, 5)$ |
| $-1$ | $2$ | $(-1, 2)$ |
| $0$ | $1$ | $(0, 1)$ |
| $1$ | $2$ | $(1, 2)$ |
| $2$ | $5$ | $(2, 5)$ |

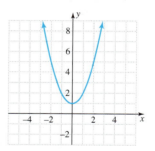

 **Calculate It!**

**Use a graphing calculator to graph the points or equations. Sketch the graph displayed on the calculator.**

**146.** $(-3, 9), (-2, 4), (-1, 1), (0, 0), (1, 1), (2, 4), (3, 9)$

**147.** $(-3, -5), (-2, -3), (-1, -1), (0, 1), (1, 3), (2, 5), (3, 7)$

**148.** $y = 4 - 2x$

**149.** $y = x^3 - 2x^2 + 4$

| SECTION 3.2 | **Graphing Linear Equations** |

## ▶ OBJECTIVES

As a result of completing this section, you will be able to

**1.** Identify a linear equation in two variables.

**2.** Plot points to graph a linear equation in two variables.

**3.** Use intercepts to graph linear equations in two variables.

**4.** Recognize and graph horizontal and vertical lines.

**5.** Solve application problems.

**6.** Troubleshoot common errors.

The median annual income for men with a bachelor's degree is given by the linear equation $y = 1443x + 38{,}843$, where $x$ is the number of years after 1990. In this section, we will graph this equation and use the equation to find important information.

## Linear Equations in Two Variables

In Chapter 2, we defined a linear equation in one variable as an equation of the form $Ax + B = C$, where $A$, $B$, and $C$ are real numbers. The defining characteristic is that the exponent of the variable is 1. A linear equation in two variables is defined in a similar manner.

> **Definition:** An equation of the form $Ax + By = C$, where $A$, $B$, and $C$ are real numbers with $A$ and $B$ not both zero is a **linear equation in two variables**.

### Objective 1 ▶

Identify a linear equation in two variables.

> **Note:** *The exponent of the variables x and y is 1.*

When the values of $A$, $B$, and $C$ are integer values with $A > 0$, the linear equation is considered to be written in the **standard form of a line**, or $Ax + By = C$.

Some examples of linear equations in two variables are

$$y = 2x - 1 \qquad \frac{x}{4} + 7y = 1 \qquad \underset{\uparrow}{3x - 5y = 15} \qquad \underset{\uparrow}{x + 4y = -8}$$
$$\qquad\qquad\qquad\qquad\qquad\text{(standard form)}\quad\text{(standard form)}$$

The last two equations are written in standard form, while the first two are not.

### Objective 1 Examples

**Determine if the equation is a linear equation in two variables. If the equation is a linear equation in two variables, write it in standard form and identify the values of $A$, $B$, and $C$.**

**1a.** $x - y = 6$  **1b.** $y = 3$  **1c.** $3x^2 + y = 9$  **1d.** $y = \dfrac{3}{2}x - 2$

**Solutions**

**INSTRUCTOR NOTE:**

Being able to write an equation in standard form will prepare students for solving systems of linear equations in two variables by the addition method.

**1a.** This equation is a linear equation in two variables because the exponents of the variables are 1. It is also written in standard form.

$$x - y = 6$$
$$1x + (-1)\,y = 6$$
$$A = 1,\, B = -1,\, C = 6$$

**1b.** This equation is a linear equation in two variables. The term with $x$ is missing, so it is understood to have a coefficient of 0.

$$y = 3$$
$$0x + 1y = 3$$
$$A = 0,\, B = 1,\, C = 3$$

> **Note:** *It is not necessary to write the coefficients of 0 or 1. They are written here for emphasis.*

**1c.** The equation $3x^2 + y = 9$ is *not* a linear equation since the exponent of $x$ is 2.

**1d.** The equation is a linear equation because the exponents of the variables are both 1. It is, however, not written in standard form. To write it in standard form, we need to clear fractions and get the variable terms on one side of the equation and the constant on the other.

$$y = \frac{3}{2}x - 2$$

| | |
|---|---|
| $2(y) = 2\left(\frac{3}{2}x - 2\right)$ | Multiply each side by the LCD, 2. |
| $2y = 3x - 4$ | Apply the distributive property and simplify. |
| $2y - 3x = 3x - 4 - 3x$ | Subtract $3x$ from each side. |
| $-3x + 2y = -4$ | Rewrite the equation with the $x$-term first. |
| $-1(-3x + 2y) = -1(-4)$ | Multiply each side by $-1$. |
| $3x - 2y = 4$ | Simplify. |

So, $A = 3$, $B = -2$, and $C = 4$.

✓ **Student Check 1**   Determine if the equation is a linear equation in two variables. If the equation is a linear equation in two variables, write it in standard form and identify the values of $A$, $B$, and $C$.

**a.** $x - 2y = -4$      **b.** $x = 4$      **c.** $x^2 + y^2 = 25$      **d.** $y = \frac{2}{3}x + 4$

## Graphing Linear Equations in Two Variables

**Objective 2** ▶

Plot points to graph a linear equation in two variables.

In Section 3.1, we learned how to graph equations in two variables. In this section, we will specifically discuss how to graph linear equations in two variables.

Consider the equation $x + y = 5$. The solutions of this equation are ordered pairs $(x, y)$ such that the sum of the $x$-coordinate and $y$-coordinate is 5. There are many ordered pairs that satisfy this condition. Some of the solutions are listed next.

| $x$ | $y$ | $(x, y)$ |
|---|---|---|
| 5 | 0 | $(5, 0)$ |
| 4 | 1 | $(4, 1)$ |
| 3 | 2 | $(3, 2)$ |
| 2 | 3 | $(2, 3)$ |
| 1 | 4 | $(1, 4)$ |
| 0 | 5 | $(0, 5)$ |
| $-1$ | 6 | $(-1, 6)$ |
| $-2.5$ | 7.5 | $(-2.5, 7.5)$ |

$5 + 0 = 5$
$4 + 1 = 5$
$3 + 2 = 5$
$2 + 3 = 5$
$1 + 4 = 5$
$0 + 5 = 5$
$-1 + 6 = 5$
$-2.5 + 7.5 = 5$

This list of solutions continues indefinitely. Since we cannot possibly list all of the solutions of this equation, we plot the points and try to recognize a pattern that enables us to connect the points and extend the graph.

Notice that the graph of this equation is a *line*. This is the reason the equation is called a *linear* equation. Any point on the graph of this line is a solution of the equation $x + y = 5$. Note that we put arrows on the ends of the graph to indicate there are infinitely many solutions.

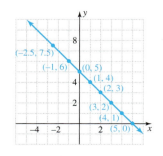

**Note:** *The graph of any linear equation in two variables is a line.*

> **Procedure:** **Graphing a Linear Equation in Two Variables by Plotting Points**
>
> **Step 1:** Make a table of at least three ordered pairs that satisfy the equation. Only two ordered pairs are required to graph a line, but a third point is good to use as a check.
>
>     **a.** To get the solutions, choose a value for $x$, substitute this in the equation and solve for $y$. (A good idea is to use a negative value, zero, and a positive value.)
>
>     **b.** We can also choose a value for $y$, substitute this in the equation, and solve for $x$.
>
> **Step 2:** Plot the three ordered pairs. These three points should lie in a straight line. If they do not, an error has been made. Recheck your work.
>
> **Step 3:** Draw the line that contains the three points. The line should extend beyond the points and have arrows at both ends to indicate that there are infinitely many solutions of the equation.

**INSTRUCTOR NOTE:**
Make sure students understand that their points do not have to be the same as someone else's. If they are comparing solutions with a friend or the back of the book, they can check that the other points lie on their line.

**Objective 2  Examples**   **Graph each linear equation.**

**2a.** $y = -x + 3$          **2b.** $y = \dfrac{4}{3}x - 4$          **2c.** $x - 2y = 4$

**Solutions**      **2a.** Let $x = -1, 0,$ and $1$. We will use the table to organize our work to obtain the solutions.

| $x$ | $y = -x + 3$ | $(x, y)$ |
|---|---|---|
| $-1$ | $y = -(-1) + 3$ <br> $y = 1 + 3$ <br> $y = 4$ | $(-1, 4)$ |
| $0$ | $y = -(0) + 3$ <br> $y = 0 + 3$ <br> $y = 3$ | $(0, 3)$ |
| $1$ | $y = -(1) + 3$ <br> $y = -1 + 3$ <br> $y = 2$ | $(1, 2)$ |

Plot the points $(-1, 4)$, $(0, 3)$, and $(1, 2)$. Since these points lie on a line, we draw a line through the points to obtain the graph of $y = -x + 3$.

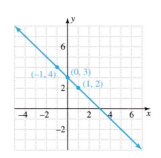

**2b.** Because the coefficient of $x$ is $\dfrac{4}{3}$, our calculations for $y$ will be easier if the values of $x$ are divisible by 3. So, we will let $x = -3, 0,$ and 3.

| $x$ | $y = \dfrac{4}{3}x - 4$ | $(x, y)$ |
|---|---|---|
| $-3$ | $y = \dfrac{4}{3}(-3) - 4$ <br> $y = -4 - 4$ <br> $y = -8$ | $(-3, -8)$ |
| $0$ | $y = \dfrac{4}{3}(0) - 4$ <br> $y = 0 - 4$ <br> $y = -4$ | $(0, -4)$ |
| $3$ | $y = \dfrac{4}{3}(3) - 4$ <br> $y = 4 - 4$ <br> $y = 0$ | $(3, 0)$ |

Plot the points $(-3, -8)$, $(0, -4)$, and $(3, 0)$. Since the points lie in a line, we draw the line through the points to obtain the graph of $y = \dfrac{4}{3}x - 4$.

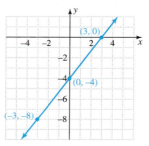

**2c.** Let $x = -2, 0, 1$.

| $x$ | $x - 2y = 4$ | $(x, y)$ |
|---|---|---|
| $-2$ | $-2 - 2y = 4$ <br> $-2 - 2y + 2 = 4 + 2$ <br> $-2y = 6$ <br> $\dfrac{-2y}{-2} = \dfrac{6}{-2}$ <br> $y = -3$ | $(-2, -3)$ |
| $0$ | $0 - 2y = 4$ <br> $-2y = 4$ <br> $\dfrac{-2y}{-2} = \dfrac{4}{-2}$ <br> $y = -2$ | $(0, -2)$ |
| $1$ | $1 - 2y = 4$ <br> $1 - 2y - 1 = 4 - 1$ <br> $-2y = 3$ <br> $\dfrac{-2y}{-2} = \dfrac{3}{-2}$ <br> $y = -\dfrac{3}{2}$ | $\left(1, -\dfrac{3}{2}\right)$ |

Plotting these points and connecting them gives us the graph of $x - 2y = 4$.

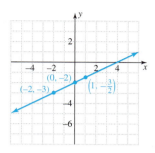

 **Student Check 2**    Graph each equation.

    **a.** $y = x - 2$                       **b.** $y = -\dfrac{1}{3}x + 6$                  **c.** $x + 3y = 9$

## Finding Intercepts

**Objective 3** ▶

Use intercepts to graph linear equations in two variables.

Another method of graphing a linear equation in two variables involves finding two important points that lie on the graph of the line. These points are called the *intercepts* of the graph.

> **Definition:** The **x-intercept** is the point on the graph where the graph intersects the $x$-axis. The **y-intercept** is the point on the graph where the graph intersects the $y$-axis.

In the graph from Example 2b, the $x$-intercept is $(3, 0)$ and the $y$-intercept is $(0, -4)$. Notice the $x$-intercept has a $y$-coordinate of zero since every point on the $x$-axis has a $y$-coordinate of zero. Similarly, the $y$-intercept has an $x$-coordinate of zero since every point on the $y$-axis has an $x$-coordinate of zero.

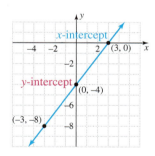

> **Procedure: Finding the Intercepts**
>
> **Step 1:** To find the $x$-intercept, replace $y$ with 0 and solve for $x$. The point will always be of the form $(a, 0)$.
>
> **Step 2:** To find the $y$-intercept, replace $x$ with 0 and solve for $y$. The point will always be of the form $(0, b)$.

 > **Note:** While we can graph a line knowing only two points, it can be helpful to find a third point as a checkpoint.

**Objective 3 Examples**  **Find the *x*- and *y*-intercepts for each equation. Use these points to graph each line.**

**3a.** $2x + 3y = 6$          **3b.** $y = 2x - 4$          **3c.** $2x - y = 0$

**Solutions**  **3a.**

|  | $x$ | $y$ | $(x, y)$ |
|---|---|---|---|
| *x*-intercept | $2x + 3(0) = 6$ $2x = 6$ $x = 3$ | $0$ | $(3, 0)$ |
| *y*-intercept | $0$ | $2(0) + 3y = 6$ $3y = 6$ $y = 2$ | $(0, 2)$ |
| Checkpoint | $1$ | $2(1) + 3y = 6$ $2 + 3y = 6$ $3y = 4$ $y = \dfrac{4}{3}$ | $\left(1, \dfrac{4}{3}\right)$ |

Plotting the points $(3, 0)$, $(0, 2)$, and $\left(1, \dfrac{4}{3}\right)$ gives us the graph of $2x + 3y = 6$.

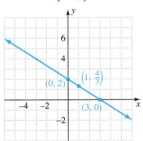

Notice that the checkpoint also lies on the graph of the line, so we know our work is correct.

**3b.**

|  | $x$ | $y$ | $(x, y)$ |
|---|---|---|---|
| *x*-intercept | $0 = 2x - 4$ $0 + 4 = 2x - 4 + 4$ $4 = 2x$ $\dfrac{4}{2} = \dfrac{2x}{2}$ $2 = x$ | $0$ | $(2, 0)$ |
| *y*-intercept | $0$ | $y = 2(0) - 4$ $y = 0 - 4$ $y = -4$ | $(0, -4)$ |
| Checkpoint | $3$ | $y = 2(3) - 4$ $y = 6 - 4$ $y = 2$ | $(3, 2)$ |

Plotting the points $(2, 0)$, $(0, -4)$, and $(3, 2)$, gives us the graph of $y = 2x - 4$.

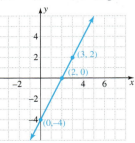

Notice that the checkpoint lies on the graph of the line, so we know our work is correct.

**3c.**

| $x$ | $y$ | $(x, y)$ |
|---|---|---|
| $2x - 0 = 0$ $2x = 0$ $\dfrac{2x}{2} = \dfrac{0}{2}$ $x = 0$ | $0$ | $(0, 0)$ |
| $0$ | $2(0) - y = 0$ $-y = 0$ $\dfrac{-y}{-1} = \dfrac{0}{-1}$ $y = 0$ | $(0, 0)$ |
| $2$ | $2(2) - y = 0$ $4 - y = 0$ $4 - y - 4 = 0 - 4$ $-y = -4$ $\dfrac{-y}{-1} = \dfrac{-4}{-1}$ $y = 4$ | $(2, 4)$ |

> The $x$- and $y$-intercepts for this graph are the same point, the origin. So, we need to find another point on the graph to have enough information to draw the line. Choose any value for $x$ and solve for $y$.

Plotting $(0, 0)$ and $(2, 4)$ gives us the graph of $2x - y = 0$.

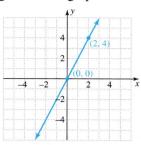

 **Note:** *Equations of the form $Ax + By = 0$ have graphs that go through the origin since $(0, 0)$ is always a solution of this type of equation.*

✓ **Student Check 3**   Find the $x$- and $y$-intercepts for each equation. Use these points to graph each line.

**a.** $4x - y = -8$ **b.** $y = 3x - 3$ **c.** $4x + y = 0$

## Horizontal and Vertical Lines

**Objective 4** ▶

Recognize and graph horizontal and vertical lines.

Horizontal and vertical lines are special cases of linear equations in two variables. Consider the graph of the *horizontal line* shown next.

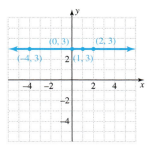

Some of the points on the line are $(0, 3)$, $(1, 3)$, $(2, 3)$, and $(-4, 3)$. The only points that this graph contains are points whose $y$-value is 3. A key characteristic for an ordered pair to be a solution of the equation of this line is that the $y$-value equals 3. Therefore, the equation of this line is $y = 3$. The equation $y = 3$ is equivalent to $0x + y = 3$. Any value of $x$ substituted in this equation is multiplied by 0, so the solutions only depend on the value of $y$ being 3.

> **Definition:** The graph of any equation of the form $y = k$, where $k$ is a real number, is a **horizontal line** through $k$ on the $y$-axis. The point $(0, k)$ is the $y$-intercept of the graph.

Now, we will examine the graph of a *vertical line,* as follows.

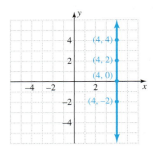

Some of the points on the line are $(4, 4)$, $(4, 2)$, $(4, 0)$, and $(4, -2)$. The only points that this graph contains are points whose $x$-value is 4. A key characteristic for an ordered pair to be a solution of the equation of this line is that the $x$-value equals 4. Therefore, the equation of this line is $x = 4$. The equation $x = 4$ is equivalent to $x + 0y = 4$. Any value of $y$ substituted in this equation is multiplied by 0, so the solutions only depend on the value of $x$ being 4.

> **Definition:** The graph of any equation of the form $x = h$, where $h$ is a real number, is a **vertical line** through $h$ on the $x$-axis. The point $(h, 0)$ is the $x$-intercept of the graph.

**Objective 4  Examples**     Graph each linear equation.

**4a.** $y = 2$          **4b.** $x = -3$          **4c.** $2x + 5 = 0$

**Solutions**   **4a.** The equation $y = 2$ can be written as $0x + y = 2$. Notice that for any $x$-value chosen, $y$ is 2. So, any ordered pair whose $y$-coordinate is 2 is a solution. If $x = 3, 0,$ and $-1$, the ordered pair solutions are

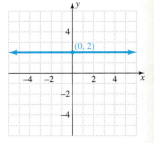

| $x$ | $y$ | $(x, y)$ |
|-----|-----|----------|
| 3 | 2 | $(3, 2)$ |
| 0 | 2 | $(0, 2)$ |
| $-1$ | 2 | $(-1, 2)$ |

The graph is a horizontal line with a $y$-intercept of $(0, 2)$. Note the graph does not have an $x$-intercept because the $y$-coordinate is never zero.

**4b.** The equation $x = -3$ can be written as
$x + 0y = -3$. Notice that for any $y$-value
chosen, $x$ is $-3$. So, any ordered pair whose
$x$-coordinate is $-3$ is a solution. If $y = 0, 1$,
and $-1$, the ordered pair solutions are

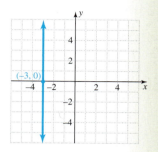

| $x$ | $y$ | $(x, y)$ |
|---|---|---|
| $-3$ | $0$ | $(3, 0)$ |
| $-3$ | $1$ | $(3, 1)$ |
| $-3$ | $-1$ | $(-3, -1)$ |

The graph is a vertical line with an $x$-intercept of $(-3, 0)$. Note the graph does
not have a $y$-intercept because the $x$-coordinate is never zero.

**4c.** The equation $2x + 5 = 0$ can be written as $2x + 0y = -5$ or as $x + 0y = -\dfrac{5}{2}$.

Notice that for any $y$-value chosen, $x$ is $-\dfrac{5}{2}$. So, any ordered pair whose

$x$-coordinate is $-\dfrac{5}{2}$ is a solution. If $y = 1, -4$, and $3$, the ordered pair
solutions are

| $x$ | $y$ | $(x, y)$ |
|---|---|---|
| $-\dfrac{5}{2}$ | $1$ | $\left(-\dfrac{5}{2}, 1\right)$ |
| $-\dfrac{5}{2}$ | $-4$ | $\left(-\dfrac{5}{2}, -4\right)$ |
| $-\dfrac{5}{2}$ | $3$ | $\left(-\dfrac{5}{2}, 3\right)$ |

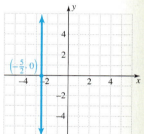

The graph is a vertical line with $x$-intercept $\left(-\dfrac{5}{2}, 0\right)$. Notice the line has no
$y$-intercept since the $x$-coordinate is never zero.

✓ **Student Check 4**  Graph each linear equation.
  **a.** $y = -1$          **b.** $x = 5$          **c.** $3y + 2 = 0$

## Applications

**Objective 5** ▶

Solve application problems.

In the application problems encountered in this section, we will be given a linear equation that models a real-world situation. We will construct the graph of the equation and
extract information from the equation.

**Objective 5 Examples**  **Use the equation to answer each question.**

**5a.** The median annual income for men with a bachelor's degree can be modeled by
the equation $y = 1443x + 38,843$, where $x$ is the number of years after 1990.
(Source: www.infoplease.com)

  **i.** Is this equation a linear equation in two variables? Why or why not?
  **ii.** Find the $y$-intercept of the equation and interpret its meaning in the context
of this problem.
  **iii.** Find the $x$-intercept of the equation and interpret its meaning in the context
of this problem.
  **iv.** Find the median annual income for the years 2000 and 2010.

**Solution**  **5a.**  **i.** The equation $y = 1443x + 38{,}843$ is a linear equation in two variables since each variable has a power of 1.

**ii.** The $y$-intercept of $y = 1443x + 38{,}843$ is found by replacing $x$ with 0 and solving for $y$.

$$y = 1443x + 38{,}843$$

$$y = 1443(0) + 38{,}843 \qquad \text{Substitute 0 for } x.$$

$$y = 0 + 38{,}843 \qquad \text{Multiply.}$$

$$y = 38{,}843 \qquad \text{Add.}$$

The $y$-intercept is the ordered pair $(0, 38{,}843)$. This means that in 0 yr after 1990, or 1990, the median annual income for men with a bachelor's degree was $38{,}843.

**iii.** The $x$-intercept of $y = 1443x + 38{,}843$ is found by replacing $y$ with 0 and solving for $x$.

$$y = 1443x + 38{,}843$$

$$0 = 1443x + 38{,}843 \qquad \text{Substitute 0 for } y.$$

$$0 - 38{,}843 = 1443x + 38{,}843 - 38{,}843 \qquad \text{Subtract 38,843 from each side.}$$

$$-38{,}843 = 1443x \qquad \text{Simplify.}$$

$$\frac{-38{,}843}{1443} = \frac{1443x}{1443} \qquad \text{Divide each side by 1443.}$$

$$-26.9 \approx x \qquad \text{Simplify.}$$

The $x$-intercept is the ordered pair $(-26.9, 0)$. This point does not make sense in the context of the problem since we cannot have a negative value for $x$.

**iv.** The median annual income for 2000 is found by replacing $x$ with 10 $(2000 - 1990 = 10)$ and the income for 2010 is found by replacing $x$ with 20 $(2010 - 1990 = 20)$.

Let $x = 10$: 

$$y = 1443(10) + 38{,}843$$

$$y = 14{,}430 + 38{,}843$$

$$y = 53{,}273$$

In 2000, the median annual income was $53{,}273.

Let $x = 20$:

$$y = 1443(20) + 38{,}843$$

$$y = 28{,}860 + 38{,}843$$

$$y = 67{,}703$$

In 2010, the median annual income was $67{,}703.

**5b.** Straight-line depreciation is the most common method of depreciating business assets. Mr. Rowell's surveying company purchased a truck for their business. On the company's tax records, Mr. Rowell must report the depreciated value of the truck. The truck's value is given by $y = -4000x + 32{,}000$, where $x$ is the age of the truck in years.

**i.** Find the $x$- and $y$-intercepts and interpret their meaning in the context of this problem.

**ii.** Find the value of the truck after 2 yr and after 4 yr.

**iii.** Use the ordered pairs found in parts (i) and (ii) to draw the graph of the equation.

**Solution**  **5b.**  **i.** The $x$- and $y$-intercepts of $y = -4000x + 32,000$ are shown next.

| x | y | Meaning |
|---|---|---|
| $0 = -4000x + 32,000$<br>$4000x = 32,000$<br>$\dfrac{4000x}{4000} = \dfrac{32,000}{4000}$<br>$x = 8$ | 0 | The $x$-intercept is $(8, 0)$. It means that when the truck is 8 yr old, its value is \$0. |
| 0 | $y = -4000(0) + 32,000$<br>$y = 32,000$ | The $y$-intercept is $(0, 32,000)$. It means that when the truck is 0 yr old, or when it is new, its value is \$32,000. |

**ii.** The value of the truck after 2 yr and after 4 yr is shown next.

| x | y | Meaning |
|---|---|---|
| 2 | $y = -4000(2) + 32,000$<br>$y = -8000 + 32,000$<br>$y = 24,000$ | The point is $(2, 24,000)$. It means that in 2 yr, the value of the truck is \$24,000. |
| 4 | $y = -4000(4) + 32,000$<br>$y = -16,000 + 32,000$<br>$y = 16,000$ | The point is $(4, 16,000)$. It means that in 4 yr, the value of the truck is \$16,000. |

**iii.** The graph of the equation is found by plotting the points $(8, 0)$, $(0, 32,000)$, $(2, 24,000)$, and $(4, 16,000)$. Because the $y$-values range from 0 to 32,000, we let each tick mark on the $y$-axis represent 4000 units. Also, the graph is drawn only in Quadrant I since it doesn't make sense for the years to be negative or for the value of the truck to be negative.

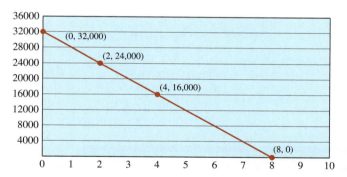

**✓ Student Check 5**  The median annual income for women with a bachelor's degree is given by the equation $y = 1100x + 27,645$, where $x$ is the number of years after 1990. (Source: www.infoplease.com)

   **a.** Find the $y$-intercept of the equation and interpret its meaning in the context of this problem.

   **b.** Find the median annual income for the years 2000 and 2010.

   **c.** Use the information in parts (a) and (b) to graph the equation.

**Objective 6 ▶**

Troubleshoot common errors.

## Troubleshooting Common Errors

Some common errors associated with graphing linear equations in two variables are illustrated next.

**Objective 6 Examples**    **A problem and an incorrect solution are given. Provide the correct solution and an explanation of the error.**

**6a.** Graph the equation $5x + y = 0$.

| Incorrect Solution | Correct Solution and Explanation |
|---|---|
| The $x$-intercept is $(0, 0)$ and the $y$-intercept is $(0, 0)$. So, the graph is 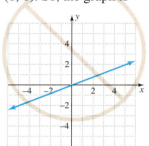 | We must have two points to draw the graph of this line. To find another point, make a substitution for $x$. If $x = 1$, then $$5x + y = 0$$ $$5(1) + y = 0$$ $$5 + y = 0$$ $$y = -5$$ We can graph the equation using the points $(0, 0)$ and $(1, -5)$. 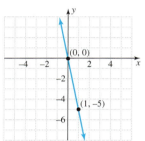 |

**6b.** Find the $x$- and $y$-intercepts of $8x - y = -16$.

| Incorrect Solution | Correct Solution and Explanation |
|---|---|
| $$8(0) - y = -16$$ $$y = -16$$ $$8x - 0 = -16$$ $$8x = -16$$ $$x = -2$$ The $x$-intercept is $(0, -16)$ and the $y$-intercept is $(-2, 0)$. | The first error is in solving the first equation. It should be $$8(0) - y = -16$$ $$-y = -16$$ $$y = 16$$ The other error was in assigning the points. The point $(0, 16)$ is the $y$-intercept and the point $(-2, 0)$ is the $x$-intercept. |

## ANSWERS TO STUDENT CHECKS

**Student Check 1**   **a.** yes; $x - 2y = -4$; $A = 1, B = -2, C = -4$   **b.** yes; $x + 0y = 4$; $A = 1, B = 0, C = 4$
**c.** no   **d.** yes; $2x - 3y = -12$; $A = 2, B = -3, C = -12$

**Student Check 2**

**a.** $y = x - 2$

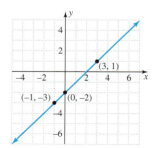

**b.** $y = -\dfrac{1}{3}x + 6$

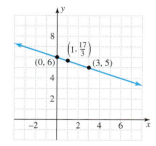

**c.** $x + 3y = 9$

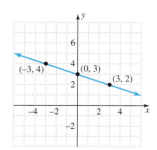

**Student Check 3**

**a.** intercepts: $(-2, 0)$ and $(0, 8)$

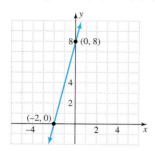

**b.** intercepts: $(1, 0)$ and $(0, -3)$

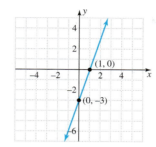

**c.** intercepts: $(0, 0)$

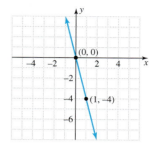

**Student Check 4**

**a.** $y = -1$

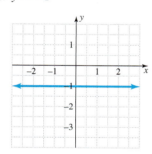

**b.** $x = 5$

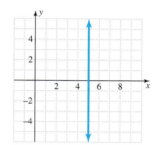

**c.** $3y + 2 = 0$

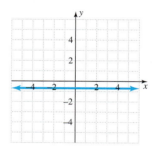

**Student Check 5** **a.** The $y$-intercept is $(0, 27{,}645)$. This means that in 1990, the median annual income for women with a bachelor's degree is $27,645. **b.** The median income in 2000 is $38,645 and was $49,645 in 2010.

**c.**

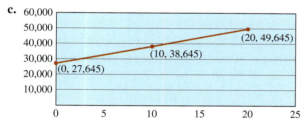

## SUMMARY OF KEY CONCEPTS

1. A linear equation in two variables is an equation of the form $Ax + By = C$, where $A$, $B$, and $C$ are real numbers with $A$ and $B$ not both zero. The exponent of the variables is one. If the values of $A$, $B$, and $C$ are not fractions and $A > 0$, this form is called the standard form. It is important to be able to write equivalent forms of a linear equation in two variables.

2. To graph an equation by plotting points, select at least three different values for $x$ or $y$, substitute into the equation, and solve for the corresponding value. These three points should lie in a straight line. If they do not, a mistake was made in your calculations or in graphing.

3. A linear equation can also be graphed by finding the $x$- and $y$-intercepts.

   **a.** To find the $x$-intercept, replace $y$ with zero and solve for $x$. This is a point of the form $(a, 0)$.

   **b.** To find the $y$-intercept, replace $x$ with zero and solve for $y$. This is a point of the form $(0, b)$.

   **c.** If both the $x$- and $y$-intercepts are $(0, 0)$, then the graph goes through the origin. Another point is needed to graph the line.

4. Horizontal lines are represented by an equation of the form $y = k$. These lines have a $y$-intercept of $(0, k)$. Equations of the form $x = h$ have graphs that are vertical lines. These lines have an $x$-intercept of $(h, 0)$.

5. When a linear equation is given that represents a real-life situation, the $x$- and $y$-intercepts have meaning. It is important to be able to state the meaning of these points. Finding other points and the graph of these models is also an important skill.

## GRAPHING CALCULATOR SKILLS

**Graphing Window**—The graphing window on the calculator comes with a standard setting, as shown. The $x$- and $y$-values range between $-10$ and $10$, with each tick mark representing 1 unit.

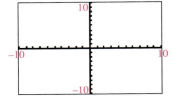
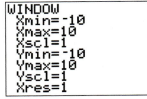

**Equation Editor**—The equation editor is what we use to enter equations into the calculator. This is accessed by pressing .

**Example:** Graph $x - 2y = 4$ on the calculator.

We must first solve the equation for $y$, because this is the only way equations can be entered into the calculator.

$$x - 2y = 4$$
$$x - 2y - x = 4 - x \qquad \text{Subtract } x \text{ from both sides.}$$
$$-2y = -x + 4 \qquad \text{Simplify.}$$
$$\frac{-2y}{-2} = \frac{-x + 4}{-2} \qquad \text{Divide both sides by } -2.$$
$$y = \frac{1}{2}x - 2 \qquad \text{Simplify.}$$

We can now graph the equation using a graphing calculator.

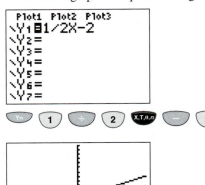

The Table Feature enables us to view ordered pairs that satisfy the equation and are, therefore, points on the graph.

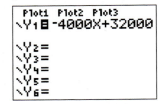

With some equations, we may have to adjust the viewing window to see the graph on the graphing calculator as this next example illustrates.

**Example:** Graph $y = -4000x + 32{,}000$.

We enter it into the equation editor. When graphed in the standard window, the graph almost looks like a vertical line (which is not the case).

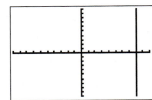

We must change the view of the window on the calculator to see the correct graph.

To change the view on the calculator, we should find the intercepts by hand to know how to change the window. From example 5b, we know the $x$-intercept is $(8, 0)$ and the $y$-intercept is $(0, 32{,}000)$. We also know that negative $x$- and $y$-values do not make sense in this problem.

To view a complete graph, we must set the window to see the intercepts. Since the graph crosses the $x$-axis at $(0, 8)$, we need the maximum $x$-value to be larger than 8. Let's use 10.

Since the graph crosses the $y$-axis at $(0, 32{,}000)$, we need the maximum $y$-value to be larger than 32,000. Let's use 35,000. The value for Yscl represents how many units each tick mark on the $y$-axis represents. If we were drawing this by hand, we might let each tick mark be 5000 units.

This is the graph of the equation in the new window.

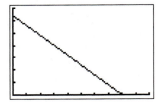

## SECTION 3.2 / EXERCISE SET

### Write About It!

Use complete sentences to explain the meaning of each term or process.

1. Graph a line by point-plotting.   Answers vary.
2. Graph a line by using the $x$-intercept and $y$-intercept.   Answers vary.
3. Which method is easier to use with the equation $y = -4x + 2$? Which method is easier to use with the equation $4x - 3y = -12$?   Answers vary.

Determine if each statement is true or false. If a statement is false, provide an explanation.

4. The equation $y = 3x - 4$ is equivalent to $3x - y = -4$.
   False, the equation is equivalent to $3x - y = 4$.
5. The graph of the equation $7x - y = 14$ passes through the point $(0, 2)$.   False, the graph goes through $(2, 0)$ and $(0, -14)$.
6. The $x$-intercept of $y = 8x - 24$ is $(0, -24)$.
   False, the $y$-intercept is $(0, -24)$.
7. The graph of the equation $x = -2$ has a $y$-intercept at $(0, -2)$.   False, the graph has no $y$-intercept.
8. The graph of the equation $y = 3$ has no $x$-intercept.   True
9. The graph of the equation $y = 5x$ passes through the origin.   True
10. The graph of the equation $2x + 3y = 0$ passes through the origin.   True

### Practice Makes Perfect!

Determine if the equation is a linear equation in two variables. (*See Objective 1.*)

11. $4x + 5y = 6$   yes
12. $3x - 2y = -12$   yes
13. $x^2 + y = 4$   no
14. $x + y^2 = 9$   no
15. $x = 2y$   yes
16. $y = -\dfrac{4}{3}x$   yes
17. $\dfrac{x}{2} - \dfrac{y}{4} = \dfrac{1}{8}$   yes
18. $x^2 + y^2 = 4$   no
19. $x = 3$   yes
20. $y = -1$   yes

Make a table of at least three ordered pairs that are solutions of the equation and use these points to graph each equation. (*See Objective 2.*)

21. $x + 4y = 8$
22. $x - 6y = 12$
23. $3x + y = -6$
24. $5x + y = 10$
25. $y = -x + 1$
26. $y = -x - 3$
27. $y = 2x - 4$
28. $y = -2x + 3$
29. $y = \dfrac{2}{3}x - 6$
30. $y = -\dfrac{2}{5}x + 4$
31. $y = -\dfrac{1}{4}x + 2$
32. $y = \dfrac{5}{6}x - 1$

Use the intercepts to graph each equation. Label the intercepts as ordered pairs on the graph. (*See Objective 3.*)

33. $x - 2y = 6$
34. $x + 6y = 6$
35. $7x - y = -7$
36. $2x + y = -4$
37. $y + 2x = 0$
38. $y - 2x = 0$

39. $y = -x + 4$
40. $y = -x - 7$
41. $y = 4x - 5$
42. $y = -6x + 3$
43. $y = \dfrac{1}{2}x + 5$
44. $y = -\dfrac{3}{4}x + 6$

Graph each equation. (*See Objective 4.*)

45. $y = 6$
46. $y = -5$
47. $x = -7$
48. $x = 4$
49. $8y - 2 = 0$
50. $6y + 3 = 0$
51. $2x + 4 = 0$
52. $3x - 9 = 0$
53. $4y + 3 = 0$
54. $2y - 7 = 0$
55. $\dfrac{1}{2}x + 5 = 0$
56. $-\dfrac{3}{4}x + 8 = 0$
57. $\dfrac{1}{3}y - 2 = 0$
58. $-\dfrac{3}{4}y + 1 = 0$

Use the equation to answer each question. (*See Objective 5.*)

59. The median annual income for men with a master's degree is given by the equation $y = 1873x + 49{,}217$, where $x$ is the number of years after 1991. (Source: www.infoplease.com)
   a. Find the $y$-intercept and interpret its meaning.
   b. Find the median annual income for men in 2001.
   c. Find the expected median annual income for men in 2012.

60. The median annual income for women with a master's degree is given by the equation $y = 1335x + 35{,}660$, where $x$ is the number of years after 1991. (Source: www.infoplease.com)
   a. Find the $y$-intercept and interpret its meaning.
   b. Find the median annual income for women in 2001.
   c. Find the expected median annual income for women in 2012.

61. The average retail price, in cents per kilowatt-hour, of electricity for residential use can be modeled by the equation $y = 0.395x + 8.154$, where $x$ is the number of years after 2001. (Source: www.eia.gov)
   a. Find the $y$-intercept and interpret its meaning.
   b. Find the average retail price of electricity for residential use in 2010.
   c. Find the expected average retail price of electricity for residential use in 2015.

62. The average domestic first purchase price of crude oil, in dollars per barrel, can be modeled by the equation $y = 6.679x + 16.173$, where $x$ is the number of years after 2000. (Source: www.eia.gov)
   a. Find the $y$-intercept and interpret its meaning.
   b. Find the average domestic first purchase price of crude oil in 2010.
   c. Find the expected average domestic first purchase price of crude oil in 2013.

Additional answers can be found in the Instructor Answer Appendix.

**Mix 'Em Up!**

**Graph each equation. Identify at least two points on the graph.**

**63.** $9x - 3y = 6$    **64.** $4x - 2y = 8$    **65.** $y - 1 = 0$

**66.** $x + 3 = 0$    **67.** $y = 3x$    **68.** $y = -2x$

**69.** $y = -\frac{1}{2}x + 5$    **70.** $y = \frac{1}{3}x - 2$    **71.** $y = -0.4x + 1$

**72.** $y = 0.25x - 1$    **73.** $y = 2.4x - 1.8$    **74.** $y = -1.5x - 4.5$

**75.** $x = 0$   *y-axis*    **76.** $y = 0$   *x-axis*    **77.** $3x - 2y = 0$

**78.** $4x + 5y = 0$    **79.** $y = -\frac{1}{5}x$    **80.** $y = \frac{5}{6}x$

**81.** $2y + 5 = 0$    **82.** $5y - 2 = 0$    **83.** $2x - 10 = 0$

**84.** $3x + 3 = 0$    **85.** $4x - 5 = 0$    **86.** $6x - 9 = 0$

**87.** $0.01x - 0.02y = 1.2$        **88.** $0.03x + 0.02y = 0.6$

**89.** $\frac{2}{3}x - \frac{1}{2}y = 1$    **90.** $\frac{3}{4}x + \frac{1}{3}y = -2$

**Use the equation to answer each question.**

**91.** The percent of high school students who are users of cigarettes is on the decline. The percent of students who are users can be modeled by the equation $y = -2.5x + 38$, where $x$ is the number of years after 1997. (Source: National Cancer Institute)

   **a.** Find the $y$-intercept of the equation and interpret its meaning.

   **b.** What percent of high school students used cigarettes in 2000?

   **c.** What percent of high school students used in cigarettes in 2010?

   **d.** Use the intercepts to graph the equation.

**92.** The expenditures (in billions) for public elementary and secondary education in the United States can be modeled by the equation $y = 20x + 285$, where $x$ is the number of years after 1997. (Source: http://nces.ed.gov/programs/projections/projections2018/tables/table_34.asp)

   **a.** Find the $y$-intercept of the equation and interpret its meaning.

   **b.** What were the expenditures for public elementary and secondary education in the United States in 2005?

   **c.** What were the expenditures for public elementary and secondary education in the United States in 2012?

   **d.** Use the information obtained in parts (a) to (c) to graph the equation.

**93.** The percent of men aged 18 yr and older who are current cigarette smokers can be modeled by the linear equation $y = -0.33x + 25.64$, where $x$ is the number of years after 1998. (Source: National Cancer Institute)

   **a.** Find the $x$-intercept of the equation and interpret its meaning. Is this realistic?

   **b.** Find the $y$-intercept of the equation and interpret its meaning.

   **c.** What percent of men aged 18 yr and older were cigarette smokers in 2005? How close is the answer to the actual estimate of 23.39%?

   **d.** What percent of men aged 18 yr and older were cigarette smokers in 2012?

   **e.** Use the intercepts to graph the equation.

   **f.** The Healthy People Campaign targets to reduce the percent of smokers. Find the year when the percent of men aged 18 yr and older who are current cigarette smokers will be reduced to 12%.

**94.** The percent of women aged 18 yr and older who are current cigarette smokers can be modeled by the linear equation $y = -0.53x + 22.12$, where $x$ is the number of years after 1998. (Source: National Cancer Institute)

   **a.** Find the $x$-intercept of the equation and interpret its meaning. Is this realistic?

   **b.** Find the $y$-intercept of the equation and interpret its meaning.

   **c.** What percent of women aged 18 yr and older were cigarette smokers in 2005? How close is the answer to the actual estimate of 18.38%?

   **d.** What percent of women aged 18 yr and older were cigarette smokers in 2012?

   **e.** Use the intercepts to graph the equation.

   **f.** The Healthy People Campaign targets to reduce the percent of smokers. Find the year when the percent of women aged 18 yr and older who are current cigarette smokers will be reduced to 12%.

**95.** The percent of U.S. adults aged 20 yr and over who are obese can be modeled by the linear equation $y = 0.71x + 19.83$, where $x$ is the number of years after 1997. (Source: National Center for Health Statistics)

   **a.** Find the $y$-intercept of the equation and interpret its meaning.

   **b.** What percent of U.S. adults were obese in 2008? How close is the answer to the actual estimate of 27.6% in 2008?

   **c.** What percent of U.S. adults were obese in 2012?

   **d.** Use the information obtained in parts (a) to (c) to graph the equation.

   **e.** What does the graph indicate about the percent of U.S. adults aged 20 yr and over who are obese?

**96.** The owner of a limousine rental company purchases a stretch limousine for $110,000. For tax purposes, the owner uses straight-line depreciation for reporting the value of the limousine. The depreciated value of the limousine is given by $y = -22,000x + 110,000$, where $x$ is the age of the limousine in years.

   **a.** Find the $x$-intercept and interpret the meaning.

   **b.** Find the $y$-intercept and interpret the meaning.

   **c.** When will the depreciated value of the limousine be $66,000?

**d.** What will be the depreciated value of the limousine when it is 3 yr old?

**e.** Use the information obtained in parts (a) to (c) to graph the equation.

 **You Be the Teacher!**

**Answer each student's question.**

**97.** What is the minimum number of points that you must find to graph a straight line? Why?
We must find two points because this allows us to determine the direction of the line.

**98.** For the equation $y = \frac{3}{7}x + 4$, how can I choose $x$-values to plug into the equation so that I get integer values for $y$?
Because the coefficient of $x$ is $\frac{3}{7}$, we need to choose values that are divisible by 7.

**99.** How do you find the $x$-intercept of the equation $x = 7$?
Write the equation as $x + 0y = 7$, replace $y$ with 0, and solve for $x$ to get (7, 0).

**100.** How do you find the $y$-intercept of the equation $y = -10$?
Write the equation as $0x + y = -10$, replace $x$ with 0, and solve for $y$ to get (0, −10).

**101.** I am trying to graph the equation $4y + x = 0$. When I find the intercepts, I can only find one point (0, 0). How can I find another point to graph the equation?
Choose any nonzero number for $x$ and substitute into the equation to solve for $y$.

**102.** I am trying to graph the equation $3x - 5y = 15$. When I find the intercepts, I only get one point (5, −3). How can I find another point to graph the equation?
The intercepts correspond to the points (0, −3) and (5, 0); they are not combined into one point.

 **Calculate It!**

**Use a graphing calculator to complete each exercise.**

**103.** Draw, by hand, the coordinate system that is defined by these window settings. Label the $x$-axis and $y$-axis with the appropriate tick marks.
Answers vary.

```
WINDOW
 Xmin=0
 Xmax=100
 Xscl=1
 Ymin=0
 Ymax=500
 Yscl=50
 Xres=1
```

**104.** Draw, by hand, the coordinate system that is defined by these window settings. Label the $x$-axis and $y$-axis with appropriate tick marks.
Answers vary.

```
WINDOW
 Xmin=-20
 Xmax=20
 Xscl=5
 Ymin=-50
 Ymax=50
 Yscl=10
 Xres=1
```

**105.** Determine the ordered pairs for the $x$-intercept and $y$-intercept of the graph if each tick mark represents one unit.
(2, 0) and (0, 6)

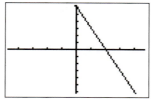

**106.** Determine the ordered pairs for the $x$-intercept and $y$-intercept of the graph if each tick mark on the $x$-axis represents one unit and each tick mark on the $y$-axis represents five units.
(5, 0) and (0, −15)

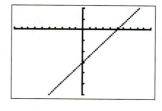

**107.** Graph $y = -2x + 20$ in the standard viewing window.

**a.** Sketch the graph shown on your calculator.
Answers vary.

**b.** Which intercept(s) can be seen on your calculator?
(10, 0)

**c.** Find the $x$-intercept and $y$-intercept by hand.
(0, 20) and (10, 0)

**d.** What viewing window will show the complete graph that contains both intercepts? State the values that you would use for Xmin, Xmax, Xscl, Ymin, Ymax, and Yscl. Answers vary.

**e.** Sketch the graph shown on your calculator using the window settings from part d. Answers vary.

**108.** Graph $y = 0.5x + 40$ in the standard viewing window.

**a.** Sketch the graph that is shown on your calculator.
Answers vary.

**b.** Which intercept(s) can be seen on your calculator?
neither one

**c.** Find the $x$ and $y$-intercepts by hand. (0, 40) and (−80, 0)

**d.** What viewing window will show the complete graph that contains both intercepts? State the values you would use for Xmin, Xmax, Xscl, Ymin, Ymax, and Yscl. Answers vary.

**e.** Sketch the graph shown on your calculator using the window settings from part d. Answers vary.

 **Think About It!**

**109.** Make a table of solutions for the equation $y = 2x + 4$. Use $x$-values of −2, −1, 0, 1, 2, and 3. Do you notice anything about how the $y$-values are changing in the table? How does this relate to the equation?

**110.** Make a table of solutions for the equation $y = -3x + 1$. Use $x$-values of −2, −1, 0, 1, 2, and 3. Do you notice anything about how the $y$-values are changing in the table? How does this relate to the equation?

## SECTION 3.3 / The Slope of a Line

▶ **OBJECTIVES**

As a result of completing this section, you will be able to

1. Use the slope formula to determine the slope of a line.
2. Use the slope-intercept form of a line to find its slope and y-intercept.
3. Graph a line given its slope and y-intercept.
4. Troubleshoot common errors.

In Section 3.2, we learned how to graph lines such as $y = 2x$ and $y = \frac{1}{2}x$, which are shown in the following graphs. In this section, we will discuss a property of lines. This property will answer the question, "How do the following graphs of the two lines compare to one another?"

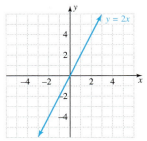 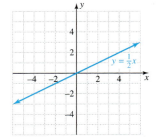

Observing the lines, we can see that the line on the left is steeper than the line on the right. A line's steepness, or *slope*, can be used to describe its graph. In this section, we will explore the slope of a line and how we can determine it.

### The Slope Formula

**Objective 1** ▶

Use the slope formula to determine the slope of a line.

In Section 3.2, we graphed the line $y = 2x - 4$. Let's explore some of its solutions and how they relate to the graph of the line.

| $x$ | $y$ | Change in $x$ | Change in $y$ | $\dfrac{\text{Change in } y}{\text{Change in } x}$ |
|---|---|---|---|---|
| 0 | $2(0) - 4 = -4$ | | | |
| 1 | $2(1) - 4 = -2$ | $1 - 0 = 1$ | $-2 - (-4) = 2$ | $\dfrac{2}{1} = 2$ |
| 2 | $2(2) - 4 = 0$ | $2 - 1 = 1$ | $0 - (-2) = 2$ | $\dfrac{2}{1} = 2$ |
| 3 | $2(3) - 4 = 2$ | $3 - 2 = 1$ | $2 - 0 = 2$ | $\dfrac{2}{1} = 2$ |

The $y$-values of the solutions in the table increase by 2 units every time the $x$-values of the solutions increase by 1 unit. This can be seen from the graph in that as we move 1 unit to the right, the graph rises 2 units.

**INSTRUCTOR NOTE:**
Point out that the slope of the line is the same regardless of which two points we use. Illustrate this with the points $(0, -4)$ and $(3, 2)$. The change in $x$ is $3 - 0 = 3$ and the change in $y$ is $2 - (-4) = 6$. The ratio of these changes is $\frac{6}{3} = 2$.

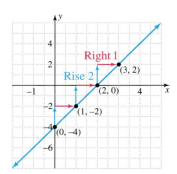

The ratio of these changes, $\dfrac{\text{Change in } y \text{ (vertical change)}}{\text{Change in } x \text{ (horizontal change)}}$, is called the slope of the line. So, the slope of the line $y = 2x - 4$ is 2.

$$\text{Slope} = \frac{\text{change in } y}{\text{change in } x} = \frac{2}{1} = 2$$

This means that for each 2 units of change in the $y$-coordinates, there is a corresponding change of 1 unit in the $x$-coordinates.

> **Note:** *The slope of the line is the same for every pair of points on the line. If the slope is not the same, then the graph is not a line.*

**Definition:** The **slope** of a line is the ratio of the change in $y$ to the change in $x$ between two points on a line. The slope is denoted by the letter $m$. It is often called "vertical change over horizontal change" or "rise over run."

$$m = \frac{\text{Change in } y}{\text{Change in } x} = \frac{\text{vertical change}}{\text{horizontal change}} = \frac{\text{rise}}{\text{run}}$$

Using a table of values to find the slope of a line can become quite cumbersome. We can also find the slope of a line by using the slope formula.

**Property: The Slope Formula**

If $(x_1, y_1)$ and $(x_2, y_2)$ are two points that lie on a line with $x_1 \neq x_2$, then the slope is

$$m = \frac{\text{change in } y}{\text{change in } x} = \frac{y_2 - y_1}{x_2 - x_1} = \frac{y_1 - y_2}{x_1 - x_2}$$

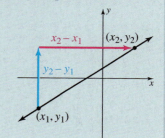

If $x_1 = x_2$, then the line through the points is vertical and the slope is undefined.

**Procedure: Using the Slope Formula to Find the Slope between Two Points**

**Step 1:** Label one point as $(x_1, y_1)$ and the otherpoint as $(x_2, y_2)$.
**Step 2:** Substitute the values into the formula

$$m = \frac{y_2 - y_1}{x_2 - x_1}$$

**Step 3:** Simplify the numerator and denominator and reduce the fraction, if necessary.

**Objective 1 Examples** / **Use the slope formula to determine the slope between each pair of points.**

**1a.** $(-1, -2)$ and $(4, 5)$      **1b.** $(5, -3)$ and $(-4, 2)$

**1c.** $(2, -3)$ and $(2, 5)$      **1d.** $(0, -2)$ and $(4, -2)$

**1e.**

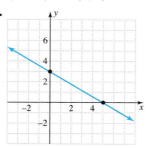

**Solutions**    **1a.** $m = \dfrac{y_2 - y_1}{x_2 - x_1}$    State the slope formula.

$m = \dfrac{5 - (-2)}{4 - (-1)}$    Let $(x_1, y_1) = (-1, -2)$ and $(x_2, y_2) = (4, 5)$.

$m = \dfrac{7}{5}$    Simplify.

Notice from the graph that if we move 7 units up from $(-1, -2)$ and 5 units to the right, we will be at the point $(4, 5)$.

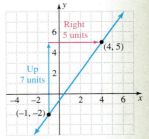

**1b.** $m = \dfrac{y_2 - y_1}{x_2 - x_1}$    State the slope formula.

$m = \dfrac{2 - (-3)}{-4 - (5)}$    Let $(x_1, y_1) = (5, -3)$ and $(x_2, y_2) = (-4, 2)$.

$m = \dfrac{5}{-9}$    Simplify.

$m = -\dfrac{5}{9}$

Notice from the graph that if we move 5 units down from $(-4, 2)$ and 9 units to the right, we will be at the point $(5, -3)$.

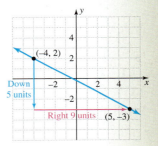

**1c.** $m = \dfrac{y_2 - y_1}{x_2 - x_1}$    State the slope formula.

$m = \dfrac{5 - (-3)}{2 - (2)}$    Let $(x_1, y_1) = (2, -3)$ and $(x_2, y_2) = (2, 5)$.

$m = \dfrac{8}{0}$ undefined    Simplify.

Notice from the graph that if we move up 8 units from the point $(2, -3)$, we will be at the point $(2, 5)$ since there is no movement left or right.

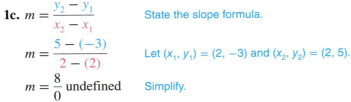

**1d.** $m = \dfrac{y_2 - y_1}{x_2 - x_1}$    State the slope formula.

$m = \dfrac{-2 - (-2)}{4 - (0)}$    Let $(x_1, y_1) = (0, -2)$ and $(x_2, y_2) = (4, -2)$.

$m = \dfrac{0}{4}$    Simplify.

$m = 0$    Simplify.

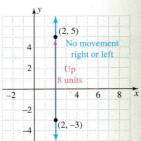

Notice from the graph that if we move right 4 units from the point $(0, -2)$, we will be at the point $(4, -2)$ since there is no movement up or down.

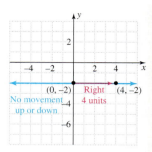

**1e.** Since we are given the graph of a line, we can find the slope by identifying two points on the graph and substituting their values in the slope formula. We can also find the ratio of the vertical change to the horizontal change.

**Method 1:** Two points on the graph are $(5, 0)$ and $(0, 3)$. So,

$$m = \frac{y_2 - y_1}{x_2 - x_1}$$

$$m = \frac{3 - 0}{0 - 5}$$

$$m = \frac{3}{-5}$$

$$m = -\frac{3}{5}$$

**Method 2:** To move from $(0, 3)$ to $(5, 0)$, we go down 3 units and right 5 units.

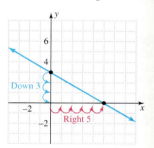

So, the slope is

$$m = \frac{\text{vertical change}}{\text{horizontal change}} = \frac{-3}{5} = -\frac{3}{5}$$

*Notice that as we examine the lines in Example 1, we find that*

- *The line in 1a "goes up" from left to right or has positive slope.*
- *The line in 1b "goes down" from left to right or has negative slope.*
- *The line in 1c is "vertical" and has an undefined slope.*
- *The line in 1d is "horizontal" and has a zero slope.*

So, we have four different possibilities for the slope of a line.

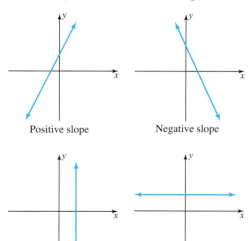

**INSTRUCTOR NOTE:**
Remind students that we must look at the line from left to right to determine if the line "goes up" or "goes down."

**✓ Student Check 1**   Use the slope formula to determine the slope of the line between each pair of points.

**a.** $(1, -6)$ and $(7, -6)$   **b.** $(-2, 5)$ and $(3, 2)$

**c.** $(4, 7)$ and $(2, -1)$   **d.** $(-4, 2)$ and $(-4, -1)$

**e.**

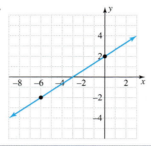

## The Slope-Intercept Form of a Line

**Objective 2** ▶

Use the slope-intercept form of a line to find its slope and $y$-intercept.

In the previous objective, we found that the slope of the line $y = 2x - 4$ is $m = \dfrac{2}{1} = 2$. We discovered this by examining the ratio of the vertical change to the horizontal change using the table on page 226. We can also confirm this by substituting two points that lie on the line into the slope formula. For instance, $(0, -4)$ and $(3, 2)$ lie on the line.

$$m = \frac{y_2 - y_1}{x_2 - x_1} \qquad \text{State the slope formula.}$$

$$m = \frac{2 - (-4)}{3 - 0} \qquad \text{Let } (x_1, y_1) = (0, -4) \text{ and } (x_2, y_2) = (3, 2).$$

$$m = \frac{6}{3} \qquad \text{Simplify.}$$

$$m = 2 \qquad \text{Simplify.}$$

The ordered pair $(0, -4)$ is the $y$-intercept of the graph of $y = 2x - 4$ since its $x$-coordinate is zero.

Note that the value of the slope and the $y$-intercept are provided in the equation, $y = 2x - 4$. The coefficient of $x$, 2, is the same as the slope of the line. The constant term, $-4$, is the same as the $y$-coordinate of the $y$-intercept of the line. The equation $y = 2x - 4$ is in what we call the *slope-intercept form* of a line.

> **Property:  The Slope-Intercept Form of a Line**
>
> The equation $y = mx + b$ is the **slope-intercept form** of a linear equation. The value $m$ is the slope of the line and the point $(0, b)$ is the $y$-intercept.
>
> $$y = \underset{\underset{\text{Slope}}{\uparrow}}{mx} + \underset{\underset{y\text{-intercept}}{\uparrow}}{b}$$

The key difference between the standard form of a linear equation and the slope-intercept form of a linear equation is that the slope-intercept form is the equation solved for $y$.

> **Procedure:  Finding the Slope and $y$-Intercept of a Linear Equation**
>
> **Step 1:** Solve the linear equation for $y$, if necessary.
> **Step 2:** The coefficient of $x$ is the slope of the line.
> **Step 3:** The constant term is the $y$-coordinate of the $y$-intercept of the line.

**Objective 2 Examples**   Write the equation in slope-intercept form, if possible. State the slope and *y*-intercept of the line from its equation. Explain how the *x*- and *y*-values change with respect to one another.

**2a.** $y = 4x + 1$        **2b.** $y = -\dfrac{2}{3}x$        **2c.** $6x - y = 2$

**2d.** $4x + 3y = 24$        **2e.** $y = 3$        **2f.** $x = 2$

**Solutions**   **2a.** The equation $y = 4x + 1$ is in slope-intercept form since it is solved for *y*.

The coefficient of *x* is 4, so the slope is $m = 4$.
The constant term is 1, so the *y*-intercept is $(0, 1)$.
The slope, $m = 4 = \dfrac{4}{1}$, means that as the *y*-values increase by 4 units, the *x*-values increase by 1 unit.

**2b.** The equation, $y = -\dfrac{2}{3}x + 0$, is in slope-intercept form since it is solved for *y*.

The coefficient of *x* is $-\dfrac{2}{3}$, so, the slope is $m = -\dfrac{2}{3}$.
The constant term is 0, so the *y*-intercept is $(0, 0)$.

The slope $m = -\dfrac{2}{3} = \dfrac{-2}{3}$ or $\dfrac{2}{-3}$. This means that as the *y*-values decrease by 2 units, the *x*-values increase by 3 units. An equivalent way to state this is that as the *y*-values increase by 2 units, the *x*-values decrease by 3 units.

**2c.** We must solve the equation for *y* to write it in the slope-intercept form.

$$6x - y = 2$$
$$6x - y - 6x = 2 - 6x \qquad \text{Subtract } 6x \text{ from each side.}$$
$$-y = -6x + 2 \qquad \text{Simplify.}$$
$$-1(-y) = -1(-6x + 2) \qquad \text{Multiply each side by } -1.$$
$$y = 6x - 2 \qquad \text{Simplify.}$$

The coefficient of *x* is 6, so the slope is $m = 6$.
The constant term is $-2$, so the *y*-intercept is $(0, -2)$.

The slope $m = 6 = \dfrac{6}{1}$. This means that as the *y*-values increase by 6 units, the *x*-values increase by 1 unit.

**2d.** We must solve the equation for *y*.

$$4x + 3y = 24$$
$$4x + 3y - 4x = 24 - 4x \qquad \text{Subtract } 4x \text{ from each side.}$$
$$3y = -4x + 24 \qquad \text{Simplify.}$$
$$\dfrac{3y}{3} = \dfrac{-4x + 24}{3} \qquad \text{Divide each side by 3.}$$
$$y = -\dfrac{4}{3}x + 8 \qquad \text{Simplify.}$$

The coefficient of *x* is $-\dfrac{4}{3}$, so the slope is $m = -\dfrac{4}{3}$.

The constant term is 8, so the *y*-intercept is $(0, 8)$.

The slope $m = -\dfrac{4}{3} = \dfrac{-4}{3}$ or $\dfrac{4}{-3}$. This means that as the *y*-values decrease by 4 units, the *x*-values increase by 3 units. An equivalent way to state this is that as the *y*-values increase by 4 units, the *x*-values decrease by 3 units.

**2e.** The equation $y = 3$ is the equation of a horizontal line. It is in slope-intercept form and is equivalent to $y = 0x + 3$.

The coefficient of $x$ is 0, so the slope is $m = 0$.

The constant term is 3, so the $y$-intercept is $(0, 3)$.

The slope $m = 0 = \dfrac{0}{1}$. This means that there is no change in the $y$-values as the $x$-values increase by 1 unit.

**2f.** The equation $x = 2$ represents a vertical line through the point $(2, 0)$. It cannot be written in slope-intercept form since there is no $y$-variable. From Example 1, we know that the slope of a vertical line is undefined. There is no $y$-intercept since the graph doesn't cross the $y$-axis.

Note that we can use the slope formula to find the slope using two points on the line, such as $(2, 0)$ and $(2, 3)$.

$$m = \frac{y_2 - y_1}{x_2 - x_1} \qquad \text{State the slope formula.}$$

$$m = \frac{3 - 0}{2 - 2} \qquad \text{Let } (x_1, y_1) = (2, 0) \text{ and } (x_2, y_2) = (2, 3).$$

$$m = \frac{3}{0} \qquad \text{Simplify.}$$

$m$ is undefined.

✓ **Student Check 2** Write the equation in slope-intercept form, if possible. State the slope and $y$-intercept of the line from its equation. Explain how the $x$- and $y$-values change with respect to one another.

**a.** $y = 7x + 8$  **b.** $y = \dfrac{1}{5}x$  **c.** $9x - y = 5$

**d.** $x + 2y = 10$  **e.** $x = -4$  **f.** $y = -2$

> **Property: Slopes of Vertical and Horizontal Lines**
>
> The slope of the equation $x = h$, where $h$ is a real number, is undefined.
> The slope of the equation $y = k$, where $k$ is a real number, is zero.

> **Note:** *To remember the slopes of horizontal and vertical lines, we can think of the following situation: If we can walk on the line, then the line has a slope. If we can't walk on the line, then the slope is undefined. Since we cannot "walk" on a vertical line, it has a slope that is undefined. We walk on horizontal "lines" every day, so the slope of a horizontal line is a real number, the real number 0.*

## Graph a Line Using Its Slope and $y$-Intercept

In Section 3.2, we graphed a line by finding at least two points on the line. We can also graph a line if we know its slope and $y$-intercept.

- The $y$-intercept $(0, b)$ is one point on the line.
- Another point can be obtained from the slope.

Since the slope is the ratio of the change in $y$ to the change in $x$, it tells us how to move from the $y$-intercept to another point on the line. Recall the numerator of the slope corresponds to the change in $y$, or the vertical movement between the two points, and the denominator of the slope corresponds to the change in $x$, or the horizontal movement between the two points.

> **Procedure: Graphing a Line Given Its Slope and $y$-Intercept**
>
> **Step 1:** Plot the $y$-intercept $(0, b)$.
> **Step 2:** Use the slope to determine another point on the line.
>  **a.** The numerator of the slope tells us how many units to go up (if positive) or down (if negative) from the $y$-intercept.
>  **b.** The denominator tells us how many units to move to the right (if positive) or left (if negative).
> **Step 3:** Draw the graph of the line through these two points.

**Note:** $y = mx + b \rightarrow b$ tells us where to begin and $m$ tells us how to move.

---

**Objective 3 Examples**  Graph the line using its slope and $y$-intercept.

**3a.** $y = 3x + 2$      **3b.** $y = -\dfrac{2}{3}x + 2$

**Solutions**  **3a.** In the equation, $y = 3x + 2$, the slope $m = 3$ and the $y$-intercept is $(0, 2)$.

**Step 1:** Plot the $y$-intercept $(0, 2)$.

**Step 2:** Since $m = 3 = \dfrac{3}{1}$, we move up 3 units as we move right 1 unit to get to another point.

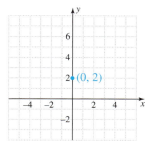

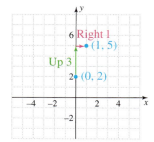

**Step 3:** Connect the points to obtain the graph of $y = 3x + 2$.

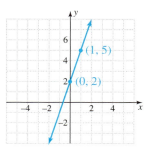

**3b.** In the equation $y = -\frac{2}{3}x + 2$, the slope $m = -\frac{2}{3}$ and the y-intercept is $(0, 2)$.

**Step 1:** Plot the y-intercept $(0, 2)$.

**Step 2:** Since $m = -\frac{2}{3} = \frac{-2}{3}$, we move down 2 units as we move right 3 units to get to another point.

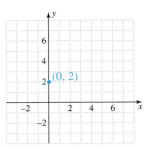

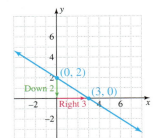

**Step 3:** Connect the points to obtain the graph of $y = -\frac{2}{3}x + 2$.

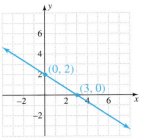

✓ **Student Check 3**   Graph the line using its slope and y-intercept.

  **a.** $y = \frac{1}{4}x + 3$                    **b.** $y = -3x + 6$

**Objective 4** ▶

Troubleshoot common errors.

## Troubleshooting Common Errors

Some common errors associated with the concept of slope are shown next.

**Objective 4 Examples**   **A problem and an incorrect solution are given. Provide the correct solution and an explanation of the error.**

**4a.** Find the slope between $(3, -2)$ and $(1, 7)$.

| Incorrect Solution | Correct Solution and Explanation |
|---|---|
| The slope is $$m = \frac{3 - 1}{-2 - 7}$$ $$m = \frac{2}{-9}$$ $$m = -\frac{2}{9}$$ | The values were substituted incorrectly. The numerator is the change in $y$ and the denominator is the change in $x$. So, the slope between these points is $$m = \frac{y_2 - y_1}{x_2 - x_1}$$ $$m = \frac{7 - (-2)}{1 - (3)}$$ $$m = \frac{9}{-2} = -\frac{9}{2}$$ |

**4b.** Find the slope of $5x + y = 10$.

| Incorrect Solution | Correct Solution and Explanation |
|---|---|
| The coefficient of $x$ is 5, so the slope $m = 5$. | The equation is not in slope-intercept form, so we cannot conclude the slope is 5. We must first solve the equation for $y$. $$5x + y = 10$$ $$5x + y - 5x = 10 - 5x$$ $$y = -5x + 10$$ The coefficient of $x$ is $-5$. So, the slope $m = -5$. |

**4c.** Graph the line $y = \dfrac{1}{2}x - 1$.

| Incorrect Solution | Correct Solution and Explanation |
|---|---|
| Begin at $(0, -1)$ and then move right 1 and up 2 since the slope is $\dfrac{1}{2}$. | The slope was used incorrectly to obtain the graph. Since the slope of $y = \dfrac{1}{2}x - 1$ is $\dfrac{1}{2}$, we move up 1 unit as we move right 2 units. |
| 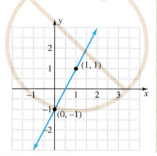 | 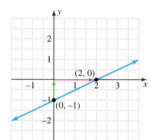 |

## ANSWERS TO STUDENT CHECKS

**Student Check 1**   **a.** $m = 0$   **b.** $m = -\dfrac{3}{5}$

**c.** $m = 4$   **d.** undefined   **e.** $m = \dfrac{2}{3}$

**Student Check 2**   **a.** $m = 7$ ($y$ increases by 7 units as $x$ increases by 1 unit); $(0, 8)$   **b.** $m = \dfrac{1}{5}$ ($y$ increases by 1 unit as $x$ increases by 5 units); $(0, 0)$   **c.** $y = 9x - 5$; $m = 9$ ($y$ increases by 9 units as $x$ increases by 1 unit); $(0, -5)$   **d.** $y = -\dfrac{1}{2}x + 5$; $m = -\dfrac{1}{2}$ ($y$ decreases by 1 unit as $x$ increases by 2 units); $(0, 5)$   **e.** can't be written in slope-intercept form, slope is undefined

**f.** $y = 0x - 2$; $m = 0$ ($y$ doesn't change as $x$ increases by 1 unit); $(0, -2)$

**Student Check 3**

**a.**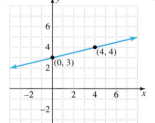

**b.**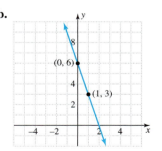

## SUMMARY OF KEY CONCEPTS

1. The slope of a line is a measure of the steepness of the line. It is defined as the ratio of the change in $y$ to the change in $x$.
2. The slope of a line can be found using its equation, using a table of solutions, using its graph, or using the slope formula.
   - To find the slope given its equation, the equation must be in slope-intercept form $y = mx + b$. The coefficient of $x$ is the slope.

- From a table of solutions, we can find the change in the $y$-values and the change in the $x$-values between the points. The ratio of these changes is the slope.
- From a graph, we can find the slope by determining the vertical change and the horizontal change between two points on the graph. The ratio of the vertical change to the horizontal change is the slope.

- If given two ordered pairs, we can find the slope by substituting the $x$ and $y$ coordinates of the two points into the formula, $m = \dfrac{y_2 - y_1}{x_2 - x_1}$, and simplifying.
3. The slope tells us the direction of the line.
   - A positive slope indicates the line goes up from left to right.

- A negative slope indicates the line goes down from left to right.
- A zero slope indicates the line is horizontal.
- An undefined slope indicates the line is vertical.
4. The graph of a line can be obtained using the slope and $y$-intercept. Plot the $y$-intercept and use the slope to move to another point on the line.

## GRAPHING CALCULATOR SKILLS

The graphing calculator can be used to verify our graphs as was illustrated in the previous section. We can also verify the slope of a line by examining the table of solutions and the graph of an equation.

**Example:** Graph $y = -\dfrac{2}{3}x + 2$ using its slope and $y$-intercept.

From the graph, we can verify that the slope is $-\dfrac{2}{3}$ since the movement between points is down 2 and right 3.

Examining the table also confirms the slope of the line. As the $y$-values decrease by 0.66667, the $x$-values increase by 1 unit or as the $y$-values decrease by 2 units, the $x$-values increase by 3 units.

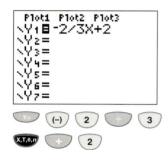

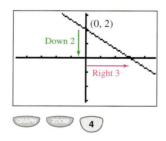

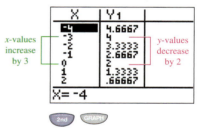

## SECTION 3.3 / EXERCISE SET

 ### Write About It!

**Use complete sentences to explain the meaning of each process.**

1. Finding the slope of line between two points   Answers vary.
2. Finding the slope of line from a linear equation   Answers vary.
3. Finding the slope of a horizontal line   Answers vary.
4. Finding the slope of a vertical line   Answers vary.

**Determine if each statement is true or false. If a statement is false, provide an explanation.**

5. The slope of the line $2x + 5y = 10$ is 2.
6. If the slope of a line is $-\dfrac{4}{3}$, we would move down 4 units and left 3 units to get to another point on the line.
7. The slope of the line $x = 5$ is 0.
   False, $x = 5$ is a vertical line and its slope is undefined.
8. The slope of the line $y = 4$ is undefined.
   False, $y = 4$ is a horizontal line and its slope is zero.
9. The slope of the line through $(3, 5)$ and $(4, -1)$ is $-\dfrac{1}{6}$.

Additional answers can be found in the Instructor Answer Appendix.

10. The slope of the line through $(-2, -3)$ and $(2, -1)$ is undefined.   False, the slope is $\dfrac{-1 - (-3)}{2 - (-2)} = \dfrac{2}{4}$.

 ### Practice Makes Perfect!

**Use the slope formula to determine the slope of the line between each pair of points. (*See Objective 1.*)**

11. $(-2, 3)$ and $(4, 5)$   $m = \dfrac{1}{3}$
12. $(1, -7)$ and $(-3, 7)$   $m = -\dfrac{7}{2}$
13. $(-6, -2)$ and $(0, 4)$   $m = 1$
14. $(5, 0)$ and $(-4, 2)$   $m = -\dfrac{2}{9}$
15. $(-4, 3)$ and $(-4, 5)$   $m = $ undefined
16. $(3, 7)$ and $(3, -2)$   $m = $ undefined
17. $(1, -8)$ and $(4, -8)$   $m = 0$
18. $(-2, 6)$ and $(5, 6)$   $m = 0$
19. $\left(-\dfrac{1}{6}, \dfrac{1}{4}\right)$ and $\left(\dfrac{1}{3}, -\dfrac{5}{4}\right)$   $m = -3$
20. $\left(-\dfrac{1}{3}, -\dfrac{5}{2}\right)$ and $\left(\dfrac{2}{3}, \dfrac{3}{2}\right)$   $m = 4$
21. $\left(\dfrac{7}{3}, \dfrac{1}{2}\right)$ and $\left(-\dfrac{2}{3}, -\dfrac{3}{2}\right)$   $m = \dfrac{2}{3}$
22. $\left(\dfrac{1}{2}, -\dfrac{2}{5}\right)$ and $\left(-\dfrac{1}{4}, \dfrac{4}{5}\right)$   $m = -\dfrac{8}{5}$

**23.** 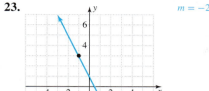    $m = -2$

**24.** 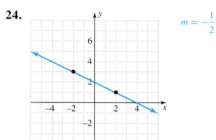    $m = -\dfrac{1}{2}$

**25.** 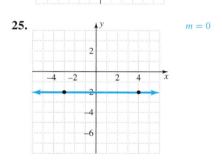    $m = 0$

**26.** 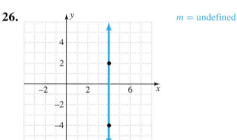    $m = $ undefined

**27.** 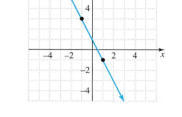    $m = \dfrac{1}{3}$

**28.**     $m = -4$

Write each equation in slope-intercept form, if necessary. State the slope and *y*-intercept of the line from its equation. Write the *y*-intercept as an ordered pair. Explain how the *x*- and *y*-values change with respect to one another. (*See Objective 2.*)

**29.** $y = 4x - 3$    **30.** $y = 3x + 5$    **31.** $y = -2x - 1$

**32.** $y = -x - 4$    **33.** $y = \dfrac{1}{2}x + \dfrac{3}{2}$    **34.** $y = -\dfrac{7}{3}x + \dfrac{1}{3}$

**35.** $y = 5x$    **36.** $y = -3x$    **37.** $4x + y = 8$

**38.** $8x + y = -24$    **39.** $7x + 4y = 0$    **40.** $9x + 2y = 0$

**41.** $x = -2$    **42.** $y = 7$    **43.** $y = -3$

**44.** $x = 1$    **45.** $6y - 8 = 0$    **46.** $3x + 5 = 0$

**47.** $2x - 1 = 0$    **48.** $7y + 3 = 0$

Graph each line using its slope and *y*-intercept. Label two points on the graph. (*See Objective 3.*)

**49.** $y = x - 2$    **50.** $y = x + 3$    **51.** $y = -x + 4$

**52.** $y = \dfrac{2}{3}x - 2$    **53.** $y = -\dfrac{3}{5}x + 4$    **54.** $y = -\dfrac{1}{3}x + 3$

**55.** $2x - 4y = -8$  **56.** $4x + 5y = 20$    **57.** $x - 2y = 0$

**58.** $2x + y = 0$

Use the given slope of a line and the point on the line to find one additional point on the line. Graph the line using the two points. (*See Objective 3.*)

**59.** $m = -4$ and $(3, 4)$    **60.** $m = -2$ and $(-4, -1)$

**61.** $m = \dfrac{3}{2}$ and $(0, 2)$    **62.** $m = \dfrac{4}{5}$ and $(0, -3)$

**63.** $m = -\dfrac{6}{7}$ and $(0, 7)$    **64.** $m = -\dfrac{3}{8}$ and $(0, -8)$

**65.** $m = 0$ and $(2, -3)$    **66.** $m = 0$ and $(-2, 5)$

**67.** $m$ is undefined and $(-1, 4)$

**68.** $m$ is undefined and $(-5, -2)$

### Mix 'Em Up!

Find the slope of each line given its equation, two points on the line, its graph, or a table of solutions.

**69.** $y = \dfrac{1}{2}x + 6$    $m = \dfrac{1}{2}$

**70.** $y = -\dfrac{1}{3}x - \dfrac{1}{4}$    $m = -\dfrac{1}{3}$    **71.** $(0, -3)$ and $(3, 5)$    $m = \dfrac{8}{3}$

**72.** $(-1, 5)$ and $(1, -5)$    **73.** $y = 14$    $m = 0$
$\quad\quad m = -5$
**74.** $x = -10$    $m = $ undefined    **75.** $(11, -4)$ and $(2, 2)$

**76.** $(-5, 1)$ and $(1, -14)$    **77.** $\left(\dfrac{1}{3}, 3\right)$ and $\left(2, \dfrac{1}{2}\right)$

**78.** $\left(\dfrac{3}{2}, -\dfrac{1}{4}\right)$ and $\left(\dfrac{1}{2}, -\dfrac{3}{4}\right)$    **79.** $3x - 5y = 4$    $m = \dfrac{3}{5}$

**80.** $8x + 3y = 7$    $m = -\dfrac{8}{3}$    **81.** $3x + 1 = 0$
$\quad\quad\quad\quad\quad\quad\quad\quad\quad\quad m = $ undefined
**82.** $4y + 3 = 0$    $m = 0$    **83.** $0.2x + 0.5y = 3.2$
$\quad\quad\quad\quad\quad\quad\quad\quad\quad\quad\quad\quad m = -0.4$
**84.** $0.45x - 0.3y = 4.8$    **85.** $(-3.5, 12.6)$ and $(1.5, 5.6)$
$\quad\quad m = 1.5$    $\quad\quad\quad\quad\quad\quad\quad m = -1.4$
**86.** $(-7.5, -0.6)$ and $(2.5, -11.6)$
$\quad\quad m = -1.1$

**87.**

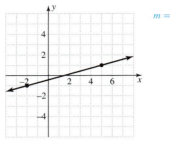

$m = \dfrac{2}{7}$

**88.**

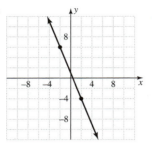

$m = -\dfrac{5}{2}$

**89.**

| X | Y₁ |
|---|---|
| -3 | 7 |
| -2 | 5.3333 |
| -1 | 3.6667 |
| 0 | 2 |
| 1 | .33333 |
| 2 | -1.333 |
| 3 | -3 |

X=3

$m = -\dfrac{5}{3}$

**90.**

| X | Y₁ |
|---|---|
| -3 | -2.75 |
| -2 | -3.5 |
| -1 | -4.25 |
| 0 | -5 |
| 1 | -5.75 |
| 2 | -6.5 |
| 3 | -7.25 |

X=3

$m = -\dfrac{3}{4}$

**Graph each described line. Label two points on the line.**

**91.** $y = 4x - 3$       **92.** $y = -2x + 5$

**93.** $m = 1$, passes through $(0, 3)$

**94.** $m = -5$, passes through $(2, -1)$

**95.** $m = 3$, $b = 6$       **96.** $m = \dfrac{1}{2}$, $b = 0$

**97.** $m = $ undefined, passes through $(3, 4)$

**98.** $m = 0$, passes through $(3, 4)$

**99.** $m = -\dfrac{1}{4}$, passes through $(0, 0)$

**100.** $m = \dfrac{3}{2}$, passes through $(0, 0)$

## You Be the Teacher!

**Each student is trying to solve the same problem: If a line that passes through the point $(-1, 1)$ has a slope of $-\dfrac{3}{4}$, specify one other point on the line. Correct each student's errors, if any.**

**101.** Lisbelle: I can rewrite the slope as $\dfrac{3}{-4}$. I also know slope = rise/run. This means that if I rise, or go up, 3 units, then I have to go left 4 units to get another point on the line. So, starting at the point $(-1, 1)$, I end up at the point $(-1, 4)$ if I rise 3 units. When I go left 4 units, I end up at the point $(-5, 4)$. So, another point on the line is $(-5, 4)$.   correct

**102.** Florida: I can write the slope as $\dfrac{3}{-4}$. I know slope = rise/run. This means that if I rise, or go right 3 units, then I have to go down 4 units to get another point on the line. So, starting at the point $(-1, 1)$, I end up at the point $(2, 1)$ when I go right 3 units. When I go down 4 units, I end up at $(2, -3)$. So, another point on the line is $(2, -3)$.   incorrect; To rise by 3 means to go up 3 units and to run $-4$ means to go left 4 units.

**103.** Lewis: I can rewrite the slope as $-\dfrac{3}{4}$. So if I start at the point $(-1, 1)$, I must go down 3 units and right 4 units to get another point on the line. This means that $(3, -2)$ is another point on the line.   correct

**104.** James: I can rewrite the slope as $-\dfrac{3}{4}$. So, if I start at the point $(-1, 1)$, I must go left 3 units and up 4 units to get another point on the line. This means that $(-4, 5)$ is another point on the line.

## Calculate It!

**Determine the slope of the line depicted in each graphing calculator screen shot.**

**105.**

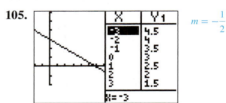

$m = -\dfrac{1}{2}$

**106.**

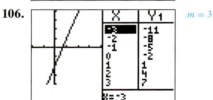

$m = 3$

## Think About It!

**Draw a graph of a line with the following characteristics, where $m$ is the slope of the line and $b$ is the $y$-value of the $y$-intercept of the line.**

**107.** $m > 0$ and $b < 0$       **108.** $m > 0$ and $b = 0$

**109.** $m > 0$ and $b > 0$       **110.** $m < 0$ and $b < 0$

**111.** $m < 0$ and $b = 0$       **112.** $m < 0$ and $b > 0$

**113.** Slope is undefined.       **114.** $m = 0$ and $b > 0$

**Determine algebraically if the ordered pair is a solution of the equation.** (*Section 3.1, Objective 1*)

**1.** $(6, 2)$; $y = |x - 4|$    yes    **2.** $(7, -4)$; $8x - 2y = 62$    no

**Identify the quadrant or axis where each point is located.** (*Section 3.1, Objective 2*)

**3.** $(-8.9, -1)$    III

**4.** $\left(\frac{7}{8}, -\frac{5}{6}\right)$    IV

**5.** $(-5, -1)$    III

**6.** $(9, 4)$    I

**Use the given graph of an equation to determine if each ordered pair is a solution of the equation.** (*Section 3.1, Objective 4*)

**7.** $(0, 3)$    no

**8.** $(2, -3)$    yes

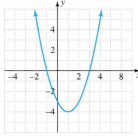

**Graph each equation.** (*Section 3.2, Objectives 2–4*)

**9.** $y = 4x + 2$    **10.** $y = -5x + 5$    **11.** $4x - y = 0$

**12.** $y = -7x$    **13.** $3x + 8 = 0$    **14.** $2y - 11 = 0$

**Use the slope formula to determine the slope of the line between each pair of points.** (*Section 3.3, Objective 1*)

**15.** $(-2, 3)$ and $(4, 7)$    $m = \frac{2}{3}$    **16.** $(-2, 8)$ and $(5, 8)$    $m = 0$

**Write each equation in slope-intercept form, if necessary. State the slope and $y$-intercept of the line from its equation. Write the $y$-intercept as an ordered pair. Explain how the $x$- and $y$-values change with respect to one another.** (*Section 3.3, Objective 2*)

**17.** $3x - 2y = 6$    $y = \frac{3}{2}x - 3$; $m = \frac{3}{2}$; $(0, -3)$; As the $x$-values increase by 2 units, the $y$-values increase by 3 units.

**18.** $5x + 3y = -15$    $y = -\frac{5}{3}x - 5$; $m = -\frac{5}{3}$; $(0, -5)$; As the $x$-values increase by 3 units, the $y$-values decrease by 5 units.

**Graph each line using its slope and $y$-intercept. Label two points on the graph.** (*Section 3.3, Objective 3*)

**19.** $y = \frac{1}{4}x - 1$    **20.** $x - 7y = 14$

---

**SECTION 3.4**    # More About Slope

The depreciated value of a truck is given by $y = -4000x + 32,000$, where $x$ is the age of the truck. How does the truck's value change each year? What is the initial value of the truck?

In this section, we will learn how to interpret the meaning of the slope and $y$-intercept in the context of applications. We will also examine the relationships between parallel and perpendicular lines.

## Interpret the Meaning of the Slope and $y$-Intercept

When a linear equation in two variables represents a real-world situation, the slope and $y$-intercept have practical meanings that relate to the situation. In the equation $y = mx + b$, we know that

- The point $(0, b)$ is the $y$-intercept. Therefore, the value $b$ is the *beginning* or **initial value**.
- The slope $m$ represents a **rate of change**. The slope tells us how the $y$-values change with respect to a change in $x$. When dealing with lines, the slope is constant. So, lines are used to describe quantities that change at a constant rate.

**Objective 1** ▶

Interpret the meaning of the slope and $y$-intercept in real-world applications.

**Objective 1 Examples**    **Interpret the meaning of the slope and $y$-intercept in the context of each situation. Use the slope and $y$-intercept to create a table of three ordered pairs that satisfy the given equation.**

**1a.** Jose's salary as an engineer can be given by $y = 3500x + 45,000$, where $x$ is the number of years he has worked for the company.

**1b.** A truck's value is given by $y = -4000x + 32,000$, where $x$ is the age of the truck in years.

**1c.** The life expectancy for men in the United States can be approximated by $y = 0.2x + 66$, where $x$ is the years after 1960. (Source: http://www.cdc.gov/nchs/data/nvsr/nvsr56/nvsr56_09.pdf)

**Solutions**

**1a.** The $x$-value represents years worked and $y$ represents salary in dollars. In $y = 3500x + 45,000$, the slope is $m = 3500 = \dfrac{3500}{1} = \dfrac{\text{change in salary}}{\text{change in years}}$ and the $y$-intercept is $(0, 45,000)$. So, Jose's initial salary is \$45,000 and his salary increases by \$3500 per year.

From this information, we can obtain the following table of solutions.

| $x$ | $y$ |
|---|---|
| 0 | 45,000 |
| $0 + 1 = 1$ | $45,000 + 3500 = 48,500$ |
| $1 + 1 = 2$ | $48,500 + 3500 = 52,000$ |

**1b.** The $x$-value represents the age of the truck in years and $y$ represents the value of the truck. In $y = -4000x + 32,000$, the slope is $m = -4000 = \dfrac{-4000}{1} = \dfrac{\text{change in value}}{\text{change in years}}$ and the $y$-intercept is $(0, 32,000)$. So, the initial value of the truck is \$32,000 and its value decreases by \$4000 per year.

From this information, we can obtain the following table of solutions.

| $x$ | $y$ |
|---|---|
| 0 | 32,000 |
| $0 + 1 = 1$ | $32,000 - 4000 = 28,000$ |
| $1 + 1 = 2$ | $28,000 - 4000 = 24,000$ |

**1c.** The $x$-value represents the years after 1960 and $y$ represents life expectancy in years. In $y = 0.2x + 66$, the slope is $m = 0.2 = \dfrac{2}{10} = \dfrac{1}{5} = \dfrac{\text{change in life expectancy}}{\text{change in years after 1960}}$ and the $y$-intercept is $(0, 66)$.

So, the life expectancy in 1960 was 66. Life expectancy increases 1 yr for every 5 yr after 1960. We can also say that life expectancy increases by 0.2 yr (2.4 months) for every year after 1960.

From this information, we can obtain the following table of solutions.

| $x$ | $y$ |
|---|---|
| 0 | 66 |
| $0 + 5 = 5$ | $66 + 1 = 67$ |
| $5 + 5 = 10$ | $67 + 1 = 68$ |

**✓ Student Check 1**

Interpret the slope and $y$-intercept in the context of each situation. Use the slope and $y$-intercept to create a table of three ordered pairs that satisfy the given equation.

**a.** The median annual income for women with a bachelor's degree is given by the equation $y = 1100x + 27,645$, where $x$ is the number of years after 1990.

**b.** A car's value is given by $y = -2500x + 15,000$, where $x$ is the age of the car.

**c.** The daily cost to rent a car is $y = 0.50x + 40$, where $x$ is the number of miles driven.

**Objective 2** ▶

Determine the slope in real-world applications.

## Slope in Real-Life Settings

Slope has many meaningful applications in real life. Slope is used to construct stairs, wheelchair ramps, highways, and roofs. It is also used to express the difficulty level of ski slopes. In these physical situations, the slope is often described in terms of rise and run.

For other applications, slope is considered a rate. For instance, it can be used to represent the growth in population, a tax rate, and other rates of change. In these situations, we must state the slope as the change in the $y$-values over the change in the $x$-values.

**Objective 2 Examples** | **Determine the slope described in each situation.**

**2a.** The guidelines for constructing a wheelchair ramp state that the maximum slope should be 1 in. vertical distance for every 12 in. of horizontal distance. Express the maximum slope of a wheelchair ramp.

**2b.** The *grade* of a highway is the measure of the incline or steepness of the road. The grade is generally expressed as a percentage. The steepest road in the United States is Canton Avenue in Pennsylvania. The road rises 37 ft of vertical distance for each 100 ft of horizontal distance. What is the grade of the road? (Source: http://www.geographylists.com/list17y.html)

**2c.** The population of a town increases by 3000 people every year.

**Solutions** | **2a.** The slope of the wheelchair ramp is

$$m = \frac{\text{vertical change}}{\text{horizontal change}} = \frac{\text{rise}}{\text{run}} = \frac{1}{12}$$

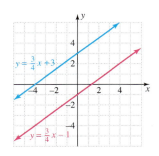

1 in.
12 in.

**2b.** The road rises 37 ft vertically for each 100 ft of horizontal change. So, the slope is

$$m = \frac{\text{vertical change}}{\text{horizontal change}} = \frac{\text{rise}}{\text{run}} = \frac{37}{100} = 37\%$$

**2c.** In real-life situations, time is generally used for the variable $x$. So, in this situation, population represents the variable $y$. Therefore, the slope is

$$m = \frac{\text{change in } y}{\text{change in } x} = \frac{3000}{1} = 3000$$

✓ **Student Check 2** | Determine the slope described in each situation.

**a.** The *pitch* of a roof is defined as the ratio of the vertical increase of the roof to its horizontal increase. If a roof rises 8 in. for each 12 in. it extends horizontally, what is the pitch?

**b.** Ski Dubai is the first indoor ski resort in the Middle East. It has five ski runs with differing difficulty levels. The longest run is 400 m with a fall of approximately 60 m. What is the slope of this run?

**c.** The value of a car decreases by $5000 each year.

## Parallel and Perpendicular Lines

**Objective 3** ▶

Determine if two lines are parallel or perpendicular.

We will now turn our discussion to the comparison of two lines. Two lines are *parallel* if they never intersect. For lines to be parallel, they must have the same slope but different $y$-intercepts. Consider the graphs of the equations $y = \frac{3}{4}x - 1$ and $y = \frac{3}{4}x + 3$. These lines have the same slope, $\frac{3}{4}$. The lines have different $y$-intercepts, so the lines are parallel.

Note that if lines have the same slope and the same $y$-intercept, the lines are the same.

> ### Definition: Parallel Lines
>
> Two nonvertical lines $y_1 = m_1 x + b_1$ and $y_2 = m_2 x + b_2$ are parallel if they have the same slope (that is, $m_1 = m_2$) and different $y$-intercepts.
>
> Vertical lines of the form $x = a_1$ and $x = a_2$ are parallel if $a_1 \neq a_2$.

Two nonvertical lines are *perpendicular* to each other if the lines form a right angle (90°) at their point of intersection. For this to happen, the graph of one line must be increasing (have positive slope) and the other line must be decreasing (have negative slope). Also, the rise of one line must be the run of the other line and vice versa. Mathematically, this means that the slopes of the lines are negative reciprocals of each other.

Consider the graphs of the equations $y = \dfrac{3}{4}x + 3$ and $y = -\dfrac{4}{3}x + 1$.

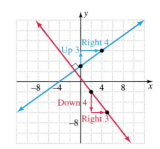

One line rises 3 units vertically and runs 4 units horizontally, so its slope is $m = \dfrac{3}{4}$.

The other line falls 4 units vertically and runs 3 units horizontally, so its slope is $m = -\dfrac{4}{3}$.

These lines have slopes that are negative reciprocals and are perpendicular to one another.

> ### Definition: Perpendicular Lines
>
> Two nonvertical lines $y_1 = m_1 x + b_1$ and $y_2 = m_2 x + b_2$ are perpendicular if their slopes are negative reciprocals—that is, if
>
> $$m_1 = -\frac{1}{m_2}$$
>
> Any vertical line of the form $x = a$ is perpendicular to any horizontal line $y = b$.

Perpendicular lines may or may not have the same $y$-intercept.

> **Note:** *The negative reciprocal means that the signs must be different and the fractions must be reciprocals of each other.*

> ### Procedure: Determining if Two Lines Are Parallel or Perpendicular
>
> **Step 1:** Write each equation in slope-intercept form.
> **Step 2:** Find the slope of each line.
> **Step 3:** Compare the slopes of the lines.
> > **a.** If the slopes are the same and the $y$-intercepts are different, then the lines are parallel.
> > **b.** If the slopes are negative reciprocals of one another, then the lines are perpendicular.

**Objective 3 Examples** / **Determine if the lines are parallel, perpendicular, or neither.**

**3a.** $y = x + 4$ and $y = -x - 3$          **3b.** $y = 2x + 3$ and $y = \dfrac{1}{2}x - 4$

**3c.** $2x + y = 4$ and $4x + 2y = 6$          **3d.** $4x - y = 8$ and $x + 4y = -4$

**Solutions**   **3a.** Both equations are in slope-intercept form.

The slope of $y = x + 4$ is $m = 1$.

The slope of $y = -x + 3$ is $m = -1$.

The slopes are not the same, so the lines are not parallel. The slopes, however, are negative reciprocals of one another since $1 = -\left(\dfrac{1}{-1}\right)$. So, the lines are perpendicular.

**3b.** Both equations are in slope-intercept form.

The slope of $y = 2x + 3$ is $m = 2$.

The slope of $y = \dfrac{1}{2}x - 4$ is $m = \dfrac{1}{2}$.

The slopes are not the same, so the lines are not parallel. The slopes are reciprocals but not opposites. Therefore, the lines are not perpendicular to one another. So, the lines are neither parallel nor perpendicular.

**3c.** We must first write each equation in slope-intercept form.

$$2x + y = 4 \qquad\qquad 4x + 2y = 6$$
$$2x + y - 2x = 4 - 2x \qquad\qquad 4x + 2y - 4x = 6 - 4x$$
$$y = -2x + 4 \qquad\qquad 2y = -4x + 6$$
$$m = -2 \qquad\qquad \frac{2y}{2} = \frac{-4x + 6}{2}$$
$$y = -2x + 3$$
$$m = -2$$

The slope of each line is $m = -2$. The lines have different $y$-intercepts. Therefore, the lines are parallel.

**3d.** We must first write each equation in slope-intercept form.

$$4x - y = 8 \qquad\qquad x + 4y = -4$$
$$4x - y - 4x = 8 - 4x \qquad\qquad x + 4y - x = -4 - x$$
$$-y = -4x + 8 \qquad\qquad 4y = -x - 4$$
$$-1(-y) = -1(-4x + 8) \qquad\qquad \frac{4y}{4} = \frac{-x - 4}{4}$$
$$y = 4x - 8 \qquad\qquad y = -\frac{1}{4}x - 1$$
$$m = 4 \qquad\qquad m = -\frac{1}{4}$$

The slopes of the lines are negative reciprocals of one another, so the lines are perpendicular.

**✓ Student Check 3**   Determine if the lines are parallel, perpendicular, or neither.

**a.** $y = -5x - 1$ and $y = \dfrac{1}{5}x + 2$          **b.** $y = 3x - 4$ and $y = -3x + 5$

**c.** $3x + 2y = 6$ and $9x + 6y = 10$          **d.** $2x + 3y = 6$ and $3x - 2y = 12$

## Graph Parallel and Perpendicular Lines

**Objective 4** ▶

Graph parallel or perpendicular lines if given the equation of one line and a point on the line parallel or perpendicular to it.

In Section 3.5, we will determine equations of lines that go through particular points that are parallel or perpendicular to a given line. It is helpful to visualize this situation before writing their equations.

> **Procedure: Graphing a Line through a Point That Is Parallel or Perpendicular to a Given Line**
>
> **Step 1:** Determine the slope of the given line.
> **Step 2:** Determine the slope of the parallel or perpendicular line.
>     **a.** The slope of the parallel line is equal to the slope of the given line.
>     **b.** The slope of the perpendicular line is equal to the negative reciprocal of the slope of the given line.
> **Step 3:** Graph the given line using its slope and $y$-intercept.
> **Step 4:** Plot the given point and use the slope in step 2 to locate another point to the line. Draw the line between these two points.

**Objective 4 Examples** | **Graph the line that satisfies each condition.**

**4a.** Graph the line $3x + y = -3$ and the line that is parallel to it that passes through the point $(1, 5)$.

**4b.** Graph the line $y = \dfrac{2}{3}x + 2$ and the line perpendicular to it that passes through $(-2, 1)$.

**Solutions** | **4a.** The slope of $3x + y = -3$ is found by writing the equation in slope-intercept form.

$$3x + y = -3$$
$$3x + y - 3x = -3 - 3x$$
$$y = -3x - 3$$

The slope of the given line is $m = -3$. So, the slope of a line parallel to it is also $m = -3$ since parallel lines have the same slope.

Graph the given equation, $y = -3x - 3$, by plotting the $y$-intercept $(0, -3)$ and then moving down 3 units vertically and right 1 unit horizontally to $(1, -6)$.

Graph the line parallel to $y = -3x - 3$ and through $(1, 5)$, by plotting $(1, 5)$. From this point, move down 3 units vertically and right 1 unit horizontally to $(2, 2)$.

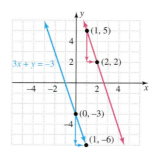

**4b.** The line $y = \dfrac{2}{3}x + 2$ has slope $m = \dfrac{2}{3}$. The line perpendicular to it has slope $m = -\dfrac{3}{2}$ since perpendicular lines have slopes that are negative reciprocals.

Graph the line, $y = \dfrac{2}{3}x + 2$, by plotting the $y$-intercept $(0, 2)$ and then rising 2 units vertically and moving right 3 units horizontally to get to $(3, 4)$.

The line perpendicular to the given line has slope of $-\dfrac{3}{2}$. So, we graph this line by plotting the point $(-2, 1)$ and then move down 3 units vertically and right 2 units horizontally to get to $(0, -2)$.

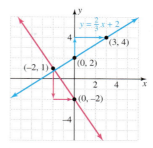

✓ **Student Check 4**  Graph the line that satisfies each condition.

**a.** Graph the line $2x - y = -4$ and the line that is parallel to it that passes through $(2, -1)$.

**b.** Graph the line $y = 5x - 5$ and the line that is perpendicular to it that passes through $(5, 1)$.

---

**Objective 5** ▶

Troubleshoot common errors.

## Troubleshooting Common Errors

A common error associated with slope and parallel and perpendicular lines is shown next.

**Objective 5 Examples** | **A problem and an incorrect solution are given. Provide the correct solution and an explanation of the error.**

Are the lines $y = 4x + 6$ and $4x + y = 0$ parallel, perpendicular, or neither?

| **Incorrect Solution** | **Correct Solution and Explanation** |
|---|---|
| The coefficient of each equation is 4, so the slopes are the same and the lines are parallel. | The second equation is not in slope-intercept form. We must rewrite $4x + y = 0$ in this form to identify its slope. $$4x + y = 0$$ $$y = -4x$$ The slope of $y = 4x + 6$ is $m = 4$. The slope of $y = -4x$ is $m = -4$. The slopes are not the same, so the lines are *not* parallel. The slopes are not negative reciprocals of one another either, so the lines are *not* perpendicular. |

---

## ANSWERS TO STUDENT CHECKS

**Student Check 1**  **a.** The median annual income for women in 1990 was \$27,645 and it increased by \$1100 per year after 1990. Three points are $(0, 27{,}645)$, $(1, 28{,}745)$, and $(2, 29{,}845)$.  **b.** The initial value of the car was \$15,000 and its value decreased by \$2500 each year. Three points are $(0, 15{,}000)$, $(1, 12{,}500)$ and $(2, 10{,}000)$  **c.** The cost of renting the car is \$40 plus \$0.50 per mile driven. Three points are $(0, 40)$, $(1, 40.50)$, and $(2, 41)$.

**Student Check 2**  **a.** $\dfrac{8}{12} = \dfrac{2}{3}$  **b.** $-\dfrac{60}{400} = -\dfrac{3}{20}$  **c.** $-\dfrac{5000}{1} = -5000$

**Student Check 3**  **a.** perpendicular  **b.** neither  **c.** parallel  **d.** perpendicular

**Student Check 4    a.**

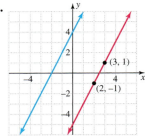    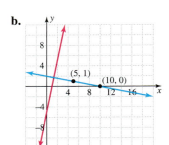

## SUMMARY OF KEY CONCEPTS

1. The slope and $y$-intercept have meaning when the linear equation represents a real-life situation. The $y$-intercept represents the initial value and the slope is a rate of change indicating how the $y$-values change as the $x$-values change.

2. Slope has many physical applications, such as the grade of a road, wheelchair ramps, and roofs. Slopes are also used to denote rates of change, such as a population rate.

3. Lines that have the same slope and different $y$-intercepts are parallel to one another. Lines that have slopes that are negative reciprocals are perpendicular to one another.

## GRAPHING CALCULATOR SKILLS

The calculator can help us determine if lines are parallel or perpendicular. It is often helpful to graph the lines in the ZDecimal or ZSquare format so that the graphs are not distorted. Because of the possible distortion, it is best to compare the line's slopes to determine how they relate to each other.

**Example:** Are $y = 2x - 3$ and $y = \frac{1}{2}x + 1$ parallel, perpendicular, or neither?

    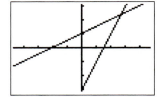

By graphing the lines on the calculator, we can verify that the lines are neither parallel nor perpendicular.

**Example:** Are $y = 2x - 3$ and $y = -\frac{1}{2}x + 1$ parallel, perpendicular, or neither?

    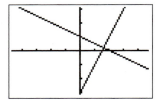

The graphs show that the lines are perpendicular to one another.

## SECTION 3.4    EXERCISE SET

### Write About It!

Interpret the meaning of the slope and $y$-intercept in the context of each situation. Use the slope and $y$-intercept to create a table of three ordered pairs that satisfy the given equation. (*See Objective 1.*)

1. The expenditure per student in fall enrollment for public elementary and secondary education in the United States can be approximated by the equation $y = 374.15x + 7672.73$, where $x$ is the number of years after 2001. (Source: http://nces.ed.gov/programs/projections/projections2018/tables/table_34.asp)

2. The total student enrollment (in thousands) in all degree-granting institutions (2-yr and 4-yr colleges and universities) can be approximated by the equation $y = 351x + 16{,}121$, where $x$ is the number of years after 2001. (Source: http://nces.ed.gov/programs/projections/projections2018/tables/table_10.asp)

3. The percentage of high school students who regularly smoke cigarettes is on the decline. The percentage of students who are smokers is given by the equation $y = -2.5x + 38$, where $x$ is the number of years after 1997. (Source: National Cancer Institute)

4. The percent of U.S. adults aged 20 yr and over who were obese can be modeled by the linear equation $y = 0.71x + 19.83$, where $x$ is the number of years after 1997. (Source: National Center for Health Statistics)

5. The owner of a limousine rental company purchases a stretch limousine for $110,000. For tax purposes, the owner uses straight-line depreciation for reporting the value of the limousine. The depreciated value of the limousine is given by $y = -22,000x + 110,000$, where $x$ is the age of the limousine in years.

6. The number of U.S. households with cable television is approximated by $y = 2.3x + 17.3$ (in millions), where $x$ is the number of years after 1977. (Source: *The World Almanac and Book of Facts 2006*)

7. The percentage of eighth-graders who reported using alcohol is given by $y = -0.7x + 27$, where $x$ is the number of years after 1993. (Source: *The World Almanac and Book of Facts 2006*)

8. The average cost, in dollars, of higher education at 2-yr institutions can be approximated by the equation $y = 334x + 5252$, where $x$ is the number of years after 1999. (Source: www.infoplease.com)

 ## Practice Makes Perfect!

**Determine the slope described in each situation. (See Objective 2.)**

9. Traversing the Cooper River in Charleston, South Carolina, the 13,200-ft Cooper River Bridge is North America's longest cable-stayed bridge.

The average grade of the bridge is 4.1%. The steepest part of the bridge occurs on a section that is three-tenths of a mile long (1584 ft) with a vertical rise of 88.7 ft. What is the grade of the bridge on this section? (Source: www .cooperriverbridge.org, http://www.cooperriverbridge.org) $\frac{887}{15840}$; 5.6%

10. Filbert Street in San Francisco, California, is one of the steepest streets in the city and in America. The street rises 63 ft for a 200-ft run. What is the grade of the street? Express the grade as a fraction and a percentage. (Source: http://wikitravel.org/en/San_Francisco) $\frac{63}{200}$, 31.5%

11. A roof rises vertically 12 in. for each 12 in. it extends horizontally. What is the pitch of the roof? 1

12. What is the pitch of a roof that rises 6 in. for each 12 in. it extends horizontally? $\frac{1}{2}$

13. The American National Standards Institute facilitates the development of national standards for the steepness of wheelchair ramps. The national standard for constructing wheelchair ramps specifies that a slope of $\frac{1}{12}$ is preferable, but a slope of $\frac{1}{8}$ is acceptable when the former is not possible. If a builder constructs a wheelchair ramp with a rise of 3 in. for every run of 24 in., does the slope of the ramp meet the guidelines? $\frac{1}{8}$; yes, meets the guidelines.

14. If a door is 15 in. above the ground, how long must the wheelchair ramp be for its slope to be $\frac{1}{12}$? Express answer in inches and feet. 180 in., 15 ft

**Determine if the two lines are parallel, perpendicular, or neither. (See Objective 3.)**

15. $y = \frac{4}{3}x - 5$ and $y = \frac{4}{3}x + 5$  parallel

16. $y = -6x + 1$ and $y = -6x$  parallel

17. $y = 9x + 3$ and $y = -9x - 7$  neither

18. $y = 7x + 2$ and $7x - y = 14$  parallel

19. $x + 3y = -6$ and $y = 3x + 4$  perpendicular

20. $2x - y = 8$ and $4x - 2y = -4$  parallel

21. $x + 5y = 10$ and $3x - 15y = 0$  neither

22. $y = 3$ and $y = 3x$  neither

23. $x = 2$ and $5x + 2 = 0$  parallel

24. $y - 5 = 0$ and $y = -2$  parallel

25. $y = 2x - 4$ and $y = -2x + 3$  neither

26. $y = -2x - 4$ and $y = \frac{1}{2}x + 5$  perpendicular

27. $y = 5x - 4$ and $y = -\frac{1}{5}x$  perpendicular

28. $3x + y = 6$ and $y = 3x - 4$  neither

29. $3x + y = 6$ and $y = -\frac{1}{3}x - 6$  neither

30. $4x + 3y = 6$ and $y = \frac{3}{4}x - 5$  perpendicular

31. $x + 7y = 21$ and $7x - y = -14$  perpendicular

32. $2x + 5y = -10$ and $5x + 2y = 20$  neither

33. $x = 5$ and $y = 2$  perpendicular

34. $y = -4$ and $y = \frac{1}{4}x$  neither

**Graph the given line and the line that is parallel to it through the given point. (See Objective 4.)**

35. $y = 2x - 4$; $(-2, 2)$     36. $y = -3x + 3$; $(0, -2)$

37. $y = \frac{1}{5}x - 1$; $(5, 5)$     38. $y = -\frac{2}{3}x + 4$; $(0, 0)$

39. $x + y = 4$; $(0, -2)$     40. $x - y = 3$; $(-3, -1)$

41. $x = 2$; $(-1, 4)$     42. $x = -3$; $(4, 2)$

43. $y = -4$; $(0, 3)$     44. $y = 1$; $(-4, -2)$

**Graph the given line and the line that is perpendicular to it through the given point. (See Objective 4.)**

45. $y = 2x - 4$; $(-2, 2)$     46. $y = -3x + 3$; $(0, -2)$

47. $y = \frac{1}{5}x - 1$; $(5, 5)$     48. $y = -\frac{2}{3}x + 4$; $(0, 0)$

49. $x + y = 4$; $(0, -2)$     50. $x - y = 3$; $(-3, -1)$

51. $x = 2$; $(-1, 4)$     52. $x = -3$; $(4, 2)$

53. $y = -4$; $(0, 3)$     54. $y = 1$; $(-4, -2)$

 ## Mix 'Em Up!

**Solve each problem.**

55. What is the pitch of a roof that rises vertically 12 in. for each 5 in. it extends horizontally? $\frac{12}{5}$

**56.** What is the pitch of a roof that rises vertically 10 in. for each 15 in. it extends horizontally? $\frac{2}{3}$

**57.** Determine if the lines are parallel, perpendicular, or neither: $y = -\frac{1}{7}x + 14$ and $y = 7x + 7$.   perpendicular

**58.** Determine if the lines are parallel, perpendicular, or neither: $y = \frac{5}{6}x - 3$ and $y = \frac{6}{5}x + \frac{1}{3}$.   neither

**59.** Graph the line that is parallel to $y = 2x + 1$ and passes through the point $(-3, 2)$.

**60.** Graph the line that is parallel to $y = \frac{2}{3}x - 4$ and passes through the point $(3, 1)$.

**61.** Graph the line that is perpendicular to $y = -5x + 1$ and passes through the point $(1, 1)$.

**62.** Graph the line that is perpendicular to $y = -6x + 5$ and passes through the point $(-1, 4)$.

**63.** Determine if the lines are parallel, perpendicular, or neither: $y = 3$ and $y = -\frac{1}{3}$.   parallel

**64.** Determine if the lines are parallel, perpendicular, or neither: $x = \frac{2}{5}$ and $x = -\frac{5}{2}$.   parallel

**Interpret the meaning of the slope and *y*-intercept in the context of each situation.**

**65.** Suzy's car is in the shop, so she is using a rental car. The total cost, in dollars, for Suzy to rent a compact car is given by $y = 149x$, where $x$ is the number of weeks the car is rented.   The *y*-intercept is (0, 0), which means the initial cost is $0. The total cost increases at a rate of $149 per week.

**66.** Mindy has a wedding cake business. Mindy's total weekly cost, in dollars, for making wedding cakes is given by $y = 20x + 50$, where $x$ is the number of cakes Mindy makes each week.

**67.** The average cost, in dollars, of higher education at 4-yr institutions can be approximated by the equation $y = 892x + 12{,}034$, where $x$ is the number of years after 1999. (Source: www.infoplease.com)

**68.** The median annual income, in dollars, for men who complete high school can be modeled by the equation $y = 627.5x + 30{,}818$, where $x$ is the number of years after 1996. (Source: www.infoplease.com)

**69.** The median annual income, in dollars, for women who complete high school can be modeled by the equation $y = 641x + 21{,}506$, where $x$ is the number of years after 1996. (Source: www.infoplease.com)

**70.** The percent of America's high school seniors who abuse alcohol can be modeled by the equation $y = -1.13x + 77.1$, where $x$ is the number of years after 2003.

 **You Be the Teacher!**

**Correct each student's errors, if any.**

**71.** Explain the steps to graph the line that is parallel to $6x - 2y = 4$ and passes through the point $(1, 2)$.

Sandee's work:

The slope of the line from the given equation is 6. From the point $(1, 2)$, go up 6 units and right 1 unit.

**72.** Graph the line that is perpendicular to $-3x + y = 4$ and passes through the point $(0, 5)$.

Aisha's work:

$$-3x + y = 4$$
$$\underline{+3x \quad + 3x}$$

$y = 3x + 4$. So, the slope of the perpendicular line is $-\frac{1}{3}$. The *y*-intercept of this line is $(0, 4)$. So, to get another point on the graph, I must go down 1 unit and right 3 units. So another point on this line is $(3, 3)$.

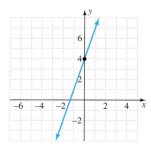

**73.** Determine if the lines are parallel, perpendicular, or neither: $y = 2x$ and $y = -\frac{1}{2}$.

Warren's work: perpendicular

**74.** Determine if the lines are parallel, perpendicular, or neither: $x = -2$ and $y = -2$.

Crystal's work: parallel   They are perpendicular to each other since $x = -2$ is vertical and $y = -2$ is horizontal.

 **Calculate It!**

**Graph each pair of equations in the ZSquare format on a graphing calculator to determine if the lines are parallel, perpendicular, or neither.**

**75.** $y = 5x - 3$ and $y = \frac{1}{5}x + 3$   neither

**76.** $y = -\frac{4}{3}x + 5$ and $y = \frac{3}{4}x + 1$   perpendicular

**77.** $y = \frac{1}{6}x + 2$ and $y = -6x - 4$   perpendicular

**78.** $y = 2x + 6$ and $y = 2x + 3$   parallel

**79.** $0.4x + 1.2y = 3.5$ and $1.8x - 0.6y = 4.9$   perpendicular

**80.** $5.6x - 3.5y = 8.9$ and $2.4x - 1.5y = 2.7$   parallel

 **Think About It!**

**Write equations of lines whose graphs are parallel and perpendicular to the graph of the given equation.**

**81.** $y = 5x - 3$

**82.** $y = -\frac{2}{3}x$

**83.** $4x + 3y = 12$

**84.** $7x - 2y = 4$

| SECTION 3.5 | **Writing Equations of Lines** |

▶ **OBJECTIVES**

As a result of completing this section, you will be able to

1. Write the equation of a line given its slope and *y*-intercept.
2. Write the equation of a line given a point and a slope.
3. Write the equation of a line given two points.
4. Write the equation of a line given a point and a relationship to another line.
5. Solve application problems.
6. Troubleshoot common errors.

The average price of a movie ticket in the United States in 1996 was \$4.42. The average price of a movie ticket in the United States in 2010 was \$7.89. Write a linear equation that models the average price of a movie ticket. To answer this question, we need to know how to use two points to write the equation of a line. (Source: http://www.natoonline.org/statisticstickets.htm)

In Sections 3.1 to 3.4, we were given the equation of a line and we had to find certain information about it. In this section, we will be given specific information about a line and will write the equation of the line that satisfies the given conditions. The answers to these types of problems will be linear equations of the form $y = mx + b$ or $Ax + By = C$.

## Use the Slope and *y*-Intercept to Write an Equation of a Line

The first case we consider when writing equations of lines is when we are given a line's slope and *y*-intercept. Recall that the slope-intercept form of a line is $y = mx + b$, where *m* is the slope and *b* is the *y*-intercept.

So, if we know the slope and *y*-intercept, we will substitute these values into the slope-intercept form to obtain the equation.

---

**Objective 1** ▶

Write the equation of a line given its slope and *y*-intercept.

**INSTRUCTOR NOTE:**
Review the information needed to graph a line:

1. two points
2. a point and a slope
3. a point and a parallel or perpendicular line

Help students see that the common thread among these criteria is a point and a slope. So, as long as we know a point and a slope, we can graph a line. This is the same information needed to write the equation. The only piece of information they may need to work to find is the slope.

---

**Procedure:  Writing the Equation of a Line Given Its Slope and *y*-Intercept**

The line's slope is the value of *m* and the *y*-coordinate of the *y*-intercept is the value *b*.

So, the equation is                $y = \underline{\quad} x + \underline{\quad}.$

---

| **Objective 1  Examples** | **Write the equation of the line that satisfies the given information.** |

**1a.** $m = 4$ and the *y*-intercept is $(0, 5)$    **1b.** $m = -\dfrac{1}{2}$ and the *y*-intercept is $(0, -4)$

**1c.** $m = 0$ and the *y*-intercept is $(0, 1)$

**1d.**

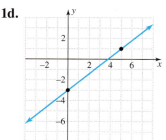

**Solutions**    **1a.** The slope is $m = 4$ and the value of $b = 5$. So, the equation is

$$y = 4x + 5$$

**1b.** The slope is $m = -\dfrac{1}{2}$ and the value of $b = -4$. So, the equation is

$$y = -\frac{1}{2}x - 4$$

**1c.** The slope is $m = 0$ and the value of $b = 1$. So, the equation is

$$y = 0x + 1 \quad \text{or} \quad y = 1$$

**1d.** The graph crosses the $y$-axis at $(0, -3)$, so $b = -3$.

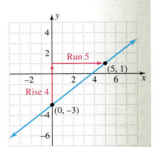

To determine the slope, we calculate the *rise* and *run* between the two points. The slope is $m = \dfrac{\text{rise}}{\text{run}} = \dfrac{4}{5}$.

So, the equation is $y = \dfrac{4}{5}x - 3$.

---

✔ **Student Check 1**    Write the equation of the line that satisfies the given information.

**a.** $m = -2$ and the $y$-intercept is $(0, 9)$    **b.** $m = \dfrac{7}{3}$ and the $y$-intercept is $(0, -1)$

**c.** $m = 0$ and the $y$-intercept is $\left(0, \dfrac{2}{3}\right)$

**d.**

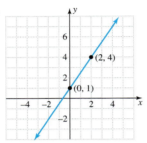

---

## Use a Point and a Slope to Write an Equation of a Line

**Objective 2** ▶

Write the equation of a line given a point and a slope.

When writing the equation of a line when the given point is not the $y$-intercept, we have to perform some work to determine the $y$-intercept and the value of $b$. There are two methods that we can use to determine the value of $b$.

- One method uses the slope-intercept form of a line $y = mx + b$.
- The other method uses another form of a line called the *point-slope form*.

The point-slope form is derived from the slope formula using $(x_1, y_1)$ as the given point and $(x, y)$ as any other point on the line.

$$m = \frac{y - y_1}{x - x_1}$$

Multiply both sides of the slope formula by the LCD $= x - x_1$ to get the point-slope form of a line.

$$y - y_1 = m(x - x_1)$$

> **Definition:** The **point-slope form** of a line is $y - y_1 = m(x - x_1)$, where $(x_1, y_1)$ is a point on the line and $m$ is the slope of the line.

To write the equation of a line given a point and a slope, we will use one of the two methods described next.

**Procedure: Writing the Equation of a Line Given a Point and a Slope**

| **Method 1:** Use the slope-intercept form of a line | **Method 2:** Use the point-slope form of a line |
|---|---|
| **Step 1:** In the equation $y = mx + b$, substitute the given point $(x, y)$ for the $x$-value and the $y$-value in the equation. Substitute the slope for $m$. | **Step 1:** In the equation $y - y_1 = m(x - x_1)$, substitute the given point $(x_1, y_1)$ and the slope for $m$. |
| **Step 2:** Solve the resulting equation for $b$. | **Step 2:** Simplify both sides of the equation. |
| **Step 3:** Write the equation in the form $y = mx + b$ by substituting the given slope and the value of $b$. | **Step 3:** Write the equation in either standard form or slope-intercept form. |
| **Step 4:** Check the equation. | **Step 4:** Check the equation. |

**Objective 2 Examples**

Write the equation of the line that has the given slope and passes through the given point. Write the answer in both slope-intercept form and standard form.

**2a.** $m = 3; (5, -2)$

**2b.** $m = -\dfrac{3}{4}; (1, 5)$

**2c.** $m = 0; (-2, 1)$

**2d.** undefined slope; $(-3, 9)$

**Solutions** **2a.**

| **Method 1** | **Method 2** |
|---|---|
| $m = 3$ and $(x, y) = (5, -2)$ | $m = 3$ and $(x_1, y_1) = (5, -2)$ |
| $y = mx + b$ | $y - y_1 = m(x - x_1)$ |
| $-2 = 3(5) + b$ | $y - (-2) = 3(x - 5)$ |
| $-2 = 15 + b$ | $y + 2 = 3x - 15$ |
| $-2 - 15 = 15 + b - 15$ | $y + 2 - 2 = 3x - 15 - 2$ |
| $-17 = b$ | $y = 3x - 17$ |
| Equation in slope-intercept form: | Equation in slope-intercept form: |
| $y = 3x - 17$ | $y = 3x - 17$ |

Equation in standard form:

| | |
|---|---|
| $y = 3x - 17$ | Begin with the slope-intercept form. |
| $y - 3x = 3x - 17 - 3x$ | Subtract 3x from each side. |
| $-3x + y = -17$ | Simplify. |
| $-1(-3x + y) = -1(-17)$ | Multiply each side by −1. |
| $3x - y = 17$ | Simplify. |

**Check:** Show that $(5, -2)$ is a solution of $y = 3x - 17$ or $3x - y = 17$.

| | |
|---|---|
| $y = 3x - 17$ | Begin with the slope-intercept form. |
| $-2 = 3(5) - 17$ | Let $x = 5$ and $y = -2$. |
| $-2 = 15 - 17$ | Simplify. |
| $-2 = -2$ | Simplify. |

Since the resulting equation is true, our work is correct.

**2b.**

| **Method 1** | **Method 2** |
|---|---|

$m = -\dfrac{3}{4}$ and $(x, y) = (1, 5)$ | $m = -\dfrac{3}{4}$ and $(x_1, y_1) = (1, 5)$

**Method 1**

$m = -\dfrac{3}{4}$ and $(x, y) = (1, 5)$

$y = mx + b$

$5 = -\dfrac{3}{4}(1) + b$

$5 = -\dfrac{3}{4} + b$

$4(5) = 4\left(-\dfrac{3}{4} + b\right)$

$20 = -3 + 4b$

$20 + 3 = -3 + 4b + 3$

$23 = 4b$

$\dfrac{23}{4} = b$

Equation in slope-intercept form:

$y = -\dfrac{3}{4}x + \dfrac{23}{4}$

**Method 2**

$m = -\dfrac{3}{4}$ and $(x_1, y_1) = (1, 5)$

$y - y_1 = m(x - x_1)$

$y - 5 = -\dfrac{3}{4}(x - 1)$

$y - 5 = -\dfrac{3}{4}x + \dfrac{3}{4}$

$4(y - 5) = 4\left(-\dfrac{3}{4}x + \dfrac{3}{4}\right)$

$4y - 20 = -3x + 3$

$4y - 20 + 20 = -3x + 3 + 20$

$4y = -3x + 23$

$y = -\dfrac{3}{4}x + \dfrac{23}{4}$

Equation in slope-intercept form:

$y = -\dfrac{3}{4}x + \dfrac{23}{4}$

**INSTRUCTOR NOTE:**
Point out that students can also use the point-slope form to write the equation in standard form.

Equation in standard form:

$y = -\dfrac{3}{4}x + \dfrac{23}{4}$    Begin with the slope-intercept form.

$4(y) = 4\left(-\dfrac{3}{4}x + \dfrac{23}{4}\right)$    Multiply each side by 4.

$4y = -3x + 23$    Simplify.

$4y + 3x = -3x + 23 + 3x$    Add $3x$ to each side.

$3x + 4y = 23$    Simplify.

**Check:** Show that $(1, 5)$ is a solution of $y = -\dfrac{3}{4}x + \dfrac{23}{4}$ or $3x + 4y = 23$.

$3x + 4y = 23$    Begin with the standard form.

$3(1) + 4(5) = 23$    Let $x = 1$ and $x = 5$.

$3 + 20 = 23$    Simplify.

$23 = 23$    Simplify.

Since the resulting equation is true, our work is correct.

**2c.**

| **Method 1** | **Method 2** |
|---|---|

**Method 1**

$m = 0$ and $(x, y) = (-2, 1)$

$y = mx + b$

$1 = 0(-2) + b$

$1 = 0 + b$

$1 = b$

Equation: $y = 0x + 1$ or $y = 1$

**Method 2**

$m = 0$ and $(x_1, y_1) = (-2, 1)$

$y - y_1 = m(x - x_1)$

$y - 1 = 0[x - (-2)]$

$y - 1 = 0$

$y - 1 + 1 = 0 + 1$

$y = 1$

Equation: $y = 1$

Recall that a line with a slope of zero is horizontal and that horizontal lines are of the form $y = k$, where $k$ is the $y$-value of a point on the line.

**2d.** A line with undefined slope is a vertical line. A vertical line is written as $x = h$, where $h$ is the $x$-value of a point on the line. Since the given point is $(-3, 9)$, $h = -3$. So, the equation of the line is $x = -3$.

✓ **Student Check 2**   Write the equation of the line that has the given slope and passes through the given point. Write the answer in both slope-intercept form and standard form.

**a.** $m = -5; (3, 1)$

**b.** $m = -\dfrac{2}{7}; (4, 5)$

**c.** $m = 0; (-6, 9)$

**d.** undefined slope; $(8, 1)$

**Note**: Using method 1 or 2 produces the same linear equation. When using the slope-intercept form, we must substitute the values of $m$ and $b$ into $y = mx + b$ to obtain the equation. When using the point-slope form, the equation is obtained through the process.

## Use Two Points to Write the Equation of a Line

**Objective 3** ▶

Write the equation of a line given two points.

In Objectives 1 and 2, we were given the slope and a point on the line. If we know only two points on the line, we can still apply the two methods discussed earlier to write the equation of the line through these points. The only difference is that now we will have to calculate the slope of the line using the slope formula. Once the slope is known, we can use method 1 or method 2 to write the equation of the line.

---

**Procedure:  Writing the Equation of a Line Given Two Points**

**Step 1:** Find the slope of the line using the slope formula, $m = \dfrac{y_2 - y_1}{x_2 - x_1}$.

**Step 2:** Use the slope and one of the given points to find the equation of the line using either the slope-intercept form of a line or the point-slope form of a line.

---

**Objective 3  Examples**   **Write the equation of the line, in slope-intercept form, that passes through the given points.**

**3a.** $(4, -3)$ and $(-2, 9)$     **3b.** $(-4, 0)$ and $(4, 12)$     **3c.** $(-2, 4)$ and $(-2, 11)$

**Solutions**   **3a.** Find the slope of the line.

$$m = \frac{y_2 - y_1}{x_2 - x_1}$$   State the slope formula.

$$m = \frac{9 - (-3)}{-2 - 4}$$   Let $(x_1, y_1) = (4, -3)$ and $(x_2, y_2) = (-2, 9)$.

$$m = \frac{12}{-6}$$   Simplify numerator and denominator.

$$m = -2$$   Simplify result.

Use the point-slope form of a line to write the equation. Either of the given points can be used in the point-slope form as shown.

| $m = -2, (x_1, y_1) = (4, -3)$ | $m = -2, (x_1, y_1) = (-2, 9)$ |
|---|---|
| $y - y_1 = m(x - x_1)$ | $y - y_1 = m(x - x_1)$ |
| $y - (-3) = -2(x - 4)$ | $y - (9) = -2[x - (-2)]$ |
| $y + 3 = -2x + 8$ | $y - 9 = -2(x + 2)$ |
| $y + 3 - 3 = -2x + 8 - 3$ | $y - 9 = -2x - 4$ |
| $y = -2x + 5$ | $y - 9 + 9 = -2x - 4 + 9$ |
| | $y = -2x + 5$ |

**Check:** Show that $(4, -3)$ and $(-2, 9)$ are solutions of the equation $y = -2x + 5$.

| $(x, y)$ | $y = -2x + 5$ | Solution? |
|----------|---------------|-----------|
| $(4, -3)$ | $-3 = -2(4) + 5$ <br> $-3 = -8 + 5$ <br> $-3 = -3$ | Yes |
| $(-2, 9)$ | $9 = -2(-2) + 5$ <br> $9 = 4 + 5$ <br> $9 = 9$ | Yes |

**3b.** Find the slope of the line.

$$m = \frac{y_2 - y_1}{x_2 - x_1}$$    Begin with the slope formula.

$$m = \frac{12 - 0}{4 - (-4)}$$    Let $(x_1, y_1) = (-4, 0)$ and $(x_2, y_2) = (4, 12)$.

$$m = \frac{12}{8}$$    Simplify the numerator and denominator.

$$m = \frac{3}{2}$$    Simplify the result.

Use the slope-intercept form of a line to write the equation. Either of the given points can be used in the slope-intercept form to find $b$.

$m = \frac{3}{2}; (x, y) = (-4, 0)$

$y = mx + b$

$0 = \frac{3}{2}(-4) + b$

$0 = -6 + b$

$0 + 6 = -6 + b + 6$

$6 = b$

Equation: $y = \frac{3}{2}x + 6$

$m = \frac{3}{2}; (x, y) = (4, 12)$

$y = mx + b$

$12 = \frac{3}{2}(4) + b$

$12 = 6 + b$

$12 - 6 = 6 + b - 6$

$6 = b$

Equation: $y = \frac{3}{2}x + 6$

**Check:** Show that $(-4, 0)$ and $(4, 12)$ are solutions of the equation $y = \frac{3}{2}x + 6$.

| $(x, y)$ | $y = \frac{3}{2}x + 6$ | Solution? |
|----------|------------------------|-----------|
| $(-4, 0)$ | $0 = \frac{3}{2}(-4) + 6$ <br> $0 = -6 + 6$ <br> $0 = 0$ | Yes |
| $(4, 12)$ | $12 = \frac{3}{2}(4) + 6$ <br> $12 = 6 + 6$ <br> $12 = 12$ | Yes |

**3c.** Find the slope of the line.

$$m = \frac{y_2 - y_1}{x_2 - x_1}$$    Begin with the slope formula.

$$m = \frac{11 - 4}{-2 - (-2)}$$    Let $(x_1, y_1) = (-2, 4)$ and $(x_2, y_2) = (-2, 11)$.

$$m = \frac{7}{0}$$    Simplify the numerator and denominator.

$$m = \text{undefined}$$    Simplify the result.

Since the slope is undefined, the line is vertical. Recall that a vertical line is represented by an equation of the form $x = h$, where $h$ is the $x$-coordinate of a point on the line. Since the given points are $(-2, 4)$ and $(-2, 11)$, $h = -2$. So, the equation of the line is $x = -2$.

> ✔ **Student Check 3**    Write the equation of the line, in slope-intercept form, that passes through the given points.
>
> **a.** $(3, -7)$ and $(1, 3)$    **b.** $(-1, 2)$ and $(5, -6)$    **c.** $(4, -1)$ and $(6, -1)$

## Use a Point and a Line to Write the Equation of a Line

**Objective 4** ▶

Write the equation of a line given a point and a relationship to another line.

The last case we will discuss is writing the equation of a line given a point and a relationship to another line. We will be given the equation of a parallel or perpendicular line. Knowing this information enables us to find the slope of the unknown line. If the lines are parallel, the slopes are the same. If the lines are perpendicular, the slopes are negative reciprocals. Once the slope is determined, the methods to find the equation of a line, as previously stated, apply.

> **Procedure: Writing the Equation of a Line Given a Point and a Relationship to Another Line**
>
> **Step 1:** Determine the slope of the given line by writing the equation in slope-intercept form.
> **Step 2:** Determine the slope of the unknown line.
>   **a.** The slope will be the same as the given line if the lines are parallel.
>   **b.** The slope will be the negative reciprocal of the given line if the lines are perpendicular.
> **Step 3:** Use the slope from step 2 and the given point to write the equation of the unknown line using either method 1 or method 2.

**Objective 4 Examples**    **Write the equation of the line that passes through the given point and is either parallel or perpendicular to the given line.**

**4a.** $(1, -5)$ and parallel to $4x - y = 8$; Write answer in slope-intercept form.
**4b.** $(-2, 4)$ and perpendicular to $y = 5x - 6$; Write answer in standard form.
**4c.** $(-2, 3)$ and parallel to $x = 5$
**4d.** $(4, -7)$ and perpendicular to $x = 5$

**Solutions**    **4a.** We first find the slope of $4x - y = 8$ by writing it in slope-intercept form.

$$4x - y = 8$$
$$4x - y - 4x = 8 - 4x \qquad \text{Subtract } 4x \text{ from each side.}$$
$$-y = -4x + 8 \qquad \text{Simplify.}$$
$$-1(-y) = -1(-4x + 8) \qquad \text{Multiply each side by } -1.$$
$$y = 4x - 8 \qquad \text{Simplify.}$$

The slope of the line $4x - y = 8$ is $m = 4$. Since the lines are parallel, the slope of the unknown line is also $m = 4$.

Now we find the equation of the line with slope $m = 4$ that passes through $(1, -5)$.

$$y - y_1 = m(x - x_1) \qquad \text{State the point-slope form.}$$
$$y - (-5) = 4(x - 1) \qquad \text{Let } m = 4, (x_1, y_1) = (1, -5).$$
$$y + 5 = 4x - 4 \qquad \text{Simplify and distribute.}$$
$$y + 5 - 5 = 4x - 4 - 5 \qquad \text{Subtract 5 from each side.}$$
$$y = 4x - 9 \qquad \text{Simplify.}$$

So, the equation of the line through $(1, -5)$ and parallel to $4x - y = 8$ is $y = 4x - 9$.

**4b.** We first find the slope of $y = 5x - 6$. Since the equation is in slope-intercept form, the slope of the line is $m = 5$. The lines are perpendicular, so the slope of the unknown line is the negative reciprocal of 5. Hence, the slope of the unknown line is $m = -\dfrac{1}{5}$.

Now we find the equation of the line with $m = -\dfrac{1}{5}$ that passes through $(-2, 4)$.

$$y - y_1 = m(x - x_1)$$

$$y - (4) = -\frac{1}{5}[x - (-2)] \qquad \text{Let } m = -\frac{1}{5}, (x_1, y_1) = (-2, 4).$$

$$y - 4 = -\frac{1}{5}(x + 2) \qquad \text{Simplify.}$$

$$y - 4 = -\frac{1}{5}x - \frac{2}{5} \qquad \text{Apply the distributive property.}$$

$$5(y - 4) = 5\left(-\frac{1}{5}x - \frac{2}{5}\right) \qquad \text{Multiply each side by 5.}$$

$$5y - 20 = -x - 2 \qquad \text{Apply the distributive property.}$$

$$5y - 20 + x = -x - 2 + x \qquad \text{Add } x \text{ to each side.}$$

$$x + 5y - 20 = -2 \qquad \text{Simplify.}$$

$$x + 5y - 20 + 20 = -2 + 20 \qquad \text{Add 20 to each side.}$$

$$x + 5y = 18 \qquad \text{Simplify.}$$

So, the equation of the line through $(-2, 4)$ and perpendicular to $y = 5x - 6$ is $x + 5y = 18$.

**4c.** We first find the slope of $x = 5$. The equation $x = 5$ represents a vertical line and, therefore, has undefined slope. Since the lines are parallel, the slope of the unknown line is also undefined.

The unknown line is a vertical line that passes through the point $(2, -3)$. So, its equation is $x = -2$.

**4d.** From part c, the slope of $x = 5$ is undefined.

Since the lines are perpendicular, the slope of the unknown line is the negative reciprocal of the slope of the given line. Because the slope of the given line is undefined, we can't find its negative reciprocal. We do know, however, that a line perpendicular to a vertical line is a horizontal line with slope $m = 0$.

So, the equation of the horizontal line through $(4, -7)$ is $y = -7$.

✓ **Student Check 4**    Write the equation of the line that goes through $(2, -3)$ that is

a. parallel to $y = \dfrac{1}{2}x + 6$

b. perpendicular to $y = \dfrac{1}{2}x + 6$

c. parallel to $x = -7$

d. perpendicular to $x = -7$

## Applications

**Objective 5** ▶

Solve application problems.

There are two types of application problems that we will examine. One type involves knowing an **initial value** (or beginning value, that is, the $y$-value that corresponds to $x = 0$) and information about how that value changes. The other type involves knowing two data points that describe a particular situation.

**Procedure:  Solving Problems Given an Initial Value and Rate of Change**

**Step 1:** Identify the initial (or beginning) value of the given quantity. This value represents $b$ in the slope-intercept form of a line.
**Step 2:** Identify how the initial value changes. This is the rate of change, or slope $m$.
**Step 3:** Write the equation in slope intercept form $y = mx + b$ by substituting the values of $m$ and $b$ into the equation.

**Procedure:  Solving Problems Given Two Data Points**

**Step 1:** Find the slope of the line between the two points.
**Step 2:** Use one data point and the slope to determine the equation by using either the slope-intercept form or the point-slope form of a line.

**Objective 5 Examples**   **Find the equation that represents each situation. Then use the equation to answer the questions.**

5a. Juanita has just been hired as a public school teacher. Her starting salary is $35,000. She will get a raise of $1500 each year she works at the school.

   **i.** Write a linear equation that represents her salary, where $x$ is the number of years worked.

   **ii.** Use the equation to determine Juanita's salary after she has worked at the school for 10 yr.

   **iii.** Use the equation to determine how long she will have to work to have a salary of $80,000.

**Solution**   5a.   **i.** The starting salary is the initial value, so $b = 35{,}000$. The raise she gets each year is the rate of change or slope.

$$m = \frac{\text{change in } y}{\text{change in } x} = \frac{\text{change in salary}}{\text{change in years}} = \frac{1500}{1} = 1500$$

The equation that models Juanita's salary is $y = 1500x + 35{,}000$, where $x$ is the number of years worked.

   **ii.** Juanita's salary after working 10 years is found by setting $x = 10$.

| | |
|---|---|
| $y = 1500x + 35{,}000$ | Begin with the model. |
| $y = 1500(10) + 35{,}000$ | Replace $x$ with 10. |
| $y = 15{,}000 + 35{,}000$ | Multiply. |
| $y = 50{,}000$ | Add. |

So, Juanita's salary after 10 yr is $50,000.

   **iii.** To determine how long she needs to work to have a salary of $80,000, we set $y = 80{,}000$.

| | |
|---|---|
| $y = 1500x + 35{,}000$ | Begin with the model. |
| $80{,}000 = 1500x + 35{,}000$ | Replace $y$ with 80,000. |
| $80{,}000 - 35{,}000 = 1500x + 35{,}000 - 35{,}000$ | Subtract 35,000 from each side. |
| $45{,}000 = 1500x$ | Simplify. |
| $\dfrac{45{,}000}{1500} = \dfrac{1500x}{1500}$ | Divide each side by 1500. |
| $30 = x$ | Simplify. |

Juanita's salary will be $80,000 after she works 30 yr at the school.

**5b.** The average price of a movie ticket in 1996 was $4.42. The average price of a movie ticket in 2010 was $7.89. (Source: http://www.natoonline.org/statisticstickets.htm)

**i.** Write a linear equation that represents the average price of a movie ticket, where $x$ is the years after 1996.

**ii.** Use the equation to predict the average price of a movie ticket in 2015.

**iii.** In what year will the average price of a movie ticket be $10.67?

**Solution** **5b.** In this problem, $x$ represents the years after 1996 and $y$ represents the average price of a movie ticket. The year 1996 corresponds to the $x$-value of 0 and the year 2010 corresponds to the $x$-value of 14 since $2010 - 1996 = 14$. So, the two ordered pairs given in the problem are $(0, 4.42)$ and $(14, 7.89)$.

**i.** To write the equation, we first find the slope.

| | |
|---|---|
| $m = \dfrac{y_2 - y_1}{x_2 - x_1}$ | Begin with the slope formula. |
| $m = \dfrac{7.89 - 4.42}{14 - 0}$ | Let $(x_1, y_1) = (0, 4.42)$ and $(x_2, y_2) = (14, 7.89)$. |
| $m = \dfrac{3.47}{14}$ | Simplify the numerator and denominator. |
| $m \approx 0.25$ | Simplify the result. |

One of the given points is $(0, 4.42)$. This is the $y$-intercept; therefore, $b = 4.42$. Knowing the slope $m = 0.25$ and $b = 4.42$, the equation that models the average price of a movie ticket is $y = 0.25x + 4.42$.

**ii.** The average price in 2015 is found by setting $x = 19$.

| | |
|---|---|
| $y = 0.25x + 4.42$ | Begin with the model. |
| $y = 0.25(19) + 4.42$ | Replace $x$ with 19. |
| $y = 4.75 + 4.42$ | Multiply. |
| $y = 9.17$ | Add. |

The average price of a movie ticket in 2015 will be $9.17.

**iii.** To find when the average price is $10.67, we set $y = 10.67$ and solve for $x$.

| | |
|---|---|
| $y = 0.25x + 4.42$ | Begin with the model. |
| $10.67 = 0.25x + 4.42$ | Replace $y$ with 10.67. |
| $10.67 - 4.42 = 0.25x + 4.42 - 4.42$ | Subtract 4.42 from each side. |
| $6.25 = 0.25x$ | Simplify. |
| $\dfrac{6.25}{0.25} = \dfrac{0.25x}{0.25}$ | Divide each side by 0.25. |
| $25 = x$ | Simplify. |

The average price of a movie ticket will be $10.67 in 25 years or in 2021.

✓ **Student Check 5** Find the equation that represents each situation. Then use the equation to answer the questions.

**a.** Ryan bought an SUV for $30,000. The value of the SUV decreases by $3000 each year.

**i.** Write a linear equation that represents the value of the SUV, where $x$ is the age in years.

**ii.** Use the equation to find the value of the SUV after 4 yr.

**iii.** When will the value of the SUV be $9000?

**b.** In Fall 2000, there were approximately 15.3 million students enrolled in post-secondary degree granting institutions. In Fall 2009, there were approximately 20.2 million students enrolled in post-secondary degree granting institutions. (Source: National Center for Education Statistics)

   **i.** Assuming that enrollment is growing linearly, write an equation that represents the number of students (in millions) enrolled in post-secondary degree granting institutions $x$ years after 2000.

   **ii.** Use the equation to estimate the enrollment in 2012.

   **iii.** When will the enrollment reach 28.26 million?

---

**Objective 6** ▶

Troubleshoot common errors.

## Troubleshooting Common Errors

Some common errors for writing equations of lines are shown next.

**Objective 6 Examples**   **A problem and an incorrect solution are given. Provide the correct solution and an explanation of the error.**

**6a.** Write the equation of the line through $(3, 0)$ with slope $m = \dfrac{2}{3}$.

| Incorrect Solution | Correct Solution and Explanation |
|---|---|
| Since the point $(3, 0)$ is given, the value of $b = 3$. So, the equation is $$y = \frac{2}{3}x + 3$$ | The point $(3, 0)$ is the $x$-intercept, *not* the $y$-intercept. So, we use the slope-intercept form to find the value of $b$. $$y = mx + b$$ $$0 = \frac{2}{3}(3) + b$$ $$0 = 2 + b$$ $$-2 = b$$ So, the equation is $y = \dfrac{2}{3}x - 2$. |

**6b.** Write the equation of the line through $(2, -1)$ and $(5, -2)$.

| Incorrect Solution | Correct Solution and Explanation |
|---|---|
| The slope is $$m = \frac{5 - 2}{-2 - (-1)} = \frac{3}{-1} = -3.$$ So, the equation is $$y - y_1 = m(x - x_1)$$ $$y - (-2) = -3(x - 5)$$ $$y + 2 = -3x + 15$$ $$y = -3x + 13$$ | The error was made in calculating the slope of the line. The slope should be $$m = \frac{\text{change in } y}{\text{change in } x} = \frac{-2 - (-1)}{5 - 2} = -\frac{1}{3}$$ We can use the point-slope form to find the equation. $$y - y_1 = m(x - x_1)$$ $$y - (-2) = -\frac{1}{3}(x - 5)$$ $$y + 2 = -\frac{1}{3}x + \frac{5}{3}$$ $$y + 2 - 2 = -\frac{1}{3}x + \frac{5}{3} - 2$$ $$y = -\frac{1}{3}x + \frac{5}{3} - \frac{6}{3}$$ $$y = -\frac{1}{3}x - \frac{1}{3}$$ |

**6c.** Write the equation of the line through $(4, -1)$ that is perpendicular to $y = 2x + 4$.

| Incorrect Solution | Correct Solution and Explanation |
|---|---|
| The slope of the given line is $m = 2$. The slope of a perpendicular line is $m = -\dfrac{1}{2}$.<br><br>The value of $b = 4$.<br>So, the equation is<br><br>$$y = -\frac{1}{2}x + 4$$ | The error was made in calculating the value of $b$. The $y$-intercept of the given line is $(0, 4)$, but this is not necessarily the $y$-intercept of the perpendicular line. We must use the slope-intercept form or point-slope form to find the equation.<br><br>Use $m = -\dfrac{1}{2}$ and $(4, -1)$ to get<br><br>$$y = mx + b$$<br>$$-1 = -\frac{1}{2}(4) + b$$<br>$$-1 = -2 + b$$<br>$$1 = b$$<br><br>So, the equation is $y = -\dfrac{1}{2}x + 1$. |

## ANSWERS TO STUDENT CHECKS

**Student Check 1**   **a.** $y = -2x + 9$   **b.** $y = \dfrac{7}{3}x - 1$
   **c.** $y = \dfrac{2}{3}$   **d.** $y = \dfrac{3}{2}x + 1$

**Student Check 2**   **a.** $y = -5x + 16$   **b.** $y = -\dfrac{2}{7}x + \dfrac{43}{7}$
   **c.** $y = 9$   **d.** $x = 8$

**Student Check 3**   **a.** $y = -5x + 8$   **b.** $y = -\dfrac{4}{3}x + \dfrac{2}{3}$
   **c.** $y = -1$   **d.** $x = -8$

**Student Check 4**   **a.** $y = \dfrac{1}{2}x - 4$   **b.** $y = -2x + 1$
   **c.** $x = 2$   **d.** $y = -3$

**Student Check 5**   **a. i.** $y = -3000x + 30{,}000$
   **ii.** \$18,000   **iii.** 7 yr
   **b. i.** $y = 0.54x + 15.3$   **ii.** 21.78 million   **iii.** 2024

## SUMMARY OF KEY CONCEPTS

1. If the slope and $y$-intercept of an equation are known, writing the equation that satisfies this information is immediate. The value of $m$ and $b$ are substituted into the slope-intercept form $y = mx + b$.

2. There are three other situations that provide enough information for an equation of a line to be written. They are:
   • a point and a slope
   • two points
   • a point and a line parallel or perpendicular
   In each of these situations, the slope must be determined. If it is not given, use the slope formula or the relationship to a given line to find it. After the slope is found, use it with one of the points in either the point-slope form or the slope-intercept form to write the equation of the line.

3. If a line is described as vertical or with undefined slope, the equation of the line will be of the form $x = h$, where $h$ is the $x$-coordinate of the given point.

4. If a line is described as horizontal or with zero slope, the equation of the line will be of the form $y = k$, where $k$ is the $y$-coordinate of the given point.

5. In application problems, we will either know the slope (how the values change) and $y$-intercept (initial value) from the problem or we will be given two points that enable us to write the equation.

## GRAPHING CALCULATOR SKILLS

The graphing calculator has the ability to calculate the equation of the line if provided enough information. At this point, it is more beneficial to use the calculator to check our work instead of allowing it to do the work for us.

**Example:** Use the calculator to verify that $y = \dfrac{3}{2}x + 6$ is the equation of the line that goes through the points $(-4, 0)$ and $(4, 12)$.

Enter the equation in the equation editor.

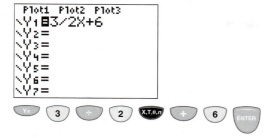

Graph the equation and use the TRACE feature to determine if the points $(-4, 0)$ and $(4, 12)$ lie on the line. Press TRACE, enter the $x$-value of the first point and press enter. Notice that this takes us to the point $(-4, 0)$ on the graph.

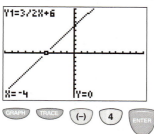

Now enter the $x$-value of the second point. Notice that the display is $x = 4$ and $y = 12$. The point is not shown on the graph since it is outside of the standard viewing window.

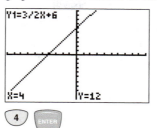

Another way to check to see if the equation contains the points as solutions is to use the TABLE feature. Press [2nd] [GRAPH] and verify that the points $(4, 12)$ and $(-4, 0)$ are in the table.

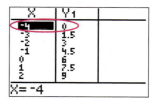

## SECTION 3.5 / EXERCISE SET

### Write About It!

**Use complete sentences in your answer to each question.**

1. How can you determine the equation of a line if you know its slope and $y$-intercept? *Answers vary.*

2. How can you determine the equation of a line if you know two points on the line? *Answers vary.*

3. Which method do you prefer to find the equation of a line—using the slope-intercept form or using the point-slope form? Why? *Answers vary.*

4. Explain how to determine the equation of the line that passes through the point $(-1, 3)$ and is parallel to $y = 4$. *Answers vary.*

5. Explain how to determine the equation of the line that passes through the point $(3, -2)$ and is perpendicular to $2x - 1 = 0$. *Answers vary.*

**Determine if each statement is true or false. If a statement is false, explain why it is false.**

6. The equation of the line with slope 7 that passes through the point $(5, 0)$ is $y = 7x + 5$.
*False, $y = 7x + 5$ is the equation of a line with slope 7 that passes through $(0, 5)$.*

7. The equation of the line that passes through the point $(2, 6)$ and is perpendicular to $y = 3x + 4$ is $y = -\frac{1}{3}x + 6$. *False, $(2, 6)$ is not the $y$-intercept of the line, so we must find $b$. The equation is $y = -\frac{1}{3}x + \frac{20}{3}$.*

8. The equation of the line that passes through the point $(3, -4)$ and is parallel to $y = 3 - 2x$ is $y = 3x - 13$.
*False, the slope of the given line is $-2$ not 3. The equation is $y = -2x + 2$.*

*Additional answers can be found in the Instructor Answer Appendix.*

### Practice Makes Perfect!

**Write the equation of the line that satisfies the given information. (See Objective 1.)**

9. $m = 4$, $b = 5$   $y = 4x + 5$

10. $m = -2$, $b = 1$   $y = -2x + 1$

11. $m = 3$, passes through $(0, -4)$   $y = 3x - 4$

12. $m = \frac{1}{2}$, passes through $(0, 7)$   $y = \frac{1}{2}x + 7$

13. $m = -\frac{2}{3}$ with $y$-intercept $(0, -10)$   $y = -\frac{2}{3}x - 10$

14. $m = \frac{4}{5}$ with $y$-intercept $(0, 20)$   $y = \frac{4}{5}x + 20$

15. $m = 0$, passes through $(0, 7)$   $y = 7$

16. $m = 0$, passes through $(0, -14)$   $y = -14$

17. $m = -\frac{1}{9}$, passes through $\left(0, -\frac{4}{9}\right)$   $y = -\frac{1}{9}x - \frac{4}{9}$

18. $m = \frac{3}{7}$, passes through $\left(0, \frac{2}{7}\right)$   $y = \frac{3}{7}x + \frac{2}{7}$

**Write the equation of the line that has the given slope and passes through the given point. Express each answer in slope-intercept form when possible. (See Objective 2.)**

19. $m = 4$; $(1, 5)$   $y = 4x + 1$

20. $m = 2$; $(4, 3)$   $y = 2x - 5$

21. $m = -5$; $(-2, -1)$   $y = -5x - 11$

22. $m = 6$; $(3, 0)$   $y = 6x - 18$

23. $m = -1$; $(-4, 0)$   $y = -x - 4$

24. $m = \frac{1}{2}$; $(4, 0)$   $y = \frac{1}{2}x - 2$

25. $m = -\frac{5}{2}$; $(-2, 0)$

26. $m = -7$; $(1, 6)$   $y = -7x + 13$

**27.** $m = 8$; $(-3, -9)$
$y = 8x + 15$

**28.** $m = -12$; $(-4, 50)$
$y = -12x + 2$

**29.** $m = 6$; $(0, 0)$ $\quad y = 6x$

**30.** $m = -1$; $(0, 0)$ $\quad y = -x$

**31.** $m = -8$; $(0, 0)$ $\quad y = -8x$

**32.** $m = 4$; $(0, 0)$ $\quad y = 4x$

**33.** $m = 0$; $(2, -5)$ $\quad y = -5$

**34.** $m = 0$; $(2, 8)$ $\quad y = 8$

**35.** $m = \dfrac{2}{3}$; $(6, -8)$ $y = \dfrac{2}{3}x - 12$

**36.** $m = \dfrac{1}{9}$; $(-9, 2)$ $y = \dfrac{1}{9}x + 3$

**37.** $m = -\dfrac{3}{5}$; $(2, -1)$

**38.** $m = -\dfrac{1}{4}$; $(6, -1)$ $y = -\dfrac{1}{4}x + \dfrac{1}{2}$

**39.** $m =$ undefined; $(9, 3)$
$x = 9$

**40.** $m =$ undefined; $(-5, 4)$
$x = -5$

Write the equation of the line that passes through the two points. Express each answer in slope-intercept form when possible. (*See Objective 3.*)

**41.** $(-4, 1)$ and $(2, 7)$
$y = x + 5$

**42.** $(-2, -5)$ and $(2, -1)$
$y = x - 3$

**43.** $(-4, -8)$ and $(2, 4)$
$y = 2x$

**44.** $(-5, 15)$ and $(1, -3)$
$y = -3x$

**45.** $(-1, -6)$ and $(5, 6)$
$y = 2x - 4$

**46.** $(-3, -14)$ and $(2, 11)$
$y = 5x + 1$

**47.** $(4, -4)$ and $(12, 0)$

**48.** $(8, -3)$ and $(-8, -15)$

**49.** $(-9, 8)$ and $(-3, 6)$

**50.** $(-11, 8)$ and $(11, -4)$

**51.** $(-2, 5)$ and $(4, 5)$
$y = 5$

**52.** $(1, -7)$ and $(-4, -7)$
$y = -7$

**53.** $(2, -1)$ and $(2, 6)$ $\quad x = 2$

**54.** $(-3, 5)$ and $(-3, 2)$
$x = -3$

Write the equation of the line that passes through the given point and is either parallel or perpendicular to the given line. Express each answer in slope-intercept form when possible. (*See Objective 4.*)

**55.** $(0, -4)$, parallel to $y = -3x + 6$ $\quad y = -3x - 4$

**56.** $(0, 3)$, parallel to $y = 7x - 3$ $\quad y = 7x + 3$

**57.** $(-7, 8)$, parallel to $2x + y = 4$ $\quad y = -2x - 6$

**58.** $(5, -2)$, parallel to $3x - y = 9$ $\quad y = 3x - 17$

**59.** $(3, -5)$, parallel to $4x + 3y = -12$ $\quad y = -\dfrac{4}{3}x - 1$

**60.** $(-6, -1)$, parallel to $5x - 6y = -30$ $\quad y = \dfrac{5}{6}x + 4$

**61.** $(-1, 4)$, parallel to $y = 3$ $\quad y = 4$

**62.** $(7, -3)$, parallel to $y = -4$ $\quad y = -3$

**63.** $(-1, 4)$, parallel to $x = 3$ $\quad x = -1$

**64.** $(2, -7)$, parallel to $x = -9$ $\quad x = 2$

**65.** $(0, -4)$, perpendicular to $y = -3x + 6$ $\quad y = \dfrac{1}{3}x - 4$

**66.** $(0, 3)$, perpendicular to $y = 7x - 3$ $\quad y = -\dfrac{1}{7}x + 3$

**67.** $(-6, 8)$, perpendicular to $2x + y = 4$ $\quad y = \dfrac{1}{2}x + 11$

**68.** $(3, -5)$, perpendicular to $3x - y = 9$ $\quad y = -\dfrac{1}{3}x - 4$

**69.** $(-8, 1)$, perpendicular to
$4x + 3y = -12$ $\quad y = \dfrac{3}{4}x + 7$

**70.** $(-5, 9)$, perpendicular to
$5x - 6y = -30$ $\quad y = -\dfrac{6}{5}x + 3$

**71.** $(-1, 4)$, perpendicular to $y = 3$ $\quad x = -1$

**72.** $(11, -2)$, perpendicular to $y = -7$ $\quad x = 11$

**73.** $(-4, 8)$, perpendicular to $x = \dfrac{1}{2}$ $\quad y = 8$

**74.** $(1, -6)$, perpendicular to $x = -4$ $\quad y = -6$

 **Mix 'Em Up!**

Write the equation of the line described. Express each answer in slope-intercept form and in standard form.

**75.** $m = -\dfrac{1}{2}$, $(-2, 4)$ $\quad y = -\dfrac{1}{2}x + 3$; $x + 2y = 6$

**76.** $m = -\dfrac{2}{3}$, $(3, -3)$ $\quad y = -\dfrac{2}{3}x - 1$; $2x + 3y = -3$

**77.** $(3.5, -4.7)$ and $(-6.3, -3.3)$ $\quad y = -\dfrac{1}{7}x - 4.2$; $x + 7y = -29.4$

**78.** $(5.3, 0)$ and $(6.2, -1.8)$ $\quad y = -2x + 10.6$; $2x + y = 10.6$

**79.** $m =$ undefined, $(-10, 12)$ $\quad x = -10$

**80.** $m = 0$, $(-15, 20)$ $\quad y = 20$

**81.** $(4, -5)$, parallel to $y = -6x + 3$ $\quad y = -6x + 19$; $6x + y = 19$

**82.** $(-1, 3)$, perpendicular to $y = -\dfrac{1}{10}x + 15$
$\quad y = 10x + 13$; $10x - y = -13$

**83.** $(5, 4)$ and $(5, -6)$ $\quad x = 5$

**84.** $(-1, -3)$ and $(2, -3)$ $\quad y = -3$

**85.** $m = 5$, $y$-intercept: $(0, -5)$ $\quad y = 5x - 5$; $5x - y = 5$

**86.** $m = -6$, $x$-intercept: $(12, 0)$ $\quad y = -6x + 72$; $6x + y = 72$

**87.** $m = 2.5$, $(-4, 7)$ $\quad y = 2.5x + 17$; $5x - 2y = -34$

**88.** $m = -1.6$, $(5, -3)$ $\quad y = -1.6x + 5$; $8x + 5y = 25$

**89.** $x$-intercept $(-4, 0)$ and $y$-intercept $(0, 5)$

**90.** $x$-intercept $(3, 0)$ and $y$-intercept $(0, 2)$ $\quad y = -\dfrac{2}{3}x + 2$; $2x + 3y = 6$

**91.** $(0, 5)$, parallel to $3x - 5y = 4$ $\quad y = \dfrac{3}{5}x + 5$; $3x - 5y = -25$

**92.** $\left(0, \dfrac{3}{5}\right)$, perpendicular to $6x - 3y = 12$

**93.** $(-2, 4)$, perpendicular to $x = -5$ $\quad y = 4$

**94.** $(-4, 0)$, parallel to $x = 10$ $\quad x = -4$

Write a linear equation to model each situation and use the model to answer each question.

**95.** Pedro has a new job as a computer programmer. His starting salary is $45,000. He will receive a raise of $2000 each year he works with the company.

   **a.** Write a linear equation that represents Pedro's salary, where $x$ is the years he has worked with the company. $\quad y = 2000x + 45{,}000$

   **b.** What is Pedro's salary after 5 yr of working for the company? $\quad \$55{,}000$

   **c.** How long does he have to work for the company to earn a salary of $71,000? $\quad$ 13 yr

**96.** Sue has a new job as a pharmaceutical sales representative. Her starting salary is $50,000. She earns a bonus based on her sales. Her bonus is 25% of her total sales for the year.

   **a.** Write a linear equation that represents Sue's yearly income, where $x$ is the total sales for the year.
   $\quad y = 0.25x + 50{,}000$

   **b.** If Sue's sales total $30,000 for the year, what is her income? $\quad \$57{,}500$

   **c.** How much does Sue need to sell to have an income of $65,000? $\quad \$60{,}000$

**97.** Shanika registers for her first semester in college. Her tuition includes fees of $400 plus $150 per credit hour.

  **a.** Write a linear equation that represents Shanika's total tuition, where $x$ is the number of credit hours Shanika will take.  $y = 150x + 400$

  **b.** What is Shanika's tuition if she registers for 12 credit hours?  $2200

  **c.** How many hours does Shanika take if her tuition is $1300?  6 credits

**98.** James joins a fitness club. There is a one-time membership fee of $300 plus a monthly charge of $35.

  **a.** Write a linear equation that represents the total cost of joining the fitness club, where $x$ is the number of months James is a member.  $y = 35x + 300$

  **b.** What is the total cost of joining the fitness club for 1 yr?  $720

  **c.** How many months has James been a member if this total cost is $1560?  3 yr

**99.** Abdul purchases a new car for $20,000. The car's value decreases by $1500 each year.

  **a.** Write a linear equation that represents the value of the car, where $x$ is the age of the car in years.
  $y = -1500x + 20,000$

  **b.** What is the car's value after 6 yr?  $11,000

  **c.** What will be the age of the car when its value is $5000?  10 yr

**100.** One of the fastest growing counties in the United States is Flagler County, Florida. In 2004, its population was approximately 69,000. In 2010, the population was approximately 96,000. (Source: U.S. Census Bureau)

  **a.** Assuming this growth is linear, write an equation that approximates the population of Flagler County where $x$ is the number of years after 2004.
  $y = 4500x + 69,000$

  **b.** What is the estimated population of Flagler County in 2015?  118,500 in 2015

  **c.** When will the population reach 150,000?  in 2022

**101.** The number of single-family, existing homes sold in Florida in January 2010 was approximately 10,700. In January 2011, the number of homes sold was approximately 12,200. (Source: http://media.living.net)

  **a.** Assuming linear growth, write an equation that approximates the number of homes sold in Florida where $x$ is the number of years after 2010.
  $y = 1500x + 10,700$

  **b.** If this trend continues, how many homes will be sold in January 2015?  18,200 homes

**102.** There were approximately 40 million Social Security beneficiaries in 1990 and approximately 61.4 million beneficiaries in 2012. (Source: www.ssa.gov)

  **a.** Write a linear equation that approximates the number of Social Security beneficiaries (in millions), where $x$ is the number of years after 1990.
  $y = 0.97x + 40$

  **b.** If this growth continues, how many beneficiaries will there be in 2015?  64.25 million beneficiaries

**103.** The following graph shows the billions of dollars spent on spectator sports in the United States for the years 1990–2003. (Source: www.infoplease.com)

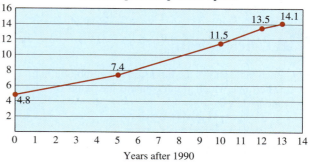

**Billions Spent on Spectator Sports**

  **a.** Use the points (0, 4.8) and (13, 14.1) to write a linear equation that models the money spent on spectator sports (in billions of dollars) $x$ years after 1990.  $y = 0.72x + 4.8$

  **b.** If this growth continues, how much will be spent on spectator sports in 2015?  approximately $22.8 billion

**104.** The median age of women at their first wedding was 20.3 in 1950. In 2010, the median age of women at their first wedding was 26.1. (Source: www.infoplease.com)

  **a.** Assuming linear growth, write an equation that represents the median age of women at their first wedding, where $x$ is the number of years after 1950.
  $y = 0.10x + 20.3$

  **b.** Use the equation to predict the median age of women on the date of their first wedding in 2015.
  26.8

## You Be the Teacher!

**Correct each student's errors, if any.**

**105.** Find the equation of the line that is parallel to $4x - 3y = 6$ and passes through the point (8, 3).

  Vivian's work:

$$4x - 3y = 6$$
$$\underline{-4x \qquad\qquad -4x}$$
$$\frac{-3y}{-3} = \frac{-4x}{-3} + \frac{6}{-3}$$

$y = -\frac{4}{3}x - 2$. So, the slope is $y = -\frac{4}{3}$. Since it passes through the point (8, 3), the equation of the line is
$y = -\frac{4}{3}x + 3.$

**106.** Find the equation of the line that passes through the points $(-3, 4)$ and $\left(0, \frac{1}{2}\right)$.

  William's work: $m = -\frac{7}{6}$ and the equation is
$y = -\frac{7}{6}x + 4$

**107.** Find the equation of the line with slope $-2$ that passes through the point $(5, -1)$.
$y - (-1) = -2(x - 5)$
$y + 1 = -2x + 10$
$y = -2x + 9$

  Desi's work: $y = -2x - 1$

**108.** Suppose $x$ = number of years after 1971. What does $x = 0$ represent? How do I figure out what $x$ is for the year 2008?

Marla's work: $x = 0$ represents 1971. For 2008, $x = 2008 - 1971 = 37$.   correct!

 **Calculate It!**

**109.** Each table shows the points on the graph of a line. Determine which graph contains the points $(0, 2)$ and $(-1, -3)$.

**a.**

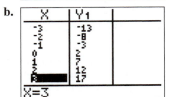

| X | Y1 |
|---|---|
| -3 | 17 |
| -2 | 12 |
| -1 | 7 |
| 0 | 2 |
| 1 | -3 |
| 2 | -8 |
| 3 | -13 |
| X=3 | |

**b.**

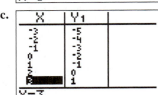

| X | Y1 |
|---|---|
| -3 | -13 |
| -2 | -8 |
| -1 | -3 |
| 0 | 2 |
| 1 | 7 |
| 2 | 12 |
| 3 | 17 |
| X=3 | |

**c.**

| X | Y1 |
|---|---|
| -3 | -5 |
| -2 | -4 |
| -1 | -3 |
| 0 | -2 |
| 1 | -1 |
| 2 | 0 |
| 3 | 1 |
| X=3 | |

**110.** Each table shows the points on the graph of a line. Determine which graph contains the points $(0, -5)$ and $(-2, 4)$.

**a.**

| X | Y1 |
|---|---|
| -4 | 3 |
| -3 | 3.5 |
| -2 | 4 |
| -1 | 4.5 |
| 0 | 5 |
| 1 | 5.5 |
| 2 | 6 |
| X=-4 | |

**b.**

| X | Y1 |
|---|---|
| -3 | -18.5 |
| -2 | -14 |
| -1 | -9.5 |
| 0 | -5 |
| 1 | -.5 |
| 2 | 4 |
| 3 | 8.5 |
| X=3 | |

**c.**

| X | Y1 |
|---|---|
| -4 | 13 |
| -3 | 8.5 |
| -2 | 4 |
| -1 | -.5 |
| 0 | -5 |
| 1 | -9.5 |
| 2 | -14 |
| X=-4 | |

---

**SECTION 3.6** / **Functions**

Suppose Greg registers for classes online. What would happen if Greg input his student ID number and another student's name came up? Besides being annoyed, Greg would have to contact the Registrar's office and inform them that his ID is not functioning correctly; it corresponds to another student.

In this section, we will learn the definition of a function and how functions apply to our everyday lives.

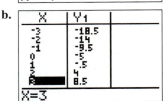

### Relations

If we look up the definition of a relation, we will find that most every description involves the words connection and/or association. This idea extends to the definition of a mathematical *relation*, as well. A relation, in mathematics, describes the connection or association between two sets of information. This description is denoted by a set of ordered pairs.

We use relations every day. When we check out at the grocery store, each item scanned corresponds to a price. When a topic in Google is entered, many search results are displayed. If we register for classes online and enter the course title, a list of class options will be shown. In each of these cases, we are pairing together a piece of information from one set to a piece of information from another set.

**Objective 1** ▶

Identify a relation and its domain and range.

**INSTRUCTOR NOTE:**
It might be helpful to introduce the domain as not only the set of all *x*-values, but as the set of all input values, the set of all first coordinates, or the set of all starting values. Likewise, range can be introduced as not only the set of all *y*-values, but as the set of all output values, the set of all second coordinates, or the set of all ending values.

The set of pairs is called a relation. The first piece of information is the *input value*, or *x*-value. The second piece of information is the *output value*, or *y*-value. The set of all *x*-coordinates of a relation is called the *domain* and the set of all *y*-coordinates of a relation is called the *range*.

> **Definition:** A **relation** is a set of ordered pairs in which the first coordinate of the ordered pairs comes from a set called the **domain** and the second coordinate of the ordered pairs comes from a set called the **range**.

Relations can be expressed in various forms: a graph, a table, a mapping, a set of ordered pairs, or an equation. Some examples are shown.

| Graph | Table | Mapping |
|-------|-------|---------|
| 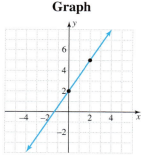 | 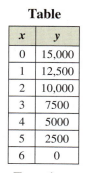 | 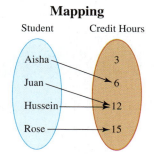 |

**Set of ordered pairs**

$\{(1, 3), (2, 4), (3, 5), (4, 6)\}$

**Equation**

$y = 0.24x + 4.42$

When we state the domain or range of a relation, it is not necessary to list values more than once.

---

**Objective 1 Examples**   **Express each described relation in the requested form. State the domain and range of each relation.**

**1a.** A class was surveyed to determine which students were enrolled in a specific number of hours. The results are shown in the following mapping. Write this relation as a set of ordered pairs.

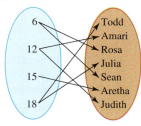

**1b.** According to T-Mobile.com, each text message sent or received in the United States costs $0.20 if you do not have a messaging plan. Express the relation between the number of text messages sent or received and the cost associated with them as an equation.

**1c.** The network TV ad revenue (in millions) for the NCAA Division I Men's Basketball Championship is given in the table. Express this relation in a graph.
(Source: http://www.internetadsales.com/march-madness-advertising-trends-report)

| Year | 2000 | 2001 | 2002 | 2003 | 2004 | 2005 | 2006 | 2007 | 2008 | 2009 |
|------|------|------|------|------|------|------|------|------|------|------|
| Revenue (in millions) | 319 | 318 | 358 | 380 | 451 | 475 | 500 | 520 | 643 | 589 |

**Solutions**

**1a.** The *x*-value of the ordered pairs represents the number of credit hours and the *y*-value represents the student. So, the set of ordered pairs for this relation is

{(6, Rosa), (6, Sean), (12, Amari), (12, Judith), (15, Aretha), (18, Todd), (18, Julia)}.

The domain of the relation is {6, 12, 15, 18} and the range of the relation is {Todd, Amari, Rosa, Julia, Sean, Aretha, Judith}.

**1b.** The *x*-value of the ordered pairs represents the number of text messages and the *y*-value represents the cost. The equation for the relation is $y = 0.20x$. Since the number of text messages sent or received can be 0, 1, 2, and so on, the domain of the relation is {0, 1, 2, 3, 4, 5, 6, . . .} and the range is {0, 0.20, 0.40, 0.60, 0.80, 1, 1.20, . . .}.

**1c.** The *x*-value of the ordered pairs represents the year and the *y*-value represents the revenue (in millions). The graph is shown.

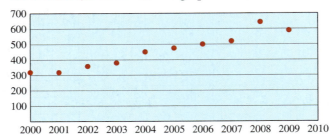

The domain is {2000, 2001, 2002, 2003, 2004, 2005, 2006, 2007, 2008, 2009}, and the range is {318, 319, 358, 380, 451, 475, 500, 520, 589, 643}.

✓ **Student Check 1** Express each described relation in the requested form. State the domain and range of each relation. For b, use only whole numbers for the input.

**a.** A class was surveyed to find out each student's declared major. The results are shown in the mapping. Write this relation as a set of ordered pairs.

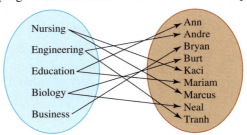

**b.** Skype allows calls from computer-to-computer or computer-to-land line or mobile phone. Calls from computer-to-computer are free but calls made from a computer to phones have an associated cost. Calls from a computer in the United States to a phone in Iraq cost approximately $0.39 per minute. Express the relation between the number of minutes and the associated cost of using Skype to call Iraq from the United States as an equation. (Source: http://www.skype.com/intl/en-us/prices/payg-rates#cc=IQ)

**c.** The percentage of Americans who are obese (BMI is greater than or equal to 30) is given in the table. Express this relation as a graph. (Source: http://apps.nccd.cdc.gov/brfss/display.asp and http://www.data360.org/index.aspx)

| Year | Percentage | Year | Percentage |
|------|-----------|------|-----------|
| 1995 | 15.9 | 2003 | 22.8 |
| 1996 | 16.8 | 2004 | 23.2 |
| 1997 | 16.6 | 2005 | 24.4 |
| 1998 | 18.3 | 2006 | 25.1 |
| 1999 | 19.8 | 2007 | 26.3 |
| 2000 | 20.1 | 2008 | 26.6 |
| 2001 | 21.1 | 2009 | 27.2 |
| 2002 | 22.2 | | |

## Functions

**Objective 2** ▶

Determine if a relation is a function.

We defined a relation as a set of ordered pairs—a correspondence between a set of inputs and outputs. A *function* is a special type of relation.

> **Definition:** A **function** is a relation in which each member of the domain (or each input) corresponds to exactly one member of the range (or an output). In other words, each input, or $x$-value, can have only one output, or $y$-value.

In the opening of this section, we mentioned a student ID that corresponds to more than one output. It corresponds to the student who was trying to register as well as to the other student whose name appeared on the screen. This was certainly due to entry error. It doesn't make sense for a student ID to correspond to more than one student. Except in the case of an error, this relation is a function. Each input (student ID) can only correspond to one output (student).

> **Procedure: Determining if a Relation Is a Function**
>
> **Step 1:** Determine how many $y$-values correspond to each $x$-value.
> 
> **a.** If there is only one corresponding $y$-value for each $x$-value, then the relation is a function.
>
> **b.** If at least one $x$-value corresponds to more than one $y$-value, then the relation is *not* a function.

**Objective 2 Examples**

Determine if each relation is a function. If not, explain why.

**2a.** $\{(-4, 0), (-3, 2), (-3, -2), (0, 4), (0, -4)\}$

**2b.** $\{(-2, -3), (-1, -3), (0, -3), (1, -3), (2, -3), (3, -3)\}$

**2c.** Let the relation be defined by the mapping, where $x$ is the math grade and $y$ is the student earning the grade.

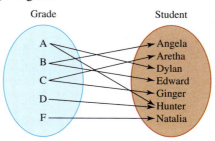

**2d.** $y = |x|$

**Solutions**

**2a.** This is *not a function* since the $x$-value of $-3$ corresponds to two $y$-values, 2 and $-2$. Also, the $x$-value of 0 corresponds to two $y$-values, $-4$ and 4.

**2b.** This *is a function* since each $x$-value corresponds to only one $y$-value.

**2c.** This is *not a function* since the grade of A, B, and C correspond to more than one student. For example, the grade of A corresponds to Dylan and Hunter. The grade of B corresponds to Angela and Edward. The grade of C corresponds to Aretha and Ginger.

**2d.** This *is a function* since each input, or $x$-value, has only one possible output, or $y$-value. When a value is substituted for $x$, there is only one possible $y$-value. Some examples are

$x = -3$; $y = |-3| = 3 \rightarrow (-3, 3)$

$x = 2$; $y = |2| = 2 \rightarrow (2, 2)$

✓ **Student Check 2**    Determine if each relation is a function. If not, explain why.

**a.** $\{(5, 1), (5, -2), (5, 6), (5, 3), (5, 0)\}$

**b.** $\{(1, 5), (-2, 5), (6, 5), (3, 5), (0, 5)\}$

**c.** Let the relation be defined by the mapping, where $x$ is the student and $y$ is the grade earned in math class.

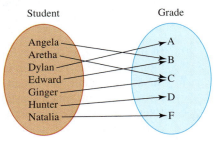

**d.** $y = x^2$

## Vertical Line Test

**Objective 3** ▶

Use the vertical line test.

We will now examine graphs of relations and see how we can use them to determine if a graph represents a function. Graphing some of the preceding examples will show us an important characteristic of the graph of a function.

Shown next are some examples of functions and their corresponding graphs.

$$\{(-2, -3), (-1, -3), (0, -3),$$
$$(1, -3), (2, -3), (3, -3)\}$$

$$y = |x|$$

If each $x$-value corresponds to only one $y$-value, we know the relation is a function. Notice that when we draw vertical lines through the graph of the relations, each vertical line touches the graph only once.

Shown next are examples of relations that are *not* functions.

$$\{(-4, 0), (-3, 2), (-3, -2),$$
$$(0, 4), (0, -4)\}$$

$$x = y^2$$

Notice that when an $x$-coordinate is paired with more than one $y$-coordinate, a vertical line can be drawn that will intersect the graph at more than one point.

These illustrations provide us a method for determining if a relation is a function by examining its graph. This method involves performing the *vertical line test*.

> **Property: Vertical Line Test**
> 1. If all vertical lines drawn on the graph of a relation intersect the graph in at most one point, then the graph is a function.
> 2. If at least one vertical line intersects the graph in more than one point, then the graph is *not* a function.

**Objective 3 Examples** | Use the vertical line test to determine if each relation is a function.

3a. 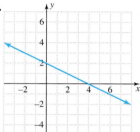    3b. 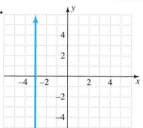    3c.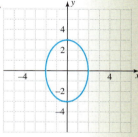

**Solutions** 

**3a.** The relation graphed *is a function* since every vertical line intersects the graph in at most one point.

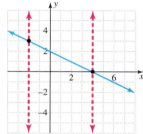

**3b.** The graph *is not a function* since a vertical line through $x = -3$ intersects infinitely many points of the graph.

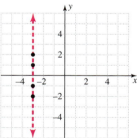

**3c.** The graph *is not a function* since any vertical line drawn between the $x$-values of $-2$ and $2$ intersects the graph in two points.

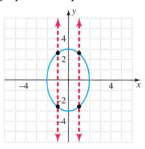

✓ **Student Check 3** Use the vertical line test to determine if each relation is a function.

a.
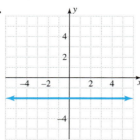

b.
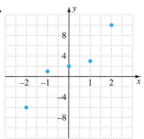

c.
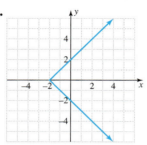

d.
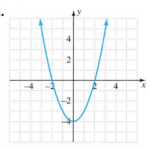

 **Note:** *All lines, except vertical lines, pass the vertical line test. Therefore, every linear equation of the form y = mx + b is a function.*

## Function Notation

**Objective 4** ▶

**Use function notation.**

An example of a linear equation that is a function is $y = 4x - 5$. This equation states a rule for writing ordered pairs where each $x$-coordinate is paired with exactly one $y$-coordinate. Function notation is a convenient way to determine this pairing of coordinates.

Suppose we want to find the $y$-value that corresponds to $x = 3$ for $y = 4x - 5$. We determine the $y$-value by replacing the variable $x$ with 3.

$$y = 4x - 5$$
$$y = 4(3) - 5$$
$$y = 12 - 5 = 7$$

So, the output value of $4x - 5$ is 7 when $x = 3$. In other words, the point $(3, 7)$ lies on the graph of $y = 4x - 5$.

**INSTRUCTOR NOTE:**

Point out that $f(x)$ is read "$f$ of $x$" and is not $f$ times $x$.

In mathematics, we use letters such as $f$, $g$, and $h$ to name functions. The symbol $f(x)$ is read "$f$ of $x$" and means "the function of $x$." This notation is called **function notation**.

> **Definition: Function Notation**
>
> The notation $f(x)$ is used to denote that $y$ is a function of $x$.
>
> $$y = f(x) \qquad \text{Name of function}$$
>
> Output value   Input value
>
> In this notation, $x$ is the input value and $f(x)$ is the output value.

We can use function notation to write an equivalent equation to $y = 4x - 5$, which is $f(x) = 4x - 5$. The equations are equivalent because $y = f(x)$.

The notation $f(3)$ means to replace $x$ with 3 and find the corresponding $y$-value. For example,

$$f(x) = 4x - 5$$
$$f(3) = 4(3) - 5 = 7$$

So, when $x = 3$, $y = 7$ or $[f(3) = 7]$ and an ordered pair solution of the equation is $(3, 7)$. Note that this is exactly what we obtain when we evaluate $y = 4x - 5$ at $x = 3$. This process is called *evaluating a function*.

> **Procedure: Evaluating a Function $f(x)$ at $x = k$**
>
> **Step 1:** If the function is given in terms of an equation, replace the variable with the number $k$ that is given in parentheses and simplify.
> **Step 2:** If the function is given in terms of a set, a mapping, or a table, find the $y$-value that corresponds to the $x$-value of $k$.
> **Step 3:** If the function is given in terms of a graph, find the ordered pair on the graph whose $x$-value is $k$. The corresponding $y$-value is the result of $f(k)$.

### Objective 4 Examples

**Evaluate each function at the given value.**

**INSTRUCTOR NOTE:**
Remind students that $y = f(x)$. Stress the importance of using parentheses in place of the variable first and then inserting the given value.

**4a.** Find $f(0)$ if $f(x) = 3x - 12$.

**4b.** Find $g(-3)$ if $g(x) = x^2 + 1$.

**4c.** Find $c(4)$ if the function $c(x)$ is given by $Y_1$.

**4d.** Find $f(2)$ if $f(x)$ is given by the following graph.

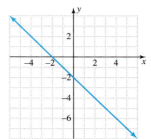

### Solutions

**4a.** $f(0) = 3(0) - 12 = 0 - 12 = -12$
**4b.** $g(-3) = (-3)^2 + 1 = 9 + 1 = 10$
**4c.** The point $(4, -8)$ is one of the ordered pairs in the table. So, $c(4) = -8$
**4d.** The point $(2, -4)$ lies on the graph of the function. So, $f(2) = -4$.

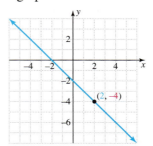

### ✓ Student Check 4

**Evaluate each function at the given value.**

**a.** Find $f(0)$ if $f(x) = 9x + 4$.

**b.** Find $g(-5)$ if $g(x) = -2x^2 - 4x + 1$.

**c.** Find $c(2)$ if the function $c(x)$ is given by $Y_1$.

**d.** Find $f(-2)$ if $f(x)$ is given by the following graph.

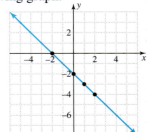

## Use Function Notation to Represent Linear Equations in Two Variables

**Objective 5** ▶

Write a linear equation in two variables in function notation.

We will now focus on a particular type of function, a *linear function*. Recall that the graph of $Ax + By = C$ or $y = mx + b$ is a line. All of these lines are also functions, except for vertical lines since they do not pass the vertical line test. Therefore, we can use function notation to write equations of non-vertical lines. When lines are written in the form $y = mx + b$, we can replace $y$ with $f(x)$ and put it in function notation.

> **Definition:** A function of the form $f(x) = mx + b$, for $m$ and $b$ real numbers is a **linear function**.

> **Procedure: Writing a Linear Equation in Two Variables in Function Notation**
>
> **Step 1:** Solve the equation for $y$, if needed.
> **Step 2:** Replace $y$ with $f(x)$.

**Objective 5 Examples**  Write each linear equation in two variables in function notation. Then, find $f(0)$, $f(2)$, and $f(-1)$ and write the corresponding ordered pairs.

**5a.** $x + 2y = 4$                                       **5b.** $y = 7$

**Solutions**    **5a.**

$$x + 2y = 4$$
$$x + 2y - x = 4 - x \qquad \text{Subtract } x \text{ from each side.}$$
$$2y = -x + 4 \qquad \text{Simplify.}$$
$$\frac{2y}{2} = \frac{-x}{2} + \frac{4}{2} \qquad \text{Divide each side by 2.}$$
$$y = -\frac{1}{2}x + 2 \qquad \text{Simplify.}$$
$$f(x) = -\frac{1}{2}x + 2 \qquad \text{Replace } y \text{ with } f(x).$$

Now we must evaluate $f(0), f(2),$ and $f(-1)$.

$$f(0) = -\frac{1}{2}(0) + 2 = 0 + 2 = 2$$

$$f(2) = -\frac{1}{2}(2) + 2 = -1 + 2 = 1$$

$$f(-1) = -\frac{1}{2}(-1) + 2 = \frac{1}{2} + 2 = \frac{1}{2} + \frac{4}{2} = \frac{5}{2}$$

The corresponding ordered pairs are $(0, 2)$, $(2, 1)$, and $\left(-1, \frac{5}{2}\right)$.

**5b.** The equation is already solved for $y$. So, in function notation, it is $f(x) = 7$. Recall that $y = 7$ is equivalent to $y = 0x + 7$. So, $f(x) = 0x + 7$.

$$f(0) = 0(0) + 7 = 0 + 7 = 7$$
$$f(2) = 0(2) + 7 = 0 + 7 = 7$$
$$f(-1) = 0(-1) + 7 = 0 + 7 = 7$$

The corresponding ordered pairs are $(0, 7)$, $(2, 7)$, and $(-1, 7)$.

✔ **Student Check 5**  Write each linear equation in two variables in function notation. Then find $f(0)$, $f(3)$, and $f(-2)$ and write the corresponding ordered pairs.

    **a.** $2x + 3y = 6$                                  **b.** $y = 1$

## Applications

Without functions, many things in life would be very chaotic. Functions enable our mail to be delivered properly since each address can only correspond to one location. Functions enable our login to different programs to work properly since each login name must correspond to exactly one person. Every social security number must correspond to exactly one person. The list of these types of situations is infinite.

In Example 6, we will see a function that relates to a real-world situation.

**Objective 6   Examples**

**According to the U.S. Department of Labor, the fastest-growing occupation for the years 2006–2016 is a network system and data communications analyst. The function $f(x) = 1343x + 45{,}657$ approximates the median salary in the United States for $x$ years of experience in this field.** (Sources: www.bls.gov and www.payscale.com)

**6a.** Find the median salary with 5 yr of experience.

**6b.** How many years of experience is required for the median salary to be $60,430?

**Solutions**

**6a.** The input $x$ represents the number of years of experience and $f(x)$, or $y$, represents the median salary. To find the salary with 5 yr of experience, we let $x = 5$.

$$f(x) = 1343x + 45{,}657$$
$$f(5) = 1343(5) + 45{,}657 = 6715 + 45{,}657 = 52{,}372$$

The median salary is $52,372 with 5 yr of experience.

**6b.** To find how many years of experience is required for the salary to be $60,430, we set $f(x) = 60{,}430$ and solve for $x$.

$$1343x + 45{,}657 = 60{,}430$$
$$1343x + 45{,}657 - 45{,}657 = 60{,}430 - 45{,}657 \qquad \text{Subtract 45,657 from each side.}$$
$$1343x = 14{,}773 \qquad \text{Divide each side by 1343.}$$
$$x = 11 \qquad \text{Simplify.}$$

The median salary is $60,430 with 11 yr of experience.

**☑ Student Check 6**

The number of cell-phone subscribers in the United States, in millions, between 1985 and 2009 can be approximated by $f(x) = 0.7082x^2 - 4.0318x + 4.8661$, where $x$ is the number of years after 1985. Use the function to determine the number of cell phone subscribers in the United States in 2015. (Source: www.infoplease.com)

## Troubleshooting Common Errors

Some common errors associated with the concept of functions are shown next.

**Objective 7   Examples**

**A problem and an incorrect solution are given. Provide the correct solution and an explanation of the error.**

**7a.** Is the relation $\{(-2, 3), (-1, 0), (0, -1), (1, 0), (2, 3)\}$ a function?

| Incorrect Solution | Correct Solution and Explanation |
|---|---|
| The relation is not a function since two $x$-values correspond to the same $y$-value. Both $-2$ and $2$ have a $y$-value of 3. | The relation *is* a function. The definition of a function is that each $x$-value have only one output. It is OK if the output values repeat. Since every value of $x$ has only one output, this relation is a function. |

**7b.** Write the linear equation $x = 5$ in function notation.

| Incorrect Solution | Correct Solution and Explanation |
|---|---|
| The equation $x = 5$ in function notation is $f(x) = 5$. | The equation $x = 5$ is a vertical line and is *not* a function. It cannot be written in function notation. |

**7c.** If $f(x) = 7x + 9$, find $f(2)$.

| Incorrect Solution | Correct Solution and Explanation |
|---|---|
| $f(2) = (7x + 9)(2) = 14x + 18$ | The notation $f(2)$ means to evaluate the function at $x = 2$. It does not mean multiplication. $$f(2) = 7(2) + 9 = 14 + 9 = 23$$ |

## ANSWERS TO STUDENT CHECKS

**Student Check 1    a.** {(Nursing, Kaci), (Nursing, Marcus), (Engineering, Andre), (Engineering, Tranh), (Education, Ann), (Education, Mariam), (Biology, Burt), (Biology, Neal), (Business, Bryan)}

Domain = {nursing, engineering, education, biology, business}

Range = {Ann, Andre, Bryan, Burt, Kaci, Mariam, Marcus, Neal, Tranh}

**b.** $y = 0.39$; Domain = {0, 1, 2, 3, . . .}, Range = {0, 0.39, 0.78, 1.17, . . .}

Domain = {1995, 1996, 1997, 1998, 1999, 2000, 2001, 2002, 2003, 2004, 2005, 2006, 2007, 2008, 2009}

Range = {15.9, 16.6, 16.8, 18.3, 19.8, 20.1, 21.1, 22.2, 22.8, 23.2, 24.4, 25.1, 26.3, 26.6, 27.2}

**Student Check 2    a.** no, the $x$-value of 5 corresponds to more than one $y$-value.    **b.** yes    **c.** yes    **d.** yes

**Student Check 3    a.** yes    **b.** yes    **c.** no    **d.** yes

**Student Check 4    a.** $f(0) = 4$    **b.** $g(-5) = -29$
   **c.** $c(2) = -6$    **d.** $f(-2) = 0$

**Student Check 5**
   **a.** $f(x) = -\dfrac{2}{3}x + 2$; $f(0) = 2$; $f(3) = 0$; $f(-2) = \dfrac{10}{3}$
   **b.** $f(x) = 1$; $f(0) = 1$; $f(3) = 1$; $f(-2) = 1$

**Student Check 6**    521.29 million cell phone subscribers.

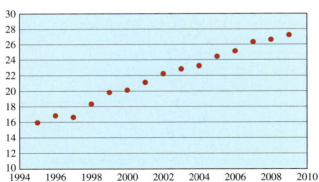

## SUMMARY OF KEY CONCEPTS

1. A relation is a set of ordered pairs. The set of $x$-values of the ordered pairs is the domain. The set of $y$-values of the ordered pairs is the range.

2. A function is a relation in which each input has only one output, that is, for every $x$ in the domain of the function, there is exactly one $y$-value. The $x$-values in the list of ordered pairs cannot repeat for the relation to be a function.

3. The vertical line test can be used to determine if a graph represents a function. If any vertical line intersects a graph in more than one point, the graph does *not* represent a function.

4. Function notation, $f(x)$, is another name for the output value $y$. It shows that $y$ is dependent on the value of $x$. When a specific value of $x$ is inside the function notation, we evaluate the function at that value.

a. In an equation, we do this by substituting the variable with the given number.

b. In a table or graph, we find the ordered pair in the table or on the graph with the given *x* value.

The *y*-value of this ordered pair is the value we need.

5. Linear equations in two variables of the form $y = mx + b$ can be written in function notation as $f(x) = mx + b$. Equations of the form $Ax + By = C$ must be solved for *y* before they can be written in function notation. Vertical lines are not functions.

6. Functions occur in many real-world situations. It is important to understand from the problem what the input and output of the function represent.

## GRAPHING CALCULATOR SKILLS

The graphing calculator can assist us in evaluating functions when the function is given in terms of an equation. Input the function into the calculator to find the requested information.

**Example:** Find $f(0)$ if $f(x) = |x - 4|$.

**Solution:** Input the equation into the calculator and use the Table or TRACE feature to find the *y*-value.

Use the Table to find the *y*-value when $x = 0$.

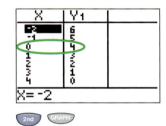

Graph the equation and use the TRACE feature to find the point that has an *x*-value of 0.

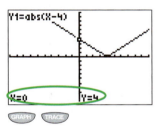

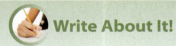

So, $f(0) = 4$.

## SECTION 3.6 / EXERCISE SET

### Write About It!

**Use complete sentences in your answer to each exercise.**

1. Define a relation and its domain and range.
   Answers vary.
2. How do we determine if a relation is a function?
   Answers vary.
3. What is the vertical line test?   Answers vary.

4. Define a function.   Answers vary.

5. What is function notation?   Answers vary.

6. What is the difference between an equation of a line and a linear equation in two variables in function notation?
   Answers vary.

**Determine if each statement is true or false. If a statement is false, explain why it is false.**

7. If $f(3) = 5$, then $(3, 5)$ is a point on the graph of the function *f*.   True

8. If $f(0) = -7$, then the *x*-intercept is $(-7, 0)$.
   False, $f(0) = -7$ corresponds to $(0, -7)$ which is the *y*-intercept.
9. If $f(8) = 0$, then the *y*-intercept is $(0, 8)$.
   False, $f(8) = 0$ corresponds to $(8,0)$ which is the *x*-intercept.
10. A relation is always a function.
   False, functions are all relations but not all relations are functions.

11. If any vertical line intersects a graph in more than one point, the graph represents a function.
   False, the vertical line can only touch one point on the graph for it to be a function.
12. A relation always has a domain and range.   True

### Practice Makes Perfect!

**Express each relation in Exercises 13–16 as a set of ordered pairs. State the domain and range of each relation. (*See Objective 1.*)**

13. The following table shows the 2009 total health expenditure per capita in the selected countries. (Source: http://www.who.int/en)

| Countries | Per Capita |
|-----------|------------|
| Canada | $4400 |
| France | $4800 |
| Italy | $3300 |
| Japan | $3300 |
| Republic of Korea | $1100 |
| United Kingdom | $3300 |
| United States | $7400 |

Additional answers can be found in the Instructor Answer Appendix.

**14.** The following table shows the number of days that selected U.S. metropolitan areas failed to meet acceptable air-quality standards in 2008. (Source: http://worldalmanac.com)

| Metropolitan Statistical Areas | Number of Days |
|---|---|
| Atlanta | 4 |
| Bakersfield | 26 |
| Baltimore | 4 |
| Detroit | 1 |
| Houston | 2 |
| Los Angeles | 28 |
| Memphis | 1 |
| New Orleans | 0 |
| New York | 1 |
| Orange County | 1 |
| Phoenix | 84 |
| Sacramento | 20 |
| Washington, DC | 3 |

**15.** The mapping below shows the number of major professional championships earned by each professional golfer. (Source: http://worldalmanac.com)

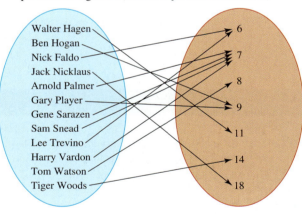

**16.** The results of the Indy Racing League (IRL) winners from 1996 to 2008 are shown in the mapping. (Source: http://worldalmanac.com)

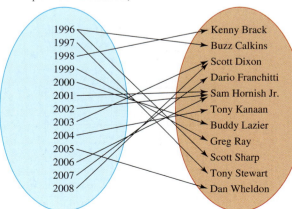

**17.** The residential average price of electricity in New Jersey was 14.52 cents per kilowatt-hour in 2009. Express the relation between the number of kilowatt-hours (in hundreds) and the cost (in dollars) associated with them as an equation. (Source: www.think-energy.net)

**18.** The residential average price of electricity in Alaska was 15.09 cents per kilowatt-hour in 2009. Express the relation between the number of kilowatt-hours (in hundreds) and the cost (in dollars) associated with them as an equation. (Source: www.think-energy.net)

**Determine if each relation is a function. If not, explain why. (*See Objective 2.*)**

**19.** $\{(2, -5), (2, -1), (2, 8), (2, 14), (2, 18)\}$
not a function; The x-value of 2 corresponds to more than one y-value.

**20.** $\{(-14, -3), (-7, -3), (-5, -3), (6, -3), (12, -3)\}$
a function

**21.** $\{(-20, 6), (-9, 6), (0, 6), (1, 6), (4, 6)\}$    a function

**22.** $\{(4, -8), (4, -1), (4, 0), (4, 7), (4, 11)\}$
not a function; The x-value of 4 corresponds to more than one y-value.

**23.** $\{(-2, 2), (-2, 4), (1, 8), (5, 12), (0, 9)\}$    not a function;
The x-value of −2 corresponds to more than one y-value.

**24.** $\{(-5, 8), (-5, -7), (1, 15), (-1, 0), (2, 10)\}$
not a function; The x-value of −5 corresponds to more than one y-value.

**25.** Assign to each student the grade earned in their physics class.

a function

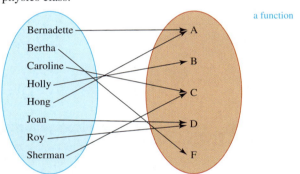

**26.** Assign to each student the grade earned in their math class.

a function

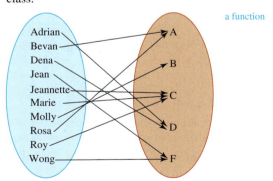

**27.** $y = 2x + 5$    a function

**28.** $y = -4x + 3$    a function

**29.** $y = |x - 1|$    a function

**30.** $y = 2 - |x|$    a function

**31.** $x = |y - 2|$    not a function, because x = 1 corresponds to y = 3 and y = 1

**32.** $x = |y + 3|$
not a function, because x = 2 corresponds to y = −1 and y = −5

**Use the vertical line test to determine if each relation is a function.** (*See Objective 3.*)

**33.**

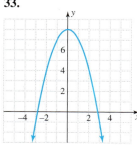

a function

**34.**

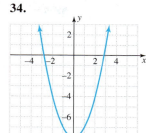

a function

**35.**

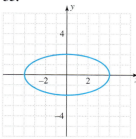

not a function

**36.**

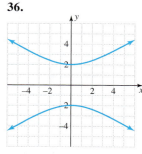

not a function

**37.**

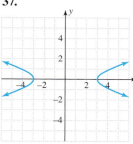

not a function

**38.**

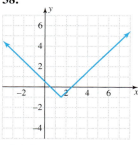

a function

**Evaluate each function at the given value.** (*See Objective 4.*)

**39.** $f(-4)$ if $f(x) = -6x + 1$    **40.** $f(0)$ if $f(x) = 2x - 7$   $-7$
    $25$
**41.** $g(5)$ if $g(x) = 2x^2 - 1$    **42.** $g(-3)$ if $g(x) = -x^2 + 2$
    $49$                                    $-7$
**43.** $h(-2)$ if $h(x) = |5x - 4|$    **44.** $h(-5)$ if $h(x) = |3x - 1|$
    $14$                                    $16$
**45.** $f(10)$ if $f(x) = -x^2 + 7x - 6$   $-36$

**46.** $f(-1)$ if $f(x) = 3x^2 - 6x + 12$    $21$

**47.** Find $c(3)$ if the function $c(x)$ is given by $Y_1$.    $-3$

| X | Y₁ | |
|---|---|---|
| -3 | 9 | |
| -2 | 7 | |
| -1 | 5 | |
| 0 | 3 | |
| 1 | 1 | |
| 2 | -1 | |
| | -3 | |
| X=3 | | |

**48.** Find $c(-2)$ if the function $c(x)$ is given by $Y_1$.    $-10$

| X | Y₁ | |
|---|---|---|
| -3 | -13 | |
| -2 | -10 | |
| -1 | -7 | |
| 0 | -4 | |
| 1 | -1 | |
| 2 | 2 | |
| | 5 | |
| X=3 | | |

**49.** Find $f(-3)$ if the function $f(x)$ is given by $Y_1$.    $-9.6$

| X | Y₁ | |
|---|---|---|
| -3 | -9.6 | |
| -2 | -7 | |
| -1 | -4.4 | |
| 0 | -1.8 | |
| 1 | .8 | |
| 2 | 3.4 | |
| | 6 | |
| X=3 | | |

**50.** Find $f(-2)$ if the function $f(x)$ is given by $Y_1$.    $9.8$

| X | Y₁ | |
|---|---|---|
| -4 | 17 | |
| -3 | 13.4 | |
| -2 | 9.8 | |
| -1 | 6.2 | |
| 0 | 2.6 | |
| 1 | -1 | |
| 2 | -4.6 | |
| X=-4 | | |

**51.** Find $x$ if $c(x) = -4$ and the function $c(x)$ is given by $Y_1$.    $0$

| X | Y₁ | |
|---|---|---|
| -3 | -13 | |
| -2 | -10 | |
| -1 | -7 | |
| 0 | -4 | |
| 1 | -1 | |
| 2 | 2 | |
| | 5 | |
| X=3 | | |

**52.** Find $x$ if $f(x) = 3.4$ and the function $f(x)$ is given by $Y_1$.    $2$

| X | Y₁ | |
|---|---|---|
| -3 | -9.6 | |
| -2 | -7 | |
| -1 | -4.4 | |
| 0 | -1.8 | |
| 1 | .8 | |
| 2 | 3.4 | |
| | 6 | |
| X=3 | | |

**53.** Find $f(-4)$ if $f(x)$ is given by the graph.    $-8$

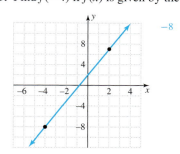

**54.** Find $f(-2)$ if $f(x)$ is given by the graph.

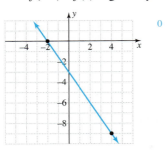

0

**55.** Find $f(2)$ if $f(x)$ is given by the graph.

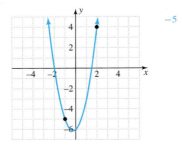

−9

**56.** Find $f(-1)$ if $f(x)$ is given by the graph.

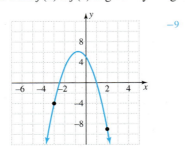

−5

**57.** Find $x$ if $f(x) = -6$ and $f(x)$ is given by the graph.

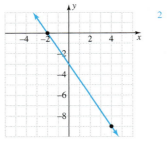

2

**58.** Find $x$ if $f(x) = 7$ and $f(x)$ is given by the graph.

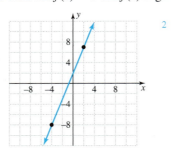

2

**Write each linear equation in two variables in function notation. Then find $f(0)$, $f(3)$, and $f(-2)$ and write the corresponding ordered pairs. (*See Objective 4.*)**

**59.** $y = 5x - 6$    **60.** $y = -7x + 3$    **61.** $4x - y = 9$

**62.** $3x + 5y = 6$    **63.** $2y - 7 = 0$    **64.** $2y + 3 = 0$

**Solve each problem. (*See Objective 5.*)**

**65.** Assume that the S&P 500 index historical prices (in dollars) can be modeled by the function $f(x) = 1.56x^4 - 35x^3 + 247x^2 - 567x + 1381$, where $x$ is the number of years after 2000.

   **a.** Use the function to find the S&P 500 index price in 2009. How close is the answer if the actual estimate was 1106?   $1005.16, about \$100.84 off

   **b.** Use the function to predict the S&P 500 index for year 2012.   $2013.16

**66.** Assume that the NASDAQ Composite historical prices (in dollars) can be modeled by the function $f(x) = 8.271x^4 - 106.302x^3 + 346.835x^2 + 34.085x + 1462.356$, where $x$ is the number of years after 2002.

   **a.** Use the function to find the NASDAQ composite price in 2006. How close is the answer if the actual estimate was $2457?   $2462.10, very close

   **b.** Use the function to predict the NASDAQ composite price for year 2010.   $3383.87

**67.** An open-top box is constructed by cutting four squares of width $x$ in. from the corners of a rectangular piece of cardboard with dimensions 12 in. by 18 in. and then folding up the sides. The volume of the box is given by the function $V(x) = 216x - 60x^2 + 4x^3$, where $0 \le x \le 6$. Evaluate $V(2)$ and interpret the answer.   224; The volume of the open-top box is 224 in.$^3$.

**68.** An open-top box is constructed by cutting four squares of width $x$ in. from the corners of a rectangular piece of cardboard with dimensions 24 in. by 30 in. and then folding up the sides. The volume of the box is given by the function $V(x) = 720x - 108x^2 + 4x^3$, where $0 \le x \le 12$. Evaluate $V(5)$ and interpret the answer.   1400; The volume of the open-top box is 1400 in.$^3$.

 **Mix 'Em Up!**

**Determine if each relation is a function. If not, explain why.**

**69.** The following table shows the 2009 total health expenditure per capita in the selected countries. (Source: http://www.who.int/en)

| Countries | Per Capita |
|---|---|
| Canada | $4400 |
| France | $4800 |
| Italy | $3300 |
| Japan | $3300 |
| Republic of Korea | $1100 |
| United Kingdom | $3300 |
| United States | $7400 |

**70.** The following table shows the number of days that selected U.S. metropolitan areas failed to meet acceptable air-quality standards in 2008. (Source: http://worldalmanac.com)

| Metropolitan Statistical Areas | Number of Days |
|---|---|
| Atlanta | 4 |
| Bakersfield | 26 |
| Baltimore | 4 |
| Detroit | 1 |
| Houston | 2 |
| Los Angeles | 28 |
| Memphis | 1 |
| New Orleans | 0 |
| New York | 1 |
| Orange County | 1 |
| Phoenix | 84 |
| Sacramento | 20 |
| Washington, DC | 3 |

a function

**71.** The mapping below shows the number of major professional championships earned by each professional golfer. (Source: http://worldalmanac.com)

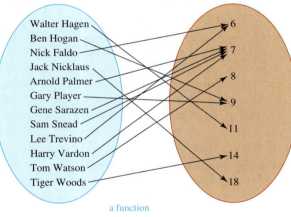

a function

**72.** The results of the Indy Racing League (IRL) winners from 1996 to 2008 are shown in the mapping. (Source: http://worldalmanac.com)

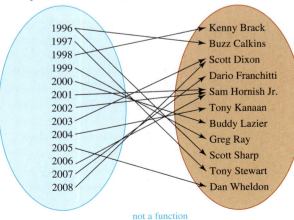

not a function

**73.** $\{(3, 9), (3, -14), (3, 4), (3, -2), (3, -8)\}$    not a function

**74.** $\{(9, 3), (-14, 3), (4, 3), (-2, 3), (-8, 3)\}$    a function

**75.** $\{(1, 3), (-2, 7), (-2, 1), (0, 5), (4, 0)\}$    not a function

**76.** $\{(6, -3), (1, -3), (0, 11), (-5, 0), (2, 0)\}$    a function

**77.** Assign to each student the grade earned in their chemistry class.

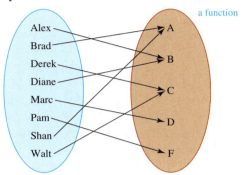

a function

**78.** Assign to each student the grade earned in their computer programming class.

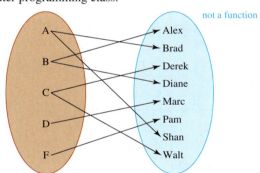

not a function

**79.** $y = 7$    a function

**80.** $x = 4$    not a function

**81.** $y = \frac{1}{2}x + 4$    a function

**82.** $y = -\frac{5}{2}x + 1$    a function

**83.**

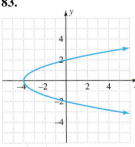

not a function

**84.**

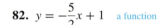

not a function

**85.**

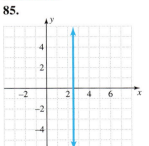

not a function

**86.**

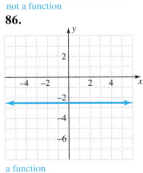

a function

**87.**

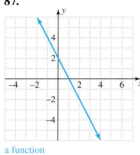

a function

**88.**

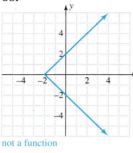

not a function

**Find the requested information.**

**89.** $f(-2)$ if $f(x) = -5x^2 + 7x + 1$   $-33$

**90.** $f(-1)$ if $f(x) = -x^2 + 6x + 18$   $11$

**91.** The function $f(x)$ is given by $Y_1$. Find **a.** $f(4)$ and **b.** $x$ if $f(x) = -10$.

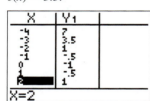

a. $-25$   b. $-1$

**92.** The function $c(x)$ is given by $Y_1$. Find **a.** $c(1)$ and **b.** $x$ if $c(x) = 3.5$.

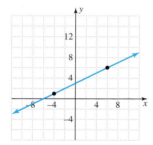

a. $-0.5$   b. $-3$

**93.** The graph of $f(x)$ is given. Find **a.** $f(6)$ and **b.** $x$ if $f(x) = 0$.

a. 6,   b. $-6$

**94.** The graph of $f(x)$ is given. Find **a.** $f(0)$ and **b.** $x$ if $f(x) = 0$.

a. 4,   b. 6

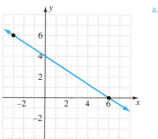

**95.** Find $f(4.2)$ if $f(x) = 1.5x - 3.2$.   3.1

**96.** Find $f(-0.8)$ if $f(x) = -2.4x + 8.4$.   10.32

**Write each linear equation in two variables in function notation. Then find $f(0)$, $f(3)$, and $f(-2)$ and write the corresponding ordered pairs.**

**97.** $9x - 2y = 10$

**98.** $7x + 14y = 4$

**Solve each problem.**

**99.** Brasil has a new job as a computer information analyst. His starting salary is $52,000. He will receive a raise of $1600 each year he works with the company.

   **a.** Write a linear function $f(x)$ that represents Brasil's salary, where $x$ is the number of years he has worked with the company.   $y = 1600x + 52,000$

   **b.** Find $f(5)$ and interpret the answer.   $60,000; If Brasil has worked with the company for 5 yr, his salary would be $60,000.

**100.** Lois has a new job as a math instructor at a local community college. Her starting salary is $41,500 with a regular teaching load of 30 credit hours. If she teaches beyond the regular load, she earns $950 per credit hour.

   **a.** Write a linear function $f(x)$ that represents Lois' yearly income, where $x$ is the number of extra credit hours she teaches.   $f(x) = 950x + 41,500$

   **b.** If Lois teaches two additional 4-credit-hour courses, what is her income?   $49,100

   **c.** Find $f(6)$ and interpret the answer.   $47,200; It means Lois has taught an additional 6 credit hours.

**101.** Lee registers for his first semester in college. His tuition includes fees of $565 plus $135 per credit hour.

   **a.** Write a linear function $f(x)$ that represents Lee's total tuition, where $x$ is the number of credit hours Lee takes during the first semester.   $f(x) = 135x + 565$

   **b.** What is Lee's tuition if he registers for 12 credit hours?   $2185

   **c.** Find $f(15)$ and interpret the answer.   $2590; It means Lee is taking 15 credits.

**102.** Aziz purchased a new car for $18,500. The car's value decreases by $1250 each year.

   **a.** Write a linear function $f(x)$ that represents the value of the car, where $x$ is the age of the car in years.   $f(x) = -1250x + 18,500$

   **b.** What is the car's value after 4 yr?   $13,500

   **c.** Find $f(8)$ and interpret the answer.   $8500; It means the value of the car after 8 yr is $8500.

 **You Be the Teacher!**

**Correct each student's errors, if any.**

**103.** Determine if each relation is a function.

   **a.** $\{(3, 1), (7, 5), (1, 2), (5, 2), (0, -2)\}$

   **b.** $\{(1, 3), (5, 7), (2, 1), (2, 5), (-2, 0)\}$

   Vivian's work: **a.** a function and **b.** not a function
   Correct!

**104.** Find $f(-3)$ if $f(x) = 12x - 5$.

   William's work:
   $f(-3) = -3(12x - 5) = -36x + 15$
   $f(-3) = 12(-3) - 5 = -36 - 5 = -41$

**105.** Find $f(0)$ and $x$ if $f(x) = 0$ from the given graph.

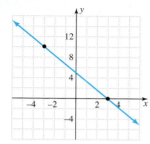

Brian's work: $f(0) = 3$ and $f(5) = 0$   $f(0) = 5$ and $f(3) = 0$

**106.** Find $f(0)$ and $x$ if $f(x) = 0$ from the given table.

| X | Y₁ | |
|---|---|---|
| -2 | -6 | |
| -1 | -4.5 | |
| 0 | -3 | |
| 1 | -1.5 | |
| 2 | 0 | |
| 3 | 1.5 | |
| 4 | 3 | |
| X= -2 | | |

Marla's work: $f(-3) = 0$ and $f(0) = 2$
$f(0) = -3$ and $f(2) = 0$

 **Calculate It!**

**Input the function in a graphing calculator. Use both the TRACE and TABLE features to find $f(0)$ and $x$ if $f(x) = 0$.**

**107.** $f(x) = 0.25x - 1.75$   $f(0) = -1.75; x = 7$

**108.** $f(x) = -0.06x^2 + 0.5x + 0.36$   $f(0) = 0.36; x = -\dfrac{2}{3}, x = 9$

**109.** $f(x) = |2x - 1| - 3$   $f(0) = -2; x = -1, 2$

**110.** The table shows the points on the graph of a function. Determine which table shows a line that contains the points $(0, -5)$ and $(-2, 4)$.

**a.**

| X | Y₁ | |
|---|---|---|
| -4 | 3 | |
| -3 | 3.5 | |
| -2 | 4 | |
| -1 | 4.5 | |
| 0 | 5 | |
| 1 | 5.5 | |
| 2 | 6 | |
| X= -4 | | |

**b.**

| X | Y₁ | |
|---|---|---|
| -3 | -18.5 | |
| -2 | -14 | |
| -1 | -9.5 | |
| 0 | -5 | |
| 1 | -.5 | |
| 2 | 4 | |
| 3 | 8.5 | |
| X=3 | | |

**c.**

| X | Y₁ | |
|---|---|---|
| -4 | 13 | |
| -3 | 8.5 | |
| -2 | 4 | |
| -1 | -.5 | |
| 0 | -5 | |
| 1 | -9.5 | |
| 2 | -14 | |
| X= -4 | | |

---

## GROUP ACTIVITY    Use a Linear Equation to Model Population Growth

1. Go to http://www.google.com/publicdata?ds=uspopulation and select a state or expand the state to select a specific county.

2. After you make your selection, a graph will appear in the main window. Use your cursor to trace the graph to see the population for each year. Write two ordered pairs of the form (year, population) for the state or county that you have selected.

3. Use the two ordered pairs from step 2 to write a linear equation that models the population of the selected state/county.

4. What is the slope of the model? What does it mean in the context of the problem?

5. What is the $y$-intercept of the model? What does it mean in the context of the problem?

6. Use the model to predict the population of your selected state/county in the year 2020.

# Linear Equations in Two Variables

**What's the big idea?** Now that you have completed Chapter 3, you should be able to see a connection between a linear equation in two variables, solutions of the equation, and the graph of the equation. You should also be able to identify important characteristics of the linear equation from any of its different forms.

## The Tools

Listed below are the key terms, skills, formulas, and properties you should know for this chapter.

The page reference is provided if you need additional help with the given topic. The Study Tips will assist in your preparation for an exam.

## Study Tips

1. Learn all of the terms, formulas, and properties. Make flash cards and have someone quiz you.
2. Rework problems from the exercises and also the ones you worked in class. Work additional problems from the review exercises.
3. Review the summaries of key concepts.
4. Work the chapter test.
5. Be sure to review the online resources for additional study materials.

## Terms

Domain   265
Function notation   270
Function   267
Horizontal line   216
Initial value   239, 256
Linear equation in two variables   209
Linear function   272
Ordered pair   194
Origin   196

Paired data   200
Parallel lines   242
Perpendicular lines   242
Point-slope form   250
Quadrants   196
Range   265
Rate of change   239
Rectangular coordinate system   196
Relation   265

Scatter plot   200
Slope   227
Slope-intercept form   230
Solution of an equation   194
Vertical line   216
$x$-axis   196
$x$-intercept   213
$y$-axis   196
$y$-intercept   213

## Formulas and Properties

- Equation of a horizontal line   216
- Equation of a vertical line   216
- Linear equation in two variables   209

- Point-slope form of a line   250
- Slope formula   227
- Slope-intercept form of a line   230

- Standard form of a line   209
- Vertical line test   269

## CHAPTER 3 / SUMMARY

How well do you know this chapter? Complete the following questions to find out. Take a look back at the section if you need help.

### SECTION 3.1 Equations and the Rectangular Coordinate System

1. A solution of an equation in two variables is a(n) ordered pair that makes the equation a(n) true statement.

2. In an ordered pair, the first coordinate tells you how to move left or right from the origin. The second coordinate tells you how far to move up or down from the y-axis.

3. To identify an ordered pair of a point on the coordinate system, determine how far left or right the point is from the origin and how far up or down the point is from the y-axis.

4. Quadrants are the four regions formed by the two axes of the rectangular coordinate system. They are labeled with Roman numerals moving in the counterclockwise direction.

5. The graph of an equation in two variables is composed of all ordered pairs that are solutions of the equation. An equation in two variables has infinitely many solutions.

6. If a point lies on the graph of an equation, it is a(n) _solution_ of the equation.

7. We used _graphs_ and _tables_ to solve application problems. From these representations, _ordered_ _pairs_ must be identified to answer questions.

## SECTION 3.2  Graphing Linear Equations

8. The standard form of a linear equation in two variables is _$Ax + By = C$_.

9. To graph a line by plotting points, _two_ points are required. However, a total of _three_ points is recommended so that one of the points serves as a check.

10. The point where a graph crosses the *x*-axis is called the _x-intercept_. To find this point, set _$y$_ = 0 and solve the resulting equation. The point where a graph crosses the *y*-axis is called the _y-intercept_. To find this point, set _$x$_ = 0 and solve the resulting equation.

11. If the value of the constant in a linear equation equals zero, then the graph of the equation will pass through the _origin_. To graph such an equation, _substitute a nonzero value for x and solve for y to get another point_.

12. The equation of a horizontal line has the form _$y = k$, where $k$ is a real number_.

13. The equation of a vertical line has the form _$x = h$, where $h$ is a real number_.

## SECTION 3.3  The Slope of a Line

14. The slope of a line measures the _steepness_ of the line.

15. The slope is defined as the ratio of the _vertical_ change to the _horizontal_ change. It is also called _rise_ over _run_.

16. The slope-intercept form of a line is _$y = mx + b$_. The *m* represents the _slope_ and *b* represents the _y-coordinate of the y-intercept_.

17. To graph a line using the slope and *y*-intercept, plot the _y-intercept_ first. From this point, use the _slope_ to move

to another point on the line. For instance, a slope of $\dfrac{4}{3}$ would mean to _move up 4 units_ and _move right 3 units_.

18. A horizontal line has a slope of _zero_ and a vertical line has a slope that is _undefined_.

19. The slope formula is _$m = \dfrac{y_2 - y_1}{x_2 - x_1}$_.

20. A positive slope corresponds to a line that is _rising_. A negative slope corresponds to a line that is _falling_. A slope of zero corresponds to a _horizontal_ line. An undefined slope corresponds to a _vertical_ line.

## SECTION 3.4  More About Slope

21. In application problems, the value of *b* represents the _initial_ value. The value *m* represents the _rate of change_.

22. Two lines are _parallel_ if they have the same _slope_.

23. Two lines are _perpendicular_ if their slopes are _negative_ reciprocals.

## SECTION 3.5  Writing Equations of Lines

24. To write the equation of a line, we must know the _slope_ and at least one _point_. The _slope-intercept_ form of a line or the _point-slope_ form of a line can be used to find the equation.

## SECTION 3.6  Functions

25. A(n) _relation_ is a set of ordered pairs. The set of *x*-values is the _domain_, and the set of *y*-values is the _range_.

26. A _function_ is a relation in which each input value corresponds to _exactly one_ output value.

27. The _vertical_ line test can be used to determine if a graph represents a function. If every vertical line touches at most _one point_ on the graph, then the graph is a function.

28. In function notation, we use the symbol _$f(x)$_ for *y*.

29. To write a linear equation in two variables in function notation, we must first solve the equation for _$y$_.

30. If $f(a) = b$, then the point _$(a, b)$_ is on the graph of $f(x)$.

---

# CHAPTER 3 / REVIEW EXERCISES

## SECTION 3.1

**Determine algebraically if the given ordered pair is a solution of the equation. (*See Objective 1.*)**

1. $(1, -1)$; $3x - 2y = 5$   
   yes
2. $(-3, -0.75)$; $x + 6y = -7.5$   
   yes
3. $(-2, 1)$; $y = |3 - x|$   
   no
4. $(4, 1)$; $y = -x^2 + 8x - 15$   
   yes
5. $(-2, 0)$; $y = \dfrac{x - 2}{x + 2}$   
   no
6. $(0, 3)$; $y = \dfrac{2x - 3}{x + 1}$   no

**Plot each ordered pair on the Cartesian coordinate system and state the quadrant or axis where each point is located. (*See Objective 2.*)**

7. $(-5, 1)$  II
8. $(2, 0)$  *x*-axis
9. $(-3, -2.5)$  III
10. $\left(\dfrac{7}{2}, 4\right)$  I
11. $(0, -4)$  *y*-axis
12. $(1, -6)$  IV

**Identify each point plotted on the coordinate system and state the quadrant or axis where each point is located. (*See Objective 2.*)**

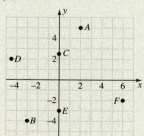

13. A  $(2, 5)$, I
14. B  $(-3, -4)$, III
15. C  $(0, 2.5)$, *y*-axis
16. D  $(-4.5, 2)$, II
17. E  $(0, -3)$, *y*-axis
18. F  $(6, -2)$, IV

Identify the quadrant or axis where each point is located.
(*See Objective 2.*)

19. $(-1, 0)$  x-axis
20. $(2, -7)$  IV
21. $(-3.2, -1.6)$  III
22. $(5.2, 0.8)$  I

Graph each equation using a table of solutions.
(*See Objective 3.*)

23. $y = 5 - 3x$
24. $y = |5 - 2x|$
25. $y = 4 - x^2$
26. $y = x^3$

Use the given graph of an equation to determine if each
ordered pair is a solution of the equation. (*See Objective 4.*)

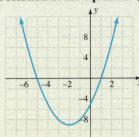

27. $(-5, 0)$  yes
28. $(0, 4)$  no
29. $(2, 4)$  no
30. $(1, 0)$  yes

Use the given graph to complete each ordered pair that is
a solution. (*See Objective 4.*)

31. ( ____, 0)  (−2, 0) and (4, 0)
32. (0, ____ )  (0, 8)
33. (1, ____ )  (1, 9)
34. ( ____, 5)  (−1, 5) and (3, 5)

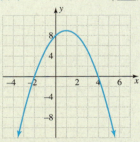

Solve each problem. (*See Objective 5.*)

35. The following graph shows the expenditures for
national health in the United States between 1995 and
2006. (Source: http://worldalmanac.com)

**National Health Expenditures in United States, in Billions (1995–2006)**

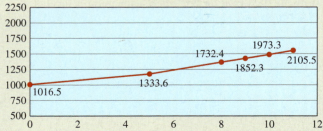

a. Write an ordered pair for each point labeled on the
graph.

b. Between what consecutive years did the expenditures
increase the least? By how much?

c. Based on the graph, how would you describe
expenditures for national health?

## SECTION 3.2

**Graph each equation. Identify at least two points on the
graph. (*See Objectives 2–4.*)**

36. $8x + y = 16$
37. $3y + 6 = 0$
38. $y + 3x = 0$
39. $5x - 10 = 0$
40. $y = 0.2x + 1.5$
41. $y = -\dfrac{2}{3}x - 1$
42. $0.6x - 0.1y = 1.2$

Use the equation to answer each question. (*See Objective 5.*)

43. The emissions of carbon monoxide, the principal air
pollutants in the United States during 1970–2008, are
on the decline. The emissions of carbon monoxide
in millions of tons can be modeled by the equation
$y = -3.36x + 213$, where $x$ is the number of years after
1970. (Source: U.S. Environmental Protection Agency, Office of
Air Quality Planning and Standards)

a. Find the $y$-intercept and interpret its meaning.

b. Find the $x$-intercept and interpret its meaning.

c. What was the emission of carbon monoxide in 2005?
The emission of carbon monoxide was about 95.4 million tons in 2005.

d. What was the emission of carbon monoxide in 2009?
The emission of carbon monoxide was about 82 million tons in 2009.

e. Use the intercepts to graph the equation.

## SECTION 3.3

**Find the slope of the line that passes through the given points
or that is defined by the given equation. (*See Objectives 1, 3,
and 4.*)**

44. $(3, -2)$ and $(3, 1)$   m = undefined
45. $y = 4$   m = 0
46. $\left(-2, -\dfrac{2}{5}\right)$ and $\left(4, \dfrac{3}{5}\right)$   $m = \dfrac{1}{6}$
47. $5x - 6y = 1$   $m = \dfrac{5}{6}$
48. $0.7x + 0.14y = 2.1$   m = −5

49. 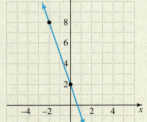   m = −3

50.

| $x$ | $y$ |
|-----|-----|
| −2 | −3 |
| −1 | −1 |
| 0 | 1 |
| 1 | 3 |
| 2 | 5 |

m = 2

Graph each line using the given slope and point. Label two
points on the line. (*See Objective 3.*)

51. $m = \dfrac{2}{7}$, $(0, -2)$
52. $m = -1$, $(0, 5)$
53. $m = -0.5$, $(2, -6)$
54. $m = $ undefined, $(-3, 1)$
55. $m = 0$, $(1, 2)$
56. $m = -4$, $(0, 0)$

## SECTION 3.4

**Interpret the meaning of the slope and *y*-intercept in the context of each situation. (*See Objective 1*.)**

**57.** Monica's car is in the shop, so she is using a rental car. The total cost in dollars for Monica to rent a compact car is given by $y = 135x$, where $x$ is the number of weeks the car is rented.

**58.** The per capita income in dollars can be modeled by the equation $y = 1155.6x + 23,507$, where $x$ is the number of years after 1995. (Source: www.infoplease.com)

**59.** What is the pitch of a roof that rises 10 in. for every horizontal move of 6 in.? (*See Objective 2.*) $\frac{5}{3}$

**60.** Determine if the lines are parallel, perpendicular, or neither: (*See Objective 3.*) $y = -\frac{2}{3}x + 1$ and $y = \frac{3}{2}x - 2.$ perpendicular

**61.** Graph the line that is parallel to $y = -x + 3$ and passes through the point $(2, -6)$. (*See Objective 4.*)

**62.** Graph the line that is perpendicular to $y = -2x + 3$ and passes through the point $(-4, -1.5)$. (*See Objective 4.*)

## SECTION 3.5

**Write the equation of each line described. Express each answer in slope-intercept form if possible. (*See Objectives 1–4.*)**

**63.** $m = -\frac{1}{4}, (-8, 1)$  $y = -\frac{1}{4}x - 1$

**64.** $(5.2, -2.3)$ and $(7.2, -4.2)$  $y = -0.95x + 2.64$

**65.** $m =$ undefined, $(-1, 2)$  $x = -1$

**66.** $(1, -2)$, parallel to $y = 4x - 1$  $y = 4x - 6$

**67.** $(2, -4)$ and $(2, 6)$  $x = 2$

**68.** $x$-intercept $(6, 0)$ and $y$-intercept $(0, -1)$  $y = \frac{1}{6}x - 1$

**69.** $(4, 1)$, perpendicular to $2x + 3y = 1$  $y = \frac{3}{2}x - 5$

**70.** $(-1, 0)$, perpendicular to $x = -5$  $y = 0$

**Solve each problem. (*See Objective 5.*)**

**71.** Amy has a new job as a graphic designer. Her starting salary is $43,500. She will receive a raise of $2160 each year she works with the company.

    **a.** Write a linear equation that represents Amy's salary where $x$ is the years she has worked with the company. $y = 2160x + 43,500$

    **b.** What is Amy's salary after 4 yr of working for the company? $52,140

    **c.** How long does she have to work for the company to earn a salary of $69,420? 12 yr

**72.** Adam registers for his first semester in college. His tuition, includes fees of $525 plus $145 per credit hour.

    **a.** Write a linear equation that represents Adam's total tuition, where $x$ is the number of credit hours Adam will take. $y = 145x + 525$

    **b.** What is Adam's tuition if he registers for 12 credit hours? $2265

    **c.** How many hours is Adam taking if his tuition is $1830? 9 credits

## SECTION 3.6

**Determine if each relation is a function. If not, explain why. (*See Objective 2.*)**

**73.** The following table shows the richest Americans in 2009 by their net worth. (Source: http://www.infoplease.com/us/statistics/richest-americans.html)  a function

| Name | Net Worth, in Billions |
|---|---|
| Bill Gates | $50 |
| Warren Buffett | $40 |
| Larry Ellison | $27 |
| Christy Walton | $21.5 |
| Jim Walton | $19.6 |
| Alice Walton | $19.3 |
| S. Robson Walter | $19 |
| Michael Bloomberg | $27.5 |
| Charles Koch | $16 |
| David Koch | $16 |

**74.** The following table shows the gas mileage of the "meanest" (i.e., most polluting) vehicles of 2010. a function

| Meanest Vehicles | Miles per Gallon |
|---|---|
| 1. Lamborghini Mucielago | 18 |
| 2. Bugatti Veyron | 8 |
| 3. Bentley Azure/Brooklands | 9 |
| 4. Maybach 57S | 10 |
| 5. Dodge Ram 2500 Mega Cab | 12 |
| 6. Ford F-250 | 12 |
| 7. Bentley Continental | 10 |
| 8. Ferrari 612 Scaglietti | 9 |
| 9. Mercedes-Benz ML 63 AMG | 11 |
| 10. Chevrolet Suburban K2500 | 13 |

(Source: http://www.infoplease.com)

**75.** $\{(2, 8), (2, -1), (1, 5), (-2, 1), (-4, 0)\}$
not a function since $x = 2$ corresponds to two different $y$-values.

**76.** Assign to each student the grade earned in their beginning algebra class. a function

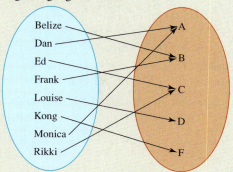

**77.** $x = 1$

**78.** $y = -\frac{1}{5}x$  a function

Use the vertical line test to determine if each relation is a function. (*See Objective 3.*)

**79.**

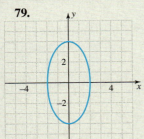

not a function

**80.**

not a function

**81.**

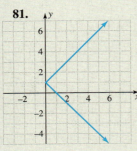

not a function

**82.**

function

Find the requested information. (*See Objective 4.*)

**83.** $f(-3)$ if $f(x) = -2x^2 + x + 3$    −18

**84.** The function $f(x)$ is given by $Y_1$. Find **a.** $f(-2)$ and **b.** $x$ if $f(x) = -14$.

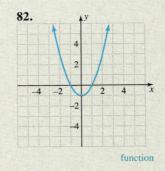

a. −19  b. 3

**85.** The graph of $f(x)$ is given below. Find **a.** $f(0)$ and **b.** $x$ if $f(x) = 4$.

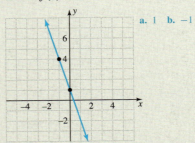

a. 1  b. −1

**86.** Find $f(1.5)$ if $f(x) = 3.4x - 0.5$.    4.6

Write each linear equation in two variables in function notation. Then find $f(0), f(3)$, and $f(-2)$ and write the corresponding ordered pairs. (*See Objective 5.*)

**87.** $2x - 7y = 21$             **88.** $y = |4x + 1|$

Solve each problem. (*See Objective 5.*)

**89.** Rikki has a new job as a math instructor at a local community college. Her starting salary is $43,200 with a regular annual teaching load of 30 credit hours. If she teaches beyond the regular load, she earns $1020 per credit hour.

   **a.** Write a linear function $f(x)$ that represents Rikki's yearly income, where $x$ is the number of extra credit hours Rikki teaches.    $f(x) = 1020x + 43{,}200$

   **b.** If Rikki teaches two 3-credit-hour courses in addition to her regular load, what is her income?    $49,320

   **c.** Find $f(9)$ and interpret the answer.
   $52,380; It means Rikki taught an additional 9 credit hours.

**90.** Wong purchases a new car for $16,400. The car's value decreases by $1325 each year.

   **a.** Write a linear function that represents the value of the car, where $x$ is the age of the car in years.
   $f(x) = -1325x + 16{,}400$

   **b.** What is the car's value after 3 yr?    $12,425

   **c.** Find $f(5)$ and interpret the answer.
   $9,775; It means the value of the car after 5 years is $9775.

## CHAPTER 3 TEST / LINEAR EQUATIONS IN TWO VARIABLES

**1.** The point $(2, -4)$ is a solution of the equation
   **a.** $y = 2x + 8$        **b.** $x = -4$
   **c.** $2x - y = 8$        **d.** $y = |x - 6|$

**2.** If $a > 0$ and $b < 0$, then the point $(a, b)$ lies in
   **a.** Quadrant I          **b.** Quadrant II
   **c.** Quadrant III        **d.** Quadrant IV

Graph each equation by creating a table of solutions.

**3.** $y = x^2 + 1$

**4.** $y = |x + 2|$

**5.** Is the point $(-1, 1)$ a solution of the equation shown in the graph?    yes

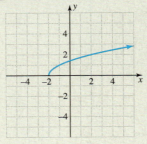

**6.** The average price (rounded to the nearest dollar) of brand name prescription drugs for certain years is shown in the table.

| Year | 1995 | 2000 | 2005 | 2006 | 2007 |
|------|------|------|------|------|------|
| Average price of brand name drugs | 40 | 65 | 98 | 107 | 120 |

   **a.** Write an ordered pair for each year, where $x$ is the number of years after 1995 and $y$ is the average price.    {(0, 40), (5, 65), (10, 98), (11, 107), (12, 120)}

   **b.** Interpret the meaning of the first and last ordered pairs.

   **c.** Plot the ordered pairs to form a scatter plot of the given data.

7. Graph each line by finding the *x*- and *y*-intercepts.

   **a.** $2x - y = 6$   (0, −6); (3, 0)   **b.** $y = \dfrac{1}{5}x + 1$   (0, 1); (−5, 0)

   **c.** $x + 3y = 0$   (0, 0)

8. The equation $x = 5$ is a __vertical__ line.

9. The equation $y = -1$ is a __horizontal__ line.

10. Suppose an airplane descends at a rate of 400 ft/min from an altitude of 8000 ft above the ground.

    **a.** Write a linear equation that represents the plane's altitude, *y*, after *x* min.   $y = -400x + 8000$

    **b.** Find the *x*-intercept and state its meaning.

    **c.** Find the *y*-intercept and state its meaning.

11. State the slope of each line.

    **a.** $y = \dfrac{4}{5}x + 3$   $m = \dfrac{4}{5}$   **b.** $3x + y = 9$   $m = -3$

    **c.** $x = 4$   $m =$ undefined   **d.** $y = 7$   $m = 0$

    **e.** Through (4, −1) and (6, 5)   $m = 3$

    **f.** Parallel to $y = 7x + 4$   $m = 7$

    **g.** Perpendicular to $y = -3x$   $m = \dfrac{1}{3}$

    **h.**

    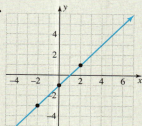
    $m = 1$

12. Graph the line $y = -\dfrac{3}{2}x + 4$ using the slope and *y*-intercept. Label the *y*-intercept and at least one other point.

13. Determine if the lines $5x + 2y = 10$ and $4x - 10y = -4$ are parallel, perpendicular, or neither.   perpendicular

14. The equation $y = 138.41x + 2961.8$ models the average undergraduate cost of attending a public 2-yr college *x* years after 1986. (Source: http://nces.ed.gov/fastfacts)

    **a.** What are the slope and *y*-intercept?
    $m = 138.41$, *y*-intercept = (0, 2961.8)

    **b.** What do they mean in the context of this problem?

    **c.** Use this model to find the cost of attending a 2-yr college in 2016.   $7114.10

15. Write the equation of the line that satisfies the given conditions.

    **a.** $m = \dfrac{3}{4}$ and passes through (0, −6)   $y = \dfrac{3}{4}x - 6$

    **b.** $m = -2$ and passes through (−7, 4)   $y = -2x - 10$

    **c.** passes through (4, −1) and (6, 5)   $y = 3x - 13$

    **d.** passes through (2, −8) and parallel to $y = 4x - 1$   $y = 4x - 16$

    **e.** passes through (3, 0) and perpendicular to $y = -3x + 9$   $y = \dfrac{1}{3}x - 1$

16. Suppose the enrollment of Einstein Community College was 16,000 in 1998 and 22,000 in 2004. If *x* is the number of years after 1998 and *y* is the college's enrollment, write a linear equation that represents Einstein's enrollment *x* years after 1998.
    $y = 1000x + 16{,}000$

17. Which of the following relations is not a function?

    **a.** {(−2, 4), (−1, 1), (0, 0), (1, 1), (2, 4)}

    **b.** $\{(x, y) \mid x =$ person and $y =$ person's Social Security number.$\}$

    **c.**       **d.**

    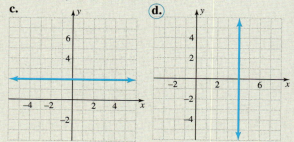

18. Find $f(0)$ and $f(-2)$ for $f(x) = x^2 - 3x + 2$.   $f(0) = 2, f(-2) = 12$

19. Write the linear equation $4x - 7y = -14$ in function notation. Then find $f(-14)$. Write the corresponding ordered pair.   $f(x) = \dfrac{4}{7}x + 2; f(-14) = -6; (-14, -6)$

20. Explain what it means for a relation to be a function. Provide a real-life example of a function.

---

## CUMULATIVE REVIEW EXERCISES / CHAPTERS 1–3

1. Evaluate each absolute value expression. (*Section 1.1, Objective 5*)

   **a.** $|6|$   6   **b.** $\left| -\dfrac{3}{4} \right|$   $\dfrac{3}{4}$   **c.** $-|-24|$   −24

2. Perform the each operation and simplify each result. (*Section 1.2, Objectives 2–7*)

   **a.** $\dfrac{15}{16} + \dfrac{7}{12} - \dfrac{17}{18}$   $\dfrac{83}{144}$

   **b.** $6\dfrac{2}{5} \cdot 3\dfrac{3}{4} \div 12$   2

   **c.** $4\dfrac{1}{5} - 1\dfrac{2}{3} + \dfrac{4}{15}$   $\dfrac{14}{5}$

3. Use the order of operations to simplify each numerical expression. (*Section 1.3, Objectives 1 and 2*)

   **a.** $-\left(-\dfrac{3}{2}\right)^3$   $\dfrac{27}{8}$   **b.** $\dfrac{1}{6}(10 - 8)^3 - \dfrac{5}{3}$   $-\dfrac{1}{3}$

   **c.** $2(7)^2 - 5(12) + 1$   39   **d.** $5[20 - 2|13 - 2(4 - 1)|]$   30

   **e.** $\dfrac{13 + \sqrt{13^2 + 5(13)(6)}}{2 \cdot 6}$   $\dfrac{13 + \sqrt{559}}{12}$

4. Evaluate each expression for the given values of the variables. (*Section 1.3, Objective 3*)

   **a.** $\dfrac{2x + 3}{x - 1}$ for $x = 0, 1, 2, 3$   −3, undefined, 7, $\dfrac{9}{2}$

   **b.** $b^2 - 4ac$ for $a = 5, b = 2, c = 15$   −296

5. The height of a baseball hit upward with an initial velocity of 112 ft/sec from an initial height of 8 ft is represented by the expression $-16t^2 + 112t + 8$, where $t$ is the number of seconds after the ball has been hit. What is the height of the ball after 4 sec? (*Section 1.3, Objective 6*)    200 ft

6. Perform the indicated operation and simplify. (*Sections 1.4 and 1.5, Objectives 1 and 2*)
   a. $5.4 - 9.2 + (-8.4) - (-11.5)$    −0.7
   b. $\dfrac{5}{6} - \dfrac{3}{4} - \left(-\dfrac{1}{4}\right)$    $\frac{1}{3}$
   c. $30 - \{7 - [4 - (2 - \sqrt{20 + 5})]\}$    30

**Write the mathematical expression needed to solve each problem and then answer any questions. (*Sections 1.4 and 1.5, Objective 3*)**

7. Sherry has $235.65 in her checking account. She deposits her paycheck of $567.45. Sherry writes a check for phone bill for $47.82, for groceries for $135.93, and then for rent for $475. What is Sherry's checking account balance?    $144.35

8. Selena wants to mix a 4% salt solution with a 30% solution to get a 25% solution. Let $x$ oz represent the volume of the 4% salt solution. If Selena wants 50 oz of 25% salt solution, write an expression that represents the volume of the 30% salt solution.    $(50 - x)$ oz

9. In a rectangle, the length is two more than four times the width. If $w$ represents the width, write an expression that represents the length of the rectangle.    $2 + 4w$

10. Find the complement and the supplement of an angle whose measure is 63°.    27°, 117°

11. Perform the indicated operation and simplify. (*Section 1.6, Objectives 1–3*)
    a. $(-5.4)(3.6)$    −19.44
    b. $(-36)\left(\dfrac{5}{16}\right)\left(-\dfrac{8}{15}\right)$    6
    c. $-(-5)^3$    125
    d. $\dfrac{-7}{0}$    undefined
    e. $6(-15)(-4)$    360
    f. $\left(\dfrac{15}{14}\right) \div \left(-\dfrac{10}{21}\right)$    $-\frac{9}{4}$

12. Evaluate each expression for the given values of the variables. (*Section 1.6, Objective 4*)
    a. $b^2 - 4ac$ for $a = -2, b = 1, c = 5$    41
    b. $\dfrac{|3 + x|}{x - 1}$ for $x = -1, 0, 1$    −1, −3, undefined

13. Apply the commutative, associative, and/or distributive properties to simplify each expression. (*Section 1.7, Objectives 2 and 3*)
    a. $13 + c - (-25)$    $c + 38$
    b. $-4(x + 3y - 8)$    $-4x - 12y + 32$
    c. $\left(x - \dfrac{11}{6}\right) + 2$    $x + \frac{1}{6}$
    d. $8\left(\dfrac{3}{4}x - \dfrac{1}{2}\right)$    $6x - 4$

14. Simplify each algebraic expression. (*Section 1.8, Objectives 3 and 4*)
    a. $3(4x - 5) - 16x$    $-4x - 15$
    b. $-(2x - 7) - 3(-2x + 5)$    $4x - 8$
    c. $4.8 - 0.3(-0.5x + 1.4)$    $0.15x + 4.38$

d. $15\left(\dfrac{2}{5}x - \dfrac{1}{3}\right) - 10\left(\dfrac{1}{2}x + \dfrac{4}{5}\right)$    $x - 13$

e. $5x^2 - 7(x + x^2) - 8x$    $-2x^2 - 15x$

15. Determine whether each of the following is an expression or an equation. (*Section 2.1, Objective 1*)
    a. $5x - 28 = 2x - 12$    $\frac{16}{3}$
    b. $x - 10 + 3x + 5$    $4x - 5$

16. Determine if the given numbers are solutions of the equation. (*Section 2.1, Objective 2*)
    a. Is $x = -2, x = 0$, or $x = 1$ a solution of $7x - 3 = -x - 19$?    $x = -2$ is a solution, $x = 0$ and $x = 1$ are not solutions
    b. Is $x = -2, x = 0$, or $x = 3$ a solution of $x - 2 = -4x + 13$?    $x = -2$ and $x = 0$ are not solutions, $x = 3$ is a solution

17. For each exercise, define the variable, write an equation, and solve the problem. (*Section 2.1, Objectives 3 and 4*)
    a. The sum of a number and 32 is the same as three times the number. Find the number.    16
    b. The difference of three times a number and $-5$ equals 7 less than the number. Find the number.    −6
    c. One angle is 14° more than another angle. Their sum is 90°. Find the measures of each angle.    38° and 52°

18. Solve each equation. (*Section 2.2, Objectives 2 and 3; Sections 2.3 and 2.4, Objectives 1 and 2*)
    a. $3(x - 2) - 28 = 5(x - 1) + 9$    −19
    b. $5(1 - x) - 7x = 13 - (2 - 4x)$    $-\frac{3}{8}$
    c. $-\dfrac{2}{5}(5x + 10) + 3 = x - 5$    $\frac{4}{3}$
    d. $-0.2y + 14.1 = 2.3y + 4.1$    4

**For each exercise: (1) assign a variable to the unknown, (2) write an equation that represents the situation, (3) solve the equation, and (4) answer the question using complete sentences. (*Section 2.2, Objective 4; Section 2.3, Objectives 3 and 4*)**

19. The cost to rent a midsize car for one day is $68 plus $0.36 per mile. This daily cost is represented by the equation $68 + 0.36x$, where $x$ is the number of miles driven. How many miles has the car been driven if the cost of the rental car is $199.40?    365 mi

20. The sum of three consecutive odd integers is 75. Find the integers.    23, 25, 27

21. The sum of the two smaller of three consecutive integers is the same as 30 less than three times the largest integer. Find the integers.    25, 26, 27

22. Solve each equation. If the equation is a contradiction, write the solution as ∅. If the equation is an identity, write the solution as ℝ. (*Section 2.4, Objectives 1–4*)
    a. $7(2y - 1) - (y - 5) = 3(y + 2) + 8$    $\frac{8}{5}$
    b. $0.7x + 1.1x = 7.2$    4
    c. $0.015x + 0.024(480 - x) = 8.82$    300
    d. $-\dfrac{1}{2}(x - 3) = \dfrac{1}{3}(2x + 9) + \dfrac{1}{6}$    $-\frac{10}{7}$

**23.** The formula $A = P(1 + r)^t$ is used to calculate the amount of money in an account at the end of $t$ years if $P$ is invested at an annual interest rate, $r$. *(Section 2.5, Objective 1)*

   **a.** Find the value of $A$ if $960 is invested for 3 yr at 1.4% annual interest.   $1000.89

   **b.** Find the value of $P$ if $2,930.88 is expected to be accumulated at the end of 2 yr at 1.6% annual interest.   $2839.30

**24.** The revenue $R$, unit price $p$, and sale level $x$ of a product are related by the formula $R = xp$. *(Section 2.5, Objective 1)*

   **a.** Find $R$ if $p = $86.50$ and $x = 40$.   $3460

   **b.** Find $p$ if $R = $3212.50$ and $x = 125$.   $25.70

**25.** Solve each formula for the specified variable. *(Section 2.5, Objective 3)*

   **a.** $V = lwh$ for $w$   $w = \dfrac{V}{lh}$   **b.** $C = 8l + 3w$ for $w$   $w = \dfrac{C - 8l}{3}$

   **c.** $5x - 9y = 18$ for $y$   **d.** $0.04x + 0.03y = 7.2$ for $x$

**26.** Find the measure of each unknown angle. *(Section 2.5, Objective 5)*

   **a.** Find the measure of an angle whose complement is 40° less than the measure of the angle.   65°

   **b.** The supplement of an angle is 12° more than three times its complement. Find the measure of the angle.   51°

**27.** Find the measure of each angle labeled in the figure. *(Section 2.5, Objectives 4 and 5)*

   **a.**   61°, 119°

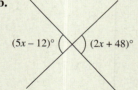
$(x + 11)°$   $(3x - 31)°$

   **b.**   88°, 88°

$(5x - 12)°$   $(2x + 48)°$

**28.** In triangle $ABC$, the measure of angle $B$ is 36° less than the measure of angle $A$. The measure of angle $C$ is 21° more than the measure of angle $A$. Find the measure of each angle in the triangle. *(Section 2.5, Objective 5)*
65°, 29°, 86°

**Write an equation that represents each situation, solve the equation, and write the answer in complete sentences.** *(Section 2.6, Objectives 1–5)*

**29.** A 46-in. LED-LCD HDTV is on sale for $1202.50. This is 35% off the original price. What is the original selling price of the TV?   $1850

**30.** A suit is on sale for $138.75. If this price is 25% off the original price, what is the original price of the suit?   $185

**31.** Sue invested $3000 in two accounts. She invested $1800 in a mutual fund that pays 2.1% annual interest and the remaining account in a money market fund that pays 1.2% annual interest. How much interest did she earn from the two accounts in 1 yr?   $52.20

**32.** Solve each inequality. Graph the solution set and write the solution set in interval notation and in set-builder notation. *(Section 2.7, Objectives 1–4)*

   **a.** $0.3a - 1.2 \geq -3.6$   **b.** $-7 < 2b - 15 < 19$

   **c.** $2x + (x - 9) \geq 8x + 16$

**33.** Determine algebraically if the given ordered pair is a solution of the equation. *(Section 3.1, Objective 1)*

   **a.** $(2, -5)$; $2x - y = 15$   (2, −5) is not a solution

   **b.** $(3, -1)$; $y = -x^2 - 3x + 1$   (3, −1) is not a solution

   **c.** $(1, 5)$; $y = |4x + 1|$   (1, 5) is a solution

**34.** Plot each ordered pair on the Cartesian coordinate system and state the quadrant or the axis where each point is located. *(Section 3.1, Objective 2)*

   **a.** $(2, -7)$   **b.** $(-3, -9)$

   **c.** $(-6, 0)$   **d.** $(5, -3)$

   **e.** $(0, 5)$

**35.** Graph each equation using a table of solutions. *(Section 3.1, Objective 3)*

   **a.** $y = 2x - 6$   **b.** $y = 3 - x^2$

   **c.** $y = |x| + 3$   **d.** $y = 4 - |x|$

**36.** Use the given graph to complete each ordered pair that is a solution. *(Section 3.1, Objective 4)*

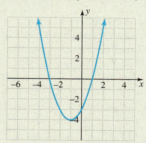

   **a.** $(\underline{\quad}, 0)$   (−3, 0) and (1, 0)   **b.** $(0, \underline{\quad})$   (0, −3)

   **c.** $(-1, \underline{\quad})$   (−1, −4)

**37.** The graph shows the average annual salary of full-time teachers in public elementary and secondary schools, 2000 to 2008. *(Section 3.1, Objective 5)* (Source: http://nces.ed.gov)

**Average Annual Salary of Teachers in Public Elementary and Secondary Schools (2002–2008)**

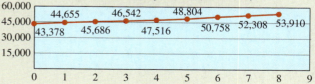

   **a.** Write an ordered pair for each point labeled on the graph.   (0, 43,378), (1, 44,655), (2, 45,686), (3, 46,542), (4, 47,516), (5, 48,804), (6, 50,758), (7, 52,308), (8, 53,910)

   **b.** Between what consecutive years did the average annual salary increase most? By how much?   Between 2007 and 2008; $1954

**38.** The table lists the annual profit (in billions of dollars) for Delta Airlines between 2000 and 2006. (*Section 3.1, Objective 5*) (Sources: *The Atlanta Journal-Constitution*, Sept. 15, 2005 and www.delta.com)

| Years After 2000 | 0 | 1 | 2 | 3 | 4 | 5 | 6 |
|---|---|---|---|---|---|---|---|
| Annual profit (in billions) | 1.23 | −1.21 | −1.27 | −0.773 | −5.2 | −1.46 | 0.058 |

    **a.** Write an ordered pair $(x, y)$ that corresponds to each year between 2000 and 2006, where $x$ is the number of years after 2000 and $y$ is the profit of Delta Airlines (in billions of dollars).

    **b.** Interpret the meaning of the first and last ordered pairs from part (a) in the context of the problem.

    **c.** In what year was the profit the highest? The lowest?

    **d.** Make a scatter plot of the data.

**39.** Graph each equation. Identify at least two points on the graph. (*Section 3.2, Objectives 2–4*)

    **a.** $2x + 7y = 14$   **b.** $y + 5 = 0$   **c.** $x − 6 = 0$

    **d.** $3x + y = 0$   **e.** $x − 0.5y = 1$   **f.** $y = −2x + 1$

    **g.** $0.4x − 0.5y = 2$

**40.** The number of full-time classroom teachers (in thousands) in elementary and secondary schools in the United States can be modeled by the equation $y = 32.5x + 3649.1$, where $x$ is the number of years after 2008. (*Section 3.2, Objective 5*) (Source: http://www .census.gov/compendia/statab/2010/tables/10s0216.pdf )

    **a.** Find the $y$-intercept and interpret its meaning. Is this realistic?

    **b.** How many full-time classroom teachers in elementary and secondary schools does the model predict in 2010?

    **c.** How many full-time classroom teachers in elementary and secondary schools does the model predict in 2014?

    **d.** Use the intercepts to graph the equation.

**41.** Find the slope of each line described. (*Section 3.3, Objectives 1, 3, 4*)

    **a.** $y = −\dfrac{2}{3}x + 4$   $−\dfrac{2}{3}$

    **b.** passes through $(2, −4)$ and $(4, 0)$  2

    **c.** $y = 6$  0       **d.** $x = −5$  undefined

    **e.** passes through $(−3.2, −1.4)$ and $(1.2, −10.2)$  −2

    **f.** $5x − 3y = 2$  $\dfrac{5}{3}$

    **g.**  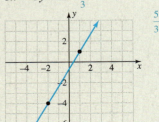  $\dfrac{5}{3}$

**42.** Graph each line described. Label two points on the line. (*Section 3.4, Objective 2*)

    **a.** $m = 2, (0, 5)$      **b.** $m = 5, (0, 3)$

    **c.** $m = −3, (0, 1)$    **d.** $m = −4, b = 3$

    **e.** $m = 0.5, b = −2$   **f.** $m =$ undefined, $(−2, 1)$

    **g.** $m = 0, (3, −4)$

**43.** The number of associate degrees (in thousands) conferred in all higher education institutions can be modeled by the equation $y = 10.93x + 73,764$, where $x$ is the number of years after 2008. Interpret the meaning of the slope and $y$-intercept in the context of the problem. Use the slope and $y$-intercept to create a table of three ordered pairs that satisfy the given equation. (*Section 3.4, Objective 1*)

**44.** Determine if each pair of lines is parallel, perpendicular, or neither. (*Section 3.4, Objective 3*)

    **a.** $y = \dfrac{3}{2}x + 1$ and $y = \dfrac{3}{2}x − 4$  parallel

    **b.** $2x + 5y = 1$ and $2x − 5y = 2$  neither

    **c.** $y = 1$ and $x = 2$  perpendicular

    **d.** $y = 0.5x$ and $y = −2x$  perpendicular

**45.** Graph the line that is parallel to $y = −3x + 2$ and passes through the point $(−4, 1)$. (*Section 3.4, Objective 4*)

**46.** Graph the line that is perpendicular to $y = 2x − 3$ and passes through the point $(−2, 3)$. (*Section 3.4, Objective 4*)

**47.** Write the equation of the line described. Express your answer in slope-intercept form and in standard form if possible. (*Section 3.5, Objectives 1–4*)

    **a.** $(−5, 6)$ and $(2, −1)$  $y = −x + 1; x + y = 1$

    **b.** $m =$ undefined, $(3, 6)$  $x = 3$

    **c.** $m = 0, (0, 12)$  $y = 12$

    **d.** $(−2, 4)$, parallel to $2x + y = 5$  $y = −2x; 2x + y = 0$

    **e.** $(3, −5)$, perpendicular to $x + 3y = 1$

    **f.** $x$-intercept $(−4, 0)$ and $y$-intercept $(0, 3)$

    **g.** $m = −4$, $x$-intercept: $(2, 0)$  $y = −4x + 8; 4x + y = 8$

**48.** Michael enrolls in a fitness club. There is a one-time membership fee of \$250 plus a monthly charge of \$40. (*Section 3.5, Objective 5*)

    **a.** Write a linear equation that represents the total cost of joining the fitness club, where $x$ is the number of months Michael is a member.  $y = 250 + 40x$

    **b.** How much money has Michael spent for his fitness club membership if he has been a member for 1 year?  \$730

    **c.** How long has Michael been a member of the club if he pays the fitness club a total of \$1450?  30 months

**49.** State the domain and range of each relation. Determine if each relation is a function. (*Section 3.6, Objectives 1–3*)

**a.** The following table shows the average number of paid vacation days per year employees receive in nine countries. (Source: http://www.infoplease.com)

| Countries | Average Number of Paid Vacation Days per Year |
|---|---|
| Italy | 42 |
| France | 37 |
| Germany | 35 |
| Brazil | 34 |
| United Kingdom | 28 |
| Canada | 26 |
| Korea | 25 |
| Japan | 25 |
| United States | 13 |

**b.** The following table shows the green score for the "greenest" vehicles in 2010. (Source: http://www.infoplease.com)

| Greenest Vehicles | Green Score |
|---|---|
| 1. Honda Civic CX | 57 |
| 2. Toyota Prius | 52 |
| 3. Honda Civic Hybrid | 51 |
| 4. Smart For Two Convertible/Coupe | 50 |
| 5. Honda Insight | 50 |
| 6. Ford Fusion Hybrid/Mercury Milan Hybrid | 47 |
| 7. Toyota Yaris | 46 |
| 8. Nissan Altima Hybrid | 46 |
| 9. Mini Cooper | 45 |
| 10. Chevrolet Cobalt XFE/Pontiac G5 XFE | 45 |

**c.** $\{(-3, 4), (-3, 1), (0, 9), (-5, 1), (2, 4)\}$

**d.** $x = 6$    D = {6}, R = $(-\infty, \infty)$, $x = 6$ is not a function

**e.** $y = 3x + 4$    D = $(-\infty, \infty)$, R = $(-\infty, \infty)$, $y = 3x + 4$ is a function

**50.** Find the requested information. (*Section 3.6, Objective 4*)

**a.** Find $f(0)$ if $f(x) = 5x - 2$.   $f(0) = -2$

**b.** Find $g(-5)$ if $g(x) = -3x^2 - x + 7$.   $g(-5) = -63$

**c.** Find $f(2)$ and $f(5)$ if $f(x)$ is given by the following graph.

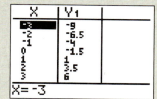

$f(2) = 3$ and $f(5) = 0$

**d.** Find $g(-1)$ and find $x$ such that $g(x) = 3.5$ if the function $g(x)$ is given by $Y_1$.

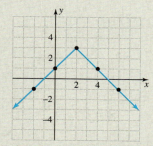

$g(-1) = -4$ and $g(2) = 3.5$

**51.** Write each linear equation in two variables in function notation. Then find $f(0), f(3),$ and $f(-2)$ and write the corresponding ordered pairs. (*Section 3.6, Objective 4*)

**a.** $4x + y = 3$        **b.** $y = 0.3x - 0.6$

**c.** $y = |3x - 2|$

**52.** Beva has a new job as a chemistry instructor at a local community college. Her starting salary is $41,500 with a regular teaching load of 30 credit hours. If she teaches beyond the regular load, she earns $950 per additional credit hour. (*Section 3.6, Objective 5*)

**a.** Write a linear function $f(x)$ that represents Beva's yearly income, where $x$ is the number of extra credit hours she teaches in an academic year.   $f(x) = 41,500 + 950x$

**b.** If Beva teaches two additional 4-credit-hour courses in an academic year, what is her income?   $49,100

**c.** Find $f(6)$ and interpret the answer.
$f(6) = $47,200$; Beva teaches 6 extra credit hours beyond the regular load and her income is $47,200.

# Systems of Linear Equations and Inequalities in Two Variables

## Study Skills

Studying math is very much like studying a foreign language. The way to learn a foreign language is to learn the alphabet, learn words, learn the rules of grammar, and so on. Learning mathematics takes a similar approach. You must learn the notation and symbols, definitions, mathematical rules and processes, and so on. If you do not have a firm foundation with the basics of mathematics, then it will be very difficult to build upon this foundation.

To learn mathematics, you must employ certain study skills. Some of these include

- Taking notes effectively.
- Reviewing notes on a regular basis.
- Studying new material as soon after class as possible.
- Seeking help on topics you do not understand before the next class meeting.
- Thinking about your performance and not comparing yourself with classmates since everyone comes from a different mathematical background.

**Question For Thought:** What study techniques have been successful for you in other courses? Can they work for you in this class? What is your biggest obstacle in studying?

## Chapter Outline

## Coming Up...

In Section 4.4, we will learn how to use systems of equations to solve this problem. As of 2010, Nintendo DS and Nintendo Wii were the two top-selling gaming units in the United States. A total of 81 million gaming units were sold. If there were 13 million more Nintendo DS units sold than Nintendo Wii units, how many of each gaming unit was sold? (Source: http://www.vgchartz.com/home.php)

> "Keep your mind on your objective, and persist until you succeed. Study, think, and plan."

—W. Clement Stone (Philanthropist, Author)

**Solving Systems of Linear Equations Graphically**

**In this chapter,** we will focus on solving systems of linear equations and inequalities. We will use our graphing skills from Chapter 3 to introduce one method of solving systems of equations. In addition, two other methods for solving systems of equations will be presented. Applications of linear systems will also be discussed. We will finish the chapter by solving linear inequalities and systems of linear inequalities in two variables.

▶ **OBJECTIVES**

As a result of completing this section, you will be able to

1. Determine if an ordered pair is a solution of a system of linear equations.
2. Determine the solution of a system of linear equations from a graph.
3. Solve a system of linear equations graphically.
4. Solve special cases of systems of linear equations graphically.
5. Determine how the lines in a system of linear equations relate, the number of solutions, and the type of system without graphing.
6. Solve applications using systems of linear equations.
7. Troubleshoot common errors.

Simone is considering two different cell phone plans but is not sure which plan is best for her. CellOne Company offers a plan that charges $60 a month for 1000 min and $0.40 for each additional minute. Phone-a-Friend Company offers a plan that charges $70 a month for 1000 min and $0.25 for each additional minute. Which plan should Simone choose?

To solve this problem, we will use a system of linear equations. We will learn how to solve the system graphically in this section.

## Solutions of Systems of Linear Equations

A **system of linear equations** is a set of two or more linear equations that must be solved together. The focus in this chapter will be on solving linear systems that contain two linear equations in two variables. Some examples of systems of linear equations in two variables are

$$\begin{cases} x + 6y = -11 \\ 2x - y = 4 \end{cases} \qquad \begin{cases} y = \frac{1}{2}x + 1 \\ x - y = 3 \end{cases} \qquad \begin{cases} x + y = 20 \\ 0.05x + 0.10(20 - x) = 1.75 \end{cases}$$

System equations are often grouped together with a brace to indicate that they belong to the system.

A **solution of a system of linear equations** is an ordered pair that satisfies both equations in the system.

**Objective 1** ▶

Determine if an ordered pair is a solution of a system of linear equations.

---

**Procedure: Determining if an Ordered Pair Is a Solution of a Linear System**

**Step 1:** Substitute the given ordered pair into each equation in the system.
**Step 2:** Simplify the resulting equations.
**Step 3:** Determine if the resulting equations are true or false.
   **a.** If both equations are true, then the ordered pair is a solution of the system.
   **b.** If one or both equations are false, then the ordered pair is not a solution of the system.

---

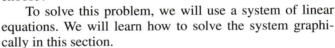

**Objective 1 Examples** Determine if the given ordered pair is a solution of the system.

**1a.** $(1, -2)$, $\begin{cases} x + 6y = -11 \\ 2x - y = 4 \end{cases}$ **1b.** $(-5, -4)$, $\begin{cases} 3x + 2y = -23 \\ 4x - y = -24 \end{cases}$

**Solutions** **1a.** Let $(x, y) = (1, -2)$.

$$x + 6y = -11 \qquad\qquad\qquad 2x - y = 4$$
$$1 + 6(-2) = -11 \qquad\qquad 2(1) - (-2) = 4$$
$$1 - 12 = -11 \qquad\qquad\qquad 2 + 2 = 4$$
$$-11 = -11 \qquad\qquad\qquad\qquad 4 = 4$$
$$\text{True} \qquad\qquad\qquad\qquad \text{True}$$

Since the ordered pair makes both equations true, $(1, -2)$ is a solution of the system.

**1b.** Let $(x, y) = (-5, -4)$.

$$3x + 2y = -23 \qquad\qquad\qquad 4x - y = -24$$
$$3(-5) + 2(-4) = -23 \qquad\qquad 4(-5) - (-4) = -24$$
$$-15 - 8 = -23 \qquad\qquad\qquad -20 + 4 = -24$$
$$-23 = -23 \qquad\qquad\qquad\qquad -16 = -24$$
$$\text{True} \qquad\qquad\qquad\qquad \text{False}$$

Since the ordered pair makes one of the equations false, $(-5, -4)$ is not a solution of the system.

✓ **Student Check 1** Determine if the given ordered pair is a solution of the system.

**a.** $(-6, 3)$, $\begin{cases} 2x + y = 15 \\ x - 4y = -18 \end{cases}$     **b.** $\left(\dfrac{1}{2}, 3\right)$, $\begin{cases} y = 8x - 1 \\ 6x - y = 0 \end{cases}$

## Use a Graph to Determine Solutions of a System

**Objective 2** ▶

Determine the solution of a system of linear equations from a graph.

As we know from Objective 1, a solution of a system of linear equations is an ordered pair that satisfies both equations. The system containing the equations $x - 3y = 6$ and $x + y = 2$ is shown in the graph. Each of these linear equations has infinitely many solutions. When we graph the two lines on the same coordinate system, we find that they share a common point. This common point is the solution of the system.

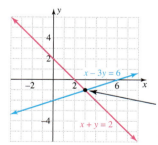

$(3, -1)$ is a point on both lines and is the solution of the system containing these two linear equations.

Graphically, a solution of a system of two linear equations is an ordered pair that the two lines have in common or the point where the two lines intersect. This is called the *point of intersection.*

> **Procedure: Determining the Solution of a System of Linear Equations from a Graph**
>
> **Step 1:** Identify the point where the two lines intersect.
> **Step 2:** Check that this point is the solution of the system by substituting the coordinates in each of the equations.

**Objective 2 Examples**   **Use the graph of the linear system to determine its solution.**

**2a.** The graph of the system

$$\begin{cases} x - y = 4 \\ 2x + y = 2 \end{cases}$$  is given.

**2b.** The graph of the system

$$\begin{cases} y = 3x - 6 \\ y = -\dfrac{5}{2}x + 5 \end{cases}$$  is given.

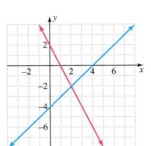     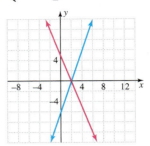

**Solutions**    **2a.** The lines intersect at the point $(2, -2)$.

**Check:**

| | |
|---|---|
| $x - y = 4$ | $2x + y = 2$ |
| $2 - (-2) = 4$ | $2(2) + (-2) = 2$ |
| $2 + 2 = 4$ | $4 - 2 = 2$ |
| $4 = 4$ | $2 = 2$ |
| True | True |

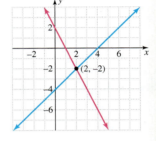

Both equations are true, so $(2, -2)$ is the solution of the system.

**2b.** The lines intersect at the point $(2, 0)$.

**Check:**

| | |
|---|---|
| $y = 3x - 6$ | $y = -\dfrac{5}{2}x + 5$ |
| $0 = 3(2) - 6$ | $0 = -\dfrac{5}{2}(2) + 5$ |
| $0 = 6 - 6$ | $0 = -5 + 5$ |
| $0 = 0$ | $0 = 0$ |
| True | True |

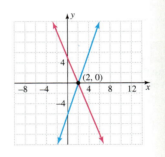

Both equations are true, so $(2, 0)$ is the solution of the system.

**✓ Student Check 2**   Use the graph of the linear system to determine its solution.

**a.** The graph of the system

$$\begin{cases} x + y = 5 \\ x - 4y = -10 \end{cases}$$  is given.

**b.** The graph of the system

$$\begin{cases} y = \dfrac{2}{3}x - 2 \\ y = -\dfrac{1}{3}x - 5 \end{cases}$$  is given.

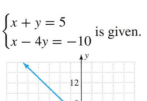
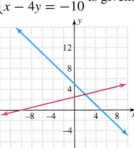     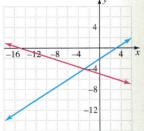

### Use Graphing to Solve Systems of Linear Equations

**Objective 3** ▶

Solve a system of linear equations graphically.

To solve a system graphically means that we need to find the point of intersection of the graphs of the equations. Since the systems will contain linear equations in two variables, we can use the methods presented in Chapter 3 to graph the equations in the systems. Recall that we can graph linear equations by plotting points, finding the $x$- and $y$-intercepts, or by using the slope and $y$-intercept.

Precision is very important when solving a system graphically. Be sure to use graph paper and a straight edge to construct the graphs. Solving a system by graphing is not as precise as other methods that we will study later in this chapter but it is a method that enables us to visualize the solution.

---

**Procedure:  Solving a System of Linear Equations Graphically**

**Step 1:** Graph each equation on the same set of axes.
**Step 2:** Identify the ordered pair of the intersection point. This is the solution of the system.
**Step 3:** Verify that the ordered pair is the solution by substituting the point in each equation.

---

**Objective 3  Examples**    **Solve each system of linear equations graphically.**

**3a.** $\begin{cases} x + y = -3 \\ x - y = 1 \end{cases}$     **3b.** $\begin{cases} 2x - y = 8 \\ y = -2 \end{cases}$

**Solutions**    **3a.** We graph the equations in the system by finding the intercepts.

| $x + y = -3$ | | |
|---|---|---|
| $x$ | $y$ | $(x, y)$ |
| $x + 0 = -3$ <br> $x = -3$ | $0$ | $(-3, 0)$ |
| $0$ | $0 + y = -3$ <br> $y = -3$ | $(0, -3)$ |

| $x - y = 1$ | | |
|---|---|---|
| $x$ | $y$ | $(x, y)$ |
| $x - 0 = 1$ <br> $x = 1$ | $0$ | $(1, 0)$ |
| $0$ | $0 - y = 1$ <br> $-y = 1$ <br> $y = -1$ | $(0, -1)$ |

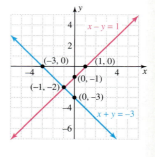

The solution of the system is the ordered pair $(-1, -2)$ since the two lines intersect at this point. We verify the solution by substituting the values into each equation in the system.

$$x + y = -3 \qquad\qquad x - y = 1$$
$$-1 + (-2) = -3 \qquad\qquad -1 - (-2) = 1$$
$$-3 = -3 \qquad\qquad\qquad -1 + 2 = 1$$
$$\qquad\qquad\qquad\qquad\qquad 1 = 1$$

$$\text{True} \qquad\qquad\qquad\qquad \text{True}$$

Since $(-1, -2)$ makes each equation in the system true, the solution set is $\{(-1, -2)\}$.

**3b.** We graph both equations in the system by using the slope and $y$-intercept.

We write $2x - y = 8$ in slope-intercept form by solving for $y$.

$$2x - y = 8$$
$$-y = -2x + 8$$
$$y = 2x - 8$$

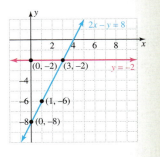

So, the slope $m = 2 = \dfrac{2}{1}$ and the $y$-intercept is $(0, -8)$. We plot $(0, -8)$ and move up 2 units and right 1 unit.

The line $y = -2$ is in slope-intercept form. It has slope $m = 0$ and $y$-intercept $(0, -2)$. It is a horizontal line through $(0, -2)$.

The solution of the system is the ordered pair $(3, -2)$ since the lines intersect at this point. We verify the solution by substituting the values into each of the equations.

| $2x - y = 8$ | $y = -2$ |
|:---:|:---:|
| $2(3) - (-2) = 8$ | $-2 = -2$ |
| $6 + 2 = 8$ | |
| $8 = 8$ | |
| True | True |

Since the ordered pair $(3, -2)$ makes each equation true, the solution set is $\{(3, -2)\}$.

---

✔ **Student Check 3**   Solve each system of linear equations graphically.

**a.** $\begin{cases} x - 3y = -6 \\ 4x - 3y = 12 \end{cases}$     **b.** $\begin{cases} y = -\dfrac{2}{5}x + 4 \\ x = -5 \end{cases}$

---

## Special Cases of Systems of Linear Equations

**Objective 4** ▶

Solve special cases of systems of linear equations graphically.

In Examples 1–3, the graphs of the lines in the system of linear equations intersected once, so there was one solution of the system of linear equations. There are two other possibilities for the graphs of the lines in a system of linear equations. The lines in the system can be parallel or they can be the same line, as shown.

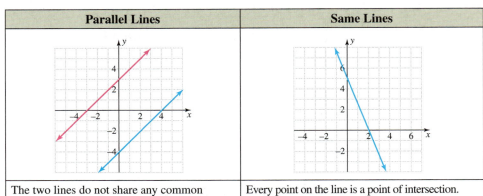

| Parallel Lines | Same Lines |
|---|---|
| The two lines do not share any common points. So, there is *no solution* of the system. | Every point on the line is a point of intersection. So, there are *infinitely many solutions* of the system. |

- A system that contains parallel lines and has no solution is called an **inconsistent system**.
- A system that contains the same lines with infinitely many solutions is called a **consistent system with dependent equations**.
- A system that contains intersecting lines with one solution is called a **consistent system with independent equations**.

---

**Procedure: Determining Solutions of Systems That Are Special Cases**

**Step 1:** Graph each equation in the system.
   **a.** If the lines are parallel, then the system has no solution. (The lines will have the same slope and different $y$-intercepts.)
   **b.** If the lines are the same, then the system has infinitely many solutions. (The lines will have the same slope and same $y$-intercept.)

**Step 2:** Write the solution set appropriately.

---

**Objective 4 Examples**  Solve each system of equations graphically.

**4a.** $\begin{cases} 2x - y = 6 \\ 4x - 2y = -8 \end{cases}$

**4b.** $\begin{cases} x + 3y = 9 \\ -2x - 6y = -18 \end{cases}$

**Solutions**  **4a.** We graph each equation using its slope and $y$-intercept by first solving each equation for $y$.

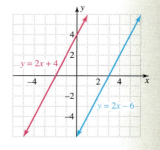

$$2x - y = 6 \qquad\qquad 4x - 2y = -8$$
$$-y = -2x + 6 \qquad -2y = -4x - 8$$
$$y = 2x - 6 \qquad\qquad y = 2x + 4$$
$$m = 2 = \frac{2}{1} \qquad\qquad m = 2 = \frac{2}{1}$$

$y$-intercept: $(0, -6)$  |  $y$-intercept: $(0, 4)$

From the slope-intercept form of the lines, we see that the lines have slope $m = 2$ and different $y$-intercepts. So, the two lines in this system are parallel, which means there is no solution of the system. We state the solution set as the empty set, $\varnothing$. The system is inconsistent.

**4b.** We graph each equation using its slope and $y$-intercept by first solving each equation for $y$.

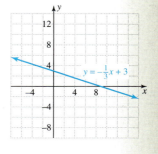

$$x + 3y = 9 \qquad\qquad -2x - 6y = -18$$
$$3y = -x + 9 \qquad\qquad -6y = 2x - 18$$
$$y = -\frac{1}{3}x + 3 \qquad\qquad y = -\frac{2}{6}x + 3$$
$$m = -\frac{1}{3} \qquad\qquad y = -\frac{1}{3}x + 3$$

$y$-intercept: $(0, 3)$  |  $m = -\frac{1}{3}$

$y$-intercept: $(0, 3)$

From the slope-intercept form of the lines, we see that the lines have slope $m = -\frac{1}{3}$ and the same $y$-intercept. Lines with the same slope and same *y*-intercept are the same line. The equations are dependent and the system is consistent.

The solution set is written as $\{(x, y) \mid x + 3y = 9\}$. This set states that every ordered pair on the graph of $x + 3y = 9$ is a solution of the system.

> **Note:** *The equations in this system are multiples of one another. If we multiply the first equation by* $-2$, *we get the second equation. This type of relationship always exists when the system of linear equations contains the same lines.*
>
> $$-2(x + 3y = 9) \text{ gives us } -2x - 6y = -18$$

☑ **Student Check 4**    Solve each system of equations graphically.

a. $\begin{cases} x + 5y = -5 \\ \dfrac{1}{15}x + \dfrac{1}{3}y = -\dfrac{1}{3} \end{cases}$      b. $\begin{cases} 6x - 8y = 24 \\ y = \dfrac{3}{4}x + 3 \end{cases}$

---

**Objective 5** ▶

Determine how the lines in a system of linear equations relate, the number of solutions, and the type of system without graphing.

## Determine the Relationship of Lines in a System of Linear Equations Without Graphing

As we have seen, there are three possibilities for the solutions of systems of linear equations in two variables. The system can have one solution, no solution, or infinitely many solutions. The number of solutions is determined by the slopes and *y*-intercepts of the lines in the system. The following table summarizes the different possibilities.

| Intersecting Lines | Parallel Lines | Same Lines |
|---|---|---|
|  | | |
| One solution | No solution | Infinitely many solutions |
| Different slopes; Same or different *y*-intercepts | Same slopes; Different *y*-intercepts | Same slopes; Same *y*-intercepts |
| Consistent system with independent equations | Inconsistent system | Consistent system with dependent equations |

---

**Objective 5 Examples**    **Determine how the lines in each system of linear equations relate, the number of solutions of the system, and the type of system without graphing.**

5a. $\begin{cases} 7x + 3y = -6 \\ 2x - 5y = 10 \end{cases}$      5b. $\begin{cases} x = \dfrac{1}{4}y + 3 \\ y = 4x - 3 \end{cases}$      5c. $\begin{cases} \dfrac{3}{2}x + \dfrac{5}{3}y = 1 \\ \dfrac{9}{5}x + 2y = \dfrac{6}{5} \end{cases}$

**Solutions**  **5a.**

| | | |
|---|---|---|
| **Slope-intercept forms** | $7x + 3y = -6$ $\\ 3y = -7x - 6$ $\\ \dfrac{3y}{3} = -\dfrac{7x}{3} - \dfrac{6}{3}$ $\\ y = -\dfrac{7}{3}x - 2$ | $2x - 5y = 10$ $\\ -5y = -2x + 10$ $\\ \dfrac{-5y}{-5} = \dfrac{-2x}{-5} + \dfrac{10}{-5}$ $\\ y = \dfrac{2}{5}x - 2$ |
| **Slopes** | $m = -\dfrac{7}{3}$ | $m = \dfrac{2}{5}$ |
| **y-intercepts** | $(0, -2)$ | $(0, -2)$ |
| **How do lines relate?** | Because the slopes are different, the two lines intersect. | |
| **Number of solutions** | One solution | |
| **Type of system** | The system is consistent with independent equations. | |

**5b.**

| | | |
|---|---|---|
| **Slope-intercept forms** | $x = \dfrac{1}{4}y + 3$ $\\ 4(x) = 4\left(\dfrac{1}{4}y + 3\right)$ $\\ 4x = y + 12$ $\\ 4x - 12 = y$ | $y = 4x - 3$ |
| **Slopes** | $m = 4$ | $m = 4$ |
| **y-intercepts** | $(0, -12)$ | $(0, -3)$ |
| **How do lines relate?** | Because the slopes are the same and the y-intercepts are different, the two lines are parallel. | |
| **Number of solutions** | No solution | |
| **Type of system** | The system is inconsistent. | |

**5c.**

| | | |
|---|---|---|
| **Slope-intercept forms** | $\dfrac{3}{2}x + \dfrac{5}{3}y = 1$ $\\ 6\left(\dfrac{3}{2}x + \dfrac{5}{3}y\right) = 6(1)$ $\\ 9x + 10y = 6$ $\\ 10y = -9x + 6$ $\\ y = -\dfrac{9}{10}x + \dfrac{6}{10}$ $\\ y = -\dfrac{9}{10}x + \dfrac{3}{5}$ | $\dfrac{9}{5}x + 2y = \dfrac{6}{5}$ $\\ 5\left(\dfrac{9}{5}x + 2y\right) = 5\left(\dfrac{6}{5}\right)$ $\\ 9x + 10y = 6$ $\\ 10y = -9x + 6$ $\\ y = -\dfrac{9}{10}x + \dfrac{6}{10}$ $\\ y = -\dfrac{9}{10}x + \dfrac{3}{5}$ |
| **Slopes** | $m = -\dfrac{9}{10}$ | $m = -\dfrac{9}{10}$ |
| **y-intercepts** | $\left(0, \dfrac{3}{5}\right)$ | $\left(0, \dfrac{3}{5}\right)$ |
| **How do lines relate?** | Because the slopes and the y-intercepts are the same, the two lines are the same. | |
| **Number of solutions** | Infinitely many solutions | |
| **Type of system** | The system is consistent with dependent equations. | |

✔ **Student Check 5**    Determine how the lines in each system of linear equations relate, the number of solutions of the system, and the type of system without graphing.

a. $\begin{cases} 6x + 2y = 10 \\ 15x + 5y = 25 \end{cases}$
b. $\begin{cases} x = 2y + 3 \\ y = \dfrac{1}{2}x - 7 \end{cases}$
c. $\begin{cases} \dfrac{3}{4}x - 2y = 8 \\ x + 6y = 12 \end{cases}$

**Objective 6 ▶**

Solve applications using systems of linear equations.

## Applications

The skills learned in this section apply to many real-life situations. We can use graphs and their equations to solve systems that represent these situations.

**Objective 6 Example**    The world's rural and urban populations are modeled by the following graph. Use the graph to approximate the year when the rural and urban populations are projected to be equal. Approximate the population at this time. (Source: http://esa.un.org/unup/)

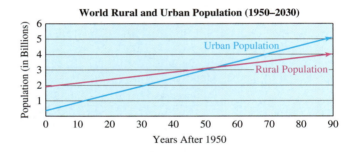

**World Rural and Urban Population (1950–2030)**

**Solution**    The point of intersection of the graph is approximately (54, 3.25). So, in 54 yr after 1950 or 2004, the urban and rural populations of the world were equal and were approximately 3.25 billion each.

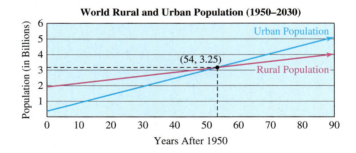

**World Rural and Urban Population (1950–2030)**

✔ **Student Check 6**    Simone is considering two different cell phone plans but is not sure which plan is best for her. CellOne Company offers a plan that charges $60 a month for 1000 min and $0.40 for each additional minute. Phone-a-Friend Company offers a plan that charges $70 a month for 1000 min and $0.25 for each additional minute. The system that is used to solve this problem is $\begin{cases} y = 0.40x + 60 \\ y = 0.25x + 70 \end{cases}$, where $x$ is the number of additional minutes. The graph of the system is given. After approximately how many minutes do the two plans have the same cost? What is this approximate cost of the plans? Based on this information, which plan should Simone choose?

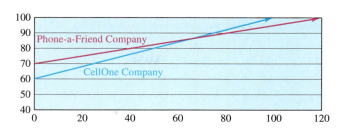

**Objective 7** ▶

Troubleshoot common errors.

## Troubleshooting Common Errors

Some common errors associated with solving systems by graphing are shown next.

**Objective 7 Examples**

**A problem and an incorrect solution are given. Provide the correct solution and an explanation of the error.**

**7a.** The graph of a system is shown. What is the solution of the system?

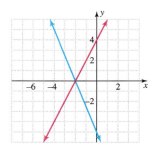

| Incorrect Solution | Correct Solution and Explanation |
|---|---|
| The solution of the system is $(0, -2)$. | The error is the order of the coordinates. Remember that the $x$-value comes first and then the $y$-value. So, the solution of the system is $(-2, 0)$. |

**7b.** Solve the system $\begin{cases} y = 3x + 7 \\ y = 4x + 6 \end{cases}$ by graphing.

| Incorrect Solution | Correct Solution and Explanation |
|---|---|
| The graph of the lines from my calculator is shown. 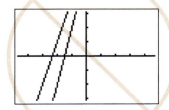 | The lines have different slopes, so they will intersect. We must adjust the viewing window to see their intersection. 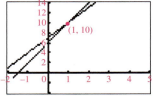 |
| Since the lines do not intersect, there is no solution. | The lines intersect at the point $(1, 10)$. So, this is the solution of the system. |

## ANSWERS TO STUDENT CHECKS

**Student Check 1**   **a.** no          **b.** yes

**Student Check 2**   **a.** $\{(2, 3)\}$    **b.** $\{(-3, -4)\}$

**Student Check 3**   **a.** $\{(6, 4)\}$    **b.** $\{(-5, 6)\}$

**Student Check 4**   **a.** $\{(x, y)\,|\,x + 5y = -5\}$

                      **b.** no solution or $\varnothing$

**Student Check 5**   **a.** same line, infinitely many solutions, consistent system with dependent equations

**b.** parallel lines, no solution, inconsistent system

**c.** intersecting lines, one solution, consistent system with independent equations

**Student Check 6**   65 min, \$85; If Simone generally talks less than 65 additional minutes, she should choose CellOne Company. Otherwise, she should select Phone-a-Friend Company.

## SUMMARY OF KEY CONCEPTS

1. A solution of a system of linear equations in two variables is an ordered pair that makes both equations in the system true. Graphically, the solution is the point where the graphs intersect.

2. There are three possible solutions of a system of linear equations in two variables.

   **a.** One ordered pair—two lines intersect in one point (Consistent system, independent equations)

   **b.** No solution—two lines are parallel and never intersect (Inconsistent system)

   **c.** Infinitely many solutions—two lines are coinciding and intersect at every point on the line (Consistent system, dependent equations)

3. The slopes and $y$-intercepts of the lines in the system determine the type of solution the system will have.

   **a.** If the slopes of the lines are different, then the lines intersect.

   **b.** If the slopes are the same and the $y$-intercepts are different, the lines are parallel.

   **c.** If the slopes and the $y$-intercepts are the same, the lines are coinciding.

## GRAPHING CALCULATOR SKILLS

The graphing calculator can be used to solve a system of equations. To solve a system on the calculator,

1. Rewrite each equation in the slope-intercept form ($y = mx + b$).
2. Enter both equations in the equation editor.
3. Graph the equations.
4. Execute the Intersect command.

**Example:** Solve the system $\begin{cases} x + y = -3 \\ x - y = 1 \end{cases}$.

**Solution:**                           $x + y = -3$          $x - y = 1$

**Step 1:** Solve each equation for $y$.        $y = -x - 3$          $-y = -x + 1$

                                                                     $y = x - 1$

**Step 2:** Enter both equations in the equation editor.

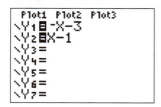

**Step 3:** Graph the equations.

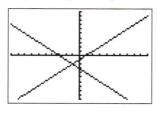

**Step 4:** Execute the Intersect command.

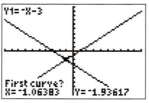

(Press Enter to accept First curve. Note the first equation is displayed in the upper left corner.)

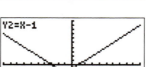

(Press Enter to accept Second curve. Note the second equation is displayed in the upper left corner.)

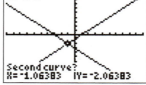

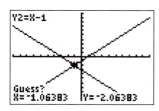

   ENTER

To input Guess, move the cursor left or right to the point of intersection and press Enter.

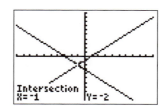

The point of intersection is displayed at the bottom of the screen. So, the solution of the system is (−1, −2). It is always important to verify the answer algebraically as well.

## SECTION 4.1 / EXERCISE SET

###  Write About It!

**Use complete sentences to answer each exercise.**

**1.** What is a linear system of equations? Answers vary.

**2.** When you graph the equations in a linear system, what part of the graph represents the solution of the system? Answers vary.

**3.** What are the steps for solving a linear system of equations by graphing? Answers vary.

**4. a.** What are the three types of solutions that a system of linear equations can have? Answers vary.

   **b.** How can you determine which of the three types of solutions that a system of linear equations has without graphing the system equations? Answers vary.

**Determine if each statement is true or false. If the statement is false, explain why.**

**5.** The system $\begin{cases} y = 4x + 5 \\ y = -4x - 3 \end{cases}$ has no solution. False, the slopes are different so the system has a solution.

**6.** The ordered pair (4, −5) is a solution of the system $\begin{cases} 2x - y = 3 \\ x + y = -1 \end{cases}$. False, (4, −5) does not satisfy the first equation.

**7.** The system $\begin{cases} 2x + y = 6 \\ 4x + 2y = 12 \end{cases}$ has no solution. False, the system has infinitely many solutions.

**8.** The system $\begin{cases} x - 2y = 4 \\ 3x - 6y = 6 \end{cases}$ has infinitely many solutions. False, the system has no solution.

###  Practice Makes Perfect!

**Determine if the given ordered pair is a solution of the system. (See Objective 1.)**

**9.** (1, 3), $\begin{cases} x - 4y = -11 \\ 2x + y = 5 \end{cases}$ yes

**10.** (4, 2), $\begin{cases} x - y = 2 \\ 3x + y = 14 \end{cases}$ yes

**11.** (−3, 5), $\begin{cases} 5x - 3y = 0 \\ x - y = -8 \end{cases}$ no

**12.** (4, −7), $\begin{cases} 4x + 3y = 5 \\ 9x - 2y = 22 \end{cases}$ no

**13.** (0, −8), $\begin{cases} y = 2x - 8 \\ y + 8 = 0 \end{cases}$ yes

**14.** (3, 5), $\begin{cases} y = -x + 8 \\ x = 3 \end{cases}$ yes

**15.** $\left(\dfrac{1}{2}, -\dfrac{1}{3}\right)$, $\begin{cases} 6x - 9y = 6 \\ 4x + 6y = 4 \end{cases}$ no

**16.** $\left(-\dfrac{2}{5}, \dfrac{4}{3}\right)$, $\begin{cases} 5x + 3y = 2 \\ 10x - 6y = -12 \end{cases}$ yes

**Use the graph of each linear system to determine its solution set. (See Objective 2.)**

**17.** $\begin{cases} 2x + y = 0 \\ x - y = -3 \end{cases}$

**18.** $\begin{cases} 2x - y = -3 \\ x - y = -3 \end{cases}$

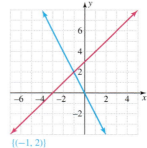

{(−1, 2)}

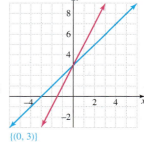

{(0, 3)}

**19.** $\begin{cases} x + 2y = 4 \\ 3x - 2y = 4 \end{cases}$

**20.** $\begin{cases} x - 4y = 4 \\ 3x - 4y = -4 \end{cases}$

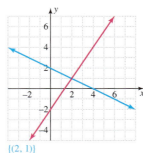

{(2, 1)}

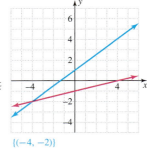

{(−4, −2)}

Additional answers can be found in the Instructor Answer Appendix.

**21.** $\begin{cases} 2x - y = 0 \\ 4x - 2y = 8 \end{cases}$

**22.** $\begin{cases} x = -1 \\ x + y = 3 \end{cases}$

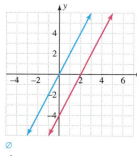

$\varnothing$

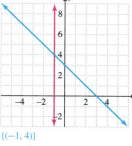

$\{(-1, 4)\}$

**23.** $\begin{cases} 2x + y = 4 \\ y = -2 \end{cases}$

**24.** $\begin{cases} x + y = 2 \\ 3x + 3y = -9 \end{cases}$

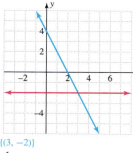

$\{(3, -2)\}$

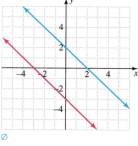

$\varnothing$

**25.** $\begin{cases} 2x - y = 0 \\ 4x - 2y = 0 \end{cases}$

**26.** $\begin{cases} x + y = 2 \\ 3x + 3y = 6 \end{cases}$

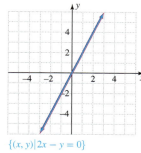

$\{(x, y)\,|\,2x - y = 0\}$

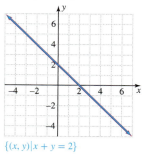

$\{(x, y)\,|\,x + y = 2\}$

**Solve each system of linear equations by graphing. (See Objective 3.)**

**27.** $\begin{cases} y = x + 1 \\ y = 2x - 2 \end{cases}$

**28.** $\begin{cases} y = -3x + 4 \\ y = x \end{cases}$

**29.** $\begin{cases} y = -\dfrac{2}{3}x + 2 \\ y = \dfrac{1}{6}x - 3 \end{cases}$

**30.** $\begin{cases} y = 4x - 1 \\ y = \dfrac{1}{2}x + 6 \end{cases}$

**31.** $\begin{cases} 3x + y = 1 \\ 2x - y = 9 \end{cases}$

**32.** $\begin{cases} x - y = 3 \\ x + 2y = -6 \end{cases}$

**33.** $\begin{cases} 5x - y = -4 \\ 3x - y = 0 \end{cases}$

**34.** $\begin{cases} 3x + 2y = -12 \\ 4x - 3y = 1 \end{cases}$

**35.** $\begin{cases} x + 5y = -10 \\ y = -3 \end{cases}$

**36.** $\begin{cases} 6x + 2y = -14 \\ y = 5 \end{cases}$

**37.** $\begin{cases} x - 4y = 8 \\ x = -4 \end{cases}$

**38.** $\begin{cases} 7x + y = 0 \\ x = 1 \end{cases}$

**39.** $\begin{cases} x = 5 \\ y = -1 \end{cases}$

**40.** $\begin{cases} x = -2 \\ y = 3 \end{cases}$

**Solve each system of linear equations by graphing. (See Objective 4.)**

**41.** $\begin{cases} 2x + y = 7 \\ \dfrac{3}{2}x + \dfrac{3}{4}y = \dfrac{15}{4} \end{cases}$ $\varnothing$

**42.** $\begin{cases} -3x + y = 6 \\ \dfrac{2}{3}x - \dfrac{2}{9}y = \dfrac{8}{9} \end{cases}$ $\varnothing$

**43.** $\begin{cases} x + 4y = -8 \\ y = -\dfrac{1}{4}x + 2 \end{cases}$ $\varnothing$

**44.** $\begin{cases} x - 6y = 3 \\ y = \dfrac{1}{6}x - \dfrac{1}{2} \end{cases}$
$\{(x, y)\,|\,x - 6y = 3\}$

**45.** $\begin{cases} 4x + 2y = 5 \\ y = -2x + \dfrac{5}{2} \end{cases}$
$\{(x, y)\,|\,4x + 2y = 5\}$

**46.** $\begin{cases} 3x - 5y = 2 \\ -15x + 25y = -10 \end{cases}$
$\{(x, y)\,|\,3x - 5y = 2\}$

**Determine how the lines in each system of linear equations relate, the number of solutions of the system, and the type of system without graphing. (See Objective 5.)**

**47.** $\begin{cases} y = x + 2 \\ y = x - 3 \end{cases}$ parallel lines, no solution, inconsistent system

**48.** $\begin{cases} y = \dfrac{1}{4}x - 1 \\ y = \dfrac{1}{4}x + 4 \end{cases}$ parallel lines, no solution, inconsistent system

**49.** $\begin{cases} 3x + y = -3 \\ 9x + 3y = -9 \end{cases}$

**50.** $\begin{cases} y = \dfrac{1}{2}x + 1 \\ x - 2y = -2 \end{cases}$ same line, infinitely many solutions, consistent system with dependent equations

**51.** $\begin{cases} x - \dfrac{1}{5}y = 4 \\ x - \dfrac{2}{10}y = -2 \end{cases}$ parallel lines, no solution, inconsistent system

**52.** $\begin{cases} 2x - y = 6 \\ 4x - 2y = 12 \end{cases}$

**53.** $\begin{cases} 6x + 2y = 4 \\ 15x + 5y = 10 \end{cases}$

**54.** $\begin{cases} x + 2y = 4 \\ 2x + 4y = 0 \end{cases}$ parallel lines, no solution, inconsistent system

**55.** $\begin{cases} y = -\dfrac{2}{3}x + 2 \\ y = -\dfrac{4}{6}x + 2 \end{cases}$

**56.** $\begin{cases} 5x - y = -10 \\ 10x - 2y = 4 \end{cases}$ parallel lines, no solution, inconsistent system

**Solve each problem. (See Objectives 2 and 6.)**

In business, a company breaks even when the revenue generated from selling a particular number of items is equal to the total cost to make the same number of items. In other words, a company breaks even when its revenue equals its cost.

**57.** A shoe factory can make sneakers for $25 per pair and the factory has fixed monthly costs of $15,000. The shoe store sells each pair of sneakers for $50. If $x$ is the number of pairs of sneakers sold in a month, the monthly cost of making sneakers is given by $C = 25x + 15,000$. The monthly revenue for selling $x$ pairs of sneakers is given by $R = 50x$. Use the system of equations

$$\begin{cases} C = 25x + 15,000 \\ R = 50x \end{cases}$$ and its graph to answer the questions that follow.

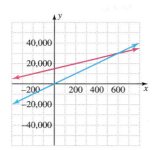

a. What is the point of intersection of the graphs? (600, 30,000)

b. What does the point of intersection mean in the context of the problem? The company will break even when 600 sneakers are sold. Both cost and revenue equal $30,000.

58. It costs a local donut shop $0.50 to make each donut and the shop has fixed monthly costs of $2000. The shop sells each donut for $1.50. If $x$ represents the number of donuts made in a month, the total monthly cost to make $x$ donuts is $C = 0.50x + 2000$. The revenue from selling $x$ donuts is $R = 1.50x$. Use the system of equations $$\begin{cases} C = 0.50x + 2000 \\ R = 1.50x \end{cases}$$ and its graph to answer the questions that follow.

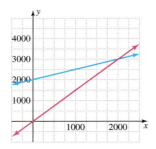

a. What is the point of intersection of the graphs? (2000, 3000)

b. What does the point of intersection mean in the context of the problem? The company will break even when 2000 donuts are sold. Both cost and revenue equal $3000.

 **Mix 'Em Up!**

**Solve each system of linear equations graphically. Then state if the system is inconsistent, consistent with dependent equations, or consistent with independent equations.**

59. $\begin{cases} y = 8x - 4 \\ y = 4x - 6 \end{cases}$   60. $\begin{cases} y = -3x + 4 \\ y = -3x + 2 \end{cases}$

61. $\begin{cases} x - y = 4 \\ 2x = 2y + 8 \end{cases}$   62. $\begin{cases} 4x - y = -3 \\ -3x + y = 6 \end{cases}$

63. $\begin{cases} x + 5y = -9 \\ y = -2x \end{cases}$   64. $\begin{cases} 3x - y = -4 \\ y = 2x \end{cases}$

65. $\begin{cases} x + 2y = -7 \\ y = -3 \end{cases}$   66. $\begin{cases} 5x - y = -4 \\ x = -2 \end{cases}$

67. $\begin{cases} x + 5y = -6 \\ 2y = -x - 3 \end{cases}$   68. $\begin{cases} 3x - y = -3 \\ 2y = -2x - 2 \end{cases}$

69. $\begin{cases} 0.25x + y = 0.25 \\ y = -0.8x + 3 \end{cases}$   70. $\begin{cases} 0.3x - 0.2y = 2.1 \\ y = 0.2x + 2.5 \end{cases}$

71. $\begin{cases} 2x - 3y = 1 \\ y = \dfrac{2}{3}x - \dfrac{1}{3} \end{cases}$   72. $\begin{cases} y = -2x \\ 2x + y = 3 \end{cases}$

73. $\begin{cases} y = -1.5x \\ y = 2.5x + 12 \end{cases}$   74. $\begin{cases} y = 1.2x - 1.6 \\ 0.2x - 0.1y = 0.72 \end{cases}$

75. $\begin{cases} y = 3x + 1 \\ x - \dfrac{1}{3}y = 2 \end{cases}$   76. $\begin{cases} x - \dfrac{1}{4}y = 2 \\ y = 2x - 2 \end{cases}$

**Solve each problem.**

77. Johanne is a college student trying to decide which cell phone company she should use. InTouch Cell offers a monthly rate of $30 plus $0.40 for each minute over 200. FriendsConnect offers a monthly rate of $60 per month plus $0.20 for each minute over 200. The monthly cost of InTouch Cell can be expressed by $y = 0.40x + 30$ and the monthly cost of FriendsConnect can be expressed by $y = 0.20x + 60$, where $x$ is the number of minutes over 200. Use the graph of the system formed by these two equations to answer the questions that follow.

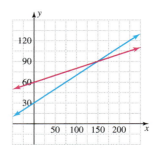

a. What is the point of intersection of this system? (150, 90)

b. What is the meaning of the intersection point in the context of the problem?

c. If Johanne expects to use 300 min each month, which plan should she choose? Why?

78. Janis is trying to decide which shipping company she should use to mail a package. You Gotta Be Shipping Me charges a flat fee of $25 plus $2 per pound. ShipItQuick charges a flat fee of $50 plus $1 per pound. For a package that weighs $x$ lb, the cost charged by You Gotta Be Shipping Me is $y = 2x + 25$. The cost charged by ShipItQuick is $y = x + 50$. Use the graph of

the system formed by these two equations to answer the questions that follow.

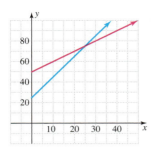

a. What is the point of intersection of this system? (25, 75).
b. What is the meaning of the intersection point in the context of this problem?
c. If Janis needs to mail a package that weighs 40 lb, which company should she use? Why?

 **You Be the Teacher!**

Correct each student's errors, if any.

**79.** Determine if (1, 3) is a solution of the system $\begin{cases} 3x - y = 0 \\ x + y = 3 \end{cases}$.

The ordered pair must be checked in both equations. Substituting (1, 3) into the second gives us $1 + 3 = 4 \neq 3$. This is false, so (1, 3) is not a solution of the system.

Janeane's work:

$3(1) - 3 = 0$

$3 - 3 = 0$.

This is true, so (1, 3) is a solution of the system.

**80.** Solve the system $\begin{cases} 4x - 2y = 4 \\ 6x + 2y = 11 \end{cases}$ graphically.

Jocelyn's work:

Equation 1: $-2y = -4x + 4$

$y = \dfrac{-4x + 4}{-2} = 2x - 2$. So, $y = 2x - 2$.

Equation 2: $2y = -6x + 11$

$y = \dfrac{-6x + 11}{2} = -3x + 5.5$. So, $y = -3x + 5.5$.

**81.** Explain how to identify the solution of the system from the graph.

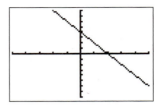

Since there is only one line, the lines are the same. There is an infinite number of solutions and the system is consistent with dependent equations.

Emelle's work:

Since there is only one line, there is no solution and the system is inconsistent.

 **Calculate It!**

Use a graphing calculator to solve each system. Round the answer to the nearest hundredth, if applicable.

**82.** $\begin{cases} 3x + y = 7 \\ y = -2x + 2 \end{cases}$ {(5, −8)} **83.** $\begin{cases} 4x - y = 3 \\ y = 2x + 5 \end{cases}$ {(4, 13)}

**84.** $\begin{cases} 2x + 3y = 6 \\ 2x - y = 4 \end{cases}$ {(2.25, 0.5)} **85.** $\begin{cases} 4x + 2y = 3 \\ 5x - y = -2 \end{cases}$ {(−0.07, 1.64)}

 **Think About It!**

Write a system of linear equations in two variables that has the given ordered pair as its solution.

**86.** (3, −2) **87.** (−4, 1) **88.** (0, 0) **89.** $\left(\dfrac{1}{2}, 3\right)$

A system of linear equations in two variables contains the given equation. Write another equation in the system that would make the system satisfy the stated condition.

**90.** $y = \dfrac{1}{3}x - 8$, inconsistent

**91.** $y = -\dfrac{5}{2}x + 2$, inconsistent

**92.** $x - 4y = 8$, consistent with dependent equations

**93.** $3x + 5y = 2$, consistent with dependent equations

---

**SECTION 4.2** / **Solving Systems of Linear Equations by Substitution**

▶ **OBJECTIVES**

As a result of completing this section, you will be able to

**1.** Solve a system of linear equations using substitution.

**2.** Solve special cases of systems of linear equations using substitution.

**3.** Determine how the lines in a system of linear equations relate, the number of solutions, and the type of system.

**4.** Troubleshoot common errors.

In Section 4.1 we solved systems by graphing. Graphing by hand, however, is not the most precise method to solve a system. Consider the system

$$\begin{cases} 2x - 3y = -3 \\ 4x + 9y = 14 \end{cases}$$

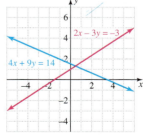

The graph of this system is shown. The point of intersection does not occur at integer values, so it is very difficult to know the exact solution based on this graph. In this section, we will learn an algebraic method to solve systems precisely.

## Substitution

**Objective 1** ▶

**Solve a system of linear equations using substitution.**

The method of substitution is a precise way to solve a system of linear equations because we will obtain a definitive answer, rather than having to approximate our answer from a graph. The goal of the **substitution method** is to substitute one equation of the system into the other equation of the system so the new equation has just one variable.

Consider the system $\begin{cases} x + 2y = 4 \\ y = x - 1 \end{cases}$. The second equation has the variable $y$ expressed in terms of $x$. When we replace the variable $y$ in the first equation with its equivalent expression, $x - 1$, from the second equation, we obtain an equation with just one variable, which we can solve for $x$. This yields

$$x + 2y = 4$$
$$x + 2(x - 1) = 4 \qquad \text{Replace } y \text{ with } x - 1.$$
$$x + 2x - 2 = 4 \qquad \text{Apply the distributive property.}$$
$$3x - 2 = 4 \qquad \text{Combine like terms.}$$
$$3x - 2 + 2 = 4 + 2 \qquad \text{Add 2 to each side.}$$
$$3x = 6 \qquad \text{Simplify.}$$
$$x = \frac{6}{3} \qquad \text{Divide each side by 3.}$$
$$x = 2 \qquad \text{Simplify.}$$

Now that we know the value of $x$, we can find the corresponding $y$-value by replacing $x$ with 2 in either of the given equations. This yields

$$y = x - 1 \qquad \text{Begin with the second equation.}$$
$$y = 2 - 1 \qquad \text{Replace } x \text{ with 2.}$$
$$y = 1 \qquad \text{Simplify.}$$

So, the solution set of the system $\begin{cases} x + 2y = 4 \\ y = x - 1 \end{cases}$ is $\{(2, 1)\}$.

The steps used to solve this system are generalized next.

**INSTRUCTOR NOTE:**
Point out that if one of the variables has a coefficient of 1 or −1, it is best to isolate this variable in order to avoid working with fractions.

**Procedure: Solving a System of Linear Equations Using Substitution**

**Step 1:** Solve one of the equations in the system for one of the variables.

**Step 2:** Substitute the expression found in step 1 into the other equation in the system.

**Step 3:** Solve the resulting equation.

**Step 4:** Substitute the value found in step 3 back into one of the equations to find the value of the other variable.

**Step 5:** The solution of the system is the ordered pair $(x, y)$.

**Step 6:** Check the solution in the system.

**Objective 1 Examples** Solve each system of linear equations using substitution.

**1a.** $\begin{cases} y = x - 4 \\ 2x + y = 5 \end{cases}$ **1b.** $\begin{cases} 3x + 2y = -2 \\ x - 6y = -14 \end{cases}$ **1c.** $\begin{cases} 2x - 3y = -3 \\ 4x + 9y = 14 \end{cases}$

**1d.** $\begin{cases} x + y = 42 \\ 0.05x + 0.10y = 3.50 \end{cases}$

**Solutions**   **1a.** The first equation is already solved for $y$.

$$y = x - 4$$

Substitute and solve.

| | |
|---|---|
| $2x + y = 5$ | Begin with the second equation. |
| $2x + (x - 4) = 5$ | Substitute $(x - 4)$ for $y$. |
| $3x - 4 = 5$ | Combine like terms. |
| $3x - 4 + 4 = 5 + 4$ | Add 4 to each side. |
| $3x = 9$ | Simplify. |
| $\dfrac{3x}{3} = \dfrac{9}{3}$ | Divide each side by 3. |
| $x = 3$ | Simplify. |

Now find the value of $y$. (For reference, both equations are shown but only one is necessary.)

| | | | | |
|---|---|---|---|---|
| $y = x - 4$ | | $2x + y = 5$ | |
| $y = 3 - 4$ | Substitute 3 for $x$. | $2(3) + y = 5$ | Substitute 3 for $x$. |
| $y = -1$ | Simplify. | $6 + y = 5$ | Simplify. |
| | | $6 + y - 6 = 5 - 6$ | Subtract 6 from each |
| | | $y = -1$ | side and simplify. |

So, the solution is the ordered pair $(3, -1)$.

Check the solution in each of the equations in the system.

| | | | |
|---|---|---|---|
| $y = x - 4$ | | $2x + y = 5$ |
| $-1 = 3 - 4$ | | $2(3) + (-1) = 5$ |
| $-1 = -1$ | | $6 - 1 = 5$ |
| | | $5 = 5$ |
| True | | True |

Since $(3, -1)$ makes both equations true, the solution set is $\{(3, -1)\}$. Because there is one solution, this system contains lines that intersect at this point. The system is consistent with independent equations. Graphing the system confirms the solution, as well.

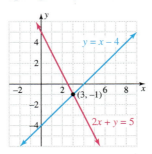

**1b.** Solve the second equation for $x$ since its coefficient is 1.

| | |
|---|---|
| $x - 6y = -14$ | |
| $x - 6y + 6y = -14 + 6y$ | Add $6y$ to each side. |
| $x = 6y - 14$ | |

Substitute and solve.

| | |
|---|---|
| $3x + 2y = -2$ | Begin with the first equation. |
| $3(6y - 14) + 2y = -2$ | Substitute $(6y - 14)$ for $x$. |

$$18y - 42 + 2y = -2 \qquad \text{Apply the distributive property.}$$
$$20y - 42 = -2 \qquad \text{Combine like terms.}$$
$$20y - 42 + 42 = -2 + 42 \qquad \text{Add 42 to each side.}$$
$$20y = 40 \qquad \text{Simplify.}$$
$$y = \frac{40}{20} \qquad \text{Divide each side by 20.}$$
$$y = 2 \qquad \text{Simplify.}$$

Now find the value of $x$.

$$x - 6y = -14 \qquad \text{Begin with the second equation.}$$
$$x - 6(2) = -14 \qquad \text{Substitute 2 for } y.$$
$$x - 12 = -14 \qquad \text{Simplify.}$$
$$x - 12 + 12 = -14 + 12 \qquad \text{Add 12 to each side.}$$
$$x = -2 \qquad \text{Simplify.}$$

So, the solution is the ordered pair $(-2, 2)$.

Check the solution.

$$
\begin{array}{c|c}
x - 6y = -14 & 3x + 2y = -2 \\
-2 - 6(2) = -14 & 3(-2) + 2(2) = -2 \\
-2 - 12 = -14 & -6 + 4 = -2 \\
-14 = -14 & -2 = -2 \\
\text{True} & \text{True}
\end{array}
$$

Because $(-2, 2)$ makes both equations in the system true, the solution set of the system is $\{(-2, 2)\}$.

**1c.** None of the coefficients on the variables is 1 or $-1$. If we solve the first equation for $x$, we will get a fraction. The fraction, however, will be cleared when we substitute into the other equation since the coefficient of $x$ is a multiple of 2.

$$2x - 3y = -3$$
$$2x - 3y + 3y = -3 + 3y \qquad \text{Add } 3y \text{ to each side.}$$
$$2x = 3y - 3 \qquad \text{Simplify.}$$
$$\frac{2x}{2} = \frac{3y - 3}{2} \qquad \text{Divide each side by 2.}$$
$$x = \frac{3y - 3}{2} \qquad \text{Simplify.}$$

Substitute and solve.

$$4x + 9y = 14 \qquad \text{Begin with the second equation.}$$
$$4\left(\frac{3y - 3}{2}\right) + 9y = 14 \qquad \text{Substitute } \left(\frac{3y - 3}{2}\right) \text{ for } x.$$
$$2(3y - 3) + 9y = 14 \qquad \text{Simplify. Note that } \frac{4}{2} = 2.$$
$$6y - 6 + 9y = 14 \qquad \text{Apply the distributive property.}$$
$$15y - 6 = 14 \qquad \text{Combine like terms.}$$
$$15y - 6 + 6 = 14 + 6 \qquad \text{Add 6 to each side.}$$
$$15y = 20 \qquad \text{Simplify.}$$
$$\frac{15y}{15} = \frac{20}{15} \qquad \text{Divide each side by 15.}$$
$$y = \frac{4}{3} \qquad \text{Simplify.}$$

Now find the value of $x$.

$$2x - 3y = -3$$

$$2x - 3\left(\frac{4}{3}\right) = -3 \qquad \text{Substitute } \frac{4}{3} \text{ for } y.$$

$$2x - 4 = -3 \qquad \text{Simplify.}$$

$$2x - 4 + 4 = -3 + 4 \qquad \text{Add 4 to each side.}$$

$$2x = 1 \qquad \text{Simplify.}$$

$$x = \frac{1}{2} \qquad \text{Divide each side by 2.}$$

So, the solution is the ordered pair $\left(\frac{1}{2}, \frac{4}{3}\right)$.

Check the solution.

$$
\begin{array}{c|c}
2x - 3y = -3 & 4x + 9y = 14 \\
2\left(\frac{1}{2}\right) - 3\left(\frac{4}{3}\right) = -3 & 4\left(\frac{1}{2}\right) + 9\left(\frac{4}{3}\right) = 14 \\
1 - 4 = -3 & 2 + 12 = 14 \\
-3 = -3 & 14 = 14 \\
\text{True} & \text{True}
\end{array}
$$

Because $\left(\frac{1}{2}, \frac{4}{3}\right)$ makes each equation in the system true, the solution set is $\left\{\left(\frac{1}{2}, \frac{4}{3}\right)\right\}$.

**1d.** Since both variables in the first equation have a coefficient of 1, we can solve for either $x$ or $y$. We will choose to solve for $y$.

$$x + y = 42$$

$$x + y - x = 42 - x \qquad \text{Subtract } x \text{ from each side.}$$

$$y = -x + 42$$

Substitute and solve.

$$0.05x + 0.10y = 3.50 \qquad \text{Begin with the second equation.}$$

$$0.05x + 0.10(-x + 42) = 3.50 \qquad \text{Substitute } (-x + 42) \text{ for } y.$$

$$0.05x - 0.10x + 4.2 = 3.50 \qquad \text{Apply the distributive property.}$$

$$-0.05x + 4.2 = 3.5 \qquad \text{Combine like terms.}$$

$$-0.05x + 4.2 - 4.2 = 3.5 - 4.2 \qquad \text{Subtract 4.2 from each side.}$$

$$-0.05x = -0.7 \qquad \text{Simplify.}$$

$$\frac{-0.05x}{-0.05} = \frac{-0.7}{-0.05} \qquad \text{Divide each side by } -0.05.$$

$$x = 14 \qquad \text{Simplify.}$$

Now find the value of $x$.

$$x + y = 42$$
$$14 + y = 42 \qquad \text{Substitute 14 for } x.$$
$$14 + y - 14 = 42 - 14 \qquad \text{Subtract 14 from each side.}$$
$$y = 28 \qquad \text{Simplify.}$$

So, the solution is the ordered pair $(14, 28)$.

Check the solution.

| | |
|---|---|
| $x + y = 42$ | $0.05x + 0.10y = 3.5$ |
| $14 + 28 = 42$ | $0.05(14) + 0.10(28) = 3.5$ |
| $42 = 42$ | $0.7 + 2.8 = 3.5$ |
| | $3.5 = 3.5$ |
| True | True |

Since $(14, 28)$ makes both equations in the system true, the solution set is $\{(14, 28)\}$.

✓ **Student Check 1**    Solve each system of linear equations using substitution.

a. $\begin{cases} 4x + y = -2 \\ y = 7x + 9 \end{cases}$    b. $\begin{cases} 5x - y = 10 \\ 3x + 4y = 6 \end{cases}$

c. $\begin{cases} 9x - 6y = 11 \\ 3x + 6y = -7 \end{cases}$    d. $\begin{cases} x + y = 50 \\ 0.25x + 0.05y = 8.50 \end{cases}$

## Special Cases of Systems of Linear Equations

**Objective 2** ▶

Solve special cases of systems of linear equations using substitution.

From Section 4.1, we know that there are two special cases of systems of linear equations—systems with no solution and systems with infinitely many solutions. Recall that a system with no solution consists of lines whose graphs are parallel; a system with infinitely many solutions consists of two lines that are the same.

**Procedure: Solving a Special Case of a System of Linear Equations Using Substitution**

**Step 1:** Solve one of the equations in the system for one of the variables.
**Step 2:** Substitute the expression found in step 1 into the other equation in the system.
**Step 3:** Solve the resulting equation.
**Step 4:** The resulting equation will be either a contradiction or an identity
   **a.** If the resulting equation is a false statement (i.e., a contradiction), then there is no solution of the system.
   **b.** If the resulting equation is a true statement (i.e., an identity), then there are infinitely many solutions.

**Objective 2  Examples**    Solve each system of linear equations using substitution.

2a. $\begin{cases} x + 5y = -1 \\ 2x + 10y = 5 \end{cases}$    2b. $\begin{cases} 9x - 3y = 27 \\ 3x - y = 9 \end{cases}$

**Solutions**    **2a.** Solve the first equation for $x$ since its coefficient is 1.

$$x + 5y = -1$$
$$x + 5y - 5y = -1 - 5y \qquad \text{Subtract } 5y \text{ from each side.}$$
$$x = -5y - 1$$

Substitute and solve.

$$2x + 10y = 5 \qquad \text{Begin with the second equation.}$$
$$2(-5y - 1) + 10y = 5 \qquad \text{Substitute } -5y - 1 \text{ for } x.$$
$$-10y - 2 + 10y = 5 \qquad \text{Apply the distributive property.}$$
$$-2 = 5 \qquad \text{Combine like terms.}$$

Because we obtained a contradiction, $-2 = 5$, in the substitution process, this system has no solution. This means that the two lines are parallel and that the system is inconsistent. So, the solution set is the empty set, or $\varnothing$.

**2b.** Solve the second equation for $y$ since its coefficient is $-1$.

$$3x - y = 9$$
$$3x - y - 3x = 9 - 3x \qquad \text{Subtract } 3x \text{ from each side.}$$
$$-y = -3x + 9 \qquad \text{Simplify.}$$
$$-1(-y) = -1(-3x + 9) \qquad \text{Multiply each side by } -1.$$
$$y = 3x - 9 \qquad \text{Simplify.}$$

Substitute and solve.

$$9x - 3y = 27 \qquad \text{Begin with the first equation.}$$
$$9x - 3(3x - 9) = 27 \qquad \text{Substitute } 3x - 9 \text{ for } y.$$
$$9x - 9x + 27 = 27 \qquad \text{Apply the distributive property.}$$
$$27 = 27 \qquad \text{Combine like terms.}$$

Because we obtained an identity, $27 = 27$, in the substitution process, this system has infinitely many solutions. The lines are the same and the system is consistent with dependent equations. The solution set consists of all ordered pairs on the line, which is denoted as $\{(x, y) | 3x - y = 9\}$.

✓ **Student Check 2**    Solve each system of linear equations using substitution.

a. $\begin{cases} 8x - 12y = -8 \\ 2x - 3y = -2 \end{cases}$    b. $\begin{cases} 10x + 5y = -20 \\ 8x + 4y = 10 \end{cases}$

**Objective 3** ▶

Determine how the lines in a system of linear equations relate, the number of solutions, and the type of system.

## Determine the Relationship of Lines in a System of Linear Equations

It is helpful to understand the equations that result in the substitution process and what they mean as far as how the lines in the system relate to one another, the number of solutions of the system, and the type of system. The following table summarizes the three possibilities.

| Intersecting Lines | Parallel Lines | Same Lines |
|---|---|---|
| After substituting and simplifying each side, the variable will remain and we will get something of the form $x = \#$ or $y = \#$. | After substituting and simplifying each side of the equation, a *false* statement or contradiction results, for example, $3 = 8$. | After substituting and simplifying each side of the equation, a *true* statement or identity results, for example, $3 = 3$. |
| One solution | No solution | Infinitely many solutions |
| Consistent system with independent equations | Inconsistent system | Consistent system with dependent equations |

**Objective 3 Examples** Determine how the lines in each system of linear equations relate, the number of solutions of the system, and the type of system from the given information. State the solution set of the system.

**3a.** $\begin{cases} \dfrac{2}{3}x - 4y = 5 \\ 4x - 24y = 30 \end{cases}$ ; Substitution process yields $30 = 30$.

**3b.** $\begin{cases} 5x - 3y = -15 \\ x + y = -3 \end{cases}$ ; Substitution process yields $y = 0$.

**3c.** $\begin{cases} y = \dfrac{1}{2}x - 7 \\ 6x - 12y = 10 \end{cases}$ ; Substitution process yields $84 = 10$.

**Solutions**

**3a.** Since the substitution process yields a true statement, or identity, this system consists of lines that are the same. There are infinitely many solutions. The system is consistent with dependent equations. The solution set is $\{(x, y) \mid 4x - 24y = 30\}$.

**3b.** Since one of the variables remains in the resulting equation, this system has intersecting lines. There is one solution. The system is consistent with independent solutions. We need to find the value of $x$ to write the solution set.

$$y = 0: x + 0 = -3$$
$$x = -3$$

So, the solution set is $\{(-3, 0)\}$.

**3c.** Since the substitution process yields a false statement, or contradiction, this system consists of parallel lines. There is no solution. The system is inconsistent. The solution set is $\varnothing$.

**✓ Student Check 3** Determine how the lines in each system of linear equations relate, the number of solutions of the system, and the type of system from the given information. State the solution set of the system.

**a.** $\begin{cases} x + 5y = 20 \\ 9x - y = -4 \end{cases}$ ; Substitution process yields $x = 0$.

**b.** $\begin{cases} \dfrac{1}{4}x + 7y = 3 \\ 2x + 56y = 8 \end{cases}$ ; Substitution process yields $24 = 8$.

**c.** $\begin{cases} y = \dfrac{1}{7}x - 6 \\ -14y = -2x + 84 \end{cases}$ ; Substitution process yields $84 = 84$.

---

**Objective 4** ▶

Troubleshoot common errors.

## Troubleshooting Common Errors

Some common errors associated with substitution are illustrated next.

**Objective 4 Examples**

**A problem and an incorrect solution are given. Provide the correct solution and an explanation of the error.**

**4a.** Solve the system $\begin{cases} y = 4x - 9 \\ 3x - y = 6 \end{cases}$.

| Incorrect Solution | Correct Solution and Explanation |
|---|---|
| $y = 4x - 9$ <br><br> $3x - 4x - 9 = 6$ <br> $-x - 9 = 6$ <br> $-x = 15$ <br> $x = -15$ <br><br> $y = 4(15) - 9$ <br> $y = 60 - 9$ <br> $y = 51$ <br><br> The solution set is $\{(-15, 51)\}$. | The error was made in the substitution step. The expression $4x - 9$ should be put in parentheses because the negative sign must be distributed. <br><br> $3x - (4x - 9) = 6$ <br> $3x - 4x + 9 = 6$ <br> $-x + 9 = 6$ <br> $-x = -3$ <br> $x = 3$ <br><br> $y = 4(3) - 9$ <br> $y = 12 - 9$ <br> $y = 3$ <br><br> The solution set is $\{(3, 3)\}$. |

**4b.** Solve the system $\begin{cases} x + 2y = 5 \\ 4x + y = 6 \end{cases}$.

| Incorrect Solution | Correct Solution and Explanation |
|---|---|
| $x = 5 - 2y$ <br><br> $5 - 2y + 2y = 5$ <br> $5 = 5$ <br><br> Since this is true, there are infinitely many solutions. The solution set is $\{(x, y) \mid x + 2y = 5\}$. | The error was made in not using the second equation for the substitution process. The expression $5 - 2y$ should be substituted for $x$ in the second equation. <br><br> $4x + y = 6$ <br> $4(5 - 2y) + y = 6$ <br> $20 - 8y + y = 6$ <br> $20 - 7y = 6$ <br> $-7y = -14$ <br> $y = 2$ <br><br> $x = 5 - 2y = 5 - 2(2) = 5 - 4 = 1$ <br> The solution set is $\{(1, 2)\}$. |

## ANSWERS TO STUDENT CHECKS

**Student Check 1**   **a.** $\{(-1, 2)\}$   **b.** $\{(2, 0)\}$
  **c.** $\left\{\left(\dfrac{1}{3}, -\dfrac{4}{3}\right)\right\}$   **d.** $\{(30, 20)\}$

**Student Check 2**   **a.** $\{(x, y)\,|\,2x - 3y = -2\}$   **b.** $\varnothing$

**Student Check 3**   **a.** intersecting lines, one solution, consistent system with independent equations, $\{(0, 4)\}$
  **b.** parallel lines, no solution, inconsistent system, $\varnothing$
  **c.** same lines, infinitely many solutions, consistent system with dependent equations, $\left\{(x, y)\,\middle|\,y = \dfrac{1}{7}x - 6\right\}$

## SUMMARY OF KEY CONCEPTS

1. The substitution method is an algebraic method that enables us to find the exact solution of a system of linear equations.

2. To solve a linear system by substitution,

   • Choose one of the equations and solve it for either $x$ or $y$.

   • Once we solve for $x$ or $y$, we substitute this value into the other equation. We will then have a simpler linear equation with one variable to solve.

   • Once we have found the value of one variable, we substitute this value back into one of the equations to find the other variable's value.

3. If, after substitution, the resulting equation is a contradiction or identity, then the solution of the system has no solution or infinitely many solutions, respectively.

4. Answers can be checked by substituting the ordered pair into both equations or by graphing.

## GRAPHING CALCULATOR SKILLS

The graphing calculator can be used to check our work. In Section 4.1, the Intersect command was shown. We can also check by substituting the $x$- and $y$-coordinates of the solution into each equation.

**Example:** Check that $\left(\dfrac{1}{2}, \dfrac{4}{3}\right)$ is the solution of $\begin{cases} 2x - 3y = -3 \\ 4x + 9y = 14 \end{cases}$.

**Solution:** Substitute the coordinates of the ordered pair into the two equations in the system.

```
2(1/2)-3(4/3)
             -3
4(1/2)+9(4/3)
             14
```

Since the ordered pair makes each equation true, the solution is correct.

## SECTION 4.2    EXERCISE SET

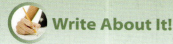

### Write About It!

**Use complete sentences to answer each exercise.**

1. Why is substitution a more precise method to find the solutions of a system of equations than graphing?   Answers vary.

2. Explain how to solve a system of linear equations using substitution.   Answers vary.

3. When using substitution to solve a system of linear equations, what will happen if there are infinitely many solutions?   Answers vary.

4. When using substitution to solve a system of linear equation, what will happen if there is no solution?   Answers vary.

5. When using substitution to solve a system of linear equations, what will happen if there is only one solution?   Answers vary.

6. Suppose that after using substitution to solve a system of linear equations, you derive $x = 0$. Does this mean that there is no solution of the system? Explain.   Answers vary.

7. When expressing the solution of a system of linear equations with infinitely many solutions, does it matter which equation is used in the set notation? Explain.   Answers vary.

8. After using substitution to solve for one of the two variables in a system of linear equations, explain how to solve for the remaining variable.   Answers vary.

Additional answers can be found in the Instructor Answer Appendix.

## Practice Makes Perfect!

**Solve each system of linear equations using substitution.** *(See Objective 1.)*

**9.** $\begin{cases} x = 2y + 1 \\ x + 3y = 6 \end{cases}$ $\{(3, 1)\}$

**10.** $\begin{cases} x = 3y + 2 \\ x + 5y = 10 \end{cases}$ $\{(5, 1)\}$

**11.** $\begin{cases} y = 5x - 3 \\ 2x + 2y = 18 \end{cases}$ $\{(2, 7)\}$

**12.** $\begin{cases} y = 4x - 1 \\ 4x + 3y = -19 \end{cases}$ $\{(-1, -5)\}$

**13.** $\begin{cases} y = \frac{2}{3}x + 9 \\ x + 6y = 24 \end{cases}$ $\{(-6, 5)\}$

**14.** $\begin{cases} y = -\frac{4}{7}x + \frac{2}{7} \\ -2x + 5y = 16 \end{cases}$ $\{(-3, 2)\}$

**15.** $\begin{cases} x - 2y = 7 \\ 3x - y = 6 \end{cases}$ $\{(1, -3)\}$

**16.** $\begin{cases} 2x - y = 5 \\ x - 5y = 4 \end{cases}$ $\left\{\left(\frac{7}{3}, -\frac{1}{3}\right)\right\}$

**17.** $\begin{cases} 2x = 4y - 6 \\ 6x - 3y = 0 \end{cases}$ $\{(1, 2)\}$

**18.** $\begin{cases} 3y = 3x + 9 \\ 4x - 5y = 10 \end{cases}$ $\{(-25, -22)\}$

**19.** $\begin{cases} 5x = 2y + 3 \\ 5x + 4y = 7 \end{cases}$

**20.** $\begin{cases} 7x + 2y = 4 \\ 2y = 3x + 1 \end{cases}$ $\left\{\left(\frac{3}{10}, \frac{19}{20}\right)\right\}$

**21.** $\begin{cases} -4x + 3y = 5 \\ 2x + 6y = -1 \end{cases}$

**22.** $\begin{cases} 3x - 5y = 2 \\ -6x + 3y = 10 \end{cases}$ $\left\{\left(-\frac{8}{3}, -2\right)\right\}$

**Solve each system of linear equations using substitution.** *(See Objective 2.)*

**23.** $\begin{cases} 4x - y = 6 \\ y = 4x - 3 \end{cases}$ $\varnothing$

**24.** $\begin{cases} x = -5y + 10 \\ 3x + 15y = 30 \end{cases}$

**25.** $\begin{cases} 2x - 4y = 6 \\ -x + 2y = -3 \end{cases}$

**26.** $\begin{cases} 2x - y = 3 \\ 14x = 7y + 28 \end{cases}$ $\varnothing$

**27.** $\begin{cases} 2x + y = 3 \\ 4x + 2y = 0 \end{cases}$ $\varnothing$

**28.** $\begin{cases} 3x + 6y = 3 \\ 5x = -10y + 5 \end{cases}$

**29.** $\begin{cases} x - 3y = 4 \\ 2x - 6y = 8 \end{cases}$

**30.** $\begin{cases} 6x = 2y + 2 \\ 9x - 3y = 6 \end{cases}$ $\varnothing$

**Determine how the lines in each system of linear equations relate, the number of solutions of the system, and the type of system from the given information. State the solution set of each system.** *(See Objective 3.)*

**31.** $\begin{cases} x + 4y = 5 \\ 2x + 8y = 1 \end{cases}$; Substitution process yields $0 = 9$.

**32.** $\begin{cases} -2x + 4y = 1 \\ 5x + 10y = 5 \end{cases}$; Substitution process yields $0 = -1$.

**33.** $\begin{cases} 6x - 3y = 3 \\ 2x = y + 1 \end{cases}$; Substitution process yields $0 = 0$.

**34.** $\begin{cases} 2x = 3y + 4 \\ 6x - 9y = 12 \end{cases}$; Substitution process yields $0 = 0$.

**35.** $\begin{cases} y = \frac{1}{3}x - 5 \\ 2x - y = 15 \end{cases}$; Substitution process yields $x = 6$.

**36.** $\begin{cases} y = -\frac{5}{3}x - 9 \\ 6x - y = -37 \end{cases}$; Substitution process yields $x = -6$.

## Mix 'Em Up!

**Solve each system of linear equations using substitution.**

**37.** $\begin{cases} 3x - 7y = -4 \\ x = 3 - 2y \end{cases}$ $\{(1, 1)\}$

**38.** $\begin{cases} 4x - 5y = 23 \\ y = 1 - 2x \end{cases}$ $\{(2, -3)\}$

**39.** $\begin{cases} y = -\frac{3}{4}x + 1 \\ 3x - 2y = 34 \end{cases}$ $\{(8, -5)\}$

**40.** $\begin{cases} y = -\frac{5}{2}x + \frac{1}{2} \\ -7x = 2y + 9 \end{cases}$ $\{(-5, 13)\}$

**41.** $\begin{cases} \frac{x}{2} + 3y = 1 \\ x = -6y + 2 \end{cases}$

**42.** $\begin{cases} \frac{x}{3} + y = 2 \\ 2x + 6y = 12 \end{cases}$

**43.** $\begin{cases} 5x + 2y = 12 \\ 2x - y = 3 \end{cases}$ $\{(2, 1)\}$

**44.** $\begin{cases} 8x = 12 - 2y \\ 4x + y = 3 \end{cases}$ $\varnothing$

**45.** $\begin{cases} 3x - 2y = 6 \\ -x + 4y = 3 \end{cases}$ $\left\{\left(3, \frac{3}{2}\right)\right\}$

**46.** $\begin{cases} \frac{x}{2} + \frac{y}{3} = -2 \\ 3x + 2y = 6 \end{cases}$ $\varnothing$

**47.** $\begin{cases} \frac{x}{5} + \frac{y}{10} = \frac{3}{5} \\ \frac{x}{4} - \frac{y}{2} = \frac{1}{4} \end{cases}$ $\left\{\left(\frac{13}{5}, \frac{4}{5}\right)\right\}$

**48.** $\begin{cases} 10y = 1 - 5x \\ 5x - 10y = -1 \end{cases}$ $\left\{\left(0, \frac{1}{10}\right)\right\}$

**49.** $\begin{cases} y = -1.8x + 1.7 \\ 0.6x - 3.5y = 25.1 \end{cases}$ $\{(4.5, -6.4)\}$

**50.** $\begin{cases} y = 6.2x - 14.5 \\ 4.8x - 2.5y = -1.2 \end{cases}$ $\{(3.5, 7.2)\}$

## You Be the Teacher!

**Answer each student's questions.**

**51.** Clarice: When I solve the system $\begin{cases} 4x + 2y = 8 \\ x + \frac{1}{2}y = 1 \end{cases}$ using substitution, all of the variables cancel out. Am I doing something wrong?

**52.** Janki: When I solve the system $\begin{cases} 9x - 15y = -15 \\ y = \frac{3}{5}x + 1 \end{cases}$ using substitution, all of the variables and constants cancel out. Am I doing something wrong?

**Correct each student's errors, if any.**

**53.** Solve the system using substitution: $\begin{cases} 4x - 2y = 10 \\ 3x + 2y = 4 \end{cases}$.

Carrel's work: I used the first equation to solve for $y$:

$4x - 2y = 10 \Rightarrow -2y = -4x + 10 \Rightarrow y = \dfrac{-4x + 10}{2} = -2x + 10$

Substituting this into the second equation for $y$:

$3x + 2(-2x + 10) = 4$

$3x - 4x + 20 = 4$

$-x + 20 = 4 \Rightarrow -x = -16 \Rightarrow x = 16$. Substituting this value into the equation I solved for $y$ gives me:

$y = -2(16) + 10 = -22$. So, my answer is $(16, -22)$.

**54.** Solve the system $\begin{cases} 2x - y = 3 \\ 3x + 2y = 1 \end{cases}$ using substitution.

Sarah's work:

I used the first equation to solve for $y$:

$2x - y = 3 \Rightarrow -y = -2x + 3 \Rightarrow y = 2x + 3$

Substituting this into the second equation for $y$:

$3x + 2(2x + 3) = 1$

$3x + 4x + 3 = 1$

$7x + 3 = 1 \Rightarrow 7x = -2 \Rightarrow x = -\dfrac{2}{7}$. Substituting this value into the equation I solved for $y$ gives me:

$y = 2\left(-\dfrac{2}{7}\right) + 3 = -\dfrac{4}{7} + 3 = \dfrac{17}{7}$.

So, my answer is $\left(-\dfrac{3}{7}, \dfrac{22}{7}\right)$.

 **Calculate It!**

**Use a graphing calculator to determine if the ordered pair is a solution of the given system.**

**55.** $\begin{cases} 3x + y = 7 \\ y = -2x + 2 \end{cases}$; $(5, -8)$    yes

**56.** $\begin{cases} 5x - y = 2 \\ y = 3x - 1 \end{cases}$; $\left(\dfrac{1}{2}, \dfrac{1}{2}\right)$    yes

**Use each substitution to solve system algebraically. Then use a graphing calculator and the Intersect command to check answers.**

**57.** $\begin{cases} 10x + 9y = -23 \\ 15x - 12y = 144 \end{cases}$    $\{(4, -7)\}$

**58.** $\begin{cases} 9x + 5y = 2 \\ 3x - 4y = 12 \end{cases}$    $\left\{\left(\dfrac{4}{3}, -2\right)\right\}$

 **Think About It!**

**59.** The expression $4x - 2$ is substituted for $y$ in the second equation of a system. If the solution of the system is $(3, 10)$, what is the equation of the other line in the system?

**60.** The expression $\dfrac{1}{3}y + 1$ is substituted for $x$ in the second equation of a system. If the solution of the system is $(-1, -6)$, what is the equation of the other line in the system?

**61.** The expression $-5x + 3$ is substituted for $y$ in the second equation of a system. If the solution set of the system is $\varnothing$, what is the equation of the other line in the system?

**62.** The expression $7y - 1$ is substituted for $x$ in the second equation of a system. If the solution set of the system is $\varnothing$, what is the equation of the other line in the system?

**63.** If one equation in a system is $2x + y = 6$ and the equation that results after substitution is $0 = 5$, what is the equation of the other line in the system?

**64.** If one equation in a system is $x - 3y = 4$ and the equation that results after substitution is $0 = -1$, what is the equation of the other line in the system?

**65.** If one equation in a system is $2x + y = 6$ and the equation that results after substitution is $0 = 0$, what is the equation of the other line in the system?

**66.** If one equation in a system is $x - 3y = 4$ and the equation that results after substitution is $0 = 0$, what is the equation of the other line in the system?

---

**SECTION 4.3**   **Solving Systems of Linear Equations by Elimination**

▶ **OBJECTIVES**

As a result of completing this section, you will be able to

**1.** Solve a system of linear equations using elimination.

**2.** Solve special cases of systems of linear equations using elimination.

**3.** Determine how the lines in a system of linear equations relate, the number of solutions, and the type of system.

**4.** Solve application problems.

**5.** Troubleshoot common errors.

As of 2010, Nintendo DS and Nintendo Wii were the two top-selling gaming units in the United States. A total of 81 million gaming units have been sold. If there have been 13 million more Nintendo DS units sold than Nintendo Wii units, how many of each gaming unit has been sold in the United States? (Source: http://www.vgchartz.com/home.php)

In this section, we will learn another method for solving systems of linear equations in two variables and how to apply these methods to solving a problem such as the one just stated.

### The Elimination Method

**Objective 1** ▶

Solve a system of linear equations using elimination.

While the substitution method is a precise algebraic method to solve a system of linear equations, it can be tedious if the coefficients of the variables in the system are not 1. In this section, we will explore another method to solve systems, the **elimination method** or **addition method**.

The goal of the elimination method is to add the two equations in the system together so that one of the variables is eliminated—that is, the terms containing that variable add to zero. This process creates an equation in just one variable that enables us to solve for one of the variables in the system.

Consider the system $\begin{cases} 3x + y = 5 \\ -3x + 3y = 7 \end{cases}$.

$$\begin{cases} 3x + y = 5 \\ -3x + 3y = 7 \end{cases}$$
$$\overline{\phantom{3x+}0x + 4y = 12}$$ Note the coefficients of $x$, 3 and $-3$, are opposites.

Add the equations in the system to eliminate $x$.

$$4y = 12$$ Simplify.

$$\frac{4y}{4} = \frac{12}{4}$$ Divide each side by 4.

$$y = 3$$ Simplify.

Now we solve for the value of $x$ by substituting the value of $y$ into one of the equations.

$$3x + y = 5$$ Choose one of the original equations.

$$3x + 3 = 5$$ Substitute 3 for $y$.

$$3x + 3 - 3 = 5 - 3$$ Subtract 3 from each side.

$$3x = 2$$ Simplify.

$$x = \frac{2}{3}$$ Divide each side by 3.

So, the solution set of the system is $\left\{ \left( \frac{2}{3}, 3 \right) \right\}$.

In the system, $\begin{cases} 4x - y = 3 \\ 5x + 2y = 7 \end{cases}$, neither the coefficients of $x$ nor $y$ are opposites. To eliminate the variable $y$, we can multiply the first equation by 2. This yields

$$\begin{cases} 2(4x - y) = 2(3) \\ 5x + 2y = 7 \end{cases}$$ Multiply the first equation by 2.

$$\begin{cases} 8x - 2y = 6 \\ 5x + 2y = 7 \end{cases}$$ The coefficients of the $y$-terms in each equation are now opposites.

$$\overline{13x + 0y = 13}$$ Add the equations to eliminate $y$.

$$13x = 13$$ Simplify.

$$x = 1$$ Divide each side by 13.

Now we solve for $y$ by substituting $x = 1$ into one of the two equations.

$$5x + 2y = 7$$ Choose one of the original equations.

$$5(1) + 2y = 7$$ Substitute 1 for $x$.

$$5 + 2y = 7$$ Simplify.

$$5 + 2y - 5 = 7 - 5$$ Subtract 5 from each side.

$$2y = 2$$ Simplify.

$$\frac{2y}{2} = \frac{2}{2}$$ Divide each side by 2.

$$y = 1$$ Simplify.

So, the solution set of the system is $\{(1, 1)\}$.

The elimination method is based on the addition property of equality, which enables us to add the equations in a system together. This property enables us to add equivalent expressions to each side of an equation and not change the solution.

> **Property: Addition Property of Equality**
>
> $$\text{If } A = B, \text{ then } A + C = B + C.$$

The steps that were used to solve the preceding systems of linear equations are generalized as follows.

> **Procedure: Solving a System of Linear Equations Using Elimination**
>
> **Step 1:** Write both equations in the system in standard form $(Ax + By = C)$.
> **Step 2:** Create a new system with opposites as coefficients on one of the variables, if necessary.
>    **a.** Choose the variable to be eliminated.
>    **b.** Find a nonzero number, $a$, and multiply one or both equations by the number that will produce coefficients of $a$ and $-a$ in the new system.
> **Step 3:** Add the equations together.
> **Step 4:** Solve the resulting equation for the variable.
> **Step 5:** Substitute this value into one of the original equations to find the value of the other variable.
> **Step 6:** Check by substituting the ordered pair into the equations in the original system.

**INSTRUCTOR NOTE:**
Remind students that they must multiply both sides of the equation by this number.

**Objective 1 Examples**    Solve each system of linear equations using elimination.

**1a.** $\begin{cases} 2x - y = 4 \\ 3x + y = 1 \end{cases}$    **1b.** $\begin{cases} 4x + 3y = 12 \\ 2x - 5y = 6 \end{cases}$    **1c.** $\begin{cases} 7x - 2y = 4 \\ 3x + 9y = -10 \end{cases}$

**Solutions**    **1a.** The coefficients of $y$, $-1$ and $1$, are opposites, so we add the equations in this form.

$$\begin{cases} 2x - y = 4 \\ 3x + y = 1 \end{cases}$$

$\qquad\qquad 5x = 5$        Add the equations.

$\qquad\qquad \dfrac{5x}{5} = \dfrac{5}{5}$        Divide each side by 5.

$\qquad\qquad x = 1$        Simplify.

Now we solve for $y$, by substituting $x = 1$ in either of the original equations.

$\qquad 2x - y = 4$        Begin with first equation.

$\qquad 2(1) - y = 4$        Substitute 1 for $x$.

$\qquad 2 - y = 4$        Simplify.

$\qquad 2 - y - 2 = 4 - 2$        Subtract 2 from each side.

$\qquad -y = 2$        Simplify.

$\qquad -1(-y) = -1(2)$        Multiply each side by $-1$.

$\qquad y = -2$        Simplify.

The solution of the system is the ordered pair $(1, -2)$.

**Check:**

$$
\begin{array}{c|c}
2x - y = 4 & 3x + y = 1 \\
2(1) - (-2) = 4 & 3(1) + (-2) = 1 \\
2 + 2 = 4 & 3 - 2 = 1 \\
4 = 4 & 1 = 1 \\
\text{True} & \text{True}
\end{array}
$$

Since $(1, -2)$ makes both equations true, the solution set is $\{(1, -2)\}$.

**1b.** Neither the coefficients of $x$ nor $y$ are opposites. We choose to eliminate the variable $x$. The coefficients of $x$, 4 and 2, divide evenly into 4. We must create an equivalent system with equations that contain $4x$ and $-4x$. The first equation contains $4x$ so nothing needs to be done to this equation. To create $-4x$ in the second equation, multiply the equation by $\left(\dfrac{-4}{2}\right) = -2$.

$$
\begin{cases} 4x + 3y = 12 \\ 2x - 5y = 6 \end{cases} \longrightarrow \begin{cases} 4x + 3y = 12 \\ -2(2x - 5y) = -2(6) \end{cases} \longrightarrow \begin{cases} 4x + 3y = 12 \\ -4x + 10y = -12 \end{cases}
$$

Now we add the equations to solve for $y$.

$$
\begin{cases} \phantom{-}4x + \phantom{1}3y = \phantom{-}12 \\ -4x + 10y = -12 \end{cases}
$$

| | |
|---|---|
| $13y = 0$ | Add the equations. |
| $\dfrac{13y}{13} = \dfrac{0}{13}$ | Divide each side by 13. |
| $y = 0$ | Simplify. |

Now we solve for $x$, by substituting $y = 0$ in either of the original equations.

| | |
|---|---|
| $2x - 5y = 6$ | Begin with the second equation. |
| $2x - 5(0) = 6$ | Substitute 0 for $y$. |
| $2x = 6$ | Simplify. |
| $\dfrac{2x}{2} = \dfrac{6}{2}$ | Divide each side by 2. |
| $x = 3$ | Simplify. |

The solution of the system is the ordered pair $(3, 0)$.

**Check:**

$$
\begin{array}{c|c}
4x + 3y = 12 & 2x - 5y = 6 \\
4(3) + 3(0) = 12 & 2(3) - 5(0) = 6 \\
12 + 0 = 12 & 6 - 0 = 6 \\
12 = 12 & 6 = 6 \\
\text{True} & \text{True}
\end{array}
$$

Since $(3, 0)$ makes both equations true, the solution set is $\{(3, 0)\}$.

**1c.** Neither the coefficients of $x$ nor $y$ are opposites. We choose to eliminate the variable $y$. The coefficients of $y$, $-2$ and 9, divide evenly into 18. We must create an equivalent system with equations that contain $18y$ and $-18y$. To create $-18y$ in the first equation, multiply the equation by $\dfrac{-18}{-2} = 9$. To create $18y$ in the second equation, multiply by $\dfrac{18}{9} = 2$.

$$
\begin{cases} 7x - 2y = 4 \\ 3x + 9y = -10 \end{cases} \longrightarrow \begin{cases} 9(7x - 2y) = 9(4) \\ 2(3x + 9y) = 2(-10) \end{cases} \longrightarrow \begin{cases} 63x - 18y = 36 \\ 6x + 18y = -20 \end{cases}
$$

Now we add the equations to solve for $x$.

$$\begin{cases} 63x - 18y = \phantom{-}36 \\ 6x + 18y = -20 \end{cases}$$

$$69x = 16 \qquad \text{Add the equations.}$$

$$\frac{69x}{69} = \frac{16}{69} \qquad \text{Divide each side by 69.}$$

$$x = \frac{16}{69} \qquad \text{Simplify.}$$

Now we solve for $y$ by substituting $x = \dfrac{16}{69}$ in either of the original equations.

$$3x + 9y = -10 \qquad \text{Begin with the second equation.}$$

$$3\left(\frac{16}{69}\right) + 9y = -10 \qquad \text{Replace } x \text{ with } \frac{16}{69}.$$

$$\frac{16}{23} + 9y = -10 \qquad \text{Simplify.}$$

$$23\left(\frac{16}{23} + 9y\right) = 23(-10) \qquad \text{Multiply each side by the LCD, 23.}$$

$$16 + 207y = -230 \qquad \text{Simplify.}$$

$$16 + 207y - 16 = -230 - 16 \qquad \text{Subtract 16 from each side.}$$

$$207y = -246 \qquad \text{Simplify.}$$

$$y = -\frac{246}{207} \qquad \text{Divide each side by 207.}$$

$$y = -\frac{82}{69} \qquad \text{Simplify the fraction.}$$

The solution of the system is the ordered pair $\left(\dfrac{16}{69}, -\dfrac{82}{69}\right)$.

**Check:**

$$7x - 2y = 4 \qquad\qquad 3x + 9y = -10$$

$$7\left(\frac{16}{69}\right) - 2\left(-\frac{82}{69}\right) = 4 \qquad 3\left(\frac{16}{69}\right) + 9\left(-\frac{82}{69}\right) = -10$$

$$\frac{112}{69} + \frac{164}{69} = 4 \qquad\qquad \frac{48}{69} - \frac{738}{69} = -10$$

$$\frac{276}{69} = 4 \qquad\qquad\qquad -\frac{690}{69} = -10$$

$$4 = 4 \qquad\qquad\qquad\qquad -10 = -10$$

$$\text{True} \qquad\qquad\qquad\qquad\quad \text{True}$$

Since $\left(\dfrac{16}{69}, -\dfrac{82}{69}\right)$ makes the equations true, the solution set is $\left\{\left(\dfrac{16}{69}, -\dfrac{82}{69}\right)\right\}$.

✓ **Student Check 1**   Solve each system of linear equations using elimination.

**a.** $\begin{cases} x + 4y = 8 \\ 3x - 4y = 4 \end{cases}$   **b.** $\begin{cases} 6x + 9y = 15 \\ 2x + 5y = 11 \end{cases}$   **c.** $\begin{cases} 7x - 3y = 6 \\ 5x + 4y = 8 \end{cases}$

**Objective 2** ▶

Solve special cases of systems of linear equations using elimination.

## Special Cases of Systems of Linear Equations

We will now see how elimination can be used to obtain solutions of special systems of linear equations.

> **Procedure: Solving a Special System of Linear Equations Using Elimination**
>
> **Step 1:** Apply the steps shown in Objective 1.
> **Step 2:** The equation that results from adding the two equations in the system together will be a contradiction or an identity.
>     **a.** If the resulting equation is a false statement (i.e., a contradiction), then there is no solution of the system and the lines are parallel.
>     **b.** If the resulting equation is a true statement (i.e., an identity), then there are infinitely many solutions and the lines are the same.

**Objective 2 Examples**    Solve each system of linear equations using elimination.

**2a.** $\begin{cases} x + 5y = -5 \\ 2x + 10y = 5 \end{cases}$      **2b.** $\begin{cases} 3x - y = 6 \\ 12x - 4y = 24 \end{cases}$

**Solutions**    **2a.** We choose to eliminate the variable $x$. The number that the coefficients of $x$, 1 and 2, divide into evenly is 2. We must create an equivalent system that contains $2x$ and $-2x$. The second equation has $2x$. To obtain $-2x$ in the first equation, we multiply it by $\dfrac{-2}{1} = -2$.

$\begin{cases} x + 5y = -5 \\ 2x + 10y = 5 \end{cases} \longrightarrow \begin{cases} -2(x + 5y) = -2(-5) \\ 2x + 10y = 5 \end{cases} \longrightarrow \begin{cases} -2x - 10y = 10 \\ 2x + 10y = 5 \end{cases}$

Now we add the equations to solve for $y$.

$\begin{cases} -2x - 10y = 10 \\ \phantom{-}2x + 10y = \phantom{1}5 \end{cases}$
$$0 = 15 \qquad \text{Add the equations.}$$

The resulting equation, $0 = 15$, is a contradiction, or a false statement. Therefore, the system has no solution. The lines are parallel and the solution set is the empty set, $\varnothing$.

**2b.** We choose to eliminate the variable $x$. The number that the coefficients of $x$, 3 and 12, divide into evenly is 12. We must create an equivalent system that contains $12x$ and $-12x$. The second equation contains $12x$. We multiply the first equation by $\dfrac{-12}{3} = -4$ to obtain $-12x$.

$\begin{cases} 3x - y = 6 \\ 12x - 4y = 24 \end{cases} \longrightarrow \begin{cases} -4(3x - y) = -4(6) \\ 12x - 4y = 24 \end{cases} \longrightarrow \begin{cases} -12x + 4y = -24 \\ 12x - 4y = 24 \end{cases}$

Now we add the equations to solve for $y$.

$\begin{cases} -12x + 4y = -24 \\ \phantom{-}12x - 4y = \phantom{-}24 \end{cases}$
$$0 = 0 \qquad \text{Add the equations.}$$

The resulting equation, $0 = 0$, is an identity, or a true statement. Therefore, the system has infinitely many solutions and the lines are the same. The solution set of the system is $\{(x, y) \mid 3x - y = 6\}$.

✓ **Student Check 2**  Solve each system of linear equations using elimination.

a. $\begin{cases} 8x - 10y = 4 \\ 12x - 15y = 6 \end{cases}$    b. $\begin{cases} 6x - 2y = -8 \\ 3x - y = -2 \end{cases}$

## Determine the Relationship of the Lines in a System of Linear Equations

**Objective 3** ▶

Determine how the lines in a system of linear equations relate, the number of solutions, and the type of system.

Just as we have seen with graphing and substitution, the elimination method for solving systems of linear equations produces one of three types of solutions. It is helpful to understand the equations that result after adding the equations in the system and what they mean as far as how the lines in the system relate to one another, the number of solutions of the system, and the type of system. The following table summarizes the three possibilities.

| Intersecting Lines | Parallel Lines | Same Lines |
|---|---|---|
| After adding the two equations, we will get something of the form $x = \#$ or $y = \#$. | After adding the equations, a *false* statement or contradiction results—for example, $0 = 8$. | After adding the equations, a *true* statement or identity results—for example, $0 = 0$. |
| One solution | No solution | Infinitely many solutions |
| Consistent system with independent equations | Inconsistent system | Consistent system with dependent equations |

**Objective 3  Examples**  Determine how the lines in each system of linear equations relate, the number of solutions of the system, and the type of system from the given information. State the solution set of the system.

**3a.** $\begin{cases} 4x + y = 12 \\ x - 5y = 3 \end{cases}$ ; The elimination process yields $21y = 0$.

**3b.** $\begin{cases} \dfrac{3}{2}x - \dfrac{7}{3}y = \dfrac{1}{6} \\ \dfrac{9}{5}x - \dfrac{14}{5}y = \dfrac{1}{5} \end{cases}$ ; The elimination process yields $0 = 0$.

**3c.** $\begin{cases} x + \dfrac{4}{3}y = 7 \\ \dfrac{1}{4}x + \dfrac{1}{3}y = 0 \end{cases}$ ; The elimination process yields $0 = 7$.

**Solutions**  **3a.** The equation $21y = 0$ can be solved for $y$. We get $y = 0$. Therefore, this system contains intersecting lines with one solution. The system is consistent with independent equations. The solution is found by replacing $y$ with 0 in one of the given equations.

$$4x + y = 12$$
$$4x + 0 = 12$$
$$4x = 12$$
$$x = 3$$

The solution set is $\{(3, 0)\}$.

**3b.** The equation $0 = 0$ is a true statement, or an identity. This means that the equations in the system are the same line and that there are infinitely many solutions. The system is consistent with dependent equations. The solution set is $\{(x, y) \mid 9x - 14y = 1\}$.

> **Note:** *Note that if we multiply the first equation by 6, we get*
>
> $$6\left(\frac{3}{2}x - \frac{7}{3}y = \frac{1}{6}\right) \longrightarrow 9x - 14y = 1$$
>
> *If we multiply the second equation by 5, we get*
>
> $$5\left(\frac{9}{5}x - \frac{14}{5}y = \frac{1}{5}\right) \longrightarrow 9x - 14y = 1$$

**3c.** The equation $0 = 7$ is a false statement, or a contradiction. This means that the system has no solution and that the lines are parallel. The system is inconsistent. The solution set is the empty set, $\varnothing$.

✓ **Student Check 3**   Determine how the lines in each system of linear equations relate, the number of solutions of the system, and the type of system from the given information. State the solution set of the system.

**a.** $\begin{cases} y = \dfrac{3}{2}x - 6 \\ 9x - 6y = 24 \end{cases}$ ; The elimination process yields $0 = -12$.

**b.** $\begin{cases} 10x + \dfrac{1}{3}y = 8 \\ 7x - y = 13 \end{cases}$ ; The elimination process yields $37x = 37$.

**c.** $\begin{cases} x - 4y = 4 \\ 5x - 20y = 20 \end{cases}$ ; The elimination process yields $0 = 0$.

## Applications

**Objective 4** ▶

Solve application problems.

Many word problems that we solve involve more than one unknown. When this is the case, systems are the best method to employ to solve the problem. A system enables us to assign a different variable for each of the unknown values.

We will assign two different variables to the two unknown values, and then use a system of linear equations in two variables to solve the problem. There are four general steps we will follow to solve such problems.

> **Procedure:  General Problem Solving Strategy**
>
> **Step 1:** Read the problem and determine the two unknown values. Assign a different variable to represent each of these values.
> **Step 2:** Read the problem and write two equations that define a relationship between the two variables.
> **Step 3:** Solve the system of equations using elimination.
> **Step 4:** Write the answer to the problem.

**Objective 4 Example**   As of 2010, Nintendo DS and Nintendo Wii were the two top-selling gaming units in the United States. A total of 81 million gaming units have been sold. If there have been 13 million more Nintendo DS units sold than Nintendo Wii units, how many of each gaming unit has been sold in the United States? (Source: http://www.vgchartz.com/home.php)

**Solution**   What are the unknowns? The unknowns are the number of Nintendo DS units and the number of Nintendo Wii units sold.

Let $x$ = number of Nintendo DS units sold.

Let $y$ = number of Nintendo Wii games sold.

What is known? The total gaming units sold was 81 million. There were 13 million more DS units sold than Wii units. The system of equations that represents this information is

$$\begin{cases} x + y = 81 \\ x = y + 13 \end{cases}$$

Solve the system using elimination.

$$\begin{array}{ll} \begin{cases} x + y = 81 \\ \underline{x - y = 13} \end{cases} & \text{Rewrite the second equation in standard} \\ & \text{form by subtracting } y \text{ from each side.} \\ 2x = 94 & \text{Add the equations.} \\ \dfrac{2x}{2} = \dfrac{94}{2} & \text{Divide each side by 2.} \\ x = 47 & \text{Simplify.} \end{array}$$

Now we solve for $y$ by replacing $x$ with 47 in either equation.

$$\begin{array}{ll} x + y = 81 & \text{Begin with the first equation.} \\ 47 + y = 81 & \text{Substitute 47 for } x. \\ 47 + y - 47 = 81 - 47 & \text{Subtract 47 from each side.} \\ y = 34 & \text{Simplify.} \end{array}$$

As of 2010, there have been 47 million Nintendo DS units sold and 34 million Nintendo Wii gaming units sold in the United States.

---

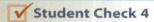 **Student Check 4**  In November 2009, the Xbox 360 and the PlayStation 3 (PS3) were the third and fourth top-selling gaming units. A total of 1,529,900 gaming units were sold. If there were 109,100 more Xbox 360 units sold than PS3 units, how many of each gaming unit was sold? (Source: http://www.1up.com/do/newsStory?cId=3177273)

---

**Objective 5** ▶

Troubleshoot common errors.

## Troubleshooting Common Errors

A common error associated with elimination is shown next.

**Objective 5  Example**  **A problem and an incorrect solution are given. Provide the correct solution and an explanation of the error.**

Solve the system $\begin{cases} 6x - 5y = 27 \\ 3x + y = 3 \end{cases}$.

| Incorrect Solution | Correct Solution and Explanation |
|---|---|
| $\begin{cases} 6x - 5y = 27 \\ 3x + y = 3 \end{cases}$ $\begin{cases} 6x - 5y = 27 \\ 15x + 5y = \phantom{0}3 \end{cases}$ $\phantom{aaaa}21x = 30$ $\phantom{aaaaa}x = \dfrac{30}{21}$ $\phantom{aaaaa}x = \dfrac{10}{7}$ | The error was made in multiplying the second equation by 5. We must distribute 5 to each side of the equation. $\begin{cases} 6x - 5y = 27 \\ 15x + 5y = 15 \end{cases}$ $\phantom{aaaa}21x = 42$ $\phantom{aaaaa}x = 2$ |

$$3\left(\frac{10}{7}\right) + y = 3$$

$$\frac{30}{7} + y = 3$$

$$30 + 7y = 21$$

$$7y = -9$$

$$y = -\frac{9}{7}$$

The solution set is $\left\{\left(\frac{10}{7}, -\frac{9}{7}\right)\right\}$.

$$3(2) + y = 3$$

$$6 + y = 3$$

$$y = -3$$

The solution set is $\{(2, -3)\}$.

## ANSWERS TO STUDENT CHECKS

**Student Check 1**  **a.** $\left\{\left(3, \frac{5}{4}\right)\right\}$   **b.** $\{(-2, 3)\}$
**c.** $\left\{\left(\frac{48}{43}, \frac{26}{43}\right)\right\}$

**Student Check 2**  **a.** $\{(x, y) | 8x - 10y = 4\}$
**b.** $\varnothing$

**Student Check 3**  **a.** Parallel lines, no solution, inconsistent system, $\varnothing$   **b.** Intersecting lines, one solution, consistent system with independent equations, $\{(1, -6)\}$   **c.** Same lines, infinitely many solutions, consistent system with dependent equations, $\{(x, y) | x - 4y = 4\}$

**Student Check 4**  There were 819,500 Xbox 360 units sold and 710,400 PS3 units sold.

## SUMMARY OF KEY CONCEPTS

1. Elimination is an algebraic method that enables us to find the exact solution of a system of linear equations.
2. To solve a linear system using elimination, the equations in the system should be written in standard form. The goal is to add the equations together so that one of the variables is eliminated. Multiply one or both equations by some nonzero number so that the coefficients of one of the variables are opposites, if needed.
3. If the resulting equation is a contradiction or an identity, then the solution of the system is no solution or infinitely many solutions, respectively.
4. Answers can be checked by substituting the ordered pair into both equations or by graphing.
5. Systems can be used to solve problems involving two unknowns. From the information given, we need to determine the two unknowns and the two equations that represent the situation.

## SECTION 4.3 / EXERCISE SET

 **Write About It!**

Use complete sentences in your answer to each exercise.

1. Explain how to solve a system of linear equations using elimination.   Answers vary.
2. How do you determine the numbers to multiply by the equations in the system in order to apply elimination?   Answers vary.
3. When solving a system of linear equations using elimination, what will happen if there are infinitely many solutions?   Answers vary.
4. When solving a system of linear equations using elimination, what will happen if there is no solution?   Answers vary.

Additional answers can be found in the Instructor Answer Appendix.

 **Practice Makes Perfect!**

Solve each system of linear equations using elimination. (*See Objective 1.*)

5. $\begin{cases} 7x + 4y = 5 \\ -7x + 2y = 1 \end{cases}$ $\left\{\left(\frac{1}{7}, 1\right)\right\}$

6. $\begin{cases} 6x - y = 9 \\ 6x + y = 0 \end{cases}$ $\left\{\left(\frac{3}{4}, -\frac{9}{2}\right)\right\}$

7. $\begin{cases} x - 4y = -2 \\ 3x + 2y = -13 \end{cases}$

8. $\begin{cases} -12x + y = 12 \\ 8x + y = 9 \end{cases}$ $\left\{\left(-\frac{3}{20}, \frac{51}{5}\right)\right\}$

9. $\begin{cases} 4x - 5y = 17 \\ 8x + 2y = -2 \end{cases}$

10. $\begin{cases} 10x - 3y = 1 \\ -2x + 6y = 16 \end{cases}$ $\{(1, 3)\}$

11. $\begin{cases} 7x + 6y = 10 \\ 3x + 9y = 12 \end{cases}$

12. $\begin{cases} 10x - 3y = -69 \\ x + \dfrac{3y}{2} = -\dfrac{3}{2} \end{cases}$ $\{(-6, 3)\}$

13. $\begin{cases} \dfrac{x}{6} + \dfrac{y}{8} = -1 \\ -3x - 2y = 4 \end{cases}$ $\{(36, -56)\}$  14. $\begin{cases} \dfrac{x}{4} - \dfrac{3y}{4} = -1 \\ 4x - 7y = 44 \end{cases}$ $\{(32, 12)\}$

**Solve each system of linear equations using elimination.** (*See Objective 2.*)

15. $\begin{cases} 4x - 6y = 5 \\ -2x + 3y = 4 \end{cases}$ $\varnothing$  16. $\begin{cases} -3x + 7y = -4 \\ 24x - 56y = 32 \end{cases}$

17. $\begin{cases} 5x - 8y = 10 \\ 10x - 16y = 20 \end{cases}$  18. $\begin{cases} 9x - 3y = 12 \\ 3x - y = -10 \end{cases}$ $\varnothing$

19. $\begin{cases} 6x + 18y = 20 \\ \dfrac{x}{6} + \dfrac{y}{2} = \dfrac{5}{9} \end{cases}$

20. $\begin{cases} 14x - 21y = 28 \\ \dfrac{x}{3} - \dfrac{y}{2} = -\dfrac{2}{3} \end{cases}$ $\varnothing$

21. $\begin{cases} -13x + 15y = -12 \\ 39x - 45y = 30 \end{cases}$  22. $\begin{cases} -12x - 11y = 13 \\ 48x + 44y = 17 \end{cases}$
$\varnothing$ $\varnothing$

**Determine how the lines in each system of linear equations relate, the number of solutions of the system, and the type of system from the given information. State the solution set of the system.** (*See Objective 3.*)

23. $\begin{cases} 7x - 2y = 14 \\ y = \dfrac{7}{2}x - 7 \end{cases}$ ; The elimination process yields $0 = 0$.

24. $\begin{cases} 2x + 5y = 12 \\ y = -0.4x + 2.4 \end{cases}$ ; The elimination process yields $0 = 0$.

25. $\begin{cases} x - 11y = 22 \\ -3x + 33y = -60 \end{cases}$ ; The elimination process yields $0 = 6$.

26. $\begin{cases} y = 1.5x - 7.5 \\ 3x - 2y = 18 \end{cases}$ ; The elimination process yields $0 = 3$.

27. $\begin{cases} 2x + 9y = 5 \\ -x + 4y = 6 \end{cases}$ ; The elimination process yields $17y = 17$.

28. $\begin{cases} 2x - 5y = 15 \\ y = -\dfrac{3}{5}x + 1 \end{cases}$ ; The elimination process yields $5x = 20$.

**Solve each problem.** (*See Objective 4.*)

29. In 2008, the two largest toy manufacturers, Mattel and Hasbro, reported a total net sales of $9.939 billion. If the reported net sales of Mattel was $2.124 billion less than twice that of Hasbro, find the net sales of each company. (Source: http://www.ita.doc.gov/td/ocg/toyoutlook_09.pdf)

30. In 2010, there were 25.8 million people, diagnosed and undiagnosed, in the United States affected by diabetes. If the number of people diagnosed with diabetes is 4.8 million more than twice the number

of people undiagnosed, find the number of diagnosed and undiagnosed cases of diabetes in the United States (Source: http://diabetes.niddk.nih.gov/dm/pubs/statistics/#fast) Diagnosed cases: 18.8 million, Undiagnosed cases: 7 million

31. In 2010, there were a total of 6.43 million homes sold or foreclosed in the United States. If the number of homes sold was 7.92 million less than four times the number of foreclosures, find the number of homes sold and foreclosed. (Sources: http://www.census.gov/const/newressales.pdf and http://www.tampabay.com/news/business/banking/florida-ranks-second-in-number-of-foreclosures-for-2010/1145229) Foreclosures: 2.87 million, Homes sold: 3.56 million

32. In May 2011, the top two automobile corporations, General Motors Corporation and Ford Motor Company, sold a total of approximately 413,000 vehicles in the United States. If the Ford Motor Company sold 250,000 less than twice the sales of General Motors, find the number of vehicles sold by each company. (Source: http://www.motorintelligence.com/m_frameset.html) General Motors: 221,000, Ford: 192,000

 **Mix 'Em Up!**

**Solve each system of linear equations using elimination.**

33. $\begin{cases} 18x + 8y = 11 \\ 9x + 4y = 5 \end{cases}$ $\varnothing$  34. $\begin{cases} 4x - 15y = -6 \\ 12x - 3y = 42 \end{cases}$ $\left\{\left(\dfrac{27}{7}, \dfrac{10}{7}\right)\right\}$

35. $\begin{cases} 2x = 5y + 8 \\ 3x - y = 1 \end{cases}$ $\left\{\left(-\dfrac{3}{13}, -\dfrac{22}{13}\right)\right\}$  36. $\begin{cases} 12x - 6 = 10y \\ -6x + 5y = -3 \end{cases}$ $\{(x, y) \mid -6x + 5y = -3\}$

37. $\begin{cases} 3x - 2y = 6 \\ 15x = 10y - 12 \end{cases}$ $\varnothing$  38. $\begin{cases} -6y = -5x + 4 \\ 20x - 3y = 19 \end{cases}$ $\left\{\left(\dfrac{34}{35}, \dfrac{1}{7}\right)\right\}$

39. $\begin{cases} 2x + 5y = 4 \\ \dfrac{x}{10} + \dfrac{y}{4} = \dfrac{1}{5} \end{cases}$ $\{(x, y) \mid 2x + 5y = 4\}$  40. $\begin{cases} 7x - 3y = 2 \\ -\dfrac{x}{4} + \dfrac{y}{7} = -\dfrac{1}{14} \end{cases}$ $\left\{\left(\dfrac{2}{7}, 0\right)\right\}$

41. $\begin{cases} \dfrac{x}{10} - y = \dfrac{5}{2} \\ 4x + 5y = 1 \end{cases}$ $\left\{\left(3, -\dfrac{11}{5}\right)\right\}$  42. $\begin{cases} 2x - 8y = 13 \\ \dfrac{x}{8} - y = 0 \end{cases}$ $\left\{\left(13, \dfrac{13}{8}\right)\right\}$

43. $\begin{cases} 0.4x - 1.4y = 5.88 \\ -1.6x + 2.5y = -13.6 \end{cases}$ $\{(3.5, -3.2)\}$  44. $\begin{cases} 4.5x + 1.2y = 5.1 \\ 1.5x = 6.4y + 18.7 \end{cases}$ $\{(1.8, -2.5)\}$

45. $\begin{cases} 3.5x = 1.2y - 11.81 \\ -0.6x + 2.4y = -3.9 \end{cases}$ $\{(-4.3, -2.7)\}$  46. $\begin{cases} 3.6x - 0.5y = 8.01 \\ 0.4x = 2.6y + 3.18 \end{cases}$ $\{(2.1, -0.9)\}$

47. $\begin{cases} 6x - 2y = 5 \\ 3x = y \end{cases}$ $\varnothing$  48. $\begin{cases} 4x = 3y - 2 \\ y = \dfrac{4}{3}x + \dfrac{2}{3} \end{cases}$
$\{(x, y) \mid 4x = 3y - 2\}$

**You Be the Teacher!**

**Answer each student's question or correct their errors, if any.**

49. Solve the system $\begin{cases} 3x + 4y = 8 \\ 5x - 3y = 1 \end{cases}$.

Cory: I'm not sure which variable I should eliminate. Which variable should I eliminate in this system and why?

**50.** Solve the system $\begin{cases} 2x - 5y = 6 \\ 6x - 2y = 1 \end{cases}$.

Andy's work:

$\begin{cases} -3(2x - 5y) = 6 \\ 6x - 2y = 1 \end{cases}$

$\begin{cases} -6x + 15y = 6 \\ \underline{6x - 2y = 1} \end{cases}$

$13y = 7$

$y = \dfrac{7}{13}$

$2x - 5y = 6$

$2x - 5\left(\dfrac{7}{13}\right) = 6$

$2x - \dfrac{35}{13} = 6$

$2x = 6 + \dfrac{35}{13}$

$2x = \dfrac{113}{13}$

$x = \dfrac{113}{26}$

So, my answer is $\left(\dfrac{113}{26}, \dfrac{7}{13}\right)$.

**51.** Solve the system $\begin{cases} 2x = 4y - 8 \\ 6x - 3y = 1 \end{cases}$.

Charlie: Please explain how to decide which method is best to solve this system.

**52.** Solve the system $\begin{cases} 14x + 5y = 86 \\ 18x - 10y = 242 \end{cases}$.

Clarissa: Please explain how to decide which method is best to solve this system.

**53.** Solve the system $\begin{cases} -12x + 15y = 10 \\ 14x + 6y = 11 \end{cases}$.

Jalen: I have decided to eliminate the $x$-variable. How do I determine the numbers to multiply each equation by to eliminate $x$?

**54.** Solve the system $\begin{cases} 6x + 28y = 25 \\ 8x - 70y = 15 \end{cases}$.

Edna: I know that 6 and 8 will each divide evenly into 48, so I write an equivalent system that contains $48x$ and $-48x$ in it. A friend told me that I should have used the least common multiple of 6 and 8, which is 24. Will I get the same solution if I use 48 instead of 24? Please explain.

 **Calculate It!**

Use elimination to solve each system algebraically. Then use a graphing calculator and the Intersect command to check the answer.

**55.** $\begin{cases} 3x - 5y = 21 \\ 4x + y = 5 \end{cases}$  {(2, −3)}

**56.** $\begin{cases} 5x + 6y = 8 \\ 2x + 3y = 4 \end{cases}$  $\left\{\left(0, \dfrac{4}{3}\right)\right\}$

**57.** $\begin{cases} 6x = 4y + 5 \\ x - y = 2 \end{cases}$  $\left\{\left(-\dfrac{3}{2}, -\dfrac{7}{2}\right)\right\}$

**58.** $\begin{cases} -9y = 10 - 10x \\ 15x - 12y = -3 \end{cases}$

---

## PIECE IT TOGETHER    SECTIONS 4.1–4.3 Review

**Determine if the given ordered pair is a solution of the system. (Section 4.1, Objective 1)**

**1.** Is $\left(\dfrac{8}{7}, 0\right)$ a solution of $\begin{cases} 7x - 8y = 8 \\ 14x - 16y = 16 \end{cases}$?   yes

**2.** Is $\left(0, -\dfrac{10}{3}\right)$ a solution of $\begin{cases} x - 3y = 10 \\ 2x - 6y = -5 \end{cases}$?   no

**Solve each system by graphing. (Section 4.1, Objective 3)**

**3.** $\begin{cases} x + y = -1 \\ 2x - y = -8 \end{cases}$

**4.** $\begin{cases} x - 4y = 8 \\ x + y = -2 \end{cases}$

**5.** $\begin{cases} x + 2y = 1 \\ 3x + 6y = -5 \end{cases}$

**6.** $\begin{cases} 3x - y = 1 \\ 3y = 9x - 3 \end{cases}$

**Solve each system using either substitution or elimination. (Sections 4.2 and 4.3, Objectives 1 and 2)**

**7.** $\begin{cases} 4x + 3y = 14 \\ x + 2y = 1 \end{cases}$  {(5, −2)}

**8.** $\begin{cases} 5x - 2y = 14 \\ 3x + y = 4 \end{cases}$  {(2, −2)}

**9.** $\begin{cases} \dfrac{x}{2} - \dfrac{y}{4} = 1 \\ x = 4y \end{cases}$  $\left\{\left(\dfrac{16}{7}, \dfrac{4}{7}\right)\right\}$

**10.** $\begin{cases} 8x - 2y = 26 \\ 2x = 3y - 1 \end{cases}$  {(4, 3)}

**11.** $\begin{cases} 2x + y = 3 \\ 4x + 2y = 0 \end{cases}$  ∅

**12.** $\begin{cases} 3x + 6y = 3 \\ 5x = -10y + 5 \end{cases}$  {(x, y) | 3x + 6y = 3}

**13.** $\begin{cases} -3x + 4y = 2 \\ -2x + 5y = 6 \end{cases}$  {(2, 2)}

**14.** $\begin{cases} 2x - 5y = 2 \\ 5x - 2y = 5 \end{cases}$  {(1, 0)}

**15.** $\begin{cases} -5x + 4y = 6 \\ 2x + 3y = 1 \end{cases}$  $\left\{\left(-\dfrac{14}{23}, \dfrac{17}{23}\right)\right\}$

**16.** $\begin{cases} 2x - 7y = -26 \\ 3x + 2y = 136 \end{cases}$  {(36, 14)}

**17.** $\begin{cases} x - 6y = 2 \\ 3x = 18y + 6 \end{cases}$  {(x, y) | x − 6y = 2}

**18.** $\begin{cases} 5x - y = 2 \\ 10x = 2y - 4 \end{cases}$  ∅

**Determine how the lines in each system of linear equations relate, the number of solutions of the system, and the type of system from the given information. State the solution set of the system. (Sections 4.2 and 4.3, Objective 3)**

**19.** $\begin{cases} y = -4x \\ 4x + y = 2 \end{cases}$; The elimination process yields $0 = -2$. The system has no solution and the lines are parallel. The system is inconsistent. The solution set is the empty set, ∅.

**20.** $\begin{cases} y = 2.5x - 7.1 \\ 6x - y = 19.7 \end{cases}$; The substitution process yields $3.5x = 12.6$.
The system contains intersecting lines with one solution. The system is consistent with independent equations. The solution set is {(3.6, 1.9)}.

**Applications of Systems of Linear Equations**

▶ **OBJECTIVES**

As a result of completing this section, you will be able to

1. Solve money applications.
2. Solve investment applications.
3. Solve mixture applications.
4. Solve distance applications.
5. Solve geometry applications.
6. Troubleshoot common errors.

Sam and Lori flew from Atlanta, Georgia, to San Francisco, California, a distance of approximately 2135 mi. The flight from Atlanta to San Francisco took 5 hr since the plane was flying against the wind. The return trip took 4.5 hr since the plane was flying with the wind. What was the speed of the plane in still air and what was the speed of the wind?

In this section, we will learn how to use a system of linear equations in two variables to solve this problem and other applications. The applications in this section are similar to the applications covered in Section 2.6. In Section 2.6, there were two unknowns in the problem and we expressed both unknowns in terms of one variable. We then used one linear equation to solve the problem.

Sometimes, it is easier to assign a different variable to each unknown. There are four general steps we will follow to solve such problems.

---

**Procedure:  General Problem Solving Strategy**

**Step 1:** Read the problem and determine the two unknown values. Assign a different variable to represent each of these values.
**Step 2:** Read the problem and write two equations that define a relationship between the two variables.
**Step 3:** Solve the system of equations using substitution or elimination.
**Step 4:** Write the answer to the problem.

---

## Money Applications

**Objective 1** ▶

Solve money applications.

Some of the applications covered in this objective require us to know how to find the value of a collection of objects. The following table illustrates how to find the total worth of a collection of books, for example, if each book in the collection has the same value.

| Number of Books | Value of Each Book | Total Worth of Collection |
|---|---|---|
| 3 | $10 | $3(10) = \$30$ |
| 7 | $15 | $7(15) = \$105$ |
| 10 | $20 | $10(20) = \$200$ |
| $x$ | $30 | $x(30) = 30x$ |

So, if we know the value of an individual item, we can determine the worth of a collection of these items by multiplying the value of the individual item by the total number of items in the collection. This concept can be extended to money as we have seen before when dealing with applications of linear equations in one variable. Suppose we have a total of 30 coins consisting of dimes and quarters. The total value of the collection of coins is provided in the following table for several different cases.

| Number of Dimes | Number of Quarters | Value of Dimes + Value of Quarters = | Total Worth of Collection |
|---|---|---|---|
| 6 | $30 - 6 = 24$ | $6(0.10) + 24(0.25) = \$0.60 + \$6.00 = \$6.60$ | |
| 20 | $30 - 20 = 10$ | $20(0.10) + 10(0.25) = \$2.00 + \$2.50 = \$4.50$ | |
| 12 | $30 - 12 = 18$ | $12(0.10) + 18(0.25) = \$1.20 + \$4.50 = \$5.70$ | |
| $x$ | $30 - x = y$ | $x(0.10) + y(0.25) = 0.10x + 0.25y$ | |

The objects in the collection can extend beyond coins. They might be a collection of tickets or a collection of hours at work, as we will see in Example 1.

| Objective 1 Examples | Solve each problem using a system of linear equations. |

**1a.** A movie theater sold 2500 tickets. Adult tickets cost $8.25 each and student tickets cost $6.00 each. If the movie theater collected a total of $17,475 for the tickets, how many adult tickets and student tickets were sold?

**Solution**  **1a.** What is unknown? The number of adult tickets and the number of student tickets sold are unknown.

Let $a$ = the number of adult tickets sold.

Let $s$ = the number of student tickets sold.

What is known? The total number of tickets sold is 2500. The total amount of money collected is $17,475. We can organize this information in a table.

| Unknowns | Number of tickets sold × Price per ticket = Total money collected | | |
|---|---|---|---|
| **Adult tickets sold** | $a$ | $8.25 | $8.25a$ |
| **Student tickets sold** | $s$ | $6.00 | $6.00s$ |
| **Totals** | 2500 | | $17,475 |

The system that represents these facts is

$$\begin{cases} a + s = 2500 & \text{A movie theater sold 2500 tickets.} \\ 8.25a + 6.00s = 17,475 & \text{The movie theater collected \$17,475.} \end{cases}$$

We solve the system using elimination since both equations are in standard form.

$$\begin{cases} -6(a + s = 2500) \\ 8.25a + 6.00s = 17,475 \end{cases} \longrightarrow \begin{cases} -6.00a - 6.00s = -15,000 & \text{Multiply the first} \\ \underline{\phantom{-}8.25a + 6.00s = \phantom{-}17,475} & \text{equation by } -6. \\ \phantom{-6.00a}2.25a = 2475 & \text{Add the equations.} \\ \phantom{-6.00a}a = 1100 & \text{Divide each side} \\ & \text{by 2.25.} \end{cases}$$

Now solve for $s$.

$$\begin{aligned} a + s &= 2500 & \text{Begin with the first equation.} \\ 1100 + s &= 2500 & \text{Substitute 1100 for } a. \\ 1100 + s - 1100 &= 2500 - 1100 & \text{Subtract 1100 from each side.} \\ s &= 1400 & \text{Simplify.} \end{aligned}$$

The movie theater sold 1100 adult tickets and 1400 student tickets.

**1b.** Monte shows his children a jar of coins that consists of only quarters and nickels. He tells them there are 300 coins totaling $51. How many quarters and nickels are in the jar?

**Solution**  **1b.** What is unknown? The number of quarters and the number of nickels in the jar are unknown.

Let $q$ = number of quarters.

Let $n$ = number of nickels.

What is known? There are 300 coins in the jar. The value of the coins is $51. We can organize the information in a table.

| Unknowns | Number of coins × Value per coin = Total value of coins | | |
|---|---|---|---|
| **Quarters** | $q$ | $0.25 | $0.25q$ |
| **Nickels** | $n$ | $0.05 | $0.05n$ |
| **Totals** | 300 | | $51 |

The system that represents these facts is

$$\begin{cases} q + n = 300 \\ 0.25q + 0.05n = 51 \end{cases}$$

There is a total of 300 coins in the jar.
The total value of the collection is $51.

We solve the system by substitution and solve the first equation for $q$.

| | |
|---|---|
| $q + n = 300$ | Begin with the first equation. |
| $q + n - n = 300 - n$ | Subtract $n$ from each side. |
| $q = 300 - n$ | Simplify. |

Now we substitute this expression in place of $q$ in the other equation.

| | |
|---|---|
| $0.25q + 0.05n = 51$ | Begin with the second equation. |
| $0.25(300 - n) + 0.05n = 51$ | Substitute $300 - n$ for $q$. |
| $75 - 0.25n + 0.05n = 51$ | Apply the distributive property. |
| $75 - 0.20n = 51$ | Combine like terms. |
| $75 - 0.20n - 75 = 51 - 75$ | Subtract 75 from each side. |
| $-0.20n = -24$ | Simplify. |
| $n = \dfrac{-24}{-0.20}$ | Divide each side by $-0.20$. |
| $n = 120$ | Simplify. |

Finally we solve for $q$.

| | |
|---|---|
| $q + n = 300$ | Begin with the first equation. |
| $q + 120 = 300$ | Replace $n$ with 120. |
| $q + 120 - 120 = 300 - 120$ | Subtract 120 from each side. |
| $q = 180$ | Simplify. |

Monte's jar contains 120 nickels and 180 quarters.

**1c.** Janna is working two jobs to help pay her way through college. She works as an office assistant and a store clerk. One week she earns $440 by working 25 hr as an office assistant and 20 hr as a store clerk. Another week she earns $390 by working 15 hr as an office assistant and 30 hr as a store clerk. What is Janna's hourly wage as an office assistant and a store clerk?

**Solution**   **1c.** What is unknown? The hourly wage Janna makes as an office assistant and a store clerk is unknown.

Let $x$ = Janna's hourly wage as an office assistant.
Let $y$ = Janna's hourly wage as a store clerk.

What is known? Janna's salary for two different weeks is given and shown in the following table.

| Unknowns | Hourly wage | Hours worked for Week 1 | Total paid | Hours worked for Week 2 | Total paid |
|---|---|---|---|---|---|
| **Office assistant wage** | $x$ | 25 | $25x$ | 15 | $15x$ |
| **Store clerk wage** | $y$ | 20 | $20y$ | 30 | $30y$ |
| **Totals** | | | 440 | | 390 |

The system that represents these facts is

$$\begin{cases} 25x + 20y = 440 \\ 15x + 30y = 390 \end{cases}$$

Week 1 earnings were $440.

Week 2 earnings were $390.

Solve the system using elimination.

$$\begin{cases} 3(25x + 20y = 440) \\ -2(15x + 30y = 390) \end{cases} \longrightarrow \begin{cases} 75x + 60y = 1320 \\ -30x - 60y = -780 \end{cases}$$

Multiply the first equation by 3.

Multiply the second equation by −2.

$$45x = 540$$   Add the equations.

$$\frac{45x}{45} = \frac{540}{45}$$   Divide each side by 45.

$$x = 12$$   Simplify.

Now we solve for $y$.

$$25x + 20y = 440$$   Begin with the first equation.

$$25(12) + 20y = 440$$   Substitute 12 for $x$.

$$300 + 20y = 440$$   Simplify.

$$300 + 20y - 300 = 440 - 300$$   Subtract 300 from each side.

$$20y = 140$$   Simplify.

$$\frac{20y}{20} = \frac{140}{20}$$   Divide each side by 20.

$$y = 7$$   Simplify.

So, Janna earns $12 an hour as an office assistant and $7 an hour as a store clerk.

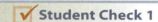 **Student Check 1**   Solve each problem using a system of linear equations.

**a.** Admission to a football game is $10 for students and $14 for general admission. If 300 tickets are sold and $3500 is collected, how many student tickets and general admission tickets are sold?

**b.** Thomas has a collection of coins consisting of nickels and dimes. If he has 500 coins totaling $40, how many nickels and dimes are in his collection?

**c.** Two families go to the movies together. One family purchases four drinks and two large popcorns for $26. The other family purchases two drinks and three large popcorns for $24. How much does each drink and popcorn cost?

## Investment Applications

**Objective 2** ▶

Solve investment applications.

Investment problems in this course deal primarily with *simple interest. Interest* is a fee paid for the use of money. Interest is calculated as a percentage of the amount of money invested. Simple interest is the most basic type of interest that can be earned.

> **Property: Simple Interest Formula**
>
> Let $P$ be the principal or initial money invested, $r$ the annual interest rate (as a decimal), and $t$ the length of the investment in years, then the earned interest $I$ is
>
> $$I = Prt$$

Suppose we invest $5000 in two different savings accounts. One account earns 5% annual interest and the other account earns 4% annual interest. The following table shows how much annual interest is earned from the two accounts if the money is invested as shown.

| | Principal × | Rate × | Time = | Interest |
|---|---|---|---|---|
| **Account 1** | $3000 | 0.05 | 1 | (3000)(0.05)(1) = $150 |
| **Account 2** | $2000 | 0.04 | 1 | (2000)(0.04)(1) = $80 |
| **Totals** | $5000 | | | $230 |

If the amount invested in each account is unknown, we can assign a variable to represent those amounts. The amount of interest earned is shown in the table.

| | Principal × | Rate × | Time = | Interest |
|---|---|---|---|---|
| **Account 1** | $x$ | 0.05 | 1 | $(x)(0.05)(1) = 0.05x$ |
| **Account 2** | $y$ | 0.04 | 1 | $(y)(0.04)(1) = 0.04y$ |
| **Totals** | $5000 | | | $0.05x + 0.04y$ |

We can use these ideas to solve word problems involving investments.

**Objective 2 Example**   Sonja invests $3000 in two different accounts. She invests part of her money in a savings account that earns 4% simple interest and the rest of her money in a money market account that earns 8% simple interest. If she earns a total of $200 interest in 1 yr, how much did she invest in each account? Use a system of linear equations to solve this problem.

**Solution**   What is unknown? The amount of money invested at 4% and at 8% is unknown.

Let $x$ = amount invested at 4%.
Let $y$ = amount invested at 8%.

What is known? The total amount invested is $3000. The total interest earned is $200. We can organize the information in a table.

| | Principal × | Rate × | Time = | Interest |
|---|---|---|---|---|
| **Savings account** | $x$ | 0.04 | 1 | $(x)(0.04)(1) = 0.04x$ |
| **Money market account** | $y$ | 0.08 | 1 | $(y)(0.08)(1) = 0.08y$ |
| **Totals** | $3000 | | | $200 |

The system that represents these facts is

$$\begin{cases} x + y = 3000 & \text{Total invested is \$3000.} \\ 0.04x + 0.08y = 200 & \text{Total interest earned is \$200.} \end{cases}$$

We solve the system using elimination but first clear the decimals from the system by multiplying the second equation by 100.

$$\begin{cases} x + y = 3000 \\ 100(0.04x + 0.08y = 200) \end{cases} \longrightarrow \begin{cases} x + y = 3000 \\ 4x + 8y = 20{,}000 \end{cases}$$

Now we multiply the first equation by $-8$ to eliminate the $y$-terms from the equations.

$$\begin{cases} -8(x + y = 3000) \\ 4x + 8y = 20{,}000 \end{cases} \longrightarrow \begin{cases} -8x - 8y = -24{,}000 \\ \underline{\phantom{-}4x + 8y = \phantom{-}20{,}000} \end{cases}$$ 

Multiply the first equation by $-8$.

$$-4x = -4000$$ 

Add the equations.

$$\frac{-4x}{-4} = \frac{-4000}{-4}$$ 

Divide each side by $-4$.

$$x = 1000$$ 

Simplify.

Next we solve for $y$.

$$x + y = 3000 \qquad \text{Begin with the first equation.}$$
$$1000 + y = 3000 \qquad \text{Replace } x \text{ with 1000.}$$
$$1000 + y - 1000 = 3000 - 1000 \qquad \text{Subtract 1000 from each side.}$$
$$y = 2000 \qquad \text{Simplify.}$$

Sonja invested $1000 into the 4% account and $2000 into the 8% account.

✓ **Student Check 2** Georgia receives an inheritance of $100,000. She invests part of the money in a certificate of deposit (CD) that earns 8% simple interest and the rest of the money in an international mutual fund that earns 12% simple interest. If Georgia earns $11,000 in interest for 1 yr, how much did she invest in each account? Use a system of linear equations to solve this problem.

## Mixture Applications

**Objective 3** ▶

Solve mixture applications.

In Section 2.6, we learned about mixture problems and discovered that to find the amount of substance in a solution, we multiply the amount of the given substance by the strength of the solution. To solve these types of applications with a system, we will assign a variable for each of the unknowns and write two equations that represent the facts in the problem.

**Objective 3 Example** A dairy farmer produces two types of milk. How many gallons of a 3.25% milk fat solution and how many gallons of a 1% milk fat solution should be mixed together to obtain 750 gal of a 2% milk fat solution?

**Solution** What is unknown? The gallons of the 3.25% milk fat solution and the gallons of the 1% milk fat solution are unknown.

Let $x$ = number of gallons of 3.25% milk fat solution.
Let $y$ = number of gallons of 1% milk fat solution.

What is known? The total mixture is 750 gal. The final mixture is 2% milk fat. We can organize the information in a table.

| Unknowns | Amount of Solution | Strength of Solution | Amount of Milk Fat |
|---|---|---|---|
| Gallons of 3.25% milk fat | $x$ | 3.25% = 0.0325 | $0.0325x$ |
| Gallons of 1% milk fat | $y$ | 1% = 0.01 | $0.01y$ |
| Final mixture | 750 | 2% = 0.02 | $0.02(750) = 15$ |

The system that represents the situation is

$$\begin{cases} x + y = 750 & \text{Total obtained is 750 gal.} \\ 0.0325x + 0.01y = 15 & \text{Mixture needs to be 2\% milk fat.} \end{cases}$$

We solve the system using substitution and solve the first equation for $y$.

$$x + y = 750$$
$$x + y - x = 750 - x \qquad \text{Subtract } x \text{ from each side.}$$
$$y = 750 - x$$

Now we solve for $x$.

| | |
|---|---|
| $0.0325x + 0.01y = 15$ | Begin with the second equation. |
| $0.0325x + 0.01(750 - x) = 15$ | Substitute $750 - x$ for $y$. |
| $0.0325x + 7.5 - 0.01x = 15$ | Apply the distributive property. |
| $0.0225x + 7.5 = 15$ | Combine like terms. |
| $0.0225x + 7.5 - 7.5 = 15 - 7.5$ | Subtract 7.5 from each side. |
| $0.0225x = 7.5$ | Simplify. |
| $x = \dfrac{7.5}{0.0225}$ | Divide each side by 0.0225. |
| $x = 333.33$ | Simplify. |

Now we solve for $y$.

| | |
|---|---|
| $x + y = 750$ | Begin with the first equation. |
| $333.33 + y = 750$ | Substitute 333.33 for $x$. |
| $333.33 + y - 333.33 = 750 - 333.33$ | Subtract 333.33 from each side. |
| $y = 416.67$ | Simplify. |

The farmer needs to mix 333.33 gal of the 3.25% milk fat solution with 416.67 gal of the 1% milk fat solution to obtain 750 gal of a 2% milk fat solution.

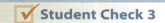 **Student Check 3**    A chemist has a 50% acid solution. She needs 30 L of a 15% acid solution. How many liters of water and how many liters of the 50% acid solution must she mix together to obtain 30 L of a 15% acid solution? Use a system of linear equations to solve this problem.

## Distance Applications

**Objective 4** ▶

**Solve distance applications.**

In Chapter 2, we learned that the total *distance* traveled equals the rate (or speed) times the time elapsed. Algebraically, we write the **distance formula** as $d = rt$, where $r$ is the rate or speed and $t$ is the time traveled. We will apply this concept to flying.

If a plane is flying *directly with the wind*, the plane will arrive at its destination sooner than it would if flying in still air since the speed of the wind increases the plane's speed. If, however, a plane flies *directly against the wind*, the plane will take longer to reach its destination since the speed of the wind decreases the speed of the plane. The following table provides some illustrations of how the speed of the wind can affect the speed of the plane.

| Speed of the Plane in Still Air | Speed of the Wind | Speed of Plane Flying **with the Wind** | Speed of Plane Flying **against the Wind** |
|---|---|---|---|
| 500 mph | 20 mph | $500 + 20 = 520$ mph | $500 - 20 = 480$ mph |
| 600 mph | $x$ mph | $600 + x$ mph | $600 - x$ mph |

Note the following:

- Flying directly *with* the wind increases the speed of the plane by the speed of the wind.
- Flying directly *against* the wind decreases the speed of the plane by the speed of the wind.

**Objective 4  Examples**    **Solve each problem using a system of linear equations.**

**4a.** Sam and Lori flew from Atlanta, Georgia (ATL), to San Francisco, California (SFO), a distance of approximately 2135 mi. The flight from Atlanta to San Francisco took 5 hr since the plane was flying against the wind. The return trip took 4.5 hr since the plane was flying with the wind. What was the speed of the plane in still air and what was the speed of the wind?

**Solution**    **4a.** What is unknown? The speed of the plane and the speed of the wind are unknown.

Let $x = $ speed of the plane.
Let $y = $ speed of the wind.

What is known? The distance between the cities is 2135 mi. The trip against the wind was 5 hr. The trip with the wind was 4.5 hr. We can organize the information in a table as shown.

|  | Rate | $\times$ | Time | = | Distance |
|---|---|---|---|---|---|
| **ATL to SFO** | $x - y$ | | 5 | | 2135 |
| **SFO to ATL** | $x + y$ | | 4.5 | | 2135 |

The system that represents the situation is

$$\begin{cases} 5(x - y) = 2135 \\ 4.5(x + y) = 2135 \end{cases}$$

We solve the system using elimination after first applying the distributive property. Then we clear decimals by multiplying the second equation by 10.

$$\begin{cases} 5x - 5y = 2135 \\ 4.5x + 4.5y = 2135 \end{cases} \longrightarrow \begin{cases} 5x - 5y = 2135 \\ 10(4.5x + 4.5y = 2135) \end{cases} \longrightarrow \begin{cases} 5x - 5y = 2135 \\ 45x + 45y = 21{,}350 \end{cases}$$

Now multiply the first equation by 9 to eliminate the $y$-terms from the system.

$$\begin{cases} 9(5x - 5y = 2135) \\ 45x + 45y = 21{,}350 \end{cases} \longrightarrow \begin{cases} 45x - 45y = 19{,}215 \\ 45x + 45y = 21{,}350 \end{cases}$$    Multiply the first equation by 9.

$$90x = 40{,}565$$    Add the equations.

$$\frac{90x}{90} = \frac{40{,}565}{90}$$    Divide each side by 90.

$$x \approx 450.7$$    Simplify.

Next solve for $y$.

$$5(x - y) = 2135$$    Begin with the first equation.

$$5(450.7 - y) \approx 2135$$    Replace $x$ with 450.7.

$$\frac{5(450.7 - y)}{5} \approx \frac{2135}{5}$$    Divide each side by 5.

$$450.7 - y \approx 427$$    Simplify.

$$450.7 - y - 450.7 \approx 427 - 450.7$$    Subtract 450.7 from each side.

$$-y \approx -23.7$$    Simplify.

$$\frac{-y}{-1} \approx \frac{-23.7}{-1}$$    Divide each side by $-1$.

$$y \approx 23.7$$    Simplify.

So, the speed of the plane is approximately 450.7 mph and the speed of the wind is approximately 23.7 mph.

**4b.** At 9 A.M., David leaves his home traveling south on his bike at an average speed of 10 mph. At 11 A.M., his wife Judy leaves their home to find David to give him an urgent message, since he left his cell phone at home. She travels in her car at an average speed of 65 mph. How long will it take for Judy to reach David?

**Solution**

**4b.** What is unknown? The time that David and Judy each travel is unknown.

> Let $x$ = time David travels.
>
> Let $y$ = time Judy travels.

What is known? David's speed is 10 mph and Judy's speed is 65 mph. The distance traveled by both David and Judy is the same. Judy's time traveling is 2 hr less than David's since she left 2 hr after David. We can organize this information in a table.

|  | Rate | × | Time | = | Distance |
|---|---|---|---|---|---|
| **David** | 10 | | $x$ | | $10x$ |
| **Judy** | 65 | | $y$ | | $65y$ |

The system that represents the situation is

$$\begin{cases} 10x = 65y & \text{Distances are the same.} \\ y = x - 2 & \text{Judy's time is 2 hr less than David's time.} \end{cases}$$

Because the second equation is solved for $y$, we solve the system using substitution.

| | |
|---|---|
| $10x = 65y$ | Begin with the first equation. |
| $10x = 65(x - 2)$ | Replace $y$ with $x - 2$. |
| $10x = 65x - 130$ | Apply the distributive property. |
| $10x - 65x = 65x - 130 - 65x$ | Subtract $65x$ from each side. |
| $-55x = -130$ | Simplify. |
| $\dfrac{-55x}{-55} = \dfrac{-130}{-55}$ | Divide each side by $-55$. |
| $x \approx 2.4$ | Simplify. |

**INSTRUCTOR NOTE:**
Remind students that 0.4 hr is equivalent to 0.4 · 60 min or 24 min.

Finally, we solve for $y$.

| | |
|---|---|
| $y = x - 2$ | Begin with the second equation. |
| $y \approx 2.4 - 2$ | Substitute 2.4 for $x$. |
| $y \approx 0.4$ | Simplify. |

It will take Judy about 0.4 hr or 24 min to reach David.

✔ **Student Check 4**  Solve each problem using a system of linear equations.

**a.** A plane can travel 1800 mi with the wind in 4 hr. The return trip against the wind takes 6 hr. What is the speed of the plane and the speed of the wind?

**b.** Cherie and her mom live 330 mi apart. They want to meet each other so that Cherie's kids can go home with her mom for a few days. Cherie leaves at 8 A.M. and travels toward her mom at an average speed of 50 mph. Her mom leaves at 8:30 A.M. and travels toward Cherie at an average speed of 65 mph. What time will they meet? (Round solutions to the nearest minute.)

## Geometry Applications

**Objective 5** ▶

Solve geometry applications.

We will use some topics from Chapter 2 to solve applications involving geometry problems. Some of the geometry problems we will solve involve perimeter of a rectangle and the relationships between complementary and supplementary angles. Recall these important facts.

- The **perimeter of a rectangle** is $2l + 2w$, where $l$ is the length and $w$ is the width.
- **Supplementary angles** are two angles whose measures add to 180°.
- **Complementary angles** are two angles whose measures add to 90°.

---

**Objective 5  Examples**      **Solve each problem using a system of linear equations.**

**5a.** The perimeter of a high school basketball court is 268 ft. The length of the court is 16 ft less than two times the width of the court. Find the dimensions of the court.

**Solution**   **5a.** What is unknown? The length and the width of the basketball court are unknown.

Let $l$ = the length of the court.
Let $w$ = the width of the court.

What is known? The perimeter of the court is 268 ft. The length is 16 ft less than two times the width. The system that represents the problem is

$$\begin{cases} 2l + 2w = 268 \\ l = 2w - 16 \end{cases}$$

Because the second equation is solved for $l$, we solve the system using substitution.

| | |
|---|---|
| $2l + 2w = 268$ | Begin with the first equation. |
| $2(2w - 16) + 2w = 268$ | Substitute $(2w - 16)$ for $l$. |
| $4w - 32 + 2w = 268$ | Apply the distributive property. |
| $6w - 32 = 268$ | Combine like terms. |
| $6w - 32 + 32 = 268 + 32$ | Add 32 to each side. |
| $6w = 300$ | Simplify. |
| $\dfrac{6w}{6} = \dfrac{300}{6}$ | Divide each side by 6. |
| $w = 50$ | Simplify. |

Next we solve for the length.

| | |
|---|---|
| $l = 2w - 16$ | Begin with the second equation. |
| $l = 2(50) - 16$ | Substitute 50 for $w$. |
| $l = 100 - 16$ | Multiply. |
| $l = 84$ | Subtract. |

So, the dimensions of the basketball court are 84 ft by 50 ft.

---

**5b.** Two angles are supplementary. The measure of one angle is 6° less than twice the measure of the other angle. Find the measure of each angle.

**Solution**   **5b.** What is unknown? The measure of each angle is unknown.

Let $x$ = the measure of one angle.
Let $y$ = the measure of its supplement.

What is known? The angles are supplementary, which means their sum is 180°. The measure of one angle is 6° less than twice the measure of the other. The system that represents the problem is

$$\begin{cases} x + y = 180 \\ x = 2y - 6 \end{cases}$$

Because the second equation is solved for $x$, we solve the system by substitution.

| | |
|---|---|
| $x + y = 180$ | Begin with the first equation. |
| $2y - 6 + y = 180$ | Substitute $2y - 6$ for $x$. |
| $3y - 6 = 180$ | Combine like terms. |
| $3y - 6 + 6 = 180 + 6$ | Add 6 to each side. |
| $3y = 186$ | Simplify. |
| $\dfrac{3y}{3} = \dfrac{186}{3}$ | Divide each side by 3. |
| $y = 62$ | Simplify. |

Now we solve for $x$.

| | |
|---|---|
| $x = 2y - 6$ | Begin with the second equation. |
| $x = 2(62) - 6$ | Substitute 62 for $y$. |
| $x = 124 - 6$ | Multiply. |
| $x = 118$ | Subtract. |

So, the measure of one angle is 118° and the measure of its supplement is 62°.

**✓ Student Check 5**   Solve each problem using a system of linear equations.

**a.** A regulation size tennis court has a perimeter of 228 ft. The length of the court is 6 ft more than twice the width. Find the dimensions of the court.

**b.** Two angles are complementary. The measure of one angle is 10° more than the measure of the other. Find the measure of each angle.

---

**Objective 6** ▶

Troubleshoot common errors.

## Troubleshooting Common Errors

Some common errors associated with applications of systems are shown next. Most of the errors are the result of setting up the equations in the system incorrectly.

**Objective 6  Examples**   **A problem and an incorrect solution are given. Provide the correct solution and an explanation of the error.**

**6a.** Micah invests money in two different accounts, a CD that earns 5% annual interest and a money market account that earns 8% annual interest. In the money market account, he invests $2000 less than 3 times the amount he invests in the CD account. If he earns a total of $3320 in yearly interest, how much does he invest in each account? Write the system that represents this situation.

| Incorrect Solution | Correct Solution and Explanation |
|---|---|
| Let $x$ = amount invested in the CD account. Let $y$ = amount invested in the money market account. $$\begin{cases} y = 2000 - 3x \\ 0.05x + 0.08y = 3320 \end{cases}$$ | The error was made in the first equation. To represent 2000 less than 3 times an amount is $3x - 2000$. So, the system should be $$\begin{cases} y = 3x - 2000 \\ 0.5x + 0.08y = 3320 \end{cases}$$ |

**6b.** A chemist needs to make a 30% iodine solution. He has a 50% iodine solution and a 10% iodine solution. How much of each should he mix together to obtain 60 mL of a 20% iodine solution? Write the system that represents this situation.

| Incorrect Solution | Correct Solution and Explanation |
|---|---|
| Let $x$ = amount of 50% iodine solution. Let $y$ = amount of 10% iodine solution. $$\begin{cases} x + y = 60 \\ 0.5x + 0.1y = 60 \end{cases}$$ | The error was made in the second equation. The left side of the equation represents the amount of pure iodine in the two solutions, so the right side should also represent the amount of pure iodine. The system should be $$\begin{cases} x + y = 60 \\ 0.5x + 0.1y = 0.2(60) \end{cases}$$ |

## ANSWERS TO STUDENT CHECKS

**Student Check 1**   **a.** 175 student tickets and 125 general tickets   **b.** 200 nickels and 300 dimes
   **c.** $3.75 per drink and $5.50 per popcorn

**Student Check 2**   $75,000 invested in the international mutual fund and $25,000 in the CD

**Student Check 3**   21 L of water and 9 L of the 50% acid solution

**Student Check 4**   **a.** The plane's speed is 375 mph and the wind speed is 75 mph.   **b.** They will meet at approximately 11:09 A.M.

**Student Check 5**   **a.** The tennis court is 78 ft long by 36 ft wide.   **b.** The angles are 50° and 40°.

## SUMMARY OF KEY CONCEPTS

While many different types of applications were solved in this section, there is really only one method that we use to solve these problems. The steps to solve an application with a system of two linear equations in two variables are

1. Determine the two unknowns in the problem. Use two different variables to represent these unknowns.
2. Write two equations that relate these unknowns to each other.
3. Once the system is set up, use substitution or elimination to solve the system.

## GRAPHING CALCULATOR SKILLS

In Section 4.2, we learned a direct method to determine if a point is a solution of a system of equations. Another way we can check if a point is a solution of a system is by using the Store command.

**Example:** Determine if $(1, -1)$ is a solution of the system
$2x + 3y = -1$
$x - 3y = 4$

**Solution:** Store the value of 1 for $x$ and store the value of $-1$ for $y$.

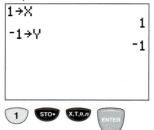

Next, enter each equation. After pressing Enter, a 1 or a 0 will be displayed. If a 1 is displayed, then the stored values of $x$ and $y$

make the equation true. If a 0 is displayed, then the stored values of $x$ and $y$ make the equation false.

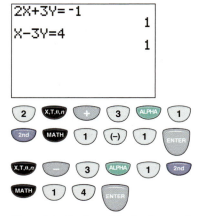

Since 1 is displayed for both equations, the point $(1, -1)$ is a solution of the given system.

**Example:** Determine if (4, 3) is a solution of the system
$$\begin{cases} 2x + 3y = -1 \\ x - 3y = 4 \end{cases}.$$

**Solution:** Store the value of 4 for $x$ and the value of 3 for $y$.

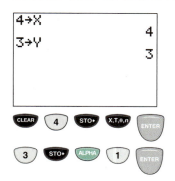

Next, enter each equation.

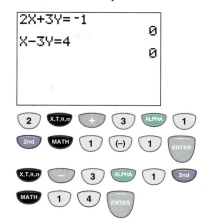

Since 0 is displayed for both equations, the point (4, 3) is not a solution of the given system.

---

## SECTION 4.4 / EXERCISE SET

 ## Write About It!

**Use complete sentences in your answer to each exercise.**

1. Explain the general strategy you should follow to solve application problems.   Answers vary.

2. This section contains applications similar to those found in Section 2.6. How does the approach to solving these problems differ in this section from the approach given in Section 2.6?

3. Give an example of a mixture application.   Answers vary.

4. In Objective 3, an explanation was given about the relationship between the wind and the net speed of a plane. Suppose that the speed of a plane in still air is 450 mph. If this plane is flying against a wind current of $x$ mph, will the net speed of the plane be $450 - x$ mph or $x - 450$ mph? Explain.

 ## Practice Makes Perfect!

**Solve each problem using a system of linear equations. Round answers to two decimal places when applicable. (See Objective 1.)**

5. Jamie's middle school put on a production of *The Wizard of Oz*. They charged students and school employees $5 per ticket, and everyone else was charged $10 per ticket. Five hundred people attended the show, and the receipts totaled $3000. How many students and school employees attended the production?
   400 students and school employees

6. Carrel's youth group raised money by having a car wash. The group charged $3 to wash a car and $5 to wash an SUV. At the end of the day, they had

washed 46 vehicles and raised $180. How many SUVs were washed by the youth group?   21 SUVs

7. Janeane takes a bag of dimes and quarters to a coin counting machine and opts to have the value of her money converted to a gift card. If 113 coins are counted and the value of the gift card is $19.55, how many dimes did she deposit in the machine?   58 dimes

8. Elise collects half-dollars and silver dollars. After several years, Elise takes her coins to the bank. She has 130 coins that total $90. How many half-dollar coins does she have?   80 half-dollars

9. Rachel buy a 2-lb bag of candied pecans and candied peanuts for $17. The candied pecans sell for $9.50 per pound, and the candied peanuts sell for $5.50 per pound. How many pounds of each type of nut are included in the bag?   1.5 lb candied pecans, 0.5 lb candied peanuts

10. Lori makes her own trail mix at a natural-foods grocery store. She mixes Spanish almonds costing $7.50 per pound with dried dates costing $9 per pound. If a 3-lb bag cost Lori $24, how many pounds of Spanish almonds and how many pounds of dried dates are included in the mixture?   2 lb Spanish almonds, 1 lb dried dates

11. Lisa works for a community food bank. She buys 100 cans of green beans and 70 cans of tuna for $92.50. The next day, she purchases 60 cans of green beans and 80 cans of tuna for $84. How much is she charged for each can of green beans and each can of tuna?   green beans: $0.40, tuna: $0.75

12. A video store has a month-long sale on movies. Pauline, who loves movies, purchases two new DVDs and three used DVDs for $46. Pauline returns the

Additional answers can be found in the Instructor Answer Appendix.

following week and purchases three new DVDs and five used DVDs for $71.50. How much does the video store charge for each new DVD and each used DVD?   new: $15.50, used: $5

**Solve each problem using a system of linear equations. (*See Objective 2.*)**

13. Behnaz invests $500 in two different accounts. She invests part of her money in a savings account that earns 5% simple interest and the rest of her money in a high-risk junk bond that earns 15% simple interest. If Behnaz's accounts earn $37.50 interest in 1 yr, how much does she invest in each account?   5%: $375, 15%: $125

14. Haazim has $1500 that he invests in two different accounts. He invests part of his money in a savings account that earns 4.5% simple interest and he invests the rest of his money in a CD that earns 9% simple interest. If Haazim's accounts earn $90 interest in 1 yr, how much does he invest in each account?   4.5%: $1000, 9%: $500

**Solve each problem using a system of linear equations. (*See Objective 3.*)**

15. A dairy farmer produces two types of milk: milk with 1% milk fat and milk with 4% milk fat. How many gallons of each type of milk should he mix to obtain 300 gal of milk with 2% milk fat?

1%: 100 gal, 4%: 200 gal

16. A dairy farmer produces two types of milk: milk with 1.5% milk fat and milk with 3.25% milk fat. How many gallons of each type of milk should he mix to obtain 420 gal of milk with 2% milk fat?   1.5%: 300 gal, 3.25%: 120 gal

17. Joe's Diner purchases whole milk (3.25% milk fat) and skim milk (0% milk fat). Some of Joe's customers order glasses of 2% lowfat milk. How many liters of whole milk and skim milk must Joe combine to obtain 26 L of 2% lowfat milk?

whole milk: 16 L, skim milk: 10 L

18. Asa likes the taste of a 1% milk fat solution. How much whole milk and skim milk must Asa mix together to obtain 13 oz of milk with 1% milk fat? (Recall that whole milk is 3.25% milk fat and skim milk is 0% milk fat.)   whole milk: 4 oz, skim milk: 9 oz

19. A chemist needs 5 mL of a solution that is 20% acid. In her lab, she has a container of solution that is 50% acid and another container of solution that is 10% acid. How much of each acid solution must the chemist combine to get the desired 5 mL of 20% acid?   50%: 1.25 mL, 10%: 3.75 mL

20. A chemist needs 12 L of a solution that is 15% acid. In his lab, he has a supply of a 30% acid solution and a 10% acid solution. How many liters of each acid solution must he combine to make 12 L of a 15% acid solution?   30%: 3 L, 10%: 9 L

21. A chemist needs 10 L of a solution that is 25% acid. In his lab, he has a 40% acid solution. How much water and how much 40% acid solution must the chemist combine to make 10 L of a 25% acid solution?   water: 3.75 L, 40%: 6.25 L

22. A chemist needs 20 mL of a solution that is 10% acid. In her lab, she has a supply of pure acid. How much water and how much pure acid must the chemist combine to make 20 mL of a 10% acid solution?   water: 18 mL, pure acid: 2 mL

**Solve each problem using a system of linear equations. Round answers to the nearest hundredth when applicable. (*See Objective 4.*)**

23. The distance between Newark Liberty International Airport in New Jersey and Denver International Airport in Colorado is approximately 1600 mi. Ming's flight from Newark to Denver takes 3.5 hr flying against the jet stream, but her return flight takes only 3 hr flying with the jet stream. What was the plane's speed in still air, and what was the speed of the jet stream?   plane: 495.24 mph, jet stream: 38.10 mph

24. Roland, who lives in Phoenix, is planning to fly 2900 mi to Honolulu during his winter break. By checking online flight times, Roland finds a flight from Phoenix to Honolulu that takes 6.5 hr traveling against the wind. He finds a return flight from Honolulu to Phoenix that takes only 5.75 hr traveling with the wind in the same type of plane. What is the speed of the plane in still air, and what is the speed of the wind?   plane: 475.25 mph, wind: 29.10 mph

**Similar to wind currents, water currents affect the rates boats travel. Suppose that a boat in still water travels 20 mph and that the speed of the water current is $x$ mph.**

**If a boat is traveling *upstream* (against the current), the net speed of the boat is *slower*: *speed of boat − speed of current* = 20 − $x$ mph**

**If a boat is traveling *downstream* (with the current), the net speed of the boat is *faster*: *speed of boat + speed of current* = 20 + $x$ mph**

25. Mitchell and his friends rent a fishing boat. They travel 36 mi upstream in 1 hr. Their return trip takes only 0.75 hr. What was the speed of the current? What was the speed of Mitchell's boat in still water?   current: 6 mph, boat: 42 mph

**26.** Ira leaves his dock and sails 60 mi upstream in 1.5 hr. It takes him only 1 hr to return to his dock. What was the speed of the current? What was the speed of Ira's boat in still water?   current: 10 mph, boat: 50 mph

**27.** Drivers A and B enter the pit row at the same time during a professional car race. Driver A stops for gas and a tire change and returns to the race immediately at an average speed of 160 mph. Driver B has mechanical problems that take 10 min to resolve. He enters the race at an average speed of 180 mph. Assuming the drivers do not make any more stops, how long will it take for Driver B to catch up to Driver A?   1 hr 20 min

**28.** A fugitive leaves the scene of an accident traveling by car at an average speed of 80 mph. The police arrive at the scene of the accident 15 min later. They travel in the same direction as the fugitive at an average speed of 90 mph. How long will it take before the police catch the fugitive?   2 hr

**Solve each problem using a system of linear equations. (See Objective 5.)**

**29.** The perimeter of a rectangle is 160 ft. If the length of the rectangle is 4 ft less than the width, find the dimensions of the rectangle.   length: 38 ft, width: 42 ft

**30.** The perimeter of a rectangle is 70 in. If the width of the rectangle is 5 in. more than 3 times the length, find the dimensions of the rectangle.   length: 7.5 in., width: 27.5 in.

**31.** Two angles are complementary. One angle is 3° less than twice the other angle. Find the measure of the larger angle.   59°

**32.** Two angles are supplementary. One angle is 6° more than 5 times the other angle. Find the measure of the smaller angle.   29°

**33.** Sandeep fences off part of his yard for his new Labrador retriever. He encloses a rectangular region with 50 yd of chain link fence. If Sandeep wants the length of the dog yard to be 5 yd longer than its width, find the dimensions of the dog yard he can build.   length: 15 yd, width: 10 yd

**34.** Chang makes wooden frames for paintings with elaborate hand-carved details. He has enough wood pieces to make a rectangular frame with a perimeter of 120 in. If the width of the frame is 10 in. less than its length, find the dimensions of the frame Chang made.   length: 35 in., width: 25 in.

**35.** Becky and Elaine are aspiring ballerinas. They are each working toward doing a perfect six-o'clock position, but neither can do it just yet. Elaine's back leg can rise 20° higher than Becky's back leg. The sum of the angles formed by the legs of both ballerinas is 280°. What angle is each ballerina able to form with her legs? Who is closer to her six-o'clock position goal?   Becky (130°); Elaine is closer (150°)

**36.** Sisters Chelsea and Cheyanne just started taking ballet classes. They are working toward executing the arabesque position in which their legs form a right angle. Chelsea, the older of the sisters, is able to raise her back leg 25° higher

than Cheyanne and still keep her balance. The sum of the angles formed by the legs of the sisters when attempting their best arabesque positions is 135°. What angle is each sister able to make?   Chelsea: 80°, Cheyanne: 55°

## Mix 'Em Up!

**Solve each problem using a system of linear equations.**

**37.** John Carpenter's *Halloween* (1978) gave Jamie Lee Curtis her big break as an actor. *Halloween* and *Halloween H20: 20 Years Later* (1998) grossed approximately $102 million at the box office in the United States. *Halloween* grossed $8 million less than *Halloween H20*. How much did each movie gross at the box office? (Source: Wikipedia)   *Halloween:* $47 million, *Halloween H20:* $55 million

**38.** Two angles are supplementary. The larger angle is 5° more than four times the smaller angle. What is the measure of each angle?   35°, 145°

**39.** Two angles are supplementary. One angle is 6° less than five times the other angle. What is the measure of each angle?   31°, 149°

**40.** Two angles are complementary. The larger angle is 21° less than twice the measure of the smaller one. What is the measure of each angle?   37°, 53°

**41.** Two angles are complementary. One angle is 28° more than three times the measure of the other one. What is the measure of each angle?   15.5°, 74.5°

**42.** Leigh and Ryan are experienced cyclists who like challenging cycling routes. Ryan gives Leigh a 15-min head start on a 50-mi route. On this route, Leigh cycles an average of 10 mph. On the same route, Ryan averages 15 mph. How long will it take Ryan to catch up to Leigh?   $\frac{1}{2}$ hour

**43.** Susan leaves her job and heads east at 55.5 mph. Vera leaves the same workplace 36 min later and heads east at 74 mph. How long will it take Vera to catch up to Susan?   1.8 hr

**44.** Paint thinner is often mixed with paint to lessen the appearance of brush strokes. A rule of thumb is to use a mixture that is 60% paint and 40% paint thinner. Jennifer has a large supply of pure paint, and a large supply of a mixture that is 50% paint (and 50% paint thinner). How much pure paint and how much paint mixture must Jennifer mix together so that she will have 3 gal of a 60% paint mixture?   pure paint: 0.6 gal, 50% paint: 2.4 gal

**45.** An angle measures 6° more than twice its complement. What is the measure of each angle?   28°, 62°

**46.** An angle measures 36° more than four times its supplement. What is the measure of each angle?   28.8°, 151.2°

**47.** A plane travels 3159.5 mi in 7.1 hr traveling with the wind. It takes 8.9 hr for the plane to travel the same distance against the wind. What is the speed of the plane in still air and what is the speed of the wind?   400 mph, 45 mph

**48.** A plane travels 1505 mi in 3.5 hr traveling with the wind. It takes 4.3 hr for the plane to travel the same distance against the wind. What is the speed of the plane in still air and what is the speed of the wind?   390 mph, 40 mph

**49.** Steven has $10,000 to invest. He invests part of the money in a savings account that pays 1.4% annual interest and the rest of the money in a CD that pays 3.1% annual interest. If he earns $268.52 interest in 1 yr, how much does he invest in each account?   $7560 at 3.1% and $2440 at 1.4%

**50.** Michelle has $6650 to invest. She invests part of the money in a stock that pays a 3.8% annual dividend and the rest of the money in a mutual fund that pays 1.6% annual interest. If she earns $228.50 in investment income in 1 yr, how much does she invest in each account?   $5550 at 3.8% and $1100 at 1.6%

**51.** One week, Danielle buys 7 cans of tuna and 6 cans of soup for $13.69. The next week, she buys 3 cans of tuna and 2 cans of soup for $5.13. What is the price of each can?   $0.85 for a can of tuna, $1.29 for a can of soup

**52.** One week, Lucia buys 6 cans of soup and 1 can of mushrooms for $15.24. The next week, she buys 9 cans of soup and 3 cans of mushrooms for $25.11. What is the price of each can?
$2.29 for a can of soup, $1.50 for a can of mushrooms

 **You Be the Teacher!**

**Explain each situation to a student.**

**53.** Explain why you should solve the applications algebraically instead of just reasoning.

**54.** Explain the process of finding the system of equations for liquid mixture applications.

**Correct each student's errors, if any.**

**55.** Jim has $10,650 to invest. He invests part of the money in a mutual fund that pays 1.1% annual interest and the rest of the money in a money market fund that pays 4.5% annual interest. If he earns $248.05 interest in 1 yr, how much did he invest in each account?

Steve's work:

Let $x$ be the amount invested at 1.1% and $y$ at 4.5%.

$$\begin{cases} 1.1x + 4.5y = 10{,}650 \\ x + y = 248.05 \end{cases}$$

I solve for $x$ and substitute into the first equation.

$$x = 248.05 - y$$

$$1.1(248.05 - y) + 4.5y = 10{,}650$$

$$2728.55 - 1.1y + 4.5y = 10{,}650$$

$$3.5y = 7921.45$$

$$y = 2263.27$$

So, $x = 248.05 - 2263.27 = -2015.22$.

**56.** Todd needs to make 266 L of a 46% antifreeze solution. How much pure antifreeze and how much of a 16% antifreeze solution should he mix together to get 266 L of a 46% antifreeze solution?

Andrew's work:

Let $x$ be the volume of the pure antifreeze solution and $y$ be the volume of a 16% antifreeze solution.

$$\begin{cases} x + y = 266(0.46) \\ 0x + 0.16y = 266 \end{cases}$$

I solve for $y$ using the second equation.

$$y = 266/0.16 = 1662.5$$

 **Calculate It!**

**Use a graphing calculator to determine if each point is a solution of the system.**

**57.** $\begin{cases} 0.3x + 0.6y = 12 \\ x + y = 25 \end{cases}$; $(10, 15)$   yes

**58.** $\begin{cases} 0.75x + 1.25y = 10.25 \\ x + y = 9 \end{cases}$; $(2, 7)$   yes

**59.** $\begin{cases} x - y = 3 \\ 2x + y = 6 \end{cases}$; $(3, 1)$   no **60.** $\begin{cases} x - 3y = 4 \\ 4x + 3y = 1 \end{cases}$; $(1, -2)$   no

---

**SECTION 4.5**    **Linear Inequalities in Two Variables**

▶ **OBJECTIVES**

**As a result of completing this section, you will be able to**

**1.** Determine if an ordered pair is a solution of a linear inequality in two variables.

**2.** Graph the solution set of a linear inequality in two variables.

**3.** Solve applications of linear inequalities in two variables.

**4.** Troubleshoot common errors.

The final grade in Tyrone's math class is based on his test average and final exam grade. The test average counts for 70% of his final grade and the final exam counts for 30% of his final grade. If Tyrone wants to have at least a 70 average in the course, what are some possible combinations of test average and final exam grades to produce the desired result?

To solve this problem, we must know how to solve linear inequalities in two variables.

## Solutions of Linear Inequalities in Two Variables

**OBJECTIVE 1** ▶

Determine if an ordered pair is a solution of a linear inequality in two variables.

At this point, we have discussed solving systems of linear equations in two variables both graphically and algebraically. In the next section, we will learn how to identify solutions of systems of linear inequalities in two variables based on their graphs. To do this, we first need to learn how to graph a single *linear inequality in two variables*.

> **Definition:** A **linear inequality in two variables** is an inequality that can be written in one of the following ways, where $A$, $B$, $C$, $m$, and $b$ are real numbers with $A$ and $B$ not both zero.
>
> $Ax + By > C$ $\qquad$ $Ax + By \geq C$ $\qquad$ $Ax + By < C$ $\qquad$ $Ax + By \leq C$
>
> $y > mx + b$ $\qquad$ $y \geq mx + b$ $\qquad$ $y < mx + b$ $\qquad$ $y \leq mx + b$

Some examples of linear inequalities in two variables are $x + y > 3$ and $y \leq \frac{1}{4}x + 5$.

Like a solution of linear equations in two variables, a **solution of a linear inequality in two variables** is an ordered pair $(x, y)$ that makes the inequality true.

> **Procedure:  Determining if an Ordered Pair Is a Solution of a Linear Inequality in Two Variables**
>
> **Step 1:** Replace the variables with the corresponding values of $x$ and $y$.
> **Step 2:** Simplify the resulting inequality.
> **Step 3:** If the ordered pair makes the inequality true, then the ordered pair is a solution. If the ordered pair makes the inequality false, then the ordered pair is not a solution.

**Objective 1  Examples**   Determine which ordered pairs are solutions of the given inequality.

**1a.** $2x + y > 6$; $(2, 8)$, $(3, 0)$, $(-1, 3)$, $(-2, -1)$, $(4, 0)$, $(6, -2)$

**1b.** $y \leq \frac{1}{2}x - 4$; $(0, 1)$, $(3, -5)$, $(8, 0)$, and $(0.1, 0.3)$

**Solutions**   **1a.**

| $(x, y)$ | $2x + y > 6$ | True or False? | Solution? |
|---|---|---|---|
| $(2, 8)$ | $2(2) + 8 > 6$ <br> $12 > 6$ | True | Yes |
| $(3, 0)$ | $2(3) + 0 > 6$ <br> $6 > 6$ | False | No |
| $(-1, 3)$ | $2(-1) + 3 > 6$ <br> $1 > 6$ | False | No |
| $(-2, -1)$ | $2(-2) + (-1) > 6$ <br> $-5 > 6$ | False | No |
| $(4, 0)$ | $2(4) + 0 > 6$ <br> $8 > 6$ | True | Yes |
| $(6, -2)$ | $2(6) + (-2) > 6$ <br> $10 > 6$ | True | Yes |

So, the ordered pairs $(2, 8)$, $(4, 0)$, and $(6, -2)$ are solutions of $2x + y > 6$.

**1b.**

| $(x, y)$ | $y \leq \dfrac{1}{2}x - 4$ | True or False? | Solution? |
|---|---|---|---|
| $(0, 1)$ | $1 \leq \dfrac{1}{2}(0) - 4$ <br> $1 \leq -4$ | False | No |
| $(3, -5)$ | $-5 \leq \dfrac{1}{2}(3) - 4$ <br> $-5 \leq -\dfrac{5}{2}$ | True | Yes |
| $(8, 0)$ | $0 \leq \dfrac{1}{2}(8) - 4$ <br> $0 \leq 0$ | True | Yes |
| $(0.1, 0.3)$ | $0.3 \leq \dfrac{1}{2}(0.1) - 4$ <br> $0.3 \leq -3.95$ | False | No |

So, $(3, -5)$ and $(8, 0)$ are solutions of $y \leq \dfrac{1}{2}x - 4$.

✔ **Student Check 1**  Determine if the ordered pair is a solution of the given inequality.

**a.** $2x - 5y > -10;\ (0, 0)$    **b.** $y \geq \dfrac{1}{3}x + 4;\ (-6, 1)$    **c.** $x < -1;\ (-2, 3)$

## Graphing Linear Inequalities in Two Variables

**Objective 2**

Graph the solution set of a linear inequality in two variables.

Now that we know what it means for an ordered pair to be a solution of a linear inequality in two variables, we will graph the solution set of a linear inequality in two variables. Just as there are infinitely many solutions of linear equations in one variable, there are infinitely many solutions of linear inequalities in two variables.

To graph a linear inequality in two variables, we begin by graphing the linear equation in two variables that is formed by replacing the inequality with an equals sign. This line is called the **boundary line** and divides the plane into two regions called **half-planes**. The region above or to the right of the line is the **upper half-plane** and the region below or to the left of the line is the **lower half-plane**. Either the upper half-plane or the lower half-plane contains solutions of the inequality. The boundary line may or may not be included in the solution set.

To determine the graph of the solution set of the inequality $2x + y > 6$, we use the boundary line, $2x + y = 6$, and the information we found in Example 1a. First, we graph the linear equation associated with this inequality: $2x + y = 6$. This equation has intercepts of $(3, 0)$ and $(0, 6)$. The graph is shown next along with the ordered pairs from Example 1a.

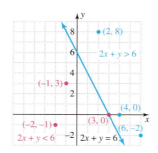

| $(x, y)$ | $2x + y > 6$ | Solution? |
|---|---|---|
| $(2, 8)$ | $2(2) + 8 > 6$ <br> $12 > 6$ | Yes |
| $(4, 0)$ | $2(4) + 0 > 6$ <br> $8 > 6$ | Yes |
| $(6, -2)$ | $2(6) + (-2) > 6$ <br> $10 > 6$ | Yes |
| $(3, 0)$ | $2(3) + 0 > 6$ <br> $6 > 6$ | No |
| $(-1, 3)$ | $2(-1) + 3 > 6$ <br> $1 > 6$ | No |
| $(-2, -1)$ | $2(-2) + (-1) > 6$ <br> $-5 > 6$ | No |

The points $(2, 8)$, $(4, 0)$, and $(6, -2)$ are solutions of $2x + y > 6$. Notice these points are all located above the line $2x + y = 6$, or in the upper half-plane. The points $(-1, 3)$ and $(-2, -1)$ are not solutions of $2x + y > 6$. Notice these points are located below the line $2x + y = 6$, or in the lower half-plane. We also found that the point $(3, 0)$ is not a solution of $2x + y > 6$. However, this point is located on the line $2x + y = 6$.

From the graph we observe the following:

- Solutions of the inequality are all located on the same side of the graph of the associated equation.
- The point $(3, 0)$ lies on the line, $2x + y = 6$, but is not a solution of the inequality.
- None of the points that lie below the line are solutions of the inequality.

---

**Property:  The Graph of a Linear Inequality in Two Variables**

The graph of the solution set of a linear inequality in two variables has three parts:

- The region, or half-plane, that contains solutions of the inequality.
- The region, or half-plane, that contains no solutions of the inequality.
- The boundary line that divides the plane into these two regions.

---

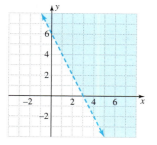

When we shade the half-plane that contains solutions of the inequality, $2x + y > 6$, we get the graph shown to the left. Notice that the half-plane that contains no solutions of $2x + y > 6$ is not shaded. The line $2x + y = 6$ is dashed since the line does not contain solutions.

**Note:**

- *If one point in a half-plane is a solution of an inequality, then all points in that half-plane are also solutions.*

- *If one point in a half-plane is not a solution of an inequality, then none of the points in that half-plane are solutions.*

These facts enable us to use a **test-point method** to graph linear inequalities in two variables. This method is described next.

---

**Procedure:  Graphing a Linear Inequality in Two Variables Using the Test-Point Method**

**Step 1:** Graph the associated linear equation in two variables by replacing the inequality symbol with an equals sign. This line forms the boundary between the region that contains solutions and the region that does not contain solutions.

  **a.** The boundary line is solid if the inequality symbol is $\leq$ or $\geq$.
     (This means that the points on the line are solutions of the inequality.)
  **b.** The boundary line is dashed if the inequality symbol is $<$ or $>$.
     (This means that the points on the line are not solutions of the inequality.)

**Step 2:** Identify a point in either one of the half-planes formed by the boundary line, but not on the boundary line, and test it in the original inequality.

  **a.** If the point makes the original inequality *true*, then shade the half-plane that contains the test point.
  **b.** If the point makes the original inequality *false*, then shade the half-plane that does not contain the test point.

---

**Note:** *The point $(0, 0)$ is the easiest point to test. Any other point in the plane will work as long as the point is not on the boundary line. If the boundary line goes through the origin, then $(0, 0)$ cannot be used as the test point.*

**Objective 2 Examples**    **Graph the solution set of each linear inequality in two variables.**

**2a.** $x - 2y > 4$    **2b.** $y \le -\dfrac{2}{3}x + 3$    **2c.** $x \ge -2$    **2d.** $y < 5$    **2e.** $y + 3x \ge 0$

**Solutions**    **2a.** We first graph the associated equation $x - 2y = 4$.

To graph $x - 2y = 4$, we find the $x$- and $y$-intercepts as shown in the table. The line is dashed since the original inequality is $<$.

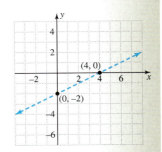

| $x$ | $y$ | $(x, y)$ |
|-----|-----|----------|
| 0 | $-2$ | $(0, -2)$ |
| 4 | 0 | $(4, 0)$ |

Next, we use $(0, 0)$ as a test point since the graph of $x - 2y = 4$ does not go through the origin.

$$x - 2y > 4 \qquad \text{Begin with the original inequality.}$$
$$0 - 2(0) > 4 \qquad \text{Replace } x \text{ and } y \text{ with 0.}$$
$$0 > 4 \qquad \text{Simplify.}$$

Since the test point makes the original inequality false, solutions of the inequality are located in the half-plane that does *not* contain $(0, 0)$. The solutions of the inequality are located in the half-plane below the boundary line, or the lower half-plane. Recall that none of the points on the boundary line are solutions since the inequality symbol is $>$. We shade the region below the boundary line to denote the graph of $x - 2y > 4$.

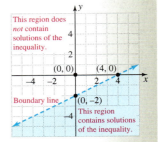

**2b.** We first graph the associated equation $y = -\dfrac{2}{3}x + 3$.

The $y$-intercept of $y = -\dfrac{2}{3}x + 3$ is $(0, 3)$ and the slope $m = -\dfrac{2}{3}$. Plot $(0, 3)$ and then move down 2 units and right 3 to get to another point on the line. The boundary line is solid since the inequality is $\le$.

Next, we use $(0, 0)$ as a test point since the graph of $y = -\dfrac{2}{3}x + 3$ does not go through the origin.

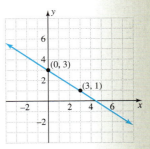

$$y \le -\dfrac{2}{3}x + 3 \qquad \text{Begin with the original inequality.}$$

$$0 \le -\dfrac{2}{3}(0) + 3 \qquad \text{Replace } x \text{ and } y \text{ with 0.}$$

$$0 \le 0 + 3 \qquad \text{Simplify.}$$
$$0 \le 3 \qquad \text{Simplify.}$$

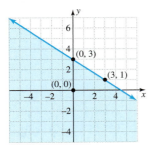

Since $(0, 0)$ makes the original inequality true, we shade the half-plane that contains $(0, 0)$. All points on the boundary line and in the lower half-plane are solutions of the inequality, $y \le -\dfrac{2}{3}x + 3$.

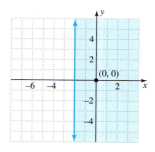

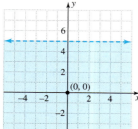

**2c.** We first graph the associated equation $x = -2$. The line $x = -2$ is a vertical line through $(-2, 0)$. The boundary line is solid since the inequality symbol is $\geq$.

Next, we test the point $(0, 0)$ since it is not on the line, $x = -2$.

$x \geq -2$     Begin with the original inequality.

$0 \geq -2$     Replace $x$ with 0.

Since $(0, 0)$ makes the inequality true, we shade the half-plane that contains $(0, 0)$. This is the upper, or right, half-plane. All points on the boundary line and in the right half-plane are solutions of the inequality, $x \geq -2$.

**2d.** We first graph the associated equation $y = 5$. The line $y = 5$ is a horizontal line through $(0, 5)$. The line is dashed since the inequality symbol is $<$.

Next, we test the point $(0, 0)$ since it is not on the line, $y = 5$.

$y < 5$     Begin with the original inequality.

$0 < 5$     Replace $y$ with 0.

Since $(0, 0)$ makes the original inequality true, we shade the half-plane that contains $(0, 0)$. Only points in the lower half-plane are solutions of the inequality; none of the points on the boundary line are solutions.

**2e.** We graph the associated equation $y + 3x = 0$. This line is equivalent to $y = -3x$, which has $y$-intercept of $(0, 0)$ and slope $m = -3$. Plot $(0, 0)$ and move down 3 units and right 1 to get to another point on the line. The boundary line is solid since the inequality is $\geq$.

Since $(0, 0)$ is on the boundary line, $y = -3x$, we must test a point other than $(0, 0)$, say, $(1, 1)$.

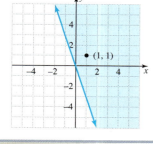

$y + 3x \geq 0$     Begin with the original inequality.

$(1) + 3(1) \geq 0$     Replace $x$ and $y$ with 1.

$4 \geq 0$     Simplify.

Since $(1, 1)$ makes the original inequality true, we shade the half-plane that contains $(1, 1)$. All points on the boundary line and in the upper half-plane are solutions of the inequality, $y + 3x \geq 0$.

☑ **Student Check 2**    Graph the solution set of each linear inequality in two variables.

    **a.** $3x + y > 6$       **b.** $y \leq 2x + 1$       **c.** $x < -2$

    **d.** $y \geq 3$           **e.** $y + 2x \geq 0$

## Applications

**Objective 3**

Solve applications of linear inequalities in two variables.

Applications of linear inequalities in two variables are very important for a branch of mathematics called *linear programming*. This field of mathematics is used to determine how to maximize or minimize different values. While this topic is beyond the scope of this class, we will explore other applications of linear inequalities.

**Objective 3 Example**

The final grade in Tyrone's math class is based on his test average and his final exam grade. The test average counts for 70% of his final grade and the final exam counts for 30% of his final grade. If Tyrone wants to have at least a 70 in the course, what are some possible combinations of test average and final exam grades that produce the desired result? (Note the final exam grade and test average cannot exceed 100.)

**Solution**   What is unknown? The test average and the final exam grade are unknown.

Let $x$ = test average and let $y$ = final exam grade.

Recall that 70% and 30% written as decimals are 0.70 and 0.30, respectively. Since the test average counts for 70% and the final counts for 30%, the final grade is computed as

$$0.70x + 0.30y$$

For the final grade to be at least 70, Tyrone must obtain a grade of 70 or higher. Therefore, we must solve the linear inequality

$$0.70x + 0.30y \geq 70$$

The graph of the boundary line $0.70x + 0.30y = 70$ is solid ($\geq$) and goes through the points $(100, 0)$ and $(70, 70)$. We graph only in the first quadrant since negative grades are not possible. We test $(0, 0)$ since it does not lie on the boundary line, $0.70x + 0.30y = 70$.

| | |
|---|---|
| $0.70x + 0.30y \geq 70$ | Begin with the original inequality. |
| $0.70(0) + 0.30(0) \geq 70$ | Replace $x$ and $y$ with 0. |
| $0 + 0 \geq 70$ | Simplify. |
| $0 \geq 70$ | Simplify. |

Since the point $(0, 0)$ makes the inequality false, we shade the portion of the plane that does not contain the origin, that is, the upper half-plane.

Any point contained in the shaded region satisfies the inequality. Points in this region give possible values for Tyrone's test average and final exam grade, provided both values are less than or equal to 100. Some possible solutions to this problem are shown on the graph in green. The final average each point yields is shown in the following table.

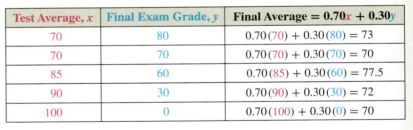

| Test Average, $x$ | Final Exam Grade, $y$ | Final Average = $0.70x + 0.30y$ |
|---|---|---|
| 70 | 80 | $0.70(70) + 0.30(80) = 73$ |
| 70 | 70 | $0.70(70) + 0.30(70) = 70$ |
| 85 | 60 | $0.70(85) + 0.30(60) = 77.5$ |
| 90 | 30 | $0.70(90) + 0.30(30) = 72$ |
| 100 | 0 | $0.70(100) + 0.30(0) = 70$ |

There are many other combinations of test averages and final exam grades that provide the desired course grade.

The points labeled in magenta $(50, 90)$, $(70, 60)$, and $(80, 30)$ are ordered pairs that do not satisfy the inequality and would, therefore, not make the desired final grade.

✓ **Student Check 3**   The final grade in Lorna's history class is based on her midterm exam grade and her final exam grade. The midterm exam counts for 40% of her grade and her final exam counts for 60% of her grade. Find three possible combinations of grades she can earn on the midterm and final exams to have at least an 80 in the class.

**Objective 4** ▶

Troubleshoot common errors.

## Troubleshooting Common Errors

Some common errors associated with linear inequalities in two variables are shown next. Be careful to draw the boundary line correctly.

**Objective 4 Examples** A problem and an incorrect solution are given. Provide the correct solution and an explanation of the error.

**4a.** Graph the solution set of $y > -x + 3$.

| Incorrect Solution | Correct Solution and Explanation |
|---|---|
| The boundary line is $y = -x + 3$. The point $(0, 0)$ makes the inequality false. | Everything is correct except for the boundary line. It should be dashed since the inequality is $>$. |

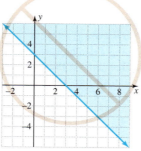

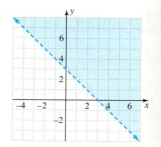

**4b.** Graph the solution set of $x - 5y < 10$.

| Incorrect Solution | Correct Solution and Explanation |
|---|---|
| The boundary line is $x - 5y = 10$. The point $(0, 0)$ makes the inequality true. The graph is | Since $(0, 0)$ makes the inequality true, the graph must include this point. We shade the half-plane that contains the origin. |

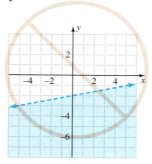

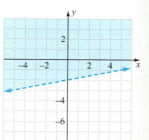

## ANSWERS TO STUDENT CHECKS

**Student Check 1** **a.** yes **b.** no **c.** yes

**Student Check 2**

**a.**

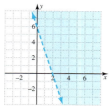

**b.**

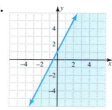

**e.**

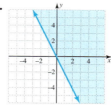

**c.**

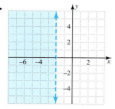

**d.**

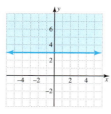

**Student Check 3** Answers vary; $(80, 80)$, $(60, 100)$, and $(70, 90)$.

## SUMMARY OF KEY CONCEPTS

1. Solutions of linear inequalities in two variables are ordered pairs that satisfy the linear inequality. There are infinitely many solutions of linear inequalities in two variables.

2. Solutions of linear inequalities can be illustrated by a graph. The solutions of a linear inequality in two variables lie in a half-plane that is formed by the graph of the associated linear equation. A single test point can be used to determine which half-plane contains solutions.

3. To use a linear inequality in two variables to solve an applied problem, graph the inequality as shown in this section. Any ordered pair in the shaded region is a solution of the problem.

## GRAPHING CALCULATOR SKILLS

To graph a linear inequality in two variables using the graphing calculator, we must first solve the inequality for $y$. If the resulting inequality is of the form $y < mx + b$, shade below the boundary line. If the resulting inequality is of the form $y > mx + b$, shade above the boundary line. If the inequality is $<$ or $>$, the boundary line will be dashed and if the inequality is $\leq$ or $\geq$, then the boundary line is solid.

**Example:** Solve $x - 2y > 4$.

**Solution:** We first solve the inequality for $y$.

$$-2y > -x + 4$$

$$y < \frac{1}{2}x - 2$$

*The inequality reverses* because we divide both sides by $-2$. Enter the associated equation. For the calculator to graph the appropriate region, we need to "tell it" which region to shade—the region above the boundary line ($>$ or $\geq$) or the region below ($<$ or $\leq$). Move the cursor to the left of $Y_1$. Press ENTER until we reach the symbol ▉ (shade above) or ▉ (shade below).

**Alternate Method for TI-84 Plus Users:** Use the APPS feature to graph inequalities.

**Example:** Solve $y > -2x - 4$.

Press ⬭ and ⬭. Scroll down in the menu and select Inequalz.

Press any key to continue. Press ⬭ ⬭ to select the $>$ symbol and then enter the remaining part of the inequality. Press ⬭

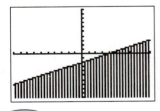

## SECTION 4.5 / EXERCISE SET

 **Write About It!**

Use complete sentences in your answer to each exercise.

1. How do you determine if an ordered pair $(x, y)$ is a solution of a linear inequality?

2. How do you determine the boundary line of a linear inequality?

3. How will you know whether the boundary line of the graph of a linear inequality should be a dotted line or a solid line?

4. How many points in a half-plane must be tested to determine if the half-plane includes solutions of a linear inequality?

 **Practice Makes Perfect!**

Determine if the ordered pair is a solution of the given inequality. (*See Objective 1.*)

5. $3x - 4y < 12$; $(1, 1)$   yes

6. $2x + y < -3$; $(0, 1)$   no

7. $y \geq 5x - 1$; $(4, -1)$   no

8. $y \leq -2x + 3$; $(0, 0)$   yes

9. $x + y < 9$; $(10, 3)$   no

10. $x - y > -5$; $(-2, -4)$   yes

11. $y > \dfrac{1}{2}x + 8$; $(6, -3)$   no

12. $y < -\dfrac{1}{3}x - 11$; $(9, 0)$   no

13. $x \geq 2y - 4$; $(0, 2)$   yes

14. $y + 2 < 0$; $(8, -1)$   no

15. $x - 3 < 0$; $(-4, 11)$   yes

16. $x + 15 > 1$; $(13, -2)$   yes

17. $4x - 2y > 1$; $\left(\dfrac{1}{2}, \dfrac{3}{4}\right)$   no

18. $x - 2y < 3$; $\left(\dfrac{2}{3}, \dfrac{1}{6}\right)$   yes

Graph the solution set of each linear inequality in two variables. (*See Objective 2.*)

19. $2x + y > 4$
20. $3x + y < 5$
21. $y \leq 4x - 8$
22. $y \geq 3x - 6$
23. $y < -2x + 12$
24. $y > -3x - 9$
25. $y \leq 5x - 15$
26. $y \geq 10x - 20$
27. $x - y < 2$
28. $y - 7x \leq 0$
29. $y < 3x$
30. $y > 2x$
31. $3x - 3y \geq 0$
32. $4x - y > -3$

33. $x > 8$
34. $x < -1$
35. $y < -4$
36. $y > 3$

Use a linear inequality in two variables to solve each problem. (*See Objective 3.*)

**For Exercises 37 and 38: Dawn's algebra course grade is based on her test average and her final exam grade. Her test average counts for 60% of her course grade and the final exam counts for 40% of her course grade.**

37. If Dawn wants to have at least an 80 in the course, (a) write a linear inequality in two variables that represents this situation, (b) graph the linear inequality, and (c) give three possible combinations of test averages and final exam grades that produce a course grade of at least 80.

38. If Dawn wants to have at least a 70 in the course, (a) write a linear inequality in two variables that represents this situation, (b) graph the linear inequality, and (c) give three possible combinations of test averages and final exam grades that produce a course grade of at least 70.

**For Exercises 39 and 40: Jenna works at a bookstore for $8 per hour and at the school library for $10.50 per hour.**

39. If Jenna wants to make at least $300 per week, (a) write a linear inequality in two variables that represents this situation; (b) graph the linear inequality; and (c) give three possible combinations of hours that Jenna could work at each job to make at least $300 per week.

40. If Jenna wants to make at least $450 per week, (a) write a linear inequality in two variables that represents this situation; (b) graph the linear inequality; and (c) give three possible combinations of hours that Jenna could work at each job to make at least $450 per week.

**For Exercises 41 and 42: Shenita charges $15 to hem a pair of pants and $25 to tailor a suit jacket.**

41. If Shenita wants to make at least $500 in one week from tailoring, (a) write a linear inequality in two variables that represents this situation; (b) graph the linear inequality; and (c) if Shenita wants to make at least $500 in one week, give three possible combinations of the number of pants she must hem and the number of suit jackets she must tailor.

Additional answers can be found in the Instructor Answer Appendix.

**42.** If Shenita wants to make at least $650 in one week from tailoring, (a) write a linear inequality in two variables that represents this situation; (b) graph the linear inequality; and (c) if Shenita wants to make at least $650 in one week, give three possible combinations of the number of pants she must hem and the number of suit jackets she must tailor.

 **Mix 'Em Up!**

**Determine if the ordered pair is a solution of the given inequality.**

**43.** $x + 3y < 6$; $(0, 0)$    yes

**44.** $x - 7y < 1$; $(10, 3)$    yes

**45.** $y > \frac{1}{2}x + 8$; $(8, -5)$    no

**46.** $y < -\frac{1}{3}x - 11$; $(-9, 0)$    no

**47.** $x > 12y$; $(1, 0)$    yes

**48.** $y > 6x$; $(0, -1)$    no

**49.** $y - 11 < 4$; $(-3, 25)$    no

**50.** $x + 5 \geq 1$; $(-2, 4)$    yes

**51.** $6x - y \leq 1$; $(0.2, -0.5)$    no

**52.** $x + 8y > 3$; $(-1.4, 0.5)$    no

**Graph the solution set of each linear inequality in two variables.**

**53.** $x + 3y < 2$

**54.** $2x - 4y \leq 3$

**55.** $5x - 2y > 0$

**56.** $3x + 2y < 0$

**57.** $y + 3 \geq 0$

**58.** $2x - 3 < 0$

**59.** $0.2x + 0.3y < 1.5$

**60.** $-0.3x + 0.4y \geq 1.2$

**For Exercises 61 and 62: Ted's biology course grade is based on his test average and his final exam grade. His test average counts for 75% of his course grade and his final exam counts for 25% of his course grade.**

**61.** If Ted wants to have at least a 90 in the course, (a) write a linear inequality in two variables that represents this situation, (b) graph the linear inequality, and (c) give three possible combinations of test averages and final exam grades that produce a course grade of at least 90.

**62.** If Ted wants to have at least an 80 in the course, (a) write a linear inequality in two variables that represents this situation, (b) graph the linear inequality, and (c) give three possible combinations of test averages and final exam grades that produce a course grade of at least 80.

**For Exercises 63 and 64: Brasil works at a grocery store for $10.50 per hour and as a student worker for $8.50 per hour.**

**63.** If Brasil wants to make at least $350 per week, (a) write a linear inequality in two variables that represents this situation, (b) graph the linear inequality, and (c) give three possible combinations of hours that Brasil could work at each job to make at least $350 per week.

**64.** If Brasil wants to make at least $450 per week, (a) write a linear inequality in two variables that represents this situation, (b) graph the linear inequality, and (c) give three possible combinations of hours that Brasil could work at each job to make at least $450 per week.

 **You Be the Teacher!**

**Correct each student's error, if any.**

**65.** Graph $4x + y < 8$.
Jamie's work: The boundary line is $4x + y = 8$ and my test point is $(0, 0)$.
$4(0) + 0 < 8$
$0 < 8$
True

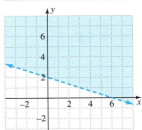

**66.** Graph $x + 3y \geq 6$.
Vince's work: The boundary line is $x + 3y > 6$ and use $(0, 0)$ as a test point.
$0 + 3(0) > 6$
$0 > 6$
False

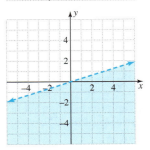

**67.** Graph $x - 4y > 0$.
Carolyn's work: The boundary line is $x - 4y = 0$.
I shade the upper half-plane since the inequality symbol is $>$.

**68.** Graph $2x \leq y$.
Brent's work:
I graph $2x = y$ and shade the lower half-plane since the symbol is $\leq$.

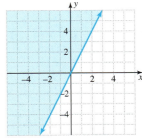

## Calculate It!

Use a graphing calculator to graph each linear inequality in two variables.

**69.** $3x + y \geq 1$      **70.** $6x + y \leq 2$

**71.** $x + 3y < 2$      **72.** $x - 5y < 4$

## Think About It!

**73.** Give an example of a linear inequality for which the point $(0, 0)$ cannot be used as the test point to determine which half-plane to shade.

**74.** Is it possible for a linear inequality in two variables to have one point as a solution? Explain.

**75.** Is it possible for a linear inequality in two variables to have only its boundary line as a solution? Explain.

**76.** Graph the inequalities in parts a–d and use your results to answer parts e–g.

    **a.** $y > -2x + 4$      **b.** $y > \frac{1}{3}x - 1$

    **c.** $y > 4x$      **d.** $y > 1$

    **e.** Which half-plane was shaded in each of these inequalities?    upper half-plane

    **f.** What can you conclude about the graph of the solution of $y > mx + b$?
    The solution set will always be the upper half-plane.

    **g.** How does the graph of $y \geq mx + b$ differ from the graph of $y > mx + b$?

**77.** Graph the inequalities in parts a–d and use your results to answer parts e–g.

    **a.** $y < -2x + 4$      **b.** $y < \frac{1}{3}x - 1$

    **c.** $y < 4x$      **d.** $y < 1$

    **e.** Which half-plane was shaded in each of these inequalities?    lower half-plane

    **f.** What can you conclude about the graph of the solution of $y < mx + b$?
    The solution set will always be the lower half-plane.

    **g.** How does the graph of $y \leq mx + b$ differ from the graph of $y < mx + b$?

---

| SECTION 4.6 | **Systems of Linear Inequalities in Two Variables** |
|---|---|

### ▶ OBJECTIVES

As a result of completing this section, you will be able to

**1.** Solve a system of linear inequalities in two variables.

**2.** Solve application problems.

**3.** Troubleshoot common errors.

The Parent Teacher Association (PTA) of a local school is selling rolls of wrapping paper and boxed chocolates for a fund-raiser. The PTA can order at most 300 items. Each roll of wrapping paper costs \$2 and each box of chocolates costs \$3. The PTA can spend no more than \$1200 on these items. From past experience, the PTA knows that they sell at least twice as many boxes of chocolates as they do rolls of wrapping paper.

How many rolls of wrapping paper and boxes of chocolates should the PTA order for their conditions to be satisfied?

Solving this problem requires us to satisfy more than one linear inequality in two variables. In this section, we will learn how to solve a system of linear inequalities.

### Solving Systems of Linear Inequalities in Two Variables

**Objective 1** ▶

Solve a system of linear inequalities in two variables.

A **system of linear inequalities** is a set of two or more linear inequalities that must be solved together. The **solution set of a system of linear inequalities in two variables** is the set of all ordered pairs that satisfies each inequality in the system.

In Section 4.1, we learned that the solution set of a system of linear equations in two variables consists of the point where the two lines intersect. Similarly, the solution set of a system of linear inequalities in two variables consists of the points where the two inequalities intersect. Graphically, this is the set of all ordered pairs where the two half-planes in the system intersect.

> **Procedure: Solving a System of Linear Inequalities in Two Variables**
>
> **Step 1:** Graph each linear inequality on the same coordinate system using the method presented in Section 4.5.
>
> **Step 2:** Identify the intersection of the shaded regions. This is the solution of the system.

| **Objective 1  Examples** | **Solve each system of linear inequalities.** |

**1a.** $\begin{cases} y > -x + 3 \\ y \le 2x - 4 \end{cases}$    **1b.** $\begin{cases} 2x + y < 6 \\ x - y \le 4 \end{cases}$    **1c.** $\begin{cases} y > -2 \\ x < 3 \end{cases}$

**Solutions**

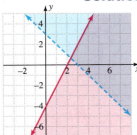

**Figure 4.4.1**

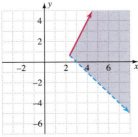

**Figure 4.4.2**

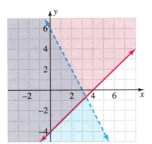

**Figure 4.4.3**

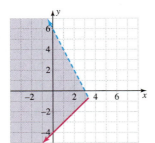

**Figure 4.4.4**

**1a.** We graph each inequality and find the intersection.

$y > -x + 3$: The boundary line is $y = -x + 3$ and is dashed. The line $y = -x + 3$ goes through $(0, 3)$ and $(3, 0)$. We test $(0, 0)$.

$$y > -x + 3$$
$$0 > -(0) + 3$$
$$0 > 3 \qquad \text{False}$$

Since $(0, 0)$ makes the original inequality false, we shade the upper half-plane that doesn't contain $(0, 0)$. The graph of the solution set of $y > -x + 3$ is shown in blue in Figure 4.4.1.

$y \le 2x - 4$: The boundary line is $y = 2x - 4$ and is solid. The line $y = 2x - 4$ goes through $(0, -4)$ and $(2, 0)$. We test $(0, 0)$.

$$y \le 2x - 4$$
$$0 \le 2(0) - 4$$
$$0 \le -4 \qquad \text{False}$$

Since $(0, 0)$ makes the original inequality false, we shade the lower half-plane that doesn't contain $(0, 0)$. The graph of the solution set of $y \le 2x - 4$ is shown in magenta in Figure 4.4.1.

The solution set of the system is the region where the two half-planes intersect as shown in Figure 4.4.2.

**1b.** We graph each inequality and find the intersection.

$2x + y < 6$: The boundary line is $2x + y = 6$ and is dashed. The line goes through $(0, 6)$ and $(3, 0)$. We test $(0, 0)$.

$$2x + y < 6$$
$$2(0) + 0 < 6$$
$$0 < 6 \qquad \text{True}$$

Since $(0, 0)$ makes the original inequality true, we shade the lower half-plane, which contains $(0, 0)$. The solution set of $2x + y < 6$ is shown in blue in Figure 4.4.3.

$x - y \le 4$: The boundary line is $x - y = 4$ and is solid. The line goes through the $(0, -4)$ and $(4, 0)$. We test $(0, 0)$.

$$x - y \le 4$$
$$0 - 0 \le 4$$
$$0 \le 4 \qquad \text{True}$$

Since $(0, 0)$ makes the original inequality true, we shade the upper half-plane, which contains $(0, 0)$. The solution set of $x - y \le 4$ is shown in magenta as shown in Figure 4.4.3.

The solution set of the system is the intersection of the two regions as shown in Figure 4.4.4.

**1c.** We graph each inequality and find the intersection.

$y > -2$: The boundary line is $y = -2$ and is dashed. The line is a horizontal line through $(0, -2)$. We test $(0, 0)$.

$$y > -2$$
$$0 > -2 \qquad \text{True}$$

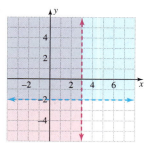

**Figure 4.4.5**

Since (0, 0) makes the inequality true, we shade the upper half-plane which contains (0, 0). The graph of $y > -2$ is shown in blue in Figure 4.4.5.

$x < 3$: The boundary line is $x = 3$ and is dashed. The line is a vertical line through (3, 0). We test (0, 0).

$$x < 3$$
$$0 < 3 \qquad \text{True}$$

Since (0, 0) makes the inequality true, we shade the half-plane that contains (0, 0). The graph of $x < 3$ is shown in magenta in Figure 4.4.5.

The solution set of the system is the intersection of the shaded regions as shown in Figure 4.4.6.

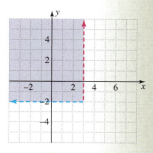

**Figure 4.4.6**

---

✔ **Student Check 1**      Solve each system of linear inequalities.

**a.** $\begin{cases} y \geq x + 5 \\ y \geq -x + 2 \end{cases}$   **b.** $\begin{cases} 3x - y < 3 \\ x + 2y > 4 \end{cases}$   **c.** $\begin{cases} x \geq -2 \\ y \leq 4 \end{cases}$

---

## Applications

**Objective 2** ▶

Solve application problems.

Systems of linear inequalities can be used to solve problems in business as well as other areas. If the variables in the problem satisfy several constraints that can be represented by inequalities, then a system is used. Key phrases such as "at most," "at least," "not more than," and "not less than" indicate that an inequality is needed.

> **Procedure: Solving an Application Involving a System of Linear Inequalities**
>
> **Step 1:** Determine the unknowns and define variables for them.
> **Step 2:** Write a system of linear inequalities to represent the constraints given in the problem.
> **Step 3:** Solve the system graphically.
> **Step 4:** Ordered pairs within the solution set satisfy all of the constraints in the problem.

---

**Objective 2 Example**      The Parent Teacher Association (PTA) of a local school is selling rolls of wrapping paper and boxed chocolates for a fund-raiser. The PTA can order at most 300 items. Each roll of wrapping paper costs \$2 and each box of chocolates costs \$3. The PTA can spend no more than \$1200 on these items. From past experience, the PTA knows that they sell at least twice as many boxes

of chocolates as they do rolls of wrapping paper. How many rolls of wrapping paper and boxes of chocolates should the PTA order for their conditions to be satisfied? Provide three specific examples that are in the solution set.

**Solution**      What is unknown? The number of rolls of wrapping paper and the number of boxes of chocolates that should be ordered are unknown.

Let $w$ = number of rolls of wrapping paper (first coordinate).

Let $c$ = number of boxes of chocolate (second coordinate).

What is known? At most 300 items can be ordered. Each roll of wrapping paper costs \$2 and each box of chocolates costs \$3. The PTA can't spend more than \$1200. The boxes of chocolates sold are at least twice as many as the rolls of

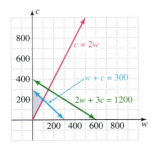

**Figure 4.4.7**

wrapping paper sold. The total number of each item sold is nonnegative. We can represent these facts in the system as shown.

$$\begin{cases} 2w + 3c \le 1200 \\ w + c \le 300 \\ c \ge 2w \\ c, w \ge 0 \end{cases}$$

The solution set of this system is shown in Figure 4.4.7.

Any ordered pair in the shaded region or on the boundary lines satisfies the system. Some possible combinations that will satisfy the PTA's requirements are stated next.

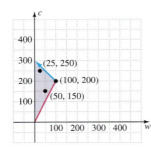

50 rolls of wrapping paper and 150 boxes of chocolates

100 rolls of wrapping paper and 200 boxes of chocolates

25 rolls of wrapping paper and 250 boxes of chocolates

✓ **Student Check 2** A motel chain plans to open a new motel. The motel will have a combination of double rooms and king-size rooms. The motel will have at most 200 rooms. Based on consumer trends, the motel knows that they should have at least 125 double rooms. They also know that they need at least twice as many double rooms as king-size rooms. Use a system of linear inequalities to find at least three combinations of double rooms and king-size rooms that will satisfy these conditions.

**Objective 3** ▶

Troubleshoot common errors.

## Troubleshooting Common Errors

A common error associated with systems of linear inequalities is shown next.

**Objective 3 Example** A problem and an incorrect solution are given. Provide the correct solution and an explanation of the error.

Solve the system $\begin{cases} y > x + 2 \\ y < \dfrac{1}{4}x - 4 \end{cases}$.

| Incorrect Solution | Correct Solution and Explanation |
|---|---|
| Each inequality is graphed here | The error was made in not extending the graphs to the left. If we adjust our scale, we get the following graph. |

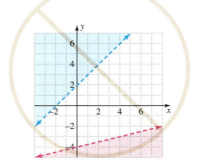

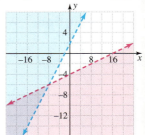

Since the graphs do not overlap, there is no solution for this system.

So, the solution set is

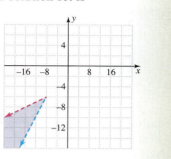

---

## ANSWERS TO STUDENT CHECKS

**Student Check 1** **a.**

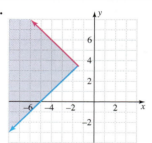

**b.**

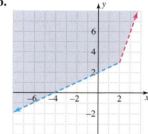

**c.**

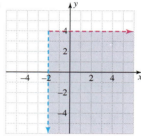

**Student Check 2**

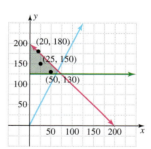

---

## SUMMARY OF KEY CONCEPTS

Solving a system of linear inequalities requires us to graph each linear inequality on the same coordinate system. The solution set for the system is the intersection of the half-planes formed by each inequality.

Be sure to draw the boundary lines correctly for each inequality. If the inequality symbol is $<$ or $>$, then the boundary line is dashed. If the inequality symbol is $\leq$ or $\geq$, then the boundary line is solid.

---

## GRAPHING CALCULATOR SKILLS

**Example:** Solve the system $\begin{cases} 2x + y < 6 \\ x - y \leq 4 \end{cases}$ on the graphing calculator.

**Solution:** Solve each inequality in the system for $y$. This gives us $\begin{cases} y < -2x + 6 \\ y \geq x - 4 \end{cases}$.

Enter each inequality in the equation editor. Since the first inequality involves a $<$ symbol, select the option to graph

below the line. The second inequality involves a $\geq$ symbol, so we select the option to graph above the line.

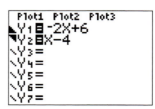

Press Graph to view the intersection of the two inequalities.

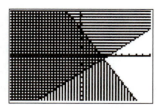

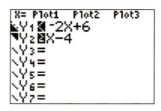

Press Graph to view the solution set of the system.

**Alternative Method for TI-84 Plus Users:** Use the APPS menu to graph the inequalities.

Press Y=, APPS, arrow down to Inequalz, and press any key to enter the program.

Select the appropriate inequality symbol by using the Alpha key and the appropriate function key.

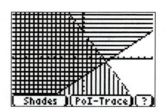

## SECTION 4.6 / EXERCISE SET

### Write About It!

**Use complete sentences in your answer to each exercise.**

1. Describe the steps to solve a system of linear inequalities.

2. Explain how the solution of a system of linear inequalities could be a straight line.

3. Explain how to determine two points $(x, y)$ that are solutions of the system $\begin{cases} y \geq 3 \\ x < 4 \end{cases}$

4. Explain how to determine two points $(x, y)$ that are solutions of the system $\begin{cases} x \geq 10 \\ y < 7 \end{cases}$.

### Practice Makes Perfect!

**Solve each system of linear inequalities. (See Objective 1.)**

5. $\begin{cases} y > x + 1 \\ y \leq 2x - 1 \end{cases}$

6. $\begin{cases} y > 3x + 3 \\ y \leq -x + 5 \end{cases}$

7. $\begin{cases} y \geq 4x - 8 \\ y > -2x + 4 \end{cases}$

8. $\begin{cases} y \geq 6x + 6 \\ y > -3x + 6 \end{cases}$

9. $\begin{cases} 5x - y > 5 \\ x - y < 7 \end{cases}$

10. $\begin{cases} 4x + y < 4 \\ -7x + y > -3 \end{cases}$

11. $\begin{cases} x \geq 5 \\ y < -4 \end{cases}$

12. $\begin{cases} x \geq -3 \\ y < 9 \end{cases}$

13. $\begin{cases} 3x + y \geq 6 \\ y \geq 2 \end{cases}$

14. $\begin{cases} 2x + y \geq 5 \\ x \geq -3 \end{cases}$

15. $\begin{cases} x \leq 2y \\ x - 4y \geq 4 \end{cases}$

16. $\begin{cases} x \leq 3y \\ x - 2y \geq -1 \end{cases}$

17. $\begin{cases} x > 2 \\ y \leq -1 \end{cases}$

18. $\begin{cases} x > -4 \\ y \leq -7 \end{cases}$

19. $\begin{cases} y \geq 10 \\ x < -8 \end{cases}$

20. $\begin{cases} y \geq -6 \\ x \leq 5 \end{cases}$

21. $\begin{cases} 3x - 4y \geq 12 \\ x < -5 \end{cases}$

22. $\begin{cases} 5x + 2y \geq 6 \\ x \geq 1 \end{cases}$

23. $\begin{cases} 2x + 3y \leq 9 \\ y \leq 5 \end{cases}$

24. $\begin{cases} 3x - 5y \leq 10 \\ y \geq -2 \end{cases}$

25. $\begin{cases} 4x + y \geq 8 \\ 4x + y \leq -1 \end{cases}$

26. $\begin{cases} 3x - 2y \geq 6 \\ -3x + 2y \geq 4 \end{cases}$

**For Exercises 27 and 28, (a) write a system of linear inequalities that represents the situation, (b) solve it by graphing, and (c) find at least three combinations of rolls of wrapping paper and boxes of chocolates that satisfy the given conditions. (See Objective 2.)**

27. The Parent Teacher Association (PTA) of a local school is selling rolls of wrapping paper and boxed chocolates for a fund-raiser. The PTA can order at most 1200 items. Each roll of wrapping paper costs $1.50 and each box of chocolates costs $4. The PTA can spend no more than $4200 on these items. From past experience, the PTA knows that they sell at least twice as many boxes of chocolates as they do rolls of wrapping paper.

28. The Parent Teacher Association (PTA) of a local school is selling rolls of wrapping paper and boxed chocolates for a fund-raiser. The PTA can order at most 750 items. Each roll of wrapping paper costs $2 and each box of chocolates costs $6.50. The PTA can spend no more than $5100 on these items. From past experience, the PTA knows that they sell at least twice as many boxes of chocolates as they do rolls of wrapping paper.

**For Exercises 29 and 30, use the following information. In the business world, the price per item, in dollars, is determined by the demand equation $y = D(x)$, where $x$ is the number of items sold. The consumer's surplus is the area of the graph of the solution set of the system of linear inequalities**

$$\begin{cases} y \le D(x) \\ y \ge k \\ x \ge 0 \end{cases}, \text{ where } k \text{ is a fixed price}$$

29. The demand equation for an item is $y = D(x) = 20 - 0.05x$ and $k = 10$. (a) Write a system of linear inequalities and solve it by graphing and (b) find the consumer's surplus.

30. The demand equation for an item is $y = D(x) = 30 - 0.45x$ and $k = 7.5$. (a) Write a system of linear inequalities and solve it by graphing and (b) find the consumer's surplus.

 ## Mix 'Em Up!

**Solve each system of linear inequalities.**

31. $\begin{cases} y \ge -\dfrac{1}{2}x \\ y \le 2x \end{cases}$

32. $\begin{cases} y > 3x \\ y < -\dfrac{1}{3}x + 2 \end{cases}$

33. $\begin{cases} y \ge 2x - 1 \\ y \le 2x + 3 \end{cases}$

34. $\begin{cases} y < -x \\ y > -x + 3 \end{cases}$

35. $\begin{cases} 3x - y \le 2 \\ x \ge -2 \end{cases}$

36. $\begin{cases} y \le 2 \\ x + 2y \ge -6 \end{cases}$

37. $\begin{cases} y \ge \dfrac{2}{3}x + 4 \\ y \ge \dfrac{1}{4}x - 1 \end{cases}$

38. $\begin{cases} y < 4x + 2 \\ y > -2x + 5 \end{cases}$

39. $\begin{cases} y \ge -2 \\ x \le 2 \end{cases}$

40. $\begin{cases} x + 2y \ge -6 \\ x + 2y \le 4 \end{cases}$

**The price per item, in dollars, is determined by the supply equation $y = S(x)$, where $x$ is the number of items produced. The producer's surplus is the area of the graph of the solution set of the system of linear inequalities**

$$\begin{cases} y \ge S(x) \\ y \le k \\ x \ge 0 \end{cases}, \text{ where } k \text{ is a fixed price}$$

41. The supply equation for an item is $y = S(x) = 25 + 0.10x$ and $k = 50$. (a) Write a system of linear inequalities and solve it by graphing and (b) find the producer's surplus.

42. The supply equation for an item is $y = S(x) = 8 + 0.004x$ and $k = 20$. (a) Write a system of linear inequalities and solve it by graphing and (b) find the producer's surplus.

 ## You Be the Teacher!

**Correct each student's error, if any.**

43. Solve $\begin{cases} x - 2y \le 2 \\ -x + y \ge 1 \end{cases}$.

Bernie's work: For the first inequality, I graphed the line $x - 2y = 2$ and shaded the lower half-plane since the inequality is $\le$. For the second inequality, I graphed the line $-x + y = 1$ and shaded the upper half-plane since the inequality is $\ge$.

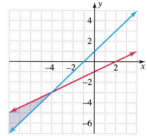

44. Solve $\begin{cases} y > 2x \\ x - 2y > 4 \end{cases}$.

Yvonne's work: For the first inequality, I graphed the line $y = 2$ as a dotted line and shaded the upper half-plane since the inequality is $>$. For the second inequality, I graphed the line $x - 2y = 4$ as a dotted line and shaded the upper half-plane since the inequality is $>$.

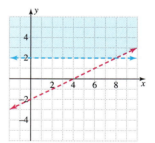

 ## Calculate It!

**Use a graphing calculator to solve each system of linear inequalities.**

45. $\begin{cases} y < 4x - 1 \\ y \ge -2x + 3 \end{cases}$

46. $\begin{cases} y \ge 3x - 6 \\ y > x \end{cases}$

47. $\begin{cases} 4x - 2y > 5 \\ -3x + y < 1 \end{cases}$

48. $\begin{cases} x - 3y \le 6 \\ 2x + y \ge -1 \end{cases}$

 ## Think About It!

49. Give an example of a system of linear inequalities that has no solution.

50. Give an example of a system of linear inequalities that has a straight line as a solution.

## GROUP ACTIVITY  The Mathematics of Choosing a Design

Mike and Amy purchased a lakefront lot for their dream home. The lot is rectangular and borders the lake on one side. Mike and Amy want to fence the other three sides of the property so that the perimeter of the fenced area is 2000 ft. They have $25,000 budgeted for this job. The cost of several types of fencing materials is shown in the table.

| Type | Cost per Linear Foot |
|---|---|
| 4-ft-high chain link | $ 6 |
| $4\frac{1}{2}$ ft wood fence | $15 |
| Vinyl fence | $30 |
| Picket fence | $20 |
| 6-ft-high cedar privacy fence | $11 |

1. Draw a diagram of the lake property and label the sides of the property with appropriate variables.

2. Choose two different fencing options, one to be used for the front of the property and the other to be used for the remaining two sides.

3. Write a system of linear equations that can be used to determine the amount of each type of fencing needed to enclose the property.

4. Solve the system using an appropriate method. Round to the nearest hundredth, if necessary.

5. Explain what the solution of the system means in the context of the problem.

6. Find the area that is enclosed by the selected fencing design.

7. Repeat the process two additional times.

8. Which design yields the largest enclosed area of the property?

# Systems of Linear Equations and Inequalities in Two Variables

**What's the big idea?** Now that we have completed Chapter 4, we should be able to solve a system of linear equations in two variables in three different ways and identify the number of solutions of the system. Systems can be used to solve applications involving two unknowns.

## The Tools

Listed below are the key terms, skills, and formulas, and properties you should know for this chapter.

The page reference is provided if you need additional help with the given topic. The Study Tips will assist in your preparation for an exam.

## Study Tips

1. Learn all of the terms, formulas, and properties. Make flash cards and have someone quiz you.
2. Rework problems from the exercises and also the ones you worked in class. Work additional problems from the review exercises.
3. Review the summaries of key concepts.
4. Work the chapter test.
5. Be sure to review the online resources for additional study materials.

## Terms

Addition method   319
Boundary line   348
Consistent system with dependent equations   299
Consistent system with independent equations   299
Elimination method   319
Half-plane   348
Inconsistent system   299
Linear inequality in two variables   347
Lower half-plane   348

Solution of a linear inequality in two variables   347
Solution of a system of linear equations   294
Solution set of a system of linear inequalities in two variables   357
Substitution method   309
System of linear equations   294
System of linear inequalities   357
Test-point method   349
Upper half-plane   348

## Formulas and Properties

- Addition property of equality   321
- Complementary angles   340
- Distance formula   337

- Perimeter of a rectangle   340
- Simple interest formula   334
- Supplementary angles   340

## CHAPTER 4 / SUMMARY

**How well do you know this chapter? Complete the following questions to find out. Take a look back at the section if you need help.**

### SECTION 4.1 Solving Systems of Linear Equations Graphically

1. A system of linear equations consists or two or more equations that must be solved simultaneously .

2. A solution of a system of linear equations in two variables is an ordered pair that satisfies both equations.

3. To solve a system of linear equations in two variables by graphing, we graph each equation on the same coordinate system and find their intersection point.

4. A system of linear equations in two variables has three possible types of solutions: one solution, no solution, or infinitely many solutions.

5. The lines in a system of linear equations will either intersect one another, be parallel to each other, or be the same line.

6. A system of linear equations that has at least one solution is called a consistent system.

7. A system of linear equations that has no solution is called an inconsistent system.

8. If the two equations in a system of linear equations are multiples of one another, then the equations are dependent. Otherwise, the equations are independent.

### SECTION 4.2 Solving Systems of Linear Equations by Substitution

9. To solve a system of linear equations using substitution, we must first solve one of the equations for one of the variables and substitute this expression into the other equation.

10. If after substitution, the resulting equation is a contradiction, then the system of linear equations is inconsistent and has no solution.

11. If after substitution, the resulting equation is an identity, then the system of linear equations is consistent with dependent equations and has infinitely many solutions.

### SECTION 4.3 Solving Systems of Linear Equations by Elimination

12. The goal of elimination is to add the equations in a system of linear equations so that one variable is eliminated.

13. For a variable to be eliminated from a system of linear equations, its coefficients must be opposites of one

another. If they are not, we must multiply one or both equations by some nonzero number.

14. If after adding the two equations in a system of linear equations together, the resulting equation is an identity, this means that the system contains two lines that are the same and there are infinitely many solutions.

15. If after adding the two equations in a system of linear equations together, the resulting equation is a contradiction, this means that the system contains two lines that are parallel and there is no solution of the system.

### SECTION 4.4 Applications of Systems of Linear Equations

16. Applications that involve two unknowns can be solved using a system of equations.

17. We assign two variables to the unknowns and write two different equations that relate the two unknowns and then solve using substitution or elimination.

### SECTION 4.5 Linear Inequalities in Two Variables

18. A linear inequality in two variables is an inequality of the form $y < mx + b$ or $Ax + By < C$ .

19. Solutions of linear inequalities in two variables are ordered pairs that make the inequality true.

20. To graph a linear inequality in two variables, we must first graph the boundary line which is formed by replacing the inequality symbol with an equals sign. If the inequality symbol is $\leq$ or $\geq$, the boundary line is solid. If the inequality symbol is $<$ or $>$, the boundary line is dashed.

21. The boundary line divides the plane into two half-planes. One half-plane contains solutions of the inequality and the other does not.

22. After the boundary line is drawn, we must use a test point in the original inequality to determine which half-plane contains solutions of the inequality. The ideal test point is $(0, 0)$ as long as the boundary line does not go through the origin.

### SECTION 4.6 Systems of Linear Inequalities in Two Variables

23. The solution set of a system of linear inequalities in two variables consists of the intersection of the half-planes formed from each linear inequality.

## CHAPTER 4 / REVIEW EXERCISES

### SECTION 4.1

**Solve each system of linear equations graphically. Then state if the system is consistent with dependent equations, consistent with independent equations, or inconsistent.** *(See Objectives 3–5.)*

1. $\begin{cases} y = 2x + 1 \\ y = 4x - 3 \end{cases}$

2. $\begin{cases} 3x - y = 2 \\ y = 3x - 2 \end{cases}$

3. $\begin{cases} x - 2y = 0 \\ 3y + x = 10 \end{cases}$

4. $\begin{cases} 2x - y = 7 \\ x = 3 \end{cases}$

5. $\begin{cases} x + y = -5 \\ y = \dfrac{1}{7}x + 3 \end{cases}$

6. $\begin{cases} y = 4x - 1 \\ x - \dfrac{1}{4}y = 2 \end{cases}$

**Solve each problem. (*See Objectives 2 and 6.*)**

7. George is trying to decide which Qwest home phone service he should use. The basic service starts at $15 per month with unlimited local calling and $0.15 per minute for long distance calls. The Home Phone Plus package starts at $30 per month with unlimited local calling and $0.05 per minute for long distance calls. If George makes $x$ minutes of long distance calls per month, the cost per month for the basic service is $y = 15 + 0.15x$. The cost per month for the Home Phone Plus package is $y = 30 + 0.05x$. Use the graph of the system formed by these two equations to answer the questions that follow.

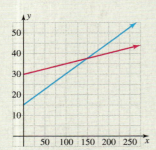

a. What is the point of intersection of this system?

b. What is the meaning of the point of intersection in the context of this problem?

c. If George makes an average of 200 min of long distance calls per month, which plan should he use? Why?

## SECTION 4.2

**Solve each system of linear equations using substitution. (*See Objectives 1 and 2.*)**

8. $\begin{cases} x + 2y = 1 \\ 7x - 5y = 26 \end{cases}$
$\{(3, -1)\}$

9. $\begin{cases} y = -\dfrac{1}{4}x + \dfrac{7}{2} \\ x + y = 2 \end{cases}$
$\{(-2, 4)\}$

10. $\begin{cases} y = -2x - 15 \\ 4x + 2y = -30 \end{cases}$
$\{(x, y) \mid y = -2x - 15\}$

11. $\begin{cases} -2x + 5y = 27 \\ y = 7 \end{cases}$
$\{(4, 7)\}$

12. $\begin{cases} x = -0.75y - 3.5 \\ 0.25x + 0.25y = -1.25 \end{cases}$
$\{(1, -6)\}$

13. $\begin{cases} y = \dfrac{5}{3}x + 2 \\ \dfrac{1}{3}x - \dfrac{1}{5}y = -\dfrac{6}{5} \end{cases}$
$\varnothing$

14. $\begin{cases} y = 2x \\ 0.3x - 0.4y = 1.5 \end{cases}$
$\{(-3, -6)\}$

## SECTION 4.3

**Solve each system using elimination. (*See Objectives 1–3.*)**

15. $\begin{cases} 3x - 7y = 4 \\ 5x - 6y = 18 \end{cases}$
$\{(6, 2)\}$

16. $\begin{cases} 4x + y = -2 \\ x + 5y = 9 \end{cases}$
$\{(-1, 2)\}$

17. $\begin{cases} 4x + y = 16 \\ y = -4x + 12 \end{cases}$
$\varnothing$

18. $\begin{cases} 3x - y = -1 \\ 2x - \dfrac{2}{3}y = -\dfrac{2}{3} \end{cases}$
$\{(x, y) \mid 3x - y = -1\}$

19. $\begin{cases} 0.7x + 0.1y = 1.7 \\ -0.9x + y = 1.2 \end{cases}$
$\{(2, 3)\}$

20. $\begin{cases} \dfrac{1}{2}x - \dfrac{2}{3}y = -2 \\ \dfrac{1}{4}x + \dfrac{5}{6}y = 6 \end{cases}$
$\{(4, 6)\}$

## SECTION 4.4

**Solve each problem. (*See Objectives 1–5.*)**

21. Two angles are supplementary. The larger angle is 28° less than three times the smaller angle. What is the measure of each angle?   52°, 128°

22. Two angles are complementary. The larger angle is 6° less than twice the measure of the smaller one. What is the measure of each angle?   32°, 58°

23. Matt and Dimitri are experienced cyclists who like challenging cycling routes. Dimitri gives Matt a 20-min head start on the route. On this route, Matt cycles an average of 8 mph. On the same route, Dimitri averages 10 mph. How long will it take Dimitri to catch up to Matt? $\left(\text{Hint: } 20 \text{ min} = \dfrac{1}{3} \text{ hr}\right)$   $\dfrac{4}{3}$ hr

24. Joseph has a 40% alcohol solution and a 20% alcohol solution. How many liters of the 40% solution and how many liters of the 20% solution must he mix together to obtain 100 L of the 36% alcohol solution?   80 L of the 40% alcohol solution and 20 L of the 20% alcohol solution

25. A plane travels 2305 mi in 4.15 hr when it is traveling with the wind. It takes 4.56 hr for the plane to travel the same distance against the wind. What is the speed of the plane in still air and what is the speed of the wind?   530 mph, 25 mph

26. Sandy has $8600 to invest. She invests part of the money in a savings account that pays 1.2% annual interest and the rest of the money in a CD that pays 2.8% annual interest. If she earns $202.40 interest in 1 yr, how much did she invest in each account?   $6200 at 2.8% and $2400 at 1.2%

27. Two families visit an amusement park. One family pays $272 for four adults and six children to be admitted. Another family pays $154 for two adults and four children to be admitted. How much does it cost for each adult and each child to be admitted to the park?   $41 for an adult and $18 for a child

28. Two families eat at a buffet restaurant. One family pays $190 for six adults and eight children. Another family pays $104.50 for two adults and seven children. How much does it cost for each adult and each child?   $19 for an adult and $9.50 for a child

## SECTION 4.5

**Determine if the ordered pair is a solution of the given inequality. (*See Objective 1.*)**

29. $3x - y < 5$; $(0, 1)$   yes

**30.** $y > -\frac{1}{2}x$; $(8, -5)$   no

**31.** $3x > 2y$; $(0, 1)$   no

**32.** $y + 5 < 7$; $(-3, 1)$   yes

**33.** $0.5x - y \geq 1.2$; $(0.3, -0.1)$   no

**34.** $7y - x < 0$; $(2, -3)$   yes

**Graph each linear inequality in two variables.**
(*See Objective 2.*)

**35.** $x - 2y \geq 3$        **36.** $6 - 2y > 0$

**37.** $x + 3 \leq 1$        **38.** $0.1x + 0.4y > 1.6$

**Use a linear inequality in two variables to solve each problem.** (*See Objective 3.*)

**39.** A geography course grade is based on a student's test average and final exam grade. The test average counts for 80% of the course grade and the final exam counts for 20% of the course grade. Write a linear inequality in two variables that will determine the test average and final exam grade for a student to earn at least a 90 in the course. Graph the linear inequality.

**40.** Gail works two jobs to put herself through college. She works at a grocery store for $11.50 per hour, and works as a student worker for $8.25 per hour. If Gail wants to make at least $450 per week, write a linear inequality in two variables that represents this situation. Graph the linear inequality.

## SECTION 4.6

**Solve each system of linear inequalities.**
(*See Objective 1.*)

**41.** $\begin{cases} y \geq -3x - 6 \\ y \leq 4x \end{cases}$        **42.** $\begin{cases} y \geq \frac{1}{2}x - 5 \\ 2y \leq x + 2 \end{cases}$

**43.** $\begin{cases} x + 2y > -4 \\ x < 2 \end{cases}$        **44.** $\begin{cases} x \geq -1 \\ y \leq 3 \end{cases}$

**Solve the problem using a system of linear inequalities.**
(*See Objective 2.*)

**45.** The price per item, in dollars, is determined by the demand equation $y = D(x)$, where $x$ is the number of items sold. The consumer's surplus is the area of the graph of the solution set of the system.

$$\begin{cases} y \leq D(x) \\ y \geq k \\ x \geq 0 \end{cases}, \text{ where } k \text{ is a fixed price}$$

The demand equation for an item is $D(x) = 25 - 0.10x$ and $k = 10$. (a) Write a system of linear inequalities and solve it by graphing and (b) find the consumer's surplus.

---

## CHAPTER 4 TEST / SYSTEMS OF LINEAR EQUATIONS AND INEQUALITIES IN TWO VARIABLES

**1.** The ordered pair that is a solution of the system $\begin{cases} 4x - y = 3 \\ y = 6x - 4 \end{cases}$ is   b

    **a.** $(2, 8)$        **b.** $\left(\frac{1}{2}, -1\right)$

    **c.** $\left(-\frac{1}{2}, -7\right)$        **d.** $(-2, -11)$

**2.** Use graphing to solve the system $\begin{cases} x - y = -2 \\ y = -\frac{2}{3}x - 3 \end{cases}$.

**3.** Without graphing each system of linear equations, complete the table.

    **a.** $\begin{cases} 7x - 4y = 8 \\ \frac{14}{8}x - y = 2 \end{cases}$

| Slope-intercept form | $y = \frac{7}{4}x - 2$ | $y = \frac{7}{4}x - 2$ |
|---|---|---|
| Slopes | $m = \frac{7}{4}$ | $m = \frac{7}{4}$ |
| $y$-intercepts | $(0, -2)$ | $(0, -2)$ |
| How do lines relate? | The lines are the same. | |

| Number of solutions | Infinitely many solutions | |
|---|---|---|
| Type of system | Consistent system with dependent equations | |
| Solution set | $\{(x, y) \mid 7x - 4y = 8\}$ | |

    **b.** $\begin{cases} y = \frac{1}{5}x + 3 \\ x - 5y = 10 \end{cases}$

| Slope-intercept form | $y = \frac{1}{5}x + 3$ | $y = \frac{1}{5}x - 2$ |
|---|---|---|
| Slopes | $m = \frac{1}{5}$ | $m = \frac{1}{5}$ |
| $y$-intercepts | $(0, 3)$ | $(0, -2)$ |
| How do lines relate? | The lines are parallel. | |
| Number of solutions | No solution | |
| Type of system | Inconsistent system | |
| Solution set | $\varnothing$ | |

**4.** The following graph shows the percentage of Americans who use the newspaper or Internet as their source for news, where $x$ is the years after 2003.

(Source: http://pewresearch.org/pubs/1066/internet-overtakes-newspapers-as-news-outlet)

**National and International News Sources for Americans (2003–2008)**

**a.** Approximate the point of intersection.

**b.** Interpret what the point of intersection means in the context of this problem.

**5.** What are the three methods to solve a system of linear equations in two variables by hand? Which method is the least precise? Why? How do you decide when to use the other two methods?

**6.** Solve each system by substitution.

**a.** $\begin{cases} x - 8y = 13 \\ 5y - 9x = -50 \end{cases}$  $\{(5, -1)\}$

**b.** $\begin{cases} -4x - y = 15 \\ 3x + 5y = -41 \end{cases}$  $\{(-2, -7)\}$

**7.** Solve each system by elimination.

**a.** $\begin{cases} 7x + 3y = -22 \\ 6x - 3y = -30 \end{cases}$  $\{(-4, 2)\}$

**b.** $\begin{cases} 3x - 2y = 19 \\ 2x + 6y = -46 \end{cases}$  $\{(1, -8)\}$

**8.** Solve each system using any method.

**a.** $\begin{cases} 24x - 4y = 24 \\ 6x = y - 6 \end{cases}$  $\varnothing$

**b.** $\begin{cases} \dfrac{x}{5} + y = \dfrac{11}{5} \\ \dfrac{3x}{5} + \dfrac{y}{2} = -\dfrac{17}{5} \end{cases}$  $\{(-9, 4)\}$

**c.** $\begin{cases} x + \dfrac{1}{4}y = y - 4 \\ \dfrac{1}{3}x + y = x + y \end{cases}$  $\left\{\left(0, \dfrac{16}{3}\right)\right\}$

**Use a system of equations to solve each problem.**

**9.** Ms. Jones invests $18,000 in two accounts, one yielding 8% interest and the other yielding 9%. She earns a total of $1490 in interest at the end of the year. How much was invested in each account?

**10.** Diane is the manager of a 10,000-seat ballpark and plans to host a special event to raise money for the Cystic Fibrosis Foundation. She plans to sell VIP seats for $100 each and regular admission seats for $37.50. If Diane wants to raise $500,000 by selling every seat, how many of seats should she designate as VIP seats? She should designate 2000 seats as VIP seats.

**11.** Dominick is on a flight from Chicago O'Hare Airport to London Heathrow Airport, a distance of approximately 3950 mi. He notices on his ticket that the travel time from Chicago to London is 7.5 hr and the return trip is 9 hr. The trip to London is with the wind and the trip from London is against the wind. What is the average speed of the plane in still air and the average speed of the wind? Round the answers to the nearest integer. The average speed of the plane is approximately 483 mph and the average speed of the wind is approximately 44 mph.

**12.** The Lincoln Memorial Reflecting Pool is the largest reflecting pool in Washington, DC. It is in the shape of a rectangle whose perimeter is 1338 m. Its length is 6 m more than 12 times its width. Find the length and width of the reflecting pool. The length is 618 m and the width is 51 m.

**13.** Using complete sentences, explain the steps used to graph the solution set of a linear inequality in two variables.

**14.** Graph the solution set of each linear inequality.

**a.** $x - 4y \geq -8$

**b.** $y > -\dfrac{3}{5}x$

**c.** $x < 2$

**d.** $y \geq 1$

**15.** Solve the system of linear inequalities.

$\begin{cases} x + y < 6 \\ y > \dfrac{1}{2}x - 4 \end{cases}$

---

## CUMULATIVE REVIEW EXERCISES / CHAPTERS 1–4

**1.** Perform each operation and simplify the result. (*Section 1.2, Objectives 2–7*)

**a.** $\dfrac{9}{10} + \dfrac{1}{2} - \dfrac{3}{5}$  $\dfrac{4}{5}$

**b.** $5\dfrac{1}{3} \cdot 2\dfrac{1}{4} \div 6$  $2$

**c.** $3\dfrac{1}{6} - 1\dfrac{3}{4} + \dfrac{7}{12}$  $2$

**2.** Use the order of operations to simplify each expression. (*Section 1.3, Objectives 1 and 2*)

**a.** $-(-2.5)^4$  $-39.0625$

**b.** $\dfrac{2}{3}(10 - 8)^2 - \dfrac{5}{6}$  $\dfrac{11}{6}$

**c.** $3(4)^2 - 6(7) + 11$   17   **d.** $4[7 - |9 - (5-3)|]$   0

**e.** $\dfrac{5 + \sqrt{5^2 - 4(2)(3)}}{2 \cdot 2}$   $\dfrac{3}{2}$

**3.** The expression $27.94x + 41.91y$ represents the weekly earnings of a mining worker, where $x$ is the number of regular hours worked in a week and $y$ is the number of overtime hours worked in a week. Find the weekly salary of a mining worker who worked 40 regular hours and 10 overtime hours. (Source: http://www.bls.gov) (*Section 1.3, Objective 6*)   $1536.70

**4.** Perform the operation and simplify. (*Sections 1.4 and 1.5, Objectives 1 and 2*)

**a.** $4.2 - 2.9 + (-1.6) - (-7.5)$   7.2

**b.** $\dfrac{5}{7} - \dfrac{1}{2} - \left(-\dfrac{13}{14}\right)$   $\dfrac{8}{7}$

**c.** $10 - \{6 - [3 - (1 - \sqrt{17 - 1})]\}$   10

**5.** Perry has $115.25 in his checking account. He deposits his paycheck of $627.85. Perry writes a check for the phone bill for $56.29, for groceries for $156.23, and then for rent for $525. What is Perry's checking account balance? (*Section 1.4, Objective 3*)   $5.58

**6.** Nate wants to mix a 10% salt solution with a 40% salt solution to get a 28% salt solution. Let $x$ represent the volume of the 10% salt solution. If Nate wants to make 80 oz of a 28% salt solution, write an expression that represents the volume of the 40% salt solution. (*Section 1.5, Objective 3*)   $80 - x$

**7.** In a rectangle, the length is six less than twice times the width. If $w$ represents the width, write an expression that represents the length of the rectangle. (*Section 1.5, Objective 3*)   $2w - 6$

**8.** Perform the indicated operation and simplify. (*Section 1.6, Objectives 1–3*)

**a.** $(-12.4)(-7.5)$   93   **b.** $(-76)\left(\dfrac{14}{19}\right)\left(-\dfrac{10}{21}\right)$   $\dfrac{80}{3}$

**c.** $(-4)^4$   256   **d.** $0 \div 16$   0

**e.** $\left(4\dfrac{1}{6}\right) \div \left(3\dfrac{3}{4}\right)$   $1\dfrac{1}{9}$

**9.** Simplify each algebraic expression. (*Section 1.8, Objectives 3 and 4*)

**a.** $-(3a - 1) - 4(-5a + 2)$   $17a - 7$

**b.** $5.8 - 0.2(-6b + 36)$   $1.2b - 1.4$

**c.** $12\left(\dfrac{1}{2}x - \dfrac{5}{6}\right) - 6\left(\dfrac{1}{3}x + 1\right)$   $4x - 16$

**d.** $12x^2 - 8(x + 2x^2) + x$   $-4x^2 - 7x$

**10.** Determine if the given numbers are solutions of the equation. (*Section 2.1, Objective 2*)

**a.** Is $x = -2$, $x = 0$, or $x = 2$ a solution of $4x - 1 = x + 5$?

**b.** Is $x = -3$, $x = 0$, or $x = 3$ a solution of $3x + 11 = -2x - 4$?

**11.** For each problem, define the variable, write an equation, and solve the problem. (*Section 2.1, Objectives 3 and 4*)

**a.** The difference of 40 and a number is the same as three times the number. Find the number.   10

**b.** The sum of three times a number and $-5$ equals 7 less than the number. Find the number.   $-1$

**c.** One angle is $34°$ more than another angle. Their sum is $90°$. Find the measure of each angle.   $28°, 62°$

**12.** Solve each equation. (*Section 2.2, Objectives 2 and 3; Sections 2.3 and 2.4, Objectives 1 and 2*)

**a.** $5(x - 1) - 14 = 2(x - 3) + 11$   {8}

**b.** $-\dfrac{2}{7}(7x + 21) + 9 = x - 6$   {3}

**c.** $-1.2y + 24.1 = 2.3y + 3.1$   {6}

**13.** Solve each equation. If the equation is a contradiction, write the solution as Ø. If the equation is an identity, write the solution as ℝ. (*Section 2.4, Objectives 1–4*)

**a.** $4(z - 2) - (z + 6) = 3(z + 2) - 20$   ℝ

**b.** $0.012x + 0.019(890 - x) = 14.46$   {350}

**c.** $-\dfrac{1}{4}(x - 3) = \dfrac{1}{2}(2x + 9) + 1$   $\left\{-\dfrac{19}{5}\right\}$

**14.** The revenue $R$, unit price $p$, and sales level $x$ of a product are related by the formula $R = xp$. (*Section 2.5, Objective 1*)

**a.** Find $R$ if $p = \$23.50$ and $x = 50$.   $1175

**b.** Find $p$ if $R = \$9480$ and $x = 395$.   $24

**15.** Solve each formula for the specified variable. (*Section 2.5, Objective 3*)

**a.** $V = \pi r^2 h$ for $h$   $h = \dfrac{V}{\pi r^2}$

**b.** $C = 2l + 5w$ for $w$   $w = \dfrac{C - 2l}{5}$

**c.** $3x - 2y = 60$ for $y$   $y = \dfrac{3}{2}x - 30$

**d.** $0.035x + 0.028y = 2.1$ for $x$   $x = 60 - \dfrac{4}{5}y$

**16.** Solve each problem. (*Section 2.5, Objective 5*)

**a.** Find the measure of an angle whose complement is $21°$ less than the measure of the angle.   $55.5°$

**b.** The supplement of an angle is $32°$ more than three times its complement. Find the measure of the angle.   $61°$

**17.** In triangle $ABC$, the measure of angle $B$ is $20°$ less than the measure of angle $A$. The measure of angle $C$ is $11°$ more than the measure of angle $A$. Find the measure of each angle in the triangle. (*Section 2.5, Objective 5*)   $63°, 43°, 74°$

Solve each problem. Write an appropriate equation, solve the equation, and write the answer in complete sentences. (*Section 2.6, Objectives 1–5*)

**18.** A suit is on sale for $110.55. If this price is 33% off the original price, what is the original price of the suit?   $165

**19.** Candice invests $3400 in two accounts. She invests $2600 in a mutual fund that pays 2.1% annual interest and the remaining amount in a money market fund that pays 1.0% annual interest. How much interest will she earn from the two accounts in 1 yr?   $62.60

**20.** Solve each inequality. For each inequality, graph the solution set, and write the solution set in interval notation, and in set-builder notation. (*Section 2.7, Objectives 1–4*)

   **a.** $0.2b - 3.8 \geq -4.2$    **b.** $-1 < 4a + 3 < 15$

**21.** Complete a table of values, plot the resulting points, and use the points to draw the graph of the equation. (*Section 3.1, Objective 3*)

   **a.** $y = -2x + 6$    **b.** $y = x^2 - 5$

   **c.** $y = 2 - |x - 1|$

**22.** Use the given graph to complete each ordered pair that is a solution. (*Section 3.1, Objective 4*)

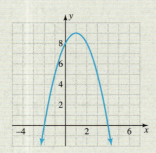

   **a.** ( _____ , 0)   (−2, 0), (4, 0)   **b.** (0, _____ )   (0, 8)

   **c.** (1, _____ )   (1, 9)      **d.** (2, _____ )   (2, 8)

**23.** The graph shows the seasonally adjusted unemployment rate in the month of September from 2000 to 2010. (Source: http://data.bls.gov) (*Section 3.1, Objective 5*)

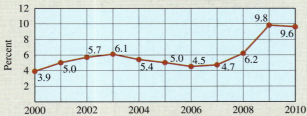

**Unemployment Rate (Seasonally Adjusted) in the Month of September 2000 to 2010**

   **a.** Write an ordered pair for each point labeled on the graph.

   **b.** Between what consecutive years did the unemployment rate increase most? By how much?

   **c.** Between what consecutive years did the unemployment rate decrease most? By how much?

**24.** The table lists the net income as reported (in millions of dollars) for Southwest Airlines between 2003 and 2009. (Source: http://phx.corporate-ir.net) (*Section 3.1, Objective 5*)

| Years After 2003 | 0 | 1 | 2 | 3 | 4 | 5 | 6 |
|---|---|---|---|---|---|---|---|
| Net income as reported (in millions) | $372 | $217 | $484 | $499 | $645 | $178 | $99 |

   **a.** Write an ordered pair $(x, y)$ that corresponds to each year between 2003 and 2009, where $x$ is the number

of years after 2003 and $y$ is the net income as reported for Southwest Airlines (in millions of dollars).

   **b.** Interpret the meaning of first and last ordered pairs from part (a) in the context of the problem.

   **c.** In what year was the net income the highest? The lowest?

   **d.** Make a scatter plot of the data.

**25.** Graph each equation. Identify at least two points on the graph. (*Section 3.2, Objectives 2–4*)

   **a.** $5x - 3y = 15$      **b.** $2y - x = 0$

   **c.** $2x - 3 = 0$       **d.** $1 - 0.5y = 0$

   **e.** $0.6x + 0.1y = 12$

**26.** The yearly cost of jet fuel and oil (in millions of dollars) for Southwest Airlines can be modeled by the equation $y = 457.7x + 1724.8$, where $x$ is the number of years after 2005. (Source: http://www.southwest.com/htm/cs/investor_relations/ar2010.html) (*Section 3.2, Objective 5*)

   **a.** Find the $y$-intercept and interpret its meaning.

   **b.** What was the yearly cost of jet fuel and oil for Southwest Airlines in 2009?

   **c.** What will the yearly cost of jet fuel and oil be for Southwest Airlines in 2013?

   **d.** Graph the equation.

**27.** Find the slope of each line (*Section 3.3, Objectives 1, 3, and 4*)

   **a.** $y = -\frac{1}{2}x + 1$   $-\frac{1}{2}$

   **b.** Passes through $(-1, -3)$ and $(-2, 0)$   $-3$

   **c.** $y = -2$   $0$

   **d.** Passes through $(-3.2, -0.4)$ and $(-3.2, -10.2)$   undefined

   **e.** $x - 4y = 1$   $\frac{1}{4}$

**28.** Use the given information to graph the line. Label two points on the line. (*Section 3.4, Objective 2*)

   **a.** $m = -3$, passes through $(0, 0)$

   **b.** $y = 3x - 5$

   **c.** $m = $ undefined, passes through $(-1, 3)$

   **d.** $m = 0$, passes through $(-3, 2)$

**29.** The average price, in dollars per barrel, of jet fuel in the United States can be modeled by $y = 8.38x + 24.02$, where $x$ is the years after 2000. Interpret the meaning of the slope and $y$-intercept in the context of the problem. (Source: http://www.airlines.org/Economics/DataAnalysis/Pages/AnnualCrudeOilandJetFuelPrices.aspx) (*Section 3.4, Objective 3*)

**30.** Determine if the lines are parallel, perpendicular, or neither. (*Section 3.4, Objective 3*)

   **a.** $3x + 7y = 1$ and $0.14y = -0.06x + 1.2$   parallel

   **b.** $y = 0.6x$ and $y = -0.6x$   neither

**31.** Write the equation of each line. Express the answer in slope-intercept form and in standard form. (*Section 3.5, Objectives 1–4*)

   **a.** passes through $(-1, 4)$ and $(-2, 0)$

   **b.** $m = 0$, passes through $(0, 12)$

   **c.** passes through $(2, -2)$, perpendicular to $x + 3y = 1$

**32.** Janis plans to enroll in a fitness club. There is a one-time membership fee of $195 plus a monthly charge of $35. (*Section 3.5, Objective 5*)

 **a.** Write a linear equation that represents the cost of joining the fitness club, where $x$ is the number of months Janis is a member.   $y = 35x + 195$

 **b.** How much money has Janis spent for her fitness club membership if she has been a member for 1 yr?   $615

 **c.** How long has Janis been a member of the club if she has paid a total of $825?   18 months

**33.** State the domain and range of each relation. Determine if the relation is a function. (*Source: http://www.eia.gov*) (*Section 3.6, Objectives 1–3*)

 **a.** The average residential natural gas prices (in dollars per thousand cubic feet) from 2004 to 2010 is shown.

| Years | Average Residential Natural Gas Prices |
|-------|----------------------------------------|
| 2004  | 10.75 |
| 2005  | 12.70 |
| 2006  | 13.73 |
| 2007  | 13.08 |
| 2008  | 13.89 |
| 2009  | 12.14 |
| 2010  | 11.20 |

 **b.** $y = 2$      **c.** $y = -2x$

**34.** Find the requested information. (*Section 3.6, Objective 4*)

 **a.** Find $f(0)$ if $f(x) = 12x - 9$.   $-9$

 **b.** Find $g(-2)$ if $g(x) = 3x^2 - 2x$.   16

 **c.** Find $f(-1)$ and $f(4)$ if $f(x)$ is given by the graph.   0, 1

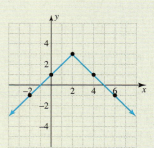

 **d.** Find $g(-3)$ and find $x$ for which $g(x) = -4$ if the function $g(x)$ is given by $Y_1$.   $-9, -1$

**35.** Write each equation in two variables in function notation. Then find $f(0)$, $f(3)$, and $f(-2)$ and write the corresponding ordered pairs. (*Section 3.6, Objectives 4 and 5*)

 **a.** $x + 2y = 1$      **b.** $y = 0.2x^2 - 0.1x$

**36.** Shawn has a new job as an English instructor at a local community college. His annual salary is $42,400 with a regular teaching load of 30 credit hours. If he teaches beyond the regular load in a year, he earns $1050 per credit hour. (*Section 3.6, Objective 6*)

 **a.** Write a linear function $f(x)$ that represents Shawn's yearly income, where $x$ is the number of additional credit hours he teaches in a year.   $f(x) = 42,400 + 1050x$

 **b.** If Shawn teaches two extra 3-credit-hour courses in a year, what is his income?   $48,700

 **c.** Find $f(2)$ and interpret the answer.

**37.** Solve each system of equations graphically. Then state if the system is consistent with dependent equations, consistent with independent equations, or inconsistent. (*Section 4.1, Objectives 3–5*)

 **a.** $\begin{cases} y = 3x - 5 \\ y = x - 1 \end{cases}$      **b.** $\begin{cases} y = 5x + 1 \\ y = 5x - 1 \end{cases}$

 **c.** $\begin{cases} x - y = 3 \\ x = 3 \end{cases}$      **d.** $\begin{cases} x - 2y = -5 \\ -x + 3y = 7 \end{cases}$

 **e.** $\begin{cases} 3x - y = 12 \\ x = \dfrac{1}{3}y + 4 \end{cases}$

**38.** Rikki needs to rent a refrigerator for her apartment. Household Rentals offers a monthly rental rate of $42 plus a one-time installation fee of $60. Rent and Go offers a monthly rental rate of $50 per month. The monthly cost of Household Rentals can be expressed by $y = 42x + 60$ and the monthly cost of Rent and Go can be expressed by $y = 50x$, where $x$ is the number of months. Use the graph of the system formed by these two equations to answer each question. (*Section 4.1, Objectives 2 and 6*)

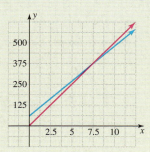

 **a.** Approximate the point of intersection of this system.

 **b.** What is the meaning of the intersection point in the context of the problem?

 **c.** If Rikki expects to sign a 12-month rental of the refrigerator, which company should she choose? Why?

**Solve each system of linear equations using substitution.** (*Section 4.2, Objectives 1 and 2*)

**39.** $\begin{cases} 7x - 5y = 16 \\ x + 3y = 6 \end{cases}$   {(3, 1)}   **40.** $\begin{cases} x - 6y = 4 \\ 4x + 2y = 3 \end{cases}$   $\left\{\left(1, -\dfrac{1}{2}\right)\right\}$

**41.** $\begin{cases} \dfrac{1}{3}x - y = 2 \\ x = 3y + 5 \end{cases}$  ∅

**Solve each system using elimination.** (*Section 4.3, Objectives 1–3*)

**42.** $\begin{cases} 3x - 10y = -56 \\ 4x + 8y = 32 \end{cases}$ {(−2, 5)}  **43.** $\begin{cases} 2x + y = 5 \\ 3y = -6x + 15 \end{cases}$ {(x, y)|2x + y = 5}

**44.** $\begin{cases} 6x - 8y = 13 \\ \dfrac{5}{6}x + \dfrac{1}{2}y = 1 \end{cases}$ $\left\{\left(\dfrac{3}{2}, -\dfrac{1}{2}\right)\right\}$  **45.** $\begin{cases} 1.5x - 0.4y = 5.8 \\ 0.5x + 2.4y = -4.4 \end{cases}$ {(3.2, −2.5)}

**Solve each problem using a system of linear equations.** (*Section 4.4, Objectives 1–5*)

**46.** Two angles are supplementary. The larger angle is 12° more than three times the smaller angle. What is the measure of each angle?   42°, 138°

**47.** Two angles are complementary. The larger angle is 24° less than twice the measure of the smaller one. What is the measure of each angle?   38°, 52°

**48.** Wendy leaves her job and heads south at 45 mph. Jay leaves the same workplace 20 min later and heads south at 60 mph. How long will it take Jay to catch up to Wendy?   1 hr

**49.** Alison needs to make 250 L of a 60% alcohol solution. How much pure alcohol and how much 50% alcohol should be mixed together to make 250 L of a 60% alcohol solution?   50 L of pure alcohol and 200 L of 50% alcohol

**50.** A plane travels 1260 mi with the wind in 2.8 hr. It takes 3.6 hr for the plane to travel the same distance against the wind. What is the speed of the plane in still air and what is the speed of the wind?   400 mph, 50 mph

**Determine if the ordered pair is a solution of the given inequality.** (*Section 4.5, Objective 1*)

**51.** $3x - 9y < 5$; (0, 2)   yes   **52.** $y \le -2x + 8$; (−2, 1)   yes

**Graph the solution set of each linear inequality in two variables.** (*Section 4.5, Objective 2*)

**53.** $x + 5y \le 10$   **54.** $4x - y \le 16$

**55.** $x + y < 0$   **56.** $x - 4 < 0$

**Use a linear inequality in two variables to solve each problem.** (*Section 4.5 Objective 3*)

**57.** Brenda works as a student worker at a college earning $8.75 an hour and at a nursing home as a nurse aide earning $11.50 per hour. If she wants to make at least $280 per week, (a) write a linear inequality in two variables that represents this situation, (b) graph the linear inequality, and (c) give three possible combinations of hours that Brenda could work at each job to make at least $280 per week.   $8.75x + 11.50y \ge 280$

**Solve each system of linear inequalities.** (*Section 4.6, Objective 1*)

**58.** $\begin{cases} y \ge -3x - 4 \\ y \le -3x + 1 \end{cases}$   **59.** $\begin{cases} x \ge 1 \\ y \le 4 \end{cases}$

**Solve the problem using a system of linear inequalities.** (*Section 4.6, Objective 2*)

**60.** The price per item, in dollars, is determined by the supply equation $y = S(x)$, where $x$ is the number of items produced. The producer's surplus is the area of the graph of the solution set of the system of linear inequalities.

$$\begin{cases} y \ge S(x) \\ y \le k \\ x \ge 0 \end{cases} \text{, where } k \text{ is a fixed price}$$

The supply equation for an item is $y = S(x) = 21 + 0.40x$ and $k = 53$. (a) Write a system of linear inequalities and solve it by graphing and (b) find the producer's surplus.   1280

# Laws of Exponents and Polynomial Operations

## Goal Setting

Having goals is essential to your success in life, in college, and in a classroom. Without goals, you do not really know what you are working toward. Take the time to write down the goals you hope to accomplish by being in this class and by being in college. Start with specific and realistic goals, such as

- I will review my class notes each day.
- I will work homework problems for an hour daily.
- I will ask questions if I do not understand.
- I will attend class each day.

You need to make sure that your goals are attainable. It is important that you review these goals often so that you stay focused on what you are working toward. Defining goals may require you to make changes in other areas of your life. Be willing to do what is necessary to accomplish your dreams.

**Question For Thought:** What do you hope to accomplish by attending college? How does this class help you reach that goal? What do you hope to accomplish in this class?

## Chapter Outline

## Coming Up...

In Section 5.2, we will work with the formula, $P = P_0 \cdot 2^{-t/m}$, which gives the amount of radioactive substance present at time $t$, where $P_0$ is the original amount of the substance and $m$ is the half-life of the substance. The half-life of a radioactive substance is the time it takes for half of a certain amount of radioactive substance to decay.

"Goals. There's no telling what you can do when you get inspired by them. There's no telling what you can do when you believe in them. There's no telling what will happen when you act upon them."

— Jim Rohn (Entrepreneur, Author, Speaker)

## SECTION 5.1 | The Product and Power Rules for Exponents

**In Chapter 1,** the concept of an exponent was defined and introduced briefly. This chapter will take a closer look at exponents and their properties. We will also examine algebraic expressions called polynomials and the basic operations performed on them.

### ▶ OBJECTIVES

As a result of completing this section, you will be able to

1. Apply the product of like bases rule for exponents.
2. Apply the power of a power rule for exponents.
3. Apply the power of a product rule for exponents.
4. Apply the power of a quotient rule for exponents.
5. Solve application problems.
6. Troubleshoot common errors.

A pizza place sells a small pizza (8 in. in diameter) for $5.45 and an extra large pizza (16 in. in diameter) for $15.95. The diameter of the extra large pizza is double that of the small pizza but the cost of the extra large pizza is nearly triple that of the small pizza. How much more pizza does a customer get when they order an extra large pizza rather than a small pizza? The answer to this question is based on knowing how to simplify the expression $(2r)^2$.

In this section, we will learn a rule for exponents that will enable us to simplify the preceding expression along with three other important rules. These rules are used quite extensively in algebra so it is very important that we become familiar with them.

### Objective 1 ▶

Apply the product of like bases rule for exponents.

**INSTRUCTOR NOTE:**
Explain the difference between the terms power and exponent. The expression $b^n$ is called a power and $n$ is the exponent. For example, some powers of 2 are $2^1$, $2^2$, and $2^3$; they are not 1, 2, and 3.

### Products of Like Bases

In Chapter 1, the definition of an exponent was presented. Recall the following.

> **Definition: Exponent**
>
> For a real number $b$ and a natural number $n$,
>
> $b^n = b \cdot b \cdot b \ldots b$, where the base $b$ is repeated $n$ times.

The exponent $n$ indicates the number of times the base $b$ is used as a factor. For example, $3 \cdot 3 \cdot 3 \cdot 3$ can be written as $3^4$. The expression $3^4$ is called an *exponential expression, or power*, and is read as "3 to the fourth" or "the fourth power of 3." In the expression $3^4$, 3 is the base and 4 is the exponent.

An exponent has the same meaning when the base is a variable. For example, $y \cdot y \cdot y$ can be written as $y^3$. Recall a few other important rules from Chapter 1.

1. A negative number raised to an even exponent is positive. For example,

$$(-5)^2 = 25$$

2. A negative number raised to an odd exponent is negative. For example,

$$(-5)^3 = -125$$

3. If a negative sign is in front of an exponential expression, the negative sign indicates the opposite of the value of the exponential expression. For example,

$$-6^4 = -(6 \cdot 6 \cdot 6 \cdot 6) = -1296$$

**4.** If there are parentheses around a negative number raised to an exponent, then the exponent is applied to the negative number. For example,

$$(-6)^4 = (-6)(-6)(-6)(-6) = 1296$$

The definition of an exponent is used to develop rules that can be used when performing operations with exponents. The first of these rules deals with multiplying exponential expressions with the same, or like, bases. The following illustration shows how to use the definition of the exponent to perform the multiplication and then states a rule that can be applied to the original problem to obtain the final result.

| **Based on Definition** | **Observed Rule** |
|---|---|
| $b^2 \cdot b^3 = (b \cdot b) \cdot (b \cdot b \cdot b)$ <br> $= b^5$ <br> Two factors of $b$ times three factors of $b$ gives a total of five factors of $b$. | The same result is obtained by adding the exponents of the like bases. <br> $b^2 \cdot b^3 = b^{2+3} = b^5$ |

As we can see from this example, when exponential expressions with like bases are multiplied the result is the base raised to the sum of its exponents.

---

**Property:  The Product of Like Bases Rule for Exponents**

For a real number $b$ and positive integers $m$ and $n$,

$$b^m \cdot b^n = b^{m+n}$$

---

It is very important to note that the product rule can be applied *only* if the base of each factor is the same.

---

**Procedure:  Multiplying Exponential Expressions with Like Bases**

**Step 1:** Keep the base the same.
**Step 2:** Add the exponents of the factors.

---

**Objective 1  Examples**   Simplify each expression by applying the product of like bases rule.

**1a.** $2^2 \cdot 2^3$     **1b.** $r^{11} \cdot r^5 \cdot r$     **1c.** $(-6)^4 \cdot (-6)^8$     **1d.** $(-7)^2 \cdot (-7)^3$
**1e.** $x^3 \cdot y^4$     **1f.** $(4y^2)(-3y^4)$     **1g.** $-5x(2x^2)(-4y^3)$

**Solutions**   **1a.**     $2^2 \cdot 2^3 = 2^{2+3}$         Apply the product of like bases rule.
                             $= 2^5$            Add the exponents.
                             $= 32$            Simplify.

**1b.**     $r^{11} \cdot r^5 \cdot r = r^{11} \cdot r^5 \cdot r^1$     Recall that $r = r^1$.
                             $= r^{11+5+1}$     Apply the product of like bases rule.
                             $= r^{17}$          Add the exponents.

**1c.** $(-6)^4 \cdot (-6)^8 = (-6)^{4+8}$     Apply the product of like bases rule.
                             $= (-6)^{12}$      Add the exponents.
                             $= 6^{12}$          A negative to an even exponent is positive.

**1d.** $(-7)^2 \cdot (-7)^3 = (-7)^{2+3}$     Apply the product of like bases rule.
                             $= (-7)^5$          Add the exponents.
                             $= -7^5$            A negative to an odd exponent is negative.

**1e.** The bases are not the same, so the expression cannot be simplified further.

**1f.**
$$(4y^2)(-3y^4) = 4(-3)y^2y^4 \qquad \text{Group coefficients and like bases.}$$
$$= -12y^{2+4} \qquad \text{Multiply the coefficients; apply the product of like bases rule.}$$
$$= -12y^6 \qquad \text{Add the exponents.}$$

**1g.**
$$-5x(2x^2)(-4y^3) = (-5)(2)(-4)x^1x^2y^3 \qquad \text{Group coefficients and like bases.}$$
$$= 40x^{1+2}y^3 \qquad \text{Multiply the coefficients; apply the product of like bases rule.}$$
$$= 40x^3y^3 \qquad \text{Add the exponents of the like bases.}$$

✓ **Student Check 1**    Simplify each expression by applying the product of like bases rule.

**a.** $m^9 \cdot m$      **b.** $5^3 \cdot 5^3$      **c.** $y^7 \cdot y^4 \cdot y^2$

**d.** $(2)^4 \cdot (3)^2$      **e.** $(-2x^4)(-8x^3)$      **f.** $(7a^4)(-5b^2)(-a^5)$

## Powers of Exponential Expressions

**Objective 2** ▶

Apply the power of a power rule for exponents.

Exponents can be applied to exponential expressions as well. The following illustration shows how to use the definition of an exponent to obtain the result as well as a rule that can be applied to the original problem to obtain the result.

| **Based on Definition** | **Observed Rule** |
|---|---|
| $(x^2)^3 = x^2 \cdot x^2 \cdot x^2$   Three factors of $x^2$ | The same result is obtained by multiplying the exponents. |
| $= x^{2+2+2}$   Apply the product rule. | |
| $= x^6$   Add the exponents. | $(x^2)^3 = x^{2 \cdot 3} = x^6$ |

As we can see from this example, when a power is raised to an exponent, the original exponents are multiplied to obtain the exponent in the final result. This brings us to the second property of exponents.

> **Property: The Power of a Power Rule for Exponents**
>
> For a real number $b$ and positive integers $m$ and $n$,
> $$(b^m)^n = b^{m \cdot n}$$

> **Procedure: Simplifying a Power of a Power**
>
> **Step 1:** Keep the base the same.
> **Step 2:** Multiply the exponents.

**Objective 2 Examples**    Simplify each expression by applying the power of a power rule.

**2a.** $(m^4)^5$      **2b.** $(2^4)^3$      **2c.** $[(-3)^4]^5$

**Solutions**

**2a.**
$$(m^4)^5 = m^{4 \cdot 5} \qquad \text{Apply the power of a power rule.}$$
$$= m^{20} \qquad \text{Keep the base and multiply the exponents.}$$

**2b.**
$$(2^4)^3 = 2^{4 \cdot 3} \qquad \text{Apply the power of a power rule.}$$
$$= 2^{12} \qquad \text{Keep the base and multiply the exponents.}$$

**2c.**
$$[(-3)^4]^5 = (-3)^{4 \cdot 5} \qquad \text{Apply the power of a power rule.}$$
$$= (-3)^{20} \qquad \text{Keep the base and multiply the exponents.}$$
$$= 3^{20} \qquad \text{A negative number raised to an even exponent is positive.}$$

✓ **Student Check 2**   Simplify each expression by applying the power of a power rule.

   **a.** $(p^6)^2$   **b.** $(10^3)^3$   **c.** $[(-2)^3]^2$

## The Power of a Product

**Objective 3** ▶

Apply the power of a product rule for exponents.

Thus far, we have simplified exponential expressions in which the base was a number, a variable, or an exponential expression. The base can also be a product, a quotient, a sum, or a difference of expressions. When the base of an exponential expression contains an operation, the entire expression in parentheses repeats when the exponent is applied to it. The following example illustrates how to simplify a power of a product.

**Based on Definition**

$$(3x^4)^2 = (3x^4)(3x^4) \quad \text{Two factors of } 3x^4$$
$$= 3 \cdot 3 \cdot x^4 \cdot x^4 \quad \text{Group coefficients and like bases.}$$
$$= 9x^8 \quad \text{Multiply the coefficients and add the exponents.}$$

**Observed Rule**

The same result is obtained by squaring 3 and squaring $x^4$.

$$(3x^4)^2 = (3)^2 (x^4)^2$$
$$= 9x^8$$

As we can see from this example, when a product is raised to an exponent, the result is the same as applying the exponent to each factor in the base. This brings us to another property of exponents.

**Property: The Power of a Product Rule for Exponents**

For real numbers $a$ and $b$ and a positive integer $n$,

$$(ab)^n = a^n b^n$$

**Procedure: Simplifying a Power of a Product**

**Step 1:** Apply the power to each factor in the parentheses.
**Step 2:** Evaluate any exponents and apply the power of a power rule, as needed.

**Objective 3 Examples**   Simplify each expression by applying the power of a product rule.

   **3a.** $(2r)^2$   **3b.** $(-6y^3)^4$   **3c.** $(3a^5 b^2)^3$

**Solutions**   **3a.**   $(2r)^2 = (2)^2(r)^2$   Apply the power of a product rule.
                         $= 4r^2$   Simplify the exponent.

   **3b.** $(-6y^3)^4 = (-6)^4(y^3)^4$   Apply the power of a product rule.
                      $= 1296y^{12}$   Simplify the exponent and apply the power of a power rule.

   **3c.** $(3a^5 b^2)^3 = (3)^3(a^5)^3(b^2)^3$   Apply the power of a product rule.
                         $= 27a^{15}b^6$   Simplify the exponent and apply the power of a power rule.

✓ **Student Check 3**   Simplify each expression by applying the power of a product rule.

   **a.** $(7p)^3$   **b.** $(-11x^4)^2$   **c.** $(2m^2 n^3)^5$

## The Power of a Quotient

Apply the power of a quotient rule for exponents.

Now we will examine what happens when a quotient is raised to an exponent.

| **Based on Definition** | | **Observed Rule** |
|---|---|---|

The same result is obtained by squaring $y$ and squaring 3.

$$\left(\frac{y}{3}\right)^2 = \left(\frac{y}{3}\right)\left(\frac{y}{3}\right) \qquad \text{Two factors of } \frac{y}{3}$$

$$= \frac{y \cdot y}{3 \cdot 3} \qquad \text{Multiply the fractions.}$$

$$= \frac{y^2}{9} \qquad \text{Simplify the numerator and denominator.}$$

$$\left(\frac{y}{3}\right)^2 = \frac{(y)^2}{(3)^2}$$

$$= \frac{y^2}{9}$$

As we can see from this example, when a quotient is raised to an exponent, the exponent is applied to both the numerator and denominator. This leads to another property.

> **Property:  The Power of a Quotient Rule for Exponents**
>
> For real numbers $a$ and $b$ ($b \neq 0$) and $n$ a positive integer,
>
> $$\left(\frac{a}{b}\right)^n = \frac{a^n}{b^n}$$

> **Procedure:  Simplifying the Power of a Quotient**
>
> **Step 1:  a.** Apply the exponent to the numerator.
> **b.** Apply the exponent to the denominator.
> **Step 2:** Simplify, if needed.

**Objective 4  Examples**     Simplify each expression by applying the power of a quotient rule.

**4a.** $\left(\dfrac{x^5}{9}\right)^2$    **4b.** $\left(-\dfrac{2a}{5b}\right)^3$    **4c.** $\left(\dfrac{3}{4s^3}\right)^5$

**Solutions    4a.**  $\left(\dfrac{x^5}{9}\right)^2 = \dfrac{(x^5)^2}{(9)^2}$    Apply the power of a quotient rule.

$$= \frac{x^{5 \cdot 2}}{81} \qquad \text{Apply the power of a power rule in the numerator and simplify } 9^2.$$

$$= \frac{x^{10}}{81} \qquad \text{Simplify the numerator by multiplying the exponents.}$$

**4b.**  $\left(-\dfrac{2a}{5b}\right)^3 = \dfrac{(-2a)^3}{(5b)^3}$    Apply the power of a quotient rule.

$$= \frac{(-2)^3(a)^3}{(5)^3(b)^3} \qquad \text{Apply the power of a power rule in the numerator and denominator.}$$

$$= \frac{-8a^3}{125b^3} \qquad \text{Simplify the numerator and denominator.}$$

$$= -\frac{8a^3}{125b^3} \qquad \text{Rewrite.}$$

**4c.** $\left(\dfrac{3}{4s^3}\right)^5 = \dfrac{(3)^5}{(4s^3)^5}$     Apply the power of a quotient rule.

$= \dfrac{243}{(4)^5(s^3)^5}$     Simplify the numerator and apply the power of a power rule in the denominator.

$= \dfrac{243}{1024s^{15}}$     Simplify the denominator.

---

✔ **Student Check 4**     Simplify each expression by applying the power of a quotient rule.

**a.** $\left(\dfrac{2}{b^4}\right)^5$      **b.** $\left(-\dfrac{6x}{7y}\right)^3$      **c.** $\left(\dfrac{5t^2}{3}\right)^4$

---

## Applications

**Objective 5** ▶

Solve application problems.

There are many applications in real life that use exponents. Exponents are very important when studying population growth, investments, the growth of bacteria, and the half-life of chemical elements, to name a few. These types of problems will be studied more closely in later math classes.

**Objective 5 Examples**     **Solve each problem.**

A pizza place sells a small pizza (8 in. in diameter) for $5.45 and an extra large pizza (16 in. in diameter) for $15.95. The diameter of the extra large pizza is double the diameter of the small pizza; however, the cost of the extra large pizza is nearly triple the cost of the small pizza.

**5a.** How much more pizza does a customer get when ordering an extra large pizza rather than a small pizza? Which pizza is the better deal?

**5b.** Is this result always true, that is, if the radius of a circle is doubled, what happens to the area?

**Solutions**     **5a.** To determine how much more pizza is in an extra large pizza, we need to compare the areas of the small and extra large pizzas. Recall, the area of a circle is given by $A = \pi r^2$. (Remember that the radius is half the diameter.)

|  **Small Pizza**  |  **Extra Large Pizza**  |
|---|---|
| The diameter of the small pizza is 8 in., so its radius is $\frac{1}{2}(8) = 4$ in. | The diameter of the extra large pizza is 16 in., so its radius is $\frac{1}{2}(16) = 8$ in. |

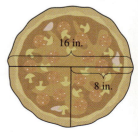

So, the area of the small pizza is

$A = \pi r^2$
$= \pi(4)^2$
$= 16\pi$ in.²

So, the area of the extra large pizza is

$A = \pi r^2$
$= \pi(8)^2$
$= 64\pi$ in.²

The area of the extra large pizza is four times as large as the small pizza. Since the extra large pizza is four times as large as the small pizza, it is the better deal since the cost is only three times as much as the small pizza.

**5b.** If $r$ represents the radius of a circle, then $2r$ represents the radius doubled.

$$A = \pi(2r)^2 \qquad \text{Replace } r \text{ with } 2r.$$

$$A = \pi(2)^2(r)^2 \qquad \text{Apply the power of a product rule.}$$

$$A = \pi \cdot 4 \cdot r^2 \qquad \text{Simplify.}$$

$$A = 4\pi r^2 \qquad \text{Rewrite using the commutative property.}$$

So, the area of a circle whose radius doubled is always four times the area of the original circle.

✓ **Student Check 5**   A room is in the shape of a square with side $x$ ft long. If each side of the square is tripled, what happens to the area of the room?

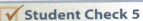**Objective 6** ▶

Troubleshoot common errors.

## Troubleshooting Common Errors

Some common errors associated with the properties of exponents are shown next.

**Objective 6  Examples**   A problem and an incorrect solution are given. Provide the correct solution and an explanation of the error.

**6a.** $y^2 \cdot y^6$

| Incorrect Solution | Correct Solution and Explanation |
|---|---|
| $y^2 \cdot y^6 = y^{12}$ | The error is that the exponents were multiplied. The product of like bases rule tells us when we multiply like bases, we must *add* exponents. $$y^2 \cdot y^6 = y^{2+6} = y^8$$ |

**6b.** $5^2 \cdot 5^3$

| Incorrect Solution | Correct Solution and Explanation |
|---|---|
| $5^2 \cdot 5^3 = 25^5$ | The error is that the bases were multiplied. The product of like bases rule tells us when we multiply like bases, we must keep the base the same. $$5^2 \cdot 5^3 = 5^5 = 3125$$ |

**6c.** $2^3 \cdot 3^2$

| Incorrect Solution | Correct Solution and Explanation |
|---|---|
| $2^3 \cdot 3^2 = 6^5$ | The error is that the product of like bases rule was applied incorrectly. The product rule applies only when the bases are the same. Since the bases are not the same, we must evaluate each expression and then multiply the results. $$2^3 \cdot 3^2 = 8 \cdot 9 = 72$$ |

**6d.** $(-4y^3)^2$

| Incorrect Solution | Correct Solution and Explanation |
|---|---|
| $(-4y^3)^2 = -4y^9$ | Two errors were made. The power of a product rule was not applied correctly since the coefficient, $-4$, was not squared. Also, the power of a power rule was not applied correctly since the exponents were not multiplied. <br><br> $$(-4y^3)^2 = (-4)^2\,(y^3)^2 = 16y^6$$ |

## ANSWERS TO STUDENT CHECKS

**Student Check 1   a.** $m^{10}$   **b.** $5^6$   **c.** $y^{13}$   **d.** $144$
   **e.** $16x^7$   **f.** $35a^9b^2$

**Student Check 2   a.** $p^{12}$   **b.** $10^9$   **c.** $2^6 = 64$

**Student Check 3   a.** $343p^3$   **b.** $121x^8$   **c.** $32m^{10}n^{15}$

**Student Check 4   a.** $\dfrac{32}{b^{20}}$   **b.** $-\dfrac{216x^3}{343y^3}$   **c.** $\dfrac{625t^8}{81}$

**Student Check 5**   The area of the enlarged room is 9 times the area of the original room.

## SUMMARY OF KEY CONCEPTS

**1.** The properties presented in this section are summarized as follows for real numbers $a$ and $b$ and positive integers $m$ and $n$.

| Product of like bases | $b^m \cdot b^n = b^{m+n}$ | $x^3 \cdot x^2 = x^5$ |
|---|---|---|
| **Power of a power** | $(b^m)^n = b^{mn}$ | $(x^3)^2 = x^6$ |
| **Power of a product** | $(ab)^n = a^n b^n$ | $(3x)^4 = 3^4x^4 = 81x^4$ |
| **Power of a quotient** | $\left(\dfrac{a}{b}\right)^n = \dfrac{a^n}{b^n},\ b \neq 0$ | $\left(\dfrac{4}{x^3}\right)^5 = \dfrac{4^5}{(x^3)^5} = \dfrac{1024}{x^{15}}$ |

**2.** Be careful when simplifying expressions with negative signs. A negative number raised to an even exponent is positive and a negative number raised to an odd exponent is negative. See the following examples.
$$(-2a)^3 = (-2)^3(a)^3 = -8a^3 \text{ and}$$
$$(-5b)^4 = (-5)^4(b)^4 = 625b^4$$

## GRAPHING CALCULATOR SKILLS

The calculator skills for this section have previously been covered, though a few examples are included here.

**Example 1:** Show that $(-6)^4 \cdot (-6)^8 = (-6)^{12}$.

**Solution:**

```
(-6)^4*(-6)^8
           2176782336
(-6)^12
           2176782336
```

**Example 2:** Show that $[(-3)^4]^5 = (-3)^{20}$.

**Solution:**

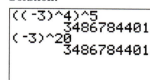

```
((-3)^4)^5
           3486784401
(-3)^20
           3486784401
```

## SECTION 5.1 / EXERCISE SET

### Write About It!

**Use complete sentences in your answer to each exercise.**

1. Explain the product of like bases rule for exponents. *Answers vary.*
2. Explain the power of a power rule for exponents. *Answers vary.*
3. Explain the power of a product rule for exponents. *Answers vary.*
4. Explain the power of a quotient rule for exponents. *Answers vary.*
5. Under what conditions do the expressions $(-2x)^n$ and $-(2x)^n$ yield the same result? Explain. *Answers vary.*
6. Under what conditions do the expressions $(-2x)^n$ and $(2x)^n$ yield the same result? Explain. *Answers vary.*

### Practice Makes Perfect!

**Simplify each expression by applying the product of like bases rule for exponents. (See Objective 1.)**

7. $r^3 \cdot r^4$   $r^7$
8. $s^5 \cdot s^2$   $s^7$
9. $m \cdot m^6$   $m^7$
10. $x \cdot x^4$   $x^5$
11. $x^2 \cdot x^3 \cdot x$   $x^6$
12. $y^3 \cdot y^5 \cdot y$   $y^9$
13. $m^3 \cdot m^5 \cdot m^4$   $m^{12}$
14. $n^4 \cdot n^5 \cdot n^6$   $n^{15}$
15. $3^{11} \cdot 3^4$   $3^{15}$
16. $4^2 \cdot 4^{10}$   $4^{12}$
17. $(-2)^5 \cdot (-2)^9$   $2^{14}$
18. $(-5)^6 \cdot (-5)^8$   $5^{14}$
19. $(3)^4 \cdot (3)^{12}$   $3^{16}$
20. $(7)^3 \cdot (7)^5$   $7^8$
21. $(5x^3) \cdot (-4x^2)$   $-20x^5$
22. $(-3y^8) \cdot (6y^3)$   $-18y^{11}$
23. $(-12a^4) \cdot (-5a^6)$   $60a^{10}$
24. $(15y^7) \cdot (-5y^{12})$   $-75y^{19}$
25. $(2x^2)(-3y^4)(-y^3)$   $6x^2y^7$
26. $(6x^5)(12y^3)(-2x^7)$   $-144x^{12}y^3$
27. $(-3a^4)(-9b^7)(-2a^9)$   $-54a^{13}b^7$
28. $(-8a^2)(-10b^8)(-a^6)$   $-80a^8b^8$

**Simplify each expression by applying the power of a power rule for exponents. (See Objective 2.)**

29. $(x^3)^2$   $x^6$
30. $(y^5)^3$   $y^{15}$
31. $(6^4)^8$   $6^{32}$
32. $(11^2)^6$   $11^{12}$
33. $(3^{12})^{13}$   $3^{156}$
34. $(6^6)^{14}$   $6^{84}$
35. $[(-4)^5]^7$   $-4^{35}$
36. $[(-5)^{10}]^{11}$   $5^{110}$

**Simplify each expression by applying the power of a product rule for exponents. (See Objective 3.)**

37. $(3x)^2$   $9x^2$
38. $(2y)^3$   $8y^3$
39. $(-10r^2)^3$   $-1000r^6$
40. $(-5m^5)^4$   $625m^{20}$
41. $(6n^6)^5$   $6^5n^{30}$
42. $(4s^7)^2$   $16s^{14}$
43. $(-2x^3y)^4$   $16x^{12}y^4$
44. $(-3x^2y^8)^3$   $-27x^6y^{24}$
45. $(7a^{10}b^8)^2$   $49a^{20}b^{16}$
46. $(8c^4d^{11})^4$   $8^4c^{16}d^{44}$

**Simplify each expression by applying the power of a quotient rule for exponents. (See Objective 4.)**

47. $\left(\dfrac{r}{3}\right)^2$   $\dfrac{r^2}{9}$
48. $\left(\dfrac{s}{4}\right)^3$   $\dfrac{s^3}{64}$
49. $\left(\dfrac{x^3}{y^2}\right)^5$   $\dfrac{x^{15}}{y^{10}}$
50. $\left(\dfrac{m^6}{n^7}\right)^8$   $\dfrac{m^{48}}{n^{56}}$
51. $-\left(\dfrac{3x}{4y}\right)^3$   $\dfrac{27x^3}{64y^3}$
52. $-\left(\dfrac{5a}{3b}\right)^2$   $-\dfrac{25a^2}{9b^2}$

Additional answers can be found in the Instructor Answer Appendix.

53. $\left(\dfrac{2}{5s^2}\right)^4$   $\dfrac{16}{625s^8}$
54. $\left(\dfrac{3}{2r^3}\right)^4$   $\dfrac{81}{16r^{12}}$
55. $\left(-\dfrac{9a^5}{5}\right)^2$   $\dfrac{81a^{10}}{25}$
56. $\left(-\dfrac{10b^6}{3}\right)^3$   $-\dfrac{1000b^{18}}{27}$
57. $\left(\dfrac{6x}{7y^5}\right)^8$   $\dfrac{6^8x^8}{7^8y^{40}}$
58. $\left(\dfrac{3x^4}{8y^{11}}\right)^5$   $\dfrac{3^5x^{20}}{8^5y^{55}}$

**Solve each problem. (See Objective 5.)**

59. A pizza shop sells a small pizza that is 6 in. in diameter and an extra large pizza whose diameter is triple that of the small pizza. Is the area of the extra large pizza triple the area of the small pizza? Explain. *No, the area of the extra large pizza will be nine times the area of the small pizza.*
60. Sliders are mini-hamburgers often served as appetizers. The diameter of a slider burger is 2 in. The diameter of a super burger is 4 times the slider's diameter. Is the area of the super burger 4 times as large as the area of the slider burger? Explain. *No, the area of the regular burger is 16 times as large as the area of the slider burger.*
61. Stephen builds a square deck that is 15 ft long. His brother, Ethan, builds a square deck with a length that is double the length of Stephen's deck. Is the area of Ethan's deck double the area of Stephen's deck? Explain. *No, Ethan's deck is four times the area of Stephen's deck.*
62. Tammy plans to fence in a square area of her yard for her dog. She is considering having the length of the dog yard be 20 ft. If she doubles this length, will the area of the dog yard also double? Explain. *No, the area of the dog yard will be four times larger.*

### Mix 'Em Up!

**Use exponent rules to simplify each expression.**

63. $(2a^3)^6$   $64a^{18}$
64. $(13y^{12})^2$   $169y^{24}$
65. $x^4 \cdot x^7 \cdot x$   $x^{12}$
66. $y^9 \cdot y \cdot y^3$   $y^{13}$
67. $(5^5)^{15}$   $5^{75}$
68. $[(-2)^7]^3$   $-2^{21}$
69. $\left(-\dfrac{2x}{y^6}\right)^7$   $-\dfrac{2^7x^7}{y^{42}}$
70. $-\left(-\dfrac{3x^5}{4y^2}\right)^3$   $\dfrac{27x^{15}}{64y^6}$
71. $(-5a^7) \cdot (-6a^8)$   $30a^{15}$
72. $(-8q^3) \cdot (7q^{13})$   $-56q^{16}$
73. $(m^5n^6)^4$   $m^{20}n^{24}$
74. $(c^4d^9)^{11}$   $c^{44}d^{99}$
75. $(-x^4)(7y^5)(-3x^{12})$   $21x^{16}y^5$
76. $(-a^{15})(-5a^2)(8b^{10})$   $40a^{17}b^{10}$
77. $(6r^8)^4$   $1296r^{32}$
78. $(-5x^6y)^3$   $-125x^{18}y^3$
79. $\left(-\dfrac{s^5}{t^3}\right)^{11}$   $-\dfrac{s^{55}}{t^{33}}$
80. $-\left(-\dfrac{7p^2}{3q}\right)^3$   $\dfrac{343p^6}{27q^3}$
81. $(-4x^5y^7)^4$   $256x^{20}y^{28}$
82. $(-9a^{11}b^7)^2$   $81a^{22}b^{14}$
83. $(3x)(14y^3)(-2x^9)^2$   $168x^{19}y^3$
84. $(-4a^{12})^2(-2b)(-9a^3)$   $288a^{27}b$
85. $\left(\dfrac{-0.2p^{12}}{q^7}\right)^4$   $\dfrac{0.0016p^{48}}{q^{28}}$
86. $\left(-\dfrac{r}{0.7s^3}\right)^3$   $-\dfrac{r^3}{0.343s^9}$
87. $(0.5a^2b)^3$   $0.125a^6b^3$
88. $(0.6x^3y^2)^2$   $0.36x^6y^4$

**Solve each problem.**

**89.** Suppose circle A has a diameter of 3 in. and circle B has a diameter of 9 in. How many times larger is the area of circle B than the area of circle A?   nine times as large

**90.** Suppose circle A has a diameter of 1 in. and circle B has a diameter of 5 in. How many times larger is the area of Circle B than the area of Circle A?   25 times as large

**91.** Nathan has a square deck that is 10 ft long. He expands his deck by doubling its length and width. How many times larger is the area of the new deck?   four times as large

**92.** Rosa has a dog pen that measures 8 ft by 8 ft. She expands the dog pen by tripling the length of each side of the pen. How many times larger is the new pen?   nine times as large

The volume $V$ of a cylinder with radius $r$ and height $h$ is $\pi r^2 h$. Express each answer in terms of $\pi$.

**93.** Suppose a cylinder has a radius of 2 in. and height of 5 in. What is the volume of the cylinder?   $20\pi$ in.³

**94.** Suppose a cylinder has a radius of 6 cm and height of 10 cm. What is the volume of the cylinder?   $360\pi$ cm³

The volume $V$ of a sphere with radius $r$ is $\frac{4}{3}\pi r^3$. Express each answer in terms of $\pi$.

**95.** Suppose a sphere has a radius of 6 in. What is the volume of the sphere?   $288\pi$ in.³

**96.** Suppose a sphere has a radius of 3 cm. What is the volume of the sphere?   $36\pi$ cm³

## You Be the Teacher!

**Correct each student's errors, if any.**

**97.** Simplify: $k^4 \cdot k^{12}$.

Rena's work: $k^4 \cdot k^{12} = k^{48}$

$k^4 \cdot k^{12} = k^{16}$; Rena incorrectly used the power of a power rule. She should have used the product of like bases rule.

**98.** Simplify: $m^{12} \cdot m^2$

Liam's work: $m^{12} \cdot m^2 = m^{24}$

$m^{12} \cdot m^2 = m^{14}$; Liam used the power of a power rule. He should have used the product of like bases rule.

**99.** Simplify: $(4a^6b^3)^3$

Ursula's work: $(4a^6b^3)^3 = 4(a^6)^3(b^3)^3 = 4a^{18}b^9$

**100.** Simplify: $(10xy^3)^5$

Ricky's work: $(10xy^3)^5 = 10(5)(x)^5(y^3)^5 = 50x^5y^{15}$

**101.** Simplify $(-2ab^2)^3$

Rob's work: $(-2ab^2)^3 = -2(3)(a)^3(b^2)^3 = -6a^3b^6$

**102.** Simplify: $(-x^2y)^5$

Wanda's work: $(-x^2y)^5 = (x^2)^5(y)^5 = x^{10}y^5$

## Calculate It!

**Use a calculator to simplify each expression.**

**103.** $3^6 \cdot 3^{12}$
$3^{18}$ or 387,420,489

**104.** $(-2)^{11} \cdot (-2)^{13}$
$(-2)^{24}$ or $2^{24}$ or 16,777,216

**105.** $[(-5)^3]^5$
$(-5)^{15}$ or $-5^{15}$, or $-30,517,578,125$

**106.** $[(-3)^2]^8$
$(-3)^{16}$ or $3^{16}$ or 43,046,721

## Think About It!

**Use the properties of exponents to simplify each expression.**

**107.** $3^{4x} \cdot 3^{x+1}$   $3^{5x+1}$

**108.** $2^{3x} \cdot 2^{4x-5}$   $2^{7x-5}$

**109.** $2^{3(a+2)} \cdot 2$   $2^{3a+7}$

**110.** $4^{5(a-1)} \cdot 4$   $4^{5a-4}$

**111.** $(7^{2x})^3$   $7^{6x}$

**112.** $(6^{4b})^5$   $6^{20b}$

**113.** $(4x^{3a})^5$   $1024x^{15a}$

**114.** $(-2x^{8b})^4$   $16x^{32b}$

---

| **SECTION 5.2** | **The Quotient Rule and Zero and Negative Exponents** |
|---|---|

### ▶ OBJECTIVES

**As a result of completing this section, you will be able to**

**1.** Apply the quotient of like bases rule for exponents.

**2.** Simplify expressions with zero as an exponent.

**3.** Simplify expressions with negative exponents.

**4.** Apply a combination of exponent properties.

**5.** Solve application problems.

**6.** Troubleshoot common errors.

Brianna and John to purchase a car borrow $25,000. The car company will finance the loan at 6% annual interest for 6 yr. To determine the monthly payment for Brianna and John to repay the loan, we

must simplify the expression $\dfrac{25,000\left(\dfrac{0.06}{12}\right)}{1-\left(1+\dfrac{0.06}{12}\right)^{-72}}$.

The graphing calculator skills will demonstrate how to simplify this expression.

In this section, we will learn how to simplify expressions with negative exponents as well as the zero exponent. We will also learn an additional property of exponents.

## Quotient of Like Bases

In Section 5.1, we learned how to simplify expressions that involve products of like bases, powers of powers, and products and quotients raised to exponents. In this section, we will learn one more property of exponents. This property deals with simplifying quotients that involve like bases. The following example will illustrate the rule that can be applied to this type of problem. Assume that the denominator is nonzero throughout this section.

<table>
<tr><th>Based on Definition</th><th>Observed Rule</th></tr>
<tr><td>

$$\frac{a^6}{a^4} = \frac{a \cdot a \cdot a \cdot a \cdot a \cdot a}{a \cdot a \cdot a \cdot a}$$

$$= \frac{\cancel{a} \cdot \cancel{a} \cdot \cancel{a} \cdot \cancel{a} \cdot a \cdot a}{\cancel{a} \cdot \cancel{a} \cdot \cancel{a} \cdot \cancel{a}}$$

$$= \frac{a \cdot a}{1}$$

$$= a^2$$

</td><td>

The same result is obtained by subtracting the exponents.

$$\frac{a^6}{a^4} = a^{6-4}$$

$$= a^2$$

</td></tr>
</table>

This example illustrates that, when exponential expressions with like bases are divided, the exponent of the quotient is the difference of the exponents in the numerator and denominator. This brings us to our last property of exponents.

---

**Property: The Quotient of Like Bases Rule**

For a real number $b$ ($b \neq 0$) and for positive integers $m$ and $n$ ($m > n$),

$$\frac{b^m}{b^n} = b^{m-n}$$

---

**Procedure: Simplifying a Quotient of Like Bases**

**Step 1:** Keep the base the same.
**Step 2:** Subtract the denominator's exponent from the numerator's exponent.
**Step 3:** Simplify the expression, if needed.

---

**Objective 1 Examples**    Simplify each expression by applying the quotient of like bases rule.

1a. $\dfrac{b^8}{b}$    1b. $\dfrac{3^5}{3^3}$    1c. $\dfrac{-4y^3}{2y^2}$    1d. $\dfrac{(-6)^9}{(-6)^4}$    1e. $\dfrac{(2x)^7}{(2x)^3}$

**Solutions**    1a.    $\dfrac{b^8}{b^1} = b^{8-1}$    Apply the quotient of like bases rule.

$= b^7$    Keep the base and subtract the exponents.

1b.    $\dfrac{3^5}{3^3} = 3^{5-3}$    Apply the quotient of like bases rule.

$= 3^2$    Keep the base and subtract the exponents.

$= 9$    Simplify.

1c.    $\dfrac{-4y^3}{2y^2} = \dfrac{-4}{2}y^{3-2}$    Divide the coefficients and apply the quotient of like bases rule.

$= -2y^1$    Keep the base and subtract the exponents.

$= -2y$    Simplify.

**1d.** $\dfrac{(-6)^9}{(-6)^4} = (-6)^{9-4}$    Apply the quotient of like bases rule.

$= (-6)^5$    Keep the base and subtract the exponents.

$= -7776$    Simplify.

**1e.** $\dfrac{(2x)^7}{(2x)^3} = (2x)^{7-3}$    Apply the quotient of like bases rule.

$= (2x)^4$    Keep the base and subtract the exponents.

$= 2^4 x^4$    Apply the power of a product rule.

$= 16x^4$    Simplify.

✔ **Student Check 1**    Simplify each expression by applying the quotient of like bases rule.

**a.** $\dfrac{y^{10}}{y}$    **b.** $\dfrac{7^8}{7^7}$    **c.** $\dfrac{-8x^6}{4x^2}$    **d.** $\dfrac{(-1)^{11}}{(-1)^5}$    **e.** $\dfrac{(9b)^7}{(9b)^4}$

## The Zero Exponent

**Objective 2** ▶

Simplify expressions with zero as an exponent.

Thus far, we have worked with exponents that are positive integers. We now investigate how to define an exponent of zero. We know that when a number is divided by itself, the result is 1. For example, $\dfrac{4}{4} = 1$. So, when we divide like exponential expressions, the result is 1.

**Based on Division Facts**

$$\dfrac{x^2}{x^2} = 1, \ (x \neq 0)$$

**Based on the Quotient Rule**

$$\dfrac{x^2}{x^2} = x^{2-2}$$

$$= x^0, \ (x \neq 0)$$

So, we have that $\dfrac{x^2}{x^2} = 1$ and also $\dfrac{x^2}{x^2} = x^0$. Therefore, it follows that

$$x^0 = 1$$

**INSTRUCTOR NOTE:**
Show students why $b \neq 0$. We know that $0^n = 0$ but the zero exponent rule would imply that $0^0 = 1$, which contradicts the previous statement.

> **Definition:  The Zero Exponent**
>
> For a nonzero real number $b$,
>
> $$b^0 = 1$$

This means that any number raised to the power of 0 is 1 as long as the number is not 0.

**Objective 2  Examples**    Use the definition of the zero exponent to simplify each expression. Assume $a$ is nonzero.

**2a.** $5^0$    **2b.** $(-3)^0$    **2c.** $-6^0$    **2d.** $(2a)^0$    **2e.** $2a^0$

**Solutions**    **2a.**    $5^0 = 1$

**2b.** $(-3)^0 = 1$

**2c.** This is the opposite of 6 raised to the power of 0.

$-6^0 = -(1)$

$= -1$

**2d.** $(2a)^0 = 1$

**2e.** This is two times $a$ raised to the power of 0.

$$2a^0 = 2(1)$$
$$= 2$$

---

✓ **Student Check 2**     Use the definition of the zero exponent to simplify each expression. Assume $x$ is nonzero.

   **a.** $4^0$     **b.** $(-20)^0$     **c.** $-9^0$     **d.** $(6x)^0$     **e.** $6x^0$

---

## Negative Exponents

**Objective 3** ▶

Simplify expressions with negative exponents.

Now that we know how to work with exponents that are whole numbers $\{0, 1, 2, \ldots\}$, we turn our attention to exponents that are negative integers.

What does it mean to raise a number to the exponent of $-3$? It is impossible to write a number as a factor "$-3$" times, so we need another way to look at this. The following illustration provides a definition of a negative exponent.

**Based on the Definition of an Exponent**

$$\frac{x}{x^4} = \frac{x}{x \cdot x \cdot x \cdot x}$$

$$= \frac{\overset{1}{\cancel{x}}}{\underset{1}{\cancel{x}} \cdot x \cdot x \cdot x}$$

$$= \frac{1}{x^3}$$

**Based on the Quotient Rule**

$$\frac{x}{x^4} = \frac{x^1}{x^4}$$

$$= x^{1-4}$$

$$= x^{-3}$$

So, we see that $\dfrac{x}{x^4} = \dfrac{1}{x^3}$ and $\dfrac{x}{x^4} = x^{-3}$. Therefore,

$$x^{-3} = \frac{1}{x^3}$$

A negative exponent is equivalent to the *reciprocal* of the base raised to the positive exponent. Recall that the reciprocal of $a$ is $\dfrac{1}{a}$.

---

**Definition: Negative Exponents**

If $b$ is a nonzero real number and $n$ is a positive integer, then

$$b^{-n} = \frac{1}{b^n}$$

---

**Procedure: Simplifying an Expression Raised to a Negative Exponent**

**Step 1:** Identify the base of the negative exponent.
**Step 2:** Rewrite the expression as 1 divided by the base to its positive exponent.
**Step 3:** Simplify the expression in the denominator.

| **Objective 3 Examples** | **Simplify each expression. Express answers with positive exponents.** |

**3a.** $4^{-2}$     **3b.** $(-5)^{-3}$     **3c.** $-3^{-4}$     **3d.** $\left(\dfrac{2}{3}\right)^{-5}$

**3e.** $(6x)^{-2}$     **3f.** $6x^{-2}$     **3g.** $\dfrac{x^{-3}}{y^{-4}}$

**Solutions**

**3a.**  $4^{-2} = \dfrac{1}{4^2}$     Apply the definition of the negative exponent with the base of 4.

$= \dfrac{1}{16}$     Simplify the denominator.

**3b.** $(-5)^{-3} = \dfrac{1}{(-5)^3}$     Apply the definition of the negative exponent with the base of $-5$.

$= \dfrac{1}{-125}$     Simplify the denominator.

$= -\dfrac{1}{125}$     Rewrite.

**3c.** Since there are no parentheses in this exponential expression, the base is 3.

$-3^{-4} = -\dfrac{1}{3^4}$     Apply the definition of the negative exponent with the base of 3.

$= -\dfrac{1}{81}$     Simplify the denominator.

**3d.** $\left(\dfrac{2}{3}\right)^{-5} = \dfrac{1}{\left(\dfrac{2}{3}\right)^5}$     Apply the definition of the negative exponent with the base of $\dfrac{2}{3}$.

$= \dfrac{1}{\dfrac{32}{243}}$     Apply the power of a quotient rule in the denominator.

$= 1 \cdot \dfrac{243}{32}$     Dividing by a fraction is the same as multiplying by its reciprocal.

$= \dfrac{243}{32}$     Simplify.

**3e.** $(6x)^{-2} = \dfrac{1}{(6x)^2}$     Apply the definition of the negative exponent with the base of $6x$.

$= \dfrac{1}{6^2 x^2}$     Apply the power of a product rule.

$= \dfrac{1}{36x^2}$     Simplify the denominator.

**3f.** Note that 6 is the coefficient of the exponential expression; it is not part of the base.

$6x^{-2} = 6 \cdot \dfrac{1}{x^2}$     Apply the definition of the negative exponent with the base $x$.

$= \dfrac{6}{x^2}$     Multiply.

**3g.** $\dfrac{x^{-3}}{y^{-4}} = \dfrac{\frac{1}{x^3}}{\frac{1}{y^4}}$    Apply the definition of the negative exponent with the base $x$ in the numerator and with the base $y$ in the denominator.

$= \dfrac{1}{x^3} \cdot \dfrac{y^4}{1}$    Rewrite as a product of the reciprocal of $\dfrac{1}{y^4}$, which is $y^4$.

$= \dfrac{y^4}{x^3}$    Multiply.

---

✓ **Student Check 3**    Simplify each expression. Express answers with positive exponents.

**a.** $7^{-3}$    **b.** $(-2)^{-4}$    **c.** $-5^{-2}$    **d.** $\left(\dfrac{1}{4}\right)^{-1}$    **e.** $(3y)^{-3}$    **f.** $3y^{-3}$    **g.** $\dfrac{a^{-5}}{b^{-2}}$

---

### Important Notes

1. Example 3(d) shows an important property when dealing with negative exponents. It illustrates that when a fraction is raised to a negative exponent, it is equivalent to raising the reciprocal of the fraction to the positive exponent.

   For example, $\left(\dfrac{2}{3}\right)^{-5} = \left(\dfrac{3}{2}\right)^{5}$. In general, we have the following rule.

   **Property:  Negative Powers of Fractions**

   If $a$ and $b$ are nonzero real numbers and $n$ is a positive integer, then

   $$\left(\dfrac{a}{b}\right)^{-n} = \left(\dfrac{b}{a}\right)^{n}$$

2. The base with a negative exponent moves from the denominator to the numerator or from the numerator to the denominator for the exponent to become positive.

   Example 3(g) shows that $\dfrac{x^{-3}}{y^{-4}} = \dfrac{y^4}{x^3}$. In general, we have the following rule.

   **Property:** If $b$ is a nonzero real number and $n$ is a positive integer, then

   $$\dfrac{1}{b^{-n}} = b^n$$

### Simplifying Exponential Expressions

**Objective 4** ▶

Apply a combination of exponent properties.

Now that we have defined exponents for all integers, the properties that have been defined apply for all integers, not just whole numbers. A summary of the rules follows.

| Product of like bases | $b^m \cdot b^n = b^{m+n}$ |
|---|---|
| Power of a power | $(b^m)^n = b^{mn}$ |
| Power of a product | $(ab)^n = a^n b^n$ |
| Power of a quotient | $\left(\dfrac{a}{b}\right)^n = \dfrac{a^n}{b^n},\ b \neq 0$ |
| Quotient of like bases | $\dfrac{b^m}{b^n} = b^{m-n},\ b \neq 0$ |
| Zero exponent | $b^0 = 1,\ b \neq 0$ |
| Negative exponent | $b^{-n} = \dfrac{1}{b^n},\ b \neq 0$ |

**Note:** *As a general rule, we should apply the other properties before rewriting negative exponents since applying the property may change the sign of the exponent.*

**Objective 4 Examples**   Simplify each expression. Write the result with positive exponents.

**4a.** $x^{-5}x^3$ 　　　　**4b.** $(y^{-3})^{-2}$ 　　　　**4c.** $(2b^{-3})^{-1}$

**4d.** $\left(\dfrac{3a}{p^3}\right)^{-4}$ 　　　**4e.** $\dfrac{5^{-2}}{5^{-4}}$ 　　　**4f.** $\left(\dfrac{12x^3y^{-2}}{2x^{-1}y^3}\right)^{-2}$

**Solutions**  **4a.**   $x^{-5}x^3 = x^{-5+3}$ 　　　Apply the product of like bases rule.

$\qquad\qquad = x^{-2}$ 　　　Add the exponents.

$\qquad\qquad = \dfrac{1}{x^2}$ 　　　Apply the definition of the negative exponent.

**4b.**   $(y^{-3})^{-2} = y^{(-3)(-2)}$ 　　　Apply the power of a power rule.

$\qquad\qquad = y^6$ 　　　Multiply the exponents.

**4c.**  $(2b^{-3})^{-1} = (2)^{-1}(b^{-3})^{-1}$ 　　　Apply the power of a product rule.

$\qquad\qquad = (2)^{-1} \cdot b^3$ 　　　Apply the power of a power rule.

$\qquad\qquad = \dfrac{1}{2} \cdot \dfrac{b^3}{1}$ 　　　Apply the definition of the negative exponent.

$\qquad\qquad = \dfrac{b^3}{2}$ 　　　Multiply.

**4d.**  $\left(\dfrac{3a}{p^3}\right)^{-4} = \dfrac{(3a)^{-4}}{(p^3)^{-4}}$ 　　　Apply the power of a quotient rule.

$\qquad\qquad = \dfrac{(3)^{-4}(a)^{-4}}{p^{-12}}$ 　　　Apply the power of a product rule in the numerator. Apply the power of a power rule in the denominator.

$\qquad\qquad = \dfrac{p^{12}}{3^4 a^4}$ 　　　Apply the definition of the negative exponent.

$\qquad\qquad = \dfrac{p^{12}}{81 a^4}$ 　　　Rewrite $3^4$ as 81.

We can also simplify this expression by applying the negative powers of fractions rule.

$\left(\dfrac{3a}{p^3}\right)^{-4} = \left(\dfrac{p^3}{3a}\right)^4$ 　　　Apply the negative powers of fractions rule.

$\qquad\qquad = \dfrac{(p^3)^4}{(3a)^4}$ 　　　Apply the power of a quotient rule.

$\qquad\qquad = \dfrac{p^{12}}{81 a^4}$ 　　　Apply the power of a power rule in the numerator. Apply the power of a product rule in the denominator.

**4e.**　　$\dfrac{5^{-2}}{5^{-4}} = 5^{-2-(-4)}$ 　　　Apply the quotient of like bases rule.

$\qquad\qquad = 5^{-2+4}$ 　　　Subtract the exponents by adding the opposite.

$\qquad\qquad = 5^2$ 　　　Simplify the exponent.

$\qquad\qquad = 25$ 　　　Simplify the exponential expression.

**4f.** Before applying the exponent of $-2$, we simplify inside the parentheses first.

$$\left(\frac{12x^3y^{-2}}{2x^{-1}y^3}\right)^{-2} = (6x^{3-(-1)}y^{-2-3})^{-2}$$  Divide the coefficients and apply the quotient of like bases rule.

$$= (6x^4y^{-5})^{-2}$$  Simplify.

$$= 6^{-2}(x^4)^{-2}(y^{-5})^{-2}$$  Apply the power of a product rule.

$$= 6^{-2}x^{-8}y^{10}$$  Simplify.

$$= \frac{1}{6^2} \cdot \frac{1}{x^8} \cdot \frac{y^{10}}{1}$$  Rewrite the negative exponents.

$$= \frac{y^{10}}{36x^8}$$  Rewrite $6^2$ as 36 and multiply the fractions.

 **Student Check 4**  Simplify each expression. Write the result with positive exponents.

**a.** $y^7y^{-4}$  **b.** $(x^{-8})^{-3}$  **c.** $(3a^{-1})^{-5}$

**d.** $\left(\frac{8b}{a^6}\right)^{-2}$  **e.** $\frac{4^{-3}}{4^{-6}}$  **f.** $\left(\frac{18a^{-4}y^3}{6a^2y^{-1}}\right)^{-4}$

## Applications

**Objective 5** ▶

Solve application problems.

There are some important formulas that involve negative exponents. Two that we will examine deal with the decay of a radioactive substance and financing a loan. At this point, we will evaluate a given formula for some specific criteria.

**Objective 5 Examples**  Solve each problem.

**5a.** The half-life of a radioactive substance is the time it takes for half of a certain amount of the radioactive substance to decay. The formula

$$P = P_0 \cdot 2^{-t/m}$$

gives the amount of radioactive material present at time $t$, where $P_0$ is the original amount of the material and $m$ is the half-life of the material.

The radioactive isotope sodium-24 is used in medical and nonmedical applications. It can be used to detect circulatory problems in patients as well as used to detect leaks in oil pipe lines. The half-life of sodium-24 is 15 hr. If a hospital has 20 g of sodium-24, how many grams will be present after 15 hr? 60 hr? (Sources: http://www.chemistryexplained.com/elements/P-T/Sodium.html and http://www.3rd1000.com/nuclear/halflife.htm)

**Solution**  **5a.** We are given $P_0 = 20$ and $m = 15$. To determine how many grams will be present in 15 hr, we replace $t$ with 15.

$$P = P_0 \cdot 2^{-t/m}$$  State the formula.

$$P = 20 \cdot 2^{-15/15}$$  Replace $P_0$ with 20, $m$ with 15, and $t$ with 15.

$$P = 20 \cdot 2^{-1}$$  Simplify the exponent.

$$P = 20 \cdot \frac{1}{2}$$  Apply the negative exponent definition.

$$P = 10$$  Multiply.

To determine how many grams will be present in 60 hr, we replace $t$ with 60.

$$P = P_0 \cdot 2^{-t/m}$$  State the formula.

$$P = 20 \cdot 2^{-60/15}$$  Replace $P_0$ with 20, $m$ with 15, and $t$ with 60.

$$P = 20 \cdot 2^{-4}$$   Simplify the exponent.

$$P = 20 \cdot \frac{1}{2^4}$$   Apply the negative exponent definition.

$$P = 20 \cdot \frac{1}{16}$$   Rewrite $2^4$ as 16.

$$P = \frac{20}{16} \quad \text{or} \quad 1.25$$   Multiply.

So, there will be 10 g of sodium-24 in 15 hr and 1.25 g in 60 hr.

**5b.** The formula to find the monthly payment to repay a loan is $P = \dfrac{A\left(\dfrac{r}{12}\right)}{1 - \left(1 + \dfrac{r}{12}\right)^{-m}}$,

where $A$ is the amount of the loan, $r$ is the annual interest rate (in decimal form), and $m$ is the number of monthly payments. Brianna and John finance a $25,000 car loan at 6% annual interest for 6 yr.

 **i.** What is the monthly payment to repay the loan?

 **ii.** How much will they pay for the car at the end of the loan?

**Solution**   **5b.** We are given $A = 25{,}000$, $r = 6\%$ or 0.06, and $m = 12(6) = 72$ (12 payments per year times 6 yr).

**i.**
$$P = \frac{A\left(\dfrac{r}{12}\right)}{1 - \left(1 + \dfrac{r}{12}\right)^{-m}}$$   State the formula.

$$P = \frac{25{,}000\left(\dfrac{0.06}{12}\right)}{1 - \left(1 + \dfrac{0.06}{12}\right)^{-72}}$$   Replace $A$ with 25,000, $r$ with 0.06, and $m$ with 72.

$$= \frac{25{,}000(0.005)}{1 - (1.005)^{-72}}$$   Simplify inside parentheses first.

$$= \frac{125}{1 - 0.6983}$$   Simplify the numerator. Simplify the exponent in the denominator.

$$= \frac{125}{0.3017}$$   Simplify the denominator.

$$= \$414.32$$   Divide.

The monthly payment to repay the loan is $414.32.

**INSTRUCTOR NOTE:**
Explain to students that the difference in the total amount paid and the original loan amount is the interest paid.

 **ii.** The total amount paid for the car is the monthly payment times the number of payments, which is

$$(\$414.32)(72) = \$29{,}831.04$$

✓ **Student Check 5**    Solve each problem.

**a.** The radioactive isotope gallium-67 can be used to locate sites of infection in a patient, for tumor imaging, and for chemotherapy for pediatric patients. Its half-life is approximately 3 days. If an imaging center has 400 MBq of gallium-67, how much will be present after 3 days? after 15 days? (Note: MBq stands for millibecquerel and is a unit used for radiation.) (Source: http://www.radiochemistry.org/nuclearmedicine/radioisotopes/ex_iso_medicine.htm)

**b.** Charles finances a $35,000 car loan.

**i.** What is the monthly payment for the car loan if it is financed for 7 yr at 4%?

**ii.** How much will Charles have paid for the car at the end of the loan?

**Objective 6** ▶

Troubleshoot common errors.

## Troubleshooting Common Errors

Some common errors associated with the quotient rule and zero and negative exponents are shown next.

**Objective 6  Examples**    A problem and an incorrect solution are given. Provide the correct solution and an explanation of the error.

**6a.** Simplify $\dfrac{b^{12}}{b^4}$.

| Incorrect Solution | Correct Solution and Explanation |
|---|---|
| $\dfrac{b^{12}}{b^4} = b^3$ | When dividing like bases, we subtract the exponents, not divide them.<br><br>$\dfrac{b^{12}}{b^4} = b^{12-4} = b^8$ |

**6b.** Evaluate $2^{-3}$.

| Incorrect Solution | Correct Solution and Explanation |
|---|---|
| $2^{-3} = -8$ | A negative exponent is equivalent to the reciprocal of the base raised to the positive exponent. It does not make the result negative.<br><br>$2^{-3} = \dfrac{1}{2^3} = \dfrac{1}{8}$ |

**6c.** Simplify $\dfrac{x^5}{x^{-2}}$.

| Incorrect Solution | Correct Solution and Explanation |
|---|---|
| <br>$\dfrac{x^5}{x^{-2}} = x^{5-2} = x^3$ | When we divide like bases, we subtract their exponents. However, the exponent in the denominator is negative. When we subtract a negative number, it becomes positive.<br><br>$\dfrac{x^5}{x^{-2}} = x^{5-(-2)}$<br>$= x^{5+2}$<br>$= x^7$ |

## ANSWERS TO STUDENT CHECKS

**Student Check 1**   **a.** $y^9$   **b.** 7   **c.** $-2x^4$   **d.** 1   **e.** $729b^3$

**Student Check 2**   **a.** 1   **b.** 1   **c.** $-1$   **d.** 1   **e.** 6

**Student Check 3**   **a.** $\dfrac{1}{343}$   **b.** $\dfrac{1}{16}$   **c.** $-\dfrac{1}{25}$   **d.** 4

   **e.** $\dfrac{1}{27y^3}$   **f.** $\dfrac{3}{y^3}$   **g.** $\dfrac{b^2}{a^5}$

**Student Check 4**   **a.** $y^3$   **b.** $x^{24}$   **c.** $\dfrac{a^5}{243}$   **d.** $\dfrac{a^{12}}{64b^2}$

   **e.** 64   **f.** $\dfrac{a^{24}}{81y^{16}}$

**Student Check 5**   **a.** 200 MBq after 3 days, 12.5 MBq after 15 days

   **b.**   **i.** Payments are $478.41 each.

      **ii.** The total amount paid for the car is $40,186.44.

## SUMMARY OF KEY CONCEPTS

1. The quotient rule applies when the bases are the same. To divide like bases, subtract their exponents. If the result contains a negative exponent, rewrite the expression with a positive exponent.

2. Any base (except 0) raised to the exponent of 0 is 1. Be careful with negative signs. If the base is a negative number, then the result is 1. If the negative sign is not in parentheses, then the result is $-1$.

$$(-b)^0 = 1 \text{ but } -b^0 = -1$$

3. The expression $b^{-n}$ is the reciprocal of $b^n$. That is,

$$b^{-n} = \frac{1}{b^n}.$$

4. The properties of exponents can be applied to zero, positive integer, and negative integer exponents.

In general, apply the properties before rewriting negative exponents.

| | |
|---|---|
| **Product of like bases** | $b^m \cdot b^n = b^{m+n}$ |
| **Power of a power** | $(b^m)^n = b^{mn}$ |
| **Power of a product** | $(ab)^n = a^n b^n$ |
| **Power of a quotient** | $\left(\dfrac{a}{b}\right)^n = \dfrac{a^n}{b^n},\ b \neq 0$ |
| **Quotient of like bases** | $\dfrac{b^m}{b^n} = b^{m-n},\ b \neq 0$ |
| **Zero exponent** | $b^0 = 1,\ b \neq 0$ |
| **Negative exponent** | $b^{-n} = \dfrac{1}{b^n},\ b \neq 0$ |

## GRAPHING CALCULATOR SKILLS

The graphing calculator skills for this section involve working with negative exponents and evaluating formulas.

**Example 1:** Simplify $7^{-3}$ and express the result as a fraction.

**Solution:**

```
7^-3
       .0029154519
Ans▶Frac
            1/343
```

**Example 2:** Simplify the expression $\dfrac{25{,}000\left(\dfrac{0.06}{12}\right)}{1-\left(1+\dfrac{0.06}{12}\right)^{-72}}$.

**Solution:** Parentheses play an important role in this example. Parentheses must be inserted around the numerator and denominator of the fraction. Because the numerator involves a product, we can enter it as 25000 * 0.06/12 instead of using another set of parentheses. Be sure to begin the denominator with an open parenthesis.

```
(25000*.06/12)/(
1-(1+.06/12)^-72
)
         414.3221973
```

## SECTION 5.2 / EXERCISE SET

### Write About It!

**Use complete sentences in the answer to each exercise.**

1. Explain the quotient of like bases rule for exponents.
   Answers vary.
2. Explain how to simplify expressions with negative exponents.    Answers vary.
3. Why does a number raised to the zero exponent equal 1, not 0?    Answers vary.
4. If a positive number is raised to a negative exponent, will the simplified answer be a positive or negative number? Explain.    Answers vary.

### Practice Makes Perfect!

**Simplify each expression by applying the quotient of like bases rule for exponents. (*See Objective 1.*)**

5. $\dfrac{b^3}{b}$    $b^2$

6. $\dfrac{a^6}{a}$    $a^5$

7. $\dfrac{a^{10}}{a^5}$    $a^5$

8. $\dfrac{b^{12}}{b^4}$    $b^8$

9. $\dfrac{6^7}{6^4}$    $216$

10. $\dfrac{3^{12}}{3^9}$    $27$

11. $\dfrac{7^8}{7^6}$    $49$

12. $\dfrac{10^{20}}{10^{12}}$    $10^8$

13. $\dfrac{-6y^5}{3y^2}$    $-2y^3$

14. $\dfrac{-10y^6}{2y^4}$    $-5y^2$

15. $\dfrac{4x^{14}}{x^4}$    $4x^{10}$

16. $\dfrac{3x^{11}}{x^3}$    $3x^8$

17. $\dfrac{a^6}{3a^4}$    $\dfrac{a^2}{3}$

18. $\dfrac{a^8}{4a^5}$    $\dfrac{a^3}{4}$

19. $\dfrac{(-12)^9}{(-12)^7}$    $144$

20. $\dfrac{(-5)^{12}}{(-5)^{10}}$    $25$

21. $\dfrac{(-7)^{10}}{(-7)^4}$    $7^6$

22. $\dfrac{(-4)^{16}}{(-4)^7}$    $-4^9$

23. $\dfrac{(3p)^9}{(3p)^6}$    $27p^3$

24. $\dfrac{(5q)^{11}}{(5q)^7}$    $625q^4$

25. $\dfrac{(2b)^3}{(2b)^2}$    $2b$

26. $\dfrac{(4b)^6}{(4b)^2}$    $256b^4$

**Use the definition of the zero exponent to simplify each expression. (*See Objective 2.*)**

27. $3^0$    $1$

28. $6^0$    $1$

29. $(-2)^0$    $1$

30. $(-5)^0$    $1$

31. $-4^0$    $-1$

32. $-8^0$    $-1$

33. $x^0$    $1$

34. $z^0$    $1$

35. $(4a)^0$    $1$

36. $(3a)^0$    $1$

37. $6b^0$    $6$

38. $8b^0$    $8$

**Simplify each expression. Write answers with positive exponents. (*See Objective 3.*)**

39. $5^{-2}$    $\dfrac{1}{25}$

40. $4^{-3}$    $\dfrac{1}{64}$

41. $(-3)^{-3}$    $-\dfrac{1}{27}$

42. $(-4)^{-4}$    $\dfrac{1}{256}$

43. $(-10)^{-3}$    $-\dfrac{1}{1000}$

44. $(-11)^{-2}$    $\dfrac{1}{121}$

45. $-6^{-4}$    $-\dfrac{1}{1296}$

46. $-7^{-5}$    $-\dfrac{1}{16,807}$

47. $\left(\dfrac{3}{5}\right)^{-4}$    $\dfrac{625}{81}$

48. $\left(\dfrac{3}{4}\right)^{-5}$    $\dfrac{1024}{243}$

49. $\left(\dfrac{1}{2}\right)^{-5}$    $32$

50. $\left(\dfrac{1}{3}\right)^{-3}$    $27$

51. $(2x)^{-4}$    $\dfrac{1}{16x^4}$

52. $(3x)^{-3}$    $\dfrac{1}{27x^3}$

53. $(-5x)^{-3}$    $-\dfrac{1}{125x^3}$

54. $(-2x)^{-5}$    $-\dfrac{1}{32x^5}$

55. $2x^{-4}$    $\dfrac{2}{x^4}$

56. $3x^{-3}$    $\dfrac{3}{x^3}$

57. $\dfrac{a^{-2}}{b^{-5}}$    $\dfrac{b^5}{a^2}$

58. $\dfrac{a^{-6}}{b^{-7}}$    $\dfrac{b^7}{a^6}$

59. $\dfrac{x}{y^{-4}}$    $xy^4$

60. $\dfrac{x^3}{y^{-2}}$    $x^3y^2$

**Simplify each expression. Write answers with positive exponents. (*See Objective 4.*)**

61. $a^8 \cdot a^{-5}$    $a^3$

62. $b^{12} \cdot b^{-1}$    $b^{11}$

63. $(x^{-4})^{-3}$    $x^{12}$

64. $(y^{-4})^{-6}$    $y^{24}$

65. $(2p^{-1})^{-3}$    $\dfrac{p^3}{8}$

66. $(5q^{-2})^{-3}$    $\dfrac{q^6}{125}$

67. $(-3r^{-1})^{-4}$    $\dfrac{r^4}{81}$

68. $(-2s^{-3})^{-5}$    $-\dfrac{s^{15}}{32}$

69. $\left(\dfrac{9x^{-4}}{y^2}\right)^{-2}$    $\dfrac{x^8y^4}{81}$

70. $\left(\dfrac{6p^{-2}}{q^3}\right)^{-4}$    $\dfrac{p^8q^{12}}{1296}$

71. $\dfrac{3^{-5}}{3^{-9}}$    $81$

72. $\dfrac{5^{-13}}{5^{-17}}$    $625$

73. $\dfrac{6^{-8}}{6^{-5}}$    $\dfrac{1}{216}$

74. $\dfrac{2^{-13}}{2^{-7}}$    $\dfrac{1}{64}$

75. $\left(\dfrac{15c^{-6}d}{3c^{-4}d^{-3}}\right)^{-3}$    $\dfrac{c^6}{125d^{12}}$

76. $\left(\dfrac{24xy^{-8}}{8x^{-2}y^{-5}}\right)^{-4}$    $\dfrac{y^{12}}{81x^{12}}$

77. $\left(-\dfrac{2p^{-7}q^{-5}}{3pq^{-2}}\right)^{-5}$    $-\dfrac{243p^{40}q^{15}}{32}$

78. $\left(-\dfrac{5xy^{-4}}{4x^{-5}y^{-7}}\right)^{-3}$    $-\dfrac{64}{125x^{18}y^9}$

79. $2^{-2} + 3^{-1}$    $\dfrac{7}{12}$

80. $4^{-1} + 6^{-1}$    $\dfrac{5}{12}$

81. $3^0 - 6^{-1}$    $\dfrac{5}{6}$

82. $5^0 + 2^{-3}$    $\dfrac{9}{8}$

83. $6^{-1} - 5^{-2}$    $\dfrac{19}{150}$

84. $2^{-2} - 3^{-1}$    $-\dfrac{1}{12}$

**Solve each problem. Round answers to two decimal places when needed. (*See Objective 5.*)**

85. Suppose $P = 100 \cdot 2^{-t}$ represents the amount, in grams, of a radioactive substance that remains after $t$ months, where the substance has a half-life of 1 month.
    a. How much substance is there originally?    100 g
    b. How much substance remains after 1 month?    50 g
    c. How much substance remains after 6 months?    1.56 g
    d. How much substance remains after 12 months?    0.02 g

86. Suppose $P = 50 \cdot 2^{-t}$ represents the amount, in grams, of a radioactive substance that remains after $t$ years, where the half-life of the substance is 1 yr.
    a. How much of the substance is there originally?    50 g
    b. How much substance remains after 1 yr?    25 g
    c. How much substance remains after 5 yr?    1.56 g
    d. How much substance remains after 10 yr?    0.05 g

Additional answers can be found in the Instructor Answer Appendix.

**For Exercises 87 and 88, the formula to find the monthly payment to repay a loan is** $P = \dfrac{A\left(\dfrac{r}{12}\right)}{1 - \left(1 + \dfrac{r}{12}\right)^{-m}}$**, where** $A$ **is the amount of the loan,** $r$ **is the interest rate (in decimal form), and** $m$ **is the number of monthly payments.**

87. Find the monthly payment needed to repay a loan of $12,000 at 8% interest for 5 yr.  $243.32

88. Find the monthly payment needed to repay a loan of $5000 at 7% interest for 4 yr.  $119.73

 **Mix 'Em Up!**

**Use the exponent rules to simplify each expression. Write answers with positive exponents.**

89. $x^{-6}x^4$   $\dfrac{1}{x^2}$

90. $x^{-10}x^5$   $\dfrac{1}{x^5}$

91. $(x^{-6})^{-2}$   $x^{12}$

92. $(x^{-5})^{-2}$   $x^{10}$

93. $\left(\dfrac{2a}{p^2}\right)^{-5}$   $\dfrac{p^{10}}{32a^5}$

94. $\left(\dfrac{4a}{p^4}\right)^{-3}$   $\dfrac{p^{12}}{64a^3}$

95. $(2y^{-5})^2\,(3y^4)$   $\dfrac{12}{y^6}$

96. $(3y^{-2})^3(5y^5)$   $\dfrac{135}{y}$

97. $(-2y^{10})(3y^{-3})$   $-6y^7$

98. $(-4y^8)(2y^{-6})$   $-8y^2$

99. $\left(-\dfrac{3a^0}{2b^{-2}}\right)^{-5}$   $-\dfrac{32}{243b^{10}}$

100. $\left(-\dfrac{5x^{-2}}{4y^0}\right)^{-3}$   $-\dfrac{64x^6}{125}$

101. $\dfrac{3^{-2}}{3^{-6}}$   $81$

102. $\dfrac{2^{-5}}{2^{-7}}$   $4$

103. $[(3a)^{-3}]^{-4}$   $531{,}441a^{12}$

104. $[(5a)^{-2}]^{-3}$   $15{,}625a^6$

105. $\dfrac{18x^{-6}}{3x^{-10}}$   $6x^4$

106. $\dfrac{5x^{-12}}{10x^{-15}}$   $\dfrac{x^3}{2}$

107. $\left(\dfrac{6b}{q^0}\right)^{-3}$   $\dfrac{1}{216b^3}$

108. $\left(\dfrac{10b^0}{q^2}\right)^{-4}$   $\dfrac{q^8}{10{,}000}$

109. $\dfrac{(-2a^3)^2}{a^{-4}}$   $4a^{10}$

110. $\dfrac{(-5b^4)^3}{b^{-11}}$   $-125b^{23}$

111. $5^{-1} + 3^{-2}$   $\dfrac{14}{45}$

112. $4^0 - 6^{-2}$   $\dfrac{35}{36}$

113. $\left(-\dfrac{x^{-5}}{6y^{-3}}\right)^{-2}$   $\dfrac{36x^{10}}{y^6}$

114. $\left(-\dfrac{7a^{-5}}{b^{-1}}\right)^{-4}$   $\dfrac{a^{20}}{2401b^4}$

**For Exercises 115 and 116, the formula**

$$P = P_0 \cdot 2^{-t/m}$$

**gives the amount of radioactive substance present at time** $t$**, where** $P_0$ **is the original amount of the substance and** $m$ **is the half-life of the substance.**

115. The half-life of a substance is 3 days. If the original amount of the substance is 12 g, how much substance remains after
    **a.** 6 days?  3 g  **b.** 12 days?  0.75 g  **c.** 18 days?  0.1875 g

116. The half-life of a substance is 10 hr. If the original amount of the substance is 24 g, how much substance remains after
    **a.** 10 hr?  12 g  **b.** 30 hr?  3 g  **c.** 50 hr?  0.75 g

**For Exercises 117 and 118, the formula to find the monthly payment to repay a loan is** $P = \dfrac{A\left(\dfrac{r}{12}\right)}{1 - \left(1 + \dfrac{r}{12}\right)^{-m}}$**, where** $A$ **is the amount of the loan,** $r$ **is the interest rate (in decimal form), and** $m$ **is the number of monthly payments.**

117. Find the monthly payment to repay a student loan of $20,000 at 5.4% interest for 10 yr.  $216.06

118. Find the monthly payment to repay a car loan of $2000 at 4% interest for 3 yr.  $59.05

 **You Be the Teacher!**

**Correct each student's errors, if any.**

119. Simplify $\dfrac{(3a)^4}{(3a)^3}$.

    Bethenny's work:   Correct

    $$\dfrac{(3a)^4}{(3a)^3} = \dfrac{81a^4}{27a^3} = \dfrac{\overset{1}{27} \cdot 3a^{4-3}}{\underset{1}{27}} = 3a^1 = 3a$$

120. Simplify: $\dfrac{a^{-3}}{a^{-7}}$.

    Luann's work:

    $$\dfrac{a^{-3}}{a^{-7}} = a^{-3-7} = a^{-10} = \dfrac{1}{a^{10}}.$$

    We must subtract the exponents. When we subtract a negative number, it becomes positive. We should get
    $\dfrac{a^{-3}}{a^{-7}} = a^{-3-(-7)} = a^{-3+7} = a^4$

**Calculate It!**

**Use a calculator for a each problem.**

121. Simplify $\left(\dfrac{3}{5}\right)^{-4}$ and express the answer as a fraction.  $\dfrac{625}{81}$

122. Simplify $\left(\dfrac{2^3}{7^0}\right)^{-2}$ and express the answer as a fraction.  $\dfrac{1}{64}$

123. Use the half-life formula given for Exercises 115 and 116. Suppose the half-life of a substance is 64 yr. If the original amount of the substance is 100 g, how much substance remains after
    **a.** 32 yr?  70.7 g  **b.** 256 yr?  6.25 g  **c.** 640 yr? 0.098 g

124. Use the monthly payment formula given for Exercises 117 and 118. Find the monthly payment to repay a car loan of $11,250 at 4.25% interest for 6 yr.  $177.29

## Think About It!

For exercises 125–128, use the properties of exponents to simplify each expression.

**125.** $\dfrac{5^{2a}}{5^{a-1}}$    $5^{a+1}$

**126.** $\dfrac{4^{x+3}}{4^{x+4}}$    $\dfrac{1}{4}$

**127.** $\dfrac{(x^{3a})^2}{x^{a-6}}$    $x^{5a+6}$

**128.** $\dfrac{(2^{4x})^3 \cdot 2^{x-1}}{2^{x+3}}$    $2^{12x-4}$

**129.** What expression needs to be squared to obtain $49a^2b^{10}$? Check your result.    $7ab^5;\ (7ab^5)^2 = 49a^2b^{10}\ (-7ab^5$ is also acceptable.)

**130.** What expression needs to be squared to obtain $144x^4y^6$? Check your result.    $12x^2y^3;\ (12x^2y^3)^2 = 144x^4y^6\ (-12x^2y^3$ is also acceptable.)

**131.** What expression needs to be cubed to obtain $-\dfrac{27r^3}{s^6}$? Check your result.    $-\dfrac{3r}{s^2};\ \left(-\dfrac{3r}{s^2}\right)^3 = -\dfrac{27r^3}{s^6}$

**132.** What expression needs to be cubed to obtain $\dfrac{125x^{12}}{y^9}$? Check your result.    $\dfrac{5x^4}{y^3};\ \left(\dfrac{5x^4}{y^3}\right)^3 = \dfrac{125x^{12}}{y^9}$

## SECTION 5.3    Scientific Notation

### OBJECTIVES

As a result of completing this section, you will be able to

1. Write numbers in scientific notation.
2. Convert numbers from scientific notation to standard notation.
3. Perform operations with numbers in scientific notation.
4. Solve application problems using scientific notation.
5. Troubleshoot common errors.

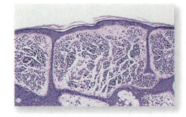

Information stored on computers is measured in bytes.

- It takes 1 byte to store a single character.
- A typewritten page takes 2000 bytes or 2 kilobytes (KB) to store.
- A floor of academic journals takes 100,000,000,000 bytes or 100 gigabytes (GB) to store.
- An academic research library takes 2,000,000,000,000 bytes or 2 terabytes (TB) to store.
- The printed collection of the U.S. Library of Congress takes 10,000,000,000,000 bytes or 10 terabytes (TB) to store. (Source: http://www2.sims.berkeley.edu/research /projects/how-much-info/datapowers.html)

Numbers that represent storage of data can get very large. Very small numbers can be used to represent things dealing with the body and things in nature. Some examples are shown.

- The thickness of human skin is 0.07 in.
- The diameter of a skin cell is 0.001181 in.
- The diameter of a red blood cell 0.000315 in.
- The diameter of an influenza virus is 0.00000472 in.
- The diameter of a water molecule is 0.0000000108 in. (Source: http://learn.genetics .utah.edu/content/begin/cells/scale/)

Very large and very small numbers are used frequently in many fields of mathematics and science. It can be very tedious to work with these numbers in standard decimal form. It is helpful to convert them to another form, called *scientific notation*, which is shorthand notation for writing extremely large and extremely small numbers. This section will discuss how to convert a number to scientific notation.

### Writing Numbers in Scientific Notation

**Objective 1** ▶

Write numbers in scientific notation.

Every digit in a number represents a power of 10. Therefore, every number can be written as some number times a power of 10. Some examples are

$$30{,}000 = 3 \times 10{,}000 = 3 \times 10^4$$
$$3000 = 3 \times 1000 = 3 \times 10^3$$
$$300 = 3 \times 100 = 3 \times 10^2$$
$$30 = 3 \times 10 = 3 \times 10^1$$
$$3 = 3 \times 1 = 3 \times 10^0$$
$$0.3 = 3 \times 0.1 = 3 \times \frac{1}{10} = 3 \times 10^{-1}$$
$$0.03 = 3 \times 0.01 = 3 \times \frac{1}{100} = 3 \times 10^{-2}$$

In these examples, the exponent of 10 represents the number of zeros in the number if the number is 10 or more. For positive values less than 1, the absolute value of the exponent of 10 represents the number of decimal places in the number. Also note the exponent of 10 is negative for positive values less than 1. This leads to the definition of scientific notation.

---

**Definition: Scientific Notation**

A number written in the form $a \times 10^n$ is in scientific notation where $1 \le a < 10$ and $n$ is an integer. The number $a$ is called the **coefficient**.

---

**Procedure: Writing a Number in Scientific Notation**

**Step 1:** Move the original decimal point to the left or right until a number between 1 and 10 is obtained. This number is the coefficient.

**Step 2:** The exponent of 10 is the number of decimal places moved in step 1. If the decimal point is moved left, the exponent of 10 is positive. If the decimal point is moved right, the exponent of 10 is negative.

**Step 3:** Multiply the coefficient obtained in step 1 by 10 raised to the exponent found in step 2.

---

**Objective 1 Examples** Write each number in scientific notation.

**1a.** The size of the printed collection of the U.S. Library of Congress is 10,000,000,000,000 bytes. (Source: http://www2.sims.berkeley.edu/research/projects/how-much-info/datapowers.html)

**1b.** The size of an average red blood cell is about 0.000315 in.

**Solutions**   **1a.** Move the decimal point until the number is between 1 and 10. So, the coefficient is 1.0.

$$1.0,000,000,000,000.0$$

The decimal point is moved 13 places to the left, so the exponent of 10 is positive.

$$10,000,000,000,000 = 1.0 \times 10^{13}$$

**1b.** Move the decimal point until the number is between 1 and 10. So, the coefficient is 3.15.

$$0.0003.15$$

The decimal point is moved 4 places to the right, so the exponent of 10 is negative.

$$0.000315 = 3.15 \times 10^{-4}$$

---

**✓ Student Check 1**   Write each number in scientific notation.

**a.** The world population is approximately 7,000,000,000.

**b.** The size of a hydrogen atom is about 0.00000005 mm in diameter.

## Writing Numbers in Standard Notation

**Objective 2** ▶

Convert numbers from scientific notation to standard notation.

To convert numbers from scientific notation to standard notation, we must understand what happens when we multiply a number by a power of 10. Consider the following examples.

$$
\begin{aligned}
1.234 \times 10^4 &= 1.234 \times 10{,}000 = 12{,}340 \\
1.234 \times 10^3 &= 1.234 \times \phantom{0}1000 = \phantom{00}1234 \\
1.234 \times 10^2 &= 1.234 \times \phantom{00}100 = \phantom{00}123.4 \\
1.234 \times 10^1 &= 1.234 \times \phantom{000}10 = \phantom{000}12.34 \\
1.234 \times 10^0 &= 1.234 \times \phantom{0000}1 = \phantom{0000}1.234 \\
1.234 \times 10^{-1} &= 1.234 \times \phantom{000}0.1 = \phantom{000}0.1234 \\
1.234 \times 10^{-2} &= 1.234 \times \phantom{00}0.01 = 0.01234
\end{aligned}
$$

When the exponent of 10 is positive, the decimal point of the coefficient is moved to the *right* the number of places indicated by the exponent. When the exponent of 10 is negative, the decimal point of the coefficient moves to the *left* the number of places indicated by the exponent. A positive exponent of 10 makes the number larger. A negative exponent of 10 makes the number smaller.

> **Procedure:  Converting a Number from Scientific Notation to Standard Notation**
>
> **Step 1:** Drop "$\times 10^n$."
> **Step 2:** Move the decimal point appropriately.
>     **a.** If $n > 0$, move the decimal point of the coefficient to the right $n$ places.
>     **b.** If $n < 0$, move the decimal point of the coefficient to the left $|n|$ places.

**Objective 2  Examples**    **Convert each number to standard notation.**

**2a.** The memory capacity of the brain is $4 \times 10^{12}$ bytes. (Source: http://www.moah.org /exhibits/archives/brains/technology.html)

**2b.** The mass of a dust particle is $7.53 \times 10^{-10}$ kg.

**Solutions**    **2a.** In $4 \times 10^{12}$, the coefficient is 4 and the exponent of 10 is 12. So, we move the decimal point in 4.0 to the right 12 times.

$$4.0 \times 10^{12} = 4{,}000{,}000{,}000{,}000$$

**2b.** In $7.53 \times 10^{-10}$, the coefficient is 7.53 and the exponent of 10 is $-10$. So, we move the decimal point in 7.53 to the left 10 times.

$$7.53 \times 10^{-10} = 0.000\,000\,000\,753$$

**✓ Student Check 2**    Convert each number to standard notation.

    **a.** The U.S. population is approximately $3.12 \times 10^8$. (Source: http://www.census.gov /population/www/popclockus.html)

    **b.** A carbon atom has a mass of $2 \times 10^{-23}$ g.

**Note:** *Helpful Method to Remember How to Convert to Standard Form*

*Think of a number line. If a number is positive, we plot it by moving to the right of zero. If a number is negative, we plot it by moving to the left of zero. This is the same movement needed to convert numbers from scientific notation to standard form. That is, if the exponent of 10 is positive, we move the decimal point of the coefficient right. If the exponent of 10 is negative, we move the decimal point of the coefficient left.*

## Operations with Numbers in Scientific Notation

**Objective 3** ▶

Perform operations with numbers in scientific notation.

It is nearly impossible to perform calculations with numbers that are very large or very small on a calculator. It is much easier to handle calculations with these types of numbers if they are written in scientific notation. This will enable us to use properties of exponents to perform the computations.

**Objective 3  Examples**   Use properties of exponents to simplify each expression. Write answers in scientific notation.

**3a.** $(3 \times 10^4)(5 \times 10^2)$      **3b.** $\dfrac{8.1 \times 10^3}{3 \times 10^9}$

**Solutions**   **3a.** $(3 \times 10^4)(5 \times 10^2) = 3 \times 5 \times 10^4 \times 10^2$   Group the coefficients and the powers of 10 together.

$= 15 \times 10^6$   Multiply the coefficients and apply the product of like bases rule.

$= 1.5 \times 10^1 \times 10^6$   Rewrite 15 in scientific notation.

$= 1.5 \times 10^7$   Apply the product of like bases rule.

**3b.**   $\dfrac{8.1 \times 10^3}{3 \times 10^9} = \dfrac{8.1}{3} \times \dfrac{10^3}{10^9}$   Divide the coefficients.

$= 2.7 \times 10^{3-9}$   Apply the quotient of like bases rule.

$= 2.7 \times 10^{-6}$   Subtract the powers.

✔ **Student Check 3**   Use properties of exponents to simplify each expression. Write answers in scientific notation.

**a.** $(7 \times 10^{-4})(6 \times 10^7)$      **b.** $\dfrac{4.5 \times 10^2}{9 \times 10^5}$

## Applications

**Objective 4** ▶

Solve application problems using scientific notation.

As we have seen, very large numbers and very small numbers exist all around us. We will convert the numbers given to scientific notation, perform operations on them, and then apply the properties of exponents to simplify the expression.

**Objective 4  Examples**   Use scientific notation and the properties of exponents to solve each problem.

**4a.** The national debt of the United States is approximately $14.6 trillion, or $14,600,000,000,000, and the population of the United States is approximately 312 million, or 312,000,000. If the debt is evenly distributed among each person in the United States, how much debt will each person owe to pay it off? (Source: http://www.brillig.com/debt_clock)

**4b.** The size of a red blood cell is about 0.000315 in. Blood has approximately 25 trillion red blood cells. If a person's blood cells are placed side by side in a line, what will be the length of the line formed by the red blood cells?

**Solutions**    **4a.** We must divide the national debt by the population.

$$\frac{14{,}600{,}000{,}000{,}000}{312{,}000{,}000} = \frac{1.46 \times 10^{13}}{3.12 \times 10^{8}}$$   Convert each number to scientific notation.

$$= \frac{1.46}{3.12} \times \frac{10^{13}}{10^{8}}$$   Divide the coefficients and the powers of 10.

$$\approx 0.47 \times 10^{5}$$   Apply the quotient of like bases rule.

$$= \$47{,}000$$   Convert to standard form.

Each person would owe approximately $47,000 to pay off the national debt.

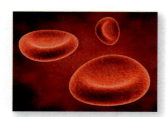

**4b.** We must multiply the number of red blood cells by the size of one red blood cell.

$$25{,}000{,}000{,}000{,}000 \times 0.000315 = (2.5 \times 10^{13})(3.15 \times 10^{-4})$$

$$= (2.5 \times 3.15) \times (10^{13} \times 10^{-4})$$

$$= 7.875 \times 10^{9}$$

$$= 7{,}875{,}000{,}000 \text{ in.}$$

So, if a person's red blood cells were put side by side, they would span a length of 7,875,000,000 in. or about 124,289.77 mi.

✓ **Student Check 4**    Use scientific notation and the properties of exponents to solve each problem.
   **a.** A standard CD holds about 700 megabytes or 700,000,000 bytes of data. How many CDs will it take to store a floor of academic journals which is about 100,000,000,000 bytes or 100 gigabytes?
   **b.** Fine hair is approximately 0.002 in. wide. The typical person has 100,000 hairs on their head. If all of the hairs of a typical person are placed side by side, how many inches will be covered by the hair?

**Objective 5 ▶**
Troubleshoot common errors.

## Troubleshooting Common Errors

Some common errors associated with scientific notation are shown next.

**Objective 5  Examples**    **A problem and an incorrect solution are given. Provide the correct solution and an explanation of the error.**

**5a.** Write the number 52,000,000 in scientific notation.

| Incorrect Solution | Correct Solution and Explanation |
|---|---|
| $52{,}000{,}000 = 52 \times 10^{6}$ | The coefficient is not between 1 and 10. The correct notation is $$52{,}000{,}000 = 5.2 \times 10^{7}$$ |

**5b.** Write the number $3.256 \times 10^{-4}$ in standard form.

| Incorrect Solution | Correct Solution and Explanation |
|---|---|
| $3.256 \times 10^{-4} = 0.00003256$ | The decimal point should only be moved four places to the left. $$3.256 \times 10^{-4} = 0.0003256$$ |

## ANSWERS TO STUDENT CHECKS

**Student Check 1   a.** $7 \times 10^9$   **b.** $5 \times 10^{-8}$
**Student Check 2   a.** 320,000,000
   **b.** 0.000 000 000 000 000 000 000 02

**Student Check 3   a.** $4.2 \times 10^4$   **b.** $5 \times 10^{-4}$
**Student Check 4   a.** Approximately $1.43 \times 10^2$ or 143 CDs
   **b.** $2 \times 10^2$ or 200 in.

## SUMMARY OF KEY CONCEPTS

1. Scientific notation is a useful way to express very large or very small numbers.
2. A number of the form $a \times 10^n$ is in scientific notation, where $1 \le a < 10$ and $n$ is an integer.
3. A number can be converted from scientific notation to standard notation by moving the decimal point in the

coefficient as indicated by the exponent of 10. A positive exponent tells us to move the decimal point $n$ places to the right. A negative exponent tells us to move $|n|$ places to the left.

4. Operations with numbers written in scientific notation can be performed using the properties of exponents.

## GRAPHING CALCULATOR SKILLS

Graphing calculators will display numbers in scientific notation if there are too many digits in the number. Understanding how to read and interpret this display is important. We should also be able to enter a number in scientific notation.

**Example 1:** Enter the number 10,000,000,000,000 on a calculator and interpret the display.

**Solution:** Notice that the display is 1E13. This means that the number entered is equivalent to $1 \times 10^{13}$. "E" represents "times 10 to the exponent of."

```
10000000000000
            1E13



```

**Example 2:** Enter the number $7.53 \times 10^{-10}$ using scientific notation on a calculator.

**Solution:** To enter the "E" on the calculator, access the EE function ( 2nd , ) on the calculator.

```
7.53E-10
        7.53E-10
```

---

## SECTION 5.3    EXERCISE SET

 **Write About It!**

Use complete sentences in your answer to each exercise.

1. Why is it useful to write numbers in scientific notation?
   Answers vary.
2. Write the name of the number $10^n$ for $n = 0, 1, 2, 3, \ldots, 12$.
3. Use an example to explain how to convert a very large number in standard notation to scientific notation.
   Answers vary.
4. Use an example to explain how to convert a very small number in standard notation to scientific notation.
   Answers vary.
5. Explain how to use the properties of exponents to simplify the expression $(8.2 \times 10^{-9})(3.2 \times 10^4)$.
   Answers vary.
6. Explain how to use the properties of exponents to simplify the expression $\dfrac{9.6 \times 10^{14}}{1.2 \times 10^{-5}}$.   Answers vary.

Additional answers can be found in the Instructor Answer Appendix.

**Practice Makes Perfect!**

Write each number in scientific notation. (*See Objective 1.*)

7. 3700   $3.7 \times 10^3$
8. 114,000   $1.14 \times 10^5$
9. 512,000   $5.12 \times 10^5$
10. 7500   $7.5 \times 10^3$
11. 0.07365   $7.365 \times 10^{-2}$
12. 0.2438   $2.438 \times 10^{-1}$
13. 0.00000001547   $1.547 \times 10^{-8}$
14. 0.000000388   $3.88 \times 10^{-7}$
15. The speed of the Space Shuttle when orbiting Earth: 17,600 mph (Source: http://www.nasa.gov)   $1.76 \times 10^4$ mph
16. The distance from the Earth to the moon: 251,000 mi (Source: http://www.britannica.com)   $2.51 \times 10^5$ mi
17. The thickness of paper: 0.0001 m (Source: http://www.nano.gov)   $1.0 \times 10^{-4}$ m
18. The diameter of a strand of hair: 0.000001 m (Source: http://www.nano.gov)   $1.0 \times 10^{-6}$ m

19. The population of China in 2009: approximately
1,331,000,000 (Source: http://data.worldbank.org/country/china)
$1.331 \times 10^9$

20. The population of Belize in 2009: approximately
333,000 (Source: http://data.worldbank.org/country/belize)
$3.33 \times 10^5$

21. The gross national income (GNI) of the United States:
$9,780,000,000,000 (Source: http://www.nationmaster.com
/graph/eco_gro_nat_inc-economy-gross-national-income)
$9.78 \times 10^{12}$

22. The gross national income (GNI) of Japan:
$4,520,000,000,000 (Source: http://www.nationmaster.com
/graph/eco_gro_nat_inc-economy-gross-national-income)
$4.52 \times 10^{12}$

**Write each number in standard notation. (See Objective 2.)**

23. $3.54 \times 10^7$  35,400,000
24. $2.36 \times 10^{12}$  2,360,000,000,000
25. $9.38 \times 10^{10}$  93,800,000,000
26. $1.03 \times 10^5$  103,000
27. $3.35 \times 10^{-3}$  0.00335
28. $5.49 \times 10^{-6}$  0.00000549
29. $6.78 \times 10^{-8}$  0.0000000678
30. $8.13 \times 10^{-10}$  0.000000000813

31. Bill Gates's net worth in 2011: $5.6 \times 10^{10}$ U.S. dollars
(Source: http://www.billgatesmicrosoft.com/networth.htm)
56,000,000,000 U.S. dollars

32. Warren Buffet's net worth in 2011: $5 \times 10^{10}$ U.S. dollars
(Source: http://www.billgatesmicrosoft.com/networth.htm)
50,000,000,000 U.S. dollars

33. Amount of blood in average adult: $5.2 \times 10^{-3}$ m³
(Source: Burdge, J. [2009]. *Chemistry*. Boston: McGraw-Hill)
0.0052 m³

34. Maximum FDA recommended daily sodium intake:
$5.291 \times 10^{-3}$ lb    0.005291 lb

35. Diameter of DNA: $5 \times 10^{-9}$ m (Source: http://www.nano.gov)
0.000000005 m

36. Mass of an electron: $9.10 \times 10^{-28}$ g (Source: Burdge, J.
[2009]. *Chemistry*. Boston: McGraw-Hill)
0.000000000000000000000000000910 g

**Use exponent rules to simplify each expression. Write each
answer in scientific notation. (See Objectives 3 and 4.)**

37. $(8.3 \times 10^5)(2.0 \times 10^3)$    $1.66 \times 10^9$
38. $(3.0 \times 10^{-4})(5.0 \times 10^{12})$    $1.5 \times 10^9$
39. $(4.0 \times 10^4)(6.0 \times 10^7)$    $2.4 \times 10^{12}$
40. $(1.3 \times 10^{-5})(7.0 \times 10^6)$    $9.1 \times 10$

41. $\dfrac{6.3 \times 10^8}{3.0 \times 10^5}$  $2.1 \times 10^3$
42. $\dfrac{8.4 \times 10^6}{2.0 \times 10^9}$  $4.2 \times 10^{-3}$
43. $\dfrac{8.0 \times 10^{12}}{4.0 \times 10^{-6}}$  $2.0 \times 10^{18}$
44. $\dfrac{2.1 \times 10^4}{2.1 \times 10^3}$  $1.0 \times 10$

45. The nanoscale is used to represent dimensions between
1 and 100 nanometers (nm). One nanometer is $10^{-9}$
m. If a sheet of paper is about $10^5$ nm thick, find the
thickness of 500 sheets in meters. Express the answer
in scientific notation. (Source: http://www.nano.gov)  $5 \times 10^{-2}$ m

46. According to the Nanoscale, there are $2.5 \times 10^7$ nm/in.
If a sheet of paper is about $10^5$ nm thick, find the
thickness of 5000 sheets in inches. Express the answer
in scientific notation. (Source: http://www.nano.gov)  $2 \times 10$ in.

## Mix 'Em Up!

**Convert numbers in standard notation to scientific
notation and numbers in scientific notation to standard
notation.**

47. 160,000  $1.6 \times 10^5$
48. 2,450,000  $2.45 \times 10^6$
49. $5.21 \times 10^{11}$  521,000,000,000
50. $7.12 \times 10^8$  712,000,000
51. 0.01498  $1.498 \times 10^{-2}$
52. 0.4317  $4.317 \times 10^{-1}$
53. 0.000006372  $6.372 \times 10^{-6}$
54. 0.000000522  $5.22 \times 10^{-7}$
55. $4.16 \times 10^{-4}$  0.000416
56. $1.033 \times 10^{-7}$  0.0000001033

57. The population of Brazil in 2009: 198,700,000 (Source:
http://www.factmonster.com)  $1.987 \times 10^8$

58. The population of European Union in 2009:
491,600,000 (Source: http://www.factmonster.com)  $4.916 \times 10^8$

59. The length of an ant: 0.004 m (Source: http://www.nano.gov)
$4.0 \times 10^{-3}$ m

60. The diameter of a large raindrop: 0.0025 m (Source:
http://www.nano.gov)  $2.5 \times 10^{-3}$ m

61. Chevron Corporation reported its first quarter profit in
2010: $4.55 \times 10^9$ U.S. dollars (Source: http://www.sfgate
.com)  4,550,000,000 U.S. dollars

62. A fast-food restaurant reported its third quarter profit in
2010: $1.26 \times 10^9$ U.S. dollars  1,260,000,000 U.S. dollars

**Use exponent rules to simplify each expression. Write each
answer in scientific notation.**

63. $(1.4 \times 10^6)(6.45 \times 10^5)$  $9.03 \times 10^{11}$
64. $\dfrac{5.6 \times 10^9}{7.0 \times 10^2}$  $8.0 \times 10^6$
65. $\dfrac{9.45 \times 10^{-13}}{4.5 \times 10^{-5}}$  $2.1 \times 10^{-8}$
66. $(8.9 \times 10^{-3})(2.5 \times 10^{15})$  $2.225 \times 10^{13}$
67. $\dfrac{3.5 \times 10^{11}}{2.5 \times 10^{-4}}$  $1.4 \times 10^{15}$
68. $(5.3 \times 10^4)(6.6 \times 10^{-9})$  $3.498 \times 10^{-4}$
69. $(4.9 \times 10^{-11})(3.7 \times 10^4)$  $1.813 \times 10^{-6}$
70. $\dfrac{8.84 \times 10^{-14}}{3.4 \times 10^3}$  $2.6 \times 10^{-17}$

71. The population of India in 2009 was approximately
1,166,000,000 and the land area is approximately
1,269,000 mi². If the land is evenly distributed among
all people in India, how much land will each person
have?  $1.09 \times 10^{-3}$ mi² per person

72. The population of Canada in 2009 was approximately
33,500,000 and the land area is approximately
3,854,000 mi². If the land is evenly distributed among all
people in Canada, how much land will each person have?
$1.15 \times 10^{-1}$ mi² per person

## You Be the Teacher!

**Explain the notation.**

73. 2.7E14  $2.7 \times 10^{14}$
74. 5.7E−13  $5.7 \times 10^{-13}$

## Calculate It!

Use a calculator to simplify each expression. Write the answer using scientific notation.

**75.** $(1.35 \times 10^5)(3.47 \times 10^8)$   $4.6845 \times 10^{13}$

**76.** $(2.35 \times 10^{-3})(5.2 \times 10^9)$   $1.222 \times 10^7$

**77.** $\dfrac{3.6 \times 10^{14}}{1.2 \times 10^{-6}}$   $3.0 \times 10^{20}$

**78.** $\dfrac{6 \times 10^{12}}{1.5 \times 10^{18}}$   $4.0 \times 10^{-6}$

---

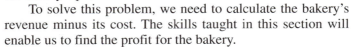

## SECTION 5.4 / Addition and Subtraction of Polynomials

### ▶ OBJECTIVES

As a result of completing this section, you will be able to

**1.** Classify polynomials and write in standard form.
**2.** Find the degree and leading coefficient of a polynomial.
**3.** Evaluate polynomials.
**4.** Add polynomials.
**5.** Subtract polynomials.
**6.** Solve application problems.
**7.** Troubleshoot common errors.

A bakery can make a donut for $0.10. It costs about $2500 to operate the bakery each month. The total operating cost can be represented by $2500 + 0.10d$, where $d$ is the number of donuts made. The bakery sells each donut for $0.75, so its revenue can be represented by $0.75d$. What expression represents the bakery's profit?

To solve this problem, we need to calculate the bakery's revenue minus its cost. The skills taught in this section will enable us to find the profit for the bakery.

### Polynomials

In Section 1.3, we discussed the concept of algebraic expressions. *Polynomials* are special types of algebraic expressions in which the variables have whole numbers as exponents. Polynomials are very important to the study of algebra.

Recall that a term is a number or a product of a number and variables raised to whole number exponents. For example, the terms in the expression $2x^2 + 3x - 4$ are $2x^2$, $3x$, and $-4$.

### Objective 1 ▶

Classify polynomials and write in standard form.

> **Definition:** A **Polynomial** is a finite sum of terms of the form $ax^n$, where $a$ is a real number and $n$ is a whole number.

An example of a polynomial is $-3x^4 + 2x^2 - 5x + 7$. This polynomial is written in **standard form**, since the powers of $x$ are in descending order.

The expressions $\dfrac{2}{x}$, $6x^{-2} + 4x^{-1}$, and $\sqrt{y^2 + 3}$ are not polynomials, since a variable occurs in the denominator, exponents are negative, and the variable occurs inside a square root, respectively.

Polynomials can have one, two, three, or more terms.

| Number of Terms | Name | Examples |
|---|---|---|
| One term | **Monomial** | $5x,\ -3y,\ 7t,\ \dfrac{2}{3}a$ |
| Two terms | **Binomial** | $x - 4,\ \dfrac{1}{2}t - 1,$ <br> $y + 2,\ -3a - 7$ |
| Three terms | **Trinomial** | $x^2 - x + 6$ <br> $4y^2 + 3y - 1$ |
| Four terms | | $-7y^3 + 9y^2 + 6y - 5$ |

> **Procedure: Classifying a Polynomial and Writing It in Standard Form**
>
> **Step 1:** Count the number of terms in the polynomial.
>   **a.** If there is one term, it is a monomial.
>   **b.** If there are two terms, it is a binomial.
>   **c.** If there are three terms, it is a trinomial.
> **Step 2:** Rearrange the terms so that the exponents on the variable are decreasing.

**Objective 1 Examples**   Classify each polynomial as a monomial, binomial, or trinomial. Write the polynomial in standard form.

**1a.** $3 - 5y^2 + 4y$       **1b.** $9 - x^2$       **1c.** $\dfrac{3}{4}a$

**Solutions**    **1a.** There are three terms so this is a trinomial. The standard form is

$$-5y^2 + 4y + 3$$

**1b.** There are two terms so this is a binomial. The standard form is

$$-x^2 + 9$$

**1c.** There is only one term so this is a monomial. It is in standard form.

✔ **Student Check 1**    Classify each polynomial as a monomial, binomial, or trinomial. Write the polynomial in standard form.

  **a.** $16t + 4t^2$       **b.** $-6 + 3h^2 + 5h$       **c.** $-8x$

## Identifying the Degree and Leading Coefficient of a Polynomial

**Objective 2** ▶

Find the degree and leading coefficient of a polynomial.

Polynomials are not only classified by the number of terms that they contain but also by their *degree*. Before we define the degree of a polynomial, we need to understand the degree of a term.

> **Definition:** The **degree of a term** is the sum of the exponents on the variables it contains.

Some terms and their degree are shown.

| Term | Degree |
|------|--------|
| $4x = 4x^1$ | 1 |
| $-3y^2$ | 2 |
| $7ab^3 = 7a^1b^3$ | $1 + 3 = 4$ |
| $5 = 5x^0$ | 0 |

The last example shows us that the degree of a constant term is 0.

Since a polynomial is made up of terms, we define its degree as follows.

> **Definition:** The **degree of a polynomial** is the largest degree of the terms in the polynomial. The **leading coefficient** is the coefficient of the term with this degree.

For example, the degree and leading coefficient of the following polynomial are shown.

Degree

$$-5y^2 + 4y + 3$$

Leading coefficient

---

**Objective 2  Examples**

**Identify the degree and the leading coefficient of each polynomial.**

**2a.** $6 - 7x + x^2$    **2b.** $-5y + 8$    **2c.** $8p - 7p^2 + 4p^3 - p^4$

**2d.** 3    **2e.** $3x^3y - 4x^2y^3 + 2xy^2$

**Solutions**

**2a.** The term with the largest degree is $x^2 = 1x^2$, so the degree of the polynomial is 2 and the coefficient of this term is the leading coefficient, which is 1.

**2b.** The term with the largest degree is $-5y = -5y^1$, so the degree of the polynomial is 1 and the coefficient of this term is the leading coefficient, which is $-5$.

**2c.** The term with the largest degree is $-p^4 = -1p^4$, so the degree of the polynomial is 4 and the coefficient of this term is the leading coefficient, which is $-1$.

**2d.** The term with the largest degree is $3 = 3x^0$, so the degree of the polynomial is 0 and the coefficient of this term is the leading coefficient, which is 3. Note also that this is a constant term.

**2e.** We need to find the degree of each term of the polynomial. The term with the largest degree is $-4x^2y^3$, so the degree of the polynomial is 5. The coefficient of this term is the leading coefficient, which is $-4$.

| Term | Degree |
|------|--------|
| $3x^3y^1$ | $3 + 1 = 4$ |
| $-4x^2y^3$ | $2 + 3 = 5$ |
| $2x^1y^2$ | $1 + 2 = 3$ |

---

✓ **Student Check 2**

Identify the degree and the leading coefficient of each polynomial.

**a.** $3y + y^2 - 4y^3$    **b.** $8x - x^2$    **c.** $5x^2 + 7x + 9x^5 - 1$

**d.** $-6$    **e.** $8r^2s^5 - 7rs^8 + 9rs$

---

## Evaluating Polynomials

**Objective 3** ▶

Evaluate polynomials.

In Section 1.3 we discussed evaluating algebraic expressions. Since polynomials are special types of algebraic expressions, we evaluate them the same way we evaluate algebraic expressions.

**Procedure:  Evaluating a Polynomial**

**Step 1:** Substitute the given number in place of the variable.
**Step 2:** Use the order of operations to simplify the resulting expression.

---

**Objective 3  Examples**

**Evaluate each polynomial for the given value.**

**3a.** $x^2 - 4$ for $x = -3$    **3b.** $4y^3 - y^2 + 6y - 8$ for $y = -1$

**3c.** $x^2 + 8xy - 4y^2$ for $x = -3$ and $y = 2$

**3d.** The height, in feet, of a ball projected upward from the top of a 100-ft building with an initial velocity of 32 ft/sec is given by $-16t^2 + 32t + 100$, where $t$ is the number of seconds after the ball is thrown. Find the height of the ball 1 sec after the ball is thrown.

**Solutions**   **3a.**

$$x^2 - 4 = (-3)^2 - 4 \qquad \text{Replace } x \text{ with } -3.$$
$$= 9 - 4 \qquad \text{Simplify the exponent.}$$
$$= 5 \qquad \text{Subtract.}$$

**3b.**  $4y^3 - y^2 + 6y - 8 = 4(-1)^3 - (-1)^2 + 6(-1) - 8$   Replace $y$ with $-1$.
$$= 4(-1) - (1) + 6(-1) - 8 \qquad \text{Simplify the exponents.}$$
$$= -4 - 1 - 6 - 8 \qquad \text{Multiply from left to right.}$$
$$= -19 \qquad \text{Add from left to right.}$$

**3c.**  $x^2 + 8xy - 4y^2 = (-3)^2 + 8(-3)(2) - 4(2)^2$   Replace the variable $x$ with $-3$ and $y$ with 2.

$$= 9 + 8(-3)(2) - 4(4) \qquad \text{Simplify the exponents.}$$
$$= 9 - 48 - 16 \qquad \text{Multiply from left to right.}$$
$$= -55 \qquad \text{Add from left to right.}$$

**3d.**  $-16t^2 + 32t + 100 = -16(1)^2 + 32(1) + 100$   Replace $t$ with 1.
$$= -16(1) + 32 + 100 \qquad \text{Simplify the exponent and multiply 32 and 1.}$$

$$= -16 + 32 + 100 \qquad \text{Multiply } -16 \text{ and 1.}$$
$$= 116 \qquad \text{Add from left to right.}$$

The ball is 116 ft above the ground after 1 sec.

✔ **Student Check 3**   Evaluate each polynomial for the given value.
  **a.** $3p^2 + 5$ for $p = -6$
  **b.** $-2x^3 + x^2 + 4x - 1$ for $x = -3$
  **c.** $5a^2 - 2ab + b^2$ for $a = 2$ and $b = -4$
  **d.** Use worked Example 3(d) to find the height of the ball after 3 sec.

## Adding Polynomials

**Objective 4** ▶

Add polynomials.

We learned in Chapter 1 that only like terms can be added together. Recall that like terms have the same variables raised to the same exponents. For example,

$$6x + 2x = (6 + 2)x = 8x$$

$$\frac{1}{2}y^2 + \frac{2}{3}y^2 = \left(\frac{1}{2} + \frac{2}{3}\right)y^2 = \left(\frac{3}{6} + \frac{4}{6}\right)y^2 = \frac{7}{6}y^2$$

Since polynomials consist of terms, we can add them by adding the like terms of each polynomial together.

| **Procedure:  Adding Polynomials** |
| --- |
| **Step 1:**  Group like terms of each polynomial together. |
| **Step 2:**  Add the coefficients of the like terms and keep the variable part the same. |
| **Step 3:**  Write the answer in standard form. |

**INSTRUCTOR NOTE:**
We can group terms vertically or horizontally as shown in Example 4.

**Objective 4  Examples**   **Perform each operation.**
  **4a.** $(6x - 4) + (2x + 3)$
  **4b.** $(x^2 - x + 6) + (x^2 + 4x - 3)$
  **4c.** $(4y^3 - 5y^2 + 6y - 7) + (y^3 - 2y^2 + 9)$

**4d.** $(6a^2 - 5ab + 4b^2) + (4a^2 - 5ab - 8b^2)$

**4e.** Add $-4x^2 + x - 7$ and $2x^2 - 3x + 1$ using a vertical format.

**Solutions**

**4a.** $(6x - 4) + (2x + 3) = (6x + 2x) + (-4 + 3)$    Group like terms.
$= 8x + (-1)$    Add like terms.
$= 8x - 1$    Simplify.

**4b.** $(x^2 - x + 6) + (x^2 + 4x - 3)$
$= (x^2 + x^2) + (-x + 4x) + (6 - 3)$    Group like terms.
$= 2x^2 + 3x + 3$    Add like terms.

**4c.** $(4y^3 - 5y^2 + 6y - 7) + (y^3 - 2y^2 + 9)$
$= (4y^3 + y^3) + (-5y^2 - 2y^2) + 6y + (-7 + 9)$    Group like terms.
$= 5y^3 + (-7y^2) + 6y + 2$    Add like terms.
$= 5y^3 - 7y^2 + 6y + 2$    Simplify.

**4d.** $(6a^2 - 5ab + 4b^2) + (4a^2 - 5ab - 8b^2)$
$= (6a^2 + 4a^2) + (-5ab - 5ab) + (4b^2 - 8b^2)$    Group like terms.
$= 10a^2 - 10ab - 4b^2$    Add like terms.

**4e.** To add vertically, line up the like terms and then add.

$(-4x^2 + x - 7)$
$+(2x^2 - 3x + 1)$
$\longrightarrow$
$\begin{array}{r} -4x^2 + x - 7 \\ 2x^2 - 3x + 1 \\ \hline -2x^2 - 2x - 6 \end{array}$

**✓ Student Check 4**   Perform each operation.
**a.** $(4p - 3) + (7p - 5)$    **b.** $(y^2 - 9y + 10) + (y^2 - y + 1)$
**c.** $(6a^3 + 2a^2 - 3a + 4) + (a^3 - 7a^2 + a)$
**d.** $(-5x^2 + 6xy - y^2) + (4x^2 + 2xy - y^2)$
**e.** Add $-2x^2 + 4x - 6$ and $3x^2 + x - 5$ using a vertical format.

## Subtracting Polynomials

**Objective 5**
Subtract polynomials.

Recall that subtraction is defined as adding the opposite:

$$a - b = a + (-b)$$

This rule also applies to subtracting polynomials. To find the opposite of a polynomial, multiply it by $-1$. For example,

Opposite of $(x - 4)$:    $-(x - 4) = -1(x - 4)$
$= -x + 4$

Opposite of $(2y^2 - 3y + 5)$:    $-(2y^2 - 3y + 5) = -1(2y^2 - 3y + 5)$
$= -2y^2 + 3y - 5$

Notice that each sign of the polynomial is *changed* when the opposite is found.

> **Procedure: Subtracting Polynomials**
> **Step 1:** Clear parentheses by finding the opposite of the polynomial being subtracted.
> **Step 2:** Add like terms.
> **Step 3:** Write the answer in standard form.

**Objective 5 Examples** **Perform each operation.**

**5a.** $(6x - 4) - (2x + 3)$
**5b.** $(x^2 - x + 6) - (x^2 + 4x - 3)$
**5c.** $(4y^3 - 5y^2 + 6y - 7) - (y^3 - 2y^2 + 9)$
**5d.** $(6a^2 - 5ab + 4b^2) - (4a^2 - 5ab - 8b^2)$
**5e.** Subtract $x^3 - 7x^2 + 4x - 6$ from $-3x^3 + 4x^2 + 5x - 1$ using a vertical format.

**Solutions** **5a.** $(6x - 4) - (2x + 3)$

| | |
|---|---|
| $= (6x - 4) - 1(2x + 3)$ | Multiply the second polynomial by $-1$. |
| $= 6x - 4 - 2x - 3$ | Apply the distributive property. |
| $= 6x - 2x - 4 - 3$ | Group like terms. |
| $= 4x - 7$ | Combine like terms. |

**5b.** $(x^2 - x + 6) - (x^2 + 4x - 3)$

| | |
|---|---|
| $= (x^2 - x + 6) - 1(x^2 + 4x - 3)$ | Multiply the second polynomial by $-1$. |
| $= x^2 - x + 6 - x^2 - 4x + 3$ | Apply the distributive property. |
| $= x^2 - x^2 - x - 4x + 6 + 3$ | Group like terms. |
| $= 0x^2 - 5x + 9$ | Combine like terms. |
| $= -5x + 9$ | Simplify. |

**5c.** $(4y^3 - 5y^2 + 6y - 7) - (y^3 - 2y^2 + 9)$

| | |
|---|---|
| $= (4y^3 - 5y^2 + 6y - 7) - 1(y^3 - 2y^2 + 9)$ | Multiply the second polynomial by $-1$. |
| $= 4y^3 - 5y^2 + 6y - 7 - y^3 + 2y^2 - 9$ | Apply the distributive property. |
| $= 4y^3 - y^3 - 5y^2 + 2y^2 + 6y - 7 - 9$ | Group like terms. |
| $= 3y^3 - 3y^2 + 6y - 16$ | Combine like terms. |

**5d.** $(6a^2 - 5ab + 4b^2) - (4a^2 - 5ab - 8b^2)$

| | |
|---|---|
| $= (6a^2 - 5ab + 4b^2) - 1(4a^2 - 5ab - 8b^2)$ | Multiply the second polynomial by $-1$. |
| $= 6a^2 - 5ab + 4b^2 - 4a^2 + 5ab + 8b^2$ | Apply the distributive property. |
| $= 6a^2 - 4a^2 - 5ab + 5ab + 4b^2 + 8b^2$ | Group like terms. |
| $= 2a^2 + 0ab + 12b^2$ | Combine like terms. |
| $= 2a^2 + 12b^2$ | Simplify. |

**5e.**
$$\begin{array}{r} -3x^3 + 4x^2 + 5x - 1 \\ -(x^3 - 7x^2 + 4x - 6) \\ \hline \end{array}$$

The second polynomial is multiplied by $-1$. Then we add.

$$\begin{array}{r} -3x^3 + 4x^2 + 5x - 1 \\ -\ x^3 + 7x^2 - 4x + 6 \\ \hline -4x^3 + 11x^2 + x + 5 \end{array}$$

✓ **Student Check 5** Perform each operation.

**a.** $(4p - 3) - (7p - 5)$
**b.** $(y^2 - 9y + 10) - (y^2 - y + 1)$
**c.** $(6a^3 + 2a^2 - 3a + 4) - (a^3 - 7a^2 + a)$
**d.** $(-5x^2 + 6xy - y^2) - (4x^2 + 2xy - y^2)$
**e.** Subtract $x^3 + 6x^2 - 3x + 10$ from $2x^3 + 9x^2 + 4x - 2$.

## Applications

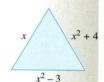

Polynomials can be used to represent real-life situations. Example 6 illustrates adding and subtracting polynomials that deal with real-life situations.

**Objective 6 Examples** | **Solve each problem.**

**6a.** One side of a triangle is three less than the square of the shortest side. The other side is four more than the square of the shortest side. Find an expression for the perimeter of the triangle.

Solution | **6a.** The perimeter is the sum of the lengths of the sides of the triangle. Let $x$ represent the length of the shortest side of the triangle. One side is $x^2 - 3$. The other side is $x^2 + 4$, as shown in the figure.

So, the perimeter is $P = x + (x^2 + 4) + (x^2 - 3)$

$$P = x^2 + x^2 + x + 4 - 3$$

$$P = 2x^2 + x + 1$$

**6b.** A bakery can make a donut for $0.10. It costs about $2500 to operate the bakery each month. The total operating cost can be represented by $2500 + 0.10d$, where $d$ is the number of donuts made. The bakery sells each donut for $0.75, so its revenue can be represented by $0.75d$.

  **i.** Write an expression that represents the bakery's profit. (Recall that profit is revenue minus cost.)

  **ii.** Find the profit from selling 7200 donuts in a month.

Solution | **6b.** **i.**

| | |
|---|---|
| Profit = Revenue − Cost | Write the profit relationship. |
| $= 0.75d - (2500 + 0.10d)$ | Substitute the revenue and cost. |
| $= 0.75d - 2500 - 0.10d$ | Apply the distributive property. |
| $= 0.65d - 2500$ | Combine like terms. |

  **ii.** To find the profit from selling 7200 donuts in a month, replace $d$ with 7200.

| | |
|---|---|
| $P = 0.65d - 2500$ | Write the expression for profit. |
| $P = 0.65(7200) - 2500$ | Replace $d$ with 7200. |
| $P = 2180$ | Simplify. |

The profit from selling 7200 donuts in a month is $2180.

✓ **Student Check 6** | Solve each problem.

**a.** Find the perimeter of the trapezoid.

$$y + 2$$
$$4y^2 + y + 5 \qquad 2y - 3$$
$$3y^2 - 2y + 1$$

**b.** The operator of a hot dog stand makes hot dogs for $0.25 each. The cost to operate his hot dog stand is $500 per month. The total monthly cost of operating the hot dog stand is $0.25h + 500$, where $h$ is the number of hot dogs made in a month. He sells hot dogs for $0.99 each. The revenue made from selling $h$ hot dogs in a month is $0.99h$. Write an expression for the profit of the hot dog stand. Find the profit from selling 3000 hot dogs in a month.

**Objective 7**

Troubleshoot common errors.

## Troubleshooting Common Errors

Some common errors associated with adding and subtracting polynomials are shown next.

**Objective 7 Examples**   A problem and an incorrect solution are given. Provide the correct solution and an explanation of the error.

**7a.** $(x^2 + 3x - 4) + (x^2 - 7x + 1)$

| Incorrect Solution | Correct Solution and Explanation |
|---|---|
| $(x^2 + 3x - 4) + (x^2 - 7x + 1)$ <br> $= x^4 - 4x - 3$ | The error was made in combining the $x^2$ terms. When we add like terms, we add their coefficients, not their exponents. <br> $(1x^2 + 3x - 4) + (1x^2 - 7x + 1)$ <br> $= 2x^2 - 7x - 3$ |

**7b.** $(x^2 + 3x - 4) - (x^2 - 7x + 1)$

| Incorrect Solution | Correct Solution and Explanation |
|---|---|
| $(x^2 + 3x - 4) - (x^2 - 7x + 1)$ <br> $= -4x - 3$ | The error was made in not distributing $-1$ to all the terms in the second polynomial. <br> $(1x^2 + 3x - 4) - (1x^2 - 7x + 1)$ <br> $= 1x^2 + 3x - 4 - 1x^2 + 7x - 1$ <br> $= 2x - 5$ |

## ANSWERS TO STUDENT CHECKS

**Student Check 1**   **a.** binomial, $4t^2 + 16t$
   **b.** trinomial, $3h^2 + 5h - 6$   **c.** monomial, $-8x$

**Student Check 2**

| | Degree | Leading Coefficient |
|---|---|---|
| **a.** | 3 | $-4$ |
| **b.** | 2 | $-1$ |
| **c.** | 5 | 9 |
| **d.** | 0 | $-6$ |
| **e.** | 9 | $-7$ |

**Student Check 3**   **a.** 113   **b.** 50   **c.** 52   **d.** 52 ft
**Student Check 4**   **a.** $11p - 8$   **b.** $2y^2 - 10y + 11$
   **c.** $7a^3 - 5a^2 - 2a + 4$   **d.** $-x^2 + 8xy - 2y^2$
   **e.** $x^2 + 5x - 11$
**Student Check 5**   **a.** $-3p + 2$   **b.** $-8y + 9$
   **c.** $5a^3 + 9a^2 - 4a + 4$   **d.** $-9x^2 + 4xy$
   **e.** $x^3 + 3x^2 + 7x - 12$
**Student Check 6**   **a.** $7y^2 + 2y + 5$   **b.** $0.74h - 500$; $1720

## SUMMARY OF KEY CONCEPTS

1. Polynomials are finite sums of terms of the form $ax^n$. If the polynomial consists of one term, it is a monomial. If it consists of two terms, it is a binomial. If it consists of three terms, it is a trinomial.

2. The degree of a term is the sum of the exponents of its variables.

3. A polynomial is written in standard form when the exponents are in descending order. The degree of a polynomial is the largest degree of all the terms in the polynomial. The coefficient of the term with the largest degree is the leading coefficient.

4. To evaluate a polynomial, replace the variable with the given value and use the order of operations to simplify the expression.

5. Polynomials are added by combining like terms. Remember that the exponent stays the same and coefficients are added.

6. To subtract polynomials, add the opposite of the polynomial. The opposite is found by multiplying the polynomial by $-1$.

## GRAPHING CALCULATOR SKILLS

The graphing calculator can be used to evaluate polynomials. We can also use the calculator to verify operations with polynomials.

**Example 1:** Find the value of $-16t^2 + 32t + 100$ for $t = 1$. There are several ways to evaluate a polynomial on a calculator.

**Method 1:** We can substitute 1 for $t$ and simply enter the resulting numerical expression on the calculator.

**Method 2:** We can enter the polynomial in the equation editor and use the table to evaluate it.

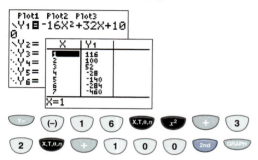

**Example 2:** Verify that $(x^2 - x + 6) - (x^2 + 4x - 3) = -5x + 9$.

**Solution:** Enter the left side of the equation in $Y_1$ and the right side of the equation in $Y_2$. Then view the table.

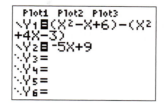

As we can see from the table, all the $Y_1$ and $Y_2$ values are the same for the given $x$ values. This means that both expressions are equivalent.

---

## SECTION 5.4   EXERCISE SET

 ### Write About It!

**Use complete sentences in your answer to each exercise.**

1. Explain what it means to write a polynomial in standard form.   *Answers vary.*
2. Explain the meaning of the degree of a polynomial.   *Answers vary.*
3. Explain how to add polynomials.   *Answers vary.*
4. Explain how to subtract polynomials.   *Answers vary.*
5. Does a trinomial have to be degree 2? Explain.   *Answers vary.*
6. Does a monomial have to be degree 1? Explain.   *Answers vary.*
7. Use an example to explain how to evaluate a polynomial for a given value of the variable.   *Answers vary.*
8. When adding or subtracting two polynomials, do they have to be of the same degree? Explain.   *Answers vary.*

 ### Practice Makes Perfect!

**Classify each polynomial as a monomial, binomial, or trinomial and write it in standard form. (*See Objective 1.*)**

9. $4x + 3$   *binomial, $4x + 3$*
10. $1 - 3x^2 - 7x$   *trinomial, $-3x^2 - 7x + 1$*
11. $5x^3$   *monomial, $5x^3$*
12. $1 - 2x$   *binomial, $-2x + 1$*
13. $6h - h^2 + 3$   *trinomial, $-h^2 + 6h + 3$*
14. $5h$   *monomial, $5h$*
15. $1 + 5y$   *binomial, $5y + 1$*
16. $3y - 4y^3 + 2$   *trinomial, $-4y^3 + 3y + 2$*

**Identify the leading coefficient and the degree of each polynomial. (*See Objective 2.*)**

17. $-6x^2 + 2x - 1$   *leading coefficient $-6$, degree 2*
18. $5x^2 - 15x + 3$   *leading coefficient 5, degree 2*
19. $2 - 3x$   *leading coefficient $-3$, degree 1*
20. $7 - x$   *leading coefficient $-1$, degree 1*
21. $4p^3 - 2p^2 + p - 6p^4 + 5$   *leading coefficient $-6$, degree 4*
22. $3p^2 + p^3 - p + 2p^4 - 6$   *leading coefficient 2, degree 4*
23. $15$   *leading coefficient 15, degree 0*
24. $-13$   *leading coefficient $-13$, degree 0*
25. $3 - h^3k + 2h^2k^4 - 4hk^2$   *leading coefficient 2, degree 6*
26. $5 - 3h^3k^4 + h^2k^2 + 11hk$   *leading coefficient $-3$, degree 7*
27. $4a^5 - a^3b + 6ab^2 - 7b^4$   *leading coefficient 4, degree 5*
28. $-2x^4y^2 + x^3y - 6x^2y^5 + y^3$   *leading coefficient $-6$, degree 7*

Additional answers can be found in the Instructor Answer Appendix.

**Evaluate each polynomial for the given value(s). (See Objective 3.)**

**29.** $10p^2 - p + 1$;   $p = 3$   88

**30.** $y^2 - 3y + 2$;   $y = -1$   6

**31.** $5h^2 - 11$;   $h = -4$   69

**32.** $h^3 - 5h^2$;   $h = 4$   −16

**33.** $8p - 8p^2$;   $p = \dfrac{1}{2}$   2

**34.** $12p^2 - 6p - 6$;   $p = \dfrac{1}{2}$   −6

**35.** $4h^3 - 2h^2 + 1$;   $h = -2$   −39

**36.** $4h^2 + 10$;   $h = -2$   26

**37.** $-y^3 + 5y - 6$;   $y = 2$   −4

**38.** $-h^4 + 5h^3 + 2h$;   $h = 3$   60

**39.** $3y^2 - 6$;   $y = \dfrac{1}{3}$   $-\dfrac{17}{3}$

**40.** $15x - 9$;   $x = -\dfrac{1}{3}$   −14

**41.** $2p^2 - pq - 3q^2$;   $p = -2, q = 3$   −13

**42.** $x^2 + 5xy - y^2$;   $x = 2, y = -1$   −7

**43.** $4a^3 + 6ab^2 - 9b^3$;   $a = \dfrac{1}{2}, b = \dfrac{1}{3}$   $\dfrac{1}{2}$

**44.** $3h^2 - 10hk - 5k^2$;   $h = -\dfrac{1}{3}, k = \dfrac{1}{5}$   $\dfrac{4}{5}$

**Add the polynomials. Write each answer in standard form. (See Objective 4.)**

**45.** $(3p + 2) + (6p + 7)$   $9p + 9$

**46.** $(6p^2 + 5) + (12p^2 + 8)$   $18p^2 + 13$

**47.** $(-p^2 + 5p - 6) + (2p^2 - 4p + 3)$   $p^2 + p - 3$

**48.** $(2p^2 + p - 5) + (p^2 - 2p + 1)$   $3p^2 - p - 4$

**49.** $4y + (6y^2 - 2y + 1)$   $6y^2 + 2y + 1$

**50.** $3y + (4y^3 - 2y^2 + 5y - 7)$   $4y^3 - 2y^2 + 8y - 7$

**51.** $(-3x^2 + 2) + (4x^3 - 2x^2 + x)$   $4x^3 - 5x^2 + x + 2$

**52.** $(10x^2 + 7) + (2x^3 - 5x^2 + 3x)$   $2x^3 + 5x^2 + 3x + 7$

**53.** $\left(\dfrac{1}{2}x^2 - \dfrac{5}{4}x + \dfrac{2}{3}\right) + \left(-\dfrac{1}{4}x^2 + \dfrac{1}{4}x - \dfrac{5}{3}\right)$   $\dfrac{1}{4}x^2 - x - 1$

**54.** $\left(\dfrac{2}{3}x^2 - \dfrac{5}{6}x - \dfrac{3}{4}\right) + \left(-\dfrac{1}{3}x^2 - \dfrac{1}{6}x + \dfrac{1}{4}\right)$   $\dfrac{1}{3}x^2 - x - \dfrac{1}{2}$

**55.**
$$5x^2 - 13x + 11$$
$$+(-2x^2 + 15x - 7)$$
$$\overline{\phantom{+}3x^2 + 2x + 4}$$

**56.**
$$x^2 + 3x - 1$$
$$+(-3x^2 + x + 9)$$
$$\overline{\phantom{+}-2x^2 + 4x + 8}$$

**57.**
$$-12y^2 + 23y - 2$$
$$+ 8y^2 \phantom{+ 23y} + 17$$
$$\overline{-4y^2 + 23y + 15}$$

**58.**
$$-14y + 32$$
$$+(-2y^2 + 3y - 19)$$
$$\overline{-2y^2 - 11y + 13}$$

**59.** $(10x^2 + 7xy + y^2) + (2x^2 - 5xy - 3y^2)$   $12x^2 + 2xy - 2y^2$

**60.** $(-a^2 + 2ab + 3b^2) + (2a^2 - 7ab - b^2)$   $a^2 - 5ab + 2b^2$

**61.** $(6h^2 + hk + 5k^2) + (-2h^2 - 4hk - k^2)$   $4h^2 - 3hk + 4k^2$

**62.** $(10x^2 + 7xy + y^2) + (2x^2 - 5xy - 3y^2)$   $12x^2 + 2xy - 2y^2$

**63.** Add $(6y^3 - 2y^2 + 3y - 1)$ and $(4y^3 + 2y^2 - 4y + 1)$.   $10y^3 - y$

**64.** Add $(5y^3 - y^2 - 4y + 3)$ and $(-6y^3 - y^2 - 4y - 3)$.   $-y^3 - 2y^2 - 8y$

**65.** $(1.6y^3 - 3.4y^2 + 1.1y - 0.9)$
$+ (0.5y^3 + 2.1y^2 - 3.9y + 2.5)$   $2.1y^3 - 1.3y^2 - 2.8y + 1.6$

**66.** $(0.5y^3 - 0.1y^2 + 2.3y + 4.7) + (-1.2y^3 + 3.5y^2$
$- 1.4y - 3.2)$   $-0.7y^3 + 3.4y^2 + 0.9y + 1.5$

**Subtract the polynomials. Write each answer in standard form. (See Objective 5.)**

**67.** $(2x + 1) - (3x - 4)$   $-x + 5$

**68.** $(5x - 2) - (4x + 3)$   $x - 5$

**69.** Subtract $(x^2 - 3x + 10)$ from $(5x + 15)$.   $-x^2 + 8x + 5$

**70.** Subtract $(x^2 - 5x - 18)$ from $(2x - 25)$.   $-x^2 + 7x - 7$

**71.** Subtract $(4x^3 - 3x^2 + 5)$ from $(x^3 - 2x - 4)$.   $-3x^3 + 3x^2 - 2x - 9$

**72.** Subtract $(6x^3 - 5x^2 + 13)$ from $(2x^3 - 15x - 24)$.   $-4x^3 + 5x^2 - 15x - 37$

**73.** $(-2y^3 + 2y^2 + 5) - (5y^3 + 2y^2 - 1)$   $-7y^3 + 6$

**74.** $(-4x^2 + 10x) - (6x + 50)$   $-4x^2 + 4x - 50$

**75.** $\left(\dfrac{1}{2}x^2 - \dfrac{5}{4}x + \dfrac{2}{3}\right) - \left(-\dfrac{1}{4}x^2 + \dfrac{1}{4}x - \dfrac{5}{3}\right)$   $\dfrac{3}{4}x^2 - \dfrac{3}{2}x + \dfrac{7}{3}$

**76.** $\left(\dfrac{2}{3}x^2 - \dfrac{5}{6}x - \dfrac{3}{4}\right) - \left(-\dfrac{1}{3}x^2 - \dfrac{1}{6}x + \dfrac{1}{4}\right)$   $x^2 - \dfrac{2}{3}x - 1$

**77.** $(0.3y^2 - 2.7y) - (1.5y^2 - 0.8y + 2.9)$   $-1.2y^2 - 1.9y - 2.9$

**78.** $(-4.5y^2 - 5.7) - (2.1y^2 + 3.1y + 0.3)$   $-6.6y^2 - 3.1y - 6$

**79.**
$$5x^2 - 12x + 7$$
$$- (-2x^2 \phantom{+} + 5x - 3)$$
$$\overline{\phantom{-}7x^2 - 17x + 10}$$

**80.**
$$-3x^2 + 5x - 4$$
$$- (2x^2 + 6x - 1)$$
$$\overline{-5x^2 - x - 3}$$

**81.**
$$-5x^2 \phantom{- 12x} + 9$$
$$- (-4x^2 + 14x - 7)$$
$$\overline{-x^2 - 14x + 16}$$

**82.**
$$-8x^2 \phantom{+ 5x} - 10$$
$$- (3x^2 - 11x - 4)$$
$$\overline{-11x^2 + 11x - 6}$$

**83.** $(10x^2 + 7xy + y^2) - (2x^2 - 5xy - 3y^2)$   $8x^2 + 12xy + 4y^2$

**84.** $(-a^2 + 2ab + 3b^2) - (2a^2 - 7ab - b^2)$   $-3a^2 + 9ab + 4b^2$

**85.** $(6h^2 + hk + 5k^2) - (-2h^2 - 4hk - k^2)$   $8h^2 + 5hk + 6k^2$

**86.** $(13x^2 - 3xy + 2y^2) - (5x^2 + 6xy - y^2)$   $8x^2 - 9xy + 3y^2$

**87.** $(10x^3 + 15x^2y - 8xy^2 + 20y^3) - (15x^3 - 30x^2y$
$- 16xy^2 + 25y^3)$   $-5x^3 + 45x^2y + 8xy^2 - 5y^3$

**88.** $(16x^3 - 13x^2y + 14xy^2 - 12y^3) - (18x^3 + 12x^2y$
$- 13xy^2 + 13y^3)$   $-2x^3 - 25x^2y + 27xy^2 - 25y^3$

**89.** $(2x^2 - 6x + 14) + (8x^2 + 11x - 3) - (10x^2 - 5x + 21)$   $10x - 10$

**90.** $(-4y^2 + 10y - 6) + (7y^2 + y + 14) - (5y^2 - 8y + 24)$   $-2y^2 + 19y - 16$

**Solve each problem. (See Objective 6.)**

**91.** The height, in feet, of an object $t$ seconds after it is thrown from the top of a 500-ft building is $h(t) = -16t^2 + 32t + 500$. Find the height of the object
**a.** 1 sec after it is thrown.   516 ft
**b.** 4 sec after it is thrown.   372 ft

**92.** The height, in feet, of an object $t$ seconds after it is thrown off the top of a 350-ft building is $h(t) = -16t^2 + 32t + 350$. Find the height of the object
**a.** 2 sec after it is thrown.   350 ft
**b.** 5 sec after it is thrown.   110 ft

**For Exercises 93–96, find an expression that represents the perimeter of the figure shown.**

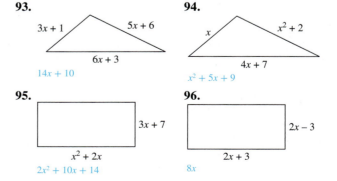

**93.**
$3x + 1$   $5x + 6$
$6x + 3$
$14x + 10$

**94.**
$x$   $x^2 + 2$
$4x + 7$
$x^2 + 5x + 9$

**95.**
$3x + 7$
$x^2 + 2x$
$2x^2 + 10x + 14$

**96.**
$2x - 3$
$2x + 3$
$8x$

**97.** A company's total revenue from selling $x$ quilts is $40x - 0.05x^2$ dollars. What is the revenue generated from selling 500 quilts? $7500

**98.** A company's total revenue from the sale of $x$ fountain pens is $20x - 0.01x^2$ dollars. What is the revenue generated from the sale of 100 fountain pens? $1900

**99.** Suppose the revenue generated by selling $x$ items is $x^3 - 8x^2 + 30x + 15$ dollars and the total cost of making $x$ items is $50x^2 - 25,000x$ dollars. Find an expression that represents the profit. Write the answer in standard form. $x^3 - 58x^2 + 25,030x + 15$ dollars

**100.** Suppose the revenue generated by selling $x$ items is $40x - 0.4x^2$ dollars and the total cost of making $x$ items is $3x + 9$ dollars. Find an expression that represents the total profit. Write the answer in standard form. $-0.4x^2 + 37x - 9$ dollars

 ## Mix 'Em Up!

**Classify each polynomial as a monomial, binomial, or trinomial and write it in standard form.**

**101.** $5 - 4x$   binomial, $-4x + 5$

**102.** $3 - 2x^2 + 11x$   trinomial, $-2x^2 + 11x + 3$

**Identify the leading coefficient and degree of each polynomial.**

**103.** $-12x^3 + 5x - 9$   leading coefficient –12, degree 3

**104.** $9x^4 - 2x^2 + 6$   leading coefficient 9, degree 4

**105.** $-2 + 4x^2y - 3x^4y^3 + 6xy^5$   leading coefficient –3, degree 7

**106.** $a^2 - ab^3 + 4ab^4 - 5b^2$   leading coefficient 4, degree 5

**Evaluate each polynomial for the given value(s).**

**107.** $-2a^2 + 5a + 7$;   $a = -3$   –26

**108.** $5b^2 - b + 9$;   $b = -2$   31

**109.** $7x^2 - xy - y^2$;   $x = -2, y = 4$   20

**110.** $3x^2 + 6xy - 10y^2$;   $x = 2, y = -1$   –10

**111.** $8x^3 + 2x - 22$;   $x = 2$   46

**112.** $-y^2 + 12$;   $y = -4$   –4

**113.** The height, in feet, of a ball projected upward from the top of an 80-ft building with an initial velocity of 64 ft/sec is given by $-16t^2 + 64t + 80$, where $t$ is the number of seconds after the ball is thrown. Find the height of the ball
   **a.** 1 sec after it is thrown.   **b.** 2 sec after it is thrown.
   128 ft                              144 ft

**114.** The height, in feet, of a ball projected upward from the top of a 400-ft building with an initial velocity of 128 ft/sec is given by $-16t^2 + 128t + 400$, where $t$ is the number of seconds after the ball is thrown. Find the height of the ball
   **a.** 4 sec after it is thrown.   **b.** 8 sec after it is thrown.
   656 ft                             400 ft

**115.** A company's total cost to make $x$ items is $10x^3 - x^2 + 50$ dollars. How much does it cost the company to make 10 items? $9950

**116.** A company's total cost to make $x$ items is $2x^2 + 3x + 12$ dollars. How much does it cost the company to make 8 items? $164

**Perform each operation. Write each answer in standard form.**

**117.** $(7x^3 + 4x^2 - 14) + (5 - 3x^2 - 8x^3)$   $-x^3 + x^2 - 9$

**118.** $(7x^4 + 3x^3 - 10) - (x^4 - 2x^3 + 9)$   $6x^4 + 5x^3 - 19$

**119.**
$$-\tfrac{2}{3}x^2 + \tfrac{6}{7}x - \tfrac{3}{4}$$
$$+\left(\tfrac{5}{3}x^2 + \tfrac{3}{7}x + \tfrac{7}{4}\right)$$
$$\overline{\phantom{-0.8x^2 + 3.7x - 5.4}}$$

**120.**
$$-\tfrac{4}{5}x^2 + \tfrac{1}{2}x - \tfrac{3}{4}$$
$$+\left(-\tfrac{1}{5}x^2 + \tfrac{1}{3}x - \tfrac{7}{4}\right)$$

**121.**
$$-0.8x^2 + 3.7x - 5.4$$
$$- (1.4x^2 + 6.8x - 3.1)$$
$-2.2x^2 - 3.1x - 2.3$

**122.**
$$4.2x^2 - 7.5x + 1.3$$
$$- (-3.2x^2 - 5.6x + 0.7)$$
$7.4x^2 - 1.9x + 0.6$

**123.** $(6x^2 + 5x - 1) - (-x^2 + x + 3) + (-4x^2 - 2x + 7)$   $3x^2 + 2x + 3$

**124.** $(3y^2 - 4y - 5) + (-7y + 10) - (-4y^2 + 9y - 18)$   $7y^2 - 20y + 23$

**125.** $(5a - 7ab + b) - (-9a + 8ab + 3b) + (-4a - 3ab + 6b)$   $10a - 18ab + 4b$

**126.** $(11c + 2cd + 5d) + (-9c + 6cd - 12d) - (4c - 10cd + d)$   $-2c + 18cd - 8d$

**127.** Subtract $(-4h^2 + 3hk - 2k^2)$ from $(-3h^2 - 7hk + 8k^2)$.   $h^2 - 10hk + 10k^2$

**128.** Add $(15x^2 - 4xy + 11y^2)$ and $(-3x^2 + 8xy - 7y^2)$.   $12x^2 + 4xy + 4y^2$

**129.** Add $(7.5a^2 + 2.4ab - 9.8b^2)$ and $(-2.4a^2 - 1.8ab + 3.2b^2)$.   $5.1a^2 + 0.6ab - 6.6b^2$

**130.** Subtract $(1.2x^2 + 5.6xy - 3.6y^2)$ from $(-1.5x^2 - 2.8xy + 4.6y^2)$.   $-2.7x^2 - 8.4xy + 8.2y^2$

**131.** The profit generated by selling $x$ items is $-0.6x^2 + 54x + 15$ dollars and the total cost is $10x + 200$ dollars. Find an expression that represents the company's revenue. Write the answer in standard form.   $-0.6x^2 + 64x + 215$ dollars

**132.** The profit generated by selling $x$ items is $84x - 0.7x^2$ dollars and the total cost is $9x + 500$ dollars. Find an expression that represents the company's total revenue. Write the answer in standard form.   $-0.7x^2 + 93x + 500$ dollars

**Find the perimeter of each figure.**

**133.**

$x^2 - 8$

$13x + 4$          $13x + 4$

$3x^2 - 15x$

$4x^2 + 11x$

**134.**
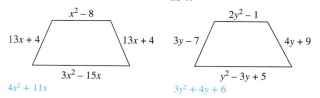
$2y^2 - 1$

$3y - 7$          $4y + 9$

$y^2 - 3y + 5$

$3y^2 + 4y + 6$

## You Be the Teacher!

**Correct each student's errors, if any.**

**135.** Subtract: $(4x^4 - 2x^2 + 3x - 1) - (3x^4 + 2x^2 - 3x + 1)$

Robin's work: $(4x^4 - 2x^2 + 3x - 1) - (3x^4 + 2x^2 - 3x + 1)$
$= 4x^4 - 2x^2 + 3x - 1 - 3x^4 + 2x^2 - 3x + 1$
$= 4x^4 - 3x^4 - 2x^2 + 2x^2 + 3x - 3x - 1 + 1$
$= x^4$

**136.** Subtract: $(3x^3 - 6x^2 + 5) - (-3x^3 + 6x^2 + 4)$

Brad's work: $(3x^3 - 6x^2 + 5) - (-3x^3 + 6x^2 + 4)$
$= 3x^3 - 6x^2 + 5 + 3x^3 + 6x^2 + 4$
$= 3x^3 + 3x^3 - 6x^2 + 6x^2 + 5 + 4$
$= 6x^3 + 9$

**137.** Evaluate: $-6x^2 - 2x + 3$ for $x = -2$.

Cameron's work:

$-6(-2)^2 - 2(-2) + 3$
$= (12)^2 + 4 + 3$
$= 144 + 7 = 151$

$-6(-2)^2 - 2(-2) + 3$
$= -6(4) + 4 + 3$
$= -24 + 7$
$= -17$

**138.** Evaluate: $x^3 - 4x^2 + 5$ for $x = -1$.

Jacquesha's work:

$(-1)^3 - 4(-1)^2 + 5$
$= -3 - 4(-2) + 5$
$= -3 + 8 + 5$
$= 5 + 5$
$= 10$

$(-1)^3 - 4(-1)^2 + 5$
$= -1 - 4(1) + 5$
$= -1 - 4 + 5$
$= -5 + 5$
$= 0$

 **Calculate It!**

Use a calculator to evaluate each polynomial for the given value.

**139.** $15x^3 - 12x^2 + 11x - 10$;   $x = 15$   48,080

**140.** $-4x^3 + 21x - 130$;   $x = -11$   4963

**141.** $8x^4 - 23x^3 + 5x^2 + 16$;   $x = -8$   44,880

**142.** $10x^3 + 16x^2 - 14x + 15$;   $x = 14$   30,395

**143.** $0.5x^4 - 1.4x^3 + 3.6x^2 + 5.2x - 7.2$;   $x = -1.8$   8.5176

**144.** $-0.2x^3 - 0.4x^2 - 2.1x + 9.7$;   $x = -3.6$   21.4072

 **Think About It!**

Write an example of a polynomial that satisfies the given conditions.

**145.** A trinomial with degree 3 and leading coefficient of $-2$.
Answers vary, $-2x^3 + 5x - 1$

**146.** A trinomial with degree 4 and leading coefficient of 7.
Answers vary, $7x^4 - 6x^3 + 3$

**147.** A monomial with degree 0.
Answers vary, 6

**148.** A monomial with degree 1.
Answers vary, $-4x$

---

## PIECE IT TOGETHER       SECTIONS 5.1–5.4 Review

Simplify each expression by applying the properties of exponents. (*See Sections 5.1 and 5.2, Objectives 1–4.*)

**1.** $(-15a^7) \cdot (-4a^{11})$   $60a^{18}$

**2.** $(-3m^6)^5$   $-243m^{30}$

**3.** $-\left(-\dfrac{2x}{3y}\right)^5$   $\dfrac{32x^5}{243y^5}$

**4.** $\dfrac{8x^{13}}{2x^5}$   $4x^8$

**5.** $\dfrac{(5p)^{10}}{(5p)^7}$   $125p^3$

**6.** $(-15)^0$   1

**7.** $\left(\dfrac{2}{3}\right)^{-2}$   $\dfrac{9}{4}$

**8.** $\left(-\dfrac{5}{4}\right)^{-3}$   $-\dfrac{64}{125}$

**9.** $\dfrac{x^5}{y^{-6}}$   $x^5y^6$

**10.** $(-2s^{-4})^{-3}$   $-\dfrac{s^{12}}{8}$

**11.** $5^{-2} + 3^{-1}$   $\dfrac{28}{75}$

Write each number in scientific or standard notation. (*See Section 5.3, Objectives 1 and 2.*)

**12.** 0.0000009547   $9.547 \times 10^{-7}$

**13.** $2.38 \times 10^5$   238,000

**14.** $3.19 \times 10^{-6}$   0.00000319

Use exponent rules to simplify each expression. Write the answer in scientific notation. (*See Section 5.3, Objectives 3 and 4.*)

**15.** $(3.0 \times 10^{-4})(6.0 \times 10^{12})$   $1.8 \times 10^9$

**16.** $\dfrac{18.0 \times 10^{15}}{6.0 \times 10^{-6}}$   $3.0 \times 10^{21}$

Identify the leading coefficient and the degree of each polynomial. (*See Section 5.4, Objective 2.*)

**17.** $12x^2 - 45x + 3$   leading coefficient 12, degree 2

**18.** $54p^3 - 20p^2 + p - 16p^4 + 35$   leading coefficient $-16$, degree 4

Evaluate each polynomial for the given values of the variable. (*See Section 5.4, Objective 3.*)

**19.** $-3h^2 + 17$;   $h = -2$   5

**20.** $h^3 - h^2$;   $h = 5$   100

**21.** $5y^2 - 9$;   $y = \dfrac{1}{2}$   $-\dfrac{31}{4}$

Perform the indicated operation. Write each answer in standard form. (*See Section 5.4, Objectives 4 and 5.*)

**22.** $(9h^2 + 2hk + 5k^2) + (-2h^2 - 4hk - k^2)$   $7h^2 - 2hk + 4k^2$

**23.** $(4.2y^3 - 3.1y^2 + 5.1y - 8.9) + (2.5y^3 + 2.8y^2 - 1.6y + 12.5)$   $6.7y^3 - 0.3y^2 + 3.5y + 3.6$

**24.** $(y^2 - 7y + 16) - (5y^2 - 4y - 23)$   $-4y^2 - 3y + 39$

**25.** $(8a^4 - 2a^2 + 4a - 7) - (2a^4 + 5a^2 - a + 6)$   $6a^4 - 7a^2 + 5a - 13$

| SECTION 5.5 | **Multiplication of Polynomials** |

► **OBJECTIVES**

As a result of completing this section, you will be able to

1. Multiply a monomial by a polynomial using the distributive property.
2. Multiply polynomials using the distributive property.
3. Multiply polynomials using the vertical format.
4. Write expressions that represent real-life situations.
5. Troubleshoot common errors.

When an airline company charges $100 per seat, they can sell 120 tickets. For each $10 increase in price, they sell one less ticket. If $x$ represents the number of increases in price, then $100 + 10x$ is the expression for the ticket price and $120 - x$ is the number of tickets sold. In this section, we will learn how to write an expression for the amount of money the airline company makes for this trip.

## Multiplying Monomials and Polynomials

**Objective 1** ►

Multiply a monomial by a polynomial using the distributive property.

In Section 5.4, we learned that we can add and subtract polynomials by combining like terms. To multiply polynomials, we use some skills that we have learned in previous sections. The tools we need are the product of like bases rule for exponents and the distributive property.

Recall the product of like bases rule for exponents enables us to multiply exponential expressions with the same base.

| Property | Example |
|----------|---------|
| $a^m a^n = a^{m+n}$ | $2x(5x) = 2 \cdot 5 \cdot x^1 \cdot x^1$ <br> $\qquad\quad = 10x^2$ |

Recall the distributive property enables us to multiply an expression by a sum or difference.

| Property | Example |
|----------|---------|
| $a(b + c) = ab + ac$ | $5(x + 4) = 5(x) + 5(4)$ <br> $\qquad\qquad = 5x + 20$ |

**Procedure:  Multiplying a Monomial by a Polynomial**

**Step 1:**  Distribute the monomial to each term of the polynomial.
**Step 2:**  Multiply the coefficients and multiply any like bases by adding the exponents.

**Objective 1  Examples**   **Multiply the expressions using the distributive property.**

**1a.** $5x(x + 4)$          **1b.** $3y^2(6y^2 - 2y + 9)$
**1c.** $-2a^3(a^3 - 8)$     **1d.** $4x^2y(3x^2y^2 - 2xy + 1)$

**Solutions**   **1a.** $5x(x + 4)$

$\quad = 5x(x) + 5x(4)$          Apply the distributive property.
$\quad = 5x^2 + 20x$             Simplify each product.

**1b.** $3y^2(6y^2 - 2y + 9)$

$\quad = 3y^2(6y^2) - 3y^2(2y) + 3y^2(9)$          Apply the distributive property.
$\quad = 3 \cdot 6 \cdot y^2 \cdot y^2 - 3 \cdot 2 \cdot y^2 \cdot y^1 + 3 \cdot 9 \cdot y^2$          Apply the commutative property.
$\quad = 18y^4 - 6y^3 + 27y^2$          Simplify each product.

**1c.** $-2a^3(a^3 - 8)$

$\quad = -2a^3(a^3) - (-2a^3)(8)$          Apply the distributive property.
$\quad = -2a^6 + 16a^3$          Simplify each product.

**1d.** $4x^2y(3x^2y^2 - 2xy + 1)$

$= 4x^2y(3x^2y^2) - 4x^2y(2xy) + 4x^2y(1)$

$= 4 \cdot 3 \cdot x^2 \cdot x^2 \cdot y \cdot y^2 - 4 \cdot 2 \cdot x^2 \cdot x \cdot y \cdot y + 4 \cdot 1x^2y$

$= 12x^4y^3 - 8x^3y^2 + 4x^2y$

---

**✓ Student Check 1**    Multiply the expressions using the distributive property.

**a.** $2y(y + 3)$    **b.** $8x^2(2x^2 - 5x + 1)$    **c.** $-7b^5(b^3 - 2)$    **d.** $3ab^3(5a^2b + ab - 1)$

---

## Multiplying Polynomials

**Objective 2** ▶

Multiply polynomials using the distributive property.

The distributive property can be used when we multiply a binomial by a polynomial. We will, in fact, have to use the distributive property twice. For instance, to multiply $(x + 3)$ by the polynomial $(x^2 + x + 6)$, we distribute the trinomial to each term in the binomial, $x$ and $3$. Then we apply the distributive property again to distribute $x$ and $3$ to each term in the trinomial, $(x^2 + x + 6)$.

$(x + 3)(x^2 + x + 6)$

$= x(x^2 + x + 6) + 3(x^2 + x + 6)$        Apply the distributive property.

$= x(x^2) + x(x) + x(6) + 3(x^2) + 3(x) + 3(6)$    Apply the distributive property.

$= x^3 + x^2 + 6x + 3x^2 + 3x + 18$        Multiply.

$= x^3 + x^2 + 3x^2 + 6x + 3x + 18$        Group like terms.

$= x^3 + 4x^2 + 9x + 18$            Add.

This multiplication results in each term of the binomial being multiplied by each term of the trinomial. So we can perform the multiplication by distributing the first term of the binomial to each term of the trinomial and then distributing the second term of the binomial to each term of the trinomial. (This is shown in the second step above.) This concept can be extended in order to multiply any two polynomials.

> **Procedure:  Multiplying Polynomials**
>
> **Step 1:** Distribute each term of the first polynomial to each term of the second polynomial.
> **Step 2:** Simplify the resulting products.
> **Step 3:** Combine like terms.
> **Step 4:** Write the polynomial in standard form.

---

**Objective 2  Examples**    Multiply the polynomials using the distributive property.

**2a.** $(x + 4)(x + 3)$        **2b.** $(2y + 4)(2y + 4)$        **2c.** $(x - 2)(x^2 + 2x + 4)$

**2d.** $(2m + 3n)(m - 5n)$    **2e.** Find the product of $(x^2 + 2x - 4)$ and $(x^2 - 8x + 3)$.

**Solutions**    **2a.** $(x + 4)(x + 3)$

$= x(x) + x(3) + 4(x) + 4(3)$        Apply the distributive property.

$= x^2 + 3x + 4x + 12$            Simplify each product.

$= x^2 + 7x + 12$                Combine like terms.

**2b.** $(2y + 4)(2y + 4)$

$= 2y(2y) + 2y(4) + 4(2y) + 4(4)$        Apply the distributive property.

$= 4y^2 + 8y + 8y + 16$            Simplify each product.

$= 4y^2 + 16y + 16$                Combine like terms.

> **Note:** *Another way to represent* $(2y + 4)(2y + 4)$ *is* $(2y + 4)^2$. *Note the base is* $(2y + 4)$ *and it is repeated as a factor two times.*

**2c.** Apply the distributive property and combine like terms.

$(x - 2)(x^2 + 2x + 4)$

$= x(x^2) + x(2x) + x(4) - 2(x^2) - 2(2x) - 2(4)$

$= x^3 + 2x^2 + 4x - 2x^2 - 4x - 8$

$= x^3 + 0x^2 + 0x - 8$

$= x^3 - 8$

**2d.** Apply the distributive property and combine like terms.

$(2m + 3n)(m - 5n)$

$= 2m(m) - 2m(5n) + 3n(m) - 3n(5n)$

$= 2m^2 - 10mn + 3mn - 15n^2$

$= 2m^2 - 7mn - 15n^2$

**2e.** $(x^2 + 2x - 4)(x^2 - 8x + 3)$

$= x^2(x^2) + x^2(-8x) + x^2(3) + 2x(x^2) + 2x(-8x) + 2x(3) - 4(x^2)$

$\quad - 4(-8x) - 4(3)$

$= x^4 - 8x^3 + 3x^2 + 2x^3 - 16x^2 + 6x - 4x^2 + 32x - 12$

$= x^4 - 6x^3 - 17x^2 + 38x - 12$

---

✓ **Student Check 2**   Multiply the polynomials using the distributive property.

**a.** $(x + 5)(x + 6)$     **b.** $(8y + 3)(8y + 3)$     **c.** $(a - 4)(a^2 + 4x + 16)$

**d.** $(4c - d)(6c + 7d)$     **e.** Find the product of $(y^2 - 5y + 7)$ and $(y^2 + 2y - 9)$.

---

## Multiplying Polynomials Vertically

**Objective 3** ▶

Multiply polynomials using the vertical format.

We can also multiply polynomials using a vertical format similar to the way we multiply numbers together. This process is illustrated in Example 3.

**Objective 3 Examples**   Perform each operation using a vertical format.

**3a.** Multiply $6x^2 - 2x + 5$ and $x - 3$.     **3b.** Multiply $4b^4 - 7b^2 + 2$ and $b - 5$.

**3c.** Multiply $(3x^2 - 2x + 1)$ and $(5x^2 + x - 3)$.

**Solutions**   **3a.** Write the product vertically.

$$
\begin{array}{r}
6x^2 - 2x + 5 \\
\times \qquad\quad x - 3 \\
\hline
-18x^2 + 6x - 15 \\
6x^3 - 2x^2 + 5x \qquad\quad \\
\hline
6x^3 - 20x^2 + 11x - 15
\end{array}
$$

Multiply $-3$ by $6x^2 - 2x + 5$.

Multiply $x$ by $6x^2 - 2x + 5$.

Line up like terms and combine them.

**3b.** Write the product vertically.

$$
\begin{array}{r}
4b^4 - 7b^2 + 2 \\
\times \qquad\quad b - 5 \\
\hline
-20b^4 \quad\; + 35b^2 \quad - 10 \\
4b^5 \qquad\; - 7b^3 \qquad\quad + 2b \qquad\quad \\
\hline
4b^5 - 20b^4 - 7b^3 + 35b^2 + 2b - 10
\end{array}
$$

Multiply $-5$ by $4b^4 - 7b^2 + 2$.

Multiply $b$ by $4b^4 - 7b^2 + 2$.

Line up like terms and combine them.

**3c.** Write the product vertically.

$$
\begin{array}{r}
3x^2 - 2x + 1 \\
\times\ \ 5x^2 +\ \ x - 3 \\
\hline
-9x^2 + 6x - 3 \\
3x^3 - 2x^2 +\ \ x \\
15x^4 - 10x^3 + 5x^2 \\
\hline
15x^4 -\ \ 7x^3 - 6x^2 + 7x - 3
\end{array}
$$

Multiply $-3$ by $3x^2 - 2x + 1$.

Multiply $x$ by $3x^2 - 2x + 1$.

Multiply $5x^2$ by $3x^2 - 2x + 1$.

Line up like terms and combine them.

 **Student Check 3** Perform each operation using a vertical format.

**a.** Multiply $7x^2 + 4x - 1$ by $2x - 1$. **b.** Multiply $9a^4 - a^2 + 5$ by $2a^2 - 3$.

**c.** Multiply $x^2 - 3x + 2$ by $5x^2 + x - 6$.

## Use Polynomials to Represent Real-Life Situations

**Objective 4** ▶

Write expressions that represent real-life situations.

In Chapter 6, we will learn how to solve real-life problems that involve multiplying polynomials. Example 4 illustrates how the expressions involved in the equations arise.

**Objective 4 Examples** **Write the product that represents the situation and find the resulting product.**

**4a.** When an airline company charges $100 per seat, they can sell 120 tickets. For each $10 increase in price, they sell one less ticket. If $x$ represents the number of increases in price, then $100 + 10x$ is the expression for the ticket price and $120 - x$ is the number of tickets sold. What simplified polynomial represents the amount of revenue the airline company earns for the trip?

**4b.** A homeowner plans to remodel his master bathroom. He wants to use an expensive tile to make a uniform border around the floor of the shower stall. If the shower stall measures 3 ft by 4 ft, what simplified polynomial represents the area of the shower floor that is not covered by the expensive tile?

**Solutions** **4a.** Revenue is calculated by multiplying the price per ticket by the number of tickets sold. To find the revenue, we must multiply the price, $(100 + 10x)$, by the number of tickets sold, $(120 - x)$.

$$
\begin{aligned}
\text{Revenue} &= (100 + 10x)(120 - x) \\
&= 100(120) + 100(-x) + 10x(120) + 10x(-x) \\
&= 12{,}000 - 100x + 1200x - 10x^2 \\
&= -10x^2 + 1100x + 12{,}000
\end{aligned}
$$

**4b.** Begin by drawing a diagram to determine the dimensions of the floor not covered by the expensive tile.

Let $x$ represent the width of the border with the expensive tile. Then the dimensions of the floor not covered by the expensive tile are $3 - 2x$ and $4 - 2x$.
The area of a rectangle is length times width. The expression that represents the area of the shower floor not covered by the expensive tile is

$$
\begin{aligned}
A &= (3 - 2x)(4 - 2x) \\
&= 3(4) + 3(-2x) - 2x(4) - 2x(-2x) \\
&= 12 - 6x - 8x + 4x^2 \\
&= 4x^2 - 14x + 12
\end{aligned}
$$

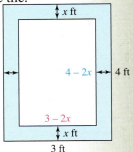

✓ **Student Check 4**   Write a product that represents the situation and find the resulting product.

**a.** An apartment manager can rent all 150 units if he charges \$400 per month rent. If he increases the price by \$50, he rents two fewer apartments. If $x$ represents the number of increases in price, then the rent for an apartment is given by $400 + 50x$ and the number of apartments rented is $150 - 2x$. What simplified polynomial represents the total rent collected each month?

**b.** A gardener has a rectangular garden that measures 5 ft by 7 ft. She wants to put a uniform border of gravel around each side of the garden. If $x$ represents the width of the border, what simplified polynomial represents the area of the garden and the border?

**Objective 5** ▶
Troubleshoot common errors.

## Troubleshooting Common Errors

Some common errors for multiplying polynomials are shown next.

**Objective 5 Examples**   **A problem and an incorrect solution are given. Provide the correct solution as well as an explanation of the error.**

**5a.** $-5x(3x^2 + 4x - 1)$

| Incorrect Solution | Correct Solution and Explanation |
|---|---|
| $-5x(3x^2 + 4x - 1)$ $= -15x^3 - 20x - 5x$ $= -15x^3 - 25x$ | When $-5x$ and $4x$ are multiplied, we get $-20x^2$. We must add the exponents of $x$ together. Also, when $-5x$ is distributed to the last term, $-1$, we get $5x$. $$-5x(3x^2 + 4x - 1) = -15x^3 - 20x^2 + 5x$$ |

**5b.** $(x - 2)(x + 7)$

| Incorrect Solution | Correct Solution and Explanation |
|---|---|
| $(x - 2)(x + 7)$ $= x + 7x - 2x - 14$ $= 6x - 14$ | The error was made in multiplying the first two terms. We should get $x \cdot x = x^2$. $$(x - 2)(x + 7) = x^2 + 7x - 2x - 14$$ $$= x^2 + 5x - 14$$ |

## ANSWERS TO STUDENT CHECKS

**Student Check 1**   **a.** $2y^2 + 6y$   **b.** $16x^4 - 40x^3 + 8x^2$
**c.** $-7b^8 + 14b^5$   **d.** $15a^3b^4 + 3a^2b^4 - 3ab^3$
**Student Check 2**   **a.** $x^2 + 11x + 30$   **b.** $64y^2 + 48y + 9$
**c.** $a^3 - 64$   **d.** $24c^2 + 22cd - 7d^2$
**e.** $y^4 - 3y^3 - 12y^2 + 59y - 63$

**Student Check 3**   **a.** $14x^3 + x^2 - 6x + 1$
**b.** $10a^6 - 29a^4 + 13a^2 - 15$
**c.** $5x^4 - 14x^3 + x^2 + 20x - 12$
**Student Check 4**   **a.** $-100x^2 + 6700x + 60,000$
**b.** $4x^2 + 24x + 35$

## SUMMARY OF KEY CONCEPTS

1. Polynomial multiplication is based on the distributive property and also uses the product rule for exponents, $x^m \cdot x^n = x^{m+n}$.

2. To multiply polynomials, distribute each term of the first polynomial to each term of the second polynomial. Combine like terms to simplify the resulting expression. Polynomials can also be multiplied using a vertical format.

3. Be very careful to apply the properties of exponents correctly. Remember that
$$x^2 \cdot x^2 = x^4 \text{ but } x^2 + x^2 = 2x^2$$

4. Polynomials can be used to represent real-life situations. These problems will be visited later in the chapter when we learn how to solve equations containing polynomials.

## GRAPHING CALCULATOR SKILLS

The graphing calculator can be used to verify polynomial multiplication. Two different methods are illustrated here.

**Example 1:** Verify that $(x + 4)(x + 3) = x^2 + 7x + 12$.

**Method 1:** Enter the product in $Y_1$ and the result in $Y_2$. If our multiplication is correct, the $y$-values will agree for each value of $x$.

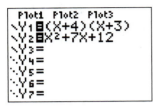

Since the $Y_1$ and $Y_2$ columns agree, the product is correct.

**Method 2:** Evaluate the product and its result at a specific value. If the two expressions agree for this input, then we have multiplied the polynomials correctly.

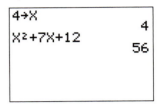

Since the two expressions yield 56 for $x = 4$, our multiplication is correct.

## SECTION 5.5 / EXERCISE SET

 **Write About It!**

**Use complete sentences in your answer to each exercise.**

1. Explain the procedure for multiplying two binomials.
   Answers vary.
2. Explain the procedure for multiplying two polynomials.
   Answers vary.

 **Practice Makes Perfect!**

**Multiply the expressions using the distributive property. (See Objective 1.)**

3. $6x(2x^2)$  $12x^3$
4. $5x^2(-7x^3)$  $-35x^5$
5. $2x(x + 5)$  $2x^2 + 10x$
6. $3x(4x - 6)$  $12x^2 - 18x$
7. $-x(x^2 + 4x - 1)$  $-x^3 - 4x^2 + x$
8. $-2x(3x^2 + 5x - 2)$  $-6x^3 - 10x^2 + 4x$
9. $4a^2b(2ab + 3)$  $8a^3b^2 + 12a^2b$
10. $2x^2y(3xy - 1)$  $6x^3y^2 - 2x^2y$
11. $3a^2(a^3 + 4a^2 + 4)$  $3a^5 + 12a^4 + 12a^2$
12. $5a^2(4a^3 - a^2 + 2a)$  $20a^5 - 5a^4 + 10a^3$
13. $-5b^3(b^5 - 7b^4)$  $-5b^8 + 35b^7$
14. $-6b^3(3b^6 - 2b^3)$  $-18b^9 + 12b^6$
15. $-10r^2s(3r^2s^2 - rs + 5)$  $-30r^4s^3 + 10r^3s^2 - 50r^2s$
16. $-11ab^2(9a^2b^2 - 2ab + 4)$  $-99a^3b^4 + 22a^2b^3 - 44ab^2$

**Multiply the polynomials using the distributive property. (See Objective 2.)**

17. $(x + 1)(x + 2)$  $x^2 + 3x + 2$
18. $(x + 5)(x + 2)$  $x^2 + 7x + 10$
19. $(x - 3)(x + 5)$  $x^2 + 2x - 15$
20. $(x - 2)(x + 6)$  $x^2 + 4x - 12$
21. $(2x - y)(x + 3y)$  $2x^2 + 5xy - 3y^2$
22. $(3x + y)(x - 4y)$  $3x^2 - 11xy - 4y^2$
23. $(3x + 5)(2x - 1)$  $6x^2 + 7x - 5$
24. $(4x - 3)(5x + 2)$  $20x^2 - 7x - 6$
25. $(3r + 2s)(r - s)$  $3r^2 - rs - 2s^2$
26. $(10a - 7b)(a - 2b)$  $10a^2 - 27ab + 14b^2$
27. $(a - 1)(a^2 - 3a + 2)$  $a^3 - 4a^2 + 5a - 2$
28. $(a + 2)(a^2 + 4a - 6)$  $a^3 + 6a^2 + 2a - 12$
29. $(a^2 + 3)(2a^2 - a + 4)$  $2a^4 - a^3 + 10a^2 - 3a + 12$
30. $(a^2 + 4)(6a^2 - 4a + 1)$  $6a^4 - 4a^3 + 25a^2 - 16a + 4$

31. $(6y - 1)(2y^2 + 3y + 1)$  $12y^3 + 16y^2 + 3y - 1$
32. $(5y + 2)(3y^2 - 4y + 2)$  $15y^3 - 14y^2 + 2y + 4$
33. $(3x - y)(9x^2 + 3xy + y^2)$  $27x^3 - y^3$
34. $(2a + 3b)(4a^2 - 6ab + 9b^2)$  $8a^3 + 27b^3$

**Multiply the polynomials using the vertical format. (See Objective 3.)**

35. $2x^2 - 3x + 1$
    $\times \quad 5x - 2$
    $10x^3 - 19x^2 + 11x - 2$
36. $3x^2 + 6x - 4$
    $\times \quad 5x + 1$
    $15x^3 + 33x^2 - 14x - 4$
37. $6x^2 + x - 4$
    $\times \quad -x + 5$
    $-6x^3 + 29x^2 + 9x - 20$
38. $-3x^2 - 5x + 2$
    $\times \quad 2x + 1$
    $-6x^3 - 13x^2 - x + 2$
39. $8x^2 \quad - 7$
    $\times \quad 2x + 3$
    $16x^3 + 24x^2 - 14x - 21$
40. $-10x^2 \quad + 4$
    $\times \quad 3x + 5$
    $-30x^3 - 50x^2 + 12x + 20$
41. $4x^2 + 3x - 1$
    $\times \quad 2x^2 - x + 6$
    $8x^4 + 2x^3 + 19x^2 + 19x - 6$
42. $-7x^2 + 2x + 5$
    $\times \quad 3x^2 - 4x - 6$
    $-21x^4 + 34x^3 + 49x^2 - 32x - 30$

43. $(2a^2 + 3a - 1)(a^2 - 2a + 4)$  $2a^4 - a^3 + a^2 + 14a - 4$
44. $(a^2 - 5a + 3)(a^2 - 3a + 1)$  $a^4 - 8a^3 + 19a^2 - 14a + 3$
45. Multiply $b^2 - 6b + 5$ by $b^2 + 4b - 2$.
    $b^4 - 2b^3 - 21b^2 + 32b - 10$
46. Multiply $3b^2 - 4b + 2$ by $b^2 - 7b + 7$.
    $3b^4 - 25b^3 + 51b^2 - 42b + 14$
47. Multiply $5y^2 + y - 1$ by $2y^2 + 3y - 2$.
    $10y^4 + 17y^3 - 9y^2 - 5y + 2$
48. Multiply $4y^2 + 5y - 3$ by $y^2 + 2y + 4$.
    $4y^4 + 13y^3 + 23y^2 + 14y - 12$
49. $(6x^2 - x - 3)(x^2 + 2x - 1)$  $6x^4 + 11x^3 - 11x^2 - 5x + 3$
50. $(7x^2 - x - 5)(8x^2 + 2x - 1)$  $56x^4 + 6x^3 - 49x^2 - 9x + 5$

**Write the product that represents the situation and find the resulting product. (See Objective 4.)**

51. When a hotel charges $140 per room, 80 rooms are rented. For each $5 increase in price, they rent one less room. If $x$ represents the number of price increases, then $140 + 5x$ represents the room charge and $80 - x$

Additional answers can be found in the Instructor Answer Appendix.

represents the number of rooms rented. What simplified polynomial represents the revenue the hotel earns? $-5x^2 + 260x + 11{,}200$

52. When a car rental company charges a daily rate of $25, 40 compact cars are rented. For each $3 increase in the daily rate, two fewer cars are rented. If $x$ represents the number of price increases, then $25 + 3x$ represents the daily rate and $40 - 2x$ represents the number of cars rented. What simplified polynomial represents the revenue the car rental company earns? $-6x^2 + 70x + 1000$

53. A rectangular garden measures 12 ft by 15 ft. A gravel path of equal width is to be laid around the garden. If $x$ is the width of the gravel path, what simplified polynomial represents the area of both the garden and the path? $4x^2 + 54x + 180$

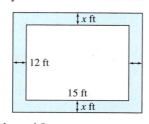

54. A picture measures 10 in. by 12 in. If the width of the frame is $x$ in., then find the simplified polynomial that represents the area of the framed picture.
$4x^2 + 44x + 120$

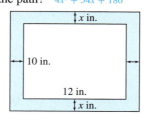

 ## Mix 'Em Up!

**Use the distributive property to multiply the expressions.**

55. $(4 - x)(x + 4)$  $-x^2 + 16$  56. $(1 - x)(2x + 3)$  $-2x^2 - x + 3$

57. $pq(3pq - 4)$  $3p^2q^2 - 4pq$  58. $ab(1 - 2ab)$  $-2a^2b^2 + ab$

59. $3a^2b(-2ab^4)$  $-6a^3b^5$  60. $-5x^3y(-3x^2y)$  $15x^5y^2$

61. $(a + 3)(a^3 - 2a + 8)$  $a^4 + 3a^3 - 2a^2 + 2a + 24$
62. $(a + 8)(a^2 - 6a + 9)$  $a^3 + 2a^2 - 39a + 72$
63. $-4n^2(3n^3 - 2n^2 + 1)$  $-12n^5 + 8n^4 - 4n^2$
64. $-6m^3(m^3 + 6m^2 + 5)$  $-6m^6 - 36m^5 - 30m^3$

65. $\left(\dfrac{2}{3}x - 2y\right)\left(\dfrac{1}{3}x + 5y\right)$  66. $\left(\dfrac{5}{4}a - 3b\right)\left(\dfrac{1}{4}a - b\right)$

67. $\left(5x + \dfrac{2}{3}\right)\left(2x - \dfrac{1}{3}\right)$  68. $\left(4x - \dfrac{2}{7}\right)\left(x + \dfrac{5}{7}\right)$

69. $\begin{array}{r} x^2 - 1.4x + 3.8 \\ \times\ \ 0.2x - 0.3 \end{array}$  70. $\begin{array}{r} x^2 + 0.9x - 1.5 \\ \times\ \ -0.3x + 0.4 \end{array}$

71. $(r + 2s)(r^2 - 2rs + 4s^2)$  $r^3 + 8s^3$

72. $(5x - y)(25x^2 + 5xy + y^2)$  $125x^3 - y^3$

73. Multiply $x^2 + 3x - 4$ by $2x^2 - 5x + 1$.
$2x^4 + x^3 - 22x^2 + 23x - 4$
74. Multiply $5x^2 - x + 6$ by $4x^2 - 3x + 2$.
$20x^4 - 19x^3 + 37x^2 - 20x + 12$

**Solve each problem.**

75. An apartment manager can rent all 100 units if he charges $600 per month rent. If he increases the price by $40, he rents two fewer apartments. If $x$ represents the number of increases in price, then the rent for an apartment is given by $600 + 40x$ dollars and the number of apartments rented is $100 - 2x$.

   a. What simplified polynomial represents the total rent collected each month? $-80x^2 + 2800x + 60{,}000$ dollars

   b. If the manager decides to increase the rent to $720 per month, how much total rent can he collect each month? $67{,}680

76. An apartment manager can rent all 120 units if he charges $700 per month rent. If he increases the price by $60, he rents one fewer apartment. If $x$ represents the number of increases in price, then the rent for an apartment is given by $700 + 60x$ and the number of apartments rented is $120 - x$.

   a. What simplified polynomial represents the total rent collected each month? $-60x^2 + 6500x + 84{,}000$

   b. If the manager decides to increase the rent to $820 per month, how much total rent can he collect each month? $96{,}760

77. Find the area of the square. $4y^2 + 4y + 1$ in.$^2$

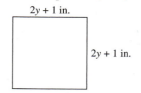

78. Find the area of the rectangle. $12x^2 - 8x$ cm$^2$

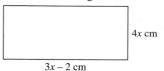

79. Find the area of the triangle. $9a^2 - 12a$ cm$^2$

80. Find the area of the triangle. $5x^2 - 2x$ mm$^2$

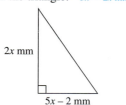

 ## You Be the Teacher!

**Correct each student's errors, if any.**

81. A rectangular swimming pool with a concrete sidewalk of uniform width that goes around it measures 50 m by 20 m. If $x$ represents the width of the sidewalk, what simplified polynomial represents the area of the pool?

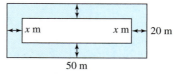

Christen's answer: $4x^2 + 140x + 1000$  $4x^2 - 140x + 1000$

**82.** A picture measures 8 in. by 12 in. A rectangular frame is placed around the picture. If the width of the frame is $x$ in., write a simplified polynomial that represents the area of the picture combined with its frame.

Jonathan's answer: $x^2 + 20x + 96$    $4x^2 + 40x + 96$

 **Calculate It!**

**Multiply the polynomials. Use a calculator to check each product.**

**83.** $(6 - x)(2x + 3)$        **84.** $(4 - x)(x + 3)$
   $-2x^2 + 9x + 18$                        $-x^2 + x + 12$
**85.** $(x^2 - 5x)(2x^2 + x - 4)$    $2x^4 - 9x^3 - 9x^2 + 20x$

**86.** $(3a^2 - a + 2)(5a^2 + 2a - 1)$    $15a^4 + a^3 + 5a^2 + 5a - 2$

---

## SECTION 5.6 — Special Products

### ▶ OBJECTIVES

As a result of completing this section, you will be able to

**1.** Simplify products of two binomials.

**2.** Simplify the square of a binomial.

**3.** Simplify the product of conjugates.

**4.** Simplify higher powers of binomials.

**5.** Troubleshoot common errors.

The area of a square in which each side is $x + 5$ ft is

$$(x + 5)(x + 5) \quad \text{or} \quad (x + 5)^2 \text{ ft}^2$$

In this section, we will learn how to simplify this special product, the square of a binomial.

$x + 5$ ft

### The Product of Binomials

In Section 5.5, we learned that polynomials can be multiplied using the distributive property. All of the problems in this section can be multiplied in the same manner. However, there are some products that repeatedly come up in algebra and it is helpful if we develop patterns to simplify these products. The first of these special products is the product of two binomials. Recall

$$(x + 3)(x + 4) = x(x) + x(4) + 3(x) + 3(4)$$
$$= x^2 + 4x + 3x + 12$$
$$= x^2 + 7x + 12$$

This product was formed by distributing $x$ to $(x + 4)$ and then $3$ to $(x + 4)$. To ensure that we get all of the terms in the product, we can remember the word FOIL.

**FOIL = First + Outer + Inner + Last**

$$(x + 3)(x + 4) = x^2 + 4x + 3x + 12$$

F O I L

#### Objective 1 ▶

Simplify products of two binomials.

---

**Procedure: Multiplying Two Binomials Using the FOIL Method**

**Step 1:** Find the product of the two first (F) terms of the binomials.

**Step 2:** Find the product of the two outer (O) terms of the binomials.

**Step 3:** Find the product of the two inner (I) terms of the binomials.

**Step 4:** Find the product of the two last (L) terms of the binomials.

**Step 5:** Simplify the products and combine like terms.

---

 **Note:** *It is very important to understand that this method applies only to the product of two binomials. This method does not extend to other types of products. Also note that this method is exactly what was covered in Section 5.5—it is simply that we are giving the process a name, the FOIL method.*

**Objective 1 Examples** **Find the product of the binomials using the FOIL method.**

**1a.** $(y - 5)(y - 7)$        **1b.** $(2x - 3)(5x + 2)$        **1c.** $(a^2 + 9)(a^2 - 4)$

**Solutions** **1a.**

$$\begin{array}{cccc} & F & O & I & L \\ (y - 5)(y - 7) = (y)(y) + (y)(-7) - 5(y) - 5(-7) \end{array}$$
$$= y^2 - 7y - 5y + 35$$
$$= y^2 - 12y + 35$$

**1b.**
$$\begin{array}{cccc} & F & O & I & L \\ (2x - 3)(5x + 2) = (2x)(5x) + (2x)(2) - 3(5x) - 3(2) \end{array}$$
$$= 10x^2 + 4x - 15x - 6$$
$$= 10x^2 - 11x - 6$$

**1c.**
$$\begin{array}{cccc} & F & O & I & L \\ (a^2 + 9)(a^2 - 4) = (a^2)(a^2) + (a^2)(-4) + 9(a^2) + 9(-4) \end{array}$$
$$= a^4 - 4a^2 + 9a^2 - 36$$
$$= a^4 + 5a^2 - 36$$

**✓ Student Check 1** Find the product of the binomials using the FOIL method.
**a.** $(x + 6)(x + 2)$      **b.** $(4y - 7)(3y + 8)$      **c.** $(a^2 - 2)(a^2 + 10)$

## The Square of a Binomial

**Objective 2** ▶

Simplify the square of a binomial.

We will examine how to simplify a product that involves the same binomial base, like $(x + 4)^2$, by applying the definition of the exponent. The power of 2 tells us to multiply the base together two times.

$$(x + 4)^2 = (x + 4)(x + 4)$$

This gives us the product of two binomials, so we can apply the FOIL method in order to multiply.

$$\begin{array}{cccc} & F & O & I & L \\ (x + 4)^2 = (x + 4)(x + 4) = (x)(x) + 4(x) + 4(x) + 4(4) \end{array}$$
$$= x^2 + 4x + 4x + 16$$
$$= x^2 + 8x + 16$$

So, $(x + 4)^2 = x^2 + 8x + 16$.

The FOIL method or distributive property will always give us the result of squaring a binomial but there is also a pattern that is worth noting. From the example, we can conclude the following about squaring a binomial.

1. The *square of a binomial* is *always* a trinomial.

2. The first term of the trinomial is the square of the first term of the binomial. $x^2$ is the first term of the trinomial and is the first term of the binomial squared:
$$(x)^2 = x^2$$

3. The middle term of the trinomial is always twice the product of the terms of the binomial. $8x$ is the middle term and is twice the product of the terms in the binomial:
$$2(x)(4) = 8x$$

4. The last term of the trinomial is the square of the last term of the binomial. 16 is the last term and is the second term of the binomial squared:
$$(4)^2 = 16$$

In mathematical notation, we write this relationship as follows.

> **Property:  The Square of a Binomial**
>
> For $a$ and $b$ real numbers,
> $$(a + b)^2 = a^2 + 2ab + b^2$$
> $$(a - b)^2 = a^2 - 2ab + b^2$$

We can verify this rule using geometry by creating a square whose length is $a + b$. The area of the large square is

$$A = (a + b)(a + b) = (a + b)^2$$

The area of the large square is also equal to the sum of the area of the four rectangles—that is,

$$A = a^2 + ab + ab + b^2 = a^2 + 2ab + b^2$$

So, we conclude that

$$(a + b)^2 = a^2 + 2ab + b^2$$

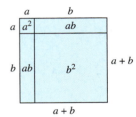

The trinomials that result from squaring a binomial are called **perfect square trinomials**.

> **Procedure:  Squaring a Binomial**
>
> **Step 1:** Square the first term.
> **Step 2:** Find twice the product of the terms in the binomial.
> **Step 3:** Square the last term.
> **Step 4:** Add the products found in steps 1 through 3.
> **Step 5:** Verify the result by FOILing.

**Objective 2  Examples**  Find each product.

**2a.** $(x + 6)^2$        **2b.** $(3a - 7b)^2$        **2c.** $(b^3 + 5)^2$

**Solutions**

**2a.** 
$$(x + 6)^2 = (x)^2 + 2(x)(6) + (6)^2$$ Apply the squaring pattern.
$$= x^2 + 12x + 36$$ Simplify.

**INSTRUCTOR NOTE:**
Point out that the last term in these trinomials is always positive.

**2b.** 
$$(3a - 7b)^2 = (3a)^2 - 2(3a)(7b) + (7b)^2$$ Apply the squaring pattern.
$$= 9a^2 - 42ab + 49b^2$$ Simplify.

**2c.** 
$$(b^3 + 5)^2 = (b^3)^2 + 2(b^3)(5) + (5)^2$$ Apply the squaring pattern.
$$= b^6 + 10b^3 + 25$$ Simplify.

**✓ Student Check 2**  Find each product.

**a.** $(y - 8)^2$        **b.** $(4x + 3y)^2$        **c.** $(a^4 - 3)^2$

## The Product of Conjugates

**Objective 3** ▶

Simplify the product of conjugates.

**Conjugates** are binomials with the same terms connected by opposite signs. For example,

$$x + 3 \text{ and } x - 3 \text{ are conjugates.}$$
$$4b + 5 \text{ and } 4b - 5 \text{ are conjugates.}$$

To find the product of conjugates, we can apply the FOIL method or distributive property.

$$(x + 3)(x - 3) = x^2 - 3x + 3x - 9 = x^2 - 9$$

$$(4b + 5)(4b - 5) = 16b^2 - 20b + 20b - 25 = 16b^2 - 25$$

In multiplying conjugates, note that the outer and inner products are opposites of one another and, therefore, add to zero. The product of the first terms minus the product of the last terms remains. Since the first and last terms are the same in conjugates, this is equivalent to the *difference of the squares* of the terms. That is,

$$
\begin{array}{ccccc}
 & \text{Square of} & & \text{Square of} & \\
 & \text{the first} & - & \text{the last} & \\
 & \text{term} & & \text{term} & \\
(x + 3)(x - 3) = & (x)^2 & - & (3)^2 & = x^2 - 9 \\
(4b + 5)(4b - 5) = & (4b)^2 & - & (5)^2 & = 16b^2 - 25
\end{array}
$$

**Property: The Product of Conjugates**

For $a$ and $b$ real numbers,

$$(a + b)(a - b) = a^2 - b^2$$

**Procedure: Finding the Product of Conjugates**

**Step 1:** Find the square of the first term.
**Step 2:** Find the square of the last term.
**Step 3:** Subtract the results from step 1 and step 2.
**Step 4:** Verify the result by FOILing.

**Note:** *The product of conjugates always results in the difference of two squares.*

---

**Objective 3 Examples** | Find each product.

**3a.** $(x + 8)(x - 8)$     **3b.** $(6a - 7b)(6a + 7b)$     **3c.** $(a^2 - 5)(a^2 + 5)$

**Solutions** | **3a.**

$$(x + 8)(x - 8) = (x)^2 - (8)^2 \qquad \text{Apply the product of conjugates property.}$$
$$= x^2 - 64 \qquad \text{Simplify.}$$

**3b.**
$$(6a - 7b)(6a + 7b) = (6a)^2 - (7b)^2 \qquad \text{Apply the product of conjugates property.}$$
$$= 36a^2 - 49b^2 \qquad \text{Simplify.}$$

**3c.**
$$(a^2 - 5)(a^2 + 5) = (a^2)^2 - (5)^2 \qquad \text{Apply the product of conjugates property.}$$
$$= a^4 - 25 \qquad \text{Simplify.}$$

---

☑ **Student Check 3** | Find each product.

    **a.** $(x + 1)(x - 1)$     **b.** $(9c - 2d)(9c + 2d)$     **c.** $(a^3 - 6)(a^3 + 6)$

## Higher Powers of Binomials

**Objective 4** ▶

Simplify higher powers of binomials.

We now examine how to raise a binomial to a larger exponent, such as 3 or 4. We begin by applying the definition of the exponent. That is, we repeat the base as a factor as indicated by the exponent.

$$(x + 2)^3 = (x + 2)(x + 2)(x + 2)$$    Repeat the base three times.

$$= (x + 2)(x + 2)^2$$    Rewrite as a product of one factor and the square of the factor.

$$= (x + 2)(x^2 + 4x + 4)$$    Apply the rule for squaring a binomial.

$$= x^3 + 4x^2 + 4x + 2x^2 + 8x + 8$$    Apply the distributive property.

$$= x^3 + 6x^2 + 12x + 8$$    Combine like terms.

$$(x + 2)^4 = (x + 2)(x + 2)(x + 2)(x + 2)$$    Repeat the base four times.

$$= (x + 2)^2(x + 2)^2$$    Rewrite as two factors which are both squares of $(x + 2)$.

$$= (x^2 + 4x + 4)(x^2 + 4x + 4)$$    Apply the rule for squaring a binomial.

$$= x^4 + 4x^3 + 4x^2 + 4x^3 + 16x^2 + 16x + 4x^2 + 16x + 16$$    Apply the distributive property.

$$= x^4 + 8x^3 + 24x^2 + 32x + 16$$    Combine like terms.

There are patterns that we could memorize to compute higher powers of binomials, but these formulas are beyond the scope of this course. For now, we can apply the rules for polynomial multiplication and special products.

> **Procedure:  Finding Higher Powers of a Binomial**
>
> **Step 1:** Rewrite as a product of the base repeated the number of times as the exponent states.
> **Step 2:** Rewrite pairs of factors as the square of the binomial and use the rule for square binomials to simplify these expressions.
> **Step 3:** Distribute the remaining polynomials to complete the multiplication.
> **Step 4:** Combine like terms and write the answer in standard form.

**Objective 4 Example**    Simplify the expression $(x - 5)^4$.

**Solution**

$$(x - 5)^4 = (x - 5)(x - 5)(x - 5)(x - 5)$$    Repeat the base four times.

$$= (x - 5)^2(x - 5)^2$$    Rewrite as two factors of $(x - 5)^2$.

$$= (x^2 - 10x + 25)(x^2 - 10x + 25)$$    Square each binomial.

$$= x^4 - 10x^3 + 25x^2 - 10x^3 + 100x^2 - 250x + 25x^2 - 250x + 625$$    Apply the distributive property.

$$= x^4 - 20x^3 + 150x^2 - 500x + 625$$    Combine like terms.

✓ **Student Check 4**    Simplify the expression $(y + 8)^3$.

## Troubleshooting Common Errors

**Objective 5** ▶

Troubleshoot common errors.

Some common errors related to special products are shown next.

**Objective 5 Examples** A problem and an incorrect solution are given. Provide the correct solution and an explanation of the error.

**5a.** $(x + 9)^2$

| Incorrect Solution | Correct Solution and Explanation |
|---|---|
| $(x + 9)^2 = x^2 + 81$ | The error was made in squaring each term in the base. Since the base is not a product, we must use the definition of the exponent or use the formula for squaring a binomial. $$(x + 9)^2 = x^2 + 18x + 81$$ |

**5b.** $(6x + 5y)(6x - 5y)$

| Incorrect Solution | Correct Solution and Explanation |
|---|---|
| $(6x + 5y)(6x - 5y)$ $= 36x^2 - 60xy - 25y^2$ | The error was made in combining the middle terms. Since the outer and inner products are opposites of one another, their sum is zero. $$(6x + 5y)(6x - 5y) = 36x^2 - 25y^2$$ |

## ANSWERS TO STUDENT CHECKS

**Student Check 1** **a.** $x^2 + 8x + 12$ **b.** $12y^2 + 11y - 56$
 **c.** $a^4 + 8a^2 - 20$

**Student Check 2** **a.** $y^2 - 16y + 64$ **b.** $16x^2 + 24xy + 9y^2$
 **c.** $a^8 - 6a^4 + 9$

**Student Check 3** **a.** $x^2 - 1$ **b.** $81c^2 - 4d^2$ **c.** $a^6 - 36$

**Student Check 4** $y^3 + 24y^2 + 192y + 512$

## SUMMARY OF KEY CONCEPTS

1. All of the products discussed in this section can be found by applying the distributive property. It is helpful to be able to apply the patterns shown in this section so that we can develop speed to simplify expressions.

2. Two binomials can be multiplied using the FOIL method. FOIL stands for *first, outer, inner*, and *last*.

3. The square of a binomial is *always* a trinomial. The middle term results from twice the product of the terms

in the binomial. The first and last terms are the squares of the first and last terms of the binomial.

4. Conjugates are expressions of the form $a + b$ and $a - b$. Their product is the difference of two squares.

5. Higher powers of binomials are found by repeating the base and then using the rules developed within this section and the last section to simplify them.

## SECTION 5.6 / EXERCISE SET

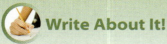 **Write About It!**

**Use complete sentences in your answer to each exercise.**

1. Explain the FOIL method.  Answers vary.

2. Is FOIL different than the distributive property? Explain.  Answers vary.

3. Explain why a trinomial results when a binomial is squared.  Answers vary.

4. Give examples of conjugates and explain what happens when they are multiplied.  Answers vary.

Additional answers can be found in the Instructor Answer Appendix.

## Practice Makes Perfect!

Find the product of the binomials using the FOIL method. Write each answer in standard form. (*See Objective 1.*)

**5.** $(x - 6)(x - 1)$   $x^2 - 7x + 6$    **6.** $(x - 2)(x - 4)$   $x^2 - 6x + 8$

**7.** $(x + 5)(x - 7)$   $x^2 - 2x - 35$    **8.** $(x + 6)(x - 9)$   $x^2 - 3x - 54$

**9.** $(3x - 2y)(5x + y)$   $15x^2 - 7xy - 2y^2$    **10.** $(2x + 3y)(4x - y)$   $8x^2 + 10xy - 3y^2$

**11.** $(a^2 + 1)(a^2 - 4)$   $a^4 - 3a^2 - 4$    **12.** $(a^2 + 9)(a^2 - 3)$   $a^4 + 6a^2 - 27$

**13.** $(4y^2 + 5)(3y^2 - 2)$   $12y^4 + 7y^2 - 10$    **14.** $(2y^2 - 1)(3y^2 + 4)$   $6y^4 + 5y^2 - 4$

**15.** $(2r^2 - 6s)(6r + 8s)$   $12r^3 + 16r^2s - 36rs - 48s^2$    **16.** $(3a^2 + 7b)(a - 3b)$   $3a^3 - 9a^2b + 7ab - 21b^2$

**17.** $\left(2x - \dfrac{3}{5}\right)\left(3x + \dfrac{2}{5}\right)$    **18.** $\left(5x - \dfrac{3}{4}\right)\left(x - \dfrac{1}{2}\right)$

**19.** $(1.5x + 0.6y)(2.4x + 1.8y)$   $3.6x^2 + 4.14xy + 1.08y^2$

**20.** $(0.8a - 1.7b)(0.6a + 2.8b)$   $0.48a^2 + 1.22ab - 4.76b^2$

**21.** $(1 - 3x)(4x + 5)$   $-12x^2 - 11x + 5$    **22.** $(6x - 5)(2 - 3x)$   $-18x^2 + 27x - 10$

Use a special product to multiply. Write each answer in standard form. (*See Objective 2.*)

**23.** $(x + 4)^2$   $x^2 + 8x + 16$    **24.** $(x + 9)^2$   $x^2 + 18x + 81$

**25.** $(2y - 1)^2$   $4y^2 - 4y + 1$    **26.** $(3y - 2)^2$   $9y^2 - 12y + 4$

**27.** $(a^2 - 2)^2$   $a^4 - 4a^2 + 4$    **28.** $(a^2 - 3)^2$   $a^4 - 6a^2 + 9$

**29.** $(4c + 5d)^2$   $16c^2 + 40cd + 25d^2$    **30.** $(5a - 6b)^2$   $25a^2 - 60ab + 36b^2$

**31.** $\left(3a - \dfrac{2}{5}\right)^2$   $9a^2 - \dfrac{12}{5}a + \dfrac{4}{25}$    **32.** $\left(2b - \dfrac{1}{3}\right)^2$   $4b^2 - \dfrac{4}{3}b + \dfrac{1}{9}$

**33.** $(1.5p - 0.6)^2$   $2.25p^2 - 1.8p + 0.36$    **34.** $(0.9p + 2.5)^2$   $0.81p^2 + 4.5p + 6.25$

**35.** $(3x^2 - 7y)^2$   $9x^4 - 42x^2y + 49y^2$    **36.** $(5a^2 + 8b)^2$   $25a^4 + 80a^2b + 64b^2$

Use a special product to multiply. Write each answer in standard form. (*See Objective 3.*)

**37.** $(x - 7)(x + 7)$   $x^2 - 49$    **38.** $(x + 4)(x - 4)$   $x^2 - 16$

**39.** $(x + 2y)(x - 2y)$   $x^2 - 4y^2$    **40.** $(a + 3b)(a - 3b)$   $a^2 - 9b^2$

**41.** $(2y + 4)(2y - 4)$   $4y^2 - 16$    **42.** $(3y + 5)(3y - 5)$   $9y^2 - 25$

**43.** $(7r - 3s)(7r + 3s)$   $49r^2 - 9s^2$    **44.** $(9p + 2q)(9p - 2q)$   $81p^2 - 4q^2$

**45.** $(1.3x - 2.5)(1.3x + 2.5)$   $1.69x^2 - 6.25$    **46.** $(0.8y + 1.5)(0.8y - 1.5)$   $0.64y^2 - 2.25$

**47.** $(a^2 + 1)(a^2 - 1)$   $a^4 - 1$    **48.** $(a^2 + 4b^2)(a^2 - 4b^2)$   $a^4 - 16b^4$

**49.** $(c^3 - 8d^3)(c^3 + 8d^3)$   $c^6 - 64d^6$    **50.** $(b^3 - 9)(b^3 + 9)$   $b^6 - 81$

**51.** $\left(x^2 - \dfrac{1}{7}\right)\left(x^2 + \dfrac{1}{7}\right)$    **52.** $\left(y^2 - \dfrac{4}{5}\right)\left(y^2 + \dfrac{4}{5}\right)$

**53.** $\left(\dfrac{3}{4}c^3 + d\right)\left(\dfrac{3}{4}c^3 - d\right)$    **54.** $\left(\dfrac{2}{9}a^3 + b\right)\left(\dfrac{2}{9}a^3 - b\right)$

Perform the indicated operation. Write each answer in standard form. (*See Objective 4.*)

**55.** $(x + 1)^3$   $x^3 + 3x^2 + 3x + 1$    **56.** $(x + 4)^3$   $x^3 + 12x^2 + 48x + 64$

**57.** $(y - 2)^3$   $y^3 - 6y^2 + 12y - 8$    **58.** $(y - 3)^3$   $y^3 - 9y^2 + 27y - 27$

**59.** $(2y - 3)^3$   $8y^3 - 36y^2 + 54y - 27$    **60.** $(3y + 1)^3$   $27y^3 + 27y^2 + 9y + 1$

**61.** $\left(b + \dfrac{1}{2}\right)^3$    **62.** $\left(c - \dfrac{1}{5}\right)^3$

**63.** $(a + 1)^4$   $a^4 + 4a^3 + 6a^2 + 4a + 1$    **64.** $(a + 2)^4$   $a^4 + 8a^3 + 24a^2 + 32a + 16$

## Mix 'Em Up!

Perform the indicated operation. Write each answer in standard form.

**65.** $(x - 12y)(x + 12y)$   $x^2 - 144y^2$    **66.** $(a + 13b)(a - 13b)$   $a^2 - 169b^2$

**67.** $(3x + 9)(2x - 5)$   $6x^2 + 3x - 45$    **68.** $(8x + 3)(9x - 4)$   $72x^2 - 5x - 12$

**69.** $(3a + 2)^2$   $9a^2 + 12a + 4$    **70.** $(5b - 7)^2$   $25b^2 - 70b + 49$

**71.** $(y + 11)^3$   $y^3 + 33y^2 + 363y + 1331$    **72.** $(a + 6)^3$   $a^3 + 18a^2 + 108a + 216$

**73.** $\left(a^2 - \dfrac{3}{7}\right)\left(a^2 + \dfrac{3}{7}\right)$    **74.** $\left(b - \dfrac{4}{3}\right)\left(b + \dfrac{4}{3}\right)$

**75.** $\left(\dfrac{1}{2}c^3 + 5d\right)\left(\dfrac{1}{2}c^3 - 5d\right)$    **76.** $\left(2a - \dfrac{1}{5}b^3\right)\left(2a + \dfrac{1}{5}b^3\right)$

**77.** $(2p - 1)^3$   $8p^3 - 12p^2 + 6p - 1$    **78.** $(2p - 1)^4$   $16p^4 - 32p^3 + 24p^2 - 8p + 1$

**79.** $(10x - 7y)(10x + 7y)$   $100x^2 - 49y^2$    **80.** $(15a - 2b)(15a + 2b)$   $225a^2 - 4b^2$

**81.** $\left(\dfrac{1}{3}x - 1\right)^2$   $\dfrac{1}{9}x^2 - \dfrac{2}{3}x + 1$    **82.** $\left(\dfrac{1}{5}x + 2\right)^2$   $\dfrac{1}{25}x^2 + \dfrac{4}{5}x + 4$

**83.** $(1.6x - 4.5)(0.2x + 1.2)$   $0.32x^2 + 1.02x - 5.4$    **84.** $(0.6x - 3.1)(0.7x - 1.9)$   $0.42x^2 - 3.31x + 5.89$

**85.** $(x^2 - 3y)(x^2 + 3y)$   $x^4 - 9y^2$    **86.** $(x^3 + 8y^2)(x^3 - 8y^2)$   $x^6 - 64y^4$

## You Be the Teacher!

Correct each student's errors, if any.

**87.** Simplify $(3x - 8)^2$.

Mark's work: $(3x - 8)^2 = 9x^2 + 64$

**88.** Simplify $(5y + 2)^3$.

Joan's work: $(5y + 2)^3 = 125y^3 + 8$

**89.** Simplify $(4x + 3)(4x - 3)$.

Antoine's work: $(4x + 3)(4x - 3)$
$$= 16x^2 - 12x + 12x - 9$$
$$= 16x^2 - 24x - 9$$

Antoine didn't combine like terms correctly. The middle terms add to zero. The result should be $16x^2 - 9$.

## Calculate It!

Use a graphing calculator to determine if each product is correct. If it is not correct, state the correct product.

**90.** $(5x - 7)^2 = 25x^2 + 49$   incorrect, $25x^2 - 70x + 49$

**91.** $(11x - 8)^2 = 121x^2 + 64$   incorrect, $121x^2 - 176x + 64$

**92.** $(4x - 5)^3 = 64x^3 - 125$   incorrect, $64x^3 - 240x^2 + 300x - 125$

**93.** $(3x + 2)^3 = 27x^3 + 8$   incorrect, $27x^3 + 54x^2 + 36x + 8$

**94.** $(3x + 7)(3x - 7) = 9x^2 + 49$   incorrect, $9x^2 - 49$

**95.** $(2x + 5)(2x - 5) = 4x^2 - 25$   correct

## Think About It!

**96.** Show how to derive the formula for $(a + b)^2$ using the FOIL method.
$(a + b)^2 = (a + b)(a + b) = a^2 + ab + ab + b^2 = a^2 + 2ab + b^2$

**97.** Show how to derive the formula for $(a - b)^2$ using the FOIL method.
$(a - b)^2 = (a - b)(a - b) = a^2 - ab - ab + b^2 = a^2 - 2ab + b^2$

**98.** Write an example of a difference of two squares. What product of conjugate produces it?
Answers vary, $36x^2 - 25 = (6x - 5)(6x + 5)$

---

| **SECTION 5.7** | **Division of Polynomials** |
| --- | --- |

### ▶ OBJECTIVES

As a result of completing this section, you will be able to

1. Divide a polynomial by a monomial.
2. Divide a polynomial by a binomial.
3. Solve application problems.
4. Troubleshoot common errors.

The area of this pool table is $4x^3y + 2x^2y$ ft$^2$ and its length is $2x^2y$ ft. How can we express the width of the table?

In this section, we will learn how to divide polynomials, which will enable us to answer this question.

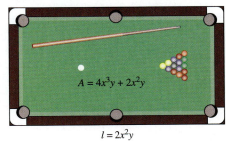

$A = 4x^3y + 2x^2y$

$l = 2x^2y$

### Dividing Polynomials by Monomials

**Objective 1 ▶**

Divide a polynomial by a monomial.

We conclude this chapter with a discussion of the last operation of polynomials, division. To divide a polynomial by a monomial, we use a basic property of fractions. The following example reviews this property.

$$\frac{14}{7} + \frac{21}{7} = \frac{14 + 21}{7} = \frac{35}{7} = 5$$

This property reminds us that to add fractions, they must have the same denominator. To use it for a property relating to division, we need to view it in reverse order.

$$\frac{14 + 21}{7} = \frac{14}{7} + \frac{21}{7} = 2 + 3 = 5$$

In this order, we see that if a sum is divided by a single term, each term in the numerator must be divided by the denominator.

This leads to the following property.

> **Property: Property of Division for Fractions**
>
> If $A$, $B$, and $C$ are monomials with $C \neq 0$, then
>
> $$\frac{A + B}{C} = \frac{A}{C} + \frac{B}{C}$$

Throughout this section, we will assume that all denominators are nonzero.

> **Procedure: Dividing a Polynomial by a Monomial**
>
> **Step 1:** Rewrite the problem so that each term in the numerator is divided by the monomial in the denominator.
> **Step 2:** Simplify each expression using properties of exponents. Recall $\frac{x^m}{x^n} = x^{m-n}$.
> **Step 3:** Check by multiplication. If $\frac{a}{b} = c$, then $bc = a$. That is, if we multiply the quotient by the divisor, we should obtain the dividend.

**Objective 1  Examples**  **Perform the operation and check by multiplying.**

**1a.** Divide $5x^2 - 10x$ by $5x$.

**1b.** $\dfrac{12y^4 - 18y^3 + 6y^2}{-6y^2}$

**1c.** Divide $9a^3 + 3a^2 - 6a$ by $9a^2$.

**Solutions**  **1a.** $\dfrac{5x^2 - 10x}{5x} = \dfrac{5x^2}{5x} - \dfrac{10x}{5x}$    Apply the property of division for fractions.

$= x - 2$    Simplify each quotient.

**Check:** $5x(x - 2) = 5x(x) - 5x(2)$

$= 5x^2 - 10x$

Since we obtained the dividend, our work is correct.

**1b.** Rewrite the problem by applying the property of division for fractions. Be careful with signs since the divisor is negative.

$$\frac{12y^4 - 18y^3 + 6y^2}{-6y^2} = \frac{12y^4}{-6y^2} - \frac{18y^3}{-6y^2} + \frac{6y^2}{-6y^2}$$

$$= -2y^2 + 3y - 1$$

**Check:** $-6y^2(-2y^2 + 3y - 1) = -6y^2(-2y^2) - 6y^2(3y) - 6y^2(-1)$

$$= 12y^4 - 18y^3 + 6y^2$$

Since we obtained the dividend, our quotient is correct.

**1c.** $\dfrac{9a^3 + 3a^2 - 6a}{9a^2} = \dfrac{9a^3}{9a^2} + \dfrac{3a^2}{9a^2} - \dfrac{6a}{9a^2}$    Apply the property of division for fractions.

$= a + \dfrac{1}{3} - \dfrac{2}{3}a^{-1}$    Simplify each quotient.

$= a + \dfrac{1}{3} - \dfrac{2}{3} \cdot \dfrac{1}{a}$    Apply the definition of a negative exponent.

$= a + \dfrac{1}{3} - \dfrac{2}{3a}$    Simplify.

**Check:** $9a^2\left(a + \dfrac{1}{3} - \dfrac{2}{3a}\right) = 9a^2(a) + 9a^2\left(\dfrac{1}{3}\right) - 9a^2\left(\dfrac{2}{3a}\right)$

$$= 9a^3 + 3a^2 - 6a$$

Since we obtained the dividend, our work is correct.

✔ **Student Check 1**    Perform the operation and check by multiplying.

**a.** Divide $8y^2 - 24y$ by $8y$.

**b.** $\dfrac{14x^4 - 21x^3 + 7x^2}{-7x^2}$

**c.** Divide $6b^5 + 15b^3 - 24b^2$ by $3b^4$.

**Objective 2** ▶

Divide a polynomial
by a binomial.

## Dividing Polynomials by Binomials

When the denominator of a quotient is not a monomial, we cannot apply the division property for fractions or the rules of exponents directly. We must use a long division process to perform the division. The long division process is exactly the process used to divide real numbers. Review the following example for the steps used in long division.

$$\begin{array}{r} 21 \\ 15\overline{)325} \\ -30 \\ \hline 25 \\ -15 \\ \hline 10 \end{array}$$

Divide: $\dfrac{32}{15} = 2$

Multiply: $2(15) = 30$

Subtract and bring down the next digit.

Divide: $\dfrac{25}{15} = 1$

Multiply: $1(15) = 15$

Subtract.

The divisor, 15, doesn't go into 10 so we are done.

We say that $\dfrac{325}{15} = 21 + \dfrac{10}{15}$. We can check our quotient: Quotient • Divisor + Remainder = Dividend. That is,

$$21(15) + 10 = 325$$
$$315 + 10 = 325 \quad \text{True}$$

This brings us to the four basic steps of long division:

*divide, multiply, subtract,* and *bring down*

The same basic steps apply to **long division of polynomials**. When applying long division to polynomials, the dividend (numerator) must be written in standard form. If a term is missing, we write it in with a coefficient of 0.

The following example illustrates how long division applies to polynomials.

$$\begin{array}{r} 2x + 3 \\ x + 4\overline{)2x^2 + 11x + 12} \\ -(2x^2 + 8x) \\ \hline 3x + 12 \\ -(3x + 12) \\ \hline 0 \end{array}$$

Divide: $\dfrac{2x^2}{x} = 2x$

Multiply: $2x(x + 4) = 2x^2 + 8x$

Subtract and bring down the next term.

Divide: $\dfrac{3x}{x} = 3$

Multiply: $3(x + 4) = 3x + 12$

Subtract and bring down.

---

**Procedure: Dividing a Polynomial by a Binomial Using Long Division**

**Step 1:** Write the numerator (dividend) in standard form writing any missing terms with a coefficient of zero.

**Step 2:** Divide the first term of the binomial in the denominator (divisor) into the first term of the polynomial in the numerator.

**Step 3:** Multiply this result by the binomial and line up like terms.

**Step 4:** Subtract the polynomials.

**Step 5:** Repeat this process until the degree of the polynomial that results from subtraction is less than the degree of the binomial.

**Step 6:** Check by multiplying: Quotient × Divisor + Remainder = Dividend.

---

**Objective 2 Examples** | **Perform each operation and check by multiplying.**

**2a.** Divide $x^2 + 2x - 35$ by $x - 5$.  **2b.** $\dfrac{x^3 + 8}{x + 2}$  **2c.** $\dfrac{7y - 6 + 3y^2}{3y + 4}$

**Solutions**   **2a.**

$$\begin{array}{r} x + 7 \\ x - 5 \overline{) x^2 + 2x - 35} \\ \underline{-(x^2 - 5x)} \\ 7x - 35 \\ \underline{-(7x - 35)} \\ 0 \end{array}$$

Divide $x^2$ by $x$, which is $x$.

Multiply $x$ by $x - 5$, which is $x^2 - 5x$.

Subtract and bring down $-35$.

Divide $7x$ by $x$ to get 7.

Multiply 7 by $x - 5$, which is $7x - 35$.

Subtract.

**Check:** $(x + 7)(x - 5) = x^2 + 2x - 35$

Because the quotient times the divisor is the dividend, our result is correct.

So, $\dfrac{x^2 + 2x - 35}{x - 5} = x + 7.$

**INSTRUCTOR NOTE:**
Remind students that in the subtraction process, the signs change.

**2b.** Write the dividend as $x^3 + 0x^2 + 0x + 8$, so that like terms can be aligned.

$$\begin{array}{r} x^2 - 2x + 4 \\ x + 2 \overline{) x^3 + 0x^2 + 0x + 8} \\ \underline{-(x^3 + 2x^2)} \\ -2x^2 + 0x \\ \underline{-(-2x^2 - 4x)} \\ 4x + 8 \\ \underline{-(4x + 8)} \\ 0 \end{array}$$

Divide $x^3$ by $x$, which is $x^2$.

Multiply $x^2$ by $x + 2$, which is $x^3 + 2x^2$.

Subtract and bring down $0x$.

Divide $-2x^2$ by $x$ to get $-2x$.

Multiply $-2x$ by $x + 2$, which is $-2x^2 - 4x$.

Subtract and bring down 8.

Divide $4x$ by $x$, which is 4.

Multiply 4 by $x + 2$, which is $4x + 8$.

Subtract.

**Check:**

$$(x^2 - 2x + 4)(x + 2) = x^3 + 2x^2 - 2x^2 - 4x + 4x + 8$$
$$= x^3 + 8$$

Because the quotient times the divisor is the dividend, our result is correct.

So, $\dfrac{x^3 + 8}{x + 2} = x^2 - 2x + 4.$

**2c.** Write the numerator in standard form, $3y^2 + 7y - 6$.

$$\begin{array}{r} y + 1 \\ 3y + 4 \overline{) 3y^2 + 7y - 6} \\ \underline{-(3y^2 + 4y)} \\ 3y - 6 \\ \underline{-(3y + 4)} \\ -10 \end{array}$$

Divide $3y^2$ by $3y$, which is $y$.

Multiply $y$ by $3y + 4$, which is $3y^2 + 4y$.

Subtract and bring down $-6$.

Divide $3y$ by $3y$ to get 1.

Multiply 1 by $3y + 4$, which is $3y + 4$.

Subtract and get $-10$.

**Check:**

$$(y + 1)(3y + 4) - 10 = 3y^2 + 4y + 3y + 4 - 10$$
$$= 3y^2 + 7y - 6$$

Because the quotient times the divisor plus the remainder is the dividend, our result is correct. So, $\dfrac{3y^2 + 7y - 6}{3y + 4} = y + 1 - \dfrac{10}{3y + 4}.$

**Student Check 2**   Perform each operation and check by multiplying.

    **a.** Divide $x^2 - x - 72$ by $x - 9$.    **b.** $\dfrac{a^3 + 125}{a + 5}$    **c.** $\dfrac{4y^2 - 4 + 10y}{4y - 2}$

## Applications

**Objective 3** ▶

Solve application problems.

The application problems we will solve involve finding an expression that represents the length of a side of a geometric figure.

> **Procedure:  Solving Applications of Division of Polynomials**
>
> **Step 1:** Write the formula for the area, volume, or perimeter given.
> **Step 2:** Write a quotient that represents the expression we need to find.
> **Step 3:** Divide using the methods from Objective 1 or 2.

**Objective 3  Example**   The area of this pool table is $4x^3y + 2x^2y$ ft$^2$ and its length is $2x^2y$ ft. Write an expression for the width of the pool table.

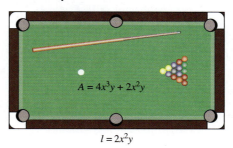

$A = 4x^3y + 2x^2y$

$l = 2x^2y$

**Solution**   We know the area of a rectangle is $A = lw$. So, to solve for $w$, we must divide the area by the length of the rectangle. That is, $w = \dfrac{A}{l}$.

$$w = \frac{A}{l}$$

$$w = \frac{4x^3y + 2x^2y}{2x^2y}$$

$$w = \frac{4x^3y}{2x^2y} + \frac{2x^2y}{2x^2y}$$

$$w = 2x + 1$$

**Check:** $(2x + 1)(2x^2y) = 4x^3y + 2x^2y$

So, the width can be represented by $2x + 1$ ft.

 **Student Check 3**   The area of a parallelogram is $9a^3b^3 - 12a^2b^2$ square units. Its height is $3a^2b$ units. Find an expression for the base of the parallelogram.

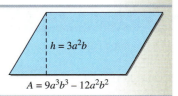

$h = 3a^2b$

$A = 9a^3b^3 - 12a^2b^2$

**Objective 4** ▶

Troubleshoot common errors.

## Troubleshooting Common Errors

Some common errors related to dividing polynomials are shown next.

**Objective 4 Examples** A problem and an incorrect solution are given. Provide the correct solution and an explanation of the error.

**4a.** $\dfrac{6x^2 + 14x}{2x}$

| Incorrect Solution | Correct Solution and Explanation |
|---|---|
| $\dfrac{6x^3 + 14x}{2x} = \dfrac{6x^3}{2x} + 14x$ <br><br> $= 3x^2 + 14x$ | We must divide $2x$ into each term in the numerator of the fraction. <br><br> $\dfrac{6x^3 + 14x}{2x} = \dfrac{6x^3}{2x} + \dfrac{14x}{2x}$ <br><br> $= 3x^2 + 7$ |

**4b.** $\dfrac{x^2 - 5x + 6}{x - 2}$

| Incorrect Solution | Correct Solution and Explanation |
|---|---|
| $\begin{array}{r} x - 7 \\ x-2\overline{)x^2 - 5x + 6} \\ \underline{x^2 - 2x} \\ -7x + 6 \\ \underline{-7x + 14} \\ 20 \end{array}$ <br><br> $\dfrac{x^2 - 5x + 6}{x - 2} = x - 7 + \dfrac{20}{x - 2}$ | We must change the signs when subtracting. <br><br> $\begin{array}{r} x - 3 \\ x-2\overline{)x^2 - 5x + 6} \\ \underline{-(x^2 - 2x)} \\ -3x + 6 \\ \underline{-(-3x + 6)} \\ 0 \end{array}$ <br><br> $\dfrac{x^2 - 5x + 6}{x - 2} = x - 3$ |

## ANSWERS TO STUDENT CHECKS

**Student Check 1** **a.** $y - 3$ **b.** $-2x^2 + 3x - 1$
**c.** $2b + \dfrac{5}{b} - \dfrac{8}{b^2}$

**Student Check 2** **a.** $x + 8$ **b.** $a^2 - 5a + 25$
**c.** $y + 3 + \dfrac{2}{4y - 2}$
**Student Check 3** $3ab^2 - 4b$

## SUMMARY OF KEY CONCEPTS

1. To divide a polynomial by a monomial, divide the monomial into each term of the polynomial. Use rules of exponents to simplify each term. Check by multiplying.
2. Dividing by a binomial requires long division. Repeat the steps involved in long division of real numbers—*divide, multiply, subtract,* and *bring down.* Continue these steps until the remainder has a degree less than the degree of the divisor. The quotient times the divisor plus the remainder should equal the dividend.
3. Division of polynomials can be used to find expressions that represent the length of a side of a geometric figure if an area, volume, or perimeter is given.

## GRAPHING CALCULATOR SKILLS

The graphing calculator can be used to check quotients for problems that involve a single variable.

**Example 1:** Verify that $\dfrac{x^3 + 8}{x + 2} = x^2 - 2x + 4$.

**Method 1:** Input the original division problem and the answer into $Y_1$ and $Y_2$, respectively. Then compare the table of values. They should agree everywhere except the value of $x$ that makes the denominator zero.

```
Plot1 Plot2 Plot3
\Y1█(X^3+8)/(X+2
)
\Y2█X²-2X+4
\Y3=
\Y4=
\Y5=
\Y6=
```

| X | Y1 | Y2 |
|---|---|---|
| -3 | 19 | 19 |
| -2 | ERR: | 12 |
| -1 | 7 | 7 |
| 0 | 4 | 4 |
| 1 | 3 | 3 |
| 2 | 4 | 4 |
| 3 | 7 | 7 |
| X=-3 | | |

**Method 2:** Use the calculator to check the multiplication. The quotient times the divisor plus the remainder should equal the dividend. Input the quotient times the divisor plus the remainder in $Y_1$ and the dividend in $Y_2$. Compare the table of values. If they agree, then the answer is correct.

**Method 1:**

Since the tables agree for $Y_1$ and $Y_2$, the answer is correct.

Since the tables agree for $Y_1$ and $Y_2$, the answer is correct.

**Method 2:**

**Example 2:** Verify that $\dfrac{3y^2 + 7y - 6}{3y + 4} = y + 1 - \dfrac{10}{3y + 4}$.

Since the tables agree for $Y_1$ and $Y_2$, the answer is correct.

---

## SECTION 5.7 / EXERCISE SET

 **Write About It!**

**Use complete sentences in your answer to each exercise.**

1. Explain how to divide a polynomial by a monomial.
   Answers vary.
2. How can you determine if you divided a polynomial by a monomial correctly?   Answers vary.
3. Explain how to divide a polynomial by a binomial.
   Answers vary.
4. Explain how to check your answer if you divided a polynomial by a binomial and ended up with a remainder.
   Answers vary.
5. How do you set up your division process if the dividend has a missing term? For example, show how you would set up $\dfrac{x^3 - 7x + 2}{x - 2}$.   Answers vary.
6. Explain how the degrees of the dividend, divisor, and remainder are related.   Answers vary.

**Practice Makes Perfect!**

**Perform the indicated operation. Check the answer using multiplication. (*See Objective 1.*)**

7. Divide $6x^2 - 12x$ by $6x$.   $x - 2$
8. Divide $4x^2 - 16x$ by $2x$.   $2x - 8$
9. Divide $10x^3 - 2x^2 + 4x$ by $2x$.   $5x^2 - x + 2$
10. Divide $7x^3 - 14x^2 + 21x$ by $7x$.   $x^2 - 2x + 3$
11. Divide $3a^4 - 12a^3 - 6a^2$ by $3a^2$.   $a^2 - 4a - 2$
12. Divide $8a^5 - 4a^4 + 16a^3$ by $4a^3$.   $2a^2 - a + 4$
13. $\dfrac{12y^5 - 16y^3 + 8y^2}{-4y^2}$   $-3y^3 + 4y - 2$
14. $\dfrac{21y^6 - 18y^5 + 9y^2}{-3y^2}$   $-7y^4 + 6y^3 - 3$

15. $\dfrac{20x^4 - 10x^3 - 5x}{5x}$   $4x^3 - 2x^2 - 1$
16. $\dfrac{45x^7 - 30x^6 + 60x^5}{15x^5}$   $3x^2 - 2x + 4$
17. $\dfrac{18y^3 - 24y^2 + 12y}{-6y}$   $-3y^2 + 4y - 2$
18. $\dfrac{24y^6 - 8y^4 + 16y^3}{8y^3}$   $3y^3 - y + 2$
19. Divide $14a^3 - 7a^2 + 28$ by $14a$.   $a^2 - \dfrac{a}{2} + \dfrac{2}{a}$
20. Divide $12a^4 - 18a^3 - 3a^2$ by $6a^3$.   $2a - 3 - \dfrac{1}{2a}$
21. Divide $5x^4 - 15x^3 + 45x^2$ by $15x^2$.   $\dfrac{x^2}{3} - x + 3$
22. Divide $10x^6 - 20x^3 - 15x$ by $10x$.   $x^5 - 2x^2 - \dfrac{3}{2}$
23. $\dfrac{3a^2 - 2a + 6}{3a^2}$   $1 - \dfrac{2}{3a} + \dfrac{2}{a^2}$
24. $\dfrac{5a^2 - 6a}{5a^2}$   $1 - \dfrac{6}{5a}$
25. $\dfrac{12x^3 - 24x^2}{12x^3}$   $1 - \dfrac{2}{x}$
26. $\dfrac{4x^5 - 16x^4 + 8}{4x^5}$
27. $\dfrac{13y^7 - 26y^6 + 26y^5}{-26y^5}$
28. $\dfrac{11y^8 - 22y^7 + 11y^6}{-22y^6}$
29. $\dfrac{6y^3 - 48y^2 - 12}{24y^2}$
30. $\dfrac{7y^4 - 63y^3 + 14}{49y^3}$

**Perform the indicated operation. Check the answer using multiplication. (*See Objective 2.*)**

31. Divide $x^2 - x - 12$ by $x + 3$.   $x - 4$
32. Divide $x^2 - 3x - 10$ by $x + 2$.   $x - 5$
33. Divide $x^2 - 16$ by $x - 4$.   $x + 4$
34. Divide $x^2 - 25$ by $x + 5$.   $x - 5$
35. $\dfrac{x^3 + 1}{x - 1}$   $x^2 + x + 1 + \dfrac{2}{x - 1}$
36. $\dfrac{x^3 + 64}{x + 4}$   $x^2 - 4x + 16$
37. $\dfrac{4a^2 - 8a}{a - 2}$   $4a$
38. $\dfrac{6a^2 + 30a}{a + 5}$   $6a$

Additional answers can be found in the Instructor Answer Appendix.

**39.** $\dfrac{2x^2 + 5x - 3}{x + 3}$   $2x - 1$

**40.** $\dfrac{3x^2 - 7x - 20}{x - 4}$   $3x + 5$

**41.** Divide $x^2 - 6x - 5$ by $x + 1$.   $x - 7 + \dfrac{2}{x + 1}$

**42.** Divide $x^2 + 3x - 4$ by $x + 2$.   $x + 1 - \dfrac{6}{x + 2}$

**43.** Divide $3y^2 - 5y + 7$ by $y - 6$.   $3y + 13 + \dfrac{85}{y - 6}$

**44.** Divide $5y^2 - 2y + 3$ by $y - 1$.   $5y + 3 + \dfrac{6}{y - 1}$

**45.** $\dfrac{2x^2 - 5x + 7}{x + 7}$

**46.** $\dfrac{3x^2 + 18x - 115}{x + 10}$

**47.** $\dfrac{6y^3 - 5y^2 - 16y + 22}{2y - 3}$

**48.** $\dfrac{8y^3 - 18y^2 + 19y - 2}{4y + 1}$

**Solve each problem. (See Objective 3.)**

**49.** The area of a rectangle is $3x^2y - 6xy^2$ square units and the length of the rectangle is $3xy$ units. Find an expression for the width of the rectangle.   $x - 2y$

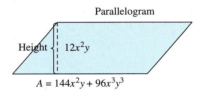
$A = 3x^2y - 6xy^2$
$3xy$

**50.** The area of a rectangle is $6xy^2 + 14x^2y$ square units and the length of the rectangle is $2xy$ units. Find an expression for the width of the rectangle.   $7x + 3y$

$A = 6xy^2 + 14x^2y$
$2xy$

**51.** The area of a parallelogram is $144x^2y + 96x^3y^3$ square units. If the height of the parallelogram is $12x^2y$ units, then find the length of its base.   $12 + 8xy^2$

Parallelogram
Height   $12x^2y$
$A = 144x^2y + 96x^3y^3$

**52.** The area of a parallelogram is $52x^2y^3 - 39xy^2$ square units. If the height of the parallelogram is $13xy^2$ units, then find the length of its base.   $4xy - 3$

Parallelogram
Height   $13xy^2$
$A = 52x^2y^3 - 39xy^2$

### Mix 'Em Up!

**Perform the indicated operation.**

**53.** Divide $6a^3 + 3a^2 - 18$ by $3a$.   $2a^2 + a - \dfrac{6}{a}$

**54.** Divide $4a^5 - 5a^4 + 6a^3$ by $a^2$.   $4a^3 - 5a^2 + 6a$

**55.** Divide $3x^2 - 11x - 4$ by $x - 4$.   $3x + 1$

**56.** Divide $6x^2 - 3x - 30$ by $2x - 5$.   $3x + 6$

**57.** $\dfrac{6x^5 - 4x^4 + 2x^3}{-2x^3}$   $-3x^2 + 2x - 1$

**58.** $\dfrac{20x^4 - 28x^3 + 12x^2}{-4x^2}$   $-5x^2 + 7x - 3$

**59.** $\dfrac{7x^3 - x^2 + 14}{x + 2}$

**60.** $\dfrac{3x^3 - 20x - 8}{x - 3}$

**61.** $\dfrac{6x^3 - x^2 - 5x + 17}{2x + 3}$   $3x^2 - 5x + 5 + \dfrac{2}{2x + 3}$

**62.** $\dfrac{5x^4 + x^3 - 10x^2 - 2x + 1}{5x + 1}$   $x^3 - 2x + \dfrac{1}{5x + 1}$

**63.** If the area of a parallelogram is $12xz^2 - 6x^2z$ square units and its height is $6xz$ units, find the length of its base.   $2z - x$

**64.** If the area of a rectangle is $4x^3y^3 - 14x^2y$ square units and its length is $2x^2y$ units, find the width of the rectangle.   $2xy^2 - 7$

### You Be the Teacher!

**Correct each student's errors, if any.**

**65.** Divide: $\dfrac{5x^2 + 2x - 3}{5x^2}$

Britney's work:
$$\dfrac{5x^2 + 2x - 3}{5x^2} = \dfrac{\cancel{5x^2} + 2x - 3}{\cancel{5x^2}} = 2x - 3$$

**66.** Divide: $\dfrac{2x^3 - 4x^2 - 1}{2x^3}$

Susan's work:
$$\dfrac{2x^3 - 4x^2 - 1}{2x^3} = \dfrac{\cancel{2x^3} - 4x^2 - 1}{\cancel{2x^3}} = -4x^2 - 1$$

**67.** Divide: $x^2 - 2x - 3$ by $x + 3$

Damien's work:
$$x + 3 \,\overline{)\, x^2 - 2x - 3}$$
← Answer $x + 1$
$$\underline{-x^2 + 3x}$$
$$x - 3$$
$$\underline{-x + 3}$$
$$0$$
$x - 5 + \dfrac{12}{x + 3}$

**68.** Divide: $x^3 + 5x^2 - 8x - 12$ by $x - 6$

Kit's work:
$$x - 6 \,\overline{)\, x^3 + 5x^2 - 8x - 12}$$
← Answer $x^2 - x - 2$
$$\underline{-x^3 - 6x^2}$$
$$-x^2 - 8x$$
$$\underline{+x^2 + 6x}$$
$$-2x - 12$$
$$\underline{+2x + 12}$$
$$0$$
$x^2 + 11x + 58 + \dfrac{336}{x - 6}$

### Calculate It!

**Use a graphing calculator to determine if each quotient is correct. If the answer is incorrect, find the correct answer.**

**69.** $\dfrac{x^4 - 3x^2}{x^4} = -3x^2$   incorrect; The correct answer is $1 - \dfrac{3}{x^2}$.

**70.** $\dfrac{5x^2 + 3x - 1}{5x^2} = 3x - 1$   incorrect; The correct answer is $1 + \dfrac{3}{5x} - \dfrac{1}{5x^2}$.

**71.** $\dfrac{x^3 + 125}{x + 5} = x^2 - 5x + 25$   correct

**72.** $\dfrac{x^3 + 1}{x + 1} = x^2 - 1$   incorrect; The correct answer is $x^2 - x + 1$.

## GROUP ACTIVITY  **The Mathematics of Finance**

The amortization formula gives the payment $P$ for repaying a loan of $A$ dollars at an annual interest rate of $r$ (as a decimal) for $t$ yr.

$$P = \dfrac{A \cdot \dfrac{r}{12}}{1 - \left(1 + \dfrac{r}{12}\right)^{-12t}}$$

1. Show why the amortization formula can be written as

$$P = \dfrac{\dfrac{r}{12}}{1 - \left(1 + \dfrac{r}{12}\right)^{-12t}} \cdot A.$$

2. Use the Internet, newspapers, a local bank, or credit union to determine the current rate for a 30-yr fixed mortgage. www.lendingtree.com once reported a 30-yr fixed rate of 4.5%.

3. Use the interest rate you found in part 2 along with $t = 30$ yr to determine the coefficient of $A$ in the amortization formula in part 1. Round the value to five decimal places.

4. Now the amortization formula in part 2 should look like $P =$ _____ $A$. This is a linear equation that we can write as $y =$ _____ $x$, where $y$ represents $P$

and $x$ represents $A$. Use this equation to complete the following table. Round answers to two decimal places.

$y = 0.00507x$

| $x$ | $y$ |
|---|---|
| 100,000 | 507 |
| 150,000 | 760.5 |
| 200,000 | 1014 |
| 250,000 | 1267.5 |
| 300,000 | 1521 |

5. What do the $y$-values represent in the table in part 4?
   The $y$-values represents the monthly payment to repay the loan.

6. Based on your financial situation, choose a loan amount you could afford. If you paid the loan off in 30 yr, what is the total amount that you would pay for the loan? How much would you pay in interest for the loan?

7. Repeat parts 2–6 with a 15-yr fixed loan.

**Note:** The payment that results from the amortization formula only represents principal plus interest. A monthly mortgage payment typically includes an amount to be paid into escrow. Escrow funds cover such items as property taxes and home insurance payments. Taxes and insurance can add several hundred dollars to the payment determined by the formula.
www.lendingtree.com reports that a 15-yr fixed rate is 3.75%.

$$\dfrac{\dfrac{r}{12}}{1 - \left(1 + \dfrac{r}{12}\right)^{-12t}} = \dfrac{\dfrac{0.0375}{12}}{1 - \left(1 + \dfrac{0.0375}{12}\right)^{-12(15)}} = 0.00727$$

$y = 0.00727x$

| $x$ | $y$ |
|---|---|
| 100,000 | 727 |
| 150,000 | 1090.50 |
| 200,000 | 1454 |
| 250,000 | 1817.50 |
| 300,000 | 2181 |

If we choose \$200,000, the total paid back is \$1454(180) = \$261,720. The total interest paid is \$261,720 − 200,000 = \$61,720.

# Laws of Exponents and Polynomial Operations

**What's the big idea?** Polynomials are very important to the study of math, chemistry, science, economics, and other fields. Polynomials are used to model real-life phenomena in these fields. Since polynomials are constructed of terms of the form $ax^n$, the properties of exponents then allow us to perform operations on the polynomials.

## The Tools

Listed below are the key terms, skills, formulas, and properties you should know for this chapter.

The page reference is provided if you need additional help with the given topic. The Study Tips will assist in your preparation for an exam.

## Study Tips

1. Learn all of the terms, formulas, and properties. Make flash cards and have someone quiz you.
2. Rework problems from the exercises and also the ones you worked in class. Work additional problems from the review exercises.
3. Review the summaries of key concepts.
4. Work the chapter test.
5. Be sure to review the online resources for additional study materials.

## Terms

| | | |
|---|---|---|
| Binomial   405 | Degree of a polynomial   406 | Polynomial   405 |
| Coefficient   399 | Leading coefficient   406 | Scientific notation   399 |
| Conjugates   426 | Monomial   405 | Standard form   405 |
| Degree of a term   406 | Perfect square trinomial   426 | Trinomial   405 |

## Formulas and Properties

- Difference of two squares   427
- Exponent   376
- FOIL   424
- Long division of polynomials   433
- Property of division for fractions   431

- Square of a binomial   426
- Negative exponent   388
- Negative powers of fractions   390
- Power of a power rule   378
- Power of a product rule   379

- Power of a quotient rule   380
- Product of conjugates   427
- Product of like bases rule   377
- Quotient of like bases rule   386
- Zero exponent   387

## CHAPTER 5 / SUMMARY

How well do you know this chapter? Complete the following questions to find out. Take a look back at the section if you need help.

### SECTION 5.1  The Product and Power Rules for Exponents

1. When exponential expressions with like bases are multiplied, the resulting power is the _sum_ of the exponents in the original expression. The base stays the _same_. This rule can be written as $x^a \cdot x^b = \underline{x^{a+b}}$.

2. When an exponential expression is raised to an exponent, the exponents are _multiplied_ to obtain the exponent in the final result. This rule can be written as $(x^a)^b = \underline{x^{ab}}$.

3. When a product is raised to an exponent, the exponent must be applied to each _factor_ in the base. This rule can be written as $(xy)^b = \underline{x^b\, y^b}$.

4. When a quotient is raised to an exponent, the exponent must be applied to the _numerator_ and _denominator_ of the base. This rule can be written as $\left(\dfrac{x}{y}\right)^b = \underline{\dfrac{x^b}{y^b}}$.

## SECTION 5.2  The Quotient Rule and Zero and Negative Exponents

**5.** When exponential expressions with like bases are divided, the resulting exponent is the _difference_ of the exponents in the original expression. The base stays the _same_. This rule can be written as $\dfrac{x^a}{x^b} = x^{a-b}$.

**6.** When a nonzero number is raised to the exponent of zero, the result is _one_.

**7.** A negative exponent is equivalent to the _reciprocal_ of the base raised to the _positive_ exponent. This can be stated as $b^{-n} = \dfrac{1}{b^n}$ for $b \neq 0$.

## SECTION 5.3  Scientific Notation

**8.** A number written in the form $a \times 10^n$ is a number written in _scientific_ _notation_ as long as $1 \leq a < 10$ and $n$ is an integer.

**9.** When a number is greater than 10 and the number is written in scientific notation, the exponent is _positive_.

**10.** When a number is between 0 and 1 the number is written in scientific notation, the exponent is _negative_.

## SECTION 5.4  Addition and Subtraction of Polynomials

**11.** A _polynomial_ consists of a finite number of terms of the form $ax^n$, where $a$ is a real number and $n$ is a whole number.

**12.** _Polynomials_ with one term are called _monomials_.

**13.** _Polynomials_ with two terms are called _binomials_.

**14.** _Polynomials_ with three terms are called _trinomials_.

**15.** When the exponents are written in descending order, the polynomial is in _standard_ _form_.

**16.** The degree of a polynomial in a single variable is the _largest_ exponent on the variable.

**17.** The _leading_ _coefficient_ is the coefficient of the term with the largest exponent.

**18.** The degree of a term with more than one variable is the _sum_ of the exponents on the variables.

**19.** The degree of a polynomial in more than one variable is the _largest_ degree of the terms in the polynomial.

**20.** To evaluate a polynomial, _substitute_ the variable with the given number and simplify.

**21.** To add polynomials, combine _like_ _terms_.

**22.** To subtract polynomials, add the _opposite_ of the second polynomial. When we find the _opposite_ of a polynomial, all of the signs are changed.

## SECTION 5.5  Multiplication of Polynomials

**23.** When a monomial is multiplied by a polynomial, we apply the _distributive_ property and the product rule for exponents.

**24.** To multiply polynomials, each term in the first polynomial must be _distributed_ to each term in the second polynomial.

## SECTION 5.6  Special Products

**25.** Two binomials can be multiplied using the _FOIL_ method. This only works for two _binomials_.

**26.** When a binomial is squared, the result is always a _trinomial_. The rule for squaring a binomial can be written as $(a + b)^2 = a^2 + 2ab + b^2$. The result of squaring a binomial is called a _perfect_ _square_ _trinomial_.

**27.** _Conjugates_ are binomials of the form $a + b$ and $a - b$.

**28.** When $(a + b)$ and $(a - b)$ are multiplied, the result is _$a^2 - b^2$_. This is called the _difference_ of two _squares_.

**29.** When we raise a binomial to a larger exponent, we must apply the definition of the _exponent_.

## SECTION 5.7  Division of Polynomials

**30.** To divide a polynomial by a monomial, we divide _each_ term of the _polynomial_ by the monomial. This can be written as $\dfrac{A + B}{C} = \dfrac{A}{C} + \dfrac{B}{C}$.

**31.** To divide a polynomial by a binomial, we use _long_ _division_.

**32.** To check division of polynomials, we can use _multiplication_. The dividend should equal the _quotient_ times the _divisor_ plus the _remainder_.

## CHAPTER 5 / REVIEW EXERCISES

### SECTION 5.1

**Use the appropriate exponent rules to simplify the expression or to answer the question. Express each answer with positive exponents. (*See Objectives 1–4.*)**

**1.** $a^2 \cdot a^{3/4}$   $a^{11/4}$

**2.** $(2b^3)^6$   $64b^{18}$

**3.** $(-4x^4)(2y^6)(-5y^8)$   $40x^4y^{14}$

**4.** $(-16a)(-2a^{10})(-3b^5)$   $-96a^{11}b^5$

**5.** $-\left(-\dfrac{2x^5}{5y^2}\right)^2$   $-\dfrac{4x^{10}}{25y^4}$

**6.** $(-3^{-2}x)^3$   $-\dfrac{x^3}{729}$

**7.** $\left(\dfrac{3x^4}{2y}\right)^3$   $\dfrac{27x^{12}}{8y^3}$

**8.** $(m^3n^2)^5$   $m^{15}n^{10}$

**9.** $(-10x^6)(2x^9)$   $-20x^{15}$

**10.** $(-15a^9)(-5a^{12})$   $75a^{21}$

**11.** $(0.5c^3d^2)^4$   $0.0625c^{12}d^8$

**12.** $\left(\dfrac{-0.2p^4}{q^3}\right)^5$   $-\dfrac{0.00032p^{20}}{q^{15}}$

**13.** Suppose circle A has a diameter of 8 in. and circle B has a diameter of 6 in. How much larger is the area of circle A than the area of circle B? Express the answer in terms of $\pi$.   $7\pi$ in.$^2$

14. Nathan has a square deck that is 12 ft long. He expands his deck by doubling the length and width of the existing deck. How much larger is the area of the new deck than the existing one?   432 ft²

## SECTION 5.2

**Use the appropriate exponent rules to simplify each expression. Express each answer with positive exponents. (*See Objectives 1–4.*)**

15. $x^{-8} x^3$   $\dfrac{1}{x^5}$

16. $(2x^4)^{-3}$   $\dfrac{1}{8x^{12}}$

17. $(3y^{-5})^{-2}$   $\dfrac{y^{10}}{9}$

18. $[(5a)^3]^{-1}$   $\dfrac{1}{125a^3}$

19. $[(2b)^{-4}]^{-2}$   $256b^8$

20. $\left(\dfrac{5p}{q^2}\right)^{-2}$   $\dfrac{q^4}{25p^2}$

21. $\left(\dfrac{9a}{b^0}\right)^{-3}$   $\dfrac{1}{729a^3}$

22. $\dfrac{12^{-3}}{12^{-2}}$   $\dfrac{1}{12}$

23. $\left(\dfrac{3c^0}{2d^{-4}}\right)^{-2}$   $\dfrac{4}{9d^8}$

24. $\left(-\dfrac{x^{-4}}{7y^3}\right)^{-3}$   $-343x^{12}\,y^9$

**Solve each problem. (*See Objective 5.*)**

25. The formula to find the amount of a substance $P$ that remains after time $t$ is $P = P_0 \cdot 2^{-t/m}$, where $P_0$ is the original amount of the substance and $m$ is its half-life. Suppose the half-life of a substance is 8 days. If the original amount of the substance is 16 g, how much substance remains after

 a. 16 days?   4 g  b. 32 days?   1 g  c. 64 days?   0.0625 g

26. The formula to find the monthly payment to repay a loan is $P = \dfrac{A\left(\dfrac{r}{12}\right)}{1 - \left(1 + \dfrac{r}{12}\right)^{-m}}$, where $A$ is the amount of the loan, $r$ is the interest rate (in decimal form), and $m$ is the number of monthly payments. Find the monthly payment to repay a student loan of $12,000 at an interest rate of 4.2% for 6 yr.   $188.84

## SECTION 5.3

**Convert each number in standard notation to scientific notation and each number in scientific notation to standard notation. (*See Objectives 1 and 2.*)**

27. 3,500,000   $3.5 \times 10^6$

28. $4.32 \times 10^9$   4,320,000,000

29. 0.0000891   $8.91 \times 10^{-5}$

30. $2.78 \times 10^{-5}$   0.0000278

31. The population of Ethiopia in 2011: 90,900,000 (Source: http://www.cia.gov)   $9.09 \times 10^7$

32. Discover Financial Services' 2008 profit: $7.088 \times 10^9$ U.S. dollars (Source: http://money.cnn.com)
7,088,000,000 U.S. dollars

**Use exponent rules to simplify each expression. Write each answer in scientific notation. (*See Objective 3.*)**

33. $(6.9 \times 10^5)(2.8 \times 10^{-12})$
1.932 × 10⁻⁶

34. $\dfrac{2.834 \times 10^{-9}}{8.72 \times 10^{-3}}$   $3.25 \times 10^{-7}$

35. The population of Japan in 2009 was approximately 127,500,000 and the land area is approximately 146,000 mi². If the area is evenly distributed among all people in Japan, how much land will each person have?
$1.145 \times 10^{-3}$ mi² per person

## SECTION 5.4

**Answer each question. (*See Objectives 1 and 2.*)**

36. Classify the polynomial, $11 - 5x + 9x^2$, as a monomial, binomial, or trinomial. Then write it in standard form.
trinomial, $9x^2 - 5x + 11$

37. Give an example of a binomial.   Answer vary; $5x + y$

38. Identify the leading coefficient and the degree of the polynomial $-21x^5 + 5x^3 - 4x + 9$.
leading coefficient −21, degree 5

39. Classify the polynomial, $13 - 2x^2 - 11x$, as a monomial, binomial, or trinomial. Then write it in standard form.   trinomial, $-2x^2 - 11x + 13$

40. Identify the leading coefficient and the degree of the polynomial $-12a^5 - 2a^3b + 14ab^2 - 2b^4$.
leading coefficient −12, degree 5

**Evaluate each polynomial for the given value(s). (*See Objectives 3 and 6.*)**

41. $-x^2 + 3x + 7$, $x = -2$   −3

42. $2y^3 - 4y - 18$, $y = 3$   24

43. $4a^2 - 3ab - b^2$, $a = -2$, $b = 3$   25

44. A company's total cost to make $x$ items is $12x^3 - 5x^2 + 100$ dollars. How much does it cost the company to make 20 items?   $94,100

45. The height, in feet, of a ball projected upward from the top of a 100-ft building with an initial velocity of 80 ft/sec is given by $-16t^2 + 80t + 100$, where $t$ is the number of seconds after the ball is thrown. Find the height of the ball

 a. 2.5 sec after it is thrown.   200 ft  b. 3 sec after it is thrown.   196 ft

**Perform each operation. Write each answer in standard form. (*See Objectives 4–6.*)**

46. $(12 - 9x^2 - 5x) - (-7 + 4x^2 + 8x)$   $-13x^2 - 13x + 19$

47. $(2x^4 + 17x^3 - 11) + (3x^4 - 12x^3 + 19)$   $5x^4 + 5x^3 + 8$

48. $-\dfrac{6}{5}x^2 + \dfrac{7}{5}x - \dfrac{4}{5}$
    $+ \left(\dfrac{1}{5}x^2 + \dfrac{3}{5}x + \dfrac{1}{5}\right)$   $-x^2 + 2x - \dfrac{3}{5}$

49. $\begin{aligned}&1.5x^2 + 2.1x - 3.1\\ -\ &(-0.8x^2 + 7.2x - 8.4)\end{aligned}$   $2.3x^2 - 5.1x + 5.3$

50. Add $(-5a^2 + 2ab - 16b^2)$ and $(-2a^2 - 4ab + 11b^2)$.
$-7a^2 - 2ab - 5b^2$

51. Subtract $(7x^2 - xy + 6y^2)$ from $(-3x^2 + 4xy - y^2)$.
$-10x^2 + 5xy - 7y^2$

52. $(6x^2 + 4xy - 3y^2) - (-5x^2 + 2xy + y^2)$
    $+ (-7x^2 - 2xy + 4y^2)$   $4x^2$

53. $(-x^3 + 8x^2 - 4x - 12) + (2x^3 + 11x - 14)$
    $- (7x^3 - x^2 + 6x - 1)$   $-6x^3 + 9x^2 + x - 25$

54. Find the perimeter of the trapezoid:   $5x^2 + 14x - 16$

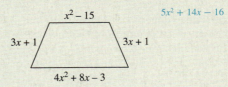

55. Suppose the profit generated by selling $x$ items is $-0.2x^2 + 124x + 75$ dollars and the total cost is $12x + 320$ dollars. Find an expression that represents the company's revenue. $-0.2x^2 + 136x + 395$ dollars

## SECTION 5.5

**Use the distributive property to multiply each expression. (*See Objectives 1–3.*)**

56. $p(-3p + 5)$  $-3p^2 + 5p$  57. $6x^2y(5xy - 2)$  $30x^3y^2 - 12x^2y$

58. $(7 - x)(x + 7)$  $-x^2 + 49$  59. $(a + 3b)(a^2 - 3ab + 9b^2)$  $a^3 + 27b^3$

60. $-4mn^2(3m^3 - 2mn^2 + n^2)$  $-12m^4n^2 + 8m^2n^4 - 4mn^4$

61. $\left(\frac{2}{3}x - 2\right)\left(\frac{1}{3}x + 5\right)$  $\frac{2}{9}x^2 + \frac{8}{3}x - 10$

62. $\left(4a + \frac{1}{3}b\right)\left(2a - \frac{5}{3}b\right)$  $8a^2 - 6ab - \frac{5}{9}b^2$

63. $\begin{array}{r} x^2 - 3.2x + 6.8 \\ \times\quad 0.3x - 0.5 \end{array}$  $0.3x^3 - 1.46x^2 + 3.64x - 3.4$

64. $(x^2 + 2x - 5)(3x^2 - 7x + 4)$  $3x^4 - x^3 - 25x^2 + 43x - 20$

**Solve each problem. (*See Objective 4.*)**

65. An apartment manager can rent all 100 units if he charges $720 per month rent. If he increases the price by $50, he rents two fewer apartments. If $x$ represents the number of increases in price, then the rent for an apartment is given by $720 + 50x$ dollars and the number of apartments rented is $100 - 2x$.

   a. What polynomial represents the total rent collected each month?  $-100x^2 + 3560x + 72,000$ dollars

   b. If the manager decides to increase the rent to $820 per month, how much total rent can he collect each month?  $78,720

## SECTION 5.6

**Perform the indicated operation. Write each answer in standard form. (*See Objectives 1–4.*)**

66. $(4x + 5)(3x - 1)$  $12x^2 + 11x - 5$  67. $(x - 12y)(x + 12y)$  $x^2 - 144y^2$

68. $(3a + 2b)^2$  $9a^2 + 12ab + 4b^2$  69. $(a + 12)^3$  $a^3 + 36a^2 + 432a + 1728$

70. $\left(c^2 - \frac{2}{5}\right)\left(c^2 + \frac{2}{5}\right)$  71. $\left(\frac{2}{3}a^3 + 5b\right)\left(\frac{2}{3}a^3 - 5b\right)$

72. $(3q + 1)^3$  $27q^3 + 27q^2 + 9q + 1$  73. $(11x - 7y)(11x + 7y)$  $121x^2 - 49y^2$

74. $(2r - 3s)(4r^2 + 6rs + 9s^2)$  $8r^3 - 27s^3$

75. $\left(\frac{3}{5}a - b\right)^2$  $\frac{9}{25}a^2 - \frac{6}{5}ab + b^2$

76. $(1.4x - 2.5)(0.1x + 1.5)$  $0.14x^2 + 1.85x - 3.75$  77. $(7a^2 - 3b)(7a^2 + 3b)$  $49a^4 - 9b^2$

## SECTION 5.7

**Perform the indicated operation. (*See Objectives 1–3.*)**

78. Divide $8a^3 + 4a^2 - 24$ by $4a$.  $2a^2 + a - \frac{6}{a}$

79. Divide $6x^2 - 7x - 20$ by $2x - 5$.  $3x + 4$

80. $\dfrac{12x^4 - 6x^3 + 14x^2}{-2x^2}$  $-6x^2 + 3x - 7$

81. $\dfrac{5x^3 + 31x^2 + 29x - 5}{x + 5}$  $5x^2 + 6x - 1$

82. $\dfrac{8x^3 - 14x^2 + 7x - 6}{2x - 3}$  $4x^2 - x + 2$

83. If the area of a parallelogram is $28xy^2 - 21x^2y$ square units and its height is $7xy$ units, find the length of its base.  $4y - 3x$

84. If the area of a triangle is $32xy^2 - 24x^2y$ square units and its height is $4xy$ units, find the length of its base.  $8y - 6x$

## CHAPTER 5 TEST / LAWS OF EXPONENTS AND POLYNOMIAL OPERATIONS

1. When simplified $(4x^5)(-2x^3)^2$ is
   a. $16x^{11}$  b. $-8x^{11}$
   c. $-8x^{10}$  d. $16x^{10}$

2. $-8^0$ is equivalent to
   a. $-8$  b. $-1$
   c. $1$  d. $0$

3. The expression $\dfrac{x^{-8}}{y^{-4}}$ is equivalent to
   a. $\dfrac{x^8}{y^4}$  b. $\dfrac{1}{x^8y^4}$
   c. $\dfrac{y^4}{x^8}$  d. $x^8y^4$

4. When simplified $\dfrac{a^6}{a^{-2}}$ is equivalent to
   a. $a^4$  b. $a^8$
   c. $\dfrac{1}{a^3}$  d. $\dfrac{1}{a^{12}}$

5. When 0.0000037 is written in scientific notation, it is
   a. $37 \times 10^5$  b. $37 \times 10^{-5}$
   c. $3.7 \times 10^6$  d. $3.7 \times 10^{-6}$

6. When simplified $(-4x^2 - 6x + 2) - (3x^2 + 2x - 7)$ is
   a. $-7x^2 - 4x + 5$  b. $-7x^2 - 8x + 9$
   c. $x^2 - 4x - 5$  d. $x^2 - 8x + 9$

7. When simplified $-3x^2(x^2 - 3x - 1)$ is
   a. $-3x^4 + 9x^3 + 3x^2$  b. $-3x^4 - 9x^3 - 3x^2$
   c. $-3x^4 - 12x^2$  d. $-3x^4 + 6x^2$

8. When simplified $(3x + 1)(3x - 1)$ is
   a. $9x^2 + 1$  b. $9x^2 - 6x - 1$
   c. $9x^2 - 1$  d. $9x^2 + 6x - 1$

9. When simplified $(x - 5)^2$ is
   a. $x^2 + 25$  b. $x^2 - 25$
   c. $x^2 + 10x + 25$  d. $x^2 - 10x + 25$

**10.** $\dfrac{8x^3 - 6x^2 - x + 5}{2x}$ is equivalent to

**a.** $4x^2 - 3x - 2 + \dfrac{5x}{2}$     **(b.)** $4x^2 - 3x - \dfrac{1}{2} + \dfrac{5}{2x}$

**c.** $-2x^2 - x + 5$     **d.** $4x^2 - 3x + 5$

**Perform each operation. Write answers using positive exponents.**

**11.** $(-6a^3)(2a^{-4})^2$   $-\dfrac{24}{a^5}$     **12.** $\left(-\dfrac{3x^5y}{2x^{-1}y^3}\right)^{-2}$   $\dfrac{4y^4}{9x^{12}}$

**13.** $-4x^0 + (-4)^0$   $-3$

**Simplify each expression.**

**14.** $(-6)^2$   $36$     **15.** $-6^2$   $-36$     **16.** $6^{-2}$   $\dfrac{1}{36}$

**17.** $-6^{-2}$   $-\dfrac{1}{36}$     **18.** $(-6)^{-2}$   $\dfrac{1}{36}$

**Complete the table for each polynomial.**

| | Polynomial | Standard Form | Type | Degree | Leading Coefficient |
|---|---|---|---|---|---|
| **19.** | $-4y - y^2 + 5$ | $-y^2 - 4y + 5$ | trinomial | 2 | $-1$ |
| **20.** | $1 - 27x^3$ | $-27x^3 + 1$ | binomial | 3 | $-27$ |
| **21.** | $9a^6$ | $9a^6$ | monomial | 6 | 9 |

**22.** Simplify: $(6x^4y^3 - 2x^3y^2 + x^2y) + (2x^4y^3 + 3x^3y^2 - 7x^2y)$
$8x^4y^3 + x^3y^2 - 6x^2y$

**23.** What is the degree of the polynomial that results in Exercise 22?   7

**24.** Evaluate the polynomial in Exercise 22 when $x = -2$ and $y = 3$.   3312

**Perform the indicated operation.**

**25.** $(2x + 5) - (3x - 7)$
$-x + 12$

**26.** $(2x + 5)(3x - 7)$
$6x^2 + x - 35$

**27.** $(5a^3 + 2b)^2$
$25a^6 + 20a^3b + 4b^2$

**28.** $(5a^3 + 2b)(5a^3 - 2b)$
$25a^6 - 4b^2$

**29.** $(3x - 5)(9x^2 + 15x + 25)$
$27x^3 - 125$

**30.** $(3x - 5)^3$
$27x^3 - 135x^2 + 225x - 125$

**31.** $\dfrac{2x^2 - 5x + 3}{x - 1}$   $2x - 3$

**32.** $\dfrac{24a^2b^3 - 12ab^2 + 4ab}{4ab}$
$6ab^2 - 3b + 1$

**Solve each problem.**

**33.** The distance traveled in 1 light-year is approximately 6 trillion miles. The North Star, Polaris, is approximately 300 light-years away from Earth. Convert each of these numbers to scientific notation and then perform an operation using these numbers that will determine how many miles away the North Star is from Earth.
$6 \times 10^{12}; 3 \times 10^2; 1.8 \times 10^{15}$

**34.** Write the polynomials that represent the perimeter and the area of a rectangle whose length is $5x^2 + 3x + 1$ units and whose width is $6x + 7$ units.
Perimeter $10x^2 + 18x + 16$; Area $30x^3 + 53x^2 + 27x + 7$

**35.** Amy owns a scarf-making business. The cost for making $x$ scarves in a month can be represented by $C(x) = 2x + 50$ dollars and the revenue for selling $x$ scarves in a month is $R(x) = 12x$ dollars. Profit is revenue minus cost.

**a.** Find a polynomial that represents the profit for Amy's business.   $10x - 50$ dollars

**b.** Find the profit for selling 30 scarves in a month.   $250

---

## CUMULATIVE REVIEW EXERCISES / CHAPTERS 1–5

**1.** Perform the indicated operation and simplify the result. (*Section 1.2, Objectives 2–7*)

**a.** $2\dfrac{1}{2} \cdot 3\dfrac{1}{5} \div 6\dfrac{1}{4}$   $1\dfrac{7}{25}$     **b.** $5\dfrac{1}{2} + 2\dfrac{3}{7} - 4\dfrac{2}{7}$   $3\dfrac{9}{14}$

**2.** The expression $29.64x + 44.46y$ represents the average weekly earnings of a utility worker, where $x$ is the number of regular hours worked in a week and $y$ is the number of overtime hours worked in a week. Find the weekly salary of a utility worker if she works 40 regular hours and 12 overtime hours. (*Section 1.3, Objective 6*)
$1719.12

**3.** Perform the indicated operation and simplify. (*Sections 1.3–1.5, Objectives 1 and 2*)

**a.** $3(5)^2 - (-4)^3 - 11 \cdot 5$   $84$

**b.** $13.2 - 8.7 + (-6.4) - (-12.5)$   $10.6$

**Write the mathematical expression needed to solve each problem.** (*Section 1.4 and 1.5, Objective 3*)

**4.** Warren mixes a 10% antifreeze solution with a 30% solution to get an 18% antifreeze solution. Let $x$ represent the volume of the 10% antifreeze solution. If Warren wants to get 120 oz of 18% antifreeze solution, write an expression that represents the volume of the 30% antifreeze solution.   $120 - x$ oz

**5.** Find the measure of each unknown angle of the given triangle.

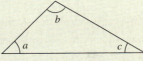

**a.** $a = 45°$, $b = 92°$   $c = 43°$   **b.** $b = 123°$, $c = 31°$
$a = 26°$

**6.** In a rectangle, the width is eight less than three times the length. If $l$ represents the length, write an expression that represents the width of the rectangle.   $3l - 8$

**7.** Perform the indicated operation and simplify. (*Section 1.6, Objectives 1–3*)

**a.** $(-5.3)(-12.2)$   $64.66$     **b.** $(-54)\left(-\dfrac{25}{18}\right)(0)$   $0$

**c.** $26 \div 0$   undefined

**8.** Simplify each algebraic expression. (*Section 1.8, Objectives 3 and 4*)

**a.** $-5(2a - 9) + 3(-a + 8)$   $-13a + 69$

**b.** $15.8 - 1.2(9b - 2)$   $18.2 - 10.8b$

**c.** $2x^2 - (4x - 3x^2) + 9x$   $5x^2 + 5x$

9. Determine whether each of the following is an expression or an equation. (*Section 2.1, Objective 1*)

   **a.** $16x - 29 = 2x - 10$   **b.** $27x - 5 - 13x + 40$
   
   equation              expression

**For each problem, define the variable, write an equation, and solve the problem. (*Section 2.1, Objective 3 and 4*)**

10. The product of a number and 4 is the same as the sum of twice the number and 14. Find the number.
    $4x = 2x + 14; x = 7$

11. The cost to rent a midsize car for one day is $62 plus $0.42 per mile. This daily cost is represented by the equation $62 + 0.42x$, where $x$ is the number of miles driven. How many miles has the car been driven if the cost of the rental car is $578.60?   1230 mi

12. Solve each equation. If the equation is a contradiction, write the solution as ∅. If the equation is an identity, write the solution as ℝ. (*Section 2.2, Objectives 2 and 3; Sections 2.3 and 2.4, Objectives 1–4*)

    **a.** $4(x - 3) - 20 = 7(x + 5) - 19$   $\{-16\}$

    **b.** $-\dfrac{1}{4}(2x + 1) + \dfrac{9}{4} = x$   $\left\{\dfrac{4}{3}\right\}$

    **c.** $-14.5x + 20.5 = 5(-2.9x + 4.3)$   ∅

13. Find the measure of an angle whose complement is 28° more than the measure of the angle. (*Section 2.5, Objective 5*)   31°

14. Bianca invested $9600 in two accounts. She invested $6500 in a certificate of deposit that pays 2.3% annual interest and the remaining amount in a money market fund that pays 1.6% annual interest. How much interest does she earn from the two accounts in 1 yr? (*Section 2.6, Objective 2*)   The interest earned is $199.10.

15. Solve each inequality. Graph the solution set, write the solution set in interval notation, and in set-builder notation. (*Section 2.7, Objectives 1–4*)

    **a.** $6x + 2(x + 22) < 11x - 16$

    **b.** $-1.6 < 0.4a + 4.3 < 1.5$

16. The following table gives the percentage of reported flight operations arriving on time as defined by the Department of Transportation (DOT) for US Airways between 2003 and 2009. (Source: http://www.transtat.bts.gov) (*Section 3.1, Objective 5*)

| Years After 2003 | 0 | 1 | 2 | 3 | 4 | 5 | 6 |
|---|---|---|---|---|---|---|---|
| Percentage of reported flight operations arriving on time | 82.0 | 75.7 | 77.8 | 76.9 | 68.7 | 80.1 | 80.9 |

   **a.** Write an ordered pair $(x, y)$ that corresponds to each year between 2003 and 2009, where $x$ is the number of years after 2003 and $y$ is the percentage of reported flight operations arriving on time as defined by DOT for US Airways.

   **b.** Interpret the meaning of first and last ordered pairs from part (a) in the context of the problem.

   **c.** In what year was the percentage of reported flight operations arriving on time for US Airways the highest? The lowest?

   **d.** Make a scatter plot of the data.

17. Graph each equation. Identify at least two points on each graph. (*Section 3.2, Objectives 2–4*)

    **a.** $5x - 2y = 10$   **b.** $x - 0.5y = 0$   **c.** $0.6x + 0.1 = 1.3$

18. The number of gallons of fuel consumed (in millions) in a year by a major airline can be modeled by the equation $y = 38x + 330.4$, where $x$ is the number of years after 2005. (*Section 3.2, Objective 5*)

    **a.** What is the $y$-intercept of the equation and what does it mean in the context of the problem?

    **b.** What was the number of gallons of fuel consumed by airline in 2010?

    **c.** What was the number of gallons of fuel consumed by airline in 2013?

    **d.** Graph the equation.

19. Find the slope of each line. (*Section 3.3, Objectives 1, 3, and 4*)

    **a.** the line that passes through the point $(-4, -1)$ and $(3, 5)$   $\dfrac{6}{7}$

    **b.** $y = 6$   0   **c.** $x + 4 = 0$   undefined

    **d.** $x - 2y = 6$   $\dfrac{1}{2}$

    **e.**   $-5$

20. The total operating revenue (in billions of dollars) for a major department store can be modeled by the equation $y = -0.281x + 12.01$, where $x$ is the number of years after 2006. Interpret the meaning of the slope and $y$-intercept in the context of the problem. (*Section 3.4, Objectives 1 and 2*)

21. Graph each line. Label two points on the line. (*Section 3.4, Objective 2*)

    **a.** $m = \dfrac{2}{3}$, passes through $(-2, 3)$

    **b.** $y = 2x - 4$

22. Determine if the lines are parallel, perpendicular, or neither. (*Section 3.4, Objective 3*)

    **a.** $x + 2y = 5$ and $2y - x = 6$   neither

    **b.** $y = 0.6x$ and $6x - 10y = 0$   parallel

23. Write the equation of each line. Express answers in slope-intercept form and in standard form. (*Section 3.5, Objectives 1–4*)

    **a.** $(3, -1)$ perpendicular to $x + 3y = 1$   $y = 3x - 10;$ $3x + y = 10$

    **b.** $(6, -5)$ parallel to $x - 6y = 7$   $y = \dfrac{1}{6}x - 6; x - 6y = 36$

    **c.** $m =$ undefined, passing through $(3, -2)$   $x = 3$

**24.** The following table shows the average residential natural gas prices (in dollars per thousand cubic feet) from 2006 to 2010. (Source: http://www.eia.gov/naturalgas/) (*Section 3.5, Objective 5*)

| Year | Average Residential Natural Gas Prices |
|------|--------------------------------------|
| 2006 | 13.73 |
| 2007 | 13.08 |
| 2008 | 13.89 |
| 2009 | 12.14 |
| 2010 | 11.19 |

  **a.** Use the points (1, 13.08) and (4, 11.19) to write a linear equation that models the average residential natural gas prices (in dollars per thousand cubic feet), where $x$ is the number of years after 2006. $y = -0.63x + 13.71$
  **b.** Find the average residential natural gas price in 2011. The average residential natural gas price in 2011 is $10.56.
  **c.** Find the average residential natural gas price in 2013. The average residential natural gas price in 2013 is $9.30.
**25.** State the domain and range of each relation. Determine if the relation is a function. (*Section 3.6, Objectives 1–3*)

  **a.** {(−4, 1), (1, −4), (0, 5), (−2, 3), (3, 1)}
  **b.** $x = 3$                    **c.** $y = -2x + 12$
**26.** Find the requested information. (*Section 3.6, Objective 4*)

  **a.** Find $f(0)$ if $f(x) = 15x - 6$.   $f(0) = -6$
  **b.** Find $g(-2)$ if $g(x) = -x^2 + 3x + 2$.   $g(-2) = -8$
  **c.** Find $f(-2)$ and solve $f(x) = 3$ if $f(x)$ is given by the following graph.   $f(-2) = -1; f(2) = 3$

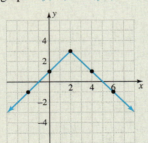

**27.** Write each equation in function notation. Then find $f(-1)$, $f(0)$, and $f(3)$ and write the corresponding ordered pairs. (*Section 3.6, Objective 4*)

  **a.** $4x + 5y = 24$           **b.** $y = 2|x - 1|$
**28.** Solve each system of equations graphically. Then state if the system is consistent with dependent equations, consistent with independent equations, or inconsistent. (*Section 4.1, Objectives 3–5*)

  **a.** $\begin{cases} y = 2x - 5 \\ y = -x + 3 \end{cases}$   **b.** $\begin{cases} x - y = 3 \\ x = -1 \end{cases}$

  **c.** $\begin{cases} 5x - y = 10 \\ x = \dfrac{1}{5}y + 2 \end{cases}$  $\{(x, y)|y = 5x - 10\}$; consistent with dependent equations

**29.** George wants to rent a 42-in. LCD TV for his apartment. Rent-Me-Cheap offers a monthly rate of $48 plus an one-time installation fee of $72. Rent to Go

offers a monthly rate of $60 per month. The monthly cost of Rent-Me-Cheap can be expressed by $y = 48x + 72$ and the monthly cost of Rent to Go can be expressed by $y = 60x$, where $x$ is the number of months. Use the graph of the system formed by these two equations to answer the questions that follow. (*Section 4.1, Objectives 2 and 6*)

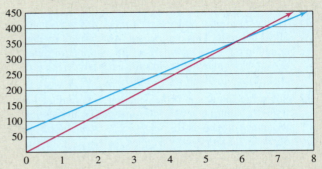

  **a.** Approximate the point of intersection of this system.   (6, 360)
  **b.** What is the meaning of the intersection point in the context of the problem?
  **c.** If George signs a 12-month rental agreement for the TV, which company should he choose? Why?

**Solve each system of linear equations using substitution. (*Section 4.2, Objectives 1 and 2*)**

**30.** $\begin{cases} 7x - y = -25 \\ x + 3y = 9 \end{cases}$  $(-3, 4)$  **31.** $\begin{cases} x - 4y = 9 \\ 5x + 2y = 23 \end{cases}$  $(5, -1)$

**Solve each system using elimination. (*Section 4.3, Objectives 1–3*)**

**32.** $\begin{cases} 4x - 7y = -59 \\ 5x + 3y = -15 \end{cases}$  $(-6, 5)$  **33.** $\begin{cases} 2x + y = -1 \\ 3y = -6x - 3 \end{cases}$

$\{(x, y)|y = -2x - 1\}$

**34.** $\begin{cases} 1.1x - 0.2y = 5.8 \\ 0.4x + 0.1y = 4.7 \end{cases}$  $(8, 15)$

**Solve each problem. (*Section 4.4, Objectives 1–5*)**

**35.** Two angles are supplementary. The larger angle is 20° more than three times the smaller angle. What is the measure of each angle?   40°, 140°

**36.** Wayne leaves his job and heads south at 50 mph. Joy leaves the same workplace 10 min later and heads south at 60 mph. How long will it take Joy to catch up to Wayne?   $\dfrac{5}{6}$ hr

**Determine if the ordered pair is a solution of the given inequality. (*Section 4.5, Objective 1*)**

**37.** $4x - 7y < 1$; (0, 2)          **38.** $y \le -3x + 2$; (−2, 1)
    (0, 2) is a solution                (−2, 1) is a solution
**Graph each linear inequality in two variables. (*Section 4.5, Objective 2*)**

**39.** $2x + 4y \le 15$     **40.** $x + 3y > 0$     **41.** $x - 4 \ge 0$

**Solve each system of linear inequalities. (*Section 4.6, Objective 1*)**

**42.** $\begin{cases} y \ge -3x + 1 \\ y \le 2x + 6 \end{cases}$   **43.** $\begin{cases} x - 2y \ge 0 \\ y \le 2 \end{cases}$

**44.** Use the appropriate exponent rules to simplify each expression. (*Section 5.1, Objectives 1–4*)

**a.** $a \cdot a^{1/2}$   $a^{3/2}$

**b.** $(-6b)(3b^3)(-4b^7)$   $72b^{11}$

**c.** $-\left(-\dfrac{x^3}{2y^5}\right)^4$   $-\dfrac{x^{12}}{16y^{20}}$

**d.** $(-5^{-2}\,xy^3)^2$   $\dfrac{x^2y^6}{625}$

**e.** $\left(\dfrac{-0.1p^2}{q^4}\right)^3$   $-\dfrac{0.001p^6}{q^{12}}$

**45.** Suppose circle A has a diameter of 9 in. and circle B has a diameter of 7 in. How much larger is the area of circle A than the area of circle B? (*Section 5.2, Objectives 1–4*)   $8\pi$ in.

**46.** Use the appropriate exponent rules to simplify each expression. Express each answer with positive exponents. (*Section 5.2, Objectives 1–4*)

**a.** $x^{-8}x^3$   $\dfrac{1}{x^5}$

**b.** $(2x^4)^{-3}$   $\dfrac{1}{8x^{12}}$

**c.** $\left(\dfrac{3x}{y^{-2}}\right)^{-3}$   $\dfrac{1}{27x^3y^6}$

**d.** $\dfrac{8^{-5}}{8^{-2}}$   $\dfrac{1}{512}$

**e.** $\left(-\dfrac{a^{-5}}{2b^2}\right)^{-4}$   $16a^{20}b^8$

**Solve each problem. (*Section 5.2, Objective 5*)**

**47.** The formula $P = P_0 \cdot 2^{-t/m}$ gives the amount of radioactive substance present at time $t$, where $P_0$ is the original amount of the substance and $m$ is the half-life of the substance. Suppose the half-life of a substance is 12 weeks. If the original amount of the substance is 10 g, how much substance remains after

**a.** 12 weeks?   5 g

**b.** 36 weeks?   1.25 g

**48.** The formula to find the monthly payment to repay a loan is $P = \dfrac{A\left(\dfrac{r}{12}\right)}{1-\left(1+\dfrac{r}{12}\right)^{-m}}$, where $A$ is the amount of the loan, $r$ is the interest rate (in decimal form), and $m$ is the number of monthly payments. Find the monthly payment for a student loan of \$15,800 at an annual interest rate of 5.1% for 8 yr.   \$200.78

**49.** Convert each number in standard notation to scientific notation and each number in scientific notation to standard notation. (*Section 5.3, Objectives 1 and 2*)

**a.** 4,820,000   $4.82 \times 10^6$

**b.** 0.0005612   $5.612 \times 10^{-4}$

**c.** $7.24 \times 10^{-6}$   0.00000724

**d.** The total number of Google searches in a day is approximately 2,960,000,000. (Source: http://www.worldometers.info/)   $2.96 \times 10^9$

**50.** Use exponent rules to simplify each expression. Write each answer in scientific notation. (*Section 5.3, Objective 3*)

**a.** $(4.2 \times 10^6)(3.5 \times 10^{-20})$   $1.47 \times 10^{-13}$

**b.** $\dfrac{7.925 \times 10^{-14}}{6.34 \times 10^{-5}}$   $1.25 \times 10^{-9}$

**51.** The population of Taiwan in 2010 was approximately 23,000,000 and the land area is approximately 13,900 mi². If the area is evenly distributed among each person in Taiwan, how much land will each person have? Write your answer in scientific notation. (Source: https://www.cia.gov/library/publications/the-world-factbook/geos/tw.html) (*Section 5.3, Objective 3*)   $6.04 \times 10^{-4}$ mi² per person

**52.** Find the requested information. (*Section 5.4, Objectives 1 and 2*)

**a.** Identify the leading coefficient and the degree of the polynomial $-2x^5 + 6x^3 - 12x + 99$.   $-2, 5$

**b.** Classify the polynomial, $-2x^2 + 1$, as a monomial, binomial, or trinomial. Then write it in standard form.   binomial, $-2x^2 + 1$

**53.** Evaluate each polynomial for the given value(s). (*Section 5.4, Objective 3*)

**a.** $-3x^2 - 5x + 10$, $x = -1$   12

**b.** $5a^2 - ab - 2b^2$, $a = -3$, $b = 2$   43

**54.** A company's total cost to make $x$ items is $24x^3 - 15x^2 + 1900$ dollars. How much does it cost the company to make 20 items? (*Section 5.4, Objectives 3 and 6*)   \$187,900

**55.** The height, in feet, of a ball projected upward from the top of a 120-ft building with an initial velocity of 80 ft/sec is given by $-16t^2 + 80t + 120$, where $t$ is the number of seconds after the ball is thrown. Find the height of the ball 3 sec after it is thrown. (*Section 5.4, Objectives 3 and 6*)   216 ft

**56.** Perform each operation. Write each answer in standard form. (*Section 5.4, Objectives 5 and 6*)

**a.** $(34 - 12x^2 - 25x) - (-27 + 6x^2 + 18x)$   $61 - 18x^2 - 43x$

**b.** $(7x^4 + 13x^3 - 16) + (2x^4 - 10x^3 + 23)$   $9x^4 + 3x^3 + 7$

**c.** $\begin{aligned} &-\tfrac{7}{6}x^2 + \tfrac{7}{3}x - \tfrac{9}{2} \\ &+\tfrac{1}{6}x^2 + \tfrac{5}{3}x + \tfrac{1}{2} \end{aligned}$   $-x^2 + 4x - 4$

**d.** Subtract $(6x^2 - 2xy + 4y^2)$ from $(-5x^2 + xy - 3y^2)$.

**e.** $(8x^2 + 12xy - 13y^2) - (-9x^2 + 4xy + y^2) + (-10x^2 - 3xy + 6y^2)$   $7x^2 + 5xy - 8y^2$

**57.** Find the perimeter of the trapezoid. (*Section 5.4, Objective 6*)   $13x^2 + 5x + 1$

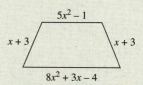

**58.** Suppose the profit generated by selling $x$ items is $-0.3x^2 + 143x + 105$ dollars and the total cost is $18x + 360$ dollars. Find an expression that represents the company's revenue. Write the answer in standard form. (*Section 5.4, Objective 6*)   $-0.3x^2 + 161x + 465$ dollars

**59.** Use the distributive property to multiply each expression. (*Section 5.5, Objectives 1–3*)

**a.** $5x^3y^2(3xy - 1)$    **b.** $(5 - 2x)(2x + 5)$

**c.** $(a + 4b)(a^2 - 4ab + 16b^2)$

**d.** $\left(\dfrac{1}{4}x - 1\right)\left(\dfrac{3}{4}x + 6\right)$    **e.**
$$\begin{array}{r} x^2 - 2.5x + 4.8 \\ \times\quad\ 0.4x - 0.6 \\ \hline \end{array}$$

**f.** $(3x^2 + 5x - 1)(2x^2 - 6x + 1)$

**Solve each problem. (*Section 5.5, Objective 4*)**

**60.** An apartment manager can rent all 100 units if he charges $750 per month rent. If he increases the price by $60, he rents three fewer apartments. If $x$ represents the number of increases in price, then the rent for an apartment is given by $750 + 60x$ dollars and the number of apartments rented is $100 - 3x$.

**a.** What polynomial represents the total rent collected each month?    $(750 + 60x)(100 - 3x) = 75,000 + 3750x - 180x^2$

**b.** If the manager decides to increase the rent to $870 per month, how much total rent can he collect each month?    $81,780

**61.** Perform the indicated operation. Write each answer in standard form. (*Section 5.6, Objectives 1–4*)

**a.** $(4x + 1)(3x - 5)$    **b.** $(b + 10)^3$
   $12x^2 - 17x - 5$          $b^3 + 30b^2 + 300b + 1000$

**c.** $\left(\dfrac{2}{3}a^3 + 5b\right)\left(\dfrac{2}{3}a^3 - 5b\right)$    $\dfrac{4}{9}a^6 - 25b^2$

**d.** $(2p - 1)^3$    $8p^3 - 12p^2 + 6p - 1$

**e.** $(11x - 5y)(11x + 5y)$    $121x^2 - 25y^2$

**f.** $(2r + 3s)(4r^2 - 6rs + 9s^2)$    $8r^3 + 27s^3$

**g.** $\left(\dfrac{2}{3}a + b\right)^2$    $\dfrac{4}{9}a^2 + \dfrac{4}{3}ab + b^2$

**h.** $(1.4x - 1.5)(0.1x + 2.5)$
   $0.14x^2 + 3.35x - 3.75$

**62.** Perform the indicated operation. (*Section 5.7, Objectives 1–3*)

**a.** Divide $12a^3 + 4a^2 - 24$ by $4a$.    $3a^2 + a - \dfrac{6}{a}$

**b.** $\dfrac{12a^4 - 48a^3 + 8a - 16}{16a^2}$    $\dfrac{3}{4}a^2 - 3a + \dfrac{1}{2a} - \dfrac{1}{a^2}$

**c.** $\dfrac{2x^3 + 20x^2 + 18x - 15}{x + 3}$    $2x^2 + 14x - 24 + \dfrac{57}{x + 3}$

**63.** If the area of a parallelogram is $32xy^3 - 16x^2y$ square units and its height is $8xy$ units, find the length of its base. (*Section 5.7, Objective 3*)    $4y^2 - 2x$ units

# Factoring Polynomials and Polynomial Equations

## Motivation

Motivation is a key element in helping us accomplish our goals like completing this class and ultimately a college degree. Something motivated you to enroll in college, so keep that vision in mind as you complete each and every course. Some courses are going to require more energy and time to complete but stay motivated through the process. Obstacles or circumstances that challenge us have a tendency to make us less motivated to finish what we started. So find what motivates you and continue to dwell on that until you achieve your goal.

**Question For Thought:** What is your motivation to complete this course? to complete your college education? How does this impact you in your daily life?

## Chapter Outline

## Coming Up...

In Section 6.6, we will solve a quadratic equation that enables us to determine how long a B.A.S.E. jumper has to open his parachute before reaching the ground after he falls off of a 1600-ft structure. (B.A.S.E. stands for buildings, antennas, structures, and earth formations.)

## SECTION 6.1 Greatest Common Factor and Grouping

**In Chapter 5** we learned how to multiply polynomials. The focus of this chapter is on the reverse process—that is, given a polynomial, find the factors that must be multiplied to get the polynomial. We will learn how to factor binomials, trinomials, and other polynomials. We will then show how factoring can be used to solve quadratic equations and their applications.

### ▶ OBJECTIVES

As a result of completing this section, you will be able to

**1.** Find the greatest common factor of a set of integers.

**2.** Find the greatest common factor of a set of terms.

**3.** Use the greatest common factor to factor a polynomial.

**4.** Use grouping to factor polynomials.

**5.** Troubleshoot common errors.

A company's profit can be represented by $-5q^2 + 200q$, where $q$ is the number of units sold in hundreds. In Section 6.6, we will solve an application involving this expression that requires us to factor it.

In Chapter 5, we began with two polynomials and multiplied them to obtain another polynomial. For instance,

$$2x(x^2 + 3) = 2x(x^2) + 2x(3) = 2x^3 + 6x$$

So, the polynomials $2x$ and $x^2 + 3$ are called **factors** of $2x^3 + 6x$ and the product $2x(x^2 + 3)$ is called the *factored form*.

In this chapter, we will be given a polynomial and must be able to find the polynomials whose product is the given polynomial. This process, called **factoring**, is the reverse of multiplication.

One of the first steps when factoring a polynomial is to determine if the terms in the polynomial have a common factor. So, we will begin our factoring techniques with finding the greatest common factor and then extend this to a method called grouping.

### The Greatest Common Factor of a Set of Integers

**Objective 1** ▶

Find the greatest common factor of a set of integers.

In the product $3 \cdot 4 = 12$, the numbers 3 and 4 are called the *factors* of 12. To *factor a number* means to write the number as a product. Often, we are interested in determining the factors that numbers have in common, specifically the largest of these factors. This is called the *greatest common factor*.

> **Definition:** The **greatest common factor** (GCF) of a set of integers is the largest integer that is a factor of each integer in the set.

Suppose we want to find the GCF of 42 and 60. Then we need to find the largest integer that is a factor of both of these numbers. There are two approaches that we can take.

First, we can list all of the factors of each number and find the largest factor that is in both lists.

Factors of 42: 1, 2, 3, 6, 7, 14, 21, 42

Factors of 60: 1, 2, 3, 4, 5, 6, 10, 12, 15, 20, 30, 60

> The common factors are 1, 2, 3, and 6 and the greatest common factor is 6.

This method works great for small integers but may not be the best method for larger integers.

Another way we can find the GCF is to use prime factorizations. Recall from Chapter 1, we used factor trees to find the prime factorization of numbers. The prime factorizations of 42 and 60 are shown.

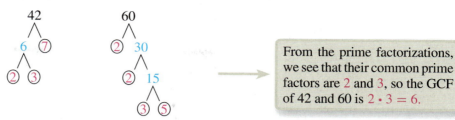

So, $42 = 2 \cdot 3 \cdot 7$.    So, $60 = 2 \cdot 2 \cdot 3 \cdot 5$.

From the prime factorizations, we see that their common prime factors are 2 and 3, so the GCF of 42 and 60 is $2 \cdot 3 = 6$.

**Note:** We can also write the factorizations horizontally as shown.

$$42 = 6 \cdot 7 = 2 \cdot 3 \cdot 7$$
$$60 = 2 \cdot 30 = 2 \cdot 2 \cdot 15 = 2 \cdot 2 \cdot 3 \cdot 5$$

**Procedure: Finding the GCF of a Set of Integers**

**Step 1:** Write each integer as a product of prime factors.
**Step 2:** Determine the prime factors that are common to each integer.
**Step 3:** The GCF is the product of these common factors. If there is no common factor, then the GCF is 1.

**Objective 1 Examples**    **Find the GCF of each set of integers.**

**1a.** 56 and 42        **1b.** 18, 54, and 81        **1c.** 48 and 35

**Solutions**    **1a.** $56 = 8 \cdot 7 = 2 \cdot 2 \cdot 2 \cdot 7$
$42 = 6 \cdot 7 = 2 \cdot 3 \cdot 7$

Since the common factors are 2 and 7, the GCF of 56 and 42 is $2 \cdot 7 = 14$.

**1b.** $18 = 2 \cdot 9 = 2 \cdot 3 \cdot 3$
$54 = 6 \cdot 9 = 2 \cdot 3 \cdot 3 \cdot 3$
$81 = 9 \cdot 9 = 3 \cdot 3 \cdot 3 \cdot 3$

Since the common factors are 3 and 3, the GCF of 18, 54, and 81 is $3 \cdot 3 = 9$.

**1c.** $48 = 4 \cdot 12 = 2 \cdot 2 \cdot 2 \cdot 2 \cdot 3$
$35 = 5 \cdot 7$

The numbers do not share any common factors. Therefore, the GCF of 48 and 35 is 1.

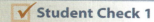 **Student Check 1**    Find the GCF of each set of integers.
**a.** 60 and 36        **b.** 75, 51, and 99        **c.** 32 and 45

## The Greatest Common Factor of a Set of Terms

**Objective 2** ▶

Find the greatest common factor of a set of terms.

We can extend the idea of the GCF of integers to finding the greatest common factor of terms. Because terms involve integer coefficients and variables raised to exponents, we first need to determine how to find the greatest common factor of variables. Since we already know how to find the GCF of the integer coefficients, we will combine these results to find the GCF of a set of terms.

To find the GCF of $x^3$ and $x^5$, we write each term in expanded form and find the product of the factors that are common.

$$x^3 = \circled{x} \cdot \circled{x} \cdot \circled{x}$$
$$x^5 = \circled{x} \cdot \circled{x} \cdot \circled{x} \cdot x \cdot x$$

The common factors are three factors of $x$.
So, the GCF is $x \cdot x \cdot x = x^3$.

**Note:** *The GCF of the variables in the terms is the common variable raised to the smallest exponent of the terms.*

> **Procedure:  Finding the GCF of a Set of Terms**
>
> **Step 1:** Find the GCF of the coefficients of the terms.
> **Step 2:** Find the GCF of the variables of the terms.
> **Step 3:** The GCF of the terms is the product of the GCF of the coefficients and the GCF of the variable factors.

**Objective 2  Examples**  Find the GCF of each set of terms.

**2a.** $4x^2$ and $10x$

**2b.** $-12a^3$ and $42a^4$

**2c.** $6x^2y$, $15xy^3$, and $20xy$

**2d.** $4(x - 2)$ and $y(x - 2)$

**Solutions**   **2a.**     $4x^2 = 2 \cdot 2 \cdot x^2$
$10x = 2 \cdot 5 \cdot x$

The greatest common factor of 4 and 10 is 2. The greatest common factor of $x^2$ and $x$ is $x$ since 1 is the smallest exponent of the variable. So, the GCF of $4x^2$ and $10x$ is $2x$.

**2b.**   $-12a^3 = -1 \cdot 2 \cdot 2 \cdot 3 \cdot a^3$
$42a^4 = 2 \cdot 3 \cdot 7 \cdot a^4$

The greatest common factor of $-12$ and 42 is $2 \cdot 3 = 6$. The greatest common factor of $a^3$ and $a^4$ is $a^3$ since 3 is the smallest exponent of the variable. So, the GCF of $-12a^3$ and $42a^4$ is $6a^3$.

**2c.**     $6x^2y = 2 \cdot 3 \cdot x^2 \cdot y$
$15xy^3 = 3 \cdot 5 \cdot x \cdot y^3$
$20xy = 2 \cdot 2 \cdot 5 \cdot x \cdot y$

The coefficients do not have a common factor, so their GCF is 1. The greatest common factor of $x^2y$, $xy^3$, and $xy$ is $xy$ since 1 is the smallest exponent of both $x$ and $y$. So, the GCF of $6x^2y$, $15xy^3$, and $20xy$ is $xy$.

**2d.** $4(x - 2) = 2 \cdot 2(x - 2)$
$y(x - 2)$

The greatest common factor of the coefficients is 1. The common variable factor, $(x - 2)$, has a smallest exponent of 1, so the GCF of the variable factors is $(x - 2)$. So, the GCF of $4(x - 2)$ and $y(x - 2)$ is $1(x - 2)$ or $(x - 2)$.

**✓ Student Check 2**   Find the GCF of each set of terms.

**a.** $6y$ and $9y^2$

**b.** $24b^5$ and $32b^4$

**c.** $8x^3y^2$, $20x^2y$, and $25x^4y^3$

**d.** $6(y + 4)$ and $z(y + 4)$

## Factor out the Greatest Common Factor from a Polynomial

**Objective 3** ▶

Use the greatest common factor to factor a polynomial.

To use the greatest common factor when factoring, we apply the distributive property in reverse by "factoring out" the GCF. Recall:

$$2x(x^2 + 3) = 2x(x^2) + 2x(3) = 2x^3 + 6x$$

When this statement is written in reverse, we have

$$2x^3 + 6x = \underbrace{2x(x^2) + 2x(3)}_{} = \underbrace{2x(x^2 + 3)}$$

$2x$ is the GCF       Factored form
of $2x^3$ and $6x$      of $2x^3 + 6x$

So, the polynomial $2x^3 + 6x$ is written as a product, or in factored form, where one of its factors is the GCF of its terms.

---

**Property:  The Distributive Property**

For $a$, $b$, and $c$ real numbers,

$$ab + ac = a(b + c)$$

---

**Note:** a represents the greatest common factor of each term of the polynomial.

---

**Procedure:  Factoring a Polynomial Using the Greatest Common Factor**

**Step 1:** Find the GCF of the terms of the polynomial.
**Step 2:** Rewrite each term of the polynomial as a product of the GCF and the remaining factor.
**Step 3:** Apply the distributive property to factor out the GCF. The GCF "$a$" is written on the outside of parentheses and the remaining factors $(b + c)$ are written on the inside of the parentheses.
**Step 4:** Check the factorization by multiplying.

---

**Objective 3  Examples**   Use the GCF to factor each polynomial.

**3a.** $6x^2 - 18x$            **3b.** $14xy^3 + 49x^2y^2 + 7xy^2$

**3c.** $-2a^3 + 8a$           **3d.** $x(x + 5) + 3(x + 5)$

**3e.** A company's profit can by represented by $-5q^2 + 200q$, where $q$ is the number of units sold in hundreds. Factor this polynomial.

**Solutions**   **3a.** The GCF of $6x^2$ and $18x$ is $6x$.

**INSTRUCTOR NOTE:**
Show students that they can use division to find the other factor. For instance: $\dfrac{6x^2}{6x} = x$ and $\dfrac{18x}{6x} = 3$.

$$6x^2 - 18x = 6x(x) - 6x(3)$$        Rewrite each term as a product involving $6x$.
$$= 6x(x - 3)$$        Apply the distributive property.

**Check:** $6x(x - 3) = 6x(x) - 6x(3) = 6x^2 - 18x$

**3b.** The GCF of $14xy^3$, $49x^2y^2$, $7xy^2$ is $7xy^2$.

$14xy^3 + 49x^2y^2 + 7xy^2$
$\quad = 7xy^2(2y) + 7xy^2(7x) + 7xy^2(1)$    Rewrite each term as a product involving $7xy^2$.
$\quad = 7xy^2(2y + 7x + 1)$    Apply the distributive property.

**Check:** $7xy^2(2y + 7x + 1) = 7xy^2(2y) + 7xy^2(7x) + 7xy^2(1)$
$\qquad\qquad\qquad\qquad = 14xy^3 + 49x^2y^2 + 7xy^2$

**3c.** The GCF of $-2a^3$ and $8a$ is $2a$. When the first term of a polynomial is negative, it is common to make the GCF negative. Therefore, we will factor out $-2a$.

$-2a^3 + 8a = -2a(a^2) + (-2a)(-4)$    Rewrite each term as a product involving $-2a$.
$\qquad\qquad = -2a(a^2 - 4)$    Apply the distributive property.

**Check:** $-2a(a^2 - 4) = -2a(a^2) + (-2a)(-4) = -2a^3 + 8a$

We could have also factored out $2a$.

$-2a^3 + 8a = 2a(-a^2) + (2a)(4)$    Rewrite each term as a product involving $2a$.
$\qquad\qquad = 2a(-a^2 + 4)$    Apply the distributive property.

**Check:** $2a(-a^2 + 4) = 2a(-a^2) + (2a)(4) = -2a^3 + 8a$

 **Note:** *The difference between the two factorizations is the signs.*

**3d.** The GCF of $x(x + 5)$ and $3(x + 5)$ is $x + 5$. Each term is already expressed as a product of $(x + 5)$ and another factor.

$x(x + 5) + 3(x + 5) = (x + 5)(x + 3)$    Apply the distributive property.

**Check:** $(x + 5)(x + 3) = (x + 5)x + (x + 5)(3) = x(x + 5) + 3(x + 5)$.

**3e.** The GCF of $-5q^2$ and $200q$ is $5q$. Since the first term is negative, we will factor out $-5q$.

$-5q^2 + 200q$
$\quad = -5q(q) + (-5q)(-40)$    Rewrite each term as a product involving $-5q$.
$\quad = -5q(q - 40)$    Apply the distributive property.

**Check:** $-5q(q - 40) = -5q(q) - 5q(-40) = -5q^2 + 200q$

 **Student Check 3**    Use the GCF to factor each polynomial.

  **a.** $5x^2 - 20x$                       **b.** $21x^2y^4 + 63x^3y^2$
  **c.** $-8a^3 + 72a^2 + 16a$             **d.** $x(x - 9) + 5(x - 9)$
  **e.** The expression $8x^2 + 20x$ represents the area of a rectangle. Factor the expression to determine an expression for the length and width of the rectangle.

## Grouping

**Objective 4** ▶

Use grouping to factor polynomials.

The grouping method of factoring applies to polynomials with four or more terms in which pairs of terms have a common factor. If, after factoring out the GCF from pairs of terms, the remaining factors are the same, the polynomial can be factored. Consider the following polynomial.

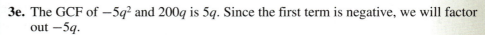

$x^3 + x^2 + 2x + 2 = (x^3 + x^2) + (2x + 2)$    Group the first two terms and the last two terms.
$\qquad\qquad\qquad = x^2(x + 1) + 2(x + 1)$    Factor out the GCF from each pair of terms.
$\qquad\qquad\qquad = (x + 1)(x^2 + 2)$    Factor out the common binomial of $(x + 1)$.

In this illustration, note that after the GCF was factored out of each pair of terms, the binomial that remained was the same, namely $(x + 1)$. If, however, the binomial is not the same after removing the GCF from the two pairs of terms, then different terms may need to be grouped together. If no grouping produces a common binomial, then we can conclude that the polynomial cannot be factored and is a **prime polynomial**.

---

**Procedure: Using Grouping to Factor a Polynomial**

**Step 1:** Group together pairs of terms.
**Step 2:** Factor out the GCF from each pair of terms.
**Step 3:** Factor out the common binomial from both terms, if possible.
  **a.** If the binomial is not the same, try rearranging the grouping of terms.
  **b.** Repeat the preceding steps.
  **c.** If this doesn't result in a common binomial factor, then the polynomial cannot be factored.
**Step 4:** Check the answer by multiplying.

---

**Objective 4 Examples** | **Use grouping to factor each polynomial. Check answer by multiplying.**

**4a.** $12x^2 + 18x + 10x + 15$      **4b.** $5x^3 - 35x^2 + x - 7$

**4c.** $3x^3 - 7x^2 - 6x + 14$       **4d.** $6x^3 + 3x^2 + 6x + 3$

**Solutions** | **4a.** $12x^2 + 18x + 10x + 15$

$= (12x^2 + 18x) + (10x + 15)$    Group together the first two and last two terms.
$= 6x(2x + 3) + 5(2x + 3)$    Factor out the GCF from each pair of terms.
$= (2x + 3)(6x + 5)$    Factor out the common binomial, $2x + 3$.

**Check:** $(2x + 3)(6x + 5) = 12x^2 + 10x + 18x + 15$

Because of the commutative property of addition, this is equivalent to the given polynomial.

**Note:** *If one grouping works, then another grouping may also work. For example, grouping the first and third terms and the second and fourth terms gives us the following.*

$12x^2 + 18x + 10x + 15$
$= (12x^2 + 10x) + (18x + 15)$    Group together the first two and last two terms.
$= 2x(6x + 5) + 3(6x + 5)$    Factor out the GCF from each pair of terms.
$= (6x + 5)(2x + 3)$    Factor out the common binomial, $6x + 5$.

**4b.** $5x^3 - 35x^2 + x - 7$

$= (5x^3 - 35x^2) + (x - 7)$    Group together the first two and last two terms.
$= 5x^2(x - 7) + 1(x - 7)$    Factor out the GCF from the first pair of terms. Write $(x - 7)$ as $1(x - 7)$.
$= (x - 7)(5x^2 + 1)$    Factor out the common binomial, $x - 7$.

**Check:** $(x - 7)(5x^2 + 1) = 5x^3 + x - 35x^2 - 7$

**4c.** Group together the first two and the last two terms. Be sure to keep the negative sign on $6x$ in the grouping. Because the negative sign is on the first term of the second pair, we need to factor out a negative GCF.

$3x^3 - 7x^2 - 6x + 14$

$\begin{aligned}
&= (3x^3 - 7x^2) + (-6x + 14) &&\text{Group together the first two and last two terms.}\\
&= x^2(3x - 7) - 2(3x - 7) &&\text{Factor out the GCF from each pair of terms.}\\
&= (3x - 7)(x^2 - 2) &&\text{Factor out the common binomial, } 3x - 7.
\end{aligned}$

**Check:** $(3x - 7)(x^2 - 2) = 3x^3 - 6x - 7x^2 + 14$

**4d.** Notice that each term in this polynomial has a common factor of 3. We factor this out first.

$$6x^3 + 3x^2 + 6x + 3 = 3(2x^3 + x^2 + 2x + 1)$$

Now we will factor $2x^3 + x^2 + 2x + 1$ by grouping.

$2x^3 + x^2 + 2x + 1$

$\begin{aligned}
&= (2x^3 + x^2) + (2x + 1) &&\text{Group together the first two and last two terms.}\\
&= x^2(2x + 1) + 1(2x + 1) &&\text{Factor out the GCF from each pair of terms.}\\
&= (2x + 1)(x^2 + 1) &&\text{Factor out the common binomial, } 2x + 1.
\end{aligned}$

So, the complete factorization is

$$\begin{aligned}
6x^3 + 3x^2 + 6x + 3 &= 3(2x^3 + x^2 + 2x + 1)\\
&= 3(2x + 1)(x^2 + 1)
\end{aligned}$$

**Check:** $3(2x + 1)(x^2 + 1) = (6x + 3)(x^2 + 1) = 6x^3 + 6x + 3x^2 + 3$

---

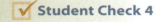 **Student Check 4**    Use grouping to factor each polynomial. Check answer by multiplying.

**a.** $8y^2 - 2y + 20y - 5$        **b.** $6x^4 - 42x^3 + x - 7$

**c.** $2x^3 - 3x^2 - 12x + 18$        **d.** $12y^3 + 8y^2 + 48y + 32$

---

**Objective 5** ▶

Troubleshoot common errors.

## Troubleshooting Common Errors

Some common errors related to factoring using the GCF and grouping are shown next.

**Objective 5  Examples**    **A problem and an incorrect solution are given. Provide the correct solution and an explanation of the error.**

**5a.** Find the GCF of $120x^3y^2$ and $50x^4y$.

| Incorrect Solution | Correct Solution and Explanation |
|---|---|
| $120 = 6 \cdot 20 = 2 \cdot 3 \cdot 2 \cdot 2 \cdot 5$<br>$50 = 5 \cdot 10 = 5 \cdot 2 \cdot 5$<br><br>The GCF of the coefficients is $2 \cdot 5 = 10$.<br>The GCF of the variables is $x^4y^2$.<br><br>So, the GCF is $10x^4y^2$. | An error was made in finding the GCF of the variables. The GCF of the variables is the smallest exponent, not the largest exponent. Therefore, the GCF of the variables is $x^3y$.<br><br>So, the GCF is $10x^3y$. |

**5b.** Factor $4x^3 - 12x$.

| Incorrect Solution | Correct Solution and Explanation |
|---|---|
| $4x^3 - 12x^2$ <br> $2x(2x^2 - 6x)$ | A common factor was factored out but not the greatest common factor. The terms in the binomial still have something in common. The GCF of the terms in the binomial is $4x^2$. <br><br> $4x^3 - 12x^2 = 4x^2(x - 3)$ |

**5c.** Factor $m^2 + 5m - 2mn - 10n$.

| Incorrect Solution | Correct Solution and Explanation |
|---|---|
| $m^2 + 5m - 2mn - 10n$ <br> $(m^2 + 5m) - (2mn - 10n)$ <br> $m(m + 5) - 2n(m - 5)$ <br><br> Since the binomials are not the same, this is prime. | Because the sign of $2mn$ is negative, we need to include the sign in the grouping. Then, factor out a negative GCF from the last two terms. <br><br> $m^2 + 5m - 2mn - 10n$ <br> $(m^2 + 5m) + (-2mn - 10n)$ <br> $m(m + 5) - 2n(m + 5)$ <br> $(m + 5)(m - 2n)$ |

**5d.** Factor $x^2 + 4x + 3x + 12$ by grouping.

| Incorrect Solution | Correct Solution and Explanation |
|---|---|
| $x^2 + 4x + 3x + 12$ <br> $(x^2 + 4x) + (3x + 12)$ <br> $x(x + 4) + 3(x + 4)$ <br> $(x + 3)(x + 4)(x + 4)$ <br> $(x + 3)(x + 4)^2$ <br><br> Since the binomials are not the same, this is prime. | The error was made in not factoring out the common binomial $(x + 4)$ correctly. <br><br> $x^2 + 4x + 3x + 12$ <br> $(x^2 + 4x) + (3x + 12)$ <br> $x(x + 4) + 3(x + 4)$ <br> $(x + 4)(x + 3)$ |

## ANSWERS TO STUDENT CHECKS

**Student Check 1** **a.** 12 **b.** 3 **c.** 1

**Student Check 2** **a.** $3y$ **b.** $8b^4$ **c.** $x^2y$ **d.** $(y + 4)$

**Student Check 3** **a.** $5x(x - 4)$ **b.** $21x^2y^2(y^2 + 3x)$
   **c.** $-8a(a^2 - 9a - 2)$ **d.** $(x - 9)(x + 5)$
   **e.** $4x$ and $2x + 5$ are the length and width of the rectangle.

**Student Check 4** **a.** $(4y - 1)(2y + 5)$ **b.** $(x - 7)(6x^3 + 1)$
   **c.** $(2x - 3)(x^2 - 6)$ **d.** $4(3y + 2)(y^2 + 4)$

## SUMMARY OF KEY CONCEPTS

1. The greatest common factor (GCF) of a set of integers is the largest integer that is a factor of each integer. We can find this by comparing the list of factors or by writing the integers as a product of prime numbers.

2. For common variables, the GCF is the variable raised to the smallest exponent on the variables. The GCF of a group of terms is the product of the GCF of the coefficients and the GCF of the common variables.

3. To factor means to express as a product. Factoring is the reverse of multiplication. The distributive property enables us to factor a polynomial by factoring out the GCF. We identify the GCF of the terms in the polynomial and rewrite each term as a product of the GCF and another factor. The factored form is the GCF times the sum of the remaining factors. Factoring can always be checked by multiplying.

4. Grouping is used to factor a polynomial with four or more terms. Pair together terms. Factor out the GCF from each pair of terms. The remaining binomial must be the same to factor further. If the binomial is the same, factor it out and multiply by the remaining factors. If it is not the same, try regrouping terms. A polynomial that cannot be factored is a prime polynomial.

## GRAPHING CALCULATOR SKILLS

The graphing calculator can be used in to find factors of integers as well as to find the GCF of a set of integers.

**Example:** Find factors of 42.

**Solution:** To find factors of 42 by hand, we divide it by 1, 2, 3, . . . . To perform this task using a calculator, we divide 42 by $x$ and then view the table. If the corresponding value in the $Y_1$ column is an integer, then $x$ is a factor of 42.

 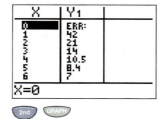

From the table, we see that factors of 42 are 1, 2, 3, 4, 6, 7, 14, 21, and 42. Some factorizations of 42 are $1 \cdot 42$, $2 \cdot 21$, $3 \cdot 14$, and $6 \cdot 7$.

**Example:** Find the GCF of 60 and 84.

**Solution:** The calculator has a built-in command called gcd which stands for the greatest common divisor, or greatest common factor. To access the command, press MATH and use the NUM menu, option 9.

So, the GCF of 60 and 84 is 12.

## SECTION 6.1 / EXERCISE SET

### Write About It!

Use complete sentences in your answer to each exercise.

1. Define the greatest common factor.
2. Explain how to find the GCF of a set of terms.
3. What does it mean to factor a polynomial?
   To factor a polynomial is to write it as a product.
4. Explain how to check if your factoring is correct.
   Check the factoring by multiplying the factors.
5. Explain how to factor by grouping.
6. Explain when the factoring is completed.   When each factor, other than the coefficient, is prime, the factorization is complete.
7. Most math operations have an inverse, which "undoes" the operation (subtraction undoes addition, etc.). What is the inverse operation for factoring? Illustrate your answer with an example.
8. What is the inverse operation for factoring out the GCF? Illustrate your answer with an example.

### Practice Makes Perfect!

**Find the GCF of each set of integers. (*See Objective 1.*)**

9. 42 and 30   6
10. 45 and 27   9
11. 54 and 24   6
12. 80, 32, and 48   16

13. 60, 45, and 90   15
14. 57, 84, and 18   3
15. 44, 66, and 110   22
16. 26, 65, and 91   13

**Find the GCF of each set of terms. (*See Objective 2.*)**

17. $6x^2$ and $3x^3$   $3x^2$
18. $4x^2$ and $8x^3$   $4x^2$
19. $-15a^2$ and $10a$   $5a$
20. $-18a^3$ and $6a$   $6a$
21. $-24b^3$ and $16b^2$   $8b^2$
22. $-36b^4$ and $24b^5$   $12b^4$
23. $12a^2b^3$ and $16a^3b^4$   $4a^2b^3$
24. $20a^5b^3$ and $12a^4b^2$   $4a^4b^2$
25. $-8ab^2$ and $14a^2b$   $2ab$
26. $-6a^3b^2$ and $15a^2b^3$   $3a^2b^2$
27. $4y, 16y^2,$ and $8y^3$   $4y$
28. $8y, 12y^2,$ and $24y^4$   $4y$
29. $6y^3, 18y^5,$ and $24$   $6$
30. $8y^4, 12y^3,$ and $16$   $4$
31. $5(x + 1)$ and $3(x + 1)$   $x + 1$
32. $6(x + 3)$ and $7(x + 3)$   $x + 3$
33. $8(y - 12)$ and $x(y - 12)$   $y - 12$
34. $10(y + 9)$ and $x^2(y + 9)$   $y + 9$
35. $6(a + 2)$ and $3(a + 2)$   $3(a + 2)$
36. $12(a + 4)$ and $14(a + 4)$   $2(a + 4)$
37. $3x(b - 3)$ and $5(b + 3)$   $1$
38. $5x(b + 7)$ and $4(b + 5)$   $1$

**Use the GCF to factor each polynomial. (*See Objective 3.*)**

39. $2x^2 - 4x$   $2x(x - 2)$
40. $3x^2 - 9x$   $3x(x - 3)$
41. $4a^4 - 16a^3$   $4a^3(a - 4)$
42. $5a^5 - 15a^3$   $5a^3(a^2 - 3)$
43. $-3y^2 + 6y$   $-3y(y - 2)$
44. $-10y^3 + 12y^2$   $-2y^2(5y - 6)$
45. $-36b^5 - 24b^3 - 12b^2$   $-12b^2(3b^3 + 2b + 1)$
46. $-48b^6 - 36b^5 + 12b^3$   $-12b^3(4b^3 + 3b^2 - 1)$

**47.** $45x^3 - 30x^2 + 15x$
$15x(3x^2 - 2x + 1)$

**48.** $18y^4 - 27y^3 - 9y^2$
$9y^2(2y^2 - 3y - 1)$

**49.** $2a^2 + 2a - 2$   $2(a^2 + a - 1)$

**50.** $5a^2 + 25a - 10$
$5(a^2 + 5a - 2)$

**51.** $-8x^3 + 20x^2 + 4x$
$-4x(2x^2 - 5x - 1)$

**52.** $-6y^3 - 12y^2 + 6y$
$-6y(y^2 + 2y - 1)$

**53.** $x(x + 3) + 2(x + 3)$
$(x + 3)(x + 2)$

**54.** $x(x - 11) + 6(x - 11)$
$(x - 11)(x + 6)$

**55.** $6(x - 2) + x(x - 2)$
$(x - 2)(x + 6)$

**56.** $5(x - 5) + x(x - 5)$
$(x - 5)(x + 5)$

**57.** $4(x + 3) - x(x + 3)$
$(x + 3)(4 - x)$

**58.** $3(x + 9) - x(x + 9)$
$(x + 9)(3 - x)$

**59.** $x^2(x + 1) - 3x(x + 1)$
$x(x + 1)(x - 3)$

**60.** $2x^2(x - 7) - x(x - 7)$
$x(x - 7)(2x - 1)$

**61.** The expression $6y^3 + 16y^2$ represents the area of a rectangle. Factor the expression to determine expressions that could represent the length and width of the rectangle.   $2y^2$ and $3y + 8$ are the length and width of the rectangle.

**62.** The expression $8y^3 + 12y^2$ represents the area of a rectangle. Factor the expression to determine expressions that could represent the length and width of the rectangle.   $4y^2$ and $2y + 3$ are the length and width of the rectangle.

**63.** A company's profit can be represented by $-20q^2 + 500q$, where $q$ is the number of units sold in thousands. Factor this polynomial.   $-20q(q - 25)$

**64.** A company's profit can be represented by $-15p^2 + 450p$, where $p$ is the number of units sold in thousands. Factor this polynomial.   $-15p(p - 30)$

**Use grouping to factor each polynomial. (*See Objective 4.*)**

**65.** $6x^2 - 2x + 3x - 1$
$(3x - 1)(2x + 1)$

**66.** $5x^2 + 2x + 20x + 8$
$(5x + 2)(x + 4)$

**67.** $4b^3 + 4b^2 + 3b + 3$
$(4b^2 + 3)(b + 1)$

**68.** $2y^3 - 6y^2 + 5y - 15$
$(2y^2 + 5)(y - 3)$

**69.** $5a - 20 - a^2 + 4a$
$(5 - a)(a - 4)$

**70.** $6x - 12 - 5x^2 + 10x$
$(6 - 5x)(x - 2)$

**71.** $r^3 + 5r^2 - 8r - 40$
$(r + 5)(r^2 - 8)$

**72.** $15s^3 - 10s^2 - 3s + 2$
$(3s - 2)(5s^2 - 1)$

**73.** $4a^3 - 20a^2 - a - 5$

**74.** $3c^3 + 6c^2 - c + 2$

**75.** $32x^2 - 56x + 12x - 21$
$(8x + 3)(4x - 7)$

**76.** $14t^2 + 21t + 10t + 15$
$(7t + 5)(2t + 3)$

**77.** $3x^3 + 12x^2 - 18x - 72$
$3(x^2 - 6)(x + 4)$

**78.** $15y^3 + 15y - 35y^2 - 35$
$5(3y - 7)(y^2 + 1)$

**79.** $8x^3 - 40x^2 - 4x + 20$
$4(2x^2 - 1)(x - 5)$

**80.** $15d^3 - 15d^2 - 5d + 5$
$5(3d^2 - 1)(d - 1)$

 **Mix 'Em Up!**

**Factor each polynomial.**

**81.** $9y^2 - 81y$   $9y(y - 9)$

**82.** $6r^2 - 21r$   $3r(2r - 7)$

**83.** $-16a^3b + 48a^2b - 8ab$
$-8ab(2a^2 - 6a + 1)$

**84.** $30c^3d - 30c^2d + 6cd$
$6cd(5c^2 - 5c + 1)$

**85.** $3x^2 - 6x + 2x - 4$
$(3x + 2)(x - 2)$

**86.** $21z^2 - 35z + 3z - 5$
$(7z + 1)(3z - 5)$

**87.** $3x^3 + 5x^2 - 6x + 10$

**88.** $9s^3 + 21s^2 - 3s + 7$

**89.** $36x^2 + 12x - 6x - 2$
$2(6x - 1)(3x + 1)$

**90.** $15t^2 - 45t + 12t - 36$
$3(5t + 4)(t - 3)$

**91.** $-16a^4 + 48a^3 + 16a^2$
$-16a^2(a^2 - 3a - 1)$

**92.** $-15b^4 - 18b^3 + 3b^2$
$-3b^2(5b^2 + 6b - 1)$

**93.** $4x^3(2x - 1) - 3x(2x - 1)$
$x(4x^2 - 3)(2x - 1)$

**94.** $6y(y + 3) - 42(y + 3)$
$6(y - 7)(y + 3)$

**95.** $4a^2 + 6a - 6$
$2(2a^2 + 3a - 3)$

**96.** $3b^3 + 5b^2 - b$
$b(3b^2 + 5b - 1)$

**97.** $6x^2 + 30x - 5x - 25$
$(6x - 5)(x + 5)$

**98.** $4z^2 - 10z - 6z + 15$
$(2z - 3)(2z - 5)$

**99.** The expression $6z^4 + 24z^2$ represents the area of a parallelogram. Factor the expression to determine expressions that could represent the base and height of the parallelogram.   Answers vary. $6z^2$ and $z^2 + 4$ are the base and height of the parallelogram.

**100.** The expression $5z^4 + 30z^2$ represents the area of a parallelogram. Factor the expression to determine expressions that could represent the base and height of the parallelogram.   Answers vary. $5z^2$ and $z^2 + 6$ are the base and height of the parallelogram.

**101.** The surface area of a circular cone is the sum of the area of the base and the area of the lateral surface, $\pi r^2 + \pi r l$, where $r$ is the radius of the base and $l$ is the slant height of the cone. Factor the expression.   $\pi r(r + l)$

**102.** The surface area of a square pyramid is given by $x^2 + 2xh$, where $x$ is the length of the square base and $h$ is the slant height of the lateral face of the pyramid. Factor the expression.   $x(x + 2h)$

 **You Be the Teacher!**

**Correct each student's errors, if any.**

**103.** Find the GCF of $x^2$ and $x^3$.

Stephanie's work:

The GCF of $x^2$ and $x^3$ is $x^3$.   $x^2$

**104.** Find the GCF of $12a^3$ and $6a^4$.

Jessica's work:

The GCF of $12a^3$ and $6a^4$ is $12a^4$.   $6a^3$

**105.** Factor $2x^2 - 6x - x + 3$.

Tammy's work:

$2x(x + 3) - 1(x + 3) = (x + 3)(2x - 1)$   $2x(x - 3) - 1(x - 3)$
$= (2x - 1)(x - 3)$

**106.** Factor $6x^3 + 15x^2 - 18x - 45$.

Danny's work:

$3x^2(2x + 5) - 9(2x + 5) = (2x + 5)(3x^2 - 9)$
$3(2x^3 + 5x^2 - 6x - 15) = 3[x^2(2x + 5) - 3(2x + 5)] = 3(2x + 5)(x^2 - 3)$

**Calculate It!**

**Find the GCF of each group of terms. Use a calculator to find the GCF of the coefficients of the given terms.**

**107.** $60x^5$ and $72x^6$   $12x^5$

**108.** $112x^8$ and $42x^3$   $14x^3$

**109.** $512x^3$ and $48x^2$   $16x^2$

**110.** $216x^6$ and $144x^5$   $72x^5$

**Think About It!**

**Write a binomial, in factored form, whose greatest common factor is given.**

**111.** $6x$

**112.** $8a$

**113.** $2y^2$

**114.** $7b^3$

**115.** $-2ab$

**116.** $-4x^2y$

## SECTION 6.2    Factoring Trinomials

▶ **OBJECTIVES**

As a result of completing this section, you will be able to

1. Factor trinomials of the form $x^2 + bx + c$.

2. Factor trinomials that have a common factor.

3. Troubleshoot common errors.

A ball is thrown upward with an initial velocity of 16 ft/sec from a height of 32 ft above the ground. The height, in feet, of the ball $t$ sec after it is thrown can be represented by $-16t^2 + 16t + 32$. In Section 6.6, we will solve an equation that enables us to determine when the ball reaches the ground. This process requires us to factor the trinomial, $-16t^2 + 16t + 32$.

As discussed in Section 6.1, factoring is the reverse of multiplication. Factoring trinomials is like solving a puzzle. There are clues to the answer provided in the given problem. This section will show us how to use the given clues to factor a trinomial.

### Factoring $x^2 + bx + c$

**Objective 1** ▶

Factor trinomials of the form $x^2 + bx + c$.

Our goal is to express a trinomial of the form $x^2 + bx + c$ as a product. This trinomial has a leading coefficient of 1. Some examples of this type of trinomial are

$$x^2 + 9x + 14 \qquad y^2 - 8y + 15$$
$$x^2 - 2x - 35 \qquad y^2 + 4y - 12$$

The products that produce these trinomials are shown.

$$(x + 7)(x + 2) = x^2 + 7x + 2x + 14 = x^2 + 9x + 14$$
$$(y - 3)(y - 5) = y^2 - 3y - 5y + 15 = y^2 - 8y + 15$$
$$(x - 7)(x + 5) = x^2 + 5x - 7x - 35 = x^2 - 2x - 35$$
$$(y + 6)(y - 2) = y^2 - 2y + 6y - 12 = y^2 + 4y - 12$$

If we reverse the preceding products, we begin with a trinomial and end with its factored form. The following table shows some important relationships between the binomial factors and the trinomial.

| Trinomial | Factored Form | Important Observations |
|---|---|---|
| $x^2 + 9x + 14$ | $= (x + 7)(x + 2)$ | 7 and 2 are factors of 14 whose sum is 9. |
| $y^2 - 8y + 15$ | $= (y - 3)(y - 5)$ | $-3$ and $-5$ are factors of 15 whose sum is $-8$. |
| $x^2 - 2x - 35$ | $= (x - 7)(x + 5)$ | $-7$ and 5 are factors of $-35$ whose sum is $-2$. |
| $y^2 + 4y - 12$ | $= (y + 6)(y - 2)$ | 6 and $-2$ are factors of $-12$ whose sum is 4. |

Observe that when the sign of the last term is *positive*, the binomial factors have the *same* sign. For example,

$$x^2 + 9x + 14 = (x + 7)(x + 2)$$
$$y^2 - 8y + 15 = (y - 3)(y - 5)$$

When the sign of the last term is *negative*, the binomial factors have *opposite* signs.

$$x^2 - 2x - 35 = (x - 7)(x + 5)$$
$$y^2 + 4y - 12 = (y + 6)(y - 2)$$

Note also that the first two terms of the binomial factors have a product equal to the first term of the trinomial.

**Property: Factoring $x^2 + bx + c$**

For $b$ and $c$ real numbers, the factored form of $x^2 + bx + c$ is

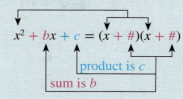

$$x^2 + bx + c = (x + \#)(x + \#)$$

product is $c$

sum is $b$

**Procedure: Factoring a Trinomial of the Form $x^2 + bx + c$**

**Step 1:** List pairs of factors of the last term of the trinomial, $c$.
**Step 2:** Determine the signs of the factors that produce the correct product.
    **a.** If the last term is positive, both factors are of the same sign.
    **b.** If the last term is negative, the factors are opposite signs.
**Step 3:** Choose the pair of factors whose sum is the middle coefficient, $b$. If no such factors exist, then the trinomial is prime and cannot be factored.
**Step 4:** Arrange the pair of factors in the binomials.
**Step 5:** Check answer by multiplying.

---

**Objective 1 Examples**

**Factor each trinomial. Check answer by multiplying.**

**1a.** $a^2 + 20a + 75$     **1b.** $h^2 - 13hk + 36k^2$     **1c.** $y^2 + 3y + 4$
**1d.** $a^4 - 9a^2 + 18$     **1e.** $x^2 - 6x - 16$     **1f.** $x^2 + 4x - 10$
**1g.** $a^2 + 7a - 30$     **1h.** $x^2 - 3xy - 54y^2$

**Solutions**

**1a.** We need factors of 75 whose sum is 20. Since the last term is positive, the signs of the factors are the same. Since the sum is positive, we consider only positive factors. The factors that satisfy the conditions are 5 and 15. So, the factorization is

| Factors of 75 | Sum of the Factors |
|---|---|
| 1, 75 | $1 + 75 = 76$ |
| 3, 25 | $3 + 25 = 28$ |
| 5, 15 | $5 + 15 = 20$ |

$$a^2 + 20a + 75 = (a + 5)(a + 15)$$

**Check:** $(a + 5)(a + 15) = a^2 + 15a + 5a + 75 = a^2 + 20a + 75$

**INSTRUCTOR NOTE:**
Remind students that when adding two negative numbers the result is negative. So, the only way for two numbers to have a positive product and a negative sum is if both numbers are negative.

**1b.** We find factors of 36 whose sum is $-13$. Since the last term is positive, the signs of the factors are the same. Since the sum is negative, we consider only negative factors.
    The factors that satisfy the conditions are $-4$ and $-9$. For the last term to have a product of $36k^2$, we use $-4k$ and $-9k$ in the factorization.

| Factors of 36 | Sum of the Factors |
|---|---|
| $-1, -36$ | $-1 + (-36) = -37$ |
| $-2, -18$ | $-2 + (-18) = -20$ |
| $-3, -12$ | $-3 + (-12) = -15$ |
| $-4, -9$ | $-4 + (-9) = -13$ |
| $-6, -6$ | $-6 + (-6) = -12$ |

$$h^2 - 13hk + 36k^2 = (h - 4k)(h - 9k)$$

**Check:** $(h - 4k)(h - 9k) = h^2 - 9hk - 4hk + 36k^2 = h^2 - 13hk + 36k^2$

**1c.** We find factors of 4 whose sum is 3. Since the sum is positive, we consider only positive factors.
    There are not any factors of 4 whose sum is 3. So, the trinomial $y^2 + 3y + 4$ is *prime*.

| Factors of 4 | Sum of the Factors |
|---|---|
| 1, 4 | $1 + 4 = 5$ |
| 2, 2 | $2 + 2 = 4$ |

**1d.** The first term is $a^4$. The factors that produce $a^4$ are $a^2$ and $a^2$. The last term is 18. We find factors of 18 whose sum is $-9$. Since the sum is negative, we consider only negative factors.

| Factors of 18 | Sum of the Factors |
|---|---|
| $-1, -18$ | $-1 + (-18) = -19$ |
| $-2, -9$ | $-2 + (-9) = -11$ |
| $-3, -6$ | $-3 + (-6) = -9$ |

The factors that satisfy the conditions are $-3$ and $-6$. So, the factorization is

$$a^4 - 9a^2 + 18 = (a^2 - 3)(a^2 - 6)$$

**Check:** $(a^2 - 3)(a^2 - 6) = a^4 - 6a^2 - 3a^2 + 18 = a^4 - 9a^2 + 18$

**INSTRUCTOR NOTE:**
Remind students how to add numbers with opposite signs. Show that two numbers with opposite signs will have a positive sum when the larger factor is positive. Also show that two numbers with opposite signs have a negative sum when the larger factor is negative.

**1e.** We find factors of $-16$ whose sum is $-6$. Since the last term is negative, the factors have opposite signs.
     The factors that satisfy the conditions are 2 and $-8$. So, the factorization is

| Factors of $-16$ | Sum of the Factors |
|---|---|
| $1, -16$ | $1 + (-16) = -15$ |
| $2, -8$ | $2 + (-8) = -6$ |
| $4, -4$ | $4 + (-4) = 0$ |
| $-1, 16$ | $-1 + 16 = 15$ |
| $-2, 8$ | $-2 + 8 = 6$ |

$$x^2 - 6x - 16 = (x + 2)(x - 8)$$

**Check:** $(x + 2)(x - 8) = x^2 - 8x + 2x - 16 = x^2 - 6x - 16$

**1f.** We find factors of $-10$ whose sum is 4. Since the last term is negative, the factors have opposite signs.
     None of the factors satisfy the required condition, so $x^2 + 4x - 10$ is *prime*.

| Factors of $-10$ | Sum of the Factors |
|---|---|
| $-1, 10$ | $-1 + 10 = 9$ |
| $-2, 5$ | $-2 + 5 = 3$ |
| $1, -10$ | $1 + (-10) = -9$ |
| $2, -5$ | $2 + (-5) = -3$ |

**1g.** We find factors of $-30$ whose sum is 7. Since the last term is negative, the factors have opposite signs.
     The factors that satisfy the conditions are $-3$ and 10. So, the factorization is

$$a^2 + 7a - 30 = (a - 3)(a + 10)$$

**Check:** $(a - 3)(a + 10) = a^2 + 10a - 3a - 30$
$$= a^2 + 7a - 30$$

| Factors of $-30$ | Sum of the Factors |
|---|---|
| $-1, 30$ | $-1 + 30 = 29$ |
| $-2, 15$ | $-2 + 15 = 13$ |
| $-3, 10$ | $-3 + 10 = 7$ |
| $-5, 6$ | $-5 + 6 = 1$ |
| $1, -30$ | $1 + (-30) = -29$ |
| $2, -15$ | $2 + (-15) = -13$ |
| $3, -10$ | $3 + (-10) = -7$ |
| $5, -6$ | $5 + (-6) = -1$ |

**1h.** We find factors of $-54$ whose sum is $-3$. Since the last term is negative, the factors have opposite signs.
     The factors 6 and $-9$ satisfy the conditions. For the last term to have a product of $-54y^2$, we use $6y$ and $-9y$ in the factorization.

$$x^2 - 3yx - 54y^2 = (x + 6y)(x - 9y)$$

**Check:** $(x + 6y)(x - 9y) = x^2 - 9xy + 6xy - 54y^2$
$$= x^2 - 3xy - 54y^2$$

| Factors of $-54$ | Sum of the Factors |
|---|---|
| $1, -54$ | $1 + (-54) = -53$ |
| $2, -27$ | $2 + (-27) = -25$ |
| $3, -18$ | $3 + (-18) = -15$ |
| $6, -9$ | $6 + (-9) = -3$ |
| $-1, 54$ | $-1 + 54 = 53$ |
| $-2, 27$ | $-2 + 27 = 25$ |
| $-3, 18$ | $-3 + 18 = 15$ |
| $-6, 9$ | $-6 + 9 = 3$ |

✔ **Student Check 1**   Factor each trinomial and check answer by multiplying.

**a.** $x^2 + 12x + 32$        **b.** $a^2 - 38ab + 72b^2$        **c.** $y^2 + 13y + 24$

**d.** $r^4 - 11r^2 + 30$      **e.** $x^2 - x - 42$              **f.** $a^2 + 12a - 45$

**g.** $x^2 + 11x - 18$        **h.** $x^2 - xy - 12y^2$

## Factoring Trinomials with Common Factors

**Objective 2** ▶

Factor trinomials that have a common factor.

One of the key steps in factoring a polynomial is to factor out any common factors first. After the common factors are factored out, the remaining polynomial may need to be factored further.

> **Procedure:  Factoring Trinomials with a Common Factor**
>
> **Step 1:** Factor out the GCF.
> **Step 2:** If the remaining trinomial has a coefficient of 1, determine if it can be factored further.
> **Step 3:** Check by multiplying.

**Objective 2  Examples**   **Factor each polynomial completely. Check the answer by multiplying.**

**2a.** $2x^2 - 12x + 16$        **2b.** $-5x^3 - 15x^2 + 20x$

**2c.** A ball is thrown upward with an initial velocity of 16 ft/sec from a height of 32 ft above the ground. The height, in feet, of the ball $t$ seconds after it is thrown can be represented by $-16t^2 + 16t + 32$.

**Solutions**   **2a.**

$$\begin{aligned} 2x^2 - 12x + 16 &= 2(x^2 - 6x + 8) \\ &= 2(x - 4)(x - 2) \end{aligned}$$

Factor out the GCF, 2.

The factors of 8 whose sum is $-6$ are $-4$ and $-2$.

**INSTRUCTOR NOTE:**
Remind students to keep the GCF in their answer.

**Check:**
$$\begin{aligned} 2(x - 4)(x - 2) &= (2x - 8)(x - 2) \\ &= 2x^2 - 4x - 8x + 16 \\ &= 2x^2 - 12x + 16 \end{aligned}$$

**2b.** The terms of the trinomial have a common factor of $5x$. Since the leading term is negative, we factor out $-5x$.

$$\begin{aligned} -5x^3 - 15x^2 + 20x &= -5x(x^2 + 3x - 4) \\ &= -5x(x - 1)(x + 4) \end{aligned}$$

Factor out the GCF, $-5x$.

The factors of $-4$ whose sum is 3 are 4 and $-1$.

**Check:**
$$\begin{aligned} -5x(x - 1)(x + 4) &= (-5x^2 + 5x)(x + 4) \\ &= -5x^3 - 20x^2 + 5x^2 + 20x \\ &= -5x^3 - 15x^2 + 20x \end{aligned}$$

**2c.** The terms of the trinomial have a GCF of 16. Since the leading term is negative, we factor out $-16$.

$$\begin{aligned} -16t^2 + 16t + 32 &= -16(t^2 - t - 2) \\ &= -16(t - 2)(t + 1) \end{aligned}$$

Factor out the GCF, $-16$.

The factors of $-2$ whose sum is $-1$ are $-2$ and 1.

**Check:**
$$\begin{aligned} -16(t - 2)(t + 1) &= (-16t + 32)(t + 1) \\ &= -16t^2 - 16t + 32t + 32 \\ &= -16t^2 + 16t + 32 \end{aligned}$$

✓ **Student Check 2**   Factor each polynomial completely. Check the answer by multiplying.

**a.** $3y^2 + 33y + 84$                **b.** $-6x^3 + 6x^2 + 12x$

**c.** The height of a ball thrown upward with an initial velocity of 32 ft/sec from a height of 128 ft above the ground can be represented by $-16t^2 + 32t + 128$, where $t$ is the number of seconds after the ball is thrown.

**Objective 3** ▶

Troubleshoot common errors.

## Troubleshooting Common Errors

Some common errors associated with factoring trinomials of the form $x^2 + bx + c$ are shown next.

**Objective 3  Examples**    A problem and an incorrect solution are given. Provide the correct solution and an explanation of the error.

**3a.** Factor $x^2 - x - 6$.

| Incorrect Solution | Correct Solution and Explanation |
|---|---|
| The factors that produce $-6$ are $-2$ and $3$. So, the factorization is $$x^2 - x - 6 = (x - 2)(x + 3)$$ | If we multiply the two binomials, we get $x^2 + x - 6$. The middle sign is wrong since $-2 + 3 = 1$, not $-1$. We need factors of $-6$ whose sum is $-1$, the middle coefficient. The factors are $-3$ and $2$. $$x^2 - x - 6 = (x - 3)(x + 2)$$ |

**3b.** Factor $y^2 + 2y + 35$.

| Incorrect Solution | Correct Solution and Explanation |
|---|---|
| The factors that produce 35 that can give us 2 are 7 and 5. So, the factorization is $$y^2 + 2y + 35 = (y + 5)(y + 7)$$ | The difference of 7 and 5 is 2 but the sum of the factors is 12. There are not any factors of 35 whose sum is 2. Therefore, $$y^2 + 2y + 35 \text{ is prime.}$$ |

**3c.** Factor $4x^3 + 8x^2 - 32x$.

| Incorrect Solution | Correct Solution and Explanation |
|---|---|
| The GCF is $4x$. When we factor $4x$ out, we get $x^2 + 2x - 8$. This factors as $$(x + 4)(x - 2)$$ | The error made was not keeping the GCF with the remaining factors. $$4x^3 + 8x^2 - 32x = 4x(x^2 + 2x - 8)$$ $$= 4x(x + 4)(x - 2)$$ |

## ANSWERS TO STUDENT CHECKS

**Student Check 1    a.** $(x + 4)(x + 8)$
**b.** $(a - 36b)(a - 2b)$   **c.** prime
**d.** $(r^2 - 5)(r^2 - 6)$   **e.** $(x - 7)(x + 6)$

**f.** $(a + 15)(a - 3)$   **g.** prime   **h.** $(x - 4y)(x + 3y)$
**Student Check 2   a.** $3(y + 7)(y + 4)$   **b.** $-6x(x - 2)(x + 1)$
**c.** $-16(t + 2)(t - 4)$

## SUMMARY OF KEY CONCEPTS

Here is a summary of the "clues" given in a trinomial of the form $x^2 + bx + c$ that will enable us to factor it.

1. The last term of the trinomial tells us the number to factor. So, begin by making a list of all the factors of the number $c$.
2. The sign of the last term tells us if the factors are going to have the same sign or different signs.
   a. If $c > 0$, the factors will have the same sign.
   b. If $c < 0$, the factors will have different signs.
3. The sign of the middle coefficient, $b$, tells us the specific signs.
   a. If the signs of the factors are the same, they are the same as the sign of the middle coefficient.
   b. If the signs of the factors are different, the sign of the middle coefficient is the sign of the larger factor.
4. The coefficient of the middle term, $b$, tells us the sum of the factors of $c$.
5. From the list of factors, we select the pair whose sum is $b$. If there are no such factors, then the trinomial is prime.
6. Always check the answer by multiplying.
7. If the leading coefficient is not 1, then look for a common factor to factor out first.

## GRAPHING CALCULATOR SKILLS

The graphing calculator can assist us in finding factors of a number and their sum. The calculator can also be used to verify our factored form.

**Example 1:** Find the factors of $-75$ whose sum is $-10$.

Recall, to factor numbers on the calculator, we can divide the number by $x$ and view the table for different values. So, to find the factors of $-75$, we enter $-75/X$ as $Y_1$. To find the sum of the factors, we enter $X + Y_1$.

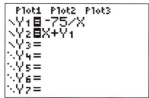

Now we view the table.

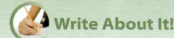

The first two columns are the factors of $-75$. The last column shows the sum of the factors. So, $-5$ and $15$ are the factors of $-75$ whose sum is 10.

**Example 2:** Verify that $x^2 - 6x - 16 = (x - 8)(x + 2)$.

Enter the trinomial in $Y_1$ and the factored form in $Y_2$. If the factored form is correct, these two expressions will have the same value when evaluated for different values of $x$. We can confirm this using the table feature.

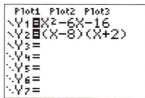

Since the values of $Y_1$ and $Y_2$ are the same for every value of $x$, our factoring is correct.

## SECTION 6.2 / EXERCISE SET

### ✏ Write About It!

**Use complete sentences in your answer to each exercise.**

1. Use an example to explain how to factor a trinomial in the form $x^2 + bx + c = 0$, where $c$ is a positive number.

2. Use an example to explain how to factor a trinomial in the form $x^2 + bx + c = 0$, where $c$ is a negative number.

3. Explain the conditions the factors of the constant term must satisfy in Exercises 7 and 8.

4. Explain the conditions the factors of the constant term must satisfy in Exercises 33 and 34.

Additional answers can be found in the Instructor Answer Appendix.

### 🎹 Practice Makes Perfect!

**Factor each trinomial completely. Check by multiplying.** (*See Objective 1.*)

5. $a^2 + 6a + 5$   $(a + 1)(a + 5)$  6. $a^2 + 7a + 6$   $(a + 1)(a + 6)$

7. $x^2 + 5x + 6$   $(x + 2)(x + 3)$  8. $x^2 + 9x + 20$   $(x + 4)(x + 5)$

9. $x^2 - 17x + 72$
   $(x - 8)(x - 9)$
10. $x^2 - 15x + 26$
   $(x - 13)(x - 2)$
11. $a^2 - 10a + 12$   prime
12. $a^2 + 12a + 6$   prime

13. $b^2 - 18b + 80$
   $(b - 8)(b - 10)$
14. $b^2 - 10b + 24$
   $(b - 4)(b - 6)$
15. $x^2 + 21xy + 90y^2$
   $(x + 6y)(x + 15y)$
16. $a^2 + 20ab + 64b^2$
   $(a + 4b)(a + 16b)$
17. $h^2 - 17hk + 70k^2$
   $(h - 7k)(h - 10k)$
18. $c^2 - 16cd + 48d^2$
   $(c - 4d)(c - 12d)$
19. $r^2 + 13r + 40$
   $(r + 5)(r + 8)$
20. $s^2 + 20s + 99$
   $(s + 9)(s + 11)$
21. $a^2 + a - 20$
   $(a - 4)(a + 5)$
22. $c^2 + 4c - 21$   $(c - 3)(c + 7)$
23. $x^2 + 12x - 45$
   $(x - 3)(x + 15)$
24. $y^2 + 8y - 9$   $(y - 1)(y + 9)$

25. $x^2 + 9xy - 10y^2$ $(x - y)(x + 10y)$
26. $a^2 - ab - 72b^2$ $(a + 8b)(a - 9b)$
27. $h^2 + 4h - 4$ prime
28. $k^2 + 10k - 14$ prime
29. $b^2 - 7b - 60$ $(b + 5)(b - 12)$
30. $r^2 + 5r - 66$ $(r - 6)(r + 11)$
31. $a^2 - 4ab - 12b^2$ $(a + 2b)(a - 6b)$
32. $c^2 - 6cd - 27d^2$ $(c + 3d)(c - 9d)$
33. $h^2 - 11h - 42$ $(h + 3)(h - 14)$
34. $k^2 - 6k - 40$ $(k + 4)(k - 10)$
35. $x^2 - 4xy - 45y^2$ $(x + 5y)(x - 9y)$
36. $s^2 + st - 56t^2$ $(s - 7t)(s + 8t)$

**Factor each trinomial completely. Check by multiplying. (*See Objective 2.*)**

37. $4x^2 - 12x - 72$ $4(x + 3)(x - 6)$
38. $3x^2 + 3x - 270$ $3(x - 9)(x + 10)$
39. $5y^2 - 15y - 20$ $5(y + 1)(y - 4)$
40. $6y^2 - 30y - 84$ $6(y + 2)(y - 7)$
41. $-4a^2 - 40a - 36$ $-4(a + 1)(a + 9)$
42. $-8b^2 + 40b + 48$ $-8(b + 1)(b - 6)$
43. $3r^3 - 21r^2 - 24r$ $3r(r - 8)(r + 1)$
44. $5s^3 - 30s^2 - 80s$ $5s(s + 2)(s - 8)$
45. $-6x^3 + 6x^2 + 72x$ $-6x(x + 3)(x - 4)$
46. $-4y^3 - 16y^2 + 84y$ $-4y(y - 3)(y + 7)$
47. $9p^4 - 9p^3 - 270p^2$ $9p^2(p + 5)(p - 6)$
48. $3q^4 - 6q^3 + 3q^2$ $3q^2(q - 1)(q - 1)$
49. $5x^3y + 5x^2y - 30xy$ $5xy(x - 2)(x + 3)$
50. $6x^3y - 36x^2y + 30xy$ $6xy(x - 1)(x - 5)$
51. $-3a^3b + 39a^2b - 126ab$ $-3ab(a - 6)(a - 7)$
52. $-2a^3b - 6a^2b + 80ab$ $-2ab(a - 5)(a + 8)$

53. The height of a ball thrown upward with an initial velocity of 96 ft/sec from a height of 640 ft above the ground can be represented by $-16t^2 + 96t + 640$ where $t$ is the number of seconds after the ball is thrown. $-16(t + 4)(t - 10)$

54. The height of a ball thrown upward with an initial velocity of 128 ft/sec from a height of 768 ft above the ground can be represented by $-16t^2 + 128t + 768$, where $t$ is the number of seconds after the ball is thrown. $-16(t + 4)(t - 12)$

55. The volume of an open box with height $x$ in. is given by the expression $x^3 + 4x^2 - 12x$ in.$^3$. $x(x + 6)(x - 2)$

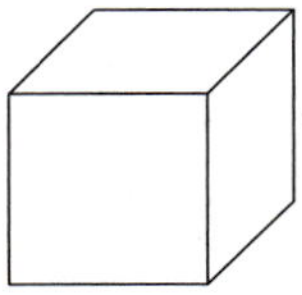
x

56. The volume of an open box with height $x$ in. is given by the expression $x^3 + 7x^2 - 8x$ in.$^3$. $x(x + 8)(x - 1)$

## Mix 'Em Up!

**Factor each trinomial completely.**

57. $x^2 - 15x + 56$ $(x - 7)(x - 8)$
58. $x^2 + 2x - 63$ $(x - 7)(x + 9)$
59. $3y^3 - 12y^2 + 12y$ $3y(y - 2)(y - 2)$
60. $-2x^3 - 4x^2 - 2x$ $-2x(x + 1)(x + 1)$
61. $p^2 - 15pq + 54q^2$ $(p - 6q)(p - 9q)$
62. $c^2 + 19cd + 60d^2$ $(c + 4d)(c + 15d)$
63. $x^2 + 4x + 7$ prime
64. $t^2 - 5t - 10$ prime
65. $-6x^4 + 24x^3 + 126x^2$ $-6x^2(x + 3)(x - 7)$
66. $4a^3 + 12a^2 - 160a$ $4a(a - 5)(a + 8)$
67. $r^2 + 16rs + 39s^2$ $(r + 3s)(r + 13s)$
68. $h^2 - 8hk - 48k^2$ $(h + 4k)(h - 12k)$
69. $3x^4 + 33x^3 + 54x^2$ $3x^2(x + 2)(x + 9)$
70. $-10x^3 + 60x^2 + 270x$ $-10x(x + 3)(x - 9)$
71. $-2s^2 - 16s + 40$ $-2(s + 10)(s - 2)$
72. $a^2 + 10ab - 9b^2$ prime
73. $x^2 + 4xy + 5y^2$ prime

74. The volume of an open box with height $x$ in. is given by the expression $x^3 + 5x^2 - 36x$ in.$^3$. $x(x + 9)(x - 4)$

75. The volume of an open box with height $x$ in. is given by the expression $x^3 + 9x^2 - 36x$ in.$^3$. $x(x + 12)(x - 3)$

76. A company's profit can be represented by $-20q^2 + 700q - 5000$, where $q$ is the number of units sold, in thousands. $-20(q - 10)(q - 25)$

77. A company's profit can be represented by $-16q^2 + 160q + 9600$, where $q$ is the number of units sold, in thousands. $-16(q - 30)(q + 20)$

## You Be the Teacher!

**Explain each situation to a student.**

78. Explain how to factor $a^2 - 6a + 8$. Find the factors of 8 whose sum is $-6$. They are $-2$ and $-4$, $(a - 2)(a - 4)$.

79. Explain how to factor $b^2 + 2b - 24$. Find the factors of $-24$ whose sum is 2. They are $-4$ and 6, $(b + 6)(b - 4)$.

**Correct each student's errors, if any.**

80. Factor $-3x^3 + 12x^2 - 36$.

Johannes's work:

$$-3x^3 + 12x - 36 = -3x(x^2 + 4x - 12)$$
$$= -3x(x - 6)(x + 2)$$

$-3x^3 + 12x^2 - 36 = -3(x^3 - 4x^2 + 12)$

81. Factor $2x^2 - 18xy + 36y^2$.

Sarah's work:

$$2x^2 - 18xy + 36y^2 = 2(x^2 - 9x + 18)$$
$$= 2(x - 6)(x - 3)$$

$2x^2 - 18xy + 36y^2 = 2(x^2 - 9xy + 18y^2)$
$= 2(x - 3y)(x - 6y)$

## Calculate It!

**Use the tables provided to write the factorization of each trinomial.**

82. $x^2 - 4x - 60$ $(x + 6)(x - 10)$

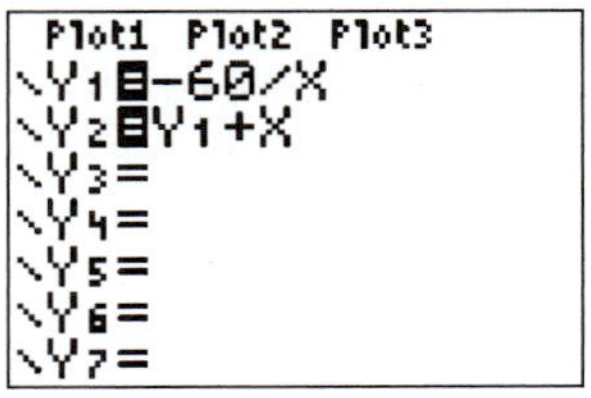
Plot1 Plot2 Plot3
\Y1=-60/X
\Y2=Y1+X
\Y3=
\Y4=
\Y5=
\Y6=
\Y7=

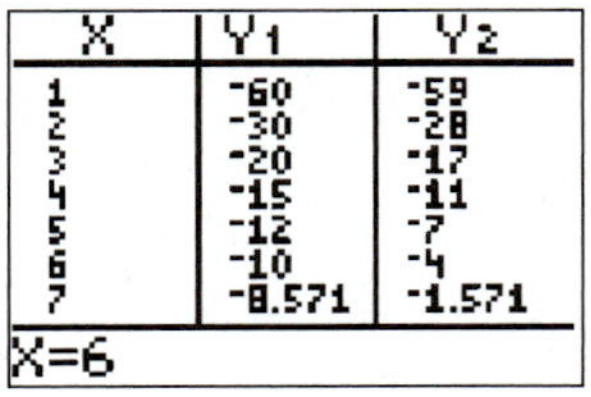
X Y1 Y2
1 -60 -59
2 -30 -28
3 -20 -17
4 -15 -11
5 -12 -7
6 -10 -4
7 -8.571 -1.571
X=6

83. $x^2 - 17x + 42$ $(x - 3)(x - 14)$

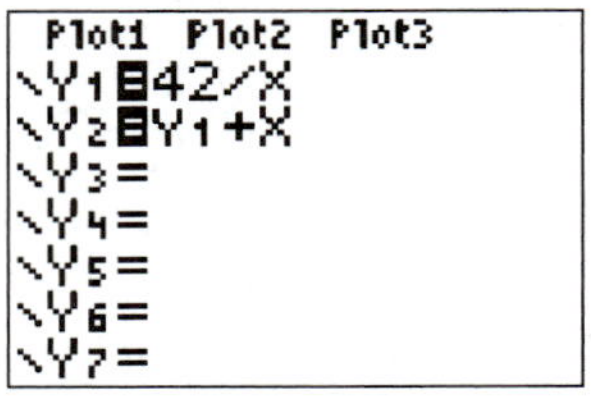
Plot1 Plot2 Plot3
\Y1=42/X
\Y2=Y1+X
\Y3=
\Y4=
\Y5=
\Y6=
\Y7=

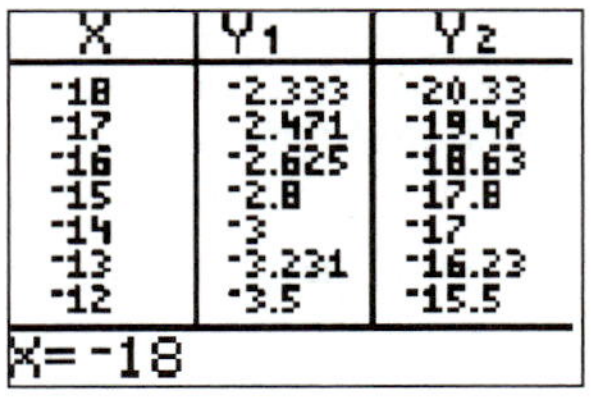
X Y1 Y2
-18 -2.333 -20.33
-17 -2.471 -19.47
-16 -2.625 -18.63
-15 -2.8 -17.8
-14 -3 -17
-13 -3.231 -16.23
-12 -3.5 -15.5
X=-18

 **Think About It!**

Write a trinomial of the form $x^2 + bx + c$ that is factorable and has the given last term. Explain how you determined the middle term of your trinomial.

**84.** 15     **85.** 28     **86.** $-12$     **87.** $-30$

Write a trinomial of the form $x^2 + bx + c$ that is prime with the given last term. Explain how you determined the middle term of your trinomial.

**88.** 14     **89.** $-8$

---

| SECTION 6.3 | More on Factoring Trinomials |

▶ **OBJECTIVES**

As a result of completing this section, you will be able to

1. Factor trinomials using trial and error.
2. Factor trinomials using grouping.
3. Factor perfect square trinomials.
4. Troubleshoot common errors.

The annual profit, in thousands of dollars, for a mobile phone company can be represented by $-4x^2 + 102x - 50$, where $x$ is the total number of mobile phones sold for the year, in thousands. In Section 6.6, we will solve an equation that enables us to determine how many phones need to be sold for the company to break even. This process requires us to factor the trinomial $-4x^2 + 102x - 50$.

This section will present two different methods for factoring trinomials of the form $ax^2 + bx + c$. Both methods will produce the same results. Your instructor may choose to present only one of these methods.

**Objective 1** ▶

Factor trinomials using trial and error.

### Factoring by Trial and Error

In Objective 1 of Section 6.2, we dealt primarily with trinomials whose leading coefficient was one. Now we will deal with trinomials whose leading coefficient is not one. Some examples are

$$6x^2 + 19x + 10 \qquad 4y^2 - 29y + 7 \qquad 12c^2 + 4c - 5$$

The products that produce these trinomials are shown. Pay particular attention to the relationship between the two binomials and the trinomial.

$$(2x + 5)(3x + 2) = 6x^2 + 4x + 15x + 10 = 6x^2 + 19x + 10$$
$$(4y - 1)(y - 7) = 4y^2 - 28y - y + 7 = 4y^2 - 29y + 7$$
$$(6c + 5)(2c - 1) = 12c^2 - 6c + 10c - 5 = 12c^2 + 4c - 5$$

By reversing the preceding statements, we get the factored form of the trinomials.

| Trinomial        Factored Form | Important Observations |
|---|---|
| $6x^2 + 19x + 10 = (2x + 5)(3x + 2)$ | $2x$ and $3x$ are factors of $6x^2$.  <br> 5 and 2 are factors of 10. <br> The sum of the outer and inner products, $4x$ and $15x$, is $19x$. |
| $4y^2 - 29y + 7 = (4y - 1)(y - 7)$ | $4y$ and $y$ are factors of $4y^2$. <br> $-1$ and $-7$ are factors of 7. <br> The sum of the outer and inner products, $-28y$ and $-y$, is $-29y$. |
| $12c^2 + 4c - 5 = (6c + 5)(2c - 1)$ | $6c$ and $2c$ are factors of $12c^2$. <br> 5 and $-1$ are factors of $-5$. <br> The sum of the outer and inner products, $-6c$ and $10c$, is $4c$. |

So we conclude the following:

1. The first terms of the binomials are factors of the first term of the trinomial.
2. The last terms of the binomials are factors of the last term of the trinomial.
3. The sum of the outer and inner products is the middle term of the trinomial.

We can use this as a guideline for factoring trinomials using the **trial-and-error method**. The process of trial and error involves arranging all possible factors of the first term of the trinomial and factors of the last term of the trinomial until we obtain the correct outer and inner products.

> **Property:** **Factoring $ax^2 + bx + c$ by Trial and Error**
>
> For $a$, $b$, and $c$ real numbers and $a \neq 0$, the factored form of $ax^2 + bx + c$ is
>
> $$\overset{\text{F}}{\phantom{a}}\ \overset{\text{O + I}}{\phantom{a}}\ \overset{\text{L}}{\phantom{a}}$$
> $$ax^2 + bx + c = (\underline{\ \ }x + \underline{\ \ })(\underline{\ \ }x + \underline{\ \ })$$

> **Procedure:** **Factoring a Trinomial of the Form $ax^2 + bx + c$ Using Trial and Error**
>
> **Step 1:** Factor out any common factors, if possible.
> **Step 2:** List the factors of the first term of the trinomial.
> **Step 3:** List the factors of the last term of the trinomial.
> **Step 4:** Determine the appropriate signs of the factors in Steps 2 and 3.
> **Step 5:** Arrange these factors in two binomials until the product produces the given trinomial.
> **Step 6:** Check by multiplying.

When trying different factors, be mindful of the following.

- If the first arrangement of factors doesn't work, obtain another combination by reversing the position of the factors of the last term.
- If the trinomial does not have a common factor, the binomial factors cannot have a common factor. So, we can eliminate any combinations in which one or both of the binomials has a common factor in this case.
- If none of the combinations work, then the trinomial is *prime*.

---

**Objective 1 Examples**   **Factor each trinomial using trial and error.**

**1a.** $3x^2 + 10x + 8$    **1b.** $2a^2 - 5a - 12$    **1c.** $15x^2y - 21xy + 6y$

**1d.** The annual profit, in thousands of dollars, for a mobile phone company can be represented by $-4x^2 + 102x - 50$, where $x$ is the total number of mobile phones sold for the year, in thousands.

**Solutions**   **1a.** We find the factors of $3x^2$ and the factors of 8. Since the last term of the trinomial is positive and the middle term is positive, we need to use only pairs of positive factors.

**INSTRUCTOR NOTE:**
Point out that because of the way the factors are selected, the first terms and last terms of the trinomials will always be correct. The difference is that they each produce different middle terms.

| Factors of $3x^2$ and 8 | | Possible Factors | |
|---|---|---|---|
| $3x, x$ | 1, 8 | $(3x + 1)(x + 8) = 3x^2 + 24x + x + 8$ $= 3x^2 + 25x + 8$ | Wrong middle term |
| | | $(3x + 8)(x + 1) = 3x^2 + 3x + 8x + 8$ $= 3x^2 + 11x + 8$ | Wrong middle term |
| | 2, 4 | $(3x + 2)(x + 4) = 3x^2 + 12x + 2x + 8$ $= 3x^2 + 14x + 8$ | Wrong middle term |
| | | $(3x + 4)(x + 2) = 3x^2 + 6x + 4x + 8$ $= 3x^2 + 10x + 8$ | Correct middle term |

So, the factorization is $3x^2 + 10x + 8 = (3x + 4)(x + 2)$.

**1b.** We find the factors of $2a^2$ and the factors of $-12$. Since the last term of the trinomial is negative, the factors of $-12$ must have opposite signs.

| Factors of $2a^2$ and $-12$ | | Possible Factors | |
|---|---|---|---|
| 2a, a | 1, −12 | $(2a + 1)(a - 12) = 2a^2 - 12a + a - 12$ $= 2a^2 - 11a - 12$ | Wrong middle term |
| | | $(2a - 12)(a + 1)$ no need to check | Common factor |
| | −1, 12 | $(2a - 1)(a + 12) = 2a^2 + 24a - a - 12$ $= 2a^2 + 23a - 12$ | Wrong middle term |
| | | $(2a + 12)(a - 1)$ no need to check | Common factor |
| | 2, −6 | $(2a + 2)(a - 6)$ no need to check | Common factor |
| | | $(2a - 6)(a + 2)$ no need to check | Common factor |
| | −2, 6 | $(2a - 2)(a + 6)$ no need to check | Common factor |
| | | $(2a + 6)(a - 2)$ no need to check | Common factor |
| | 3, −4 | $(2a + 3)(a - 4) = 2a^2 - 8a + 3a - 12$ $= 2a^2 - 5a - 12$ | Correct middle term |
| | | $(2a - 4)(a + 3)$ no need to check | Common factor |
| | −3, 4 | $(2a - 3)(a + 4) = 2a^2 + 8a - 3a - 12$ $= 2a^2 + 5a - 12$ | Wrong middle term |
| | | $(2a + 4)(a - 3)$ no need to check | Common factor |

So, the factorization is $2a^2 - 5a - 12 = (2a + 3)(a - 4)$.

**1c.** The terms in this trinomial have a GCF of $3y$. We factor this out first.

$$15x^2y - 21xy + 6y = 3y(5x^2 - 7x + 2)$$

Now we use trial and error to factor $5x^2 - 7x + 2$. Since the last term is positive, the signs of its factors are the same. To obtain a negative sum, the factors of the last term are both negative.

| Factors of $5x^2$ and 2 | | Possible Factors | |
|---|---|---|---|
| 5x, x | −1, −2 | $(5x - 1)(x - 2) = 5x^2 - 10x - x + 2$ $= 5x^2 - 11x + 2$ | Wrong middle term |
| | | $(5x - 2)(x - 1) = 5x^2 - 5x - 2x + 2$ $= 5x^2 - 7x + 2$ | Correct middle term |

So, the factorization is

$$15x^2y - 21xy + 6y = 3y(5x^2 - 7x + 2)$$
$$= 3y(5x - 2)(x - 1)$$

**1d.** The terms in this trinomial have a GCF of 2. Since the first term is negative, we factor out $-2$.

$$-4x^2 + 102x - 50 = -2(2x^2 - 51x + 25)$$

Now we use trial and error to factor $2x^2 - 51x + 25$.

| Factors of $2x^2$ and $25$ | | Possible Factors | |
|---|---|---|---|
| $2x, x$ | $-1, -25$ | $(2x - 1)(x - 25) = 2x^2 - 50x - x + 25$ $= 2x^2 - 51x + 25$ | Correct middle term |
| | | $(2x - 25)(x - 1) = 2x^2 - 2x - 25x + 25$ $= 2x^2 - 27x + 25$ | Wrong middle term |
| | $-5, -5$ | $(2x - 5)(x - 5) = 2x^2 - 10x - 5x + 25$ $= 2x^2 - 15x + 25$ | Wrong middle term |

So, the factorization is

$$-4x^2 + 102x - 50 = -2(2x^2 - 51x + 25)$$
$$= -2(2x - 1)(x - 25)$$

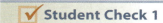 **Student Check 1**   Factor each polynomial using trial and error.

    **a.** $5x^2 + 21x + 4$          **b.** $7x^2 - 11x - 6$          **c.** $12xy^3 + 2xy^2 - 30xy$

    **d.** The profit, in thousands, of a toy manufacturer is $-9x^2 + 129x + 90$, where $x$ is in hundreds.

## Factoring by Grouping

**Objective 2** ▶

**Factor trinomials using grouping.**

Factoring by trial and error is a method that works but it can be a lengthy and tedious process, especially when the first term of the trinomial and/or the last term have several factors. The method illustrated in this objective provides us with a very systematic approach to factoring trinomials.

This process expands the middle term of the trinomial to two terms, specifically the outer and inner products that would result from multiplying the two binomial factors together. This expanded polynomial now has four terms, which we can factor by grouping. This method is basically FOIL in reverse.

We will use the products that we examined at the beginning of this section to illustrate this concept. It is important for us to determine how the outer and inner products relate to the terms in the trinomial.

| Trinomial and Its Factored Form | Observations |
|---|---|
| $6x^2 + 19x + 10$ $6x^2 + 4x + 15x + 10$ $(2x + 5)(3x + 2)$ | The coefficients of the outer and inner terms are 4 and 15. Note that $4(15) = 60$ and $4 + 15 = 19$. Also note that $ac = 6(10) = 60$. |
| $12c^2 + 4c - 5$ $12c^2 - 6c + 10c - 5$ $(6c + 5)(2c - 1)$ | The coefficients of the outer and inner terms are $-6$ and 10. Note that $-6(10) = -60$ and $-6 + 10 = 4$. Also note that $ac = 12(-5) = -60$. |
| $4y^2 - 29y + 7$ $4y^2 - 28y - y + 7$ $(4y - 1)(y - 7)$ | The coefficients of the outer and inner terms are $-28$ and $-1$. Note that $-28(-1) = 28$ and $-28 + (-1) = -29$. Also note that $ac = 4(7) = 28$. |

From the observations, we can conclude that the coefficients of the outer and inner terms have a product equal to $ac$ and have a sum equal to the middle coefficient of the trinomial. We summarize this as follows.

> **Property: Factoring $ax^2 + bx + c$ by Grouping**
> The coefficients of the outer and inner terms are factors of the number $a \cdot c$, whose sum is $b$.
>
> $$ax^2 + bx + c = ax^2 + \underbrace{b_1x + b_2x}_{\substack{b_1 \cdot b_2 = a \cdot c \\ b_1 + b_2 = b}} + c = \underbrace{(\_\_x + \_\_)(\_\_x + \_\_)}_{\text{Obtained by grouping}}$$

> **Procedure: Factoring a Trinomial of the Form $ax^2 + bx + c$ Using Grouping**
>
> **Step 1:** Factor out any common factors.
> **Step 2:** Find the product of the leading coefficient and the constant term; that is, $a \cdot c$.
> **Step 3:** List the factors of this number to find the pair of factors whose sum is $b$, the middle coefficient of the trinomial. If there is no such pair, the trinomial is prime.
> **Step 4:** Replace the middle term of the trinomial with a sum that uses the factors from Step 3.
> **Step 5:** Factor by grouping.
> **Step 6:** Check by multiplying.

**Objective 2 Examples**   Factor each trinomial by grouping. Check by multiplying.

**2a.** $15x^2 - 16x + 4$    **2b.** $7y^2 + 41y - 6$    **2c.** $-16x^3 + 26x^2 + 12x$

**Solutions**   **2a.** The terms of this trinomial contain no common factors other than 1. This trinomial is in the form $ax^2 + bx + c$, where $a = 15$ and $c = 4$. The product $a \cdot c = 15(4) = 60$.

We find the factors of 60 whose sum is $b = -16$, the middle coefficient. Since the product is positive and the sum is negative, both factors of 60 must be negative. The factors whose sum produces the correct middle coefficient are $-6$ and $-10$.

| $a \cdot c = 60$ | Sum of Factors |
|---|---|
| $-1, -60$ | $-1 + (-60) = -61$ |
| $-2, -30$ | $-2 + (-30) = -32$ |
| $-3, -20$ | $-3 + (-20) = -23$ |
| $-4, -15$ | $-4 + (-15) = -19$ |
| $-6, -10$ | $-6 + (-10) = -16$ |

$$
\begin{aligned}
15x^2 - 16x + 4 &= 15x^2 - 6x - 10x + 4 && \text{Replace } -16x \text{ with } -6x - 10x. \\
&= (15x^2 - 6x) + (-10x + 4) && \text{Group together first two and last two terms.} \\
&= 3x(5x - 2) - 2(5x - 2) && \text{Factor out the GCF from each pair of terms.} \\
&= (5x - 2)(3x - 2) && \text{Factor out the common binomial, } (5x - 2).
\end{aligned}
$$

So, the factorization is $15x^2 - 16x + 4 = (5x - 2)(3x - 2)$.

**Check:** $(5x - 2)(3x - 2) = 15x^2 - 10x - 6x + 4$
$$= 15x^2 - 16x + 4$$

**2b.** The terms of this trinomial contain no common factors other than 1. This trinomial is in the form $ax^2 + bx + c$, where $a = 7$ and $c = -6$. The product $a \cdot c = 7(-6) = -42$.

We find the factors of $-42$ whose sum is the middle coefficient, $b = 41$. The signs of the factors of $-42$ are opposite since it is negative. The factors whose sum produces the correct middle coefficient are $-1$ and $42$.

| $a \cdot c = -42$ | Sum of Factors |
|---|---|
| $-1, 42$ | $-1 + 42 = 41$ |
| $1, -42$ | $1 + (-42) = -41$ |
| $-2, 21$ | $-2 + 21 = 19$ |
| $2, -21$ | $2 + (-21) = -19$ |
| $-3, 14$ | $-3 + 14 = 11$ |
| $3, -14$ | $3 + (-14) = -11$ |
| $-6, 7$ | $-6 + 7 = 1$ |
| $6, -7$ | $6 + (-7) = -1$ |

$$
\begin{aligned}
7y^2 + 41y - 6 &= 7y^2 - y + 42y - 6 &&\text{Replace } 41y \text{ with } -y + 42y.\\
&= (7y^2 - y) + (42y - 6) &&\text{Group together first two and last two terms.}\\
&= y(7y - 1) + 6(7y - 1) &&\text{Factor out the GCF from each pair of terms.}\\
&= (7y - 1)(y + 6) &&\text{Factor out the common binomial, } (7y - 1).
\end{aligned}
$$

So, the factorization is $7y^2 + 41x - 6 = (7y - 1)(y + 6)$.

**Check:** $(7y - 1)(y + 6) = 7y^2 + 42y - y - 6$
$\qquad\qquad\qquad\qquad = 7y^2 + 41y - 6$

**Note:** *The order in which we insert the two terms will not affect the factorization. For instance, we could replace $41y$ with $42y + (-1y)$ or $42y - y$.*

$$
\begin{aligned}
7y^2 + 41y - 6 &= 7y^2 + 42y - y - 6 &&\text{Replace } 41y \text{ with } -y + 42y.\\
&= (7y^2 + 42y) + (-y - 6) &&\text{Group together first two and last two terms.}\\
&= 7y(y + 6) - 1(y + 6) &&\text{Factor out the GCF from each pair of terms.}\\
&= (y + 6)(7y - 1) &&\text{Factor out the common binomial, } (7y - 1).
\end{aligned}
$$

**2c.** The terms in the trinomial have a common factor of $-2x$. We factor this out.

$$-16x^3 + 26x^2 + 12x = -2x(8x^2 - 13x - 6)$$

Now, we factor $8x^2 - 13x - 6$, which is of the form $ax^2 + bx + c$, where $a = 8$ and $c = -6$. The product $a \cdot c = 8(-6) = -48$.

We find the factors of $-48$ whose sum is $b = -13$. The signs of the factors of $-48$ are opposite since it is negative. The factors whose sum produces the correct middle coefficient are $3$ and $-16$.

| $a \cdot c = -48$ | Sum of Factors |
|---|---|
| $-1, 48$ | $-1 + 48 = 47$ |
| $1, -48$ | $1 + (-48) = -47$ |
| $-2, 24$ | $-2 + 24 = 22$ |
| $2, -24$ | $2 + (-24) = -22$ |
| $-3, 16$ | $-3 + 16 = 13$ |
| $3, -16$ | $3 + (-16) = -13$ |
| $-4, 12$ | $-4 + 12 = 8$ |
| $4, -12$ | $4 + (-12) = -8$ |
| $-6, 8$ | $-6 + 8 = 2$ |
| $6, -8$ | $6 + (-8) = -2$ |

$$
\begin{aligned}
8x^2 - 13x - 6 &= 8x^2 + 3x - 16x - 6 &&\text{Replace } -13x \text{ with } 3x - 16x.\\
&= (8x^2 + 3x) + (-16x - 6) &&\text{Group together first two and last two terms.}\\
&= x(8x + 3) - 2(8x + 3) &&\text{Factor out the GCF from each pair of terms.}\\
&= (8x + 3)(x - 2) &&\text{Factor out the common binomial, } (8x + 3).
\end{aligned}
$$

So, the factorization of the original trinomial is

$$-16x^3 + 26x^2 + 12x = -2x(8x^2 - 13x - 6)$$
$$= -2x(8x + 3)(x - 2)$$

**Check:** $-2x(8x + 3)(x - 2) = (-16x^2 - 6x)(x - 2)$
$$= -16x^3 + 32x^2 - 6x^2 + 12$$
$$= -16x^3 + 26x^2 + 12x$$

✓ **Student Check 2**  Factor each trinomial using grouping. Check by multiplying.
  **a.** $4x^2 + 13x + 3$      **b.** $6x^2 - x - 12$      **c.** $-40y^3 - 44y^2 + 32y$

## Perfect Square Trinomials

**Objective 3** ▶

Factor perfect square trinomials.

Recall that **perfect square trinomials** are trinomials whose factored form is obtained from squaring a binomial. In Section 5.6, we learned the rules for squaring binomials.

$(a + b)^2 = a^2 + 2ab + b^2$      Example: $(3x + 4)^2 = 9x^2 + 24x + 16$

$(a - b)^2 = a^2 - 2ab + b^2$      Example: $(x - 5)^2 = x^2 - 10x + 25$

So, in factored form, we have

$a^2 + 2ab + b^2 = (a + b)^2$      Example: $9x^2 + 24x + 16 = (3x + 4)^2$

$a^2 - 2ab + b^2 = (a - b)^2$      Example: $x^2 - 10x + 25 = (x - 5)^2$

Recall that some examples of perfect squares are $1, 4, 9, 16, 25, \ldots$, and $x^2, x^4, x^6, \ldots$.

**INSTRUCTOR NOTE:**
Explain that the last term must be positive because the last term is a square. Two real numbers squared is always positive.

> **Procedure: Identifying and Factoring a Perfect Square Trinomial**
>
> **Step 1:** Is the first term of the trinomial a perfect square? If so, write it as $a^2$.
> **Step 2:** Is the last term of the trinomial a perfect square and positive? If so, write it as $b^2$.
> **Step 3:** Is the middle term twice the product of $a$ and $b$ or $2ab$? If so, the trinomial is a perfect square trinomial. If not, the trinomial is not a perfect square trinomial.
> **Step 4:** Factor the trinomial as $(a + b)^2$ or $(a - b)^2$, as appropriate.

**Objective 3  Examples**  Determine if each trinomial is a perfect square trinomial and factor, if possible.
  **3a.** $y^2 - 8y + 16$     **3b.** $4x^2 + 20x + 25$     **3c.** $4a^2 + 8a + 1$

**Solutions**  **3a.** The first term is a perfect square: $y^2 = (y)^2$
The last term is a perfect square: $16 = (4)^2$
The middle term is twice the product of $a \cdot b$: $8y = 2(y)(4)$
So, $y^2 - 8y + 16$ is a perfect square trinomial. It is factored as

$$y^2 - 8y + 16 = (y - 4)^2$$

**Note:** We can also factor $y^2 - 8y + 16$ using the techniques of Section 6.1. We find the factors of $16$ whose sum is $-8$. The factors are $-4$ and $-4$.
So, $y^2 - 8y + 16 = (y - 4)(y - 4)$ or $(y - 4)^2$.

**3b.** The first term is a perfect square: $4x^2 = (2x)^2$
The last term is a perfect square: $25 = (5)^2$
The middle term is twice the product of $a \cdot b$: $20x = 2(2x)(5)$
So, $4x^2 + 20x + 25$ is a perfect square trinomial. It is factored as

$$4x^2 + 20x + 25 = (2x + 5)^2$$

> **Note:** *We can also factor* $4x^2 + 20x + 25$ *using the techniques presented in this section.*
>
> $$4x^2 + 20x + 25 \qquad a \cdot c = 4 \cdot 25 = 100$$
>
> *We find the factors of* $100$ *whose sum is* $20$. *The factors are* $10$ *and* $10$.
>
> | | |
> |---|---|
> | $4x^2 + 20x + 25 = 4x^2 + 10x + 10x + 25$ | Replace 20x with 10x + 10x. |
> | $\qquad\qquad\quad = (4x^2 + 10x) + (10x + 25)$ | Group together first two and last two terms. |
> | $\qquad\qquad\quad = 2x(2x + 5) + 5(2x + 5)$ | Factor out the GCF from each pair of terms. |
> | $\qquad\qquad\quad = (2x + 5)(2x + 5)$ | Factor out the common binomial, (2x + 5). |
> | $\qquad\qquad\quad = (2x + 5)^2$ | Write in exponential form. |

**3c.** The first term is a perfect square: $4a^2 = (2a)^2$

The last term is a perfect square: $1 = (1)^2$

The middle term is *not* twice the product of $a \cdot b$: $8a \neq 2(2a)(1)$.

Since the trinomial is not a perfect square trinomial, we use trial and error or grouping to factor it. If we use grouping, we need to find factors of $a \cdot c = 4(1) = 4$ whose sum is the middle coefficient, 8. Since the factors of 4 are 1 and 4 or 2 and 2, this polynomial can't be factored since the sum of the factors is not 8. So, $4a^2 + 8a + 1$ is prime.

---

✔ **Student Check 3**    Determine if each trinomial is a perfect square trinomial and factor, if possible.

     **a.** $x^2 + 6x + 9$          **b.** $16x^2 - 24x + 9$          **c.** $y^2 + 4y + 16$

---

**Objective 4** ▶

Troubleshoot common errors.

## Troubleshooting Common Errors

Some common errors related to factoring trinomials are shown next.

**Objective 4 Examples**    **A problem and an incorrect solution are given. Provide the correct solution and an explanation of the error.**

**4a.** Factor $10r^2 + 47r - 15$ by trial and error.

| Incorrect Solution | Correct Solution and Explanation |
|---|---|
| The first terms are factors of $10r^2$ and the last terms are factors of $-15$. So, factors of $10r^2$ are $5r$ and $2r$ and factors of $-15$ are $-5$ and 3. <br><br> $10r^2 + 47r - 15$ <br> $(5r + 3)(2r - 5)$ <br>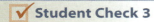 | While the first term and last term of the product are $10r^2$ and $-15$, the middle term is not $47r$. If we multiply the binomials together, we get $10r^2 - 19r - 15$. <br><br> The factors of $10r^2$ also include $10r$ and $r$. The factors of $-15$ also include $-3, 5; -1, 15;$ and $1, -15$. If we try all possible arrangements, the one that works is <br><br> $$10r^2 + 47r - 15$$ $$(10r - 3)(r + 5)$$ <br> Note that the outer and inner products are $50r - 3r = 47r$. |

**4b.** Factor $18y^2 - 17y + 4$ by grouping.

| Incorrect Solution | Correct Solution and Explanation |
|---|---|
| To use grouping, we need to find the factors of $18 \cdot 4 = 72$ which add to 17. This is 8 and 9. $$18y^2 - 17y + 4$$ $$18y^2 - 8y + 9y + 4$$ $$2y(9y - 4) + 1(9y + 4)$$ $$(9y - 4)(2y + 1)$$ | The factors of 72 need to add to $-17$. Also, the binomial after grouping in the incorrect solution is not the same. The factors that we need are $-8$ and $-9$. $$18y^2 - 17y + 4$$ $$18y^2 - 8y - 9y + 4$$ $$2y(9y - 4) - 1(9y - 4)$$ $$(9y - 4)(2y - 1)$$ |

## ANSWERS TO STUDENT CHECKS

**Student Check 1  a.** $(5x + 1)(x + 4)$  **b.** $(7x + 3)(x - 2)$
**c.** $2xy(2y - 3)(3y + 5)$  **d.** $-3(3x + 2)(x - 15)$

**Student Check 2  a.** $(4x + 1)(x + 3)$  **b.** $(3x + 4)(2x - 3)$
**c.** $-4y(5y + 8)(2y - 1)$

**Student Check 3  a.** $(x + 3)^2$  **b.** $(4x - 3)^2$  **c.** prime

## SUMMARY OF KEY CONCEPTS

1. Two methods can be used to factor a trinomial. We can use the trial-and-error method or the grouping method.

2. The trial-and-error method requires us to try every combination of factors of the first term with factors of the last term until the correct middle term is obtained.

3. Grouping is a methodical process that requires us to find factors of the product of the leading coefficient and the last term that adds to the middle coefficient.

4. Perfect square trinomials are special trinomials that can be factored as a binomial squared. Memorizing the pattern is helpful but not absolutely necessary. These trinomials can be factored using the methods of factoring trinomials.

5. Every factoring problem can be checked by multiplying.

## SECTION 6.3 / EXERCISE SET

 ### Write About It!

Use complete sentences in your answer to each exercise.

1. Explain the steps to factor a trinomial of the form $ax^2 + bx + c$, $a \neq 1$, by trial and error.

2. Explain the steps to factor a trinomial of the form $ax^2 + bx + c$, $a \neq 1$, by grouping.

3. When factoring a trinomial, how do you determine the signs of the last terms in the binomial factors?

4. What is a perfect square trinomial?
   A perfect square trinomial is a binomial squared.

5. Find the missing factor of $6x^2 + 7x - 3 = (3x - 1)$ ( ? ) and explain how you arrived at your answer.

6. Use an example to show how to factor a trinomial which can be written as a perfect square trinomial.
   Answers vary; $9x^2 + 42x + 49 = (3x + 7)^2$

7. Explain what types of numbers you need to find to factor Exercises 17 and 18 by trial and error.

8. Explain what types of numbers you need to find to factor Exercises 33 and 34 by grouping. In Exercise 33, we need to find the factors of 3 and 4 that add to 7. In Exercise 34, we need to find the factors of the product of 2 and 6 that add to 13.

Additional answers can be found in the Instructor Answer Appendix.

 ### Practice Makes Perfect!

Factor each trinomial completely using trial and error. Check by multiplying. (*See Objective 1.*)

9. $3x^2 + 13x + 4$
   $(3x + 1)(x + 4)$

10. $2y^2 + 5y + 2$
   $(2y + 1)(y + 2)$

11. $4s^2 - 29s + 7$
   $(4s - 1)(s - 7)$

12. $5t^2 - 17t + 6$
   $(5t - 2)(t - 3)$

13. $7a^2 - 41a - 6$
   $(7a + 1)(a - 6)$

14. $7h^2 - 2h - 5$
   $(7h + 5)(h - 1)$

15. $3x^2 + 20x - 7$
   $(3x - 1)(x + 7)$

16. $5m^2 + m - 4$
   $(5m - 4)(m + 1)$

17. $4k^2 - 16k + 15$
   $(2k - 3)(2k - 5)$

18. $9x^2 + 15x + 4$
   $(3x + 1)(3x + 4)$

19. $6y^2 - 17y - 14$
   $(2y - 7)(3y + 2)$

20. $15a^2 - a - 6$
   $(3a - 2)(5a + 3)$

21. $10t^2 + 11t - 8$
   $(2t - 1)(5t + 8)$

22. $21b^2 + 20b - 9$
   $(3b - 1)(7b + 9)$

23. $2y^2 + 3y + 5$  prime

24. $4s^2 + 8s + 5$  prime

25. $3m^2 - 10m - 7$  prime

26. $6a^2 + 7a - 2$  prime

27. $16k^2 - 4k - 6$
   $2(4k - 3)(2k + 1)$

28. $12x^2 - 48x + 45$
   $3(2x - 3)(2x - 5)$

29. $-7b^2 + b + 6$
   $-(7b + 6)(b - 1)$

30. $-4y^2 + 16y - 15$
   $-(2y - 3)(2y - 5)$

31. $8h^3 - 12h^2 - 8h$
   $4h(2h + 1)(h - 2)$

32. $15x^3 - 70x^2 + 40x$
   $5x(3x - 2)(x - 4)$

**Factor each trinomial completely using grouping. Check by multiplying.** (*See Objective 2.*)

**33.** $3y^2 + 7y + 4$
$(3y + 4)(y + 1)$

**34.** $2a^2 + 13a + 6$
$(2a + 1)(a + 6)$

**35.** $3m^2 - 8m + 5$
$(3m - 5)(m - 1)$

**36.** $2t^2 - 11t + 12$
$(2t - 3)(t - 4)$

**37.** $5k^2 - 28k - 12$
$(5k + 2)(k - 6)$

**38.** $7m^2 - 4m - 3$
$(7m + 3)(m - 1)$

**39.** $11h^2 + 19h - 6$
$(11h - 3)(h + 2)$

**40.** $6b^2 + 7b - 10$
$(b + 2)(6b - 5)$

**41.** $4x^2 + 33x + 8$
$(4x + 1)(x + 8)$

**42.** $10y^2 + 13y + 4$
$(2y + 1)(5y + 4)$

**43.** $9r^2 - 29r + 6$
$(9r - 2)(r - 3)$

**44.** $8m^2 - 38m + 9$
$(4m - 1)(2m - 9)$

**45.** $15u^2 - 14u - 8$
$(5u + 2)(3u - 4)$

**46.** $12s^2 - 11s - 15$
$(3s - 5)(4s + 3)$

**47.** $4a^2 + 20a - 11$
$(2a + 11)(2a - 1)$

**48.** $18s^2 + 17s - 15$
$(9s - 5)(2s + 3)$

**49.** $8x^2 + 13x + 6$
prime

**50.** $6y^2 + 13y - 12$
prime

**51.** $-5a^2 - 13a - 6$
$-(5a + 3)(a + 2)$

**52.** $24x^3 - 6x^2 - 9x$
$3x(4x - 3)(2x + 1)$

**53.** $36h^2k - 6hk - 20k$
$2k(6h - 5)(3h + 2)$

**54.** $60p^2q - 27pq - 60q$
$3q(4p - 5)(5p + 4)$

**55.** The annual operating income, in millions of dollars, of an airline company from 2007 to 2009 can be modeled by the expression $2120x^2 - 4505x + 530$, where $x$ is the number of years after 2007. $265(8x - 1)(x - 2)$

**56.** The net nonoperating income, in millions of dollars, of an airline company from 2007 to 2009 be modeled by the expression $-256x^2 + 704x - 288$, where $x$ is the number of years after 2007. $-32(2x - 1)(4x - 9)$

**Factor each perfect square trinomial.** (*See Objective 3.*)

**57.** $x^2 + 4x + 4$
$(x + 2)^2$

**58.** $m^2 + 10m + 25$
$(m + 5)^2$

**59.** $b^2 - 6b + 9$
$(b - 3)^2$

**60.** $y^2 - 14y + 49$
$(y - 7)^2$

**61.** $25r^2 + 20r + 4$
$(5r + 2)^2$

**62.** $48h^2 + 24h + 3$
$3(4h + 1)^2$

**63.** $18u^2 - 60u + 50$
$2(3u - 5)^2$

**64.** $36a^2 + 84a + 49$
$(6a + 7)^2$

## Mix 'Em Up!

**Use either trial and error or grouping to factor each trinomial completely.**

**65.** $2x^2 + 15x + 18$
$(2x + 3)(x + 6)$

**66.** $2y^2 + 25y + 50$
$(2y + 5)(y + 10)$

**67.** $12t^2 - 5t - 2$
$(4t + 1)(3t - 2)$

**68.** $16s^2 + 34s - 15$
$(8s - 3)(2s + 5)$

**69.** $81b^2 - 36b + 4$ $(9b - 2)^2$

**70.** $25c^2 + 40c + 16$
$(5c + 4)^2$

**71.** $40u^2 - 2u - 21$
$(4u - 3)(10u + 7)$

**72.** $33x^2 - 49x - 10$
$(3x - 5)(11x + 2)$

**73.** $-6a^2 + 29a + 5$
$-(6a + 1)(a - 5)$

**74.** $-5y^2 + 7y + 6$
$-(5y + 3)(y - 2)$

**75.** $4x^2 + 6x - 28$
$2(2x + 7)(x - 2)$

**76.** $18t^2 + 12t - 48$
$6(3t - 4)(t + 2)$

**77.** $5u^2 - 2u + 3$ prime

**78.** $6r^2 - 5r + 4$ prime

**79.** $45p^2 + 120p + 80$
$5(3p + 4)^2$

**80.** $12q^2 - 84q + 147$
$3(2q - 7)^2$

**81.** $10a^2b - 29ab - 21b$
$b(5a + 3)(2a - 7)$

**82.** $32c^2 - 52cd + 6d^2$
$2(8c - d)(2c - 3d)$

**83.** $52x^2 - 34xy + 4y^2$
$2(13x - 2y)(2x - y)$

**84.** $96p^2q - 156pq + 45q$
$3q(4p - 5)(8p - 3)$

**85.** $18a^2c - 84abc + 98b^2c$
$2c(3a - 7b)^2$

**86.** $150x^2z + 120xyz + 24y^2z$
$6z(5x + 2y)^2$

**87.** The profit, in millions, of a video game company is given by the expression $-60x^2 + 555x + 450$, where $x$ is the number of units sold, in thousands. $-15(4x + 3)(x - 10)$

**88.** The profit, in millions, of a software company is given by the expression $-150x^2 + 1425x + 4500$, where $x$ is the number of units sold, in thousands. $-75(2x + 5)(x - 12)$

## You Be the Teacher!

**Correct each student's errors, if any.**

**89.** Factor $3y^2 + 11y - 4$.

Tearra's work:

$3y^2 + 11y - 4 = (3y + 1)(y - 4)$ $(3y - 1)(y + 4)$

**90.** Factor $-6x^2 + 31x - 35$.

Charlene's work:

$6x^2 - 31x + 35 = (3x - 5)(2x - 7)$
$-(6x^2 - 31x + 35) = -(3x - 5)(2x - 7)$

**91.** Factor $5x^2 + 13x - 6$ by grouping.

Darlene's work:

$5x^2 - 15x + 2x - 6 = 5x(x - 3) + 2(x - 3)$
$= (x - 3)(5x + 2)$
$5x^2 + 15x - 2x - 6 = 5x(x + 3) - 2(x + 3) = (5x - 2)(x + 3)$

**92.** Factor $3r^2 + 12r + 12$ by grouping.

Coppola's work:

$3r^2 + 6r + 6r + 12 = 3r(r + 2) + 6(r + 2)$
$= 3(r + 2)^2(r + 2)$
$= 3(r + 2)^3$
$3r^2 + 6r + 6r + 12 = 3r(r + 2) + 6(r + 2) = 3(r + 2)(r + 2) = 3(r + 2)^2$

## Calculate It!

**93.** To factor $33x^2 - 49x - 10$ by grouping, what number do we need to factor? Use a calculator to list the factors of this number. Find the pair of factors that produce the correct middle term. Complete your factoring by hand.
$(3x - 5)(11x + 2)$

**94.** Use a calculator to verify which, if any, of the following factorizations of $40b^2 - 11b - 21$ is correct.

**a.** $(10b - 7)(4b + 3)$       **b.** $(10b + 7)(4b - 3)$
Neither factorization is correct.

## Think About It!

**Write a trinomial, $a \neq 1$, that satisfies the given conditions and factor it.**

**95.** $a \cdot c = -36$ and $b = -5$   $6x^2 - 5x - 6 = (2x - 3)(3x + 2)$

**96.** $a \cdot c = -20$ and $b = 19$   $4x^2 + 19x - 5 = (4x - 1)(x + 5)$

**97.** $a \cdot c = 144$ and $b = 24$   $9x^2 + 24x + 16 = (3x + 4)^2$

**98.** $a \cdot c = 324$ and $b = 36$   $81x^2 + 36x + 4 = (9x + 2)^2$

**For what values of $b$ is the trinomial factorable?**

**99.** $2x^2 + bx + 5$
$b = -11, -7, 7, 11$

**100.** $6x^2 + bx + 2$
$b = -13, -8, -7, 7, 8, 13$

**101.** $5x^2 + bx - 4$
$b = -19, -8, -1, 1, 8, 19$

**102.** $3x^2 + bx - 7$
$b = -20, -4, 4, 20$

/ **Factoring Binomials**

▶ **OBJECTIVES**

As a result of completing this section, you will be able to

1. Factor the difference of two squares.
2. Factor the sum and difference of two cubes.
3. Factor binomials with higher exponents.
4. Troubleshoot common errors.

**Objective 1** ▶

Factor the difference of two squares.

The west face of Mount Thor in Canada is home to the Earth's greatest vertical drop. The cliff is approximately 4096 ft high. The expression $-16t^2 + 4096$ represents the height, in feet, of an object dropped off the top of Mount Thor $t$ sec after it was dropped. In Section 6.6, we will solve an equation that determines how long it takes for a falling object to reach the ground. This process requires us to factor $-16t^2 + 4096$. (Source: http://en.wikipedia.org/wiki/Extremes_on_Earth#Greatest_vertical_drop)

So far in this chapter, we have learned how to factor trinomials. We will now finish our journey into factoring by learning how to factor three types of binomials.

## The Difference of Two Squares

In Section 5.6, we learned that the product of two conjugates results in the *difference of two squares*. That is,

$$(a - b)(a + b) = a^2 - b^2$$

An example is

$$(x - 3)(x + 3) = (x)^2 - (3)^2 = x^2 - 9$$

When we reverse this statement, we have a method that enables us to factor the difference of two squares. The difference of two squares can be factored into the product of two conjugates. The factored form is

$$x^2 - 9 = (x)^2 - (3)^2 = (x - 3)(x + 3)$$

> **Property: The Difference of Two Squares**
> $$a^2 - b^2 = (a - b)(a + b)$$

It is very helpful to be able to recognize the difference of two squares. To do this, we must be able to identify perfect squares. Some perfect squares are

$$1, 4, 9, 16, 25, \ldots, \quad \text{and} \quad x^2, x^4, x^6, \ldots$$

Some examples of the difference of two squares are

$$x^2 - 4 \qquad 16a^2 - 25 \qquad 49r^2 - 81s^2$$

Notice that each term in the binomial is a perfect square and the terms are connected by subtraction.

> **Procedure: Factoring the Difference of Two Squares**
>
> **Step 1:** Factor out any common factor.
> **Step 2:** Rewrite the first term as $a^2$, where $a$ is what must be squared to obtain the first term.
> **Step 3:** Rewrite the second term as $b^2$, where $b$ is what must be squared to obtain the second term.
> **Step 4:** Factor as $(a - b)(a + b)$.
> **Step 5:** Check by multiplying.

**Objective 1 Examples**   Use the difference of two squares property to factor each binomial.

**1a.** $y^2 - 49$   **1b.** $100 - x^2$   **1c.** $2x^3 - 72x$   **1d.** $49r^2 - \dfrac{1}{9}s^2$   **1e.** $x^2 + 4$

**1f.** The west face of Mount Thor in Canada is home to the Earth's greatest vertical drop. The cliff is approximately 4096 ft high. The expression $-16t^2 + 4096$ represents the height, in feet, of an object dropped off the top of Mount Thor $t$ sec after it was dropped. Factor $-16t^2 + 4096$. (Source: http://en.wikipedia.org/wiki/Extremes_on_Earth#Greatest_vertical_drop)

**Solutions**  **1a.**
$$y^2 - 49 = (y)^2 - (7)^2 \qquad \text{Write as } a^2 - b^2.$$
$$= (y - 7)(y + 7) \qquad \text{Apply the difference of two squares property.}$$

**1b.**
$$100 - x^2 = (10)^2 - (x)^2 \qquad \text{Write as } a^2 - b^2.$$
$$= (10 - x)(10 + x) \qquad \text{Apply the difference of two squares property.}$$

**1c.** The terms are not perfect squares but they have a common factor, $2x$.
$$2x^3 - 72x = 2x(x^2 - 36) \qquad \text{Factor out the common factor, } 2x.$$
$$= 2x(x^2 - 6^2) \qquad \text{Write the binomial as } a^2 - b^2.$$
$$= 2x(x - 6)(x + 6) \qquad \text{Apply the difference of two squares property.}$$

**1d.**
$$49r^2 - \frac{1}{9}s^2 = (7r)^2 - \left(\frac{1}{3}s\right)^2 \qquad \text{Write as } a^2 - b^2.$$
$$= \left(7r - \frac{1}{3}s\right)\left(7r + \frac{1}{3}s\right) \quad \text{Apply the difference of two squares property.}$$

**1e.** Each term is a perfect square, but this binomial is the **sum of squares**, not the difference of squares. This binomial cannot be factored as the product of conjugates.

A common answer for the factorization of this binomial is $(x + 2)^2$. But when we square this binomial, we get

$$(x + 2)^2 = (x + 2)(x + 2) = x^2 + 4x + 4$$

This is not the same as the given binomial, $x^2 + 4$. So, $(x + 2)^2$ is not the factorization.

We can think of $x^2 + 4$ as $x^2 + 0x + 4$. Using the rules from Section 6.2, we find the factors of 4 whose sum is 0. No such factors exist. So, $x^2 + 4$ is a *prime* polynomial.

 **Note:** *The sum of squares $a^2 + b^2$ cannot be factored unless there is a GCF.*

**1f.** The terms in the binomial, $-16t^2 + 4096$, have a common factor, $-16$.
$$-16t^2 + 4096 = -16(t^2 - 256) \qquad \text{Factor out the GCF, } -16.$$
$$= -16(t^2 - 16^2) \qquad \text{Write the binomial as } a^2 - b^2.$$
$$= -16(t - 16)(t + 16) \qquad \text{Apply the difference of two squares property.}$$

✓ **Student Check 1**   Use the difference of squares property to factor each binomial.
  **a.** $x^2 - 64$    **b.** $25 - y^2$    **c.** $3x^3 - 3x$    **d.** $9a^2 - 4b^2$    **e.** $4y^2 + 25$
  **f.** B.A.S.E. jumping is an activity whereby people fall off structures with only a brief period before they must open a parachute to survive. If a person free-falls off a 1600-ft structure, his height can be represented by $-16t^2 + 1600$, where $t$ is the number of seconds after the fall.

## The Sum and Difference of Two Cubes

**Objective 2** ▶

Factor the sum and difference of two cubes.

We will now factor binomials that are the sum and difference of two cubes; that is, binomials of the form $a^3 + b^3$ and $a^3 - b^3$. Again, it will be helpful to identify perfect cubes. Perfect cubes come from cubing integers and powers of $x$. A list of some perfect cubes is shown.

> **Note:** *Perfect Cubes:*
>
> $$1, 8, 27, 64, 125, 216 \ldots, \quad \text{and} \quad x^3, x^6, x^9, x^{12}, x^{15} \ldots$$

In Section 5.5, we found the following product.

$$(x + 2)(x^2 - 2x + 4) = x(x^2) - x(2x) + x(4) + 2(x^2) - 2(2x) + 2(4)$$
$$= x^3 - 2x^2 + 4x + 2x^2 - 4x + 8$$
$$= x^3 + 8$$

The result of the product of this binomial and trinomial is the *sum of two cubes*,

$$x^3 + 8 = x^3 + 2^3$$

If we reverse the statement, we obtain the factorization of $x^3 + 8$.

$$x^3 + 8 = x^3 + 2^3 = (x + 2)(x^2 - 2x + 4)$$

Notice the binomial in the factorization, $(x + 2)$, consists of the terms that we cube to get $x^3$ and $8$. The trinomial results from a special pattern, which is shown. The pattern also applies to the difference of two cubes.

---

**Property: Sum and Difference of Two Cubes**

$$a^3 + b^3 = (a + b)(a^2 - ab + b^2)$$
$$a^3 - b^3 = (a - b)(a^2 + ab + b^2)$$

---

Some observations:

1. The sign of the binomial factor is the same as the given binomial.
2. The first term of the trinomial factor, $a^2$, is the first term of the binomial factor squared.
3. The middle term of the trinomial factor, $ab$, is the product of the terms in the binomial. Its sign is always the opposite of the sign in the binomial factor.
4. The last term of the trinomial factor, $b^2$, is the last term of the binomial factor squared. Its sign is always positive.

> **Note:** *Helpful hint to remember the signs of the factors is the word SOAP.*
>
> S = same sign
>
> O = opposite sign
>
> AP = always positive
>
> $$\overset{\text{Same}}{\phantom{x}} \quad \overset{\text{Always positive}}{\phantom{x}}$$
> $$x^3 + 8 = (x + 2)(x^2 - 2x + 4)$$
> $$\underset{\text{Opposite}}{\phantom{x}}$$

---

**Procedure: Factoring the Sum or Difference of Two Cubes**

**Step 1:** Factor out any common factor.

**Step 2:** Rewrite the first term as $a^3$, where $a$ is what must be cubed to obtain the first term.

**Step 3:** Rewrite the second term as $b^3$, where $b$ is what must be cubed to obtain the second term.

**Step 4:** Apply the sum and difference of two cubes property.
   **a.** If the binomial is a sum of cubes, the first factor is $(a + b)$ and the second factor is $(a^2 - ab + b^2)$.
   **b.** If the binomial is a difference of cubes, the first factor is $(a - b)$ and the second factor is $(a^2 + ab + b^2)$.

**Step 5:** Check by multiplying.

**Objective 2 Examples** | Factor each binomial.

**2a.** $x^3 - 125$     **2b.** $8r^3 - s^3$     **2c.** $m^3 + 216n^3$     **2d.** $54y^3 + 128$

**Solutions**

**2a.**

$$x^3 - 125 = (x)^3 - (5)^3$$    Write as $a^3 - b^3$.

$$= (x - 5)[x^2 + (x)(5) + 5^2]$$    Apply the property with $a = x$ and $b = 5$.

$$= (x - 5)(x^2 + 5x + 25)$$    Simplify.

**Check:** $(x - 5)(x^2 + 5x + 25) = x^3 + 5x^2 + 25x - 5x^2 - 25x - 125$
$$= x^3 - 125$$

**2b.**

$$8r^3 - s^3 = (2r)^3 - (s)^3$$    Write as $a^3 - b^3$.

$$= (2r - s)[(2r)^2 + (2r)(s) + s^2]$$    Apply the property with $a = 2r$ and $b = s$.

$$= (2r - s)(4r^2 + 2rs + s^2)$$    Simplify.

**2c.**

$$m^3 + 216n^3 = m^3 + (6n)^3$$    Write as $a^3 + b^3$.

$$= (m + 6n)[m^2 - (m)(6n) + (6n)^2]$$    Apply the property with $a = m$ and $b = 6n$.

$$= (m + 6n)(m^2 - 6mn + 36n^2)$$    Simplify.

**2d.** The terms in the binomial are not perfect cubes. They do, however, have a common factor, 2.

$$54y^3 + 128 = 2(27y^3 + 64)$$    Factor out the GCF, 2.

$$= 2[(3y)^3 + (4)^3]$$    Write the binomial as $a^3 + b^3$.

$$= 2(3y + 4)[(3y)^2 - (3y)(4) + 4^2]$$    Apply the property with $a = 3y$ and $b = 4$.

$$= 2(3y + 4)(9y^2 - 12y + 16)$$    Simplify.

 **Student Check 2** | Factor each binomial.

**a.** $y^3 - 64$     **b.** $27a^3 - b^3$     **c.** $x^3 + 8y^3$     **d.** $-24m^3 - 3$

## Factoring Binomials with Exponents Greater Than 3

**Objective 3** ▶

Factor binomials with higher exponents.

Binomials containing exponents greater than 3 may not initially appear to satisfy the patterns we have discussed thus far. But if we can write the binomial as the difference of two squares, sum of two cubes, or difference of two cubes, then we can factor the polynomial.

> **Procedure: Factoring Binomials with Higher Exponents**
>
> **Step 1:** Factor out common factors, if applicable.
> **Step 2:** Determine if the binomial can be expressed as $a^2 - b^2$, $a^3 - b^3$, or $a^3 + b^3$.
>     **a.** If all of the exponents are even, try applying the difference of squares property.
>     **b.** If all of the exponents are multiples of 3, try applying the sum or difference of cubes property.
> **Step 3:** Make sure that any resulting factors cannot be factored any further.
> **Step 4:** Check by multiplying.

**Objective 3 Examples** | Factor each binomial completely.

**3a.** $x^4 - 81$     **3b.** $y^6 + 64$     **3c.** $125r^3s^9 - 216t^{15}$

**Solutions**   **3a.** Both terms are perfect squares since $x^4 = (x^2)^2$ and $81 = 9^2$, so we can apply the difference of two squares property.

$$x^4 - 81 = (x^2)^2 - 9^2 \qquad \text{Write as } a^2 - b^2.$$
$$= (x^2 - 9)(x^2 + 9) \qquad \text{Apply the difference of two squares property.}$$
$$= (x - 3)(x + 3)(x^2 + 9) \qquad \text{Apply the difference of two squares property on } x^2 - 9.$$

**3b.** The terms in this binomial are both perfect squares and perfect cubes.

$$y^6 + 64 = (y^3)^2 + 8^2 \quad \text{or} \quad y^6 + 64 = (y^2)^3 + 4^3$$

We cannot factor the sum of squares over the real numbers, so we must factor the binomial as the sum of cubes.

$$y^6 + 64 = (y^2)^3 + 4^3 \qquad \text{Write as } a^3 + b^3.$$
$$= (y^2 + 4)[(y^2)^2 - (y^2)(4) + 4^2] \qquad \text{Apply the property with } a = y^2 \text{ and } b = 4.$$
$$= (y^2 + 4)(y^4 - 4y^2 + 16) \qquad \text{Simplify.}$$

**3c.** All the powers in this binomial are multiples of 3 and the coefficients are perfect cubes, so this is the difference of two cubes.

$$125r^3s^9 - 216t^{15}$$
$$= (5rs^3)^3 - (6t^5)^3 \qquad \text{Write as } a^3 - b^3.$$
$$= (5rs^3 + 6t^5)[(5rs^3)^2 - (5rs^3)(6t^5) + (6t^5)^2] \qquad \text{Apply the property with } a = 5rs^3 \text{ and } b = 6t^5.$$
$$= (5rs^3 + 6t^5)(25r^2s^6 - 30rs^3t^5 + 36t^{10}) \qquad \text{Simplify by applying the properties of exponents.}$$

 **Student Check 3**   Factor each binomial completely.

  **a.** $x^4 - 16$    **b.** $y^6 + 1$    **c.** $16m^4n^{10} - 25p^6$

**Objective 4** ▶
Troubleshoot common errors.

## Troubleshooting Common Errors

Some common errors with applying the factoring rules for binomials are shown next.

**Objective 4  Examples**   **A problem and an incorrect solution are given. Provide the correct solution and an explanation of the error.**

**4a.** Factor $m^2 - 36$.

| Incorrect Solution | Correct Solution and Explanation |
|---|---|
| $m^2 - 36$ <br> $(m - 6)(m - 6)$ <br> $(m - 6)^2$ | If we square $(m - 6)$, we get $m^2 - 12m + 36$. This is not the given binomial. The binomial is a difference of two squares and factors as the product of conjugates. <br><br> $$m^2 - 36 = (m - 6)(m + 6)$$ |

**4b.** Factor $4y^2 + 49$.

| Incorrect Solution | Correct Solution and Explanation |
|---|---|
| ~~$4y^2 + 49$<br>$(2y + 7)(2y + 7)$<br>$(2y + 7)^2$~~ | If we square $(2y + 7)$, we get $4y^2 + 28y + 49$. This is not the given binomial. The binomial is a sum of two squares, not the difference of two squares. It cannot be factored over the real numbers.<br><br>$4y^2 + 49$ is prime. |

**4c.** Factor $a^3 - 27$.

| Incorrect Solution | Correct Solution and Explanation |
|---|---|
| ~~$a^3 - 27$<br>$(a - 3)^3$~~ | When we cube $(a - 3)$, we get $a^3 - 9a^2 + 27a - 27$. This does not yield $a^3 - 27$. The binomial is a difference of two cubes and is factored in the form $(a - b)(a^2 + ab + b^2)$.<br><br>$a^3 - 27 = (a - 3)(a^2 + 3a + 9)$ |

## ANSWERS TO STUDENT CHECKS

**Student Check 1** **a.** $(x - 8)(x + 8)$ **b.** $(5 - y)(5 + y)$
**c.** $3x(x - 1)(x + 1)$ **d.** $(3a - 2b)(3a + 2b)$ **e.** prime
**f.** $-16(t - 10)(t + 10)$

**Student Check 2** **a.** $(y - 4)(y^2 + 4y + 16)$
**b.** $(3a - b)(9a^2 + 3ab + b^2)$ **c.** $(x + 2y)(x^2 - 2xy + 4y^2)$
**d.** $-3(2m + 1)(4m^2 - 2m + 1)$

**Student Check 3** **a.** $(x - 2)(x + 2)(x^2 + 4)$
**b.** $(y^2 + 1)(y^4 - y^2 + 1)$
**c.** $(4m^2n^5 - 5p^3)(4m^2n^5 + 5p^3)$

## SUMMARY OF KEY CONCEPTS

1. The difference of two squares can be factored as a product of conjugates: $a^2 - b^2 = (a - b)(a + b)$. The sum of two squares is prime.

2. The sum and difference of two cubes can be factored as
$$a^3 + b^3 = (a + b)(a^2 - ab + b^2)$$
$$a^3 - b^3 = (a - b)(a^2 + ab + b^2)$$

3. Binomials with larger exponents may be factored into the difference of two squares, difference of two cubes, or the sum of two cubes. The key is to rewrite each term of the binomial as a quantity that is squared or cubed.

## GRAPHING CALCULATOR SKILLS

The graphing calculator can be used to verify our factorization as described in previous sections. It can also be used to generate a list of perfect squares and perfect cubes that might help in identifying the values of $a$ and $b$ in the factoring properties.

**Example 1:** List the perfect squares.

Enter $x$ squared in the equation editor. Then view the table. The values in $Y_1$ are the perfect squares. The corresponding $x$-value is the number that must be squared to obtain the value in $Y_1$.

```
Plot1 Plot2 Plot3
\Y1▧X²
\Y2=
\Y3=
\Y4=
\Y5=
\Y6=
\Y7=
```

| X | Y1 |
|---|---|
| 0 | 0 |
| 1 | 1 |
| 2 | 4 |
| 3 | 9 |
| 4 | 16 |
| 5 | 25 |
| 6 | 36 |

X=6

**Example 2:** List the perfect cubes.

Enter $x$ cubed in the equation editor. Then view the table. The values in $Y_1$ are the perfect cubes. The corresponding $x$-value is the number that must be cubed to obtain the value in $Y_1$.

```
Plot1 Plot2 Plot3
\Y1▧X^3
\Y2=
\Y3=
\Y4=
\Y5=
\Y6=
\Y7=
```

| X | Y1 |
|---|---|
| 0 | 0 |
| 1 | 1 |
| 2 | 8 |
| 3 | 27 |
| 4 | 64 |
| 5 | 125 |
| 6 | 216 |

X=0

# SECTION 6.4 / EXERCISE SET

## Write About It!

Use complete sentences in your answer to each exercise.

1. Explain how to factor a binomial that is the difference of two squares.

2. Explain how to factor a binomial that is the difference of two cubes.

3. Explain how to factor a binomial that is the sum of two cubes.

4. Use an example to show how to factor a binomial that can be written as a difference of two squares.   Answers vary; $49x^2 - 64y^2 = (7x)^2 - (8y)^2 = (7x + 8y)(7x - 8y)$

5. Use the example, $125a^3 + 27b^3$, to explain how to factor a sum of two cubes.   Answers vary; $125a^3 + 27b^3 = (5a)^3 + (3b)^3 = (5a + 3b)(25a^2 - 15ab + 9b^2)$

6. Use the example, $64a^3 - 27b^3$, to explain how to factor a difference of two cubes.   Answers vary; $64a^3 - 27b^3 = (4a)^3 - (3b)^3 = (4a - 3b)(16a^2 + 12ab + 9b^2)$

7. Explain why $x^2 + 9$ is prime.   We cannot write $x^2 + 9$ as a product of two binomials.

8. Explain how to factor the binomial: $a^4 - b^4$.   $a^4 - b^4 = (a^2)^2 - (b^2)^2 = (a^2 + b^2)(a^2 - b^2) = (a^2 + b^2)(a + b)(a - b)$

## Practice Makes Perfect!

Use the difference of two squares property to factor each binomial, if possible. (*See Objective 1.*)

9. $x^2 - 1$   $(x + 1)(x - 1)$
10. $x^2 - 4$   $(x + 2)(x - 2)$
11. $y^2 - 36$   $(y + 6)(y - 6)$
12. $y^2 - 9$   $(y + 3)(y - 3)$
13. $2x^3 - 8x$   $2x(x + 2)(x - 2)$
14. $5x^3 - 125x$   $5x(x + 5)(x - 5)$
15. $9a^2 - 4$   $(3a + 2)(3a - 2)$
16. $16a^2 - 1$   $(4a + 1)(4a - 1)$
17. $25b^2 - 49$   $(5b + 7)(5b - 7)$
18. $64b^2 - 25$   $(8b + 5)(8b - 5)$
19. $121 - 4x^2$   $(11 + 2x)(11 - 2x)$
20. $81 - 49x^2$   $(9 + 7x)(9 - 7x)$
21. $8t^2 - 72$   $8(t + 3)(t - 3)$
22. $7t^2 - 112$   $7(t + 4)(t - 4)$
23. $9m^2 - 256$   $(3m + 16)(3m - 16)$
24. $121m^2 - 4$   $(11m + 2)(11m - 2)$
25. $3x^5 - 48x^3$   $3x^3(x + 4)(x - 4)$
26. $2x^7 - 18x^5$   $2x^5(x + 3)(x - 3)$
27. $9a^2 + 4$   prime
28. $25 + 49b^2$   prime
29. $121y^2 - 144$   $(11y + 12)(11y - 12)$
30. $81y^2 - 64$   $(9y + 8)(9y - 8)$
31. $64a^2 - 9b^2$   $(8a + 3b)(8a - 3b)$
32. $25a^2 - 16b^2$   $(5a + 4b)(5a - 4b)$
33. $m^2n^2 - 121$   $(mn + 11)(mn - 11)$
34. $m^2n^2 - 256$   $(mn + 16)(mn - 16)$
35. $9x^2 - \dfrac{1}{4}$   $\left(3x + \dfrac{1}{2}\right)\left(3x - \dfrac{1}{2}\right)$
36. $4y^2 - \dfrac{9}{25}$   $\left(2y + \dfrac{3}{5}\right)\left(2y - \dfrac{3}{5}\right)$
37. $\dfrac{1}{9}a^2 - \dfrac{4}{49}$
38. $\dfrac{1}{4}b^2 - \dfrac{16}{25}$

Use the difference of two cubes property to factor each binomial, if possible. (*See Objective 2.*)

39. $x^3 - 8$   $(x - 2)(x^2 + 2x + 4)$
40. $x^3 - 1$   $(x - 1)(x^2 + x + 1)$
41. $a^3 - 64$   $(a - 4)(a^2 + 4a + 16)$
42. $a^3 - 125$   $(a - 5)(a^2 + 5a + 25)$
43. $2t^4 - 54t$   $2t(t - 3)(t^2 + 3t + 9)$
44. $3t^4 - 24t$   $3t(t - 2)(t^2 + 2t + 4)$
45. $125y^3 - 1$   $(5y - 1)(25y^2 + 5y + 1)$
46. $64y^3 - 27$   $(4y - 3)(16y^2 + 12y + 9)$
47. $1 - 8t^3$   $(1 - 2t)(1 + 2t + 4t^2)$
48. $8 - 27t^3$   $(2 - 3t)(4 + 6t + 9t^2)$
49. $m^3 - 216n^3$   $(m - 6n)(m^2 + 6mn + 36n^2)$
50. $8m^3 - 1$   $(2m - 1)(4m^2 + 2m + 1)$
51. $\dfrac{1}{27}s^3 - 1$
52. $\dfrac{1}{8}t^3 - 125$

Use the sum of two cubes property to factor each binomial, if possible. (*See Objective 2.*)

53. $y^3 + 27$   $(y + 3)(y^2 - 3y + 9)$
54. $y^3 + 1$   $(y + 1)(y^2 - y + 1)$
55. $a^3 + 64$   $(a + 4)(a^2 - 4a + 16)$
56. $a^3 + 125$   $(a + 5)(a^2 - 5a + 25)$
57. $5t^4 + 320t$   $5t(t + 4)(t^2 - 4t + 16)$
58. $-3t^4 - 24t$   $-3t(t + 2)(t^2 - 2t + 4)$
59. $1000 + 125b^2$   $125(8 + b^2)$
60. $8 + 216b^2$   $8(1 + 27b^2)$
61. $8m^3 + n^3$   $(2m + n)(4m^2 - 2mn + n^2)$
62. $m^3 + 27n^3$   $(m + 3n)(m^2 - 3mn + 9n^2)$
63. $a^3 + \dfrac{1}{27}$
64. $b^3 + \dfrac{8}{125}$

Factor each binomial completely. (*See Objective 3.*)

65. $x^4 - 16$   $(x^2 + 4)(x + 2)(x - 2)$
66. $x^4 - 625$   $(x^2 + 25)(x + 5)(x - 5)$
67. $2y^6 + 250$   $2(y^2 + 5)(y^4 - 5y^2 + 25)$
68. $3p^6 + 24$   $3(p^2 + 2)(p^4 - 2p^2 + 4)$
69. $256 - y^6$   $(16 + y^3)(16 - y^3)$
70. $1 - 16y^6$   $(1 + 4y^3)(1 - 4y^3)$
71. $r^4 - \dfrac{81}{16}$
72. $s^4 - \dfrac{1}{625}$
73. $16a^4 - 81$   $(4a^2 + 9)(2a + 3)(2a - 3)$
74. $1 - 256a^4$   $(1 + 16a^2)(1 + 4a)(1 - 4a)$

## Mix 'Em Up!

Factor each binomial completely.

75. $9y^3 - 36y$   $9y(y + 2)(y - 2)$
76. $36m^2 - 81n^2$   $9(2m + 3n)(2m - 3n)$
77. $256 - a^2$   $(16 + a)(16 - a)$
78. $4m^2 - 121$   $(2m + 11)(2m - 11)$
79. $169 - b^2$   $(13 + b)(13 - b)$
80. $625 - c^2$   $(25 + c)(25 - c)$
81. $8a^3 - 125$   $(2a - 5)(4a^2 + 10a + 25)$
82. $4y^3 - 324$   $4(y^3 - 81)$
83. $r^2 + 100s^2$   prime
84. $s^4 + 16t^4$   prime
85. $a^3 + 1000$   $(a + 10)(a^2 - 10a + 100)$
86. $8y^3 + 64$   $8(y + 2)(y^2 - 2y + 4)$
87. $16x^4 - 81$   $(4x^2 + 9)(2x + 3)(2x - 3)$
88. $256x^4 - 1$   $(16x^2 + 1)(4x + 1)(4x - 1)$
89. $2y^6 - 54$   $2(y^2 - 3)(y^4 + 3y^2 + 9)$
90. $5y^6 + 40$   $5(y^2 + 2)(y^4 - 2y^2 + 4)$
91. $5x^4y^4 - 405z^4$   $5(x^2y^2 + 9z^2)(xy + 3z)(xy - 3z)$
92. $7p^4q^4 - 7r^4$   $7(p^2q^2 + r^2)(pq + r)(pq - r)$
93. $64r^2 - 49s^2$   $(8r + 7s)(8r - 7s)$
94. $100m^2 - 49n^2$   $(10m + 7n)(10m - 7n)$
95. $0.16a^2 - 0.25b^2$   $(0.4a - 0.5b)(0.4a + 0.5b)$
96. $0.01c^2 - 0.09d^2$   $(0.1c - 0.3d)(0.1c + 0.3d)$
97. $8x^3 - \dfrac{1}{125}y^3$
98. $27x^3 + \dfrac{1}{8}y^3$
99. $\dfrac{1}{16}x^4 - \dfrac{1}{81}$
100. $\dfrac{1}{625}y^4 - \dfrac{1}{16}$

## You Be the Teacher!

**Correct each student's errors, if any.**

**101.** Factor $x^2 - 64x$.

Sam's work: $x^2 - 64x = (x + 8)(x - 8)$    $x^2 - 64x = x(x - 64)$

**102.** Factor $y^2 + 25$.

Cameron's work: $y^2 + 25 = (y + 5)^2$
$y^2 + 25$ is a sum of two squares with no common factor, so it is prime.

**103.** Factor $x^3 + 64$.

Hyunh's work: $x^3 + 64 = (x + 4)(x^2 + 4x + 16)$
$x^3 + 64 = (x + 4)(x^2 - 4x + 16)$

**104.** Factor $y^3 - 8$.

Dawn's work: $y^3 - 8 = (y - 2)^3$
$y^3 - 8 = (y - 2)(y^2 + 2y + 4)$

## Calculate It!

**Factor each binomial and use a calculator to verify it.**

**105.** $x^3 - 8$
$(x - 2)(x^2 + 2x + 4)$

**106.** $x^3 + 27$
$(x + 3)(x^2 - 3x + 9)$

**107.** $81x^4 - 16$
$(9x^2 + 4)(3x - 2)(3x + 2)$

**108.** $10,000x^4 - 1$
$(100x^2 + 1)(10x - 1)(10x + 1)$

---

## PIECE IT TOGETHER / SECTIONS 6.1–6.4 Review

**Factor each polynomial completely.**

**1.** $3x^3 + 15x^2 - 18x - 90$
$3(x^2 - 6)(x + 5)$

**2.** $16s^3 - 48s^2 - s + 3$
$(4s + 1)(4s - 1)(s - 3)$

**3.** $x^2 + 13x + 36$
$(x + 4)(x + 9)$

**4.** $y^2 - y - 30$    $(y - 6)(y + 5)$

**5.** $a^2 + 14a + 7$    prime

**6.** $b^2 - 19b + 60$
$(b - 4)(b - 15)$

**7.** $5y^2 + 15y - 20$    $5(y + 4)(y - 1)$

**8.** $6y^2 + 30y - 84$
$6(y - 2)(y + 7)$

**9.** $p^3q^2 + 2p^2q - 35p$
$p(pq - 5)(pq + 7)$

**10.** $5r^3s^2 - 14r^2s - 24r$
$r(rs - 4)(5rs + 6)$

**11.** $75x^2 - 65x + 10$
$5(3x - 2)(5x - 1)$

**12.** $10y^2 - 13y + 4$
$(2y - 1)(5y - 4)$

**13.** $4r^2 - 20r + 25$    $(2r - 5)^2$

**14.** $50h^2 + 20h + 2$
$2(5h + 1)^2$

**15.** $4y^2 - \frac{49}{25}$    $\left(2y + \frac{7}{5}\right)\left(2y - \frac{7}{5}\right)$

**16.** $\frac{1}{9}a^2 + \frac{4}{15}$    prime

**17.** $x^2 - 9y^2$    $(x + 3y)(x - 3y)$

**18.** $1000a^3 + b^3$
$(10a + b)(100a^2 - 10ab + b^2)$

**19.** $8x^3 - 125y^3$
$(2x - 5y)(4x^2 + 10xy + 25y^2)$

**20.** $81a^4 - 1$
$(9a^2 + 1)(3a + 1)(3a - 1)$

---

## SECTION 6.5 · Solving Quadratic Equations and Other Polynomial Equations by Factoring

### OBJECTIVES

As a result of completing this section, you will be able to

**1.** Solve quadratic equations by factoring.

**2.** Solve equations of degree 3 or higher by factoring.

**3.** Troubleshoot common errors.

B.A.S.E. jumping is an activity whereby people leap off structures with only a brief period before they must open a parachute to survive. If a person free-falls from a 1600-ft structure, his height can be represented by $-16t^2 + 1600$, where $t$ is the number of seconds after the fall. The jumper must open the parachute well before reaching the ground, to give it time to open fully and slow the descent. If the jumper does not open his parachute, when does he reach the ground?

To answer this question, we must solve the equation $-16t^2 + 1600 = 0$. This is an example of a *quadratic equation*. We will learn the steps needed to solve this type of equation in this section.

### Objective 1 ▶

Solve quadratic equations by factoring.

**INSTRUCTOR NOTE:**
The word quadratic comes from a Latin word "quadratus," which means squared. So, while quad usually signifies a four-sided figure, quadratic equations arose from geometric problems involving squares.

## Solve Quadratic Equations by Factoring

So far in this chapter, we have learned how to factor different types of polynomials. Now we are going to use our knowledge of factoring to solve a new type of equation called a quadratic equation.

> **Definition:** A **quadratic equation** is an equation that can be written in the form $ax^2 + bx + c = 0$, where $a$, $b$, and $c$ are real numbers and $a \neq 0$. This form is called the **standard form** of a quadratic equation.

Some examples of quadratic equations in standard form are

$$2x^2 + 7x - 4 = 0 \qquad y^2 + 5y + 6 = 0 \qquad h^2 - 4 = 0$$

Some examples of quadratic equations that are not in standard form are

$$a^2 - a = 12 \qquad t^2 = 5t \qquad (x - 3)(x + 1) = 8$$

There are numerous ways to solve quadratic equations, but in this section we are going to focus on just one method—solving quadratic equations by factoring. Other methods will be discussed in Chapter 9.

Suppose two numbers are multiplied and the product is zero. Then at least one of the numbers has to be zero. For example,

$$-4 \cdot 0 = 0 \quad \text{and} \quad 0 \cdot \frac{1}{2} = 0$$

The only way for a product to equal zero is if one of the factors is zero. This concept is called the *zero products property*.

> **Property:  Zero Products Property**
>
> If $a$ and $b$ are real numbers and $a \cdot b = 0$, then
>
> $$a = 0 \quad \text{or} \quad b = 0$$

This property provides a method for solving quadratic equations. It states that if one side of the equation is expressed in factored form and the other side of the equation is zero, the solutions are found by setting each factor equal to zero.

> **Procedure:  Solving Quadratic Equations by Factoring**
>
> **Step 1:** Write the quadratic equation in standard form, if necessary.
> **Step 2:** Factor the resulting polynomial.
> **Step 3:** Use the zero products property to solve the equation.
> **Step 4:** Check the solutions in the original equation.

**Objective 1  Examples**  **Solve each quadratic equation by factoring.**

**1a.** $(x + 1)(x - 3) = 0$      **1b.** $4y(y - 5) = 0$      **1c.** $x^2 - 4x - 5 = 0$

**1d.** $x^2 = 49$      **1e.** $4x^2 + 20x + 25 = 0$      **1f.** $3y^2 + 4 = 7y$

**1g.** $12 - 5x = 2x^2$      **1h.** $(x - 1)(x + 2) = -2$

**Solutions**  **1a.** We can apply the zero products property to solve the equation since the left side of the equation is in factored form and the right side is zero.

$$(x + 1)(x - 3) = 0$$

$$x + 1 = 0 \qquad \text{or} \qquad x - 3 = 0 \qquad \text{Apply the zero products property.}$$

$$x + 1 - 1 = 0 - 1 \qquad x - 3 + 3 = 0 + 3 \qquad \text{Solve each equation.}$$

$$x = -1 \qquad\qquad x = 3$$

**INSTRUCTOR NOTE:**
Inform students that we generally list solutions in order from smallest to largest in a solution set.

The solutions are $-1$ and $3$, and the solution set is $\{-1, 3\}$. Check solution by replacing the variable with its corresponding value.

$x = -1$:

$$(x + 1)(x - 3) = 0$$
$$(-1 + 1)(-1 - 3) = 0$$
$$(0)(-4) = 0$$
$$0 = 0 \quad \text{True}$$

$x = 3$:

$$(x + 1)(x - 3) = 0$$
$$(3 + 1)(3 - 3) = 0$$
$$(4)(0) = 0$$
$$0 = 0 \quad \text{True}$$

**1b.**  $4y(y-5) = 0$  — The equation is factored and equal to zero.

$4y = 0$  or  $y - 5 = 0$  — Apply the zero products property.

$\dfrac{4y}{4} = \dfrac{0}{4}$  $y - 5 + 5 = 0 + 5$  — Solve each equation.

$y = 0$  $y = 5$

The solutions are 0 and 5, and the solution set is $\{0, 5\}$.

**Check:**

$y = 0$:

$$4y(y-5) = 0$$
$$4(0)(0-5) = 0$$
$$(0)(-5) = 0$$
$$0 = 0 \quad \text{True}$$

$y = 5$:

$$4y(y-5) = 0$$
$$4(5)(5-5) = 0$$
$$20(0) = 0$$
$$0 = 0 \quad \text{True}$$

**1c.**  $x^2 - 4x - 5 = 0$  — The equation is in standard form.

$(x-5)(x+1) = 0$  — Factor the trinomial on the left side.

$x - 5 = 0$  or  $x + 1 = 0$  — Apply the zero products property.

$x - 5 + 5 = 0 + 5$  $x + 1 - 1 = 0 - 1$  — Solve each equation.

$x = 5$  $x = -1$

The solutions are 5 and $-1$, and the solution set is $\{-1, 5\}$.

**Check:**

$x = 5$:

$$x^2 - 4x - 5 = 0$$
$$(5)^2 - 4(5) - 5 = 0$$
$$25 - 20 - 5 = 0$$
$$5 - 5 = 0$$
$$0 = 0 \quad \text{True}$$

$x = -1$:

$$x^2 - 4x - 5 = 0$$
$$(-1)^2 - 4(-1) - 5 = 0$$
$$1 + 4 - 5 = 0$$
$$5 - 5 = 0$$
$$0 = 0 \quad \text{True}$$

**1d.**  $x^2 = 49$  — The equation is not in standard form.

$x^2 - 49 = 49 - 49$  — Subtract 49 from each side.

$x^2 - 49 = 0$  — Simplify.

$(x-7)(x+7) = 0$  — Factor the binomial on the left side.

$x - 7 = 0$  or  $x + 7 = 0$  — Apply the zero products property.

$x - 7 + 7 = 0 + 7$  $x + 7 - 7 = 0 - 7$  — Solve each equation.

$x = 7$  $x = -7$

The solutions are 7 and $-7$, and the solution set is $\{-7, 7\}$. We can write the solution set as $\{\pm 7\}$ to represent both the positive and negative value of 7.

**Check:**

$x = 7$:

$$x^2 = 49$$
$$(7)^2 = 49$$
$$49 = 49 \quad \text{True}$$

$x = -7$:

$$x^2 = 49$$
$$(-7)^2 = 49$$
$$49 = 49 \quad \text{True}$$

**1e.**

$$4x^2 + 20x + 25 = 0$$ The equation is in standard form.

$$(2x + 5)(2x + 5) = 0$$ Factor the trinomial on the left side.

$$2x + 5 = 0$$ Apply the zero products property.

$$2x + 5 - 5 = 0 - 5$$ Subtract 5 from each side.

$$2x = -5$$ Simplify.

$$x = -\frac{5}{2}$$

The solution is $-\frac{5}{2}$, and the solution set is $\left\{-\frac{5}{2}\right\}$.

**Check:**

$$x = -\frac{5}{2}: \qquad 4x^2 + 20x + 25 = 0$$

$$4\left(-\frac{5}{2}\right)^2 + 20\left(-\frac{5}{2}\right) + 25 = 0$$

$$4\left(\frac{25}{4}\right) - 50 + 25 = 0$$

$$25 - 50 + 25 = 0$$

$$0 = 0 \quad \text{True}$$

**INSTRUCTOR NOTE:**
Remind students that they can factor the trinomial with trial and error or grouping.

**1f.**

$$3y^2 + 4 = 7y$$ The equation is not in standard form.

$$3y^2 + 4 - 7y = 7y - 7y$$ Subtract $7y$ from each side.

$$3y^2 - 7y + 4 = 0$$ Simplify and write in standard form.

$$3y^2 - 3y - 4y + 4 = 0$$ Replace $-7y$ with $-3y - 4y$.

$$3y(y - 1) - 4(y - 1) = 0$$ Factor by grouping.

$$(y - 1)(3y - 4) = 0$$

$$y - 1 = 0 \qquad \text{or} \qquad 3y - 4 = 0$$ Apply the zero products property.

$$y - 1 + 1 = 0 + 1 \qquad 3y - 4 + 4 = 0 + 4$$ Solve each equation.

$$y = 1 \qquad\qquad 3y = 4$$

$$\frac{3y}{3} = \frac{4}{3}$$

$$y = \frac{4}{3}$$

The solutions are 1 and $\frac{4}{3}$, and the solution set is $\left\{1, \frac{4}{3}\right\}$.

**Check:**

$$y = 1: \qquad 3y^2 + 4 = 7y \qquad\qquad\qquad y = \frac{4}{3}: \qquad 3y^2 + 4 = 7y$$

$$3(1)^2 + 4 = 7(1) \qquad\qquad\qquad\qquad 3\left(\frac{4}{3}\right)^2 + 4 = 7\left(\frac{4}{3}\right)$$

$$3(1) + 4 = 7 \qquad\qquad\qquad\qquad\qquad 3\left(\frac{16}{9}\right) + 4 = \frac{28}{3}$$

$$3 + 4 = 7 \qquad\qquad\qquad\qquad\qquad\qquad \frac{48}{9} + \frac{36}{9} = \frac{28}{3}$$

$$7 = 7 \quad \text{True} \qquad\qquad\qquad\qquad\qquad \frac{84}{9} = \frac{28}{3}$$

$$\frac{28}{3} = \frac{28}{3} \quad \text{True}$$

**1g.** We need to perform operations on each side of the equation to make one side equal to zero. We can either make the left side of the equation zero or the right side of the equation zero.

| **Method 1** | **Method 2** |
|---|---|
| Make the right side of equation zero. | Make the left side of equation zero. |

**Method 1**

$$12 - 5x = 2x^2$$
$$12 - 5x - 2x^2 = 2x^2 - 2x^2$$
$$-2x^2 - 5x + 12 = 0$$
$$-1(-2x^2 - 5x + 12) = -1(0)$$
$$2x^2 + 5x - 12 = 0$$
$$2x^2 - 3x + 8x - 12 = 0$$
$$x(2x - 3) + 4(2x - 3) = 0$$
$$(2x - 3)(x + 4) = 0$$
$$2x - 3 = 0 \quad \text{or} \quad x + 4 = 0$$
$$x = \frac{3}{2} \qquad\qquad x = -4$$

**Method 2**

$$12 - 5x = 2x^2$$
$$12 - 5x - 12 + 5x = 2x^2 - 12 + 5x$$
$$0 = 2x^2 + 5x - 12$$
$$0 = 2x^2 - 3x + 8x - 12$$
$$0 = x(2x - 3) + 4(2x - 3)$$
$$0 = (2x - 3)(x + 4)$$
$$2x - 3 = 0 \quad \text{or} \quad x + 4 = 0$$
$$x = \frac{3}{2} \qquad\qquad x = -4$$

The solutions are $\frac{3}{2}$ and $-4$, and the solution set is $\left\{ -4, \frac{3}{2} \right\}$.

**Check:**

$x = \frac{3}{2}$:

$$12 - 5x = 2x^2$$
$$12 - 5\left(\frac{3}{2}\right) = 2\left(\frac{3}{2}\right)^2$$
$$12 - \frac{15}{2} = 2\left(\frac{9}{4}\right)$$
$$\frac{24}{2} - \frac{15}{2} = \frac{18}{4}$$
$$\frac{9}{2} = \frac{9}{2} \quad \text{True}$$

$x = -4$:

$$12 - 5x = 2x^2$$
$$12 - 5(-4) = 2(-4)^2$$
$$12 + 20 = 2(16)$$
$$32 = 32 \quad \text{True}$$

**Note:** *Both methods provide the same solutions, but the second method provides a positive coefficient on the squared term in fewer steps. In general, we want to write the equation in standard form in a way that makes the coefficient of the squared term positive.*

**1h.** The left side of the equation is in factored form, but we cannot apply the zero products property because the right side of the equation is *not* zero. So, we first multiply the binomials on the left.

| | |
|---|---|
| $(x - 1)(x + 2) = -2$ | The equation is not in standard form. |
| $x^2 + 2x - x - 2 = -2$ | Multiply the two binomials on the left side. |
| $x^2 + x - 2 = -2$ | Combine like terms on the left side. |

Now that the left side of the equation has been simplified, we can write the equation in standard form.

$$x^2 + x - 2 + 2 = -2 + 2 \qquad \text{Add 2 to each side.}$$
$$x^2 + x = 0 \qquad \text{Simplify.}$$
$$x(x + 1) = 0 \qquad \text{Factor.}$$
$$x = 0 \quad \text{or} \quad x + 1 = 0 \qquad \text{Apply the zero products property.}$$
$$x + 1 - 1 = 1 - 1 \qquad \text{Solve the resulting equations.}$$
$$x = -1$$

The solutions are 0 or $-1$ and the solution set is $\{-1, 0\}$.

**Check:**

$x = 0$:

$$(x - 1)(x + 2) = -2$$
$$(0 - 1)(0 + 2) = -2$$
$$(-1)(2) = -2$$
$$-2 = -2 \quad \text{True}$$

$x = -1$:

$$(x - 1)(x + 2) = -2$$
$$(-1 - 1)(-1 + 2) = -2$$
$$(-2)(1) = -2$$
$$-2 = -2 \quad \text{True}$$

✔ **Student Check 1**    Solve each quadratic equation by factoring.

   **a.** $(x - 10)(x + 8) = 0$      **b.** $2b(4b + 3) = 0$      **c.** $y^2 - 15y + 36 = 0$

   **d.** $r^2 = 36$                  **e.** $9a^2 - 12a + 4 = 0$      **f.** $4a^2 - 5 = -8a$

   **g.** $7x = 3x^2 - 20$        **h.** $(x - 4)(x - 2) = 8$

Recall that the degree of a polynomial is the largest degree of the terms of the polynomial. So, the degree of each of the equations in Example 1 is 2. Note that each of the equations in Example 1 has two solutions except for 1e. Example 1e has 1 solution that repeats. The degree of an equation determines the maximum number of solutions of that equation.

**Note:** *The degree of a polynomial equation determines the maximum number of solutions of the equation.*

## Solving Polynomial Equations

**Objective 2** ▶

Solve equations of degree 3 or higher by factoring.

**Polynomial equations** are equations that involve polynomials. A quadratic equation is a special case of a polynomial equation whose degree is 2. The following table shows examples of polynomial equations, the type of equation, its degree, and its maximum number of solutions.

| Example | Type of Equation | Degree | Maximum Number of Solutions |
|---|---|---|---|
| $x^2 - x - 6 = 0$ | Quadratic | 2 | 2 |
| $2y^3 - 18y = 0$ | Cubic | 3 | 3 |
| $a^4 - 5a^2 + 4 = 0$ | Quartic | 4 | 4 |

We now learn how to solve polynomial equations that have a degree of 3 or higher by factoring. The process of solving these equations is similar to the process we used to solve quadratic equations. We must write the polynomial equation in standard form, factor the polynomial, and apply the zero products property.

**Procedure: Solving Polynomial Equations of Degree 3 or Higher**

**Step 1:** Write the equation in standard form, that is "polynomial = 0."
**Step 2:** Factor the polynomial.
**Step 3:** Apply the zero products property and set each factor equal to zero.
**Step 4:** Solve the resulting equations.
**Step 5:** Check each solution in the original equation.

**Objective 2 Examples** / **Solve each equation by factoring.**

**2a.** $(x - 2)(4x - 1)(x + 6) = 0$ **2b.** $2y^3 - 18y = 0$

**2c.** $(3x - 5)(5x^2 + 9x - 2) = 0$ **2d.** $a^4 - 5a^2 + 4 = 0$

**Solutions** **2a.** We set each factor equal to zero and solve.

$$(x - 2)(4x - 1)(x + 6) = 0$$

$$
\begin{array}{ccc}
x - 2 = 0 & 4x - 1 = 0 & x + 6 = 0 \\
x - 2 + 2 = 0 + 2 & 4x - 1 + 1 = 0 + 1 & x + 6 - 6 = 0 - 6 \\
x = 2 & 4x = 1 & x = -6 \\
& \dfrac{4x}{4} = \dfrac{1}{4} & \\
& x = \dfrac{1}{4} &
\end{array}
$$

The solution set is $\left\{ -6, \dfrac{1}{4}, 2 \right\}$. Check each solution.

$x = 2$:

$$(x - 2)(4x - 1)(x + 6) = 0$$
$$(2 - 2)[4(2) - 1](2 + 6) = 0$$
$$(0)(7)(8) = 0$$
$$0 = 0$$

$x = \dfrac{1}{4}$:

$$(x - 2)(4x - 1)(x + 6) = 0$$
$$\left(\dfrac{1}{4} - 2\right)\left[4\left(\dfrac{1}{4}\right) - 1\right]\left(\dfrac{1}{4} + 6\right) = 0$$
$$\left(-\dfrac{7}{4}\right)(0)\left(\dfrac{25}{4}\right) = 0$$
$$0 = 0$$

$x = -6$:

$$(x - 2)(4x - 1)(x + 6) = 0$$
$$(-6 - 2)[4(-6) - 1](-6 + 6) = 0$$
$$(-8)(-25)(0) = 0$$
$$0 = 0$$

Since each solution makes the equation true, our solution set is correct.

**2b.**

$$2y^3 - 18y = 0$$
$$2y(y^2 - 9) = 0 \qquad \text{Factor out the common factor.}$$

$$2y(y - 3)(y + 3) = 0 \qquad \text{Apply the difference of two squares property.}$$

$$
\begin{array}{lll}
2y = 0 & y - 3 = 0 & y + 3 = 0 \qquad \text{Apply the zero products property.} \\
\dfrac{2y}{2} = \dfrac{0}{2} & y - 3 + 3 = 0 + 3 & y + 3 - 3 = 0 - 3 \qquad \text{Solve the resulting} \\
y = 0 & y = 3 & y = -3 \qquad \text{equations.}
\end{array}
$$

The solution set is $\{-3, 0, 3\}$. Check each solution.

$y = 0$:

$$2y^3 - 18y = 0$$
$$2(0)^3 - 18(0) = 0$$
$$2(0) - 0 = 0$$
$$0 - 0 = 0$$
$$0 = 0$$

$y = 3$:

$$2y^3 - 18y = 0$$
$$2(3)^3 - 18(3) = 0$$
$$2(27) - 54 = 0$$
$$54 - 54 = 0$$
$$0 = 0$$

$y = -3$:

$$2y^3 - 18y = 0$$
$$2(-3)^3 - 18(-3) = 0$$
$$2(-27) + 54 = 0$$
$$-54 + 54 = 0$$
$$0 = 0$$

Since each solution makes the equation true, our solution set is correct.

**2c.**   $(3x - 5)(5x^2 + 9x - 2) = 0$

$(3x - 5)(5x - 1)(x + 2) = 0$   Factor the trinomial.

$3x - 5 = 0$   or   $5x - 1 = 0$   or   $x + 2 = 0$   Apply the zero products property.

$3x = 5$          $5x = 1$          $x = -2$   Solve each equation.

$x = \dfrac{5}{3}$          $x = \dfrac{1}{5}$

The solution set is $\left\{ -2, \dfrac{1}{5}, \dfrac{5}{3} \right\}$. Each solution checks.

**2d.**   $a^4 - 5a^2 + 4 = 0$

$(a^2 - 4)(a^2 - 1) = 0$   Factor the trinomial.

$(a - 2)(a + 2)(a - 1)(a + 1) = 0$   Factor the difference of squares.

$a - 2 = 0$   or   $a + 2 = 0$   or   $a - 1 = 0$   or   $a + 1 = 0$   Apply the zero products property.

$a = 2$          $a = -2$          $a = 1$          $a = -1$   Solve each equation.

The solution set is $\{-2, -1, 1, 2\}$. We can also write the four solutions as $\{\pm 1, \pm 2\}$. Each solution checks.

✓ **Student Check 2**   Solve each equation by factoring.

**a.** $(a - 1)(3a + 6)(a - 5) = 0$    **b.** $16x - 4x^3 = 0$

**c.** $(y + 5)(y^2 - 3y + 2) = 0$    **d.** $b^4 - 13b^2 + 36 = 0$

**Objective 3** ▶
Troubleshoot common errors.

## Troubleshooting Common Errors

Some common errors related to solving quadratic equations and polynomial equations are shown next.

**Objective 3  Examples**   **A problem and an incorrect solution are given. Provide the correct solution and an explanation of the error.**

**3a.** Solve $3x(x - 4) = 0$.

| Incorrect Solution | Correct Solution and Explanation |
|---|---|
| $3x(x - 4) = 0$ <br> $3x = 0$   or   $x - 4 = 0$ <br> $x = -3$         $x = 4$ <br> The solution set is $\{-3, 4\}$. | The error was made in solving the equation $3x = 0$. To solve this equation, we must divide both sides by 3 to get $x = \dfrac{0}{3} = 0$. So, the solution set is $\{0, 4\}$. |

**3b.** Solve $x^2 = 9x$.

| Incorrect Solution | Correct Solution and Explanation |
|---|---|
| $x^2 = 9x$ <br> $x^2 - 9x = 0$ <br> $(x - 3)(x + 3) = 0$ <br> $x - 3 = 0$ or $x + 3 = 0$ <br> $x = 3 \qquad x = -3$ <br> The solution set is $\{-3, 3\}$. | The binomial $x^2 - 9$ factors as $(x - 3)(x + 3)$ but $x^2 - 9x$ is not the difference of two squares. We must factor out a common factor of $x$ first. <br><br> $x^2 = 9x$ <br> $x^2 - 9x = 0$ <br> $x(x - 9) = 0$ <br> $x = 0$ or $x - 9 = 0$ <br> $\qquad\qquad x = 9$ <br><br> So, the solution set is $\{0, 9\}$. |

**3c.** Solve $(x - 6)(x + 3) = -8$.

| Incorrect Solution | Correct Solution and Explanation |
|---|---|
| $(x - 6)(x + 3) = -8$ <br> $x - 6 = -8$ or $x + 3 = -8$ <br> $x = -2 \qquad x = -11$ <br> The solution set is $\{-11, -2\}$. | We cannot apply the zero products property since the right side is not equal to zero. So, we must first write the equation in standard form. <br><br> $(x - 6)(x + 3) = -8$ <br> $x^2 + 3x - 6x - 18 = -8$ <br> $x^2 - 3x - 10 = 0$ <br> $(x - 5)(x + 2) = 0$ <br> $x - 5 = 0$ or $x + 2 = 0$ <br> $x = 5 \qquad x = -2$ <br><br> So, the solution set is $\{-2, 5\}$. |

**3d.** Solve $4x^2 - 5x - 6 = 0$.

| Incorrect Solution | Correct Solution and Explanation |
|---|---|
| $4x^2 - 5x - 6 = 0$ <br> $(4x - 3)(x + 2) = 0$ <br> $4x - 3 = 0$ or $x + 2 = 0$ <br> $x = \dfrac{3}{4} \qquad x = -2$ <br> The solution set is $\left\{-2, \dfrac{3}{4}\right\}$. | The error was made in assigning the signs of the binomial factors. Because the middle term of the trinomial is negative, the larger of the outer and inner product must be negative, so the negative sign should be on 2. <br><br> $4x^2 - 5x - 6 = 0$ <br> $(4x + 3)(x - 2) = 0$ <br> $4x + 3 = 0$ or $x - 2 = 0$ <br> $x = -\dfrac{3}{4} \qquad x = 2$ <br><br> The solution set is $\left\{-\dfrac{3}{4}, 2\right\}$. |

**3e.** Solve $2x^3 - 18x = 0$.

| Incorrect Solution | Correct Solution and Explanation |
|---|---|
| $2x^3 - 18x = 0$ $$\frac{2x^3 - 18x}{2x} = \frac{0}{2x}$$ $x^2 - 9 = 0$ $(x - 3)(x + 3) = 0$ $x - 3 = 0$ or $x + 3 = 0$ $x = 3 \qquad x = -3$ The solution set is $\{-3, 3\}$. | The degree of the equation is 3, so we should have 3 solutions. The error was made in dividing both sides by a variable. We cannot divide by a variable because we might be dividing by zero. This also loses one of the solutions. We should begin by factoring out the common factor. $2x^3 - 18x = 0$ $2x(x^2 - 9) = 0$ $2x(x - 3)(x + 3) = 0$ $2x = 0$ or $x - 3 = 0$ or $x + 3 = 0$ $x = 0 \qquad x = 3 \qquad x = -3$ The solution set is $\{-3, 0, 3\}$. |

## ANSWERS TO STUDENT CHECKS

**Student Check 1**    **a.** $\{-8, 10\}$    **b.** $\left\{-\dfrac{3}{4}, 0\right\}$    **c.** $\{3, 12\}$

**d.** $\{-6, 6\}$    **e.** $\dfrac{2}{3}$    **f.** $\left\{-\dfrac{5}{2}, \dfrac{1}{2}\right\}$    **g.** $\left\{-\dfrac{5}{3}, 4\right\}$

**h.** $\{0, 6\}$

**Student Check 2**    **a.** $\{-2, 1, 5\}$    **b.** $\{-2, 0, 2\}$

**c.** $\{-5, 1, 2\}$    **d.** $\{-3, -2, 2, 3\}$

## SUMMARY OF KEY CONCEPTS

1. The zero products property provides the basis for solving polynomial equations. It states that if the product of factors is zero, then one of those factors must equal zero.

2. To solve a quadratic equation, write the equation in standard form; that is, make one side of the equation zero and factor the other side. Lastly, set each factor equal to 0 and solve the resulting equations.

3. To solve an equation with degree of 3 or higher, write the equation in the form "polynomial = 0." Factor the resulting polynomial. Lastly, set each factor equal to 0 and solve the resulting equations.

## GRAPHING CALCULATOR SKILLS

The graphing calculator can be used to verify solutions of quadratic equations.

**Example:** Verify that the solutions of $x^2 - 8x - 20 = 0$ are $x = 10$ or $x = 2$.

**Solution:** To verify the solutions, the equation must be in standard form. Then enter the polynomial from the equation into $Y_1$.

Then view the table. For the solutions to be correct, the $Y_1$ column should equal zero for the given $x$-values. Press 2nd Graph and view the $x$-values of 2 and 10.

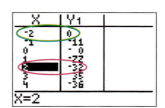 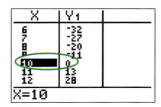

Notice that the $Y_1$ value for $x = 2$ is $-32$. This means that 2 is *not* a solution of the equation.

From the table, we see that the $Y_1$ value for $x = -2$ is 0. So, $-2$ is a solution. The $Y_1$ value for $x = 10$ is 0, so 10 is a solution of the equation as well.

## Write About It!

**Use complete sentences in your answer to each exercise.**

1. Explain the zero products property. *If the product of two numbers is zero, then at least one of the numbers has to be zero.*
2. What is a quadratic equation?
3. Explain how to solve a quadratic equation by using the zero products property.
4. Explain how the zero products property can be used to solve polynomial equations of degree 3 or higher.
5. Explain the difference in solving the equations $(x - 5)(x - 1) = 12$ and $(x - 5)(x - 1) = 0$.
6. Explain the difference in solving the equations $4x(x - 5)(x - 1) = 0$ and $4(x - 5)(x - 1) = 0$.

## Practice Makes Perfect!

**Solve each equation. (See Objective 1.)**

7. $(x + 2)(x - 5) = 0$ $\{-2, 5\}$
8. $(x - 1)(x + 4) = 0$ $\{-4, 1\}$
9. $x(x - 3) = 0$ $\{0, 3\}$
10. $x(x + 1) = 0$ $\{-1, 0\}$
11. $(10 - x)(x + 7) = 0$ $\{-7, 10\}$
12. $(4 - x)(x + 2) = 0$ $\{-2, 4\}$
13. $(2x - 1)(3x + 6) = 0$
14. $(3x + 4)(4x - 1) = 0$
15. $5(6 - x) = 0$ $\{6\}$
16. $3(1 - x) = 0$ $\{1\}$
17. $-4(x + 12)(2x - 3) = 0$
18. $-2(4x - 5)(x + 5) = 0$
19. $x(3x - 6)(x + 6) = 0$ $\{-6, 0, 2\}$
20. $x(5x - 10)(x + 3) = 0$ $\{-3, 0, 2\}$
21. $2x(x - 7) = 0$ $\{0, 7\}$
22. $3x(x - 4) = 0$ $\{0, 4\}$
23. $(x + 7)(x - 1)(x + 3) = 0$ $\{-7, -3, 1\}$
24. $(x + 8)(x - 9)(x + 1) = 0$ $\{-8, -1, 9\}$
25. $x^2 - x - 6 = 0$ $\{-2, 3\}$
26. $x^2 - 2x - 8 = 0$ $\{-2, 4\}$
27. $x^2 - 2x + 1 = 0$ $\{1\}$
28. $x^2 - 6x + 9 = 0$ $\{3\}$
29. $x^2 + 7x + 10 = 0$ $\{-5, -2\}$
30. $x^2 + 10x + 24 = 0$ $\{-6, -4\}$
31. $2x^2 - x - 3 = 0$ $\left\{-1, \frac{3}{2}\right\}$
32. $3x^2 + 5x - 2 = 0$ $\left\{-2, \frac{1}{3}\right\}$
33. $x^2 = 25$ $\{-5, 5\}$
34. $x^2 = 16$ $\{-4, 4\}$
35. $(x - 6)(x + 7) = -42$ $\{-1, 0\}$
36. $(x + 3)(x - 8) = -24$ $\{0, 5\}$
37. $x^2 + 4x = 32$ $\{-8, 4\}$
38. $x^2 - 3x = 18$ $\{-3, 6\}$
39. $6x^2 + 20 = 23x$ $\left\{\frac{4}{3}, \frac{5}{2}\right\}$
40. $3x^2 - 19x = 14$ $\left\{-\frac{2}{3}, 7\right\}$
41. $(x - 5)(x + 3) = -7$ $\{-2, 4\}$
42. $(x + 7)(x + 2) = 24$ $\{-10, 1\}$

**Solve each polynomial equation. (See Objective 2.)**

43. $2x^3 - 18x = 0$ $\{-3, 0, 3\}$
44. $3x^3 - 12x = 0$ $\{-2, 0, 2\}$
45. $(2x - 1)(x^2 - 6x - 16) = 0$ $\left\{-2, \frac{1}{2}, 8\right\}$
46. $(4x + 3)(x^2 + 5x - 14) = 0$ $\left\{-7, -\frac{3}{4}, 2\right\}$
47. $x^4 - 5x^2 + 4 = 0$ $\{-2, -1, 1, 2\}$
48. $x^4 - 20x^2 + 64 = 0$ $\{-4, -2, 2, 4\}$

49. $5x^4 - 15x^3 - 90x^2 = 0$ $\{-3, 0, 6\}$
50. $4x^4 - 8x^3 + 4x^2 = 0$ $\{0, 1\}$
51. $5x^3 = 5x$ $\{-1, 0, 1\}$
52. $2x^3 = 32x$ $\{-4, 0, 4\}$
53. $(49x^2 - 1)(4x^2 - 9) = 0$
54. $(16x^2 - 25)(9x^2 - 4) = 0$

## Mix 'Em Up!

**Solve each equation.**

55. $4x^2 - 5x - 6 = 0$ $\left\{-\frac{3}{4}, 2\right\}$
56. $5x^2 + 3x - 2 = 0$ $\left\{-1, \frac{2}{5}\right\}$
57. $(x + 9)^2 = 0$ $\{-9\}$
58. $(2x - 1)^2 = 0$ $\left\{\frac{1}{2}\right\}$
59. $(x + 2)(x + 11) = 22$ $\{-13, 0\}$
60. $(2x - 3)(x - 4) = 12$
61. $(3 - x)(10 + 3x) = 0$
62. $(12 - 5x)(6 - x) = 0$
63. $-4x^2 + x + 5 = 0$
64. $(x^2 - 9)(x^2 - 36) = 0$ $\{-6, -3, 3, 6\}$
65. $(x^2 - 100)(x^2 - 1) = 0$ $\{-10, -1, 1, 10\}$
66. $4x^2 + 10x - 24 = 0$
67. $4x^2 = 16$ $\{-2, 2\}$
68. $5x^2 = 245$ $\{-7, 7\}$
69. $3x^3 - 363x = 0$ $\{-11, 0, 11\}$
70. $4x^4 - 324x^2 = 0$ $\{-9, 0, 9\}$
71. $4x^4 - 37x^2 + 9 = 0$
72. $9x^4 - 13x^2 + 4 = 0$
73. $10x^2 + 12x = 16$ $\left\{-2, \frac{4}{5}\right\}$
74. $12x^2 + 14 = 34x$ $\left\{\frac{1}{2}, \frac{7}{3}\right\}$
75. $3x(x + 2)(x - 1) = 0$ $\{-2, 0, 1\}$
76. $2x(x - 5)(x - 4) = 0$ $\{0, 4, 5\}$
77. $25x^2 + 9 = 0$ $\varnothing$
78. $16x^2 + 1 = 0$ $\varnothing$
79. $7x^3 - 63x = 0$ $\{-3, 0, 3\}$
80. $5x^3 - 125x = 0$ $\{-5, 0, 5\}$
81. $6x^2 + 38x + 40 = 0$
82. $20x^2 + 94x + 18 = 0$
83. $(2x - 3)(x^2 + x - 20) = 0$ $\left\{-5, \frac{3}{2}, 4\right\}$
84. $(7x + 2)(x^2 + 3x - 18) = 0$ $\left\{-6, -\frac{2}{7}, 3\right\}$

## You Be the Teacher!

**Correct each student's errors, if any.**

85. Solve: $(x - 3)(x - 1) = 3$.

    Bruce's work:

    $(x - 3)(x - 1) = 3$

    $x - 3 = 3$ or $x - 1 = 3$

    $x = 6$      $x = 4$

    $(x - 3)(x - 1) = 3$
    $x^2 - 4x + 3 = 3$
    $x^2 - 4x = 0$
    $x(x - 4) = 0$
    $x = 0$ or $x = 4$

86. Solve: $5x(x - 4) = 0$.

    Jordan's work:

    $5x(x - 4) = 0$

    $5x = 0$ or $2x - 4 = 0$

    $x = 5$      $x = 2$

    $5x(x - 4) = 0$
    $x = 0$ or $x - 4 = 0$
    $x = 0$ or      $x = 4$

87. Solve: $2(x + 1) = 0$.

    Chris's work:

    $2(x + 1) = 0$

    $x = 2$ or $x + 1 = 0$

    $x = -1$

    $2(x + 1) = 0$
    $x + 1 = 0$
    $x = -1$

*Additional answers can be found in the Instructor Answer Appendix.*

**88.** Give an example of a quadratic equation
with 1 solution.

Taylor's answer: $x^2 - 9 = 0$

Answers vary. Taylor's equation has two solutions, $-3$ and $3$. If we set a perfect square trinomial equal to zero, we will have one solution for an answer. $x^2 + 2x + 1 = 0$.

### Calculate It!

Use a graphing calculator to determine if the solutions are correct. If they are not, state the correct solutions.

**89.** $x^2 + 3x - 54 = 0, x = -6, x = -9$   $\{-9, 6\}$

**90.** $4x^2 + 7x - 2 = 0, x = \dfrac{1}{4}, x = 2$   $\left\{-2, \dfrac{1}{4}\right\}$

### Think About It!

Write a polynomial equation that has the given solutions.

**91.** $x = 2$ and $x = 5$
$x^2 - 7x + 10 = 0$

**92.** $x = -1$ and $x = 4$
$x^2 - 3x - 4 = 0$

**93.** $x = 0, x = -3$   $x^2 + 3x = 0$   **94.** $x = 0, x = \dfrac{1}{2}$   $2x^2 - x = 0$

**95.** $x = 4$ (occurs twice as a solution)   $x^2 - 8x + 16 = 0$

**96.** $x = -2$ (occurs twice as a solution)   $x^2 + 4x + 4 = 0$

**97.** $x = 1, x = 3,$ and $x = -4$   **98.** $x = 0, x = \dfrac{2}{5}, x = -6$
$(x - 1)(x - 3)(x + 4) = 0$           $x(5x - 2)(x + 6) = 0$

---

## SECTION 6.6  Applications of Quadratic Equations

### ▶ OBJECTIVES

As a result of completing this section, you will be able to use quadratic equations to

1. Solve applications involving area.
2. Solve applications involving consecutive integers.
3. Solve applications involving revenue and profit.
4. Solve applications using the Pythagorean theorem.
5. Solve applications given a model.
6. Troubleshoot common errors.

Suppose that when an airline charges $150 per seat, they can sell 100 tickets. For each $5 decrease in price, they sell 10 additional tickets. How much should the airline charge for each ticket if they want to generate $15,000 in revenue?

This and other real-world situations can be modeled by quadratic equations. In physics, quadratic equations can be used to determine the exact speed or position of moving objects. In business, quadratic models can be used to determine information about a company's revenue and profit. We will study these and other applications in this section.

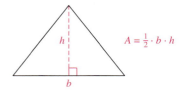

### Applications Involving Area

Some area formulas involve the multiplication of two measurements. For example,

Area of a rectangle $=$ length $\times$ width          Area of a triangle $= \dfrac{1}{2} \times$ base $\times$ height

#### ▶ Objective 1

Solve applications involving area.

When we multiply the two measurements in these area formulas, we get square units. For example, if a rectangle has length 3 in. and width 4 in., its area is

$$A = lw$$
$$A = (3 \text{ in.})(4 \text{ in.})$$
$$A = 12 \text{ in.}^2$$

So, it is not surprising that quadratic equations can often be used to model applications involving area.

We can use the following general problem solving strategy to guide us in solving the applications in this section.

---

**Procedure:  Using the General Problem Solving Strategy**

**Step 1: Read** the question carefully to determine what is known and what is unknown. Assign a variable to the unknown, as needed. Make note of any formulas that apply and draw a picture of the situation, if possible.

**Step 2: Translate** the words into a mathematical equation.

**Step 3: Solve** the equation.

**Step 4: Check** the answer.

**Step 5: Answer** the original question and make sure the answer is reasonable in the context of the problem.

---

**Objective 1  Examples**    Solve each problem.

**1a.** Find the dimensions of a rectangle that has a length of $2x + 1$ units, a width of $x - 3$ units, and an area of 22 units$^2$.

**Solution**    **1a.** What is unknown? The length is $2x + 1$.

The width is $x - 3$.

$2x + 1$ units

| Area = 22 units$^2$ | $x - 3$ units |

What is known? The area is 22 units$^2$ and $A = lw$.

Translate and solve: Use the area formula to write the equation.

| | |
|---|---|
| $(2x + 1)(x - 3) = 22$ | Apply the area formula. |
| $2x^2 - 6x + x - 3 = 22$ | Multiply the binomials. |
| $2x^2 - 5x - 3 = 22$ | Combine like terms. |
| $2x^2 - 5x - 3 - 22 = 22 - 22$ | Subtract 22 from each side. |
| $2x^2 - 5x - 25 = 0$ | Simplify. |
| $(2x + 5)(x - 5) = 0$ | Factor the resulting trinomial. |
| $2x + 5 = 0 \quad$ or $\quad x - 5 = 0$ | Apply the zero products property. |
| $2x = -5 \qquad\qquad x = 5$ | Solve each equation. |
| $x = -\dfrac{5}{2}$ | |

Now we need to find the value of the length and width. Replace the variable in the expressions for the length and width with the possible solutions.

| $x = -\dfrac{5}{2}$ | | $x = 5$ | |
|---|---|---|---|
| $l = 2x + 1$ | $w = x - 3$ | $l = 2x + 1$ | $w = x - 3$ |
| $= 2\left(-\dfrac{5}{2}\right) + 1$ | $= -\dfrac{5}{2} - \dfrac{6}{2}$ | $= 2(5) + 1$ | $= 5 - 3$ |
| $= -5 + 1$ | $= -\dfrac{11}{2}$ | $= 10 + 1$ | $= 2$ |
| $= -4$ | | $= 11$ | |

The value $-\dfrac{5}{2}$ doesn't make sense in the context of this situation since it makes the length and width of the rectangle negative. So, the length of the rectangle is 11 units and the width is 2 units.

**1b.** Jentezen plans to pour concrete in a uniform width around his rectangular pool to form a walkway. If his pool measures 20 ft by 50 ft and he has enough concrete to cover 800 ft², how wide can the walkway be?

**Solution**

**1b.** What is unknown? The width of the walkway is unknown. Let $x$ represent the width of the walkway.

What is known? The pool measures 20 ft × 50 ft. The area of the walkway is 800 ft². The formula that applies is $A = lw$.

Translate and solve: We need an expression that represents the area of the walkway. The area of the walkway is the area of the pool and walkway combined minus the area of the pool.

area of the pool − area of pool = area of the
and walkway                      walkway

$$(50 + 2x)(20 + 2x) - (50)(20) = 800 \qquad \text{Write the equation.}$$
$$1000 + 100x + 40x + 4x^2 - 1000 = 800 \qquad \text{Find each product.}$$
$$4x^2 + 140x = 800 \qquad \text{Combine like terms.}$$
$$4x^2 + 140x - 800 = 800 - 800 \qquad \text{Subtract 800 from each side.}$$
$$4x^2 + 140x - 800 = 0 \qquad \text{Simplify.}$$
$$4(x^2 + 35x - 200) = 0 \qquad \text{Factor out the GCF, 4.}$$
$$4(x + 40)(x - 5) = 0 \qquad \text{Factor the trinomial.}$$
$$x + 40 = 0 \quad \text{or} \quad x - 5 = 0 \qquad \text{Apply the zero products property.}$$
$$x = -40 \qquad x = 5 \qquad \text{Solve each equation.}$$

Since $x$ represents the width of the walkway, the negative value doesn't make sense in this context. Therefore, the walkway should be 5 ft wide.

**✓ Student Check 1**

**a.** Find the length of the base and height of a triangle if the base is $4x$ units, the height is $x + 2$ units, and the area is 30 units².

**b.** Kathleen built a rectangular koi pond in her backyard that measures 5 ft by 4 ft. She purchased gravel to spread uniformly around the pond for decoration. If Kathleen has enough gravel to cover 52 ft², how wide can she make the gravel border?

## Applications Involving Consecutive Integers

**Objective 2** ▶

Solve applications involving consecutive integers.

In Section 2.3, we learned that **consecutive integers** are integers that are successive, or follow one another without interruption. We also learned how to express consecutive integers algebraically as shown.

**Property: Consecutive Integer Representation**

Consecutive integers: $x, x + 1, x + 2, \ldots$, where $x$ is the first integer.
Consecutive odd integers: $x, x + 2, x + 4, \ldots$, where $x$ is the first odd integer.
Consecutive even integers: $x, x + 2, x + 4, \ldots$, where $x$ is the first even integer.

**Objective 2 Examples**    **Solve each problem.**

**2a.** The product of two consecutive integers is 90. Find the integers.

**Solution**    **2a.** What is unknown? The two consecutive integers are unknown.

Let $x$, $x + 1$ represent the two integers.

What is known? The product of the two integers is 90.

Translate and solve: The product of the integers is $x(x + 1)$ and is equal to 90.

$$x(x + 1) = 90 \quad \text{Write the equation.}$$
$$x^2 + x = 90 \quad \text{Multiply the expressions on the left.}$$
$$x^2 + x - 90 = 90 - 90 \quad \text{Subtract 90 from each side.}$$
$$x^2 + x - 90 = 0 \quad \text{Write in standard form.}$$
$$(x + 10)(x - 9) = 0 \quad \text{Factor.}$$
$$x + 10 = 0 \quad \text{or} \quad x - 9 = 0 \quad \text{Apply the zero products property.}$$
$$x = -10 \qquad\qquad x = 9 \quad \text{Solve each equation.}$$

We have two possible values for $x$.

If $x = -10$, then $x + 1 = -10 + 1 = -9$.  |  If $x = 9$, then $x + 1 = 9 + 1 = 10$.
$-10$ and $-9$ are consecutive integers with | 9 and 10 are consecutive integers
a product of 90.  |  with a product of 90.

So, there are two pairs of integers that solve this problem. The integers are either $-10$ and $-9$ or 9 and 10.

**2b.** The product of two consecutive positive even integers is 14 more than their sum. Find the integers.

**Solution**    **2b.** What is unknown? The two consecutive positive even integers are unknown.

Let $x$ and $x + 2$ represent the consecutive even integers.

What is known? The product of two integers is 14 more than their sum.

Translate and solve: The product of the integers is $x(x + 2)$. The sum of the integers is $x + x + 2$.

product of two integers is 14 more than their sum

$$x(x + 2) = x + x + 2 + 14 \quad \text{Write the equation.}$$
$$x^2 + 2x = 2x + 16 \quad \text{Simplify each side.}$$
$$x^2 + 2x - 2x - 16 = 2x + 16 - 2x - 16 \quad \text{Subtract 2x and 16 from each side.}$$
$$x^2 - 16 = 0 \quad \text{Write the equation in standard form.}$$
$$(x - 4)(x + 4) = 0 \quad \text{Factor.}$$
$$x - 4 = 0 \quad \text{or} \quad x + 4 = 0 \quad \text{Apply the zero products property.}$$
$$x = 4 \qquad\qquad x = -4 \quad \text{Solve each equation.}$$

Because the integers are positive, the value $-4$ must be discarded. So, if 4 is one integer, the next even integer is $x + 2 = 4 + 2 = 6$. So, the integers are 4 and 6.

**✓ Student Check 2**    Solve each problem.

**a.** The product of two consecutive integers is 72. Find the integers.

**b.** The product of two consecutive odd integers is 23 more than their sum. Find the integers.

## Applications Involving Revenue and Profit

**Objective 3** ▶

Solve applications involving revenue and profit.

In Chapter 5, we learned that **revenue** is calculated by multiplying the price of an item by the number of items sold and that **profit** is the difference of revenue and cost. We can use our knowledge of revenue, profit, and quadratic equations to solve applications involving these quantities.

**Objective 3  Examples** | Solve each problem.

**3a.** The annual profit, in thousands of dollars, for a mobile phone company can be represented by $p = -4x^2 + 102x - 50$, where $x$ is the total number of mobile phones sold for the year, in thousands. Find the number of phones the company needs to sell to break even; that is, find when profit is equal to zero dollars.

**Solution**  **3a.**

| | |
|---|---|
| $-4x^2 + 102x - 50 = p$ | Begin with the profit model. |
| $-4x^2 + 102x - 50 = 0$ | Replace $p$ with 0. |
| $-2(2x^2 - 51x + 25) = 0$ | Factor out the GCF. |
| $-2(2x - 1)(x - 25) = 0$ | Factor the trinomial. |
| $2x - 1 = 0$  or  $x - 25 = 0$ | Apply the zero products property. |
| $x = \dfrac{1}{2}$     $x = 25$ | Solve each equation. |
| $x = 0.5$ | |

So, the company will break even when they sell 0.5 thousand (or 500) phones or when they sell 25 thousand (or 25,000) phones.

**3b.** Suppose that when an airline charges $150 per seat, they can sell 100 tickets. For each $5 decrease in price, they sell 10 additional tickets. If $x$ represents the number of price decreases, then $150 - 5x$ is the expression for the ticket price and $100 + 10x$ is the number of tickets sold. How much should the airline charge for each ticket if they want to generate $18,200 in revenue?

**Solution**  **3b.** What is unknown? The price the airline should charge for each ticket is unknown. This is represented by the quantity $150 - 5x$.

What is known? The revenue generated is $18,200.

Translate and solve: Since revenue = price times items sold, we have the following.

**price × items sold = revenue**

| | |
|---|---|
| $(150 - 5x)(100 + 10x) = 18{,}200$ | Write the equation. |
| $15{,}000 + 1500x - 500x - 50x^2 = 18{,}200$ | Simplify the product. |
| $-50x^2 + 1000x + 15{,}000 = 18{,}200$ | Combine like terms. |
| $-50x^2 + 1000x + 15{,}000 - 18{,}200 = 18{,}200 - 18{,}200$ | Subtract 18,200 from each side. |
| $-50x^2 + 1000x - 3200 = 0$ | Simplify. |
| $-50(x^2 - 20x + 64) = 0$ | Factor out the GCF, −50. |
| $-50(x - 16)(x - 4) = 0$ | Factor the trinomial. |
| $x - 16 = 0$  or  $x - 4 = 0$ | Apply the zero products property. |
| $x = 16$     $x = 4$ | Solve each equation. |

We have two possible solutions, $x = 16$ or $x = 4$. To find the corresponding price the airline should charge for each of these values, substitute the values into the expression $150 - 5x$.

If $x = 16$, then the price is $150 - 5(16) = 70$ or $70.

If $x = 4$, then the price is $150 - 5(4) = 130$ or $130.

So, the airline could charge $70 per ticket or $130 per ticket in order to generate $18,200 in revenue.

✔ **Student Check 3**    Solve each problem.

a. The profit, in hundreds of dollars, for a company is given by $p = -x^2 + 23x - 60$, where $x$ is the number of units sold. Find how many units need to be sold for the company to break even.

b. When a boutique charges \$30 for one of its designer shirts, they can sell 25 shirts. For each \$10 increase in price, they sell five fewer shirts. If $x$ represents the number of price increases, then $30 + 10x$ represents the price of the shirt and $25 - 5x$ represents the number of shirts sold. What price should the boutique charge for a designer shirt if they want to generate \$800 in revenue?

## Applications Involving the Pythagorean Theorem

**Objective 4** ▶

Solve applications using the Pythagorean theorem.

The Greek mathematician Pythagoras is credited with discovering a special relationship that exists between the lengths of the sides of a *right triangle* (i.e., a triangle with a 90° angle). This special relationship is called the *Pythagorean theorem*.

---

**Property: Pythagorean Theorem**

For any right triangle, where $a$ and $b$ are the lengths of the legs and $c$ is the length of the hypotenuse (the side opposite the right angle),

$$a^2 + b^2 = c^2$$

$$(\text{leg})^2 + (\text{other leg})^2 = (\text{hypotenuse})^2$$

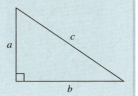

---

**Objective 4  Examples**    Solve each problem.

**4a.** A 13-ft ladder leans against the side of a house. If the distance from the ground to the bottom of the ladder is 5 ft, how far up the house does the ladder reach?

**Solution**    **4a.** What is unknown? The distance from the ground to where the ladder touches the house is unknown. Let $x$ represent this distance.

What is known? The length of the ladder is 13 ft. The distance from the house to the ladder is 5 ft.

Translate and solve: Since we are dealing with a right triangle, we can use the Pythagorean theorem.

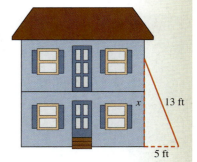

| | |
|---|---|
| $a^2 + b^2 = c^2$ | State the Pythagorean theorem. |
| $x^2 + 5^2 = 13^2$ | Replace $a$, $b$, and $c$ with $x$, 5, and 13, respectively. |
| $x^2 + 25 = 169$ | Simplify. |
| $x^2 + 25 - 169 = 169 - 169$ | Subtract 169 from each side. |
| $x^2 - 144 = 0$ | Simplify. |
| $(x - 12)(x + 12) = 0$ | Factor. |
| $x - 12 = 0$  or  $x + 12 = 0$ | Apply the zero products property. |
| $x = 12$ \qquad $x = -12$ | Solve each equation. |

Since $x$ represents a length, the only possible answer is 12. So, the ladder touches the house at a height of 12 ft.

**4b.** Donna and Jerry leave their home at the same time one morning. Jerry travels north, and Donna travels east. When they are 10 mi apart, Donna has traveled 2 mi more than Jerry. How many miles have Donna and Jerry traveled?

**Solution**   **4b.** What is unknown? The distance that Jerry and Donna traveled is unknown. Let $x$ represent the distance Jerry traveled. Then $x + 2$ represents the distance Donna traveled.

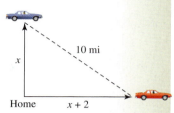

What is known? The distance between the two is 10 mi.

Translate and solve: We are dealing with a right triangle, so we use the Pythagorean theorem to get an equation.

| | |
|---|---|
| $a^2 + b^2 = c^2$ | State the Pythagorean theorem. |
| $x^2 + (x + 2)^2 = 10^2$ | Replace $a$, $b$, and $c$, with $x$, $x + 2$, and 10, respectively. |
| $x^2 + x^2 + 4x + 4 = 100$ | Square the binomial. |
| $2x^2 + 4x + 4 = 100$ | Combine like terms. |
| $2x^2 + 4x + 4 - 100 = 100 - 100$ | Subtract 100 from each side. |
| $2x^2 + 4x - 96 = 0$ | Simplify. |
| $2(x^2 + 2x - 48) = 0$ | Factor out the GCF. |
| $2(x + 8)(x - 6) = 0$ | Factor the polynomial. |
| $x + 8 = 0$  or  $x - 6 = 0$ | Apply the zero products property. |
| $x = -8$   $x = 6$ | Solve each equation. |

Since $x$ represents a distance, the only possible answer is 6. So, Jerry traveled 6 mi and Donna traveled $x + 2 = 6 + 2 = 8$ mi.

**✓ Student Check 4**   Solve each problem.

**a.** A 20-ft ladder leans against the side of a house. If distance from the house to the ladder is 12 ft, what is the distance the ladder reaches on the house?

**b.** One leg of a triangle is 1 unit longer than the other leg. If the hypotenuse measures 5 units, then what is the length of each leg?

**Objective 5** ▶

Solve applications given a model.

## Applications Involving a Model

Now we will use a quadratic model to solve some application problems.

**Objective 5 Examples**   Solve each problem.

**5a.** A B.A.S.E. jumper leaps from a 1600-ft structure. His height, in feet, can be represented by $s = -16t^2 + 1600$, where $t$ is the number of seconds after he has jumped. The jumper must open his parachute before he reaches the ground. If the jumper does not open his parachute, how long will it take before he reaches the ground?

**Solution**   **5a.** If the jumper reaches the ground, his height above the ground will be 0 ft. So, we replace $s$ with 0 and solve the resulting equation.

| | |
|---|---|
| $s = -16t^2 + 1600$ | Begin with the given model. |
| $0 = -16t^2 + 1600$ | Replace $s$ with 0. |
| $0 = -16(t^2 - 100)$ | Factor out the common factor, $-16$. |
| $0 = -16(t - 10)(t + 10)$ | Factor the binomial. |
| $t - 10 = 0$  or  $t + 10 = 0$ | Apply the zero products property. |
| $t = 10$   $t = -10$ | Solve. |

Since $t$ represents the number of seconds after the jumper leaps off the building, only the positive answer makes sense for this problem. So, the jumper would reach the ground in 10 sec. He has less than 10 sec to open his parachute after jumping.

**5b.** A ball is thrown upward with an initial velocity of 16 ft/sec from a height of 32 ft above the ground. The height, in feet, of the ball $t$ sec after it is thrown can be represented by $s = -16t^2 + 16t + 32$. How long will it take for the ball to reach the ground?

**Solution**   **5b.** When the ball reaches the ground, its height above the ground is 0 ft.

| | |
|---|---|
| $s = -16t^2 + 16t + 32$ | Begin with the given model. |
| $0 = -16t^2 + 16t + 32$ | Replace $s$ with 0. |
| $0 = -16(t^2 - t - 2)$ | Factor out the common factor, $-16$. |
| $0 = -16(t - 2)(t + 1)$ | Factor the trinomial. |
| $t - 2 = 0$   or   $t + 1 = 0$ | Apply the zero products property. |
| $t = 2$        $t = -1$ | Solve. |

The negative value doesn't make sense in this context. So, the ball reaches the ground 2 sec after it was thrown.

✓ **Student Check 5**   Solve each problem.

**a.** Suppose a penny is dropped from the top of the Empire State Building, which is approximately 1296 ft high. The quadratic equation $s = -16t^2 + 1296$ gives the distance from the ground (in feet) that the penny will be after $t$ sec. How long will it take the penny to reach the ground?

**b.** The height of a ball thrown upward with an initial velocity of 32 ft/sec from a height of 128 ft above the ground can be represented by $-16t^2 + 32t + 128$, where $t$ is the number of seconds after the ball is thrown. How long will it take the ball to reach the ground?

**Objective 6** ▶

Troubleshoot common errors.

## Troubleshooting Common Errors

Some of the common errors associated with applications of quadratic equations are shown. Most of the problems occur from setting up the equation incorrectly. Others come from simplifying the equation incorrectly.

**Objective 6  Examples**   **A problem and an incorrect solution are given. Provide the correct solution and provide an explanation of the error.**

**6a.** The product of two consecutive odd integers is 63. Write an equation that can be used to find the integers.

| Incorrect Solution | Correct Solution and Explanation |
|---|---|
| $x(x + 1) = 63$ | Because the integers are consecutive odd integers, the equation is $x(x + 2) = 63$ |

**6b.** The sides of a right triangle are consecutive integers. Find the length of each side.

| Incorrect Solution | Correct Solution and Explanation |
|---|---|
| The sides are $x$, $x + 1$, and $x + 2$. So, the equation is $$x^2 + (x + 1)^2 = (x + 2)^2$$ $$x^2 + x^2 + 1 = x^2 + 4$$ $$2x^2 + 1 = x^2 + 4$$ $$2x^2 + 1 - x^2 - 4 = x^2 + 4 - x^2 - 4$$ $$x^2 - 3 = 0$$ Can't factor so there are no solutions. | The binomials were not squared correctly. Remember that $$(a + b)^2 = a^2 + 2ab + b^2$$ So, we get $$x^2 + (x + 1)^2 = (x + 2)^2$$ $$x^2 + x^2 + 2x + 1 = x^2 + 4x + 4$$ $$2x^2 + 2x + 1 = x^2 + 4x + 4$$ $$x^2 - 2x - 3 = 0$$ $$(x - 3)(x + 1) = 0$$ $$x = 3 \quad \text{or} \quad x = -1$$ So, the sides of the right triangle are 3, 4, and 5 units. |

## ANSWERS TO STUDENT CHECKS

**Student Check 1   a.** The base is 12 units and the height is 5 units.   **b.** The width of the gravel border is 2 ft.

**Student Check 2   a.** The integers are 8 and 9 or −9 and −8.   **b.** The integers are 5 and 7 or −5 and −3.

**Student Check 3   a.** The company will break even after selling 3 units or 20 units.   **b.** The shirts should be sold for $40 each.

**Student Check 4   a.** The ladder reaches 16 ft up the house.   **b.** The legs of the triangle are 3 units and 4 units.

**Student Check 5   a.** The penny will reach the ground in 9 sec.   **b.** The ball will reach the ground in 4 sec.

## SUMMARY OF KEY CONCEPTS

1. When solving applications involving area, identify the shape of the object whose area is being examined. Next, write down the area formula for that shape. Draw a picture to visualize the problem.

2. Consecutive integers are integers that follow one another without interruption.
   a. Consecutive integers can be expressed as $x$, $x + 1$, . . . .
   b. Consecutive even integers can be expressed as $x$, $x + 2$, . . . , where $x$ is an even integer.
   c. Consecutive odd integers can be expressed as $x$, $x + 2$, . . . , where $x$ is an odd integer.

3. When solving applications involving revenue, remember the formula: revenue = price × quantity. When solving applications involving profit, remember the formula: profit = revenue − cost.

4. The Pythagorean theorem defines a relationship between the sides of a right triangle. The Pythagorean theorem is $(\text{leg})^2 + (\text{other leg})^2 = (\text{hypotenuse})^2$.

5. If a model is given, replace the appropriate variable with the known value.

## GRAPHING CALCULATOR SKILLS

The graphing calculator can help us solve equations. One method is to use the Table feature. If we input the two sides of the equation into $Y_1$ and $Y_2$, respectively, we can use the Table feature to determine the value of $x$ for which the two sides are equal.

**Example:** Solve $(150 - 5x)(100 + 10x) = 18{,}200$ using the Table feature.

**Solution:** Enter each side of the equation into the equation editor. Examine the table for the value of $x$ that makes $Y_1 = Y_2$.

```
Plot1 Plot2 Plot3
\Y1日(150-5X)(100
+10X)
\Y2日18200
\Y3=
\Y4=
\Y5=
·.Y6=
```

| X | Y1 | Y2 |
|---|---|---|
| 0 | 15000 | 18200 |
| 1 | 15950 | 18200 |
| 2 | 16800 | 18200 |
| 3 | 17550 | 18200 |
| 4 | 18200 | 18200 |
| 5 | 18750 | 18200 |
| 6 | 19200 | 18200 |

X=4

| X | Y1 | Y2 |
|---|---|---|
| 12 | 19800 | 18200 |
| 13 | 19550 | 18200 |
| 14 | 19200 | 18200 |
| 15 | 18750 | 18200 |
| 16 | 18200 | 18200 |
| 17 | 17550 | 18200 |
| 18 | 16800 | 18200 |

X=16

From the table, we see the solutions are 4 and 16.

## SECTION 6.6 / EXERCISE SET

### Write About It!

**Use complete sentences in your answer to each exercise.**

1. Explain the first two steps in solving an application problem.

2. Give examples of consecutive integers.
   *Answers vary; 21, 23, 25, . . . or 75, 76, 77, . . .*

3. Explain what expressions represent consecutive odd and consecutive even integers.

4. What is revenue? What is a profit?

### Practice Makes Perfect!

**Solve each problem involving area. (*See Objective 1.*)**

5. Allison wants to section off a rectangular portion of her yard for a flowerbed. She wants the flowerbed to be 8 ft longer than it is wide. If the area of the flowerbed will be 48 ft², find the dimensions of the flowerbed. *4 ft by 12 ft*

6. A county park builds rectangular pool. The length of the pool is 10 ft more than twice its width. If the area of the pool is 1000 ft², find the dimensions of the pool. *20 ft by 50 ft*

7. Janna wants to mat her wedding picture before framing it. Her picture measures 14 in. by 20 in. and she wants the mat in uniform width around the picture. How wide can the mat be if the area of the picture and mat is 1360 in.²? *10 in.*

8. Mona wants to professionally frame her son's first school photo. The photo is 8 in. by 10 in. She wants to mat the picture with an even width around the picture. How wide should the mat be if the area of the photo and the frame is 360 in.²? *5 in.*

9. Jensen wants to pour concrete in a uniform width around his pool that measures 30 ft by 40 ft. If he has enough concrete to cover an area of 624 ft², then what should be the width of the concrete walkway that is created? *4 ft*

10. Persia wants to put mulch in a uniform width around a rectangular reflection pond that she built in her backyard. If the pond measures 8 ft by 12 ft and she has enough mulch to cover 300 ft², then what should be the width of the mulch border? *5 ft*

**Solve each problem involving consecutive integers. (*See Objective 2.*)**

11. The product of two positive consecutive integers is 132. Find the integers. *11, 12*

12. The product of two positive consecutive integers is 20. Find the integers. *4, 5*

13. The product of two consecutive even integers is 34 more than their sum. Find the integers. *6 and 8 or −6 and −4*

14. The product of two consecutive even integers is 2 more than their sum. Find the integers. *2 and 4 or −2 and 0*

15. The product of two consecutive odd integers is equal to 3 times the larger integer. Find the integers.
    *3 and 5 or −2 and 0*

16. The product of two consecutive odd integers equals 9 more than 6 times the larger integer. Find the integers.
    *7 and 9 or −3 and −1*

**Solve each problem involving revenue and profit. (*See Objective 3.*)**

17. A luxury kennel provides accommodations for cats. When the kennel charges a daily rate of $45, they can board 10 cats. For each $2 decrease in price, 1 more cat is boarded. If $x$ represents the number of price decreases, then $45 - 2x$ represents the daily boarding rate and $10 + x$ represents the number of cats boarded. How many cats do they need to board to make $375 in revenue? *25 cats*

18. Ashley monograms T-shirts. When she charges $5 per name, she monograms 50 T-shirts. For every $1 increase in price, five fewer T-shirts are monogrammed. If $x$ represents the number of price increases, then $5 + x$ represents the price to monogram a name and $50 - 5x$ represents the number of T-shirts monogrammed. How much should Ashley charge to monogram a name to earn $250 in revenue? *$5 or $10*

19. When an airline charges $200 per seat, they can sell 50 tickets. For every $2 decrease in price, five more tickets are sold. If $x$ represents the number of price decreases, then $200 - 2x$ represents the ticket price and $50 + 5x$ represents the number of passengers. How many passengers are needed for the airline to earn $18,000 in revenue? *100 or 450*

20. When an online shoe seller charges $100 for a pair of wedge sandals, 50 pairs are sold. For each $5 decrease in price, 10 more pairs are sold. If $x$ represents the number of price decreases, then $100 - 5x$ represents the price of each pair of sandals and $50 + 10x$ represents the number of pairs of sandals sold. How much should the online shoe seller charge for each pair of sandals to earn $5000 in revenue? *$25 or $100*

21. The profit, in millions, of a video game company is given by $p = -60x^2 + 555x + 450$, where $x$ is the number of units sold, in thousands. How many units need to be sold for the company to break even? *10,000 units*

22. The profit, in millions, of a software company is given by $p = -150x^2 + 1425x + 4500$, where $x$ is the number of units sold, in thousands. How many units need to be sold for the company to break even? *12,000 units*

**Find the lengths of the sides of each right triangle. (*See Objective 4.*)**

23.

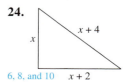

*4, 3, and 5*

24.

*6, 8, and 10*    $x + 2$

**25.**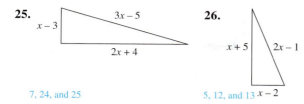

7, 24, and 25

**26.** $x + 5$  $2x - 1$  $x - 2$

5, 12, and 13

**Solve each problem. (See Objective 4.)**

**27.** A 15-ft ladder leans against the side of a house. If the distance from the bottom of the ladder to the house is 3 ft shorter than the distance from the ground to the top of the ladder, then what is the distance from the ground to the top of the ladder?   12 ft

**28.** The length of a ladder leaning against a house is 3 ft more than twice the distance from the bottom of the ladder to the house. If the distance from the bottom of the house to the top of the ladder is 12 ft, what is the distance from the bottom of the ladder to the house?   5 ft

**29.** Heidi and Spencer leave a restaurant in two separate cars at the same time. Heidi travels south and Spencer travels west. When they are 20 mi apart, Heidi has traveled 4 mi less than Spencer. How many miles have Heidi and Spencer each traveled?
Heidi traveled 12 mi and Spencer traveled 16 mi.

**30.** Theodore and Wallace leave their dorm room at the same time. Theodore travels north and Wallace travels west. When they are 30 mi apart, Theodore has traveled 6 mi more than the distance Wallace has traveled. How far did Wallace travel?   Wallace traveled 18 mi.

**Solve each problem. (See Objective 5.)**

**31.** A penny is dropped from a 144-ft bridge. The distance between the penny and the ground can be approximated by $s = -16t^2 + 144$ (in feet), where $t$ is the time (in seconds) after the penny is dropped. How long will it take the penny to hit the ground?   3 sec

**32.** How long will it take the penny from Exercise 31 to be 80 ft from the ground?   2 sec

**33.** The distance of a ball thrown directly upwards from the ground is $s = -16t^2 + 120t$ (in feet), where $t$ is the time (in seconds) after the ball is thrown. When will the ball be 104 ft in the air?   1 and 6.5 sec

**34.** A toy rocket is projected upward from the ground with an initial velocity of 96 ft/sec. The distance of the rocket from the ground is $s = -16t^2 + 96t$ (in feet), where $t$ is the time (in seconds) after the rocket is launched. When will the rocket reach 80 ft?   1 and 5 sec

**35.** A sky diver is dropped from a plane at 3600 ft. His distance from the ground can be approximated by $s = -16t^2 + 3600$ (in feet), where $t$ is the time (in seconds) after the diver is dropped. How long will it take the diver to reach the ground if his parachute doesn't open?   15 sec

**36.** How long will it take the diver from Exercise 35 to be 2000 ft from the ground?   10 sec

**37.** Young performs at birthday parties for children. The revenue generated from performing at $x$ birthday parties

is $R = x^2 + 250x$ (in dollars). How many birthday parties must Young perform at to earn $2600 in revenue?   10 parties

**38.** The total cost of producing $x$ items is given by $C = 3x^2 - 22x + 10$ (in dollars). How many items can be produced at a total cost of $90?   10 items

**39.** Joyce's Bakery makes creative birthday cakes. The profit generated from the sale of $x$ cakes is $P = 100x - x^2$ (in dollars). How many cakes must be sold for Joyce's Bakery to have a profit of $2500?   50 cakes

**40.** Leroy is a popular deejay. He determined that his revenue from hosting $x$ parties is approximately $R = x^2 + 244x$ (in dollars). How many parties does Leroy have to host to earn $1500 in revenue?   6 parties

**41.** The height of a ball thrown upward with an initial velocity of 96 ft/sec from a height of 640 ft above the ground can be represented by $S = -16t^2 + 96t + 640$, where $t$ is the number of seconds after the ball is thrown. How long will it take the ball to reach the ground?   10 sec

**42.** The height of a ball thrown upward with an initial velocity of 128 ft/sec from a height of 768 ft above the ground can be represented by $S = -16t^2 + 128t + 768$, where $t$ is the number of seconds after the ball is thrown. How long will it take the ball to reach the ground?   12 sec

 **You Be the Teacher!**

**Correct each student's errors, if any.**

**43.** The product of two positive odd consecutive integers is 1 less than twice their sum. Find the integers.

Chris's work:

$x =$ 1st odd integer
$x + 1 =$ 2nd odd integer
$x(x + 1) = 2(2x + 1) - 1$
$x^2 + x = 4x + 1 - 1$
$x^2 - 3x = 0$
$x(x - 3) = 0$
  $x = 0$  or  $x = 3$
  $x + 1 = 1$  or  $x = 4$

$x(x + 2) = 2(2x + 2) - 1$
$x^2 + 2x = 4x + 4 - 1$
$x^2 - 2x - 3 = 0$
$(x - 3)(x + 1) = 0$
$x = 3$  or  $x = -1$
The two consecutive positive odd integers are 3 and 5.

**44.** Set up the equation that models the situation: "When a local theater charges $6 per seat, they can sell 200 tickets. For every $1 increase in price, 5 less tickets are sold. How many tickets are needed to earn $1800?"

Jordan's work:

$x =$ number of tickets
$(6 + x)(5x - 200) = 1800$
$(6 + x)(200 - 5x) = 1800$

 **Calculate It!**

**Use the table feature in a graphing calculator to solve each equation.**

**45.** $(10 + x)(120 - 3x) = 1443$   3, 27

**46.** $(15 + x)(200 - 5x) = 3570$   6, 19

## GROUP ACTIVITY        The Mathematics of Home Entertainment

The table shows consumer spending on video home systems (VHS) and universal media disks (UMD), digital video discs (DVD), blue-ray discs (BD) and high definition TVs (HDTV), and digital movies. (Source: http://www.degonline.org)

**U.S. Consumer Spending for Home Entertainment Rentals and Purchases (in billions)**

| Year | VHS/UMD | DVD | BD/HDTV | Digital | Total |
|------|---------|------|---------|---------|-------|
| 1999 | $12.2 | $ 1.1 | $0.0 | $0.6 | $13.9 |
| 2000 | $11.4 | $ 2.4 | $0.0 | $0.7 | $14.5 |
| 2001 | $10.9 | $ 5.3 | $0.0 | $0.7 | $16.9 |
| 2002 | $ 9.6 | $ 8.6 | $0.0 | $0.7 | $19.0 |
| 2003 | $ 6.9 | $13.1 | $0.0 | $0.7 | $20.7 |
| 2004 | $ 4.4 | $16.7 | $0.0 | $0.7 | $21.8 |
| 2005 | $ 2.1 | $18.9 | $0.0 | $0.8 | $21.7 |
| 2006 | $ 0.4 | $20.2 | $0.0 | $1.0 | $21.6 |
| 2007 | $ 0.1 | $19.7 | $0.3 | $1.3 | $21.4 |
| 2008 | $ 0.1 | $18.4 | $0.9 | $1.6 | $21.0 |
| 2009 | $ 0.0 | $15.8 | $1.5 | $2.1 | $19.4 |
| 2010 | $ 0.0 | $14.0 | $2.3 | $2.5 | $18.8 |

The following graph shows a scatter plot of the data presented in the table for three of the categories.

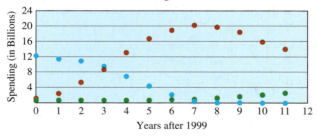

1. Determine which data each scatter plot represents. Describe the trend shown by each scatter plot.

2. Connect the points in a smooth curve for each scatter plot to form three graphs. Approximate the three points of intersection for the graphs and interpret the results.

3. A model for the red graph is
   $$y = -0.0411x^3 + 0.3327x^2 + 2.4591x + 0.2187.$$
   A model for the blue graph is
   $$y = 0.0305x^3 - 0.4126x^2 - 0.212x + 12.334$$
   and a model for the green graph is
   $$y = 0.0267x^2 - 0.1408x + 0.7643.$$
   Write equations in standard form that can be used to determine the point of intersection for the following graphs. Do not solve the equations.
   a. the blue and green graphs
   b. the red and green graphs
   c. the blue and red graphs

4. The equations in step 3 are examples of polynomial equations that cannot be solved by factoring. So, use a graphing calculator to approximate the point of intersection for the green and red graphs and interpret the results.

# Factoring Polynomials and Polynomial Equations

> **What's the big idea?** The ability to factor polynomials provides us with the foundation for solving polynomial equations. This skill is utilized with a basic property of real numbers; that is, if a product is equal to zero, then one of the factors must be zero. Once we know how to solve polynomial equations, we can then solve applications that are modeled by this type of equation.

## The Tools

Listed below are the key terms, skills, formulas, and properties you should know for this chapter.

The page reference is provided if you need additional help with the given topic. The Study Tips will assist in your preparation for an exam.

## Study Tips

1. Learn all of the terms, formulas, and properties. Make flash cards and have someone quiz you.
2. Rework problems from the exercises and also the ones you worked in class. Work additional problems from the review exercises.
3. Review the summaries of key concepts.
4. Work the chapter test.
5. Be sure to review the online resources for additional study materials.

## Terms

Consecutive integers   497
Factor   450
Factoring   450
Greatest common factor   450

Perfect square trinomial   473
Polynomial equation   489
Prime polynomial   455
Quadratic equation   484

Revenue and profit   499
Standard form   484
Sum of squares   478
Trial-and-error method   468

## Formulas and Properties

- Area formulas   495
- Difference of two cubes property   479

- Difference of two squares property   477
- Distributive property   453

- Pythagorean theorem   500
- Sum of two cubes property   479
- Zero products property   485

## CHAPTER 6 / SUMMARY

✎ **How well do you know this chapter? Complete the following questions to find out. Take a look back at the section if you need help.**

### SECTION 6.1 Greatest Common Factor and Grouping

1. To factor a polynomial means to write the polynomial as a(n) _product_ .

2. The _greatest_ _common_ _factor_ of a set of integers is the largest factor that is common to each integer.

3. The GCF of a set of variable terms is the common variable raised to the _smallest_ exponent of the terms.

4. The GCF of a set of terms is the _product_ of the GCF of the _coefficients_ and the GCF of the _variable_ _expressions_ .

5. To factor using the GCF, we apply the _distributive_ _property_ in reverse. That is, $ab + ac = $ _$a(b + c)$_ .

6. The _grouping_ method applies to factoring polynomials with four or more terms.

7. All factoring can be checked by _multiplying_ .

### SECTION 6.2 Factoring Trinomials

8. To factor a trinomial of the form $x^2 + bx + c$, we must find the factors of _c_ whose sum is _b_ .

9. If the last term of the trinomial is positive, the signs in the binomial factors are the _same_ .

10. If the last term of the trinomial is negative, the signs in the binomial factors are _different_ .

11. Before factoring a trinomial, we should factor out any _common_ _factor_ , if necessary.

## SECTION 6.3 More on Factoring Trinomials

**12.** There are two methods for factoring trinomials of the form $ax^2 + bx + c$, $a \neq 1$. We can use <u>trial</u> and <u>error</u> or the method of <u>grouping</u>.

**13.** A <u>perfect</u> <u>square</u> trinomial factors as a binomial squared.

## SECTION 6.4 Factoring Binomials

**14.** A binomial of the form $a^2 - b^2$ can be factored as <u>$(a - b)(a + b)$</u>. An example is <u>Answers vary; $x^2 - 16$</u>.

**15.** A binomial of the form $a^3 - b^3$ can be factored as <u>$(a - b)(a^2 + ab + b^2)$</u>. An example is <u>Answers vary; $x^3 - 8$</u>.

**16.** A binomial of the form $a^3 + b^3$ can be factored as <u>$(a + b)(a^2 - ab + b^2)$</u>. An example is <u>Answers vary; $x^3 + 27$</u>.

**17.** The sum of squares <u>does</u> <u>not</u> <u>factor</u> over the real numbers. An example is <u>Answers vary; $x^2 + 4$</u>.

## SECTION 6.5 Solving Quadratic Equations and Other Polynomial Equations by Factoring

**18.** The zero products property states that if a product is zero, then one of the <u>factors</u> must be <u>zero</u>.

**19.** A quadratic equation is an equation of the form <u>$ax^2 + bx + c = 0$</u>, where $a \neq 0$.

**20.** To solve a quadratic equation by factoring, write the equation in <u>standard</u> <u>form</u>, <u>factor</u> the polynomial, set each <u>factor</u> equal to <u>zero</u>, and <u>solve</u>.

**21.** Polynomial equations are equations that involve <u>polynomials</u>.

**22.** A polynomial equation with degree 3 is called a(n) <u>cubic</u> equation. A polynomial equation with degree 4 is called a(n) <u>quartic</u> equation.

**23.** The degree of an equation determines the maximum number of <u>solutions</u> of the equation.

## SECTION 6.6 Applications of Quadratic Equations

**24.** Quadratic equations can be used to solve problems involving <u>area</u> of rectangles and triangles.

**25.** Consecutive integers can be represented as <u>$x, x + 1, x + 2, \ldots$</u>. Consecutive odd integers can be represented as <u>$x, x + 2, x + 4, \ldots$</u>. Consecutive even integers can be represented as <u>$x, x + 2, x + 4, \ldots$</u>.

**26.** <u>Revenue</u> is calculated by multiplying the price of an item by the number of items sold and <u>profit</u> is the difference of revenue and cost.

**27.** The Pythagorean theorem is <u>$a^2 + b^2 = c^2$</u>, where $a$ and $b$ represent the <u>legs</u> and $c$ represents the <u>hypotenuse</u>.

# CHAPTER 6 / REVIEW EXERCISES

## SECTION 6.1

**Find the GCF of the terms in each set.** (*See Objectives 1 and 2.*)

**1.** $12x^3$ and $28x$   $4x$

**2.** $6a^2$ and $15$   $3$

**3.** $-20y^2$, $36y^4$, and $28y$   $4y$

**4.** $-44ab^3$ and $33a^2b^2$   $11ab^2$

**5.** $112r^4s^2$ and $84r^3$   $28r^3$

**6.** $10xy$, $16xy^2$, and $18x^3y$   $2xy$

**Factor each polynomial.** (*See Objectives 3 and 4.*)

**7.** $2x^2 - 6x + 5x - 15$   $(2x + 5)(x - 3)$

**8.** $3x^2 + 10x + 3x + 10$   $(x + 1)(3x + 10)$

**9.** $7x^2 + 28x - 3x - 12$   $(7x - 3)(x + 4)$

**10.** $12b^3 - 60b^2$   $12b^2(b - 5)$

**11.** $4c^2(3c - 2d) - 5d(3c - 2d)$   $(4c^2 - 5d)(3c - 2d)$

**12.** $7ac - 21ad + 2bc - 6bd$   $(7a + 2b)(c - 3d)$

**13.** $-10p^2 - 35p$   $-5p(2p + 7)$

**14.** $-3a^2 + 15a - 6$   $-3(a^2 - 5a + 2)$

**15.** $3x^2(x + 1) - (x - 1)$   prime

**16.** $6x^2 + 21xy - 10xy - 35y^2$   $(3x - 5y)(2x + 7y)$

**17.** The surface area of a cylindrical can is given as $2\pi r^2 + 2\pi rh$, where $r$ is the radius of the base and $h$ is the height of the cylinder.   $2\pi r(r + h)$

**18.** The volume of a box is given by $2x^2 + 24x$ in.$^3$.   $2x(x + 12)$

## SECTION 6.2

**Factor each trinomial.** (*See Objectives 1 and 2.*)

**19.** $x^2 + 3x - 28$   $(x - 4)(x + 7)$

**20.** $6x^3 + 12x^2 - 90x$   $6x(x - 3)(x + 5)$

**21.** $4y^2 + y + 3$   prime

**22.** $4a^3b + 44a^2b^2 + 96ab^3$   $4ab(a + 3b)(a + 8b)$

**23.** $-x^4 + 9x^3 - 20x^2$   $-x^2(x - 4)(x - 5)$

**24.** $x^2 + 16xy + 60y^2$   $(x + 6y)(x + 10y)$

**25.** $-5x^2y + 10xy - 30y$   $-5y(x^2 - 2x + 6)$

**26.** $a^3 + 5a^2b - 84ab^2$   $a(a + 12b)(a - 7b)$

## SECTION 6.3

**Use either trial and error or grouping to factor each trinomial completely.** (*See Objectives 1–3.*)

**27.** $4a^2 - 28a + 49$   $(2a - 7)^2$

**28.** $3x^2 - 13x - 10$   $(3x + 2)(x - 5)$

**29.** $-3y^2 - 2y + 21$   $-(3y - 7)(y + 3)$

**30.** $8t^2 + 2t - 45$   $(2t + 5)(4t - 9)$

**31.** $16x^2 - 12x + 9$   prime

**32.** $9a^2 + 30ab + 25b^2$   $(3a + 5b)^2$

**33.** $-14h^2 + 28h - 7$   $-7(2h^2 - 4h + 1)$

**34.** $4x^2 + x + 1$   prime

**35.** $26x^2 - 6xy - 20y^2$   $2(13x + 10y)(x - y)$

**36.** $18x^2y^4 - 2z^6$   $2(3xy^2 + z^3)(3xy^2 - z^3)$

**37.** $-108a^2 - 9ab + 42b^2$   $-3(12a - 7b)(3a + 2b)$

**38.** $p^3q^6 + 8r^9$   $(pq^2 + 2r^3)(p^2q^4 - 2pq^2r^3 + 4r^6)$

**39.** $-12h^4 + 12h^2 - 3$   $-3(2h^2 - 1)^2$

**40.** $25u^2 - 10uv + v^2$   $(5u - v)^2$

**41.** The profit, in thousands of dollars, of a video game company is given by the expression $-60x^2 + 456x + 192$, where $x$ is the number of units sold, in hundreds.   $-12(5x + 2)(x - 8)$

**42.** The profit, in millions of dollars, of a footwear company is given by the expression $-150x^2 + 1150x + 3500$, where $x$ is the number of units sold, in thousands.   $-50(3x + 7)(x - 10)$

## SECTION 6.4

**Factor each binomial completely. (*See Objectives 1–3.*)**

**43.** $9y^4 - 72y$
$9y(y-2)(y^2+2y+4)$

**44.** $8a^2 - 50b^2$
$2(2a-5b)(2a+5b)$

**45.** $64x^3y + 125y$
$y(4x+5)(16x^2-20x+25)$

**46.** $32r^4s^4 - 2r^4$
$2(4r^2s^2+t^2)(2rs-t)(2rs+t)$

**47.** $x^6y^6 + z^6$

**48.** $a^4 + 100b^4$   prime

**49.** $a^3 - 1000b^3c^3$

**50.** $343x^3 - 8$

**51.** $625y^4 - 16z^4$
$(25y^2+4z^2)(5y+2z)(5y-2z)$

**52.** $a^{15}b^{12} - 216c^6$

**53.** $\dfrac{1}{8}a^3 + b^3c^3$

**54.** $x^4y^4 - \dfrac{1}{16}z^4$

**55.** $x^3 + 27y^5$   prime

**56.** $0.36r^6 - 0.49s^8$
$(0.6r^3-0.7s^4)(0.6r^3+0.7s^4)$

## SECTION 6.5

**Solve each equation. (*See Objectives 1 and 2.*)**

**57.** $5x^2 - 8x = 4$   $\left\{-\dfrac{2}{5}, 2\right\}$

**58.** $(3x+4)^2 = 0$   $\left\{-\dfrac{4}{3}\right\}$

**59.** $(x+6)(x+5) = 30$   $\{-11, 0\}$

**60.** $-2x^2 + 11x - 14 = 0$

**61.** $6x^3 + 15x^2 - 9x = 0$

**62.** $9x^2 = 25$   $\left\{-\dfrac{5}{3}, \dfrac{5}{3}\right\}$

**63.** $7(x+4)(x-12) = 0$   $\{-4, 12\}$

**64.** $(4x-1)(x^2+6x-16) = 0$   $\left\{-8, \dfrac{1}{4}, 2\right\}$

**65.** $75x^3 - 3x = 0$   $\left\{-\dfrac{1}{5}, 0, \dfrac{1}{5}\right\}$

**66.** $x^4 - 5x^2 + 4 = 0$
$\{-2, -1, 1, 2\}$

## SECTION 6.6

**Solve each problem. (*See Objectives 1–5.*)**

**67.** George wants to section off a rectangular portion of his yard for a vegetable garden. He wants the garden to be 7 ft longer than it is wide. If the area of the garden will be 144 ft², find the dimensions of the garden.   9 ft by 16 ft

**68.** Shutter wants to pour concrete in a uniform width around his backyard pool that measures 35 ft by 40 ft. If he has enough concrete to cover an area of 850 ft², what should be the width of the concrete walkway that is created?   5 ft

**69.** The product of two positive consecutive integers is 240. Find the integers.   15, 16

**70.** The product of two positive consecutive even integers is equal to 6 more than 3 times the sum of the two integers. Find the integers.   6, 8

**Find the lengths of the sides of each right triangle.**

**71.**

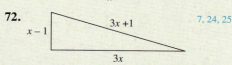

8, 15, and 17

**72.**

7, 24, 25

**73.** A 25-ft ladder leans against the side of a house. If the distance from the bottom of the ladder to the house is 5 ft shorter than the distance from the ground to the top of the ladder, then what is the distance from the ground to the top of the ladder?   20 ft

**74.** Anna and Bob leave a restaurant in two separate cars at the same time. Anna travels north and Bob travels east. When they are 26 mi apart, Anna has traveled 14 mi less than Bob. How many miles have Anna and Bob each traveled?   Anna traveled 10 mi and Bob traveled 24 mi.

**75.** A sky diver is dropped from a plane at 2500 ft. His distance from the ground can be approximated by $s = -16t^2 + 2500$ (in feet), where $t$ is the time (in seconds) after the diver is dropped. How long will it take the diver to land on the ground if his parachute doesn't open?   12.5 sec

**76.** Wing plays the piano for weddings. The revenue generated from performing at $x$ weddings is $R = x^2 + 160x$ (in dollars). At how many weddings must Wing perform to receive $2625 in revenue?   15

---

# CHAPTER 6 TEST  /  FACTORING POLYNOMIALS AND POLYNOMIAL EQUATIONS

**1.** The greatest common factor of $4x^5 - 8x^4 + 12x^3$ is
  **a.** 4
  **b.** $4x$
  **(c.)** $4x^3$
  **d.** $4x^5$

**2.** The complete factorization of $x^2 + 2x - 3$ is
  **a.** $(x-1)(x-3)$
  **(b.)** $(x-1)(x+3)$
  **c.** $(x+1)(x-3)$
  **d.** $(x+1)(x+3)$

**3.** One of the prime factors of $6t^2 - 19t - 20$ is
  **a.** $t+5$
  **b.** $2t+5$
  **(c.)** $6t+5$
  **d.** $t+1$

**4.** The prime factorization of $16x^2 + 48x + 36$ is
  **a.** $(8x+12)(2x+3)$
  **(b.)** $4(2x+3)^2$
  **c.** $4(2x-3)^2$
  **d.** $(4x+9)^2$

**5.** The prime factorization of $16x^2 - 25$ is
  **a.** $(8x-5)(2x+5)$
  **b.** $(8x-5)(8x+5)$
  **c.** $(16x-25)(16x+25)$
  **(d.)** $(4x-5)(4x+5)$

**6.** One of the prime factors of $8x^3 + 27$ is
  **(a.)** $4x^2 - 6x + 9$
  **b.** $4x^2 - 6x - 9$
  **c.** $4x^2 + 6x + 9$
  **d.** $2x^2 - 6x + 3$

**7.** The prime factorization of $ab^2 + ac + b^2d + cd$ is
  **a.** $(a+c)(d+b^2)$
  **(b.)** $(a+d)(b^2+c)$
  **c.** $a(b^2+c) + (b^2+c)$
  **d.** $(a+d)^2(b^2+c)$

**8.** The solution set for $x^2 = 16$ is
  **a.** $\{4\}$
  **b.** $\{-4\}$
  **c.** $\{16\}$
  **(d.)** $\{-4, 4\}$

**9.** The solution set for $6x^2 + x = 2$ is
  **a.** $\left\{-\dfrac{3}{2}, 2\right\}$
  **b.** $\left\{-\dfrac{1}{2}, \dfrac{2}{3}\right\}$
  **c.** $\left\{-\dfrac{3}{2}, \dfrac{1}{2}\right\}$
  **(d.)** $\left\{-\dfrac{2}{3}, \dfrac{1}{2}\right\}$

**Factor each polynomial completely.**

**10.** $5x^2 + 10x$   $5x(x + 2)$

**11.** $20 - 5y - 4p + yp$   $(4 - y)(5 - p)$

**12.** $r^3 + r^2 - 4r - 4$   $(r + 1)(r - 2)(r + 2)$

**13.** $3a^2 - 3a - 18$   $3(a - 3)(a + 2)$

**14.** $9a^2 + 6a - 8$   $(3a + 4)(3a - 2)$

**15.** $10a^2 - 35a - 20$   $5(2a + 1)(a - 4)$

**16.** $9x^2 + 30x + 25$   $(3x + 5)^2$

**17.** $81p^2 - 49$   $(9p - 7)(9p + 7)$

**18.** $4a^2 + 25$   prime

**19.** $125x^3 - 27y^3$   $(5x - 3y)(25x^2 + 15xy + 9y^2)$

**20.** $81t^4 - 16$   $(3t - 2)(3t + 2)(9t^2 + 4)$

**21.** $64y^6 + 1$   $(4y^2 + 1)(16y^4 - 4y^2 + 1)$

**Solve each equation.**

**22.** $(x - 5)(x + 6) = 0$   $\{-6, 5\}$

**23.** $4x^2 = 36$   $\{-3, 3\}$

**24.** $a^2 + a = 56$   $\{-8, 7\}$

**25.** $6z^2 = 5z + 6$   $\left\{-\dfrac{2}{3}, \dfrac{3}{2}\right\}$

**26.** $(y + 1)(y - 3) = 5$   $\{-2, 4\}$

**27.** $(x - 4)(x + 3)(x + 5) = 0$   $\{-5, -3, 4\}$

**28.** $x^4 - 29x^2 + 100 = 0$   $\{-5, -2, 2, 5\}$

**Solve each problem.**

**29.** Find the lengths of the sides of a right triangle if the legs are consecutive even integers and the hypotenuse is 10 units.   The legs are 6 and 8 units.

**30.** The Grand Canyon Skywalk is a glass walkway that stands about 3600 ft above the floor of the Grand Canyon. The distance an object would be above the floor of the canyon if dropped from the skywalk is given by $s = -16t^2 + 3600$ ft after $t$ sec. How long will it take for a rock dropped from the skywalk to reach the floor of the canyon? (Source: http://www.grandcanyonskywalk.com /skywalk.html)   It will take the rock 15 sec to reach the canyon floor.

# CUMULATIVE REVIEW EXERCISES / CHAPTERS 1–6

**1.** Perform the indicated operation and simplify. (*Sections 1.3–1.6, Objectives 1 and 2*)

a. $-(-3)^4 + (-2)^5 - 8 \cdot 9$   $-185$

b. $15.1 - 8.6 + (-4.3) - (-7.8)$   $10$

c. $(-7.3)(0)(-11.2)$   $0$

**2.** Simplify each algebraic expression. (*Section 1.8, Objectives 3 and 4*)

a. $-12(3a - 4) + 5(-a + 9)$   $-41a + 93$

b. $14.2 - 1.2(2b - 2.5)$   $17.2 - 2.4b$

c. $6x^2 - (-5x^2 + 16x - 5) + 12x - 11$   $11x^2 - 4x - 6$

**3.** Solve each equation. If the equation is a contradiction, write the solution as ∅. If the equation is an identity, write the solution as $\mathbb{R}$. (*Section 2.2, Objectives 2 and 3; Section 2.3, Objectives 1 and 2; Section 2.4, Objectives 1–4*)

a. $-(2x - 9) - 21 = 5(x - 1)$   $\{-1\}$

b. $\dfrac{2}{3}(2x - 5) + \dfrac{5}{6} = x$   $\left\{\dfrac{15}{2}\right\}$

c. $-4.5x + 20.5 = 3(-1.5x + 4.3)$   ∅

**4.** Ellen invests $5400 in two accounts. She invests $4100 in a certificate of deposit that pays 1.6% annual interest and the remaining amount in a money market fund that pays 1.9% annual interest. How much interest does she earn from the two accounts in 1 yr? (*Section 2.6, Objective 2*)   $90.30

**5.** Solve each inequality. For each inequality, graph the solution set, write the solution set in interval notation and in set-builder notation. (*Section 2.7, Objectives 1–4*)

a. $9x + 3(x - 10) < 8x - 10$

b. $-2.4 < 0.6a + 1.8 < 3.9$

**6.** Graph each line. Label two points on the line. (*Section 3.2, Objective 2*)

a. $m = \dfrac{2}{5}$, passes through $(-10, 6)$

b. $y = 2x + 7$

**7.** The U.S. Bureau of Labor Statistics conducted a Consumer Expenditure Survey to find out how many U.S. households own desktops and laptops. The percent of computer ownership from 2005 to 2008 can be modeled by the equation $y = 2.45x + 68.6$, where $x$ is the number of years after 2005. (Source: http://www.bls.gov) (*Section 3.2, Objective 5*)

a. Find the $y$-intercept and interpret its meaning in the context of the problem.

b. Based on the model, what is the percent of computer ownership in 2011?   The percent of computer ownership is predicted to be 83.3 in 2011.

c. Based on the model, what is the percent of computer ownership in 2013?   The percent of computer ownership is predicted to be 88.2 in 2013.

d. Graph the equation.

**8.** Write the equation of each line. Express each answer in slope-intercept form and in standard form. (*Section 3.5, Objectives 1–4*)

a. $(10, -5)$, perpendicular to $x - 2y = 9$

b. $(2, -1)$, parallel to $6x - y = 2$   $y = 6x - 13, 6x - y = 13$

c. $m = 0$, passes through $(3, -4)$   $y = -4$

**9.** Find the requested information. (*Section 3.6, Objective 4*)

a. Find $f(0)$ if $f(x) = 5x^2 - x + 3$.   $f(0) = 3$

b. Find $g(-3)$ if $g(x) = -x^2 + 2x$.   $g(-3) = -15$

c. Find $f(3)$ and find all $x$ such that $f(x) = -1$ if $f(x)$ is given by the graph.   $f(3) = 2; x = -2, x = 6$

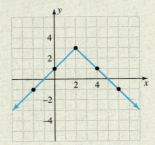

10. Solve each system of equations graphically. Then state if the system is consistent with dependent equations, consistent with independent equations, or inconsistent. (*Section 4.1, Objectives 3 and 4*)

    a. $\begin{cases} y = x - 4 \\ y = -3x \end{cases}$
    b. $\begin{cases} x - y = 2 \\ y = -3 \end{cases}$

11. Solve each system of linear equations using substitution. (*Section 4.2, Objective 1*)

    a. $\begin{cases} 3x - 2y = -8 \\ x + 5y = 3 \end{cases}$ $(-2, 1)$
    b. $\begin{cases} 4x - y = 15 \\ 2x + 7y = -45 \end{cases}$ $(2, -7)$

12. Solve the systems using the elimination method. (*Section 4.3, Objective 1*)

    a. $\begin{cases} 3x - 5y = -13 \\ -2x + 9y = 20 \end{cases}$ $(-1, 2)$
    b. $\begin{cases} \dfrac{2}{3}x - \dfrac{1}{4}y = 2 \\ \dfrac{5}{6}x + \dfrac{1}{2}y = \dfrac{5}{2} \end{cases}$ $(3, 0)$

13. Graph each linear inequality in two variables. (*Section 4.5, Objective 2*)

    a. $x + 3y \le 12$
    b. $x + 4y \ge 0$

14. Solve each system of linear inequalities. (*Section 4.6, Objective 1*)

    a. $\begin{cases} y \ge -4x \\ y \le 3x - 7 \end{cases}$
    b. $\begin{cases} x - 2y \ge 0 \\ y \le -2 \end{cases}$

15. Use the appropriate exponent rules to simplify each expression. (*Section 5.1, Objectives 1–4*)

    a. $(5x^2y^4)^3$ $125x^6y^{12}$
    b. $\left(-\dfrac{2x}{3y^3}\right)^5$ $-\dfrac{32x^5}{243y^{15}}$
    c. $(2^{-1}ab^6)^5$ $\dfrac{a^5b^{30}}{32}$

16. Use the appropriate exponent rules to simplify each expression. Express all answers with positive exponents. (*Section 5.2, Objectives 1–4*)

    a. $(3^{-1}x^5)^{-3}$ $\dfrac{27}{x^{15}}$
    b. $\left(\dfrac{5x^2}{y^{-4}}\right)^{-3}$ $\dfrac{1}{125x^6y^{12}}$
    c. $\left(-\dfrac{2c^{-3}}{d^4}\right)^{-3}$ $-\dfrac{c^9d^{12}}{8}$

17. The formula $P = P_0 \cdot 2^{-t/m}$ gives the amount of substance, $P$, present at time $t$, where $P_0$ is the original amount of the substance and $m$ is its half-life. Suppose the half-life of a substance is 12 weeks. If the original amount of the substance is 10 g, how much substance will be left after (*Section 5.2, Objective 5*)

    a. 12 weeks? 5 g
    b. 36 weeks? 1.25 g

18. Use exponent rules to simplify each expression. Write answer in scientific notation. (*Section 5.3, Objective 3*)

    a. $(4.2 \times 10^6)(3.5 \times 10^{-20})$ $1.47 \times 10^{-13}$
    b. $\dfrac{7.925 \times 10^{-14}}{6.34 \times 10^{-5}}$ $1.25 \times 10^{-9}$

19. The Consumer Assistance to Recycle and Save Act (CARS), also known as the "Cash for Clunkers" Program, was signed into law June 24, 2009, to provide incentives to motorists to trade in less fuel-efficient vehicles for more fuel-efficient ones. When the program ended, there were about 700,000 car sales with rebate applications totaling $2.877 billion. If the rebate was evenly distributed among the car buyers, how much would each buyer receive? Write answer in scientific notation. (1 billion $= 10^9$) (Source: http://www.cars.gov /carsreport/index.html) (*Section 5.3, Objective 3*) $4.11 \times 10^3$

20. A company's total cost to make $x$ items is $24x^3 - 15x^2 + 1900$ dollars. How much does it cost the company to make 18 items? (*Section 5.4, Objectives 3 and 6*) $137,008

21. The height, in feet, of a ball projected upward from the top of a 240-ft building with an initial velocity of 128 ft/sec is given by $-16t^2 + 128t + 240$, where $t$ is the number of seconds after the ball is thrown. Find the height of the ball. (*Section 5.4, Objectives 3 and 6*)

    a. 3 sec after it is thrown 480 ft
    b. 4 sec after it is thrown 496 ft

22. Perform each operation. Write answers in standard form. (*Section 5.4, Objective 5*)

    a. $(21 - 15x^2 - 9x) + (-16 + 10x^2 + 28x)$
    b. $(32x^3 + 19x^2 - 16x) - (20x^3 + 7x^2 - 22x)$
    c. Subtract $(4x^2 - 6xy + 9y^2)$ from $(-3x^2 + 4xy - 13y^2)$.
    d. $(5x^2 + 11xy - 3y^2) - (-6x^2 + 14xy + y^2) + (-12x^2 - 18xy - 6y^2)$ $-x^2 - 21xy - 10y^2$

23. Suppose the profit generated by selling $x$ items is $-0.6x^2 + 98x + 213$ dollars and the total cost is $28x + 460$ dollars. Find an expression that represents the company's revenue. Write answer in standard form. (*Section 5.4, Objectives 4 and 6*) $-0.6x^2 + 126x + 673$ dollars

24. Simplify each product. (*Section 5.5, Objectives 1–3*)

    a. $6x^2y(2xy^2 - 8)$
    b. $(13 - 4x)(3x + 1)$
    c. $(3a - 2b)(9a^2 + 6ab + 4b^2)$ $27a^3 - 8b^3$
    d. $\left(\dfrac{1}{5}x - 3\right)\left(\dfrac{3}{5}x + 2\right)$ $\dfrac{3}{25}x^2 - \dfrac{7}{5}x - 6$
    e. $(x^2 + 6x - 5)(3x^2 - 4x + 1)$ $3x^4 + 14x^3 - 38x^2 + 26x - 5$

25. Perform the indicated operation. Write answers in standard form. (*Section 5.6, Objectives 1–4*)

    a. $(5x + 3)(4x - 1)$
    b. $(5b + 12)^3$
    c. $(p - 3)^3$
    d. $(15x - y)(15x + y)$ $225x^2 - y^2$
    e. $(2r + 3s)(4r^2 - 6rs + 9s^2)$ $8r^3 + 27s^3$

26. Perform the indicated operation. (*Section 5.7, Objectives 1 and 2*)

    a. Divide $15x^3 + 6x^2 - 24x + 7$ by $3x$. $5x^2 + 2x - 8 + \dfrac{7}{3x}$
    b. $\dfrac{5x^3 - 18x^2 + 23x - 12}{x - 4}$ $5x^2 + 2x + 31 + \dfrac{112}{x - 4}$

27. Factor each trinomial completely. (*Section 6.2, Objectives 1 and 2; Section 6.3, Objectives 1 and 2*)

    a. $x^3 + 3x^2 - 108x$
    b. $10y^3 - 25y^2 - 60y$
    c. $3x^2 + 4x + 8$ prime
    d. $-7x^4 + 91x^2 - 252$
    e. $-9x^3 + 36x^2 - 36x$ $-9x(x - 2)^2$

28. The volume of an open box with height $x$ inches is given by the expression $x^3 - 7x^2 - 60x$ in.$^3$. Factor this trinomial. (*Section 6.2, Objectives 1 and 2*) $x(x - 12)(x + 5)$

**29.** Factor each trinomial completely. (*Section 6.2, Objectives 1 and 2; Section 6.3, Objectives 1 and 2*)

**a.** $x^2 - 15x + 54$     **b.** $3y^3 - 18y^2 + 27y$   $3y(y-3)^2$

**c.** $-12x^3 - 54x^2 + 30x$    **d.** $a^2 - 16ab + 48b^2$

**e.** $2x^2y + 42xy + 216y$     **f.** $x^2 + 6xy + 7y^2$   prime
    $2y(x+12)(x+9)$

**30.** A company's profit can be represented by $-14q^2 + 350q + 8400$, where $q$ is the number of units sold in thousands. Factor this polynomial. (*Section 6.3, Objectives 1 and 2*)   $-14(q-40)(q+15)$

**31.** Use either trial and error or grouping to factor each trinomial completely. (*Section 6.3, Objectives 1 and 2*)

**a.** $5p^2 + 24p + 27$      **b.** $5pq^2 - 6pq - 27p$

**c.** $30x^2 - x - 7$       **d.** $16x^2 + 56xy + 49y^2$

**e.** $117a^2b^2 + 3ab - 12$    **f.** $3p^2 + 4pq + 7q^2$   prime
   $3(13ab-4)(3ab+1)$

**32.** The profit, in millions of dollars, of a video game company is given by $p = -60x^2 + 555x + 450$, where $x$ is the number of units sold in thousands. How many units need to be sold for the company to break even? (*See Section 6.6, Objective 3*)   10,000 units

**33.** Factor completely. (*Section 6.4, Objectives 1–3*)

**a.** $5x^3 - 45x$   $5x(x-3)(x+3)$ **b.** $50m^2 - 32n^2$

**c.** $54x^3 - 128y^3$      **d.** $2y^3 + 250z^3$

**e.** $r^2 + 64s^2$   prime     **f.** $81a^4 - 16$

**g.** $3c^6 - 24$         **h.** $x^4y^4 - 16z^4$

**i.** $\dfrac{1}{4}a^2 - \dfrac{9}{25}b^2$   $\left(\dfrac{1}{2}a - \dfrac{3}{5}b\right)\left(\dfrac{1}{2}a + \dfrac{3}{5}b\right)$

**34.** Solve each equation. (*Section 6.5, Objectives 1 and 2*)

**a.** $4x^2 - 5x = 6$       **b.** $5x^2 - 2 = 3x$

**c.** $4x^2 + 4x + 1 = 0$     **d.** $(x+2)(x+11) = 22$   $\{-13, 0\}$

**e.** $(2-x)(3+5x) = 0$   **f.** $-3x^2 + 17x - 20 = 0$

**g.** $x^4 - 10x^2 + 9 = 0$    **h.** $3x(x+2)(x-5) = 0$

**i.** $4x^2 + 25 = 0$       **j.** $48x^2 + 116x + 70 = 0$
   no real solution

**35.** Solve each problem. (*Section 6.6, Objectives 1–5*)

**a.** Patty wants to section off a rectangular portion of her yard for a flowerbed. She wants the flowerbed to be 3 ft longer than it is wide. If the area of the flowerbed will be 54 ft$^2$, find the dimensions of the flowerbed.   9 ft by 6 ft

**b.** The product of two positive consecutive integers is 156. Find the integers.   12 and 13

**c.** The product of two consecutive odd integers equals 12 more than 9 times the larger integer. Find the integers.   $-1$ and $-3$, 10 and 12

**d.** When an express bus company charges $40 per seat from New York City to Philadelphia, they can sell 90 tickets. For every $2 decrease in price, 5 more tickets are sold. If $x$ represents the number of price decreases, then $40 - 2x$ represents the ticket price and $90 + 5x$ represents the number of passengers. How many passengers are needed for the express bus company to earn $2400?   150 passengers

**e.** Find the lengths of the sides of the right triangle.

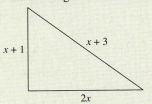

          4, 3, and 5

**f.** The profit, in millions, of a furniture company is given by $p = -40x^2 + 464x + 192$, where $x$ is the number of units sold in thousands. How many units need to be sold for the company to break even?   12,000 units

**g.** Kelly and Sue leave their dorm room at the same time. Kelly travels north and Sue travels west. When they are 30 mi apart, Kelly has traveled 6 mi more than the distance Sue has traveled. How far did Sue travel?   18 mi

**h.** The total cost of producing $x$ items is given by $C = 12x^2 - 540x + 6000$ dollars, where $x$ is the number of units sold. How many items can be produced at a total cost of $1800?   10 items or 35 items

**i.** The height of a ball thrown upward with an initial velocity of 128 ft/sec from a height of 144 ft above the ground can be represented by $s = -16t^2 + 128t + 144$, where $t$ is the number of seconds after the ball is thrown. How long will it take the ball to reach the ground?   9 sec

# Rational Expressions and Equations

## Learning Strategies

Albert Einstein is probably the epitome of an incredible learner. Could you ever be like him? What gave him the capacity to acquire all of his knowledge? While you and I are not Albert Einstein, there are things that we can do to improve our learning.

- Find out what kind of learner you are—visual, auditory, kinesthetic, and the like. The Internet offers many different learning styles inventories that you can complete to determine this information. One is also provided in Chapter S.
- As soon as class ends, take a few moments to write down a summary of the main points that were presented in class.
- It has been proven that most people forget up to 40% of new material within a few hours of being exposed to it. Therefore, you should find the time to review class notes within a couple hours of attending class.
- Review your notes frequently over the next few days so that the concepts get transferred to your long-term memory.
- Distribute your studying over several sessions rather than one long cram session. Studies have shown that students who distribute their practice over several sessions remember 67% more than those who learn by cramming. (Source: John J. Donovan and David J. Radosevich, "A Meta-Analytic Review of the Distribution of Practice Effect: Now You See It, Now You Don't," *Journal of Applied Psychology*, 84 [1999]: 795–805)

**Question For Thought:** What type of learner are you? How long after new material is presented do you review it?

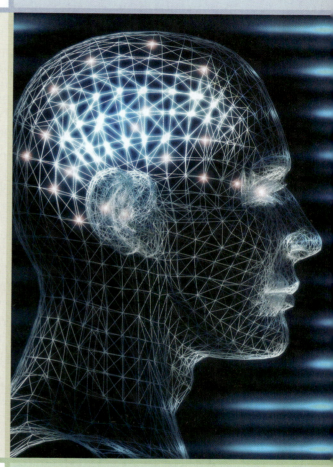

## Chapter Outline

## Coming Up...

In Section 7.7, we will learn how to find the speed of a flight to Denver if we know that an airplane flies 1573 mi from Denver to Philadelphia in the same amount of time another airplane flies 1090 mi from New York to Orlando and that the speed of the Denver flight was 220 mph faster than the New York flight.

"Imagination is more important than knowledge. For while knowledge defines all we currently know and understand, imagination points to all we might yet discover and create."

— Albert Einstein (Mathematician and scientist)

## SECTION 7.1 / Rational Expressions

**Chapters 5 and 6** introduced us to polynomials, operations on polynomials, and factoring polynomials. In this chapter, our focus will be on quotients of polynomials and equations involving them. It is critical that you have a good understanding of the prior chapters as the skills you learned will be used quite extensively.

### ▶ OBJECTIVES

As a result of completing this section, you will be able to

1. **Evaluate rational expressions.**
2. **Find the values that make a rational expression undefined.**
3. **Simplify rational expressions.**
4. **Write equivalent forms of rational expressions that include negative signs.**
5. **Troubleshoot common errors.**

The body mass index (BMI) provides an indication of whether a person is underweight, at a normal weight, overweight, or obese. It is calculated using the expression $\dfrac{703w}{h^2}$, where $w$ is a person's weight in pounds and $h$ is a person's height in inches. Find the BMI of a person who is 5 ft 8 in. tall and weighs 175 lb.

To answer this question, we must evaluate the expression $\dfrac{703w}{h^2}$ at the given values. In this section, we will learn how to work with expressions of this form.

### Evaluating Rational Expressions

**Objective 1 ▶**

Evaluate rational expressions.

In Chapter 1, we learned that a rational number is the quotient, or ratio, of two integers. Some examples of rational numbers are $\dfrac{2}{3}$, $-5$, and $0.4$. Recall that $-5$ can be written as $-\dfrac{5}{1}$ and $0.4$ can be written as $\dfrac{4}{10}$.

A *rational expression* is a quotient, or ratio, of two polynomials. Some examples of rational expressions are as follows:

$$\frac{3}{4} \qquad \frac{6xy^3}{12x^2y} \qquad \frac{y^2 + 5y - 1}{2y^3 - 3y^2 + y} \qquad -\frac{a+2}{a^3+8} \qquad b^2 - 4$$

Note that $b^2 - 4$ is a rational expression since it can be written as $\dfrac{b^2-4}{1}$. The fraction $\dfrac{\sqrt{x+2}}{x-1}$ is not a rational expression since the numerator is not a polynomial.

> **Definition:** If $N$ and $D$ are polynomials, then the expression $\dfrac{N}{D}$ is a **rational expression** as long as $D \neq 0$.

The definition of a rational expression has a restriction that the denominator cannot equal zero. Recall the following facts from Chapter 1.

- When the *numerator* of a fraction is zero, the value of the fraction is zero. For instance, $\dfrac{0}{5} = 0$.

- When the *denominator* of a fraction is zero, the value of the fraction is undefined. For example, $\dfrac{5}{0}$ is undefined.

**INSTRUCTOR NOTE:**
Remind students that a negative number divided by a positive number is negative.

Since division by zero is undefined, we must make the stated restriction for a rational expression to be defined.

Also, recall that when the numerator or denominator of a fraction is negative, we can make the entire fraction negative. For instance,

$$\frac{-2}{3} = -\frac{2}{3} \quad \text{and} \quad \frac{2}{-3} = -\frac{2}{3}$$

Throughout this course, we have evaluated algebraic expressions. Evaluating a rational expression is no different.

> **Procedure: Evaluating a Rational Expression**
>
> **Step 1:** Replace each occurrence of the variable with the given value.
> **Step 2:** Use the order of operations to simplify the numerator and denominator.
> **Step 3:** Simplify the answer to lowest terms.

**Objective 1 Examples**    **Evaluate each expression for the given values.**

**1a.** $\dfrac{x + 3}{x - 1}$ for $x = -3, 0, 1$     **1b.** $\dfrac{3x^2 - 1}{4x + 3}$ for $x = -2, 0, \dfrac{1}{2}$

**1c.** The body mass index (BMI) provides an indication of whether a person is underweight, at a normal weight, overweight, or obese. It is calculated using the expression $\dfrac{703w}{h^2}$, where $w$ is a person's

| BMI | Weight Status |
|---|---|
| Below 18.5 | Underweight |
| 18.5–24.9 | Normal |
| 25.0–29.9 | Overweight |
| 30.0 and Above | Obese |

weight in pounds and $h$ is a person's height in inches. Find the BMI of a person who is 5 ft 8 in. tall and weighs 175 lb. Based on the chart, what can you conclude about the person's weight?

**Solutions**   **1a.**

| $x$ | $\dfrac{x + 3}{x - 1}$ |
|---|---|
| $-3$ | $\dfrac{-3 + 3}{-3 - 1} = \dfrac{0}{-4} = 0$ |
| $0$ | $\dfrac{0 + 3}{0 - 1} = \dfrac{3}{-1} = -3$ |
| $1$ | $\dfrac{1 + 3}{1 - 1} = \dfrac{4}{0}$ undefined |

**1b.**

| $x$ | $\dfrac{3x^2 - 1}{4x + 3}$ |
|---|---|
| $-2$ | $\dfrac{3(-2)^2 - 1}{4(-2) + 3} = \dfrac{3(4) - 1}{-8 + 3} = \dfrac{12 - 1}{-5} = \dfrac{11}{-5} = -\dfrac{11}{5}$ |
| $0$ | $\dfrac{3(0)^2 - 1}{4(0) + 3} = \dfrac{3(0) - 1}{0 + 3} = \dfrac{0 - 1}{3} = \dfrac{-1}{3} = -\dfrac{1}{3}$ |
| $\dfrac{1}{2}$ | $\dfrac{3\left(\dfrac{1}{2}\right)^2 - 1}{4\left(\dfrac{1}{2}\right) + 3} = \dfrac{3\left(\dfrac{1}{4}\right) - 1}{2 + 3} = \dfrac{\dfrac{3}{4} - \dfrac{4}{4}}{5} = \dfrac{-\dfrac{1}{4}}{5} = -\dfrac{1}{4} \cdot \dfrac{1}{5} = -\dfrac{1}{20}$ |

**1c.** The given height is 5 ft 8 in., or 68 in., and the weight is 175 lb.

$$\frac{703w}{h^2} = \frac{703(175)}{(68)^2} \qquad \text{Replace } w \text{ with 175 and } h \text{ with 68.}$$

$$= \frac{123{,}025}{4624} \qquad \text{Simplify the numerator and denominator.}$$

$$\approx 26.61 \qquad \text{Divide.}$$

So, a person who is 5 ft 8 in. tall and weighs 175 lb has a BMI of 26.61. Based on the chart, this person is considered overweight.

✔ **Student Check 1** Evaluate each rational expression for the given values.

**a.** $\dfrac{x-5}{x+4}$ for $x = -4, 0, 5$      **b.** $\dfrac{2x+3}{x^2+1}$ for $x = -2, 0, 1$

**c.** A fitness club charges a one-time membership fee of $120 when a new member signs a contract. A new member is charged a monthly fee of $48, and the one-time fee is divided equally among the monthly payments. If $x$ represents the number of months in a new member's contract, the expression $\dfrac{120 + 48x}{x}$ represents the monthly payment. What is the monthly payment if a new member signs a contract for 6 months? 12 months? 24 months?

## Values That Make a Rational Expression Undefined

**Objective 2** ▶

Find the values that make a rational expression undefined.

In Example 1a, we found that $\dfrac{x+3}{x-1}$ is undefined for $x = 1$. This is because $x = 1$ makes the denominator of the rational expression zero. When working with rational expressions, it is important to identify the values that make the expression undefined.

> **Definition:** A rational expression is **undefined** for values that make its denominator equal to 0.

> **Procedure: Finding the Values That Make a Rational Expression Undefined**
>
> **Step 1:** Set the denominator equal to 0.
> **Step 2:** Solve the resulting equation.

**Objective 2 Examples** Find the value(s) that make each rational expression undefined.

**2a.** $\dfrac{2x-7}{6x}$      **2b.** $\dfrac{y-2}{y^2-3y-10}$      **2c.** $\dfrac{a^2+a-6}{a^2+4}$

**Solutions** **2a.**

$$6x = 0 \qquad \text{Set the denominator equal to zero.}$$

$$\frac{6x}{6} = \frac{0}{6} \qquad \text{Divide each side by 6.}$$

$$x = 0 \qquad \text{Simplify.}$$

So, the expression $\dfrac{2x-7}{6x}$ is undefined for $x = 0$.

**2b.**
$$y^2 - 3y - 10 = 0 \qquad \text{Set the denominator equal to zero.}$$
$$(y - 5)(y + 2) = 0 \qquad \text{Factor.}$$
$$y - 5 = 0 \quad \text{or} \quad y + 2 = 0 \qquad \text{Set each factor equal to zero.}$$
$$y = 5 \qquad\qquad y = -2 \qquad \text{Solve each equation.}$$

So, the expression $\dfrac{y - 2}{y^2 - 3y - 10}$ is undefined for $y = 5$ and $y = -2$.

**INSTRUCTOR NOTE:**
Help students understand
that $a^2 \geq 0$ for all values of $a$.
So, $a^2 + 4 \geq 0$ for all values of $a$.

**2c.** To find the values that make the expression undefined, we must solve

$$a^2 + 4 = 0$$

Because $a^2 + 4$ is the sum of squares, we can't solve this equation by factoring. So, we will substitute values into $a^2 + 4$.

From the table, note that the value of $a^2 + 4$ is always greater than or equal to 4 and is never equal to 0. There are no values of $a$ that make the rational expression undefined.

| $a$ | $a^2 + 4$ |
|---|---|
| $-2$ | $(-2)^2 + 4 = 4 + 4 = 8$ |
| $-1$ | $(-1)^2 + 4 = 1 + 4 = 5$ |
| $0$ | $(0)^2 + 4 = 0 + 4 = 4$ |
| $1$ | $(1)^2 + 4 = 1 + 4 = 5$ |
| $2$ | $(2)^2 + 4 = 4 + 4 = 8$ |

So, the rational expression $\dfrac{a^2 + a - 6}{a^2 + 4}$ is defined for all values of $a$.

✔ **Student Check 2**   Find the value(s) that make each rational expression undefined.

**a.** $\dfrac{y + 1}{2y + 10}$
**b.** $\dfrac{h - 4}{h^2 - 4h + 3}$
**c.** $\dfrac{9x - 7}{2}$

## Simplifying Rational Expressions

**Objective 3** ▶

Simplify rational expressions.

In Chapter 1, we learned that a fraction is in **lowest terms** if the common factor of the numerator and denominator is 1. For example, $\dfrac{26}{39}$ is not in lowest terms because the common factors of 26 and 39 are 13 and 1. To simplify the fraction, we factor the numerator and denominator and divide out the common factor, 13.

$$\frac{26}{39} = \frac{2 \cdot 13}{3 \cdot 13} = \frac{2}{3} \cdot \frac{13}{13} = \frac{2}{3} \cdot 1 = \frac{2}{3}$$

This process can be extended to rational expressions.

> **Property: Fundamental Property of Rational Expressions**
> If $N$, $D$, and $C$ are polynomials such that $D$ and $C$ are not zero,
>
> $$\frac{N \cdot C}{D \cdot C} = \frac{N}{D} \cdot \frac{C}{C} = \frac{N}{D} \cdot 1 = \frac{N}{D}$$

The key to simplifying rational expressions is in recognizing quotients that are equivalent to 1 or $-1$.

- A quotient has a value of 1 when an expression is divided by itself. For example, assuming the denominator is not zero,

$$\frac{3x}{3x} = 1 \qquad \frac{y + 4}{y + 4} = 1 \qquad \frac{2a - 5}{2a - 5} = 1$$

- A quotient has a value of $-1$ when an expression and its opposite are divided. For example, assuming the denominator is not zero,

$$\frac{5}{-5} = -1 \qquad \frac{-2y}{2y} = -1 \qquad \frac{t-2}{2-t} = -1$$

> Recall: $2 - t = -t + 2$
> $\qquad\quad = -1(t-2)$
> $\qquad\quad = -(t-2)$

---

**Property: Dividing Opposites**

The expressions $x - y$ and $y - x$ are **opposites** of one another. As long as $x \neq y$,

$$\frac{x-y}{y-x} = -1$$

---

**Procedure: Simplifying a Rational Expression**

**Step 1:** Factor the numerator and denominator.
**Step 2:** Divide the numerator and denominator by the common factors.
**Step 3:** Simplify.

---

**Objective 3 Examples**    Simplify each rational expression. Assume that the denominators are not zero.

**3a.** $\dfrac{4x^2 + 12x}{7x + 21}$      **3b.** $\dfrac{x^2 - 4x - 12}{x^2 + 6x + 8}$      **3c.** $\dfrac{4y^2 - 25}{2y^2 + y - 10}$

**3d.** $\dfrac{a^3 - 27}{a^2 + ab - 3a - 3b}$      **3e.** $\dfrac{24x^3 - 54x}{24x^2 + 28x - 12}$      **3f.** $\dfrac{5t - 10}{4 - 2t}$

**Solutions**    **3a.** $\dfrac{4x^2 + 12x}{7x + 21} = \dfrac{4x(x+3)}{7(x+3)}$      Factor the numerator and denominator.

$$= \frac{4x}{7} \cdot \frac{x+3}{x+3} \qquad \text{Apply the fundamental property of rational expressions.}$$

$$= \frac{4x}{7} \cdot 1 \qquad \text{Divide out the common factor, } x + 3.$$

$$= \frac{4x}{7} \qquad \text{Simplify.}$$

We can also show the steps as

$$\frac{4x^2 + 12x}{7x + 21} = \frac{4x\cancel{(x+3)}}{7\cancel{(x+3)}} = \frac{4x}{7}$$

The numerator and denominator of the rational expression are divided by their common factor, $x + 3$. We write a "1" above and below this common factor to show that $\dfrac{x+3}{x+3} = 1$.

**3b.** $\dfrac{x^2 - 4x - 12}{x^2 + 6x + 8} = \dfrac{(x-6)\cancel{(x+2)}}{(x+4)\cancel{(x+2)}}$      Factor the numerator and denominator.

$$= \frac{x-6}{x+4} \qquad \text{Divide out the common factor, } x + 2.$$

**3c.**
$$\frac{4y^2 - 25}{2y^2 + y - 10} = \frac{(2y - 5)\overset{1}{(\cancel{2y + 5})}}{\overset{}{(\cancel{2y + 5})}(y - 2)}$$
Factor the numerator and denominator.

$$= \frac{2y - 5}{y - 2}$$
Divide out the common factor, $2y + 5$.

**3d.**
$$\frac{a^3 - 27}{a^2 + ab - 3a - 3b} = \frac{\overset{1}{(\cancel{a - 3})}(a^2 + 3a + 9)}{(a + b)\underset{1}{(\cancel{a - 3})}}$$
Factor the numerator and denominator.

$$= \frac{a^2 + 3a + 9}{a + b}$$
Divide out the common factor, $a - 3$.

**3e.**
$$\frac{24x^3 - 54x}{24x^2 + 28x - 12} = \frac{6x(4x^2 - 9)}{4(6x^2 + 7x - 3)}$$
Factor out the GCF in the numerator and denominator.

$$= \frac{\overset{3}{\cancel{6}}x\overset{1}{(\cancel{2x + 3})}(2x - 3)}{\underset{2}{\cancel{4}}(3x - 1)\underset{1}{(\cancel{2x + 3})}}$$
Factor the numerator and denominator further.

$$= \frac{3x(2x - 3)}{2(3x - 1)}$$
Divide out the common factors, $2x + 3$ and 2.

**3f.**
$$\frac{5t - 10}{4 - 2t} = \frac{5\overset{1}{(\cancel{t - 2})}}{2\underset{-1}{(\cancel{2 - t})}}$$
Factor the numerator and denominator.

$$= \frac{5}{2}(-1)$$
Apply the property of dividing opposites.

$$= -\frac{5}{2}$$
Simplify.

---

✓ **Student Check 3**   Simplify each rational expression. Assume that the denominators are not zero.

**a.** $\dfrac{4x - 12}{8x - 24}$     **b.** $\dfrac{a^2 - a - 20}{a^2 + a - 12}$     **c.** $\dfrac{4y^2 + 23y - 6}{y^2 - 36}$

**d.** $\dfrac{3x^2 - xy + 6x - 2y}{x^3 + 8}$     **e.** $\dfrac{32b^2 + 8b - 40}{10b^2 + 10b - 20}$     **f.** $\dfrac{3r^2 - 9r}{6 - 2r}$

---

## Negative Rational Expressions

**Objective 4** ▶

Write equivalent forms of rational expressions that include negative signs.

When working with rational expressions, it is helpful to recognize equivalent forms especially when negative signs are involved. When a rational expression is negative, we can write it in two equivalent forms.

> **Property: Rule of Negative Rational Expressions**
>
> If $N$ and $D$ are polynomials, and $D \neq 0$, then
>
> $$-\frac{N}{D} = \frac{-N}{D} = \frac{N}{-D}$$

**Objective 4 Examples**   Use the rule of negative rational expressions to write two equivalent forms of each rational expression.

**4a.** $-\dfrac{4x + 1}{x - 2}$     **4b.** $\dfrac{-3x + 5}{x - 1}$

**Solutions**

**4a.** $-\dfrac{4x+1}{x-2} = \dfrac{-(4x+1)}{x-2}$  Apply $-\dfrac{N}{D}=\dfrac{-N}{D}$.

$= \dfrac{-4x-1}{x-2}$  Distribute.

$-\dfrac{4x+1}{x-2} = \dfrac{4x+1}{-(x-2)}$  Apply $-\dfrac{N}{D}=\dfrac{N}{-D}$.

$= \dfrac{4x+1}{-x+2}$  Distribute.

$= \dfrac{4x+1}{2-x}$  Apply the commutative property.

So,

$$-\dfrac{4x+1}{x-2} = \dfrac{-4x-1}{x-2} = \dfrac{4x+1}{2-x}$$

**4b.** Because the first term in the numerator is negative, we can factor out $-1$ from the numerator.

$$\dfrac{-3x+5}{x-1} = \dfrac{-1(3x-5)}{x-1} = \dfrac{-(3x-5)}{x-1}$$

Now we can write the two equivalent expressions.

$\dfrac{-(3x-5)}{x-1} = -\dfrac{3x-5}{x-1}$  Apply $\dfrac{-N}{D}=-\dfrac{N}{D}$.

$\dfrac{-(3x-5)}{x-1} = \dfrac{3x-5}{-(x-1)}$  Apply $\dfrac{-N}{D}=\dfrac{N}{-D}$.

$= \dfrac{3x-5}{-x+1}$  Distribute.

$= \dfrac{3x-5}{1-x}$  Apply the commutative property.

So,

$$\dfrac{-3x+5}{x-1} = -\dfrac{3x-5}{x-1} = \dfrac{3x-5}{1-x}$$

**✓ Student Check 4** Use the rule of negative rational expressions to write two equivalent forms of each rational expression.

**a.** $-\dfrac{3x-1}{6x+1}$

**b.** $\dfrac{-x-12}{x+7}$

---

**Objective 5** ▶

Troubleshoot common errors.

## Troubleshooting Common Errors

Some common errors associated with rational expressions are shown next.

**Objective 5 Examples** A problem and an incorrect solution are given. Provide the correct solution and an explanation of the error.

**5a.** Find the value of $\dfrac{3x+2}{x-1}$ for $x=1$.

| Incorrect Solution | Correct Solution and Explanation |
|---|---|
| $\dfrac{3x+2}{x-1} = \dfrac{3(1)+2}{1-1}$ $= \dfrac{3+2}{0} = \dfrac{5}{0} = 0$ | The process was correct but division by zero is undefined. So, $\dfrac{5}{0}$ is undefined. It is not equal to zero. Therefore, $\dfrac{3x+2}{x-1}$ is undefined for $x=1$. |

**5b.** Find the values that make $\dfrac{x+1}{x^2-4}$ undefined.

| Incorrect Solution | Correct Solution and Explanation |
|---|---|
| The expression $\dfrac{x+1}{x^2-4}$ is undefined for $x=-1$ and $x=2$ since $x=-1$ makes the numerator zero and $x=2$ makes the denominator zero. | Rational expressions are undefined when the denominator is equal to zero, not the numerator. We must set the denominator equal to zero and solve. $$x^2-4=0$$ $$(x-2)(x+2)=0$$ $$x-2=0 \quad \text{or} \quad x+2=0$$ $$x=2 \qquad\qquad x=-2$$ So, it is undefined for $x=2$ or $x=-2$. |

**5c.** Simplify $\dfrac{x^2+x}{x}$.

| Incorrect Solution | Correct Solution and Explanation |
|---|---|
| $\dfrac{x^2+\cancel{x}}{\cancel{x}}=x^2+1$ | The error is that a common term was "cancelled out." We can only simplify fractions by dividing out common factors. So, we factor the numerator and divide out the common factor, $x$. $$\dfrac{x^2+x}{x}=\dfrac{x(x+1)}{x}=x+1$$ |

## ANSWERS TO STUDENT CHECKS

**Student Check 1**  **a.** undefined; $-\dfrac{5}{4}$; 0  **b.** $-\dfrac{1}{5}$; 3; $\dfrac{5}{2}$
**c.** \$68, \$58, and \$53

**Student Check 2**  **a.** $y=-5$  **b.** $h=3$ or $h=1$
**c.** defined for all real numbers

**Student Check 3**  **a.** $\dfrac{1}{2}$  **b.** $\dfrac{a-5}{a-3}$  **c.** $\dfrac{4y-1}{y-6}$

**d.** $\dfrac{3x-y}{x^2-2x+4}$  **e.** $\dfrac{16b+20}{5b+10}$  **f.** $-\dfrac{3r}{2}$

**Student Check 4**  **a.** $\dfrac{1-3x}{6x+1}$ or $\dfrac{3x-1}{-6x-1}$

**b.** $-\dfrac{x+12}{x+7}$ or $\dfrac{x+12}{-x-7}$

## SUMMARY OF KEY CONCEPTS

1. A rational expression is a quotient of two polynomials, in which the denominator cannot be zero.
2. Rational expressions can be evaluated by substituting the given value in place of the variable.
3. Rational expressions are undefined for the values that make the denominator equal to zero since division by zero is undefined.
4. To simplify a rational expression, factor the numerator and denominator and divide out any common factors.
5. If a rational expression has a negative sign in front of it, $-1$ can be distributed through the numerator *or* the denominator, but not both.

## GRAPHING CALCULATOR SKILLS

We can use the Store command in the calculator to evaluate rational expressions. When entering rational expressions in the calculator, parentheses are important to preserve the order of operations.

**Example:** Use the Store command to evaluate the expression $\dfrac{x+1}{4x-3}$ for $x=3$.

**Solution:** Enter the function in the equation editor.

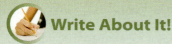

On the home screen, store the value 3 for the variable $x$.

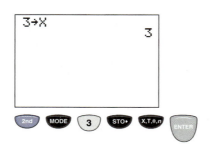

Now evaluate the expression $Y_1$. Use the Fraction command to put the answer in fraction form.

So, the value of the expression $\dfrac{x+1}{4x-3}$ at $x=3$ is $\dfrac{4}{9}$.

## SECTION 7.1 EXERCISE SET

 **Write About It!**

**Use complete sentences in your answer to each exercise.**

1. What is a rational expression? *A rational expression is a quotient or ratio of two polynomials, provided the denominator is not equal to zero.*
2. How can you determine the values that make a rational expression undefined?
3. Give three examples of rational expressions.
4. Give three examples of fractions that are not rational expressions.
5. Is every rational expression a polynomial? Explain.
6. Is every polynomial a rational expression? Explain.
7. Give an example of a rational expression that is undefined for $x = 0$. Explain how you came up with your answer.
8. Give an example of a rational expression that is undefined for $x = 2$ and $x = -2$. Explain how you came up with your answer.
9. Given an example of a rational expression that is defined for all real numbers. Explain how you came up with your answer.
10. Is the expression $\dfrac{4x+6}{x}$ in simplified form? Why or why not?
11. Is the expression $\dfrac{4x^2+6x}{x}$ in simplified form? Why or why not?
12. Is the expression $\dfrac{x^2-9}{x-3}$ in simplified form? Why or why not?

**Practice Makes Perfect!**

**Evaluate each rational expression for $x = 2, -1,$ and $-3$. Give each answer as a fraction in lowest terms, when applicable. (*See Objective 1.*)**

13. $\dfrac{2x+1}{x+2}$   $\dfrac{5}{4}, -1, 5$

14. $\dfrac{x-1}{3x-3}$   $\dfrac{1}{3}, \dfrac{1}{3}, \dfrac{1}{3}$

15. $\dfrac{x^2}{3-2x}$   $-4, \dfrac{1}{5}, 1$

16. $\dfrac{x^3}{1-4x}$   $-\dfrac{8}{7}, \dfrac{1}{5}, \dfrac{27}{13}$

17. $\dfrac{x^3+5x-3}{x^2+4}$   $\dfrac{15}{8}, -\dfrac{9}{5}, -\dfrac{45}{13}$

18. $\dfrac{x^2-6x+1}{x^3-2x}$   $-\dfrac{7}{4}, 8, -\dfrac{4}{3}$

19. $\dfrac{1}{4x-3}$   $\dfrac{1}{5}, -\dfrac{1}{7}, -\dfrac{1}{15}$

20. $\dfrac{3}{2x+1}$   $\dfrac{3}{5}, -3, -\dfrac{3}{5}$

21. $\dfrac{x}{3-2x}$   $-2, -\dfrac{1}{5}, -\dfrac{1}{3}$

22. $\dfrac{2x}{1-3x}$   $-\dfrac{4}{5}, \dfrac{1}{2}, -\dfrac{3}{5}$

23. A fitness club charges a one-time membership fee of $108 when a new member signs a contract. A new member is charged a monthly fee of $45, and the one-time fee is divided equally among the monthly payments. If $x$ represents the number of months in the new member's contract, the expression $\dfrac{108+45x}{x}$ represents the monthly payment, in dollars. What is the monthly payment if the new member signs a contract for 6 months? 18 months?   $63; $51

24. A buyers' club charges a one-time membership fee of $48 when a new member signs a contract. A new member is charged a monthly fee of $10, and the one-time fee is divided equally among the monthly payments. If $x$ represents the number of months in the new member's contract, the expression $\dfrac{48+10x}{x}$ represents the monthly payment, in dollars. What is the monthly payment if the new member signs a contract for 6 months? 12 months?   $18; $14

25. The total cost, in dollars, of producing $x$ items is given by $C = 12x^2 - 540x + 6000$, and the average cost per item is $\dfrac{12x^2 - 540x + 6000}{x}$, where $x$ is the number of units produced. What is the average cost for each item when 10 items are produced? 12 items are produced?   $180; $104

*Additional answers can be found in the Instructor Answer Appendix.*

**26.** A company's total cost, in dollars, to make $x$ items is $C = 4x^2 - 15x + 1900$ and the average cost per item is $\dfrac{4x^2 - 15x + 1900}{x}$. What is the average cost of each item when 20 items are made? 25 items are made? $160; $161

**Find the value(s) that make each rational expression undefined. (*See Objective 2.*)**

**27.** $\dfrac{x+3}{x-1}$  $x = 1$

**28.** $\dfrac{x-6}{x+7}$  $x = -7$

**29.** $\dfrac{2x-4}{3x+6}$  $x = -2$

**30.** $\dfrac{5x+5}{4x-8}$  $x = 2$

**31.** $\dfrac{3x+4}{6}$  none

**32.** $\dfrac{7-4x}{3}$  none

**33.** $\dfrac{3}{x^2-9}$  $x = -3$ or $x = 3$

**34.** $\dfrac{5}{x^2-25}$  $x = -5$ or $x = 5$

**35.** $\dfrac{x^2-4}{2}$  none

**36.** $\dfrac{x^2-16}{4}$  none

**37.** $\dfrac{x^2+3x-10}{2x^2+5x-3}$  $x = -3$ or $x = \frac{1}{2}$

**38.** $\dfrac{x^2-7x-18}{3x^2-13x-10}$  $x = -\frac{2}{3}$ or $x = 5$

**39.** $\dfrac{x^2-4x+5}{11}$  none

**40.** $\dfrac{x^2-9}{9}$  none

**Simplify each rational expression. Assume all denominators are not zero. (*See Objective 3.*)**

**41.** $\dfrac{2x}{6x^2}$  $\frac{1}{3x}$

**42.** $\dfrac{8a^3}{10a}$  $\frac{4a^2}{5}$

**43.** $\dfrac{6x}{2x^2+4}$  $\frac{3x}{x^2+2}$

**44.** $\dfrac{5x}{10x-15}$  $\frac{x}{2x-3}$

**45.** $\dfrac{4x+8}{2x}$  $\frac{2x+4}{x}$

**46.** $\dfrac{8x+24}{4x}$  $\frac{2x+6}{x}$

**47.** $\dfrac{x^2+5x+4}{x^2+x-12}$  $\frac{x+1}{x-3}$

**48.** $\dfrac{x^2+4x-12}{x^2-5x+6}$  $\frac{x+6}{x-3}$

**49.** $\dfrac{6x}{6x+10}$  $\frac{3x}{3x+5}$

**50.** $\dfrac{7x}{7x-14}$  $\frac{x}{x-2}$

**51.** $\dfrac{x^2+x-2}{x^2+8x-9}$  $\frac{x+2}{x+9}$

**52.** $\dfrac{x+2}{x^2-8x-20}$  $\frac{1}{x-10}$

**53.** $\dfrac{3x^2-3x}{x^2+2x}$  $\frac{3x-3}{x+2}$

**54.** $\dfrac{4x^3-20x}{16x^2+4x}$  $\frac{x^2-5}{4x+1}$

**55.** $\dfrac{x^3-1}{x^2-6x+5}$  $\frac{x^2+x+1}{x-5}$

**56.** $\dfrac{x^2+7x+12}{x^3+27}$  $\frac{x+4}{x^2-3x+9}$

**57.** $\dfrac{6xy+2ay-3x-a}{8y^3-1}$

**58.** $\dfrac{a^3-125}{2ab-10b+3a-15}$

**59.** $\dfrac{16-x^2}{x+4}$  $4-x$

**60.** $\dfrac{x-5}{25-x^2}$  $\frac{-1}{x+5}$

**Use the rule of negative rational expressions to write two equivalent forms of each rational expression. (*See Objective 4.*)**

**61.** $-\dfrac{x-3}{2x+4}$

**62.** $-\dfrac{2x-3}{4}$

**63.** $-\dfrac{-3x+4}{x-1}$

**64.** $-\dfrac{-x+8}{2x-1}$

**65.** $\dfrac{-3x+1}{x-4}$

**66.** $\dfrac{-4x+8}{x-2}$

**67.** $\dfrac{-4x-9}{2x+1}$

**68.** $\dfrac{-2x+15}{3x-12}$

## Mix 'Em Up!

**Solve each problem.**

**69.** Evaluate $\dfrac{x^2-3x}{2x+1}$ for $x = -2, 0, 3$.  $-\frac{10}{3}, 0, 0$

**70.** Evaluate $\dfrac{3y}{y+1}$ for $y = -2, 1, 3$.  $6, \frac{3}{2}, \frac{9}{4}$

**71.** Is $\dfrac{\sqrt{x}+1}{x+3}$ a rational expression? Explain.

**72.** Is $\dfrac{1}{\sqrt{x}+5}$ a rational expression? Explain.

**73.** Is $\dfrac{x^2}{2x-1}$ a rational expression? Explain.  yes

**74.** Is $\dfrac{5x^2}{2x-3}$ a rational expression? Explain.  yes

**75.** A company's, profit in dollars, can be represented by $-20q^2 + 500q$, where $q$ is the number of items sold. The average profit per item is given by $\dfrac{-20q^2 + 500q}{q}$. What is the average profit per item when 10 items are sold? 20 items are sold?  $300; $100

**76.** A company's profit, in dollars, can be represented by $-15p^2 + 450p$, where $p$ is the number of items sold. The average profit per item is given by $\dfrac{-15p^2 + 450p}{p}$. What is the average profit per item when 20 items are sold? 25 items are sold?  $150; $75

**77.** The volume in cubic inches of an open box with height $x$ in. is given by the expression $x^3 + 8x^2 + 15x$. The area in square inches of the base of the box is given by $\dfrac{x^3 + 8x^2 + 15x}{x}$. What is the area of the base when the height is 2 in.? 6 in.?  35 in.²; 99 in.²

**78.** The volume in cubic inches of an open box with height $x$ in. is given by the expression $x^3 + 4x^2 - 12x$. The area in square inches of the base of the box is given by $\dfrac{x^3 + 4x^2 - 12x}{x}$. What is the area of the base when the height is 4 in.? 6 in.?  20 in.²; 48 in.²

**Find the value(s) that make each rational expression undefined. Then simplify each rational expression.**

**79.** $\dfrac{x+5}{4x-2}$  $x = \frac{1}{2}$

**80.** $\dfrac{2x-6}{3x+5}$  $x = -\frac{5}{3}$

**81.** $\dfrac{3x}{2x^2+7x}$  $x = 0, -\frac{7}{2}; \frac{3}{2x+7}$

**82.** $\dfrac{4x}{3x^2-x}$  $x = 0, \frac{1}{3}; \frac{4}{3x-1}$

**83.** $\dfrac{3x + 6}{2x^2 + 4x}$ $\quad x = 0, -2; \dfrac{3}{2x}$

**84.** $\dfrac{-2x + 10}{3x^2 - 15x}$ $\quad x = 0, 5; -\dfrac{2}{3x}$

**85.** $\dfrac{x^2 - 81}{x^2 + 8x - 9}$ $\quad x = 1, -9; \dfrac{x - 9}{x - 1}$

**86.** $\dfrac{x^2 + 7x + 10}{x^2 - 4}$ $\quad x = 2, -2; \dfrac{x + 5}{x - 2}$

**87.** $\dfrac{12x}{20x - 8}$ $\quad x = \dfrac{2}{5}; \dfrac{3x}{5x - 2}$

**88.** $\dfrac{16x}{8x - 24}$ $\quad x = 3; \dfrac{2x}{x - 3}$

**89.** $\dfrac{x^3 + 8}{x^2 - 3x - 10}$

**90.** $\dfrac{x^3 - 64}{x^2 + 3x - 28}$

**91.** $\dfrac{2x^2 + 8x}{x^2 - 7x}$ $\quad x = 0, 7; \dfrac{2x + 8}{x - 7}$

**92.** $\dfrac{-6x^2 + 6x}{2x^2 + 10x}$

Use the rule of negative rational expressions to write two equivalent forms of each rational expression.

**93.** $\dfrac{-9x + 18}{x - 10}$

**94.** $\dfrac{-2x + 7}{x + 1}$

**95.** $-\dfrac{x^2 + 3x}{4x - 1}$

**96.** $-\dfrac{x^2 - x}{-3x + 2}$

 **You Be the Teacher!**

Correct each student's errors, if any.

**97.** Simplify $\dfrac{6x - 12}{6x}$.

Victor's work:

$$\dfrac{\overset{1}{\cancel{6}}x - 12}{\underset{1}{\cancel{6}}x} = \dfrac{1 - 12}{1} = -11 \qquad \dfrac{6x - 12}{6x} = \dfrac{6(x - 2)}{6x} = \dfrac{x - 2}{x}$$

**98.** Simplify $\dfrac{x^2 - 5x}{25 - x^2}$.

Justin's work:

$$\dfrac{x^2 - 5x}{25 - x^2} = \dfrac{\cancel{x^2} - 5x}{25 - \cancel{x^2}}$$
$$= \dfrac{-5x}{25}$$
$$= -\dfrac{x}{5}$$

$$\dfrac{x^2 - 5x}{25 - x^2} = \dfrac{x\overset{1}{\cancel{(x - 5)}}}{\underset{-1}{\cancel{(5 - x)}}(5 + x)}$$
$$= -\dfrac{x}{x + 5}$$

**99.** Simplify $\dfrac{81 - x^2}{x + 9}$.

Hillary's work:

$$\dfrac{81 - x^2}{x + 9} = \dfrac{x^2 - 81}{x + 9} = \dfrac{\cancel{(x + 9)}(x - 9)}{\cancel{x + 9}} = x - 9$$

**100.** Simplify $\dfrac{x^2 + 3x - 10}{x^2 + 9x + 20}$.

Will's work:

$$\dfrac{x^2 + 3x - 10}{x^2 + 9x + 20} = \dfrac{\cancel{x^2} + 3\cancel{x} - \cancel{10}}{\cancel{x^2} + 9\cancel{x_3} + \cancel{20_2}} = \dfrac{1}{5}$$

**101.** Use the rule of negative rational expressions to write an equivalent form of the expression $\dfrac{-4x - 4}{5x + 1}$.

Janice's work:

$$\dfrac{-4x - 4}{5x + 1} = -\dfrac{4x - 4}{5x + 1} \qquad \dfrac{-4x - 4}{5x + 1} = -\dfrac{4x + 4}{5x + 1}$$

**102.** Is $\dfrac{3x^4 - 12x^3}{4}$ a rational expression? Explain.

Marlon's explanation: I do not think it is a rational expression because of the 4 in the denominator. I think there need to be variables in the top and the bottom for it to be rational. *Since both the numerator and the denominator contain polynomials, it is a rational expression.*

 **Calculate It!**

Use the Store feature in a calculator to evaluate each rational expression for $x = -10, 0,$ and $10$. Give each answer as a fraction in lowest terms, when applicable.

**103.** $\dfrac{x^2 + 3x - 4}{2x + 8}$ $\quad -\dfrac{11}{2}, -\dfrac{1}{2}, \dfrac{9}{2}$

**104.** $\dfrac{3x^2 - 15}{2x + 5}$ $\quad -19, -3, \dfrac{57}{5}$

**105.** $-\dfrac{2x + 3}{x - 1}$ $\quad -\dfrac{17}{11}, 3, -\dfrac{23}{9}$

**106.** $-\dfrac{3x}{7 - x}$ $\quad \dfrac{30}{17}, 0, 10$

**SECTION 7.2** **Multiplication and Division of Rational Expressions**

▶ **OBJECTIVES**

As a result of completing this section, you will be able to

1. Multiply rational expressions.
2. Divide rational expressions.
3. Solve problems requiring conversion of units.
4. Troubleshoot common errors.

The distance for most races and marathons is given in kilometers (K or km) rather than miles. How many miles are in a 5K race? Once we know the conversion rate between kilometers and miles, we can use multiplication of rational expressions to answer this question.

## Multiplying Rational Expressions

Multiplying rational expressions is very similar to multiplying numeric fractions. Recall

$$\frac{5}{3} \cdot \frac{2}{7} = \frac{5 \cdot 2}{3 \cdot 7} = \frac{10}{21}$$

We multiply the numerators together, multiply the denominators together, and then express in the lowest terms.

**Objective 1** ▶

Multiply rational expressions.

---

**Property:  Multiplying Rational Expressions**

If $\dfrac{A}{B}$ and $\dfrac{C}{D}$ are rational expressions, where $B$ and $D$ are not zero, then

$$\frac{A}{B} \cdot \frac{C}{D} = \frac{A \cdot C}{B \cdot D}$$

---

**Procedure:  Multiplying Rational Expressions**

**Step 1:** Factor each numerator and denominator, if possible.
**Step 2:** Divide out common factors.
**Step 3:** The result is the quotient of the product of the remaining factors in the numerators and the product of the remaining factors in the denominators.

---

**Objective 1  Examples** **Multiply the rational expressions. Write answers in lowest terms.**

**1a.** $\dfrac{12a^2}{5} \cdot \dfrac{10}{3a^3}$   **1b.** $\dfrac{x^2 + 5x + 6}{x + 3} \cdot \dfrac{3}{x + 2}$   **1c.** $-\dfrac{2y}{y^2 - 16} \cdot \dfrac{3y^2 - 10y - 8}{12y^2 + 8y}$

**Solutions** **1a.** $\dfrac{12a^2}{5} \cdot \dfrac{10}{3a^3} = \dfrac{2 \cdot 2 \cdot 3 \cdot a^2}{5} \cdot \dfrac{2 \cdot 5}{3 \cdot a^2 \cdot a}$    Factor. The common factors are $3a^2$ and 5.

$\qquad\qquad = \dfrac{2 \cdot 2 \cdot 2 \cdot 1}{1 \cdot a}$    Divide out the common factors and multiply the rational expressions.

$\qquad\qquad = \dfrac{8}{a}$    Simplify.

We can also multiply first and then simplify.

$\dfrac{12a^2}{5} \cdot \dfrac{10}{3a^3} = \dfrac{(12a^2)(10)}{(5)(3a^3)}$    Multiply the rational expressions.

$\qquad\qquad = \dfrac{120a^2}{15a^3}$    Simplify the products.

$\qquad\qquad = \dfrac{15a^2(8)}{15a^2(a)}$    Factor out the common factor of the numerator and denominator.

$\qquad\qquad = \dfrac{8}{a}$    Apply the fundamental property of rational expressions.

**1b.** $\dfrac{x^2 + 5x + 6}{x + 3} \cdot \dfrac{3}{x + 2}$

$= \dfrac{(x + 3)(x + 2)}{x + 3} \cdot \dfrac{3}{x + 2}$     Factor. The common factors are $x + 3$ and $x + 2$.

$= \dfrac{1 \cdot 3}{1}$     Divide out the common factors and multiply the remaining factors.

$= \dfrac{3}{1}$     Simplify the products.

$= 3$     Divide.

**1c.** $-\dfrac{2y}{y^2 - 16} \cdot \dfrac{3y^2 - 10y - 8}{12y^2 + 8y}$

$= -\dfrac{2y}{(y - 4)(y + 4)} \cdot \dfrac{(3y + 2)(y - 4)}{2 \cdot 2y(3y + 2)}$     Factor. The common factors are $2y$, $3y + 2$, and $y - 4$.

$= -\dfrac{1}{1 \cdot (y + 4) \cdot 2}$     Divide out the common factors and multiply the rational expressions.

$= -\dfrac{1}{2(y + 4)}$     Simplify.

---

✓ **Student Check 1**    Multiply the rational expressions. Write answers in lowest terms.

**a.** $-\dfrac{12x^3}{5y^2} \cdot \dfrac{y^3}{6x^4}$    **b.** $\dfrac{x^2 - 6x + 5}{4x + 4} \cdot \dfrac{4}{x - 5}$    **c.** $-\dfrac{10h}{h^2 + 6h + 9} \cdot \dfrac{2h^2 + 11h + 15}{4h^2 + 10h}$

---

## Dividing Rational Expressions

**Objective 2** ▶

Divide rational expressions.

In Chapter 1, we learned that dividing by a number is the same as multiplying by its reciprocal. Recall that the *reciprocal* of $\dfrac{a}{b}$ is $\dfrac{b}{a}$, where $a \neq 0$ and $b \neq 0$. For example, we know that

$$36 \div 9 = 4, \text{ but we also know that } 36 \times \dfrac{1}{9} = \dfrac{36}{1} \times \dfrac{1}{9} = \dfrac{36}{9} = 4.$$

So, division by 9 is the same as multiplication by $\dfrac{1}{9}$. Notice the reciprocal of 9 or $\dfrac{9}{1}$ is $\dfrac{1}{9}$.

The same rule holds true when dividing rational expressions: dividing by a rational expression is the same as multiplying by the reciprocal of the expression.

---

**Property: Dividing Rational Expressions**

If $\dfrac{A}{B}$ and $\dfrac{C}{D}$ are rational expressions, where $B$, $C$, and $D$ are not zero, then

$$\dfrac{A}{B} \div \dfrac{C}{D} = \dfrac{A}{B} \cdot \dfrac{D}{C} = \dfrac{A \cdot D}{B \cdot C}$$

---

**Procedure: Dividing Rational Expressions**

**Step 1:** Convert the division to multiplication by the reciprocal of the second fraction.

**Step 2:** Perform the steps to multiply the expressions.

**Step 3:** Express the answer in lowest terms.

**Objective 2 Examples** **Divide the rational expressions. Write answers in lowest terms.**

**2a.** $\dfrac{4a^7}{3} \div \dfrac{8a^8}{21b}$  **2b.** $\dfrac{-3y^2 + 3y}{5y^2 + 10y} \div \dfrac{y^2 + 8y - 9}{15}$

**2c.** $\dfrac{4a^2 - a - 5}{a^3 + 1} \div \dfrac{16a^2 - 25}{-2a^3 + 2a^2 - 2a}$

**Solutions**  **2a.** $\dfrac{4a^7}{3} \div \dfrac{8a^8}{21b} = \dfrac{4a^7}{3} \cdot \dfrac{21b}{8a^8}$  Multiply by the reciprocal of $\dfrac{8a^8}{21b}$.

$= \dfrac{2 \cdot 2 \cdot a^7}{3} \cdot \dfrac{7 \cdot 3b}{2 \cdot 2 \cdot 2 \cdot a^7 \cdot a}$  Factor and identify the common factors.

$= \dfrac{1 \cdot 7b}{1 \cdot 2a}$  Divide out the common factors and multiply the rational expressions.

$= \dfrac{7b}{2a}$  Simplify.

**2b.** $\dfrac{-3y^2 + 3y}{5y^2 + 10y} \div \dfrac{y^2 + 8y - 9}{15}$

$= \dfrac{-3y^2 + 3y}{5y^2 + 10y} \cdot \dfrac{15}{y^2 + 8y - 9}$  Multiply by the reciprocal of $\dfrac{y^2 + 8y - 9}{15}$.

$= \dfrac{-3y(y - 1)}{5y(y + 2)} \cdot \dfrac{3 \cdot 5}{(y + 9)(y - 1)}$  Factor and identify the common factors.

$= \dfrac{-3 \cdot 3 \cdot 1}{1(y + 2)(y + 9)}$  Divide out the common factors and multiply the rational expressions.

$= \dfrac{-9}{(y + 2)(y + 9)}$  Simplify.

$= -\dfrac{9}{(y + 2)(y + 9)}$  Apply the rule of negative rational expressions.

**2c.** $\dfrac{4a^2 - a - 5}{a^3 + 1} \div \dfrac{16a^2 - 25}{-2a^3 + 2a^2 - 2a}$

$= \dfrac{4a^2 - a - 5}{a^3 + 1} \cdot \dfrac{-2a^3 + 2a^2 - 2a}{16a^2 - 25}$  Multiply by the reciprocal of the second fraction.

$= \dfrac{(4a - 5)(a + 1)}{(a + 1)(a^2 - a + 1)} \cdot \dfrac{-2a(a^2 - a + 1)}{(4a - 5)(4a + 5)}$  Factor and identify the common factors.

$= \dfrac{1(-2a)}{1(4a + 5)}$  Divide out the common factors and multiply the expressions.

$= \dfrac{-2a}{4a + 5}$ or $-\dfrac{2a}{4a + 5}$  Simplify.

✔ **Student Check 2**  Divide the rational expressions. Write answers in lowest terms.

**a.** $\dfrac{3a^3}{14b^2} \div \dfrac{9a^2}{12b}$  **b.** $\dfrac{3x^2 + 10x - 8}{6x^3 - 4x^2} \div \dfrac{x^2 + 7x + 12}{2x^2}$

**c.** $\dfrac{8y^3 - 27}{2y^2 + 5y + 3} \div \dfrac{12y^2 + 18y + 27}{4y^2 - 9}$

## Unit Conversion

There are many real-life situations that require us to convert between different units of a quantity. When cooking, we may need to convert a measurement from liters to cups. In the medical field, we may need to convert a quantity from pounds to kilograms. In the section opener, we stated an example of the need to convert a distance from kilometers to miles.

To convert from one unit to another, we will use the fact that when equal quantities are divided, the result is 1. For example, we know that 60 min is equal to 1 hr, so

$$\frac{60 \text{ min}}{1 \text{ hr}} = 1$$

We use this fact to convert 5 hr to minutes, for example.

$$5 \text{ hr} = 5 \text{ hr} \times 1 = 5 \cancel{\text{hr}} \times \frac{60 \text{ min}}{1 \cancel{\text{hr}}} = 5 \times 60 \text{ min} = 300 \text{ min}$$

---

**Property: Rule for Converting Units**

If $a$ and $b$ are equal quantities, then $\dfrac{a}{b} = 1$.

---

**Procedure: Converting a Unit to Another Unit**

**Step 1:** Write the given unit as a product of itself and 1.
**Step 2:** Replace 1 with a quotient that involves equivalent units. The numerator represents the new units and the denominator represents the given units.
**Step 3:** Divide out the common factor and multiply the remaining fractions.

---

**Objective 3  Examples**    Answer each question.

**3a.** The dosage of medicine that a patient should receive is often based on the patient's weight in kilograms (kg). If a patient weights 165 pounds (lb), what is his weight in kilograms? (Note: 1 lb = 0.45 kg.)

**Solution**    **3a.** 165 lb = 165 lb  · 1

$$= \frac{165 \cancel{\text{lb}}}{1} \cdot \frac{0.45 \text{ kg}}{1 \cancel{\text{lb}}}$$

$$= \frac{165(0.45 \text{ kg})}{1 \cdot 1}$$

$$= 74.25 \text{ kg}$$

**3b.** The Willis Tower (formerly known as the Sears Tower) is the tallest building in the United States with a height of 527 m. What is the height of the Willis Tower in feet? (Note: 1 m = 3.28 ft.) (Source: Wikipedia)

**Solution**    **3b.** 527 m = 527 m  · 1

$$= \frac{527 \cancel{\text{m}}}{1} \cdot \frac{3.28 \text{ ft}}{1 \cancel{\text{m}}}$$

$$= \frac{527(3.28 \text{ ft})}{1 \cdot 1}$$

$$= 1728.56 \text{ ft}$$

**3c.** The distance for most races and marathons is given in kilometers (K or km) rather than miles. How many miles are in a 5K race? (Note: 1 km = 0.62 mi.)

**Solution**   **3c.** $5\,K = 5\text{ km} \cdot 1$

$$= \frac{5\text{ km}}{1} \cdot \frac{0.62\text{ mi}}{1\text{ km}}$$

$$= \frac{5(0.62\text{ mi})}{1 \cdot 1}$$

$$= 3.1\text{ mi}$$

✓ **Student Check 3**   Use the conversion equivalents given in Example 3 to answer each question.

  **a.** An average-size cat weighs 3.3 kg. How many pounds does an average cat weigh?

  **b.** The length of an Olympic-size pool is 50 m. What is this length in feet?

  **c.** The New York Marathon is 42.2 km. What is this distance in miles?

**Objective 4** ▶

Troubleshoot common errors.

## Troubleshooting Common Errors

A common error related to dividing rational expressions is shown.

**Objective 4  Example**   **A problem and an incorrect solution are given. Provide the correct solution and an explanation of the error.**

Perform the operation $\dfrac{x^2 + 4x + 3}{x - 3} \div (x^2 - 9)$.

| Incorrect Solution | Correct Solution and Explanation |
|---|---|
| $\dfrac{x^2 + 4x + 3}{x - 3} \div (x^2 - 9)$ $= \dfrac{x^2 + 4x + 3}{x - 3} \cdot \dfrac{(x^2 - 9)}{1}$ $= \dfrac{(x + 3)(x + 1)}{\cancel{x-3}} \cdot \dfrac{\cancel{(x-3)}(x + 3)}{1}$ $= (x + 3)^2 (x + 1)$ | We must multiply by the reciprocal of $x^2 - 9$, which is $\dfrac{1}{x^2 - 9}$. $\dfrac{x^2 + 4x + 3}{x - 3} \div (x^2 - 9)$ $= \dfrac{x^2 + 4x + 3}{x - 3} \cdot \dfrac{1}{(x^2 - 9)}$ $= \dfrac{(x + 3)(x + 1)}{x - 3} \cdot \dfrac{1}{(x - 3)(x + 3)}$ $= \dfrac{x + 1}{(x - 3)^2}$ |

## ANSWERS TO STUDENT CHECKS

**Student Check 1**   **a.** $-\dfrac{2y}{5x}$   **b.** $\dfrac{x - 1}{x + 1}$   **c.** $-\dfrac{5}{h + 3}$

**Student Check 2**   **a.** $\dfrac{2a}{7b}$   **b.** $\dfrac{1}{x + 3}$   **c.** $\dfrac{(2y - 3)^2}{3(y + 1)}$

**Student Check 3**   **a.** 7.33 lb   **b.** 164 ft   **c.** 26.164 mi

## SUMMARY OF KEY CONCEPTS

1. The key to multiplying rational expressions is factoring. We begin by factoring each of the numerators and denominators. Then we divide out any common factors and multiply the remaining factors.

2. Division of rational expressions is just like dividing numerical fractions. We first convert the division to multiplication by the reciprocal of the second rational expression and follow the steps for multiplying fractions.

3. The key to unit conversions is multiplying the expression to be converted by a form of 1. The form of 1 involves a ratio of two equivalent quantities.

## GRAPHING CALCULATOR SKILLS

The graphing calculator can be used to verify products or quotients.

**Example:** Verify that $\dfrac{x^2 - 36}{x + 1} \div \dfrac{2x + 12}{x - 1} = \dfrac{(x - 6)(x - 1)}{2(x + 1)}$.

**Solution:** Enter the original problem in $Y_1$ and the result in $Y_2$. Be very careful about the use of parentheses to enter these expressions. If the numerator or denominator has a sum or product, the entire expression must be entered in parentheses.

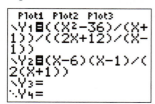

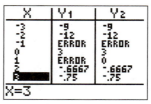

View the Table to compare the $y$-values. If the $y$-values agree for each $x$-value, then the answer is correct.

Note: The original problem and the final answer may have different values for which the expressions are undefined. The $y$-value has an ERROR message for the values at which it is undefined.

## SECTION 7.2 / EXERCISE SET

###  Write About It!

**Use complete sentences in your answer to each exercise.**

1. Explain the process of multiplying rational expressions.

2. Explain how to convert division by a polynomial to multiplication.   *Convert division by a polynomial to multiplication of 1 over the polynomial.*

3. Explain the process of dividing rational expressions.

4. Explain the two general rules that are used when converting units.

###  Practice Makes Perfect!

**Multiply the rational expressions. Write answers in simplest terms.** (*See Objective 1.*)

5. $\dfrac{4a^3}{3} \cdot \dfrac{9}{12a^5}$   $\dfrac{1}{a^2}$

6. $\dfrac{16a^7}{9} \cdot \dfrac{18}{a^2}$   $32a^5$

7. $\dfrac{6a}{b^4} \cdot \dfrac{3b^5}{14a^2}$   $\dfrac{9b}{7a}$

8. $\dfrac{4a^4}{3b} \cdot \dfrac{9b^2}{a^3}$   $12ab$

9. $\dfrac{3xy^2}{16y} \cdot \dfrac{4x}{6x^2y}$   $\dfrac{1}{8}$

10. $\dfrac{12x^3y}{5x} \cdot \dfrac{6y^2}{12xy^4}$   $\dfrac{6x}{5y}$

11. $\dfrac{x^2 - 8x + 15}{x + 1} \cdot \dfrac{2}{x - 3}$

12. $\dfrac{x - 2}{5} \cdot \dfrac{x + 7}{x^2 - 3x + 2}$

13. $\dfrac{x^2 - 1}{x + 3} \cdot \dfrac{4x + 12}{x + 1}$   $4(x - 1)$

14. $\dfrac{x^2 - 4}{x - 4} \cdot \dfrac{3x - 12}{x + 2}$   $3(x - 2)$

15. $\dfrac{-5x}{(x - 5)^2} \cdot \dfrac{2x - 10}{15x^3}$

16. $\dfrac{-2x^3}{(x + 6)^3} \cdot \dfrac{4x + 24}{6x^2}$

17. $\dfrac{2x + 1}{x - 3} \cdot \dfrac{x^2 - 9}{2x^2 + 7x + 3}$   $1$

18. $\dfrac{3x - 2}{x + 1} \cdot \dfrac{x^2 - 1}{3x^2 - 5x + 2}$   $1$

19. $\dfrac{x^3 - 8}{x + 3} \cdot \dfrac{4}{x - 2}$   $\dfrac{4(x^2 + 2x + 4)}{x + 3}$

20. $\dfrac{2}{x^3 + 1} \cdot \dfrac{x + 1}{x + 2}$

21. $\dfrac{x - 3}{2x^2 + 4x} \cdot 4x$   $\dfrac{2(x - 3)}{x + 2}$

22. $\dfrac{x - 1}{3x^3 - 6x^2} \cdot 6x^2$

23. $5x^3 \cdot \dfrac{2x + 3}{15x^2 - 5x}$   $\dfrac{x^2(2x + 3)}{3x - 1}$

24. $10x^2 \cdot \dfrac{3x + 5}{4x^5 - 8x^3}$   $\dfrac{5(3x + 5)}{2x(x^2 - 2)}$

*Additional answers can be found in the Instructor Answer Appendix.*

**Divide the rational expressions. Write answers in simplest terms.** (*See Objective 2.*)

**25.** $\dfrac{2a^4}{5} \div \dfrac{10a^3}{3}$   $\dfrac{3a}{25}$

**26.** $\dfrac{3a^6}{14} \div \dfrac{21a^7}{9}$   $\dfrac{9}{98a}$

**27.** $\dfrac{12a^2}{5b} \div \dfrac{3ab}{20}$   $\dfrac{16a}{b^2}$

**28.** $\dfrac{6a^3}{25b^2} \div \dfrac{12a^2b}{5}$   $\dfrac{a}{10b^3}$

**29.** $\dfrac{2xy^4}{x^3} \div \dfrac{6x^2y}{5x}$   $\dfrac{5y^3}{3x^3}$

**30.** $\dfrac{8x^2y^3}{5y} \div \dfrac{16xy^4}{15y}$   $\dfrac{3x}{2y}$

**31.** $\dfrac{x+3}{2x} \div \dfrac{x^2-9}{8}$   $\dfrac{4}{x(x-3)}$

**32.** $\dfrac{x-6}{3x} \div \dfrac{x^2-36}{x^2}$   $\dfrac{x}{3(x+6)}$

**33.** $\dfrac{x^2}{x^2-1} \div \dfrac{4x}{x-1}$   $\dfrac{x}{4(x+1)}$

**34.** $\dfrac{x^2-49}{7x^3} \div \dfrac{x+7}{14x}$   $\dfrac{2(x-7)}{x^2}$

**35.** $\dfrac{x^2+2x-15}{x-2} \div (x-3)$   $\dfrac{x+5}{x-2}$

**36.** $\dfrac{x^2+5x+6}{x+6} \div (x+2)$   $\dfrac{x+3}{x+6}$

**37.** $\dfrac{x^2-6x+1}{x^2-2x} \div \dfrac{x-1}{x}$   $\dfrac{x^2-6x+1}{(x-2)(x+1)}$

**38.** $\dfrac{x^2-8x+10}{x^3-4x^2} \div \dfrac{x+1}{x^2}$   $\dfrac{x^2-8x+10}{(x-4)(x+1)}$

**39.** $\dfrac{x^2-5x-14}{2x-2} \div \dfrac{x^2-6x-7}{x-1}$   $\dfrac{x+2}{2(x+1)}$

**40.** $\dfrac{3x+12}{x^2-2x-3} \div \dfrac{x+4}{x^2-x-2}$   $\dfrac{3(x-2)}{x-3}$

**41.** $\dfrac{x^2-3x}{x^2-1} \div \dfrac{2x^2-6x}{x^2+2x-3}$   $\dfrac{x+3}{2(x+1)}$

**42.** $\dfrac{3x^2-3x}{x^2-4} \div \dfrac{x^2+6x-7}{x^2-x-6}$   $\dfrac{3x(x-3)}{(x-2)(x+7)}$

**Use the given conversion equivalents to answer each question.** (*See Objective 3.*)

**43.** A dosage of medicine is based on a patient's weight in kilograms. If a patient weighs 180 lb, what is his weight in kilograms? (1 lb = 0.45 kg)   81 kg

**44.** A dosage of medicine is based on a patient's weight in kilograms. If a patient weighs 210 lb, what is his weight in kilograms? (1 lb = 0.45 kg)   94.5 kg

**45.** Joanie watched a British television show on which a woman claimed to have lost 3 stone (st) in weight in 5 months. How many pounds did the woman lose? (1 st = 14 lb)   42 lb

**46.** Danielle has an interview with a London modeling agency. Her weight is 135 lb, but she wants to know her weight in stones for the modeling agency. How many stones does Danielle weigh? (1 st = 14 lb)   9.6 st

**47.** The height of the Eiffel Tower in Paris is 324 m. What is its height in feet? (1 m = 3.28 ft) (Source: http://www.tour-eiffel.com/)   1062.7 ft

**48.** The height of the Ferris wheel at Hershey Park in Pennsylvania is approximately 100 ft. What is its height in meters? (1 m = 3.28 ft) (Source: http://www.hersheypark.com/rides/detail.php?q=yes&id=34)   30.5 m

**49.** Rose saves $200 to spend during her trip to Barcelona. How many Euros will Rose be able to spend? (1 U.S. dollar = 0.694 Euros)   138.8 Euros

**50.** When Eliza returns from her trip to Russia, she has 624 rubles. How many U.S. dollars does Eliza have? (1 U.S. dollar = 24.71 Russian rubles.)   25.25 U.S. dollars

## Mix 'Em Up!

**Perform each operation and simplify the answer.**

**51.** $\dfrac{6x^2y^4}{5} \cdot \dfrac{21x}{20y^6}$   $\dfrac{63x^3}{50y^2}$

**52.** $\dfrac{12a^2}{35b} \div \dfrac{4a^{12}}{7b^5}$   $\dfrac{3b^4}{5a^{10}}$

**53.** $\dfrac{x^2+3x}{4} \cdot \dfrac{2x-10}{x^3}$

**54.** $\dfrac{2x-1}{2x^2-3x+1} \cdot \dfrac{4x-4}{x+3}$

**55.** $\dfrac{x^2-4x-12}{x^2} \div \dfrac{x^2+4x+4}{3x}$   $\dfrac{3(x-6)}{x(x+2)}$

**56.** $\dfrac{x^2-1}{x+3} \div \dfrac{x^2-5x+4}{x^2-9}$   $\dfrac{(x+1)(x-3)}{x-4}$

**57.** $\dfrac{x^3+16}{x-2} \div (x+2)$   $\dfrac{x^3+16}{(x-2)(x+2)}$

**58.** $\dfrac{x+4}{3x^2-10x-8} \cdot (6x+4)$   $\dfrac{2(x+4)}{x-4}$

**59.** $\dfrac{x^3+8}{x^2-2x-8} \div \dfrac{2x^2-4x+8}{x^2-x-12}$   $\dfrac{x+3}{2}$

**60.** $\dfrac{x^3-1}{x^2-6x+5} \div \dfrac{3x^2+3x+3}{x^2-25}$   $\dfrac{x+5}{3}$

**61.** $\dfrac{x^3+3x^2+5x+15}{x^2-9} \cdot \dfrac{x^2-2x-3}{x^3+5x}$   $\dfrac{x+1}{x}$

**62.** $\dfrac{x^3-4x^2+x-4}{x^2-3x-4} \cdot \dfrac{x^2-6x-7}{2x^3+2x}$   $\dfrac{x-7}{2x}$

**63.** $\dfrac{8x^2-20x}{2x^2+x-15} \div \dfrac{6x^3+2x^2}{3x^2+10x+3}$   $\dfrac{2}{x}$

**64.** $\dfrac{35x^2+41x+12}{7x^2+18x+8} \div \dfrac{25x^2-1}{5x^2+9x-2}$   $\dfrac{5x+3}{5x+1}$

**65.** $\dfrac{x^2+8x+15}{x^2+x-20} \cdot \dfrac{2x^2-x-28}{2x^2+x-21}$   $\dfrac{x+3}{x-3}$

**66.** $\dfrac{2x^2 + 11x + 9}{3x^2 + 4x - 4} \cdot \dfrac{3x^2 - 8x + 4}{2x^2 + 5x - 18}$    $\dfrac{x+1}{x+2}$

**67.** $\dfrac{x^4 - 16}{5x^2 - 10x} \div \dfrac{x^3 - 3x^2 + 4x - 12}{x^2 - 3x}$    $\dfrac{x+2}{5}$

**68.** $\dfrac{3x^3 - 2x^2 + 3x - 2}{18x^2 - 12x} \div \dfrac{x^4 - 1}{5x^2 + 5x}$    $\dfrac{5}{6(x-1)}$

**Use the given conversion equivalents to answer each question.**

**69.** The distance between Frankfurt and Würzburg is 217 km. What is this distance in miles? (Note: 1 km = 0.62 mi.)    134.54 mi

**70.** The distance between London and Paris is 340 km. What is this distance in miles? (Note: 1 km = 0.62 mi.)
210.8 mi

**71.** How many minutes are in 0.7 hr? (Note: 1 hr = 60 min.)
42 min

**72.** How many minutes are in 1.2 hr? (Note: 1 hr = 60 min.)
72 min

**73.** Convert 45 kg to pounds. (Note: 1 lb = 0.45 kg.)    100 lb

**74.** Convert 98 kg to pounds. (Note: 1 lb = 0.45 kg.)    217.78 lb

 ## You Be the Teacher!

**Correct each student's errors, if any.**

**75.** Convert 6 ft into meters. (Note: 1 m = 3.28 ft.)

Marissa's work:

$$\frac{x \text{ m}}{3.28 \text{ ft}} = \frac{6 \text{ ft}}{1 \text{ m}}$$

$$x = 6 \cdot 3.28 = 19.68 \text{ m}$$

6 ft = 6 ft · 1

$= \dfrac{6 \text{ ft} \cdot 1 \text{ m}}{3.28 \text{ ft}}$

$= \dfrac{6}{3.28} \text{ m}$

$= 1.829 \text{ m}$

**76.** Convert 0.88 hr into minutes. (Note: 1 hr = 60 min.)

Rozalynn's work:

$$\frac{x \text{ min}}{1 \text{ hr}} = \frac{60 \text{ min}}{0.88 \text{ hr}}$$

0.88 hr = 0.88 hr · 1

$= \dfrac{0.88 \text{ hr} \cdot 60 \text{ min}}{1 \text{ hr}}$

$= 0.88 \cdot 60 \text{ min}$

$= 52.8 \text{ min}$

**77.** Simplify $\dfrac{x+3}{2x^3} \div 4x^2$.    $\dfrac{x+3}{2x^3} \div 4x^2 = \dfrac{x+3}{2x^3} \cdot \dfrac{1}{4x^2} = \dfrac{x+3}{8x^5}$

Rupert's work:

$$\frac{x+3}{2x^3} \div 4x^2 = \frac{x+3}{2x^3} \cdot \frac{4x^2}{1}$$

$$= \frac{x+3}{2x^3} \cdot \frac{\overset{2}{\cancel{4x^2}}}{1} = \frac{(x+3) \cdot 2}{x \cdot 1} = \frac{2x+6}{x}$$

**78.** Simplify $\dfrac{x^2 - 9}{2x + 8} \div \dfrac{x+3}{4x}$.

Norman's work:

$$\frac{x^2 - 9}{2x + 8} \div \frac{x+3}{4x} = \frac{(x+3)(x-3)}{2\cancel{(x+4)}} \div \frac{\cancel{x+4}}{4x}$$

$$= \frac{(x+3)(x-3)}{2} \div \frac{1}{4x}$$

$$= \frac{(x+3)(x-3)}{\underset{1}{\cancel{2}}} \cdot \frac{\overset{2}{\cancel{4}}x}{1}$$

$$= \frac{(x^2 - 9) \cdot 2x}{1} = 2x^3 - 18x$$

 ## Calculate It!

**Use a graphing calculator to verify each product and quotient.**

**79.** $\dfrac{25x^2 - 49}{2x + 6} \div \dfrac{5x^2 + 3x - 14}{x^2 + 5x + 6} = \dfrac{5x + 7}{2}$

**80.** $\dfrac{3x^2 - 12}{2x^2 + x - 10} \div \dfrac{x^3 + 8}{4x + 10} = \dfrac{6}{x^2 - 2x + 4}$

**81.** $\dfrac{x^2 - 9}{x + 2} \cdot \dfrac{x^2 + 2x}{3x + 9} = \dfrac{x^2 - 3x}{3}$

**82.** $\dfrac{x^2 - 16}{x^2 - 4x} \cdot \dfrac{4x - 8}{5x + 20} = \dfrac{4(x - 2)}{5x}$

---

**SECTION 7.3**    **Addition and Subtraction of Rational Expressions with Like Denominators and the Least Common Denominator**

▶ **OBJECTIVES**

As a result of completing this section, you will be able to

**1.** Add rational expressions with like denominators.

**2.** Subtract rational expressions with like denominators.

**3.** Find the least common denominator of rational expressions.

**4.** Write equivalent rational expressions.

**5.** Troubleshoot common errors.

Have you ever heard the expression that you cannot compare apples to oranges? This saying lends itself perfectly to fractions. In order to compare two fractions, they must be the same kind of "fruit." Think of the denominator as the type of fruit. We learned in Chapter 1 that fractions must have a common denominator before we can compare them, or add them together. The same is true for rational expressions.

## Adding Rational Expressions with Like Denominators

**Objective 1** ▶

Add rational expressions with like denominators.

As we know, we can add fractions when their denominators are the same. For example,

$$\frac{2}{7} + \frac{3}{7} = \frac{2+3}{7} = \frac{5}{7}$$

This is also true with rational expressions.

---

**Property:** **Adding Rational Expressions**

If $\dfrac{A}{C}$ and $\dfrac{B}{C}$ are rational expressions with $C \neq 0$, then

$$\frac{A}{C} + \frac{B}{C} = \frac{A+B}{C}$$

---

**Procedure:** **Adding Rational Expressions with Like Denominators**

**Step 1:** Add the numerators and place the sum over the like denominator.
**Step 2:** Simplify, if possible.

---

**Objective 1 Examples**    Add the rational expressions. Write each answer in lowest terms.

**1a.** $\dfrac{3}{x} + \dfrac{5}{x}$    **1b.** $\dfrac{2}{5a} + \dfrac{8}{5a}$    **1c.** $\dfrac{2x}{x^2+6x+5} + \dfrac{2}{x^2+6x+5}$

**Solutions**    **1a.**

$$\frac{3}{x} + \frac{5}{x} = \frac{3+5}{x}$$    Add the numerators and place over the denominator.

$$= \frac{8}{x}$$    Simplify the numerator.

**1b.**

$$\frac{2}{5a} + \frac{8}{5a} = \frac{2+8}{5a}$$    Add the numerators and place over the denominator.

$$= \frac{10}{5a}$$    Simplify the numerator.

$$= \frac{5 \cdot 2}{5 \cdot a}$$    Factor.

$$= \frac{2}{a}$$    Simplify.

**1c.**

$$\frac{2x}{x^2+6x+5} + \frac{2}{x^2+6x+5} = \frac{2x+2}{x^2+6x+5}$$    Add the numerators and place over the denominator.

$$= \frac{2(x+1)}{(x+5)(x+1)}$$    Factor the numerator and the denominator.

$$= \frac{2}{x+5}$$    Simplify.

---

✔ **Student Check 1**    Add the rational expressions. Write each answer in lowest terms.

**a.** $\dfrac{11}{2y} + \dfrac{3}{2y}$    **b.** $\dfrac{1}{x+1} + \dfrac{5}{x+1}$    **c.** $\dfrac{2x+5}{6x^2+13x-5} + \dfrac{4x-7}{6x^2+13x-5}$

## Subtracting Rational Expressions with Like Denominators

**Objective 2** ▶

Subtract rational expressions with like denominators.

Subtracting fractions is similar to adding fractions in that we must have like denominators. If we have like denominators, we subtract the numerators and put the result over the common denominator. For example,

$$\frac{2}{5} - \frac{3}{5} = \frac{2-3}{5}$$

---

**Property: Subtracting Rational Expressions**

If $\dfrac{A}{C}$ and $\dfrac{B}{C}$ are rational expressions with $C \neq 0$, then

$$\frac{A}{C} - \frac{B}{C} = \frac{A-B}{C}$$

---

**Note:** *When we subtract rational expressions we must apply the negative sign to the entire expression in the second numerator. This may require us to apply the distributive property.*

---

**Procedure: Subtracting Rational Expressions with Like Denominators**

**Step 1:** Subtract the numerators and place the difference over the like denominator.
**Step 2:** Simplify, if possible.

---

**Objective 2 Examples**    Subtract the rational expressions. Write each answer in lowest terms.

**2a.** $\dfrac{5t}{t-6} - \dfrac{6t}{t-6}$    **2b.** $\dfrac{6x^2 - 4x}{5x+10} - \dfrac{5x^2 - 6x}{5x+10}$    **2c.** $\dfrac{4x}{x^2 + x - 12} - \dfrac{3x-4}{x^2 + x - 12}$

**Solutions**    **2a.**

$$\frac{5t}{t-6} - \frac{6t}{t-6} = \frac{5t - 6t}{t-6}$$    Subtract the numerators and place over the denominator.

$$= \frac{-t}{t-6}$$    Simplify the numerator.

$$= -\frac{t}{t-6}$$    Apply the rule of negative rational expressions.

**2b.** $\dfrac{6x^2 - 4x}{5x+10} - \dfrac{5x^2 - 6x}{5x+10} = \dfrac{6x^2 - 4x - (5x^2 - 6x)}{5x+10}$    Subtract the numerators and place over the denominator.

$$= \frac{6x^2 - 4x - 5x^2 + 6x}{5x+10}$$    Distribute $-1$ to the numerator of the second fraction.

$$= \frac{x^2 + 2x}{5x+10}$$    Simplify the numerator.

$$= \frac{x(x+2)}{5(x+2)}$$    Factor the numerator and denominator.

$$= \frac{x}{5}$$    Simplify.

**2c.** $\dfrac{4x}{x^2 + x - 12} - \dfrac{3x - 4}{x^2 + x - 12} = \dfrac{4x - (3x - 4)}{x^2 + x - 12}$   Subtract the numerators and place over the denominator.

$\qquad\qquad\qquad\qquad\quad = \dfrac{4x - 3x + 4}{x^2 + x - 12}$   Distribute $-1$ to the numerator of the second fraction.

$\qquad\qquad\qquad\qquad\quad = \dfrac{x + 4}{(x + 4)(x - 3)}$   Simplify the numerator and factor the denominator.

$\qquad\qquad\qquad\qquad\quad = \dfrac{1}{x - 3}$   Simplify.

✓ **Student Check 2**   Subtract the rational expressions. Write each answer in lowest terms.

**a.** $\dfrac{2a}{a - 4} - \dfrac{7a}{a - 4}$   **b.** $\dfrac{3b}{b^2 - 5b} - \dfrac{2b + 5}{b^2 - 5b}$   **c.** $\dfrac{5m^2 + m}{m^2 - 2m - 3} - \dfrac{4m^2 - 4m}{m^2 - 2m - 3}$

## The Least Common Denominator

**Objective 3** ▶

Find the least common denominator of rational expressions.

Recall that if fractions have unlike denominators, we can find their least common denominator (LCD) and convert each fraction to an equivalent fraction with the LCD in order to add or subtract them.

Consider the rational numbers $\dfrac{5}{12}$ and $\dfrac{7}{30}$. To find the **least common denominator** of these fractions, we begin with the prime factorization of each denominator.

$12 = 4 \cdot 3 = 2 \cdot 2 \cdot 3$
$30 = 6 \cdot 5 = 2 \cdot 3 \cdot 5$

The LCD is a number that 12 and 30 divide into, that is, it is a number that contains the factors from each number. So, the LCD of 12 and 30 is

Factors of 30

$LCD = 2 \cdot 2 \cdot 3 \cdot 5 = 60$

Factors of 12

The LCD is the product of the common factors of the denominators and the other factors from each of the denominators. That is, the LCD contains all of the factors from each denominator.

We can also find the LCD using the exponents of the factors in the denominators. Rewrite the prime factorization using exponents.

$12 = 4 \cdot 3 = 2 \cdot 2 \cdot 3 = 2^2 \cdot 3$
$30 = 6 \cdot 5 = 2 \cdot 3 \cdot 5$

$LCD = 2^2 \cdot 3 \cdot 5 = 60$

So, the LCD is the product of all the different factors raised to the largest exponent that the factor occurs in any one denominator. So, if a factor occurs three times in one denominator and two times in another denominator, it must occur three times in the LCD.

---

**Procedure:  Finding the Least Common Denominator**

**Step 1:** Write each denominator in factored form.
**Step 2:** "Build" the LCD by forming the product of all factors from the first denominator.
**Step 3:** Include any additional factors from the other denominators that have not been included in Step 2.
**Step 4:** The product of the factors from Steps 2 and 3 is the LCD. Leave the LCD in factored form.

**Note:** *The least common denominator is the same as the least common multiple.*

**Objective 3 Examples** | **Determine the least common denominator of the given rational expressions.**

**3a.** $\dfrac{1}{3ab}, \dfrac{2}{15a^2}$     **3b.** $\dfrac{x}{x+1}, \dfrac{4-x}{x^2+2x+1}$     **3c.** $\dfrac{5}{t}, \dfrac{t}{t-1}$

**3d.** $\dfrac{7}{2y+18}, \dfrac{y}{y^2-81}, \dfrac{y+2}{y^2-10y+9}$     **3e.** $\dfrac{x+1}{x-6}, \dfrac{2}{6-x}$

**Solutions** **3a.**

$$3ab = 3 \cdot a \cdot b$$
$$15a^2 = 3 \cdot 5 \cdot a \cdot a$$

$\longrightarrow$   $\text{LCD} = 3 \cdot a \cdot b \cdot 5 \cdot a = 15a^2b$

When we include the factors of $15a^2$ in the LCD, we only need to include the factors of 5 and $a$ since the factors of 3 and $a$ are already included from the first denominator.

**3b.**

$$x + 1 = (x+1)$$
$$x^2 + 2x + 1 = (x+1)(x+1)$$

$\longrightarrow$   $\text{LCD} = (x+1)(x+1) = (x+1)^2$

The exponent of $(x+1)$ is 1 in the first denominator and 2 in the second denominator, so its exponent is 2 in the LCD.

**3c.**

$$t = (t)$$
$$t + 1 = (t+1)$$

$\longrightarrow$   $\text{LCD} = (t)(t+1)$

Since the denominators do not share any common factors, the LCD is the product of the denominators.

**3d.**

$$2y + 18 = 2(y+9)$$
$$y^2 - 81 = (y+9)(y-9)$$
$$y^2 - 10y + 9 = (y-9)(y-1)$$

$\longrightarrow$   $\text{LCD} = 2(y+9)(y-9)(y-1)$

**3e.** The denominators $x - 6$ and $6 - x$ are opposites of one another. When the denominators are opposite, the LCD can be either one of the denominators. That is, it can be $x - 6$ or $6 - x$, since we can go from one expression to the other by multiplying by $-1$.

$$x - 6 = (x-6) \qquad\qquad \text{LCD} = (x-6)$$
$$\text{or}$$
$$6 - x = -x + 6 = -1(x-6) \longrightarrow \text{LCD} = -1(x-6) = 6 - x$$

**✓ Student Check 3**   Determine the least common denominator of the given rational expressions.

**a.** $\dfrac{5}{18xy^2}, \dfrac{1}{4xy}$     **b.** $\dfrac{x}{x+4}, \dfrac{-2}{x^2+8x+16}$     **c.** $\dfrac{a}{a+6}, \dfrac{7}{a-2}$

**d.** $\dfrac{-1}{3y-21}, \dfrac{y}{y^2-49}, \dfrac{y+2}{2y^2+9y-35}$     **e.** $\dfrac{3}{x-1}, \dfrac{5}{1-x}$

**Objective 4** ▶

**Write equivalent rational expressions.**

## Equivalent Rational Expressions

Now that we know how to find the LCD of rational expressions, we will write equivalent forms of rational expressions. This is required before we can add or subtract rational expressions with unlike denominators.

Converting a fraction to an equivalent fraction is a skill that was presented in Chapter 1. Recall that we can write the fraction $\frac{3}{8}$ as a fraction with a denominator of 16 by multiplying the fraction by $\frac{2}{2}$, a form of 1.

$$\frac{3}{8} = \frac{3}{8} \cdot \frac{2}{2} = \frac{6}{16}$$

This skill is based on the *fundamental property of rational expressions*, which states that for $N$, $C \neq 0$,

$$\frac{N \cdot C}{D \cdot C} = \frac{N}{D}$$

We use this property to write equivalent forms of rational expressions as well. The identity property of multiplication states that any number multiplied by 1 is unchanged in value. Therefore, we can multiply a rational expression $\frac{N}{D}$ by a form of 1 to obtain an equivalent expression.

> **Property:  Equivalent Rational Expressions**
>
> For $D$, $C \neq 0$,
>
> $$\frac{N}{D} = \frac{N}{D} \cdot 1 = \frac{N}{D} \cdot \frac{C}{C} = \frac{N \cdot C}{D \cdot C}$$

Some examples of **equivalent rational expressions** are as follows.

| Equivalent Fractions | | | |
|---|---|---|---|
| $\dfrac{2}{3}$ | $\dfrac{2}{3} \cdot \dfrac{5}{5} = \dfrac{10}{15}$ | $\dfrac{2}{3} \cdot \dfrac{x}{x} = \dfrac{2x}{3x}$ | |
| $\dfrac{7}{x+2}$ | $\dfrac{7}{x+2} \cdot \dfrac{2}{2} = \dfrac{14}{2x+4}$ | $\dfrac{7}{x+2} \cdot \dfrac{2x-3}{2x-3} = \dfrac{14x-21}{2x^2+x-6}$ | |

We see that $\frac{2}{3}, \frac{10}{15}$, and $\frac{2x}{3x}$ are equivalent rational expressions since we obtain the last two from the first by multiplying by 1 in the form of $\frac{5}{5}$ and $\frac{x}{x}$, respectively. Also, $\frac{7}{x+2}, \frac{14}{2x+4}$, and $\frac{14x-21}{2x^2+x-6}$ are equivalent rational expressions since we obtain the last two from the first by multiplying by 1 in the form of $\frac{2}{2}$ and $\frac{2x-3}{2x-3}$, respectively.

> **Procedure:  Finding an Equivalent Form of a Rational Expression**
>
> **Step 1:** Multiply the numerator and denominator of a rational expression by the same nonzero value.
>
> **Step 2:** Simplify each product.

**Objective 4  Examples**   **Find the numerator that will make the two rational expressions equivalent.**

**4a.** $\dfrac{7}{3x} = \dfrac{?}{6x^2}$

**4b.** $\dfrac{3a}{4b} = \dfrac{?}{24ab^2}$

**4c.** $\dfrac{x+2}{x-1} = \dfrac{?}{x^2+3x-4}$

**4d.** $\dfrac{5x}{3x+6} = \dfrac{?}{3x^2-12}$

**Solutions**

**4a.** Write the prime factorization of the two denominators. Determine the factors that are not in the original denominator. These factors are used in the form of 1.

| Original denominator | $3x = 3 \cdot x$ |
|---|---|
| Denominator in the equivalent fraction | $6x^2 = 2 \cdot 3 \cdot x \cdot x$ |
| What do we multiply the original denominator by to obtain the second denominator? | $2x$ |

We must multiply by 1 in the form of $\dfrac{2x}{2x}$. So, the equivalent fraction with the given denominator is

$$\frac{7}{3x} = \frac{7}{3x} \cdot 1 = \frac{7}{3x} \cdot \frac{2x}{2x} = \frac{7(2x)}{(3x)(2x)} = \frac{14x}{6x^2}$$

**4b.**

| Original denominator | $4b = 2 \cdot 2 \cdot b$ |
|---|---|
| Denominator in the equivalent fraction | $24ab^2 = 2 \cdot 2 \cdot 2 \cdot 3 \cdot a \cdot b \cdot b$ |
| What do we multiply the original denominator by to obtain the second denominator? | $2 \cdot 3 \cdot a \cdot b = 6ab$ |

We must multiply by 1 in the form of $\dfrac{6ab}{6ab}$. So, the equivalent fraction with the given denominator is

$$\frac{3a}{4b} = \frac{3a}{4b} \cdot 1 = \frac{3a}{4b} \cdot \frac{6ab}{6ab} = \frac{3a(6ab)}{(4b)(6ab)} = \frac{18a^2b}{24ab^2}$$

**4c.**

| Original denominator | $x - 1$ |
|---|---|
| Denominator in the equivalent fraction | $x^2 + 3x - 4 = (x - 1)(x + 4)$ |
| What do we multiply the original denominator by to obtain the second denominator? | $(x + 4)$ |

We must multiply by 1 in the form of $\dfrac{x + 4}{x + 4}$. So, the equivalent fraction with the given denominator is

$$\frac{x + 2}{x - 1} = \frac{x + 2}{x - 1} \cdot 1 = \frac{x + 2}{x - 1} \cdot \frac{(x + 4)}{(x + 4)} = \frac{(x + 2)(x + 4)}{(x - 1)(x + 4)} = \frac{x^2 + 6x + 8}{x^2 + 3x - 4}$$

**4d.**

| Original denominator | $3x + 6 = 3(x + 2)$ |
|---|---|
| Denominator in the equivalent fraction | $3x^2 - 12 = 3(x^2 - 4) = 3(x - 2)(x + 2)$ |
| What do we multiply the original denominator by to obtain the second denominator? | $(x - 2)$ |

We must multiply by 1 in the form of $\dfrac{x - 2}{x - 2}$. So, the equivalent fraction with the given denominator is

$$\frac{5x}{3x + 6} = \frac{5x}{3(x + 2)} \cdot 1 = \frac{5x}{3(x + 2)} \cdot \frac{(x - 2)}{(x - 2)} = \frac{(5x)(x - 2)}{3(x + 2)(x - 2)} = \frac{5x^2 - 10x}{3x^2 - 12}$$

✓ **Student Check 4** Find the numerator that will make the two rational expressions equivalent.

**a.** $\dfrac{6}{7y} = \dfrac{?}{35y^2}$

**b.** $\dfrac{3a}{4b} = \dfrac{?}{36ab^3}$

**c.** $\dfrac{x - 3}{x - 5} = \dfrac{?}{x^2 - 9x + 20}$

**d.** $\dfrac{2b + 1}{4b + 32} = \dfrac{?}{4b^2 - 256}$

**Objective 5** ▶

Troubleshoot common errors.

## Troubleshooting Common Errors

Some common errors for adding and subtracting rational expressions, finding the LCD, and writing equivalent rational expressions are shown next.

**Objective 5 Examples**  A problem and an incorrect solution are given. Provide the correct solution and an explanation of the error.

**5a.** $\dfrac{6}{x} + \dfrac{1}{x}$

| Incorrect Solution | Correct Solution and Explanation |
|---|---|
| $\dfrac{6}{x} + \dfrac{1}{x} = \dfrac{7}{2x}$ | When we add rational expressions, we do not add their denominators. We add the numerators and place their sum over the common denominator. $$\dfrac{6}{x} + \dfrac{1}{x} = \dfrac{6+1}{x} = \dfrac{7}{x}$$ |

**5b.** $\dfrac{y+5}{y+4} - \dfrac{y+3}{y+4}$

| Incorrect Solution | Correct Solution and Explanation |
|---|---|
| $\dfrac{y+5}{y+4} - \dfrac{y+3}{y+4} = \dfrac{y+5-y+3}{y+4}$ $= \dfrac{8}{y+4}$ | When we subtract two rational expressions, the negative sign must be applied to the entire second numerator, not just the first term of the numerator. $$\dfrac{y+5}{y+4} - \dfrac{y+3}{y+4} = \dfrac{y+5-(y+3)}{y+4}$$ $$= \dfrac{y+5-y-3}{y+4}$$ $$= \dfrac{2}{y+4}$$ |

**5c.** Determine the numerator that will make the expressions equivalent: $\dfrac{4x}{x-1} = \dfrac{?}{x^2-1}$

| Incorrect Solution | Correct Solution and Explanation |
|---|---|
| $\dfrac{4x}{x-1}\left(\dfrac{x}{x}\right) = \dfrac{4x^2}{x^2-1}$ | If we multiply $x$ by $x-1$, we get $x^2-x$, not $x^2-1$. The second denominator $x^2-1$ is equal to $(x-1)(x+1)$, so we must multiply the initial fraction by 1 in the form of $\dfrac{x+1}{x+1}$. $$\dfrac{4x}{x-1}\left(\dfrac{x+1}{x+1}\right) = \dfrac{4x^2+4x}{x^2-1}$$ |

**5d.** Find the LCD of $\dfrac{4}{x}$ and $\dfrac{3}{x+5}$.

| Incorrect Solution | Correct Solution and Explanation |
|---|---|
| The LCD is $x+5$ since $x$ is included in the expression $x+5$. | The LCD consists of factors of each denominator. In the second denominator, $x+5$, the $x$ is a term and not a factor. So, the LCD is $x(x+5)$. |

## ANSWERS TO STUDENT CHECKS

**Student Check 1**   **a.** $\dfrac{7}{y}$   **b.** $\dfrac{6}{x+1}$   **c.** $\dfrac{2}{2x+5}$

**Student Check 2**   **a.** $-\dfrac{5a}{a-4}$   or   $\dfrac{5a}{4-a}$   **b.** $\dfrac{1}{b}$

   **c.** $\dfrac{m^2+5m}{(m-3)(m+1)}$   or   $\dfrac{m(m+5)}{(m-3)(m+1)}$

**Student Check 3**   **a.** $36xy^2$   **b.** $(x+4)(x+4)$   or   $(x+4)^2$
   **c.** $(a+6)(a-2)$   **d.** $3(y-7)(y+7)(2y-5)$
   **e.** $x-1$   or   $1-x$

**Student Check 4**   **a.** $30y$   **b.** $27a^2b^2$   **c.** $x^2-7x+12$
   **d.** $2b^2-15b-8$

## SUMMARY OF KEY CONCEPTS

1. If the rational expressions being added have like denominators, then add the numerators and place the sum over the common denominator. Simplify the result, if possible.

2. If the rational expressions being subtracted have like denominators, then subtract the numerators and place the difference over the common denominator. Remember to distribute $-1$ to all terms in the numerator of the second fraction. Simplify the result, if possible.

3. To find the LCD of rational expressions, factor all of the denominators. Build the LCD by forming the product of all the factors from the first denominator. Multiply this by any factors from the other denominators that have not yet been included.

4. To convert a rational expression to an equivalent fraction in which the denominator of the equivalent fraction is known, we have to determine what factor was multiplied by the original denominator to obtain the new denominator. Rewrite the original denominator and the new denominator in terms of their prime factors. Identify the factors that are missing from the original denominator and multiply the original numerator by this factor to obtain an equivalent fraction.

## GRAPHING CALCULATOR SKILLS

We can use the graphing calculator to find the least common multiple of two numbers. This can assist us in finding the least common denominator for some rational expressions.

**Example:** Find the coefficient of the least common denominator of the two rational expressions: $\dfrac{5}{14x^4y^5}, \dfrac{x}{104y^4}$

**Solution:** We need to find the least common multiple (lcm) of 14 and 104. The lcm is found by pressing Math and accessing the Num menu and selecting option 8.

So, the LCD of the rational expressions is $728x^4y^5$.

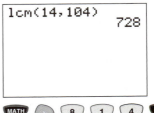

## SECTION 7.3   EXERCISE SET

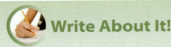 **Write About It!**

Use complete sentences in your answer to each exercise.

1. Explain how to add or subtract rational expressions with the like denominators.   *You add or subtract the numerators and place the result over the like denominator.*

2. Explain the fundamental property of rational expressions.

3. What does it mean for two rational expressions to be equivalent?

4. Explain how to create a rational expression that is equivalent to a given rational expression.

5. Explain the process for finding the least common denominator of rational expressions.

6. Explain why $\dfrac{1}{-x-1}$ and $-\dfrac{1}{x-1}$ are not equivalent.

*Additional answers can be found in the Instructor Answer Appendix.*

 **Practice Makes Perfect!**

**Add or subtract the rational expressions. Write each answer in lowest terms. (*See Objectives 1 and 2.*)**

**7.** $\dfrac{4}{3x} + \dfrac{2}{3x}$    $\dfrac{2}{x}$

**8.** $\dfrac{4}{5x} + \dfrac{1}{5x}$    $\dfrac{1}{x}$

**9.** $\dfrac{3x}{x-1} - \dfrac{2x}{x-1}$    $\dfrac{x}{x-1}$

**10.** $\dfrac{6x}{x+2} - \dfrac{x}{x+2}$    $\dfrac{5x}{x+2}$

**11.** $\dfrac{y}{y-1} + \dfrac{2y+3}{y-1}$    $\dfrac{3y+3}{y-1}$

**12.** $\dfrac{3y}{4y+1} + \dfrac{y+1}{4y+1}$    $1$

**13.** $\dfrac{2x}{(2x-1)(x-4)} - \dfrac{1}{(2x-1)(x-4)}$    $\dfrac{1}{x-4}$

**14.** $\dfrac{4x}{(4x-5)(x+2)} - \dfrac{5}{(4x-5)(x+2)}$    $\dfrac{1}{x+2}$

**15.** $\dfrac{y}{y^2-1} + \dfrac{1}{y^2-1}$    $\dfrac{1}{y-1}$

**16.** $\dfrac{y}{y^2-25} + \dfrac{5}{y^2-25}$    $\dfrac{1}{y-5}$

**17.** $\dfrac{3x-2}{x+1} - \dfrac{x-3}{x+1}$    $\dfrac{2x+1}{x+1}$

**18.** $\dfrac{6x-7}{x+7} - \dfrac{3x-6}{x+7}$    $\dfrac{3x-1}{x+7}$

**19.** $\dfrac{4x}{x^2+x-6} - \dfrac{3x+2}{x^2+x-6}$    $\dfrac{1}{x+3}$

**20.** $\dfrac{5x-8}{(x-2)(x-7)} - \dfrac{6x-10}{(x-2)(x-7)}$    $-\dfrac{1}{x-7}$

**21.** $\dfrac{3x-5}{(x-3)(x+1)} - \dfrac{4x-8}{(x-3)(x+1)}$    $-\dfrac{1}{x+1}$

**22.** $\dfrac{6x+7}{x^2-3x-4} - \dfrac{5x+11}{x^2-3x-4}$    $\dfrac{1}{x+1}$

**23.** $\dfrac{x+1}{x-3} + \dfrac{2x-7}{x-3}$    $\dfrac{3(x-2)}{x-3}$

**24.** $\dfrac{4x-3}{2x+1} + \dfrac{8x+9}{2x+1}$    $6$

**Find the least common denominator of the given rational expressions. (*See Objective 3.*)**

**25.** $\dfrac{1}{5a}, \dfrac{3}{15a}$    $15a$

**26.** $\dfrac{3}{10a}, \dfrac{5}{20a}$    $20a$

**27.** $\dfrac{7}{2x}, \dfrac{3}{6x^2}$    $6x^2$

**28.** $\dfrac{2}{5x}, \dfrac{-6}{20x^3}$    $20x^3$

**29.** $\dfrac{3}{4x}, \dfrac{1}{3x^2}$    $12x^2$

**30.** $\dfrac{10}{7x^2}, \dfrac{11}{6x}$    $42x^2$

**31.** $\dfrac{3}{xy^2}, \dfrac{3}{x^3y}$    $x^3y^2$

**32.** $\dfrac{a+1}{2a^3b}, \dfrac{b}{7a}$    $14a^3b$

**33.** $\dfrac{10}{3ab^3}, \dfrac{6}{5a^5b^4}$    $15a^5b^4$

**34.** $\dfrac{9}{12x^3y^2}, \dfrac{4}{8xy^2}$    $24x^3y^2$

**35.** $\dfrac{x}{x(x-5)}, \dfrac{2}{x}$    $x(x-5)$

**36.** $\dfrac{2x}{x(3x-4)}, \dfrac{x^2}{3x-4}$    $x(3x-4)$

**37.** $\dfrac{4}{x+2}, \dfrac{5}{x^2+2x}$    $x(x+2)$

**38.** $\dfrac{1}{5-x}, \dfrac{x+1}{5x-x^2}$    $x(5-x)$

**39.** $\dfrac{x+1}{x-3}, \dfrac{x}{x-1}$    $(x-3)(x-1)$

**40.** $\dfrac{2x-1}{x+5}, \dfrac{4x}{x-6}$    $(x+5)(x-6)$

**41.** $\dfrac{3}{x^2+7x+10}, \dfrac{5}{x+2}$    $(x+2)(x+5)$

**42.** $\dfrac{8}{x^2+3x-18}, \dfrac{6}{x-3}$    $(x-3)(x+6)$

**43.** $\dfrac{6x}{x^2+2x-35}, \dfrac{4x}{x+7}$    $(x-5)(x+7)$

**44.** $\dfrac{2x}{x^2-6x+5}, \dfrac{x^2}{x-1}$    $(x-5)(x-1)$

**45.** $\dfrac{3x}{x^2-9}, \dfrac{4x}{x^2-6x+9}$    $(x+3)(x-3)(x-3)$

**46.** $\dfrac{2x}{x^2-25}, \dfrac{x}{x^2-x-20}$    $(x+4)(x-5)(x+5)$

**47.** $\dfrac{2x}{4x^2-9}, \dfrac{5x}{6x^2-5x-6}$    $(2x+3)(2x-3)(3x+2)$

**48.** $\dfrac{-9}{6x^2-13x-5}, \dfrac{8}{2x^2+3x-20}$    $(3x+1)(2x-5)(x+4)$

**Find the numerator that makes each equation true. (*See Objective 4.*)**

**49.** $\dfrac{1}{2x} = \dfrac{?}{6x}$    $3$

**50.** $\dfrac{2}{5x} = \dfrac{?}{15x}$    $6$

**51.** $\dfrac{2}{x^2} = \dfrac{?}{4x^2}$    $8$

**52.** $\dfrac{3}{y} = \dfrac{?}{4y}$    $12$

**53.** $\dfrac{4x}{y} = \dfrac{?}{xy}$    $4x^2$

**54.** $\dfrac{3x^2}{y^2} = \dfrac{?}{2y^2}$    $6x^2$

**55.** $\dfrac{x-1}{y} = \dfrac{?}{3y}$    $3(x-1)$

**56.** $\dfrac{x+4}{y} = \dfrac{?}{5y}$    $5(x+4)$

**57.** $\dfrac{2a}{b} = \dfrac{?}{3ab}$    $6a^2$

**58.** $\dfrac{3}{4b} = \dfrac{?}{8ab}$    $6a$

**59.** $\dfrac{2x}{x-1} = \dfrac{?}{(x-1)(x+2)}$    $2x(x+2)$

**60.** $\dfrac{x^2}{x+3} = \dfrac{?}{(x+3)(x-5)}$    $x^2(x-5)$

**61.** $\dfrac{x+3}{x-1} = \dfrac{?}{3x-3}$    $3(x+3)$

**62.** $\dfrac{x-6}{x+7} = \dfrac{?}{2x+14}$    $2(x-6)$

**63.** $\dfrac{x}{x+1} = \dfrac{?}{3x+3}$    $3x$

**64.** $\dfrac{y}{4-y} = \dfrac{?}{8-2y}$    $2y$

**65.** $\dfrac{1-2x}{x} = \dfrac{?}{x^2}$    $x(1-2x)$

**66.** $\dfrac{3+5x}{x} = \dfrac{?}{2x}$    $2(3+5x)$

**67.** $\dfrac{3x}{1-2x} = \dfrac{?}{4x^2-2x}$    $-6x^2$

**68.** $\dfrac{5x}{3-4x} = \dfrac{?}{12x^2-9x}$    $-15x^2$

**69.** $\dfrac{x}{x+3} = \dfrac{?}{x^2-9}$    $x(x-3)$

**70.** $\dfrac{2}{x+5} = \dfrac{?}{x^2-25}$    $2(x-5)$

**71.** $\dfrac{t^3+t}{t-2} = \dfrac{?}{2t-4}$    $2t(t^2+1)$

**72.** $\dfrac{t^2-7t}{3t+5} = \dfrac{?}{12t+20}$    $4t^2-28t$

**73.** $\dfrac{4a-5}{3a+12} = \dfrac{?}{6ab+24b}$    $2b(4a-5)$

**74.** $\dfrac{2a-1}{5a-6} = \dfrac{?}{20ab-24b}$    $4b(2a-1)$

## Mix 'Em Up!

**Add or subtract the rational expressions. Write each answer in lowest terms.**

**75.** $\dfrac{-3x}{3x+8}+\dfrac{19x}{3x+8}$ $\quad\frac{16x}{3x+8}$

**76.** $\dfrac{8x}{x+7}-\dfrac{9x}{x+7}$ $\quad-\frac{x}{x+7}$

**77.** $\dfrac{6a+5}{9a-1}+\dfrac{3a-2}{9a-1}$ $\quad\frac{9a+3}{9a-1}$

**78.** $\dfrac{5b-6}{2b-15}+\dfrac{b-4}{2b-15}$ $\quad\frac{6b-10}{2b-15}$

**79.** $\dfrac{5x+3}{2x+7}-\dfrac{3x-4}{2x+7}$ $\quad1$

**80.** $\dfrac{22x-9}{12x-5}-\dfrac{10x-4}{12x-5}$ $\quad1$

**81.** $\dfrac{3x}{9x^2-4}-\dfrac{2}{9x^2-4}$ $\quad\frac{1}{3x+2}$

**82.** $\dfrac{x}{x^2-36}+\dfrac{6}{x^2-36}$ $\quad\frac{1}{x-6}$

**83.** $\dfrac{2x}{4x^2-1}+\dfrac{1}{4x^2-1}$ $\quad\frac{1}{2x-1}$

**84.** $\dfrac{x}{x^2-100}-\dfrac{10}{x^2-100}$ $\quad\frac{1}{x+10}$

**85.** $\dfrac{3x+7}{(x-2)(x+5)}-\dfrac{6x+22}{(x-2)(x+5)}$ $\quad-\frac{3}{x-2}$

**86.** $\dfrac{4y-7}{y^2+7y-18}-\dfrac{2y-3}{y^2+7y-18}$ $\quad\frac{2}{y+9}$

**Find the least common denominator of the rational expressions.**

**87.** $\dfrac{x}{x-9},\dfrac{3}{9x-x^2}$ $\quad x(x-9)$

**88.** $\dfrac{1}{x-12},\dfrac{5x}{12x-x^2}$ $\quad x(x-12)$

**89.** $\dfrac{x}{4x+3},\dfrac{3}{x+1}$ $\quad(4x+3)(x+1)$

**90.** $\dfrac{5}{x+3},\dfrac{x}{x^2-2x-15}$ $\quad(x+3)(x-5)$

**91.** $\dfrac{6x}{25x^2-4},\dfrac{-x}{5x^2+13x-6}$ $\quad(5x+2)(5x-2)(x+3)$

**92.** $\dfrac{5x}{4x+6},\dfrac{3}{4x^2+12x+9}$ $\quad2(2x+3)^2$

**93.** $\dfrac{x}{15x-6},\dfrac{7}{25x^2-20x+4}$ $\quad3(5x-2)^2$

**94.** $\dfrac{7}{6x^2-5x-4},\dfrac{-5}{2x^2-9x-5}$ $\quad(3x-4)(2x+1)(x-5)$

**Find the numerator that makes the two rational expressions equivalent.**

**95.** $\dfrac{x+1}{x-6}=\dfrac{?}{x^2-6x}$ $\quad x(x+1)$

**96.** $\dfrac{2x}{x-4}=\dfrac{?}{2x^2-7x-4}$ $\quad2x(2x+1)$

**97.** $\dfrac{x-1}{3x+6}=\dfrac{?}{12x^2+24x}$ $\quad4x(x-1)$

**98.** $\dfrac{3x}{x+1}=\dfrac{?}{x^2+4x+3}$ $\quad3x(x+3)$

**99.** $\dfrac{3}{4a}=\dfrac{?}{12a^2}$ $\quad9a$

**100.** $\dfrac{2}{xy}=\dfrac{?}{3x^2y}$ $\quad6x$

**101.** $\dfrac{5x}{7-2x}=\dfrac{?}{8x^2-28x}$ $\quad-20x^2$

**102.** $\dfrac{x}{1-3x}=\dfrac{?}{15x^2-5x}$ $\quad-5x^2$

## You Be the Teacher!

**Correct each student's errors, if any.**

**103.** Add $\dfrac{2}{5x}+\dfrac{3}{5x}$.

Ricky's work:

$$\dfrac{2}{5x}+\dfrac{3}{5x}=\dfrac{5}{5x}=\dfrac{\cancel{5}}{\cancel{5}x}=x$$

$$\dfrac{2}{5x}+\dfrac{3}{5x}=\dfrac{5}{5x}=\dfrac{\cancel{5}}{\cancel{5}x}=\dfrac{1}{x}$$

**104.** Subtract $\dfrac{4x+3}{2x-5}-\dfrac{6x+1}{2x-5}$.

Rodney's work:

$$\dfrac{4x+3}{2x-5}-\dfrac{6x+1}{2x-5}=\dfrac{4x+3-6x+1}{2x-5}=\dfrac{-2x+4}{2x-5}$$

**105.** Find the least common denominator of $\dfrac{3}{x+2}$ and $\dfrac{1}{x}$.

Caridee's work:

The least common denominator of $\dfrac{3}{x+2}$ and $\dfrac{1}{x}$ is $x+2$. $\quad x(x+2)$

**106.** Find the least common denominator of $\dfrac{3}{x+1}$ and $\dfrac{x}{x^2-1}$.

Meg's work:

The least common denominator of $\dfrac{3}{x+1}$ and $\dfrac{x}{x^2-1}$ is $x^2-1$. $\quad$ correct

# Addition and Subtraction of Rational Expressions with Unlike Denominators

▶ **OBJECTIVES**

As a result of completing this section, you will be able to

1. Add rational expressions with unlike denominators.

2. Subtract rational expressions with unlike denominators.

3. Troubleshoot common errors.

Two rectangles are placed side by side. The length of one rectangle is $\dfrac{4}{x+1}$ and the length of the other rectangle is $\dfrac{2}{x-4}$. What is the total length formed by the two rectangles? To answer this question, we must simplify $\dfrac{4}{x+1} + \dfrac{2}{x-4}$. These are rational expressions with unlike denominators.

## Adding Rational Expressions with Unlike Denominators

To add fractions with unlike denominators, we can use their LCD. For example, the LCD of $\dfrac{7}{6}$ and $\dfrac{5}{22}$ is 66. So, we multiply $\dfrac{7}{6}$ by 1 in the form of $\dfrac{11}{11}$ and we multiply $\dfrac{5}{22}$ by 1 in the form of $\dfrac{3}{3}$.

**Objective 1** ▶

Add rational expressions with unlike denominators.

$$
\begin{aligned}
\frac{7}{6} + \frac{5}{22} &= \frac{7}{6} \cdot \frac{11}{11} + \frac{5}{22} \cdot \frac{3}{3} \\
&= \frac{77}{66} + \frac{15}{66} \\
&= \frac{92}{66} \\
&= \frac{46}{33}
\end{aligned}
$$

We add rational expressions with unlike denominators in this same way. We will apply the skills we learned in Section 7.3 to determine the LCD, rewrite equivalent rational expressions, and convert rational expressions to an equivalent fraction using the LCD. Then we will be able to add the expressions as we did in Section 7.3.

---

**Procedure:  Adding Rational Expressions with Unlike Denominators**

**Step 1:** Determine the LCD of the rational expressions.

**Step 2:** Convert each rational expression to an equivalent fraction with the LCD as its denominator.

**Step 3:** Add the rational expressions.

**Step 4:** Simplify, if possible.

---

**Objective 1 Examples**   Add the rational expressions. Write each answer in lowest terms.

**1a.** $\dfrac{3}{x} + \dfrac{5}{x^2}$

**1b.** $\dfrac{4}{x+1} + \dfrac{2}{x-4}$

**1c.** $\dfrac{5b}{b^2-9} + \dfrac{6}{2b^2-b-15}$

**1d.** $\dfrac{3}{x-1} + \dfrac{2}{1-x}$

---

**Solutions**   **1a.** $\dfrac{3}{x} + \dfrac{5}{x^2} = \dfrac{3}{x} \cdot \dfrac{x}{x} + \dfrac{5}{x^2}$      Write equivalent fractions with the LCD, $x^2$.

$\qquad\qquad = \dfrac{3x}{x^2} + \dfrac{5}{x^2}$      Simplify the product.

$\qquad\qquad = \dfrac{3x+5}{x^2}$      Add the numerators and place over the LCD.

**1b.** $\dfrac{4}{x+1} + \dfrac{2}{x-4} = \dfrac{4}{x+1} \cdot \dfrac{x-4}{x-4} + \dfrac{2}{x-4} \cdot \dfrac{x+1}{x+1}$     Write equivalent fractions with the LCD, $(x+1)(x-4)$.

$$= \dfrac{4x-16}{(x+1)(x-4)} + \dfrac{2x+2}{(x+1)(x-4)}$$     Simplify each product.

$$= \dfrac{4x-16+2x+2}{(x+1)(x-4)}$$     Add the numerators and place over the LCD.

$$= \dfrac{6x-14}{(x+1)(x-4)}$$     Simplify.

**INSTRUCTORS NOTE:**
Inform students that they should determine if the numerator and denominator of the final result have any common factors to be sure the answer is in lowest terms.

**1c.** Factor the denominators to determine the LCD.

$$b^2 - 9 = (b-3)(b+3)$$
$$2b^2 - b - 15 = (2b+5)(b-3)$$     $\longrightarrow$     $LCD = (b-3)(b+3)(2b+5)$

Multiply each expression by a form of 1 that contains the factor needed for the original denominator to become the LCD. Then add the numerators of the like denominators.

$$\dfrac{5b}{b^2-9} + \dfrac{6}{2b^2-b-15} = \dfrac{5b}{(b-3)(b+3)} \cdot \dfrac{2b+5}{2b+5} + \dfrac{6}{(2b+5)(b-3)} \cdot \dfrac{b+3}{b+3}$$

$$= \dfrac{5b(2b+5)}{(b-3)(b+3)(2b+5)} + \dfrac{6(b+3)}{(b-3)(b+3)(2b+5)}$$

$$= \dfrac{10b^2+25b}{(b-3)(b+3)(2b+5)} + \dfrac{6b+18}{(b-3)(b+3)(2b+5)}$$

$$= \dfrac{10b^2+25b+6b+18}{(b-3)(b+3)(2b+5)}$$

$$= \dfrac{10b^2+31b+18}{(b-3)(b+3)(2b+5)}$$

**1d.**     $\dfrac{3}{x-1} + \dfrac{2}{1-x} = \dfrac{3}{x-1} + \dfrac{2}{-(x-1)}$     Rewrite $1-x$ as $-(x-1)$.

$$= \dfrac{3}{x-1} + \dfrac{-2}{x-1}$$     Apply the rule of negative rational expressions.

$$= \dfrac{3-2}{x-1}$$     Add the numerators and place over the LCD.

$$= \dfrac{1}{x-1}$$     Simplify.

✔ **Student Check 1**     Add the rational expressions. Write each answer in lowest terms.

**a.** $\dfrac{x^2}{x^2-7x} + \dfrac{2x}{x-7}$     **b.** $\dfrac{5}{x+6} + \dfrac{3}{x+2}$

**c.** $\dfrac{6y}{y^2-4} + \dfrac{4}{3y^2+7y+2}$     **d.** $\dfrac{6}{a-5} + \dfrac{9}{5-a}$

## Subtracting Rational Expressions with Unlike Denominators

**Objective 2** ▶

Subtract rational expressions with unlike denominators.

Subtracting rational expressions with unlike denominators is similar to adding rational expressions with unlike denominators. After the LCD is determined and each rational expression has been converted to an equivalent rational expression, we subtract the numerators. So, the key is that we subtract the entire expression in the numerator of the rational expression being subtracted. To subtract this numerator is to add the opposite of it.

> **Procedure: Subtracting Rational Expressions with Unlike Denominators**
> **Step 1:** Determine the LCD of the rational expressions.
> **Step 2:** Convert each rational expression to an equivalent fraction with the LCD as its denominator.
> **Step 3:** Subtract the rational expressions.
> **Step 4:** Simplify, if possible.

**Objective 2 Examples**

**Subtract the rational expressions. Write each answer in lowest terms.**

**2a.** $\dfrac{2a}{a^2 - 3a} - \dfrac{1}{2a - 6}$    **2b.** $\dfrac{4}{y - 1} - \dfrac{7}{4y}$

**2c.** $2 - \dfrac{3x}{x + 4}$    **2d.** $\dfrac{x}{3x^2 + 19x - 14} - \dfrac{2}{x^2 + 15x + 56}$

**Solutions**

**2a.** Factor the denominators to determine the LCD.

$$a^2 - 3a = a(a - 3)$$
$$2a - 6 = 2(a - 3) \quad \longrightarrow \quad \text{LCD} = 2a(a - 3)$$

$$\dfrac{2a}{a^2 - 3a} - \dfrac{1}{2a - 6} = \dfrac{2a}{a(a - 3)} \cdot \dfrac{2}{2} - \dfrac{1}{2(a - 3)} \cdot \dfrac{a}{a} \qquad \text{Write equivalent fractions with the LCD, } 2a(a - 3).$$

$$= \dfrac{4a}{2a(a - 3)} - \dfrac{a}{2a(a - 3)} \qquad \text{Simplify each product.}$$

$$= \dfrac{4a - a}{2a(a - 3)} \qquad \text{Subtract the numerators and place over the LCD.}$$

$$= \dfrac{3a}{2a(a - 3)} \qquad \text{Simplify.}$$

$$= \dfrac{3}{2(a - 3)} \qquad \text{Divide out the common factor, } a.$$

**2b.**

$$\dfrac{4}{y - 1} - \dfrac{7}{4y} = \dfrac{4}{y - 1} \cdot \dfrac{4y}{4y} - \dfrac{7}{4y} \cdot \dfrac{y - 1}{y - 1} \qquad \text{Write equivalent fractions with the LCD, } 4y(y - 1).$$

$$= \dfrac{4(4y)}{4y(y - 1)} - \dfrac{7(y - 1)}{4y(y - 1)} \qquad \text{Simplify each product.}$$

$$= \dfrac{16y - 7(y - 1)}{4y(y - 1)} \qquad \text{Subtract the numerators and place over the LCD.}$$

$$= \dfrac{16y - 7y + 7}{4y(y - 1)} \qquad \text{Apply the distributive property in the numerator.}$$

$$= \dfrac{9y + 7}{4y(y - 1)} \qquad \text{Simplify.}$$

**2c.**

$$2 - \dfrac{3x}{x + 4} = \dfrac{2}{1} \cdot \dfrac{x + 4}{x + 4} - \dfrac{3x}{x + 4} \qquad \text{Write equivalent fractions with the LCD, } x + 4.$$

$$= \dfrac{2(x + 4)}{x + 4} - \dfrac{3x}{x + 4} \qquad \text{Simplify the product.}$$

$$= \dfrac{2x + 8 - 3x}{x + 4} \qquad \text{Subtract the numerators and place over the LCD.}$$

$$= \dfrac{-x + 8}{x + 4} \qquad \text{Simplify.}$$

$$= -\dfrac{x - 8}{x + 4} \qquad \text{Recall } \dfrac{-N}{D} = -\dfrac{N}{D} \text{ and that } -x + 8 = -(x - 8).$$

**2d.** Factor the denominators to determine the LCD.

$$3x^2 + 19x - 14 = (3x - 2)(x + 7)$$
$$x^2 + 15x + 56 = (x + 8)(x + 7) \qquad \longrightarrow \qquad \text{LCD} = (3x - 2)(x + 8)(x + 7)$$

$$\frac{x}{3x^2 + 19x - 14} - \frac{2}{x^2 + 15x + 56}$$

$$= \frac{x}{(3x - 2)(x + 7)} \cdot \frac{x + 8}{x + 8} - \frac{2}{(x + 8)(x + 7)} \cdot \frac{3x - 2}{3x - 2}$$

$$= \frac{x(x + 8)}{(3x - 2)(x + 8)(x + 7)} - \frac{2(3x - 2)}{(3x - 2)(x + 8)(x + 7)}$$

$$= \frac{x^2 + 8x - 2(3x - 2)}{(3x - 2)(x - 8)(x + 7)}$$

$$= \frac{x^2 + 8x - 6x + 4}{(3x - 2)(x - 8)(x + 7)}$$

$$= \frac{x^2 + 2x + 4}{(3x - 2)(x - 8)(x + 7)}$$

✔ **Student Check 2**  Subtract the rational expressions. Write each answer in lowest terms.

**a.** $\dfrac{2}{a} - \dfrac{4}{3a^2}$ 

**b.** $\dfrac{3y}{2y + 1} - \dfrac{2}{5y}$

**c.** $6 - \dfrac{2}{x + 3}$ 

**d.** $\dfrac{3x}{x^2 + 5x + 6} - \dfrac{4}{2x^2 + 3x - 2}$

---

**Objective 3** ▶

Troubleshoot common errors.

## Troubleshooting Common Errors

A common error associated with subtracting rational expressions with unlike denominators is shown.

**Objective 3 Example**  **A problem and an incorrect solution are given. Provide the correct solution and an explanation of the error.**

Simplify $\dfrac{3}{x + 5} - \dfrac{6}{x - 1}$.

| Incorrect Solution | Correct Solution and Explanation |
|---|---|
| $\dfrac{3}{x + 5} - \dfrac{6}{x - 1}$ | The negative sign was not distributed to the numerator of the second expression. |
| $= \dfrac{3}{x + 5} \cdot \dfrac{x - 1}{x - 1} - \dfrac{6}{x - 1} \cdot \dfrac{x + 5}{x + 5}$ | $\dfrac{3}{x + 5} - \dfrac{6}{x - 1}$ |
| $= \dfrac{3x - 3 - 6x + 30}{(x - 1)(x + 5)}$ | $= \dfrac{3}{x + 5} \cdot \dfrac{x - 1}{x - 1} - \dfrac{6}{x - 1} \cdot \dfrac{x + 5}{x + 5}$ |
| $= \dfrac{-3x + 27}{(x - 1)(x + 5)}$ | $= \dfrac{3x - 3 - 6x - 30}{(x - 1)(x + 5)}$ |
| | $= \dfrac{-3x - 33}{(x - 1)(x + 5)}$ |

## ANSWERS TO STUDENT CHECKS

**Student Check 1**    **a.** $\dfrac{3x}{x-7}$    **b.** $\dfrac{8x+28}{(x+2)(x+6)}$

   **c.** $\dfrac{18y^2+10y-8}{(y-2)(y+2)(3y+1)}$    **d.** $-\dfrac{3}{a-5}$

**Student Check 2**    **a.** $\dfrac{6a-4}{3a^2}$    **b.** $\dfrac{15y^2-4y-2}{5y(2y+1)}$

   **c.** $\dfrac{6x+16}{x+3}$    **d.** $\dfrac{6x^2-7x-12}{(x+2)(x+3)(2x-1)}$

## SUMMARY OF KEY CONCEPTS

1. If the rational expressions being added or subtracted do not have the same denominator, then convert each expression to an equivalent fraction with the LCD as its denominator. Then add or subtract the numerators and put the result over the LCD. Simplify, if possible.

2. When subtracting rational expressions, be sure to apply the negative sign to the entire numerator of the expression being subtracted.

## GRAPHING CALCULATOR SKILLS

The graphing calculator can help us determine if we worked a problem correctly.

**Example:** Verify that $\dfrac{2x}{x^2-3x}-\dfrac{1}{x-3}=\dfrac{1}{x-3}$.

**Solution:** Enter the original problem in $Y_1$ and the simplified expression in $Y_2$. Compare the $y$-values of these expressions. If they agree, then we performed the operation correctly. The only values for which the two quantities may not agree are the values where the expressions are undefined.

```
Plot1 Plot2 Plot3
\Y1◘(2X)/(X²-3X)
-1/(X-3)
\Y2◘1/(X-3)
\Y3=
\Y4=
\Y5=
\Y6=
```

| X | Y₁ | Y₂ |
|---|------|------|
| -3 | -.1667 | -.1667 |
| -2 | -.2 | -.2 |
| -1 | -.25 | -.25 |
| 0 | ERROR | -.3333 |
| 1 | -.5 | -.5 |
| 2 | -1 | -1 |
| 3 | ERROR | ERROR |

X= -3

Since the two columns agree, except for the values where the expressions are undefined, our answer is correct.

## SECTION 7.4 / EXERCISE SET

### Write About It!

Use complete sentences in your answer to each exercise.

1. Explain how to add or subtract rational expressions with different denominators.

2. What is a key step in subtracting rational expressions?

### Practice Makes Perfect!

**Add or subtract the rational expressions. Write each answer in lowest terms. (See Objectives 1 and 2.)**

3. $\dfrac{5}{x}+\dfrac{1}{x^2}$    $\dfrac{5x+1}{x^2}$

4. $\dfrac{3}{x^3}+\dfrac{2}{x^2}$    $\dfrac{2x+3}{x^3}$

5. $\dfrac{1}{3x}-\dfrac{1}{6x^2}$    $\dfrac{2x-1}{6x^2}$

6. $\dfrac{4}{5x}-\dfrac{2}{15x^3}$    $\dfrac{12x^2-2}{15x^3}$

7. $\dfrac{1}{x}-\dfrac{2}{x(x+2)}$    $\dfrac{1}{x+2}$

8. $\dfrac{1}{x}-\dfrac{9}{x(x+9)}$    $\dfrac{1}{x+9}$

9. $\dfrac{100}{r-5}+\dfrac{50}{r}$    $\dfrac{50(3r-5)}{r(r-5)}$

10. $\dfrac{20}{r+1}+\dfrac{40}{r}$    $\dfrac{20(3r+2)}{r(r+1)}$

11. $\dfrac{4}{x-1}+\dfrac{3x}{(x-1)(x+2)}$

12. $\dfrac{2}{x+4}+\dfrac{x}{(x+4)(x-2)}$

13. $\dfrac{4x}{x+1}+\dfrac{x^2}{x^2+6x+5}$

14. $\dfrac{x}{x-6}+\dfrac{x-1}{x^2-5x-6}$

15. $\dfrac{2x}{x^2-9}-\dfrac{1}{x+3}$    $\dfrac{1}{x-3}$

16. $\dfrac{5}{x^2-1}-\dfrac{3}{x+1}$

17. $\dfrac{x}{x+3}+\dfrac{4}{x+1}$

18. $\dfrac{2x}{x-1}+\dfrac{x}{x-4}$

19. $\dfrac{3x}{x-2}-\dfrac{x}{x-3}$

20. $\dfrac{6x}{x+9}-\dfrac{x}{x-4}$

21. $\dfrac{3x}{x-2}-\dfrac{2-x}{x}$

22. $\dfrac{2x}{x-3}-\dfrac{3-x}{x}$

23. $\dfrac{2x}{x+3}-\dfrac{5}{4}$    $\dfrac{3(x-5)}{4(x+3)}$

24. $\dfrac{x}{3x-1}-\dfrac{2}{7}$    $\dfrac{x+2}{7(3x-1)}$

25. $\dfrac{3}{x}+\dfrac{2}{x+5}$    $\dfrac{5(x+3)}{x(x+5)}$

26. $\dfrac{x+2}{x-2}+\dfrac{3}{x}$    $\dfrac{(x-1)(x+6)}{x(x-2)}$

Additional answers can be found in the Instructor Answer Appendix.

**27.** $5 - \dfrac{2}{x-2}$    $\dfrac{5x-12}{x-2}$

**28.** $2 - \dfrac{5}{x+4}$    $\dfrac{2x+3}{x+4}$

**29.** $4 + \dfrac{6}{x-5}$    $\dfrac{4x-14}{x-5}$

**30.** $8 + \dfrac{6}{x+1}$    $\dfrac{2(4x+7)}{x+1}$

**31.** $\dfrac{4}{x^2-1} - \dfrac{5}{2x^2-x-1}$    $\dfrac{3x-1}{(x+1)(x-1)(2x+1)}$

**32.** $\dfrac{6}{9x^2-1} - \dfrac{3}{6x^2+13x-5}$    $\dfrac{3(x+9)}{(3x+1)(3x-1)(2x+5)}$

**33.** $\dfrac{5}{2x^2-5x-3} + \dfrac{2}{4x^2-9x-9}$    $\dfrac{24x+17}{(2x+1)(x-3)(4x+3)}$

**34.** $\dfrac{2}{x^2-x-20} + \dfrac{3}{3x^2-14x-5}$    $\dfrac{9x+14}{(x+4)(x-5)(3x+1)}$

 ## Mix 'Em Up!

**Add or subtract the rational expressions. Write each answer in lowest terms.**

**35.** $\dfrac{3x}{x-2} + \dfrac{4x}{2-x}$   $-\dfrac{x}{x-2}$

**36.** $\dfrac{x+1}{x-3} + \dfrac{2x}{x+3}$

**37.** $\dfrac{5}{x-10} + \dfrac{2}{x+2}$

**38.** $\dfrac{4x-1}{x+1} - \dfrac{3x}{x-1}$   $\dfrac{x^2-8x+1}{(x+1)(x-1)}$

**39.** $\dfrac{3x}{x^2-8x-9} + \dfrac{2x-1}{x-9}$   $\dfrac{2x^2+4x-1}{(x-9)(x+1)}$

**40.** $\dfrac{3x}{x^2-3x-18} - \dfrac{3x^2}{x-6}$   $\dfrac{-3x(x^2+3x-1)}{(x-6)(x+3)}$

**41.** $5 + \dfrac{1}{x+1}$   $\dfrac{5x+6}{x+1}$

**42.** $4 + \dfrac{2x}{x-3}$   $\dfrac{6(x-2)}{x-3}$

**43.** $\dfrac{4x}{x-1} + \dfrac{3x-6}{1-x}$   $\dfrac{x+6}{x-1}$

**44.** $\dfrac{2x^2}{x^2-6x-40} + \dfrac{5x+3}{x+4}$

**45.** $\dfrac{8}{x^2-9} - \dfrac{5}{x^2+6x+9}$   $\dfrac{3(x+13)}{(x+3)^2(x-3)}$

**46.** $\dfrac{3}{x^2-8x+16} + \dfrac{1}{x^2-16}$   $\dfrac{4(x+2)}{(x-4)^2(x+4)}$

**47.** $\dfrac{6}{x^2-3x+2} - \dfrac{4}{x^2-5x+6}$   $\dfrac{2(x-7)}{(x-1)(x-2)(x-3)}$

**48.** $\dfrac{1}{x^2+x-42} + \dfrac{2}{x^2-5x-6}$   $\dfrac{3(x+5)}{(x+7)(x-6)(x+1)}$

**49.** $\dfrac{450}{r-10} + \dfrac{300}{r+3}$

**50.** $\dfrac{150}{r+40} + \dfrac{60}{r+20}$

 ## You Be the Teacher!

**Correct each student's errors, if any.**

**51.** Add $\dfrac{7a}{3} + \dfrac{7a}{4}$.

Barry's work:

$$\dfrac{7a}{3} + \dfrac{7a}{4} = \dfrac{7a}{7} = \dfrac{\boldsymbol{7}a}{\boldsymbol{7}} = a$$   $\dfrac{7a}{3} + \dfrac{7a}{4} = \dfrac{28a+21a}{12} = \dfrac{49a}{12}$

**52.** Subtract $\dfrac{x+3}{4x+1} - \dfrac{2x-1}{4x-1}$.

Darren's work:

$$\dfrac{2x}{4x+1} - \dfrac{x}{4x-1} = \dfrac{2x}{4x+1} + \dfrac{x}{4x+1} = \dfrac{2x+x}{4x+1} = \dfrac{3x}{4x+1}$$

 ## Calculate It!

**Use a graphing calculator to verify that each operation has been performed correctly.**

**53.** $\dfrac{3x-1}{x+3} + \dfrac{-4x+2}{x+3} = \dfrac{-x+1}{x+3}$

**54.** $\dfrac{6x-4}{2x-5} - \dfrac{3x}{2x-5} = \dfrac{3x-4}{2x-5}$

**55.** $\dfrac{2x+4}{x^2+6x+9} - \dfrac{x}{x+3} = -\dfrac{x^2+x-4}{(x+3)^2}$

**56.** $\dfrac{-5x}{x^2+2x-8} + \dfrac{3x}{x-2} = \dfrac{x(3x+7)}{(x+4)(x-2)}$

---

## PIECE IT TOGETHER    SECTIONS 7.1–7.4 Review

**Find the value(s) that make each rational expression undefined.** (*Section 7.1, Objective 2*)

**1.** $\dfrac{x-1}{2x-10}$   $x=5$

**2.** $\dfrac{2x^2+5x-3}{x^2+3x-10}$   $x=-5$ or $x=2$

**Simplify each rational expression.** (*Section 7.1, Objective 3*)

**3.** $\dfrac{10x}{5x^2+15}$   $\dfrac{2x}{x^2+3}$

**4.** $\dfrac{x^2-5x+6}{x^2+4x-12}$   $\dfrac{x-3}{x+6}$

**5.** $\dfrac{4x^2-25}{6x^2+13x-5}$   $\dfrac{2x-5}{3x-1}$

**6.** $\dfrac{x^3-125}{x^2-25}$   $\dfrac{x^2+5x+25}{x+5}$

**Use the rule of negative rational expressions to write two equivalent forms of the rational expression.** (*Section 7.1, Objective 4*)

**7.** $-\dfrac{-5x-1}{x+1}$   $\dfrac{5x+1}{x+1}$ or $\dfrac{-5x-1}{-x-1}$

**Perform each operation with the rational expressions. Write each answer in lowest terms.** (*Section 7.2, Objectives 1 and 2; Section 7.3, Objectives 1 and 2; Section 7.4, Objectives 1 and 2*)

**8.** $\dfrac{x^2-9}{x-9} \cdot \dfrac{5x-45}{x+3}$   $5(x-3)$

**9.** $\dfrac{-6x}{(x+1)^2} \cdot \dfrac{3x+3}{18x^2}$   $-\dfrac{1}{x(x+1)}$

**10.** $\dfrac{x^3 + 8}{4x - 20} \cdot \dfrac{x - 5}{x + 2}$  $\dfrac{x^2 - 2x + 4}{4}$  **11.** $\dfrac{3x + 2}{x + 2} \cdot \dfrac{x^2 - 4}{3x^2 + 5x + 2}$  $\dfrac{x - 2}{x + 1}$  **17.** $\dfrac{y + 10}{y + 5} + \dfrac{2y + 5}{y + 5}$  $3$

**12.** $\dfrac{25a^2}{6b^3} \div \dfrac{5ab^2}{18}$  $\dfrac{15a}{b^5}$  **13.** $\dfrac{x^2 - 36}{5x^3} \div \dfrac{x - 6}{10x}$  $\dfrac{2(x + 6)}{x^2}$  **18.** $\dfrac{4x}{x^2 + x - 12} - \dfrac{3x + 3}{x^2 + x - 12}$  $\dfrac{1}{x + 4}$

**14.** $\dfrac{x^2 - 2x - 8}{x - 4} \div (x + 2)$  $1$  **15.** $\dfrac{x^2 - 3x - 10}{x^3 - 5x^2} \div \dfrac{x + 2}{x^2}$  $1$  **19.** $\dfrac{5x}{x^2 - 49} - \dfrac{2}{x + 7}$  $\dfrac{3x + 14}{(x - 7)(x + 7)}$

**16.** $\dfrac{x^2 + 4x - 12}{3x + 6} \div \dfrac{x^2 + x - 30}{2x + 4}$  $\dfrac{2(x - 2)}{3(x - 5)}$  **20.** $\dfrac{4}{x^2 - 1} - \dfrac{5}{2x^2 - x - 1}$  $\dfrac{3x - 1}{(2x + 1)(x + 1)(x - 1)}$

---

## SECTION 7.5    Complex Fractions

Larry and Tim watch football together on Sunday. They order 1 pepperoni pizza and 1 combination pizza. During the first quarter, Larry eats $\dfrac{1}{2}$ of the pepperoni pizza and Tim eats $\dfrac{3}{4}$ of the combination pizza. Jose stops by to watch the last quarter of the game. The three guys split the leftover pizza evenly between them. How much of the leftover pizza does each person get? To answer this question, we can simplify the expression $\dfrac{2 - \dfrac{1}{2} - \dfrac{3}{4}}{3}$.

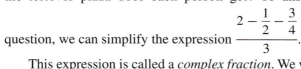

This expression is called a *complex fraction*. We will learn how to simplify complex fractions using two different methods.

### Simplifying Complex Fractions (Method 1)

**Objective 1** ▶

Simplify complex fractions using division (method 1).

A **complex fraction** is a fraction that contains fractions in its numerator and/or denominator. Some examples of complex fractions are

$$\dfrac{\dfrac{1}{3}}{\dfrac{2}{5}} \qquad \dfrac{2 - \dfrac{1}{4}}{\dfrac{1}{8}} \qquad \dfrac{\dfrac{6}{x}}{x + \dfrac{1}{x}}$$

Numerator of the complex fraction
Denominator of the complex fraction

Complex fractions are not simplified forms of fractions. Our goal is to simplify a complex fraction as a rational number or rational expression. That is, we want to write the complex fraction so that it does not have fractions in its numerator or denominator.

> **Definition: Simplest Form of a Complex Fraction**
>
> A complex fraction is in **simplest form** when it is in the form $\dfrac{A}{B}$, where $A$ and $B$ are polynomials ($B \neq 0$) that have no common factors.

Two methods will be illustrated in this section. The first method uses division. The goal of this method is to rewrite the complex fraction as a single fraction over a single fraction. Then we apply the rules for dividing rational expressions. Recall

$$\dfrac{\dfrac{A}{B}}{\dfrac{C}{D}} = \dfrac{A}{B} \div \dfrac{C}{D} = \dfrac{A}{B} \cdot \dfrac{D}{C}$$

---

**Procedure: Simplifying a Complex Fraction Using Division (Method 1)**

**Step 1:** Simplify the numerator and/or denominator of the complex fraction, if necessary.

**Step 2:** Write as a division problem.

**Step 3:** Divide the expressions by multiplying the first rational expression by the reciprocal of the second rational expression.

**Step 4:** Simplify the resulting rational expression.

---

**Objective 1 Examples** | Simplify each complex fraction using method 1.

**1a.** $\dfrac{\dfrac{1}{x}}{\dfrac{3}{x^2}}$　　**1b.** $\dfrac{\dfrac{x}{x^2 - 3x + 2}}{\dfrac{x + 2}{x - 1}}$　　**1c.** $\dfrac{\dfrac{1}{5} - \dfrac{2}{3}}{\dfrac{3}{4} + \dfrac{1}{3}}$　　**1d.** $\dfrac{\dfrac{2}{5a} + \dfrac{8}{5a}}{2a}$　　**1e.** $\dfrac{3}{\dfrac{1}{x} + \dfrac{2}{x^2}}$

**Solutions**　**1a.**

$\dfrac{\dfrac{1}{x}}{\dfrac{3}{x^2}} = \dfrac{1}{x} \div \dfrac{3}{x^2}$ 　　Write as a division problem.

$= \dfrac{1}{x} \cdot \dfrac{x^2}{3}$ 　　Multiply by the reciprocal of $\dfrac{3}{x^2}$.

$= \dfrac{x}{3}$ 　　Divide out the common factor, $x$, and multiply.

**1b.**

$\dfrac{\dfrac{x}{x^2 - 3x + 2}}{\dfrac{x + 2}{x - 1}} = \dfrac{x}{(x - 2)(x - 1)} \div \dfrac{x + 2}{x - 1}$ 　　Write as a division problem and factor $x^2 - 3x + 2$.

$= \dfrac{x}{(x - 2)(x - 1)} \cdot \dfrac{x - 1}{x + 2}$ 　　Multiply by the reciprocal of $\dfrac{x + 2}{x - 1}$.

$= \dfrac{x}{(x - 2)(x + 2)}$ 　　Divide out the common factor, $(x - 1)$, and multiply the remaining factors.

**1c.**

$\dfrac{\dfrac{1}{5} - \dfrac{2}{3}}{\dfrac{3}{4} + \dfrac{1}{3}} = \dfrac{\dfrac{1}{5} \cdot \dfrac{3}{3} - \dfrac{2}{3} \cdot \dfrac{5}{5}}{\dfrac{3}{4} \cdot \dfrac{3}{3} + \dfrac{1}{3} \cdot \dfrac{4}{4}}$ 　　Convert each fraction in the numerator to an equivalent fraction with the LCD, 15.

Convert each fraction in the denominator to an equivalent fraction with the LCD, 12.

$= \dfrac{\dfrac{3}{15} - \dfrac{10}{15}}{\dfrac{9}{12} + \dfrac{4}{12}}$ 　　Simplify each product.

$= \dfrac{-\dfrac{7}{15}}{\dfrac{13}{12}}$ 　　Simplify the numerator and denominator of the complex fraction.

$= -\dfrac{7}{15} \cdot \dfrac{12}{13}$ 　　Write as a multiplication problem.

$= -\dfrac{7 \cdot 3 \cdot 4}{3 \cdot 5 \cdot 13}$ 　　Multiply and factor.

$= -\dfrac{28}{65}$ 　　Simplify by dividing out the common factor, 3.

**1d.**  $\dfrac{\dfrac{2}{5a}+\dfrac{8}{5a}}{2a} = \dfrac{\dfrac{10}{5a}}{\dfrac{2a}{1}}$          Combine the terms in the numerator of the complex fraction.

$= \dfrac{10}{5a}\cdot\dfrac{1}{2a}$          Rewrite as a multiplication problem.

$= \dfrac{10}{10a^2}$          Multiply.

$= \dfrac{1}{a^2}$          Divide out the common factor, 10.

**1e.**  $\dfrac{3}{\dfrac{1}{x}+\dfrac{2}{x^2}} = \dfrac{\dfrac{3}{1}}{\dfrac{1}{x}\cdot\dfrac{x}{x}+\dfrac{2}{x^2}}$          Write the numerator as a fraction. Convert the terms in the denominator to equivalent fractions with the LCD, $x^2$.

$= \dfrac{\dfrac{3}{1}}{\dfrac{x}{x^2}+\dfrac{2}{x^2}}$          Simplify the product in the denominator.

$= \dfrac{\dfrac{3}{1}}{\dfrac{x+2}{x^2}}$          Add the rational expressions in the denominator.

$= \dfrac{3}{1}\cdot\dfrac{x^2}{x+2}$          Rewrite as a multiplication problem.

$= \dfrac{3x^2}{x+2}$          Multiply.

✔ **Student Check 1**    Simplify each complex fraction using method 1.

**a.** $\dfrac{\dfrac{3}{2a}}{\dfrac{a}{4}}$     **b.** $\dfrac{\dfrac{3x}{x^2-6x+8}}{\dfrac{6x}{x-4}}$     **c.** $\dfrac{\dfrac{2}{7}+\dfrac{4}{5}}{\dfrac{1}{10}-\dfrac{3}{14}}$     **d.** $\dfrac{\dfrac{3}{y}-\dfrac{1}{y}}{\dfrac{1}{y^2}}$     **e.** $\dfrac{\dfrac{6}{b}}{\dfrac{1}{b}+\dfrac{4}{b^2}}$

## Simplifying Complex Fractions (Method 2)

**Objective 2** ▶

Simplify complex fractions using multiplication by the LCD (method 2).

A second way that we can simplify complex fractions is by multiplying the numerator and denominator of the complex fraction by the LCD of all the fractions in the expression. This operation will eliminate the fractions within the complex fraction.

To illustrate how method 2 produces the same answer as method 1, we will simplify the same complex fractions from Example 1 in Example 2.

> **Procedure:  Simplifying a Complex Fraction Using Multiplication by the LCD (Method 2)**
>
> **Step 1:** Determine the LCD of all the fractions in the numerator and denominator of the complex fraction.
> **Step 2:** Multiply the numerator and denominator of the complex fraction by the LCD.
> **Step 3:** Simplify the resulting products.

**Objective 2 Examples**   **Simplify each complex fraction using method 2.**

**2a.** $\dfrac{\dfrac{1}{x}}{\dfrac{3}{x^2}}$   **2b.** $\dfrac{\dfrac{x}{x^2 - 3x + 2}}{\dfrac{x + 2}{x - 1}}$   **2c.** $\dfrac{\dfrac{1}{5} - \dfrac{2}{3}}{\dfrac{3}{4} + \dfrac{1}{3}}$   **2d.** $\dfrac{\dfrac{2}{5a} + \dfrac{8}{5a}}{2a}$   **2e.** $\dfrac{3}{\dfrac{1}{x} + \dfrac{2}{x^2}}$

**Solutions**   **2a.**

$$\frac{\dfrac{1}{x}}{\dfrac{3}{x^2}} = \frac{\left(\dfrac{1}{x}\right)(x^2)}{\left(\dfrac{3}{x^2}\right)(x^2)}$$

Multiply the numerator and denominator of the complex fraction by the LCD, $x^2$. Divide out the common factors.

$$= \frac{x}{3}$$

Multiply the remaining factors.

**2b.** Since $x^2 - 3x + 2 = (x - 2)(x - 1)$, the LCD of the fractions in the complex fraction is $(x - 2)(x - 1)$. Note that $(x - 2)(x - 1) = \dfrac{(x - 2)(x - 1)}{1}$.

$$\frac{\dfrac{x}{x^2 - 3x + 2}}{\dfrac{x + 2}{x - 1}} = \frac{\dfrac{x}{(x - 2)(x - 1)} \cdot \dfrac{(x - 2)(x - 1)}{1}}{\dfrac{x + 2}{x - 1} \cdot \dfrac{(x - 2)(x - 1)}{1}}$$

Multiply the numerator and denominator of the complex fraction by the LCD.

$$= \frac{x}{(x - 2)(x + 2)}$$

Divide out the common factors and multiply the remaining factors.

**2c.**

$$\frac{\dfrac{1}{5} - \dfrac{2}{3}}{\dfrac{3}{4} + \dfrac{1}{3}} = \frac{\left(\dfrac{1}{5} - \dfrac{2}{3}\right)60}{\left(\dfrac{3}{4} + \dfrac{1}{3}\right)60}$$

Multiply the numerator and denominator of the complex fraction by the LCD, 60.

$$= \frac{\dfrac{1}{5} \cdot 60 - \dfrac{2}{3} \cdot 60}{\dfrac{3}{4} \cdot 60 + \dfrac{1}{3} \cdot 60}$$

Apply the distributive property.

$$= \frac{12 - 40}{45 + 20}$$

Simplify each product.

$$= \frac{-28}{65}$$

Simplify the numerator and denominator.

$$= -\frac{28}{65}$$

Apply the rule of negative rational expressions.

**2d.**

$$\frac{\dfrac{2}{5a} + \dfrac{8}{5a}}{2a} = \frac{\left(\dfrac{2}{5a} + \dfrac{8}{5a}\right)5a}{(2a)(5a)}$$

Multiply the numerator and denominator of the complex fraction by the LCD, $5a$.

$$= \frac{\dfrac{2}{5a} \cdot 5a + \dfrac{8}{5a} \cdot 5a}{10a^2}$$

Apply the distributive property in the numerator and simplify the denominator.

$$= \frac{2 + 8}{10a^2}$$

Simplify each product in the numerator.

$$= \frac{10}{10a^2}$$

Add the terms in the numerator.

$$= \frac{1}{a^2}$$

Simplify by dividing out the common factor, 10.

**2e.**

$$\frac{3}{\dfrac{1}{x}+\dfrac{2}{x^2}}=\frac{3\cdot x^2}{\left(\dfrac{1}{x}+\dfrac{2}{x^2}\right)\cdot x^2}$$

Multiply the numerator and denominator of the complex fraction by the LCD, $x^2$.

$$=\frac{3x^2}{\dfrac{1}{x}\cdot x^2+\dfrac{2}{x^2}\cdot x^2}$$

Apply the distributive property in the denominator and simplify the numerator.

$$=\frac{3x^2}{x+2}$$

Simplify each product in the denominator.

---

☑ **Student Check 2**   Simplify each complex fraction using method 2.

**a.** $\dfrac{\dfrac{3}{2a}}{\dfrac{a}{4}}$   **b.** $\dfrac{\dfrac{3x}{x^2-6x+8}}{\dfrac{6x}{x-4}}$   **c.** $\dfrac{\dfrac{2}{7}+\dfrac{4}{5}}{\dfrac{1}{10}-\dfrac{3}{14}}$   **d.** $\dfrac{\dfrac{3}{y}-\dfrac{1}{y}}{\dfrac{1}{y^2}}$   **e.** $\dfrac{\dfrac{6}{b}}{\dfrac{1}{b}+\dfrac{4}{b^2}}$

---

## Applications

**Objective 3** ▶

Solve applications of complex fractions.

There are problems from real-life situations that can be solved by simplifying complex fractions as illustrated.

**Objective 3 Example**   Larry and Tim watch football together on Sunday. They order 1 pepperoni pizza and 1 combination pizza. During the first quarter, Larry eats $\dfrac{1}{2}$ of the pepperoni pizza and Tim eats $\dfrac{3}{4}$ of the combination pizza. Jose stops by to watch the last quarter of the game. The three friends split the leftover pizza evenly between them. How much of the leftover pizza does each person get?

**Solution**   Larry and Tim began with two whole pizzas. Since Larry eats $\dfrac{1}{2}$ of a pizza and Tim eats $\dfrac{3}{4}$ of a pizza, $2-\dfrac{1}{2}-\dfrac{3}{4}$ pizza remains. We must divide this expression by 3 to determine how much of the leftover pizza each person receives.

$$\frac{2-\dfrac{1}{2}-\dfrac{3}{4}}{3}=\frac{\left(2-\dfrac{1}{2}-\dfrac{3}{4}\right)8}{(3)8}$$

Multiply the numerator and denominator of the complex fraction by the LCD, 8.

$$=\frac{(2)8-\left(\dfrac{1}{2}\right)8-\left(\dfrac{3}{4}\right)8}{24}$$

Apply the distributive property and simplify the denominator.

$$=\frac{16-4-6}{24}$$

Simplify each product.

$$=\frac{6}{24}$$

Combine the terms in the numerator.

$$=\frac{1}{4}$$

Simplify.

So, each person gets $\dfrac{1}{4}$ of a pizza from the leftover pizza.

✓ **Student Check 3** Lisa has a half-cup of fertilizer that she wants to use on three plants. How much fertilizer can she use for each plant?

**Objective 4** ▶

Troubleshoot common errors.

## Troubleshooting Common Errors

A common error related to complex fractions is shown next.

**Objective 4  Example**  **A problem and an incorrect solution are given. Provide the correct solution and an explanation of the error.**

Simplify $\dfrac{xy}{\frac{x}{y}}$.

| Incorrect Solution | Correct Solution and Explanation |
|---|---|
| $\dfrac{xy}{\frac{x}{y}} = \dfrac{xy}{x} \cdot \dfrac{1}{y} = 1$ | The expression is $xy$ divided by $\dfrac{x}{y}$, not $\dfrac{xy}{x}$ divided by $y$. $$\dfrac{xy}{\frac{x}{y}} = \dfrac{xy}{1} \cdot \dfrac{y}{x} = y^2$$ |

## ANSWERS TO STUDENT CHECKS

**Student Check 1**  **a.** $\dfrac{6}{a^2}$  **b.** $\dfrac{1}{2(x-2)}$  **c.** $-\dfrac{19}{2}$

  **d.** $2y$  **e.** $\dfrac{6b}{b+4}$

**Student Check 2**  **a.** $\dfrac{6}{a^2}$  **b.** $\dfrac{1}{2(x-2)}$  **c.** $-\dfrac{19}{2}$

  **d.** $2y$  **e.** $\dfrac{6b}{b+4}$

**Student Check 3**  **a.** $\dfrac{1}{6}$ cup

## SUMMARY OF KEY CONCEPTS

1. There are two methods to simplify a complex fraction. Method 1 requires us to combine the numerator and/or denominator of the complex fraction into a single fraction so that we have one fraction divided by another fraction. Then we multiply the numerator of the complex fraction by the reciprocal of the denominator.

2. Method 2 requires us to eliminate the fractions from the numerator and/or denominator of the complex fraction initially. This is done by multiplying the numerator and the denominator of the complex fraction by the LCD of all the fractions within the complex fraction. Combine any like terms and simplify the fraction.

## GRAPHING CALCULATOR SKILLS

When entering complex fractions in the calculator, it is important that we use parentheses around the fractions.

**Example:** Simplify

$$\dfrac{\frac{1}{3}}{\frac{2}{5}}$$

**Solution:** We must enter the numerator and denominator of the complex fraction in parentheses.

```
(1/3)/(2/5)
         .8333333333
Ans▸Frac
               5/6
```

Recall to convert the answer to fraction form, press Math, 1, Enter.

## SECTION 7.5 / EXERCISE SET

### Write About It!

Use complete sentences in your answer to each exercise.

1. What is a complex fraction?

2. Explain why $\dfrac{\dfrac{1}{a}+\dfrac{1}{b}}{\dfrac{1}{c}+\dfrac{1}{d}} \neq \dfrac{c+d}{a+b}$.

3. Explain how to simplify a complex fraction using division.

4. Explain how to simplify a complex fraction using multiplication by the LCD.

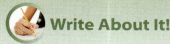

### Practice Makes Perfect!

Simplify each complex fraction using method 1, and then method 2. (*See Objectives 1 and 2.*)

5. $\dfrac{\dfrac{5}{x}}{\dfrac{6}{x^2}}$   $\dfrac{5x}{6}$

6. $\dfrac{\dfrac{2}{y^3}}{\dfrac{1}{y^2}}$   $\dfrac{2}{y}$

7. $\dfrac{\dfrac{10}{3y}}{\dfrac{5}{6y^2}}$   $4y$

8. $\dfrac{\dfrac{2}{3x^3}}{\dfrac{6}{5x^5}}$   $\dfrac{5x^2}{9}$

9. $\dfrac{\dfrac{1}{2}+\dfrac{1}{5}}{\dfrac{3}{5}-\dfrac{2}{5}}$   $\dfrac{7}{2}$

10. $\dfrac{\dfrac{2}{3}-\dfrac{1}{3}}{\dfrac{1}{4}+\dfrac{1}{2}}$   $\dfrac{4}{9}$

11. $\dfrac{7-\dfrac{1}{2}}{6+\dfrac{3}{4}}$   $\dfrac{26}{27}$

12. $\dfrac{2-\dfrac{4}{9}}{5+\dfrac{6}{5}}$   $\dfrac{70}{279}$

13. $\dfrac{3+\dfrac{x}{y}}{2-\dfrac{x}{y}}$   $\dfrac{3y+x}{2y-x}$

14. $\dfrac{4+\dfrac{x}{4y}}{3-\dfrac{y}{3x}}$   $\dfrac{3x(16y+x)}{4y(9x-y)}$

15. $\dfrac{\dfrac{x}{x+3}}{\dfrac{x^2}{x-1}}$   $\dfrac{x-1}{x(x+3)}$

16. $\dfrac{\dfrac{x-1}{x+2}}{\dfrac{x}{x+2}}$   $\dfrac{x-1}{x}$

17. $\dfrac{\dfrac{2y+1}{y-2}}{\dfrac{3y}{y^2-4}}$   $\dfrac{(2y+1)(y+2)}{3y}$

18. $\dfrac{\dfrac{y+4}{y^2}}{\dfrac{y^2-16}{y}}$   $\dfrac{1}{y(y-4)}$

19. $\dfrac{\dfrac{3x}{x+1}}{6}$   $\dfrac{x}{2(x+1)}$

20. $\dfrac{\dfrac{2x}{x+3}}{12}$   $\dfrac{x}{6(x+3)}$

21. $\dfrac{\dfrac{3}{2a}+\dfrac{4}{2a}}{6a}$   $\dfrac{7}{12a^2}$

22. $\dfrac{\dfrac{4}{7a}+\dfrac{3}{7a}}{14a}$   $\dfrac{1}{14a^2}$

23. $\dfrac{1}{\dfrac{1}{2x}+\dfrac{3}{x}}$   $\dfrac{2x}{7}$

24. $\dfrac{2}{\dfrac{1}{x}-\dfrac{3}{x^2}}$   $\dfrac{2x^2}{x-3}$

25. $\dfrac{\dfrac{x}{x+3}}{x-5}$   $\dfrac{x(x-5)}{x+3}$

26. $\dfrac{\dfrac{3x}{2x-1}}{x+6}$   $\dfrac{3x(x+6)}{2x-1}$

27. $\dfrac{\dfrac{500}{r+3}}{\dfrac{40}{r^2+8r+15}}$   $\dfrac{25(r+5)}{2}$

28. $\dfrac{\dfrac{100}{r^2-10r+25}}{\dfrac{75}{r-5}}$   $\dfrac{4}{3(r-5)}$

29. $\dfrac{\dfrac{2}{5x}-\dfrac{3}{x}}{\dfrac{3}{x^2}+\dfrac{4}{x}}$   $-\dfrac{13x}{5(3+4x)}$

30. $\dfrac{\dfrac{1}{3x}+\dfrac{3}{12x}}{\dfrac{2}{x^3}-\dfrac{1}{x^2}}$   $\dfrac{7x^2}{12(2-x)}$

31. $\dfrac{1+\dfrac{4}{x}}{3-\dfrac{1}{x^2}}$   $\dfrac{x(x+4)}{3x^2-1}$

32. $\dfrac{5+\dfrac{x}{2}}{4-\dfrac{x}{3}}$   $\dfrac{3(10+x)}{2(12-x)}$

33. $\dfrac{\dfrac{x^2+x-6}{x^2+6x+5}}{\dfrac{x^2+4x+3}{x^2+3x-10}}$   $\dfrac{(x-2)(x-2)}{(x+1)(x+1)}$

34. $\dfrac{\dfrac{x^2-5x}{x^2+3x+2}}{\dfrac{x^2-3x-10}{x^2-3x-4}}$

35. $\dfrac{\dfrac{x^2+3x}{x^2-6x-7}}{\dfrac{2x^2+x}{x^2+8x+7}}$   $\dfrac{(x+3)(x+7)}{(x-7)(2x+1)}$

36. $\dfrac{\dfrac{2x^2-6x}{x^2-x-12}}{\dfrac{5x-15}{x^2+x-6}}$   $\dfrac{2x(x-2)}{5(x-4)}$

37. $\dfrac{1+\dfrac{x-1}{x+3}}{4-\dfrac{2}{x+3}}$   $\dfrac{x+1}{2x+5}$

38. $\dfrac{5-\dfrac{x}{3x+1}}{6+\dfrac{3}{3x+1}}$   $\dfrac{14x+5}{9(2x+1)}$

Write a complex fraction that represents each situation, and then simplify the complex fraction. (*See Objective 3.*)

39. Quinn and her daughter Lauri each order a pizza at a restaurant. Quinn eats $\dfrac{1}{3}$ of her pizza and Lauri eats $\dfrac{1}{4}$ of her pizza. Their leftover pizza will be split between Quinn's two sons. How much pizza will each son get?

$\dfrac{17}{24}$ of a pizza

Additional answers can be found in the Instructor Answer Appendix.

**40.** A recipe for 12 people calls for $\dfrac{3}{4}$ cup of flour. How much flour should be used if the recipe is made for 3 people (i.e., only $\dfrac{1}{4}$ of the recipe is made)? $\dfrac{3}{16}$ of a cup

**41.** What is the average of $\dfrac{1}{3}$ and $\dfrac{1}{4}$? $\dfrac{7}{24}$

**42.** What is the average of $\dfrac{1}{3}$ and $\dfrac{1}{5}$? $\dfrac{4}{15}$

**43.** During the month of January, Leslie lost $\dfrac{1}{2}$ lb the first week, 1 lb the second week, $\dfrac{3}{4}$ lb the third week, and $\dfrac{1}{4}$ lb the last week. What is the average number of pounds that Leslie lost each week in January? $\dfrac{5}{8}$ lb

**44.** Suzy eats $\dfrac{1}{6}$ of an apple pie and Sienna eats $\dfrac{1}{3}$ of the same pie. If the remaining pie is to be split equally among three people, what fraction of the pie will each person receive? $\dfrac{1}{6}$ of a pie

 **Mix 'Em Up!**

**Simplify each complex fraction using method 1, and then method 2.**

**45.** $\dfrac{\dfrac{12}{2a^2}}{\dfrac{21}{3a^5}}$ $\dfrac{6a^3}{7}$

**46.** $\dfrac{\dfrac{5}{15b^8}}{\dfrac{20}{2b^3}}$ $\dfrac{1}{30b^5}$

**47.** $\dfrac{5-\dfrac{d}{c}}{3+\dfrac{c}{d}}$ $\dfrac{d(5c-d)}{c(3d+c)}$

**48.** $\dfrac{6-\dfrac{3a}{2b}}{1-\dfrac{a}{4b}}$ $6$

**49.** $\dfrac{\dfrac{3x}{2x+3}}{\dfrac{15x^2}{4x^2-9}}$ $\dfrac{2x-3}{5x}$

**50.** $\dfrac{\dfrac{9y^2-1}{18y^2}}{\dfrac{3y-1}{6y}}$ $\dfrac{3y+1}{3y}$

**51.** $\dfrac{\dfrac{1}{x}+\dfrac{1}{2}}{6+3x}$ $\dfrac{1}{6x}$

**52.** $\dfrac{\dfrac{1}{y}-\dfrac{2}{7}}{4y-14}$ $-\dfrac{1}{14y}$

**53.** $\dfrac{\dfrac{10xy}{2x+y}}{\dfrac{12xy}{2x^2-xy-y^2}}$ $\dfrac{5(x-y)}{6}$

**54.** $\dfrac{\dfrac{14rs}{r^2-8rs+16s^2}}{\dfrac{21rs}{r-4s}}$ $\dfrac{2}{3(r-4s)}$

**55.** $\dfrac{\dfrac{2}{x}+\dfrac{3}{y}}{\dfrac{4}{x^2}-\dfrac{9}{y^2}}$ $\dfrac{xy}{2y-3x}$

**56.** $\dfrac{\dfrac{5}{r}-\dfrac{2}{s}}{\dfrac{25}{r^2}-\dfrac{4}{s^2}}$ $\dfrac{rs}{5s+2r}$

**57.** $\dfrac{\dfrac{x^2-9}{x^2+2x+4}}{\dfrac{x^2-5x+6}{x^3-8}}$ $x+3$

**58.** $\dfrac{\dfrac{x^2-25}{x^2-x+1}}{\dfrac{x^2+6x+5}{x^3+1}}$ $x-5$

**59.** $\dfrac{\dfrac{x^2+4x}{x^2+5x+4}}{\dfrac{3x^2+x}{3x^2+16x+5}}$ $\dfrac{x+5}{x+1}$

**60.** $\dfrac{\dfrac{3x^2-6x}{x^2+4x}}{\dfrac{6x-12}{x^2+3x-4}}$ $\dfrac{x-1}{2}$

**61.** $\dfrac{1-\dfrac{x+2}{x-1}}{6+\dfrac{5}{x-1}}$ $-\dfrac{3}{6x-1}$

**62.** $\dfrac{1-\dfrac{2x}{4x-1}}{2+\dfrac{1}{4x-1}}$ $\dfrac{2x-1}{8x-1}$

**Write a complex fraction that represents each situation, and then simplify the complex fraction.**

**63.** Wing ate $\dfrac{1}{2}$ of a cherry pie and Nam ate $\dfrac{1}{3}$ of the same pie. If the remaining pie is to be split equally among two friends, what fraction of the pie will each person receive?

**64.** A recipe for 10 people calls for $\dfrac{5}{4}$ cup of flour. How much flour should be used if the recipe is made for 6 people?

**65.** What is the average of $\dfrac{7}{3}$, $\dfrac{11}{15}$, and $\dfrac{2}{5}$? $\dfrac{52}{45}$

**66.** What is the average of $\dfrac{8}{5}$, $\dfrac{7}{10}$, and $\dfrac{3}{2}$? $\dfrac{19}{15}$

**For Exercises 67 and 68, a person can do a job alone in $x$ hours and another person can do the same job in $y$ hours. When these two people work together to do the same job, it will take them $\dfrac{1}{\dfrac{1}{x}+\dfrac{1}{y}}$ hours.**

**67.** Shan can paint her bedroom in 4 hr. Her brother Hong can paint the same room in 6 hr. If they work together, how long will it take them to paint the room? 2.4 hr

**68.** Sam can deliver newspapers to the customers on his assigned route in 2 hr. If Sam is sick, his wife delivers the newspapers in 3.6 hr. If they work together, how long will it take them to deliver the newspapers?

 **You Be the Teacher!**

**Correct each student's errors, if any.**

**69.** Simplify $\dfrac{\dfrac{2x}{x+1}}{x^2}$.

Ian's work:

$$\dfrac{\dfrac{2x}{x+1}}{x^2}=\dfrac{\dfrac{2x}{x+1}}{\dfrac{x^2}{1}}=\dfrac{2x}{x+1}\cdot\dfrac{1}{x^2}=\dfrac{2\cancel{x}}{x+1}\cdot\dfrac{1}{\cancel{x}^2_x}$$

$$=\dfrac{2}{x(x+1)}=\dfrac{2}{x^2+x}\quad\text{correct}$$

**70.** Simplify $\dfrac{\dfrac{3x}{x+1}}{2x}$.

Vivienne's work:

$$\dfrac{\dfrac{3x}{x+1}}{2x} = \dfrac{\dfrac{3x}{x+1}}{\dfrac{2x}{1}} = \dfrac{3x}{x+1} \cdot \dfrac{2x}{1} = \dfrac{6x^2}{x+1}$$

**71.** Simplify $\dfrac{1 - \dfrac{3}{x}}{2 + \dfrac{4}{3x}}$.

Mary Clarice's work:

$$\dfrac{1 - \dfrac{3}{x}}{2 + \dfrac{4}{3x}} = \dfrac{\dfrac{-2}{x}}{\dfrac{6}{3x}} = -\dfrac{2}{x} \cdot \dfrac{3x}{6} = \dfrac{6x}{6x} = 1$$

**72.** Simplify $\dfrac{1 - \dfrac{x}{2x+1}}{\dfrac{3x}{2x+1}}$.

Sebastian's work:

$$\dfrac{1 - \dfrac{x}{2x+1}}{\dfrac{2x+1}{3x}} = \dfrac{\dfrac{2x+1}{2x+1} - \dfrac{x}{2x+1}}{\dfrac{2x+1}{3x}}$$

$$= \dfrac{\dfrac{2x+1-x}{2x+1}}{\dfrac{2x+1}{3x}} = \dfrac{\dfrac{x+1}{2x+1}}{\dfrac{2x+1}{3x}}$$

$$= \dfrac{\dfrac{x+1}{\cancel{2x+1}}}{\dfrac{\cancel{2x+1}}{3x}} = \dfrac{x+1}{3x}$$

**73.** Simplify $\dfrac{\dfrac{2}{a} + \dfrac{1}{b}}{\dfrac{4}{a^2} - \dfrac{1}{b^2}}$.

Karen's work:

$$\dfrac{\dfrac{2}{a} + \dfrac{1}{b}}{\dfrac{4}{a^2} - \dfrac{1}{b^2}} = \dfrac{\dfrac{\cancel{2}^{1}}{\cancel{a}} + \dfrac{\cancel{1}^{1}}{\cancel{b}}}{\cancel{\dfrac{4}{a^2}}_{\frac{2}{a}} - \cancel{\dfrac{1}{b^2}}_{\frac{1}{b}}} = \dfrac{1}{\dfrac{2}{a} - \dfrac{1}{b}} = \dfrac{a-b}{2}$$

**74.** Simplify $\dfrac{a+b}{\dfrac{1}{a} + \dfrac{1}{b}}$.

Joanna's work:

$$\dfrac{a+b}{\dfrac{1}{a} + \dfrac{1}{b}} = \dfrac{a}{\dfrac{1}{a}} + \dfrac{b}{\dfrac{1}{b}} = a \cdot a + b \cdot b = a^2 + b^2$$

 **Calculate It!**

**Use a calculator to simplify each complex fraction. State each answer as a fraction in lowest terms.**

**75.** $\dfrac{\dfrac{10}{7}}{\dfrac{3}{5}}$    $\dfrac{50}{21}$

**76.** $\dfrac{\dfrac{7}{6}}{\dfrac{10}{9}}$    $\dfrac{21}{20}$

**77.** $\dfrac{1 + \dfrac{1}{4}}{2 - \dfrac{1}{3}}$    $\dfrac{3}{4}$

**78.** $\dfrac{\dfrac{1}{2} + \dfrac{1}{5}}{\dfrac{3}{5} - \dfrac{2}{3}}$    $-\dfrac{21}{2}$

| SECTION 7.6 | Solving Rational Equations |

**OBJECTIVES**

As a result of completing this section, you will be able to

1. Solve rational equations.
2. Solve rational equations for a specific variable.
3. Troubleshoot common errors.

Suppose Dionne travels 165 mi to visit a friend. If she wants her trip to take 2 hr, how fast would she have to drive? To answer this question, we need to solve the equation $2 = \dfrac{165}{s}$, where $s$ is in mph.

This is an example of a *rational equation*. We will learn how to solve this type of equation in this section.

## Rational Equations

**Objective 1** ▶

Solve rational equations.

A **rational equation** is an equation that contains a rational expression. Some examples are

$$\frac{4}{x} = \frac{3}{x+1} \qquad \frac{1}{x-2} + \frac{1}{x+2} = \frac{3}{x^2-4} \qquad \frac{2}{m} + 4 = \frac{3}{2}$$

Recall that in Chapter 2, we solved linear equations that contained fractions. To clear fractions, we applied the multiplication property of equality by multiplying each side of the equation by the LCD. An example is shown.

$$\frac{1}{3}x + \frac{2}{9} = \frac{1}{4}x$$

$$36\left(\frac{1}{3}x + \frac{2}{9}\right) = 36\left(\frac{1}{4}x\right) \qquad \text{Multiply each side by the LCD, 36.}$$

$$36\left(\frac{1}{3}x\right) + 36\left(\frac{2}{9}\right) = 9x \qquad \text{Apply the distributive property and simplify.}$$

$$12x + 8 = 9x \qquad \text{Simplify each product.}$$

$$3x = -8 \qquad \text{Subtract } 9x \text{ and 8 from each side.}$$

$$x = -\frac{8}{3} \qquad \text{Divide each side by 3.}$$

When we solve rational equations, the LCD will involve algebraic expressions. The multiplication property of equality states that we must multiply both sides of an equation by a *nonzero number*. So, we must determine the values of the variable that make the LCD equal to zero. These values must be excluded from the solution set since they will make the denominator of the rational expression zero and the rational expression undefined.

**INSTRUCTOR NOTE:**
Tell students that the resulting equation will either be linear or quadratic in this section.

**Procedure:  Solving Rational Equations**

**Step 1:** Determine the LCD of the rational expressions in the equation.
**Step 2:** Multiply each side of the equation by the LCD.
**Step 3:** Solve the resulting equation.
**Step 4:** Check the solutions in the original equation. Write the solution set.

**Note:** *If one of the solutions is a value that makes the LCD zero, then it is an extraneous solution and must be discarded from the solution set.*

**Objective 1 Examples** Solve each rational equation.

**1a.** $\dfrac{3}{x} + 5 = 8$  **1b.** $\dfrac{1}{m-3} + 2 = \dfrac{1}{m-3}$  **1c.** $\dfrac{4}{x^2} + \dfrac{5}{x} = -1$

**1d.** $\dfrac{2}{x-2} - \dfrac{1}{x+2} = \dfrac{3}{x^2-4}$  **1e.** $\dfrac{r+2}{r+8} = \dfrac{2}{r}$

**1f.** $\dfrac{-4x}{3x^2+5x+2} + \dfrac{x}{3x+2} = \dfrac{-4}{x+1}$

**Solutions** **1a.**

$$\dfrac{3}{x} + 5 = 8$$

$$x\left(\dfrac{3}{x} + 5\right) = x(8)$$ Multiply each side by the LCD, $x$.

$$(x)\dfrac{3}{x} + (x)(5) = 8x$$ Apply the distributive property.

$$3 + 5x = 8x$$ Simplify.

$$3 + 5x - 5x = 8x - 5x$$ Subtract $5x$ from each side.

$$3 = 3x$$ Simplify.

$$\dfrac{3}{3} = \dfrac{3x}{3}$$ Divide each side by 3.

$$1 = x$$ Simplify.

**Check:** Replace $x$ with 1 in the original equation.

$$\dfrac{3}{x} + 5 = 8$$

$$\dfrac{3}{1} + 5 = 8$$

$$3 + 5 = 8$$

$$8 = 8$$

Since 1 makes the equation true and does not make the denominator in the original equation zero, it is the solution of the equation. Therefore, the solution set is $\{1\}$.

**1b.**

$$\dfrac{1}{m-3} + 2 = \dfrac{1}{m-3}$$

$$(m-3)\left(\dfrac{1}{m-3} + 2\right) = (m-3)\left(\dfrac{1}{m-3}\right)$$ Multiply each side by the LCD, $m-3$.

$$(m-3)\left(\dfrac{1}{m-3}\right) + 2(m-3) = 1$$ Apply the distributive property and simplify the right side.

$$1 + 2m - 6 = 1$$ Simplify each product.

$$2m - 5 = 1$$ Combine like terms.

$$2m - 5 + 5 = 1 + 5$$ Add 5 to each side.

$$2m = 6$$ Simplify.

$$\dfrac{2m}{2} = \dfrac{6}{2}$$ Divide each side by 2.

$$m = 3$$ Simplify.

**Check:** Replace $m$ with 3 in the original equation.

$$\frac{1}{m-3} + 2 = \frac{1}{m-3}$$

$$\frac{1}{3-3} + 2 = \frac{1}{3-3}$$

$$\frac{1}{0} + 2 = \frac{1}{0}$$

undefined

The proposed solution, 3, is the value that makes the LCD equal to zero and the rational expression undefined. So, 3 is an *extraneous* solution and must be excluded from the solution set. Since the only possible solution is rejected, this equation has no solution. The solution set is the empty set, or ∅.

**1c.**         $$\frac{4}{x^2} + \frac{5}{x} = -1$$

$$x^2\left(\frac{4}{x^2} + \frac{5}{x}\right) = x^2(-1) \qquad \text{Multiply each side by the LCD, } x^2.$$

$$x^2\left(\frac{4}{x^2}\right) + x^2\left(\frac{5}{x}\right) = -x^2 \qquad \text{Apply the distributive property.}$$

$$4 + 5x = -x^2 \qquad \text{Simplify each product.}$$

The resulting equation is quadratic, so we must write the equation in standard form and solve it by factoring.

$$4 + 5x + x^2 = -x^2 + x^2 \qquad \text{Add } x^2 \text{ to each side.}$$
$$x^2 + 5x + 4 = 0 \qquad \text{Simplify and write in standard form.}$$
$$(x + 1)(x + 4) = 0 \qquad \text{Factor.}$$
$$x + 1 = 0 \quad \text{or} \quad x + 4 = 0 \qquad \text{Apply the zero products property.}$$
$$x = -1 \qquad\qquad x = -4 \qquad \text{Solve each equation.}$$

**Check:** Replace $x$ with $-4$ and then $-1$ in the original equation.

$x = -4$:

$$\frac{4}{x^2} + \frac{5}{x} = -1$$

$$\frac{4}{(-4)^2} + \frac{5}{(-4)} = -1$$

$$\frac{4}{16} - \frac{5}{4} = -1$$

$$\frac{1}{4} - \frac{5}{4} = -1$$

$$-\frac{4}{4} = -1$$

$$-1 = -1$$

$x = -1$:

$$\frac{4}{x^2} + \frac{5}{x} = -1$$

$$\frac{4}{(-1)^2} + \frac{5}{(-1)} = -1$$

$$\frac{4}{1} - 5 = -1$$

$$4 - 5 = -1$$

$$-1 = -1$$

**INSTRUCTOR NOTE:**
Discuss the values that must be excluded from the solution set.

Both $-4$ and $-1$ make the equation true and are, therefore, solutions of the equation. So, the solution set is $\{-4, -1\}$.

 **Note:** *Neither $-4$ nor $-1$ makes the denominator in the original equation equal to zero. The value of $0$ makes the rational expression undefined and could not be a solution.*

**1d.** The denominators are $x - 2$, $x + 2$, and $x^2 - 4 = (x - 2)(x + 2)$, so the LCD is $(x - 2)(x + 2)$. Multiply each side by the LCD.

$$\frac{2}{x - 2} - \frac{1}{x + 2} = \frac{3}{x^2 - 4}$$

$$(x - 2)(x + 2)\left(\frac{2}{x - 2} - \frac{1}{x + 2}\right) = (x - 2)(x + 2)\left(\frac{3}{(x - 2)(x + 2)}\right)$$

$$(x - 2)(x + 2)\left(\frac{2}{x - 2}\right) - (x - 2)(x + 2)\left(\frac{1}{x + 2}\right) = 3$$

$$2(x + 2) - 1(x - 2) = 3$$

$$2x + 4 - x + 2 = 3$$

$$x + 6 = 3$$

$$x + 6 - 6 = 3 - 6$$

$$x = -3$$

**Check:** Replace $x$ with $-3$ in the original equation.

$$\frac{2}{x - 2} - \frac{1}{x + 2} = \frac{3}{x^2 - 4}$$

$$\frac{2}{-3 - 2} - \frac{1}{-3 + 2} = \frac{3}{(-3)^2 - 4}$$

$$\frac{2}{-5} - \frac{1}{-1} = \frac{3}{9 - 4}$$

$$-\frac{2}{5} + 1 = \frac{3}{5}$$

$$-\frac{2}{5} + \frac{5}{5} = \frac{3}{5}$$

$$\frac{3}{5} = \frac{3}{5}$$

The value $-3$ makes the equation a true statement, so it is a solution. The solution set is $\{-3\}$.

**Note:** *The value $-3$ does not make the denominator in the original equation equal to zero. The values 2 or $-2$ make the rational expression undefined and could not be solutions.*

**1e.**

$$\frac{r + 2}{r + 8} = \frac{2}{r}$$

$$r(r + 8)\frac{r + 2}{r + 8} = r(r + 8)\frac{2}{r} \qquad \text{Multiply each side by the LCD, } r(r + 8).$$

$$r(r + 2) = 2(r + 8) \qquad \text{Simplify each product.}$$

$$r^2 + 2r = 2r + 16 \qquad \text{Apply the distributive property.}$$

Note that the resulting equation is quadratic, so we write the equation in standard form and solve it by factoring.

$$r^2 + 2r - 2r - 16 = 2r + 16 - 2r - 16$$    Subtract $2r$ and 16 from each side.

$$r^2 - 16 = 0$$    Simplify.

$$(r - 4)(r + 4) = 0$$    Factor.

$$r - 4 = 0 \quad \text{or} \quad r + 4 = 0$$    Apply the zero products property.

$$r = 4 \qquad\qquad r = -4$$    Solve each equation.

**Check:** Replace $r$ with 4 and then $-4$ in the original equation.

$r = 4$:

$$\frac{r + 2}{r + 8} = \frac{2}{r}$$

$$\frac{4 + 2}{4 + 8} = \frac{2}{4}$$

$$\frac{6}{12} = \frac{1}{2}$$

$$\frac{1}{2} = \frac{1}{2}$$

$r = -4$:

$$\frac{r + 2}{r + 8} = \frac{2}{r}$$

$$\frac{-4 + 2}{-4 + 8} = \frac{2}{-4}$$

$$\frac{-2}{4} = -\frac{1}{2}$$

$$-\frac{1}{2} = -\frac{1}{2}$$

The values 4 and $-4$ make the equation true and are, therefore, solutions of the equation. The solution set is $\{-4, 4\}$.

**Note:** *Neither 4 nor $-4$ makes the denominator in the original equation equal to zero. The values 0 or $-8$ make the rational expression undefined and could not be solutions.*

**1f.** The denominators are $3x^2 + 5x + 2 = (3x + 2)(x + 1)$, $(3x + 2)$, and $(x + 1)$. So, the LCD is $(3x + 2)(x + 1)$.

$$\frac{-4x}{3x^2 + 5x + 2} + \frac{x}{3x + 2} = \frac{-4}{x + 1}$$

$$(3x + 2)(x + 1)\left(\frac{-4x}{(3x + 2)(x + 1)} + \frac{x}{3x + 2}\right) = (3x + 2)(x + 1)\left(\frac{-4}{x + 1}\right)$$

$$(3x + 2)(x + 1)\frac{-4x}{(3x + 2)(x + 1)} + (3x + 2)(x + 1)\frac{x}{3x + 2} = -4(3x + 2)$$

$$-4x + x(x + 1) = -12x - 8$$

$$-4x + x^2 + x = -12x - 8$$

$$x^2 - 3x = -12x - 8$$

$$x^2 - 3x + 12x + 8 = -12x - 8 + 12x + 8$$

$$x^2 + 9x + 8 = 0$$

$$(x + 8)(x + 1) = 0$$

$$x + 8 = 0 \quad \text{or} \quad x + 1 = 0$$

$$x = -8 \qquad\qquad x = -1$$

**Check:** Replace $x$ with $-8$ and then $-1$ in the original equation.

$x = -8$:

$$\frac{-4x}{3x^2 + 5x + 2} + \frac{x}{3x + 2} = \frac{-4}{x + 1}$$

$$\frac{-4(-8)}{3(-8)^2 + 5(-8) + 2}$$

$$+ \frac{-8}{3(-8) + 2} = \frac{-4}{-8 + 1}$$

$$\frac{32}{3(64) - 40 + 2} + \frac{-8}{-24 + 2} = \frac{-4}{-7}$$

$$\frac{32}{154} + \frac{-8}{-22} = \frac{4}{7}$$

$$\frac{16}{77} + \frac{4}{11} = \frac{4}{7}$$

$$\frac{16}{77} + \frac{28}{77} = \frac{44}{77}$$

$$\frac{44}{77} = \frac{44}{77}$$

$x = -1$:

$$\frac{-4x}{3x^2 + 5x + 2} + \frac{x}{3x + 2} = \frac{-4}{x + 1}$$

$$\frac{-4(-1)}{3(-1)^2 + 5(-1) + 2}$$

$$+ \frac{-1}{3(-1) + 2} = \frac{-4}{-1 + 1}$$

$$\frac{4}{3(1) - 5 + 2} + \frac{-1}{-3 + 2} = \frac{-4}{0}$$

undefined

The value $-1$ makes the expression undefined and must be excluded from the solution set. The value $-8$ makes the equation true, and is, therefore, a solution of the equation. So, the solution set is $\{-8\}$.

✓ **Student Check 1**    Solve each rational equation.

**a.** $\dfrac{4}{x} = 1 + \dfrac{2}{x}$    **b.** $\dfrac{2}{y + 7} - 1 = \dfrac{2}{y + 7}$    **c.** $1 + \dfrac{2}{x} = \dfrac{8}{x^2}$

**d.** $\dfrac{3}{x} + \dfrac{1}{x - 2} = \dfrac{6}{x^2 - 2x}$    **e.** $\dfrac{a - 7}{a + 3} = \dfrac{4}{a}$    **f.** $\dfrac{x}{2x + 1} - \dfrac{13 + x}{2x^2 - 3x - 2} = \dfrac{-3}{x - 2}$

## Solving Rational Equations for a Specific Variable

**Objective 2** ▶

Solve rational equations for a specific variable.

The skills used for solving rational equations provide the framework for rewriting formulas for specific variables. When the formula contains a rational expression, we must multiply each side of the equation by the LCD to enable us to solve for a different variable.

**Objective 2  Examples**    Solve each rational equation for the specified variable.

**2a.** $a = \dfrac{GM}{r^2}$ for $M$ (acceleration of gravity)    **2b.** $a = \dfrac{f - v}{t}$ for $v$

**2c.** $\dfrac{1}{a} + \dfrac{1}{b} = \dfrac{1}{t}$ for $b$

**Solutions**    **2a.**    $a = \dfrac{GM}{r^2}$    Identify the variable to be isolated.

$r^2(a) = r^2\left(\dfrac{GM}{r^2}\right)$    Multiply each side by the LCD, $r^2$.

$ar^2 = GM$    Simplify each side.

$\dfrac{ar^2}{G} = \dfrac{GM}{G}$    Divide each side by $G$.

$\dfrac{ar^2}{G} = M$    Simplify.

**2b.**

$$a = \frac{f - v}{t}$$  Identify the variable to be isolated.

$$t(a) = t\left(\frac{f - v}{t}\right)$$  Multiply each side by the LCD, $t$.

$$at = f - v$$  Simplify.

$$at - f = f - v - f$$  Subtract $f$ from each side.

$$at - f = -v$$  Simplify.

$$-1(at - f) = -1(-v)$$  Multiply each side by $-1$.

$$-at + f = v$$  Simplify.

**2c.**

$$\frac{1}{a} + \frac{1}{b} = \frac{1}{t}$$  Identify the variable to be isolated.

$$abt\left(\frac{1}{a} + \frac{1}{b}\right) = abt\left(\frac{1}{t}\right)$$  Multiply each side by the LCD, $abt$.

$$abt\left(\frac{1}{a}\right) + abt\left(\frac{1}{b}\right) = ab$$  Apply the distributive property.

$$bt + at = ab$$  Simplify each product.

$$bt + at - bt = ab - bt$$  Subtract $bt$ from each side.

$$at = b(a - t)$$  Factor out $b$ from the terms on the right side.

$$\frac{at}{a - t} = \frac{b(a - t)}{a - t}$$  Divide each side by $a - t$.

$$\frac{at}{a - t} = b$$  Simplify.

✔ **Student Check 2**    Solve each rational equation for the specified variable.

**a.** $d = \dfrac{m}{v}$ for $m$ (density)    **b.** $r^2 = \dfrac{V}{ph}$ for $p$    **c.** $\dfrac{1}{a} + \dfrac{1}{b} = \dfrac{1}{t}$ for $t$

**Objective 3** ▶
Troubleshoot common errors.

## Troubleshooting Common Errors

Some common errors related to rational equations are shown next.

**Objective 3 Examples**    A problem and an incorrect solution are given. Provide the correct solution and an explanation of the error.

**3a.** Solve $\dfrac{x}{x - 2} + 3 = \dfrac{2}{x - 2}$.

| Incorrect Solution | Correct Solution and Explanation |
|---|---|
| $$\frac{x}{x - 2} + 3 = \frac{2}{x - 2}$$ $$(x - 2)\left(\frac{x}{x - 2}\right) + 3(x - 2) = \left(\frac{2}{x - 2}\right)(x - 2)$$ $$x + 3x - 6 = 2$$ $$4x - 6 = 2$$ $$4x - 6 + 6 = 2 + 6$$ $$4x = 8$$ $$x = 2$$ So, the solution set is $\{2\}$. | The error was made in not excluding the solution 2 from the solution set. This value makes the denominator in the original equation zero and, therefore, the rational expression undefined. So, the solution set is the empty set, $\varnothing$. |

**3b.** Solve $\dfrac{4}{x+5} - \dfrac{3}{x-2} = \dfrac{1}{x^2+3x-10}$.

| Incorrect Solution | Correct Solution and Explanation |
|---|---|
| $\dfrac{4}{x+5} - \dfrac{3}{x-2} = \dfrac{1}{x^2+3x-10}$ $\dfrac{4}{x+5} - \dfrac{3}{x-2} = \dfrac{1}{(x+5)(x-2)}$ $(x+5)(x-2)\left(\dfrac{4}{x+5} - \dfrac{3}{x-2}\right)$ $= (x+5)(x-2)\left[\dfrac{1}{(x+5)(x-2)}\right]$ $4x - 8 - 3x + 15 = 1$ $x + 7 = 1$ $x = -6$ | The error was made in writing the equation that results after the fractions are cleared. The $-3$, rather than 3, should be distributed to $(x+5)$. $$4x - 8 - 3x - 15 = 1$$ $$x - 23 = 1$$ $$x = 24$$ So, the solution set is $\{24\}$. |

## ANSWERS TO STUDENT CHECKS

**Student Check 1** **a.** $\{2\}$ **b.** $\varnothing$ **c.** $\{-4, 2\}$
**d.** $\{3\}$ **e.** $\{-1, 12\}$ **f.** $\{-5\}$

**Student Check 2** **a.** $m = vd$ **b.** $p = \dfrac{V}{r^2 h}$ **c.** $t = \dfrac{ab}{b+a}$

## SUMMARY OF KEY CONCEPTS

1. The method to solve a rational equation involves clearing fractions from the equation by multiplying by the LCD. Once the fractions are cleared, solve the resulting equation. It is essential to check the solutions in the original equation to make sure that the proposed solution is not extraneous.

2. We can solve formulas for a specific variable as well. If the formula has a fraction, clear the fractions by multiplying by the LCD and then isolate the specified variable.

## GRAPHING CALCULATOR SKILLS

We can use the Table feature of a graphing calculator to check our possible answers when solving rational equations.

**Example:** Solve $x + \dfrac{6}{x-2} = \dfrac{3x}{x-2} + 1$.

**Solution:** We first solve the equation by hand.

$$(x-2)\left(x + \dfrac{6}{x-2}\right) = \left(\dfrac{3x}{x-2} + 1\right)(x-2)$$

$$(x-2)x + (x-2)\dfrac{6}{x-2} = \dfrac{3x}{x-2}(x-2) + 1(x-2)$$

$$x^2 - 2x + 6 = 3x + x - 2$$
$$x^2 - 2x + 6 = 4x - 2$$
$$x^2 - 6x + 8 = 0$$
$$(x-4)(x-2) = 0$$
$$x - 4 = 0 \qquad x - 2 = 0$$
$$x = 4 \qquad x = 2$$

We now need to check our answers in the original equation.

Enter two functions in the calculator. The first function $Y_1$ is the left side of the equation and the second function $Y_2$ is the right side of the equation.

```
Plot1 Plot2 Plot3
\Y1■X+6/(X-2)
\Y2■(3X)/(X-2)+1
\Y3=
\Y4=
\Y5=
\Y6=
```

Next we examine the table of values.

```
X    Y1     Y2
0    -3     1
1    -5     -2
2    ERROR  ERROR
3    9      10
4    7      7
5    7.5    6
6           5.5
X=4
```

Since the two $y$-values are equal at $x = 4$, this confirms 4 is a solution. At $x = 2$, the $y$-values have the ERROR message. This means that 2 makes the rational expression undefined. Therefore, 2 cannot be a solution.

## SECTION 7.6 / EXERCISE SET

### Write About It!

**Use complete sentences in your answer to each exercise.**

1. What is a rational equation?

2. Explain the steps to solve a rational equation.

3. Why must we check our answers when solving rational equations?

4. Explain the difference between a rational expression and a rational equation.

5. Explain the difference in the use of the LCD in $\dfrac{4}{x+5} + \dfrac{5}{(x-3)(x+5)}$ and $\dfrac{4}{x+5} + \dfrac{5}{(x-3)(x+5)} = 1$.

6. Explain how to solve $\dfrac{1}{a} + \dfrac{2}{b} = \dfrac{3}{c}$ for $a$.

### Practice Makes Perfect!

**Solve each rational equation. (See Objective 1.)**

7. $\dfrac{4}{x} = 8$ $\left\{\dfrac{1}{2}\right\}$

8. $\dfrac{5}{x} = 10$ $\left\{\dfrac{1}{2}\right\}$

9. $\dfrac{16}{2x+3} = 4$ $\left\{\dfrac{1}{2}\right\}$

10. $\dfrac{8}{3x+2} = 1$ $\{2\}$

11. $\dfrac{x}{3} = \dfrac{x+1}{4}$ $\{3\}$

12. $\dfrac{y}{5} = \dfrac{y-2}{6}$ $\{-10\}$

13. $\dfrac{y-4}{3} + \dfrac{1}{2} = \dfrac{y}{3}$ $\varnothing$

14. $\dfrac{a+3}{4} - \dfrac{2}{5} = \dfrac{a}{4}$ $\varnothing$

15. $\dfrac{4}{3a+5} = -2$ $\left\{-\dfrac{7}{3}\right\}$

16. $\dfrac{4}{2y-8} = 5$ $\left\{\dfrac{22}{5}\right\}$

17. $\dfrac{9}{2x} + 2 = 12$ $\left\{\dfrac{9}{20}\right\}$

18. $\dfrac{5}{3x} + 3 = 7$ $\left\{\dfrac{5}{12}\right\}$

19. $1 - \dfrac{5}{x} = -\dfrac{4}{x^2}$ $\{4, 1\}$

20. $2 - \dfrac{1}{x} = \dfrac{6}{x^2}$ $\left\{-\dfrac{3}{2}, 2\right\}$

21. $5 + \dfrac{3}{t+1} = -\dfrac{2}{t+1}$ $\{-2\}$

22. $\dfrac{3}{t+2} = \dfrac{6}{t+2} - 2$ $\left\{-\dfrac{1}{2}\right\}$

23. $\dfrac{3}{x+3} = \dfrac{4}{x-2}$ $\{-18\}$

24. $\dfrac{1}{x-5} = \dfrac{2}{x+4}$ $\{14\}$

25. $\dfrac{3}{x+1} - \dfrac{x}{x^2-1} = \dfrac{2}{x-1}$ $\varnothing$

26. $\dfrac{4}{x+2} + \dfrac{3x}{x^2-4} = \dfrac{7}{x-2}$ $\varnothing$

27. $\dfrac{x^2}{x+9} = \dfrac{49}{x+9}$ $\{-7, 7\}$

28. $\dfrac{3x^2}{x+10} = \dfrac{12}{x+10}$ $\{-2, 2\}$

29. $\dfrac{x^2}{x^2+7x+6} = \dfrac{1}{x+1}$ $\{-2, 3\}$

30. $\dfrac{x^2}{2x^2+5x-12} = \dfrac{-x}{2x-3}$ $\{-2, 0\}$

31. $\dfrac{5}{2x-1} + \dfrac{3x}{2x^2+7x-4} = \dfrac{4}{x+4}$ $\varnothing$

32. $\dfrac{4x}{3x^2-13x-10} - \dfrac{1}{x-5} = \dfrac{2}{3x+2}$ $\{8\}$

33. $\dfrac{x^2-10}{2x^2+5x-7} = \dfrac{x}{2x+1}$ $\left\{-\dfrac{5}{4}, -2\right\}$

34. $\dfrac{x^2-1}{3x^2-13x-6} = \dfrac{2x}{6x+1}$ $\left\{-\dfrac{1}{3}, \dfrac{1}{9}\right\}$

**Solve each rational equation for the specified variable. (See Objective 2.)**

35. $A = \dfrac{bh}{2}$ for $h$ (area of triangle) $h = \dfrac{2A}{b}$

36. $A = \dfrac{bh}{2}$ for $b$ (area of triangle) $b = \dfrac{2A}{h}$

37. $P = \dfrac{f}{a}$ for $a$ (pressure) $a = \dfrac{f}{P}$

38. $P = \dfrac{f}{a}$ for $f$ (pressure) $f = aP$

39. $a = \dfrac{v-v_0}{t}$ for $t$ (acceleration) $t = \dfrac{v-v_0}{a}$

40. $a = \dfrac{v-v_0}{t}$ for $v$ (acceleration) $v = v_0 + at$

41. $F = \dfrac{mv^2}{r}$ for $r$ (force) $r = \dfrac{mv^2}{F}$

42. $F = \dfrac{mv^2}{r}$ for $m$ (force) $m = \dfrac{Fr}{v^2}$

43. $u = \dfrac{P^2L}{2AE}$ for $A$ (mechanics, strain energy) $A = \dfrac{P^2L}{2uE}$

44. $u = \dfrac{P^2L}{2AE}$ for $E$ (mechanics, strain energy) $E = \dfrac{P^2L}{2uA}$

45. $G = \dfrac{E}{2(I+v)}$ for $I$ (engineering, shear stress) $I = \dfrac{E}{2G} - v$

46. $G = \dfrac{E}{2(I+v)}$ for $v$ (engineering, shear stress) $v = \dfrac{E}{2G} - I$

47. $\dfrac{1}{a} + \dfrac{2}{b} = \dfrac{3}{c}$ for $b$ $b = \dfrac{2ac}{3a-c}$

48. $\dfrac{4}{r} - \dfrac{5}{s} = \dfrac{1}{t}$ for $r$ $r = \dfrac{4st}{s+5t}$

### Mix 'Em Up!

**Solve each rational equation.**

49. $\dfrac{a}{3} = \dfrac{3a-1}{7}$ $\left\{\dfrac{3}{2}\right\}$

50. $y = \dfrac{2y-5}{6}$ $\left\{-\dfrac{5}{4}\right\}$

51. $\dfrac{x-5}{2} - \dfrac{1}{6} = \dfrac{4x}{3}$ $\left\{-\dfrac{16}{5}\right\}$

52. $\dfrac{4y+1}{5} + \dfrac{7}{10} = \dfrac{3y}{2}$ $\left\{\dfrac{9}{7}\right\}$

53. $\dfrac{4a}{10a+3} = \dfrac{2}{5}$ $\varnothing$

54. $\dfrac{2a}{6a-5} = \dfrac{1}{3}$ $\varnothing$

55. $\dfrac{5}{7x} - 1 = 14$ $\left\{\dfrac{1}{21}\right\}$

56. $\dfrac{3}{4x} + 2 = 11$ $\left\{\dfrac{1}{12}\right\}$

Additional answers can be found in the Instructor Answer Appendix.

**57.** $\dfrac{2x^2 - 3x - 2}{x^2 - 6} = \dfrac{2x + 1}{x}$ **58.** $\dfrac{x^2 + 2x}{2x^2 - 5x - 3} = \dfrac{2x}{4x + 3}$

**59.** $3 + \dfrac{13}{x} = \dfrac{10}{x^2}$ $\left\{-5, \dfrac{2}{3}\right\}$ **60.** $2 + \dfrac{7}{x} = -\dfrac{3}{x^2}$ $\left\{-\dfrac{1}{2}, -3\right\}$

**61.** $3 + \dfrac{4}{t - 2} = -\dfrac{1}{t - 2}$ $\left\{\dfrac{1}{3}\right\}$ **62.** $\dfrac{2}{t + 5} = \dfrac{1}{t + 5} - 4$ $\left\{-\dfrac{21}{4}\right\}$

**63.** $\dfrac{1}{3x + 2} = \dfrac{2}{5x - 1}$ $\{-5\}$ **64.** $\dfrac{5}{2x - 7} = \dfrac{3}{x + 3}$ $\{36\}$

**65.** $\dfrac{3}{2x + 3} - \dfrac{2x}{4x^2 - 9} = \dfrac{2}{2x - 3}$ $\varnothing$

**66.** $\dfrac{4}{x - 5} + \dfrac{2x}{x^2 - 25} = \dfrac{6}{x + 5}$ $\varnothing$

**Solve each rational equation for the specified variable.**

**67.** $A = \dfrac{1}{2}h(a + b)$ for $b$ (area of trapezoid) $b = \dfrac{2A - ha}{h}$

**68.** $A = \dfrac{1}{2}h(a + b)$ for $h$ (area of trapezoid) $h = \dfrac{2A}{a + b}$

**69.** $\dfrac{PV}{T} = k$ for $V$ (combined gas law) $V = \dfrac{kT}{P}$

**70.** $\dfrac{PV}{T} = k$ for $T$ (combined gas law) $T = \dfrac{PV}{k}$

**71.** $\dfrac{1}{R} = \dfrac{1}{R_1} + \dfrac{1}{R_2}$ for $R$ (resistors in parallel) $R = \dfrac{R_1 R_2}{R_1 + R_2}$

**72.** $\dfrac{1}{R} = \dfrac{1}{R_1} + \dfrac{1}{R_2}$ for $R_1$ (resistors in parallel) $R_1 = \dfrac{RR_2}{R_2 - R}$

**73.** $m = \dfrac{y_2 - y_1}{x_2 - x_1}$ for $y_2$ (slope formula) $y_2 = mx_2 - mx_1 + y_1$

**74.** $m = \dfrac{y_2 - y_1}{x_2 - x_1}$ for $x_2$ (slope formula) $x_2 = \dfrac{y_2 - y_1 + mx_1}{m}$

## You Be the Teacher!

**Correct each student's errors, if any.**

**75.** Solve $\dfrac{3}{x + 1} = \dfrac{3}{x + 4}$.

Baxter's work:

$(x + 1)(x + 4)\dfrac{3}{x + 1} = \dfrac{3}{x + 4}(x + 1)(x + 4)$

$3(x + 4) = 3(x + 1)$

$3x + 12 = 3x + 3$

$0 = -9$

So, $x = 0$, $x = -9$.

**76.** Solve $\dfrac{x^2}{x - 1} = \dfrac{1}{x - 1}$.

Lydia's work:

$(x - 1)\dfrac{x^2}{x - 1} = \dfrac{1}{x - 1}(x - 1)$

$x^2 = 1$

$x^2 - 1 = 0$

$(x - 1)(x + 1) = 0$

$x - 1 = 0 \quad x + 1 = 0$

$x = 1 \quad\quad x = -1$

So, the solutions are 1 and $-1$.

**77.** Solve $\dfrac{2}{x + 2} - \dfrac{5}{3x - 1} = \dfrac{4x}{3x^2 + 5x - 2}$.

Holly's work:

$\dfrac{2}{x + 2} - \dfrac{5}{3x - 1} = \dfrac{4x}{(3x - 1)(x + 2)}$

$(3x - 1)(x + 2)\dfrac{2}{x + 2} - \dfrac{5}{3x - 1}(3x - 1)(x + 2)$

$= \dfrac{4x}{(3x - 1)(x + 2)}(3x - 1)(x + 2)$

$2(3x - 1) - 5(x + 2) = 4x$

$6x - 2 - 5x + 10 = 4x$

$x + 8 = 4x$

$8 = 3x$

$x = \dfrac{8}{3}$

So, the solution is $\dfrac{8}{3}$.

**78.** Solve $\dfrac{2}{x + 1} - \dfrac{1}{2x - 1} = \dfrac{4x}{2x^2 + x - 1}$.

Eddie's work:

$\dfrac{2}{x + 1} - \dfrac{1}{2x - 1} = \dfrac{4x}{2x^2 + x - 1}$

$(2x - 1)(x + 1)\dfrac{2}{x + 1} - \dfrac{1}{2x - 1}(2x - 1)(x + 1)$

$= \dfrac{4x}{(2x - 1)(x + 1)}(2x - 1)(x + 1)$

$2(2x - 1) - (x + 1) = 4x$

$4x - 2 - x + 1 = 4x$

$3x - 1 = 4x$

$x = -1$

So, the solution is $-1$.

## Calculate It!

**Solve each rational equation. Use a graphing calculator to check the answers.**

**79.** $\dfrac{3}{x} + \dfrac{5}{x^2} = 2$ $\left\{-1, \dfrac{5}{2}\right\}$ **80.** $\dfrac{2}{x} + \dfrac{1}{x^2} = 3$ $\left\{-\dfrac{1}{3}, 1\right\}$

**81.** $\dfrac{x}{x + 5} - \dfrac{3}{x - 5} = 1$ $\left\{\dfrac{5}{4}\right\}$ **82.** $\dfrac{2x}{x + 6} - \dfrac{2}{x - 6} = 1$ $\{2, 12\}$

**Proportions and Applications of Rational Equations**

Dolls are one of the most popular toys among young girls. However, some people are concerned that these toys display an unrealistic body image to girls. An 11.5-in.-tall doll has a waist measurement of 3 in. What would be the waist measurement of a 5-ft-tall doll with the same proportions?

In this section, we use ratios and proportions to answer this and other real-life questions.

## Proportions

A **ratio** relates the measurement of two quantities. Ratios can be expressed in two ways: (1) with a colon, or (2) as a fraction. For example, suppose that a fragrance spa gives away one free lotion sample for every five people who enter the store. The ratio of free lotion samples to people who enter the store can be expressed as 1:5 (read "1 to 5") or the ratio can be expressed as the fraction $\frac{1}{5}$. We will express ratios as fractions in this section.

A **proportion** is an equation with two ratios set equal to each other. A proportion is an equation of the form $\frac{a}{b} = \frac{c}{d}$, where $b$ and $d \neq 0$. When a proportion includes variables, it is a rational equation. Therefore, we can apply the methods for solving rational equations to solve proportions. When we multiply a proportion by its LCD, it is equivalent to the equation that results from cross multiplication.

**Objective 1** ▶

Solve proportions.

$$\frac{a}{b} = \frac{c}{d}$$

$$bd\left(\frac{a}{b}\right) = bd\left(\frac{c}{d}\right) \qquad \text{Multiply by the LCD, } bd.$$

$$ad = bc \qquad \text{Simplify.}$$

$$\frac{a}{b} \diagup\!\!\!\!= \diagdown \frac{c}{d}$$

$$ad = bc$$

cross products

**Property: Cross Products**

If $b$ and $d \neq 0$, and $\frac{a}{b} = \frac{c}{d}$, then the **cross products** are $ad$ and $bc$. The cross products are equal. So,

$$ad = bc$$

For example, since $\frac{2}{3} = \frac{4}{6}$, then

$$2 \cdot 6 = 3 \cdot 4$$
$$12 = 12$$

**Procedure: Solving a Proportion**

**Step 1:** Multiply each side of the equation by the LCD or cross multiply.
**Step 2:** Solve for the indicated variable.
**Step 3:** Check the answer in the original equation.

**Objective 1 Examples** Solve each proportion.

**1a.** $\dfrac{4}{5} = \dfrac{x}{10}$  **1b.** $\dfrac{x+2}{4} = \dfrac{2}{3}$  **1c.** $\dfrac{1}{2x} = \dfrac{5}{9}$

**Solutions**

**1a.** **Method 1:** Multiply by the LCD.

$$\frac{4}{5} = \frac{x}{10}$$

$$10\left(\frac{4}{5}\right) = 10\left(\frac{x}{10}\right) \quad \text{Multiply each side by 10.}$$

$$2 \cdot 4 = 1 \cdot x \quad \text{Divide out the common factors.}$$

$$8 = x \quad \text{Simplify.}$$

**Method 2:** Cross multiply.

$$\frac{4}{5} = \frac{x}{10}$$

$$4(10) = 5x \quad \text{Cross multiply.}$$

$$40 = 5x \quad \text{Simplify.}$$

$$\frac{40}{5} = \frac{5x}{5} \quad \text{Divide each side by 5.}$$

$$8 = x \quad \text{Simplify.}$$

So, the solution set is {8}. We can check by substituting 8 into the original equation.

**1b.** **Method 1:** Multiply by the LCD.

$$\frac{x+2}{4} = \frac{2}{3}$$

$$12\left(\frac{x+2}{4}\right) = 12\left(\frac{2}{3}\right) \quad \text{Multiply each side by 12.}$$

$$3(x+2) = 4 \cdot 2 \quad \text{Divide out the common factors.}$$

$$3x + 6 = 8 \quad \text{Distribute.}$$

$$3x + 6 - 6 = 8 - 6 \quad \text{Subtract 6 from each side.}$$

$$3x = 2 \quad \text{Simplify.}$$

$$x = \frac{2}{3} \quad \text{Divide each side by 3.}$$

**Method 2:** Cross multiply.

$$\frac{x+2}{4} = \frac{2}{3}$$

$$3(x+2) = 2(4) \quad \text{Cross multiply.}$$

$$3x + 6 = 8 \quad \text{Simplify.}$$

$$3x + 6 - 6 = 8 - 6 \quad \text{Subtract 6 from each side.}$$

$$3x = 2 \quad \text{Simplify.}$$

$$x = \frac{2}{3} \quad \text{Divide each side by 3.}$$

So, the solution set is $\left\{\frac{2}{3}\right\}$. We can check by substituting $\frac{2}{3}$ into the original equation.

**1c.** **Method 1:** Multiply by the LCD.

$$\frac{1}{2x} = \frac{5}{9}$$

$$18x\left(\frac{1}{2x}\right) = 18x\left(\frac{5}{9}\right) \quad \text{Multiply each side by 18x.}$$

$$9(1) = 2x(5) \quad \text{Divide out common factors.}$$

$$9 = 10x \quad \text{Simplify.}$$

$$\frac{9}{10} = x \quad \text{Divide each side by 10.}$$

**Method 2:** Cross multiply.

$$\frac{1}{2x} = \frac{5}{9}$$

$$1(9) = 2x(5) \quad \text{Cross multiply.}$$

$$9 = 10x \quad \text{Simplify.}$$

$$\frac{9}{10} = x \quad \text{Divide each side by 10.}$$

So, the solution set is $\left\{\frac{9}{10}\right\}$. We can check by substituting $\frac{9}{10}$ into the original equation.

✔ **Student Check 1** Solve each proportion.

**a.** $\dfrac{3}{4} = \dfrac{x}{6}$   **b.** $\dfrac{x-3}{5} = \dfrac{x}{2}$   **c.** $\dfrac{3}{x} = \dfrac{7}{10}$

## Applications

**Objective 2** ▶

Use proportions in various real-world applications.

Proportions can be used to solve problems from the real world. Recall the fragrance spa store that gives away one free lotion sample for every five people who enter the store. Suppose we want to determine the number of free lotion samples that the fragrance spa has given away after 60 people have entered the store. We can write the following proportion to determine that information.

Number of free samples → $\dfrac{1}{5} = \dfrac{x}{60}$ ← Number of free samples
Number of people who entered store → $\phantom{\dfrac{1}{5} = }$ ← Number of people who entered store

To solve, we cross multiply to get

$$1(60) = x(5)$$
$$60 = 5x$$
$$\frac{60}{5} = x$$
$$12 = x$$

So, the fragrance spa has given away 12 free lotion samples after 60 people have entered the store.

---

**Procedure: Writing a Proportion**

**Step 1:** Define the ratio so that the numerator represents one quantity and the denominator represents the other quantity.

**Step 2:** Write the proportion, setting the known ratio equal to the unknown ratio. Use a variable to represent the unknown quantity.

**Step 3:** Solve the proportion.

---

**Objective 2 Examples** / **Use a proportion to solve each problem.**

**2a.** Jennifer's favorite chocolate chip cookie recipe calls for 1 cup of brown sugar and it yields 36 cookies. If Jennifer wants to make only 24 cookies, how much brown sugar should she use?

**Solution**

**2a.** Let the ratio be cups of brown sugar to the number of servings in the recipe.

Cups of brown sugar → $\dfrac{1}{36} = \dfrac{x}{24}$   Write the proportion, where $x$ is the number
Number of servings → $\phantom{\dfrac{1}{36} = }$   of cups needed to make 24 cookies.

$$1(24) = x(36)$$   Cross multiply.

$$24 = 36x$$   Simplify each product.

$$\frac{24}{36} = x$$   Divide each side by 36.

$$\frac{2}{3} = x$$   Simplify.

So, Jennifer should use $\dfrac{2}{3}$ cup of brown sugar to make 24 cookies.

**2b.** When Iason returns to the United States from Greece, he has 245 Euros. If the conversion rate is 1 Euro = \$1.50, how much money in U.S. dollars does Iason have?

**Solution** **2b.** Let the ratio be Euros to U.S. dollars.

$$\text{Euros} \rightarrow \frac{245}{x} = \frac{1}{1.50} \qquad \text{U.S. dollars} \rightarrow$$

Write the proportion, where $x$ is the number of U.S. dollars for 245 Euros.

$$245(1.50) = x(1) \qquad \text{Cross multiply.}$$

$$367.5 = x \qquad \text{Simplify.}$$

So, Iason has the equivalent of $367.50.

**2c.** An 11.5-in.-tall doll has a waist measurement of 3 in. What would be the waist measurement of a 5-ft-tall doll with the same proportions?

**Solution** **2c.** Let the ratio be height to waist measurement.

$$\text{Height} \rightarrow \frac{11.5}{3} = \frac{60}{x} \qquad \text{Waist} \rightarrow$$

Write the proportion, where $x$ is the waist measurement of a 60-in.-tall doll.

$$3x\left(\frac{11.5}{3}\right) = 3x\left(\frac{60}{x}\right) \qquad \text{Multiply each side by the LCD, } 3x.$$

$$11.5x = 180 \qquad \text{Simplify.}$$

$$x = \frac{180}{11.5} \qquad \text{Divide each side by 11.5.}$$

$$x \approx 15.65 \qquad \text{Simplify.}$$

So, a 5-ft.-tall doll would have a waist measurement of 15.65 in.

✔ **Student Check 2** Use a proportion to solve each problem.

**a.** A punch recipe that serves 40 calls for $1\frac{3}{4}$ cups lemon juice. How many cups of lemon juice is needed to serve 100 people?

**b.** How many U.S. dollars are in 3000 Pakistani rupee if the conversion rate is 1 U.S. dollar = 61.2 Pakistan rupee?

**c.** If 1500 people were surveyed in a small town and 850 of them were Republicans, how many in the town of 10,000 are Republicans? Round to the nearest integer.

## Similar Triangles

**Objective 3** ▶

Use proportions to solve similar triangle problems.

**Similar triangles** have the same shape, but different sizes. Corresponding angles of similar triangles are equal, and the lengths of corresponding sides of similar triangles are proportional. Triangles $ABC$ and $LMN$ are similar triangles.

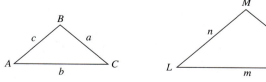

The measure of angle $A$ is equal to the measure of angle $L$, the measure of angle $B$ is equal to the measure of angle $M$, and the measure of angle $C$ is equal to the measure of angle $N$. Since the sides of the similar triangles are proportional, we can write

$$\frac{a}{l} = \frac{b}{m} = \frac{c}{n}$$

**Objective 3 Example**  Triangles *DEF* and *JKL* are similar. Find the missing length, *l*. Note the figures are not drawn to scale.

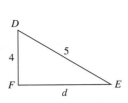

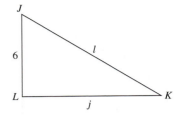

**Solution**

$$\frac{4}{6} = \frac{5}{l}$$   Write a proportion with ratios of corresponding sides of the triangle.

$$4l = 30$$   Cross multiply.

$$l = \frac{30}{4}$$   Divide each side by 4.

$$l = 7.5$$   Simplify.

✓ **Student Check 3**  If $d = 7$, use the result of Example 3 to find the missing length *j* in triangle *JKL*.

## Work Problems

**Objective 4** ▶
Solve work problems.

Now we will solve problems from other applications of rational equations. The first of these involves work applications. Work applications involve finding the amount of time it takes two individuals to complete a task when working together.

For example, if Jodi can do a job in 6 hr, then she completes $\frac{1}{6}$ of the job in 1 hr. If Kyle can do the same job in 4 hr, then he completes $\frac{1}{4}$ of the job in 1 hr. To determine the time it will take for Jodi and Kyle to complete the job if they work together, we add up the portion of the job that can be completed in 1 hr by each of them working alone.

$$\frac{1}{4} + \frac{1}{6} = \frac{3}{12} + \frac{2}{12}$$
$$= \frac{5}{12}$$

So, $\frac{5}{12}$ represents the portion of the job that can be completed in 1 hr. Since $\frac{5}{12} = \frac{1}{\frac{12}{5}}$,

it will take Jodi and Kyle $\frac{12}{5} = 2.4$ hr to complete the job if they work together. This leads to the following property.

> **Property: Work Equation**
>
> Work equations are of the form
>
> $$\frac{1}{a} + \frac{1}{b} = \frac{1}{t}$$
>
> where *a* is the time for one person to complete the job working alone, *b* is the time for another person to complete the job working alone, and *t* is the time for the two people working together to complete the job.

**Objective 4 Example**

Suppose it takes Clarice 5 hr to rake and bag the leaves in her yard when she works alone. Her son, Carrel, can complete the job by himself in 3 hr. How long would it take Clarice and Carrel to complete the job if they worked together?

**Solution**

What is unknown? The time it takes for Clarice and Carrel to complete the job working together. Let $x$ represent this time in hours.

What is known? Clarice can complete the job in 5 hr. Carrel can complete the job in 3 hr.

So, we can apply the property to write the equation.

$$\frac{1}{a} + \frac{1}{b} = \frac{1}{t}$$   State the work equation.

$$\frac{1}{5} + \frac{1}{3} = \frac{1}{x}$$   Replace the times working alone, 5 hr and 3 hr, and the time working together, $x$.

$$15x\left(\frac{1}{5} + \frac{1}{3}\right) = 15x\left(\frac{1}{x}\right)$$   Multiply each side by the LCD, 15x.

$$15x\left(\frac{1}{5}\right) + 15x\left(\frac{1}{3}\right) = 15$$   Apply the distributive property.

$$3x + 5x = 15$$   Simplify each product.

$$8x = 15$$   Combine like terms.

$$x = \frac{15}{8}$$   Divide each side by 8.

$$x \approx 1.89$$   Approximate the solution.

So, together they can complete the job in approximately 1.89 hr.

**✓ Student Check 4**

Helena cleans newly constructed homes for a real estate company. Alone, Helena can clean a 2400 ft² home in 4 hr. Her partner Kenneth can clean the same size home in 6 hr. How long would it take them to clean the home if they worked together?

## Motion Problems

**Objective 5** ▶

Solve motion problems.

A *rate* is a ratio of distance to time. A rate that we are familiar with is the speed of a car, usually given as $\frac{miles}{hour}$ or $\frac{kilometers}{hour}$, depending on the country where one lives. We know that rate (speed) is equal to distance (miles) divided by time (hours), that is, $r = \frac{d}{t}$. A more familiar form is $d = rt$, that is distance is equal to rate times time. This is true no matter what unit is being used for distance or time.

> **Property: Distance-Rate-Time Formula**
>
> If an object travels a distance $d$ at rate $r$ for time $t$, then $d = rt$. Solving for $r$, we obtain the formula $r = \frac{d}{t}$. Solving for $t$, we obtain the formula $t = \frac{d}{r}$.

**Objective 5 Example**

An airplane flies 1573 mi from Denver to Philadelphia in the same amount of time another airplane flies 1090 mi from New York to Orlando. If the speed of the Denver flight is 220 mph faster than the New York flight, find the speed of the Denver flight.

**Solution**

What is unknown? The speed of the each flight is unknown. Let $x$ represent the speed of the New York flight. Then $x + 220$ is the speed of the Denver flight.

What is known? The total distance from Denver to Philadelphia is 1573 mi. The total distance from New York to Orlando is 1090 mi.

We can use a table to summarize the information. We complete the distance and rate columns with the known and unknown information. We then use the

distance-rate-time formula to complete the time column. So, time is distance divided by rate.

| | Distance (mi) = | Rate (mph) | × | Time (hr) |
|---|---|---|---|---|
| Denver flight | 1573 | $x + 220$ | | $\dfrac{1573}{x + 220}$ |
| New York flight | 1090 | $x$ | | $\dfrac{1090}{x}$ |

Since both flights take the same amount of time, we write an equation by setting the expressions for time equal.

$$\frac{1573}{x + 220} = \frac{1090}{x}$$ Begin with the model.

$$1573x = 1090(x + 220)$$ Cross multiply.

$$1573x = 1090x + 239{,}800$$ Apply the distributive property.

$$1573x - 1090x = 1090x + 239{,}800 - 1090x$$ Subtract 1090x from each side.

$$483x = 239{,}800$$ Simplify.

$$x \approx 496.48$$ Divide each side by 483.

The New York flight traveled at approximately 496.48 mph, so the speed of the Denver flight was approximately 496.48 + 220 or 716.48 mph.

✓ **Student Check 5** Courtney ran 3.5 mi in the same amount of time it took Tammy to run 6 mi. If Courtney ran 2 mph slower than Tammy, how fast did each woman run?

**Objective 6** ▶
Troubleshoot common errors.

## Troubleshooting Common Errors

Some common errors associated with applications of rational equations are shown next.

**Objective 6 Examples**  **A problem and an incorrect solution are given. Provide the correct solution and an explanation of the error.**

**6a.** Marice can complete a job in 4 hr working alone and Gwen can complete the job in 3 hr working alone. How long can they complete the job working together?

| Incorrect Solution | Correct Solution and Explanation |
|---|---|
| 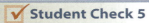 $\dfrac{4 + 3}{2} = 3.5 \text{ hr}$ | We must find the portions that they each can do in an hour and set the sum equal to the portion they can do together in an hour. Let $x$ represent the time working together to complete the job. $$\frac{1}{4} + \frac{1}{3} = \frac{1}{x}$$ $$12x\left(\frac{1}{4} + \frac{1}{3}\right) = 12x\left(\frac{1}{x}\right)$$ $$3x + 4x = 12$$ $$7x = 12$$ $$x = \frac{12}{7} \approx 1.7 \text{ hr}$$ |

**6b.** The triangles are similar. Find the value of $x$. Note the figures are not drawn to scale.

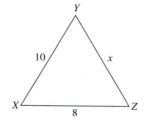

| Incorrect Solution | Correct Solution and Explanation |
|---|---|
| $$\dfrac{b}{8} = \dfrac{7}{x}$$ $$bx = 56$$ $$x = \dfrac{56}{b}$$ | We must use a known ratio in our equation. $$\dfrac{5}{10} = \dfrac{7}{x}$$ $$5x = 70$$ $$x = 14$$ |

## ANSWERS TO STUDENT CHECKS

**Student Check 1   a.** $4.5$  **b.** $-2$  **c.** $\dfrac{30}{7}$

**Student Check 2   a.** $4.38$ cups  **b.** $49.02$ U.S. dollars
  **c.** $5667$ Republicans

**Student Check 3**   $10.5$

**Student Check 4**   It would take them 2.4 hr to clean the home.

**Student Check 5**   Courtney ran 2.8 mph and Tammy ran 4.8 mph.

## SUMMARY OF KEY CONCEPTS

1. A ratio relates the measures of two quantities and can be expressed as a fraction. A proportion is an equation in which two ratios are equal.

2. Proportions are solved by multiplying both sides of the equation by the LCD or by cross multiplying.

3. Proportions can be used to solve many real-world problems. To solve a problem, set up a proportion with the known ratio equal to the unknown ratio. In these ratios, let the numerators represent one quantity and the denominators represent the other quantity.

4. Similar triangles are triangles in which the lengths of corresponding sides are proportional. To find a missing side, set up a proportion.

5. In work problems, we either find how long it will take one person working alone or two people working together to complete a job. We must represent the portion of the job that each person can do in one hour. We then add the individual portions and set equal to $\dfrac{1}{x}$, the portion of the job they can complete together.

6. To solve motion problems, use the formula $d = rt$. Make a chart and complete the columns. One of the columns is the given information. Variables are used to represent one of the other columns. The remaining column is obtained by applying the $d = rt$ formula. Write an appropriate equation and solve.

## SECTION 7.7 / EXERCISE SET

 **Write About It!**

**Use complete sentences in your answer to each exercise.**

1. Explain the meaning of a ratio.

2. Explain the meaning of a proportion.

3. What is the difference between a ratio and a proportion?

4. Explain how to solve a proportion.

Additional answers can be found in the Instructor Answer Appendix.

5. What does it mean for two triangles to be similar?

6. Explain how to set up a work problem.

 **Practice Makes Perfect!**

**Solve each proportion. (See Objective 1.)**

7. $\dfrac{2}{3} = \dfrac{x}{6}$   4

8. $\dfrac{1}{4} = \dfrac{x}{12}$   3

9. $\dfrac{1}{6} = \dfrac{x}{20}$   $\dfrac{10}{3}$

**10.** $\dfrac{1}{5} = \dfrac{x}{12}$  $\dfrac{12}{5}$   **11.** $\dfrac{x+1}{4} = \dfrac{3}{5}$  $\dfrac{7}{5}$   **12.** $\dfrac{x-4}{3} = \dfrac{2}{5}$  $\dfrac{26}{5}$

**13.** $\dfrac{\frac{1}{2}}{12} = \dfrac{3}{x}$  72   **14.** $\dfrac{\frac{2}{3}}{8} = \dfrac{2}{x}$  24   **15.** $\dfrac{3.65}{2} = \dfrac{x}{10}$  18.25

**16.** $\dfrac{12}{2.88} = \dfrac{100}{x}$  24   **17.** $\dfrac{100}{x+3} = \dfrac{250}{x+5}$  $-\dfrac{5}{3}$

**18.** $\dfrac{300}{x+1} = \dfrac{400}{x-4}$  $-16$

**Use proportions to solve each problem. (*See Objective 2.*)**

Use the Country Biscuits recipe for Exercises 19–22.

> Country Biscuits
>   2 cups all-purpose flour
>   $\dfrac{3}{4}$ cup shortening
>   1 cup milk
>   1 tsp salt
>   3 tsp baking powder
>
> Directions: Combine the ingredients in a bowl. On a floured surface, roll out the dough and cut out 12 biscuits. Bake on a cookie sheet for 375° F until lightly browned. (Makes 12 biscuits.)

**19.** How much flour should you use to increase the recipe to 20 biscuits?  $\dfrac{10}{3}$ cups

**20.** How much shortening should you use if you want to make 20 biscuits?  $\dfrac{5}{4}$ cups

**21.** If you only have $\dfrac{1}{2}$ cup of flour, approximately how many biscuits can you make?  3

**22.** If you only have $\dfrac{1}{3}$ cup of milk, approximately how many biscuits can you make?  4

**23.** How many U.S. dollars are in 15 Saudi riyals? (Note: 1 U.S. dollar = 3.75 Saudi riyals.)  4 U.S. dollars

**24.** How many U.S. dollars are in 1500 South Korean won? (Note: 1 U.S. dollar = 1056.15 South Korean won.)  1.42 U.S. dollars

**25.** How many Swiss francs are in 10 U.S. dollars? (Note: 1 U.S. dollar = 0.8238 Swiss francs.)  8.24 Swiss francs

**26.** How many Japanese yen are in 15,000 U.S. dollars? (Note: 1 U.S. dollar = 79.19 Japanese yen.)  1,187,850 Japanese yen

**27.** If 1 in. represents 120 mi on a map, how many inches represent the distance between Los Angeles, California, and Denver, Colorado, which is 1020 mi?  8.5 in.

**28.** If 1 in. represents 40 mi on a map, how many inches represent the distance between Boston, Massachusetts, and Baltimore, Maryland, which is 408 mi?  10.2 in.

**29.** If 1 cm represents 210 km on a map, how many centimeters represent the distance between Philadelphia, Pennsylvania, and Chicago, Illinois, which is 1224.3 km?  5.83 cm

**30.** If 1 cm represents 210 km on a map, how many centimeters represent the distance between New Orleans, Louisiana, and San Diego, California, which is 2927.4 km?  13.94 cm

**31.** An object weighing 1 lb on Earth weighs 0.166 lb on the moon. Find Cynthia's weight on the moon if she weighs 125 lb on Earth.  20.75 lb

**32.** An object weighing 1 lb on Earth weighs 0.377 lb on Mars. Find Andrew's weight on Mars if he weighs 180 lb on Earth.  67.86 lb

**33.** An object weighing 1 lb on Earth weighs 0.067 lb on Pluto. Find Louise's weight on Earth if she weighs 6.03 lb on Pluto.  90 lb

**34.** An object weighing 1 lb on Earth weighs 2.364 lb on Jupiter. Find Mark's weight on Earth if he weighs 531.9 lb on Jupiter.  225 lb

**Use similar triangles *ABC* and *XYZ* for Exercises 35 and 36. Note the figures are not drawn to scale. (*See Objective 3.*)**

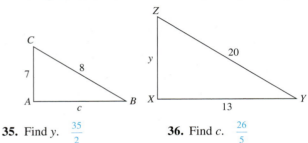

**35.** Find $y$.  $\dfrac{35}{2}$   **36.** Find $c$.  $\dfrac{26}{5}$

**Use similar triangles *ABC* and *XYZ* for Exercises 37 and 38. Note the figures are not drawn to scale (*See Objective 3.*)**

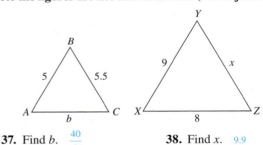

**37.** Find $b$.  $\dfrac{40}{9}$   **38.** Find $x$.  9.9

**Solve each work problem. (*See Objective 4.*)**

**39.** Lynn and Paul enjoy working on jigsaw puzzles. On average, it takes Lynn 6 hr to complete a 1000-piece puzzle. It takes Paul 10 hr to complete a 1000-piece puzzle on average. How long will it take Lynn and Paul to complete a 1000-piece puzzle working together?  $\dfrac{15}{4}$ hr

**40.** Lindsey distributes a town newspaper by foot. It takes her 3 hr to deliver the paper to the customers in her assigned area. Jamie can deliver newspapers in the same area in 2 hr. How long will it take Lindsey and Jamie to distribute the newspapers if they work together?  $\dfrac{6}{5}$ hr

**Solve each motion problem. (*See Objective 5.*)**

**41.** It takes Margie 4 hr to drive to a conference. If she drives at a constant rate of 50 mph, how far away is the conference?  200 mi

**42.** How long will it take Howie to drive 264 mi if he is traveling a constant rate of 60 mph?   4.4 hr

**43.** Cassie runs 3 mi in the time that it takes Tran to walk 1 mi. Cassie runs 2 mph faster than Tran walks. At what speed does each person travel?   Cassie runs at 3 mph and Tran walks at 1 mph.

**44.** Esther runs 5 mi in the same time that it takes Janice to run 6 mi. If Esther runs 1 mph slower than Janice, find the speed that each woman runs.   Esther runs at 5 mph and Janice runs at 6 mph.

**45.** An airplane flies 1960 mi in the same amount of time it takes a second plane to fly 1610 mi. If the speed of the first airplane is 100 mph faster than the speed of the second airplane, find the speed of each airplane.
The first airplane flies at 560 mph and the second airplane flies at 460 mph.

**46.** An airplane flies 1350 mi in the same amount of time it takes a second plane to fly 1200 mi. If the speed of the first airplane is 60 mph faster than the speed of the second airplane, find the speed of each airplane.
The first airplane flies at 540 mph and the second airplane flies at 480 mph.

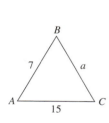 **Mix 'Em Up!**

**Solve each problem.**

**47.** A supermarket has a special on chunk white albacore tuna—three cans for $2.85. How much will it cost to purchase seven cans of tuna?   $6.65

**48.** Mandy and Bridgette together can decorate a wedding venue in 1 hr. Mandy decorates a similar size venue alone in 3 hr. How long will it take Bridgette to decorate the wedding venue alone?   1.5 hr

**Triangles *ABC* and *STU* are similar. Use the information in the diagram for Exercises 49 and 50. Note that figures are not drawn to scale.**

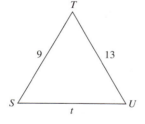

**49.** Find *a*.   $\frac{91}{9}$     **50.** Find *t*.   $\frac{135}{7}$

**51.** An order of six chicken nuggets from a popular fast food restaurant has 250 cal. Jillian is trying to cut back on her calories and only eats four of the nuggets. How many calories does she consume?   $\frac{500}{3}$ cal

**52.** If 1 in. represents 120 mi on a map, how many inches represent the distance between Danbury, Connecticut, and Miami, Florida, which is 1338 mi?   11.15 in.

**53.** An airplane flies 1320 mi in the same amount of time it takes a second plane to fly 1100 mi. If the speed of the first airplane is 80 mph faster than the speed of the second airplane, find the speed of each airplane.
The first airplane flies at 480 mph and the second airplane flies at 400 mph.

**54.** If 1 cm represents 210 km on a map, how many centimeters represent the distance between Buffalo, New York, and Des Moines, Iowa, which is 1375.5 km?   6.55 cm

**55.** An object weighing 1 lb on Earth weighs 0.378 lb on Mercury. Find Frank's weight on Mercury if he weighs 195 lb on Earth.   73.71 lb

**56.** An object weighing 1 lb on Earth weighs 1.125 lb on Neptune. Find Lucia's weight on Earth if she weighs 117 lb on Neptune.   104 lb

**57.** Olivia walks 5 mi in the same time that Carol jogs 8 mi. If Carol jogs 2 mph faster than Olivia walks, then how fast does Carol jog?   $\frac{16}{3}$ mph

**58.** A pump can fill a tank in 2 hr. A second pump can fill the same tank in 4 hr. How long will it take to fill the tank if both pumps are used?   $\frac{4}{3}$ hr

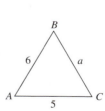 **You Be the Teacher!**

**Determine if the student provides the correct equation to answer the given question. If not, provide the correct equation. Then solve the equation to answer the question.**

**59.** A recipe that serves 20 people calls for 3 cups of cheese. How many cups of cheese is needed if the recipe is being increased to serve 25 people?

Jacqueline's work:

$\frac{20}{3} = \frac{25}{x}$   correct; $x = \frac{15}{4}$ cups

**60.** How many Hong Kong dollars are in 30 U.S. dollars? (Note: 1 U.S. dollar = 7.8125 Hong Kong dollars.)

Helen's work:

$\frac{x}{30} = \frac{1}{7.8125}$

**61.** Annette can clean an office building in 5 hr. It takes her work partner 9 hr to clean the same building. How long would it take for Annette and her partner to clean the building together?

Scott's work:

Let $x$ = number of hours worked, $\frac{x}{5} + \frac{x}{9} = 14$

**62.** Use the similar triangles *ABC* and *STU* to find *a*. Note the figures are not drawn to scale.

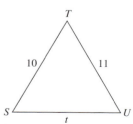

Suzanne's work:

$\frac{5}{t} = \frac{a}{11}$

## GROUP ACTIVITY   The Mathematics of Drug Dosage Calculations

The following proportion is used to calculate the amount of medication for oral, intramuscular, intravenous, or subcutaneous injections to be administered when given a dosage, stock dose, and stock volume.

$$\frac{\text{Required dose}}{\text{Stock dose}} = \frac{\text{Required volume}}{\text{Stock volume}}$$

The *required dose* is what has been prescribed or needed by the patient. The *stock dose* is the dose of the drug available. The *stock volume* is the amount of the drug that is available. The *required volume* is the amount of medication that should be administered. Note: Units for the required dose and stock dose must be the same.

1. Dr. Rawls has prescribed Carolyn 15 mg of Stemetil to treat her vertigo. The hospital has 2 mL of solution on hand, which contains 25 mg of Stemetil. How much of the drug should be administered to Carolyn?   1.2 mL

2. Many drug dosages are based on a patient's weight in kilograms (kg) instead of pounds (lb). If 1 kg is equivalent to 2.2 lb, use a proportion to determine the weight, in kg, of patients who weigh 35 lb, 100 lb, and 150 lb.
   15.88 kg; 45.36 kg; 68.04 kg

3. Dr. Dunagan has prescribed that Cory take amoxicillin at 35 mg per kg in 3 equal doses over a 24-hr period. Cory weighs 26.4 lb. The pharmacist has 125 mg in 5 mL suspension available. How many mL must be administered in each dose? (Round to two decimal points.)

   a. Determine Cory's weight in kilograms.   12 kg

   b. Determine the dose prescribed in a 24-hr period.
      420 mg

   c. Determine the amount required in a single dose.
      140 mg/dose

   d. Use the formula to calculate the mL to be administered.   5.6 mL

   e. If the doctor prescribed this medication for 10 days, what is the total amount that the pharmacist needs to provide to the patient?   56 mL

# Rational Expressions and Equations

**What's the big idea?** Now that you have completed Chapter 7, you should understand that the quotient of polynomials is a rational expression. You should be able to simplify rational expressions and to perform operations on rational expressions. Factoring is a key skill that is used to reduce rational expressions, multiply and divide rational expressions, and to find the LCD. The LCD is used in adding and subtracting rational expressions, simplifying complex fractions, and in solving rational equations.

## The Tools

Listed below are the key terms, skills, formulas, and properties you should know for this chapter.

The page reference is provided if you need additional help with the given topic. The Study Tips will assist in your preparation for an exam.

## Study Tips

1. Learn all of the terms, formulas, and properties. Make flash cards and have someone quiz you.
2. Rework problems from the exercises and also the ones you worked in class. Work additional problems from the review exercises.
3. Review the summaries of key concepts.
4. Work the chapter test.
5. Be sure to review the online resources for additional study materials.

## Terms

Complex fraction 549
Cross products 568
Equivalent rational
 expressions 537
Least common denominator 535

Lowest terms 517
Opposites 518
Proportion 568
Ratio 568
Rational equation 558

Rational expression 514
Similar triangles 571
Simplest form 549
Undefined rational expression 516

## Formulas and Properties

- Adding rational expressions 533
- Distance-rate-time formula 573
- Dividing rational expressions 526
- Dividing opposites 518

- Fundamental property of rational expressions 517
- Multiplying rational expressions 525

- Negative rational expressions 519
- Subtracting rational expressions 534

## CHAPTER 7 / SUMMARY

How well do you know this chapter? Complete the following questions to find out. Take a look back at the section if you need help.

### SECTION 7.1 Rational Expressions

1. A rational expression is a(n) _quotient_ of two _polynomials_.
2. To evaluate a rational expression, substitute the given _value_ in place of the _variable_ and simplify.
3. When the numerator of a rational expression is zero, the value of the expression is _zero_. When the denominator of a rational expression is zero, the expression is _undefined_.

4. To find the values that make a rational expression undefined, set the _denominator_ equal to _zero_ and solve.
5. To simplify a rational expression, _factor_ the numerator and denominator and divide out any common _factors_.
6. The expressions $x - y$ and _y - x_ are opposites. The quotient of opposites is _-1_.
7. The expression $-\dfrac{A}{B}$ equivalent to $\dfrac{-A}{B}$ and $\dfrac{A}{-B}$.

## SECTION 7.2 Multiplication and Division of Rational Expressions

**8.** To multiply rational expressions, _factor_ each numerator and denominator, _divide_ out common factors, and _multiply_ the remaining factors in the numerator and denominator.

**9.** To divide rational expressions, convert the division to a product of the first expression and the _reciprocal_ of the second expression. That is, $\dfrac{A}{B} \div \dfrac{C}{D} = \dfrac{A}{B} \cdot \dfrac{D}{C}$.

**10.** To convert units, we use the fact that if $a$ and $b$ are equal quantities, then $\dfrac{a}{b} = \underline{1}$, and the fact that multiplying a quantity by 1 _does not_ change the quantity.

## SECTION 7.3 Addition and Subtraction of Rational Expressions with Like Denominators and the Least Common Denominator

**11.** To add or subtract rational expressions, they must have the _same_ denominators.

**12.** The _least common denominator_ is the product of all different factors raised to the _largest_ power that the factor occurs in any one denominator.

**13.** When we multiply a rational expression by a form of 1, we obtain a(n) _equivalent_ rational expression.

## SECTION 7.4 Addition and Subtraction of Rational Expressions with Unlike Denominators

**14.** To add or subtract rational expressions with different denominators, we must find the _LCD_ and convert each fraction to a(n) _equivalent_ fraction with the _LCD_ as its denominator. Then we can add or subtract the expressions.

## SECTION 7.5 Complex Fractions

**15.** A(n) _complex_ fraction is a fraction that contains fractions in its numerator and/or denominator.

**16.** There are two methods to simplify a complex fraction. We may either _combine the fractions in the numerators and denominators and multiply the reciprocal_ or multiply the numerator and denominator of the complex fraction by the _LCD of all the fractions_.

## SECTION 7.6 Solving Rational Equations

**17.** A(n) _rational_ equation is an equation that contains a rational expression.

**18.** To solve rational equation, multiply both sides by the _LCD_ and then solve the resulting equation.

**19.** A solution that makes the LCD equal to _zero_ or the rational expression _undefined_ is a(n) _extraneous_ solution and must be excluded from the solution set.

## SECTION 7.7 Proportions and Applications of Rational Equations

**20.** A(n) _ratio_ relates the measurement of two quantities and can be expressed with a(n) _colon_ or as a(n) _fraction_.

**21.** A(n) _proportion_ is an equation with two _ratios_ set equal to each other.

**22.** _Similar triangles_ are triangles with the same _shape_ but different _sizes_. Corresponding angles are _equal_ and the lengths of corresponding sides are _proportional_.

**23.** Work applications involve finding the time it takes for two individuals to complete a task when _working together_.

**24.** To solve motion problems, use the formula _$d = rt$_.

---

## CHAPTER 7 / REVIEW EXERCISES

### SECTION 7.1

**Write each rational expression in lowest terms. (See Objective 3.)**

**1.** $\dfrac{18x^2}{12x^5}$   $\dfrac{3}{2x^3}$

**2.** $\dfrac{25a^6}{45a^2}$   $\dfrac{5a^4}{9}$

**3.** $\dfrac{8x-4}{12x^2-6x}$   $\dfrac{2}{3x}$

**4.** $\dfrac{x^2-4}{x^2+5x-14}$   $\dfrac{x+2}{x+7}$

**5.** $\dfrac{x^3+1}{x^2-1}$   $\dfrac{x^2-x+1}{x-1}$

**6.** $\dfrac{x^2-7x+12}{x^2-16}$   $\dfrac{x-3}{x+4}$

**7.** $\dfrac{18x}{6x+24}$   $\dfrac{3x}{x+4}$

**8.** $\dfrac{-8x^2-8x}{2x^2+6x}$   $\dfrac{-4(x+1)}{x+3}$

**Use the rule of negative rational expressions to write an equivalent form of each rational expression. (See Objective 4.)**

**9.** $\dfrac{2x-9}{-x-10}$   $\dfrac{-2x+9}{x+10}$ or $-\dfrac{2x-9}{x+10}$

**10.** $\dfrac{-3x-1}{x-4}$

**11.** $-\dfrac{-x^2+7x}{x^2-5}$

**12.** $-\dfrac{2x^2-x}{-3x^2+2}$

### SECTION 7.2

**Perform each operation and simplify the answer. (See Objectives 1 and 2.)**

**13.** $\dfrac{14x^3y^2}{5} \cdot \dfrac{20x}{21y^5}$   $\dfrac{8x^4}{3y^3}$

**14.** $\dfrac{24a^3}{25b} \div \dfrac{6a^{15}}{5b^8}$   $\dfrac{4b^7}{5a^{12}}$

**15.** $\dfrac{2x^2+6x}{6} \cdot \dfrac{3x+3}{x^3}$   $\dfrac{(x+3)(x+1)}{x^2}$

**16.** $\dfrac{6x+4}{3x^2-13x-10} \cdot \dfrac{3x-15}{12x-12}$   $\dfrac{1}{2(x-1)}$

**17.** $\dfrac{x^2-3x-10}{4x} \div \dfrac{x^2-25}{x^2}$   $\dfrac{x(x+2)}{4(x+5)}$

**18.** $\dfrac{2x^2-8}{x+7} \div \dfrac{x^2-5x-14}{x^2-49}$   $x(x-2)$

**19.** $\dfrac{x^3+1}{x-1} \div (x+1)$   $\dfrac{x^2-x+1}{x-1}$

**20.** $\dfrac{2x-3}{2x^2+3x-9} \cdot (2x+2)$   $\dfrac{2(x+1)}{x+3}$

**21.** $\dfrac{x^2 + x - 12}{x^3 - 27} \div \dfrac{x^2 + 8x + 16}{2x^2 + 6x + 18}$  $\dfrac{2}{x+4}$

**22.** $\dfrac{x^2 - 1}{x^2 - 2x - 3} \div \dfrac{x^3 - 1}{2x - 6}$  $\dfrac{2}{x^2 + x + 1}$

**23.** $\dfrac{x^3 + 125}{x^2 + 3x - 10} \div \dfrac{x^3 - 5x^2 + 25x}{x^2 + 2x}$  $\dfrac{x+2}{x-2}$

**24.** $\dfrac{x^3 - 2x^2 - 7x + 14}{x^2 - 4} \cdot \dfrac{x^2 + x - 2}{x^3 - 7x}$  $\dfrac{x-1}{x}$

**25.** $\dfrac{x^4 - 81}{6x^2 - 18x} \div \dfrac{x^3 + 4x^2 + 9x + 36}{2x^2 + 8x}$  $\dfrac{x+3}{3}$

**26.** $\dfrac{x^3 - 1}{x^4 - 1} \div \dfrac{x^2 - 2x - 3}{x^3 - 3x^2 + x - 3}$  $\dfrac{x^2 + x + 1}{(x+1)^2}$

**Use the given conversion equivalents to solve each problem. (*See Objective 3.*)**

**27.** The distance between London and Manchester is 321.7 km. What is this distance in miles? (Note: 1 km = 0.62 mi.)  199.5 mi

**28.** The distance between Paris and Rome is 892.7 mi. What is this distance in kilometers? (Note: 1 km = 0.62 mi.)  1439.8 km

**29.** How many minutes are in 0.35 hr? (Note: 1 hr = 60 min.)  21 min

**30.** How many minutes are in 1.5 hr? (Note: 1 hr = 60 min.)  90 min

**31.** Convert 28.8 kg to pounds. (Note: 1 lb = 0.45 kg.)  64 lb

**32.** Convert 150 lb to kilograms. (Note: 1 lb = 0.45 kg.)  67.5 kg

### SECTION 7.3

**Find the least common denominator of each pair of rational expressions. (*See Objective 3.*)**

**33.** $\dfrac{1}{x+3}, \dfrac{5}{-6x - 2x^2}$  $2x(x+3)$

**34.** $\dfrac{x}{4x+1}, \dfrac{2}{x-3}$  $(4x+1)(x-3)$

**35.** $\dfrac{5}{x+7}, \dfrac{x}{3x^2 + 19x - 14}$  $(x+7)(3x-2)$

**36.** $\dfrac{5}{6x+4}, \dfrac{6}{9x^2 + 12x + 4}$  $2(3x+2)^2$

**Find the numerator that makes each equation true. (*See Objective 4.*)**

**37.** $\dfrac{7}{2x} = \dfrac{?}{10x^2}$  $35x$

**38.** $\dfrac{x-9}{x+1} = \dfrac{?}{x^2 + x}$  $x(x-9)$

**39.** $\dfrac{x}{2x+5} = \dfrac{?}{2x^2 + 9x + 10}$  $x(x+2)$

**40.** $\dfrac{x}{1 - 6x} = \dfrac{?}{18x^2 - 3x}$  $-3x^2$

### SECTION 7.4

**Add or subtract the rational expressions. Write each answer in lowest terms. (*See Objectives 1 and 2.*)**

**41.** $\dfrac{5x}{x-6} - \dfrac{9x}{6-x}$  $\dfrac{14x}{x-6}$

**42.** $\dfrac{x-4}{x-2} + \dfrac{x}{x+2}$

**43.** $\dfrac{x}{5x+2} + \dfrac{2x}{2x+5}$

**44.** $\dfrac{7}{12x^3} - \dfrac{5}{18x^2}$  $\dfrac{21 - 10x}{36x^3}$

**45.** $\dfrac{3}{x-12} + \dfrac{2}{x+8}$

**46.** $\dfrac{2x-5}{x+1} - \dfrac{3x+1}{x+1}$

**47.** $\dfrac{5x-4}{x^2 - 6x - 7} + \dfrac{3x}{x-7}$

**48.** $\dfrac{3x}{2x^2 - 5x - 3} - \dfrac{x^2}{2x+1}$

**49.** $2 + \dfrac{2x}{5x+3}$  $\dfrac{12x + 6}{5x + 3}$

**50.** $\dfrac{5}{x^2 - 6x + 9} - \dfrac{4}{x^2 - 9}$

### SECTION 7.5

**Simplify each complex fraction using method 1 and method 2. (*See Objectives 1 and 2.*)**

**51.** $\dfrac{\dfrac{18}{3x^2}}{\dfrac{42}{8x^4}}$  $\dfrac{8x^2}{7}$

**52.** $\dfrac{\dfrac{7}{35y^6}}{\dfrac{32}{4y^3}}$  $\dfrac{1}{40y^3}$

**53.** $\dfrac{6 - \dfrac{a}{b}}{5 + \dfrac{b}{a}}$  $\dfrac{a(6b - a)}{b(5a + b)}$

**54.** $\dfrac{\dfrac{15x}{3x+2}}{\dfrac{5x^2}{9x^2 - 4}}$  $\dfrac{3(3x-2)}{x}$

**55.** $\dfrac{\dfrac{5}{2x} + \dfrac{1}{2}}{35 + 7x}$  $\dfrac{1}{14x}$

**56.** $\dfrac{\dfrac{24ab}{9a^2 - 6ab + b^2}}{\dfrac{18ab}{3a - b}}$

**57.** $\dfrac{\dfrac{7}{x} - \dfrac{3}{y}}{\dfrac{49}{x^2} - \dfrac{9}{y^2}}$  $\dfrac{xy}{7y + 3x}$

**58.** $\dfrac{\dfrac{3}{r} + \dfrac{1}{s}}{\dfrac{9}{r^2} - \dfrac{1}{s^2}}$  $\dfrac{rs}{3s - r}$

**59.** $\dfrac{\dfrac{x^2 - 16}{x^2 - 3x + 9}}{\dfrac{x^2 + 7x + 12}{x^3 + 27}}$  $x - 4$

**60.** $\dfrac{\dfrac{2x^2 - 14x}{x^2 + 4x + 3}}{\dfrac{5x^2 - 2x}{5x^2 + 3x - 2}}$  $\dfrac{2(x - 7)}{x + 3}$

**61.** $\dfrac{\dfrac{5x^2 + 5x}{4x^2 - x}}{\dfrac{45x + 30}{12x^2 + 5x - 2}}$  $\dfrac{x + 1}{3}$

**62.** $\dfrac{1 - \dfrac{x}{x-3}}{2 + \dfrac{5}{x-3}}$  $\dfrac{-3}{2x - 1}$

**Write a complex fraction that represents each situation, and then simplify the complex fraction. (*See Objective 3.*)**

**63.** Fong ate $\dfrac{1}{3}$ of an apple pie and Wang ate $\dfrac{1}{2}$ of the same pie. If the remaining pie is to be split equally among two friends, what fraction of the pie will each person receive?  $\dfrac{1}{12}$ of a pie

**64.** A recipe for 8 people calls for $\dfrac{3}{2}$ cup of flour. How much flour should be used if the recipe is made for 6 people?  $\dfrac{9}{8}$ of a cup

**65.** What is the average of $\dfrac{5}{4}$, $\dfrac{5}{12}$, and $\dfrac{4}{5}$?  $\dfrac{37}{45}$

**66.** Meredith can paint her bedroom in 9 hr. Her brother Johnson can paint the same room in 3 hr. If they work together, how long will it take them to paint the room?  2.25 hr

### SECTION 7.6

**Solve each rational equation. (*See Objective 1.*)**

**67.** $\dfrac{b}{5} = \dfrac{2b - 3}{6}$  $\left\{\dfrac{15}{4}\right\}$

**68.** $\dfrac{y-7}{5} - \dfrac{14}{15} = \dfrac{2y}{3}$  $\{-5\}$

**69.** $\dfrac{2x-5}{3} - 1 = \dfrac{2x}{3}$  ∅  **70.** $\dfrac{6x}{8x-1} = \dfrac{3}{4}$  ∅

**71.** $\dfrac{2}{3x} + 5 = 9$  $\left\{\dfrac{1}{6}\right\}$  **72.** $\dfrac{3x^2-20}{x^2-8} = \dfrac{3x+2}{x}$  $\{-2, 4\}$

**73.** $5 + \dfrac{18}{x} = \dfrac{8}{x^2}$  $\left\{\dfrac{2}{5}, -4\right\}$  **74.** $\dfrac{2}{t-6} = \dfrac{3}{t-6} - 5$  $\left\{\dfrac{31}{5}\right\}$

**75.** $\dfrac{1}{4x-3} = \dfrac{3}{6x-7}$  $\left\{\dfrac{1}{3}\right\}$  **76.** $\dfrac{3}{3x+1} - \dfrac{2x}{9x^2-1} = \dfrac{2}{3x-1}$  $\{5\}$

**Solve each rational equation for the specified variable. (See Objective 2.)**

**77.** $A = \dfrac{1}{2}h(y_0 + 2y_1 + y_2)$ for $y_1$ (trapezoidal rule)

**78.** $A = \dfrac{1}{2}h(y_0 + 2y_1 + y_2)$ for $h$ (trapezoidal rule)

**79.** $\dfrac{1}{R} = \dfrac{1}{R_1} + \dfrac{1}{R_2} + \dfrac{1}{R_3}$ for $R_2$ (resistors in parallel)

**80.** $t = \dfrac{d}{x+y}$ for $x$   $x = \dfrac{d-ty}{t}$

## SECTION 7.7

**Solve each problem. (See Objectives 1–5.)**

**81.** The Weis supermarket has a special on Campbell's chunky roasted beef tips and vegetables soup—three cans for $5.97. How much will it cost to purchase eight cans of soup?   $15.92

**82.** Working together, Warren and Diane together can mow a lawn in $\dfrac{3}{4}$ hr. Warren can mow a lawn of similar size alone in 1 hr. How long will it take Diane to mow a lawn of similar size alone?   3 hr

**Triangles *ABC* and *STU* are similar. Use the information in the diagrams for Exercises 83 and 84. Note the figures are not drawn to scale.**

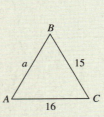
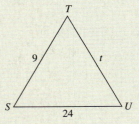

**83.** Find *a*.  6   **84.** Find *t*.  $\dfrac{45}{2}$

**85.** If 1 in. represents 120 mi on a map, how many inches represent the distance between Harrisburg, Pennsylvania, and Atlanta, Georgia, which is 720 mi?  6 in.

**86.** If 1 cm represents 210 km on a map, how many centimeters represent the distance between Topeka, Kansas, and Sacramento, California, which is 2730 km?  13 cm

**87.** An object weighing 1 lb on Earth weighs 0.378 lb on Mercury. Find Edmond's weight on Mercury if he weighs 165 lb on Earth.  62.37 lb

**88.** An airplane flies 1122 mi in the same amount of time it takes a second plane to fly 946 mi. If the speed of the first airplane is 80 mph faster than the speed of the second airplane, find the speed of each airplane.  The first airplane flies at 510 mph and the second airplane flies at 430 mph.

**89.** Renee walks 4 mi in the same time that Jane jogs 7.2 mi. If Jane jogs 2 mph faster than Renee walks, then how fast does Jane jog?  $\dfrac{5}{2}$ mph

**90.** A pump fills a tank in 3 hr. A second pump fills the same tank in 2 hr. How long will it take to fill the tank if both pumps are used?  $\dfrac{6}{5}$ hr

## CHAPTER 7 TEST / RATIONAL EXPRESSIONS AND EQUATIONS

**1.** When simplified completely, $\dfrac{2t-1}{t+2} \cdot \dfrac{t^2+2t}{2t^2+t-1}$ is equivalent to

  **a.** $\dfrac{t}{t+1}$  **b.** $\dfrac{t(2t-1)}{2t^2+t+1}$

  **c.** $\dfrac{t^2+2t}{(t+2)(t+1)}$  **d.** $\dfrac{t}{t-1}$

**2.** When simplified completely, $\dfrac{x}{x^2-1} \div \dfrac{x-1}{x^2-3x-4}$ is equivalent to

  **a.** $\dfrac{x-1}{x}$  **b.** $\dfrac{x(x-4)}{(x-1)^2}$

  **c.** $\dfrac{x}{x-1}$  **d.** $\dfrac{x}{(x+1)^2(x-4)}$

**3.** When simplified completely, $\dfrac{x^2}{x+4} + \dfrac{4x}{x+4}$ is equivalent to

  **a.** $x$  **b.** $\dfrac{4x^3}{x+4}$

  **c.** $\dfrac{5x}{x+4}$  **d.** $\dfrac{x^2+4x}{x+4}$

**4.** When simplified completely, $\dfrac{5-2x}{x+7} - \dfrac{x+3}{x+7}$ is equivalent to

  **a.** $\dfrac{-3x+2}{0}$  **b.** $\dfrac{-3x+8}{x+7}$

  **c.** $\dfrac{-x+8}{x+7}$  **d.** $\dfrac{-3x+2}{x+7}$

**5.** The solution set for $\dfrac{6}{3x} - \dfrac{2}{x} = 4$ is

  **a.** $\{0\}$  **b.** $\left\{\dfrac{1}{12}\right\}$

  **c.** $\{12\}$  **d.** ∅

**6.** The solution set for $\dfrac{x}{x-1} = \dfrac{2x}{x+1}$ is

  **a.** $\{0, 3\}$  **b.** $\{3\}$

  **c.** $\{0\}$  **d.** ∅

**7.** $\dfrac{x-7}{(x-1)(x+3)}$ is undefined for the $x$ value(s)

  **a.** 0  **b.** 7, −3, and 1

  **c.** −1 and 3  **d.** 1 and −3

**8.** Find the value of $\dfrac{2x^2 - 4x + 5}{3x + 1}$ for $x = -1$.  $-\dfrac{11}{2}$

**9.** Write each expression in lowest terms.

**a.** $\dfrac{5 - 3x}{9x^2 - 25}$   **b.** $\dfrac{2x^2 - 5x - 7}{x^2 + 5x + 4}$   **c.** $\dfrac{8a^3 - 27}{4a^2 - 9}$

**10.** What numerator makes the statement

$$\dfrac{6y}{5y + 10} = \dfrac{?}{5y^2 - 20} \text{ true?} \quad 6y^2 - 12y$$

**11.** Find the LCD of each pair of rational expressions.

**a.** $\dfrac{3}{8x^2}, \dfrac{5}{10x^3}$  $40x^3$   **b.** $\dfrac{4}{b}, -\dfrac{2}{b+5}$  $b(b+5)$

**c.** $\dfrac{x+5}{x^2 + 6x - 7}, \dfrac{3x - 1}{2x^2 + 3x - 5}$  $(x+7)(x-1)(2x+5)$

**12.** Perform each indicated operation.

**a.** $\dfrac{4}{x - 2} - \dfrac{6}{2 - x}$  $\dfrac{10}{x-2}$   **b.** $\dfrac{2a}{a + 1} - \dfrac{5}{3a}$  $\dfrac{6a^2 - 5a - 5}{3a(a + 1)}$

**c.** $\dfrac{3}{2x + 8} - \dfrac{x}{x^2 - 16}$  $\dfrac{x - 12}{2(x - 4)(x + 4)}$

**13.** Simplify each complex fraction.

**a.** $\dfrac{\dfrac{12a^3}{5b^2}}{\dfrac{21a^4}{10b}}$  $\dfrac{8}{7ab}$   **b.** $\dfrac{\dfrac{1}{a} - \dfrac{1}{b}}{\dfrac{1}{a} + \dfrac{1}{b}}$  $\dfrac{b - a}{a + b}$

**14.** Solve each equation.

**a.** $\dfrac{5}{x} + \dfrac{3}{2} = \dfrac{x + 1}{x}$  $\{-8\}$   **b.** $1 - \dfrac{4}{x} - \dfrac{21}{x^2} = 0$  $\{-3, 7\}$

**c.** $\dfrac{x}{x + 3} + \dfrac{6}{x + 1} = \dfrac{6}{x^2 + 4x + 3}$  $\{-4\}$

**15.** Solve each problem.

**a.** The average weight of a newborn baby is 3.2 kg. Given that 1 lb = 0.45 kg, convert the average weight of a newborn baby to pounds.  7.1 lb

**b.** The triangles are similar. Use the information in the diagram to find the length of $a$. Note the figures are not drawn to scale.  $a = \dfrac{33}{5}$

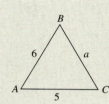
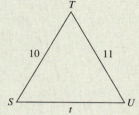

**c.** Mary and Sue have a maid service. Mary cleans an average-size house in 3 hr when she works alone. Sue cleans an average-size house in 4 hr when she works alone. How long will it take the two of them to clean an average-size house if they work together?  1.7 hr

**d.** Todd runs 10 mi in the same time that Andrea walks 8 mi. Todd runs 1 mph faster than Andrea walks. How fast does Todd run?  5 mph

---

## CUMULATIVE REVIEW EXERCISES / CHAPTERS 1–7

**1.** Perform each operation and simplify each result. (*Section 1.2, Objectives 2–7*)

**a.** $\dfrac{8}{15} + \dfrac{7}{12}$   **b.** $3\dfrac{1}{4} - 1\dfrac{5}{6}$   **c.** $12\dfrac{3}{5} \cdot 4\dfrac{2}{7}$

**2.** Perform the indicated operation and simplify. (*Sections 1.4 and 1.5, Objectives 1 and 2*)

**a.** $2 - 12 + (-23) - (-10)$  $-23$

**b.** $3.1 - (-4.5) + (-6.7) + 7.2$  8.1

**c.** $17 - \{8 - [4 - (1 - \sqrt{25 - 16})]\}$  15

**d.** $-(-3)^3 + (-14) - |5 - 9|$  9

**3.** The lowest recorded temperature in Alaska was $-62°C$ at the Prospect Creek Camp weather station on January 23, 1971. The highest recorded temperature in Nevada was 53°C at the Lake Havasu City weather station on June 29, 1994. What is the difference between the highest and lowest temperatures? (*Section 1.5, Objective 3*)  115°C

**4.** Evaluate each expression for the given values of the variables. (*Section 1.6, Objective 4*)

**a.** $\sqrt{(x_1 - x_2)^2 + (y_1 - y_2)^2}$ for $x_1 = -5, x_2 = 7,$ $y_1 = 13, y_2 = 8$  13

**b.** $\dfrac{|x + 1|}{2x - 4}$ for $x = -1, 0, 2$  $0, -\dfrac{1}{4},$ undefined

**5.** Solve each equation. (*Section 2.2, Objectives 2 and 3; Sections 2.3 and 2.4, Objectives 1 and 2*)

**a.** $-3(x - 5) + 8 = 4(x - 1) - 22$  $\{7\}$

**b.** $5.2x - 6.3 + 1.7x = 9.4x - 12.3$  $\{2.4\}$

**c.** $\dfrac{1}{7}(28x - 14) + 12 = x + 5$  $\left\{-\dfrac{5}{3}\right\}$

**6.** Solve each equation. If the equation is a contradiction, write the solution as $\varnothing$. If the equation is an identity, write the solution as $\mathbb{R}$. (*Section 2.4, Objectives 1–4*)

**a.** $6(2a + 1) - 2(a + 4) = 5(2a + 1) - 7$  $\mathbb{R}$

**b.** $\dfrac{2}{5}x + 3 = \dfrac{3}{10}x$  $\{-30\}$

**c.** $0.015x + 0.024(950 - x) = 20.28$  $\{280\}$

**d.** $\dfrac{3}{2}(x + 1) = \dfrac{1}{3}(5x - 2) + \dfrac{5}{6}$  $\{8\}$

**7.** Solve each formula for the specified variable. (*Section 2.5, Objective 3*)

**a.** $S = 2\pi r^2 + 2\pi rh$ for $h$   **b.** $C = 4l + 8w$ for $l$

**c.** $x + 5y = 10$ for $y$   **d.** $0.06x + 0.03y = 4.2$ for $x$

**8.** A coat is on sale for $144. If this price is 60% off the original price, what is the original price of the coat? (*Section 2.6, Objective 1*)  $360

9. Determine if the lines are parallel, perpendicular, or neither. (*Section 3.4, Objective 3*)

   **a.** $y = \frac{3}{2}x + 5$ and $y = \frac{3}{2}x - 1$.   parallel

   **b.** $y = 2$ and $x = 5$   **c.** $y = 0.25x$ and $y = -4x$
   perpendicular                perpendicular

10. Write the equation of each line described. Express each answer in slope-intercept form and in standard form. (*Section 3.5, Objectives 2 and 4*)

    **a.** passes through $(-8, 5)$ perpendicular to $4x - 3y = 1$

    **b.** passes through $(-1, 4)$ parallel to $6x + y = 5$

    **c.** $m =$ undefined, passes through $(3, -4)$
    $y = -6x - 2; 6x + y = -2$
    $x = 3$

11. Ming has a new job as a History instructor at a college. Her starting salary is $43,500 with a yearly teaching load of 30 credit-hours. If she teaches beyond the regular load, she earns $1050 per credit-hour. (*Section 3.6, Objectives 4–6*)

    **a.** Write a linear function $f(x)$ that represents Ming's yearly income, where $x$ is the number of extra credit-hours taught in a year.  $f(x) = 1050x + 43,500$

    **b.** If Ming has a 39 credit-hour teaching load in one school year, what is her income?  $52,950

    **c.** Find $f(8)$ and interpret the answer.
    $f(8) = $51,900$; Ming's income for teaching 8 additional credit-hours is $51,900.

12. Solve each system of linear equations using substitution. (*Section 4.2, Objectives 1 and 2*)

    **a.** $\begin{cases} 2x - 3y = 16 \\ x + 7y = -9 \end{cases}$  $\{(5, -2)\}$  **b.** $\begin{cases} \frac{1}{2}x - y = -4 \\ \frac{1}{4}x + 3y = 5 \end{cases}$  $\{(-4, 2)\}$

13. Solve each system of linear equations using elimination. (*Section 4.3, Objectives 1–3*)

    **a.** $\begin{cases} 4x - 3y = 78 \\ -5x + 2y = 40 \end{cases}$  **b.** $\begin{cases} \frac{2}{3}x - \frac{1}{4}y = -6 \\ \frac{5}{6}x + \frac{1}{2}y = -1 \end{cases}$  $\{(-6, 8)\}$
    $\left\{-\frac{276}{7}, -\frac{550}{7}\right\}$

14. Solve each system of linear inequalities. (*Section 4.6, Objective 1*)

    **a.** $\begin{cases} y \geq 5x - 5 \\ y \leq -4x + 8 \end{cases}$  **b.** $\begin{cases} x \geq 2 \\ x - 3y \leq 0 \end{cases}$

15. Use the appropriate exponent rules to simplify each expression. (*Section 5.1, Objectives 1–4*)

    **a.** $(5^{-1}x^{-3}y^4)^2$  $\frac{y^8}{25x^6}$  **b.** $\left(-\frac{4x^{-1}}{y^3}\right)^3$  $-\frac{64}{x^3y^9}$

16. Use the appropriate exponent rules to simplify each expression. Express each answer with positive exponents. (*Section 5.2, Objectives 1–4*)

    **a.** $(4^{-1}x^6)^{-2}$  **b.** $\left(\frac{3x^2}{y^{-7}}\right)^{-3}$  **c.** $\left(-\frac{c^{-2}}{2d^5}\right)^2$

17. Perform each operation. Write each answer in standard form. (*Section 5.4, Objectives 4 and 5*)

    **a.** $(18 - 5x^2 - 29x) + (-12 + 25x^2 + 37x)$  $20x^2 + 8x + 6$

    **b.** Subtract $(4x^2 - 9xy + 6y^2)$ from $(-3x^2 + 12xy - 15y^2)$
    $-7x^2 + 21xy - 21y^2$

    **c.** $(9x^2 + xy - 3y^2) - (-6x^2 + 11xy + 3y^2) + (-2x^2 - 18xy - 6y^2)$  $13x^2 - 28xy - 12y^2$

18. Suppose the profit generated by selling $x$ items is $-0.4x^2 + 86x + 198$ dollars and the total cost to manufacture $x$ items is $23x + 420$ dollars. Find an expression that represents the company's revenue. Write the answer in standard form. (*Section 5.4, Objectives 5 and 6*)  $-0.4x^2 + 109x + 618$

19. Perform the indicated operation. Write each answer in standard form. (*Section 5.5, Objectives 1 and 2; Section 5.6, Objectives 1, 3, and 4*)

    **a.** $3ab(a^2 - 6ab + 5b^2)$  **b.** $(7x - 3)(4x + 5)$

    **c.** $(a + 2)^4$  **d.** $(2b - 1)^3$

    **e.** $(12x - 7y)(12x + 7y)$  $144x^2 - 49y^2$

    **f.** $(3r - 2s)(9r^2 + 6rs + 4s^2)$  $27r^3 - 8s^3$

20. Perform the indicated operation. (*Section 5.7, Objectives 1 and 2*)

    **a.** Divide $27x^3 - 12x^2 + 24x + 6$ by $6x$  $\frac{9x^2}{2} - 2x + 4 + \frac{1}{x}$

    **b.** $\frac{6x^3 - 18x^2 + 21x - 10}{x + 3}$  $6x^2 - 36x + 129 - \frac{397}{x + 3}$

21. Factor each trinomial completely. (*Section 6.1, Objectives 1 and 2*)

    **a.** $x^3 + 3x^2 - 10x$  **b.** $10y^3 + 25y^2 - 60y$
    $x(x + 5)(x - 2)$

    **c.** $3x^2 - 4x + 8$  prime  **d.** $-7a^4 - 91a^2 - 252$

    **e.** $-9b^3 + 36b^2 - 36b$  $-9b(b - 2)^2$

22. The volume in cubic inches of an open box with height $x$ in. is given by the expression $x^3 + 7x^2 - 60x$. Factor this trinomial. (*Section 6.1, Objective 3*)
    $x(x - 5)(x + 12)$

23. Factor each trinomial completely. (*Section 6.2, Objectives 1 and 2*)

    **a.** $3y^3 - 24y^2 + 48y$  **b.** $-12x^3 + 54x^2 + 30x$

    **c.** $a^2 - 15ab + 36b^2$  **d.** $2x^2y + 34xy + 144y$

    **e.** $x^2 + 7xy + 8y^2$  prime

24. A company's profit can be represented by $-4q^2 + 140q - 1000$, where $q$ is the thousands of units sold. Factor this polynomial. (*Section 6.2, Objective 2*)
    $-4(q - 10)(q - 25)$

25. Use either trial and error or grouping to factor the trinomials completely. (*Section 6.3, Objectives 1–3*)

    **a.** $35a^2 + 31a + 6$  **b.** $40ab^2 - 35ab - 90a$

    **c.** $12x^2 - 2xy - 30y^2$  **d.** $9x^2 + 12xy + 4y^2$

    **e.** $90a^2b^2c + 3abc - 3c$  **f.** $3p^2 + 4pq - 7q^2$
    $3c(6ab - 1)(5ab + 1)$      $(3p + 7q)(p - q)$

26. Factor completely. (*Section 6.4, Objectives 1–4*)

    **a.** $98a^2 - 18b^2$  **b.** $5x^3 + 40y^3$

    **c.** $y^3 - 1000z^3$  **d.** $a^2 + 16b^2$  prime

    **e.** $a^6 - 64$  **f.** $x^4 - 81y^4$
    $(a + 2)(a^2 - 2a + 4)(a - 2)(a^2 + 2a + 4)$   $(x^2 + 9y^2)(x - 3y)(x + 3y)$

27. Solve each equation. (*Section 6.5, Objectives 1 and 2*)

    **a.** $(x + 2)(x - 8) = 24$  **b.** $(5 - x)(4 + x) = 0$

    **c.** $x^4 - 5x^2 + 4 = 0$  **d.** $x^3 - 3x^2 - 10x = 0$
    $\{-2, -1, 1, 2\}$            $\{-2, 0, 5\}$

28. Solve each problem. (*Section 6.6, Objectives 1–5*)

    **a.** The profit, in millions of dollars, of a toy company is given by $p = -12x^2 + 180x + 3000$, where $x$ is the number of units sold, in thousands. How many units need to be sold for the company to break even? (*Section 6.3, Objectives 1–3*)  25,000 units

**b.** The product of two consecutive odd integers is 143. Find the integers.   11 and 13 or −13 and −11

**c.** When an express bus company charges \$40 per seat from Boston to Princeton, they can sell 90 tickets. For every \$1 decrease in price, 2 more tickets are sold. If $x$ represents the number of price decreases, then $40 - x$ represents the ticket price and $90 + 2x$ represents the number of passengers. How many passengers are needed for the express bus company to earn \$3300?   110 passengers

**d.** The height of a ball thrown upward with an initial velocity of 48 ft/sec from a height of 160 ft above the ground can be represented by $s = -16t^2 + 48t + 160$, where $t$ is the time in seconds after the ball is thrown. How long will it take the ball to reach the ground?   5 sec

**29.** Evaluate each expression for the given. (*Section 7.1, Objective 1*)

**a.** $\dfrac{x^2 - 3x}{x + 2}$ for $x = -2, 0, 3$   **b.** $\dfrac{4x}{x - 1}$ for $x = -1, 0, 1$

**30.** Is the given expression a rational expression? Explain. (*Section 7.1, Objective 1*)

**a.** $\dfrac{5x^2}{3x - 2}$   yes   **b.** $\dfrac{\sqrt{x + 3}}{x + 1}$
no, because of the radical

**31.** A company's revenue in dollars can be represented by $40q^2 + 600q$, where $q$ is the number of items sold. If the average revenue per item is given by $\dfrac{40q^2 + 600q}{q}$, what is the average revenue per item when 10 items are sold? 20 items are sold? (*Section 7.1, Objective 1*)
\$1000; \$1400

**32.** The volume in cubic inches of an open box with height $x$ in. is given by the expression $x^3 + 5x^2 - 24x$. If the area of the base of the box is given by $\dfrac{x^3 + 5x^2 - 24x}{x}$, what is the area of the base when the height is 5 in.? 6 in.? (*Section 7.1, Objective 1*)   26 in.² ; 42 in.²

**33.** Find the value(s) that make each rational expression undefined. Then simplify each rational expression. (*Section 7.1, Objectives 2 and 3*)

**a.** $\dfrac{2x}{x^2 + 5x}$   $x = 0, -5; \dfrac{2}{x + 5}$   **b.** $\dfrac{x^2 - 1}{x^2 + 5x - 6}$

**c.** $\dfrac{x^3 - 8}{x^2 + 3x - 10}$   **d.** $\dfrac{3x^2 - 6x}{x^3 - 4x}$ $x = 0, -2, 2; \dfrac{3}{x + 2}$

**34.** Use the rule of negative rational expressions to write two equivalent forms of each rational expression. (*Section 7.1, Objective 4*)

**a.** $\dfrac{-3x + 6}{x + 4}$   **b.** $-\dfrac{1 - x}{2x - 3}$

**35.** Perform each operation and simplify the answer. (*Section 7.2, Objectives 1 and 2*)

**a.** $\dfrac{x^2 - 3x}{4} \cdot \dfrac{2x - 10}{x^3 - 9x}$   $\dfrac{x - 5}{2(x + 3)}$

**b.** $\dfrac{x^2 + 4x - 12}{6x^2} \div \dfrac{x^2 - 4x + 4}{3x}$   $\dfrac{x + 6}{2x(x - 2)}$

**c.** $\dfrac{x^3 - 1}{x + 1} \div (x - 1)$   $\dfrac{x^2 + x + 1}{x + 1}$

**d.** $\dfrac{x^3 + 4x - 7x^2 - 28}{x^2 - 49} \cdot \dfrac{x^2 + 10x + 21}{x^3 + 4x}$   $\dfrac{x + 3}{x}$

**e.** $\dfrac{8x^3 + 4x^2}{2x^2 + 7x + 3} \div \dfrac{2x^2 + 10x}{x^2 + 8x + 15}$   $2x$

**f.** $\dfrac{x^4 - 1}{x^2 + 2x + 1} \cdot \dfrac{x^2 - 4x - 5}{x^3 + x - 5x^2 - 5}$   $x - 1$

**36.** Use the given conversion equivalents to answer each question. (*Section 7.2, Objective 3*)

**a.** The distance between London and Rylstone in the United Kingdom is 224 mi. What is this distance in inches on the map? (Note: 1 in. = 18 mi.)   $\dfrac{112}{9}$ or $12\dfrac{4}{9}$ in.

**b.** How many minutes are in 0.8 hr? (Note: 1 hr = 60 min.)   48 min

**c.** Convert 2.25 kg to pounds. (Note: 1 lb = 0.45 kg.)
5 lb

**37.** Find the numerator that makes each equation true. (*Section 7.3, Objective 4*)

**a.** $\dfrac{1}{3a} = \dfrac{?}{18a^2b}$   $6ab$   **b.** $\dfrac{7}{xy} = \dfrac{?}{6x^2y}$   $42x$

**c.** $\dfrac{x + 2}{x - 3} = \dfrac{?}{2x^2 - 6x}$ $2x^2 + 4x$   **d.** $\dfrac{x}{1 - 2x} = \dfrac{?}{6x^2 - 3x}$ $-3x^2$

**38.** Find the least common denominator of the rational expressions. (*Section 7.3, Objective 3*)

**a.** $\dfrac{x}{6x + 15}, \dfrac{1}{x}$   $6x^2 + 15x$   **b.** $\dfrac{5x}{2x + 3}, \dfrac{3}{x - 4}$
$(2x + 3)(x - 4)$

**c.** $\dfrac{5}{x - 3}, \dfrac{x}{x^2 + 4x - 21}$
$(x + 7)(x - 3)$   **d.** $\dfrac{5x}{x^2 - 16}, \dfrac{7x}{5x^2 + 19x - 4}$
$(x + 4)(x - 4)(5x - 1)$

**e.** $\dfrac{7}{x^3 - 8}, \dfrac{-5}{3x^2 + 6x + 12}$   $3(x - 2)(x^2 + 2x + 4)$

**39.** Add or subtract the rational expressions. Write each answer in lowest terms. (*Section 7.4, Objectives 1 and 2*)

**a.** $\dfrac{x}{x - 4} - \dfrac{2x}{x - 4}$   $-\dfrac{x}{x - 4}$   **b.** $\dfrac{4}{x - 6} + \dfrac{5}{x + 3}$

**c.** $\dfrac{4x}{x - 2} + \dfrac{3x - 6}{2 - x}$   **d.** $\dfrac{2x^2}{x^2 - 25} - \dfrac{2x}{x - 5}$

**e.** $4 + \dfrac{3x}{x - 1}$   $\dfrac{7x - 4}{x - 1}$   **f.** $\dfrac{4x^2}{x^2 + 7x - 30} - \dfrac{2x + 1}{x + 10}$

**g.** $\dfrac{6}{x^2 - 9} - \dfrac{7}{x^2 - 6x + 9}$   $\dfrac{-x - 39}{(x + 3)(x - 3)^2}$

**h.** $\dfrac{3}{x^2 + 3x + 2} + \dfrac{4}{x^2 + 7x + 6}$   $\dfrac{7x + 26}{(x + 2)(x + 1)(x + 6)}$

**i.** $\dfrac{250}{x - 8} + \dfrac{100}{x + 3}$   $\dfrac{350x - 50}{(x - 8)(x + 3)}$

**40.** Simplify each complex fraction using method 1, and then method 2. (*Section 7.5, Objectives 1 and 2*)

**a.** $\dfrac{\dfrac{18}{3x^6}}{\dfrac{24}{5x^4}}$   **b.** $\dfrac{4 - \dfrac{a}{b}}{2 + \dfrac{a}{b}}$   **c.** $\dfrac{\dfrac{4y^2 - 9}{20y^2}}{\dfrac{2y - 3}{5y}}$

**d.** $\dfrac{\dfrac{1}{a} - \dfrac{2}{3}}{4a - 6}$    **e.** $\dfrac{\dfrac{8xy}{2x - y}}{\dfrac{6xy}{2x^2 + 5xy - 3y^2}}$    **f.** $\dfrac{\dfrac{4}{x} - \dfrac{3}{y}}{\dfrac{16}{x^2} - \dfrac{9}{y^2}}$

**g.** $\dfrac{\dfrac{x^2 - 16}{x^2 + 3x + 9}}{\dfrac{x^2 + x - 12}{x^3 - 27}}$   $x - 4$    **h.** $\dfrac{1 - \dfrac{2x}{5x - 2}}{2 + \dfrac{1}{5x - 2}}$   $\dfrac{3x - 2}{10x - 3}$

**41.** Write a complex fraction that represents each situation, and then simplify the complex fraction. (*Section 7.5, Objective 3*)

   **a.** Shan ate $\dfrac{1}{2}$ of a cherry pie and Nam ate $\dfrac{1}{4}$ of the same pie. If the remaining pie is to be split equally among two friends, what fraction of the pie will each person receive?   $\dfrac{1}{8}$ of the pie

   **b.** What is the average of $\dfrac{5}{4}, \dfrac{7}{12},$ and $\dfrac{5}{6}$?   $\dfrac{8}{9}$

**42.** A person can do a job alone in $x$ hr and another can do the same job in $y$ hr. If they work together to do the same job, it will take them $\dfrac{1}{\dfrac{1}{x} + \dfrac{1}{y}}$ hr. (*Section 7.5, Objective 3*)   $\dfrac{117}{7}, 14$

   **a.** Sharon can paint her basement in 5 hr. Her brother Erik can paint the same basement in 3 hr. If they work together, how long will it take them to paint the basement?   $\dfrac{15}{8}$ hr

   **b.** Laurn delivers newspapers to the customers on his assigned route in 2 hr. His wife delivers the newspapers on the same route in 4 hr. If they work together, how long will it take them to deliver the newspapers?   $\dfrac{4}{3}$ hr

**43.** Solve each rational equation. (*Section 7.6, Objective 1*)

   **a.** $\dfrac{a}{3} = \dfrac{a - 5}{4}$   $\{-15\}$    **b.** $\dfrac{3y - 2}{4} + \dfrac{5}{6} = \dfrac{2y}{3}$   $\{-4\}$

   **c.** $\dfrac{2a}{8a - 5} = \dfrac{1}{4}$   $\varnothing$    **d.** $\dfrac{x^2 - 5x + 1}{x^2 - 4} = \dfrac{x + 2}{x}$   $\varnothing$

   **e.** $1 = \dfrac{2}{x} + \dfrac{35}{x^2}$   $\{-5, 7\}$    **f.** $6 - \dfrac{1}{t - 2} = \dfrac{5}{t - 2}$   $\{3\}$

   **g.** $\dfrac{3}{7x - 5} = \dfrac{2}{4x - 3}$   $\left\{\dfrac{1}{2}\right\}$

   **h.** $\dfrac{3}{2x - 3} - \dfrac{2x}{4x^2 - 9} = \dfrac{2}{2x + 3}$   $\varnothing$

**44.** Solve each rational equation for the specified variable. (*Section 7.6, Objective 2*)

   **a.** $\dfrac{1}{R} = \dfrac{1}{R_1} + \dfrac{1}{R_2} + \dfrac{1}{R_3}$ for $R_1$ (resistors in parallel)

   **b.** $m = \dfrac{y_2 - y_1}{x_2 - x_1}$ for $x_1$ (slope formula)

**45.** Solve each problem. (*Section 7.7, Objectives 1–5*)

   **a.** A supermarket has a special on chunky soup—three cans for $3.87. How much will it cost to purchase seven cans of chunky soup?   $9.03

   **b.** Triangles *ABC* and *STU* are similar. Use the information in the diagram to find $a$ and $t$. Note the figures are not drawn to scale.

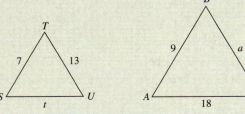

   **c.** If 1 in. represents 120 mi on a map, how many inches represent the distance between Las Vegas, Nevada, and Denver, Colorado, which is 750 mi?   6.25 in.

   **d.** An airplane flies 1500 mi in the same amount of time it takes a second plane to fly 1350 mi. If the speed of the first airplane is 50 mph faster than the speed of the second airplane, find the speed of each airplane.   500 mph and 450 mph

   **e.** An object weighing 1 lb on Earth weighs 1.125 lb on Neptune. Find Lucia's weight on Earth if she weighs 144 lb on Neptune.   128 lb

# Radical Expressions and Equations

## Refocus

Do you need to refocus your efforts to be successful? Maybe it has been a long term for you and it is a struggle to stay motivated and attentive to your studies. Don't give up. The light is at the end of the tunnel. Take a moment to remember why you are attending college and why you are enrolled in this class. Keep sight of your short-term and long-term goals. It will be worth it!

"What comes first, the compass or the clock? Before one can truly manage time (the clock), it is important to know where you are going, what your priorities and goals are, in which direction you are headed (the compass). Where you are headed is more important than how fast you are going. Rather than always *focusing* on what's urgent, learn to focus on what is really important" (author unknown, emphasis added).

**Question For Thought:** What are the things that distract you from accomplishing tasks or drain you of your time and energy? (Examples include Facebook, Web surfing, gaming, talking to co-workers, watching TV, and the like.) How can you refocus your time and energy toward your studies?

## Chapter Outline

## Coming Up...

In Section 8.6, we are going to learn how to use a formula that contains rational exponents to find the wind chill temperature on top of Mount Everest, where the wind speed can reach 115 mph and the temperature is −15°F.

"One reason so few of us achieves what we truly want is that we never direct our focus; we never concentrate our power. Most people dabble their way through life, never deciding to master anything in particular. In fact, I believe most people fail in life simply because they major in minor things."

—Anthony Robbins (Author, Life Strategist)

## SECTION 8.1 / Radical Expressions

**This chapter** deals with roots and radicals. Square roots were introduced in Chapter 1. We will review these as well as introduce cube roots and higher roots. The chapter will explore operations with these types of expressions as well as solving equations with radical expressions. The last section will show how roots are connected to exponents.

### OBJECTIVES

As a result of completing this section, you will be able to

1. Evaluate square root expressions.
2. Evaluate cube root expressions.
3. Evaluate *n*th roots.
4. Find the approximate value of irrational square roots.
5. Simplify radical expressions containing variables.
6. Troubleshoot common errors.

### Objective 1 ▶

Evaluate square root expressions.

In Chapters 1 and 5, we worked with exponential expressions. For instance, we know that $6^2 = 36$ and that $4^3 = 64$. In this section, we begin with numbers such as 36 and 64 and perform operations on them that will take us back to the base of the exponent.

The operation that takes us back to the base involves something called **radicals**. The word radical comes from the Latin word *radix*, which means "root"; that is, a radical takes us back to the source of the number.

## Square Roots

In Section 1.1, we briefly introduced the concept of the square root of a number. Recall that

The square of 5 is $5^2 = 25$.

The square of $-5$ is $(-5)^2 = 25$.

The reverse operation of squaring a number is finding the *square root* of a number. For example,

A square root of 25 is 5 because $5^2 = 25$.

A square root of 25 is $-5$ because $(-5)^2 = 25$.

> **Definition:** The **square root** of a real number $a$ is a real number $b$ if $b^2 = a$.

We know that $2^2 = 4$ and $(-2)^2 = 4$, so we say that 2 and $-2$ are square roots of 4. Also, $6^2 = 36$ and $(-6)^2 = 36$, so the square roots of 36 are 6 and $-6$.

When we use the notation $\sqrt{\phantom{a}}$, we get only one square root because the symbol denotes the **principal** or **positive square root**. For example,

$$\sqrt{4} = \sqrt{2^2} = 2 \quad \text{and} \quad \sqrt{36} = \sqrt{6^2} = 6$$

To denote a **negative square root**, we use the notation $-\sqrt{\phantom{a}}$. For example,

$$-\sqrt{4} = -2 \quad \text{and} \quad -\sqrt{36} = -6$$

The symbol $\sqrt{\phantom{a}}$ is called a **radical sign**. The number inside the radical is called the **radicand**. An expression that contains a radical is called a **radical expression**.

> **Property:** Square Root Notation
>
> If $a$ is a real number and $a > 0$, $\sqrt{a}$ denotes the *positive square root* of $a$ and $-\sqrt{a}$ denotes the *negative square root* of $a$.
>
> $$\sqrt{a} = b \text{ if and only if } b^2 = a \text{ and } b > 0$$
>
> Note that since $b^2 = a$, $\sqrt{a} = \sqrt{b^2} = b$, for $b > 0$.

**INSTRUCTOR NOTE:**
Point out the difference between $-\sqrt{4}$ and $\sqrt{-4}$.

Note that if $a = 0$, $\sqrt{0} = 0$ since $0^2 = 0$. For $a < 0$, consider the following expression. $\sqrt{-4}$ represents the number that must be squared to obtain $-4$.

When we square 2, we get $2^2 = 4$.

When we square $-2$, we get $(-2)^2 = 4$.

In fact, there is no real number that we can square to get $-4$. Recall from our study of exponents, that when a negative real number is squared, the result is always positive. So, we say that $\sqrt{-4}$ is not a real number.

> **Property: Square Root of a Negative Real Number**
>
> If $a$ is a real number and $a < 0$, $\sqrt{a}$ is *not a real number*.

Because of the relationship between squares and square roots, it is very helpful to remember the perfect squares: 1, 4, 9, 16, 25, 36, 49, and so on.

---

**Objective 1 Examples** | Simplify each expression.

| Problems | Solutions |
|---|---|
| **1a.** $\sqrt{64}$ | $\sqrt{64} = 8$ because $8^2 = 64$.<br>Also note: $\sqrt{64} = \sqrt{8^2} = 8$. |
| **1b.** $\sqrt{\dfrac{9}{121}}$ | $\sqrt{\dfrac{9}{121}} = \dfrac{3}{11}$ because $\left(\dfrac{3}{11}\right)^2 = \dfrac{9}{121}$.<br>Also note: $\sqrt{\dfrac{9}{121}} = \sqrt{\left(\dfrac{3}{11}\right)^2} = \dfrac{3}{11}$. |
| **1c.** $-\sqrt{49}$ | $-\sqrt{49} = -7$ because $7^2 = 49$.<br>The negative sign in front of the radical expression denotes the negative square root.<br>Also note: $-\sqrt{49} = -\sqrt{7^2} = -7$. |
| **1d.** $\sqrt{0.36}$ | $\sqrt{0.36} = 0.6$ because $(0.6)^2 = 0.36$.<br>Also note: $\sqrt{0.36} = \sqrt{(0.6)^2} = 0.6$. |
| **1e.** $\sqrt{-16}$ | $\sqrt{-16}$ is *not a real number* since there is no real number we can square to obtain $-16$.<br>Note: The square root of a negative number is actually a complex number. We will study these more in Chapter 9. |

---

✔ **Student Check 1** | Simplify each expression.

    **a.** $\sqrt{169}$     **b.** $\sqrt{\dfrac{49}{81}}$     **c.** $-\sqrt{\dfrac{1}{4}}$     **d.** $\sqrt{0.25}$     **e.** $\sqrt{-100}$

---

## Cube Roots

**Objective 2** ▶

**Evaluate cube root expressions.**

The square root enables us to find the number that we must square to obtain a given number. If we want to find the number that we cube to obtain a number, we use the *cube root*. Recall the following facts.

The cube of 2 is $2^3 = 8$.

The cube of $-2$ is $(-2)^3 = -8$.

**INSTRUCTOR NOTE:**
Begin by reviewing some examples
of cubing different values. Note
that any negative number cubed is
negative and any positive number
cubed is positive.

Remind students of perfect cubes.

The reverse operation of cubing a number is finding the *cube root* of a number. For example,

The cube root of 8 is 2 because $2^3 = 8$.

The cube root of $-8$ is $-2$ because $(-2)^3 = -8$.

---

**Property:  Cube Root**

The **cube root** of a real number $a$ is a real number $b$ if $b^3 = a$. The cube root of $a$ is denoted by $\sqrt[3]{a}$. The number 3 is called the **index** of the radical expression.

$$\sqrt[3]{a} = b \text{ if and only if } b^3 = a.$$

Note that since $a = b^3$, $\sqrt[3]{a} = \sqrt[3]{b^3} = b$, for all $b$.

---

For example,

$$\sqrt[3]{8} = 2 \text{ because } 2^3 = 8 \rightarrow \sqrt[3]{2^3} = 2$$

$$\sqrt[3]{-27} = -3 \text{ because } (-3)^3 = -27 \rightarrow \sqrt[3]{(-3)^3} = -3$$

Notice that, unlike the square root of a negative number, the cube root of a negative number is a real number. Recall that when we raise a negative number to an odd exponent, the result is a negative number. So, the cube root of a real number $a$ is defined for all real values of $a$.

Because of the relationship between cubes and cube roots, it is very helpful to remember the perfect cubes: 1, 8, 27, 64, 125, 216, and so on.

---

**Objective 2  Examples**    Simplify each expression.

| Problems | Solutions |
|---|---|
| **2a.**  $\sqrt[3]{64}$ | $\sqrt[3]{64} = 4$ because $4^3 = 64$. <br> Also note: $\sqrt[3]{64} = \sqrt[3]{4^3} = 4$. |
| **2b.**  $\sqrt[3]{-\dfrac{27}{125}}$ | $\sqrt[3]{-\dfrac{27}{125}} = -\dfrac{3}{5}$ because $\left(-\dfrac{3}{5}\right)^3 = -\dfrac{27}{125}$. <br><br> Also note: $\sqrt[3]{-\dfrac{27}{125}} = \sqrt[3]{\left(-\dfrac{3}{5}\right)^3} = -\dfrac{3}{5}$. |
| **2c.**  $-\sqrt[3]{216}$ | $-\sqrt[3]{216} = -6$ because $6^3 = 216$. <br> Also note: $-\sqrt[3]{216} = -\sqrt[3]{6^3} = -6$. |

---

✓ **Student Check 2**    Simplify each expression.

**a.**  $\sqrt[3]{343}$          **b.**  $\sqrt[3]{\dfrac{8}{27}}$          **c.**  $\sqrt[3]{-1000}$

---

## *n*th Roots

**Objective 3** ▶

Evaluate *n*th roots.

As we have seen, square roots and cube roots enable us to find the number that we must square or cube to obtain a given number. This concept can be extended to fourth roots, fifth roots, . . . , *n*th roots, where $n = 2, 3, 4, 5, \ldots$.

> **Property: The *n*th Root**
>
> The ***n*th root** of a real number $a$ is a real number $b$ if $b^n = a$ for $n = 2, 3, 4, 5, \ldots$ We write
>
> $$\sqrt[n]{a} = b \text{ if and only if } b^n = a.$$
>
> The number $n$ is called the *index* of the radical and $a$ is called the *radicand*. Note that the index 2 is usually omitted for square roots.

Note that if an index is *even*, such as $\sqrt{\phantom{x}}$, $\sqrt[4]{\phantom{x}}$, $\sqrt[6]{\phantom{x}}$, and the like, the radicand must be nonnegative for the root to be a real number because any real number raised to an even exponent is nonnegative. For example, $\sqrt[4]{16} = 2$ because $2^4 = 16$, but $\sqrt[4]{-16}$ is not a real number because no number raised to an exponent of 4 will give us $-16$.

If the index is *odd*, such as $\sqrt[3]{\phantom{x}}$, $\sqrt[5]{\phantom{x}}$, $\sqrt[7]{\phantom{x}}$, and the like, the radicand can be a negative number for the root to be defined because a negative number raised to an odd exponent is negative. For example, $\sqrt[7]{-128} = -2$ because $(-2)^7 = -128$.

The following diagrams show the relationship between exponents and roots. When a number is raised to an exponent of $n$, we obtain a specific number. When we take the $n$th root of that "new" number, we get the original number. This is based on the mathematical concept of inverses, which is beyond the scope of this course.

| Problems | Solutions |
|---|---|
| **Objective 3 Examples** | **Simplify each expression.** |

| Problems | Solutions |
|---|---|
| **3a.** $\sqrt[4]{16}$ | $\sqrt[4]{16} = 2$ because $2^4 = 16$.<br>Also note: $\sqrt[4]{16} = \sqrt[4]{2^4} = 2$. |
| **3b.** $\sqrt[4]{-81}$ | $\sqrt[4]{-81}$ is not a real number since there is no real number that we can raise to the fourth to obtain $-81$. |
| **3c.** $\sqrt[5]{-32}$ | $\sqrt[5]{-32} = -2$ because $(-2)^5 = -32$.<br>Also note: $\sqrt[5]{-32} = \sqrt[5]{(-2)^5} = -2$. |

✔ **Student Check 3**  Simplify each expression.

  **a.** $\sqrt[4]{625}$  **b.** $\sqrt[4]{-1}$  **c.** $\sqrt[5]{-243}$

## Approximating Square Roots

**Objective 4** ▶

Find the approximate value of irrational square roots.

Thus far, all of the roots we have evaluated resulted in a rational number (except in the case of a nonreal number). The roots are rational numbers because we evaluated square roots of perfect squares, cube roots of perfect cubes, fourth roots of powers of 4, and the like. Recall a **perfect square** is obtained when we raise a rational number to an exponent of 2, a **perfect cube** is obtained when we raise a rational number to an exponent of 3, and so on.

We will now examine square roots of numbers that are not perfect squares, such as $\sqrt{3}, \sqrt{5}, \sqrt{7}$, and the like. In this case, the result will be an irrational number (except in the case of a nonreal number). Recall from Chapter 1 that an irrational number is a real number that cannot be written as the quotient of two integers. An irrational number is a nonrepeating, nonterminating decimal. We can use the square root key, $\boxed{\sqrt{\phantom{x}}}$, on a calculator to approximate the value of irrational square roots.

**Objective 4 Examples**    **Classify each number as rational or irrational. If the number is irrational, decide what two integers it is between, and then use a calculator to approximate its value to the nearest hundredth.**

| Problems | Solutions |
|---|---|
| **4a.** $\sqrt{3}$ | The radicand 3 is not a perfect square, so $\sqrt{3}$ is irrational. The number 3 lies between the perfect squares of 1 and 4, so $\sqrt{1} < \sqrt{3} < \sqrt{4}$ or $1 < \sqrt{3} < 2$. Using the calculator, we find $$\sqrt{3} \approx 1.73$$ |
| **4b.** $\sqrt{110}$ | The radicand 110 is not a perfect square, so $\sqrt{110}$ is irrational. The number 110 lies between the perfect squares 100 and 121, so $\sqrt{100} < \sqrt{110} < \sqrt{121}$ or $10 < \sqrt{110} < 11$. Using the calculator, we find $$\sqrt{110} \approx 10.49$$ |
| **4c.** $\sqrt{\dfrac{1}{16}}$ | The radicand $\dfrac{1}{16}$ is a perfect square since it is $\left(\dfrac{1}{4}\right)^2$. So, $\sqrt{\dfrac{1}{16}}$ is rational. $$\sqrt{\frac{1}{16}} = \sqrt{\left(\frac{1}{4}\right)^2} = \frac{1}{4}$$ |
| **4d.** $-\sqrt{144}$ | The radicand 144 is a perfect square since it is $12^2$. So, $-\sqrt{144}$ is rational. $$-\sqrt{144} = -\sqrt{12^2} = -12$$ |

✓ **Student Check 4**    Classify each number as rational or irrational. If the number is irrational, decide what two integers it is between, and then use a calculator to approximate its value to the nearest hundredth.

   **a.** $\sqrt{10}$        **b.** $-\sqrt{45}$        **c.** $\sqrt{\dfrac{2}{9}}$        **d.** $\sqrt{256}$

## Radical Expressions Containing Variables

**Objective 5** ▶

Simplify radical expressions containing variables.

We have evaluated only radicals that contained real numbers in their radicands. We will now evaluate radical expressions that contain variables in their radicands. We will first examine some special roots.

We have already shown that $\sqrt{a^2} = a$, for $a \geq 0$. For instance,

$$\sqrt{(4)^2} = \sqrt{16} = 4$$

The same value

This, however, does not hold true for $a < 0$. For instance,

$$\sqrt{(-4)^2} = \sqrt{16} = 4$$

Not the same value

In this case, $\sqrt{(-4)^2} = |-4| = 4$.

To be mathematically precise, we state the following.

> **Property: Square Root of $a^2$**
>
> For any real number $a$,
> $$\sqrt{a^2} = |a|$$

A common mistake is to state that $\sqrt{a^2}$ is $a$, but this only holds true when $a \geq 0$.
- If we are not given any restriction on what the variable can be, then we must use the absolute value symbols when we simplify $\sqrt{a^2}$.
- If, however, we assume that the variable represents a positive real number, then we do not need to include the absolute value symbols.

The following illustration shows how the result differs based on our assumptions about the variable $x$.

| If $x$ represents *any* real number, then | If $x$ is a *positive* real number, then |
|---|---|
| $\sqrt{4x^2} = \sqrt{(2x)^2}$ $= \|2x\| = 2\|x\|$ | $\sqrt{4x^2} = \sqrt{(2x)^2}$ $= 2x$ |

**INSTRUCTOR NOTE:**
Help students see that squaring a variable expression is the opposite of taking the square root of the variable expression.
For instance, $(y^4)^2 = y^{4 \cdot 2}$ but $\sqrt{y^8} = y^{8/2}$.

Since we have already stated that $\sqrt[3]{a^3} = a$ for all real numbers $a$, it is not necessary to use the absolute value symbols when we simplify this type of expression.

In general, to simplify a radical expression that contains variables, write the expression in the radical as a base to an exponent equal to the index and use the fact that $\sqrt[n]{a^n} = a$. We will assume that all variables represent positive real numbers so that we do not have to use the absolute value for roots with even indices. For example,

We know that $y^8 = (y^4)^2$, so $\sqrt{y^8} = \sqrt{(y^4)^2} = y^4$.

We know that $x^6 = (x^2)^3$, so $\sqrt[3]{x^6} = \sqrt[3]{(x^2)^3} = x^2$.

> **Procedure: Simplifying a Radical Expression Containing Variables**
>
> **Step 1:** Write the radicand as an expression raised to the index. Recall the rules for exponents:
> **a.** $(a^m)^n = a^{mn}$
> **b.** $a^n b^n = (ab)^n$
> **Step 2:** Apply the rules for roots:
> **a.** $\sqrt[n]{a^n} = a$, for $a \geq 0$ if $n$ is even
> **b.** $\sqrt[n]{a^n} = a$, for any real number $a$ if $n$ is odd
> **Step 3:** Check by raising the result to the index. If this yields the radicand, the answer is correct.

**Objective 5 Examples** Simplify each expression. Assume the variables represent positive real numbers.

**5a.** $\sqrt{x^2}$      **5b.** $\sqrt{a^{12}}$      **5c.** $\sqrt{81y^6}$      **5d.** $\sqrt{25x^4 y^{10}}$

**5e.** $\sqrt[3]{-8a^3}$      **5f.** $\sqrt[3]{\dfrac{27x^9}{125}}$      **5g.** $\sqrt[4]{16r^8 t^4}$

**Solutions**

**5a.** $\sqrt{x^2} = \sqrt{(x)^2}$     Write $x^2$ as an expression squared.

        $= x$            Apply the property $\sqrt[n]{a^n} = a$.

**Check:** $(x)^2 = x^2$

**5b.** $\sqrt{a^{12}} = \sqrt{(a^6)^2}$     Write $a^{12}$ as an expression squared.

         $= a^6$           Apply the property $\sqrt[n]{a^n} = a$.

**Check:** $(a^6)^2 = a^{12}$

**5c.** $\sqrt{81y^6} = \sqrt{(9y^3)^2}$      Write $81y^6$ as an expression squared.

$\phantom{\sqrt{81y^6}} = 9y^3$      Apply the property $\sqrt[n]{a^n} = a$.

**Check:** $(9y^3)^2 = 81y^6$

**5d.** $\sqrt{25x^4y^{10}} = \sqrt{(5x^2y^5)^2}$      Write $25x^4y^{10}$ as an expression squared.

$\phantom{\sqrt{25x^4y^{10}}} = 5x^2y^5$      Apply the property $\sqrt[n]{a^n} = a$.

**Check:** $(5x^2y^5)^2 = 25x^4y^{10}$

**5e.** $\sqrt[3]{-8a^3} = \sqrt[3]{(-2a)^3}$      Write $-8a^3$ as an expression cubed.

$\phantom{\sqrt[3]{-8a^3}} = -2a$      Apply the property $\sqrt[n]{a^n} = a$.

**Check:** $(-2a)^3 = -8a^3$

**5f.** $\sqrt[3]{\dfrac{27x^9}{125}} = \sqrt[3]{\left(\dfrac{3x^3}{5}\right)^3}$      Write $\dfrac{27x^9}{125}$ as an expression cubed.

$\phantom{\sqrt[3]{\dfrac{27x^9}{125}}} = \dfrac{3x^3}{5}$      Apply the property $\sqrt[n]{a^n} = a$.

**Check:** $\left(\dfrac{3x^3}{5}\right)^3 = \dfrac{27x^9}{125}$

**5g.** $\sqrt[4]{16r^8t^4} = \sqrt[4]{(2r^2t)^4}$      Write $16r^8t^4$ as an expression to the fourth.

$\phantom{\sqrt[4]{16r^8t^4}} = 2r^2t$      Apply the property $\sqrt[n]{a^n} = a$.

**Check:** $(2r^2t)^4 = 16r^8t^4$

✓ **Student Check 5**    Simplify each expression. Assume the variables represent positive real numbers.

**a.** $\sqrt{y^{10}}$      **b.** $\sqrt[3]{b^3}$      **c.** $\sqrt{49x^4}$      **d.** $\sqrt{100x^2y^8}$

**e.** $\sqrt[3]{64b^6}$      **f.** $\sqrt[3]{-\dfrac{8a^{12}}{b^3}}$      **g.** $\sqrt[4]{81x^4y^{12}}$

**Objective 6** ▶

Troubleshoot common errors.

## Troubleshooting Common Errors

Some common errors associated with $n$th roots are shown next.

**Objective 6 Examples**    **A problem and an incorrect solution are given. Provide the correct solution and an explanation of the error.**

**6a.** Simplify $\sqrt{-25}$.

| Incorrect Solution | Correct Solution and Explanation |
|---|---|
| $\sqrt{-25} = -5$ | The square root of a negative number is not a real number. When we square $-5$, we get 25 not $-25$. So, $\sqrt{-25}$ is not a real number. |

**6b.** Simplify $\sqrt[3]{-64}$.

| Incorrect Solution | Correct Solution and Explanation |
|---|---|
| $\sqrt[3]{-64}$ is not a real number. | The cube root of a negative number is a real number. Since $(-4)^3 = -64$, $\sqrt[3]{-64} = -4$. |

## ANSWER TO STUDENT CHECKS

**Student Check 1   a.** 13   **b.** $\dfrac{7}{9}$   **c.** $-\dfrac{1}{2}$   **d.** 0.5
**e.** not a real number

**Student Check 2   a.** 7   **b.** $\dfrac{2}{3}$   **c.** $-10$

**Student Check 3   a.** 5   **b.** not a real number   **c.** $-3$

**Student Check 4   a.** irrational, between 3 and 4, 3.16
**b.** irrational, between $-6$ and $-7$, $-6.71$   **c.** irrational, between 0 and 1, 0.47   **d.** rational, 16

**Student Check 5   a.** $y^5$   **b.** $b$   **c.** $7x^2$   **d.** $10xy^4$
**e.** $4b^2$   **f.** $-\dfrac{2a^4}{b}$   **g.** $3xy^3$

## SUMMARY OF KEY CONCEPTS

1. The square root of a real number $a$, $a \geq 0$, is the number $b$ such that $b^2 = a$. In symbols, $\sqrt{a} = b$, where $b \geq 0$.

2. The cube root of a real number $a$ is the number $b$ such that $b^3 = a$. In symbols, $\sqrt[3]{a} = b$.

3. The $n$th root of a real number $a$ is the number $b$ such that $b^n = a$. In symbols, $\sqrt[n]{a} = b$. If $n$ is even, "$a$ must be nonnegative" for $\sqrt[n]{a}$ to be a real number. The number $a$ is called the radicand and $n$ is called the index. Some of the perfect powers are shown.

| $b$ | $b^2$ | $b^3$ | $b^4$ | $b^5$ |
|---|---|---|---|---|
| 1 | 1 | 1 | 1 | 1 |
| 2 | 4 | 8 | 16 | 32 |
| 3 | 9 | 27 | 81 | 243 |
| 4 | 16 | 64 | 256 | |
| 5 | 25 | 125 | 625 | |
| 6 | 36 | 216 | | |
| 7 | 49 | 343 | | |
| 8 | 64 | | | |
| 9 | 81 | | | |
| 10 | 100 | | | |
| 11 | 121 | | | |
| 12 | 144 | | | |
| 13 | 169 | | | |
| 14 | 196 | | | |
| 15 | 225 | | | |

4. The square root of numbers that are not perfect squares are irrational numbers. The calculator must be used to approximate these square roots.

5. To simplify a radical expression that contains variables, write the radicand as an expression raised to the index and use the fact that $\sqrt[n]{a^n} = a$. We will assume that the variable represents a positive real number so that we do not have to use the absolute value for roots with even indices. Some of the perfect powers of variable expressions are shown.

| | Perfect squares | Perfect cubes | Perfect fourths | Perfect fifths |
|---|---|---|---|---|
| $b$ | $b^2$ | $b^3$ | $b^4$ | $b^5$ |
| $b^2$ | $b^4$ | $b^6$ | $b^8$ | $b^{10}$ |
| $b^3$ | $b^6$ | $b^9$ | $b^{12}$ | $b^{15}$ |
| $b^4$ | $b^8$ | $b^{12}$ | $b^{16}$ | $b^{20}$ |
| $b^5$ | $b^{10}$ | $b^{15}$ | $b^{20}$ | $b^{25}$ |

## GRAPHING CALCULATOR SKILLS

We can use the graphing calculator to simplify square roots, cube roots, and radicals with an index greater than 3.

**Example:** Simplify $\sqrt{2704}$.

**Solution:**

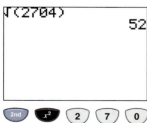

**Example:** Simplify $\sqrt[3]{1728}$.

**Solution:**

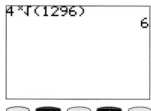

**Example:** Simplify $\sqrt[4]{1296}$.

**Solution:**
To simplify radicals with indices greater than 3, we use the $x$th-root function in the calculator. Enter the index first.

## SECTION 8.1    EXERCISE SET

### Write About It!

Use complete sentences in your answer to each exercise.

1. If $a$ is a positive number, will $-\sqrt{a}$ be positive, negative, or neither? Explain.
   If $a > 0$, $\sqrt{a}$ is positive. So, $-\sqrt{a}$ will be negative.

2. If $a$ is a positive number, will $\sqrt{-a}$ be positive, negative, or neither? Explain.

3. If $n$ is even, what values of $x$ will result in a real number for $\sqrt[n]{x}$?

4. If $n$ is odd, what values of $x$ will result in a real number for $\sqrt[n]{x}$?

5. What does the expression $\sqrt[3]{x^3}$ simplify to if we don't know if $x$ is positive or negative?
   When the index is odd, the expression $\sqrt[n]{x^n} = x$, so $\sqrt[3]{x^3} = x$.

6. What does the expression $\sqrt[4]{x^4}$ simplify to if we don't know if $x$ is positive or negative?
   When the index is even, the expression $\sqrt[n]{x^n} = |x|$, so $\sqrt[4]{x^4} = |x|$.

7. Explain how to simplify the radical $\sqrt[3]{8x^{12}y^3}$.

8. Explain how to simplify the radical $\sqrt[4]{\dfrac{81a^{12}}{b^8}}$.

### Practice Makes Perfect!

Simplify each expression. (*See Objective 1.*)

9. $\sqrt{16}$    4
10. $\sqrt{36}$    6
11. $\sqrt{\dfrac{1}{4}}$    $\dfrac{1}{2}$
12. $\sqrt{\dfrac{4}{49}}$    $\dfrac{2}{7}$
13. $\sqrt{2.25}$    1.5
14. $\sqrt{4.41}$    2.1
15. $-\sqrt{64}$    $-8$
16. $-\sqrt{100}$    $-10$
17. $\sqrt{-121}$    not a real number
18. $\sqrt{-169}$    not a real number
19. $-\sqrt{0.0049}$    $-0.07$
20. $-\sqrt{0.0016}$    $-0.04$
21. $-\sqrt{\dfrac{169}{225}}$    $-\dfrac{13}{15}$
22. $-\sqrt{\dfrac{64}{25}}$    $-\dfrac{8}{5}$
23. $-\sqrt{625}$    $-25$
24. $-\sqrt{400}$    $-20$

Simplify each expression. (*See Objective 2.*)

25. $\sqrt[3]{27}$    3
26. $\sqrt[3]{125}$    5
27. $\sqrt[3]{-64}$    $-4$
28. $\sqrt[3]{-8}$    $-2$
29. $-\sqrt[3]{1000}$    $-10$
30. $-\sqrt[3]{64}$    $-4$
31. $\sqrt[3]{0.008}$    0.2
32. $\sqrt[3]{0.729}$    0.9
33. $\sqrt[3]{\dfrac{1}{8}}$    $\dfrac{1}{2}$
34. $\sqrt[3]{\dfrac{64}{343}}$    $\dfrac{4}{7}$
35. $-\sqrt[3]{-343}$    7
36. $-\sqrt[3]{-216}$    6

Simplify each expression. (*See Objective 3.*)

37. $\sqrt[4]{81}$    3
38. $\sqrt[4]{625}$    5
39. $-\sqrt[4]{-1}$    not a real number
40. $-\sqrt[4]{-16}$    not a real number
41. $\sqrt[5]{32}$    2
42. $\sqrt[5]{243}$    3

Classify each number as rational or irrational. If the number is irrational, use a calculator to approximate its value to the nearest hundredth. (*See Objective 4.*)

43. $\sqrt{5}$    irrational, 2.24
44. $\sqrt{6}$    irrational, 2.45
45. $-\sqrt{81}$    rational, $-9$
46. $-\sqrt{49}$    rational, $-7$
47. $\sqrt{\dfrac{9}{4}}$    rational, $\dfrac{3}{2}$
48. $\sqrt{\dfrac{1}{25}}$
49. $\sqrt{99}$    irrational, 9.95
50. $\sqrt{-121}$    not a real number
51. $\sqrt{-144}$    not a real number
52. $\sqrt{\dfrac{1}{5}}$    irrational, 0.45

Simplify each expression. Assume the variables represent positive real numbers. (*See Objective 5.*)

53. $\sqrt{a^2}$    $a$
54. $\sqrt{b^2}$    $b$
55. $\sqrt{100x^4}$    $10x^2$
56. $\sqrt{900x^8}$    $30x^4$
57. $\sqrt{25x^4y^8}$    $5x^2y^4$
58. $\sqrt{81x^2y^6}$    $9xy^3$
59. $\sqrt[3]{64a^6}$    $4a^2$
60. $\sqrt[3]{1000b^9}$    $10b^3$
61. $\sqrt[3]{-125r^3s^{12}}$    $-5rs^4$
62. $\sqrt[3]{8c^{15}d^6}$    $2c^5d^2$
63. $\sqrt[3]{-\dfrac{64r^6}{s^{18}}}$
64. $\sqrt[3]{\dfrac{343a^3}{b^9}}$    $\dfrac{7a}{b^3}$
65. $\sqrt[4]{16u^4v^8}$    $2uv^2$
66. $\sqrt[4]{81a^{12}b^{16}}$    $3a^3b^4$
67. $\sqrt[4]{\dfrac{x^8y^{16}}{256}}$    $\dfrac{x^2y^4}{4}$
68. $\sqrt[4]{\dfrac{625c^8}{d^4}}$    $\dfrac{5c^2}{d}$

### Mix 'Em Up!

Simplify each expression. Assume the variables represent positive real numbers.

69. $\sqrt{-9}$    not a real number
70. $-\sqrt{196}$    $-14$
71. $\sqrt{\dfrac{81}{100}}$    $\dfrac{9}{10}$
72. $\sqrt{18}$    4.24
73. $\sqrt{16a^4}$    $4a^2$
74. $\sqrt[3]{-\dfrac{1}{27}}$    $-\dfrac{1}{3}$
75. $\sqrt[4]{\dfrac{625}{81}}$    $\dfrac{5}{3}$
76. $\sqrt{\dfrac{25a^4}{9b^2}}$    $\dfrac{5a^2}{3b}$
77. $\sqrt[3]{\dfrac{0.125u^6}{v^3}}$
78. $\sqrt[3]{0.343r^3s^6}$    $0.7rs^2$
79. $\sqrt[5]{243x^5y^{15}}$    $3xy^3$
80. $\sqrt[5]{1024r^{20}s^{10}}$    $4r^4s^2$
81. $\sqrt{64x^{14}}$    $8x^7$
82. $\sqrt{-900c^2}$    not a real number
83. $-\sqrt{10,000z^8}$    $-100z^4$
84. $\sqrt[4]{81c^8d^{12}}$    $3c^2d^3$
85. $\sqrt[3]{-\dfrac{8a^{12}}{125}}$
86. $\sqrt[3]{125r^{18}s^6}$    $5r^6s^2$
87. $\sqrt{0.64x^4}$    $0.8x^2$
88. $\sqrt{0.0121y^6}$    $0.11y^3$
89. $\sqrt[4]{16x^4y^8}$    $2xy^2$
90. $\sqrt[3]{\dfrac{1000c^{15}}{27d^6}}$    $\dfrac{10c^5}{3d^2}$

### You Be the Teacher!

Correct each student's errors, if any.

91. Simplify $-\sqrt{-9}$.
    Angela's work: $-\sqrt{-9} = \sqrt{9} = 3$.
    $\sqrt{-9}$ is not a real number, so $-\sqrt{-9}$ is not a real number.

**92.** Simplify $-\sqrt[3]{-8}$.

Robert's work: $-\sqrt[3]{-8}$ is not a real number.

$-\sqrt[3]{-8} = -(-2) = 2$

**93.** Simplify $\sqrt{16b^4}$.

Masum's work: $\sqrt{16b^4} = 8b^2$   The product rule was applied but $\sqrt{16} = 4$ not 8, so $\sqrt{16b^4} = 4b^2$.

**94.** Simplify $\sqrt[4]{\dfrac{16x^8}{y^4}}$.

Estrella's work: $\sqrt[4]{\dfrac{16x^8}{y^4}} = \dfrac{4x^2}{y^4}$   $\sqrt[4]{\dfrac{16x^8}{y^4}} = \dfrac{\sqrt[4]{16x^8}}{\sqrt[4]{y^4}} = \dfrac{2x^2}{y}$

## Calculate It!

Use a calculator to simplify each radical. Classify each number as rational or irrational and approximate values to the nearest hundredth when applicable.

**95.** $\sqrt[3]{-343}$   rational, $-7$

**96.** $\sqrt[3]{-729}$   rational, $-9$

**97.** $\sqrt{122}$   irrational, 11.05

**98.** $\sqrt{227}$   irrational, 15.07

**99.** $\sqrt[5]{248,832}$   rational, 12

**100.** $\sqrt[3]{-800}$   irrational, $-9.28$

**101.** $\sqrt[5]{-1000}$   irrational, $-3.98$

**102.** $\sqrt[4]{28,561}$   rational, 13

---

## SECTION 8.2  /  Simplifying Radicals and the Distance Formula

### ▶ OBJECTIVES

**As a result of completing this section, you will be able to**

**1.** Use the product rule to simplify square roots.

**2.** Simplify square roots of variable expressions.

**3.** Use the quotient rule to simplify square roots.

**4.** Simplify higher roots.

**5.** Use the distance formula.

**6.** Troubleshoot common errors.

One of the key skills in mathematics is having the ability to rewrite expressions in equivalent forms. For instance, we have learned how to rewrite fractions in their simplest form by dividing out common factors from their numerators and denominators. We have also learned how to rewrite polynomials in their factored form. Radicals can also be simplified. Simplifying radicals is a skill we need when we solve quadratic equations in Chapter 9.

In this section, we will discuss how to rewrite radicals in their simplest form.

### The Product Rule for Square Roots

In Section 8.1, we discussed how to evaluate roots. For instance, we know that $\sqrt{25}$ is a radical expression that is not in its simplest form, since $\sqrt{25} = 5$. When the radicand is a perfect square, the radical can be simplified fairly easily. We can also simplify square roots when the radicand has a perfect square as one of its factors. Before this is illustrated, we need to learn the product rule for radicals. Consider the following expressions.

### Objective 1 ▶

Use the product rule to simplify square roots.

$$\sqrt{4 \cdot 4} = \sqrt{16} = 4 \qquad \sqrt{4} \cdot \sqrt{4} = 2 \cdot 2 = 4 \quad \rightarrow \quad \sqrt{4 \cdot 4} = \sqrt{4} \cdot \sqrt{4} = 4$$

$$\sqrt{4 \cdot 25} = \sqrt{100} = 10 \qquad \sqrt{4} \cdot \sqrt{25} = 2 \cdot 5 = 10 \quad \rightarrow \quad \sqrt{4 \cdot 25} = \sqrt{4} \cdot \sqrt{25} = 10$$

These examples illustrate an important property of radicals. They show that the square root of a product is the same as the product of the square roots.

> **Property:  Product Rule for Square Roots:** For $a$ and $b$ positive real numbers,
> $$\sqrt{a \cdot b} = \sqrt{a} \cdot \sqrt{b}$$

For a square root to be in its simplest form, the radicand cannot contain any perfect square factors. Our goal is to extract all perfect squares from the radical.

Here are some examples of square roots that are in their simplest form and some that are not.

| In Simplest Form | Not in Simplest Form |
|---|---|
| $\sqrt{3},\, 4\sqrt{6}$ | $\sqrt{12},\, 2\sqrt{72}$ |

The radical $\sqrt{12}$ is not in simplest form since $12 = 4 \cdot 3$. The factor 4 is a perfect square. The radical, $2\sqrt{72}$, is not in simplest form since $72 = 36 \cdot 2$. The factor 36 is a perfect square.

There are two methods that we can apply to simplify square roots. One method relies on listing pairs of factors for the radicand. The other method relies on expressing the radicand in its prime factorization.

---

**Procedure: Simplifying Square Roots Using the Product Rule**

| **Method 1** | **Method 2** |
|---|---|
| **Step 1:** List the factors of the radicand. | **Step 1:** Rewrite the radicand as a product of its prime factors. |
| **Step 2:** Select the pair of factors for which one factor is a perfect square. If there is more than one perfect square factor, use the pair with the larger perfect square. | **Step 2:** Apply the product rule. |
|  | **Step 3:** Extract any perfect squares. |
| **Step 3:** Rewrite the radicand as the product of numbers you found in step 2. Write the perfect square first. |  |
| **Step 4:** Apply the product rule. |  |
| **Step 5:** Extract any perfect squares. |  |

---

**Objective 1 Examples**  Simplify each radical.

**1a.** $\sqrt{12}$ **1b.** $\sqrt{80}$ **1c.** $2\sqrt{72}$ **1d.** $-4\sqrt{54}$

**Solutions**

**1a. Method 1:** List the factors of 12.

1, 12

2, 6

3, 4    4 is a perfect square factor.

$$\sqrt{12} = \sqrt{4 \cdot 3} \quad \text{Rewrite 12 as } 4 \cdot 3.$$
$$= \sqrt{4}\sqrt{3} \quad \text{Apply the product rule.}$$
$$= 2\sqrt{3} \quad \text{Write } \sqrt{4} \text{ as 2.}$$

**Method 2:** Find the prime factorization of 12.

$$12 = 2 \cdot 6$$
$$= 2 \cdot 2 \cdot 3$$
$$= 2^2 \cdot 3$$

$$\sqrt{12} = \sqrt{2^2 \cdot 3} \quad \text{Replace 12 with } 2^2 \cdot 3.$$
$$= \sqrt{2^2}\sqrt{3} \quad \text{Apply the product rule.}$$
$$= 2\sqrt{3} \quad \text{Write } \sqrt{2^2} \text{ as 2.}$$

**1b. Method 1:** List the factors of 80.

1, 80

2, 40

4, 20    4 and 16 are both perfect square

5, 16    factors. We will use 16 since it

8, 10    is larger.

$$\sqrt{80} = \sqrt{16 \cdot 5} \quad \text{Rewrite 80 as } 16 \cdot 5.$$
$$= \sqrt{16}\sqrt{5} \quad \text{Apply the product rule.}$$
$$= 4\sqrt{5} \quad \text{Write } \sqrt{16} \text{ as 4.}$$

**Method 2:** Find the prime factorization of 80.

$$80 = 4 \cdot 20$$
$$= 2 \cdot 2 \cdot 4 \cdot 5$$
$$= 2 \cdot 2 \cdot 2 \cdot 2 \cdot 5$$
$$= 2^4 \cdot 5$$

$$\sqrt{80} = \sqrt{2^4 \cdot 5} \quad \text{Replace 80 with } 2^4 \cdot 5.$$
$$= \sqrt{2^4}\sqrt{5} \quad \text{Apply the product rule.}$$
$$= 2^2\sqrt{5} \quad \text{Write } \sqrt{2^4} \text{ as } 2^2.$$
$$= 4\sqrt{5} \quad \text{Simplify } 2^2.$$

**1c. Method 1:** List the factors of 72.

1, 72

2, 36    36, 4, and 9 are perfect square

3, 24    factors. We will use 36 since it is

4, 18    the largest.

6, 12

8, 9

$$2\sqrt{72} = 2\sqrt{36 \cdot 2}$$
$$= 2\sqrt{36}\sqrt{2}$$
$$= 2(6)\sqrt{2}$$
$$= 12\sqrt{2}$$

**Method 2:** Find the prime factorization of 72.

$$72 = 2 \cdot 36$$
$$= 2 \cdot 6 \cdot 6$$
$$= 2 \cdot 2 \cdot 3 \cdot 2 \cdot 3$$
$$= 2^2 \cdot 3^2 \cdot 2$$

$$2\sqrt{72} = 2\sqrt{2^2 \cdot 3^2 \cdot 2}$$
$$= 2\sqrt{2^2 \cdot 3^2}\sqrt{2}$$
$$= 2(2)(3)\sqrt{2}$$
$$= 12\sqrt{2}$$

**1d. Method 1:** List the factors of 54.

1, 54

2, 27

3, 18

6, 9    9 is a perfect square factor.

$$-4\sqrt{54} = -4\sqrt{9 \cdot 6}$$
$$= -4\sqrt{9}\sqrt{6}$$
$$= -4(3)\sqrt{6}$$
$$= -12\sqrt{6}$$

**Method 2:** Find the prime factorization of 54.

$$54 = 6 \cdot 9$$
$$= 2 \cdot 3 \cdot 3 \cdot 3$$
$$= 2 \cdot 3 \cdot 3^2$$

$$-4\sqrt{54} = -4\sqrt{3^2 \cdot 2 \cdot 3}$$
$$= -4\sqrt{3^2}\sqrt{2 \cdot 3}$$
$$= -4(3)\sqrt{6}$$
$$= -12\sqrt{6}$$

✓ **Student Check 1**    Simplify each radical.

**a.** $\sqrt{18}$      **b.** $\sqrt{75}$      **c.** $7\sqrt{98}$      **d.** $-3\sqrt{76}$

**Note:** *In Example 1c, we will obtain the same result if we use a different pair of factors.*

$$2\sqrt{72} = 2\sqrt{9 \cdot 8}$$    Rewrite 72 as 9 • 8.
$$= 2\sqrt{9}\sqrt{8}$$    Apply the product rule.
$$= 2(3)\sqrt{8}$$    Write $\sqrt{9}$ as 3.
$$= 6\sqrt{8}$$    Multiply the coefficients.
$$= 6\sqrt{4 \cdot 2}$$    Rewrite 8 as 4 • 2.
$$= 6\sqrt{4}\sqrt{2}$$    Apply the product rule.
$$= 6(2)\sqrt{2}$$    Write $\sqrt{4}$ as 2.
$$= 12\sqrt{2}$$    Multiply the coefficients.

*If we do not use the largest perfect square factor, the radicand still contains a perfect square factor. Therefore, the steps to simplify have to be repeated.*

## Simplifying Square Roots Containing Variable Expressions

**Objective 2** ▶

Simplify square roots of variable expressions.

As discussed in Section 8.1, perfect squares result from squaring an expression. We will examine what types of expressions come from squaring powers of a variable. Recall the power rule for exponents states that $(a^m)^n = a^{mn}$.

$$(x)^2 = x^2$$
$$(x^2)^2 = x^4$$
$$(x^3)^2 = x^6$$
$$(x^4)^2 = x^8$$

So, the expressions $x^2, x^4, x^6$, and $x^8$ are perfect squares. We can conclude the following.

- A variable raised to an *even* exponent is a *perfect square*.
- A variable raised to an *odd* exponent is *not* a perfect square.

We will use these facts to simplify the expression $\sqrt{x^5}$, $x \geq 0$. To simplify $\sqrt{x^5}$, we must rewrite the radicand, $x^5$, as a perfect square times another factor. Recall that the product rule of exponents is $a^m \cdot a^n = a^{m+n}$. So, the factorizations of $x^5$ are

$$x^5 = x \cdot x^4$$
$$x^5 = x^2 \cdot x^3$$

Since each factorization includes $x$ raised to an even exponent, $x^4$ and $x^2$, both factorizations contain a perfect square as a factor. Note, however, that the second factorization, $x^2 \cdot x^3$, still has a perfect square as a factor since $x^3 = x^2 \cdot x$. So, to simplify $\sqrt{x^5}$, we use the largest perfect square factor, $x^4$, to save some steps. We get

$$\sqrt{x^5} = \sqrt{x^4 \cdot x}$$
$$= \sqrt{x^4} \cdot \sqrt{x}$$
$$= x^2\sqrt{x}$$

---

**Procedure: Simplifying Expressions of the Form $\sqrt{x^n}$**

**Step 1:** Rewrite the radicand as a product involving a perfect square factor.
    **a.** If $n$ is even, $x^n$ is a perfect square.
    **b.** If $n$ is odd, write $x^n$ as $x^{n-1}x$.
**Step 2:** Apply the product rule. Put all perfect square factors in one radical and the remaining factors in another radical.
**Step 3:** Simplify any square roots.

---

**Note:** *For a square root to be completely simplified, the radicand cannot contain any exponents greater than or equal to 2.*

For the purpose of this section, we assume that all variables represent positive real numbers so that $\sqrt{x^n}$ is a real number. Therefore, we do not need to write roots with absolute values.

---

**Objective 2 Examples**    Simplify each radical completely.

    **2a.** $\sqrt{y^7}$      **2b.** $\sqrt{24x^6}$      **2c.** $\sqrt{14a^3}$      **2d.** $\sqrt{32x^4y^9}$

**Solutions**    **2a.** $\sqrt{y^7} = \sqrt{y^6 \cdot y}$     Rewrite $y^7$ as $y^6 \cdot y$.

                $= \sqrt{y^6}\sqrt{y}$     Apply the product rule.

                $= y^3\sqrt{y}$     Write $\sqrt{y^6}$ as $y^3$.

**2b.**  $\sqrt{24x^6} = \sqrt{4 \cdot 6x^6}$      Rewrite 24 as 4 · 6. Note: $x^6$ is a perfect square.

$\qquad\qquad = \sqrt{4x^6}\sqrt{6}$      Apply the product rule.

$\qquad\qquad = 2x^3\sqrt{6}$       Write $\sqrt{4x^6}$ as $2x^3$.

**2c.**  $\sqrt{14a^3} = \sqrt{14a^2 \cdot a}$      Rewrite $a^3$ as $a^2 \cdot a$.

$\qquad\qquad = \sqrt{a^2}\sqrt{14a}$      Apply the product rule.

$\qquad\qquad = a\sqrt{14a}$       Write $\sqrt{a^2}$ as $a$.

**2d.** $\sqrt{32x^4y^9} = \sqrt{4^2 \cdot 2x^4y^8y}$     Rewrite 32 as $4^2 \cdot 2$ and $y^9$ as $y^8 \cdot y$.

$\qquad\qquad = \sqrt{4^2x^4y^8}\sqrt{2y}$      Apply the product rule.

$\qquad\qquad = 4x^2y^4\sqrt{2y}$      Write $\sqrt{4^2x^4y^8}$ as $4x^2y^4$.

✔ **Student Check 2**   Simplify each radical completely.

   **a.** $\sqrt{t^9}$      **b.** $\sqrt{48y^{10}}$      **c.** $\sqrt{10x^7}$      **d.** $\sqrt{63a^8b^5}$

### The Quotient Rule for Square Roots

**Objective 3** ▶

Use the quotient rule to simplify square roots.

When the radicand is a quotient, we can simplify it by applying the square root to the numerator and denominator. Consider the following expressions.

$$\sqrt{\frac{36}{4}} = \sqrt{9} = 3 \qquad \frac{\sqrt{36}}{\sqrt{4}} = \frac{6}{2} = 3 \qquad \longrightarrow \qquad \sqrt{\frac{36}{4}} = \frac{\sqrt{36}}{\sqrt{4}} = 3$$

These examples illustrate another important property of square roots. The square root of a quotient is the same as the quotient of the square roots.

> **Property:  Quotient Rule for Square Roots**
>
> For $a$ and $b$ positive real numbers,
> $$\sqrt{\frac{a}{b}} = \frac{\sqrt{a}}{\sqrt{b}}$$

For a square root to be in its simplest form, there cannot be any fractions within the square root. There also cannot be any square roots in the denominator of a fraction.

> **Procedure:  Simplifying the Square Root of a Quotient**
>
> **Step 1:** Rewrite the expression as the square root of the numerator divided by the square root of the denominator.
> **Step 2:** Simplify each square root completely.

**Objective 3  Examples**   Simplify each radical completely.

   **3a.** $\sqrt{\dfrac{4}{9}}$      **3b.** $\sqrt{\dfrac{x^2}{25}}$      **3c.** $\sqrt{\dfrac{20}{9y^4}}$      **3d.** $\sqrt{\dfrac{3a^5}{16}}$

**Solutions**   **3a.** $\sqrt{\dfrac{4}{9}} = \dfrac{\sqrt{4}}{\sqrt{9}}$      Apply the quotient rule.

$\qquad\qquad\qquad = \dfrac{2}{3}$       Simplify the square roots.

**3b.** $\sqrt{\dfrac{x^2}{25}} = \dfrac{\sqrt{x^2}}{\sqrt{25}}$       Apply the quotient rule.

$= \dfrac{x}{5}$       Write $\sqrt{x^2}$ as $x$ and $\sqrt{25}$ as 5.

**3c.** $\sqrt{\dfrac{20}{9y^4}} = \dfrac{\sqrt{20}}{\sqrt{9y^4}}$       Apply the quotient rule.

$= \dfrac{\sqrt{4 \cdot 5}}{3y^2}$       Apply the product rule to write $\sqrt{9y^4}$ as $3y^2$ and write 20 as $4 \cdot 5$.

$= \dfrac{2\sqrt{5}}{3y^2}$       Write $\sqrt{4}$ as 2.

**3d.** $\sqrt{\dfrac{3a^5}{16}} = \dfrac{\sqrt{3a^5}}{\sqrt{16}}$       Apply the quotient rule.

$= \dfrac{\sqrt{3a^4 \cdot a}}{4}$       Write $3a^5$ as $3a^4 \cdot a$ and $\sqrt{16}$ as 4.

$= \dfrac{\sqrt{a^4}\sqrt{3a}}{4}$       Apply the product rule in the numerator.

$= \dfrac{a^2\sqrt{3a}}{4}$       Write $\sqrt{a^4}$ as $a^2$.

---

✔ **Student Check 3**   Simplify each radical completely.

**a.** $\sqrt{\dfrac{81}{49}}$      **b.** $\sqrt{\dfrac{y^4}{36}}$      **c.** $\sqrt{\dfrac{44}{9x^2}}$      **d.** $\sqrt{\dfrac{15b^3}{4}}$

### Simplifying Higher Roots

**Objective 4** ▶

Simplify higher roots.

Now we will simplify radical expressions that involve higher roots (cube roots, fourth roots, etc.). The rules that were presented for square roots also apply to higher roots.

> **Property: Product Rule for Roots:** For $a$ and $b$ nonnegative real numbers and $n = 2, 3, 4, \ldots,$
>
> $$\sqrt[n]{a \cdot b} = \sqrt[n]{a} \cdot \sqrt[n]{b}.$$

> **Property: Quotient Rule for Roots:** For $a$ and $b$ nonnegative real numbers with $b \neq 0$ and $n = 2, 3, 4, \ldots,$
>
> $$\sqrt[n]{\dfrac{a}{b}} = \dfrac{\sqrt[n]{a}}{\sqrt[n]{b}}.$$

For a cube root to be completely simplified, the radicand cannot contain any exponents greater than or equal to 3. For a fourth root to be completely simplified, the radicand cannot contain any exponents greater than or equal to 4, and so on. Recall the following from Section 8.1 Summary of Key Concepts:

Perfect cubes are $x^3$, $x^6$, $x^9$, $x^{12}$, and so on. Note that the exponent of a perfect cube is a multiple of 3.

Perfect fourths are $x^4$, $x^8$, $x^{12}$, $x^{16}$, and so on. Note that the exponent of a perfect fourth is a multiple of 4.

This pattern continues for perfect fifths, perfect sixths, and so on.

When a factor in a radicand can be written as an expression raised to the index, a root can be extracted.

---

**Procedure:  Simplifying a Higher Root**

**Step 1:** If the radicand is a quotient, apply the quotient rule.
**Step 2:** Rewrite each radicand as a product of an expression raised to the index and another factor. Method 1 or 2 from Objective 1 can be used.
**Step 3:** Apply the product rule.
**Step 4:** Simplify the radicals, if possible.

---

**Objective 4 Examples**    Simplify each radical completely.

**4a.** $\sqrt[3]{54}$    **4b.** $\sqrt[3]{24b^3}$    **4c.** $\sqrt[4]{80a^7b^8}$    **4d.** $\sqrt[3]{\dfrac{192x^5}{27}}$

**Solutions**    **4a. Method 1:** List the factors of 54.

1, 54
2, 27                    27 is a perfect cube.
3, 18
6, 9

$$\sqrt[3]{54} = \sqrt[3]{27 \cdot 2} \qquad \text{Rewrite 54 as } 27 \cdot 2.$$
$$= \sqrt[3]{27}\sqrt[3]{2} \qquad \text{Apply the product rule.}$$
$$= 3\sqrt[3]{2} \qquad \text{Write } \sqrt[3]{27} \text{ as 3.}$$

**4b. Method 2:** Find the prime factorization of 24.

$$24 = 4 \cdot 6$$
$$= 2 \cdot 2 \cdot 2 \cdot 3$$
$$= 2^3 \cdot 3$$

$$\sqrt[3]{24b^3} = \sqrt[3]{2^3 \cdot 3b^3} \qquad \text{Replace 24 with } 2^3 \cdot 3.$$
$$= \sqrt[3]{2^3b^3}\sqrt[3]{3} \qquad \text{Apply the product rule.}$$
$$= 2b\sqrt[3]{3} \qquad \text{Write } \sqrt[3]{2^3b^3} \text{ as } 2b.$$

**4c. Method 1:** List the factors of 80.

Factors of 80:
1, 80
2, 40
4, 20
5, 16            16 is a perfect fourths.
8, 10

Note: $a^7 = a^4 \cdot a^3$ and $b^8 = (b^2)^4$

$$\sqrt[4]{80a^7b^8} = \sqrt[4]{16 \cdot 5a^4a^3b^8} \qquad \text{Rewrite each factor as shown.}$$
$$= \sqrt[4]{16a^4(b^2)^4}\sqrt[4]{5a^3} \qquad \text{Apply the product rule.}$$
$$= 2ab^2\sqrt[4]{5a^3} \qquad \text{Simplify.}$$

**4d. Method 1:**

Factors of 192:

1, 192

2, 96

3, 64    64 is a perfect cube.

4, 48

6, 32

8, 24

12, 16

Note: $x^5 = x^3 \cdot x^2$ and 27 is a perfect cube.

$$\sqrt[3]{\frac{192x^5}{27}} = \frac{\sqrt[3]{192x^5}}{\sqrt[3]{27}} \qquad \text{Apply the quotient rule.}$$

$$= \frac{\sqrt[3]{64 \cdot 3x^3x^2}}{\sqrt[3]{27}} \qquad \text{Write each factor as shown above.}$$

$$= \frac{\sqrt[3]{64x^3}\sqrt[3]{3x^2}}{3} \qquad \text{Apply the product rule.}$$

$$= \frac{4x\sqrt[3]{3x^2}}{3} \qquad \text{Simplify.}$$

---

✓ **Student Check 4**  Simplify each radical completely.

**a.** $\sqrt[3]{128}$  **b.** $\sqrt[3]{48a^6}$  **c.** $\sqrt[3]{\dfrac{40y^8}{343}}$  **d.** $\sqrt[4]{162x^6y^4}$

---

## The Distance Formula

**Objective 5** ▶

Use the distance formula.

The distance formula provides a way for us to measure the distance between two ordered pairs on the coordinate system. The distance formula is based on the Pythagorean theorem, $a^2 + b^2 = c^2$.

**Property: Distance Formula**

If $(x_1, y_1)$ and $(x_2, y_2)$ are two points, then the distance between them is

$$d = \sqrt{(x_2 - x_1)^2 + (y_2 - y_1)^2}$$

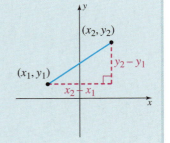

**Procedure: Finding the Distance Between Two Points**

**Step 1:** Label one point $(x_1, y_1)$ and the other point $(x_2, y_2)$.
**Step 2:** Substitute the values into the distance formula.
**Step 3:** Simplify the expressions inside the radical.
**Step 4:** Simplify the radical by applying the product rule.

**Objective 5   Examples**

**Find the distance between the points. Express each answer as a simplified radical.**

**5a.** $(-3, 5)$ and $(3, 8)$          **5b.** $\left(4, \sqrt{7}\right)$ and $(-7, 0)$

**Solutions**

**5a.** $d = \sqrt{(x_2 - x_1)^2 + (y_2 - y_1)^2}$          State the distance formula.

$d = \sqrt{[3 - (-3)]^2 + (8 - 5)^2}$          Let $(x_1, y_1) = (-3, 5)$ and $(x_2, y_2) = (3, 8)$.

$d = \sqrt{(6)^2 + (3)^2}$          Simplify within parentheses and brackets.

$d = \sqrt{36 + 9}$          Simplify each exponential expression.

$d = \sqrt{45}$          Combine the numbers inside the radical.

$d = \sqrt{9 \cdot 5}$          Write 45 as 9 · 5.

$d = 3\sqrt{5}$          Write $\sqrt{9}$ as 3.

**5b.** $d = \sqrt{(x_2 - x_1)^2 + (y_2 - y_1)^2}$          State the distance formula.

$d = \sqrt{(-7 - 4)^2 + \left(0 - \sqrt{7}\right)^2}$          Let $(x_1, y_1) = \left(4, \sqrt{7}\right)$ and $(x_2, y_2) = (-7, 0)$.

$d = \sqrt{(-11)^2 + \left(-\sqrt{7}\right)^2}$          Simplify within parentheses.

$d = \sqrt{121 + 7}$          Simplify each exponential expression.

$d = \sqrt{128}$          Combine the numbers inside the radical.

$d = \sqrt{64 \cdot 2}$          Write 128 as 64 · 2.

$d = 8\sqrt{2}$          Write $\sqrt{64}$ as 8.

✓ **Student Check 5**

**Find the distance between the points. Express each answer as a simplified radical.**

**a.** $(-4, 2)$ and $(4, 6)$          **b.** $\left(5, \sqrt{3}\right)$ and $(-2, 0)$

**Objective 6** ▶

Troubleshoot common errors.

## Troubleshooting Common Errors

Some common errors associated with simplifying radicals are shown next.

**Objective 6   Examples**

**A problem and an incorrect solution are given. Provide the correct solution and an explanation of the error.**

**6a.** Simplify $\sqrt{48}$.

| Incorrect Solution | Correct Solution and Explanation |
|---|---|
| $\sqrt{48} = \sqrt{4 \cdot 12}$ $= \sqrt{4}\sqrt{12}$ $= 2\sqrt{12}$ | The error was made in not simplifying $\sqrt{12}$ further. $$\sqrt{48} = \sqrt{4 \cdot 12}$$ $$= \sqrt{4}\sqrt{12}$$ $$= 2\sqrt{12}$$ $$= 2\sqrt{4 \cdot 3}$$ $$= 2\sqrt{4} \cdot \sqrt{3}$$ $$= 2(2)\sqrt{3}$$ $$= 4\sqrt{3}$$ Note that if we use the largest perfect square factor of 48, we obtain the simplified form immediately. $$\sqrt{48} = \sqrt{16 \cdot 3}$$ $$= \sqrt{16}\sqrt{3}$$ $$= 4\sqrt{3}$$ |

**6b.** Simplify $\sqrt[3]{54}$.

| Incorrect Solution | Correct Solution and Explanation |
|---|---|
| 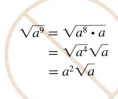 | We must find a perfect cube factor of 54 not a perfect square factor. Also, the index was omitted in the last two lines. $$\sqrt[3]{54} = \sqrt[3]{27 \cdot 2}$$ $$= \sqrt[3]{27}\sqrt[3]{2}$$ $$= 3\sqrt[3]{2}$$ |

**6c.** Simplify $\sqrt{a^9}$.

| Incorrect Solution | Correct Solution and Explanation |
|---|---|
| $$\sqrt{a^9} = \sqrt{a^8 \cdot a}$$ $$= \sqrt{a^4}\sqrt{a}$$ $$= a^2\sqrt{a}$$ | When the perfect square is extracted, it should not be written inside the radical. That is, $\sqrt{a^8} = a^4$ not $\sqrt{a^4}$. $$\sqrt{a^9} = \sqrt{a^8 \cdot a}$$ $$= \sqrt{a^8}\sqrt{a}$$ $$= \sqrt{(a^4)^2}\sqrt{a}$$ $$= a^4\sqrt{a}$$ |

## ANSWERS TO STUDENT CHECKS

**Student Check 1**   **a.** $3\sqrt{2}$   **b.** $5\sqrt{3}$   **c.** $49\sqrt{2}$
   **d.** $-6\sqrt{19}$

**Student Check 2**   **a.** $t^4\sqrt{t}$   **b.** $4y^5\sqrt{3}$   **c.** $x^3\sqrt{10x}$
   **d.** $3a^4b^2\sqrt{7b}$

**Student Check 3**   **a.** $\dfrac{9}{7}$   **b.** $\dfrac{y^2}{6}$   **c.** $\dfrac{2\sqrt{11}}{3x}$   **d.** $\dfrac{b\sqrt{15b}}{2}$

**Student Check 4**   **a.** $4\sqrt[3]{2}$   **b.** $2a^2\sqrt[3]{6}$   **c.** $\dfrac{2y^2\sqrt[3]{5y^2}}{7}$
   **d.** $3xy\sqrt[4]{2x^2}$

**Student Check 5**   **a.** $4\sqrt{5}$   **b.** $2\sqrt{13}$

## SUMMARY OF KEY CONCEPTS

1. Simplify a radical by writing it as a product of an expression raised to the index and another factor. Use the largest perfect power. Apply the product rule.
   - 1, 4, 9, 16, 25, 36, and so on are perfect squares.
   - 1, 8, 27, 64, 125, and so on are perfect cubes.
   - 1, 16, 81, 256, and so on are perfect fourths.

2. Variables raised to exponents that are multiples of the index are perfect squares, cubes, and so on.
   - $x^2, x^4, x^6, x^8$, and so on are perfect squares.
   - $x^3, x^6, x^9, x^{12}$, and so on are perfect cubes.
   - $x^4, x^8, x^{12}, x^{16}$, and so on are perfect fourth powers.

3. If the radical contains a fraction, use the quotient rule to apply the root to the numerator and the denominator.

4. The distance formula enables us to calculate the distance between two points on the coordinate system. The distance is equal to the square root of the sum of the squares of the horizontal and vertical distances between the points.

## GRAPHING CALCULATOR SKILLS

We can use the graphing calculator to learn perfect powers, find factors of numbers, and check our answers.

**Example:** Find the perfect squares, cubes, and fourths.

**Solution:** Enter $x$ raised to the appropriate exponent and view the table. The values in the Y columns are perfect squares, cubes, and fourths, respectively.

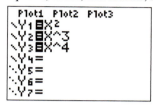

**Example:** Find factors of 54.

**Solution:** Enter $\dfrac{54}{x}$ in the equation editor. Then view the table. If both the X and $Y_1$ values are integers, then the pair of numbers are factors.

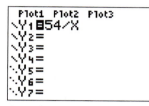

The factors of 54 are 1 and 54, 2 and 27, 3 and 18, and 6 and 9.

**Example:** Show that $\sqrt{54} = 3\sqrt{6}$ and that $\sqrt[3]{54} = 3\sqrt[3]{2}$.

**Solution:**

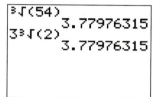

---

# SECTION 8.2 / EXERCISE SET

## Write About It!

Use complete sentences in your answer to each exercise.

1. What does it mean for the square root of a number to be in simplest form? *The radicand doesn't contain any perfect square factors.*

2. How can you tell if the square root of a variable term is in simplest form? *The exponent of the variable remaining in the radicand is less than the index.*

3. Explain the product rule for square roots.

4. Explain the quotient rule for square roots.

## Practice Makes Perfect!

Simplify each radical completely. (*See Objective 1.*)

5. $\sqrt{8}$  $2\sqrt{2}$  6. $\sqrt{24}$  $2\sqrt{6}$  7. $\sqrt{20}$  $2\sqrt{5}$

8. $\sqrt{27}$  $3\sqrt{3}$  9. $2\sqrt{45}$  $6\sqrt{5}$  10. $3\sqrt{48}$  $12\sqrt{3}$

11. $5\sqrt{60}$  $10\sqrt{15}$  12. $-3\sqrt{52}$  $-6\sqrt{13}$  13. $-4\sqrt{44}$  $-8\sqrt{11}$

14. $\sqrt{108}$  $6\sqrt{3}$  15. $\sqrt{28}$  $2\sqrt{7}$  16. $7\sqrt{75}$  $35\sqrt{3}$

17. $-5\sqrt{96}$  $-20\sqrt{6}$  18. $-4\sqrt{56}$  $-8\sqrt{14}$

Simplify each radical completely. Assume that all variables represent positive real numbers. (*See Objective 2.*)

19. $\sqrt{y^4}$  $y^2$  20. $\sqrt{y^8}$  $y^4$  21. $\sqrt{16y^{16}}$  $4y^8$

22. $\sqrt{25y^{12}}$  $5y^6$  23. $\sqrt{12y^8}$  $2y^4\sqrt{3}$  24. $\sqrt{24y^{14}}$  $2y^7\sqrt{6}$

25. $\sqrt{4y^{15}}$  $2y^7\sqrt{y}$  26. $\sqrt{9y^{19}}$  $3y^9\sqrt{y}$  27. $\sqrt{96y^3}$  $4y\sqrt{6y}$

28. $\sqrt{72y^5}$  $6y^2\sqrt{2y}$

Simplify each radical completely. Assume that all variables represent positive real numbers. (*See Objective 3.*)

29. $\sqrt{\dfrac{4}{25}}$  $\dfrac{2}{5}$  30. $\sqrt{\dfrac{9}{16}}$  $\dfrac{3}{4}$  31. $\sqrt{\dfrac{12}{49}}$  $\dfrac{2\sqrt{3}}{7}$

32. $\sqrt{\dfrac{18}{121}}$  $\dfrac{3\sqrt{2}}{11}$  33. $\sqrt{\dfrac{x^2}{169}}$  $\dfrac{x}{13}$  34. $\sqrt{\dfrac{x^4}{144}}$  $\dfrac{x^2}{12}$

*Additional answers can be found in the Instructor Answer Appendix.*

**35.** $\sqrt{\dfrac{4a^2}{9}}$ $\dfrac{2a}{3}$ **36.** $\sqrt{\dfrac{81a^4}{25}}$ $\dfrac{9a^2}{5}$ **37.** $\sqrt{\dfrac{45a^7}{81}}$ $\dfrac{a^3\sqrt{5a}}{3}$

**38.** $\sqrt{\dfrac{18a^9}{225}}$ $\dfrac{a^4\sqrt{2a}}{5}$

**Simplify each radical completely. Assume that all variables represent positive real numbers. (See Objective 4.)**

**39.** $\sqrt[3]{81}$ $3\sqrt[3]{3}$ **40.** $\sqrt[3]{40}$ $2\sqrt[3]{5}$ **41.** $\sqrt[3]{16a^4}$ $2a\sqrt[3]{2a}$

**42.** $\sqrt[3]{125a^5}$ $5a\sqrt[3]{a^2}$ **43.** $\sqrt[4]{80a^6}$ $2a\sqrt[4]{5a^2}$ **44.** $\sqrt[4]{162a^7}$ $3a\sqrt[4]{2a^3}$

**45.** $\sqrt[4]{16a^9b^8}$ $2a^2b^2\sqrt[4]{a}$ **46.** $\sqrt[4]{96r^{10}s^4}$ **47.** $\sqrt[4]{405x^7y^8}$ $3xy^2\sqrt[4]{5x^3}$

**48.** $\sqrt[4]{112c^{12}d^{11}}$ $2c^3d^2\sqrt[4]{7d^3}$ **49.** $\sqrt[3]{\dfrac{160x^7}{125}}$ **50.** $\sqrt[3]{\dfrac{27a^8}{64}}$ $\dfrac{3a^2\sqrt[3]{a^2}}{4}$

**Find the distance between the points. Express each answer as a simplified radical. (See Objective 5.)**

**51.** $(4, -1)$ and $(-5, -13)$ $15$

**52.** $(10, 0)$ and $(30, -15)$ $25$

**53.** $(3, -11)$ and $(-12, -3)$ $17$

**54.** $(-5, 13)$ and $(-17, 29)$ $20$

**55.** $(6, -4\sqrt{5})$ and $(-1, -5\sqrt{5})$ $3\sqrt{6}$

**56.** $(9, -3\sqrt{2})$ and $(12, -6\sqrt{2})$ $3\sqrt{3}$

**57.** $(-9, 5\sqrt{3})$ and $(-13, 3\sqrt{3})$ $2\sqrt{7}$

**58.** $(-1, 9\sqrt{15})$ and $(4, 10\sqrt{15})$ $2\sqrt{10}$

**59.** $\left(2, \dfrac{3}{2}\right)$ and $\left(\dfrac{5}{4}, \dfrac{5}{2}\right)$ $\dfrac{5}{4}$

**60.** $\left(\dfrac{4}{5}, -\dfrac{1}{5}\right)$ and $\left(\dfrac{12}{5}, \dfrac{14}{5}\right)$ $\dfrac{17}{5}$

 **Mix 'Em Up!**

**Simplify each radical completely. Assume that all variables represent positive real numbers.**

**61.** $6\sqrt{50}$ $30\sqrt{2}$ **62.** $\sqrt[4]{128b^8}$ $2b^2\sqrt[4]{8}$ **63.** $\sqrt{y^{21}}$ $y^{10}\sqrt{y}$

**64.** $\sqrt{12c^{10}}$ $2c^5\sqrt{3}$ **65.** $\sqrt[3]{16x^8}$ $2x^2\sqrt[3]{2x^2}$ **66.** $\sqrt{\dfrac{121a^5}{25}}$ $\dfrac{11a^2\sqrt{a}}{5}$

**67.** $\sqrt{-75x^2}$ not a real number **68.** $\sqrt[3]{-24x^{16}}$ $-2x^5\sqrt[3]{3x}$ **69.** $\sqrt{150x}$ $5\sqrt{6x}$

**70.** $\sqrt[4]{-32r^{12}}$ not a real number **71.** $\sqrt{\dfrac{225a^4}{16}}$ $\dfrac{15a^2}{4}$ **72.** $7\sqrt{147}$ $49\sqrt{3}$

**73.** $\sqrt[4]{144b^7}$ $2b\sqrt[4]{9b^3}$ **74.** $\sqrt{216}$ $6\sqrt{6}$ **75.** $\sqrt[3]{\dfrac{1000x^{11}}{y^6}}$ $\dfrac{10x^3\sqrt[3]{x^2}}{y^2}$

**76.** $7\sqrt[5]{64a^{13}b^5}$ $14a^2b\sqrt[5]{2a^3}$ **77.** $5\sqrt[4]{162r^9t^{14}}$ $15r^2t^3\sqrt[4]{2rt^2}$ **78.** $6\sqrt[4]{\dfrac{80x^{10}}{81y^8}}$ $\dfrac{4x^2\sqrt[4]{5x^2}}{y^2}$

**79.** $\sqrt[5]{-729c^5d^3}$ $-3c\sqrt[5]{3d^3}$ **80.** $\sqrt[5]{\dfrac{486u^7}{v^{15}}}$ $\dfrac{3u\sqrt[5]{2u^2}}{v^3}$

**Find the distance between the points. Express each answer as a simplified radical.**

**81.** $(-10, 0)$ and $(14, -10)$ $26$

**82.** $(6, -2)$ and $(46, 7)$ $41$

**83.** $(5, -\sqrt{6})$ and $(13, -3\sqrt{6})$ $2\sqrt{22}$

**84.** $(8, -8\sqrt{5})$ and $(1, -7\sqrt{5})$ $3\sqrt{6}$

**85.** $(3, -10\sqrt{2})$ and $(7, -8\sqrt{2})$ $2\sqrt{6}$

**86.** $(9, 9\sqrt{3})$ and $(18, 10\sqrt{3})$ $2\sqrt{21}$

**87.** $\left(\dfrac{1}{2}, \dfrac{2}{3}\right)$ and $\left(-\dfrac{1}{4}, \dfrac{5}{3}\right)$ $\dfrac{5}{4}$

**88.** $\left(\dfrac{1}{3}, \dfrac{7}{6}\right)$ and $\left(-\dfrac{1}{3}, \dfrac{29}{12}\right)$ $\dfrac{17}{12}$

**89.** $\left(1, \dfrac{11}{6}\right)$ and $\left(\dfrac{1}{2}, \dfrac{5}{2}\right)$ $\dfrac{5}{6}$

**90.** $\left(\dfrac{1}{4}, \dfrac{1}{12}\right)$ and $\left(-\dfrac{37}{12}, -\dfrac{2}{3}\right)$ $\dfrac{41}{12}$

**91.** $(-0.04, 0.02)$ and $(0.05, -0.38)$ $0.41$

**92.** $(3.5, -7)$ and $(5.5, -8.5)$ $2.5$

 **You Be the Teacher!**

**Correct each student's errors, if any.**

**93.** Simplify $\sqrt[3]{24}$.

Jose's work:

$\sqrt[3]{24} = \sqrt[3]{4 \cdot 6} = 2\sqrt[3]{6}$

The index is 3 so we need to find a perfect cube factor not a perfect square factor. $\sqrt[3]{24} = \sqrt[3]{8 \cdot 3} = 2\sqrt[3]{3}$

**94.** Simplify $\sqrt{162x^4}$.

Maria's work:

$\sqrt{162x^4} = \sqrt{81x^4 \cdot 2}$
$= \sqrt{9x^2 \cdot 2} = 3x\sqrt{2}$

When we simplify the $\sqrt{81x^4}$, we get $9x^2$ not $\sqrt{9x^2}$. $\sqrt{162x^4} = \sqrt{81x^4 \cdot 2} = 9x^2\sqrt{2}$

**95.** Find the distance between $(6, -1)$ and $(4, 3)$.

Rita's work:

$d = \sqrt{(4 - 6) + [3 - (-1)]}$
$= \sqrt{-2 + 4} = \sqrt{2}$

**96.** Find the distance between the points $(3, -1)$ and $(11, -16)$.

Glenda's work:

$\sqrt{(11 - 3)^2 + (-16 + 1)^2}$
$= \sqrt{(8)^2 + (-15)^2}$
$= \sqrt{64 + 225}$
$= 8 + 15$
$= 23$

**Calculate It!**

**Use a calculator to determine if each radical is simplified correctly. If it is not simplified correctly, find the correct form.**

**97.** $\sqrt{720} = 8\sqrt{14}$ no, $12\sqrt{5}$ **98.** $\sqrt{675} = 15\sqrt{3}$ yes

**99.** $\sqrt[3]{162} = 9\sqrt[3]{6}$ no, $3\sqrt[3]{6}$ **100.** $\sqrt[3]{320} = 4\sqrt[3]{5}$ yes

## SECTION 8.3 / **Adding and Subtracting Radical Expressions**

▶ **OBJECTIVES**

As a result of completing this section, you will be able to

**1.** Add or subtract like radicals.

**2.** Add or subtract radicals that require simplification.

**3.** Troubleshoot common errors.

Now that we know how to simplify radicals, we will learn how to perform operations on radical expressions. In this section, we will learn how to add and subtract radicals. This skill is needed to find the perimeter of the given triangle. The perimeter is the sum of the lengths of the sides of the triangle. Therefore, $P = 3\sqrt{2} + 3\sqrt{2} + 6$. The first objective will provide instructions on how to simplify this expression.

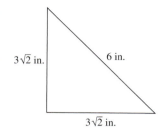

### Adding or Subtracting Like Radicals

**Objective 1** ▶

Add or subtract like radicals.

To add expressions together, they must be of the same form. Recall we can add only variable expressions that have like terms. For instance,

$$2x + 4x = (2 + 4)x = 6x$$

This same rule applies to radicals. We can add radicals only if they are *like radicals*. For instance,

$$2\sqrt{5} + 4\sqrt{5} = (2 + 4)\sqrt{5} = 6\sqrt{5}$$

Notice that when we add or subtract like radicals, we add or subtract the coefficients of the radicals and keep the radical the same.

> **Definition:**  **Like radicals** are radicals with the same index and the same radicand.

| Examples of Like Radicals | Examples of Unlike Radicals |
|---|---|
| $2\sqrt{5}$ and $4\sqrt{5}$ | $2\sqrt{3}$ and $3\sqrt{5}$ |
| $-\sqrt{6}$ and $\sqrt{6}$ | $\sqrt{6}$ and $\sqrt[3]{6}$ |

Note that $2\sqrt{3} + 3\sqrt{5}$ cannot be simplified further because the radicands are different. Note that $\sqrt{6} + \sqrt[3]{6}$ cannot be simplified further because their indices are different.

**Objective 1  Examples**  **Perform each operation.**

**1a.** $\sqrt{6} + \sqrt{6}$       **1b.** $\sqrt{15} - 3\sqrt{15}$

**1c.** $-7\sqrt[3]{2} + 3\sqrt[3]{2} - 6\sqrt[3]{2}$       **1d.** $4\sqrt[3]{5} + 3\sqrt{5}$

**1e.** Find the perimeter of the figure.

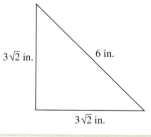

**Solutions**  **1a.**   $1\sqrt{6} + 1\sqrt{6} = (1 + 1)\sqrt{6}$          Add the coefficients.

$= 2\sqrt{6}$          Keep the radical the same.

**1b.**  $1\sqrt{15} - 3\sqrt{15} = (1 - 3)\sqrt{15}$          Subtract the coefficients.

$= -2\sqrt{15}$          Keep the radical the same.

**1c.** $-7\sqrt[3]{2} + 3\sqrt[3]{2} - 6\sqrt[3]{2} = (-7 + 3 - 6)\sqrt[3]{2}$     Combine the coefficients.

$= -10\sqrt[3]{2}$     Keep the radical the same.

**1d.** These radicals are not alike and cannot be combined. Though their radicands are the same, the indices are different. This expression is in its simplest form.

**1e.** $P = 3\sqrt{2} + 3\sqrt{2} + 6$     Add the three lengths.

$= (3 + 3)\sqrt{6} + 6$     Add the two like radical terms.

$= 6\sqrt{6} + 6$ in.     The constant term, 6, cannot be combined with the radical expressions.

✓ **Student Check 1**    Perform each operation.

**a.** $9\sqrt{2} + 7\sqrt{2}$        **b.** $\sqrt{10} + \sqrt{10}$        **c.** $\sqrt{7} - 5\sqrt{7}$

**d.** $-3\sqrt[3]{5} + 8\sqrt[3]{5} - 2\sqrt[3]{5}$        **e.** $6\sqrt{15} - 4\sqrt{5}$

**f.** Find the perimeter of a rectangle with length $3\sqrt{7}$ ft and width $5\sqrt{7}$ ft.

## Adding or Subtracting Radicals Not in Simplified Form

**Objective 2** ▶

Add or subtract radicals that require simplification.

Some radical expressions may not appear to be like radicals. Before we state that we can't combine such radicals, we need to determine if the radical expressions can be simplified by extracting any roots.

> **Procedure: Adding or Subtracting Radicals That Require Simplification**
>
> **Step 1:** Use the product rule to simplify the radicals, if possible.
> **Step 2:** After simplifying, add or subtract any like radicals.

**Objective 2 Examples**    **Perform each operation and express each answer in simplest radical form. Assume variables represent positive real numbers.**

**2a.** $\sqrt{12} + \sqrt{48} - \sqrt{75}$     **2b.** $3\sqrt{20} - 6\sqrt{45}$     **2c.** $\sqrt[3]{16} + 5\sqrt[3]{54}$

**2d.** $\sqrt{16} - 3x\sqrt{4x} + 2\sqrt{x^3}$     **2e.** $\sqrt{\dfrac{24}{25}} + \sqrt{\dfrac{54}{49}}$

**Solutions**    **2a.** $\sqrt{12} + \sqrt{48} - \sqrt{75} = \sqrt{4 \cdot 3} + \sqrt{16 \cdot 3} - \sqrt{25 \cdot 3}$    Factor each radicand.

$= \sqrt{4}\sqrt{3} + \sqrt{16}\sqrt{3} - \sqrt{25}\sqrt{3}$    Apply the product rule.

$= 2\sqrt{3} + 4\sqrt{3} - 5\sqrt{3}$    Simplify $\sqrt{4}, \sqrt{16}$, and $\sqrt{25}$.

$= (2 + 4 - 5)\sqrt{3}$    Add the like radicals.

$= 1\sqrt{3}$    Simplify.

$= \sqrt{3}$

**2b.**    $3\sqrt{20} - 6\sqrt{45} = 3\sqrt{4 \cdot 5} - 6\sqrt{9 \cdot 5}$    Factor each radicand.

$= 3\sqrt{4}\sqrt{5} - 6\sqrt{9}\sqrt{5}$    Apply the product rule.

$= 3(2)\sqrt{5} - 6(3)\sqrt{5}$    Simplify $\sqrt{4}$ and $\sqrt{9}$.

$= 6\sqrt{5} - 18\sqrt{5}$    Simplify $3(2)$ and $6(3)$.

$= (6 - 18)\sqrt{5}$    Subtract the like radicals.

$= -12\sqrt{5}$    Simplify.

**2c.** $\sqrt[3]{16} + 5\sqrt[3]{54} = \sqrt[3]{8 \cdot 2} + 5\sqrt[3]{27 \cdot 2}$    Factor each radicand.

$\qquad = \sqrt[3]{8}\sqrt[3]{2} + 5\sqrt[3]{27}\sqrt[3]{2}$    Apply the product rule.

$\qquad = 2\sqrt[3]{2} + 5(3)\sqrt[3]{2}$    Simplify $\sqrt[3]{8}$ and $\sqrt[3]{27}$.

$\qquad = 2\sqrt[3]{2} + 15\sqrt[3]{2}$    Simplify 5(3).

$\qquad = (2 + 15)\sqrt[3]{2}$    Add the like radicals.

$\qquad = 17\sqrt[3]{2}$    Simplify.

**2d.** $\sqrt{16} - 3x\sqrt{4x} + 2\sqrt{x^3}$

$\qquad = \sqrt{16} - 3x\sqrt{4}\sqrt{x} + 2\sqrt{x^2 \cdot x}$    Apply the product rule to the middle term. Factor $x^3$.

$\qquad = 4 - 3x(2)\sqrt{x} + 2\sqrt{x^2}\sqrt{x}$    Simplify $\sqrt{16}$ and $\sqrt{4}$. Apply the product rule to the last term.

$\qquad = 4 - 6x\sqrt{x} + 2x\sqrt{x}$    Simplify 3x(2) and $\sqrt{x^2}$.

$\qquad = 4 + (-6x + 2x)\sqrt{x}$    Add like radicals.

$\qquad = 4 - 4x\sqrt{x}$    Simplify.

We cannot combine the constant term, 4, with the radical expression, $-4x\sqrt{x}$, since they are not like terms.

**2e.** $\sqrt{\dfrac{24}{25}} + \sqrt{\dfrac{54}{49}} = \dfrac{\sqrt{24}}{\sqrt{25}} + \dfrac{\sqrt{54}}{\sqrt{49}}$    Apply the quotient rule.

$\qquad = \dfrac{\sqrt{4 \cdot 6}}{5} + \dfrac{\sqrt{9 \cdot 6}}{7}$    Factor the radicand in each numerator and simplify $\sqrt{25}$ and $\sqrt{49}$.

$\qquad = \dfrac{\sqrt{4}\sqrt{6}}{5} + \dfrac{\sqrt{9}\sqrt{6}}{7}$    Apply the product rule.

$\qquad = \dfrac{2\sqrt{6}}{5} + \dfrac{3\sqrt{6}}{7}$    Simplify $\sqrt{4}$ and $\sqrt{9}$.

$\qquad = \dfrac{2\sqrt{6}}{5} \cdot \dfrac{7}{7} + \dfrac{3\sqrt{6}}{7} \cdot \dfrac{5}{5}$    Convert each fraction to an equivalent fraction with a denominator of 35.

$\qquad = \dfrac{14\sqrt{6}}{35} + \dfrac{15\sqrt{6}}{35}$    Simplify each product.

$\qquad = \dfrac{(14 + 15)\sqrt{6}}{35}$    Add the numerators by combining like the radicals.

$\qquad = \dfrac{29\sqrt{6}}{35}$    Simplify.

✓ **Student Check 2**   Perform each operation and express each answer in simplest radical form. Assume variables represent positive real numbers.

**a.** $\sqrt{24} + \sqrt{54} - \sqrt{6}$     **b.** $5\sqrt{40} - 8\sqrt{90}$     **c.** $\sqrt[3]{24} + 9\sqrt[3]{81}$

**d.** $\sqrt{36} - 7y\sqrt{16y} + 4\sqrt{y^3}$     **e.** $\sqrt{\dfrac{12}{121}} + \sqrt{\dfrac{27}{16}}$

**Objective 3** ▶

Troubleshoot common errors.

## Troubleshooting Common Errors

Some common errors associated with adding like radicals are shown next.

**Objective 3 Examples**  A problem and an incorrect solution are given. Provide the correct solution and an explanation of the error.

**3a.** Simplify $\sqrt{10} + \sqrt{10}$.

| Incorrect Solution | Correct Solution and Explanation |
|---|---|
| $\sqrt{10} + \sqrt{10} = \sqrt{20}$ <br> $= \sqrt{4 \cdot 5}$ <br> $= 2\sqrt{5}$ | When we add like radicals, we add the coefficients of the radicals and keep the radical the same. <br><br> $\sqrt{10} + \sqrt{10} = 1\sqrt{10} + 1\sqrt{10}$ <br> $= (1 + 1)\sqrt{10}$ <br> $= 2\sqrt{10}$ |

**3b.** Simplify $3 + 2\sqrt{18}$.

| Incorrect Solution | Correct Solution and Explanation |
|---|---|
| $3 + 2\sqrt{18} = 3 + 2\sqrt{9 \cdot 2}$ <br> $= 3 + 2(3)\sqrt{2}$ <br> $= 3 + 6\sqrt{2}$ <br> $= 9\sqrt{2}$ | The constant term, 3, is not a radical term. <br><br> $3 + 2\sqrt{18} = 3 + 2\sqrt{9 \cdot 2}$ <br> $= 3 + 2(3)\sqrt{2}$ <br> $= 3 + 6\sqrt{2}$ |

## ANSWERS TO STUDENT CHECKS

**Student Check 1**  **a.** $16\sqrt{2}$  **b.** $2\sqrt{10}$  **c.** $-4\sqrt{7}$
  **d.** $3\sqrt[3]{5}$  **e.** can't combine  **f.** $16\sqrt{7}$ ft

**Student Check 2**  **a.** $4\sqrt{6}$  **b.** $-14\sqrt{10}$  **c.** $29\sqrt[3]{3}$
  **d.** $6 - 24y\sqrt{y}$  **e.** $\dfrac{41\sqrt{3}}{44}$

## SUMMARY OF KEY CONCEPTS

1. Like radicals are radicals with the same index and same radicand.
2. Like radicals can be added or subtracted by adding or subtracting their coefficients and keeping the radical the same.
3. If the radicals are unlike, try simplifying them according to the methods shown in Section 8.2. After simplifying the radicals, combine like terms, if possible.

## GRAPHING CALCULATOR SKILLS

We can use the graphing calculator to verify that we have performed an operation correctly.

**Example:** Verify that $\sqrt{12} + 2\sqrt{48} - \sqrt{75} = 5\sqrt{3}$.

**Solution:**

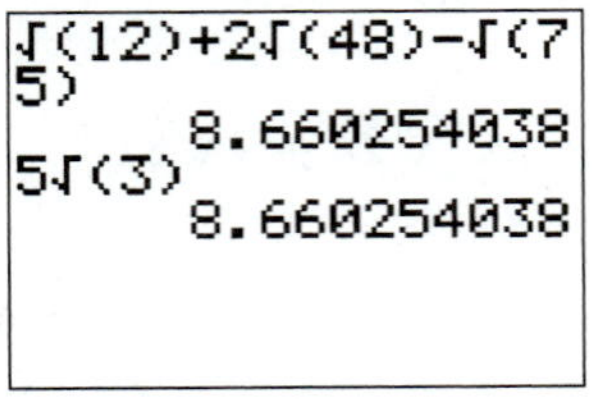

```
√(12)+2√(48)-√(7
5)
        8.660254038
5√(3)
        8.660254038
```

The two radical expressions have the same decimal value and are, therefore, equivalent.

## SECTION 8.3 / EXERCISE SET

### Write About It!

**Use complete sentences in your answer to each exercise.**

1. Can the sum of two square root expressions always be simplified? Explain.   No, if the radicands are different after simplifying each radical, they cannot be combined.
2. When are two radical expressions considered like terms?   Radical expressions are like terms when their indices and radicands are the same.
3. Explain how to add or subtract radical expressions.
4. Explain two conditions in which radical expressions cannot be simplified when added or subtracted.
   If the index of the radical expressions is different or if the radicands are different, the expressions cannot be combined.

### Practice Makes Perfect!

**Perform each operation and write each answer in simplest form. (See Objective 1.)**

5. $2\sqrt{3} + 4\sqrt{3}$   $6\sqrt{3}$
6. $5\sqrt{2} + 3\sqrt{2}$   $8\sqrt{2}$
7. $7\sqrt{5} - \sqrt{5}$   $6\sqrt{5}$
8. $6\sqrt{3} - 7\sqrt{3}$   $-\sqrt{3}$
9. $\sqrt{30} - 2\sqrt{30}$   $-\sqrt{30}$
10. $\sqrt{26} - 10\sqrt{26}$   $-9\sqrt{26}$
11. $4\sqrt[3]{2} + 8\sqrt[3]{2}$   $12\sqrt[3]{2}$
12. $10\sqrt[3]{4} + 2\sqrt[3]{4}$   $12\sqrt[3]{4}$
13. $2x\sqrt{3} + 5x\sqrt{3}$   $7x\sqrt{3}$
14. $6x\sqrt{5} + x\sqrt{5}$   $7x\sqrt{5}$
15. $x\sqrt{2} - 3x\sqrt{2}$   $-2x\sqrt{2}$
16. $6x\sqrt{6} - x\sqrt{6}$   $5x\sqrt{6}$
17. $\sqrt{14} - \sqrt{7}$   cannot be combined
18. $\sqrt{21} - \sqrt{3}$   cannot be combined
19. $3\sqrt{15} - 2\sqrt{15} + 3\sqrt{15}$   $4\sqrt{15}$
20. $12\sqrt{10} - \sqrt{10} + 6\sqrt{10}$   $17\sqrt{10}$
21. $4\sqrt{11} - 5\sqrt{11} + 2\sqrt{22}$   $-\sqrt{11} + 2\sqrt{22}$
22. $10\sqrt{5} - 12\sqrt{5} + 6\sqrt{10}$   $-2\sqrt{5} + 6\sqrt{10}$
23. Find the perimeter of the triangle.   $24\sqrt{3} + 12\sqrt{6}$ in.

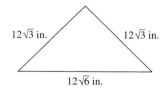

12√3 in.   12√3 in.

12√6 in.

24. Find the perimeter of the triangle.   $3\sqrt{5} + \sqrt{15}$ cm

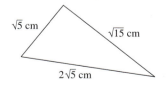

√5 cm   √15 cm

2√5 cm

**Perform each operation and write each answer in simplest form. (See Objective 2.)**

25. $3\sqrt{5} + \sqrt{20}$   $5\sqrt{5}$
26. $4\sqrt{6} + \sqrt{54}$   $7\sqrt{6}$
27. $2\sqrt{7} - \sqrt{28}$   $0$
28. $6\sqrt{11} - \sqrt{99}$   $3\sqrt{11}$

29. $12\sqrt{3} - 3\sqrt{4}$   $12\sqrt{3} - 6$
30. $6\sqrt{5} - 7\sqrt{16}$   $6\sqrt{5} - 28$
31. $4\sqrt{28} - 2\sqrt{175}$   $-2\sqrt{7}$
32. $\sqrt{117} - 3\sqrt{52}$   $-3\sqrt{13}$
33. $\sqrt[3]{32} + \sqrt[3]{16}$   $2\sqrt[3]{4} + 2\sqrt[3]{2}$
34. $\sqrt[3]{54} + \sqrt[3]{81}$   $3\sqrt[3]{2} + 3\sqrt[3]{3}$
35. $2\sqrt{18} + 5\sqrt{45}$   $6\sqrt{2} + 15\sqrt{5}$
36. $6\sqrt{45} - 3\sqrt{54}$   $18\sqrt{5} - 9\sqrt{6}$
37. $3\sqrt{12x^5} + 4x\sqrt{27x^3}$   $18x^2\sqrt{3x}$
38. $-\sqrt{180y^2} + 2\sqrt{80y^2}$   $2y\sqrt{5}$
39. $\sqrt{18a^3b} - \sqrt{12a^3b}$   $3a\sqrt{2ab} - 2a\sqrt{3ab}$
40. $\sqrt{21x^5y^2} - x\sqrt{28x^3y^2}$   $x^2y\sqrt{21x} - 2x^2y\sqrt{7x}$
41. Find the perimeter of the rectangle.

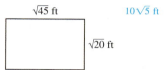

√45 ft   10√5 ft

√20 ft

42. Find the perimeter of the rectangle.

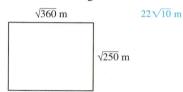

√360 m   22√10 m

√250 m

### Mix 'Em Up!

**Perform each operation and write each answer in simplest form.**

43. $\sqrt{5} + 9\sqrt{5}$   $10\sqrt{5}$
44. $6\sqrt{32} - 5\sqrt{18}$   $9\sqrt{2}$
45. $\sqrt{125} - 6\sqrt{5}$   $-\sqrt{5}$
46. $\sqrt{24y^2} + 3y\sqrt{6}$   $5y\sqrt{6}$
47. $4x\sqrt{3} - 6x\sqrt{3}$   $-2x\sqrt{3}$
48. $8\sqrt{17} + 3\sqrt{17}$   $11\sqrt{17}$
49. $2\sqrt{48} + 2\sqrt{27}$   $14\sqrt{3}$
50. $5\sqrt{18} - 6\sqrt{2}$   $9\sqrt{2}$
51. $10\sqrt[3]{x^2} - 6\sqrt[3]{x^2}$   $4\sqrt[3]{x^2}$
52. $3\sqrt{24} - 5\sqrt{6}$   $\sqrt{6}$
53. $\sqrt{44} - 3\sqrt{11}$   $-\sqrt{11}$
54. $2\sqrt[3]{128x} - 4\sqrt[3]{16x}$   $0$
55. $10x\sqrt{2} - 4\sqrt{18x^2}$   $-2x\sqrt{2}$
56. $\sqrt{75x^2} + \sqrt{12x^2}$   $7x\sqrt{3}$

### You Be the Teacher!

**Correct each student's errors, if any.**

57. Add $\sqrt{32} + \sqrt{18}$.

   Chris's work:

   $$\sqrt{32} + \sqrt{18} = \sqrt{50} = \sqrt{25 \cdot 2} = 5\sqrt{2}$$
   $\sqrt{32} + \sqrt{18} = \sqrt{16 \cdot 2} + \sqrt{9 \cdot 2} = 4\sqrt{2} + 3\sqrt{2} = 7\sqrt{2}$

58. Add $\sqrt{54} + \sqrt{96}$.

   Yolanda's work:

   $$\sqrt{54} + \sqrt{96} = \sqrt{150} = \sqrt{25 \cdot 6} = 5\sqrt{6}$$
   $\sqrt{54} + \sqrt{96} = \sqrt{9 \cdot 6} + \sqrt{16 \cdot 6} = 3\sqrt{6} + 4\sqrt{6} = 7\sqrt{6}$

Additional answers can be found in the Instructor Answer Appendix.

**59.** Subtract $\sqrt{12} - \sqrt{6}$.

Linda's work:

$$\sqrt{12} - \sqrt{6} = \sqrt{2 \cdot 6} - \sqrt{6} = 2\sqrt{6} - \sqrt{6} = \sqrt{6}$$

$\sqrt{12} - \sqrt{6} = \sqrt{2 \cdot 2 \cdot 3} - \sqrt{6} = 2\sqrt{3} - \sqrt{6}$

**60.** Subtract $3\sqrt{45} - 4\sqrt{80}$.

Seth's work:

$$3\sqrt{45} - 4\sqrt{80}$$
$$= 3\sqrt{9 \cdot 5} - 4\sqrt{16 \cdot 5}$$
$$= 3\sqrt{5} - 16\sqrt{5}$$
$$= -13\sqrt{5}$$

$3\sqrt{45} - 4\sqrt{80} = 3\sqrt{9 \cdot 5} - 4\sqrt{16 \cdot 5} = 3 \cdot 3\sqrt{5} - 16\sqrt{5} = -7\sqrt{5}$

## Calculate It!

Use a calculator to determine if each expression is simplified correctly. If the radical is not simplified correctly, provide the correct form.

**61.** $\sqrt{48} - 5\sqrt{12} = 6\sqrt{3}$    no, $-6\sqrt{3}$

**62.** $2\sqrt{63} + \sqrt{7} = 6\sqrt{7}$    no, $7\sqrt{7}$

**63.** $\sqrt{125} + 2\sqrt{45} = 11\sqrt{5}$    yes

**64.** $2\sqrt{112} + 3\sqrt{175} = 23\sqrt{7}$    yes

---

## PIECE IT TOGETHER    SECTIONS 8.1–8.3 Review

Use a calculator to evaluate each radical. Classify the number as rational or irrational and approximate its value to the nearest hundredth when applicable. (*Section 8.1, Objective 4*)

**1.** $\sqrt{48}$    irrational, 6.93
**2.** $\sqrt[4]{38{,}416}$    rational, 14

Simplify each radical completely. Assume variables represent positive real numbers. (*Section 8.1, Objective 5; Section 8.2, Objectives 1–4*)

**3.** $\sqrt{121x^6y^4}$    $11x^3y^2$
**4.** $\sqrt[3]{-125p^6q^3}$    $-5p^2q$

**5.** $\sqrt[3]{\dfrac{125r^6}{s^{12}}}$    $\dfrac{5r^2}{s^4}$
**6.** $\sqrt[3]{-\dfrac{8x^3}{27}}$    $-\dfrac{2x}{3}$

**7.** $5\sqrt{18}$    $15\sqrt{2}$
**8.** $\sqrt{y^{13}}$    $y^6\sqrt{y}$

**9.** $\sqrt{72x^3}$    $6x\sqrt{2x}$
**10.** $5\sqrt{120y^5}$    $10y^2\sqrt{30y}$

**11.** $\sqrt[3]{128}$    $4\sqrt[3]{2}$
**12.** $\sqrt[4]{1875z^{15}}$    $5z^3\sqrt[4]{3z^3}$

Find the distance between the points. Write each answer in simplest form. (*Section 8.2, Objective 5*)

**13.** $(12, -3\sqrt{2})$ and $(9, -6\sqrt{2})$    $3\sqrt{3}$

**14.** $\left(\dfrac{1}{6}, 1\right)$ and $\left(\dfrac{3}{2}, 2\right)$    $\dfrac{5}{3}$  **15.** $(0, 2)$ and $\left(\dfrac{1}{2}, \dfrac{8}{3}\right)$    $\dfrac{5}{6}$

**16.** $(0.7, -0.6)$ and $(6.7, 0.5)$    6.1

Perform the indicated operation and write each answer in simplest form. (*Section 8.3, Objectives 1 and 2*)

**17.** $3\sqrt{11} - 2\sqrt{11} + 3\sqrt{11}$    $4\sqrt{11}$
**18.** $6\sqrt{5} - 7\sqrt{16}$    $6\sqrt{5} - 28$
**19.** $3\sqrt{54} - 5\sqrt{6}$    $4\sqrt{6}$
**20.** $3\sqrt{28x^5} + 4x\sqrt{63x^3}$    $18x^2\sqrt{7x}$

---

## SECTION 8.4    Multiplying and Dividing Radical Expressions

### ▶ OBJECTIVES

As a result of completing this section, you will be able to

**1.** Multiply radical expressions.
**2.** Simplify expressions of the form $(\sqrt[n]{a})^n$, $a > 0$.
**3.** Use the distributive property to multiply radical expressions.
**4.** Divide radical expressions.
**5.** Rationalize the denominator.
**6.** Rationalize the denominator using conjugates.
**7.** Troubleshoot common errors.

### Objective 1 ▶

Multiply radical expressions.

Multiplying and dividing radical expressions is used when working with right triangles. This skill is also needed in Chapter 9 for solving quadratic equations with the quadratic formula. For now, consider how the area of this right triangle is calculated. Recall the area of a triangle is given by $A = \dfrac{1}{2}bh$. The area of this triangle is

$$A = \frac{1}{2}(3\sqrt{2})(3\sqrt{2})$$

In this section, we will learn how to simplify this product.

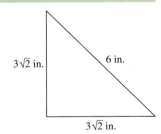

### Multiplying Radical Expressions

In Section 8.2, the product rule for radicals was introduced. The rule stated that, for $a$ and $b$ positive real numbers,

$$\sqrt[n]{a \cdot b} = \sqrt[n]{a} \cdot \sqrt[n]{b}$$

This property can also be used to multiply radicals.

**INSTRUCTOR NOTE:**
Point out that the indices are the same which is what enables us to multiply the expressions.

> **Property:** **Product Rule for Roots:** For $a$ and $b$ positive real numbers and $n = 2, 3, 4, \ldots$,
>
> $$\sqrt[n]{a} \cdot \sqrt[n]{b} = \sqrt[n]{a \cdot b}$$

> **Procedure:** **Multiplying Radical Expressions**
>
> **Step 1:** Form the $n$th root of the product of the radicands.
> **Step 2:** Simplify the product.
> **Step 3:** Simplify the radical, if necessary.

**Objective 1 Examples** Simplify each product and express each answer in simplest radical form. Assume all variables represent positive real numbers.

**1a.** $\sqrt{2} \cdot \sqrt{6}$ \qquad\qquad **1b.** $\left(3\sqrt{7x}\right)\left(2\sqrt{8x}\right)$

**1c.** Find the area of the right triangle.

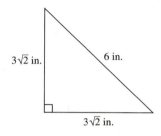

**1d.** $\sqrt{3xy^3} \cdot \sqrt{6xy}$ \qquad\qquad **1e.** $\sqrt[3]{4a^2} \cdot \sqrt[3]{2a}$

**Solutions** **1a.**

$$\sqrt{2} \cdot \sqrt{6} = \sqrt{2 \cdot 6}$$ \quad Apply the product rule to multiply the radicals.

$$= \sqrt{12}$$ \quad Simplify the radicand.

$$= \sqrt{4 \cdot 3}$$ \quad Factor 12 as $4 \cdot 3$.

$$= \sqrt{4} \cdot \sqrt{3}$$ \quad Apply the product rule to simplify the radical.

$$= 2\sqrt{3}$$ \quad Write $\sqrt{4}$ as 2.

**1b.** $\left(3\sqrt{7x}\right)\left(2\sqrt{8x}\right) = 3 \cdot 2\sqrt{7x}\sqrt{8x}$ \quad Apply the commutative property.

$$= 6\sqrt{7x \cdot 8x}$$ \quad Multiply the coefficients. Apply the product rule to multiply the radicals.

$$= 6\sqrt{56x^2}$$ \quad Simplify the radicand.

$$= 6\sqrt{4 \cdot 14x^2}$$ \quad Factor 56 as $4 \cdot 14$. Note that $x^2$ is a perfect square.

$$= 6\sqrt{4x^2} \cdot \sqrt{14}$$ \quad Apply the product rule to simplify the radical.

$$= 6(2x)\sqrt{14}$$ \quad Write $\sqrt{4x^2}$ as $2x$.

$$= 12x\sqrt{14}$$ \quad Multiply $6(2x)$.

**1c.**

$$A = \frac{1}{2}bh$$    State the area formula.

$$= \frac{1}{2}(3\sqrt{2})(3\sqrt{2})$$    Substitute $3\sqrt{2}$ for $b$ and $h$.

$$= \frac{1}{2}(3)(3)\sqrt{2} \cdot \sqrt{2}$$    Apply the commutative property.

$$= \frac{9}{2}\sqrt{2 \cdot 2}$$    Multiply the coefficients. Apply the product rule to multiply the radicals.

$$= \frac{9}{2}\sqrt{4}$$    Simplify the radicand.

$$= \frac{9}{2}(2)$$    Write $\sqrt{4}$ as 2.

$$= 9$$    Multiply $\frac{9}{2}(2)$.

So, the area of the triangle is 9 in.$^2$.

**1d.** $\sqrt{3xy^3} \cdot \sqrt{6xy} = \sqrt{(3xy^3)(6xy)}$    Apply the product rule to multiply the radicals.

$$= \sqrt{18x^2y^4}$$    Simplify the radicand.

$$= \sqrt{9 \cdot 2x^2(y^2)^2}$$    Write 18 as $9 \cdot 2$ and $y^4 = (y^2)^2$.

$$= \sqrt{9x^2(y^2)^2} \cdot \sqrt{2}$$    Apply the product rule to simplify the radical.

$$= 3xy^2\sqrt{2}$$    Write $\sqrt{9x^2(y^2)^2}$ as $3xy^2$.

**1e.** $\sqrt[3]{4a^2} \cdot \sqrt[3]{2a} = \sqrt[3]{(4a^2)(2a)}$    Apply the product rule to multiply the radicals.

$$= \sqrt[3]{8a^3}$$    Simplify the radicand.

$$= 2a$$    Write $\sqrt[3]{8a^3}$ as $2a$.

---

✓ **Student Check 1**    Simplify each product and express each answer in simplest radical form. Assume variables represent positive real numbers.

**a.** $\sqrt{8} \cdot \sqrt{3}$         **b.** $(2\sqrt{6})(5\sqrt{15})$

**c.** Find the area of the triangle with a base and a height of $4\sqrt{5}$ in.

**d.** $\sqrt{2x^3y} \cdot \sqrt{10xy}$         **e.** $\sqrt[3]{3b^4} \cdot \sqrt[3]{9b^2}$

---

## Simplifying Expressions of the Form $\left(\sqrt[n]{a}\right)^n$, $a > 0$

In Section 8.1, we learned that $\sqrt[n]{a^n} = a$, for $a > 0$. Now that we know how to multiply radicals, we will see that this expression is the same as $\left(\sqrt[n]{a}\right)^n$ for $a > 0$. Consider the following examples.

$$(\sqrt{5})^2 = \sqrt{5} \cdot \sqrt{5}$$    Apply the definition of the exponent.

$$= \sqrt{5 \cdot 5}$$    Apply the product rule.

$$= \sqrt{25}$$    Simplify the radicand.

$$= 5$$    Simplify $\sqrt{25}$.

$$(\sqrt[3]{2})^3 = \sqrt[3]{2} \cdot \sqrt[3]{2} \cdot \sqrt[3]{2}$$    Apply the definition of the exponent.

$$= \sqrt[3]{2 \cdot 2 \cdot 2}$$    Apply the product rule.

$$= \sqrt[3]{8}$$    Simplify the radicand.

$$= 2$$    Simplify $\sqrt[3]{8}$.

So, note that $(\sqrt{5})^2 = 5$ and $(\sqrt[3]{2})^3 = 2$.

This shows us that when the index and exponent of a radical expression are the same, the result is the radicand provided the radical expression is defined.

**Property:** For $a > 0$,
$$\left(\sqrt[n]{a}\right)^n = a \text{ or } \underbrace{\left(\sqrt[n]{a}\right)\left(\sqrt[n]{a}\right)\left(\sqrt[n]{a}\right)...\left(\sqrt[n]{a}\right)}_{n \text{ times}} = a$$

So, when we simplify expressions of the form $\left(\sqrt[n]{a}\right)^n$, for $a > 0$, note the following.

- The result is simply the radicand, $a$.
- The $n$th root and raising to an exponent of $n$ are reverse operations of one another.

That is, when both of the operations are performed on a positive number $a$, then it is as if nothing was done to the number $a$ since we arrive back at that value.

**Note:** *This property is fundamental to solving equations containing radicals that we will cover in Section 8.5.*

**Objective 2 Examples** **Simplify each expression. Assume all variable expressions are positive numbers.**

**2a.** $\left(\sqrt{10}\right)^2$    **2b.** $\left(\sqrt{3a}\right)^2$    **2c.** $\left(2\sqrt{5}\right)^2$    **2d.** $\left(\sqrt{x+2}\right)^2$    **2e.** $\left(\sqrt[3]{7x-1}\right)^3$

**Solutions** **2a.**    $\left(\sqrt{10}\right)^2 = \sqrt{10} \cdot \sqrt{10}$     Apply the definition of the exponent.

$= \sqrt{100}$     Apply the product rule.

$= 10$     Simplify.

**INSTRUCTORS NOTE:**
Remind students that if we don't state that the variable expressions are positive real numbers, $\left(\sqrt{x+2}\right)^2 = |x+2|$.

**2b.**    $\left(\sqrt{3a}\right)^2 = \sqrt{3a} \cdot \sqrt{3a}$     Apply the definition of the exponent.

$= \sqrt{9a^2}$     Apply the product rule.

$= 3a$     Simplify.

**2c.**    $\left(2\sqrt{5}\right)^2 = (2)^2\left(\sqrt{5}\right)^2$     Square each factor in the base.

$= 4(5)$     Simplify each exponential expression.

$= 20$     Multiply the resulting factors.

**2d.**   $\left(\sqrt{x+2}\right)^2 = x + 2$     Apply the property: $\left(\sqrt[n]{a}\right)^n = a$.

**2e.** $\left(\sqrt[3]{7x-1}\right)^3 = 7x - 1$     Apply the property: $\left(\sqrt[n]{a}\right)^n = a$.

✓ **Student Check 2**   Simplify each expression. Assume all variable expressions are positive real numbers.

    **a.** $\left(\sqrt{18}\right)^2$    **b.** $\left(\sqrt{6h}\right)^2$    **c.** $\left(7\sqrt{2}\right)^2$    **d.** $\left(\sqrt{y-5}\right)^2$    **e.** $\left(\sqrt[3]{2a+9}\right)^3$

## More on Multiplying Radical Expressions

**Objective 3** ▶

Use the distributive property to multiply radical expressions.

Now that we know how to multiply basic radical expressions, we will multiply radical expressions that require the use of the distributive property. This will include multiplying two binomials that involve radicals as well as some special cases of this type—squaring a binomial and multiplying conjugates. Recall conjugates are binomials of the form $a + b$ and $a - b$.

> **Procedure: Multiplying Radical Expressions Using the Distributive Property**
>
> **Step 1:** Apply the distributive property to multiply the radical expressions. If the product involves two binomials, use the FOIL method to multiply the expressions.
> **Step 2:** Simplify each resulting radical expression, if necessary.
> **Step 3:** Combine like terms.

**Objective 3 Examples** Simplify each product and express each answer in simplest radical form. Assume all variables represent positive real numbers.

**3a.** $\sqrt{6}(2 + \sqrt{3})$     **3b.** $(2\sqrt{3} + \sqrt{5})(3\sqrt{3} - 2\sqrt{5})$     **3c.** $(2 + \sqrt{7})^2$

**3d.** $(2 + \sqrt{7})(2 - \sqrt{7})$     **3e.** $(\sqrt[3]{16} + 2\sqrt[3]{x^2})(\sqrt[3]{4} + \sqrt[3]{x^2})$

**Solutions** **3a.**

$$\sqrt{6}(2 + \sqrt{3}) = \sqrt{6}(2) + \sqrt{6}(\sqrt{3}) \qquad \text{Apply the distributive property.}$$
$$= 2\sqrt{6} + \sqrt{6 \cdot 3} \qquad \text{Apply the product rule.}$$
$$= 2\sqrt{6} + \sqrt{18} \qquad \text{Simplify the radicand.}$$
$$= 2\sqrt{6} + \sqrt{9 \cdot 2} \qquad \text{Factor 18 as } 9 \cdot 2.$$
$$= 2\sqrt{6} + \sqrt{9}\sqrt{2} \qquad \text{Apply the product rule to simplify the radical.}$$
$$= 2\sqrt{6} + 3\sqrt{2} \qquad \text{Write } \sqrt{9} \text{ as 3.}$$

The result contains terms that are unlike radicals that cannot be combined further.

**3b.** Apply the distributive property. Then simplify each product and combine any like terms.

$$(2\sqrt{3} + \sqrt{5})(3\sqrt{3} - 2\sqrt{5}) = (2\sqrt{3})(3\sqrt{3}) - (2\sqrt{3})(2\sqrt{5})$$
$$+ (\sqrt{5})(3\sqrt{3}) - (\sqrt{5})(2\sqrt{5})$$
$$= 6\sqrt{9} - 4\sqrt{15} + 3\sqrt{15} - 2\sqrt{25}$$
$$= 6(3) - \sqrt{15} - 2(5)$$
$$= 18 - \sqrt{15} - 10$$
$$= 8 - \sqrt{15}$$

**3c.** A binomial can be squared in two ways. We can repeat the base twice and multiply the binomials or we can apply the rule for squaring a binomial. Recall $(a + b)^2 = a^2 + 2ab + b^2$.

**Method 1:**

$$(2 + \sqrt{7})^2 = (2 + \sqrt{7})(2 + \sqrt{7})$$
$$= 2(2) + 2\sqrt{7} + 2\sqrt{7} + \sqrt{7}\sqrt{7}$$
$$= 4 + 4\sqrt{7} + \sqrt{49}$$
$$= 4 + 4\sqrt{7} + 7$$
$$= 11 + 4\sqrt{7}$$

**Method 2:**

$$(2 + \sqrt{7})^2 = (2)^2 + 2(2)\sqrt{7} + (\sqrt{7})^2$$
$$= 4 + 4\sqrt{7} + 7$$
$$= 11 + 4\sqrt{7}$$

**3d.** This expression is the product of conjugates. We can multiply the binomials by applying the FOIL method or by using the product of conjugates property. Recall $(a + b)(a - b) = a^2 - b^2$.

**Method 1:**

$(2 + \sqrt{7})(2 - \sqrt{7})$

$= 2(2) - 2\sqrt{7} + \sqrt{7}(2) - \sqrt{7}(\sqrt{7})$

$= 4 - 2\sqrt{7} + 2\sqrt{7} - 7$

$= -3$

**Method 2:**

$(2 + \sqrt{7})(2 - \sqrt{7}) = (2)^2 - (\sqrt{7})^2$

$= 4 - 7$

$= -3$

 **Note:** *The product of conjugates doesn't contain a radical!*

**3e.**

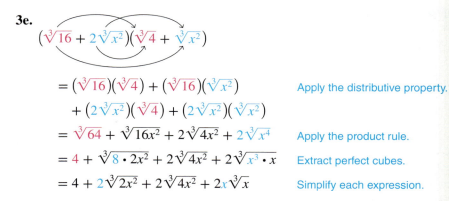

$$= (\sqrt[3]{16})(\sqrt[3]{4}) + (\sqrt[3]{16})(\sqrt[3]{x^2})$$ Apply the distributive property.

$$+ (2\sqrt[3]{x^2})(\sqrt[3]{4}) + (2\sqrt[3]{x^2})(\sqrt[3]{x^2})$$

$$= \sqrt[3]{64} + \sqrt[3]{16x^2} + 2\sqrt[3]{4x^2} + 2\sqrt[3]{x^4}$$ Apply the product rule.

$$= 4 + \sqrt[3]{8 \cdot 2x^2} + 2\sqrt[3]{4x^2} + 2\sqrt[3]{x^3 \cdot x}$$ Extract perfect cubes.

$$= 4 + 2\sqrt[3]{2x^2} + 2\sqrt[3]{4x^2} + 2x\sqrt[3]{x}$$ Simplify each expression.

None of the terms are like radicals, so the expression cannot be simplified any further.

✓ **Student Check 3**  Simplify each product and express each answer in simplest radical form. Assume all variables represent positive real numbers.

**a.** $\sqrt{3}(4 - \sqrt{15})$  **b.** $(7\sqrt{2} + \sqrt{6})(2\sqrt{2} - 3\sqrt{6})$  **c.** $(3 + \sqrt{5})^2$

**d.** $(3 + \sqrt{5})(3 - \sqrt{5})$  **e.** $(\sqrt[3]{4} + 3\sqrt[3]{x})(\sqrt[3]{6} + \sqrt[3]{x^2})$

## Dividing Radical Expressions

**Objective 4** ▶

Divide radical expressions.

In Section 8.2, the quotient rule for radicals was introduced. The rule states that, for $a$ and $b$ positive real numbers,

$$\sqrt[n]{\frac{a}{b}} = \frac{\sqrt[n]{a}}{\sqrt[n]{b}}$$

This property also provides a method for dividing radical expressions as shown next.

> **Property: Quotient Rule for Roots**
>
> For $a$ and $b$ positive real numbers,
>
> $$\frac{\sqrt[n]{a}}{\sqrt[n]{b}} = \sqrt[n]{\frac{a}{b}}$$

Dividing radical expressions enables us to remove the radical from the denominator when the radicand in the denominator evenly divides the radicand in the numerator. For instance,

$$\frac{\sqrt{8}}{\sqrt{2}} = \sqrt{\frac{8}{2}} = \sqrt{4} = 2$$

When the denominator's radicand does not divide evenly into the numerator's radicand, we have to apply another technique to remove the radical from the denominator. This is illustrated in Objective 5.

> **Procedure: Dividing Radical Expressions Using the Quotient Rule**
>
> **Step 1:** Rewrite the quotient as the $n$th root of the quotient of the radicands in the numerator and denominator.
> **Step 2:** Simplify the quotient.
> **Step 3:** Simplify the radical, if necessary.

**Objective 4 Examples** Simplify each quotient and express each answer in simplest radical form. Assume all variables represent positive real numbers.

**4a.** $\dfrac{\sqrt{15}}{\sqrt{5}}$ **4b.** $\dfrac{\sqrt{48y^3}}{\sqrt{3y}}$ **4c.** $\dfrac{\sqrt{21a^3b^7}}{\sqrt{3ab}}$ **4d.** $\dfrac{\sqrt[3]{54x^4}}{\sqrt[3]{2x}}$

**Solutions**

**INSTRUCTOR NOTE:**
Remind students of the quotient of like bases rule, $\dfrac{a^m}{a^n} = a^{m-n}$.

**4a.**
$$\frac{\sqrt{15}}{\sqrt{5}} = \sqrt{\frac{15}{5}} \qquad \text{Apply the quotient rule.}$$
$$= \sqrt{3} \qquad \text{Simplify the radicand.}$$

**4b.**
$$\frac{\sqrt{48y^3}}{\sqrt{3y}} = \sqrt{\frac{48y^3}{3y}} \qquad \text{Apply the quotient rule.}$$
$$= \sqrt{16y^2} \qquad \text{Simplify the radicand.}$$
$$= 4y \qquad \text{Write } \sqrt{16y^2} \text{ as } 4y.$$

**4c.**
$$\frac{\sqrt{21a^3b^7}}{\sqrt{3ab}} = \sqrt{\frac{21a^3b^7}{3ab}} \qquad \text{Apply the quotient rule.}$$
$$= \sqrt{7a^2b^6} \qquad \text{Simplify the radicand.}$$
$$= \sqrt{7a^2(b^3)^2} \qquad \text{Write } b^6 \text{ as } (b^3)^2.$$
$$= \sqrt{a^2(b^3)^2} \cdot \sqrt{7} \qquad \text{Apply the product rule.}$$
$$= ab^3\sqrt{7} \qquad \text{Write } \sqrt{a^2(b^3)^2} \text{ as } ab^3.$$

**4d.**
$$\frac{\sqrt[3]{54x^4}}{\sqrt[3]{2x}} = \sqrt[3]{\frac{54x^4}{2x}} \qquad \text{Apply the quotient rule.}$$
$$= \sqrt[3]{27x^3} \qquad \text{Simplify the radicand.}$$
$$= 3x \qquad \text{Write } \sqrt[3]{27x^3} \text{ as } 3x.$$

✔ **Student Check 4** Simplify each quotient and express each answer in simplest radical form. Assume all variables represent positive real numbers.

**a.** $\dfrac{\sqrt{35}}{\sqrt{7}}$ **b.** $\dfrac{\sqrt{50b^5}}{\sqrt{2b}}$ **c.** $\dfrac{\sqrt{30x^6y^5}}{\sqrt{10x^2y}}$ **d.** $\dfrac{\sqrt[3]{48a^7}}{\sqrt[3]{2a}}$

## Rationalizing the Denominator

**Objective 5** ▶

Rationalize the denominator.

Now we will examine quotients in which the radicand in the denominator does not divide evenly into the numerator. In this case, a radical remains in the denominator or a fraction remains within the radical. This is a violation of being in simplest radical form. For example,

$$\frac{\sqrt{2}}{\sqrt{3}} = \sqrt{\frac{2}{3}}$$

In this case, the quotient rule for radicals does not eliminate the radical from the denominator. So, we need another way to simplify this expression. To eliminate $\sqrt{3}$ from the denominator, we must multiply it by a radical expression that will make the radicand in the denominator a perfect square. If we multiply $\sqrt{3}$ by $\sqrt{3}$, we get $\sqrt{9}$ or 3. So that the value of $\dfrac{\sqrt{2}}{\sqrt{3}}$ does not change, we must multiply it by a form of 1, or $\dfrac{\sqrt{3}}{\sqrt{3}}$. Thus,

$$\frac{\sqrt{2}}{\sqrt{3}} = \frac{\sqrt{2}}{\sqrt{3}} \cdot \frac{\sqrt{3}}{\sqrt{3}} = \frac{\sqrt{6}}{\sqrt{9}} = \frac{\sqrt{6}}{3}$$

Note that the denominator of the original fraction, $\sqrt{3}$, is irrational. After we multiply the original fraction by a form of 1, its denominator becomes 3, a rational number. The value of this expression is much easier to calculate now that its denominator is rational.

The process of removing the radical from the denominator of a fraction is called **rationalizing the denominator**. To "rationalize" means to make rational. So, when we rationalize the denominator, we are going through a process that changes the irrational denominator into a rational denominator. Our goal then is to rewrite the original radical expression as an equivalent expression that does not contain a radical in the denominator.

We use the following facts to rationalize a denominator.

- Multiplying a fraction by a form of 1 does not change the value of the expression. Recall that

$$\frac{a}{b} \cdot 1 = \frac{a}{b} \cdot \frac{c}{c} = \frac{a \cdot c}{b \cdot c}$$

- $\sqrt[n]{a^n} = a$, provided $a$ is a positive real number.

---

**Procedure:  Rationalizing a Denominator**

**Step 1:** Simplify the radical in the denominator, if possible.

**Step 2:** Multiply the fraction by a form of 1 that makes the radicand in the denominator a perfect square if the index is 2, a perfect cube if the index is 3, and so on.

**Step 3:** Simplify the product by multiplying the numerators together and the denominators together.

**Step 4:** Simplify the resulting radical expressions.

---

**Note:** *The denominator should no longer contain a radical expression after performing the steps to rationalize it.*

---

**Objective 5  Examples**   Rationalize the denominator in each expression. Assume all variables represent positive real numbers.

**5a.** $\dfrac{1}{\sqrt{2}}$     **5b.** $\dfrac{\sqrt{7}}{\sqrt{5y}}$     **5c.** $\dfrac{\sqrt{5}}{\sqrt{48}}$     **5d.** $\dfrac{3}{\sqrt[3]{2}}$

**Solutions**   **5a.** $\dfrac{1}{\sqrt{2}} = \dfrac{1}{\sqrt{2}} \cdot \dfrac{\sqrt{2}}{\sqrt{2}}$     Multiply the given expression by $\dfrac{\sqrt{2}}{\sqrt{2}}$.

$= \dfrac{1 \cdot \sqrt{2}}{\sqrt{2} \cdot \sqrt{2}}$     Multiply the numerator and denominator by $\sqrt{2}$.

$= \dfrac{\sqrt{2}}{\sqrt{4}}$     Apply the product rule for radicals.

$= \dfrac{\sqrt{2}}{2}$     Write $\sqrt{4}$ as 2.

**5b.** $\dfrac{\sqrt{7}}{\sqrt{5y}} = \dfrac{\sqrt{7}}{\sqrt{5y}} \cdot \dfrac{\sqrt{5y}}{\sqrt{5y}}$   Multiply the given expression by $\dfrac{\sqrt{5y}}{\sqrt{5y}}$.

$\phantom{\dfrac{\sqrt{7}}{\sqrt{5y}}} = \dfrac{\sqrt{7} \cdot \sqrt{5y}}{\sqrt{5y} \cdot \sqrt{5y}}$   Multiply the numerator and denominator by $\sqrt{5y}$.

$\phantom{\dfrac{\sqrt{7}}{\sqrt{5y}}} = \dfrac{\sqrt{35y}}{\sqrt{25y^2}}$   Apply the product rule for radicals.

$\phantom{\dfrac{\sqrt{7}}{\sqrt{5y}}} = \dfrac{\sqrt{35y}}{5y}$   Write $\sqrt{25y^2}$ as $5y$.

**5c.** $\dfrac{\sqrt{5}}{\sqrt{48}} = \dfrac{\sqrt{5}}{\sqrt{16 \cdot 3}}$   Simplify the radical in the denominator.

$\phantom{\dfrac{\sqrt{5}}{\sqrt{48}}} = \dfrac{\sqrt{5}}{4\sqrt{3}} \cdot \dfrac{\sqrt{3}}{\sqrt{3}}$   Multiply by $\dfrac{\sqrt{3}}{\sqrt{3}}$.

$\phantom{\dfrac{\sqrt{5}}{\sqrt{48}}} = \dfrac{\sqrt{5} \cdot \sqrt{3}}{4\sqrt{3} \cdot \sqrt{3}}$   Multiply the numerator and denominator by $\sqrt{3}$.

$\phantom{\dfrac{\sqrt{5}}{\sqrt{48}}} = \dfrac{\sqrt{15}}{4\sqrt{9}}$   Apply the product rule for radicals.

$\phantom{\dfrac{\sqrt{5}}{\sqrt{48}}} = \dfrac{\sqrt{15}}{4(3)}$   Write $\sqrt{9}$ as 3.

$\phantom{\dfrac{\sqrt{5}}{\sqrt{48}}} = \dfrac{\sqrt{15}}{12}$   Simplify the denominator.

**5d.** We need to multiply the denominator by a radical expression that will make its radicand a perfect cube. We know that $2^3 = 8$ is a perfect cube. So, if we multiply 2 by 4, we get 8. Therefore, we need to multiply the fraction by 1 in the form of $\dfrac{\sqrt[3]{4}}{\sqrt[3]{4}}$.

$\dfrac{3}{\sqrt[3]{2}} = \dfrac{3}{\sqrt[3]{2}} \cdot \dfrac{\sqrt[3]{4}}{\sqrt[3]{4}}$   Multiply the given expression by $\dfrac{\sqrt[3]{4}}{\sqrt[3]{4}}$.

$\phantom{\dfrac{3}{\sqrt[3]{2}}} = \dfrac{3 \cdot \sqrt[3]{4}}{\sqrt[3]{2} \cdot \sqrt[3]{4}}$   Multiply the numerator and denominator by $\sqrt[3]{4}$.

$\phantom{\dfrac{3}{\sqrt[3]{2}}} = \dfrac{3\sqrt[3]{4}}{\sqrt[3]{8}}$   Apply the product rule for radicals.

$\phantom{\dfrac{3}{\sqrt[3]{2}}} = \dfrac{3\sqrt[3]{4}}{2}$   Write $\sqrt[3]{8}$ as 2.

✓ **Student Check 5**   Rationalize the denominator in each expression. Assume all variables represent positive real numbers.

**a.** $\dfrac{1}{\sqrt{3}}$   **b.** $\dfrac{\sqrt{11}}{\sqrt{6x}}$   **c.** $\dfrac{\sqrt{3}}{\sqrt{50}}$   **d.** $\dfrac{4}{\sqrt[3]{9}}$

## Rationalizing the Denominator Using Conjugates

If the denominator of a fraction contains a sum or difference of square roots, then the factor needed to rationalize the denominator involves the conjugate of the denominator. Recall from Example 3d that the product of conjugates does not contain a radical. For example, the expression $\dfrac{2}{5 + \sqrt{7}}$ has a denominator that contains a sum. In order to rationalize the denominator, we multiply the numerator and denominator by $5 - \sqrt{7}$, which is the conjugate of $5 + \sqrt{7}$.

$$\frac{3}{5 + \sqrt{7}} = \frac{3}{5 + \sqrt{7}} \cdot \frac{5 - \sqrt{7}}{5 - \sqrt{7}}$$   Multiply the original expression by $\dfrac{5 - \sqrt{7}}{5 - \sqrt{7}}$.

$$= \frac{3(5 - \sqrt{7})}{(5 + \sqrt{7})(5 - \sqrt{7})}$$   Multiply the numerator and denominator by $5 - \sqrt{7}$.

$$= \frac{15 - 3\sqrt{7}}{25 - 5\sqrt{7} + 5\sqrt{7} - \sqrt{49}}$$   Apply the distributive property in the numerator and denominator.

$$= \frac{15 - 3\sqrt{7}}{25 - 7}$$   Simplify the denominator.

$$= \frac{3(5 - \sqrt{7})}{18}$$   Factor out the GCF of 3 from the numerator and combine like terms in the denominator.

$$= \frac{5 - \sqrt{7}}{6}$$   Simplify the quotient by dividing the numerator and denominator by 3.

**Note:** *Multiplying the numerator and denominator by the conjugate of the denominator produces an equivalent expression that does not contain a radical in the denominator.*

Recall the following important facts about conjugates.

- $a + b$ and $a - b$ are conjugates.
- $(a + b)(a - b) = a^2 - b^2$

**Procedure:  Rationalizing a Denominator Using Conjugates**

**Step 1:** Multiply the given fraction by a form of 1, which is the conjugate of the denominator divided by itself.

**Step 2:** Simplify the product in the numerator and denominator by applying the distributive property.

**Step 3:** Simplify the resulting expressions in the numerator and denominator.

**Objective 6  Examples**   **Rationalize the denominator and express each answer in simplest radical form.**

**6a.** $\dfrac{6}{\sqrt{2} + \sqrt{3}}$       **6b.** $\dfrac{3 - \sqrt{2}}{7 + \sqrt{2}}$

**Solutions**     **6a.** The conjugate of the denominator is $\sqrt{2} - \sqrt{3}$. So, we multiply the fraction by 1 in the form of $\dfrac{\sqrt{2} - \sqrt{3}}{\sqrt{2} - \sqrt{3}}$.

$$\dfrac{6}{\sqrt{2} + \sqrt{3}} = \dfrac{6}{\sqrt{2} + \sqrt{3}} \cdot \dfrac{\sqrt{2} - \sqrt{3}}{\sqrt{2} - \sqrt{3}} \qquad \text{Multiply the given fraction by } \dfrac{\sqrt{2} - \sqrt{3}}{\sqrt{2} - \sqrt{3}}.$$

$$= \dfrac{6(\sqrt{2} - \sqrt{3})}{(\sqrt{2} + \sqrt{3})(\sqrt{2} - \sqrt{3})} \qquad \text{Multiply the numerator and denominator by } \sqrt{2} - \sqrt{3}.$$

$$= \dfrac{6\sqrt{2} - 6\sqrt{3}}{(\sqrt{2})^2 - (\sqrt{3})^2} \qquad \text{Simplify each product.}$$

$$= \dfrac{6\sqrt{2} - 6\sqrt{3}}{2 - 3} \qquad \text{Simplify the denominator.}$$

$$= \dfrac{6\sqrt{2} - 6\sqrt{3}}{-1} \qquad \text{Combine the terms in the denominator.}$$

$$= -6\sqrt{2} + 6\sqrt{3} \qquad \text{Divide each term in the numerator by } -1.$$

**6b.** The conjugate of the denominator is $7 - \sqrt{2}$. So, we multiply the fraction by 1 in the form of $\dfrac{7 - \sqrt{2}}{7 - \sqrt{2}}$.

$$\dfrac{3 - \sqrt{2}}{7 + \sqrt{2}} = \dfrac{3 - \sqrt{2}}{7 + \sqrt{2}} \cdot \dfrac{7 - \sqrt{2}}{7 - \sqrt{2}} \qquad \text{Multiply the given fraction by } \dfrac{7 - \sqrt{2}}{7 - \sqrt{2}}.$$

$$= \dfrac{(3 - \sqrt{2})(7 - \sqrt{2})}{(7 + \sqrt{2})(7 - \sqrt{2})} \qquad \text{Multiply the numerator and denominator by } 7 - \sqrt{2}.$$

$$= \dfrac{21 - 3\sqrt{2} - 7\sqrt{2} - \sqrt{4}}{(7)^2 - (\sqrt{2})^2} \qquad \text{Simplify each product.}$$

$$= \dfrac{21 - 10\sqrt{2} - 2}{49 - 2} \qquad \text{Combine like terms in the numerator and simplify } \sqrt{4}. \text{ Simplify the denominator.}$$

$$= \dfrac{19 - 10\sqrt{2}}{47} \qquad \text{Combine like terms in the numerator and denominator.}$$

✔ **Student Check 6**     Rationalize the denominator and express each answer in simplest radical form.

**a.** $\dfrac{8}{\sqrt{6} - \sqrt{2}}$     **b.** $\dfrac{5 + \sqrt{6}}{4 + \sqrt{6}}$

**Objective 7** ▶
Troubleshoot common errors.

## Troubleshooting Common Errors

Some common errors associated with multiplying and dividing radical expressions are shown next.

**Objective 7 Examples**   A problem and an incorrect solution are given. Provide the correct solution and an explanation of the error.

**7a.** Simplify $(4\sqrt{3})^2$.

| Incorrect Solution | Correct Solution and Explanation |
|---|---|
| $(4\sqrt{3})^2 = 4(3) = 12$ | The square was not applied to the coefficient of 4. Recall $(ab)^n = a^n b^n$.<br><br>$(4\sqrt{3})^2 = (4)^2(\sqrt{3})^2 = 16(3) = 48$ |

**7b.** Simplify $(5 + \sqrt{6})^2$.

| Incorrect Solution | Correct Solution and Explanation |
|---|---|
| $(5 + \sqrt{6})^2 = 25 + 6 = 31$ | This is a binomial squared, so we cannot just square each term. We multiply the two binomials using the FOIL method or by using the squaring pattern.<br><br>$(5 + \sqrt{6})^2 = (5 + \sqrt{6})(5 + \sqrt{6})$<br>$= 25 + 5\sqrt{6} + 5\sqrt{6} + (\sqrt{6})^2$<br>$= 25 + 10\sqrt{6} + 6$<br>$= 31 + 10\sqrt{6}$ |

## ANSWERS TO STUDENT CHECKS

**Student Check 1**   **a.** $2\sqrt{6}$   **b.** $30\sqrt{10}$   **c.** 40 in.$^2$
   **d.** $2x^2y\sqrt{5}$   **e.** $3b^2$

**Student Check 2**   **a.** 18   **b.** $6h$   **c.** 98   **d.** $y - 5$
   **e.** $2a + 9$

**Student Check 3**   **a.** $4\sqrt{3} - 3\sqrt{5}$   **b.** $10 - 38\sqrt{3}$
   **c.** $14 + 6\sqrt{5}$   **d.** 4   **e.** $2\sqrt[3]{3} + \sqrt[3]{4x^2} + 3\sqrt[3]{6x} + 3x$

**Student Check 4**   **a.** $\sqrt{5}$   **b.** $5b^2$   **c.** $x^2y^2\sqrt{3}$
   **d.** $2a^2\sqrt[3]{3}$

**Student Check 5**   **a.** $\dfrac{\sqrt{3}}{3}$   **b.** $\dfrac{\sqrt{66x}}{6x}$   **c.** $\dfrac{\sqrt{6}}{10}$
   **d.** $\dfrac{4\sqrt[3]{3}}{3}$

**Student Check 6**   **a.** $2(\sqrt{6} + \sqrt{2})$ or $2\sqrt{6} + 2\sqrt{2}$
   **b.** $\dfrac{14 - \sqrt{6}}{10}$

## SUMMARY OF KEY CONCEPTS

1. To multiply radical expressions, multiply the coefficients if possible, multiply the radicands, and then find the appropriate root of the result.

2. When an $n$th root of an expression is raised to the exponent $n$, $(\sqrt[n]{a})^n$, where $a > 0$, the result is the radicand $a$.

3. We can apply the distributive property to multiply radicals.

4. The special products of squaring a binomial and multiplying conjugates can be used for the appropriate expressions involving radicals.

5. Divide radical expressions by dividing the radicands and then find the appropriate root of the quotient.

6. If a fraction has a radical in the denominator, we must rationalize it. If there is a single term in the denominator, multiply the fraction by a form of 1 that makes the radicand in the denominator a perfect square if the index is 2, a perfect cube if the index is 3, and so on. If the denominator has a binomial, multiply the fraction by a form of 1 that is the conjugate of the denominator divided by itself.

## GRAPHING CALCULATOR SKILLS

We can use a graphing calculator to check our answers. Enter the given expression and the result to determine if the two values are the same. If both values are the same, then the expressions are equivalent.

**Examples:** Show that the expressions are equivalent.

**a.** $(2 + \sqrt{7})(2 - \sqrt{7}) = -3$

**b.** $(2 + \sqrt{7})^2 = 11 + 4\sqrt{7}$

**c.** $\dfrac{1}{\sqrt{2}} = \dfrac{\sqrt{2}}{2}$

**Solution:**

```
(2+√(7))(2-√(7))
                -3
```

**Solution:**

```
(2+√(7))²
        21.58300524
11+4√(7)
        21.58300524
```

**Solution:**

```
1/√(2)
        .7071067812
√(2)/2
        .7071067812
```

So, each of the three expressions has been simplified correctly.

---

## SECTION 8.4  EXERCISE SET

### Write About It!

Use complete sentences in your answer to each exercise.

1. Explain the process for multiplying two square root expressions.

2. Why does multiplying a square root expression by itself remove the radical? Assume the radicand is a positive real number. *A square root expression multiplied by itself produces an expression of the form $(\sqrt{x})^2$ which simplifies to $x$.*

3. Explain why the product of conjugates involving square roots does not contain any radicals.

4. Explain the process for dividing square root expressions.

5. State what the fraction $\dfrac{3}{\sqrt{3}}$ must be multiplied by to rationalize the denominator. Why does this work?

6. Explain how to rationalize the denominator of the radical expression $\dfrac{6}{\sqrt{5} + \sqrt{3}}$.

### Practice Makes Perfect!

Simplify each product and express each answer in simplest radical form. Assume all variables represent positive real numbers. (*See Objective 1.*)

7. $\sqrt{3} \cdot \sqrt{10}$  $\sqrt{30}$

8. $\sqrt{5} \cdot \sqrt{2}$  $\sqrt{10}$

9. $\sqrt{8} \cdot \sqrt{3}$  $2\sqrt{6}$

10. $\sqrt{12} \cdot \sqrt{5}$  $2\sqrt{15}$

11. $(2\sqrt{7})(4\sqrt{6})$  $8\sqrt{42}$

12. $(9\sqrt{8})(3\sqrt{2})$  $108$

13. $(6\sqrt{15})(\sqrt{5})$  $30\sqrt{3}$

14. $(8\sqrt{11})(\sqrt{8})$  $16\sqrt{22}$

15. $\sqrt{2y} \cdot \sqrt{8y}$  $4y$

16. $\sqrt{2x} \cdot \sqrt{32x}$  $8x$

17. $\sqrt{10a} \cdot \sqrt{5a^3}$  $5a^2\sqrt{2}$

18. $\sqrt{12h^5} \cdot \sqrt{6h}$  $6h^3\sqrt{2}$

19. $\sqrt{5x^5} \cdot \sqrt{60xy^2}$  $10x^3y\sqrt{3}$

20. $\sqrt{6cd^3} \cdot \sqrt{15c^2d}$  $3cd^2\sqrt{10c}$

Additional answers can be found in the Instructor Answer Appendix.

21. $\sqrt[3]{4ab} \cdot \sqrt[3]{18a^2b^7}$  $2ab^2\sqrt[3]{9b^2}$

22. $\sqrt[3]{25xy} \cdot \sqrt[3]{10x^2y^4}$

23. $\sqrt[3]{18rt} \cdot \sqrt[3]{60rt^2}$  $6t\sqrt[3]{5r^2}$

24. $\sqrt[3]{32u^4v} \cdot \sqrt[3]{6u^2v}$

**Find the area of each figure.**

25.

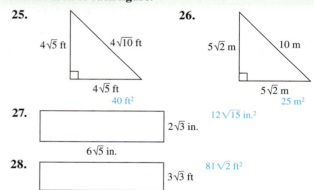

$4\sqrt{5}$ ft    $4\sqrt{10}$ ft

$4\sqrt{5}$ ft

$40$ ft$^2$

26.

$5\sqrt{2}$ m    $10$ m

$5\sqrt{2}$ m

$25$ m$^2$

27.

$2\sqrt{3}$ in.

$6\sqrt{5}$ in.

$12\sqrt{15}$ in.$^2$

28.

$3\sqrt{3}$ ft

$9\sqrt{6}$ ft

$81\sqrt{2}$ ft$^2$

**Simplify each expression. Assume all radicands are positive real numbers. (*See Objective 2.*)**

29. $\sqrt{6} \cdot \sqrt{6}$  $6$

30. $\sqrt{9} \cdot \sqrt{9}$  $9$

31. $(\sqrt{25})^2$  $25$

32. $(\sqrt{36})^2$  $36$

33. $(\sqrt{7})^2$  $7$

34. $(\sqrt{8})^2$  $8$

35. $(2\sqrt{3})^2$  $12$

36. $(5\sqrt{2})^2$  $50$

37. $(\sqrt{6x})^2$  $6x$

38. $(\sqrt{10y})^2$  $10y$

39. $(\sqrt{x-4})^2$  $x-4$

40. $(\sqrt{y+3})^2$  $y+3$

41. $(\sqrt{x^2 + 2x - 1})^2$  $x^2 + 2x - 1$

42. $(\sqrt{x^2 - 5x + 3})^2$  $x^2 - 5x + 3$

43. $(\sqrt[3]{4x-1})^3$  $4x-1$

44. $(\sqrt[3]{5x+3})^3$  $5x+3$

**Simplify each product and express each answer in simplest radical form. Assume all variables represent positive real numbers. (*See Objective 3.*)**

45. $2(1 + \sqrt{5})$  $2 + 2\sqrt{5}$

46. $6(2 + \sqrt{11})$  $12 + 6\sqrt{11}$

47. $3(\sqrt{7} + \sqrt{2})$  $3\sqrt{7} + 3\sqrt{2}$

48. $5(\sqrt{3} + \sqrt{10})$

49. $\sqrt{2}(\sqrt{5} + \sqrt{3})$  $\sqrt{10} + \sqrt{6}$

50. $\sqrt{3}(\sqrt{11} + \sqrt{7})$

**51.** $\sqrt{10}(\sqrt{2} + \sqrt{5})$    $2\sqrt{5} + 5\sqrt{2}$    **52.** $\sqrt{12}(\sqrt{6} + \sqrt{7})$    $6\sqrt{2} + 2\sqrt{21}$

**53.** $(9 + \sqrt{7})(2 + \sqrt{7})$    $25 + 11\sqrt{7}$    **54.** $(4 - \sqrt{5})(3 - \sqrt{5})$    $17 - 7\sqrt{5}$

**55.** $(6\sqrt{2} + 4)(\sqrt{2} + 8)$    $44 + 52\sqrt{2}$    **56.** $(3\sqrt{6} - 1)(\sqrt{6} + 10)$    $8 + 29\sqrt{6}$

**57.** $(7\sqrt{3} + \sqrt{5})(2\sqrt{3} + 4\sqrt{5})$    $62 + 30\sqrt{15}$

**58.** $(2\sqrt{11} + \sqrt{7})(3\sqrt{11} - 6\sqrt{7})$    $24 - 9\sqrt{77}$

**59.** $(1 + \sqrt{3})^2$    $4 + 2\sqrt{3}$    **60.** $(2 + \sqrt{5})^2$    $9 + 4\sqrt{5}$

**61.** $(5 - \sqrt{6})^2$    $31 - 10\sqrt{6}$    **62.** $(8 - \sqrt{2})^2$    $66 - 16\sqrt{2}$

**63.** $(\sqrt{5} + \sqrt{7})^2$    $12 + 2\sqrt{35}$    **64.** $(\sqrt{8} - \sqrt{6})^2$    $14 - 8\sqrt{3}$

**65.** $(3\sqrt{7} + 2)^2$    $67 + 12\sqrt{7}$    **66.** $(4\sqrt{10} + 8)^2$    $224 + 64\sqrt{10}$

**67.** $(\sqrt{x} + 3)^2$    $x + 9 + 6\sqrt{x}$    **68.** $(\sqrt{y} + 4)^2$    $y + 16 + 8\sqrt{y}$

**69.** $(1 + \sqrt{3})(1 - \sqrt{3})$    $-2$

**70.** $(2 + \sqrt{5})(2 - \sqrt{5})$    $-1$

**71.** $(\sqrt{2} + \sqrt{3})(\sqrt{2} - \sqrt{3})$    $-1$

**72.** $(\sqrt{5} + \sqrt{7})(\sqrt{5} - \sqrt{7})$    $-2$

**73.** $(\sqrt{8} - \sqrt{6})(\sqrt{8} + \sqrt{6})$    $2$

**74.** $(\sqrt{11} - \sqrt{3})(\sqrt{11} + \sqrt{3})$    $8$

**75.** $(3\sqrt{7} + 2)(3\sqrt{7} - 2)$    $59$

**76.** $(4\sqrt{10} + 8)(4\sqrt{10} - 8)$    $96$

**77.** $(\sqrt{x} + 3)(\sqrt{x} - 3)$    $x - 9$

**78.** $(\sqrt{y} + 4)(\sqrt{y} - 4)$    $y - 16$

**79.** $(\sqrt[3]{6} + \sqrt[3]{y^2})(\sqrt[3]{9} - \sqrt[3]{y})$    $3\sqrt[3]{2} + \sqrt[3]{9y^2} - \sqrt[3]{6y} - y$

**80.** $(\sqrt[3]{25} - \sqrt[3]{2a^2})(\sqrt[3]{5} - \sqrt[3]{4a})$    $5 - \sqrt[3]{10a^2} - \sqrt[3]{100a} + 2a$

**81.** $(\sqrt[3]{12} + \sqrt[3]{3b})(\sqrt[3]{2} - \sqrt[3]{9b^2})$    $2\sqrt[3]{3} + \sqrt[3]{6b} - 3\sqrt[3]{4b^2} - 3b$

**82.** $(\sqrt[3]{9x^2} - \sqrt[3]{4})(\sqrt[3]{6x} - \sqrt[3]{10})$    $3x\sqrt[3]{2} - 2\sqrt[3]{3x} - \sqrt[3]{90x^2} + 2\sqrt[3]{5}$

**Simplify each quotient and express each answer in simplest radical form. Assume all variables represent positive real numbers. (See Objective 4.)**

**83.** $\dfrac{\sqrt{6}}{\sqrt{2}}$    $\sqrt{3}$    **84.** $\dfrac{\sqrt{15}}{\sqrt{3}}$    $\sqrt{5}$    **85.** $\dfrac{\sqrt{12}}{\sqrt{3}}$    $2$

**86.** $\dfrac{\sqrt{18}}{\sqrt{2}}$    $3$    **87.** $\dfrac{\sqrt{54}}{\sqrt{3}}$    $3\sqrt{2}$    **88.** $\dfrac{\sqrt{72}}{\sqrt{6}}$    $2\sqrt{3}$

**89.** $\dfrac{\sqrt{75a^2}}{\sqrt{5a}}$    $\sqrt{15a}$    **90.** $\dfrac{\sqrt{84b^2}}{\sqrt{12b}}$    $\sqrt{7b}$    **91.** $\dfrac{\sqrt{200x^3}}{\sqrt{2x}}$    $10x$

**92.** $\dfrac{\sqrt{27y^3}}{\sqrt{3y}}$    $3y$    **93.** $\dfrac{\sqrt[3]{80x^5}}{\sqrt[3]{10x}}$    $2x\sqrt[3]{x}$    **94.** $\dfrac{\sqrt[3]{96a^5}}{\sqrt[3]{3a^2}}$    $2a\sqrt[3]{4}$

**Rationalize each denominator. Assume all variables represent positive real numbers. (See Objective 5.)**

**95.** $\dfrac{2}{\sqrt{2}}$    $\sqrt{2}$    **96.** $\dfrac{5}{\sqrt{5}}$    $\sqrt{5}$    **97.** $\dfrac{\sqrt{7}}{\sqrt{6}}$    $\dfrac{\sqrt{42}}{6}$

**98.** $\dfrac{\sqrt{3}}{\sqrt{2}}$    $\dfrac{\sqrt{6}}{2}$    **99.** $\dfrac{21}{\sqrt[3]{49}}$    $3\sqrt[3]{7}$    **100.** $\dfrac{2}{\sqrt[3]{18}}$    $\dfrac{\sqrt[3]{12}}{3}$

**101.** $\dfrac{3}{\sqrt{5x}}$    $\dfrac{3\sqrt{5x}}{5x}$    **102.** $\dfrac{9}{\sqrt{3y}}$    $\dfrac{3\sqrt{3y}}{y}$    **103.** $\dfrac{\sqrt{8}}{\sqrt{7x}}$    $\dfrac{2\sqrt{14x}}{7x}$

**104.** $\dfrac{\sqrt{10}}{\sqrt{6y}}$    $\dfrac{\sqrt{15y}}{3y}$

**Rationalize each denominator. (See Objective 6.)**

**105.** $\dfrac{1}{2 - \sqrt{5}}$    $-2 - \sqrt{5}$    **106.** $\dfrac{1}{3 - \sqrt{2}}$    $\dfrac{3 + \sqrt{2}}{7}$

**107.** $\dfrac{7}{9 + \sqrt{3}}$    $\dfrac{63 - 7\sqrt{3}}{78}$    **108.** $\dfrac{4}{8 + \sqrt{6}}$    $\dfrac{16 - 2\sqrt{6}}{29}$

**109.** $\dfrac{10}{\sqrt{6} + \sqrt{7}}$    $10\sqrt{7} - 10\sqrt{6}$    **110.** $\dfrac{6}{\sqrt{2} + \sqrt{10}}$    $\dfrac{3\sqrt{10} - 3\sqrt{2}}{4}$

**111.** $\dfrac{3 + \sqrt{5}}{2 - \sqrt{5}}$    $-11 - 5\sqrt{5}$    **112.** $\dfrac{6 - \sqrt{3}}{4 + \sqrt{3}}$    $\dfrac{27 - 10\sqrt{3}}{13}$

**113.** $\dfrac{9 - \sqrt{7}}{1 + \sqrt{7}}$    $\dfrac{-8 + 5\sqrt{7}}{3}$    **114.** $\dfrac{2 - \sqrt{11}}{8 - \sqrt{11}}$    $\dfrac{5 - 6\sqrt{11}}{53}$

 **Mix 'Em Up!**

**Perform the indicated operation and write each answer in simplest form. Rationalize denominators when necessary. Assume all radicands represent positive real numbers.**

**115.** $\sqrt{10x} \cdot \sqrt{2x}$    $2x\sqrt{5}$    **116.** $\sqrt{3y} \cdot \sqrt{6y}$    $3y\sqrt{2}$

**117.** $(9 + \sqrt{3})^2$    $84 + 18\sqrt{3}$    **118.** $(3 - \sqrt{7})^2$    $16 - 6\sqrt{7}$

**119.** $\dfrac{\sqrt{14y}}{\sqrt{7y}}$    $\sqrt{2}$    **120.** $\dfrac{\sqrt{90x}}{\sqrt{5x}}$    $3\sqrt{2}$

**121.** $\dfrac{8}{\sqrt[3]{54}}$    $\dfrac{4\sqrt[3]{4}}{3}$    **122.** $\dfrac{20}{\sqrt[3]{75}}$    $\dfrac{4\sqrt[3]{45}}{3}$

**123.** $(\sqrt{5} - 2)(\sqrt{5} + 2)$    $1$    **124.** $(\sqrt{3} - 4)(\sqrt{3} + 4)$    $-13$

**125.** $(\sqrt[3]{2} - \sqrt[3]{b^2})(\sqrt[3]{4} - \sqrt[3]{b})$    $2 - \sqrt[3]{4b^2} - \sqrt[3]{2b} + b$

**126.** $(\sqrt[3]{25} + \sqrt[3]{x})(\sqrt[3]{5} + \sqrt[3]{x^2})$    $5 + \sqrt[3]{5x} + \sqrt[3]{25x^2} + x$

**127.** $(9\sqrt{2})^2$    $162$    **128.** $(7\sqrt{5})^2$    $245$

**129.** $\dfrac{1}{\sqrt{7}}$    $\dfrac{\sqrt{7}}{7}$    **130.** $\dfrac{3}{\sqrt{8}}$    $\dfrac{3\sqrt{2}}{4}$

**131.** $(2\sqrt{6} - \sqrt{7})(4\sqrt{6} + 3\sqrt{7})$    $27 + 2\sqrt{42}$

**132.** $(5\sqrt{11} - \sqrt{3})(2\sqrt{11} + 6\sqrt{3})$    $92 + 28\sqrt{33}$

**133.** $(\sqrt{a^2 + 3a - 2})^2$    $a^2 + 3a - 2$    **134.** $(\sqrt{b^2 - 4b + 3})^2$    $b^2 - 4b + 3$

**135.** $(\sqrt[3]{2x + 5})^3$    $2x + 5$    **136.** $(\sqrt[3]{3x + 7})^3$    $3x + 7$

**137.** $\dfrac{9}{5 - \sqrt{3}}$    $\dfrac{45 + 9\sqrt{3}}{22}$    **138.** $\dfrac{4}{6 + \sqrt{2}}$    $\dfrac{12 - 2\sqrt{2}}{17}$

**139.** $\dfrac{9 + \sqrt{2}}{4 - \sqrt{2}}$    $\dfrac{38 + 13\sqrt{2}}{14}$    **140.** $\dfrac{1 - \sqrt{13}}{5 + \sqrt{13}}$    $\dfrac{3 - \sqrt{13}}{2}$

**You Be the Teacher!**

**Correct each student's errors, if any.**

**141.** Simplify $(\sqrt{2} + \sqrt{5})^2$.

Lily's work:

$$(\sqrt{2} + \sqrt{5})^2 = (\sqrt{2} + \sqrt{5})(\sqrt{2} + \sqrt{5})$$
$$= \sqrt{4} + \sqrt{10} + \sqrt{10} + \sqrt{25}$$
$$= 2 + \sqrt{20} + 5$$
$$= 7 + 2\sqrt{5}$$

**142.** Simplify $(3\sqrt{2})^2$.

Mike's work:

$$(3\sqrt{2})^2 = 3\sqrt{2^2}$$
$$= 3 \cdot 2$$
$$= 6$$

The square must be applied to the coefficient of 3.

$$(3\sqrt{2})^2 = 3^2\sqrt{2^2}$$
$$= 9 \cdot 2$$
$$= 18$$

 **Calculate It!**

**Use a calculator to determine if each expression is simplified correctly.**

**143.** Is $(\sqrt{5} + 2)(\sqrt{5} - 2) = 1$?   yes

**144.** Is $(\sqrt{5} + 2)^2 = 9 + 4\sqrt{5}$?   yes

---

**SECTION 8.5** / **Solving Square Root Equations and Applications Involving Square Roots**

▶ **OBJECTIVES**

**As a result of completing this section, you will be able to**

**1.** Solve square root equations by applying the squaring principle.

**2.** Solve square root equations by applying the squaring principle twice.

**3.** Solve applications of square root equations.

**4.** Troubleshoot common errors.

In Chapter 7, we discussed how to compute a person's BMI (body mass index). Another formula of importance is Mosteller's formula to calculate a person's body surface area (BSA). This value is often used in calculating drug dosages.

A person's BSA in square meters is given by, $\text{BSA} = \sqrt{\dfrac{hw}{3131}}$,

where $h$ is a person's height in inches and $w$ is a person's weight in pounds. The average BSA for women is 1.6 m². For a woman who is 5 ft tall, what weight would make her have an average BSA?

To answer this question, we must solve the equation $\sqrt{\dfrac{60w}{3131}} = 1.6$. We will learn how to solve this type of equation in this section.

### The Squaring Principle and Square Root Equations

**Objective 1** ▶

Solve square root equations by applying the squaring principle.

Now that we know how to work with square root expressions, we will solve equations that contain these types of expressions. Some examples of square root equations are

$$\sqrt{x - 5} = 3 \qquad \sqrt{4x - 1} = \sqrt{3x}$$

To solve these equations, our goal is to produce an equivalent equation that doesn't contain radicals. We will use the property that we learned in Section 8.4 to accomplish this. Recall that for $a > 0$,

$$\left(\sqrt[n]{a}\right)^n = a$$

For instance,

$$\left(\sqrt{x - 5}\right)^2 = x - 5$$

This property together with the fact that, what we do to one side of an equation, we must do to the other, brings us to the *squaring principle*.

**Property: The Squaring Principle**

If $a = b$, then

$$a^2 = b^2$$

This property tells us that if two quantities are equal, then the squares of the quantities are equal as well. The squared equation that results does not necessarily have the same solutions as the original equation. Consider the following illustration.

| **Original Equation** | $x = -4$ | Solution Set: $\{-4\}$ |
|---|---|---|
| **Squared Equation** | $(x)^2 = (-4)^2$ <br> $x^2 = 16$ <br> $x^2 - 16 = 0$ <br> $(x - 4)(x + 4) = 0$ <br> $x = 4 \quad \text{or} \quad x = -4$ | Solution Set: $\{-4, 4\}$ <br><br> Note: This solution set contains a solution, 4, that is *not* a solution of the original equation. This is an **extraneous solution.** |

This illustration shows us that when we apply the squaring principle, we obtain solutions of the original equation but we can also get solutions that do not satisfy the original equation. That is, we may obtain *extraneous* solutions when we apply the squaring principle. Therefore, we must *always* check the proposed solutions in the original equation. Recall we also encountered extraneous solutions when we solved rational equations.

> **Procedure:  Solving Square Root Equations**
>
> **Step 1:** Isolate the square root expression on one side of the equation.
> **Step 2:** Apply the squaring principle and square each side of the equation.
> **Step 3:** Solve the resulting equation.
> **Step 4:** Check the possible answer(s) in the original equation. If the solution makes the equation false, exclude this solution from the solution set.
> **Step 5:** State the solution set.

**Objective 1  Examples**   Solve each equation.

**1a.** $\sqrt{x - 5} = 3$      **1b.** $\sqrt{x + 1} - 3 = 5$      **1c.** $\sqrt{a} + 7 = 4$

**1d.** $\sqrt{5x - 2} = \sqrt{3x}$      **1e.** $\sqrt{y - 4} = y - 10$

**Solutions**   **1a.**

$$\sqrt{x - 5} = 3$$

$$\left(\sqrt{x - 5}\right)^2 = (3)^2 \qquad \text{Square each side of the equation.}$$

$$x - 5 = 9 \qquad \text{Simplify: } (\sqrt{x - 5})^2 = x - 5 \text{ and } 3^2 = 9.$$

$$x - 5 + 5 = 9 + 5 \qquad \text{Add 5 to each side.}$$

$$x = 14 \qquad \text{Simplify.}$$

**Check:**

$$\sqrt{x - 5} = 3 \qquad \text{Begin with the original equation.}$$

$$\sqrt{14 - 5} = 3 \qquad \text{Replace } x \text{ with 14.}$$

$$\sqrt{9} = 3 \qquad \text{Simplify the radicand.}$$

$$3 = 3 \qquad \text{Simplify the radical.}$$

The value 14 makes the equation true. So, the solution set is $\{14\}$.

**1b.**

$$\sqrt{x + 1} - 3 = 5$$

$$\sqrt{x + 1} - 3 + 3 = 5 + 3 \qquad \text{Add 3 to each side.}$$

$$\sqrt{x + 1} = 8 \qquad \text{Simplify.}$$

$$\left(\sqrt{x + 1}\right)^2 = (8)^2 \qquad \text{Square each side of the equation.}$$

$$x + 1 = 64 \qquad \text{Simplify: } (\sqrt{x + 1})^2 = x + 1 \text{ and } 8^2 = 64.$$

$$x + 1 - 1 = 64 - 1 \qquad \text{Subtract 1 from each side of the equation.}$$

$$x = 63 \qquad \text{Simplify.}$$

**Check:** $\sqrt{x + 1} - 3 = 5$       Begin with the original equation.

$\sqrt{63 + 1} - 3 = 5$       Replace $x$ with 63.

$\sqrt{64} - 3 = 5$       Simplify the radicand.

$8 - 3 = 5$       Simplify the radical.

$5 = 5$       Simplify.

The value 63 makes the equation true. So, the solution set is $\{63\}$.

**1c.**       $\sqrt{a} + 7 = 4$

$\sqrt{a} + 7 - 7 = 4 - 7$       Subtract 7 from each side.

$\sqrt{a} = -3$       Simplify.

$(\sqrt{a})^2 = (-3)^2$       Square each side of the equation.

$a = 9$       Simplify: $(\sqrt{a})^2 = a$ and $(-3)^2 = 9$.

**Check:**       $\sqrt{a} + 7 = 4$       Begin with the original equation.

$\sqrt{9} + 7 = 4$       Replace $a$ with 9.

$3 + 7 = 4$       Simplify the radical.

$10 = 4$       Simplify.

The value 9 makes the equation false and, therefore, must be excluded from the solution set. It is an extraneous solution. So, the solution set is the empty set, or $\varnothing$.

> **Note:** We can conclude that the previous equation has no solution without going through all of the steps to solve and check. After we isolate the radical, we obtain the equation, $\sqrt{a} = -3$. This statement is always false, because the square root of $a$ represents the nonnegative square root of $a$.

**1d.**       $\sqrt{5x - 2} = \sqrt{3x}$

$(\sqrt{5x - 2})^2 = (\sqrt{3x})^2$       Square each side.

$5x - 2 = 3x$       Simplify: $(\sqrt{5x - 2})^2 = 5x - 2$ and $(\sqrt{3x})^2 = 3x$.

$5x - 2 + 2 = 3x + 2$       Add 2 to each side.

$5x = 3x + 2$       Simplify.

$5x - 3x = 3x + 2 - 3x$       Subtract $3x$ from each side.

$2x = 2$       Simplify.

$\dfrac{2x}{2} = \dfrac{2}{2}$       Divide each side by 2.

$x = 1$       Simplify.

**Check:**       $\sqrt{5x - 2} = \sqrt{3x}$       Begin with the original equation.

$\sqrt{5(1) - 2} = \sqrt{3(1)}$       Replace $x$ with 1.

$\sqrt{5 - 2} = \sqrt{3}$       Simplify each product

$\sqrt{3} = \sqrt{3}$       Simplify.

The value 1 makes the equation true. So, the solution set is $\{1\}$.

**INSTRUCTOR NOTE:**
Remind students that the square of a binomial is a trinomial.

**1e.** $\sqrt{y-4} = y - 10$

$\left(\sqrt{y-4}\right)^2 = (y-10)^2$      Square each side.

$y - 4 = y^2 - 20y + 100$      Simplify each side.

To solve the resulting quadratic equation, we must write the equation in standard form, that is, get one side of the equation equal to 0. Then we will solve the equation by factoring.

$y - 4 - y = y^2 - 20y + 100 - y$      Subtract $y$ from each side.

$-4 = y^2 - 21y + 100$      Simplify.

$-4 + 4 = y^2 - 21y + 100 + 4$      Add 4 to each side.

$0 = y^2 - 21y + 104$      Simplify.

$0 = (y - 13)(y - 8)$      Factor.

$y - 13 = 0$   or   $y - 8 = 0$      Set factor equal to 0.

$y = 13$          $y = 8$      Solve the resulting equations.

**Check:** Replace $y$ with 13.      Replace $y$ with 8.

$\sqrt{y-4} = y - 10$          $\sqrt{y-4} = y - 10$

$\sqrt{13 - 4} = 13 - 10$          $\sqrt{8 - 4} = 8 - 10$

$\sqrt{9} = 3$               $\sqrt{4} = -2$

$3 = 3$   True          $2 = -2$   False

The only value that makes the equation true is 13. So, the solution set is $\{13\}$.

✓ **Student Check 1**    Solve each equation.

     **a.** $\sqrt{2x+3} = 5$      **b.** $\sqrt{y+3} - 4 = 13$      **c.** $\sqrt{a} + 10 = 9$

     **d.** $\sqrt{3x-1} = \sqrt{2x}$      **e.** $\sqrt{z+2} = z - 10$

**Objective 2** ▶

Solve square root equations by applying the squaring principle twice.

## Equations That Require Squaring Twice

Some equations that contain more than one radical may require applying the squaring principle twice.

**Objective 2 Examples**    Solve each equation.

     **2a.** $\sqrt{x+1} - \sqrt{x} = 1$      **2b.** $\sqrt{2a+3} - \sqrt{a+1} = 1$

**Solutions**    **2a.** We must isolate one of the radicals on one side of the equation.

$\sqrt{x+1} - \sqrt{x} = 1$

$\sqrt{x+1} - \sqrt{x} + \sqrt{x} = 1 + \sqrt{x}$      Add $\sqrt{x}$ to each side to isolate $\sqrt{x+1}$.

$\sqrt{x+1} = 1 + \sqrt{x}$      Simplify.

$\left(\sqrt{x+1}\right)^2 = \left(1 + \sqrt{x}\right)^2$      Square each side.

$x + 1 = 1 + 2\sqrt{x} + x$      Recall that $(a+b)^2 = a^2 + 2ab + b^2$ and $(\sqrt{x+1})^2 = x + 1$.

Now we must isolate the radical $2\sqrt{x}$ on one side of the equation.

| | |
|---|---|
| $x + 1 - 1 = 1 + 2\sqrt{x} + x - 1$ | Subtract 1 from each side. |
| $x = 2\sqrt{x} + x$ | Simplify. |
| $x - x = 2\sqrt{x} + x - x$ | Subtract $x$ from each side. |
| $0 = 2\sqrt{x}$ | Simplify. |
| $(0)^2 = \left(2\sqrt{x}\right)^2$ | Square each side. |
| $0 = 4x$ | Simplify: $(2\sqrt{x})^2 = (2)^2(\sqrt{x})^2 = 4x$. |
| $\dfrac{0}{4} = \dfrac{4x}{4}$ | Divide each side by 4. |
| $0 = x$ | Simplify. |

| | |
|---|---|
| **Check:** $\sqrt{x+1} - \sqrt{x} = 1$ | Begin with the original equation. |
| $\sqrt{0+1} - \sqrt{0} = 1$ | Replace $x$ with 0. |
| $\sqrt{1} - 0 = 1$ | Simplify each radical expression. |
| $1 - 0 = 1$ | Simplify the radical. |
| $1 = 1$ | Simplify. |

The value 0 makes the equation true. So, the solution set is $\{0\}$.

**2b.** We must isolate one of the radicals on one side of the equation.

| | |
|---|---|
| $\sqrt{2a+3} - \sqrt{a+1} = 1$ | |
| $\sqrt{2a+3} - \sqrt{a+1} + \sqrt{a+1} = 1 + \sqrt{a+1}$ | Add $\sqrt{a+1}$ to each side. |
| $\sqrt{2a+3} = 1 + \sqrt{a+1}$ | Simplify. |
| $\left(\sqrt{2a+3}\right)^2 = \left(1 + \sqrt{a+1}\right)^2$ | Square each side. |
| $2a + 3 = 1 + 2\sqrt{a+1} + a + 1$ | Simplify. |

Now we must isolate the radical $2\sqrt{a+1}$ on one side of the equation.

| | |
|---|---|
| $2a + 3 = 2\sqrt{a+1} + a + 2$ | Simplify the right side. |
| $2a + 3 - 2 = 2\sqrt{a+1} + a + 2 - 2$ | Subtract 2 from each side. |
| $2a + 1 = 2\sqrt{a+1} + a$ | Simplify. |
| $2a + 1 - a = 2\sqrt{a+1} + a - a$ | Subtract $a$ from each side. |
| $a + 1 = 2\sqrt{a+1}$ | Simplify. |
| $(a+1)^2 = \left(2\sqrt{a+1}\right)^2$ | Square each side. |
| $a^2 + 2a + 1 = 4(a+1)$ | Simplify each side. |
| $a^2 + 2a + 1 = 4a + 4$ | Apply the distributive property. |
| $a^2 + 2a + 1 - 4a = 4a + 4 - 4a$ | Subtract $4a$ from each side. |
| $a^2 - 2a + 1 = 4$ | Simplify. |
| $a^2 - 2a + 1 - 4 = 4 - 4$ | Subtract 4 from each side. |
| $a^2 - 2a - 3 = 0$ | Simplify. |
| $(a-3)(a+1) = 0$ | Solve by factoring. |
| $a - 3 = 0 \quad \text{or} \quad a + 1 = 0$ | Set each factor equal to 0. |
| $a = 3 \qquad\qquad a = -1$ | Solve each equation. |

**Check:** Replace $a$ with 3.

$$\sqrt{2a + 3} - \sqrt{a + 1} = 1$$
$$\sqrt{2(3) + 3} - \sqrt{3 + 1} = 1$$
$$\sqrt{6 + 3} - \sqrt{4} = 1$$
$$\sqrt{9} - 2 = 1$$
$$3 - 2 = 1$$
$$1 = 1$$

Replace $a$ with $-1$.

$$\sqrt{2a + 3} - \sqrt{a + 1} = 1$$
$$\sqrt{2(-1) + 3} - \sqrt{-1 + 1} = 1$$
$$\sqrt{-2 + 3} - \sqrt{0} = 1$$
$$\sqrt{1} - 0 = 1$$
$$1 - 0 = 1$$
$$1 = 1$$

Since the values 3 and $-1$ both make the original equation true, the solution set is $\{-1, 3\}$.

✔ **Student Check 2**   Solve each equation.

a. $\sqrt{3x - 2} - \sqrt{x} = 2$       b. $\sqrt{3x - 5} - \sqrt{x + 2} = 1$

## Applications

**Objective 3** ▶

Solve applications of square root equations.

Some applications of square root equations deal with right triangles. Recall the *Pythagorean theorem* shows the relationship between the sides of a right triangle.

> **Property:  Pythagorean Theorem**
>
> For a right triangle with legs $a$ and $b$ and hypotenuse $c$,
>
>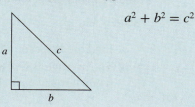
>
> $$a^2 + b^2 = c^2$$

We will also solve some other applications for which we are given a model.

**Objective 3  Examples**   Solve each problem.

**3a.** Find the exact length of the unknown side of the right triangle.

5 ft    15 ft
$b$

**Solution**   **3a.**

$$a^2 + b^2 = c^2$$        State the Pythagorean theorem.
$$5^2 + b^2 = 15^2$$        Replace $a$ with 5 and $c$ with 15.
$$25 + b^2 = 225$$        Simplify: $5^2 = 25$ and $15^2 = 225$.
$$25 + b^2 - 25 = 225 - 25$$        Subtract 25 from each side.
$$b^2 = 200$$        Simplify.
$$b = \sqrt{200}$$        Solve for $b$.
$$b = \sqrt{100 \cdot 2}$$        Factor 200 as $100 \cdot 2$.
$$b = 10\sqrt{2} \text{ ft}$$        Apply the product rule.

So, the value of the other leg of the right triangle is $10\sqrt{2}$ ft.

**3b.** Shirlene plans to have her house painted. One side of her house is lined with hydrangeas. So that her hydrangeas will not be damaged, the bottom of the painter's ladder will have to be 8 ft from the house. If the ladder must reach a point 20 ft from the ground, how long of a ladder will be needed to paint her house?

**Solution** **3b.** We can use the Pythagorean theorem since the ladder forms the hypotenuse of the triangle formed with the house and ground. Let $x$ represent the length of the ladder.

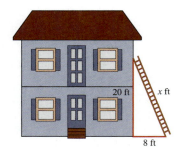

| | |
|---|---|
| $a^2 + b^2 = c^2$ | State the Pythagorean theorem. |
| $20^2 + 8^2 = x^2$ | Replace $a$ with 20, $b$ with 8, and $c$ with $x$. |
| $400 + 64 = x^2$ | Simplify: $20^2 = 400$ and $8^2 = 64$. |
| $464 = x^2$ | Simplify the left side. |
| $\sqrt{464} = x$ | Solve for $x$. |
| $\sqrt{16 \cdot 29} = x$ | Factor 464 as $16 \cdot 29$. |
| $4\sqrt{29} = x$ | Apply the product rule. |
| $x \approx 21.54$ | Approximate the value. |

So, a ladder that is approximately 22 ft is needed to paint Shirlene's house.

**3c.** Mosteller's formula is used to calculate a person's body surface area (BSA). This value is often used in calculating drug dosages. A person's BSA in square meters is given by $BSA = \sqrt{\dfrac{hw}{3131}}$, where $h$ is a person's height in inches and $w$ is a person's weight in pounds.

   **i.** Find the BSA for a female who is 5 ft 6 in. tall and weighs 165 lb.
  **ii.** The average BSA for women is 1.6 m². Find the weight needed for a woman to have an average BSA if she is 5 ft tall.

**Solution** **3c.** **i.**

| | |
|---|---|
| $BSA = \sqrt{\dfrac{hw}{3131}}$ | State the BSA formula. |
| $BSA = \sqrt{\dfrac{(66)(165)}{3131}}$ | Replace $h$ with 66 (5 ft 6 in. is $5(12) + 6 = 60 + 6 = 66$ in.) and $w$ with 165. |
| $BSA = \sqrt{\dfrac{10{,}890}{3131}}$ | Simplify (66)(165). |
| $BSA = \sqrt{3.478122}$ | Simplify the quotient. |
| $BSA \approx 1.86$ m² | Approximate the value. |

So, a 5 ft 6 in. tall woman who weighs 165 lb has a BSA of approximately 1.86 m².

  **ii.** Let $BSA = 1.6$, $h = 60$ in. and solve for $w$.

| | |
|---|---|
| $BSA = \sqrt{\dfrac{hw}{3131}}$ | State the BSA formula. |
| $1.6 = \sqrt{\dfrac{60w}{3131}}$ | Replace BSA with 1.6 and $h$ with 60. |
| $(1.6)^2 = \left(\sqrt{\dfrac{60w}{3131}}\right)^2$ | Square each side. |

$$2.56 = \frac{60w}{3131}$$    Simplify: $(1.6)^2 = 2.56$ and $\left(\sqrt{\dfrac{60w}{3131}}\right)^2 = \dfrac{60w}{3131}$.

$$2.56(3131) = \left(\frac{60w}{3131}\right)(3131)$$    Multiply each side by 3131.

$$8015.36 = 60w$$    Simplify.

$$\frac{8015.36}{60} = \frac{60w}{60}$$    Divide each side by 60.

$$133.59 \approx w$$    Simplify.

So, a woman 5 ft tall should weigh approximately 134 lb to have an average BSA.

**✓ Student Check 3**   Solve each problem.

**a.** Find the length of the missing side of a right triangle if one of the legs is 8 ft and the hypotenuse is 12 ft.

**b.** One of the ropes that is used to tie down a tent needs to be replaced. Determine the length of the rope needed if the tent is 6 ft tall and the tie down stake is placed 4 ft from the tent.

**c. i.** Find the BSA of a person who is 5 ft 4 in. and weighs 140 lb.

  **ii.** The average BSA of a man is 1.9 m². Find the weight needed for a man to have average BSA if he is 5 ft 8 in. tall.

**Objective 4 ▶**

Troubleshoot common errors.

## Troubleshooting Common Errors

Some common errors associated with solving square root equations are shown next.

**Objective 4  Examples**   **A problem and an incorrect solution are given. Provide the correct solution and an explanation of the error.**

**4a.** Solve $\sqrt{3x - 5} + 2 = 7$.

| Incorrect Solution | Correct Solution and Explanation |
|---|---|
| $\sqrt{3x - 5} + 2 = 7$ <br> $\left(\sqrt{3x - 5} + 2\right)^2 = 7^2$ <br> $3x - 5 + 4 = 49$ <br> $3x - 1 = 49$ <br> $3x = 50$ <br> $x = \dfrac{50}{3}$ <br> $\left\{\dfrac{50}{3}\right\}$ | We must isolate the radical expression before squaring each side. When we square a binomial, we cannot simply square the first term and square the last term. <br><br> $\sqrt{3x - 5} + 2 = 7$ <br> $\sqrt{3x - 5} + 2 - 2 = 7 - 2$ <br> $\sqrt{3x - 5} = 5$ <br> $\left(\sqrt{3x - 5}\right)^2 = 5^2$ <br> $3x - 5 = 25$ <br> $3x = 30$ <br> $x = 10$ <br><br> The value 10 makes the equation true, so the solution set is $\{10\}$. |

**4b.** Solve $\sqrt{z+2}+4=z$.

| Incorrect Solution | Correct Solution and Explanation |
|---|---|
| $\sqrt{z+2}+4=z$ $\sqrt{z+2}=z-4$ $\left(\sqrt{z+2}\right)^2=(z-4)^2$ $z+2=z^2-8z+16$ $0=z^2-9z+14$ $0=(z-2)(z-7)$ $z=2 \quad \text{or} \quad z=7$ $\{2,7\}$ | The steps to solve the equation were performed correctly. However, the solutions were not checked in the original equation. **Check:** $z=2$: $\sqrt{z+2}+4=z$ $\sqrt{2+2}+4=2$ $\sqrt{4}+4=2$ $2+4=2$ $6=2 \quad \text{False}$ **Check:** $z=7$: $\sqrt{z+2}+4=z$ $\sqrt{7+2}+4=7$ $\sqrt{9}+4=7$ $3+4=7$ $7=7 \quad \text{True}$ So, the solution set is $\{7\}$. |

## ANSWERS TO STUDENT CHECKS

**Student Check 1 a.** $\{11\}$ **b.** $\{286\}$ **c.** $\varnothing$ **d.** $\{1\}$ **e.** $\{14\}$   **Student Check 3 a.** $4\sqrt{5}$ ft **b.** $2\sqrt{13}$ ft **c. i.** $1.69$ m$^2$
**Student Check 2 a.** $\{9\}$ **b.** $\{7\}$                              **ii.** $166.2$ lb

## SUMMARY OF KEY CONCEPTS

1. There are five main steps that we must follow to solve square root equations:

   **a.** Isolate the radical on one side of the equation.

   **b.** Square each side of the equation. Make sure to square a binomial correctly, when needed.

   **c.** Solve the resulting equation.

   **d.** Check each answer in the original equation. It is possible that extraneous solutions may result, so do not skip the checking step.

   **e.** State the solution set.

2. When solving equations that have more than one radical, we can still use the five steps for solving square root equations. However, we may need to repeat the first and second steps until all of the radicals are eliminated.

3. When solving applications involving the Pythagorean theorem, we often must take the square root in order to eliminate the square. In other words, if $x^2=A$ for a positive number $x$, then $x=\sqrt{A}$. Other real-life models can be represented by radical expressions. Thus, we may have to solve a radical equation to obtain specific information about the model.

## GRAPHING CALCULATOR SKILLS

As we have seen in previous sections, we can use a graphing calculator to check the answers derived when solving square root equations.

**Example:** Determine if 13 or 8 is a solution of $\sqrt{y-4}=y-10$.

**Solution:** First, enter the left side of the original equation as $Y_1$ and the right side of the original equation as $Y_2$. (Notice that we use the variable X in the calculator, rather than $y$.)

```
Plot1  Plot2  Plot3
\Y1■√(X-4)
\Y2■X-10
\Y3=
\Y4=
\Y5=
\Y6=
\Y7=
```

Next, go to the Table. Enter the value 13. Notice that $Y_1$ (the left side of the equation) and $Y_2$ (the right side of the equation) are the same when $x = 13$. Therefore, 13 is a solution of the equation. Enter the value 8. Notice that $Y_1$ is not equal to $Y_2$. So, 8 is not a solution of the equation.

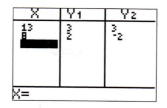

---

## SECTION 8.5 / EXERCISE SET

### Write About It!

**Use complete sentences in your answer to each exercise.**

1. What is an extraneous solution of a square root equation? *An extraneous solution is a solution of the squared equation but not the original square root equation.*

2. Explain how to solve a square root equation.

3. Explain how to solve a square root equation with more than one radical. *Isolate one square root expression and then square each side. Then isolate the square root expression and square again.*

4. Explain how square roots can be used to solve problems involving the Pythagorean theorem. *We replace the known sides into the Pythagorean theorem. The resulting equation will be of the form $x^2 = a$, so $x = \sqrt{a}$.*

### Practice Makes Perfect!

**Solve each equation. (See Objective 1.)**

5. $\sqrt{x + 1} = 3$ {8}

6. $\sqrt{x - 6} = 4$ {22}

7. $\sqrt{x - 5} = 7$ {144}

8. $\sqrt{x + 3} = 8$ {25}

9. $\sqrt{2x + 9} = \sqrt{5x}$ {3}

10. $\sqrt{6x - 5} = \sqrt{x}$ {1}

11. $\sqrt{x + 2} + 4 = x$ {7}

12. $\sqrt{x + 8} - 2 = x$ {1}

13. $\sqrt{x^2 - 4x + 8} = x + 4$

14. $\sqrt{x^2 - 4x - 4} = x + 2$ {−1}

15. $\sqrt{3x - 2} = x - 2$ {6}

16. $\sqrt{4x - 3} = x - 6$ {13}

17. $\sqrt{3 - 5x} + 10 = 12$ $\left\{-\dfrac{1}{5}\right\}$

18. $6 - \sqrt{1 - 4x} = 5$ {0}

19. $4\sqrt{x + 3} - 8 = 2x$ {−2}

20. $2\sqrt{x + 1} - 2 = 3x$ {0}

**Solve each equation. (See Objective 2.)**

21. $\sqrt{2x + 1} - \sqrt{2x} = 1$ {0}

22. $\sqrt{x} - \sqrt{x - 3} = 1$ {4}

23. $\sqrt{3x + 1} - \sqrt{x} = 3$ {16}

24. $\sqrt{x - 5} + \sqrt{x - 1} = 2$ {5}

25. $2 + \sqrt{x - 4} = \sqrt{2x - 1}$ {5, 13}

26. $2 + \sqrt{x - 2} = \sqrt{3x - 2}$ {2, 6}

27. $\sqrt{x + 2} = \sqrt{3x - 2}$ {9}

28. $\sqrt{x + 1} = \sqrt{6x - 2}$ {1}

29. $2 + \sqrt{4 - x} = \sqrt{x}$ {4}

30. $3 + \sqrt{9 - x} = \sqrt{x}$ {9}

31. $\sqrt{2x + 5} - 1 = \sqrt{2x}$ {2}

32. $\sqrt{x + 1} + 1 = \sqrt{3x}$ {3}

**Use the Pythagorean theorem to find the missing length of each triangle. (See Objective 3.)**

33.
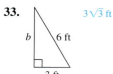
$3\sqrt{3}$ ft
$b$   6 ft
3 ft

34.
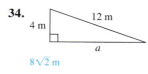
12 m
4 m
$a$
$8\sqrt{2}$ m

*Additional answers can be found in the Instructor Answer Appendix.*

35.
4 in.
$x$
7 in.
$\sqrt{65}$ in.

36.
$2\sqrt{14}$ cm
$b$
9 cm
5 cm

37.
10 m
15 m
$a$
$5\sqrt{5}$ m

38.
3 ft
$y$
8 ft
$\sqrt{73}$ ft

39. A 10-ft ladder leans against a house. If the base of the ladder is 3 ft from the house, then how far from the ground is the top of the ladder? $\sqrt{91}$ ft or 9.54 ft

40. A 12-ft ladder leans against a house. If the base of the ladder is 6 ft from the house, then how far from the ground is the top of the ladder? $6\sqrt{3}$ ft or 10.39 ft

41. Emma wants to paint the trim on her second-floor windows. To avoid the bushes under her window, the base of the ladder must be 5 ft from the house. How long must the ladder be if she needs the ladder to reach 30 ft from the ground? $5\sqrt{37}$ ft or 30.41 ft

42. Fred wants to fix the gutter on his house. If the base of the ladder must be 2 ft from his house, then how long must the ladder be to reach 25 ft from the ground? $\sqrt{629}$ ft or 25.08 ft

**The radius $r$ of a sphere given its surface area $S$ can be found by using the formula, $r = \sqrt{\dfrac{S}{4\pi}}$. Round to one decimal place.**

43. Find the radius of a sphere if its surface area is 400 ft². 5.6 ft

44. Find the radius of a sphere if its surface area is 650 m². 7.2 m

**The speed $s$, in miles per hour, of a car before an accident can be calculated by measuring the distance $d$, in feet, of the skid marks after the accident by the formula $s = \sqrt{30kd}$, where $k$ is the coefficient of friction which depends on the road surface.**

| Road Surface | $k$ |
|---|---|
| Dry tar | 1.0 |
| Wet tar | 0.5 |
| Dry concrete | 0.8 |
| Wet concrete | 0.4 |

**45.** Find the speed of a car before an accident if its skid marks are 145 ft on a wet tar road. 46.6 mph

**46.** Find the speed of a car before an accident if its skid marks are 280 ft on a dry concrete road. 82.0 mph

**47.** Find the speed of a car before an accident if its skid marks are 160 ft on a dry tar road. 69.3 mph

**48.** Find the speed of a car before an accident if its skid marks are 450 ft on a wet concrete road. 73.5 mph

The period $T$, in seconds, of a simple pendulum is the time for the pendulum to complete a full cycle and it can be approximated by $T = 2\pi\sqrt{\dfrac{L}{g}}$, where $L$ is the length of the pendulum and $g$ is the acceleration due to gravity.

**49.** Find the exact value of the period of a simple pendulum if its length is 16 ft and $g$ is 32 ft/sec². $\pi\sqrt{2}$ sec

**50.** Find the exact value of the period of a simple pendulum if its length is 48 ft and $g$ is 32 ft/sec². $\pi\sqrt{6}$ sec

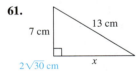 **Mix 'Em Up!**

Solve each equation.

**51.** $4 - \sqrt{x} = 2$  {4}   **52.** $3 - \sqrt{2x} = 1$  {2}

**53.** $\sqrt{x - 6} = 4$  {22}   **54.** $\sqrt{5 - x} = 3$  {−4}

**55.** $1 + \sqrt{x} = \sqrt{2x + 1}$   **56.** $\sqrt{5x} - \sqrt{5x - 3} = 1$  $\left\{\dfrac{4}{5}\right\}$
{0, 4}

**57.** $\sqrt{2x - 3} = x - 3$  {6}   **58.** $2\sqrt{x + 1} = x - 2$  {8}

**59.** $\sqrt{1 - 2x} = 5$  {−12}   **60.** $\sqrt{2x - 3} = 1$  {2}

**Find the length of the unknown side of each right triangle.**

**61.**

7 cm   13 cm
x
$2\sqrt{30}$ cm

**62.**
13 in.   a
10 in.
$\sqrt{269}$ in.

**63.**
8 ft   c
$\sqrt{185}$ ft   11 ft

**64.**
x   20 m
12 m   16 m

The period $T$, in seconds, of a simple pendulum can be approximated by $T = 2\pi\sqrt{\dfrac{L}{g}}$, where $L$ is the length of the pendulum and $g$ is the acceleration due to gravity.

**65.** Find the exact value of the period of a simple pendulum if its length is 2.8 m and $g$ is 9.8 m/sec². $\dfrac{2\pi\sqrt{14}}{7}$ sec

**66.** Find the exact value of the period of a simple pendulum if its length is 12.25 m and $g$ is 9.8 m/sec². $\sqrt{5}\pi$ sec

The speed $s$, in miles per hour, of a car before an accident can be calculated by the formula $s = \sqrt{30kd}$, where $k$ is the coefficient of friction and $d$ is the length, in feet, of the skid marks.

| Road Surface and Condition | $k$ |
|---|---|
| Dry tar | 1.0 |
| Wet tar | 0.5 |
| Dry concrete | 0.8 |
| Wet concrete | 0.4 |

**67.** Find the speed of a car before an accident if its skid marks are 190 ft on a dry concrete road. 67.5 mph

**68.** Find the speed of a car before an accident if its skid marks are 90 ft on a wet tar road. 36.7 mph

 **You Be the Teacher!**

**Correct each student's errors, if any.**

**69.** Solve: $\sqrt{5x + 4} = x - 2$

Phoebe's work:

$\sqrt{5x + 4} = x - 2$
$(\sqrt{5x + 4})^2 = (x - 2)^2$
$5x + 4 = x^2 + 4$
$0 = x^2 - 5x$
$0 = x(x - 5)$
$x = 0$   or   $x = 5$

So, the solution set is $\{0, 5\}$.

When we square $(x - 2)$, we get a trinomial.

$\sqrt{5x + 4} = x - 2$
$(\sqrt{5x + 4})^2 = (x - 2)^2$
$5x + 4 = x^2 - 4x + 4$
$0 = x^2 - 9x$
$0 = x(x - 9)$
$x = 0$   or   $x = 9$

Only $x = 9$ checks, so the solution set is $\{9\}$.

**70.** Solve: $\sqrt{x^2 - 5x + 3} = x - 2$

Hallie's work:

$\sqrt{x^2 - 5x + 3} = x - 2$
$(\sqrt{x^2 - 5x + 3})^2 = (x - 2)^2$
$x^2 - 5x + 3 = x^2 - 4x + 4$
$-5x + 3 = -4x + 4$
$-x = 1 \Rightarrow x = -1$

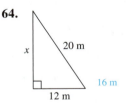 **Calculate It!**

**Use a calculator to determine which of the values is a solution of the given equation.**

**71.** $\sqrt{x - 2} = x - 8$; $x = 6$ and $x = 11$  {11}

**72.** $\sqrt{2x - 1} = x - 2$; $x = 1$ and $x = 5$  {5}

**SECTION 8.6** / **Rational Exponents**

The compound interest formula is given by $A = P\left(1 + \dfrac{r}{n}\right)^{nt}$, where $A$ is the amount of money in an account in $t$ years if $P$ dollars are invested at an annual interest rate ($r$), compounded $n$ times a year. How much money will there be in an account after 6 months $\left(\dfrac{1}{2} \text{ of a year}\right)$ if \$3000 is invested into an account that pays 5% interest compounded annually?

To answer this question, we must evaluate the right side of the equation.

$$A = 3000\left(1 + \frac{0.05}{1}\right)^{1(1/2)} = 3000(1.05)^{1/2}$$

Notice that this equation has an exponent that is a fraction. This is called a **fractional exponent**, or a **rational exponent**. In this section, we will learn how to deal with these types of exponents. We will also discover how rational exponents are related to radicals.

## Expressions of the Form $b^{1/n}$

**Objective 1** ▶

Evaluate expressions of the form $b^{1/n}$.

Until this point, we have evaluated exponential expressions with only integer exponents.

- Positive integer exponents indicate how many times a factor is repeated. For example, $4^3 = 4 \cdot 4 \cdot 4$.
- A nonzero number raised to the exponent of 0 is one. For example, $4^0 = 1$.
- A negative integer exponent requires us to take the reciprocal of the base.

For example, $4^{-3} = \dfrac{1}{4^3}$.

We will now learn how to evaluate exponential expressions with fractional exponents, such as, $4^{1/2}$ or $8^{2/3}$.

Using a calculator, we can determine the value of the following expressions. It is important that the fractional exponent be entered in parentheses.

$$4^{1/2} = 2 \qquad\qquad \sqrt{4} = 2$$
$$9^{1/2} = 3 \qquad\qquad \sqrt{9} = 3$$
$$16^{1/2} = 4 \qquad\qquad \sqrt{16} = 4$$
$$8^{1/3} = 2 \qquad\qquad \sqrt[3]{8} = 2$$
$$27^{1/3} = 3 \qquad\qquad \sqrt[3]{27} = 3$$

The value of each exponential expression is also the same as the value of a radical expression.

This example illustrates the definition of $b^{1/n}$:

$$4^{1/2} = \sqrt{4}$$
$$9^{1/2} = \sqrt{9}$$
$$16^{1/2} = \sqrt{16}$$
$$8^{1/3} = \sqrt[3]{8}$$
$$27^{1/3} = \sqrt[3]{27}$$

Rational exponents are simply another way to write a radical expression. Note that when the exponent is of the form $1/n$, the value $n$ represents the index of the equivalent radical.

> **Definition: $b^{1/n}$**
>
> For $n$ a positive integer greater than 1 and $\sqrt[n]{b}$ a real number,
> $$b^{1/n} = \sqrt[n]{b}$$

Recall that we defined the $n$th root of a number $b$ as $\sqrt[n]{b} = a$ if and only if $a^n = b$. This definition also applies to rational exponents. For example, we know that

$$\left(\sqrt[3]{8}\right)^3 = 2 \text{ because } 2^3 = 8$$

This also follows if we write $\sqrt[3]{8} = 8^{1/3}$.

$$(8^{1/3})^3 = 8^{(1/3)(3)} = 8^1 = 8$$

So, we have that

$$\left(\sqrt[n]{b}\right)^n = (b^{1/n})^n = b^{(1/n)(n)} = b^1 = b$$

> **Procedure: Simplifying Expressions of the Form $b^{1/n}$**
>
> **Step 1:** Convert the expression to a radical of the form $\sqrt[n]{b}$.
>   **a.** The denominator of the fractional exponent becomes the index of the radical.
>   **b.** The base of the fractional exponent becomes the radicand of the radical.
> **Step 2:** Simplify, if possible.

**Objective 1 Examples** | Rewrite each expression as a radical expression and simplify, if possible. Assume all variables represent positive real numbers.

**1a.** $81^{1/2}$    **1b.** $(-8)^{1/3}$    **1c.** $y^{1/4}$    **1d.** $-144^{1/2}$    **1e.** $(-144)^{1/2}$
**1f.** $(32x^{10})^{1/5}$

**Solutions**

**1a.** $81^{1/2} = \sqrt{81} = 9$

**1b.** $(-8)^{1/3} = \sqrt[3]{-8} = -2$

**1c.** $y^{1/4} = \sqrt[4]{y}$

**1d.** Since there are no parentheses, the base of the exponent is 144. We must find the opposite of the square root of 144.
$$-144^{1/2} = -\sqrt{144} = -12$$

**1e.** The base of the exponent is $-144$. We must find the square root of $-144$.
$$(-144)^{1/2} = \sqrt{-144} \text{ not a real number}$$

**1f.** $(32x^{10})^{1/5} = \sqrt[5]{32x^{10}} = 2x^2$

**✓ Student Check 1** | Rewrite each expression as a radical expression and simplify, if possible. Assume all variables represent positive real numbers.

   **a.** $25^{1/2}$          **b.** $(-64)^{1/3}$          **c.** $x^{1/5}$
   **d.** $-36^{1/2}$          **e.** $(-36)^{1/2}$          **f.** $(81x^{12})^{1/4}$

## Expressions of the Form $b^{m/n}$

**Objective 2** ▶

Evaluate expressions of the form $b^{m/n}$.

To be thorough in our discussion of rational exponents, we need to define rational exponents for numerators other than one. In other words, we will define what it means to raise an expression to exponents such as $\frac{2}{3}, \frac{3}{2}, \frac{3}{4}$, and so on. These exponents are defined based on a property of exponents. Recall that $(b^m)^n = b^{mn}$.

So, we can think of $16^{3/4}$ as $16^{(1/4)(3)} = (16^{1/4})^3 = (\sqrt[4]{16})^3 = 2^3 = 8$.

---

**Definition:** $b^{m/n}$

If $m$ and $n$ are positive integers greater than 1 and $\sqrt[n]{b}$ is a real number, then

$$b^{m/n} = (\sqrt[n]{b})^m \quad \text{or} \quad \sqrt[n]{b^m}$$

The exponent $m$ can be applied after the root is simplified or before the root is taken.

---

**Procedure:  Evaluating an Expression of the Form $b^{m/n}$**

**Step 1:** Convert the expression to radical form $(\sqrt[n]{b})^m$.
    **a.** The base of the exponent is the radicand of the radical.
    **b.** The denominator of the fractional exponent is the index of the radical.
    **c.** The numerator of the fractional exponent is the exponent of the radical expression.
**Step 2:** Simplify the root and then apply the exponent.

---

**Objective 2  Examples**

**Rewrite each expression as a radical and simplify, if possible. Assume all bases with variables represent positive real numbers.**

**2a.** $16^{5/4}$     **2b.** $-25^{3/2}$     **2c.** $(-64)^{4/3}$     **2d.** $\left(\dfrac{1}{8}\right)^{2/3}$     **2e.** $(x-3)^{5/6}$

**Solutions**

**2a.** $16^{5/4} = (\sqrt[4]{16})^5 = 2^5 = 32$

**2b.** $-25^{3/2} = -(\sqrt{25})^3 = -5^3 = -125$

**2c.** $(-64)^{4/3} = (\sqrt[3]{-64})^4 = (-4)^4 = 256$

**2d.** $\left(\dfrac{1}{8}\right)^{2/3} = \left(\sqrt[3]{\dfrac{1}{8}}\right)^2 = \left(\dfrac{1}{2}\right)^2 = \dfrac{1}{4}$

**2e.** $(x-3)^{5/6} = (\sqrt[6]{x-3})^5 = \sqrt[6]{(x-3)^5}$

---

**✓ Student Check 2**

Rewrite each expression as a radical and simplify, if possible. Assume all variables represent positive real numbers.
    **a.** $9^{3/2}$            **b.** $-81^{3/4}$          **c.** $(-27)^{2/3}$
    **d.** $\left(\dfrac{1}{32}\right)^{3/5}$      **e.** $(x+2)^{6/7}$

---

## Expressions of the Form $b^{-m/n}$

**Objective 3 ▶**

Evaluate expressions of the form $b^{-m/n}$.

The rules for rational exponents also apply when the exponents are negative. We must use the definition of a negative exponent as well. Recall that,

$$x^{-n} = \frac{1}{x^n}$$

So, using the preceding two definitions together we get the following.

**Definition:** $b^{-m/n}$

For $b \neq 0$,

$$b^{-m/n} = \frac{1}{b^{m/n}}$$

**Procedure:** Evaluating Expressions of the Form $b^{-m/n}$

**Step 1:** Rewrite the expression with a positive exponent.
**Step 2:** Apply the definition of $b^{1/m}$ or $b^{m/n}$.
**Step 3:** Simplify.

**Objective 3 Examples** | Rewrite each expression with a positive exponent and simplify.

**3a.** $16^{-1/2}$ **3b.** $(-8)^{-4/3}$ **3c.** $-81^{-3/4}$

**Solutions** **3a.** $16^{-1/2} = \dfrac{1}{16^{1/2}} = \dfrac{1}{\sqrt{16}} = \dfrac{1}{4}$

**3b.** $(-8)^{-4/3} = \dfrac{1}{(-8)^{4/3}} = \dfrac{1}{\left(\sqrt[3]{-8}\right)^4} = \dfrac{1}{(-2)^4} = \dfrac{1}{16}$

**3c.** $-81^{-3/4} = \dfrac{1}{-81^{3/4}} = \dfrac{1}{-\left(\sqrt[4]{81}\right)^3} = \dfrac{1}{-(3)^3} = \dfrac{1}{-27} = -\dfrac{1}{27}$

**✓ Student Check 3** | Rewrite each expression with a positive exponent and simplify.
**a.** $32^{-1/5}$ **b.** $(-27)^{-2/3}$ **c.** $-25^{-3/2}$

## Apply Exponent Rules to Expressions with Rational Exponents

**Objective 4** ▶

Use exponent rules to simplify expressions with rational exponents.

The exponent rules that we learned in Chapter 5 work not only for integer exponents, but also for rational exponents. Here is a review of these rules.

**Property: Rules of Exponents**

For rational numbers $m$ and $n$, and real numbers $a$ and $b$ such that the following values exist,

**1.** $b^m \cdot b^n = b^{m+n}$ **2.** $(b^m)^n = b^{mn}$ **3.** $\dfrac{b^m}{b^n} = b^{m-n}, b \neq 0$

**4.** $(ab)^m = a^m b^m$ **5.** $\left(\dfrac{a}{b}\right)^m = \dfrac{a^m}{b^m}, b \neq 0$ **6.** $b^0 = 1, b \neq 0$

**7.** $b^{-n} = \dfrac{1}{b^n}, b \neq 0$

**Objective 4 Examples** | Simplify each expression. Assume all variables represent positive real numbers. Write each answer with positive exponents.

**4a.** $4^{4/3} \cdot 4^{2/3}$ **4b.** $x^{3/5} \cdot x^{-4/5}$ **4c.** $y^{1/2} \cdot y^{1/4}$ **4d.** $\dfrac{8^{1/5}}{8^{6/5}}$

**4e.** $(a^4)^{3/2}$ **4f.** $(8x^6)^{2/3}$ **4g.** $\left(\dfrac{3b^{1/2}c^{-1/3}}{b^2}\right)^2$ **4h.** $x^{1/3}(x^{1/3} + 2x^{2/3})$

**Solutions**

**4a.**

$$4^{4/3} \cdot 4^{2/3} = 4^{4/3+2/3}$$     Apply the product of like bases rule.

$$= 4^{6/3}$$     Add the exponents.

$$= 4^2$$     Simplify the exponent.

$$= 16$$     Simplify.

**4b.**

$$x^{3/5} \cdot x^{-4/5} = x^{3/5+(-4/5)}$$     Apply the product of like bases rule.

$$= x^{-1/5}$$     Add the exponents.

$$= \frac{1}{x^{1/5}}$$     Apply the definition of a negative exponent.

**4c.**

$$y^{1/2} \cdot y^{1/4} = y^{1/2+1/4}$$     Apply the product of like bases rule.

$$= y^{2/4+1/4}$$     Convert the exponents to a fraction with the LCD, 4.

$$= y^{3/4}$$     Add the exponents.

**4d.**

$$\frac{8^{1/5}}{8^{6/5}} = 8^{1/5-6/5}$$     Apply the quotient of like bases rule.

$$= 8^{-5/5}$$     Subtract the exponents.

$$= 8^{-1}$$     Simplify the exponent.

$$= \frac{1}{8^1}$$     Apply the definition of a negative exponent.

$$= \frac{1}{8}$$     Simplify.

**4e.**

$$(a^4)^{3/2} = a^{4(3/2)}$$     Apply the power of a power rule.

$$= a^6$$     Multiply the exponents.

**4f.**

$$(8x^6)^{2/3} = (8)^{2/3} (x^6)^{2/3}$$     Apply the power of a product rule.

$$= 4x^4$$     Simplify each exponent.

**4g.**

$$\left(\frac{3b^{1/2}c^{-1/3}}{b^2}\right)^2 = \frac{(3b^{1/2}c^{-1/3})^2}{(b^2)^2}$$     Apply the power of a quotient rule.

$$= \frac{(3)^2(b^{1/2})^2(c^{-1/3})^2}{b^4}$$     Apply the power of a product rule.

$$= \frac{9b^1 c^{-2/3}}{b^4}$$     Simplify $3^2$ and multiply the exponents.

$$= 9b^{1-4}c^{-2/3}$$     Subtract the exponents of $b$.

$$= 9b^{-3}c^{-2/3}$$     Simplify.

$$= \frac{9}{b^3 c^{2/3}}$$     Rewrite with positive exponents.

**4h.**

$$x^{1/3}(x^{1/3} + 2x^{2/3}) = x^{1/3}(x^{1/3}) + x^{1/3}(2x^{2/3})$$     Apply the distributive property.

$$= x^{1/3+1/3} + 2x^{1/3+2/3}$$     Apply the product of like bases rule.

$$= x^{2/3} + 2x^{3/3}$$     Add the exponents.

$$= x^{2/3} + 2x$$     Simplify.

---

✓ **Student Check 4**     Simplify each expression. Assume all variables represent positive real numbers. Write each answer with positive exponents.

**a.** $7^{1/6} \cdot 7^{5/6}$     **b.** $y^{3/7} \cdot y^{-6/7}$     **c.** $x^{1/3} \cdot x^{1/6}$     **d.** $\dfrac{3^{2/3}}{3^{5/3}}$

**e.** $(a^4)^{5/4}$     **f.** $(9x^6)^{3/2}$     **g.** $\left(\dfrac{2x^{2/3}y^{-1/2}}{x^2}\right)^3$     **h.** $x^{1/4}(x^{1/4} - 3x^{3/4})$

## Simplify Radical Expressions Using Rational Exponents

**Objective 5** ▶

Use rational exponents to simplify radical expressions.

Radical expressions can often be simplified if we write the radical as an exponential expression. In these problems we begin with a radical expression, convert it to an exponential expression, apply exponent properties, and then write the result in radical form.

**Objective 5 Examples** | Use rational exponents to simplify each expression. Write each answer in radical form, if possible. Assume that variables represent positive real numbers.

**5a.** $\sqrt[3]{x^6}$      **5b.** $\sqrt[6]{x^3}$      **5c.** $\sqrt[4]{9}$      **5d.** $\sqrt[3]{2} \cdot \sqrt{2}$

**5e.** $\dfrac{\sqrt[4]{b^3}}{\sqrt[3]{b^2}}$      **5f.** $\left(\sqrt[3]{c^2}\right)^6$

**Solutions**

**5a.**
$$\sqrt[3]{x^6} = x^{6/3} \qquad \text{Apply the definition of } b^{m/n}.$$
$$= x^2 \qquad \text{Simplify the exponent.}$$

**5b.**
$$\sqrt[6]{x^3} = x^{3/6} \qquad \text{Apply the definition of } b^{m/n}.$$
$$= x^{1/2} \qquad \text{Simplify the exponent.}$$
$$= \sqrt{x} \qquad \text{Apply the definition of } b^{1/n}.$$

**5c.**
$$\sqrt[4]{9} = \sqrt[4]{3^2} \qquad \text{Rewrite 9 as } 3^2.$$
$$= 3^{2/4} \qquad \text{Apply the definition of } b^{m/n}.$$
$$= 3^{1/2} \qquad \text{Simplify the exponent.}$$
$$= \sqrt{3} \qquad \text{Apply the definition of } b^{1/n}.$$

**5d.** Because these radicals do not have the same index, we cannot multiply them using the product rule for radicals.
$$\sqrt[3]{2} \cdot \sqrt{2} = 2^{1/3} \cdot 2^{1/2} \qquad \text{Convert each expression to exponential form.}$$
$$= 2^{1/3 + 1/2} \qquad \text{Apply the product of like bases rule.}$$
$$= 2^{2/6 + 3/6} \qquad \text{Write the fractions with a common denominator.}$$
$$= 2^{5/6} \qquad \text{Add the exponents.}$$
$$= \sqrt[6]{2^5} \qquad \text{Apply the definition of } b^{m/n}.$$
$$= \sqrt[6]{32} \qquad \text{Simplify the radicand.}$$

**5e.** Because the indices are different, we must convert each radical to its exponential form and then apply the rules for exponents.
$$\frac{\sqrt[4]{b^3}}{\sqrt[3]{b^2}} = \frac{b^{3/4}}{b^{2/3}} \qquad \text{Convert each radical to an exponential form.}$$
$$= b^{3/4 - 2/3} \qquad \text{Apply the quotient of like bases rule.}$$
$$= b^{9/12 - 8/12} \qquad \text{Write the fractions with a common denominator.}$$
$$= b^{1/12} \qquad \text{Subtract the exponents.}$$
$$= \sqrt[12]{b} \qquad \text{Apply the definition of } b^{1/n}.$$

**5f.**
$$\left(\sqrt[3]{c^2}\right)^6 = \left(c^{2/3}\right)^6 \qquad \text{Rewrite } \sqrt[3]{c^2} \text{ as } c^{2/3}.$$
$$= c^{(2/3)(6)} \qquad \text{Apply the power of a power rule.}$$
$$= c^4 \qquad \text{Multiply the exponents.}$$

✓ **Student Check 5** Use rational exponents to simplify each expression. Write each answer in radical form, if possible. Assume that variables represent positive real numbers.

**a.** $\sqrt{x^8}$

**b.** $\sqrt[8]{x^2}$

**c.** $\sqrt[4]{4}$

**d.** $\sqrt[3]{5} \cdot \sqrt[4]{5}$

**e.** $\dfrac{\sqrt[5]{y^6}}{\sqrt[3]{y^2}}$

**f.** $\left(\sqrt[5]{z^3}\right)^{15}$

## Applications

**Objective 6** ▶

Solve applications with rational exponents.

Rational exponents are used in many formulas. Any formula that contains a radical can be written with rational exponents. Formulas that involve exponents are also applications of rational exponents when those exponents are rational. There are many formulas from physics, meteorology, oceanography, and business that contain rational exponents or radicals. Example 6 illustrates two of these applications.

**Objective 6 Examples** Solve each problem.

**6a.** The compound interest formula is given by $A = P\left(1 + \dfrac{r}{n}\right)^{nt}$, where $A$ is the amount of money in an account in $t$ years if $P$ dollars is invested at an annual interest rate $(r)$, compounded $n$ times a year. How much money is in an account after 6 months if \$3000 is invested at 5% interest compounded annually?

**Solution**

**6a.**

$A = P\left(1 + \dfrac{r}{n}\right)^{nt}$     State the formula.

$A = 3000\left(1 + \dfrac{0.05}{1}\right)^{1(1/2)}$     Let $P = 3000$, $r = 0.05$, $n = 1$, and $t = \dfrac{1}{2}$.

$A = 3000(1.05)^{1/2}$     Simplify the expression inside the parentheses. Simplify the power.

$A = 3000(1.024695)$     Apply the exponent.

$A = \$3074.09$     Multiply by 3000.

There will be approximately \$3074.09 in the account after 6 months.

**6b.** The wind chill temperature, WC, in degrees Fahrenheit, can be determined by the air temperature, $T$, in °F, and the wind speed, $V$, in mph, by the given formula.

$$\text{WC} = 35.74 + 0.6215T - 35.75V^{0.16} + 0.4275TV^{0.16}$$

Use the formula to determine the wind chill if the air temperature is 45° and the wind speed is 15 mph. Round the answer to two decimal places.

**INSTRUCTOR NOTE:**
Remind students that the power $0.16 = \dfrac{16}{100} = \dfrac{4}{25}$. So, this is an application of rational exponents.

**Solution** **6b.** Replace $T$ with 45 and $V$ with 15.

$\text{WC} = 35.74 + 0.6215T - 35.75V^{0.16} + 0.4275TV^{0.16}$

$\text{WC} = 35.74 + 0.6215(45) - 35.75(15)^{0.16} + 0.4275(45)(15)^{0.16}$

$\text{WC} \approx 35.74 + 27.9675 - 35.75(1.54232) + 0.4274(45)(1.54232)$

$\text{WC} \approx 35.74 + 27.9675 - 55.13794 + 29.66344$

$\text{WC} \approx 38.23$

So, the wind chill temperature is approximately 38.23°F when the air temperature is 45°F and the wind speed is 15 mph.

✓ **Student Check 6**    Solve each problem.

  **a.** How much money is in an account after 3 months if $5000 is invested at 4% annual interest compounded quarterly ($n = 4$)?

  **b.** Use the wind chill temperature formula to calculate the wind chill when the air temperature is 20°F and the wind speed is 30 mph. Round the answer to two decimal places.

**Objective 7** ▶

Troubleshoot common errors.

## Troubleshooting Common Errors

Some common errors associated with rational exponents are illustrated next.

**Objective 7 Examples**    A problem and an incorrect solution are given. Provide the correct solution and explanation of the error.

**7a.** Simplify $(-27)^{2/3}$.

| Incorrect Solution | Correct Solution and Explanation |
|---|---|
| $(-27)^{2/3} = (\sqrt[3]{-27})^2 = -3^2 = -9$ | The correct radical form was used. The error was not applying the exponent of 2 to $-3$. $$(-27)^{2/3} = (\sqrt[3]{-27})^2 = (-3)^2 = 9$$ |

**7b.** Simplify $5^{1/4} \cdot 5^{3/4}$.

| Incorrect Solution | Correct Solution and Explanation |
|---|---|
| $5^{1/4} \cdot 5^{3/4} = 25^{4/4} = 25^1 = 25$ | When we multiply like bases, the bases stay the same. $$5^{1/4} \cdot 5^{3/4} = 5^{4/4} = 5^1 = 5$$ |

## ANSWERS TO STUDENT CHECKS

**Student Check 1**    **a.** 5    **b.** $-4$    **c.** $\sqrt[5]{x}$    **d.** $-6$
  **e.** not a real number    **f.** $3x^3$

**Student Check 2**    **a.** 27    **b.** $-27$    **c.** 9    **d.** $\dfrac{1}{8}$
  **e.** $\sqrt[7]{(x+2)^6}$

**Student Check 3**    **a.** $\dfrac{1}{2}$    **b.** $\dfrac{1}{9}$    **c.** $-\dfrac{1}{125}$

**Student Check 4**    **a.** 7    **b.** $\dfrac{1}{y^{3/7}}$    **c.** $x^{1/2}$    **d.** $\dfrac{1}{3}$    **e.** $a^5$
  **f.** $27x^9$    **g.** $\dfrac{8}{x^4 y^{3/2}}$    **h.** $x^{(1/2)} - 3x$

**Student Check 5**    **a.** $x^4$    **b.** $\sqrt[4]{x}$    **c.** $\sqrt{2}$    **d.** $\sqrt[12]{5^7}$
  **e.** $y^{8/15}$    **f.** $z^9$

**Student Check 6**    **a.** $5050.00    **b.** 1.30°F

## SUMMARY OF KEY CONCEPTS

1. Radical expressions can be converted to exponential form, or vice versa, by using the formulas

$$b^{1/n} = \sqrt[n]{b}$$

$$b^{m/n} = (\sqrt[n]{b})^m \text{ or } \sqrt[n]{b^m}$$

$$b^{-m/n} = (\sqrt[n]{b})^{-m} = \frac{1}{(\sqrt[n]{b})^m}$$

2. All of the exponent rules are defined for rational exponents. For rational numbers $m$ and $n$, and real numbers $a$ and $b$ such that the following values exist,

   **a.** Power of like bases rule: $b^m \cdot b^n = b^{m+n}$

   **b.** Power of a power rule: $(b^m)^n = b^{mn}$

   **c.** Quotient of like bases rule: $\dfrac{b^m}{b^n} = b^{m-n}, b \neq 0$

   **d.** Power of a product rule: $(ab)^m = a^m b^m$

   **e.** Power of a quotient rule: $\left(\dfrac{a}{b}\right)^m = \dfrac{a^m}{b^m}, b \neq 0$

   **f.** Zero exponent: $b^0 = 1, b \neq 0$

   **g.** Negative exponent: $b^{-n} = \dfrac{1}{b^n}, b \neq 0$

3. Radical expressions can often be simplified if we convert them to exponential form. This enables us to reduce the exponents and, thereby, simplify the expression. After the exponents are simplified, the expression be converted back to radical form.

## GRAPHING CALCULATOR SKILLS

We can use the graphing calculator to simplify expressions with rational exponents. The key is to put parentheses around the entire exponent.

**Example:** Simplify $(-32)^{6/5}$.

**Solution:**

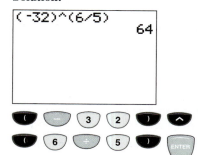

**Example:** Simplify $\sqrt[4]{12}$ using rational exponents.

**Solution:** We convert $\sqrt[4]{12}$ to $12^{1/4}$ and then follow the previous steps to enter it on the calculator. We get $12^{1/4} \approx 1.86$.

## SECTION 8.6 / EXERCISE SET

 **Write About It!**

Use complete sentences in your answer to each exercise.

1. Explain how a radical expression can be written in exponential form.   Answers vary.

2. Explain a situation in which the exponential form of a radical is more useful than the radical form.   Answers vary.

3. Use an example to explain how to apply the rules of exponents to a product of two terms with rational exponents.   Answers vary.

4. Use an example to explain how to apply the rules of exponents to a quotient of two terms with rational exponents.   Answers vary.

 **Practice Makes Perfect!**

Rewrite each expression as a radical expression and simplify, if possible. Assume all variables represent positive real numbers. (*See Objective 1.*)

5. $49^{1/2}$   7
6. $64^{1/2}$   8
7. $125^{1/3}$   5
8. $64^{1/3}$   4
9. $x^{1/3}$   $\sqrt[3]{x}$
10. $y^{1/3}$   $\sqrt[3]{y}$
11. $r^{1/2}$   $\sqrt{r}$
12. $s^{1/5}$   $\sqrt[5]{s}$
13. $-16^{1/2}$   $-4$
14. $-49^{1/2}$   $-7$
15. $(-81)^{1/2}$   not a real number
16. $(-121)^{1/2}$   not a real number
17. $(16x^8)^{1/4}$   $2x^2$
18. $(27y^{15})^{1/3}$   $3y^5$
19. $(32a^{20})^{1/5}$   $2a^4$
20. $(-343b^{12})^{1/3}$   $-7b^4$

Rewrite each expression as a radical and simplify, if possible. Assume all variables represent positive real numbers. (*See Objective 2.*)

21. $25^{3/2}$   125
22. $49^{3/2}$   343
23. $100^{5/2}$   100,000
24. $121^{1/2}$   11
25. $10,000^{3/4}$   1000
26. $169^{1/2}$   13

27. $-16^{3/4}$   $-8$
28. $-625^{3/4}$   $-125$
29. $-10,000^{1/4}$   $-10$
30. $-1296^{1/4}$   $-6$
31. $(-81)^{3/4}$   not a real number
32. $(-256)^{3/4}$   not a real number
33. $(-64)^{2/3}$   16
34. $(-125)^{2/3}$   25
35. $(-343)^{2/3}$   49
36. $(-512)^{2/3}$   64
37. $\left(\dfrac{1}{32}\right)^{2/5}$   $\dfrac{1}{4}$
38. $\left(\dfrac{1}{243}\right)^{3/5}$   $\dfrac{1}{27}$
39. $(x+1)^{2/3}$   $\sqrt[3]{(x+1)^2}$
40. $(2x+3)^{3/4}$   $\sqrt[4]{(2x+3)^3}$
41. $(3x+5)^{4/5}$   $\sqrt[5]{(3x+5)^4}$
42. $(6x-7)^{6/7}$   $\sqrt[7]{(6x-7)^6}$

Rewrite each expression with a positive exponent and simplify, if possible. (*See Objective 3.*)

43. $27^{-1/3}$   $\dfrac{1}{3}$
44. $125^{-1/3}$   $\dfrac{1}{5}$
45. $(-32)^{-3/5}$   $-\dfrac{1}{8}$
46. $(-243)^{-3/5}$   $-\dfrac{1}{27}$
47. $(-625)^{-3/4}$   not a real number
48. $(-256)^{-3/4}$   not a real number
49. $-81^{-3/2}$   $-\dfrac{1}{729}$
50. $-125^{-4/3}$   $-\dfrac{1}{625}$

Simplify each expression. Write each answer with positive exponents. Assume all variables represent positive real numbers. (*See Objective 4.*)

51. $3^{1/2} \cdot 3^{1/4}$   $3^{3/4}$
52. $5^{1/3} \cdot 5^{1/6}$   $5^{1/2}$
53. $y^{1/2} \cdot y^{-1/3}$   $y^{1/6}$
54. $x^{2/3} \cdot x^{-1/2}$   $x^{1/6}$
55. $a^{3/4} \cdot a$   $a^{7/4}$
56. $x \cdot x^{1/2}$   $x^{3/2}$
57. $\dfrac{4^{1/3}}{4^{4/3}}$   $\dfrac{1}{4}$
58. $\dfrac{5^{1/2}}{5^{3/2}}$   $\dfrac{1}{5}$
59. $\dfrac{6^{1/5}}{6^{6/5}}$   $\dfrac{1}{6}$
60. $\dfrac{7^{3/5}}{7^{8/5}}$   $\dfrac{1}{7}$
61. $(a^4)^{3/4}$   $a^3$
62. $(b^3)^{5/3}$   $b^5$
63. $(16x^8)^{3/2}$   $64x^{12}$
64. $(25y^6)^{3/2}$   $125y^9$
65. $(8a^6)^{5/3}$   $32a^{10}$
66. $(125b^{12})^{4/3}$   $625b^{16}$
67. $\left(\dfrac{4a^{1/3}b^{-1/2}}{a^2}\right)^3$   $\dfrac{64}{a^5 b^{3/2}}$
68. $\left(\dfrac{2c^{-3/5}d^{-2/5}}{c}\right)^5$   $\dfrac{32}{c^8 d^2}$
69. $\left(\dfrac{3x^{-3/10}y^{-1/5}}{y^2}\right)^5$   $\dfrac{243}{x^{3/2}y^{11}}$
70. $\left(\dfrac{6r^{-5/6}s^{-1/3}}{s}\right)^3$   $\dfrac{216}{r^{5/2}s^4}$
71. $x^{1/6}(x^{1/6} - 4x^{5/6})$   $x^{1/3} - 4x$
72. $y^{1/2}(y^{1/2} - 7y^{3/4})$   $y - 7y^{5/4}$
73. $a^{2/3}(a^{1/3} + 2a^{1/6})$   $a + 2a^{5/6}$
74. $b^{4/5}(b^{1/5} + 8b^{3/10})$   $b + 8b^{11/10}$

Additional answers can be found in the Instructor Answer Appendix.

**Use rational exponents to simplify each expression. Write each answer in radical form, if possible. Assume that variables represent positive real numbers.** (*See Objective 5.*)

75. $\sqrt{x^{10}}$    $x^5$

76. $\sqrt{y^{16}}$    $y^8$

77. $\sqrt[6]{a^2}$    $\sqrt[3]{a}$

78. $\sqrt[8]{b^6}$    $\sqrt[4]{b^3}$

79. $\sqrt[10]{r^4}$    $\sqrt[5]{r^2}$

80. $\sqrt[4]{49}$    $\sqrt{7}$

81. $\sqrt[6]{8}$    $\sqrt{2}$

82. $\sqrt[6]{125}$    $\sqrt{5}$

83. $\sqrt[5]{2} \cdot \sqrt{2}$    $\sqrt[10]{128}$

84. $\sqrt[5]{3} \cdot \sqrt{3}$    $\sqrt[10]{243}$

85. $\sqrt[5]{x^3} \cdot \sqrt{x}$    $\sqrt[10]{x^{11}}$

86. $\sqrt{b} \cdot \sqrt[4]{b^3}$    $\sqrt[4]{b^5}$

87. $\dfrac{\sqrt[4]{x^5}}{\sqrt[3]{x}}$    $\sqrt[12]{x^{11}}$

88. $\dfrac{\sqrt[3]{y^4}}{\sqrt[5]{y^2}}$    $\sqrt[15]{y^{14}}$

89. $\dfrac{\sqrt{a}}{\sqrt[3]{a}}$    $\sqrt[6]{a}$

90. $\dfrac{\sqrt{b}}{\sqrt[3]{b^2}}$    $\dfrac{1}{\sqrt[6]{b}}$

91. $\left(\sqrt[3]{a}\right)^9$    $x^3$

92. $\left(\sqrt[4]{b^3}\right)^8$    $b^6$

**Solve each problem.** (*See Objective 6.*)

**For Exercises 93 and 94, use the compound interest formula** $A = P\left(1 + \dfrac{r}{n}\right)^{nt}$, **where** $A$ **is the amount of money in an account in** $t$ **years if** $P$ **dollars is invested at an annual interest rate** $r$ **compounded** $n$ **times a year.**

93. How much money is in an account after 4 months if $2000 is invested at 3% annual interest compounded quarterly ($n = 4$)?    $2020.02

94. How much money is an account after 6 months if $4500 is invested at 2% annual interest compounded monthly ($n = 12$)?    $4545.19

**For Exercises 95 and 96, use the wind chill formula,** $WC = 35.74 + 0.6215T - 35.75V^{0.16} + 0.4275TV^{0.16}$, **where** $T$ **is the air temperature in °F and** $V$ **is the wind speed in mph. Round each answer to two decimal places.**

95. What is the wind chill temperature when the air temperature is 32°F and the wind speed is 20 mph?    19.99 °F

96. What is the wind chill temperature when the air temperature is 0°F and the wind speed is 25 mph?    −24.09°F

 **Mix 'Em Up!**

**Simplify each expression. Write answers with positive exponents, when applicable. Assume all variables represent positive real numbers.**

97. $\sqrt[4]{(x^2 - 1)^9}$    $(x^2 - 1)^{9/4}$

98. $\sqrt[3]{(y^2 + 1)^5}$    $(y^2 + 1)^{5/3}$

99. $625^{3/4}$    125

100. $\sqrt[3]{5} \cdot \sqrt{5}$    $3125^{1/6}$

101. $\sqrt[4]{2} \cdot \sqrt[3]{2}$    $128^{1/12}$

102. $\left(-\dfrac{1}{32}\right)^{4/5}$    $\dfrac{1}{16}$

103. $\left(\dfrac{1}{243}\right)^{2/5}$    $\dfrac{1}{9}$

104. $81^{1/4}$    3

105. $\left(\dfrac{9}{25}\right)^{-1/2}$    $\dfrac{5}{3}$

106. $1000^{-1/3}$    $\dfrac{1}{10}$

107. $a^{-1/5} \cdot a^{2/3}$    $a^{7/15}$

108. $y^{1/4} \cdot y^{1/6}$    $y^{5/12}$

109. $\left(\dfrac{3a^{-1/6}b^{-2/3}}{b^3}\right)^3$

110. $\left(\dfrac{2r^{-5/6}s^{-7/12}}{r}\right)^6$

111. $(-625)^{1/4}$    not a real number

112. $(-169)^{1/2}$    not a real number

113. $(32x^5)^{3/5}$    $8x^3$

114. $(27y^{15})^{2/3}$    $9y^{10}$

115. $x^{1/3}(x^{2/3} - 4x^{1/6})$    $x - 4x^{1/2}$

116. $y^{3/2}(y^{1/2} + 6y^{1/3})$    $y^2 + 6y^{11/6}$

117. $\dfrac{7^{1/4}}{7^{1/2}}$    $\dfrac{1}{7^{1/4}}$

118. $\dfrac{9^{1/2}}{9^{1/3}}$    $9^{1/6}$

119. $(8a^6)^{7/3}$    $128a^{14}$

120. $(27b^{12})^{5/3}$    $243b^{20}$

**Use rational exponents to simplify each expression. Write each answer in radical form, if possible. Assume that variables represent positive real numbers.**

121. $\sqrt{x} \cdot \sqrt[3]{x^5}$    $\sqrt[6]{x^{13}}$

122. $\sqrt{y^3} \cdot \sqrt[4]{y}$    $\sqrt[4]{y^7}$

123. $\dfrac{\sqrt[4]{s^3}}{\sqrt[3]{s^5}}$    $\dfrac{1}{\sqrt[12]{s^{11}}}$

124. $\dfrac{\sqrt[3]{t^2}}{\sqrt[4]{t}}$    $\sqrt[12]{t^5}$

125. $(-81)^{-1/4}$    not a real number

126. $(-625)^{-5/4}$    not a real number

127. $-100^{-3/2}$    $-\dfrac{1}{1000}$

128. $-121^{-3/2}$    $-\dfrac{1}{1331}$

129. $\left(\sqrt{x^3}\right)^4$    $x^6$

130. $\left(\sqrt[3]{b^5}\right)^6$    $b^{10}$

**Solve each problem.**

131. How much money is in an account after 3 months if $3600 is invested at 2.5% interest compounded annually ($n = 1$)?    $3622.29

132. How much money is in an account after 8 months if $6400 is invested at 3% annual interest compounded quarterly ($n = 4$)?    $6528.80

133. On top of Mount Everest, the wind speed can reach 115 mph and the temperature on a good day is −15°F. Use the wind chill formula to determine the wind chill temperature in these conditions.    −63.66°F.

134. The formula $WS = 3.13d^{1/2}$ gives the wave speed, in meters per second (mps), of surface waves in water that is $d$ meters deep. The fastest surface waves are those caused by tsunamis. The average ocean depth is 4000 m (about 2.5 mi). Estimate the wave speed of a tsunami.    197.96 mps (about 443 mph)

 **You Be the Teacher!**

**Correct each student's errors, if any.**

135. Simplify. $\sqrt[3]{x} \cdot \sqrt{x} = \sqrt[6]{x}$.

    Phil's work: $\sqrt[3]{x} \cdot \sqrt{x} = \sqrt[6]{x}$.

136. Simplify $\dfrac{\sqrt{x}}{\sqrt[4]{x}} = \sqrt[4]{x}$.

    Sophia's work: $\dfrac{\sqrt{x}}{\sqrt[4]{x}} = \sqrt[4]{x}$.

 **Calculate It!**

**Use a calculator to evaluate each expression. Round each answer to two decimal places when necessary.**

137. $15,625^{4/5}$    2264.94

138. $1296^{3/2}$    46,656

139. $729^{-1/6}$    0.33

140. $1024^{-1/10}$    0.5

**For Exercises 141–144, state the rational expression needed to simplify each radical and state its value rounded to two decimal places.**

141. $\sqrt[5]{20}$    $20^{1/5}$, 1.82

142. $\sqrt[7]{42}$    $42^{1/7}$, 1.71

143. $\sqrt[6]{54}$    $54^{1/6}$, 1.94

144. $\sqrt[4]{200}$    $200^{1/4}$, 3.76

## GROUP ACTIVITY   On a Clear Day

**Suppose it's a clear day and we have a view with no obstructions. How far would we be able to see to the horizon? We can use the Pythagorean theorem and other concepts in this chapter to find this distance.**

1. How many feet are in a mile?   5280 ft = 1 mi

   **a.** Convert 15 ft into miles.   $\dfrac{1}{352}$ mi   or   0.00284 mi

   **b.** Express the quantity $h$ ft in miles.   $\dfrac{h}{5280}$ mi

2. In the given diagram, $h$ represents a person's height (in feet) above the Earth, $r$ is the radius (in miles) of the Earth, and $d$ is the distance a person can see to the horizon (in miles). Replace $h$ with the quantity that represents $h$ ft in miles. Use the Pythagorean Theorem to write an equation that relates $r$, $d$, and $h$. Do not simplify the equation.   $\left(\dfrac{h}{5280}+r\right)^2 = r^2 + d^2$

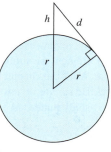

3. The radius of Earth is about 3960 mi. The distance (in miles) that a person can see to the horizon is given by $d = 1.225\sqrt{h}$, where $h$ is the person's height (in feet) above the Earth.

   **a.** How far can you see on a clear day with no obstructions? Round to two decimal places.   Answers will vary based on the person's height. A person 5 ft tall can see $d \approx 2.74$ mi

   **b.** How far can you see if you are in a plane that is 4000 ft above the Earth? Round to two decimal places.   77.48 mi

4. Use the formula $d = 1.225\sqrt{h}$ for $h$, where $d$ is in miles and $h$ is in feet to find how high above the Earth a person would need to be to see 7 mi to the horizon.   32.65 ft

5. The highest peak in the United States is Mount McKinley in Alaska at 20,320 ft. above the Earth's surface. How far can you see to the horizon if you are on top of Mount McKinley? Round to two decimal places.   174.62 mi

# Chapter 8 / REVIEW

# Radical Expressions and Equations

> **What's the big idea?** Now that you have completed Chapter 8, you should be able to see a connection between radical expressions and exponents. You should also be able to simplify radicals, perform operations with radicals, and solve equations containing radical expressions.

## The Tools

Listed below are the key terms, skills, formulas, and properties you should know for this chapter.

The page reference is provided if you need additional help with the given topic. The Study Tips will assist in your preparation for an exam.

## Study Tips

1. Learn all of the terms, formulas, and properties. Make flash cards and have someone quiz you.
2. Rework problems from the exercises and also the ones you worked in class. Work additional problems from the review exercises.
3. Review the summaries of key concepts.
4. Work the chapter test.
5. Be sure to review the online resources for additional study materials.

## Terms

Cube root   590
Extraneous solution   629
Index   590
Like radicals   609
Negative square root   588
nth root   591

Perfect cube   591
Perfect square   591
Principal or positive square
   root   588
Radical   588
Radical expression   588

Radical sign   588
Radicand   588
Rational or fractional exponent   639
Rationalizing the denominator   621
Square root   588

## Formulas and Properties

- $\sqrt[n]{a^n}$   617
- $b^{1/n}$   640
- $b^{m/n}$   641
- $b^{-m/n}$   642
- Cube root   590
- nth root   591

- Product rule for roots   602
- Product rule for square
   roots   597
- Pythagorean theorem   633
- Quotient rule for roots   602

- Quotient rule for square roots   601
- Rules of exponents   642
- Square root notation   588
- Square root of $a^2$   593
- Squaring principle   628

## CHAPTER 8 / SUMMARY

How well do you know this chapter? Complete the following questions to find out. Take a look back at the section if you need help.

### SECTION 8.1 Radical Expressions

1. The square root of a real number $A$ is a real number $b$ if $\underline{b^2 = A}$.

2. The symbol $\underline{\sqrt{\phantom{x}}}$ denotes the principal or $\underline{positive}$ square root.

3. The number inside the radical is called the $\underline{radicand}$.

4. If $A < 0$, then the $\sqrt{A}$ is $\underline{not}$ $\underline{a}$ $\underline{real}$ $\underline{number}$.

5. The cube root of a real number $A$ is a real number $b$ if $\underline{b^3 = A}$.

6. The nth root of a real number $A$ is a real number $b$ if $\underline{b^n = A}$.

7. If $n$ is an even number, the nth root of $A$ is a real number as long as $A$ is $\underline{nonnegative}$.

8. $\sqrt{a^2} = \underline{a}$ for $a \geq 0$ and $\sqrt{a^2} = \underline{|a|}$ for $a < 0$.

## SECTION 8.2 Simplifying Radicals and the Distance Formula

**9.** The product rule for square roots states that for $a$ and $b$ real numbers, $\sqrt{ab} =$ <u>$\sqrt{a}\sqrt{b}$ where $a, b \geq 0$</u>.

**10.** For a square root to be in its simplest form, the radicand cannot contain any <u>perfect</u> <u>square</u> factors or any exponents greater than or equal to <u>2</u>.

**11.** A variable raised to a(n) <u>even</u> exponent is a perfect square. An example is <u>$x^4$ (answers vary)</u>.

**12.** The quotient rule for square roots states that for $a$ and $b$ real numbers, $\sqrt{\dfrac{a}{b}} = \dfrac{\sqrt{a}}{\sqrt{b}}$ <u>where $a, b \geq 0$</u>.

**13.** The product rule for roots states that for $a$ and $b$ positive real numbers and $n = 2, 3, 4 \ldots$, $\sqrt[n]{ab} =$ <u>$\sqrt[n]{a}\sqrt[n]{b}$</u> and the quotient rule for roots states that $\sqrt[n]{\dfrac{a}{b}} = $ <u>$\dfrac{\sqrt[n]{a}}{\sqrt[n]{b}}$</u>.

## SECTION 8.3 Adding and Subtracting Radical Expressions

**14.** Like radicals are radicals with the same <u>radicand</u> and the same <u>index</u>.

**15.** To add like radicals, combine the <u>coefficients</u> of the like radical expressions then keep the <u>radicand</u> the same.

**16.** Radicals should be <u>simplified</u> before adding.

## SECTION 8.4 Multiplying and Dividing Radical Expressions

**17.** We can multiply radicals with the same <u>index</u>. In this case, we multiply the <u>radicands</u> and simplify the product.

**18.** For $a \geq 0$, $\left(\sqrt[n]{a}\right)^n = $ <u>$a$</u>.

**19.** We can multiply radical expressions by applying the <u>distributive</u> property and the <u>FOIL</u> method.

**20.** The product of <u>conjugates</u> doesn't contain a radical.

**21.** We can divide radicals with the same <u>index</u>. In this case, divide the <u>radicands</u> and simplify the quotient.

**22.** The process of removing a radical from the denominator of a fraction is called <u>rationalizing the denominator</u>.

**23.** To remove the radical from the denominator, we must multiply the fraction by a form of <u>one</u> that makes the denominator a perfect square, cube, and so on, if the <u>index</u> is 2, 3, and so on.

**24.** To remove the radical from the denominator if the denominator is a binomial, we must multiply by a form of one that involves the <u>conjugate</u> of the denominator.

## SECTION 8.5 Solving Square Root Equations and Applications Involving Square Roots

**25.** The squaring principle states that if $a = b$, then <u>$a^2 = b^2$</u>.

**26.** When we apply the squaring principle, we obtain the solutions of the original equation but we may also obtain <u>extraneous</u> solutions. Therefore, we must <u>check</u> the proposed solutions in the original equation.

**27.** To apply the squaring principle, the radical expression must be <u>isolated</u> on one side of the equation.

**28.** If an equation contains more than one <u>radical</u> expression, we may have to apply the squaring principle <u>twice</u>.

**29.** The Pythagorean theorem states that if $a$ and $b$ are legs of a right triangle and $c$ is a hypotenuse, then <u>$a^2 + b^2 = c^2$</u>.

## SECTION 8.6 Rational Exponents

**30.** An exponent with a fraction is called a(n) <u>rational</u> exponent.

**31.** For $n$ a positive integer greater than 1 and $\sqrt[n]{b}$ a real number, $b^{1/n} = $ <u>$\sqrt[n]{b}$</u>.

**32.** If $m$ and $n$ are positive integers greater than 1 and $\sqrt[n]{b}$ is a real number, then $b^{m/n} = $ <u>$\sqrt[n]{b^m}$</u>.

**33.** For $b \neq 0$, $b^{-m/n} = $ <u>$\dfrac{1}{b^{m/n}}$</u>.

**34.** The exponent rules work for integer exponents but also for <u>rational</u> exponents.

## CHAPTER 8 / REVIEW EXERCISES

### SECTION 8.1

**Evaluate each expression. Approximate irrational numbers to the nearest hundredth. Assume all variables represent positive real numbers. (*See Objectives 1–5.*)**

**1.** $\sqrt{-49}$   not a real number

**2.** $\sqrt{\dfrac{121}{100}}$   $\dfrac{11}{10}$

**3.** $\sqrt{144r^8}$   $12r^4$

**4.** $\sqrt{49s^{24}}$   $7s^{12}$

**5.** $\sqrt{28}$   5.29

**6.** $\sqrt[3]{216}$   6

**7.** $\sqrt{625m^4n^2}$   $25m^2n$

**8.** $-\sqrt{2500}$   $-50$

**9.** $-\sqrt{-900}$   not a real number

**10.** $-\sqrt[3]{\dfrac{27}{64}}$   $-\dfrac{3}{4}$

**11.** $\sqrt[4]{\dfrac{256}{81a^4}}$   $\dfrac{4}{3a}$

**12.** $\sqrt{\dfrac{49r^6}{121s^2}}$   $\dfrac{7r^3}{11s}$

**13.** $\sqrt[4]{64c^{12}d^{20}}$   $2.83c^3d^5$

**14.** $\sqrt[3]{-\dfrac{27x^{15}}{1000}}$   $-\dfrac{3x^5}{10}$

**15.** $\sqrt[3]{343r^9s^{12}}$   $7r^3s^4$

**16.** $\sqrt[5]{-243x^{15}y^{25}}$   $-3x^3y^5$

### SECTION 8.2

**Simplify each radical. Assume all variables represent positive real numbers. (*See Objectives 1–4.*)**

**17.** $3\sqrt{98}$   $21\sqrt{2}$

**18.** $\sqrt{288}$   $12\sqrt{2}$

**19.** $\sqrt{x^{17}}$   $x^8\sqrt{x}$

**20.** $\sqrt[3]{32y^{14}}$   $2y^4\sqrt[3]{4y^2}$

**21.** $\sqrt{96r^{21}}$   $4r^{10}\sqrt{6r}$

**22.** $\sqrt{-48a^4}$   not a real number

**23.** $\sqrt[3]{-40y^{20}}$   $-2y^6\sqrt[3]{5y^2}$

**24.** $\sqrt[3]{54x^{29}}$   $3x^9\sqrt[3]{2x^2}$

**25.** $\sqrt{\dfrac{4b^6}{9}}$   $\dfrac{2b^3}{3}$

26. $\sqrt[3]{-3000c^4}$    27. $5\sqrt[5]{160a^{22}b^{10}}$   28. $8\sqrt[4]{1250r^{10}t^{23}}$
    $-10c\sqrt[3]{3c}$     $10a^4b^2\sqrt[5]{5a^2}$        $40r^2t^5\sqrt[4]{2r^2t^3}$

29. $3\sqrt[4]{\dfrac{112x^{17}}{81y^{12}}}$   30. $\sqrt[3]{-96c^{13}d^5}$   31. $\sqrt[5]{\dfrac{729u^9}{v^{10}}}$   $\dfrac{3u\sqrt[5]{3u^4}}{v^2}$
                $-2c^2d\sqrt[3]{3c^3}$

## SECTION 8.3

**Perform the indicated operation and write each answer in simplest form. Assume all variables represent positive real numbers. (*See Objectives 1 and 2.*)**

32. $\sqrt{7} + 11\sqrt{7}$   $12\sqrt{7}$   33. $5\sqrt{98} - 2\sqrt{18}$   $29\sqrt{2}$

34. $7x\sqrt{12} - 6\sqrt{27x^2}$   35. $15\sqrt{50} + 9\sqrt{32}$   $111\sqrt{2}$
    $-4x\sqrt{3}$

36. $\sqrt{24} - \sqrt{2}$         37. $3\sqrt{12} - 2\sqrt{48} + 3\sqrt{18}$
    $2\sqrt{6} - \sqrt{2}$            $-2\sqrt{3} + 9\sqrt{2}$

38. $7\sqrt[3]{8x^4} - 5x\sqrt[3]{64x}$   39. $12\sqrt[3]{16} - 5\sqrt[3]{54}$   $9\sqrt[3]{2}$
    $-6x\sqrt[3]{x}$

40. $\sqrt{90} - \sqrt{40}$   $\sqrt{10}$   41. $5\sqrt[4]{16x^8y^5} - 2\sqrt[4]{81x^8y^5}$
                        $4x^2y\sqrt[4]{y}$

42. $15b\sqrt{24a^5} - 4\sqrt{54a^5b^2}$   $18a^2b\sqrt{6a}$

43. $\sqrt[3]{250a^9b} - 10a^3\sqrt[3]{54b}$   $-25a^3\sqrt[3]{2b}$

44. $\sqrt[5]{32m^5n} + \sqrt[5]{243m^5n}$   $5m\sqrt[5]{n}$

45. Find the perimeter of the triangle.   $9\sqrt{2}$ cm

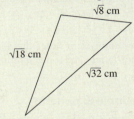

$\sqrt{8}$ cm

$\sqrt{18}$ cm

$\sqrt{32}$ cm

46. Find the perimeter of the rectangle.   $22\sqrt{2}$ in.

$\sqrt{72}$ in.

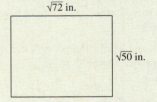

$\sqrt{50}$ in.

## SECTION 8.4

**Perform the indicated operation and write each answer in simplest form. Rationalize denominators when necessary. Assume all variables represent positive real numbers. (*See Objectives 1–6.*)**

47. $\sqrt{42x} \cdot \sqrt{6x}$   48. $(4 - \sqrt{7})^2$   49. $\dfrac{\sqrt{15w}}{\sqrt{3w}}$   $\sqrt{5}$
    $6x\sqrt{7}$       $23 - 8\sqrt{7}$

50. $\dfrac{6}{\sqrt[3]{56}}$   $\dfrac{3\sqrt[3]{49}}{7}$   51. $(3\sqrt{7} + 1)(2\sqrt{7} + 5)$   $47 + 17\sqrt{7}$

52. $(7\sqrt{3})^2$   $147$   53. $\dfrac{8}{\sqrt{12}}$   $\dfrac{4\sqrt{3}}{3}$

54. $(3\sqrt{2} - 4\sqrt{5})(\sqrt{2} - \sqrt{5})$   $26 - 7\sqrt{10}$

55. $(\sqrt{7} - \sqrt{x})(\sqrt{7} + \sqrt{x})$   $7 - x$

56. $(\sqrt{5} + 2)^2$   57. $(\sqrt[3]{5} - \sqrt[3]{a})(\sqrt[3]{25} - \sqrt[3]{a^5})$
    $9 + 4\sqrt{5}$        $5 - \sqrt[3]{25a} - a\sqrt[3]{5a^2} + a^2$

58. $(9\sqrt{2})^2$   $162$   59. $(\sqrt{2x + 9})^2$   60. $\dfrac{6}{\sqrt{18}}$   $\sqrt{2}$
              $2x + 9$

61. $(\sqrt{c^2 - 6c + 5})^2$         62. $(\sqrt[3]{4x + 1})^3$
    $c^2 - 6c + 5$                   $4x + 1$

63. $\dfrac{6}{4 + \sqrt{7}}$     64. $\dfrac{x - 2}{\sqrt{x} - \sqrt{2}}$
                $\sqrt{x} + \sqrt{2}$

## SECTION 8.5

**Solve each equation, or find the missing length. (*See Objectives 1 and 3.*)**

65. $7 - \sqrt{x} = 5$   {4}   66. $12 + \sqrt{x} = 15$   {9}

67. $\sqrt{x - 10} = 3$   {19}   68. $\sqrt{4 - 3x} = 7$   {−15}

69. $\sqrt{x + 27} = x - 3$   {9}   70. $\sqrt{x + 12} = 2x + 3$   $\left\{\dfrac{1}{4}\right\}$

71. $\sqrt{x^2 - 3x} = 2$   {−1, 4}   72. $\sqrt{x^2 - 5x} = 6$   {−4, 9}

73. Find $x$:   $10\sqrt{2}$     74. Find $x$:   $3\sqrt{10}$

                15                  $x$

  5                  3

        $x$                    9

## SECTION 8.6

**Simplify each expression. Write each answer with positive exponents, when applicable. (*See Objectives 1–5.*)**

75. $\sqrt[4]{xy}$   $(xy)^{1/4}$   76. $\sqrt[3]{(x^2 + 4)^5}$   77. $125^{2/3}$   25
                     $(x^2 + 4)^{5/3}$

78. $\left(\dfrac{49}{16}\right)^{1/2}$   $\dfrac{7}{4}$   79. $b^{-3/5} \cdot b^{1/3}$   $\dfrac{1}{b^{4/15}}$   80. $\dfrac{10^{3/4}}{10^{2/5}}$   $10^{7/20}$

81. $\sqrt[4]{2} \cdot \sqrt{2}$   $8^{1/4}$   82. $\sqrt{x} \cdot \sqrt[5]{x^3}$   83. $\dfrac{\sqrt[3]{r^5}}{\sqrt[4]{r^7}}$   $\dfrac{1}{r^{1/12}}$
                     $x^{11/10}$

84. $(\sqrt{x})^6$   $x^3$   85. $(\sqrt[3]{y^2})^6$   $y^4$   86. $8000^{-2/3}$   $\dfrac{1}{400}$

87. $\left(\dfrac{3r^{-1/6}s^{-2/3}}{r^{1/3}}\right)^6$   $\dfrac{729}{r^3s^4}$   88. $(-49)^{1/4}$
                          not a real number

89. $(125y^{12})^{1/3}$   90. $x^{1/4}(x^{3/4} - 5x^{1/2})$
    $5y^4$               $x - 5x^{3/4}$

91. $(1000y^9)^{4/3}$   92. $(64a^{12})^{4/3}$   93. $(-16)^{-3/4}$
    $10{,}000y^{12}$     $256a^{16}$       not a real number

94. $-(-27)^{-2/3}$   $-\dfrac{1}{9}$

**For Exercises 95 and 96, use the compound interest formula** $A = P\left(1 + \dfrac{r}{n}\right)^{nt}$, **where $A$ is the amount of money in an account in $t$ years if $P$ dollars is invested at an annual interest rate $r$ compounded $n$ times a year.**

95. How much money is in an account after 8 months if $1600 is invested at 2% interest compounded annually ($n = 1$)?   $1621.26

96. How much money is in an account after 18 months if $2800 is invested at 3% annual interest compounded quarterly ($n = 4$)?   $2928.39

# CHAPTER 8 TEST / RADICAL EXPRESSIONS AND EQUATIONS

1. $-\sqrt{25a^{16}} =$
   a. $5a^4$
   b. $-5a^4$
   c. $5a^8$
   d. $-5a^8$ ✓

2. $\sqrt{-256} =$
   a. $-16$
   b. $16$
   c. $-128$
   d. not a real number ✓

3. $\sqrt[3]{-729} =$
   a. $9$
   b. $-9$ ✓
   c. not a real number
   d. $243$

4. $-\sqrt[4]{16x^{12}} =$
   a. $-2x^3$ ✓
   b. $2x^3$
   c. $-4x^3$
   d. $4x^3$

5. The best approximation for $\sqrt[7]{12{,}516}$ is
   a. $3.849$ ✓
   b. $1788$
   c. $111.875$
   d. $783.125$

6. $\sqrt{75} - \sqrt{12} + \sqrt{27} =$
   a. $\sqrt{60}$
   b. $6\sqrt{3}$ ✓
   c. $2\sqrt{15}$
   d. $10\sqrt{3}$

7. $(\sqrt{5} - 1)(2\sqrt{5} + 7) =$
   a. $-7 + 7\sqrt{5}$
   b. $3 - 5\sqrt{5}$
   c. $3 + 5\sqrt{5}$ ✓
   d. $-7 - 5\sqrt{5}$

8. $(4\sqrt{3} - 6\sqrt{2})^2 =$
   a. $120 - 48\sqrt{6}$ ✓
   b. $-24 - 48\sqrt{6}$
   c. $120$
   d. $-24$

9. $\dfrac{5\sqrt{12}}{10\sqrt{2}} =$
   a. $\sqrt{3}$
   b. $\dfrac{\sqrt{6}}{2}$ ✓
   c. $3$
   d. $12$

10. The solution set of $\sqrt{m^2 + 5m - 8} = m + 1$ is
    a. $\{3\}$ ✓
    b. $\left\{\dfrac{9}{5}\right\}$
    c. $\{-3\}$
    d. $\{2\}$

11. The solution set of $\sqrt{x + 6} = x$ is
    a. $\{-2, 3\}$
    b. $\{-2\}$
    c. $\{3\}$ ✓
    d. $\varnothing$

12. $625^{3/4} =$
    a. $15$
    b. $25$
    c. $125$ ✓
    d. $5$

13. $4^{3/5} \cdot 4^{7/5} =$
    a. $8$
    b. $16$ ✓
    c. $256$
    d. $4$

14. $\dfrac{16^{-5/4}}{16^{-3/4}} =$
    a. $8$
    b. $4$
    c. $\dfrac{1}{8}$
    d. $\dfrac{1}{4}$ ✓

15. $(32k^{10})^{1/5} =$
    a. $2k^2$ ✓
    b. $32k^2$
    c. $32k^{50}$
    d. $2k^{50}$

**Simplify each radical. Assume all variables represent positive real numbers.**

16. $\sqrt{50x^4y^3}$  $\quad 5x^2y\sqrt{2y}$

17. $\sqrt{\dfrac{49a^6}{16}}$  $\quad \dfrac{7a^3}{4}$

18. $\sqrt[3]{40}$  $\quad 2\sqrt[3]{5}$

19. $\sqrt[3]{\dfrac{8}{27b^6}}$  $\quad \dfrac{2}{3b^2}$

20. $\sqrt[4]{162a^7}$  $\quad 3a\sqrt[4]{2a^3}$

**Perform the operation and express each answer in simplified radical form. Rationalize the denominators when necessary.**

21. $\sqrt{6x} \cdot \sqrt{2x}$  $\quad 2x\sqrt{3}$

22. $\sqrt[3]{4y^2} \cdot \sqrt[3]{6y^2}$  $\quad 2y\sqrt[3]{3y}$

23. $(\sqrt{4x - 9})^2$  $\quad 4x - 9$

24. $\dfrac{\sqrt{18b^3}}{\sqrt{2b}}$  $\quad 3b$

25. $\dfrac{1}{\sqrt{3}}$  $\quad \dfrac{\sqrt{3}}{3}$

26. $\dfrac{6\sqrt[3]{2}}{\sqrt[3]{5}}$  $\quad \dfrac{6\sqrt[3]{50}}{5}$

27. $\dfrac{\sqrt{3}}{\sqrt{5} + \sqrt{2}}$  $\quad \dfrac{\sqrt{15} - \sqrt{6}}{3}$

28. $\sqrt{48} - 5\sqrt{300}$  $\quad -46\sqrt{3}$

29. $2\sqrt[3]{16} - 4\sqrt[3]{54}$  $\quad -8\sqrt[3]{2}$

30. $(\sqrt{5} + 6)^2$  $\quad 41 + 12\sqrt{5}$

31. $(\sqrt{5} + 6)(\sqrt{5} - 6)$  $\quad -31$

**Solve each radical equation.**

32. $\sqrt{x + 2} = 3$  $\quad \{7\}$

33. $\sqrt{5x + 2} + 4 = 3$  $\quad \varnothing$

34. $\sqrt{2x + 1} = x - 1$  $\quad \{4\}$

35. A right triangle has legs that are each $4\sqrt{5}$ units. Find the simplest radical expression that represents the length of the hypotenuse. Then find the perimeter and area of the triangle.

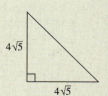

The hypotenuse is $4\sqrt{10}$ units. The perimeter is $8\sqrt{5} + 4\sqrt{10}$ units. The area is 40 square units.

## CUMULATIVE REVIEW EXERCISES / CHAPTERS 1–8

**1.** Use the order of operations to simplify each numerical expression. (*Section 1.3, Objectives 1 and 2*)

**a.** $(-2)^5$  $-32$

**b.** $-\left(-\dfrac{3}{5}\right)^3$  $\dfrac{27}{125}$

**c.** $\dfrac{1}{6}(12 - 8)^2 - \dfrac{5}{3}$  1

**d.** $2(9)^2 - 6(8) + 12$  126

**e.** $\dfrac{-16 + \sqrt{16^2 - 4(12)(-3)}}{2(12)}$  $\dfrac{1}{6}$

**2.** Evaluate each expression for the given value of the variables. (*Section 1.3, Objective 3*)

**a.** $1.4x + 2.6y$ for $x = 5$ and $y = 2$  12.2

**b.** $b^2 - 4ac$ for $a = 6, b = 9, c = 2$  33

**3.** Perform the indicated operation and simplify. (*Sections 1.4 and 1.5, Objectives 1 and 2*)

**a.** $2 - 12 + (-23) - (-10)$  $-23$

**b.** $3.1 - (-4.5) + (-6.7) + 7.2$  8.1

**c.** $-(-3)^3 + (-14) - |5 - 9|$  9

**4.** Perform the indicated operation and simplify. (*Section 1.6, Objectives 1–3*)

**a.** $(-4.5)(1.8)$  $-8.1$

**b.** $(-12)\left(\dfrac{-11}{20}\right)$  6.6

**c.** $-(-5)^3$  125

**5.** Evaluate each expression for the given values of the variables. (*Section 1.6, Objective 4*)

**a.** $\sqrt{(x_1 - x_2)^2 + (y_1 - y_2)^2}$ for $x_1 = -1, x_2 = 3,$ $y_1 = 13, y_2 = 16$  5

**b.** $\dfrac{|x - 1|}{2x - 4}$ for $x = -1, 0, 1, 2$

**6.** Solve each equation. If the equation is a contradiction, write the solution as ∅. If the equation is an identity, write the solution as ℝ. (*Section 2.2, Objective 2; Section 2.3, Objective 1; Section 2.4, Objective 2*)

**a.** $5(3a + 4) - 6(a + 1) = 3(3a + 1) - 7$  ∅

**b.** $0.014x + 0.021(870 - x) = 14.63$  {520}

**7.** The following table shows the average fare, excluding transportation taxes, per revenue passenger flying AirTran Airways between 2005 and 2009. The average fare is calculated as total passenger revenue divided by total number of passengers. (Source: http://investor .airtran.com/phoenix.zhtml?c=64267&p=irol-reportsAnnual) (*Section 3.1, Objective 5*)

| Years after 2005 | 0 | 1 | 2 | 3 | 4 |
|---|---|---|---|---|---|
| Average fare | 83.93 | 90.51 | 92.47 | 98.04 | 87.05 |

**a.** Write an ordered pair $(x, y)$ that corresponds to each year between 2005 and 2009, where $x$ is the number of years after 2005 and $y$ is the average fare, excluding transportation taxes, per revenue passenger flying AirTran Airways.

**b.** Interpret the meaning of first and last ordered pairs from part (a) in the context of the problem.

**c.** In what year was the average fare per revenue passenger the highest? The lowest?

**d.** Make a scatter plot of the data.

**8.** Write the equation of the line described. Express each answer in slope-intercept form and in standard form. (*Section 3.5, Objective 4*)

**a.** passes through the point $(-6, 1)$, perpendicular to $3x + 5y = 15$  $y = \dfrac{5}{3}x + 11; 5x - 3y = -33$

**b.** passes through the point $(-1, -4)$, parallel to $2x + y = 0$  $y = -2x - 6; 2x + y = -6$

**9.** Solve each system of equations graphically. Then state if the system is consistent with independent equations, consistent with dependent equations, or inconsistent. (*Section 4.1, Objective 3*)

**a.** $\begin{cases} 2x - y = -8 \\ x + 5y = 7 \end{cases}$

**b.** $\begin{cases} y = 5x + 2 \\ -5x + y = -5 \end{cases}$

**10.** Solve each system of linear inequalities. (*Section 4.6, Objective 1*)

**a.** $\begin{cases} y \geq 2 \\ y \leq -2x + 4 \end{cases}$

**b.** $\begin{cases} x < -2 \\ x - y > 0 \end{cases}$

**11.** Use the appropriate exponent rules to simplify each expression. (*Section 5.1, Objectives 3 and 4*)

**a.** $(-2^{-1}a^{-5}b^2)^3$  $-\dfrac{b^6}{8a^{15}}$

**b.** $\left(\dfrac{3x^2}{y^{-3}}\right)^2$  $9x^4y^6$

**12.** Use the appropriate exponent rules to simplify each expression. Express all answers with positive exponents. (*Section 5.2, Objectives 3 and 4*)

**a.** $(2^{-1}p^2)^{-5}$

**b.** $\left(\dfrac{4x^{-2}}{y^5}\right)^{-3}$

**c.** $\left(-\dfrac{a^{-5}}{3d^2}\right)^3$

**13.** Perform each operation. Write answers in standard form. (*Section 5.5, Objectives 1–3; Section 5.6, Objectives 1 and 3; Section 5.7, Objectives 1 and 2*)

**a.** $(5p^2 + 2pq - 8q^2) - (-2p^2 + 14pq + q^2) + (3p^2 - 6pq - 4q^2)$  $10p^2 - 18pq - 13q^2$

**b.** $7xy(2x^2 - xy + 3y^2)$

**c.** $(4x - 9y)(4x + 9y)$

**d.** $(a - 2b)(a + 2b)(a^2 + 4b^2)$  $a^4 - 16b^4$

**e.** Divide $28x^3 - 12x^2 + 24x + 16$ by $4x$  $7x^2 - 3x + 6 + \dfrac{4}{x}$

**f.** $\dfrac{4x^3 - 10x^2 + 21x - 18}{x - 2}$  $4x^2 - 2x + 17 + \dfrac{16}{x - 2}$

**14.** Factor each trinomial completely. (*Sections 6.1 and 6.2, Objectives 1 and 2; Section 6.3, Objectives 1–3; Section 6.4, Objectives 1–4*)

**a.** $50x^2y - 40xy^2 + 8y^3$

**b.** $a^4 - 26a^2 + 25$

**c.** $36x^2 - 3x - 5$

**d.** $27a^3 + 8b^3$

**e.** $25a^2 + b^2$  prime

**f.** $1000p^3 - q^3$

**g.** $c^6 - 1$  $(c - 1)(c^2 + c + 1)(c + 1)(c^2 - c + 1)$

**15.** The profit $p$, in millions of dollars, of a furniture company is given by $p = -6x^2 + 138x + 2520$, where $x$ is the number of units sold in thousands. How many units need to be sold for the company to break even? (*Section 6, Objectives 1–3*)  35 units

**16.** Solve each equation. (*Section 6.5, Objectives 1 and 2*)

a. $2x^2 = 9x + 35$   $\left\{-\dfrac{5}{2}, 7\right\}$   b. $(x^2 - 1)(x^2 - 12) = -24$

c. $x^3 - 3x^2 - 10x = 0$   $\{0, -2, 5\}$

**17.** The product of two consecutive odd integers is 99. Find the integers. (*Section 6.6, Objectives 1–5*)   9 and 11 or −11 and −9

**18.** Evaluate each expression for the given values. (*Section 7.1, Objective 1*)

a. $\dfrac{x^2 + x}{x - 2}$ for $x = -1, 0, 2$

b. $\dfrac{4x}{2x - 3}$ for $x = -1, 0, 1.5$

**19.** A company's revenue can be represented by $R = 0.1q^2 + 60q$, where $q$ is the number of items sold. If the average revenue per item is given by $\dfrac{R = 0.1q^2 + 60q}{q}$, what is the average revenue per item when 100 items are sold? 200 items are sold? (*Section 7.1, Objective 1*)   $70; $80

**20.** Find the values that make each expression undefined. Then simplify each expression. (*Section 7.1, Objectives 2 and 3*)

a. $\dfrac{x + 4}{3x^2 + 10x - 8}$   b. $\dfrac{2x^2 + 6x}{x^3 - 9x}$

**21.** Write two equivalent forms of each rational expression. (*Section 7.1, Objective 4*)

a. $\dfrac{x + 5}{-x + 7}$   b. $-\dfrac{2 - x}{7x + 3}$

**22.** Perform each operation and simplify the answer. (*Section 7.2, Objectives 1 and 2*)

a. $\dfrac{2x^2 + 4x}{x^2 + x - 12} \cdot \dfrac{3 - x}{x^2 + 2x}$   $-\dfrac{2}{x + 4}$

b. $\dfrac{x^2 + 6x - 27}{x^2 + 2x} \div \dfrac{x^2 - 81}{x^2 - 9x}$   $\dfrac{x - 3}{x + 2}$

c. $\dfrac{x^2 - 9}{x^3 + 3x^2 + 9x + 27} \div \dfrac{x^2 - 5x + 6}{x^4 + 5x^2 - 36}$   $x + 2$

d. $\dfrac{x^4 - 16}{x^2 - 4} \cdot \dfrac{x^2 - 3x + 2}{x^4 + 3x^2 - 4}$   $\dfrac{x - 2}{x + 1}$

**Use the given conversion equivalents to solve each problem.** (*Section 7.2, Objective 3*)

**23.** The distance between San Francisco, California, and Pittsburgh, Pennsylvania, is 2592 mi. What is this distance in inches on a map for which 1 inch = 360 mi?   7.2 in.

**24.** Convert 13.5 kilograms to pounds. (Use 1 lb = 0.45 kg.)   30 lb

**25.** Add or subtract the rational expressions. Write answers in lowest terms. (*Section 7.4, Objectives 1 and 2*)

a. $\dfrac{4x - 7}{x - 3} - \dfrac{x - 8}{3 - x}$   5   b. $\dfrac{3x^2}{x^2 - 2x - 24} - \dfrac{x + 1}{x - 6}$

c. $\dfrac{4}{x - 3} + \dfrac{2x}{x^2 + 4x - 21}$   d. $\dfrac{6}{x^2 - 1} - \dfrac{7}{x^2 + 2x + 1}$

e. $\dfrac{3}{x^3 + 8} - \dfrac{4}{2x^2 - 4x + 8}$   $\dfrac{-2x - 1}{(x + 2)(x^2 - 2x + 4)}$

**26.** Simplify each complex fraction. (*Section 7.5, Objectives 1 and 2*)

a. $\dfrac{\dfrac{8a^2b}{28x^4}}{\dfrac{6ab^2}{14x^6}}$   $\dfrac{2ax^2}{3b}$   b. $\dfrac{\dfrac{1}{a} + \dfrac{3}{4}}{6a + 8}$   $\dfrac{1}{8a}$   c. $\dfrac{\dfrac{4}{x} + \dfrac{3}{y}}{\dfrac{16}{x^2} - \dfrac{9}{y^2}}$

**27.** What is the average of $\dfrac{7}{10}, \dfrac{3}{2}$, and $\dfrac{2}{5}$? (*Section 7.5, Objective 3*)   $\dfrac{13}{15}$

**28.** A person can do a job alone in $x$ hr and another can do the same job in $y$ hr. Working together, they can do the job in $\dfrac{1}{\dfrac{1}{x} + \dfrac{1}{y}}$. (*Section 7.5, Objective 3*)

a. Aaron can paint his basement in 5 hr. His brother John can paint the same basement in 3 hr. How long it will take them to paint the basement working together?   $\dfrac{15}{8}$ hr

b. Carol can deliver newspapers in her assigned route in 2 hr. Her husband can deliver the newspapers on the same route in 5 hr. How long it will take them to deliver the newspapers working together?   $\dfrac{10}{7}$ hr

**29.** Solve each rational equation. (*Section 7.6, Objective 1*)

a. $2 + \dfrac{7}{x} = \dfrac{9}{x^2}$   $\left\{-\dfrac{9}{2}, 1\right\}$   b. $\dfrac{2}{5x - 7} = \dfrac{3}{3x - 2}$   $\left\{\dfrac{17}{9}\right\}$

c. $\dfrac{6}{2x^2 - x - 3} - \dfrac{5}{4x^2 - 9} = \dfrac{4}{2x^2 + 5x + 3}$   $\{25\}$

**30.** Solve each equation for the specified variable. (*Section 7.6, Objective 2*)

a. $\dfrac{1}{a} = \dfrac{2}{b} + \dfrac{3}{c} + \dfrac{4}{d}$ for $b$   $b = \dfrac{2acd}{cd - 3ad - 4ac}$

b. $x_m = \dfrac{x_1 + x_2}{2}$ for $x_1$ (midpoint formula)   $x_1 = 2x_m - x_2$

**31.** An airplane can travel 1350 mi in the same amount of time it takes a second plane to fly 1500 mi. If the speed of the first airplane is 50 mph slower than the speed of the second airplane, find the speed of each airplane. (*Section 7.7, Objectives 2 and 5*)   450 mph, 500 mph

**32.** Simplify each radical. Assume all variables represent positive real numbers. (*Section 8.1, Objectives 1–3; Section 8.2, Objectives 1–4*)

a. $\sqrt{-16x^4}$   not a real number   b. $\sqrt[3]{-\dfrac{2000y^{10}}{27}}$   $-\dfrac{10y^3\sqrt[3]{2y}}{3}$

c. $\sqrt{28a^6b^2}$   $2a^3b\sqrt{7}$   d. $\sqrt[3]{-216c^6d^{10}}$   $-6c^2d^3\sqrt[3]{d}$

e. $\sqrt[5]{-32x^{16}y^{20}}$   $-2x^3y^4\sqrt[5]{x}$   f. $\sqrt[4]{\dfrac{81u^9}{v^8}}$   $\dfrac{3u^2\sqrt[4]{u}}{v^2}$

**33.** Perform the indicated operation and write answers in simplest form. Assume all variables represent positive real numbers. (*Section 8.3, Objectives 1 and 2*)

a. $2\sqrt{24} + \sqrt{54}$   $7\sqrt{6}$   b. $6x\sqrt{98x} - 5\sqrt{18x^3}$   $27x\sqrt{2x}$

c. $7\sqrt{27} - 4\sqrt{75} + 2\sqrt{48}$   $9\sqrt{3}$

**d.** $6\sqrt[4]{81x^8y^{10}} + 5\sqrt[4]{16x^8y^{10}}$   $28x^2y^2\sqrt[4]{y}$

**e.** $\sqrt[3]{250a^6b} + 10a^3\sqrt[3]{54b}$   **f.** $\sqrt{900a^3b} - a\sqrt{400ab}$

**34.** Find the perimeter of the triangle. (*Section 8.3, Objective 2*)   $9\sqrt{5}$ cm

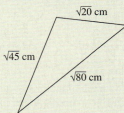

$\sqrt{20}$ cm

$\sqrt{45}$ cm

$\sqrt{80}$ cm

**35.** Perform the indicated operation and write answers in simplest form. Rationalize denominators when necessary. Assume all variables represent positive real numbers. (*Section 8.4, Objectives 1–6*)

**a.** $\sqrt{42a} \cdot \sqrt{7ab^3}$   $7ab\sqrt{6b}$   **b.** $(2 - \sqrt{5})^2$   $9 - 4\sqrt{5}$

**c.** $\dfrac{5w}{\sqrt{12w}}$   $\dfrac{5\sqrt{3w}}{6}$   **d.** $\dfrac{8}{\sqrt[3]{56a^2}}$   $\dfrac{4\sqrt[3]{49a}}{7a}$

**e.** $(4\sqrt{3} - 1)(2\sqrt{3} + 5)$   **f.** $(2\sqrt{5})^2$   $20$
  $19 + 18\sqrt{3}$

**g.** $(\sqrt{5} + \sqrt{x})^2$   $5 + 2\sqrt{5x} + x$

**h.** $(\sqrt[3]{4} - \sqrt[3]{a^2})(\sqrt[3]{16} - \sqrt[3]{a^4})$   $4 - 2\sqrt[3]{2a^2} - a\sqrt[3]{4a} + a^2$

**i.** $(\sqrt[3]{x^3 + 8x + 1})^3$   **j.** $\dfrac{15}{4 - \sqrt{7}}$   $\dfrac{5(4 + \sqrt{7})}{3}$
  $x^3 + 8x + 1$

**36.** Solve each equation, or find the missing length. (*Section 8.5, Objectives 1 and 3*)

**a.** $5 - \sqrt{x} = 2$   $\{9\}$     **b.** $\sqrt{x + 10} = x - 2$   $\{6\}$

**c.** $\sqrt{x + 14} = 3x - 2$   $\{2\}$   **d.** $\sqrt{x^2 - 24x} = 5$   $\{-1, 25\}$

**e.** Find $x$:     $5\sqrt{3}$ m

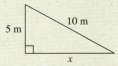

10 m

5 m

$x$

**37.** Simplify each expression. Write answers with positive exponents, when applicable. (*Section 8.6, Objectives 4 and 5*)

**a.** $\sqrt{(x^2 + 1)^5}$   $(x^2 + 1)^{5/2}$   **b.** $32^{-3/5}$   $\dfrac{1}{8}$

**c.** $a^{-2/3} \cdot a^{2/3}$   $1$     **d.** $\dfrac{7^{2/5}}{7^{3/4}}$   $\dfrac{1}{7^{7/20}}$

**e.** $\sqrt[3]{5} \cdot \sqrt{5}$   $5^{5/6}$     **f.** $\dfrac{\sqrt[4]{a^5}}{\sqrt[3]{a^8}}$   $\dfrac{1}{a^{17/12}}$

**g.** $\left(\dfrac{2^{1/5}r^{-2/5}s^{-1/3}}{r^{2/15}}\right)^{15}$   $\dfrac{8}{r^8s^5}$   **h.** $y^{1/3}(y^{1/6} - 5y^{-1/3})$
  $y^{1/2} - 5$

**For Exercises 38–39, use the compound interest formula** $A = P\left(1 + \dfrac{r}{n}\right)^{nt}$**, where** $A$ **is the amount of money in an account in** $t$ **years if** $P$ **dollars is invested at an annual interest rate** $r$ **compounded** $n$ **times a year. (*Section 8.6, Objective 6*)**

**38.** How much money is in an account after 6 months if $1000 is invested at 1% interest compounded annually ($n = 1$)?   $1004.99

**39.** How much money is in an account after 15 months if $2000 is invested at 2% annual interest compounded quarterly ($n = 4$)?   $2050.50

# Quadratic Equations

## Reflection

As you come to the end of this book and perhaps the end of this course, take a few moments to reflect on the past weeks. Think about what you will walk away with as you complete this course. Certainly that should include some math, but we hope it might also include a deeper understanding of what it takes to be successful in a math course and in college. What methods have you used that aided you in successfully understanding and retaining the material? What methods did not work? What concepts do you understand well? What topics are not as clear?

Chances are this is not the last math course required for your program of study (talk to your instructor or advisor to be sure). It is our hope that you will not take just math skills from this course but also something that can help you succeed in other classes. So, before you forget, write down the answers to the following questions.

- What worked in this course that you want to continue doing as you take other classes?
- What did you not do in this course that you want to start doing as you take other classes?
- What did you determine that you need stop doing as you take other classes?

**Question For Thought:** What will you continue doing, begin doing, and stop doing to allow you to achieve your goals?

## Chapter Outline

## Coming Up...

In Section 9.3, we will learn how to use quadratic equations to determine when the height, $h$ in feet, of a basketball, thrown upward by a player at the free-throw line reaches 10 ft when $h = -16t^2 + 22t + 7$.

> "Men go abroad to wonder at the heights of mountains, at the huge waves of the sea, at the long courses of the rivers, at the vast compass of the ocean, at the circular motions of the stars; and they pass by themselves without wondering."
>
> — St. Augustine (Philosopher and Theologian)

## SECTION 9.1 — Solve Quadratic Equations by the Square Root Property

**In Chapter 6,** quadratic equations were introduced. We solved the equations by factoring. Factoring, however, does not solve every quadratic equation. This chapter will present several additional methods that can be used to solve quadratic equations.

### ▶ OBJECTIVES

As a result of completing this section, you will be able to

1. Solve quadratic equations by the square root property.
2. Solve application problems.
3. Troubleshoot common errors.

According to the *Guinness Book of World Records*, the highest commercial bungee jump is off the Bloukrans River Bridge in South Africa. The jump takes place from a platform below the roadway of the bridge and is 216 m (710 ft) above the ground. A jumper is in free-fall until the elasticity of the cord affects the rate of fall. While the jumper is free-falling, his height, in feet, above the valley floor is $h = -16t^2 + 710$, where $t$ is the number of seconds after the jumper begins to fall. How many seconds will it take for the jumper to be 410 ft above the valley floor?

To answer this question, we must solve the equation $-16t^2 + 710 = 410$. In this section, we will learn how to solve equations of this type.

### The Square Root Property

**Objective 1** ▶

Solve quadratic equations by the square root property.

In Chapter 6, we learned that a **quadratic equation** is an equation of the form $ax^2 + bx + c = 0$, where $a$, $b$, and $c$ are real numbers with $a \neq 0$. We also learned how to solve these equations by applying the zero products property. This property is reviewed with the following equation.

$$x^2 = 36$$
$$x^2 - 36 = 0 \qquad \text{Subtract 36 from each side to set one side equal to 0.}$$
$$(x - 6)(x + 6) = 0 \qquad \text{Factor } x^2 - 36 \text{ as the difference of two squares.}$$
$$x - 6 = 0 \quad \text{or} \quad x + 6 = 0 \qquad \text{Set each factor equal to 0.}$$
$$x = 6 \quad \text{or} \quad x = -6 \qquad \text{Solve the resulting equations.}$$
$$\{-6, 6\} \quad \text{or} \quad \{\pm 6\} \qquad \text{Write the solution set.}$$

So, when the equation is of the form $x^2 = k$, where $k$ is a positive perfect square as just shown, there are two solutions of the equation. The solutions are the positive and negative square roots of $k$.

In the equation $x^2 = 36$, the value of $x$ must be a number whose square is 36. Therefore,

$$x = \sqrt{36} = 6 \quad \text{or} \quad x = -\sqrt{36} = -6$$

This result is true for *any* positive value of $k$. For example, in the equation $x^2 = 3$, the value of $x$ must be a number whose square is 3. Therefore,

$$x = \sqrt{3} \quad \text{or} \quad x = -\sqrt{3}$$

This leads to an important property for solving quadratic equations.

**INSTRUCTOR NOTE:**
We can also state this as:
if $x^2 = k$, then $x = \pm\sqrt{k}$.

> **Property: The Square Root Property**
>
> If $x^2 = k$, where $k \geq 0$, then
>
> $$x = \sqrt{k} \quad \text{or} \quad x = -\sqrt{k}$$

> **Note:** *If k < 0, then the equation has no real solutions since any real number squared is nonnegative.*

For example, the equation $x^2 = -4$ has no real solution since there is no real number whose square is $-4$. We will study this type of equation later in this chapter.

> **Procedure:  Solving a Quadratic Equation Using the Square Root Property**
>
> **Step 1:** Isolate the squared term on one side of the equation and the constant on the other side.
> **Step 2:** Remove the square by applying the square root property.
> **Step 3:** Solve each of the resulting equations. There are two real solutions if $k > 0$, one real solution if $k = 0$, and no real solutions if $k < 0$.

**Objective 1 Examples** Solve each equation by applying the square root property. Express radicals in simplest form.

**1a.** $x^2 = 49$    **1b.** $y^2 = 10$    **1c.** $a^2 + 3 = 7$    **1d.** $2y^2 - 5 = 9$
**1e.** $(x + 3)^2 = 25$    **1f.** $(3a - 4)^2 - 2 = 6$    **1g.** $(y - 2)^2 = -49$

**Solutions**   **1a.** $x^2 = 49$

$x = \sqrt{49}$  or  $x = -\sqrt{49}$       Apply the square root property.
$x = 7$         $x = -7$         Simplify each radical expression.

**Check:**  $x = 7$:                    $x = -7$:

$$x^2 = 49 \qquad\qquad\qquad x^2 = 49$$
$$(7)^2 = 49 \qquad\qquad\qquad (-7)^2 = 49$$
$$49 = 49 \qquad\qquad\qquad 49 = 49$$

Both solutions check, so the solution set is $\{-7, 7\}$ or $\{\pm 7\}$.

**1b.** $y^2 = 10$

$y = \sqrt{10}$  or  $y = -\sqrt{10}$       Apply the square root property.

**Check:**  $y = \sqrt{10}$:                $y = -\sqrt{10}$:

$$y^2 = 10 \qquad\qquad\qquad y^2 = 10$$
$$(\sqrt{10})^2 = 10 \qquad\qquad (-\sqrt{10})^2 = 10$$
$$10 = 10 \qquad\qquad\qquad 10 = 10$$

Both solutions check, so the solution set is $\{-\sqrt{10}, \sqrt{10}\}$ or $\{\pm\sqrt{10}\}$.

**1c.**      $a^2 + 3 = 7$
$a^2 + 3 - 3 = 7 - 3$                Subtract 3 from each side.
$a^2 = 4$                Simplify.
$a = \sqrt{4}$  or  $a = -\sqrt{4}$       Apply the square root property.
$a = 2$         $a = -2$         Simplify each radical expression.

**Check:**  $a = 2$:                    $a = -2$:

$$a^2 + 3 = 7 \qquad\qquad\qquad a^2 + 3 = 7$$
$$(2)^2 + 3 = 7 \qquad\qquad\qquad (-2)^2 + 3 = 7$$
$$4 + 3 = 7 \qquad\qquad\qquad 4 + 3 = 7$$
$$7 = 7 \qquad\qquad\qquad 7 = 7$$

Both solutions check, so the solution set is $\{-2, 2\}$ or $\{\pm 2\}$.

**1d.** $2y^2 - 5 = 9$

$2y^2 - 5 + 5 = 9 + 5$      Add 5 to each side.

$2y^2 = 14$      Simplify.

$\dfrac{2y^2}{2} = \dfrac{14}{2}$      Divide each side by 2.

$y^2 = 7$      Simplify.

$y = \sqrt{7}$   or   $y = -\sqrt{7}$      Apply the square root property.

**Check:**   $y = \sqrt{7}$:

$$2y^2 - 5 = 9$$
$$2(\sqrt{7})^2 - 5 = 9$$
$$2(7) - 5 = 9$$
$$14 - 5 = 9$$
$$9 = 9$$

$y = -\sqrt{7}$:

$$2y^2 - 5 = 9$$
$$2(-\sqrt{7})^2 - 5 = 9$$
$$2(7) - 5 = 9$$
$$14 - 5 = 9$$
$$9 = 9$$

Both solutions check, so the solution set is $\{-\sqrt{7}, \sqrt{7}\}$ or $\{\pm\sqrt{7}\}$.

**INSTRUCTOR NOTE:**
Emphasize the fact that in the square root property, *x* doesn't just have to be a single variable. It can be any expression squared, as long as the other side is a constant.

**1e.** $(x + 3)^2 = 25$

$x + 3 = \sqrt{25}$   or   $x + 3 = -\sqrt{25}$      Apply the square root property.

$x + 3 = 5$          $x + 3 = -5$      Simplify each radical expression.

$x + 3 - 3 = 5 - 3$    $x + 3 - 3 = -5 - 3$      Subtract 3 from each side.

$x = 2$            $x = -8$      Simplify.

**Check:**   $x = 2$:

$$(x + 3)^2 = 25$$
$$(2 + 3)^2 = 25$$
$$(5)^2 = 25$$
$$25 = 25$$

$x = -8$:

$$(x + 3)^2 = 25$$
$$(-8 + 3)^2 = 25$$
$$(-5)^2 = 25$$
$$25 = 25$$

Both solutions check, so the solution set is $\{-8, 2\}$.

**1f.** $(3a - 4)^2 - 2 = 6$

$(3a - 4)^2 - 2 + 2 = 6 + 2$      Add 2 to each side.

$(3a - 4)^2 = 8$      Simplify.

$3a - 4 = \sqrt{8}$   or   $3a - 4 = -\sqrt{8}$      Apply the square root property.

$3a - 4 = 2\sqrt{2}$        $3a - 4 = -2\sqrt{2}$      Simplify each radical expression.

$3a - 4 + 4 = 2\sqrt{2} + 4$   $3a - 4 + 4 = -2\sqrt{2} + 4$      Add 4 to each side.

$3a = 4 + 2\sqrt{2}$        $3a = 4 - 2\sqrt{2}$      Simplify.

$a = \dfrac{4 + 2\sqrt{2}}{3}$        $a = \dfrac{4 - 2\sqrt{2}}{3}$      Divide each side by 3.

**Check:**   $a = \dfrac{4 + 2\sqrt{2}}{3}$:

$$(3a - 4)^2 - 2 = 6$$

$$\left[3\left(\dfrac{4 + 2\sqrt{2}}{3}\right) - 4\right]^2 - 2 = 6$$

$$(4 + 2\sqrt{2} - 4)^2 - 2 = 6$$

$$(2\sqrt{2})^2 - 2 = 6$$

$$4(2) - 2 = 6$$

$$8 - 2 = 6$$

$$6 = 6$$

$a = \dfrac{4 - 2\sqrt{2}}{3}$:

$$(3a - 4)^2 - 2 = 6$$

$$\left[3\left(\dfrac{4 - 2\sqrt{2}}{3}\right) - 4\right]^2 - 2 = 6$$

$$(4 - 2\sqrt{2} - 4)^2 - 2 = 6$$

$$(-2\sqrt{2})^2 - 2 = 6$$

$$4(2) - 2 = 6$$

$$8 - 2 = 6$$

$$6 = 6$$

Both solutions check, so the solution set is $\left\{\dfrac{4 - 2\sqrt{2}}{3}, \dfrac{4 + 2\sqrt{2}}{3}\right\}$. These solutions are the exact solutions. We can approximate the solutions, as well.

$$\dfrac{4 + 2\sqrt{2}}{3} \approx 2.28 \quad \text{or} \quad \dfrac{4 - 2\sqrt{2}}{3} \approx 0.39$$

**1g.** $(y - 2)^2 = -49$

$$y - 2 = \sqrt{-49} \quad \text{or} \quad y - 2 = -\sqrt{-49} \qquad \text{Apply the square root property.}$$

Since $\sqrt{-49}$ is not a real number, there are no real solutions of this equation.

✓ **Student Check 1**   Solve each equation by applying the square root property. Express all radicals in simplest form.

**a.** $x^2 = 64$   **b.** $y^2 = 21$   **c.** $a^2 + 7 = 42$   **d.** $3y^2 - 8 = 22$

**e.** $(x + 7)^2 = 25$   **f.** $(5a - 7)^2 + 2 = 14$   **g.** $(y - 6)^2 = -1$

## Applications

**Objective 2** ▶

Solve application problems.

Quadratic equations model many real-life situations. Some of these include area of geometric figures, height of objects, and investment-related problems. For each of these types for problems, there are formulas that we use to set up the equations that solve the problem.

**Objective 2 Examples**   **Solve each problem.**

**2a.** The base of the Great Pyramid of Giza in Egypt is a square. If the area of the base of the pyramid is 570,780.25 ft$^2$, how long is each side of the base of the pyramid? (Source: http://www.plim.org/greatpyramid.html)

**Solution**   **2a.** $s^2 = A$                                                State the area formula.

$s^2 = 570{,}780.25$                                Replace $A$ with 570,780.25.

$s = \sqrt{570{,}780.25} \quad \text{or} \quad s = -\sqrt{570{,}780.25}$   Apply the square root property.

$s = 755.5 \qquad\qquad\qquad s = -755.5$   Simplify each radical expression.

Since $s$ represents the length of the side of the base of the pyramid, it cannot be a negative value. So, the length of each side of the base of the pyramid is 755.5 ft.

**2b.** According to the *Guinness Book of World Records*, the highest commercial bungee jump is off of the Bloukrans River Bridge in South Africa. The jump takes place from a platform below the roadway of the bridge and is 710 ft above the valley floor. A jumper is in free-fall until the elasticity of the cord affects the rate of fall. While the jumper is free-falling, his height, in feet, above the valley floor is $h = -16t^2 + 710$, where $t$ is the number of seconds after the jumper begins to fall. How many seconds will it take for the jumper to be 410 ft above the valley floor? Approximate the solution to the nearest hundredth.

**Solution** **2b.**

| | |
|---|---|
| $-16t^2 + 710 = h$ | State the given model. |
| $-16t^2 + 710 = 410$ | Replace $h$ with 410. |
| $-16t^2 + 710 - 710 = 410 - 710$ | Subtract 710 from each side. |
| $-16t^2 = -300$ | Simplify. |
| $\dfrac{-16t^2}{-16} = \dfrac{-300}{-16}$ | Divide each side by $-16$. |
| $t^2 = 18.75$ | Simplify the quotient. |
| $t = \sqrt{18.75}$ or $t = -\sqrt{18.75}$ | Apply the square root property. |
| $t \approx 4.33$ $\quad\quad$ $t \approx -4.33$ | Approximate the values. |

Since $t$ represents time, only the positive value makes sense in the context of the problem. The jumper will be 410 ft above the valley floor in approximately 4.33 sec after he jumps off the bridge.

**2c.** At what rate will $1000 need to be invested in a 2-yr CD to have $1060.90 once the CD matures? Use the formula $A = P(1 + r)^2$, where $P$ is the amount invested and $r$ is the annual interest rate.

**Solution** **2c.**

| | |
|---|---|
| $P(1 + r)^2 = A$ | State the given model. |
| $1000(1 + r)^2 = 1060.90$ | Replace $P$ with 1000 and $A$ with 1060.90. |
| $\dfrac{1000(1 + r)^2}{1000} = \dfrac{1060.90}{1000}$ | Divide each side by 1000. |
| $(1 + r)^2 = 1.0609$ | Simplify. |
| $1 + r = \sqrt{1.0609}$ or $1 + r = -\sqrt{1.0609}$ | Apply the square root property. |
| $1 + r = 1.03$ $\quad\quad$ $1 + r = -1.03$ | Simplify the radical. |
| $1 + r - 1 = 1.03 - 1$ $\quad$ $1 + r - 1 = -1.03 - 1$ | Subtract 1 from each side. |
| $r = 0.03$ $\quad\quad\quad$ $r = -2.03$ | Simplify. |

Since $r$ represents the rate, only the positive value makes sense in the context of the problem. So, the money should be invested at $r = 0.03$ or 3% annual interest rate.

✓ **Student Check 2** Solve each problem.

**a.** A square city block has an area of 435,600 ft². What is the length of the side of the block?

**b.** In 2009, the Burj Dubai became the tallest building in the world reaching 2717 ft. If a penny is dropped from the top of this building, its height, in feet, $t$ seconds after it is dropped is given by $h = -16t^2 + 2717$. How many seconds will it take for the penny to reach the ground? (Source: http://www.burjdubaiskyscraper.com/facts.html)

**c.** At what rate will $25,000 need to be invested in a 2-yr CD to have $28,090 once the CD matures? Use the formula $A = P(1 + r)^2$, where $P$ is the amount invested and $r$ is the annual interest rate.

| **Objective 3** ▶ | **Troubleshooting Common Errors** |
|---|---|
| Troubleshoot common errors. | Some common errors associated with the square root property are shown next. |

**Objective 3 Examples** | **A problem and an incorrect solution are given. Provide the correct solution and an explanation of the error.**

**3a.** Solve $(x - 2)^2 = 9$.

| **Incorrect Solution** | **Correct Solution and Explanation** |
|---|---|
| $(x - 2)^2 = 9$ <br> $x - 2 = \sqrt{9}$ <br> $x - 2 = 3$ <br> $x = 5$ <br> The solution set is $\{5\}$. | When the square root property is applied, we obtain two equations. <br><br> $(x - 2)^2 = 9$ <br> $x - 2 = \sqrt{9}$  or  $x - 2 = -\sqrt{9}$ <br> $x - 2 = 3$ $\qquad$ $x - 2 = -3$ <br> $x = 5$ $\qquad\qquad$ $x = -1$ <br> The solution set is $\{-1, 5\}$. |

**3b.** Solve $2x^2 - 1 = 9$.

| **Incorrect Solution** | **Correct Solution and Explanation** |
|---|---|
| $2x^2 - 1 = 9$ <br> $2x^2 = 10$ <br> $2x = \sqrt{10}$  or  $2x = -\sqrt{10}$ <br> $x = \dfrac{\sqrt{10}}{2}$ $\qquad$ $x = -\dfrac{\sqrt{10}}{2}$ <br> The solution set is $\left\{ -\dfrac{\sqrt{10}}{2}, \dfrac{\sqrt{10}}{2} \right\}$. | We must divide each side by 2 to isolate the squared variable. <br><br> $2x^2 - 1 = 9$ <br> $2x^2 = 10$ <br> $x^2 = 5$ <br> $x = \sqrt{5}$  or  $x = -\sqrt{5}$ <br> The solution set is $\{-\sqrt{5}, \sqrt{5}\}$. |

## ANSWERS TO STUDENT CHECKS

**Student Check 1** **a.** $\{-8, 8\}$ **b.** $\{-\sqrt{21}, \sqrt{21}\}$ **Student Check 2** **a.** 660 ft **b.** 13.03 sec **c.** 6%

**c.** $\{-\sqrt{35}, \sqrt{35}\}$ **d.** $\{-\sqrt{10}, \sqrt{10}\}$

**e.** $\{-12, -2\}$ **f.** $\left\{ \dfrac{7 - 2\sqrt{3}}{5}, \dfrac{7 + 2\sqrt{3}}{5} \right\}$

**g.** no real solution

## SUMMARY OF KEY CONCEPTS

1. When an equation is of the form $x^2 = k$, where $k \geq 0$, the square root property can be applied. The property enables us to remove the square from the expression on the left and set that expression equal to the positive and negative square root of the number on the right side of the equation.

2. If a squared expression is equal to a negative number, the equation doesn't have any *real* solutions.

3. To solve application problems involving quadratic equations, use the appropriate mathematical formula. Substitute the known values into the model and use the square root property to solve the resulting equation.

## GRAPHING CALCULATOR SKILLS

We can use a graphing calculator to approximate the solutions of quadratic equations and to check solutions.

**Example 1:** Show that $\dfrac{4 + 2\sqrt{2}}{3} \approx 2.28$ and $\dfrac{4 - 2\sqrt{2}}{3} \approx 0.39$.

**Solution:** Be sure to enter parentheses around the numerator of the fraction. Also, close the parentheses after the radicand.

```
(4+2√(2))/3
            2.276142375
(4-2√(2))/3
            .3905242918
```

**Example 2:** Show that $\dfrac{4 + 2\sqrt{2}}{3}$ is a solution of $(3x - 4)^2 = 8$.

**Solution:** Store the solution for the value of $x$.

Enter the expression on the left side of the equation. If the result agrees with the right side of the equation, then the value entered is a solution of the equation.

```
(4+2√(2))/3→X
            2.276142375
(3X-4)²
                      8
```

## SECTION 9.1   EXERCISE SET

### Write About It!

**Use complete sentences in your answer to each exercise.**

1. Explain how to apply the square root property to solve a quadratic equation.

2. Does every equation of the form $x^2 = k$ have 2 real solutions? Explain. *Every equation of the form $x^2 = k$ has 2 real solutions only if $k > 0$.*

3. Give an example of an equation that would be easier to solve using the square root property rather than by factoring.

4. Give an example of an equation that can be solved using the square root property that has 3 and $-3$ as its solutions. *Since we have solutions $x = 3$ or $x = -3$, we can square each equation and get $x^2 = 9$.*

### Practice Makes Perfect!

**Solve each equation by applying the square root property. Express radicals in simplest form. (See Objective 1.)**

5. $x^2 = 4$   $\{-2, 2\}$
6. $x^2 = 9$   $\{-3, 3\}$
7. $y^2 = 100$   $\{-10, 10\}$
8. $y^2 = 81$   $\{-9, 9\}$
9. $x^2 = 14$   $\{-\sqrt{14}, \sqrt{14}\}$
10. $x^2 = 6$   $\{-\sqrt{6}, \sqrt{6}\}$
11. $a^2 = 12$   $\{-2\sqrt{3}, 2\sqrt{3}\}$
12. $a^2 = 18$   $\{-3\sqrt{2}, 3\sqrt{2}\}$
13. $m^2 = -25$   *no real solutions*
14. $m^2 = -16$   *no real solutions*
15. $n^2 + 10 = 9$   *no real solutions*
16. $n^2 + 11 = 7$   *no real solutions*
17. $3x^2 - 2 = 25$   $\{-3, 3\}$
18. $4x^2 - 3 = 1$   $\{-1, 1\}$
19. $5a^2 + 3 = 18$   $\{-\sqrt{3}, \sqrt{3}\}$
20. $6a^2 + 5 = 17$   $\{-\sqrt{2}, \sqrt{2}\}$
21. $(x + 3)^2 = 25$   $\{-8, 2\}$
22. $(x + 5)^2 = 9$   $\{-8, -2\}$
23. $(2a + 1)^2 = 36$   $\left\{-\dfrac{7}{2}, \dfrac{5}{2}\right\}$
24. $(3a + 2)^2 = 100$   $\left\{-4, \dfrac{8}{3}\right\}$
25. $(y - 2)^2 = -36$   *no real solutions*
26. $(y - 4)^2 = -4$   *no real solutions*

27. $(2a - 4)^2 = 24$   $\{2 - \sqrt{6}, 2 + \sqrt{6}\}$
28. $(b + 2)^2 = 75$   $\{-2 - 5\sqrt{3}, -2 + 5\sqrt{3}\}$
29. $(4a - 6)^2 = 32$
30. $(5a + 1)^2 = 20$

**Solve each problem. (See Objective 2.)**

31. The Pyramid Arena in Memphis, Tennessee, is the third largest pyramid in the world. Its square base covers approximately 360,000 ft². How long is each side of the base of the pyramid? (Source: http://www.pyramidarena.com/overview.html)   600 ft

32. The base of the Washington Monument sits on 282.24 m². If the base is a square, how long is each side of the base of the monument? (Source: http://www.enchantedlearning.com)   16.8 m

33. The area of a circle is $144\pi$ in². Find the radius of the circle.   12 in.

34. The area of a circle is $62{,}500\pi$ ft². Find the radius of the circle.   250 ft

35. A group of friends organized bungee jumps from the Moyie Spring Bridge in Idaho. While the jumper is free-falling, his height, in feet, above the ground is $h = -16t^2 + 464$ ft, where $t$ is the number of seconds after the jumper begins to fall. How many seconds will it take for the jumper to be 64 ft above the ground?   5 sec

36. A group of friends organized bungee jumps from the Rio Grande High Bridge in New Mexico. While the jumper is free-falling, his height, in feet, above the ground is $h = -16t^2 + 670$ ft, where $t$ is the number of seconds after the jumper begins to fall. How many seconds will it take for the jumper to be 474 ft above the ground?   3.5 sec

*Additional answers can be found in the Instructor Answer Appendix.*

**Use the formula $A = P(1 + r)^2$, where $P$ is the amount invested and $r$ is the annual interest rate, to answer Exercises 37 and 38.**

**37.** At what rate will $1000 need to be invested in a 2-yr CD to have $1102.50 once the CD matures?   5%

**38.** At what rate will $2500 need to be invested in a 2-yr CD to have $2704 once the CD matures?   4%

 **Mix 'Em Up!**

**Solve each problem. Express radicals in simplest form.**

**39.** $5x^2 + 9 = 9$   {0}          **40.** $6x^2 - 2 = -2$   {0}

**41.** $a^2 = 20$   $\{-2\sqrt{5}, 2\sqrt{5}\}$     **42.** $b^2 = 48$   $\{-4\sqrt{3}, 4\sqrt{3}\}$

**43.** $(2x - 5)^2 = 25$   {0, 5}     **44.** $(4x + 6)^2 = 36$   {-3, 0}

**45.** $a^2 = 1.44$   {-1.2, 1.2}     **46.** $b^2 = 0.64$   {-0.8, 0.8}

**47.** $4x^2 - 1 = 8$   $\left\{-\frac{3}{2}, \frac{3}{2}\right\}$     **48.** $3x^2 + 10 = 49$   $\{-\sqrt{13}, \sqrt{13}\}$

**49.** $(5x + 4)^2 = 4$   $\left\{-\frac{6}{5}, -\frac{2}{5}\right\}$     **50.** $(3x - 2)^2 = 25$   $\left\{-1, \frac{7}{3}\right\}$

**51.** $(x - 6)^2 = -4$   no real solutions     **52.** $2x^2 + 5 = 3$   no real solutions

**53.** $(2a + 3)^2 = 72$          **54.** $(7b - 1)^2 = 52$

**55.** Alan's backyard is a square with an area of 16,900 ft². What are the dimensions of his backyard?   130 ft by 130 ft

**56.** The surface area, $S$, of a sphere is calculated using the formula $S = 4\pi r^2$, where $r$ is the radius of the sphere. If the surface area of a sphere is $32,400\pi$ ft², what is the radius of the sphere?   90 ft

 **You Be the Teacher!**

**Correct each student's errors, if any.**

**57.** Solve the equation $a^2 = 16$.

Ginger's work: $a = 4$

**58.** Solve the equation $(x + 1)^2 = 9$.

Adam's work: $x = 2$

**59.** Solve the equation $4t^2 + 1 = 9$.

Todd's work:

$$\sqrt{4t^2 + 1} = \pm\sqrt{9}$$

$$2t + 1 = \pm 3$$

$$2t = -1 \pm 3$$

$$t = \frac{-1 \pm 3}{2} \Rightarrow t = \frac{-1 + 3}{2} = \frac{2}{2} = 1 \quad \text{or} \quad t = \frac{-1 - 3}{2} = \frac{-4}{2} = -2$$

**60.** Solve the equation $(y - 3)^2 = -1$.

Desiree's work:

$$\sqrt{(y - 3)^2} = \pm\sqrt{-1}$$

$$y - 3 = \pm -1$$

$$y = 3 \pm -1 \Rightarrow y = 3 + (-1) = 2 \quad \text{or} \quad y = 3 - (-1) = 4$$

Since $\sqrt{(-1)}$ is not a real number, there are no real solutions of the equation.

 **Calculate It!**

**Use a calculator to evaluate each expression. Round each answer to two decimal places.**

**61.** $\dfrac{3 - \sqrt{10}}{7}$   −0.02          **62.** $\dfrac{8 + \sqrt{11}}{5}$   2.26

**Use a calculator to show each expression is a solution of the given equation.**

**63.** $\dfrac{\sqrt{6} - 2}{3}$; $(3x + 2)^2 = 6$

**64.** $\dfrac{\sqrt{3} + 5}{6}$; $(6x - 5)^2 = 3$

---

**SECTION 9.2**   **Solve Quadratic Equations by Completing the Square**

**► OBJECTIVES**

As a result of completing this section, you will be able to

**1.** Create a perfect square trinomial.

**2.** Use completing the square to solve a quadratic equation of the form $x^2 + bx + c = 0$.

**3.** Use completing the square to solve a quadratic equation of the form $ax^2 + bx + c = 0$, $a \neq 0$, $a \neq 1$.

**4.** Troubleshoot common errors.

The expression $x^2 + 4x$ can be represented geometrically as a square with each side having length $x$ units and four rectangles $x$ units wide and 1 unit long. If the four rectangles are combined with the square, another square is almost formed. It lacks one piece from making the shape a perfect square. In this section, we will discuss what value is needed to make the expression $x^2 + 4x$ a perfect square.

### Perfect Square Trinomials

In Section 5.6, we learned about perfect square trinomials. A **perfect square trinomial** is a trinomial that results from squaring a binomial. Here are some examples.

$$\left.\begin{array}{l} (x + 3)^2 = x^2 + 6x + 9 \\ (x - 6)^2 = x^2 - 12x + 36 \\ (x - 10)^2 = x^2 - 20x + 100 \end{array}\right\} \begin{array}{l} \text{Perfect} \\ \text{square} \\ \text{trinomials} \end{array}$$

**► Objective 1 ►**

Create a perfect square trinomial.

Now we examine the relationship between the constant term and the coefficient of $x$.

| Constant Term | Coefficient of $x$ | Relationship |
|:---:|:---:|:---:|
| 9 | 6 | $9 = 3^2$ and $3 = \dfrac{6}{2} \rightarrow 9 = \left(\dfrac{6}{2}\right)^2$ |
| 36 | $-12$ | $36 = (-6)^2$ and $-6 = \dfrac{-12}{2} \rightarrow 36 = \left(\dfrac{-12}{2}\right)^2$ |
| 100 | $-20$ | $100 = (-10)^2$ and $-10 = \dfrac{-20}{2} \rightarrow 100 = \left(\dfrac{-20}{2}\right)^2$ |

This illustration shows us that, in perfect square trinomials *whose leading coefficient is 1*, the constant term is half of the coefficient of $x$, squared. We will use this fact to create a perfect square trinomial from a binomial expression. This process is called *completing the square*. Completing the square can be used as a method to solve quadratic equations, which will be shown later in this section.

> **Procedure: Completing the Square**
>
> For $x^2 + bx$ to become a perfect square trinomial,
>
> **Step 1:** Find the constant term of the trinomial by squaring half of $b$, $\left(\dfrac{b}{2}\right)^2$.
>
> **Step 2:** Add the number found in step 1 to the expression $x^2 + bx$.
>
> **Step 3:** The trinomial can be factored as $\left(x + \dfrac{b}{2}\right)^2$.
>
> $$x^2 + bx + c \text{ is a perfect square trinomial if } c = \left(\dfrac{b}{2}\right)^2.$$

**INSTRUCTOR NOTE:**

Completing the square is used in later courses to write parabolas, circles, and other conic sections in standard form.

**Objective 1 Examples**   **Make each binomial a perfect square trinomial by completing the square and then express it in factored form.**

**1a.** $x^2 + 4x$        **1b.** $y^2 - 8y$        **1c.** $x^2 - 3x$

**Solutions**   **1a.** In the binomial $x^2 + 4x$, the value of $b = 4$. So, $c = \left(\dfrac{4}{2}\right)^2 = (2)^2 = 4$.

The binomial $x^2 + 4x$ can be made into a perfect square trinomial by adding 4 to it.

$$x^2 + 4x + 4 = (x + 2)(x + 2) = (x + 2)^2$$

We can see from the diagram at the right that it takes 4 units to "complete the square" formed by $x^2$ and $4x$.

**1b.** In the binomial $y^2 - 8y$, the value of $b = -8$. So, $c = \left(\dfrac{-8}{2}\right)^2 = (-4)^2 = 16$.

The binomial $y^2 - 8y$ can be made into a perfect square trinomial be adding 16 to it.

$$y^2 - 8y + 16 = (y - 4)(y - 4) = (y - 4)^2$$

**1c.** In the binomial $x^2 - 3x$, the value of $b = -3$. So, $c = \left(\dfrac{-3}{2}\right)^2 = \left(-\dfrac{3}{2}\right)^2 = \dfrac{9}{4}$.

The binomial $x^2 - 3x$ can be made into a perfect square trinomial by adding $\dfrac{9}{4}$ to it.

$$x^2 - 3x + \dfrac{9}{4} = \left(x - \dfrac{3}{2}\right)\left(x - \dfrac{3}{2}\right) = \left(x - \dfrac{3}{2}\right)^2$$

✔ **Student Check 1**   Make each binomial a perfect square trinomial by completing the square and then express it in factored form.

**a.** $x^2 + 12x$   **b.** $y^2 - 2y$   **c.** $x^2 - 5x$

**Note:** *In Example 1, parts a–c, it is important to note that when we complete the square of a binomial, we are changing the value of the given expression. The value is changed from a binomial to a trinomial. The two expressions are not equivalent.*

## Completing the Square (Coefficient of 1 on $x^2$)

**Objective 2** ▶

Use completing the square to solve a quadratic equation of the form $x^2 + bx + c = 0$.

In Section 9.1, we used the square root property to solve quadratic equations. This method applies to equations that can be written as an expression squared equal to some number. Since perfect square trinomials can be expressed as a binomial squared, our goal is to make one side of the equation a perfect square trinomial. This will enable us to factor the perfect square trinomial and apply the square root property to solve the equation; that is, we will rewrite the equation in a form that we already know how to solve.

Consider the equation $x^2 - 6x + 2 = 0$. The trinomial $x^2 - 6x + 2$ cannot be factored, so we cannot solve the equation by factoring. We must use another method to solve this equation. We will complete the square to create a perfect square trinomial so that the square root property can be applied. To do this, we must get the binomial $x^2 - 6x$ on a side by itself and then complete the square. The addition property of equality requires that we add the number that completes the square to both sides of the equation to produce an equivalent equation.

$$x^2 - 6x + 2 = 0$$

$$x^2 - 6x + 2 - 2 = 0 - 2 \qquad \text{Subtract 2 from each side.}$$

$$x^2 - 6x = -2 \qquad \text{Simplify.}$$

$$x^2 - 6x + 9 = -2 + 9 \qquad \text{Complete the square by adding } \left(-\frac{6}{2}\right)^2 = 9 \text{ to each side.}$$

$$(x - 3)^2 = 7 \qquad \text{Factor } x^2 - 6x + 9 \text{ and simplify the right side.}$$

$$x - 3 = \sqrt{7} \quad \text{or} \quad x - 3 = -\sqrt{7} \qquad \text{Apply the square root property.}$$

$$x = 3 + \sqrt{7} \qquad x = 3 - \sqrt{7} \qquad \text{Add 3 to each side.}$$

So, the solution set of the equation is $\{3 - \sqrt{7}, 3 + \sqrt{7}\}$. We have now solved an equation that could not be solved by factoring. The steps used are summarized as follows.

**Procedure:** **Solving an Equation of the Form $x^2 + bx + c = 0$ by Completing the Square**

**Step 1:** Move the constant term to the right side of the equation, if necessary. (The variable terms must be on a side by themselves.)

**Step 2:** Complete the square by finding half of the coefficient of $x$ (or $b$) and squaring it. Add this number to each side of the equation.

**Step 3:** Factor the perfect square trinomial and simplify the right side of the equation.

**Step 4:** Apply the square root property to solve the equation.

**Step 5:** Solve the resulting equations, if possible.

**Objective 2 Examples**   Solve each equation by completing the square.

**2a.** $x^2 + 8x - 5 = 0$   **2b.** $y^2 - y - 4 = 0$

**Solutions**  **2a.**
$$x^2 + 8x - 5 = 0$$
$$x^2 + 8x - 5 + 5 = 0 + 5 \qquad \text{Add 5 to each side.}$$
$$x^2 + 8x = 5 \qquad \text{Simplify.}$$
$$x^2 + 8x + 16 = 5 + 16 \qquad \text{Add } \left(\frac{8}{2}\right)^2 = 16 \text{ to each side.}$$
$$(x + 4)^2 = 21 \qquad \text{Factor and simplify.}$$
$$x + 4 = \sqrt{21} \quad \text{or} \quad x + 4 = -\sqrt{21} \qquad \text{Apply the square root property.}$$
$$x = -4 + \sqrt{21} \qquad\qquad x = -4 - \sqrt{21} \qquad \text{Subtract 4 from each side.}$$

**Check:**  $x = -4 + \sqrt{21}$:   $\bigg|$   $x = -4 - \sqrt{21}$:

$$x^2 + 8x - 5 = 0 \qquad\qquad\qquad\qquad\qquad x^2 + 8x - 5 = 0$$
$$\left(-4 + \sqrt{21}\right)^2 + 8\left(-4 + \sqrt{21}\right) - 5 = 0 \quad \bigg| \quad \left(-4 - \sqrt{21}\right)^2 + 8\left(-4 - \sqrt{21}\right) - 5 = 0$$
$$16 - 8\sqrt{21} + 21 - 32 + 8\sqrt{21} - 5 = 0 \quad \bigg| \quad 16 + 8\sqrt{21} + 21 - 32 - 8\sqrt{21} - 5 = 0$$
$$37 - 32 - 5 = 0 \qquad\qquad\qquad\qquad\qquad 37 - 32 - 5 = 0$$
$$0 = 0 \qquad\qquad\qquad\qquad\qquad\qquad 0 = 0$$

Both solutions make the equation true, so the solution set is $\left\{-4 - \sqrt{21},\ -4 + \sqrt{21}\right\}$. We can write this more compactly as $\left\{-4 \pm \sqrt{21}\right\}$.

**2b.**
$$y^2 - y - 4 = 0$$
$$y^2 - y - 4 + 4 = 0 + 4 \qquad \text{Add 4 to each side.}$$
$$y^2 - y = 4 \qquad \text{Simplify.}$$
$$y^2 - y + \frac{1}{4} = 4 + \frac{1}{4} \qquad \text{Add } \left(\frac{-1}{2}\right)^2 = \frac{1}{4} \text{ to each side.}$$

Factor and simplify:
$$\left(y - \frac{1}{2}\right)^2 = \frac{17}{4} \qquad 4 + \frac{1}{4} = \frac{16}{4} + \frac{1}{4} = \frac{17}{4}$$

$$y - \frac{1}{2} = \sqrt{\frac{17}{4}} \quad \text{or} \quad y - \frac{1}{2} = -\sqrt{\frac{17}{4}} \qquad \text{Apply the square root property.}$$

$$y - \frac{1}{2} = \frac{\sqrt{17}}{2} \qquad y - \frac{1}{2} = -\frac{\sqrt{17}}{2} \qquad \text{Simplify each radical expression.}$$

$$y = \frac{1}{2} + \frac{\sqrt{17}}{2} \qquad y = \frac{1}{2} - \frac{\sqrt{17}}{2} \qquad \text{Add } \frac{1}{2} \text{ to each side.}$$

**Check:**  $y = \frac{1}{2} + \frac{\sqrt{17}}{2}$:   $\bigg|$   $y = \frac{1}{2} - \frac{\sqrt{17}}{2}$:

$$y^2 - y - 4 = 0 \qquad\qquad\qquad\qquad\qquad y^2 - y - 4 = 0$$
$$\left(\frac{1}{2} + \frac{\sqrt{17}}{2}\right)^2 - \left(\frac{1}{2} + \frac{\sqrt{17}}{2}\right) - 4 = 0 \quad \bigg| \quad \left(\frac{1}{2} - \frac{\sqrt{17}}{2}\right)^2 - \left(\frac{1}{2} - \frac{\sqrt{17}}{2}\right) - 4 = 0$$
$$\frac{1}{4} + \frac{\sqrt{17}}{2} + \frac{17}{4} - \frac{1}{2} - \frac{\sqrt{17}}{2} - 4 = 0 \quad \bigg| \quad \frac{1}{4} - \frac{\sqrt{17}}{2} + \frac{17}{4} - \frac{1}{2} + \frac{\sqrt{17}}{2} - 4 = 0$$

$$\frac{1}{4} + \frac{17}{4} - \frac{2}{4} - \frac{16}{4} = 0 \qquad\qquad \frac{1}{4} + \frac{17}{4} - \frac{2}{4} - \frac{16}{4} = 0$$

$$0 = 0 \qquad\qquad\qquad\qquad 0 = 0$$

Since both solutions make the equation true, the solution set is

$$\left\{ \frac{1}{2} - \frac{\sqrt{17}}{2}, \frac{1}{2} + \frac{\sqrt{17}}{2} \right\}.$$

> **Note:** *The solutions can be written in several different ways as shown.*
>
> $$\left\{ \frac{1}{2} \pm \frac{\sqrt{17}}{2} \right\}, \left\{ \frac{1 - \sqrt{17}}{2}, \frac{1 + \sqrt{17}}{2} \right\}, \text{ or } \left\{ \frac{1 \pm \sqrt{17}}{2} \right\}.$$

✔ **Student Check 2**   Solve each equation by completing the square.

   **a.** $a^2 + 10a + 20 = 0$         **b.** $x^2 - 7x - 12 = 0$

## Completing the Square (Coefficient Other Than 1 on $x^2$)

**Objective 3** ▶

Use completing the square to solve a quadratic equation of the form $ax^2 + bx + c = 0$, $a \neq 0$, $a \neq 1$.

In the previous examples, the coefficient of the squared term is 1. This must be the case to complete the square. The following example illustrates this fact.

Although $(2x - 3)^2 = 4x^2 - 12x + 9$ is a perfect square trinomial, the relationship between the constant term and the coefficient of $x$ that we discussed earlier *does not* exist.

$$\text{Note: } 9 \neq \left( \frac{-12}{2} \right)^2 = (-6)^2 = 36$$

So, solving quadratic equations that have a coefficient other than 1 on the squared term by completing the square requires us to rewrite the equation with a coefficient of 1 on the squared term.

For instance, to solve $4x^2 - 24x + 4 = 0$, we must obtain an equivalent equation with a coefficient of 1 on $x^2$. To do this, we divide each side by 4, the coefficient of $x^2$.

$$\frac{4x^2}{4} - \frac{24x}{4} + \frac{4}{4} = 0 \rightarrow x^2 - 6x + 1 = 0$$

This gives us an equation that we can solve using the methods shown in Objective 2.

**INSTRUCTOR NOTE:**
Remind students that this is the same procedure as before but with an additional first step.

> **Procedure: Solving an Equation of the Form** $ax^2 + bx + c = 0$, $a \neq 0$, $a \neq 1$ **by Completing the Square**
>
> **Step 1:** Divide each side of the equation by $a$, the coefficient of the squared term.
> **Step 2:** Move the constant term to the side opposite the variable terms.
> **Step 3:** Complete the square by finding half of the coefficient of $x$ and squaring it. Add this number to both sides of the equation.
> **Step 4:** Factor the perfect square trinomial and simplify the right side of the equation.
> **Step 5:** Apply the square root property.
> **Step 6:** Solve the resulting equations, if possible.

**Objective 3  Examples**   **Solve each equation by completing the square.**

   **3a.** $2x^2 - 24x + 18 = 0$         **3b.** $3x^2 - 15x + 27 = 0$

**Solutions**   **3a.**

$$2x^2 - 24x + 18 = 0$$

$$\frac{2x^2 - 24x + 18}{2} = \frac{0}{2}$$   Divide each side by 2.

$$x^2 - 12x + 9 = 0$$   Simplify.

$$x^2 - 12x = -9$$   Subtract 9 from each side.

$$x^2 - 12x + 36 = -9 + 36$$   Add $\left(-\frac{12}{2}\right)^2 = 36$ to each side.

$$(x - 6)^2 = 27$$   Factor and simplify.

$$x - 6 = \sqrt{27} \quad \text{or} \quad x - 6 = -\sqrt{27}$$   Apply the square root property.

$$x - 6 = 3\sqrt{3} \qquad\qquad x - 6 = -3\sqrt{3}$$   Simplify each radical.

$$x = 6 + 3\sqrt{3} \qquad\qquad x = 6 - 3\sqrt{3}$$   Add 6 to each side.

The solution set is $\{6 - 3\sqrt{3}, 6 + 3\sqrt{3}\}$ or $\{6 \pm 3\sqrt{3}\}$. The solutions can be checked by substituting these values into the original equation.

**3b.**

$$3x^2 - 15x + 27 = 0$$

$$\frac{3x^2 - 15x + 27}{3} = \frac{0}{3}$$   Divide each side by 3.

$$x^2 - 5x + 9 = 0$$   Simplify.

$$x^2 - 5x = -9$$   Subtract 9 from each side.

$$x^2 - 5x + \frac{25}{4} = -9 + \frac{25}{4}$$   Add $\left(\frac{-5}{2}\right)^2 = \frac{25}{4}$ to each side.

$$\left(x - \frac{5}{2}\right)^2 = -\frac{11}{4}$$   Factor and simplify:

$$-9 + \frac{25}{4} = -\frac{36}{4} + \frac{25}{4} = -\frac{11}{4}$$

$$x - \frac{5}{2} = \sqrt{-\frac{11}{4}} \quad \text{or} \quad x - \frac{5}{2} = -\sqrt{-\frac{11}{4}}$$   Apply the square root property.

Because the square root of a negative number is not a real number, there are no real solutions of this equation.

☑ **Student Check 3**   Solve each equation by completing the square.
   **a.** $4x^2 + 8x - 20 = 0$      **b.** $2y^2 - y - 3 = 0$

**Objective 4** ▶
Troubleshoot common errors.

## Troubleshooting Common Errors

Some common errors associated with completing the square are shown next.

**Objective 4 Examples**   **A problem and an incorrect solution are given. Provide the correct solution and an explanation of the error.**

**4a.** Solve $x^2 - 8x + 9 = 0$ by completing the square.

| Incorrect Solution | Correct Solution and Explanation |
|---|---|
| $x^2 - 8x + 9 = 0$ <br> $x^2 - 8x = -9$ <br> $x^2 - 8x + 16 = -9$ <br> $(x - 4)^2 = -9$ <br> $x - 4 = \sqrt{-9} \quad \text{or} \quad x - 4 = -\sqrt{-9}$ <br> No real solutions | We must add 16 to both sides to make an equivalent equation. <br> $x^2 - 8x + 16 = -9 + 16$ <br> $(x - 4)^2 = 7$ <br> $x - 4 = \sqrt{7} \quad \text{or} \quad x - 4 = -\sqrt{7}$ <br> $x = 4 + \sqrt{7} \qquad\qquad x = 4 - \sqrt{7}$ <br> $\{4 - \sqrt{7}, 4 + \sqrt{7}\}$ |

**4b.** Solve $4x^2 + 4x - 12 = 0$ by completing the square.

| Incorrect Solution | Correct Solution and Explanation |
|---|---|
| $4x^2 + 4x - 12 = 0$ <br> $4x^2 + 4x = 12$ <br> $4x^2 + 4x + 4 = 12 + 4$ <br> $(2x + 2)^2 = 16$ <br> $2x + 2 = \sqrt{16}$ or $2x + 2 = -\sqrt{16}$ <br> $2x + 2 = 4 \qquad\qquad 2x + 2 = -4$ <br> $2x = 2 \qquad\qquad\quad 2x = -6$ <br> $x = 1 \qquad\qquad\qquad x = -3$ <br> $\{-3, 1\}$ | We should divide each side by 4 first to get a coefficient of 1 on $x^2$. <br><br> $4x^2 + 4x - 12 = 0$ <br> $\dfrac{4x^2 + 4x - 12}{4} = \dfrac{0}{4}$ <br> $x^2 + x - 3 = 0$ <br> $x^2 + x = 3$ <br> $x^2 + x + \dfrac{1}{4} = 3 + \dfrac{1}{4}$ <br> $\left(x + \dfrac{1}{2}\right)^2 = \dfrac{13}{4}$ <br> $x + \dfrac{1}{2} = \sqrt{\dfrac{13}{4}}$ or $x + \dfrac{1}{2} = -\sqrt{\dfrac{13}{4}}$ <br> $x + \dfrac{1}{2} = \dfrac{\sqrt{13}}{2} \qquad x + \dfrac{1}{2} = -\dfrac{\sqrt{13}}{2}$ <br> $x = -\dfrac{1}{2} + \dfrac{\sqrt{13}}{2} \qquad x = -\dfrac{1}{2} - \dfrac{\sqrt{13}}{2}$ |

## ANSWERS TO STUDENT CHECKS

**Student Check 1**   **a.** $x^2 + 12x + 36 = (x + 6)^2$

    **b.** $y^2 - 2y + 1 = (y - 1)^2$

    **c.** $x^2 - 5x + \dfrac{25}{4} = \left(x - \dfrac{5}{2}\right)^2$

**Student Check 2**   **a.** $\{-5 - \sqrt{5}, -5 + \sqrt{5}\}$

    **b.** $\left\{\dfrac{7 - \sqrt{97}}{2}, \dfrac{7 + \sqrt{97}}{2}\right\}$

**Student Check 3**   **a.** $\{-1 - \sqrt{6}, -1 + \sqrt{6}\}$

    **b.** $\left\{-1, \dfrac{3}{2}\right\}$

## SUMMARY OF KEY CONCEPTS

1. A perfect square trinomial results from squaring a binomial. When the leading coefficient is 1, the constant term is half the coefficient of $x$, squared.

2. To create a perfect square trinomial from a binomial of the form $x^2 + bx$, we add the number $c = \left(\dfrac{b}{2}\right)^2$.

3. Completing the square can be used to solve any quadratic equation. To apply this method, we must do the following.

    **a.** If the leading coefficient is not 1, divide each side by the given coefficient.

    **b.** Move the constant term to the side opposite the variable terms.

    **c.** Find the number that completes the square and add it to both sides of the equation.

    **d.** Factor the trinomial as a binomial squared and simplify the right side.

    **e.** Apply the square root property and solve.

## SECTION 9.2 / EXERCISE SET

### Write About It!

Use complete sentences in your answer to each exercise.

1. What is the purpose of completing the square when solving a quadratic equation? *The purpose is to apply the square root property to solve the equation.*
2. Why might the method of completing the square be better for solving quadratic equations than factoring?
3. Explain how to solve a quadratic equation by completing the square
   a. if the coefficient of the squared term is 1.
   b. if the coefficient of the squared term is not 1.
4. What binomial would require the number 144 to complete its square? Explain.

### Practice Makes Perfect!

Make each binomial a perfect square trinomial and express it in factored form. (*See Objective 1.*)

5. $x^2 + 6x$
   $x^2 + 6x + 9 = (x + 3)^2$
6. $x^2 + 10x$
   $x^2 + 10x + 25 = (x + 5)^2$
7. $x^2 - 8x$
   $x^2 - 8x + 16 = (x - 4)^2$
8. $x^2 - 16x$
   $x^2 - 16x + 64 = (x - 8)^2$
9. $y^2 + 2y$
   $y^2 + 2y + 1 = (y + 1)^2$
10. $y^2 + 14y$
    $y^2 + 14y + 49 = (y + 7)^2$
11. $a^2 + 3a$
12. $a^2 + 5a$
13. $b^2 - b$   $b^2 - b + \frac{1}{4} = \left(b - \frac{1}{2}\right)^2$
14. $x^2 - 11x$
15. $b^2 + 50b$
    $b^2 + 50b + 625 = (b + 25)^2$
16. $b^2 + 30b$
    $b^2 + 30b + 225 = (b + 15)^2$

Solve each equation by completing the square.
(*See Objective 2.*)

17. $x^2 + 12x - 13 = 0$
    $\{-13, 1\}$
18. $x^2 + 4x - 12 = 0$
    $\{-6, 2\}$
19. $x^2 - 10x + 8 = 0$
    $\{5 - \sqrt{17}, 5 + \sqrt{17}\}$
20. $x^2 - 14x + 20 = 0$
    $\{7 - \sqrt{29}, 7 + \sqrt{29}\}$
21. $x^2 - 4x + 6 = 0$
    no real solutions
22. $y^2 + 2y + 5 = 0$
    no real solutions
23. $y^2 - 3y - 2 = 0$
24. $a^2 - 5a - 8 = 0$
25. $a^2 - 36a - 18 = 0$
    $\{18 - 3\sqrt{38}, 18 + 3\sqrt{38}\}$
26. $a^2 - 18a + 20 = 0$
    $\{9 - \sqrt{61}, 9 + \sqrt{61}\}$
27. $b^2 - 22b + 14 = 0$
    $\{11 - \sqrt{107}, 11 + \sqrt{107}\}$
28. $b^2 - 20b + 10 = 0$
    $\{10 - 3\sqrt{10}, 10 + 3\sqrt{10}\}$
29. $y^2 + 8y - 12 = 0$
    $\{-4 - 2\sqrt{7}, -4 + 2\sqrt{7}\}$
30. $y^2 + 16y - 8 = 0$
    $\{-8 - 6\sqrt{2}, -8 + 6\sqrt{2}\}$

Solve each equation by completing the square.
(*See Objective 3.*)

31. $6x^2 - 24x - 72 = 0$
    $\{-2, 6\}$
32. $4x^2 + 32x - 80 = 0$
    $\{-10, 2\}$
33. $3x^2 - 18x - 24 = 0$
    $\{3 - \sqrt{17}, 3 + \sqrt{17}\}$
34. $2x^2 + 8x - 14 = 0$
    $\{-2 - \sqrt{11}, -2 + \sqrt{11}\}$
35. $8a^2 - 16a + 8 = 0$  $\{1\}$
36. $2y^2 + 4y - 5 = 0$
37. $6y^2 + 12y - 1 = 0$
38. $2x^2 - 3x - 2 = 0$  $\left\{-\frac{1}{2}, 2\right\}$
39. $3x^2 + x - 4 = 0$
    $\left\{-\frac{4}{3}, 1\right\}$
40. $5x^2 + x - 6 = 0$
    $\left\{-\frac{6}{5}, 1\right\}$

### Mix 'Em Up!

Make each binomial a perfect square trinomial and express it in factored form.

41. $x^2 - 14x$
    $x^2 - 14x + 49 = (x - 7)^2$
42. $a^2 - 13a$
43. $b^2 + 0.6b$
    $b^2 + 0.6b + 0.09 = (b + 0.3)^2$
44. $y^2 - 1.6y$
    $y^2 - 1.6y + 0.64 = (y - 0.8)^2$
45. $a^2 + 15a$
46. $a^2 - 7a$
47. $x^2 + \frac{2}{3}x$
48. $y^2 - \frac{2}{5}y$

Solve each equation by completing the square.

49. $3x^2 - 42x - 72 = 0$
    $\{7 - \sqrt{73}, 7 + \sqrt{73}\}$
50. $2y^2 - 20y + 36 = 0$
    $\{5 - \sqrt{7}, 5 + \sqrt{7}\}$
51. $a^2 - 6a = 0$   $\{0, 6\}$
52. $3b^2 - 2b = 0$   $\left\{0, \frac{2}{3}\right\}$
53. $x^2 - 5x + 8 = 0$
    no real solutions
54. $x^2 - x + 4 = 0$
    no real solutions
55. $5x^2 - 15x + 1 = 0$
56. $3x^2 - 4x - 3 = 0$
57. $2x^2 + 9x - 5 = 0$   $\left\{-5, \frac{1}{2}\right\}$
58. $4x^2 - 12x + 9 = 0$
    $\left\{\frac{3}{2}\right\}$

### You Be the Teacher!

Correct each student's errors, if any.

59. Solve $2x^2 - 2x - 4 = 0$ by completing the square.

    Clinton's work:
    $$2x^2 - 2x - 4 = 0$$
    $$\frac{2x^2 - 2x - 4}{2} = \frac{0}{2}$$
    $$x^2 - 2x - 2 = 0$$
    $$x^2 - 2x = 2$$
    $$x^2 - 2x + 1 = 2 + 1$$
    $$(x - 1)^2 = 3 \Rightarrow x - 1 = \pm\sqrt{3} \Rightarrow x = 1 \pm \sqrt{3}$$

60. Solve $x^2 + x - 3 = 0$ by completing the square.

    Gabriella's work:
    $$x^2 + x - 3 = 0$$
    $$x^2 + x = 3$$
    $$x^2 + 1x = 3$$
    $$x^2 + 1x + \frac{1}{2} = 3 + \frac{1}{2}$$
    $$\left(x + \frac{1}{2}\right)^2 = \frac{7}{2}$$
    $$x + \frac{1}{2} = \pm\sqrt{\frac{7}{2}}$$
    $$x = -\frac{1}{2} \pm \sqrt{\frac{7}{2}}$$

Additional answers can be found in the Instructor Answer Appendix.

**Answer each student's question.**

**61.** Libby: I always have trouble factoring when the coefficient of $x$ is not an even number. Please help me see the pattern of how to factor when completing the square if the coefficient of $x$ is an odd number. (Hint: You may want to show Libby how to make perfect square trinomials from $x^2 + x$, $x^2 + 3x$, $x^2 + 5x$, and so on, and then factor each resulting trinomial. A pattern of how to factor should evident.)

**62.** Conchita: I was trying to solve $x^2 - 10x + 24 = 0$ by completing the square. But I notice that I can solve this equation by factoring instead. Does it matter which method I use? How do I know which method I should use?

 **Calculate It!**

Use a calculator to show that the given expressions are solutions of the equation.

**63.** $\dfrac{5 + \sqrt{13}}{6}$ and $\dfrac{5 - \sqrt{13}}{6}$; $3x^2 - 5x + 1 = 0$

**64.** $\dfrac{1 + \sqrt{31}}{6}$ and $\dfrac{1 - \sqrt{31}}{6}$; $6x^2 - 2x = 5$

---

| SECTION 9.3 | **Solve Quadratic Equations by the Quadratic Formula** |

▶ **OBJECTIVES**

As a result of completing this section, you will be able to

**1.** Simplify quotients.

**2.** Solve quadratic equations by the quadratic formula.

**3.** Use the discriminant to determine the types of solutions.

**4.** Solve application problems.

**5.** Troubleshoot common errors.

The height of a basketball, in feet, thrown upward by a player at the free throw line is given by $h = -16t^2 + 22t + 7$, where $t$ is the number of seconds after the ball is thrown. When will the height of the ball be equal to 10 ft, the height of the basketball hoop? To answer this question, we must solve the equation $-16t^2 + 22t + 7 = 10$.

In this section, we will learn another method that enables us to solve this quadratic equation.

**Objective 1** ▶

Simplify quotients.

## Simplifying Quotients

Solutions of quadratic equations are often quotients that need to be simplified. Simplifying the quotients that result from solving these equations involves simplifying radicals and simplifying fractions.

---

**Procedure: Simplifying Quotients**

**Step 1:** Simplify any radical expression in the fraction.
**Step 2:** Factor out any common factor from the numerator.
**Step 3:** Divide the numerator and denominator by their common factor. If there is no common factor between the numerator and denominator, then the fraction cannot be simplified.

---

**Objective 1 Examples** Simplify each quotient.

**1a.** $\dfrac{4 + 8\sqrt{3}}{4}$  **1b.** $\dfrac{6 - \sqrt{24}}{2}$  **1c.** $\dfrac{5 + \sqrt{20}}{5}$  **1d.** $\dfrac{6 - \sqrt{18}}{9}$

**Solutions** **1a.** $\dfrac{4 + 8\sqrt{3}}{4} = \dfrac{4(1 + 2\sqrt{3})}{4}$  Factor out the common factor of 4 from the numerator.

$= 1 + 2\sqrt{3}$  Divide the numerator and denominator by the common factor, 4.

**1b.** $\dfrac{6 - \sqrt{24}}{2} = \dfrac{6 - \sqrt{4 \cdot 6}}{2}$

$\phantom{1b.} = \dfrac{6 - 2\sqrt{6}}{2}$      Simplify the radical.

$\phantom{1b.} = \dfrac{2(3 - \sqrt{6})}{2}$      Factor out the common factor of 2 from the numerator.

$\phantom{1b.} = 3 - \sqrt{6}$      Divide the numerator and denominator by the common factor.

**1c.** $\dfrac{5 + \sqrt{20}}{5} = \dfrac{5 + \sqrt{4 \cdot 5}}{5}$

$\phantom{1c.} = \dfrac{5 + 2\sqrt{5}}{5}$      Simplify the radical.

There is no common factor in the numerator so the fraction can't be simplified.

**1d.** $\dfrac{6 - \sqrt{18}}{9} = \dfrac{6 - \sqrt{9 \cdot 2}}{9}$

$\phantom{1d.} = \dfrac{6 - 3\sqrt{2}}{9}$      Simplify the radical.

$\phantom{1d.} = \dfrac{3(2 - \sqrt{2})}{3 \cdot 3}$      Factor out the common factor of 3 from the numerator.

$\phantom{1d.} = \dfrac{2 - \sqrt{2}}{3}$      Divide the numerator and denominator by the common factor, 3.

✓ **Student Check 1**    Simplify each quotient.

     **a.** $\dfrac{6 + 18\sqrt{5}}{6}$      **b.** $\dfrac{5 - \sqrt{75}}{5}$      **c.** $\dfrac{3 + \sqrt{8}}{3}$      **d.** $\dfrac{4 - \sqrt{72}}{10}$

## The Quadratic Formula

**Objective 2** ▶

Solve quadratic equations by the quadratic formula.

Thus far, we have learned three methods for solving quadratic equations. These include the zero products property (factoring), the square root property, and completing the square.

| Factoring | Square Root Property | Completing the Square |
|---|---|---|
| Is an efficient method but it *does not* solve every equation. It solves only equations that can be factored. | Applies only to equations of the form $x^2 = k$. | Solves *every* quadratic equation but can be somewhat tedious, especially if the leading coefficient is not 1 and the coefficient of the $x$-term is not even. |

We can use the technique of completing the square on the standard form of a quadratic equation ($ax^2 + bx + c = 0$, where $a \neq 0$) to develop a formula to find solutions of any quadratic equation. The formula that is derived by completing the square on the standard form of a quadratic equation is called the *quadratic formula*. The derivation of the quadratic formula follows.

### Derivation of the Quadratic Formula

$ax^2 + bx + c = 0$

$\dfrac{ax^2 + bx + c}{a} = \dfrac{0}{a}$      Divide each side by $a$ to get a coefficient of 1 on $x^2$.

$$x^2 + \frac{b}{a}x + \frac{c}{a} = 0 \qquad \text{Simplify.}$$

$$x^2 + \frac{b}{a}x = -\frac{c}{a} \qquad \text{Subtract } \frac{c}{a} \text{ from each side.}$$

$$x^2 + \frac{b}{a}x + \frac{b^2}{4a^2} = -\frac{c}{a} + \frac{b^2}{4a^2} \qquad \text{Add } \left(\frac{1}{2} \cdot \frac{b}{a}\right)^2 = \left(\frac{b}{2a}\right)^2 = \frac{b^2}{4a^2} \text{ to each side.}$$

$$\left(x + \frac{b}{2a}\right)^2 = \frac{b^2 - 4ac}{4a^2}$$

Factor the left side and simplify the right side.

$$-\frac{c}{a} + \frac{b^2}{4a^2} = -\frac{4ac}{4a \cdot a} + \frac{b^2}{4a^2} = \frac{b^2 - 4ac}{4a^2}$$

$$x + \frac{b}{2a} = \pm\sqrt{\frac{b^2 - 4ac}{4a^2}} \qquad \text{Apply the square root property. Remember that } x = \sqrt{k} \text{ or } x = -\sqrt{k} \text{ can be written as } x = \pm\sqrt{k}.$$

$$x + \frac{b}{2a} = \pm\frac{\sqrt{b^2 - 4ac}}{\sqrt{4a^2}} \qquad \text{Apply the quotient rule for radicals.}$$

$$x = -\frac{b}{2a} \pm \frac{\sqrt{b^2 - 4ac}}{2a} \qquad \text{Simplify the radical.}$$

$$x = \frac{-b \pm \sqrt{b^2 - 4ac}}{2a} \qquad \text{Add the fractions.}$$

So, the solutions are $x = \dfrac{-b + \sqrt{b^2 - 4ac}}{2a}$ or $x = \dfrac{-b - \sqrt{b^2 - 4ac}}{2a}$.

---

**Property: The Quadratic Formula**

If $a$, $b$, and $c$ are real numbers, a quadratic equation written in the form $ax^2 + bx + c = 0$ ($a \neq 0$), has solutions

$$x = \frac{-b \pm \sqrt{b^2 - 4ac}}{2a}$$

---

**Procedure: Solving a Quadratic Equation Using the Quadratic Formula**

**Step 1:** Write the equation in standard from; that is, all terms on one side equal to 0.
**Step 2:** Identify the values of $a$, $b$, and $c$ from the standard form.
**Step 3:** Substitute the values of $a$, $b$, and $c$ into the quadratic formula.
**Step 4:** Use the order of operations to simplify the expression.
**Step 5:** Simplify the quotient.

---

**Objective 2 Examples** Solve each equation using the quadratic formula. If the solutions are irrational numbers, approximate the values to two decimal places.

**2a.** $x^2 - 6x + 9 = 0$     **2b.** $y^2 + 4y = 7$     **2c.** $4h^2 = 3h + 5$

**2d.** $6x^2 + x = 0$     **2e.** $\frac{1}{3}y^2 + 4 = 0$

**Solutions** **2a.** The equation is in standard form, $1x^2 - 6x + 9 = 0$. So, $a = 1$, $b = -6$, and $c = 9$.

$$x = \frac{-b \pm \sqrt{b^2 - 4ac}}{2a}$$      State the quadratic formula.

$$x = \frac{-(-6) \pm \sqrt{(-6)^2 - 4(1)(9)}}{2(1)}$$      Substitute the values as stated.

$$x = \frac{6 \pm \sqrt{36 - 36}}{2}$$      Simplify the numerator and denominator.

$$x = \frac{6 \pm \sqrt{0}}{2}$$      Simplify the radicand.

$$x = \frac{6 \pm 0}{2}$$      Simplify the radical expression.

$$x = \frac{6 + 0}{2} \quad \text{or} \quad x = \frac{6 - 0}{2}$$      Write the two solutions.

$$x = \frac{6}{2} \qquad\qquad x = \frac{6}{2}$$      Simplify each expression.

$$x = 3 \qquad\qquad x = 3$$

So, the solution set is $\{3\}$. We can check the solution in the original equation as well.

**2b.** We first write the equation in standard form by subtracting 7 from each side of the equation.

$$y^2 + 4y = 7$$
$$1y^2 + 4y - 7 = 0$$
$$a = 1, b = 4, c = -7$$

**INSTRUCTOR NOTE:**
Point out the difference between the exact solutions and the approximate solutions.

$$y = \frac{-b \pm \sqrt{b^2 - 4ac}}{2a}$$      State the quadratic formula.

$$y = \frac{-(4) \pm \sqrt{(4)^2 - 4(1)(-7)}}{2(1)}$$      Substitute the values as stated.

$$y = \frac{-4 \pm \sqrt{16 + 28}}{2}$$      Simplify the numerator and denominator.

$$y = \frac{-4 \pm \sqrt{44}}{2}$$      Simplify the radicand.

$$y = \frac{-4 \pm \sqrt{4 \cdot 11}}{2}$$

$$y = \frac{-4 \pm 2\sqrt{11}}{2}$$      Simplify the radical expression.

$$y = \frac{2(-2 \pm \sqrt{11})}{2}$$

$$y = -2 \pm \sqrt{11}$$      Simplify the quotient.

$$y = -2 + \sqrt{11} \quad \text{or} \quad y = -2 - \sqrt{11}$$      Write the two solutions.

So, the solution set is $\{-2 - \sqrt{11}, -2 + \sqrt{11}\}$. We can check the solutions in the original equation as well. The solutions are irrational, so the approximate values of the solutions are

$$-2 + \sqrt{11} \approx 1.32 \quad \text{or} \quad -2 - \sqrt{11} \approx -5.32$$

**2c.** We subtract $3h$ and 5 from each side to write the equation in standard form.

$$4h^2 = 3h + 5$$
$$4h^2 - 3h - 5 = 0$$
$$a = 4, b = -3, c = -5$$

$$h = \frac{-b \pm \sqrt{b^2 - 4ac}}{2a}$$
State the quadratic formula.

$$h = \frac{-(-3) \pm \sqrt{(-3)^2 - 4(4)(-5)}}{2(4)}$$
Substitute the values as stated.

$$h = \frac{3 \pm \sqrt{9 + 80}}{8}$$
Simplify the numerator and denominator.

$$h = \frac{3 \pm \sqrt{89}}{8}$$
Simplify the radicand.

$$h = \frac{3 + \sqrt{89}}{8} \quad \text{or} \quad h = \frac{3 - \sqrt{89}}{8}$$
Write the two solutions.

So, the solution set is $\left\{ \dfrac{3 - \sqrt{89}}{8}, \dfrac{3 + \sqrt{89}}{8} \right\}$. We can check the solutions in the original equation as well. The solutions are irrational, so the approximate values of the solutions are

$$\frac{3 - \sqrt{89}}{8} \approx -0.80 \quad \text{or} \quad \frac{3 + \sqrt{89}}{8} \approx 1.55$$

**2d.** The equation is in standard form, $6x^2 + 1x + 0 = 0$. So, $a = 6$, $b = 1$, and $c = 0$.

$$x = \frac{-b \pm \sqrt{b^2 - 4ac}}{2a}$$
State the quadratic formula.

$$x = \frac{-(1) \pm \sqrt{(1)^2 - 4(6)(0)}}{2(6)}$$
Substitute the values as stated.

$$x = \frac{-1 \pm \sqrt{1 - 0}}{12}$$
Simplify the numerator and denominator.

$$x = \frac{-1 \pm \sqrt{1}}{12}$$
Simplify the radicand.

$$x = \frac{-1 \pm 1}{12}$$
Simplify the radical expression.

$$x = \frac{-1 + 1}{12} \quad \text{or} \quad x = \frac{-1 - 1}{12}$$
Write the two solutions.

$$x = \frac{0}{12} \qquad\qquad x = \frac{-2}{12}$$
Simplify each expression.

$$x = 0 \qquad\qquad x = -\frac{1}{6}$$

So, the solution set is $\left\{ -\dfrac{1}{6}, 0 \right\}$. We can check the solutions in the original equation as well.

**2e.** The equation is in standard form and we can use fractional values in the quadratic formula. We can, however, multiply each side by the LCD of 3 to remove the fractions.

$$3\left( \frac{1}{3}y^2 + 4 \right) = 3(0)$$
$$y^2 + 12 = 0$$
$$1y^2 + 0y + 12 = 0$$
$$a = 1, b = 0, c = 12$$

$$y = \frac{-b \pm \sqrt{b^2 - 4ac}}{2a}$$
State the quadratic formula.

$$y = \frac{-(0) \pm \sqrt{(0)^2 - 4(1)(12)}}{2(1)}$$
Substitute the values as stated.

$$y = \frac{0 \pm \sqrt{0 - 48}}{2}$$
Simplify the numerator and denominator.

$$y = \frac{\pm\sqrt{-48}}{2}$$
Simplify the radicand.

Because the number inside the radical is negative, there are no real solutions of this equation.

 **Note:** *If we used the values of* $a = \dfrac{1}{3}, b = 0,$ *and* $c = 4,$ *we would have obtained the same solutions.*

✔ **Student Check 2**     Solve each equation using the quadratic formula.

**a.** $x^2 - 2x - 8 = 0$         **b.** $y^2 + 3y = 9$         **c.** $5h^2 = 4h + 1$
**d.** $7x^2 + 2x = 0$         **e.** $y^2 + 9 = 0$

### The Discriminant of a Quadratic Equation

**Objective 3** ▶

Use the discriminant to determine the types of solutions.

From the previous examples and in previous sections, we know that a quadratic equation will have either two real solutions, one real solution, or no real solutions. We can actually predict which type of solutions a quadratic equation will have if we know a particular number. This number is called the *discriminant* of the quadratic equation.

The **discriminant** is the number that is inside the radical of the quadratic formula.

$$y = \frac{-b \pm \sqrt{b^2 - 4ac}}{2a}$$

The discriminant

The value of this number determines the number and nature of the solutions of a quadratic equation.

> **Property: Discriminant**
> If $ax^2 + bx + c = 0$ $(a \neq 0)$, then the discriminant is $b^2 - 4ac$.
>
> | Sign of the Discriminant | Types of Solutions |
> |---|---|
> | Positive | Two nonrepeating real solutions |
> | Zero | One repeating real solution |
> | Negative | No real solutions |

In Section 9.4, we will study equations that have a negative discriminant in more depth.

**Objective 3 Examples**     Find the discriminant of each quadratic equation and use the information to determine the type of solutions.

**3a.** $5x^2 + x - 3 = 0$         **3b.** $x^2 + x = -1$         **3c.** $4x^2 - 12x + 9 = 0$

**Solutions**     **3a.** The equation is in standard form, so $a = 5$, $b = 1$, and $c = -3$.

$$b^2 - 4ac = (1)^2 - 4(5)(-3)$$
$$= 1 + 60$$
$$= 61$$

Since the discriminant is positive, the equation $5x^2 + x - 3 = 0$ has two nonrepeating real solutions.

**3b.** We first write the equation in standard form by adding 1 to each side.

$$x^2 + x = -1$$
$$1x^2 + 1x + 1 = 0$$
$$a = 1, b = 1, c = 1$$

$$b^2 - 4ac = (1)^2 - 4(1)(1)$$
$$= 1 - 4$$
$$= -3$$

Since the discriminant is negative, the equation has no real solutions.

**3c.** The equation is in standard form, so $a = 4$, $b = -12$, and $c = 9$.

$$b^2 - 4ac = (-12)^2 - 4(4)(9)$$
$$= 144 - 144$$
$$= 0$$

Since the discriminant is zero, the equation has one repeating real solution.

> **✓ Student Check 3**  Find the discriminant of each quadratic equation and use the information to determine the type of solutions.
>
>   **a.** $2x^2 - 5x = 0$    **b.** $x^2 + 16 = 8x$    **c.** $6x^2 - x + 5 = 0$

## Applications

**Objective 4** ▶
Solve application problems.

In Chapter 6, we solved quadratic applications by factoring. We will solve similar types of applications but will use the quadratic formula to solve the problems.

**Objective 4 Example**  The height, in feet, of a basketball thrown upward by a player at the free-throw line is given by $h = -16t^2 + 22t + 7$, where $t$ is the number of seconds after the ball is thrown. To the nearest tenth of a second, when will the ball reach the height of the basketball hoop, which is 10 ft from the floor?

**Solution**  We want to find when the height is 10 ft. So, replace the value of $h$ with 10 and solve using the quadratic formula.

| | |
|---|---|
| $-16t^2 + 22t + 7 = h$ | Begin with the model. |
| $-16t^2 + 22t + 7 = 10$ | Replace $h$ with 10. |
| $-16t^2 + 22t + 7 - 10 = 10 - 10$ | Subtract 10 from each side. |
| $-16t^2 + 22t - 3 = 0$ | Simplify. |
| $a = -16, b = 22, c = -3$ | Identify $a$, $b$, and $c$. |

$$t = \frac{-b \pm \sqrt{b^2 - 4ac}}{2a} \qquad \text{State the quadratic formula.}$$

$$t = \frac{-(22) \pm \sqrt{(22)^2 - 4(-16)(-3)}}{2(-16)} \qquad \text{Substitute the values of } a, b, \text{ and } c.$$

$$t = \frac{-22 \pm \sqrt{484 - 192}}{-32} \qquad \text{Simplify the numerator and denominator.}$$

$$t = \frac{-22 \pm \sqrt{292}}{-32} \qquad \text{Simplify the radicand.}$$

$$t = \frac{-22 + \sqrt{292}}{-32} \quad \text{or} \quad t = \frac{-22 - \sqrt{292}}{-32} \qquad \text{Write the two solutions.}$$

$$t \approx 0.2 \text{ sec} \qquad \text{or} \quad t \approx 1.2 \text{ sec} \qquad \text{Approximate the solutions.}$$

It takes 0.2 sec for the ball to reach 10 ft as the ball goes up and 1.2 sec for the ball to reach 10 ft as the ball comes down.

> **✓ Student Check 4**  The height, in feet, of a ball thrown upward from a height of 20 ft is $h = -16t^2 + 32t + 20$, where $t$ is the number of seconds after the ball is thrown. How many seconds will it take for the ball to reach a height of 30 ft?

**Objective 5** ▶

Troubleshoot common errors.

## Troubleshooting Common Errors

Some common errors associated with the quadratic formula and simplifying quotients are shown next.

**Objective 5  Examples**    **A problem and an incorrect solution are given. Provide the correct solution and an explanation of the error.**

**5a.** Simplify $\dfrac{2 + \sqrt{32}}{2}$.

| Incorrect Solution | Correct Solution and Explanation |
|---|---|
| $$\frac{2 + \sqrt{32}}{2} = 1 + \sqrt{32}$$ $$= 1 + 4\sqrt{2}$$ | A common term was canceled out instead of a common factor being divided out. We should factor the numerator and divide the numerator and denominator by their common factor. $$\frac{2 + \sqrt{32}}{2} = \frac{2 + 4\sqrt{2}}{2}$$ $$= \frac{2(1 + 2\sqrt{2})}{2}$$ $$= 1 + 2\sqrt{2}$$ |

**5b.** Solve $7x^2 + 4x = 3$ using the Quadratic Formula.

| Incorrect Solution | Correct Solution and Explanation |
|---|---|
| $7x^2 + 4x = 3 \rightarrow a = 7, b = 4, c = 3$ $$x = \frac{-b \pm \sqrt{b^2 - 4ac}}{2a}$$ $$x = \frac{-(4) \pm \sqrt{(4)^2 - 4(7)(3)}}{2(7)}$$ $$x = \frac{-4 \pm \sqrt{16 - 84}}{14}$$ $$x = \frac{-4 \pm \sqrt{-68}}{14}$$ There are no real solutions since the number in the radical is negative. | The equation was not put in standard form before identifying $a$, $b$, and $c$. We should subtract 3 from both sides first. $7x^2 + 4x - 3 = 0 \rightarrow a = 7, b = 4, c = -3$ $$x = \frac{-b \pm \sqrt{b^2 - 4ac}}{2a}$$ $$x = \frac{-(4) \pm \sqrt{(4)^2 - 4(7)(-3)}}{2(7)}$$ $$x = \frac{-4 \pm \sqrt{16 + 84}}{14}$$ $$x = \frac{-4 \pm \sqrt{100}}{14}$$ $$x = \frac{-4 \pm 10}{14}$$ $$x = \frac{-4 + 10}{14} \quad \text{or} \quad x = \frac{-4 - 10}{14}$$ $$x = \frac{6}{14} \qquad\qquad x = \frac{-14}{14}$$ $$x = \frac{3}{7} \qquad\qquad x = -1$$ |

## ANSWERS TO STUDENT CHECKS

**Student Check 1**   **a.** $1 + 3\sqrt{5}$   **b.** $1 - \sqrt{3}$

**c.** $\dfrac{3 + 2\sqrt{2}}{3}$   **d.** $\dfrac{2 - 3\sqrt{2}}{5}$

**Student Check 2**   **a.** $\{-2, 4\}$

**b.** $\left\{\dfrac{-3 - 3\sqrt{5}}{2}, \dfrac{-3 + 3\sqrt{5}}{2}\right\}$

**c.** $\left\{-\dfrac{1}{5}, 1\right\}$   **d.** $\left\{-\dfrac{2}{7}, 0\right\}$

**e.** no real solutions

**Student Check 3**   **a.** 25, two nonrepeating real solutions

**b.** 0, one repeating real solution

**c.** −119, no real solutions

**Student Check 4**   0.4 sec or 1.6 sec

## SUMMARY OF KEY CONCEPTS

1. When simplifying a quotient in which the numerator has two terms with one being a radical, simplify the radical first. Then factor out a common factor from the terms in the numerator, if possible. Divide the numerator and denominator by their common factor.

2. We now know four different methods for solving quadratic equations.

   **a.** Zero products property (factoring)—doesn't solve all equations.

   **b.** Square root property—solves equations of the form $x^2 = k$, where $k$ is a real number.

   **c.** Completing the square—solves all equations but can be very tedious.

   **d.** The quadratic formula—solves all equations that are in standard form.

3. Quadratic equations have at most two real solutions. The types of solutions can be determined by the discriminant, $b^2 - 4ac$.

   **a.** If the discriminant is positive, there are two real solutions.

   **b.** If it is zero, there is one real solution.

   **c.** If it is negative, there are no real solutions.

4. Application problems can be solved by setting up a quadratic equation that models the situation.

## GRAPHING CALCULATOR SKILLS

We can use the graphing calculator to determine the number of solutions of a quadratic equation and to approximate the solutions of a quadratic equation.

**Example:** Determine how many solutions there are to the following equations: (a) $x^2 - 2x - 3 = 0$ and (b) $x^2 + 4x + 7 = 0$.

**Solution:** We can graph the quadratic equation to see how many times the graph crosses the $x$-axis. If the graph crosses the $x$-axis two times, the equation has two real solutions. If the graph crosses the $x$-axis one time, the equation has one real solution. If the graph doesn't cross the $x$-axis, the equation has no real solutions.

**a.** Graph $y = x^2 - 2x - 3$.

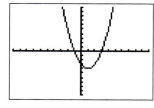

The graph crosses the $x$-axis two times. So, the equation $x^2 - 2x - 3 = 0$ has two real solutions.

**b.** Graph $y = x^2 + 4x + 7$.

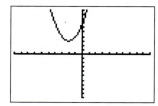

The graph doesn't cross the $x$-axis. So, the equation $x^2 + 4x + 7 = 0$ doesn't have any real solutions.

**Example:** The solutions of $2x^2 - 5x - 6 = 0$ are $\dfrac{5 \pm \sqrt{73}}{4}$. Approximate these solutions.

**Solution:** Enter the two values into the calculator being careful to enter parentheses around the expression in the numerator and the radicand.

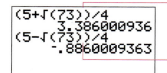

Note the use of double parentheses before entering the denominator. One pair of parentheses closes the radical and the other ends the numerator.

## SECTION 9.3 / EXERCISE SET

### Write About It!

**Use complete sentences in your answer to each exercise.**

1. State the quadratic formula.

2. Do you prefer to solve quadratic equations by completing the square or by using the quadratic formula? Explain.

3. Give an example of an equation that is easier to solve by the quadratic formula than by completing the square. Explain your answer.

4. Give an example of an equation that is easier to solve by completing the square rather than by the quadratic formula. Explain your answer.

### Practice Makes Perfect!

**Simplify each quotient. (See Objective 1.)**

5. $\dfrac{3 - 6\sqrt{2}}{12}$    $\dfrac{1 - 2\sqrt{2}}{4}$

6. $\dfrac{2 - 4\sqrt{3}}{8}$    $\dfrac{1 - 2\sqrt{3}}{4}$

7. $\dfrac{10 - 5\sqrt{6}}{5}$    $2 - \sqrt{6}$

8. $\dfrac{9 - 3\sqrt{2}}{3}$    $3 - \sqrt{2}$

9. $\dfrac{4 + 2\sqrt{6}}{4}$    $\dfrac{2 + \sqrt{6}}{2}$

10. $\dfrac{6 - 3\sqrt{7}}{6}$    $\dfrac{2 - \sqrt{7}}{2}$

11. $\dfrac{5 + \sqrt{75}}{10}$    $\dfrac{1 + \sqrt{3}}{2}$

12. $\dfrac{3 + \sqrt{18}}{6}$    $\dfrac{1 + \sqrt{2}}{2}$

13. $\dfrac{8 - \sqrt{32}}{8}$    $\dfrac{2 - \sqrt{2}}{2}$

14. $\dfrac{2 - \sqrt{24}}{2}$    $1 - \sqrt{6}$

15. $\dfrac{12 - \sqrt{52}}{2}$    $6 - \sqrt{13}$

16. $\dfrac{16 - \sqrt{128}}{8}$    $2 - \sqrt{2}$

17. $\dfrac{12 - 4\sqrt{56}}{6}$    $\dfrac{6 - 4\sqrt{14}}{3}$

18. $\dfrac{16 - 2\sqrt{48}}{4}$    $4 - 2\sqrt{3}$

**Solve each equation using the quadratic formula. (See Objective 2.)**

19. $2x^2 - 7x - 9 = 0$

20. $3x^2 - 4x - 4 = 0$

21. $3x^2 + 2x - 8 = 0$

22. $2x^2 - 5x - 12 = 0$

23. $x^2 = 7x - 5$

24. $x^2 = 8x - 2$

25. $-4y^2 + 5y - 1 = 0$

26. $-5y^2 + 3y + 2 = 0$

27. $6x^2 + 3x - 4 = 0$

28. $7x^2 - 6x - 4 = 0$

29. $10x^2 - x = 0$

30. $8x^2 - 3x = 0$

31. $5x^2 = 4x + 10$

32. $4x^2 = 8x + 9$

33. $x = 7x^2 + 3$    no real solutions

34. $5x = 2x^2 + 9$    no real solutions

35. $t^2 + 10t - 8 = 0$    $\{-5 - \sqrt{33}, -5 + \sqrt{33}\}$

36. $t^2 - 14t + 12 = 0$    $\{7 - \sqrt{37}, 7 + \sqrt{37}\}$

37. $r^2 + 1 = 0$    no real solutions

38. $r^2 + 9 = 0$    no real solutions

**Solve each equation using the quadratic formula. Give solutions in exact form and also approximate the solutions to the nearest hundredth. (See Objective 2.)**

39. $9x^2 - 30x + 10 = 0$

40. $2x^2 - 8x + 3 = 0$

41. $3x^2 + 2x = 7$

42. $2x^2 + 5x = 1$

**Find the discriminant of each equation and use the information to determine the type of solutions. (See Objective 3.)**

43. $5x^2 + 10x = 0$    100; two nonrepeating real solutions

44. $4x^2 + 12x + 13 = 0$    −64; no real solutions

45. $x^2 - 8x + 25 = 0$    −36; no real solutions

46. $2x^2 - 7x = 0$    49; two nonrepeating real solutions

47. $9x^2 - 12x + 4 = 0$    0; one repeating real solution

48. $2x^2 - 5x + 13 = 0$    −79; no real solutions

49. $3x^2 - 4x = 0$    16; two nonrepeating real solutions

50. $x^2 + 8x + 16 = 0$    0; one repeating real solution

**Solve each problem. (See Objective 4.)**

The height (in meters) of a ball projected directly upward is modeled by the equation $h = -9.8t^2 + 40t$, where $t$ is the number of seconds after the ball is released. Use this equation to answer Exercises 51 and 52. Round answers to the nearest hundredth.

51. When will the ball be 5 m above the ground?    0.13 sec or 3.95 sec

52. When will the ball reach 10 m above the ground?    0.27 sec or 3.81 sec

A cake company makes wedding cakes. Their monthly profit (in dollars) from the sale of $x$ cakes can be modeled by $P = -x^2 + 100x - 300$. Use this equation to answer Exercises 53 and 54.

53. How many cakes must the company sell to have a profit of $1800?    30 or 70 cakes

54. How many cakes must the company sell to have a profit of $2100?    40 or 60 cakes

An open box is made from a rectangular sheet of cardboard whose length is 2 in. longer than the width by cutting a square of length $h$ inches from each corner and folding the cardboard as shown. The volume of the open box can be modeled by $V = h(x + 2 - 2h)(x - 2h)$ in.³. Find the dimensions of each box whose height and volume are given in Exercises 55 and 56.

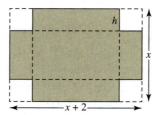

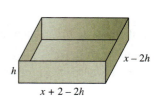

55. The height is 8 in. and the volume is 1344 in.³.    12 in. by 14 in. by 8 in.

56. The height is 10 in. and the volume is 1680 in.³.    12 in. by 14 in. by 10 in.

**57.** If the hypotenuse of a right triangle is 6 m and one leg is 2 m shorter than the other leg, find the exact length of each leg. Then approximate the length of each leg to the nearest hundredth of a meter.
$1 + \sqrt{17} \approx 5.12$ m and $-1 + \sqrt{17} \approx 3.12$ m

**58.** If the hypotenuse of a right triangle is 4 ft and one leg is 2 ft longer than the other leg, find the exact length of each leg. Then approximate the length of each leg to the nearest hundredth of a foot.
$-1 + \sqrt{7} \approx 1.65$ ft and $1 + \sqrt{7} \approx 3.65$ ft

## Mix 'Em Up!

**Simplify each quotient.**

**59.** $\dfrac{15 - \sqrt{45}}{5}$   $\dfrac{15 - 3\sqrt{5}}{5}$   **60.** $\dfrac{3 - \sqrt{81}}{6}$   $-1$

**61.** $\dfrac{10 - \sqrt{50}}{10}$   $\dfrac{2 - \sqrt{2}}{2}$   **62.** $\dfrac{16 - 4\sqrt{42}}{6}$   $\dfrac{8 - 2\sqrt{42}}{3}$

**63.** $\dfrac{18 - 3\sqrt{72}}{4}$   $\dfrac{9 - 9\sqrt{2}}{2}$   **64.** $\dfrac{8 + 5\sqrt{24}}{4}$   $\dfrac{4 + 5\sqrt{6}}{2}$

**Determine the type of solutions of each equation by finding the discriminant. Then solve the equation and give the exact solutions and approximate the solutions to the nearest hundredth, if needed.**

**65.** $7 - 3x = 2x^2$        **66.** $11 - x = x^2$

**67.** $10x^2 + x - 2 = 0$    **68.** $6x^2 - 3x = 0$

**69.** $h^2 - 3h + 9 = 0$     **70.** $r^2 + r + 3 = 0$
    $-27$; no real solutions           $-11$; no real solutions
**71.** $5x^2 + 10x = 0$       **72.** $2x^2 - 7x = 0$   $49$; two nonrepeating real solutions; $\{0, 3.5\}$

**73.** $4x^2 + 12x + 13 = 0$  **74.** $x^2 - 8x + 25 = 0$
    $-64$; no real solutions           $-36$; no real solutions
**75.** $9x^2 - 12x + 4 = 0$   **76.** $x^2 + 8x + 16 = 0$
                                $0$; one repeating real solution; $\{-4\}$
**77.** $5 + 4x = 3x^2$        **78.** $10x = 2x^2 + 9$

**Solve each problem. Round answers to two decimal places.**

**79.** The height (in feet) of a model rocket $t$ seconds after it is launched from ground level is given by the equation $h = -16t^2 + 150t$. Find when the model rocket is 20 ft from the ground.   $0.14$ sec or $9.24$ sec

**80.** The height (in feet) of a model rocket $t$ seconds after it is launched from ground level is given by the equation $h = -16t^2 + 54t$. Find when the model rocket is 10 ft from the ground.   $0.20$ sec or $3.18$ sec

## You Be the Teacher!

**Correct each student's errors, if any.**

**81.** Solve the equation $x^2 - 3x - 4 = 0$ using the quadratic formula.   The value of $b$ should be placed in parentheses in the radical expression.

Josh's work:

$x^2 - 3x - 4 = 0$

$x = \dfrac{3 \pm \sqrt{-3^2 - 4(-4)}}{2(1)}$

$= \dfrac{3 \pm \sqrt{-9 + 16}}{2}$

$= \dfrac{3 \pm \sqrt{7}}{2}$

$x^2 - 3x - 4 = 0$

$x = \dfrac{-(-3) \pm \sqrt{(-3)^2 - 4(1)(-4)}}{2(1)}$

$= \dfrac{3 \pm \sqrt{9 + 16}}{2}$

$= \dfrac{3 \pm \sqrt{25}}{2}$

$= \dfrac{3 + 5}{2}$ or $\dfrac{3 - 5}{2}$

$= 4$ or $-1$

**82.** Solve the equation $4x^2 + 16x = 13$ using the quadratic formula.

Matt's work:

$x = \dfrac{-16 \pm \sqrt{16^2 - 4(4)(13)}}{4}$

$x = \dfrac{-16 \pm \sqrt{48}}{4}$

$x = -4 \pm \sqrt{12}$

$x = -4 \pm 2\sqrt{3}$

## Calculate It!

**Use a calculator to determine if the given expressions are the same.**

**83.** $\dfrac{12 + \sqrt{48}}{4}$ and $3 + \sqrt{3}$   the same   **84.** $\dfrac{15 - \sqrt{18}}{3}$ and $5 - \sqrt{2}$   the same

**85.** $\dfrac{10 - \sqrt{60}}{5}$ and $2 - 2\sqrt{15}$   not the same   **86.** $\dfrac{3 + \sqrt{7}}{3}$ and $1 + \sqrt{7}$   not the same

**Determine if the values are solutions of the equation.**

**87.** $\dfrac{-2 + \sqrt{11}}{3}$ and $\dfrac{-2 - \sqrt{11}}{3}$; $9x^2 + 12x = 7$   yes

**88.** $\dfrac{7 + \sqrt{21}}{5}$ and $\dfrac{7 - \sqrt{21}}{5}$; $25x^2 + 28 = 70x$   yes

---

## PIECE IT TOGETHER      SECTIONS 9.1–9.3 Review

**Solve each equation by applying the best method— factoring, square root property, completing the square, or the quadratic formula.**

**1.** $x^2 - 3x = 10$
    $\{-2, 5\}$
**2.** $x^2 + x = 12$   $\{-4, 3\}$
**3.** $(3x + 2)^2 = 28$
**4.** $(2x - 1)^2 = 76$
**5.** $4x^2 + 36 = 0$
    no real solutions
**6.** $25x^2 = 1$   $\left\{-\dfrac{1}{5}, \dfrac{1}{5}\right\}$
**7.** $(5x - 2)^2 = -16$
    no real solutions
**8.** $(2x + 7)^2 = -49$
    no real solutions

**9.** $0.3x^2 + 0.7x + 1.5 = 0$   **10.** $0.6x^2 - 0.5x - 2.1 = 0$
    no real solutions
**11.** $9x^2 - 60x + 98 = 0$   **12.** $4x^2 + 36x + 88 = 0$
                                    no real solutions
**13.** $8x^2 + 19x - 15 = 0$   **14.** $14x^2 + 9x + 1 = 0$   $\left\{-\dfrac{1}{2}, -\dfrac{1}{7}\right\}$
**15.** $\left(x - \dfrac{2}{3}\right)^2 = \dfrac{50}{9}$   **16.** $\left(x + \dfrac{3}{4}\right)^2 = \dfrac{75}{16}$
**17.** $2x^2 - 3x = 0$   $\left\{0, \dfrac{3}{2}\right\}$   **18.** $5x^2 + 4x = 0$   $\left\{-\dfrac{4}{5}, 0\right\}$
**19.** $x^2 + 6x - 3 = 0$   **20.** $x^2 - 4x - 3 = 0$
    $\{-3 - 2\sqrt{3}, -3 + 2\sqrt{3}\}$      $\{2 - \sqrt{7}, 2 + \sqrt{7}\}$

/ **Complex Numbers and Quadratic Equations with Complex Solutions**

### OBJECTIVES

As a result of completing this section, you will be able to

1. Write square roots of negative numbers as complex numbers.
2. Add and subtract complex numbers.
3. Multiply complex numbers.
4. Divide complex numbers.
5. Solve quadratic equations with complex solutions.
6. Troubleshoot common errors.

In Section 8.1, we discussed square roots such as $\sqrt{-16}$, $\sqrt{-8}$, and $\sqrt{-7}$. These types of numbers were also seen when we solved some quadratic equations. Up until now, we stated that these types of numbers are not real numbers since there is no real number that, when squared, equals a negative number. While it is true that they are not real, we will introduce some notation that represents these types of expressions using a different number system.

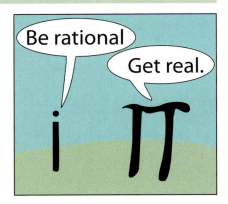

### Objective 1 ▶

Write square roots of negative numbers as complex numbers.

## Complex Numbers

Centuries ago, mathematicians realized that the square root of negative numbers arose when solving certain types of equations. They actually referred to these types of solutions as "imaginary" or "ghostly." Mathematicians devised a new number system to include numbers like $\sqrt{-16}$ and $\sqrt{-8}$. This new number system is called the *complex number* system. The **imaginary unit**, represented by the symbol $i$, is included in this system. The imaginary unit is a number that can be squared to obtain a negative number.

> **Definition: Imaginary Unit, $i$**
> The number $i$ is the number whose square is $-1$.
>
> $$i = \sqrt{-1} \quad \text{and} \quad i^2 = -1$$

We use $i$ to write numbers like $\sqrt{-16}$ as the product of a real number and $i$. Since $i = \sqrt{-1}$, $\sqrt{-16} = \sqrt{-1 \cdot 16} = \sqrt{-1} \cdot \sqrt{16} = i(4)$ or $4i$.

> **Definition:** Any number of the form $a + bi$, where $a$ and $b$ are real numbers, is a **complex number**. This form is the standard form of a complex number, where $a$ is the real part and $b$ is the imaginary part of the complex number.
>
> If $a = 0$, then $a + bi = 0 + bi = bi$. This number is *purely imaginary*.
> If $b = 0$, then $a + bi = a + 0(i) = a$. This number is *purely real*.

Some examples of complex numbers are $3 + 2i$, $-1 - 5i$, $0 + i\sqrt{7}$, and $-2 + 0i$.

> **Note:** *The number* $3 + 2i$ *represents* $3 + \sqrt{-4}$ *since* $\sqrt{-4} = \sqrt{-1 \cdot 4} = \sqrt{-1} \cdot \sqrt{4} = 2i$.

The set of complex numbers includes all of the real numbers since any real number $a$ can be written as $a + 0i$. We can visualize the complex number system as shown.

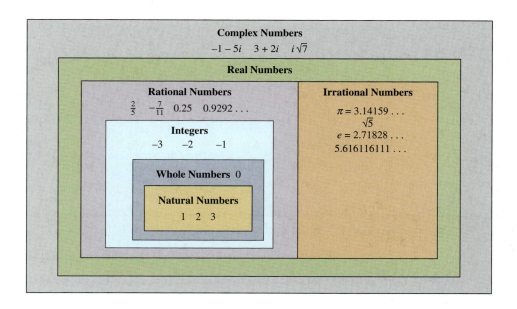

**Procedure:  Writing the Square Root of a Negative Number in Terms of $i$**

**Step 1:** Rewrite $\sqrt{-c}$ as $\sqrt{-1 \cdot c}$, where $c > 0$.
**Step 2:** Apply the product rule: $\sqrt{-1 \cdot c} = \sqrt{-1}\sqrt{c} = i\sqrt{c}$.
**Step 3:** Simplify $\sqrt{c}$ as necessary.

**Objective 1  Examples**   Write each expression in terms of $i$.

**1a.** $\sqrt{-36}$   **1b.** $\sqrt{-8}$   **1c.** $\sqrt{-7}$   **1d.** $4 + \sqrt{-25}$   **1e.** $\dfrac{-3 + \sqrt{-18}}{4}$

**Solutions**   **1a.**   $\sqrt{-36} = \sqrt{-1 \cdot 36}$   Rewrite the radicand as a product involving $-1$.

$= \sqrt{-1}\sqrt{36}$   Apply the product rule.

$= i(6)$   Simplify each radical.

$= 6i$

**1b.**   $\sqrt{-8} = \sqrt{-1 \cdot 4 \cdot 2}$   Rewrite the radicand as a product involving $-1$ and a perfect square.

$= \sqrt{-1 \cdot 4}\sqrt{2}$   Apply the product rule.

$= i(2)\sqrt{2}$   Simplify each radical.

$= 2i\sqrt{2}$

**1c.**   $\sqrt{-7} = \sqrt{-1 \cdot 7}$   Rewrite the radicand as a product involving $-1$.

$= \sqrt{-1}\sqrt{7}$   Apply the product rule.

$= i\sqrt{7}$   Simplify each radical.

**1d.** $4 + \sqrt{-25} = 4 + \sqrt{-1 \cdot 25}$   Rewrite the radicand as a product involving $-1$.

$= 4 + i(5)$   Apply the product rule and simplify.

$= 4 + 5i$   Write in standard form.

**1e.** $\dfrac{-3 + \sqrt{-18}}{4} = \dfrac{-3 + \sqrt{-1 \cdot 9 \cdot 2}}{4}$   Rewrite the radicand as a product involving −1 and a perfect square.

$= \dfrac{-3 + i(3)\sqrt{2}}{4}$   Apply the product rule and simplify.

$= \dfrac{-3 + 3i\sqrt{2}}{4}$   Simplify.

$= -\dfrac{3}{4} + \dfrac{3i\sqrt{2}}{4}$   Divide each term in the numerator by 4.

$= -\dfrac{3}{4} + \dfrac{3\sqrt{2}}{4}i$   Write in standard form.

---

✔ **Student Check 1**   Write each expression in terms of $i$.

**a.** $\sqrt{-49}$   **b.** $\sqrt{-20}$   **c.** $\sqrt{-11}$   **d.** $6 + \sqrt{-4}$   **e.** $\dfrac{-5 + \sqrt{-24}}{6}$

---

## Adding and Subtracting Complex Numbers

**Objective 2** ▶

Add and subtract complex numbers.

Any operation that can be performed on real numbers can also be performed on complex numbers; that is, we can add, subtract, multiply, and divide complex numbers.

We perform operations on complex numbers the same way we perform operations on polynomials. To add or subtract complex numbers, we combine like terms. Like terms are the real parts and the imaginary parts of the complex numbers.

> **Procedure: Adding or Subtracting Complex Numbers**
>
> **Step 1:** Clear the parentheses. If subtracting, distribute −1 to the complex number being subtracted.
> **Step 2:** Group the real parts and imaginary parts.
> **Step 3:** Combine like terms.

---

**Objective 2 Examples**   **Perform the operation on the complex numbers.**

**2a.** $(4 - 5i) + (7 - 9i)$   **2b.** $(4 - 5i) - (7 - 9i)$   **2c.** $2i + (7 - i) - (3 + i)$

**Solutions**   **2a.**   $(4 - 5i) + (7 - 9i) = 4 - 5i + 7 - 9i$   Clear parentheses.

$= 4 + 7 - 5i - 9i$   Group real parts and imaginary parts.

$= 11 - 14i$   Add the real parts and the imaginary parts.

**2b.**   $(4 - 5i) - 1(7 - 9i) = 4 - 5i - 7 + 9i$   Clear parentheses.

$= 4 - 7 - 5i + 9i$   Group real parts and imaginary parts.

$= -3 + 4i$   Add the real parts and the imaginary parts.

**2c.**   $2i + (7 - i) - 1(3 + i) = 2i + 7 - i - 3 - i$   Clear parentheses.

$= 7 - 3 + 2i - i - i$   Group real parts and imaginary parts.

$= 4 + 0i$   Add the real parts and the imaginary parts.

$= 4$   Simplify.

✓ **Student Check 2**    Perform the operation on the complex numbers.

     **a.** $(2 + 3i) + (8 - 5i)$     **b.** $(2 + 3i) - (8 - 5i)$     **c.** $4 + (8 + 7i) - (4 + 6i)$

## Multiplying Complex Numbers

**Objective 3** ▶

Multiply complex numbers.

We multiply complex numbers just like we multiply polynomials. We use the distributive property and the FOIL method. The key is to express the product in the form $a + bi$. After multiplying, we must replace all instances of $i^2$ with $-1$.

> **Procedure: Multiplying Complex Numbers**
>
> **Step 1:** Write any radicals in terms of $i$, if necessary.
> **Step 2:** Apply the distributive property or product of like bases rule.
> **Step 3:** Rewrite $i^2$ as $-1$ and combine like terms.
> **Step 4:** Simplify.
> **Step 5:** Write the answer in standard form, $a + bi$.

**Objective 3 Examples**    **Multiply the complex numbers and write each answer in standard form.**

   **3a.** $\sqrt{-25} \cdot \sqrt{-4}$      **3b.** $3i(2 + 6i)$      **3c.** $(4 - 5i)(7 - 9i)$

   **3d.** $(4 - 5i)(4 + 5i)$      **3e.** $(1 - 3i)^2$

**Solutions**    **3a.** $\sqrt{-25} \cdot \sqrt{-4} = (5i)(2i)$      Simplify each radical.

$$= 10i^2 \qquad \text{Multiply.}$$
$$= 10(-1) \qquad \text{Replace } i^2 \text{ with } -1.$$
$$= -10 \qquad \text{Multiply.}$$

   **3b.** $\qquad 3i(2 + 6i) = (3i)(2) + (3i)(6i)$      Distribute.

$$= 6i + 18i^2 \qquad \text{Simplify each product.}$$
$$= 6i + 18(-1) \qquad \text{Rewrite } i^2 \text{ as } -1.$$
$$= 6i - 18 \qquad \text{Simplify.}$$
$$= -18 + 6i \qquad \text{Write in standard form.}$$

   **3c.** $(4 - 5i)(7 - 9i) = 4(7) - 4(9i) - 5i(7) - 5i(-9i)$      Distribute.

$$= 28 - 36i - 35i + 45i^2 \qquad \text{Simplify each product.}$$
$$= 28 - 71i + 45(-1) \qquad \text{Rewrite } i^2 \text{ as } -1.$$
$$= 28 - 71i - 45 \qquad \text{Simplify.}$$
$$= -17 - 71i \qquad \text{Combine like terms.}$$

   **3d.** $(4 - 5i)(4 + 5i) = 4(4) + 4(5i) - 5i(4) - 5i(5i)$      Distribute.

$$= 16 + 20i - 20i - 25i^2 \qquad \text{Simplify.}$$
$$= 16 - 25(-1) \qquad \text{Rewrite } i^2 \text{ as } -1 \text{ and combine like terms.}$$
$$= 16 + 25 \qquad \text{Simplify.}$$
$$= 41 \qquad \text{Combine like terms.}$$

**3e.** Square the complex number by either repeating the base or by applying the formula for squaring a binomial.

| **Method 1** | **Method 2** |
|---|---|
| $(1 - 3i)^2 = (1 - 3i)(1 - 3i)$ | $(1 - 3i)^2 = (1)^2 - 2(1)(3i) + (3i)^2$ |
| $= 1 - 3i - 3i + 9i^2$ | $= 1 - 6i + 9i^2$ |
| $= 1 - 6i + 9(-1)$ | $= 1 - 6i - 9$ |
| $= 1 - 6i - 9$ | $= -8 - 6i$ |
| $= -8 - 6i$ | |

✓ **Student Check 3**    Multiply the complex numbers and write each answer in standard form.

**a.** $\sqrt{-49} \cdot \sqrt{-36}$    **b.** $7i(3 + 2i)$    **c.** $(2 - 3i)(4 - i)$

**d.** $(2 - 3i)(2 + 3i)$    **e.** $(4 - 5i)^2$

## Dividing Complex Numbers

**Objective 4** ▶

Divide complex numbers.

The goal in dividing complex numbers is to write the result as a complex number in standard form $a + bi$. That is, we must remove the complex number from the denominator. This will involve the product of conjugates.

Recall that expressions of the form $3x - 2$ and $3x + 2$ are conjugates of one another. Complex numbers also have conjugates. For example, the numbers $4 + 5i$ and $4 - 5i$ are *complex conjugates*.

**Definition:** The **complex conjugate** of $a + bi$ is $a - bi$.

Some other examples of complex conjugates are shown.

| **Complex Number** | **Complex Conjugate** |
|---|---|
| $-2 - 7i$ | $-2 + 7i$ |
| $5i = 0 + 5i$ | $-5i = 0 - 5i$ |
| $1 + \sqrt{11}i$ | $1 - \sqrt{11}i$ |
| $\dfrac{3}{5} - \dfrac{\sqrt{2}}{5}i$ | $\dfrac{3}{5} + \dfrac{\sqrt{2}}{5}i$ |
| $4 = 4 + 0i$ | $4 = 4 - 0i$ |

Note from Example 3d that the product of complex conjugates, $(4 - 5i)(4 + 5i)$, is a real number, 41. Recall that the product of conjugates results in the difference of two squares. We can use this fact to simplify the product $(4 - 5i)(4 + 5i)$ as follows.

$$(4 - 5i)(4 + 5i) = (4)^2 - (5i)^2$$
$$= 16 - 25i^2$$
$$= 16 - 25(-1)$$
$$= 16 + 25$$
$$= 41$$

We generalize this relationship as follows.

**Property: Product of Complex Conjugates**

$$(a + bi)(a - bi) = a^2 + b^2$$

We will use the fact that the product of complex conjugates is a real number to divide complex numbers.

---

**Procedure: Dividing Complex Numbers**

**Step 1:** If necessary, write the square root of a negative number as an imaginary number.

**Step 2:** Multiply the numerator and the denominator by the complex conjugate of the denominator.

**Step 3:** Simplify the result.

---

**Objective 4 Examples**    **Divide the complex numbers and write each answer in standard form.**

**4a.** $\dfrac{\sqrt{-36}}{\sqrt{-4}}$      **4b.** $\dfrac{5 - 3i}{4i}$      **4c.** $\dfrac{-2 + i}{4 + 3i}$

**Solutions**    **4a.** $\dfrac{\sqrt{-36}}{\sqrt{-4}} = \dfrac{6i}{2i}$     Rewrite the numerator and denominator as complex numbers.

$= \dfrac{6i(-2i)}{2i(-2i)}$     Multiply the numerator and denominator by the complex conjugate of $2i$, $-2i$.

$= \dfrac{-12i^2}{-4i^2}$     Simplify each product.

$= \dfrac{-12(-1)}{-4(-1)}$     Rewrite $i^2$ as $-1$.

$= \dfrac{12}{4}$     Simplify the numerator and denominator.

$= 3$     Divide.

**4b.** $\dfrac{5 - 3i}{4i} = \dfrac{(5 - 3i)(-4i)}{(4i)(-4i)}$     Multiply the numerator and denominator by the complex conjugate of $4i$, $-4i$.

$= \dfrac{-20i + 12i^2}{-16i^2}$     Simplify each product.

$= \dfrac{-20i + 12(-1)}{-16(-1)}$     Rewrite $i^2$ as $-1$.

$= \dfrac{-20i - 12}{16}$     Simplify.

$= \dfrac{-20i}{16} - \dfrac{12}{16}$     Divide each term in the numerator by 16.

$= -\dfrac{12}{16} - \dfrac{20}{16}i$     Write in standard form.

$= -\dfrac{3}{4} - \dfrac{5}{4}i$     Simplify the real and imaginary parts.

**4c.** $\dfrac{-2 + i}{4 + 3i} = \dfrac{(-2 + i)(4 - 3i)}{(4 + 3i)(4 - 3i)}$     Multiply the numerator and denominator by the complex conjugate of $4 + 3i$, $4 - 3i$.

$= \dfrac{-8 + 6i + 4i - 3i^2}{16 - 9i^2}$     Simplify each product.

$= \dfrac{-8 + 10i - 3(-1)}{16 - 9(-1)}$     Rewrite $i^2$ as $-1$.

$= \dfrac{-8 + 10i + 3}{16 + 9}$     Simplify.

$$= \frac{-5 + 10i}{25} \qquad \text{Combine like terms.}$$

$$= \frac{-5}{25} + \frac{10}{25}i \qquad \text{Write in standard form.}$$

$$= -\frac{1}{5} + \frac{2}{5}i \qquad \text{Simplify the real and imaginary parts.}$$

✔ **Student Check 4**    Divide the complex numbers and write each answer in standard form.

a. $\dfrac{\sqrt{-64}}{\sqrt{-16}}$    b. $\dfrac{4 + 3i}{5i}$    c. $\dfrac{1 + 2i}{4 + i}$

## Quadratic Equations with Complex Solutions

**Objective 5** ▶

Solve quadratic equations with complex solutions.

In Sections 9.1 through 9.3, we encountered solutions of quadratic equations that have no real solutions. This came about since the solution contained the square root of a negative number. Now that we know how to write these types of numbers as complex numbers, we can write the nonreal solutions of quadratic equations as complex numbers.

> **Procedure: Solving Quadratic Equations with Complex Solutions**
>
> **Step 1:** Use an appropriate method to solve the equation. Because there are complex solutions, factoring is not an option.
> **Step 2:** Rewrite the square root of the negative number in terms of $i$.
> **Step 3:** Write each solution in the form $a + bi$.

**Objective 5 Examples**    Solve each quadratic equation for its complex solutions.

**5a.** $(x - 2)^2 = -9$        **5b.** $3y^2 - 4y + 5 = 0$

**Solutions**    **5a.** Since the equation is of the form, $x^2 = k$, we can apply the square root property.

$$(x - 2)^2 = -9$$

$x - 2 = \sqrt{-9}$  or  $x - 2 = -\sqrt{-9}$    Apply the square root property.

$x - 2 = 3i \qquad\qquad x - 2 = -3i$    Rewrite $\sqrt{-9}$ as $3i$.

$x = 2 + 3i \qquad\qquad x = 2 - 3i$    Add 2 to each side.

The solution set is $\{2 - 3i, 2 + 3i\}$ or $\{2 \pm 3i\}$. The solutions can be checked by substituting them into the original equation.

**5b.** Since the equation is in standard form, we can use the quadratic formula to solve it with $a = 3$, $b = -4$, and $c = 5$.

$$y = \frac{-b \pm \sqrt{b^2 - 4ac}}{2a} \qquad \text{State the quadratic formula.}$$

$$y = \frac{-(-4) \pm \sqrt{(-4)^2 - 4(3)(5)}}{2(3)} \qquad \text{Substitute the values for } a, b, \text{ and } c.$$

$$y = \frac{4 \pm \sqrt{16 - 60}}{6} \qquad \text{Simplify the numerator and denominator.}$$

$$y = \frac{4 \pm \sqrt{-44}}{6}$$  Simplify the radicand.

$$y = \frac{4 \pm 2i\sqrt{11}}{6}$$  Simplify:
$$\sqrt{-44} = \sqrt{-4} \cdot \sqrt{11} = 2i\sqrt{11}$$

$$y = \frac{2(2 \pm i\sqrt{11})}{6}$$  Factor out a common factor of 2 in the numerator.

$$y = \frac{2 \pm i\sqrt{11}}{3}$$  Divide out the common factor of 2 from the numerator and denominator.

$$y = \frac{2 + i\sqrt{11}}{3} \quad \text{or} \quad y = \frac{2 - i\sqrt{11}}{3}$$  Write the two solutions.

$$y = \frac{2}{3} + \frac{\sqrt{11}}{3}i \quad \text{or} \quad y = \frac{2}{3} - \frac{\sqrt{11}}{3}i$$  Write in standard form.

So, the solution set is $\left\{ \dfrac{2 - i\sqrt{11}}{3}, \dfrac{2 + i\sqrt{11}}{3} \right\}$. We can also write the solutions in

the standard form of a complex number to get $\left\{ \dfrac{2}{3} - \dfrac{\sqrt{11}}{3}i, \dfrac{2}{3} + \dfrac{\sqrt{11}}{3}i \right\}$.

✓ **Student Check 5**    Solve each quadratic equation for its complex solutions.
  **a.** $(x + 4)^2 = -81$     **b.** $2y^2 - 6y + 7 = 0$

---

**Objective 6** ▶
Troubleshoot common errors.

## Troubleshooting Common Errors

Some common errors associated with complex numbers and quadratic equations are shown next.

**Objective 6  Examples**    **A problem and an incorrect solution are given. Provide the correct solution and an explanation of the error.**

**6a.** Simplify $(4 + 2i) + (3 - 7i)$.

| Incorrect Solution | Correct Solution and Explanation |
|---|---|
| $(4 + 2i) + (3 - 7i)$ <br> $= 12 - 28i + 6i - 14i^2$ <br> $= 12 - 22i - 14(-1)$ <br> $= 12 - 22i + 14$ <br> $= 26 - 22i$ | This is the *sum* of the two complex numbers, not the product. We combine the real and imaginary parts. <br><br> $(4 + 2i) + (3 - 7i) = 4 + 3 + 2i - 7i$ <br> $= 7 - 5i$ |

**6b.** Simplify $(5 - 6i)^2$.

| Incorrect Solution | Correct Solution and Explanation |
|---|---|
| $(5 - 6i)^2 = 25 + 36i^2$ <br> $= 25 + 36(-1)$ <br> $= 25 - 36$ <br> $= -11$ | Since the base is a binomial, we must apply the squaring pattern or multiply the base by itself. <br><br> $(5 - 6i)^2 = 25 - 60i + 36i^2$ <br> $= 25 - 60i - 36$ <br> $= -11 - 60i$ |

**6c.** Solve $(x - 1)^2 = -4$ for its complex solutions.

| Incorrect Solution | Correct Solution and Explanation |
|---|---|
| ~~$(x - 1)^2 = -4$~~ <br> ~~$x - 1 = \sqrt{-4}$~~ <br> ~~$x - 1 = 2i$~~ <br> ~~$x = 1 + 2i$~~ <br> ~~The solution set is $\{1 + 2i\}$.~~ | We must also set the expression equal to the negative square root of $-4$. <br><br> $(x - 1)^2 = -4$ <br> $x - 1 = \sqrt{-4}$  or  $x - 1 = -\sqrt{-4}$ <br> $x - 1 = 2i$      $x - 1 = -2i$ <br> $x = 1 + 2i$      $x = 1 - 2i$ <br><br> The solution set is $\{1 - 2i, 1 + 2i\}$. |

## ANSWERS TO STUDENT CHECKS

**Student Check 1**   **a.** $7i$   **b.** $2i\sqrt{5}$   **c.** $i\sqrt{11}$
   **d.** $6 + 2i$   **e.** $-\dfrac{5}{6} + \dfrac{\sqrt{6}}{3}i$

**Student Check 2**   **a.** $10 - 2i$   **b.** $-6 + 8i$   **c.** $8 + i$

**Student Check 3**   **a.** $-42$   **b.** $-14 + 21i$   **c.** $5 - 14i$
   **d.** $13$   **e.** $-9 - 40i$

**Student Check 4**   **a.** $2$   **b.** $\dfrac{3}{5} - \dfrac{4}{5}i$   **c.** $\dfrac{6}{17} + \dfrac{7}{17}i$

**Student Check 5**   **a.** $\{-4 - 9i, -4 + 9i\}$
   **b.** $\left\{ \dfrac{3}{2} - \dfrac{\sqrt{5}}{2}i, \dfrac{3}{2} + \dfrac{\sqrt{5}}{2}i \right\}$

## SUMMARY OF KEY CONCEPTS

1. A complex number is a number that can be written in the form $a + bi$, where $i^2 = -1$ and $i = \sqrt{-1}$. The expression $\sqrt{-c} = i\sqrt{c}$, for $c > 0$.

2. We add and subtract complex numbers just like we add and subtract polynomials.

3. We also multiply complex numbers like we multiply polynomials, by using the distributive property and FOIL. Rewrite $i^2$ as $-1$ and write the answer in the form $a + bi$.

4. Quadratic equations have complex solutions when the radicand, the discriminant, is negative. In this case, we want to express the solutions in terms of the imaginary unit, $i$.

## GRAPHING CALCULATOR SKILLS

We can use the calculator to express square roots of negative numbers in terms of $i$. We can also perform operations with complex numbers using a calculator.

**Example:** Simplify $4 + \sqrt{-25}$.

**Solution:** Change the calculator MODE from REAL to $a + bi$.

**Example:** Use the calculator to perform each operation:
(a) $(4 - 5i) - (7 - 9i)$ and (b) $(4 - 5i)(7 - 9i)$

**Solution:** To enter the imaginary unit $i$, press (2nd) (.).

```
(4-5i)-(7-9i)
              -3+4i
(4-5i)(7-9i)
            -17-71i
```

# SECTION 9.4 / EXERCISE SET

## Write About It!

Use complete sentences in your answer to each exercise.

1. What is a complex number?

2. Is every imaginary number a complex number? Explain.

3. Is every complex number an imaginary number? Explain.

4. Is every real number a complex number? Explain.

5. Use an example to explain how to write the square root of a negative number as a complex number.

6. Explain how to add and subtract complex numbers.

7. Explain how to multiply complex numbers.

8. Explain why the product of $(a + bi)(a - bi)$ is always a real number.

## Practice Makes Perfect!

**Write each expression as a complex number.** *(See Objective 1.)*

9. $\sqrt{-25}$   $5i$
10. $\sqrt{-4}$   $2i$
11. $\sqrt{-100}$   $10i$
12. $\sqrt{-64}$   $8i$
13. $\sqrt{-12}$   $2\sqrt{3}i$
14. $\sqrt{-18}$   $3\sqrt{2}i$
15. $\sqrt{-3}$   $\sqrt{3}i$
16. $\sqrt{-2}$   $\sqrt{2}i$
17. $3 + \sqrt{-1}$   $3 + i$
18. $7 + \sqrt{-9}$   $7 + 3i$
19. $2 + \sqrt{-80}$   $2 + 4i\sqrt{5}$
20. $3 - \sqrt{-75}$   $3 - 5i\sqrt{3}$
21. $\dfrac{-1 + \sqrt{-20}}{2}$   $-\dfrac{1}{2} + i\sqrt{5}$
22. $\dfrac{-3 + \sqrt{-32}}{4}$   $-\dfrac{3}{4} + i\sqrt{2}$
23. $\dfrac{10 - \sqrt{-18}}{6}$   $\dfrac{5}{3} - \dfrac{\sqrt{2}}{2}i$
24. $\dfrac{6 - \sqrt{-24}}{8}$   $\dfrac{3}{4} - \dfrac{\sqrt{6}}{4}i$

**Add or subtract the complex numbers and write each answer in standard form.** *(See Objective 2.)*

25. $(3 - 2i) + (-4 - 6i)$   $-1 - 8i$
26. $(5 - 5i) + (-2 - 3i)$   $3 - 8i$
27. $(10 + 11i) + (6 + 4i)$   $16 + 15i$
28. $(7 + 9i) + (11 + 5i)$   $18 + 14i$
29. $(12 + 8i) + (10 - 7i)$   $22 + i$
30. $(6 - 3i) - (2 + 5i)$   $4 - 8i$
31. $(14 + 5i) - (7 - 3i)$   $7 + 8i$
32. $(-4 + 2i) + (4 - 3i)$   $-i$
33. $6 + (3 - 2i) - (5 + 4i)$   $4 - 6i$
34. $10 + (10 + 6i) - (14 - 3i)$   $6 + 9i$
35. $4i - (2 - 9i) + (6 + 3i)$   $4 + 16i$
36. $-3i - (11 - 9i) + (20 + 6i)$   $9 + 12i$

**Multiply the complex numbers and write each answer in standard form.** *(See Objective 3.)*

37. $i(10 + 3i)$   $-3 + 10i$
38. $3i(4 + 2i)$   $-6 + 12i$
39. $-4i(2 - 3i)$   $-12 - 8i$
40. $-6i(10 - 5i)$   $-30 - 60i$

Additional answers can be found in the Instructor Answer Appendix.

41. $(1 - 3i)(2 + 5i)$   $17 - i$
42. $(6 - 4i)(7 + 3i)$   $54 - 10i$
43. $(5 - 2i)(5 + 2i)$   $29$
44. $(4 - 3i)(4 + 3i)$   $25$
45. $(2 - 2i)^2$   $-8i$
46. $(3 - 3i)^2$   $-18i$
47. $(7 + 5i)^2$   $24 + 70i$
48. $(10 + 4i)^2$   $84 + 80i$

**Divide the complex numbers and write each answer in standard form.** *(See Objective 4.)*

49. $\dfrac{1 + 3i}{i}$   $3 - i$
50. $\dfrac{-4 + 5i}{i}$   $5 + 4i$
51. $\dfrac{4 - 8i}{-2i}$   $4 + 2i$
52. $\dfrac{-6 + 9i}{-3i}$   $-3 - 2i$
53. $\dfrac{2 - 5i}{1 + 2i}$   $-\dfrac{8}{5} - \dfrac{9}{5}i$
54. $\dfrac{-3 + 2i}{4 - 2i}$   $-\dfrac{4}{5} + \dfrac{1}{10}i$
55. $\dfrac{10i}{2 + i}$   $2 + 4i$
56. $\dfrac{-20i}{1 - 7i}$   $\dfrac{14}{5} - \dfrac{2}{5}i$

**Solve each quadratic equation for its complex solutions.** *(See Objective 5.)*

57. $(x - 4)^2 = -9$   $\{4 - 3i, 4 + 3i\}$
58. $(x - 1)^2 = -4$   $\{1 - 2i, 1 + 2i\}$
59. $(x - 2)^2 = -28$   $\{2 - 2i\sqrt{7}, 2 + 2i\sqrt{7}\}$
60. $(x + 1)^2 = -75$   $\{-1 - 5i\sqrt{3}, -1 + 5i\sqrt{3}\}$
61. $(2x - 3)^2 = -16$
62. $(3x + 1)^2 = -1$
63. $x^2 + 3x + 3 = 0$
64. $x^2 + x + 2 = 0$
65. $6y^2 - 2y + 3 = 0$
66. $2y^2 - y + 4 = 0$
67. $2x(x - 3) = -10$
68. $5x(x - 2) = -8$
69. $7x - 8 = 2x^2$   $\left\{\dfrac{7 - \sqrt{15}i}{4}, \dfrac{7 + \sqrt{15}i}{4}\right\}$
70. $3x - 2 = 5x^2$   $\left\{\dfrac{3 - \sqrt{31}i}{10}, \dfrac{3 + \sqrt{31}i}{10}\right\}$

## Mix 'Em Up!

**Perform each operation on the complex numbers and write each answer in standard form.**

71. $\sqrt{-169} + (2 - \sqrt{-25})$   $2 + 8i$
72. $(6 - \sqrt{-121}) - (1 + \sqrt{-81})$   $5 - 20i$
73. $(2 - 8i) - (4 - i)$   $-2 - 7i$
74. $(9 - 4i) + (12 + 6i)$   $21 + 2i$
75. $(12 + 2i) + (10 - 7i)$   $22 - 5i$
76. $(20 + 13i) - (1 - i)$   $19 + 14i$
77. $-2i(4 - 5i)$   $-10 - 8i$
78. $6i(-4i)$   $24$
79. $(7 - i)^2$   $48 - 14i$
80. $(3 + 2i)^2$   $5 + 12i$
81. $(2 + 3i)(7 - i)$   $17 + 19i$
82. $(1 + i)(1 - 6i)$   $7 - 5i$
83. $\left(\dfrac{1}{2} - \dfrac{2}{5}i\right)\left(\dfrac{1}{2} + \dfrac{3}{5}i\right)$
84. $\left(\dfrac{2}{5} - \dfrac{1}{3}i\right) - \left(\dfrac{6}{5} - \dfrac{2}{3}i\right)$   $-\dfrac{4}{5} + \dfrac{1}{3}i$
85. $(1.2 - 0.5i)(2.4 + 1.2i)$   $3.48 + 0.24i$
86. $(7.9 + 3.1i) - (-4.2 - 8.7i)$   $12.1 + 11.8i$
87. $(-3i)(-9i)$   $-27$
88. $2i(5 - 3i)$   $6 + 10i$
89. $(5.2 - 1.4i) + (2.6 + 4.6i)$   $7.8 + 3.2i$
90. $(2.5 + 0.3i)(1.6 + 1.5i)$   $3.55 + 4.23i$

**91.** $\dfrac{6 - 9i}{-3i}$  $3 + 2i$  **92.** $\dfrac{-12 - 8i}{4i}$  $-2 + 3i$

**93.** $\dfrac{-2 + 3i}{6 - 2i}$  $-\dfrac{9}{20} + \dfrac{7}{20}i$  **94.** $\dfrac{4 - 2i}{5 + 3i}$  $\dfrac{7}{17} - \dfrac{11}{17}i$

**95.** $\dfrac{15}{3 + i}$  $\dfrac{9}{2} - \dfrac{3}{2}i$  **96.** $\dfrac{5i}{-2 + 4i}$  $1 - \dfrac{1}{2}i$

**97.** $\left(\dfrac{3}{2} + \dfrac{5}{2}i\right) + \left(\dfrac{1}{4} - \dfrac{1}{2}i\right)$ **98.** $\left(\dfrac{2}{3} - \dfrac{1}{2}i\right)\left(\dfrac{1}{3} + \dfrac{1}{2}i\right)$  $\dfrac{17}{36} + \dfrac{1}{6}i$
$\dfrac{7}{4} + 2i$

**Solve each equation for its complex solutions.**

**99.** $1 - x = -x^2$  **100.** $5x^2 - 6x = -3$

**101.** $2x^2 + 18 = 0$  **102.** $5x^2 + 20 = 0$  $\{-2i, 2i\}$
$\{-3i, 3i\}$

**103.** $x^3 + 2x^2 + 19x = 0$  **104.** $x^3 - 4x^2 + 31x = 0$
$\{0, -1 - 3i\sqrt{2}, -1 + 3i\sqrt{2}\}$  $\{0, 2 - 3i\sqrt{3}, 2 + 3i\sqrt{3}\}$

**105.** $x^4 + 3x^2 - 4 = 0$  **106.** $x^4 + 5x^2 - 36 = 0$
$\{-2i, 2i, -1, 1\}$  $\{-3i, 3i, -2, 2\}$

 **You Be the Teacher!**

**Correct each student's errors, if any.**

**107.** Simplify $\dfrac{3 - \sqrt{-27}}{3}$.

Marty's work:

$$\dfrac{3 - \sqrt{-27}}{3} = \dfrac{3 + \sqrt{27}}{3} = \dfrac{3 + 3\sqrt{3}}{3} = 1 + \sqrt{3}$$

$\dfrac{3 - \sqrt{-27}}{3} = \dfrac{3 - \sqrt{27}i}{3} = \dfrac{3 - 3\sqrt{3}i}{3} = 1 - \sqrt{3}i$

**108.** Simplify $\dfrac{2 + \sqrt{-8}}{2}$.

Lois's work:

$$\dfrac{2 + \sqrt{-8}}{2} = 1 + \sqrt{-8} = 1 + i\sqrt{8} = 1 + 2i\sqrt{2}$$

$\dfrac{2 + \sqrt{-8}}{2} = \dfrac{2 + \sqrt{8}i}{2} = \dfrac{2 + 2\sqrt{2}i}{2} = 1 + \sqrt{2}i$

**109.** Simplify: $(5 + 2i)(3 - i)$

Dottie's work:

$$(5 + 2i)(3 - i) = 15 - 5i + 6i - 2i^2 = 15 + i - 2i^2$$

$(5 + 2i)(3 - i) = 15 - 5i + 6i - 2i^2 = 15 + i - 2i^2 = 17 + i$

**110.** Simplify $(-2 - 3i)^2$.

Fred's work:

$$(-2 - 3i)^2 = (-2)^2 + (-3i)^2 = 4 + 9i^2 = 4 + 9(-1)$$
$$= 4 - 9 = -5$$

$(-2 - 3i)^2 = (-2)^2 + 2(-2)(-3i) + (-3i)^2 = 4 + 12i + 9i^2$
$= 4 + 12i + 9(-1) = 4 + 12i - 9 = -5 + 12i$

 **Calculate It!**

**Use a calculator to perform each operation.**

**111.** $(116 - 15i) - (243 + 112i)$  $-127 - 127i$

**112.** $(201 - 24i) + (16 + 15i)$  $217 - 9i$

**113.** $(15 - 3i)^2$  $216 - 90i$  **114.** $(11 - 5i)^2$  $96 - 110i$

**115.** $(21 + 10i)(19 + 9i)$  **116.** $\dfrac{3 + 7i}{5 + 2i}$  $1 + i$
$309 + 379i$

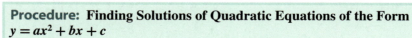

## SECTION 9.5 | Solve Quadratic Equations by Graphical Methods

▶ **OBJECTIVES**

As a result of completing this section, you will be able to

**1.** Find solutions of the equation $y = ax^2 + bx + c$.

**2.** Graph the equation $y = ax^2 + bx + c$.

**3.** Use the graph of $y = ax^2 + bx + c$ to solve the equation $ax^2 + bx + c = 0$.

**4.** Troubleshoot common errors.

Many architectural objects have a special shape, a U-shape. Recall from Section 3.1 that the graph of the equation $y = x^2 + 1$ is a U-shape. These types of curves result from graphing quadratic equations. In this section, we will study the graph of a quadratic equation.

### Solutions of Equations of the Form $y = ax^2 + bx + c$

In Chapter 3, we found that *solutions* of equations in two variables are ordered pairs $(x, y)$ that satisfy the equation. We also found that there are infinitely many solutions of equations in two variables. We are not going to be able to find every solution, so our goal is to find a few solutions that will then enable us to graph the equation.

**Objective 1** ▶

Find solutions of the equation $y = ax^2 + bx + c$.

> **Procedure: Finding Solutions of Quadratic Equations of the Form $y = ax^2 + bx + c$**
>
> **Step 1:** Make a table and assign values to $x$, such as $-2, -1, 0, 1$, or $2$.
> **Step 2:** Substitute the value of $x$ in the equation and solve for the corresponding $y$-value.
> **Step 3:** Solutions are of the form $(x, y)$.

**Objective 1  Examples**  Complete a table to find solutions of each quadratic equation.

**1a.** $y = x^2 - 2x$       **1b.** $y = x^2 - 4x + 3$       **1c.** $f(x) = x^2 - 4$

**Solutions**   **1a.**

| $x$ | $y = x^2 - 2x$ | $(x, y)$ |
|---|---|---|
| $-2$ | $y = (-2)^2 - 2(-2) = 4 + 4 = 8$ | $(-2, 8)$ |
| $-1$ | $y = (-1)^2 - 2(-1) = 1 + 2 = 3$ | $(-1, 3)$ |
| $0$ | $y = (0)^2 - 2(0) = 0 - 0 = 0$ | $(0, 0)$ |
| $1$ | $y = (1)^2 - 2(1) = 1 - 2 = -1$ | $(1, -1)$ |
| $2$ | $y = (2)^2 - 2(2) = 4 - 4 = 0$ | $(2, 0)$ |
| $3$ | $y = (3)^2 - 2(3) = 9 - 6 = 3$ | $(3, 3)$ |

**1b.**

| $x$ | $y = x^2 - 4x + 3$ | $(x, y)$ |
|---|---|---|
| $-2$ | $y = (-2)^2 - 4(-2) + 3 = 4 + 8 + 3 = 15$ | $(-2, 15)$ |
| $-1$ | $y = (-1)^2 - 4(-1) + 3 = 1 + 4 + 3 = 8$ | $(-1, 8)$ |
| $0$ | $y = (0)^2 - 4(0) + 3 = 0 - 0 + 3 = 3$ | $(0, 3)$ |
| $1$ | $y = (1)^2 - 4(1) + 3 = 1 - 4 + 3 = 0$ | $(1, 0)$ |
| $2$ | $y = (2)^2 - 4(2) + 3 = 4 - 8 + 3 = -1$ | $(2, -1)$ |
| $3$ | $y = (3)^2 - 4(3) + 3 = 9 - 12 + 3 = 0$ | $(3, 0)$ |
| $4$ | $y = (4)^2 - 4(4) + 3 = 16 - 16 + 3 = 3$ | $(4, 3)$ |

**1c.** Recall that $f(x)$ is another way to represent $y$. So, we can write the equation as $y = x^2 - 4$.

| $x$ | $y = x^2 - 4$ | $(x, y)$ |
|---|---|---|
| $-2$ | $y = (-2)^2 - 4 = 4 - 4 = 0$ | $(-2, 0)$ |
| $-1$ | $y = (-1)^2 - 4 = 1 - 4 = -3$ | $(-1, -3)$ |
| $0$ | $y = (0)^2 - 4 = 0 - 4 = -4$ | $(0, -4)$ |
| $1$ | $y = (1)^2 - 4 = 1 - 4 = -3$ | $(1, -3)$ |
| $2$ | $y = (2)^2 - 4 = 4 - 4 = 0$ | $(2, 0)$ |
| $3$ | $y = (3)^2 - 4 = 9 - 4 = 5$ | $(3, 5)$ |

**Student Check 1**   Complete a table to find solutions of each quadratic equation.

**a.** $y = x^2 + 4x$       **b.** $y = x^2 - 2x - 8$       **c.** $f(x) = x^2 - 1$

## Graphing $y = ax^2 + bx + c$

**Objective 2** ▶

Graph the equation
$y = ax^2 + bx + c$.

The graph of the equation $y = ax^2 + bx + c$ is obtained from plotting several solutions and connecting the points to determine its shape. Using the solutions from Example 1a, we can get a general idea of the shape of the graph of an equation of the form $y = ax^2 + bx + c$. Plotting the solutions we found in the example enables us to see the general shape. Connecting the points with a smooth curve gives us the graph of $y = x^2 - 2x$.

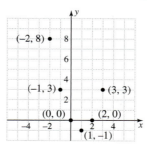
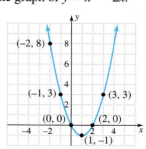

The graph of a quadratic equation $y = ax^2 + bx + c$ is a **parabola**. Parabolas have some important characteristics.

1. Each parabola is a smooth curve.
2. Each parabola has a highest or lowest point, called the **vertex**. In the preceding graph, the point $(1, -1)$ is the vertex.
3. Parabolas are *symmetric* about their vertex. If a parabola is folded vertically in half through its vertex, the left side and the right side of the parabola coincide; that is, the two halves are mirror images. If two points on a parabola have $x$-values that are the same distance from the vertex, then the points have the same $y$-value.
4. Parabolas open up if the value of $a$ is positive. Parabolas open down if the value of $a$ is negative.

The key points on the graph of a parabola are as follows.

| | |
|---|---|
| **x-intercepts** | Recall *x-intercepts* are found by replacing $y$ with 0 and solving for $x$. So, the $x$-intercepts are found by solving the equation $$ax^2 + bx + c = 0$$ Since this type of equation has either 2, 1, or 0 real solutions, the graph will have either 2 $x$-intercepts, 1 $x$-intercept, or no $x$-intercepts. |
| **y-intercept** | The *y-intercept* is found by replacing $x$ with 0 and solving for $y$. In other words, the $y$-intercept is found when we simplify $$y = a(0)^2 + b(0) + c = c$$ This shows that the $y$-intercept of $y = ax^2 + bx + c$ is the point $(0, c)$. A quadratic equation of the form $y = ax^2 + bx + c$ has only one $y$-intercept. |
| **Vertex** | The *vertex* is located halfway between the $x$-intercepts provided there are two $x$-intercepts. The following discussion provides a formula for finding the vertex of a quadratic equation. |

In Example 1a, the vertex of the graph is $(1, -1)$. The $x$-values of the $x$-intercepts are 0 and 2. The number halfway between these values is $\dfrac{0 + 2}{2} = \dfrac{2}{2} = 1$, which is the $x$-value of the vertex.

The graph of the equation from Example 1b, $y = x^2 - 4x + 3$, is shown. The vertex is $(2, -1)$. The $x$-values of the $x$-intercepts are 1 and 3. The number halfway between these values is $\dfrac{1 + 3}{2} = \dfrac{4}{2} = 2$, which is the $x$-value of the vertex.

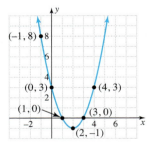

So, one way to calculate the $x$-value of the vertex is to solve for the $x$-intercepts and then find the midpoint between these numbers. There is also a formula for finding the $x$-value of the vertex that is not dependent on solving for the $x$-intercepts.

> **Property: The Vertex Formula**
>
> The $x$-value of the vertex of the graph of $y = ax^2 + bx + c$ is $x = -\dfrac{b}{2a}$.
>
> The $y$-value is found by substituting this value of $x$ into the equation.

So, for the equation $y = x^2 - 4x + 3$, $a = 1$ and $b = -4$. Substituting these values in the vertex formula gives us

$$x = -\frac{b}{2a} = -\frac{-4}{2(1)} = \frac{4}{2} = 2$$

To find the $y$-value of the vertex, we replace $x$ with 2 in the given equation.

$$y = x^2 - 4x + 3$$
$$= (2)^2 - 4(2) + 3$$
$$= 4 - 8 + 3$$
$$= -4 + 3$$
$$= -1$$

So, we find that the vertex is $(2, -1)$, which agrees with the vertex we found using the graph.

---

**Procedure: Graphing an Equation of the Form $y = ax^2 + bx + c$**

**Step 1:** Find the $y$-intercept. The $y$-intercept is $(0, c)$.
**Step 2:** Find the $x$-intercepts. The $x$-intercepts are solutions of the equation $ax^2 + bx + c = 0$. Use techniques from Sections 9.1–9.3 to solve this equation.
**Step 3:** Find the vertex point.
**Step 4:** Plot additional points, if needed.
**Step 5:** Connect the points with a smooth curve.

---

**Objective 2 Examples**

**Graph each equation by finding the intercepts, the vertex, and additional points, if needed.**

**2a.** $y = x^2 + 4x - 5$      **2b.** $y = -x^2 + 1$
**2c.** $f(x) = x^2 - 6x + 9$      **2d.** $y = x^2 + 2x + 4$

**Solutions**

**2a.**

| $x$-intercepts | $y$-intercept | Vertex |
|---|---|---|
| $y = x^2 + 4x - 5$ | $y = x^2 + 4x - 5$ | $x = -\dfrac{b}{2a} = -\dfrac{4}{2(1)} = -\dfrac{4}{2} = -2$ |
| $x^2 + 4x - 5 = 0$ | $y = (0)^2 + 4(0) - 5$ | |
| $(x + 5)(x - 1) = 0$ | $y = 0 - 5$ | $y = (-2)^2 + 4(-2) - 5$ |
| $x + 5 = 0$   or   $x - 1 = 0$ | $y = -5$ | $y = 4 - 8 - 5$ |
| $x = -5$      $x = 1$ | | $y = -9$ |
| $(-5, 0)$   and   $(1, 0)$ | $(0, -5)$ | $(-2, -9)$ |

Since $a = 1 > 0$, the parabola opens upward. We plot the intercepts and vertex and get the following graph.

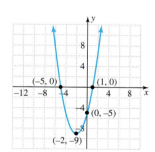

---

**Note:** We are not restricted to plotting just these particular points to obtain the graph. We can choose any $x$-value and solve for $y$ to find additional points on the graph.

**2b.**

| x-intercepts | y-intercept | Vertex |
|---|---|---|
| $y = -x^2 + 1$ | $y = -x^2 + 1$ | $x = -\dfrac{b}{2a} = -\dfrac{0}{2(-1)} = -\dfrac{0}{-2} = 0$ |
| $-x^2 + 1 = 0$ | $y = -(0)^2 + 1$ | |
| $-x^2 = -1$ | $y = 0 + 1$ | $y = -(0)^2 + 1 = 0 + 1 = 1$ |
| $x^2 = 1$ | $y = 1$ | |
| $x = \sqrt{1}$  or  $x = -\sqrt{1}$ | | |
| $x = 1$          $x = -1$ | | |
| $(-1, 0)$  and  $(1, 0)$ | $(0, 1)$ | $(0, 1)$ |

Since $a = -1 < 0$, the parabola opens downward. We plot the intercepts and vertex and get the following graph.

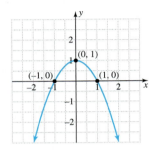

**2c.** We can write the function as $y = x^2 - 6x + 9$.

| x-intercepts | y-intercept | Vertex |
|---|---|---|
| $y = x^2 - 6x + 9$ | $y = x^2 - 6x + 9$ | $x = -\dfrac{b}{2a} = -\dfrac{-6}{2(1)} = \dfrac{6}{2} = 3$ |
| $x^2 - 6x + 9 = 0$ | $y = (0)^2 - 6(0) + 9$ | |
| $(x - 3)(x - 3) = 0$ | $y = 0 + 9$ | $y = (3)^2 - 6(3) + 9$ |
| $x - 3 = 0$  or  $x - 3 = 0$ | $y = 9$ | $y = 9 - 18 + 9$ |
| $x = 3$          $x = 3$ | | $y = 0$ |
| $(3, 0)$ | $(0, 9)$ | $(3, 0)$ |

Since the x-intercept and the vertex are the same, we need to find an additional point to graph the equation. We can either substitute another x-value in the function to find y or we can use symmetry. Because of symmetry, the point $(6, 9)$ is also on the graph. We can confirm this by substituting $x = 6$ into the equation.

$$y = x^2 - 6x + 9$$
$$y = (6)^2 - 6(6) + 9$$
$$y = 36 - 36 + 9$$
$$y = 9$$

Since $a = 1 > 0$, the parabola opens upward. Plotting the points $(0, 9)$, $(3, 0)$, and $(6, 9)$ gives us the following graph.

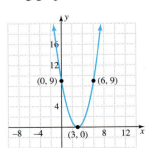

**2d.**

| | $x$-intercepts | $y$-intercept | Vertex |
|---|---|---|---|
| | $y = x^2 + 2x + 4$ <br> $1x^2 + 2x + 4 = 0$ <br> $a = 1, b = 2, c = 4$ <br><br> $x = \dfrac{-b \pm \sqrt{b^2 - 4ac}}{2a}$ <br><br> $x = \dfrac{-(2) \pm \sqrt{(2)^2 - 4(1)(4)}}{2(1)}$ <br><br> $x = \dfrac{-2 \pm \sqrt{4 - 16}}{2}$ <br><br> $x = \dfrac{-2 \pm \sqrt{-12}}{2}$ <br><br> Since the solutions are complex numbers, there are no $x$-intercepts of the graph. | $y = x^2 + 2x + 4$ <br> $y = (0)^2 + 2(0) + 4$ <br> $y = 0 + 4$ <br> $y = 4$ <br><br> $(0, 4)$ | $x = -\dfrac{b}{2a} = -\dfrac{2}{2(1)} = -\dfrac{2}{2} = -1$ <br> $y = (-1)^2 + 2(-1) + 4$ <br> $y = 1 - 2 + 4$ <br> $y = 3$ <br><br> $(-1, 3)$ |

Since there are no $x$-intercepts, we need to find an additional point to graph the equation. Because of symmetry, the point $(-2, 4)$ also lies on the graph. We can confirm this by substituting $x = -2$ into the equation.

$$y = x^2 + 2x + 4$$
$$y = (-2)^2 + 2(-2) + 4$$
$$y = 4$$

Since $a = 1 > 0$, the parabola opens upward. Plotting the points $(0, 4)$, $(-1, 3)$, and $(-2, 4)$ gives us the graph as shown.

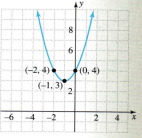

✓ **Student Check 2**   Graph each quadratic equation by finding the intercepts, the vertex, and additional points, if needed.

**a.** $y = -x^2 + 4$    **b.** $y = x^2 + 6x$    **c.** $y = x^2 - 4x + 4$    **d.** $y = x^2 + x + 3$

Example 2 illustrates how the $x$-intercepts of the graph of $y = ax^2 + bx + c$ relate to the solutions of the equation $ax^2 + bx + c = 0$. The three possibilities are summarized in the following table.

| Two Real Solutions | One Real Solution | No Real Solutions |
|---|---|---|
| If the equation has two real solutions, the graph of the equation intersects the $x$-axis twice, or has two $x$-intercepts. (See Example 2a.) | If the equation has one real solution, the graph of the equation intersects the $x$-axis once, or has one $x$-intercept. (See Example 2c.) | If the equation has no real solutions, the graph of the equation does not intersect the $x$-axis, or has no $x$-intercepts. (See Example 2d.) |

## Solving a Quadratic Equation Graphically

**Objective 3** ▶

Use the graph of $y = ax^2 + bx + c$ to solve the equation $ax^2 + bx + c = 0$.

As already noted, the $x$-intercepts are important parts of the graph of a quadratic equation. The solutions of an equation $ax^2 + bx + c = 0$ are the $x$-coordinates of the $x$-intercepts of the graph of $y = ax^2 + bx + c$. The graph of $y = x^2 + 4x - 5$ is shown. Notice the $x$-coordinates of the $x$-intercepts are $-5$ and $1$. These values are exactly the solutions of the equation $x^2 + 4x - 5 = 0$. (See Example 2a.)

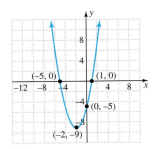

**Property:** **The Graph of $y = ax^2 + bx + c$ and Solutions of $ax^2 + bx + c = 0$**

If the point $(r, 0)$ is on the graph of $y = ax^2 + bx + c$, then $x = r$ is a solution of the equation $ax^2 + bx + c = 0$.

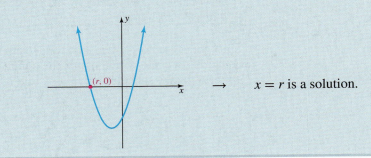

$\rightarrow$    $x = r$ is a solution.

**Procedure:** **Solving the Equation $ax^2 + bx + c = 0$ by Graphical Methods**

**Step 1:** Use the graph of the equation $y = ax^2 + bx + c$ to identify the $x$-coordinates of the $x$-intercepts.

**Step 2:** The values from step 1 are the solutions of the equation. There can be 2 solutions, 1 solution, or no solutions.

**Objective 3  Example**  Find the solution set of $2x^2 + x - 3 = 0$ using the graph of $y = 2x^2 + x - 3$.

**Solution**    The $x$-coordinates of the $x$-intercepts are $-\dfrac{3}{2}$ and $1$.

Therefore, the solution set of $2x^2 + x - 3 = 0$ is

$$\left\{ -\frac{3}{2}, 1 \right\}.$$

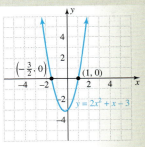

✓ **Student Check 3**    Find the solution set of $x^2 - 8x + 15 = 0$ using the graph of $y = x^2 - 8x + 15$.

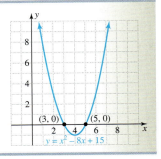

**Objective 4** ▶

Troubleshoot common errors.

## Troubleshooting Common Errors

A common error associated with finding the vertex of a quadratic equation is shown.

**Objective 4 Example** | A problem and an incorrect solution are given. Provide the correct solution and an explanation of the error.

Find the vertex of $y = 2x^2 - 4x + 3$.

| Incorrect Solution | Correct Solution and Explanation |
|---|---|
| $x = -\dfrac{b}{2a} = -\dfrac{4}{2(2)} = -1$ <br><br> $y = 2(-1)^2 - 4(-1) + 3$ <br> $y = 2(1) + 4 + 3$ <br> $y = 9$ <br> The vertex is $(-1, 9)$. | The wrong value was substituted for $b$. The value of $b = -4$. <br><br> $x = -\dfrac{b}{2a} = -\dfrac{-4}{2(2)} = 1$ <br><br> $y = 2(1)^2 - 4(1) + 3$ <br> $y = 2(1) - 4 + 3$ <br> $y = -2 + 3$ <br> $y = 1$ <br><br> Therefore, the vertex is $(1, 1)$. |

## ANSWERS TO STUDENT CHECKS

**Student Check 1**

**a.**

| $x$ | $y$ | $(x, y)$ |
|---|---|---|
| $-4$ | $0$ | $(-4, 0)$ |
| $-2$ | $-4$ | $(-2, -4)$ |
| $0$ | $0$ | $(0, 0)$ |
| $2$ | $12$ | $(2, 12)$ |
| $4$ | $32$ | $(4, 32)$ |

**b.**

| $x$ | $y$ | $(x, y)$ |
|---|---|---|
| $-4$ | $16$ | $(-4, 16)$ |
| $-2$ | $0$ | $(-2, 0)$ |
| $0$ | $-8$ | $(0, -8)$ |
| $2$ | $-8$ | $(2, -8)$ |
| $4$ | $0$ | $(4, 0)$ |

**c.**

| $x$ | $y$ | $(x, y)$ |
|---|---|---|
| $-4$ | $15$ | $(-4, 15)$ |
| $-2$ | $3$ | $(-2, 3)$ |
| $0$ | $-1$ | $(0, -1)$ |
| $2$ | $3$ | $(2, 3)$ |
| $4$ | $15$ | $(4, 15)$ |

**Student Check 2**

**a.** 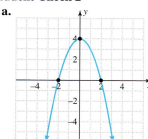  **b.** 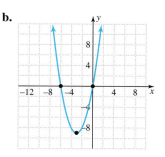  **c.** 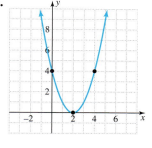  **d.**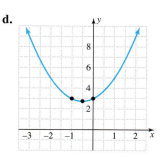

**Student Check 3** $\{3, 5\}$

## SUMMARY OF KEY CONCEPTS

1. Solutions of $y = ax^2 + bx + c$ are ordered pairs $(x, y)$ that satisfy the equation. Solutions can be found by substituting values for $x$ and solving for the corresponding $y$-value.

2. The graph of $y = ax^2 + bx + c$ is a parabola. Key points on the parabola are the $x$-intercepts, the $y$-intercept, and the vertex.

   **a.** The $x$-intercepts are found by replacing $y$ with $0$ and solving for $x$.

   **b.** The $y$-intercept is found by replacing $x$ with $0$ and solving for $y$.

   **c.** The vertex can be found by using the vertex formula.

   The $x$-value of the vertex is $x = -\dfrac{b}{2a}$ and the corresponding $y$-value is found by substituting this $x$-value into the equation and solving for $y$.

   **d.** If $a > 0$, the parabola opens upward. If $a < 0$, the parabola opens downward.

3. The solutions of the equation $ax^2 + bx + c = 0$ can be found from the graph of $y = ax^2 + bx + c$. The solutions are the $x$-coordinates of the $x$-intercepts of the graph.

## GRAPHING CALCULATOR SKILLS

The graphing calculator can be used to find ordered pairs that are solutions of the equation $y = ax^2 + bx + c$ and also to find the graph of the equation. The calculator can also be used to find the vertex of a parabola, but this skill is beyond the scope of this class. For the purpose of this class, we will use the calculator to confirm the points we find by hand.

**Example:** Graph the equation $y = x^2 - 4x + 3$ and use it to find solutions of $x^2 - 4x + 3 = 0$.

**Solution:** Enter the equation into the equation editor and view the table and graph.

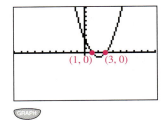

The solutions of $x^2 - 4x + 3 = 0$ can be found by identifying the $x$-intercepts of the graph of $y = x^2 - 4x + 3$. Recall that the $x$-intercepts are the points whose $y$-values are zero. From the table and graph, we find the $x$-intercepts are $(1, 0)$ and $(3, 0)$. Therefore, the solutions of $x^2 - 4x + 3 = 0$ are $x = 1$ or $x = 3$.

## SECTION 9.5 / EXERCISE SET

 **Write About It!**

Use complete sentences in your answer to each exercise.

1. How can we find solutions of the equation $y = ax^2 + bx + c$?

2. Explain how to find the solutions of the quadratic equation $ax^2 + bx + c = 0$.

3. What is a parabola? *The graph of the function $y = ax^2 + bx + c$ is a parabola.*

4. What are the key points on the graph of a parabola?

5. Explain the difference between the solutions of a quadratic equation and the $x$-intercepts of the graph of the quadratic equation. *The solutions of a quadratic equation are the $x$-values of the $x$-intercepts of the graph of the quadratic equation.*

6. What is the vertex formula?

7. How many times does the graph of the quadratic equation $y = 2x^2 + 2x + 3$ intersect the $x$-axis? What does this indicate about the solutions of the equation $2x^2 + 2x + 3 = 0$?

8. If the graph of the quadratic equation does not intersect the $x$-axis, explain how to complete the graph.

 **Practice Makes Perfect!**

Find solutions $(x, y)$ of each quadratic equation using $x = -2, -1, 0, 1, 2,$ and 3. (*See Objective 1.*)

9. $y = x^2 - 3x$          10. $y = x^2 - x$

11. $y = x^2 - 2x + 1$     12. $y = x^2 - x - 6$

13. $y = x^2 + 2x - 8$     14. $y = x^2 - 2x - 3$

Additional answers can be found in the Instructor Answer Appendix.

15. $y = x^2 - 6x + 9$     16. $y = x^2 - 4x + 4$

17. $y = -2x^2 + x - 1$    18. $y = -3x^2 + 2x - 2$

19. $y = -x^2 + 8$         20. $y = -x^2 + 6$

**Graph each equation by finding the intercept(s), vertex, and additional points if needed.** (*See Objective 2.*)

21. $y = x^2 + 4x - 12$    22. $y = x^2 - 2x - 3$

23. $y = x^2 + 4x - 32$    24. $y = x^2 - 2x - 24$

25. $y = -x^2 - x + 6$     26. $y = -x^2 - 7x - 10$

27. $y = x^2 + 4x$         28. $y = x^2 + 8x$

29. $y = -x^2 + 9$         30. $y = -x^2 + 16$

31. $y = -x^2 + 5x$        32. $y = -x^2 + 7x$

33. $y = 2x^2 - 3x - 5$    34. $y = -2x^2 + 5x + 12$

35. $y = x^2 + x + 4$      36. $y = -x^2 + x - 1$

**Find the solution set of each equation using the given graph.** (*See Objective 3.*)

37. $3x^2 - 15x - 18 = 0$   38. $2x^2 - 12x + 16 = 0$

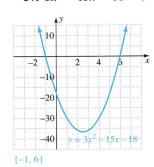

$\{-1, 6\}$

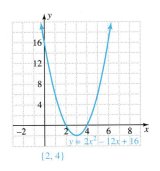

$\{2, 4\}$

**39.** $-x^2 - 7x + 18 = 0$

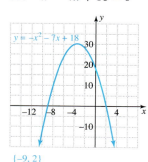

$\{-9, 2\}$

**40.** $-x^2 - 7x - 12 = 0$

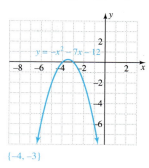

$\{-4, -3\}$

**41.** $-x^2 - 7x - 6 = 0$

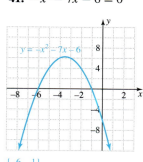

$\{-6, -1\}$

**42.** $-x^2 - x + 6 = 0$

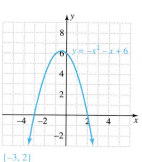

$\{-3, 2\}$

**52.** $x^2 - x - 30 = 0$

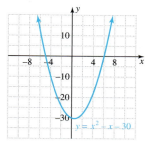

$\{-5, 6\}$

**53.** $2x^2 + x - 6 = 0$

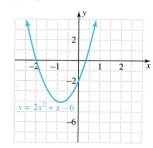

$\{-2, 1.5\}$

**54.** $4x^2 - 14x + 6 = 0$

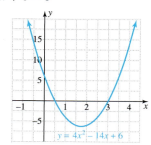

$\{0.5, 3\}$

 **Mix 'Em Up!**

**Find solutions $(x, y)$ of each quadratic equation using $x = -1, 0,$ and 1.**

**43.** $y = 6x^2 - 2x + 1$ $\quad\{(-1, 9), (0, 1), (1, 5)\}$

**44.** $y = -x^2 + 12$ $\quad\{(-1, 11), (0, 12), (1, 11)\}$

**45.** $y = 4x^2 - 3x + 2$ $\quad\{(-1, 9), (0, 2), (1, 3)\}$

**46.** $y = -3x^2 + 5$ $\quad\{(-1, 2), (0, 5), (1, 2)\}$

**Graph each equation by finding the intercept(s), vertex, and additional points if needed.**

**47.** $y = x^2 + 6x - 40$

**48.** $y = x^2 - 10x + 9$

**49.** $y = x^2 + 12x - 28$

**50.** $y = -x^2 - 4x + 21$

**Find the solution set of each equation using the given graph.**

**51.** $x^2 - 6x - 27 = 0$

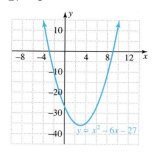

$\{-3, 9\}$

**You Be the Teacher!**

**Answer each student's question.**

**55.** Paul: When I try to solve the equation $4x^2 - 3x - 5 = 0$ using the graph, the $x$-intercepts are not integers. How can I get an exact answer?

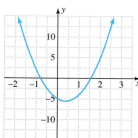

You can always use the quadratic formula to solve the equation for $x$.

**56.** Joel: When I use this graph to solve $-x^2 - 3 = 0$, there are no $x$-intercepts. So, how do I find the solutions of this equation?

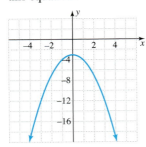

There are no real solutions of this equation based on the graph. When solving, the solution set is $\{-i\sqrt{3}, i\sqrt{3}\}$.

## Calculate It!

Find the solutions of each quadratic equation by hand. Then use a graphing calculator to verify the solutions.

**57.** $y = -3x^2 + 5x - 6$ at $x = -2, -1$, and 2
$\{(-2, -28), (-1, -14), (2, -8)\}$
**58.** $y = -x^2 + 2x + 3$ at $x = -2, -1$, and 3
$\{(-2, -5), (-1, 0), (3, 0)\}$

## Think About It!

**59.** Use the graph of Exercise 47 to solve the equation $-x^2 - 7x - 6 = 4$.   $\{-2, -5\}$
**60.** Use the graph of Exercise 38 to solve the equation $2x^2 - 12x + 16 = 6$.   $\{1, 5\}$

---

 **GROUP ACTIVITY**       **Connecting Graphs and Equations**

---

The goal of this activity is to help you understand the relationship between a quadratic equation, its solutions, and its graph.

Match the equation with its solution set and associated graph.

**1.** $x^2 - 8x + 11 = 0$   b, iv
**2.** $(3x - 1)(x + 2) = 48$   h, iii
**3.** $x^2 - 4 = 0$   c, vii
**4.** $2x^2 + x = 15$   a, ii
**5.** $5x^2 + 3 = 7x$   i, v
**6.** $-2x^2 + 8x = 0$   e, viii
**7.** $x^2 - 4x = -4$   d, vi
**8.** $(x + 2)^2 + 10 = 1$   g, i
**9.** $-x^2 - 4 = 0$   f, ix

**a.** $\left\{-3, \dfrac{5}{2}\right\}$
**b.** $\{4 - \sqrt{5}, 4 + \sqrt{5}\}$
**c.** $\{-2, 2\}$
**d.** $\{2\}$
**e.** $\{0, 4\}$
**f.** $\{-2i, 2i\}$
**g.** $\{-2 + 3i, -2 - 3i\}$
**h.** $\left\{-5, \dfrac{10}{3}\right\}$
**i.** $\left\{\dfrac{7}{10} - \dfrac{\sqrt{11}}{10}i, \dfrac{7}{10} + \dfrac{\sqrt{11}}{10}i\right\}$

**i.**
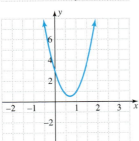

**ii.**
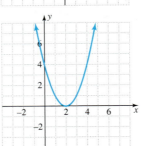

**iii.**
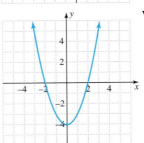

**iv.**
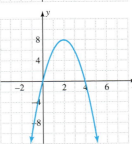

**v.**
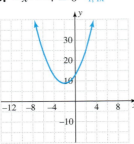

**vi.**
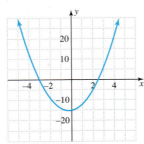

**vii.**
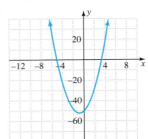

**viii.**
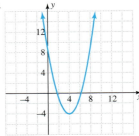

**ix.**
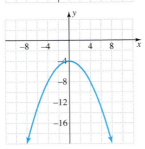

**INSTRUCTOR NOTE:**
To make this project more interactive, copy the problems shown here and randomly distribute an equation, a solution set, or a graph to each student (or pair of students). The goal is for students to find their counterparts that have the graph, equation, or solution set that matches the form they have been given.

# Quadratic Equations

**What's the big idea?** Now that you have completed Chapter 9, you should be able to solve a quadratic equation using various methods. You should also be able to see how the solutions of a quadratic equation relate to its graph. Lastly, you should be able to apply quadratic equations to real-life situations.

## The Tools

Listed below are the key terms, skills, formulas, and properties you should know for this chapter.

The page reference is provided if you need additional help with the given topic. The Study Tips will assist in your preparation for an exam.

## Study Tips

1. Learn all of the terms, formulas, and properties. Make flash cards and have someone quiz you.
2. Rework problems from the exercises and also the ones you worked in class. Work additional problems from the review exercises.
3. Review the summaries of key concepts.
4. Work the chapter test.
5. Be sure to review the online resources for additional study materials.

## Terms

Complex conjugate   688
Complex number   684
Discriminant   678

Imaginary unit, $i$   684
Parabola   696
Perfect square trinomial   665

Quadratic equation   658
Vertex   696

## Formulas and Properties

- Completing the square   666
- Discriminant   678
- Product of complex conjugates   688

- Quadratic formula   675
- Square root property   658

- Vertex formula   696

## CHAPTER 9 / SUMMARY

How well do you know this chapter? Complete the following questions to find out! Take a look back at the section if you need help.

### SECTION 9.1 Solve Quadratic Equations by the Square Root Property

1. The square root property states that if $x^2 = k$, where $k$ is a real number, then $x = \underline{\sqrt{k}}$ or $x = \underline{-\sqrt{k}}$.
   a. If $k > 0$, the equation has <u>two</u> solutions.
   b. If $k = 0$, the equation has <u>one</u> solution.
   c. If $k < 0$, the equation has <u>no</u> solutions.

2. To apply the square root property, the squared expression must be <u>isolated</u> to one side and equal to a(n) <u>constant</u>.

### SECTION 9.2 Solve Quadratic Equations by Completing the Square

3. A(n) <u>perfect</u> <u>square</u> <u>trinomial</u> results from squaring a binomial.

4. In a perfect square trinomial with a leading coefficient of 1, the <u>middle</u> term is half of the <u>coefficient</u> of $x$, <u>squared</u>. In symbols, $c = \left(\dfrac{b}{2}\right)^2$.

5. When we add the number that makes a binomial a perfect square trinomial, we are <u>completing</u> the <u>square</u>.

6. To solve an equation by completing the square,
   - The coefficient of the squared term must be <u>one</u>.
   - If it is not, we <u>divide</u> both sides of the equation by the coefficient.
   - Move the <u>constant</u> term to the side opposite the variable terms.
   - Find the number that <u>completes</u> the <u>square</u> and <u>add</u> it to both sides of the equation.
   - <u>factor</u> the perfect square trinomial and simplify the other side.
   - Apply the <u>square</u> <u>root</u> <u>property</u> and solve the equation.

## SECTION 9.3 Solve Quadratic Equations by the Quadratic Formula

**7.** The _quadratic_ _formula_ is derived from completing the square. If $ax^2 + bx + c = 0$, then $x = \dfrac{-b \pm \sqrt{b^2 - 4ac}}{2a}$.

**8.** The _discriminant_ of a quadratic equation determines the number and types of solutions. Its value is _$b^2 - 4ac$_.

   **a.** If this value is positive, there is/are _two_ _solutions_.

   **b.** If this value is zero, there is/are _one_ _solution_.

   **c.** If this value is negative, there is/are _no_ _real_ _solutions_.

## SECTION 9.4 Complex Numbers and Quadratic Equations with Complex Solutions

**9.** The _imaginary_ _unit_ is defined as the number whose square is equal to $-1$. We write it as _$i$_.

$$i = \sqrt{-1} \quad \text{and} \quad i^2 = -1.$$

**10.** A complex number is a number of the form _$a + bi$_, where _$a$_ and _$b$_ are real numbers.

**11.** We add and subtract complex numbers by _adding_ like terms.

**12.** We multiply complex numbers by applying the _distributive_ property and the _FOIL_ method.

**13.** To divide complex numbers, multiply the numerator and denominator by the _conjugate_ of the denominator. If the denominator is $a + bi$, then multiply by _$a - bi$_.

**14.** A quadratic equation with nonreal solutions has _complex_ solutions.

## SECTION 9.5 Solve Quadratic Equations by Graphical Methods

**15.** Solutions of equations of the form $y = ax^2 + bx + c$ are _ordered_ _pairs_ that satisfy the equation. There are _infinitely_ many solutions.

**16.** The graph of an equation of the form $y = ax^2 + bx + c$ is a(n) _parabola_.

**17.** The highest or lowest point on the _parabola_ is called the _vertex_.

**18.** The other key points are the $x$- and $y$-_intercepts_.

**19.** The graph of a quadratic equation can have 2, 1, or 0 _$x$_-intercepts but only 1 _$y$_-intercept.

**20.** To find the vertex, we use the vertex formula $x = \dfrac{-b}{2a}$. After we know this value, we find its corresponding _$y$_ value.

**21.** The graph of the parabola opens _up_ if $a > 0$ and opens _down_ if $a < 0$.

**22.** The solutions of the equation $ax^2 + bx + c = 0$ are the _$x$_-values of the _$x$_-intercepts of the graph of $y = ax^2 + bx + c$. That is, if $x = r$ is a solution of the equation, the point _$(r, 0)$_ lies on the graph of the equation.

---

# CHAPTER 9 / REVIEW EXERCISES

## SECTION 9.1

**Solve each equation. Express radicals in simplest form. (See Objective 1.)**

**1.** $x^2 = 28$  $\{-2\sqrt{7}, 2\sqrt{7}\}$   **2.** $3y^2 + 2 = 2$  $\{0\}$

**3.** $9x^2 - 5 = 20$  $\left\{-\dfrac{5}{3}, \dfrac{5}{3}\right\}$   **4.** $2b^2 = 12.5$  $\{-2.5, 2.5\}$

**5.** $(2a - 1)^2 - 6 = 18$   **6.** $2y^2 = 98$  $\{-7, 7\}$

**7.** $(5b - 3)^2 + 4 = 40$   **8.** $(2x - 3)^2 = 49$  $\{-2, 5\}$

**9.** $(c - 2.4)^2 = 1.69$  $\{1.1, 3.7\}$   **10.** $(y - 3)^2 = -25$  not a real number

**Solve each problem. (See Objective 2.)**

**11.** At what rate does \$2000 need to be invested in a 2-yr CD to have \$2101.25 once the CD matures?  2.5%

**12.** Judy's backyard is a square with an area of 22,500 ft². What are the dimensions of her backyard?  150 ft by 150 ft

## SECTION 9.2

**Make the binomial a perfect square trinomial and express it in factored form. (See Objective 1.)**

**13.** $a^2 - 20a$   **14.** $b^2 + 3.6b$   **15.** $x^2 + 11x$   **16.** $y^2 - \dfrac{6}{5}y$

**Solve by completing the square. (See Objectives 2 and 3.)**

**17.** $2x^2 - 16x - 8 = 0$  $\{4 - 2\sqrt{5}, 4 + 2\sqrt{5}\}$   **18.** $x^2 - 8x + 20 = 0$  no real solutions

**19.** $4y^2 + 12y + 3 = 0$   **20.** $6x^2 - 4x - 1 = 0$

**21.** $x^2 + 5x - 24 = 0$  $\{-8, 3\}$   **22.** $3x^2 + 2x = 0$  $\left\{-\dfrac{2}{3}, 0\right\}$

**23.** $x^2 + 1.2x - 1.08 = 0$  $\{-1.8, 0.6\}$   **24.** $x^2 - 16x + 64 = 0$  $\{8\}$

## SECTION 9.3

**Simplify each quotient. (See Objective 1.)**

**25.** $\dfrac{2 - \sqrt{28}}{2}$  $1 - \sqrt{7}$   **26.** $\dfrac{15 + \sqrt{45}}{3}$  $5 + \sqrt{5}$

**27.** $\dfrac{6 + \sqrt{32}}{12}$  $\dfrac{3 + 2\sqrt{2}}{6}$   **28.** $\dfrac{5 - \sqrt{160}}{10}$  $\dfrac{5 - 4\sqrt{10}}{10}$

**Solve each equation using the quadratic formula. (See Objective 2.)**

**29.** $4x^2 = 17 - 4x$   **30.** $15x^2 - x = 2$  $\left\{-\dfrac{1}{3}, \dfrac{2}{5}\right\}$

**31.** $3y^2 - 2y + 3 = 0$  no real solutions   **32.** $3a^2 + 7a = 0$  $\left\{-\dfrac{7}{3}, 0\right\}$

**33.** $4x^2 + 49 = 0$  no real solutions   **34.** $9y^2 - 6y + 1 = 0$  $\left\{\dfrac{1}{3}\right\}$

**35.** $y^2 + 0.7y - 3.3 = 0$  $\{-2.2, 1.5\}$

**36.** $0.6a^2 + 0.3a - 8.4 = 0$  $\{-4, 3.5\}$

**Find the discriminant of each quadratic equation and use the information to determine the types of solutions. (See Objective 3.)**

**37.** $3x^2 - 4x + 12 = 0$  $-128$; no real solutions   **38.** $2y^2 = 6y + 11$  124; two nonrepeating real solutions

**39.** $4x^2 + 9 = 12x$  0; one repeating real solution   **40.** $4y^2 - 3y = 0$  9; two nonrepeating real solutions

**Solve each problem. (*See Objective 4.*)**

41. If the height (in feet) of an object launched from the ground after $t$ seconds is modeled by the equation $h = -16t^2 + 64t$, then when will the object be 64 ft from the ground?    2 sec

42. If the height (in feet) of an object from the ground after $t$ seconds is modeled by the equation $h = -16t^2 + 112t$, then when will the object be 160 ft from the ground?
    2 sec or 5 sec

## SECTION 9.4

**Write each expression as a complex number.**
**(*See Objective 1.*)**

43. $\sqrt{-225}$    15i    44. $\sqrt{-180}$    45. $\sqrt{-15}$    $i\sqrt{15}$
        $6i\sqrt{5}$

46. $\sqrt{-52}$    $2i\sqrt{13}$    47. $8 - \sqrt{-18}$    48. $\dfrac{-8 + \sqrt{-68}}{12}$
                                        $8 - 3i\sqrt{2}$

**Perform the operation on the complex numbers.**
**(*See Objectives 2–4.*)**

49. $(24 - 9i) - (11 - 7i)$    50. $(17 + 10i) + (-18 - 21i)$
    $13 - 2i$                          $-1 - 11i$

51. $\left(\dfrac{5}{3} + \dfrac{1}{2}i\right) + \left(\dfrac{7}{6} - \dfrac{3}{4}i\right)$    $\dfrac{17}{6} - \dfrac{1}{4}i$

52. $(12.7 - 5.1i) - (7.6 + 2.8i)$    $5.1 - 7.9i$

53. $-3i(8 + 5i)$    $15 - 24i$    54. $4i(-7i)$    28

55. $(2 - 9i)^2$    $-77 - 36i$    56. $\dfrac{-2 + 3i}{2i}$    $\dfrac{3}{2} + i$

57. $\dfrac{4 - 5i}{-1 - 3i}$    $\dfrac{11}{10} + \dfrac{17}{10}i$    58. $(5 + 6i)(2 - 3i)$    $28 - 3i$

59. $(1.7 - 0.6i)(0.8 + 1.3i)$    60. $\left(\dfrac{3}{4} - \dfrac{1}{2}i\right)\left(\dfrac{3}{4} + \dfrac{1}{2}i\right)$    $\dfrac{13}{16}$
        $2.14 + 1.73i$

**Solve each quadratic equation for its complex solutions.**
**(*See Objective 5.*)**

61. $x^2 + 1 = x$    62. $2y^2 + 14 = 0$
                                $\{i\sqrt{7}, -i\sqrt{7}\}$

63. $52 - 4x = -2x^2$    64. $x^2 + 6x + 14 = 0$
    $\{1 - 5i, 1 + 5i\}$        $\{-3 - i\sqrt{5}, -3 + i\sqrt{5}\}$

65. $(x^2 - 100)(x^2 + 9) = 0$    66. $(4x^2 + 1)(x^2 + 9) = 0$
    $\{-3i, 3i, -10, 10\}$            $\left\{-3i, 3i, -\dfrac{1}{2}i, \dfrac{1}{2}i\right\}$

## SECTION 9.5

**Find three solutions of each quadratic equation.**
**(*See Objective 1.*)**

67. $y = -3x^2 - x + 1$    68. $y = x^2 - 5x + 3$
    Answers vary. $\{(-2, -9), (0, 1), (1, -3)\}$    Answers vary. $\{(-1, 9), (0, 3), (2, -3)\}$

**Graph each equation by finding the intercepts, vertex, and additional points if needed. (*See Objective 2.*)**

69. $y = -x^2 - 4x + 12$    70. $y = x^2 - 4x + 5$

**Find the solutions of each equation by using the given graph. (*See Objective 3.*)**

71. $x^2 + 6x + 9 = 0$    $\{-3\}$    72. $-x^2 + 5x + 6 = 0$    $\{-1, 6\}$

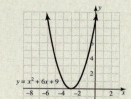

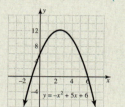

---

## CHAPTER 9 TEST / QUADRATIC EQUATIONS

1. What are the four methods for solving a quadratic equation? Which method(s) can be used to solve any quadratic equation?

2. What determines the number and type of solutions of a quadratic equation? What expression is used to calculate this number? State the different cases.

3. Applying the square root property for solving $(x + 5)^2 + 2 = 9$ gives us the solutions
   a. $\{-4\}$    b. $\{-10, -4\}$
   c. $\{-5 + \sqrt{7}\}$    (d.) $\{-5 - \sqrt{7}, -5 + \sqrt{7}\}$

4. If \$5000 is invested in a savings account for 2 yr, the amount in the account is represented by $5000(1 + r)^2$, where $r$ is the annual interest rate. Find the interest rate that is required for an investment of \$5000 to grow to \$6050 in 2 yr.    10%

5. The following trinomial that is a perfect square trinomial is
   a. $x^2 - 2x + 4$    b. $x^2 + 6x - 9$
   (c.) $x^2 + 12x + 36$    d. $4x^2 + 10x + 25$

6. Add the number that makes the binomial $x^2 - 20x$ a perfect square trinomial and express it in factored form.
   $x^2 - 20x + 100 = (x - 10)^2$

7. Solve the equation $x^2 - 2x - 7 = 0$ by completing the square.    $\{1 - 2\sqrt{2}, 1 + 2\sqrt{2}\}$

8. Solve the equation $5x^2 + 20x + 10 = 0$ by completing the square.    $\{-2 - \sqrt{2}, -2 + \sqrt{2}\}$

9. The approximate value of $\dfrac{4 + \sqrt{5}}{2}$ is
   (a.) 3.1    b. 4.2
   c. 5.1    d. 5.6

10. Simplify the quotient $\dfrac{6 - \sqrt{12}}{10}$.    $\dfrac{3 - \sqrt{3}}{5}$

11. Solve each equation using the quadratic formula.
    a. $y^2 + 16 = 8y$    $\{4\}$
    b. $2a^2 - 3a = 4$    $\left\{\dfrac{3 - \sqrt{41}}{4}, \dfrac{3 + \sqrt{41}}{4}\right\}$
    c. $b^2 + b + 5 = 0$

12. If the discriminant of a quadratic equation is 33, the equation
    a. has no real solutions
    b. has one real solution
    (c.) has two real solutions
    d. can't be determined from this information

13. Write each number as a complex number.
    a. $\sqrt{-64}$    8i
    b. $4 + \sqrt{-9}$    $4 + 3i$

**14.** When simplified, $3i(5 + 2i) - 4i$ is

    **a.** $-6 + 11i$    **b.** $-6 + 4i$    **c.** $5i$    **d.** $17i$

**15.** Perform the operations with complex numbers. Write each answer in standard form.

    **a.** $(7 + 3i) - (-2 - i)$   $9 + 4i$

    **b.** $(-2 + 10i) + (-5 - 5i) + (12 + 2i)$   $5 + 7i$

    **c.** $4i(8i)$   $-32$      **d.** $3i(-7 + 4i)$   $-12 - 21i$

    **e.** $(9 + 4i)(9 - 4i)$   $97$    **f.** $(9 + 4i)^2$   $65 + 72i$

    **g.** $\dfrac{4}{3i}$   $0 - \dfrac{4}{3}i$      **h.** $\dfrac{13}{1 + 5i}$   $\dfrac{1}{2} - \dfrac{5}{2}i$

**16.** The graph of $y = ax^2 + bx + c$ is given. The solution set of $ax^2 + bx + c = 0$ is

    **a.** $\{-3, 2\}$   **b.** $\{-2, 3\}$   **c.** $\{6\}$   **d.** $\{-6\}$

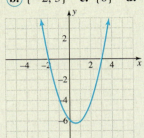

**17.** The vertex of $y = 2x^2 - 8x + 3$ is

    **a.** $(2, -5)$   **b.** $(2, 11)$   **c.** $(-2, 27)$   **d.** $(-2, 11)$

**18.** Graph each quadratic equation by finding the intercepts, vertex, and additional points, if necessary.

    **a.** $y = x^2 - 4x - 12$      **b.** $y = (x - 2)^2$

    **c.** $y = 2x^2 - 6x + 5$

**19.** The Devil's Pool is a swimming area located atop Victoria Falls along the Zambezi River, which is located on the border of Zambia and Zimbabwe. Swimmers can literally look over the edge of the falls. One swimmer was trying to take a picture over the edge and dropped his camera. The height, in meters, of the camera is represented by $h = -16t^2 + 128$, where $t$ is the number of seconds after the camera is dropped. How many seconds (to the nearest tenth) will it take for the camera to reach the bottom of the falls?   2.8 sec

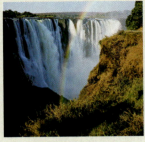

**20.** The national unemployment rate in January of each of the years 2004–2010 are shown in the table. The unemployment rate can be modeled by the equation $y = 0.39x^2 - 1.69x + 6.61$, where $x$ is the years after 2004. Assuming the unemployment rate continues to grow in this manner, in what year will the unemployment rate be 25? (Source: http://www .google.com/publicdata/home) in approximately 9.4 yr after 2004, or during the year 2013

| Year | Unemployment Rate |
|------|-------------------|
| 2004 | 6.3 |
| 2005 | 5.7 |
| 2006 | 5.1 |
| 2007 | 5 |
| 2008 | 5.4 |
| 2009 | 8.5 |
| 2010 | 10.6 |

---

## CUMULATIVE REVIEW EXERCISES / CHAPTERS 1–9

**1.** Simplify each algebraic expression. (*Section 1.8, Objectives 3 and 4*)

    **a.** $12(3x - 2) - 36x$   $-24$   **b.** $-(2x - 18) - 6(3x + 1)$   $-20x + 12$

    **c.** $7.4 - 1.8(-0.5x + 1.5)$   $0.9x + 4.7$

    **d.** $-20\left(\dfrac{3}{10}x + \dfrac{1}{2}\right) - 12\left(\dfrac{1}{6}x - \dfrac{2}{3}\right)$   $-8x - 2$

    **e.** $y^2 - 2(-3y + y^2) - 6y$   $-y^2$

**2.** Solve each equation. (*Section 2.3, Objective 2; Section 2.4, Objective 2*)

    **a.** $5(a + 4) - 8(a - 1) = 3(a + 2) - 9$   $\dfrac{31}{6}$

    **b.** $0.014x + 0.021(1200 - x) = 22.68$   $360$

**3.** In triangle $ABC$, the measure of angle $B$ is 28° less than the measure of angle $A$. The measure of angle $C$ is 40° more than the measure of angle $A$. Find the measure of each angle in the triangle. (*Section 2.5, Objective 6*)   56°, 28°, and 96°

**4.** Paul needs to make 80 L of a 28.8% salt solution. How much 45% salt solution and how much 18% salt solution should be mixed to get 80 L of a 28.8% solution? (*Section 2.6, Objective 3*)   32 L of the 45% salt solution and 48 L of the 18% salt solution

**5.** The average fare, excluding transportation taxes, per revenue passenger flying AirTran Airways can be modeled by the equation $y = 1.377x + 87.646$, where $x$ is the number of years after 2005. (Source: http://phx .corporate-ir.net) (*Section 3.2, Objective 5*)

    **a.** Find the $y$-intercept of the equation and interpret its meaning in the context of the problem.

    **b.** What was the average fare, excluding transportation taxes, per revenue passenger flying AirTran Airways in 2010?

    **c.** In what year is the average fare, excluding transportation taxes, per revenue passenger flying AirTran Airways $100?   in about 9 yr or 2014

**6.** Find the slope of each line. (*Section 3.3, Objectives 1 and 2*)

    **a.** $5x - 3y = 6$   $\dfrac{5}{3}$

    **b.** passes through $(-1, -5)$ and $(0, 4)$   $9$

    **c.** $x = 3$   undefined

**7.** Find the information from the given graph. (*Section 3.6, Objective 4*)

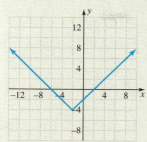

**a.** Find $f(-6)$ and $f(0)$.    $f(-6) = 0$ and $f(0) = -2$

**b.** Find $x$ if $f(x) = 2$ and $f(x) = 4$.    $f(-8) = 2, f(4) = 2,$ $f(-10) = 4, f(6) = 4$

**8.** Solve each system using elimination. Then state if the system is consistent with independent equations, consistent with dependent equations, or inconsistent. (*Section 4.3, Objectives 1 and 2*)

**a.** $\begin{cases} 2x - y = 8 \\ 3x + 5y = -1 \end{cases}$    **b.** $\begin{cases} 7y = 5x + 2 \\ -5x + 7y = -5 \end{cases}$

**9.** The price per item is determined by the demand equation $y = D(x)$, where $x$ is the number of items sold. The consumer's surplus is the area of the solution of the following system of linear inequalities. (*Section 4.6, Objective 2*)

$$\begin{cases} y \le D(x) \\ y \ge k \\ x \ge 0 \end{cases}$$, where $k$ is a fixed price

The supply equation for an item is $y = D(x) = 21 - 0.40x$ and $k = 11$.

**a.** Set up a system of linear inequalities.

**b.** Find the consumer's surplus.    125

**10.** Use the exponent rules to simplify each expression. Express each answer with positive exponents. (*Section 5.2, Objective 4*)

**a.** $(-3c^{-2}d^4)^{-3}$    $-\dfrac{c^6}{27d^{12}}$

**b.** $\left(\dfrac{5a^{-4}}{d^3}\right)^4$    $\dfrac{625}{a^{16}d^{12}}$

**c.** $\left(\dfrac{2x^2}{y^{-5}}\right)^5$    $32x^{10}y^{25}$

**d.** $(-4^{-1}p^5q^{-1})^2$    $\dfrac{p^{10}}{16q^2}$

**11.** Perform each operation. Write answers in standard form. (*Section 5.4, Objective 4; Section 5.5, Objectives 1 and 2; Section 5.7, Objective 2*)

**a.** $(7x^2 - xy - 4y^2) + (-x^2 + 9xy + 3y^2)$    $6x^2 + 8xy - y^2$

**b.** $8x^2y^3(x^2 - 2xy + 5y^2)$    **c.** $(x - 3y)(4x + 5y)$

**d.** $(2a - 3b)(2a + 3b)(4a^2 + 9b^2)$    $16a^4 - 81b^4$

**e.** $\dfrac{x^3 + 2x^2 + 14}{x + 3}$    $x^2 - x + 3 + \dfrac{5}{x + 3}$

**12.** Factor each polynomial completely. (*Section 6.1, Objective 3; Section 6.2, Objectives 1 and 2; Section 6.3, Objectives 1–3; Section 6.4, Objectives 1–3*)

**a.** $10x^2y + 18xy^2 - 4xy$    **b.** $p^4 - 17p^2 + 16$

**c.** $16a^2 - 24ab + 9b^2$    **d.** $u^3 - 8v^3$
   $(u - 2v)(u^2 + 2uv + 4v^2)$

**e.** $100r^2 + s^2$    prime    **f.** $p^3 + 64q^3$
   $(p + 4q)(p^2 - 4pq + 16q^2)$

**g.** $x^8 - y^8$    $(x^4 + y^4)(x^2 + y^2)(x + y)(x - y)$

**13.** Solve each equation. (*Section 6.5, Objectives 1 and 2*)

**a.** $2x^2 + 9x = 35$    **b.** $(x - 3)(x + 1) = 5$    $\{-2, 4\}$

**c.** $x^3 - 15x^2 + 44x = 0$    $\{0, 4, 11\}$

**14.** Perform each operation and simplify the result. (*Section 7.2, Objectives 1 and 2; Section 7.4, Objectives 1 and 2*)

**a.** $\dfrac{4x^2 - 6x}{4x^2 - 9} \cdot \dfrac{2x^2 - 7x - 15}{x^3 - 25x}$    $\dfrac{2}{x + 5}$

**b.** $\dfrac{x^2 + 5x - 24}{x^3 - 27} \div \dfrac{x^2 - 64}{x^3 + 3x^2 + 9x}$    $\dfrac{x}{x - 8}$

**c.** $\dfrac{4x - 3}{x - 4} + \dfrac{x + 9}{4 - x}$    3

**d.** $\dfrac{3x^2}{x^2 + 3x - 18} - \dfrac{x - 1}{x + 6}$

**e.** $\dfrac{4}{x + 8} + \dfrac{2x}{x^2 + 4x - 32}$    **f.** $\dfrac{2}{x^3 - 1} - \dfrac{1}{2x^2 + 2x + 2}$

**15.** Simplify each complex fraction using method 1 or method 2. (*Section 7.5, Objectives 1 and 2*)

**a.** $\dfrac{\dfrac{2}{x} + 6}{3x + 1}$    $\dfrac{2}{x}$

**b.** $\dfrac{\dfrac{1}{x} - \dfrac{5}{y}}{\dfrac{1}{x^2} - \dfrac{25}{y^2}}$    $\dfrac{xy}{5x + y}$

**16.** Solve each rational equation. (*Section 7.6, Objective 1*)

**a.** $\dfrac{5}{4x - 7} = \dfrac{3}{2 - 3x}$    $\left\{\dfrac{31}{27}\right\}$

**b.** $\dfrac{4}{2x^2 - 5x - 3} - \dfrac{5}{x^2 - 9} = \dfrac{6}{2x^2 + 7x + 3}$    $\left\{\dfrac{25}{12}\right\}$

**17.** Solve each rational equation for the specified variable. (*Section 7.6, Objective 2*)

**a.** $\dfrac{1}{a} = \dfrac{2}{b} + \dfrac{3}{c} + \dfrac{4}{d}$ for $c$    **b.** $y_m = \dfrac{y_1 + y_2}{2}$ for $y_2$

**18.** An airplane flies 504 mi in the same amount of time it takes a second plane to fly 570 mi. If the speed of the first airplane is 55 mph slower than the speed of the second airplane, find the speed of each airplane. (*Section 7.7, Objective 5*)    475 mph and 420 mph

**19.** Simplify each radical. Assume all variables represent positive real numbers. (*Section 8.2, Objectives 1–4*)

**a.** $\sqrt{16x^4y^7}$    $4x^2y^3\sqrt{y}$    **b.** $\sqrt[4]{-\dfrac{9x^8}{y^4}}$    not a real number

**c.** $\sqrt{48a^{10}b^6}$    $4a^5b^3\sqrt{3}$    **d.** $\sqrt[3]{-125c^9d^{14}}$
   $-5c^3d^4\sqrt[3]{d^2}$

**e.** $\sqrt[5]{\dfrac{486u^9}{v^{20}}}$    $\dfrac{3u\sqrt[5]{2u^4}}{v^4}$

**20.** Perform the indicated operation and write each answer in simplest form. Rationalize the denominator when necessary. Assume all variables represent positive real numbers. (*Section 8.3, Objectives 1 and 2; Section 8.4, Objectives 1–6*)

**a.** $2x^2\sqrt{12x} + 3\sqrt{75x^5}$    $19x^2\sqrt{3x}$

**b.** $12\sqrt{20} - 8\sqrt{45} + 6\sqrt{80}$    $24\sqrt{5}$

**c.** $6\sqrt[4]{81x^7y^{12}} - 5x\sqrt[4]{16x^3y^{12}}$    $8xy^3\sqrt[4]{x^3}$

**d.** $\sqrt[3]{250x^6y} + 12x^3\sqrt[3]{54y}$    $5x^2\sqrt[3]{2y} + 36x^3\sqrt[3]{2y}$

**e.** $\sqrt{45a} \cdot \sqrt{10a^3b^5}$    $15a^2b^2\sqrt{2b}$

**f.** $(3 - \sqrt{x})^2$ $\quad 9 - 6\sqrt{x} + x$ **g.** $\dfrac{8}{\sqrt[3]{40a}}$ $\quad \dfrac{4\sqrt[3]{25a^2}}{5a}$

**h.** $(2\sqrt{6} - 1)(5\sqrt{6} + 4)$ **i.** $\dfrac{6}{\sqrt{5} + \sqrt{2}}$ $\quad 2(\sqrt{5} - \sqrt{2})$

$56 + 3\sqrt{6}$

**21.** Solve each equation. (*Section 8.5, Objective 1*)

**a.** $12 + \sqrt{x} = 8$ $\quad \varnothing$ **b.** $\sqrt{x - 4} = x - 6$ $\quad \{8\}$

**22.** Simplify each expression using the properties of exponents. Write each answer with positive exponents, when applicable. (*Section 8.6, Objective 4*)

**a.** $a^{-2/5} \cdot a^{2/5}$ $\quad 1$ **b.** $\dfrac{x^{2/5}}{x^{3/4}}$ $\quad \dfrac{1}{x^{7/20}}$

**c.** $\sqrt[3]{5x^2} \cdot \sqrt{5x}$ $\quad 5^{5/6}x^{7/6}$ **d.** $\dfrac{\sqrt[4]{b^3}}{\sqrt[3]{b^7}}$ $\quad \dfrac{1}{b^{19/12}}$

**e.** $\left(\dfrac{2^{1/5}r^{-2/5}s^{-1/3}}{s^{2/15}}\right)^{15}$ $\quad \dfrac{8}{r^6 s^7}$ **f.** $a^{1/5}(a^{1/5} - 2a^{-1/5})$ $\quad a^{2/5} - 2$

**23.** Solve each equation. Express radicals in simplest form. (*Section 9.1, Objectives 1 and 2*)

**a.** $x^2 + 2x + 5 = 5$ $\quad \{-2, 0\}$ **b.** $y^2 = 24$ $\quad \{-2\sqrt{6}, 2\sqrt{6}\}$

**c.** $(2x + 7)^2 = 49$ $\quad \{-7, 0\}$ **d.** $9x^2 - 5 = 20$ $\quad \left\{-\dfrac{5}{3}, \dfrac{5}{3}\right\}$

**e.** $(x - 8)^2 = -1$ **f.** $(3a - 5)^2 = 72$

no real solutions

**24.** Marie's backyard is a square with an area of 12,100 ft². What are the dimensions of her backyard? (*Section 9.1, Objectives 1 and 2*) $\quad$ 110 ft by 110 ft

**25.** The surface area of a sphere is given by $S = 4\pi r^2$, where $r$ is the radius of the sphere. Find the radius of a sphere whose surface area is $2304\pi$ ft². (*Section 9.1, Objective 2*) $\quad$ 24 ft

**26.** Make each binomial a perfect square trinomial and express the trinomial in factored form. (*Section 9.2, Objective 1*)

**a.** $x^2 - 16x$ $\quad x^2 - 16x + 64; (x - 8)^2$

**b.** $y^2 + 9y$ $\quad y^2 + 9y + \dfrac{81}{4}; \left(y + \dfrac{9}{2}\right)^2$

**c.** $a^2 + 1.8a$ $\quad a^2 - 1.8a + 0.81; (a + 0.9)^2$

**d.** $x^2 - \dfrac{2}{5}x$ $\quad x^2 - \dfrac{2}{5}x + \dfrac{1}{25}; \left(x - \dfrac{1}{5}\right)^2$

**27.** Solve each equation by completing the square. (*Section 9.2, Objectives 2 and 3*)

**a.** $x^2 - 10x + 13 = 0$ **b.** $y^2 - 14y = 0$ $\quad \{0, 14\}$

**c.** $2x^2 + 6x + 9 = 0$ **d.** $9x^2 + 6x + 26 = 0$

no real solutions $\qquad\qquad$ no real solutions

**28.** Determine the type of solutions of each equation by finding the discriminant. Then solve each equation and approximate the solutions to the nearest hundredth. (*Section 9.3, Objectives 2 and 3*)

**a.** $x^2 + 4x = 1$ **b.** $6x^2 + 5x + 1 = 0$

**c.** $2x^2 - 3x = 0$ **d.** $x^2 - 12x + 40 = 0$

two real solutions, 0 or 1.5 $\qquad$ no real solutions

**29.** If the height (in feet) of an object launched from the ground after $t$ seconds is modeled by the equation $h = -32t^2 + 150t$, when will the object be 18 ft from the ground? (*Section 9.3, Objective 4*) $\quad$ 0.12 sec or 4.56 sec

**30.** Perform the indicated operation on the complex numbers. Write each answer in standard form. (*Section 9.4, Objectives 1–4*)

**a.** $\sqrt{-144} + (7 - \sqrt{-49})$ $\quad 7 + 5i$

**b.** $(3 - 7i) - (-5 + 13i)$ **c.** $-3i(10 - 7i)$ $\quad -21 - 30i$

**d.** $(5 - 3i)^2$ $\quad 16 - 30i$ **e.** $(8 + 5i)(6 - 4i)$ $\quad 68 - 2i$

**f.** $(1.2 - 0.5i)(1.5 + 0.6i)$ **g.** $\dfrac{10 - 12i}{-6i}$ $\quad 2 + \dfrac{5}{3}i$

$2.1 - 0.03i$

**h.** $\dfrac{5 - i}{3 + 2i}$ $\quad 1 - i$ **i.** $\dfrac{-4 + 2i}{1 - 3i}$ $\quad -1 - i$

**31.** Solve each quadratic equation for its complex solutions. (*Section 9.4, Objective 5*)

**a.** $2x^2 + 50 = 0$ $\quad \{-5i, 5i\}$ **b.** $x^3 - 2x^2 + 19x = 0$

$\{0, 1 - 3i\sqrt{2}, 1 + 3i\sqrt{2}\}$

**c.** $x^4 - 3x^2 - 4 = 0$

$\{-2, 2, -i, i\}$

**32.** Find the intercepts and vertex and graph the equation $y = x^2 + 12x - 28$. (*Section 9.5, Objective 2*)

**33.** Find the solutions of $y = 5x^2 - 4x - 1$ for $x = 0, 1,$ and $-2$. (*Section 9.5, Objective 1*) $\quad (-2, 27), (0, -1), (1, 0)$

**34.** Use the given graph to solve the equation. (*Section 9.5, Objective 3*)

**a.** $x^2 + 6x - 40 = 0$ **b.** $x^2 - 10x + 9 = 0$

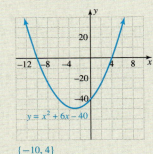
$y = x^2 + 6x - 40$

$\{-10, 4\}$

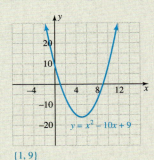
$y = x^2 - 10x + 9$

$\{1, 9\}$

# Instructor Answer Appendix

### CHAPTER 1

## Section 1.1

**9.** False, for example $\frac{1}{3} = 0.333333 \ldots$ and is rational.

**14.** natural, whole, integer, $\frac{10}{1}$ rational, real

**15.** $\frac{5}{2}$ rational, real

**21.** natural, whole, integer, $\frac{9}{1}$ rational, real

**41.**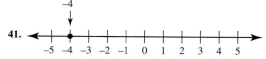

**42.**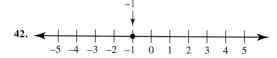

**43.**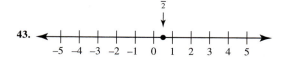

**44.**

**45.**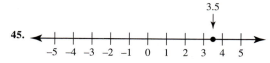

**46.**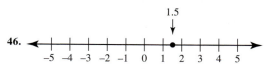

**47.**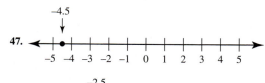

**48.**

**49.**

**50.**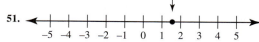

**51.**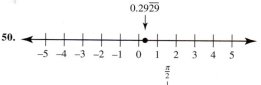

**52.**

**53.**

**54.**

**55.**

**56.**

**95.** natural, whole, integer, rational, real, $\frac{96}{1}$

**96.** natural, whole, integer, rational, real, $\frac{4160}{1}$

**105.** integer, rational, real, $-\frac{1293}{1}$

**107.**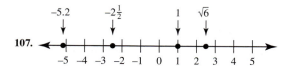

**108.**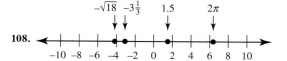

**109.**

**110.**

**126.**

**Section 1.2**

**30.** Democrats: $\dfrac{8}{17}$; Republications: $\dfrac{9}{17}$

**104.** $\dfrac{2}{5} \times \dfrac{3}{10} = \dfrac{2 \cdot 3}{5 \cdot 10} = \dfrac{3}{25}$

**105.** $2\dfrac{1}{4} \times 6\dfrac{3}{5} = \dfrac{9}{4} \times \dfrac{33}{5} = \dfrac{297}{20} = 14\dfrac{17}{20}$

**106.** LCD = 24

$\dfrac{3}{8} - \dfrac{1}{12} = \dfrac{3 \times 3}{8 \times 3} - \dfrac{1 \times 2}{12 \times 2} = \dfrac{9}{24} - \dfrac{2}{24} = \dfrac{7}{24}$

**Section 1.3**

**45.**

| $x$ | $3x + 2$ |
|---|---|
| 0 | 2 |
| $\dfrac{1}{3}$ | 3 |
| 2 | 8 |
| 4 | 14 |

**46.**

| $x$ | $4x + 1$ |
|---|---|
| $\dfrac{1}{4}$ | 2 |
| 1 | 5 |
| 2 | 9 |
| 4 | 17 |

**47.**

| $x$ | $\dfrac{3x - 2}{x}$ |
|---|---|
| 1 | 1 |
| 2 | 2 |
| 3 | $\dfrac{7}{3}$ |
| 4 | $\dfrac{5}{2}$ |

**48.**

| $x$ | $\dfrac{5x - 2}{x}$ |
|---|---|
| 1 | 3 |
| 2 | 4 |
| 3 | $\dfrac{13}{3}$ |
| 4 | $\dfrac{9}{2}$ |

**49.**

| $x$ | $\dfrac{x + 3}{x - 1}$ |
|---|---|
| 2 | 5 |
| 3 | 3 |
| 4 | $\dfrac{7}{3}$ |
| 5 | 2 |

**50.**

| $x$ | $\dfrac{x + 1}{x - 2}$ |
|---|---|
| 3 | 4 |
| 4 | $\dfrac{5}{2}$ |
| 5 | 2 |
| 6 | $\dfrac{7}{4}$ |

**123.**

| $x$ | $5x - 3$ |
|---|---|
| 1 | 2 |
| 2 | 7 |
| 3 | 12 |
| 4 | 17 |

**124.**

| $x$ | $\dfrac{3x + 2}{x + 1}$ |
|---|---|
| 0 | 2 |
| 1 | $\dfrac{5}{2}$ |
| 2 | $\dfrac{8}{3}$ |
| 3 | $\dfrac{11}{4}$ |

**152.** $-\dfrac{2}{3} \cdot 3 + 5 = -\dfrac{2}{3} \cdot \dfrac{3}{1} + 5 = -2 + 5 = 3$

**155.** $-3 \cdot 2^2 + 28 \div 7 \cdot 4 + 36 = -3 \cdot 4 + 4 \cdot 4 + 36$
$= -12 + 16 + 36 = 40$

**156.** $38 - 2^2 - 45 \div 5 \cdot 3 = 38 - 4 - 9 \cdot 3$
$= 38 - 4 - 27 = 7$

**Piece It Together  Sections 1.1–1.4**

**1.**

**17.**

| $x$ | $\dfrac{2}{5}x + 3$ |
|---|---|
| 0 | 3 |
| 1 | $\dfrac{17}{5}$ |
| 5 | 5 |
| 10 | 7 |

**Section 1.5**

**93.** $\dfrac{1}{5} - \left(-1\dfrac{3}{4}\right) = \dfrac{1}{5} - \left(-\dfrac{7}{4}\right) = \dfrac{1}{5} + \dfrac{7}{4}$

$= \dfrac{1 \times 4}{5 \times 4} + \dfrac{7 \times 5}{4 \times 5} = \dfrac{4}{20} + \dfrac{35}{20} = \dfrac{39}{20}$

**94.** $-\dfrac{1}{4} - (-3) = -\dfrac{1}{4} + 3 = -\dfrac{1}{4} + \dfrac{3}{1} = -\dfrac{1}{4} + \dfrac{3 \times 4}{1 \times 4}$

$= -\dfrac{1}{4} + \dfrac{12}{4} = \dfrac{11}{4}$

**Section 1.6**

**77.**

| $x$ | $2x^2 - 5x + 1$ |
|---|---|
| $-2$ | 19 |
| 0 | 1 |
| $\dfrac{1}{2}$ | $-1$ |
| 2 | $-1$ |

**78.**

| $x$ | $3x^2 + x - 2$ |
|---|---|
| $-2$ | 8 |
| 0 | $-2$ |
| $\dfrac{1}{2}$ | $-\dfrac{3}{4}$ |
| 2 | 12 |

**79.**

| $x$ | $-2x^2 + 4x + 3$ |
|---|---|
| $-2$ | $-13$ |
| $-1$ | $-3$ |
| $\dfrac{1}{2}$ | $\dfrac{9}{2}$ |
| 2 | 3 |

**80.**

| $x$ | $-x^2 - 3x + 4$ |
|---|---|
| $-2$ | 6 |
| $-1$ | 6 |
| $\dfrac{1}{2}$ | $\dfrac{9}{4}$ |
| 2 | $-6$ |

**83.**

| $x$ | $\dfrac{|x - 2|}{x + 1}$ |
|---|---|
| $-2$ | $-4$ |
| $-1$ | Undefined |
| 0 | 2 |
| 1 | $\dfrac{1}{2}$ |
| 2 | 0 |

**84.**

| $x$ | $\dfrac{|x - 2|}{x + 2}$ |
|---|---|
| $-2$ | Undefined |
| $-1$ | 3 |
| 0 | 1 |
| 1 | $\dfrac{1}{3}$ |
| 2 | 0 |

**91. a.**

| Date | Closing Price | Daily Change in Price | Percent Change in Price |
|---|---|---|---|
| 10/8/2010 | $48.75 | $0.87 | 0.0182 or 1.82% |
| 10/7/2010 | $47.88 | −$0.29 | −0.0060 or −0.60% |
| 10/6/2010 | $48.17 | $0.07 | 0.0015 or 0.15% |
| 10/5/2010 | $48.10 | $0.47 | 0.0099 or 0.99% |
| 10/4/2010 | $47.63 | n/a | n/a |

**92. a.**

| Date | Closing Price | Weekly Change in Price | Percent Change in Price |
|---|---|---|---|
| 10/8/2010 | $54.41 | $1.05 | 0.0197 or 1.97% |
| 10/1/2010 | $53.36 | −$0.72 | −0.0133 or −1.33% |
| 9/24/2010 | $54.08 | $1.07 | 0.0202 or 2.02% |
| 9/17/2010 | $53.01 | $1.04 | 0.0200 or 2.00% |
| 9/10/2010 | $51.97 | n/a | n/a |

**143.**

| $x$ | $\dfrac{|x-2|}{x+3}$ |
|---|---|
| −3 | Undefined |
| −2 | 4 |
| 0 | $\dfrac{2}{3}$ |
| 1 | $\dfrac{1}{4}$ |
| 2 | 0 |

**144.**

| $x$ | $\dfrac{|2x-3|}{x}$ |
|---|---|
| −2 | $-\dfrac{7}{2}$ |
| −1 | −5 |
| 0 | Undefined |
| 1 | 1 |
| 2 | $\dfrac{1}{2}$ |

**146. a.**

| Date | Closing Price | Weekly Change in Price | Percent Change in Price |
|---|---|---|---|
| 10/8/2010 | $11,006.48 | $176.80 | 0.0163 or 1.63% |
| 10/1/2010 | $10,829.68 | −$30.58 | −0.0028 or −0.28% |
| 9/24/2010 | $10,860.26 | $252.41 | 0.0238 or 2.38% |
| 9/17/2010 | $10,607.85 | $145.08 | 0.0139 or 1.39% |
| 9/10/2010 | $10,462.77 | n/a | n/a |

## CHAPTER 1  REVIEW EXERCISES

### Section 1.1

**9.**

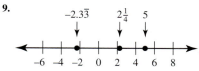

**10.**

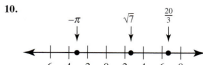

### Section 1.3

**61.** $\dfrac{28}{42} = \dfrac{2}{3}$

### Section 1.6

**108.**

| $x$ | $y$ |
|---|---|
| −4 | 15 |
| −3 | 13 |
| −2 | 11 |
| −1 | 9 |
| 0 | 7 |

**109.**

| $x$ | $y$ |
|---|---|
| −2 | $\dfrac{3}{5}$ |
| −1 | $\dfrac{1}{4}$ |
| 0 | $-\dfrac{1}{3}$ |
| 1 | $-\dfrac{3}{2}$ |
| 2 | −5 |

**110.**

| $x$ | $y$ |
|---|---|
| −2 | 1 |
| −1 | 4 |
| 0 | 5 |
| 1 | 4 |
| 2 | 1 |

**111.**

| $x$ | $y$ |
|---|---|
| −2 | 2 |
| −1 | −1 |
| 0 | −2 |
| 1 | −1 |
| 2 | 2 |

**112.**

| $x$ | $y$ |
|---|---|
| −2 | 7 |
| −1 | 3 |
| 0 | 1 |
| 1 | 1 |
| 2 | 3 |

**113.**

| $x$ | $y$ |
|---|---|
| −4 | −21 |
| −2 | −11 |
| 0 | 1 |
| 2 | 9 |
| 4 | 19 |

## CHAPTER 1  TEST

**9. e.** identity property of multiplication,

## CHAPTER 2

### Section 2.1

**17.**

| | |
|---|---|
| $x = -2$ | Is a solution |
| $x = -1$ | Is not a solution |
| $x = 0$ | Is not a solution |
| $x = 1$ | Is not a solution |
| $x = 2$ | Is not a solution |

**18.**

| | |
|---|---|
| $x = -2$ | Is not a solution |
| $x = -1$ | Is a solution |
| $x = 0$ | Is not a solution |
| $x = 1$ | Is not a solution |
| $x = 2$ | Is not a solution |

**19.**

| | |
|---|---|
| $x = -2$ | Is not a solution |
| $x = -1$ | Is not a solution |
| $x = 0$ | Is a solution |
| $x = 1$ | Is not a solution |
| $x = 2$ | Is not a solution |

**20.**

| | |
|---|---|
| $x = -2$ | Is not a solution |
| $x = -1$ | Is not a solution |
| $x = 0$ | Is not a solution |
| $x = 1$ | Is not a solution |
| $x = 2$ | Is a solution |

**21.**

| | |
|---|---|
| $x = -2$ | Is not a solution |
| $x = -1$ | Is a solution |
| $x = 0$ | Is not a solution |
| $x = 1$ | Is not a solution |
| $x = 2$ | Is a solution |

**22.**

| | |
|---|---|
| $x = -2$ | Is a solution |
| $x = -1$ | Is not a solution |
| $x = 0$ | Is not a solution |
| $x = 1$ | Is a solution |
| $x = 2$ | Is not a solution |

**23.**

| $x = -2$ | Is not a solution |
|---|---|
| $x = -1$ | Is a solution |
| $x = 0$ | Is not a solution |
| $x = 1$ | Is a solution |
| $x = 2$ | Is not a solution |

**24.**

| $x = -2$ | Is not a solution |
|---|---|
| $x = -1$ | Is not a solution |
| $x = 0$ | Is a solution |
| $x = 1$ | Is not a solution |
| $x = 2$ | Is a solution |

**25.**

| $x = -2$ | Is not a solution |
|---|---|
| $x = -1$ | Is not a solution |
| $x = 0$ | Is not a solution |
| $x = 1$ | Is not a solution |
| $x = 2$ | Is not a solution |

**26.**

| $x = -2$ | Is a solution |
|---|---|
| $x = -1$ | Is a solution |
| $x = 0$ | Is a solution |
| $x = 1$ | Is a solution |
| $x = 2$ | Is a solution |

**65.**

| $x = -4$ | Is not a solution |
|---|---|
| $x = -1$ | Is not a solution |
| $x = 0$ | Is not a solution |
| $x = 4$ | Is not a solution |
| $x = \dfrac{5}{14}$ | Is a solution |

**66.**

| $x = -7$ | Is not a solution |
|---|---|
| $x = -2$ | Is not a solution |
| $x = 0$ | Is not a solution |
| $x = 1$ | Is not a solution |
| $x = -\dfrac{11}{4}$ | Is a solution |

**67.**

| $x = -4$ | Is a solution |
|---|---|
| $x = -2$ | Is not a solution |
| $x = 0$ | Is not a solution |
| $x = 5$ | Is a solution |
| $x = \dfrac{1}{2}$ | Is not a solution |

**68.**

| $x = -2$ | Is a solution |
|---|---|
| $x = -1$ | Is a solution |
| $x = 0$ | Is not a solution |
| $x = 2$ | Is not a solution |
| $x = \dfrac{1}{3}$ | Is not a solution |

**89.**

| $x = -1$ | Is not a solution |
|---|---|
| $x = 0$ | Is a solution |
| $x = 1$ | Is a solution |

**90.**

| $x = -2$ | Is a solution |
|---|---|
| $x = 0$ | Is not a solution |
| $x = 2$ | Is not a solution |

**91.**

| $x = 0$ | Is not a solution |
|---|---|
| $x = \dfrac{1}{2}$ | Is a solution |
| $x = 1$ | Is not a solution |

**92.**

| $x = -1$ | Is not a solution |
|---|---|
| $x = -\dfrac{1}{2}$ | Is a solution |
| $x = \dfrac{1}{2}$ | Is not a solution |

## Section 2.2

**57.** Zuckerberg's net worth was $6.9 billion and Moskovitz's net worth was $1.4 billion.

## Section 2.3

**74.** $\left\{\dfrac{20}{9}\right\}$

## Section 2.4

**25.** $\left\{-\dfrac{18}{7}\right\}$

## Section 2.7

## Section 2.5

**44.** $w = \dfrac{P - 2l}{2}$    **46.** $y = -\dfrac{1}{2}x + 2$

**51.** $x = \dfrac{3}{7}y - 3$    **52.** $y = \dfrac{3}{5}x + 3$

**54.** $P = \dfrac{A}{1 + rt}$

## Section 2.6

**4.** False, we must compute $\dfrac{75 - 50}{50}$. We divide by the original amount.

**40.** 2 hr 10 min from home and 1 hr 50 min back home; $\dfrac{1}{3}$ mi total distance

**73.** Let $x$ be the price of the car without tax.

$$x + 0.08x = 52{,}629.48$$
$$1.08x = 52{,}629.48$$
$$x = \dfrac{52{,}629.48}{1.08}$$
$$x = 48{,}731$$

The price of the car without tax is $48,731.

**74.** Let $x$ be the original selling price of the TV.

$$x - 0.40x = 209.99$$
$$0.60x = 209.99$$
$$x = \dfrac{209.99}{0.60}$$
$$x = 349.98$$

The original selling price of the TV is $349.98.

**75.** Average percent $= \dfrac{34.06 - 30.17}{30.17}$

$$= \dfrac{3.89}{30.17}$$
$$= 0.128936$$
$$\approx 12.9\%$$

The average percent increase is about 12.9%.

**76.** Let $x$ be the amount invested in the stock and $13{,}850 - x$ in the mutual fund account.

$$0.045x + 0.038(13850 - x) = 581.95$$
$$0.045x + 526.30 - 0.038x = 581.95$$
$$0.007x = 55.65$$
$$x = 7950$$
$$13{,}850 - x = 5900$$

Patel has invested $7950 in the stock and $5900 in the mutual fund.

**77.** Let $x$ be the number of liters of pure acid and $210 - x$ be the number of liters of 40% solution.

$$1.00x + 0.40(210 - x) = 0.60(210)$$
$$0.60x + 84 = 126$$
$$0.60x = 42$$
$$x = 70$$
$$210 - x = 140$$

Hong needs 70 L of pure acid and 140 L of 40% solution.

| Inequality | Graph | Interval Notation | Set-Builder Notation |
|---|---|---|---|
| **37.** $x > -4$ | $-9\ -8\ -7\ -6\ -5\ -4\ -3\ -2\ -1\ \ 0\ \ 1$ | $(-4, \infty)$ | $\{x \mid x > -4\}$ |
| **38.** $y \geq -9$ | $-14\ -13\ -12\ -11\ -10\ -9\ -8\ -7\ -6\ -5\ -4$ | $[-9, \infty)$ | $\{y \mid y \geq -9\}$ |

| Inequality | Graph | Interval Notation | Set-Builder Notation |
|---|---|---|---|
| 39. $x \geq 7$ | 2 3 4 5 6 7 8 9 10 11 12 | $[7, \infty)$ | $\{x \mid x \geq 7\}$ |
| 40. $a \geq -11$ | −16 −15 −14 −13 −12 −11 −10 −9 −8 −7 −6 | $[-11, \infty)$ | $\{a \mid a \geq -11\}$ |
| 41. $y < 0$ | −5 −4 −3 −2 −1 0 1 2 3 4 5 | $(-\infty, 0)$ | $\{y \mid y < 0\}$ |
| 42. $a > 0$ | −5 −4 −3 −2 −1 0 1 2 3 4 5 | $(0, \infty)$ | $\{a \mid a > 0\}$ |
| 43. $x \leq -4$ | −9 −8 −7 −6 −5 −4 −3 −2 −1 0 1 | $(-\infty, -4]$ | $\{x \mid x \leq -4\}$ |
| 44. $y \leq -1$ | −6 −5 −4 −3 −2 −1 0 1 2 3 4 | $(-\infty, -1]$ | $\{y \mid y \leq -1\}$ |
| 45. $y > -1$ | −6 −5 −4 −3 −2 −1 0 1 2 3 4 | $(-1, \infty)$ | $\{y \mid y > -1\}$ |
| 46. $a \leq 3$ | −2 −1 0 1 2 3 4 5 6 7 8 | $(-\infty, 3]$ | $\{a \mid a \leq 3\}$ |
| 47. $x > -3$ | −8 −7 −6 −5 −4 −3 −2 −1 0 1 2 | $(-3, \infty)$ | $\{x \mid x > -3\}$ |
| 48. $a \geq -2$ | −7 −6 −5 −4 −3 −2 −1 0 1 2 3 | $[-2, \infty)$ | $\{a \mid a \geq -2\}$ |
| 49. $y > 4$ | −1 0 1 2 3 4 5 6 7 8 9 | $(4, \infty)$ | $\{y \mid y > 4\}$ |
| 50. $x \geq -10$ | −15 −14 −13 −12 −11 −10 −9 −8 −7 −6 −5 | $[-10, \infty)$ | $\{x \mid x \geq -10\}$ |
| 51. $a < -8$ | −13 −12 −11 −10 −9 −8 −7 −6 −5 −4 −3 | $(-\infty, -8)$ | $\{a \mid a < -8\}$ |
| 52. $y \geq 4$ | −1 0 1 2 3 4 5 6 7 8 9 | $[4, \infty)$ | $\{y \mid y \geq 4\}$ |
| 53. $a \leq 2$ | −3 −2 −1 0 1 2 3 4 5 6 7 | $(-\infty, 2]$ | $\{a \mid a \leq 2\}$ |
| 54. $a > -3$ | −8 −7 −6 −5 −4 −3 −2 −1 0 1 2 | $(-3, \infty)$ | $\{a \mid a > -3\}$ |
| 55. $x < -27$ | −32 −31 −30 −29 −28 −27 −26 −25 −24 −23 −22 | $(-\infty, -27)$ | $\{x \mid x < -27\}$ |
| 56. $a \geq 20$ | 15 16 17 18 19 20 21 22 23 24 25 | $[20, \infty)$ | $\{a \mid a \geq 20\}$ |
| 57. $y < -2$ | −7 −6 −5 −4 −3 −2 −1 0 1 2 3 | $(-\infty, -2)$ | $\{y \mid y < -2\}$ |
| 58. $a > -1$ | −6 −5 −4 −3 −2 −1 0 1 2 3 4 | $(-1, \infty)$ | $\{a \mid a > -1\}$ |
| 59. $a \leq 2$ | −3 −2 −1 0 1 2 3 4 5 6 7 | $(-\infty, 2]$ | $\{a \mid a \leq 2\}$ |
| 60. $x \geq 2$ | −3 −2 −1 0 1 2 3 4 5 6 7 | $[2, \infty)$ | $\{x \mid x \geq 2\}$ |

| Inequality | Graph | Interval Notation | Set-Builder Notation |
|---|---|---|---|
| **61.** $x \le \dfrac{13}{4}$ | –2 –1 0 1 2 3 4 5 6 7 8 ; $\dfrac{13}{4}$ | $\left(-\infty, \dfrac{13}{4}\right]$ | $\left\{x \middle\| x \le \dfrac{13}{4}\right\}$ |
| **62.** $x < \dfrac{4}{17}$ | –5 –4 –3 –2 –1 1 2 3 4 5 ; $\dfrac{4}{17}$ | $\left(-\infty, \dfrac{4}{17}\right)$ | $\left\{x \middle\| x < \dfrac{4}{17}\right\}$ |
| **63.** $x < 18$ | 13 14 15 16 17 18 19 20 21 22 23 | $(-\infty, 18)$ | $\left\{x \middle\| x < 18\right\}$ |
| **64.** $x \le -\dfrac{5}{17}$ | –5 –4 –3 –2 –1 0 1 2 3 4 5 ; $-\dfrac{5}{17}$ | $\left(-\infty, -\dfrac{5}{17}\right]$ | $\left\{x \middle\| x \le -\dfrac{5}{17}\right\}$ |
| **65.** $y \le \dfrac{19}{3}$ | 1 2 3 4 5 6 7 8 9 10 12 ; $\dfrac{19}{3}$ | $\left(-\infty, \dfrac{19}{3}\right]$ | $\left\{y \middle\| y \le \dfrac{19}{3}\right\}$ |
| **66.** $a < -\dfrac{11}{6}$ | –5 –4 –3 –1 0 1 2 3 4 5 ; $-\dfrac{11}{6}$ | $\left(-\infty, -\dfrac{11}{6}\right)$ | $\left\{a \middle\| a < -\dfrac{11}{6}\right\}$ |
| **67.** $x < 100$ | 95 96 97 98 99 100 101 102 103 104 105 | $(-\infty, 100)$ | $\left\{x \middle\| x < 100\right\}$ |
| **68.** $a > 140$ | 135 136 137 138 139 140 141 142 143 144 145 | $(140, \infty)$ | $\left\{a \middle\| a > 140\right\}$ |
| **69.** $-13 < a < 8$ | –20 –16 –12 –8 –4 0 4 8 12 16 20 | $(-13, 8)$ | $\left\{a \middle\| -13 < a < 8\right\}$ |
| **70.** $4 \le b \le 13$ | –12 –8 –4 0 4 8 16 20 24 28 ; 13 | $[4, 13]$ | $\left\{b \middle\| 4 \le b \le 13\right\}$ |
| **71.** $-2 \le x \le 5$ | –4 –3 –2 –1 0 1 2 3 4 5 6 | $[-2, 5]$ | $\left\{x \middle\| -2 \le x \le 5\right\}$ |
| **72.** $-2 < y < 4$ | –4 –3 –2 –1 0 1 2 3 4 5 6 | $(-2, 4)$ | $\left\{y \middle\| -2 < y < 4\right\}$ |
| **73.** $-7 \le x \le -3$ | –10 –9 –8 –7 –6 –5 –4 –3 –2 –1 0 | $[-7, -3]$ | $\left\{x \middle\| -7 \le x \le -3\right\}$ |
| **74.** $-11 \le y < 7$ | –20 –16 –8 –4 0 4 8 12 16 20 ; –11 ; 7 | $[-11, 7)$ | $\left\{y \middle\| -11 < y < 7\right\}$ |
| **75.** $-3 \le a \le -1$ | –7 –6 –5 –4 –3 –2 –1 0 1 2 3 | $[-3, -1]$ | $\left\{a \middle\| -3 \le a \le -1\right\}$ |
| **76.** $-3 < b < 2$ | –5 –4 –3 –2 –1 0 1 2 3 4 5 | $(-3, 2)$ | $\left\{b \middle\| -3 < b < 2\right\}$ |
| **77.** $-\dfrac{7}{6} < a < -\dfrac{1}{2}$ | –5 –4 –3 –2 0 1 2 3 4 5 ; $-\dfrac{7}{6}$ ; $-\dfrac{1}{2}$ | $\left(-\dfrac{7}{6}, -\dfrac{1}{2}\right)$ | $\left\{a \middle\| -\dfrac{7}{6} < a < -\dfrac{1}{2}\right\}$ |

| Inequality | Graph | Interval Notation | Set-Builder Notation |
|---|---|---|---|
| 78. $\dfrac{7}{8} \le b \le \dfrac{13}{4}$ | number line graph | $\left(\dfrac{7}{8}, \dfrac{13}{4}\right)$ | $\left\{ b \,\middle|\, \dfrac{7}{8} \le b \le \dfrac{13}{4} \right\}$ |
| 79. $-2 < x < 7$ | number line graph | $(-2, 7)$ | $\{x \mid -2 < x < 7\}$ |
| 80. $-19 \le x \le 1$ | number line graph | $[-19, 1]$ | $\{x \mid -19 \le x \le 1\}$ |
| 81. $x \ge 3$ | number line graph | $[3, \infty)$ | $\{x \mid x \ge 3\}$ |
| 82. $x \le 2$ | number line graph | $(-\infty, 2]$ | $\{x \mid x \le 2\}$ |
| 83. $x > -6$ | number line graph | $(-6, \infty)$ | $\{x \mid x > -6\}$ |
| 84. $x < -7$ | number line graph | $(-\infty, -7)$ | $\{x \mid x < -7\}$ |
| 85. $y > -\dfrac{1}{21}$ | number line graph | $\left(-\dfrac{1}{21}, \infty\right)$ | $\left\{ y \,\middle|\, y > -\dfrac{1}{21} \right\}$ |
| 86. $y < \dfrac{2}{21}$ | number line graph | $\left(-\infty, \dfrac{2}{21}\right)$ | $\left\{ y \,\middle|\, y < \dfrac{2}{21} \right\}$ |
| 87. $x < -19$ | number line graph | $(-\infty, -19)$ | $\{x \mid x < -19\}$ |
| 88. $y > 5$ | number line graph | $(5, \infty)$ | $\{y \mid y > 5\}$ |
| 95. $x > 24$ | number line graph | $(24, \infty)$ | $\{x \mid x > 24\}$ |
| 96. $y \le 2$ | number line graph | $(-\infty, 2]$ | $\{y \mid y \le 2\}$ |
| 97. $x > 2$ | number line graph | $(2, \infty)$ | $\{x \mid x > 2\}$ |
| 98. $y \ge -2$ | number line graph | $[-2, \infty)$ | $\{y \mid y \ge -2\}$ |
| 99. $a \le 27.5$ | number line graph | $(-\infty, 27.5]$ | $\{a \mid a \le 27.5\}$ |
| 100. $b > -20$ | number line graph | $(-20, \infty)$ | $\{b \mid b > -20\}$ |
| 101. $y \le 64$ | number line graph | $(-\infty, 64]$ | $\{y \mid y \le 64\}$ |

| Inequality | Graph | Interval Notation | Set-Builder Notation |
|---|---|---|---|
| **102.** $x < 11$ | 6 7 8 9 10 11 12 13 14 15 16 | $(-\infty, 11)$ | $\{x \mid x < 11\}$ |
| **103.** $x \geq -\dfrac{27}{5}$ | −7 −6 $-\frac{27}{5}$ −4 −3 −2 −1 0 1 2 3 | $\left[-\dfrac{27}{5}, \infty\right)$ | $\left\{x \mid x \geq -\dfrac{27}{5}\right\}$ |
| **104.** $y \geq -1$ | −6 −5 −4 −3 −2 −1 0 1 2 3 4 | $[-1, \infty)$ | $\{y \mid y \geq -1\}$ |
| **105.** $-5 < a < 9$ | −8 −6 −5 −2 0 2 4 6 8 9 10 12 | $(-5, 9)$ | $\{a \mid -5 < a < 9\}$ |
| **106.** $3 < b \leq 4$ | −1 0 1 2 3 4 5 6 7 8 9 | $(3, 4]$ | $\{b \mid 3 < b \leq 4\}$ |
| **107.** $-2 \leq x \leq 4$ | −4 −3 −2 −1 0 1 2 3 4 5 6 | $[-2, 4]$ | $\{x \mid -2 \leq x \leq 4\}$ |
| **108.** $-7 < y \leq 1$ | −8 −7 −6 −5 −4 −3 −2 −1 0 1 2 | $(-7, 1]$ | $\{y \mid -7 < y \leq 1\}$ |
| **109.** $0 < a < 7$ | −2 −1 0 1 2 3 4 5 6 7 8 | $(0, 7)$ | $\{a \mid 0 < a < 7\}$ |
| **110.** $-4 \leq b < -2$ | −8 −7 −6 −5 −4 −3 −2 −1 0 1 2 | $[-4, -2)$ | $\{b \mid -4 \leq b < -2\}$ |
| **111.** $-\infty < x < \infty$ | −∞ −4 −3 −2 −1 0 1 2 3 4 ∞ | $(-\infty, \infty)$ | $\{x \mid -\infty < x < \infty\}$ |
| **112.** $-\infty < x < \infty$ | −∞ −4 −3 −2 −1 0 1 2 3 4 ∞ | $(-\infty, \infty)$ | $\{x \mid -\infty < x < \infty\}$ |
| **113.** $\varnothing$ | | $\varnothing$ | $\varnothing$ |
| **114.** $\varnothing$ | | $\varnothing$ | $\varnothing$ |
| **115.** $x \geq -8$ | −13 −12 −11 −10 −9 −8 −7 −6 −5 −4 −3 | $[-8, \infty)$ | $\{x \mid x \geq -8\}$ |
| **116.** $y \leq \dfrac{6}{11}$ | −5 −4 −3 −2 −1 0 $\frac{6}{11}$ 1 2 3 4 5 | $\left(-\infty, \dfrac{6}{11}\right]$ | $\left\{y \mid y \leq \dfrac{6}{11}\right\}$ |
| **117.** $4 \leq x \leq 6$ | −1 0 1 2 3 4 5 6 7 8 9 | $[4, 6]$ | $\{x \mid 4 \leq x \leq 6\}$ |
| **118.** $-3 < y < 1$ | −6 −5 −4 −3 −2 −1 0 1 2 3 4 | $(-3, 1)$ | $\{y \mid -3 < y < 1\}$ |
| **119.** $-\dfrac{15}{2} < x < \dfrac{19}{2}$ | −10 −8 $-\frac{15}{2}$ −6 −4 −2 0 1 4 6 8 $\frac{19}{2}$ 10 | $\left(-\dfrac{15}{2}, \dfrac{19}{2}\right)$ | $\left\{x \mid -\dfrac{15}{2} < x < \dfrac{19}{2}\right\}$ |
| **120.** $-\dfrac{28}{5} \leq y \leq \dfrac{2}{5}$ | −7 −6 $-\frac{28}{5}$ −5 −4 −3 −2 −1 0 $\frac{2}{5}$ 1 2 3 | $\left[-\dfrac{28}{5}, \dfrac{2}{5}\right]$ | $\left\{y \mid -\dfrac{28}{5} \leq y \leq \dfrac{2}{5}\right\}$ |
| **121.** $-4 < x < 64$ | −8 −4 0 8 16 24 32 40 48 56 64 72 | $(-4, 64)$ | $\{x \mid -4 < x < 64\}$ |

| Inequality | Graph | Interval Notation | Set-Builder Notation |
|---|---|---|---|
| **122.** $-5 \le y < 80$ | | $[-5, 80)$ | $\{y \mid -5 \le y < 80\}$ |

## CHAPTER 2  GROUP ACTIVITY

**3.** $0.93p = 286.8$, where p is the U.S. population; The U.S. population is approximately 308.4 million.

**4.** 312 million (as of 8/5/11); The number of wireless subscriber connections with reach 312 million in 2011.

## CHAPTER 2  REVIEW EXERCISES

### Section 2.1

**5.**

| | |
|---|---|
| $x = -4$ | Is not a solution |
| $x = -1$ | Is not a solution |
| $x = 0$ | Is not a solution |
| $x = 4$ | Is not a solution |
| $x = -\dfrac{4}{5}$ | Is a solution |

**6.**

| | |
|---|---|
| $x = -6$ | Is a solution |
| $x = -2$ | Is not a solution |
| $x = 0$ | Is not a solution |
| $x = 3$ | Is a solution |
| $x = \dfrac{1}{2}$ | Is not a solution |

### Section 2.7

| Inequality | Graph | Interval Notation | Set-Builder Notation |
|---|---|---|---|
| **97.** $a \ge -14$ | | $[-14, \infty)$ | $\{a \mid a \ge -14\}$ |
| **98.** $x > 3$ | | $(3, \infty)$ | $\{x \mid x > 3\}$ |
| **99.** $a \le 16$ | | $(-\infty, 16]$ | $\{a \mid a \le 16\}$ |
| **100.** $x \ge -1$ | | $[-1, \infty)$ | $\{x \mid x \ge -1\}$ |
| **101.** $-1 < a < 7$ | | $(-1, 7)$ | $\{a \mid -1 < a < 7\}$ |
| **102.** $-3 \le x \le 4$ | | $[-3, 4]$ | $\{x \mid -3 \le x \le 4\}$ |
| **103.** $x < -6$ | | $(-\infty, -6)$ | $\{x \mid x < -6\}$ |
| **104.** $x > 9$ | | $(9, \infty)$ | $\{x \mid x > 9\}$ |
| **105.** $y \ge -\dfrac{1}{4}$ | | $\left[-\dfrac{1}{4}, \infty\right)$ | $\left\{y \mid y \ge -\dfrac{1}{4}\right\}$ |
| **106.** $2 \le x \le 11$ | | $[2, 11]$ | $\{x \mid 2 \le x \le 11\}$ |
| **107.** $-9.8 < y < -0.4$ | | $(-9.8, -0.4)$ | $\{y \mid -9.8 < y < -0.4\}$ |
| **108.** $4 < x < 50$ | | $(4, 50)$ | $\{x \mid 4 < x < 50\}$ |

## CHAPTER 2 TEST

**5. a.** $x + (x - 4007) = 112,499$
**b.** There were approximately 58,253 deaths in the Vietnam War and approximately 54,246 deaths in the Korean War.
**18.** To solve a linear inequality and linear equation, we can add or subtract the same number from both sides of the equation or inequality. We can also multiply or divide both sides by a positive number. The difference is that when we multiply or divide both sides of a linear inequality by a negative number, we must also reverse the inequality relationship.

## CHAPTERS 1 AND 2 CUMULATIVE REVIEW EXERCISES

**3. a.**

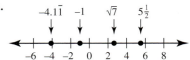

**3. b.**

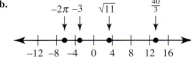

**43. c.** $y = \dfrac{3x - 24}{8} = \dfrac{3}{8}x - 3$    **d.** $x = -\dfrac{5}{2}y + 225$

**51. a.** $a \leq 14$; $(-\infty, 14]$; $\{a \mid a \leq 14\}$

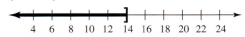

**b.** $-3 < a < 5$; $(-3, 5)$; $\{a \mid -3 < a < 5\}$

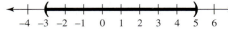

**c.** $x < -10$; $(-\infty, -10)$; $\{x \mid x < -10\}$

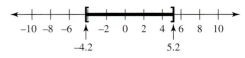

**d.** $-4.2 \leq y \leq 5.2$; $[-4.2, 5.2]$; $\{y \mid -4.2 \leq y \leq 5.2\}$

## CHAPTER 3

### Section 3.1

**29.–44.**

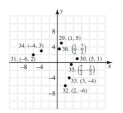
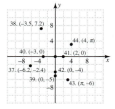

**65.**

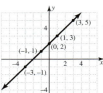

**66.**

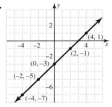

**67.**

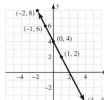

**68.**

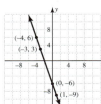

**69.**

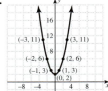

**70.**

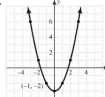

**71.**

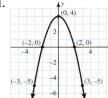

**72.**

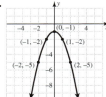

**73.**

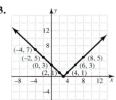

**74.**

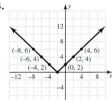

**93. a.** (0, 4.7), (1, 5.8), (2, 6.0), (3, 5.5), (4, 5.1), (5, 4.6), (6, 4.6), (7, 5.8), (8, 9.3), (9, 9.6)
**b.** The point (0, 4.7) means that in 0 yr after 2001, or in 2001, the unemployment rate was 4.7%. The point (9, 9.6) means that 9 yr after 2001, or in 2010, the unemployment rate was 9.6%.
**c.** The unemployment rate was the highest in 2010 and the lowest in 2006 and 2007.
**d.**

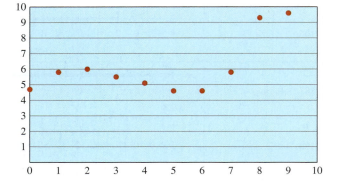

**94. a.** (0, 11.3), (1, 35), (2, 0.7), (3, 10), (4, 11), (5, 11.5), (6, 14)
  **b.** The point (0, 11.3) means that in 0 yr after 2003, or in 2003, Manning's salary was $11.3 million. The point (6, 14) means that in 6 yr after 2003, or in 2009, Manning's salary was $14 million.
  **c.** Manning's salary was the highest in 2004 and the lowest in 2005.
  **d.**

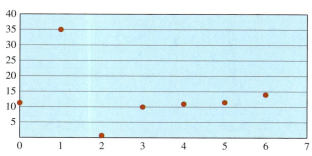

**95. a.** (0, 20.71), (1, 21.38), (2, 22.31), (3, 23.19), (4, 23.96), (5, 25.39), (6, 26.77), (7, 27.35)
  **b.** The national average hourly salary for nurses in 1998 was $20.71.
  **c.** The national average hourly salary for nurses in 2002 was $23.96.
  **d.** The average hourly salary has increased $6.64 from 1998 to 2005.
  **e.** The hourly salaries of nurses increased from 1998 to 2005.
**96. a.** (aaa.com, 3.77), (expedia.com, 15.37), (hotels.com, 6.08), (hotwire.com, 6.21), (kayak.com, 5.08), (orbitz.com, 8.38), (priceline.com, 10.72), (travelocity.com, 9.89), (tripadvisor.com, 12)
  **b.** The expedia.com site had the largest number of visitors, 15.37 million, in August 2010.
  **c.** The aaa.com site had the least number of visitors, 3.77 million, in August 2010.
**109.–116.**

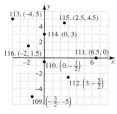

**125.**

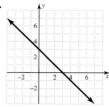

**126.**

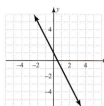

**127.**

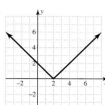

**128.**

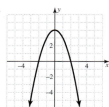

**129.**

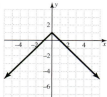

**130.**

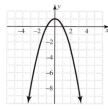

**141. a.** (0, 31.8), (1, 33.3), (2, 36.7), (3, 31.4), (4, 28.3), (5, 31.2), (6, 33.3), (7, 34.4), (8, 35.9), (9, 36.7)
  **b.** The expenditures were the greatest 2 yr after 1998, or in 2000 and in 9 yr after 1998, or 2007.
  **c.** The expenditures decreased by $5.3 billion between 2000 and 2001 and $3.3 billion between 2001 and 2002.
  **d.** The expenditures for U.S. airline transportation increased from 1998 to 2000, decreased from 2000 to 2002, and increased from 2002 to 2007.
**142. a.** (0, 5.67), (1, 5.81), (2, 6.23), (3, 6.78), (4, 7.88), (5, 7.93), (6, 8.68), (7, 9.23), (8, 10.70), (9, 12.31), (10, 13.58)
  **b.** The greatest amount was $13.58 trillion in 2010 and the least was $5.67 trillion in 2000.
  **c.** The debt increased by $1.61 trillion between 2008 and 2009.
  **d.** The public debt increased from 2000 to 2010.
**143.** Always begin in the upper right corner and then rotate counterclockwise.
**144.** Bernadette did not take the absolute value of the number. The table should be

| $x$ | $y = |x - 4|$ | $(x, y)$ |
|---|---|---|
| $-1$ | $|-1 - 4| = |-5| = 5$ | $(-1, 5)$ |
| $0$ | $|0 - 4| = |-4| = 4$ | $(0, 4)$ |
| $1$ | $|1 - 4| = |-3| = 3$ | $(1, 3)$ |
| $2$ | $|2 - 4| = |-2| = 2$ | $(2, 2)$ |
| $3$ | $|3 - 4| = |-1| = 1$ | $(3, 1)$ |

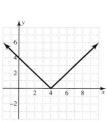

**145.** Valerie did not apply the negative sign after she squared the number.

| $x$ | $y = -x^2 + 1$ | $(x, y)$ |
|---|---|---|
| $-2$ | $-(-2)^2 + 1 = -4 + 1 = -3$ | $(-2, -3)$ |
| $-1$ | $-1(-1)^2 + 1 = -1 + 1 = 0$ | $(-1, 0)$ |
| $0$ | $-(0)^2 + 1 = 0 + 1 = 1$ | $(0, 1)$ |
| $1$ | $-(1)^2 + 1 = -1 + 1 = 0$ | $(1, 0)$ |
| $2$ | $-(2)^2 + 1 = -4 + 1 = -3$ | $(2, -3)$ |

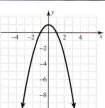

**146.**

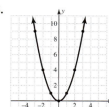

**147.**

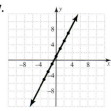

**148.**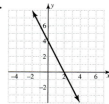

**149.**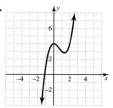

**33.**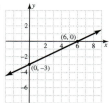

**34.**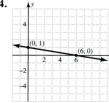

## Section 3.2

**21.**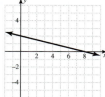

**22.**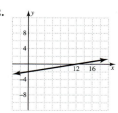

**35.**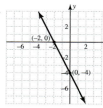

**36.**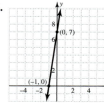

**23.**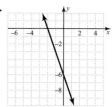

**24.**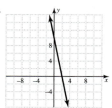

**37.**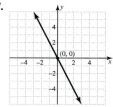

**38.**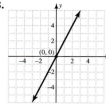

**25.**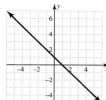

**26.**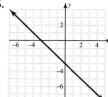

**39.**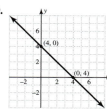

**40.**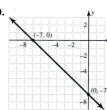

**27.**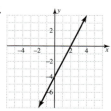

**28.**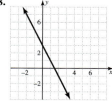

**41.**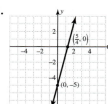

**42.**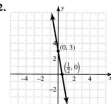

**29.**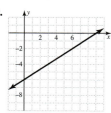

**30.**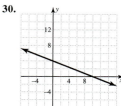

**43.**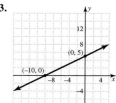

**44.**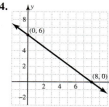

**31.**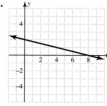

**32.**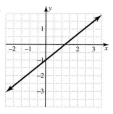

**45.**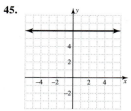

**46.**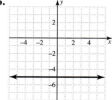

**47.**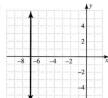

**48.**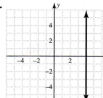

**49.**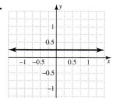

**50.**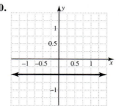

**51.**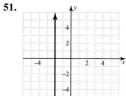

**52.**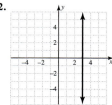

**53.**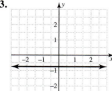

**54.**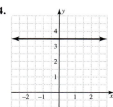

**55.**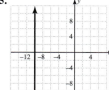

**56.**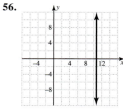

**57.**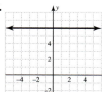

**58.**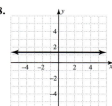

**59. a.** *y*-intercept is (0, 49,217). The median annual income for men with a master's degree is $49,217 in 1991.
  **b.** The median annual income for men in 2001 is $67,947.
  **c.** The median annual income for men in 2012 is $88,550.
**60. a.** *y*-intercept is (0, 35,660). The median annual income for women with a master's degree is $35,660 in 1991.
  **b.** The median annual income for women in 2001 is $49,010.
  **c.** The median annual income for women in 2012 is $63,695.
**61. a.** *y*-intercept is (0, 8.154). The average retail price of electricity for residential use was 8.154 cents per kilowatt-hour in 2001.
  **b.** The average retail price of electricity for residential use was 11.709 cents per kilowatt-hour in 2010.
  **c.** The expected average retail price of electricity for residential use will be 13.684 cents per kilowatt-hour in 2015.

**62. a.** *y*-intercept is (0, 16.173). The average domestic first purchase price of crude oil was $16.173 per barrel in 2000.
  **b.** The average domestic first purchase price of crude oil was $82.963 per barrel in 2010.
  **c.** The expected domestic first purchase price of crude oil will be $103 per barrel in 2013.

**63.**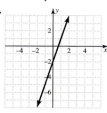

**64.**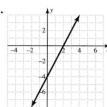

**65.**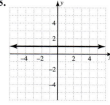

**66.**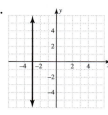

**67.**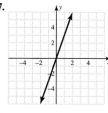

**68.**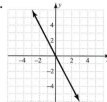

**69.**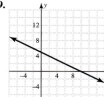

**70.**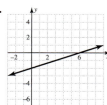

**71.**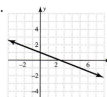

**72.**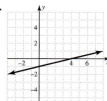

**73.**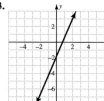

**74.**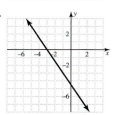

**77.**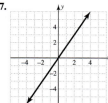

**78.**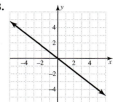

**79.**

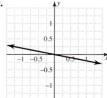

**80.**

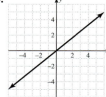

**81.**

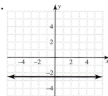

**82.**

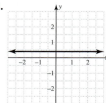

**83.**

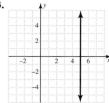

**84.**

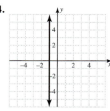

**85.**

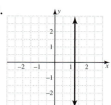

**86.**

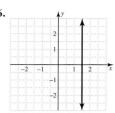

**87.**

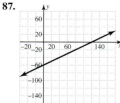

**88.**

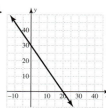

**89.**

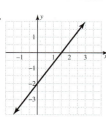

**90.**

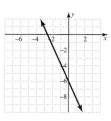

**91. a.** The *y*-intercept is (0, 38). It means 38% of high school students were users of cigarettes in 1997.
  **b.** 30.5% of high school students used cigarettes in 2000.
  **c.** 5.5% of high school students used in cigarettes in 2010.
  **d.**

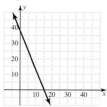

**92. a.** The *y*-intercept is (0, 285). It means that the expenditures for public elementary and secondary education in the United States were $285 billion in 1997.
  **b.** The expenditures for public elementary and secondary education in the United States were $445 billion in 2005.
  **c.** The expenditures for public elementary and secondary education in the United States were $585 billion in 2012.
  **d.**

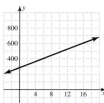

**93. a.** The *x*-intercept is (77.7, 0). It means about 77.7 yr after 1998, or in 2076, 0% of men aged 18 yr and older will be cigarette smokers. No, it is not realistic because there will most likely always be smokers.
  **b.** The *y*-intercept is (0, 25.64). It means 25.64% of the men aged 18 yr and older were cigarette smokers in 1998.
  **c.** 23.33% in 2005 and is very close to the actual estimate.
  **d.** 21.02% in 2012
  **e.**

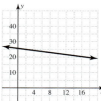

  **f.** About 41 yr after 1998 or in year 2039.

**94. a.** The *x*-intercept is (41.7, 0). It means about 41.7 yr after 1998, or in 2040, 0% of women aged 18 yr and older will be cigarette smokers. No, it is not realistic.
  **b.** The *y*-intercept is (0, 22.12). It means 22.12% of the women aged 18 yr and older were cigarette smokers in 1998.
  **c.** 18.41% in 2005 and is very close to the actual estimate.
  **d.** 14.17% in 2012
  **e.**

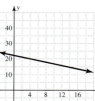

  **f.** About 19 yr after 1998, or in 2017. It is possible.

**95. a.** The *y*-intercept is (0, 19.83). It means 19.83% of the U.S. adults aged 20 yr and over were obese in 1997.
  **b.** 27.64% in 2008. It is very close to the actual estimate.
  **c.** 30.48% in 2012
  **d.**

  **e.** The percent of U.S. adults aged 20 years and over who are obese is increasing.

**96. a.** The *x*-intercept is (5, 0). It means in 5 yr, the depreciated value of the limo is $0.
  **b.** The *y*-intercept is (0, 110,000). The initial value of the limo is $110,000.

**c.** In 2 yr, the depreciated value of the limo is $66,000.
**d.** The depreciated value of the limo is $44,000 in 3 yr.
**e.**

**109.** The *y*-values increase by 2 units for each unit increase in *x*. The coefficient of *x* is also 2.

| *x* | *y* |
|----|----|
| −2 | 0 |
| −1 | 2 |
| 0 | 4 |
| 1 | 6 |
| 2 | 8 |

**110.** The *y*-values decrease by 3 units for each unit increase in *x*. The coefficient on *x* is also −3.

| *x* | *y* |
|----|----|
| −2 | 7 |
| −1 | 4 |
| 0 | 1 |
| 1 | −2 |
| 2 | −5 |

## Section 3.3

**5.** False, the slope is the coefficient of *x* only if the equation is in slope-intercept form.

**6.** False, a slope of $-\dfrac{4}{3}$ means to move down 4 units and right 3 units.

**9.** False, the slope is the change in *y* over the change in *x*, so the slope is $-\dfrac{6}{1}$ or −6.

**29.** $m = 4$; $(0, -3)$ The *x*-values increase by 1 unit, the *y*-values increase by 4 units or the *x*-values decrease by 1 unit, the *y*-values decrease by 4 units.

**30.** $m = 3$; $(0, 5)$ The *x*-values increase by 1 unit, the *y*-values increase by 3 units or the *x*-values decrease by 1 unit, the *y*-values decrease by 3 units.

**31.** $m = -2$; $(0, -1)$ The *x*-values increase by 1 unit, the *y*-values decrease by 2 units or the *x*-values decrease by 1 unit, the *y*-values increase by 2 units.

**32.** $m = -1$; $(0, -4)$ The *x*-values increase by 1 unit, the *y*-values decrease by 1 unit or the *x*-values decrease by 1 unit, the *y*-values increase by 1 unit.

**33.** $m = \dfrac{1}{2}$; $\left(0, \dfrac{3}{2}\right)$ The *x*-values increase by 2 units, the *y*-values increase by 1 unit or the *x*-values decrease by 2 units, the *y*-values decrease by 1 unit.

**34.** $m = -\dfrac{7}{3}$; $\left(0, \dfrac{1}{3}\right)$ The *x*-values increase by 3 units, the *y*-values decrease by 7 units or the *x*-values decrease by 3 units, the *y*-values increase by 7 units.

**35.** $m = 5$; $(0, 0)$ The *x*-values increase by 1 unit, the *y*-values increase by 5 units or the *x*-values decrease by 1 unit, the *y*-values decrease by 5 units.

**36.** $m = -3$; $(0, 0)$ The *x*-values increase by 1 unit, the *y*-values decrease by 3 units or the *x*-values decrease by 1 unit, the *y*-values increase by 3 units.

**37.** $y = -4x + 8$; $m = -4$; $(0, 8)$ The *x*-values increase by 1 unit, the *y*-values decrease by 4 units or the *x*-values decrease by 1 unit, the *y*-values increase by 4 units.

**38.** $y = -8x - 24$; $m = -8$; $(0, -24)$ The *x*-values increase by 1 unit, the *y*-values decrease by 8 units or the *x*-values decrease by 1 unit, the *y*-values increase by 8 units.

**39.** $y = -\dfrac{7}{4}x$; $m = -\dfrac{7}{4}$; $(0, 0)$ The *x*-values increase by 4 units, the *y*-values decrease by 7 units or the *x*-values decrease by 4 units, the *y*-values increase by 7 units.

**40.** $y = -\dfrac{9}{2}x$; $m = -\dfrac{9}{2}$; $(0, 0)$ The *x*-values increase by 2 units, the *y*-values decrease by 9 units or the *x*-values decrease by 2 units, the *y*-values increase by 9 units.

**41.** vertical line, cannot be written in slope-intercept form, slope is undefined

**42.** $y = 0x + 7$, $m = 0$, no change in the *y*-value as the *x*-value increases by 1 unit

**43.** $y = 0x - 3$, $m = 0$, no change in the *y*-value as the *x*-value increases by 1 unit

**44.** vertical line, cannot be written in slope-intercept form, slope is undefined

**45.** $y = \dfrac{4}{3}$, $m = 0$, no change in the *y*-value as the *x*-value increases by 1 unit

**46.** vertical line, cannot be written in slope-intercept form, slope is undefined

**47.** vertical line, cannot be written in slope-intercept form, slope is undefined

**48.** $m = -\dfrac{3}{7}$, $m = 0$, no change in the *y*-value as the *x*-value increases by 1 unit

**49.**

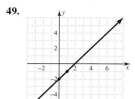

**50.**

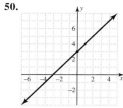

**51.**

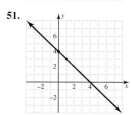

**52.**

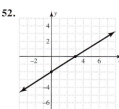

**53.**

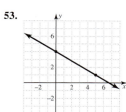

**54.**

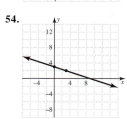

**55.**

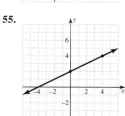

**56.**

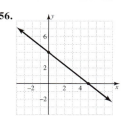

**57.**

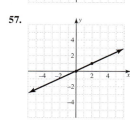

**58.**

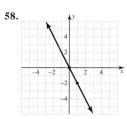

**59.**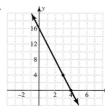

**60.**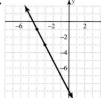

**61.**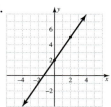

**62.**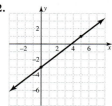

**63.**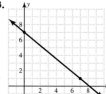

**64.**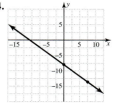

**65.**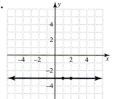

**66.**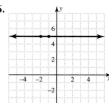

**67.**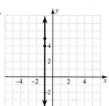

**68.**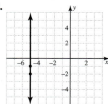

**75.** $m = -\dfrac{2}{3}$

**76.** $m = -\dfrac{5}{2}$

**77.** $m = -\dfrac{3}{2}$

**78.** $m = \dfrac{1}{2}$

**91.**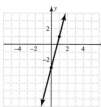

**92.**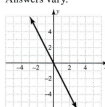

**93.**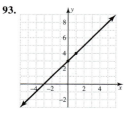

**94.**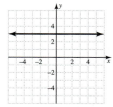

**95.**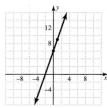

**96.**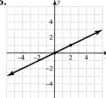

**97.**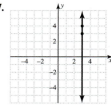

**98.**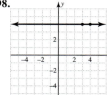

**99.**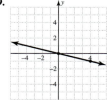

**100.**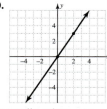

**104.** incorrect; The numerator of the slope tells us how many units to move up or down and the denominator tells us how many units to go right or left. So, we must go down 3 and right 4.

**107.** Answers vary.

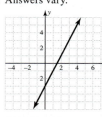

**108.** Answers vary.

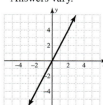

**109.** Answers vary.

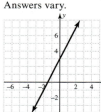

**110.** Answers vary.

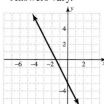

**111.** Answers vary.

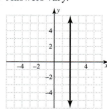

**112.** Answers vary.

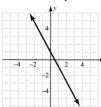

**113.** Answers vary.

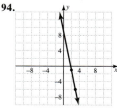

**114.** Answers vary.

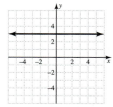

## Piece It Together  Sections 3.1–3.3

**9.**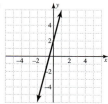
**10.**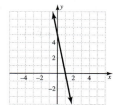

**11.**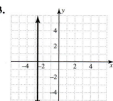
**12.**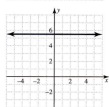

**13.**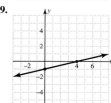
**14.**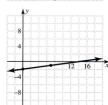

**19.**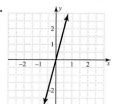
**20.**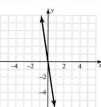

**35.**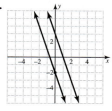
**36.**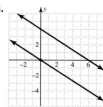

**37.**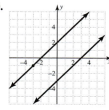
**38.**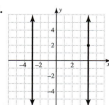

**39.**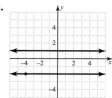
**40.**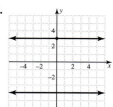

**41.**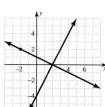
**42.**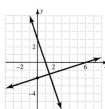

**43.**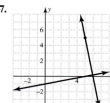
**44.**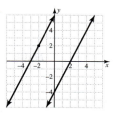

**45.**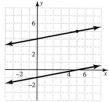
**46.**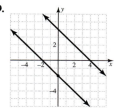

**47.**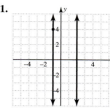
**48.**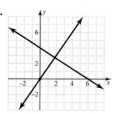

## Section 3.4

**1.** The *y*-intercept denotes the expenditure per student in fall enrollment for public elementary and secondary education was $7672.73 in 2001. The slope means that expenditures are increasing at a rate of $374.15 per year.
The ordered pairs: (4, 9169.33), (9, 11,040.08), (11, 11,788.38)

**2.** The *y*-intercept denotes there were 16,121,000 students enrolled in all degree-granting institution in 2001. The slope means that the enrollment is increasing at a rate of 351,000 per year.
The ordered pairs: (4, 17525), (9, 19280), (11, 19982)

**3.** The *y*-intercept denotes that 38% of high school students smoked in 1997. The slope means that the percent of students who are smokers is decreasing at a rate of 2.5% annually.
The ordered pairs: (8, 18), (13, 5.5), (15, 0.5)

**4.** The *y*-intercept denotes that 19.83% of U.S. adults aged 20 yr and over were obese in 1997. The slope means that the percent of U.S. adults aged 20 yr and over who were obese is increasing at a rate of 0.71% annually.
The ordered pairs: (8, 25.51), (13, 29.06), (15, 30.48)

**5.** The initial value of the limo was $110,000 and depreciates at a rate of $22,000 annually.
The ordered pairs: (2, 66,000), (4, 22,000), (5, 0)

**6.** The number of U.S. households with cable television in 1977 was $17.3 million and increases at a rate of $2.3 million annually.
The ordered pairs: (8, 81.7), (13, 93.2), (15, 97.8)

**7.** The percent of eighth-graders who reported using alcohol in 1993 was 27% and decreases at a rate of 0.7% annually.
The ordered pairs: (12, 18.6), (17, 15.1), (19, 13.7)

**8.** The cost of higher education at 2-yr institutions in 1999 was $5252 and increases at a rate of $334 annually.
The ordered pairs: (12, 7256), (17, 8926), (19, 9594)

**49.** 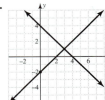    **50.**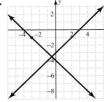

**51.** 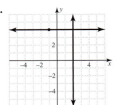    **52.**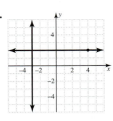

**53.** 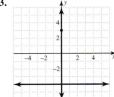    **54.**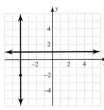

**59.** 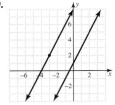    **60.**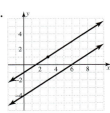

**61.** 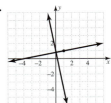    **62.**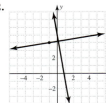

**66.** The $y$-intercept is $(0, 50)$, which means the initial cost is $50. The total cost increases at a rate of $20 per week.

**67.** The $y$-intercept is $(0, 12{,}034)$ which means the cost of higher education at 4-yr institutions was $12,034 in 1999. The cost increases at a rate of $892 per year.

**68.** The $y$-intercept is $(0, 30{,}818)$ which means the median annual income for men who completed high school was $30,818 in 1996. The income increases at a rate of $627.50 per year.

**69.** The $y$-intercept is $(0, 21{,}506)$ which means the median annual income for women who completed high school was $21,506 in 1996. The income increases at a rate of $641 per year.

**70.** The $y$-intercept is $(0, 77.1)$ which means 77.1% of America's high school seniors abused alcohol in 2003. The percent decreases at a rate of 1.13% per year.

**71.** We must write the equation in slope-intercept form first.
$$6x - 2y = 4$$
$$-2y = -6x + 4$$
$$y = \frac{-6}{-2}x + \frac{4}{-2}$$
$$y = 3x - 2$$
slope $= 3$
From the point $(1, 2)$, go up 3 units and right 1 unit.

**72.** The $y$-intercept of the perpendicular line is $(0, 5)$ and its slope is $-\frac{1}{3}$. So, to get another point on the graph, I must go down 1 unit and right 3 units. So another point on this line is $(3, 4)$.

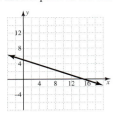

**73.** $y = -\frac{1}{2}$ is a horizontal line and $m = 0$. $y = 2x$ is not a horizontal line and $m = 2$. The lines are neither parallel nor perpendicular.

**81.** Answers vary; A parallel line is $y = 5x + 1$ and a perpendicular line is $y = -\frac{1}{5}x + 7$.

**82.** Answers vary; A parallel line is $y = -\frac{2}{3}x - 4$ and a perpendicular line is $y = \frac{3}{2}x + 2$.

**83.** Answers vary; A parallel line is $y = -\frac{4}{3}x - 1$ and a perpendicular line is $y = \frac{3}{4}x + 5$.

**84.** Answers vary; A parallel line is $y = \frac{7}{2}x + 8$ and a perpendicular line is $y = -\frac{2}{7}x + 1$.

## Section 3.5

**25.** $y = -\frac{5}{2}x - 5$    **37.** $y = -\frac{3}{5}x + \frac{1}{5}$    **47.** $y = \frac{1}{2}x - 6$

**48.** $y = \frac{3}{4}x - 9$    **49.** $y = -\frac{1}{3}x + 5$    **50.** $y = -\frac{6}{11}x + 2$

**89.** $y = \frac{5}{4}x + 5;\ -5x + 4y = 20$

**92.** $y = -\frac{1}{2}x + \frac{3}{5};\ 5x + 10y = 6$

**105.** $y = \frac{4}{3}x - 2$. So, the slope is $\frac{4}{3}$. Since it passes through the point $(8, 3)$, the equation of the line is $y = \frac{4}{3}x - \frac{23}{3}$.

**106.** $m = \dfrac{\frac{1}{2} - 4}{0 + 3} = \dfrac{-\frac{7}{2}}{3} = -\dfrac{7}{6}$
$$y = -\frac{7}{6}x + \frac{1}{2}$$

## Section 3.6

**13.** Relation = {(Canada, 4400), (France, 4800), (Italy, 3300), (Japan, 3300), (Republic of Korea, 1100), (United Kingdom, 3300), (United States, 7400)}; Domain = {Canada, France, Italy, Japan, Republic of Korea, United Kingdom, United States}; Range = {1100, 3300, 4400, 4800, 7400}

**14.** Relation = {(Atlanta, 4), (Bakersfield, 26), (Baltimore, 4), (Detroit, 1), (Houston, 2), (Los Angeles, 28), (Memphis, 1), (New Orleans, 0), (New York, 1), (Orange County, 1), (Phoenix, 84), (Sacramento, 20), (Washington, 3)}; Domain = {Atlanta, Bakersfield, Baltimore, Detroit, Houston, Los Angeles, Memphis, New Orleans, New York, Orange County, Phoenix, Sacramento, Washington}; Range = {0, 1, 2, 3, 4, 20, 26, 28, 84}

**15.** Relation = {(Walter Hagen, 11), (Ben Hogan, 9), (Nick Faldo, 6), (Jack Nicklaus, 18), (Arnold Palmer, 7), (Gary Player, 9), (Gene Sarazen, 7), (Sam Snead, 7), (Lee Trevino, 6), (Harry Vardon, 7), (Tom Watson, 8), (Tiger Woods, 14)}; Domain = {Walter Hagen, Ben Hogan, Nick Faldo, Jack Nicklaus, Arnold Palmer, Gary Player, Gene Sarazen, Sam Snead, Lee Trevino, Harry Vardon, Tom Watson, Tiger Woods}; Range = {6, 7, 8, 9, 11, 14, 18}

**16.** Relation = {(1996, Buzz Calkins), (1996, Scott Sharp), (1997, Tony Stewart), (1998, Kenny Brack), (1999, Greg Ray), (2000, Buddy Lazier), (2001, Sam Hornish Jr.), (2002, Sam Hornish Jr.), (2003, Scott Dixon), (2004, Tony Kanaan), (2005, Dan Wheldon), (2006, Sam Hornish Jr.), (2007, Dario Franchitti), (2008, Scott Dixon)}; Domain = {1996, 1997, 1998, 1999, 2000, 2001, 2002, 2003, 2004, 2005, 2006, 2007, 2008};
Range = {Kenny Brack, Buzz Calkins, Scott Dixon, Dario Franchitti, Sam Hornish Jr., Tony Kanaan, Buddy Lazier, Greg Ray, Scott Sharp, Tony Stewart, Dan Wheldon}

**17.** The cost $y = 14.52x$, where $x$ is the number of kilowatt-hours in hundreds; Domain = {0, 1, 2, 3, . . .}; Range = {0, 14.52, 29.04, 43.56, . . .}

**18.** The cost $y = 15.09x$, where $x$ is the number of kilowatt-hours in hundreds; Domain = {0, 1, 2, 3, . . .}; Range = {0, 15.09, 30.18, 45.27, . . .}

**59.** $f(x) = 5x - 6$;
$f(0) = -6, f(3) = 9, f(-2) = -16$;
$(0, -6), (3, 9), (-2, -16)$

**60.** $f(x) = -7x + 3$;
$f(0) = 3, f(3) = -18, f(-2) = 17$;
$(0, 3), (3, -18), (-2, 17)$

**61.** $f(x) = 4x - 9$;
$f(0) = -9, f(3) = 3, f(-2) = -17$;
$(0, -9), (3, 3), (-2, -17)$

**62.** $f(x) = -\dfrac{3}{5}x + \dfrac{6}{5}$;
$f(0) = \dfrac{6}{5}, f(3) = -\dfrac{3}{5}, f(-2) = \dfrac{12}{5}$;
$\left(0, \dfrac{6}{5}\right), \left(3, -\dfrac{3}{5}\right), \left(-2, \dfrac{12}{5}\right)$

**63.** $f(x) = \dfrac{7}{2}$;
$f(0) = \dfrac{7}{2}, f(3) = \dfrac{7}{2}, f(-2) = \dfrac{7}{2}$;
$\left(0, \dfrac{7}{2},\right), \left(3, \dfrac{7}{2}\right), \left(-2, \dfrac{7}{2}\right)$

**64.** $f(x) = -\dfrac{3}{2}$;
$f(0) = -\dfrac{3}{2}, f(3) = -\dfrac{3}{2}, f(-2) = -\dfrac{3}{2}$;
$\left(0, -\dfrac{3}{2}\right), \left(3, -\dfrac{3}{2}\right), \left(-2, -\dfrac{3}{2}\right)$

**69.** Relation = {(Canada, 4400), (France, 4800), (Italy, 3300), (Japan, 3300), (Republic of Korea, 1100), (United Kingdom, 3300), (United States, 7400)}; Domain = {Canada, France, Italy, Japan, Republic of Korea, United Kingdom, United States}; Range = {1100, 3300, 4400, 4800, 7400}

**97.** $f(x) = \dfrac{9}{2}x - 5$;
$f(0) = -5, f(3) = \dfrac{17}{2}, f(-2) = -14$;
$(0, -5), \left(3, \dfrac{17}{2}\right), (-2, -14)$

**98.** $f(x) = -\dfrac{1}{2}x + \dfrac{2}{7}$;
$f(0) = \dfrac{2}{7}, f(3) = -\dfrac{17}{14}, f(-2) = \dfrac{9}{7}$;
$\left(0, \dfrac{2}{7}\right), \left(3, -\dfrac{17}{14}\right), \left(-2, \dfrac{9}{7}\right)$

## CHAPTER 3 REVIEW EXERCISES

### Section 3.1

**7.–12.**

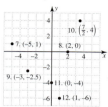

**23.**

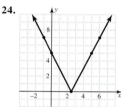

**24.**

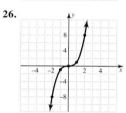

**25.**

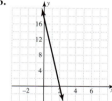

**26.**

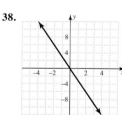

**35. a.** (0, 1016.5), (5, 1333.6), (8, 1732.4), (9, 1852.3), (10, 1973.3), (11, 2105.5)
   **b.** The expenditures increased the least between 1995 and 2000, by \$337.1 billon in 5 yr.
   **c.** The expenditures for national health are increasing.

### Section 3.2

**36.**

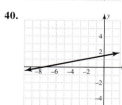

**37.**

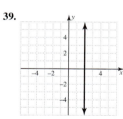

**38.**

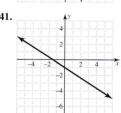

**39.**

**40.**

**41.**

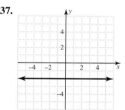

**42.**

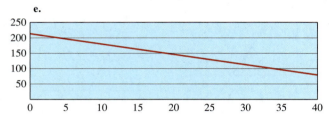

**43. a.** The $y$-intercept is $(0, 213)$. The emission of carbon monoxide was about 213 metric tons in 1970.

**b.** The $x$-intercept is $(63, 0)$. In 2033, the emission of carbon monoxide is 0 metric tons. No, it is not realistic.

**e.**

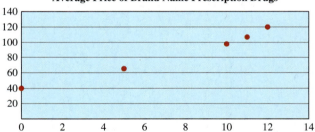

## Section 3.3

**51.**

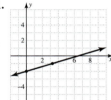

**52.**

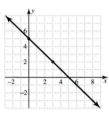

**53.**

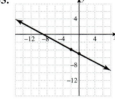

**54.**

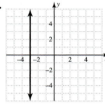

**55.**

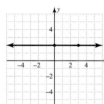

**56.**

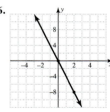

## Section 3.4

**57.** The $y$-intercept is $(0, 0)$, which means the initial cost is $0. The total cost increases at a rate of $135 per week.

**58.** The $y$-intercept is $(0, 23{,}507)$, which means the per capital personal income was $23,507 in 1995. The income increases at a rate of $1155.60 per year.

**61.**

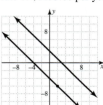

**62.**

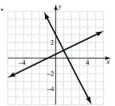

## Section 3.6

**77.** This is not a function since $(1, 0)$ and $(1, 2)$, to name two points, both satisfy the equation. So, $x = 1$ corresponds to more than one $y$-value.

**87.** $f(x) = \dfrac{2}{7}x - 3$;

$f(0) = -3, f(3) = -\dfrac{15}{7}, f(-2) = -\dfrac{25}{7}$;

$(0, -3), \left(3, -\dfrac{15}{7}\right), \left(-2, -\dfrac{25}{7}\right)$

**88.** $f(x) = |4x + 1|$;

$f(0) = 1, f(3) = 13, f(-2) = 7$;

$(0, 1), (3, 13), (-2, 7)$

## CHAPTER 3 TEST

**3.**

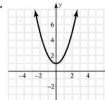

**4.**

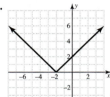

**6. b.** In 1995, the average price of brand name prescription drugs was approximately $40. In 2007, the average price of brand name prescription drugs was approximately $120.

**c.**

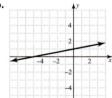

**7. a.**

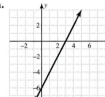

**b.**

**c.**

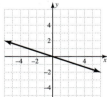

**10. b.** The $x$-intercept is $(20, 0)$. This means that it will take 20 min for the plane to reach an elevation of 0 ft, or to reach the ground.

**c.** The $y$-intercept is $(0, 8000)$. This means that the initial elevation of the plane was 8000 ft above the ground.

**12.**

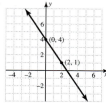

**14. b.** The average undergraduate cost of attending a public 2-yr college was $2961.80 in 1986. The cost is increasing by $138.41 each year after 1986.

**20.** Answers vary. A relation is a function if every $x$-value in the domain of the relation corresponds to exactly one output value. A real-life example is the set of ordered pairs whose $x$-value is student and whose $y$-value is student ID number.

# CHAPTERS 1–3 CUMULATIVE REVIEW EXERCISES

**25. c.** $y = \dfrac{5}{9}x - 2$         **d.** $x = -\dfrac{3}{4}y + 180$

**32. a.**

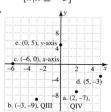

$[-8, \infty)$
$\{a \mid a \geq -8\}$

**b.**

$(4, 17)$
$\{b \mid 4 < b < 17\}$

**c.**

$(-\infty, -5]$
$\{x \mid x \leq -5\}$

**34.**

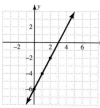
e. (0, 5), $y$-axis
c. (−6, 0), $x$-axis
d. (5, −3)
a. (2, −7), QIV
b. (−3, −9), QIII

**35. a.**

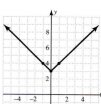

**b.**

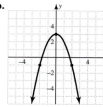

**c.**

**d.**

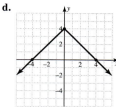

**38. a.** (0, 1.23), (1, −1.21), (2, −1.27), (3, −0.773), (4, −5.2), (5, −1.46), (6, 0.058).
   **b.** Annual profit was $1.23 billion in 2000 and $0.058 billion in 2006.
   **c.** In 2000, the profit was the highest and in 2004, the profit was the lowest.
   **d.**

**Annual Profits for Delta Airlines, in Billions (2000–2006)**

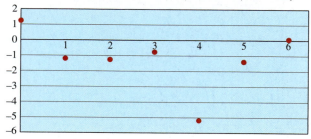

**39. a.**

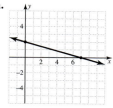

**b.**

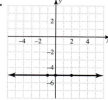

**c.**

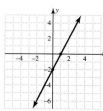

**d.**

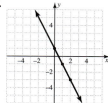

**e.**

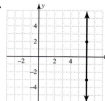

**f.**

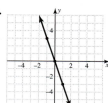

**g.**

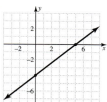

**40. a.** (0, 3649.1); The number of full-time classroom teachers in elementary and secondary schools in the United States was approximately 3,649,100 in 2008.

**b.** There were approximately 3,714,100 full-time classroom teachers in elementary and secondary schools in 2010.

**c.** There were approximately 3,844,100 full-time classroom teachers in elementary and secondary schools in 2014.

**d.**

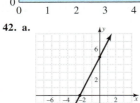

**42. a.**

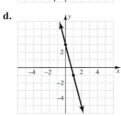

**b.**

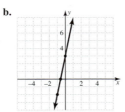

**c.**

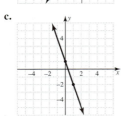

**d.**

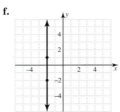

**e.**

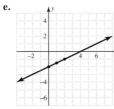

**f.**

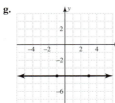

**g.**

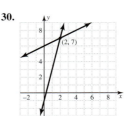

**43.** $m = 10.93$, $(0, 73764)$; The number of associate degrees conferred in all higher education institutions is increasing at a rate of 10,930 per year. In 2008, the number of associate degrees conferred was approximately 73,764,000.

| $x$ | $y$ (in thousands) |
|---|---|
| 0 | 73,764 |
| 4 | 73,807.72 |
| 8 | 73,851.44 |

**45.**

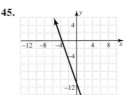

**46.**

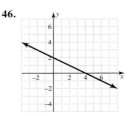

**47. e.** $y = 3x - 14$; $3x - y = 14$

    **f.** $y = \dfrac{3}{4}x + 3$; $3x - 4y = -12$

**49. a.** D = {Italy, France, Germany, Brazil, United Kingdom, Canada, Korea, Japan, U.S.}, R = {13, 25, 26, 28, 34, 35, 37, 42}, The relation is a function.

    **b.** D = {Honda Civic CX, Toyota Prius, Honda Civic Hybrid, Smart For Two Convertible/Coupe, Honda Insight, Ford Fusion Hybrid/mercury Milan Hybrid, Toyota Yaris, Nissan Altima Hybrid, Mini Cooper, Chevrolet Cobalt XFE/Pontiac G5 XFE}, R = {45, 46, 47, 50, 51, 52, 57}, The relation is a function.

    **c.** D = {−5, −3, 0, 2}
    R = {1, 4, 9}, The relation is not a function.

**51. a.** $f(x) = -4x + 3$;
    $f(0) = 3, f(3) = -9$,
    $f(-2) = 11$; $(0, 3)$,
    $(3, -9), (-2, 11)$

    **b.** $f(x) = 0.3x - 0.6$;
    $f(0) = -0.6, f(3) = 0.3$,
    $f(-2) = -1.2$; $(0, -0.6)$,
    $(3, 0.3), (-2, -1.2)$

    **c.** $f(x) = |3x - 2|$;
    $f(0) = 2, f(3) = 7$,
    $f(-2) = 8$; $(0, 2), (3, 7)$,
    $(-2, 8)$

# CHAPTER 4

## Section 4.1

**27.**

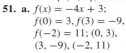

**28.**

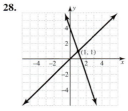

**29.**

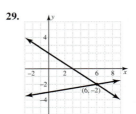

**30.**

**31.**

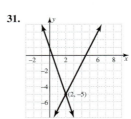

**32.**

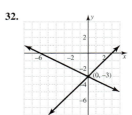

**33.**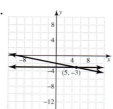

**34.**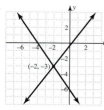

**35.**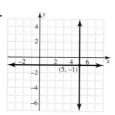

**36.**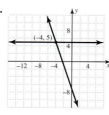

**37.**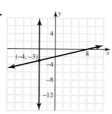

**38.**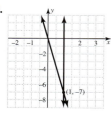

**39.**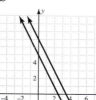

**40.**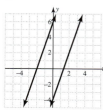

**41.** ∅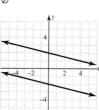

**42.** ∅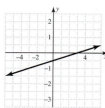

**43.** ∅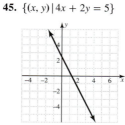

**44.** $\{(x, y) \mid x - 6y = 3\}$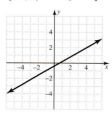

**45.** $\{(x, y) \mid 4x + 2y = 5\}$

**46.** $\{(x, y) \mid 3x - 5y = 2\}$

**49.** same line, infinitely many solutions, consistent system with dependent equations

**52.** same line, infinitely many solutions, consistent system with dependent equations

**53.** same line, infinitely many solutions, consistent system with dependent equations

**55.** same line, infinitely many solutions, consistent system with dependent equations

**59.** $\left\{\left(-\frac{1}{2}, -8\right)\right\}$, consistent system with independent equations

**60.** ∅, inconsistent system

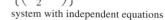

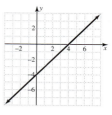
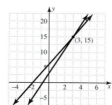
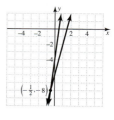

**61.** $\{(x, y) \mid x - y = 4\}$, consistent system with dependent equations

**62.** $\{(3, 15)\}$, consistent system with independent equations

**63.** $\{(1, -2)\}$, consistent system with independent equations

**64.** $\{(-4, -8)\}$, consistent system with independent equations

**65.** $\{(-1, -3)\}$, consistent system with independent equations

**66.** $\{(-2, -6)\}$, consistent system with independent equations

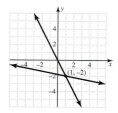
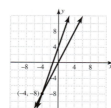
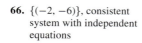
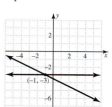
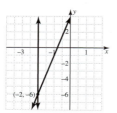
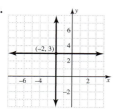
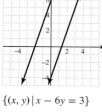
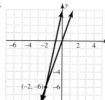

**67.** $\{(-1, -1)\}$, consistent system with independent equations

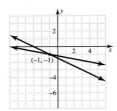

**68.** $\{(-1, 0)\}$, consistent system with independent equations

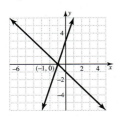

**69.** $\{(5, -1)\}$, consistent system with independent equations

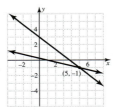

**70.** $\left\{\left(10, \dfrac{9}{2}\right)\right\}$, consistent system with independent equations

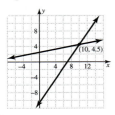

**71.** $\{(x, y) \mid 2x - 3y = 1\}$, consistent system with dependent equations

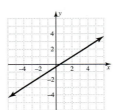

**72.** $\varnothing$, inconsistent system

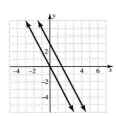

**73.** $\left\{\left(-3, \dfrac{9}{2}\right)\right\}$, consistent system with independent equations

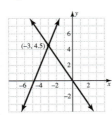

**74.** $\left\{\left(7, \dfrac{34}{5}\right)\right\}$, consistent system with independent equations

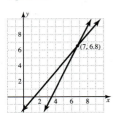

**75.** $\varnothing$, inconsistent system

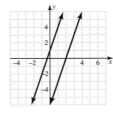

**76.** $\{(3, 4)\}$, consistent system with independent equations

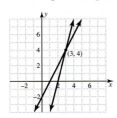

**77. b.** Both plans will have the same cost of $90 if Johanne expects to use 350 min.
  **c.** She should use the InTouch Cell plan because it will cost her $70.
**78. b.** The cost of shipping a 25-lb package will be the same by both companies at $75.
  **c.** She should use ShipItQuick to mail a 40-lb package because it will cost her $90.
**80.** After solving each equation for $y$, we must graph the equations on the same coordinate system.

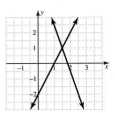

**86.** Answers vary, $\begin{cases} x + y = 1 \\ x - y = 5 \end{cases}$.

**87.** Answers vary, $\begin{cases} x + y = -3 \\ x - y = -5 \end{cases}$.

**88.** Answers vary, $\begin{cases} 2x + 3y = 0 \\ x - y = 0 \end{cases}$.

**89.** Answers vary, $\begin{cases} 2x + y = 4 \\ 4x - 3y = -7 \end{cases}$.

**90.** Answers vary, $y = \dfrac{1}{3}x + 2$.

**91.** Answers vary, $y = -\dfrac{5}{2}x - 1$.

**92.** Answers vary, $2x - 8y = 16$.

**93.** Answers vary, $-9x - 15y = -6$.

## Section 4.2

**19.** $\left\{\left(\dfrac{13}{15}, -\dfrac{2}{3}\right)\right\}$    **21.** $\left\{\left(-\dfrac{11}{10}, \dfrac{1}{5}\right)\right\}$

**24.** $\{(x, y) \mid x = -5y + 10\}$    **25.** $\{(x, y) \mid -x + 2y = -3\}$

**28.** $\{(x, y) \mid 3x + 6y = 3\}$    **29.** $\{(x, y) \mid x - 3y = 4\}$

**31.** The system has no solution and the lines are parallel. The system is inconsistent. The solution set is the empty set, $\varnothing$.
**32.** The system has no solution and the lines are parallel. The system is inconsistent. The solution set is the empty set, $\varnothing$.
**33.** The equations give the same line and there are infinitely many solutions. The system is consistent with dependent solutions. The solution set is $\{(x, y) \mid 2x = y + 1\}$.
**34.** The equations give the same line and there are infinitely many solutions. The system is consistent with dependent solutions. The solution set is $\{(x, y) \mid 2x = 3y + 4\}$.
**35.** The system contains intersecting lines with one solution. The system is consistent with independent equations. The solution set is $\{(6, -3)\}$.
**36.** The system contains intersecting lines with one solution. The system is consistent with independent equations. The solution set is $\{(-6, 1)\}$.
**41.** $\{(x, y) \mid x = -6y + 2\}$    **42.** $\{(x, y) \mid 2x + 6y = 12\}$
**51.** When the variables cancel out, there is either no solution or infinitely many solutions. In this case, the remaining statement is false, so there is no solution.
**52.** When this happens, the lines are the same. So, there are infinitely many solutions.

**53.** $4x - 2y = 10 \Rightarrow -2y = -4x + 10 \Rightarrow y = \dfrac{-4x + 10}{2} = 2x - 5$

$3x + 2(2x - 5) = 4 \Rightarrow 3x + 4x - 10 = 4 \Rightarrow 7x = 14 \Rightarrow x = 2$

$y = 2(2) - 5 = 4 - 5 = -1$

So, the answer is $\{(2, -1)\}$.

**54.** $2x - y = 3 \Rightarrow -y = -2x + 3 \Rightarrow y = 2x - 3$

$3x + 2(2x - 3) = 1 \Rightarrow 3x + 4x - 6 = 1 \Rightarrow 7x = 7 \Rightarrow x = 1$

$y = 2(1) - 3 = 2 - 3 = -1$

So, the answer is $\{(1, -1)\}$.

**59.** Answers vary, $2x - y = -4$.

**60.** Answers vary, $5x + y = -11$.

**61.** Answers vary though $m = -5$ and $b \neq 3$; $y = -5x - 6$.

**62.** Answers vary though $m = \dfrac{1}{7}$ and $b \neq \dfrac{1}{7}$; $y = \dfrac{1}{7}x + 2$.

**63.** $2x + y = 1$    **64.** $x - 3y = 5$

**65.** Answers vary, $4x + 2y = 12$.

**66.** Answers vary, $3x - 9y = 12$.

## Section 4.3

**7.** $\left\{ \left( -4, -\dfrac{1}{2} \right) \right\}$    **9.** $\left\{ \left( \dfrac{1}{2}, -3 \right) \right\}$    **11.** $\left\{ \left( \dfrac{2}{5}, \dfrac{6}{5} \right) \right\}$

**16.** $\{(x, y) \,|\, -3x + 7y = -4\}$

**17.** $\{(x, y) \,|\, \{5x - 8y = 10\}$

**19.** $\{(x, y) \,|\, \{6x + 18y = 20\}$

**23.** The equations give the same line and there are infinitely many solutions. The system is consistent with dependent solutions. The solution set is $\{(x, y) \,|\, \{7x - 2y = 14\}$.

**24.** The equations give the same line and there are infinitely many solutions. The system is consistent with dependent solutions. The solution set is $\{(x, y) \,|\, 2x + 5y = 12\}$.

**25.** The system has no solution and the lines are parallel. The system is inconsistent. The solution set is the empty set, $\varnothing$.

**26.** The system has no solution and the lines are parallel. The system is inconsistent. The solution set is the empty set, $\varnothing$.

**27.** The system contains intersecting lines with one solution. The system is consistent with independent equations. The solution set is $\{(-2, 1)\}$.

**28.** The system contains intersecting lines with one solution. The system is consistent with independent equations. The solution set is $\left\{ \left( 4, -\dfrac{7}{5} \right) \right\}$.

**29.** Mattel reported net sales of $5.918 billion and Hasbro reported net sales of $4.021 billion.

**49.** It doesn't matter which variable you choose to eliminate. If you choose to eliminate $x$, you must find the number that both 3 and 5 divide into evenly. If you choose to eliminate $y$, you must find the number that both 4 and 3 divide into. In either case, we will multiply both equations by a nonzero number.

**50.** Andy made a mistake by not multiplying the right side of the first equation by $-3$. The answer should be

$\begin{cases} -3(2x - 5y) = -3(6) \\ 6x - 2y = 1 \end{cases}$

$\begin{cases} -6x + 15y = -18 \\ \underline{\quad 6x - 2y = 1 \quad} \\ \phantom{aaaa}13y = -17 \\ \phantom{aaaa}y = -\dfrac{17}{13} \end{cases}$

$2x - 5y = 6$

$2x - 5\left( -\dfrac{17}{13} \right) = 6$

$2x + \dfrac{85}{13} = 6$

$2x = 6 - \dfrac{85}{13}$

$2x = -\dfrac{7}{13}$

$x = -\dfrac{7}{26}$

So, the answer is $\left\{ \left( -\dfrac{7}{26}, -\dfrac{17}{13} \right) \right\}$.

**51.** We can solve for $x$ by dividing both sides by 2. So, the substitution method is better to use to solve this system.

**52.** Since none of the variables have coefficients of 1 or $-1$ and because each equation is in standard form, it is best to use the elimination method to solve this system.

**53.** We must find the number that both 12 and 14 divide into evenly. We can write the prime factorization of each number to get $12 = 2 \cdot 2 \cdot 2 \cdot 3$ and $14 = 2 \cdot 7$. So, the number that both 12 and 14 divide into is $2 \cdot 2 \cdot 3 \cdot 7 = 84$. We multiply the first equation by $\dfrac{-84}{-12} = 7$ and the second equation by $\dfrac{84}{14} = 6$.

**54.** Yes, you will get the same solution using either 24 or 48. As long as your new system was obtained by multiplying both sides of each equation by the appropriate number, your result will be the same.

**58.** $\left\{ \left( \dfrac{-147}{15}, -12 \right) \right\}$

## Piece It Together  Sections 4.1–4.3

**3.**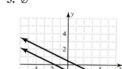

**4.**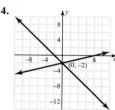

**5.** $\varnothing$    **6.** $\{(x, y) \,|\, 3x - y = 1\}$

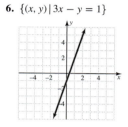

## Section 4.4

**2.** Answers vary. In this section, we assign two variables to the two unknowns. In Section 2.6, we assigned one variable to one unknown and then wrote the second unknown in terms of this same variable.

**4.** Answers vary. $450 - x$, because flying against the wind decreases the speed of the plane by the speed of the wind. This is translated as $450 - x$.

**53.** Answers vary. It is important to be able to explain how you derived your answer to someone else so that they can replicate your process.

**54.** Answers vary. For one equation, you will have the sum of the amounts of liquid being mixed together equal to the total amount of liquid desired. The second equation will represent the amount of pure liquid (e.g., milk fat, acid, gasoline, etc.) in each quantity.

**55.**
$$x + y = 10{,}650$$
$$0.011x + 0.045y = 248.05$$
$$x = 10{,}650 - y \Rightarrow 0.011(10{,}650 - y) + 0.045y = 248.05$$
$$117.15 - 0.011y + 0.045y = 248.05$$
$$0.034y = 130.9$$
$$y = 3850$$
$$x = 10{,}650 - 3850 = 6800$$
Jim invests $6800 at 1.1% and $3850 at 4.5%.

**56.**
$$x + y = 266$$
$$1.00x + 0.16y = 0.46(266)$$
$$x = 266 - y \Rightarrow (266 - y) + 0.16y = 122.36$$
$$266 - y + 0.16y = 122.36$$
$$-0.84y = -143.64$$
$$y = 171$$
$$x = 266 - 171 = 95$$
Todd needs 95 L pure antifreeze and 171 L of 16% antifreeze.

## Section 4.5

1. Answers vary. Substitute the coordinates in for $x$ and $y$. If a true inequality results, then the ordered pair is a solution. If a false inequality results, then the ordered pair is not a solution.
2. Answers vary. Replace the inequality with an equal sign. The graph of this equation is the boundary line.
3. Answers vary. If the inequality is strictly greater than or less than, then the boundary line with be dotted. If the inequality is greater than or equal to or less than or equal to, then the boundary line will be solid.
4. Answers vary. Just one point in a half-plane must be tested to determine if it includes solutions of a linear inequality.

**19.**
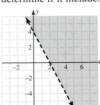

**20.**
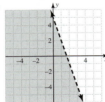

**21.**
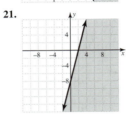

**22.**
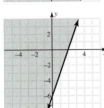

**23.**
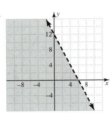

**24.**
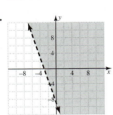

**25.**
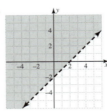

**26.**
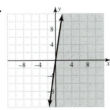

**27.**

**28.**
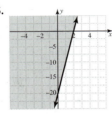

**29.**
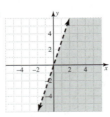

**30.**
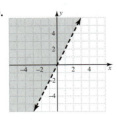

**31.**
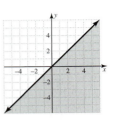

**32.**
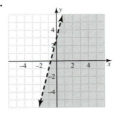

**33.**
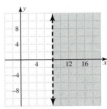

**34.**
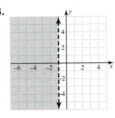

**35.**
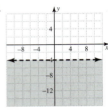

**36.**
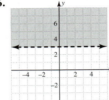

**37. a.** $0.6x + 0.4y \geq 80$

**b.**
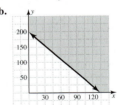

**c.** Answers vary. Three possible combinations are (80, 90), (90, 70), and (80, 80).

**38. a.** $0.6x + 0.4y \geq 70$

**b.**
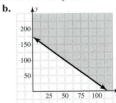

**c.** Answers vary. Three possible combinations are (70, 80), (80, 70), and (100, 60).

**39. a.** $8x + 10.50y \geq 300$

**b.**

**c.** Answers vary. bookstore: 10 hr & library: 25 hr, bookstore: 12 hr & library: 20 hr, bookstore: 20 hr & library: 15 hr

**40. a.** $8x + 10.50y \geq 450$
**b.**
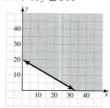

**c.** Answers vary. bookstore: 20 hr & library: 28 hr, bookstore: 25 hr & library: 24 hr, bookstore: 12 hr & library: 34 hr

**41. a.** $15x + 25y \geq 500$
**b.**
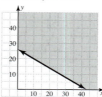

**c.** Answers vary. 20 pants & 15 jackets, 5 pants & 20 jackets, 15 pants & 15 jackets

**42. a.** $15x + 25y \geq 650$
**b.**
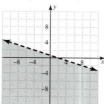

**c.** Answers vary. 20 pants & 15 jackets, 5 pants & 24 jackets, 15 pants & 18 jackets

**53.**
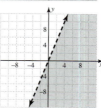

**54.**
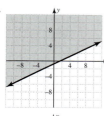

**55.**
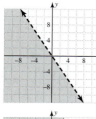

**56.**
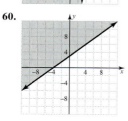

**57.**
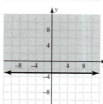

**58.**
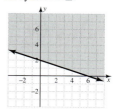

**59.**

**60.**

**61. a.** $0.75x + 0.25y \geq 90$
**b.**
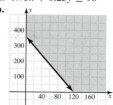

**c.** Answers vary. Three possible combinations are (95, 80), (88, 98), and (90, 90).

**62. a.** $0.75x + 0.25y \geq 80$
**b.**
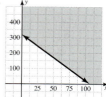

**c.** Answers vary. Three possible combinations are (80, 90), (78, 90), and (90, 70).

**63. a.** $10.50x + 8.5y \geq 350$
**b.**
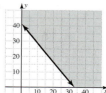

**c.** Answers vary. grocery store: 24 hr & school: 12 hr, grocery store: 22 hr & school: 14 hr, grocery store: 20 hr & school: 17 hr

**64. a.** $10.50x + 8.5y \geq 450$
**b.**
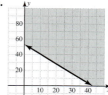

**c.** Answers vary. grocery store: 25 hr & school: 25 hr, grocery store: 30 hr & school: 20 hr, grocery store: 24 hr & school: 27 hr

**65.** Jamie shaded the correct half-plane and has the correct boundary line. The mistake was made in drawing the boundary line solid. It should be dashed since the inequality is $<$.

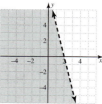

**66.** Vince shaded the correct half-plane and has the correct boundary line. The mistake is that the boundary line should be solid since the symbol is $\geq$.

**67.** Since $y$ is not isolated on the left side of the inequality, Carolyn cannot shade based on the inequality symbol. In this form, we must use a test point. We use (4, 2) since the boundary line goes through (0, 0).

$$4 - 4(2) > 0$$
$$4 - 8 > 0$$
$$-4 > 0$$
False

We shade the half-plane that doesn't contain (4, 2).

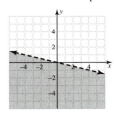

**68.** Since $y$ is not isolated on the left side of the inequality, Brent cannot shade based on the inequality symbol. We use the point (3, 1) since the boundary line goes through (0, 0).

$$2(3) \le 1$$
$$6 \le 1$$

Since this is false, we shade the half-plane that doesn't contain (3, 1).

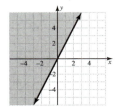

**69.**

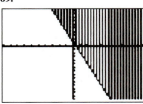

**70.**

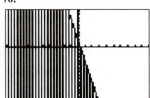

**71.**

**72.**

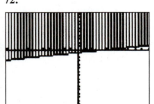

**73.** Answers vary, $x + y > 0$ or any inequality of the form $Ax + By > 0$, $Ax + By \ge 0$, $Ax + By < 0$, and $Ax + By \le 0$.

**74.** Answers vary. No, the solution set will be a half-plane and there are infinitely many points in a half-plane.

**75.** Answers vary. No, if the boundary line is part of the solution, it will be in addition to a half-plane.

**76. a.**

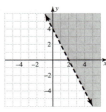

**b.**

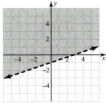

**c.**

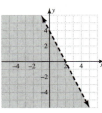

**d.**

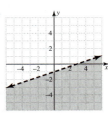

**e.** upper half-plane

**f.** The solution set will always be the upper half-plane.

**g.** Both will have the upper half-plane as solutions but $y \ge mx + b$ will have a solid boundary line and $y > mx + b$ will have a dashed boundary line.

**77. a.**

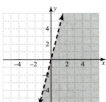

**b.**

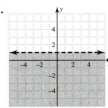

**c.**

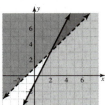

**d.**

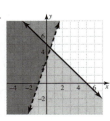

**e.** lower half-plane

**f.** The solution set will always be the lower half-plane.

**g.** Both will have the lower half-plane as solutions but $y \le mx + b$ will have a solid boundary line and $y < mx + b$ will have a dashed boundary line.

## Section 4.6

**1.** Answers vary. First graph each linear inequality. The overlapping portions of the graphs represent the solution of the system of linear inequalities.

**2.** Answers vary. If the only difference between the inequalities is that one is $\ge$ and the other is $\le$, then graphs of the inequalities will intersect at the boundary line.

**3.** Answers vary. Graph the system and then choose any two points that lie in the solution space. Two points that are in the solution set of this system are (3, 4) and (0, 5).

**4.** Answers vary. Graph the system and then choose any two points that lie in the solution space. Two points that are in the solution set of this system are (11, 6) and (12, 5).

**5.**

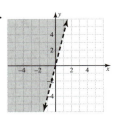

**6.**

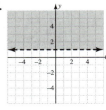

**7.**

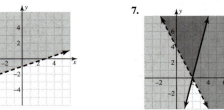

**8.**

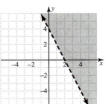

**9.**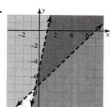

**10.**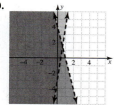

**23.**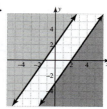

**24.**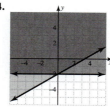

**11.**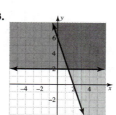

**12.**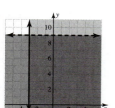

**25.**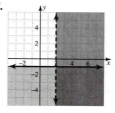

**26.**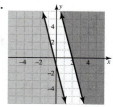

**13.**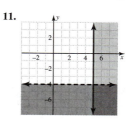

**14.**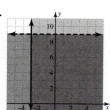

**27. a.** $x + y \leq 1200$; $1.5x + 4y \leq 4200$; $y \geq 2x$

**b.**

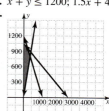

**c.** Answers vary. (0, 1050), (400, 800), (240, 960)

**15.**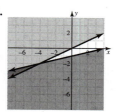

**16.**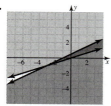

**28. a.** $x + y \leq 750$, $2x + 6.5y \leq 5100$, $y \geq 2x$;

**b.**

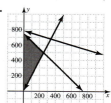

**c.** 250 rolls and 500 boxes; 0 rolls and 750 boxes; 150 rolls and 320 boxes (Answers vary.)

**17.**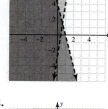

**18.**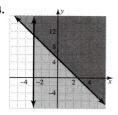

**29. a.**    **b.** 1000

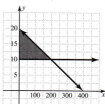

**19.**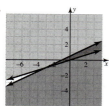

**20.**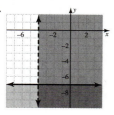

**30. a.**    **b.** 562.5

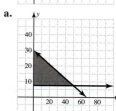

**21.**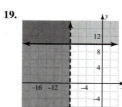

**22.**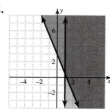

**31.**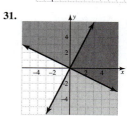

**32.**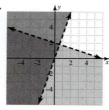

**33.**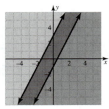

**34.**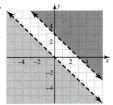

**35.**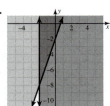

**36.**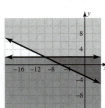

**37.**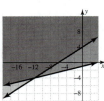

**38.**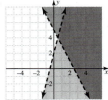

**39.**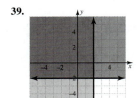

**40.**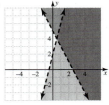

**41. a.**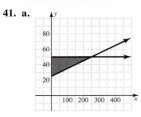

**b.** 3125

**42. a.**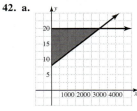

**b.** 18,000

**43.**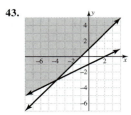

**44.**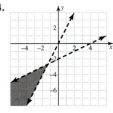

**45.**

**46.**

**47.**

**48.**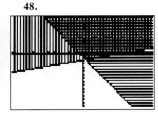

**49.** Answers vary, $x > 3$ and $x < 3$.
**50.** Answers vary, $y \geq 1$ and $y \leq 1$.

## CHAPTER 4 GROUP ACTIVITY

**1.**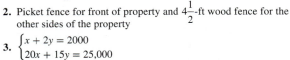

**2.** Picket fence for front of property and $4\frac{1}{2}$-ft wood fence for the other sides of the property

**3.** $\begin{cases} x + 2y = 2000 \\ 20x + 15y = 25{,}000 \end{cases}$

**4.** $x = 800$ and $y = 600$

**5.** If Mike and Amy choose these materials, the enclosed property would be 800 ft by 600 ft.

**6.** The area of the enclosed property would be 480,000 ft$^2$.

**7.** Option 2: Vinyl fencing in front and chain link on other sides

System: $\begin{cases} x + 2y = 2000 \\ 30x + 6y = 25{,}000 \end{cases}$

Solution: $x = 703.7$ and $y = 648.15$
Area: 456,103.16 ft$^2$

Option 3: $4\frac{1}{2}$-ft wood fence in front and privacy fence on other two sides

System: $\begin{cases} x + 2y = 2000 \\ 15x + 11y = 25{,}000 \end{cases}$

Solution: $x = 1473.68$ and $y = 263.16$
Area: 387,813.63 ft$^2$

**8.** The design that yields the largest enclosed area is the first option, picket fence in the front and the $4\frac{1}{2}$-ft wood fence on the other two sides.

## CHAPTER 4 REVIEW EXERCISES

### Section 4.1

**1.** consistent system with independent equations

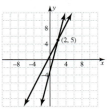

**2.** infinitely many solutions, consistent system with dependent equations

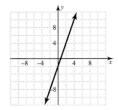

**3.** consistent system with independent equations

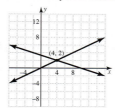

**4.** consistent system with independent equations

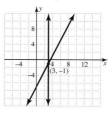

**5.** consistent system with independent equations

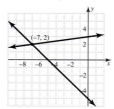

**6.** inconsistent system

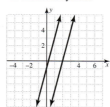

**7. a.** $(150, 37.5)$
   **b.** The cost of 150 min of long distance calls will be the same for both plans at \$37.50.
   **c.** The $x$-value 200 is to the right of the intersection point. The Home Phone Plus graph (magenta) is lower than the Qwest (blue) for this $x$-value. So the cost for Home Phone Plus is less than the cost for Qwest. He should use Qwest.

## Section 4.5

**35.** $x - 2y \geq 3$

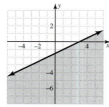

**36.** $6 - 2y > 0$

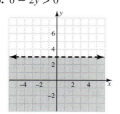

**37.** $x + 3 \leq 1$

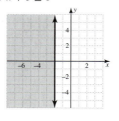

**38.** $0.1x + 0.4y > 1.6$

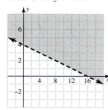

**39.** $0.8x + 0.2y \geq 90$

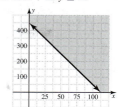

**40.** $11.5x + 8.25y \geq 450$

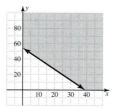

## Section 4.6

**41.**

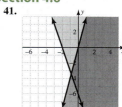

**42.**

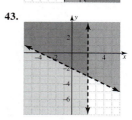

**43.**

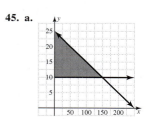

**44.**

**45. a.**

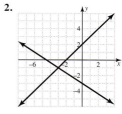

   **b.** 1125

## CHAPTER 4 TEST

**1.** b

**2.**
The solution is $\{(-3, -1)\}$.

**4. a.** $(5, 32)$
   **b.** In 2008 (5 yr after 2003), about 32% of Americans used the newspaper to obtain their news and 32% of Americans used the Internet to obtain their news.

**5.** Answers vary. The three methods are graphing, substitution, and elimination. Graphing is the least precise since it is impossible to determine exact fractional solutions when I graph by hand. I would use substitution if one of the equations in the system is already solved for one of the variables. I would use elimination for almost every other system as long as I first put the equations in standard form.

**6. a.** $\{(5, -1)\}$  **b.** $\{(-2, -7)\}$  **7. a.** $\{(-4, 2)\}$  **b.** $\{(1, -8)\}$

**9.** Ms. Jones invested \$13,000 in the 8% account and \$5000 in the 9% account.

**13.** Answers vary. The first step is to graph the boundary line, which is the equation formed by replacing the inequality symbol with an equals sign. Draw the line solid if the inequality is ≤ or ≥ and dashed if the inequality is < or >. Test a point not on the boundary line in the original inequality. If the point makes the inequality true, shade the half-plane that contains the point. If the point makes the inequality false, shade the half-plane that doesn't contain the point.

**14. a.**

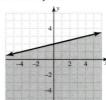

**b.**

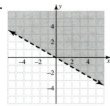

**c.**

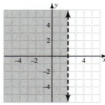

**d.**

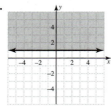

**15.**

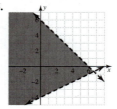

# CHAPTERS 1–4   CUMULATIVE REVIEW EXERCISES

**10. a.** $x = -2$ is not a solution; $x = 0$ is not a solution; $x = 2$ is a solution
  **b.** $x = -3$ is a solution; $x = 0$ is not a solution; $x = 3$ is not a solution

**20. a.** $[-2, \infty)$; $\{b \mid b \geq -2\}$

  **b.** $(-1, 3)$; $\{a \mid -1 < a < 3\}$

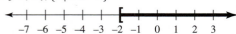

**21. a.**

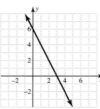

**b.**

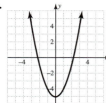

**c.**

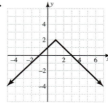

**23. a.** (2000, 3.9), (2001, 5.0), (2002, 5.7), (2003, 6.1), (2004, 5.4), (2005, 5.0), (2006, 4.5), (2007, 4.7), (2008, 6.2), (2009, 9.8), (2010, 9.6)
  **b.** The rate increased most between 2008 and 2009. It increased by 3.6%.
  **c.** The rate decreased most between 2003 and 2004. It decreased by 0.7%.

**24. a.** (0, 372), (1, 217), (2, 484), (3, 499), (4, 645), (5, 178), (6, 99)
  **b.** In 2003, the net income as reported by Southwest Airlines was $372 million. In 2009, the net income as reported by Southwest Airlines was $99 million.
  **c.** The net income as reported by Southwest Airlines was $645 million in 2007. The net income as reported by Southwest Airlines was $99 million in 2009.
  **d.**

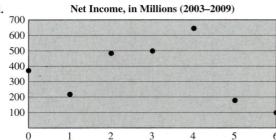

Net Income, in Millions (2003–2009)

**25. a.**

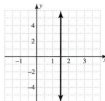

**b.**

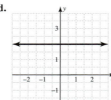

**c.**

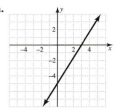

**d.**

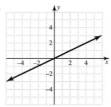

**e.**

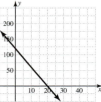

**26. a.** (0, 1724.8), The $y$-intercept means the yearly cost of jet fuel and oil for Southwest Airlines in 2005 was about $1724.8 million.
  **b.** The yearly cost of jet fuel and oil for Southwest Airlines in 2009 was about $3555.6 million.
  **c.** The yearly cost of jet fuel and oil for Southwest Airlines in 2013 was about $5386.4 million.
  **d.**

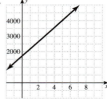

**28. a.** 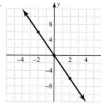   **b.**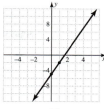

**c.** 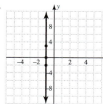   **d.**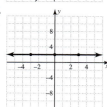

**29.** The $y$-intercept is $(0, 24.02)$ means that in 2000, the average price of jet fuel was about $24.02 per barrel. The slope is 8.38 and means that the average price of jet fuel increased by about $8.38 per barrel per year.

**31. a.** $y = 4x + 8$ or $-4x + y = 8$   **b.** $y = 12$
**c.** $y = 3x - 8$ or $3x - y = 8$

**33. a.** Domain = {2004, 2005, 2006, 2007, 2008, 2009, 2010}; Range = {10.75, 12.70, 13.73, 13.08, 13.89, 12.14, 11.20}; The relation is a function.
**b.** Domain = $(-\infty, \infty)$; Range = {2}; The relation is a function.
**c.** Domain = $(-\infty, \infty)$; Range = $(-\infty, \infty)$; The relation is a function.

**35. a.** $f(x) = -\dfrac{1}{2}x + \dfrac{1}{2}$; $\left(0, \dfrac{1}{2}\right), \left(-2, \dfrac{3}{2}\right), (3, -1)$
**b.** $f(x) = 0.2x^2 - 0.1x$; $(0, 0), (-2, 1), (3, 1.5)$

**36. c.** $f(2) = \$44{,}500$. It means that if Shawn teaches 2 extra credit hours, he will earns $44,500.

**37.** 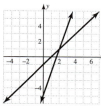   **a.** {(2, 1)}, one solution, consistent system with independent equations

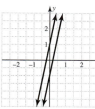   **b.** parallel lines, no solution, inconsistent system

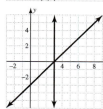   **c.** {(3, 0)}, one solution, consistent system with independent equations

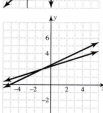   **d.** {(-1, 2)}, one solution, consistent system with independent equations

**e.** infinite number of solutions, consistent system with dependent equations

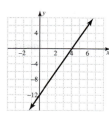

**38. a.** (8, 400)
**b.** Both plans will have the same cost of $400 if Rikki expects to rent the refrigerator for 8 months.
**c.** Since Rikki expects to sign a 12-month rental, she should use Household Rentals because it will cost her $564.

**53.** 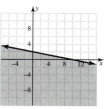   **54.**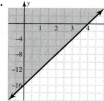

**55.** 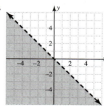   **56.**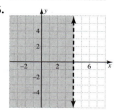

**57. b.**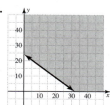

**c.** Answers vary. Student worker: 10 hr & nurse aide: 17 hr; student worker: 15 hr & nurse aide: 13 hr; student worker: 9 hr & nurse aide: 18 hr

**58.** 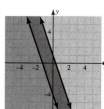   **59.**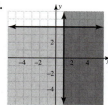

**60. a.** $\begin{cases} y \geq S(x) \\ y \leq k \\ x \geq 0 \end{cases} = \begin{cases} y \geq 21 + 0.40x \\ y \leq 53 \\ x \geq 0 \end{cases}$

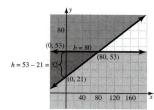

# CHAPTER 5

## Section 5.1

**99.** $(4a^6b^3)^3 = 4^3(a^6)^3(b^3)^3 = 64a^{18}b^9$; Ursula did not apply the exponent to the coefficient of the product in the base.

**100.** $(10xy^3)^5 = 10^5(x)^5(y^3)^5 = 100{,}000x^5y^{15}$; Ricky multiplied the exponent and the coefficient instead of raising the coefficient to the fifth.

**101.** $(-2ab^2)^3 = (-2)^3(a)^3(b^2)^3 = -8a^3b^6$; Rob multiplied the exponent of 3 and the coefficient $-2$ instead of raising $-2$ to the third.

**102.** $(-x^2y)^5 = (-1)^5(x^2)^5(y)^5 = -x^{10}y^5$; The coefficient of the base is $-1$. When $-1$ is raised to the fifth, the result is $-1$. A negative base to an odd exponent is negative.

## Section 5.3

**2.** one, ten, one hundred, one thousand, ten thousand, one hundred thousand, one million, ten million, one hundred million, one billion, ten billion, one hundred billion, one trillion

## Section 5.4

**119.** $x^2 + \dfrac{9}{7}x + 1$    **120.** $-x^2 + \dfrac{5}{6}x - \dfrac{5}{2}$

**135.** $(4x^4 - 2x^2 + 3x - 1) - (3x^4 + 2x^2 - 3x + 1)$
$= 4x^4 - 2x^2 + 3x - 1 - 3x^4 - 2x^2 + 3x - 1$
$= 4x^4 - 3x^4 - 2x^2 - 2x^2 + 3x + 3x - 1 - 1$
$= x^4 - 4x^2 + 6x - 2$

**136.** $(3x^3 - 6x^2 + 5) - (-3x^3 + 6x^2 + 4)$
$= 3x^3 - 6x^2 + 5 + 3x^3 - 6x^2 - 4$
$= 3x^3 + 3x^3 - 6x^2 - 6x^2 + 5 - 4$
$= 6x^3 - 12x^2 + 1$

## Section 5.5

**65.** $\dfrac{2}{9}x^2 + \dfrac{8}{3}xy - 10y^2$    **66.** $\dfrac{5}{16}a^2 - 2ab + 3b^2$

**67.** $10x^2 - \dfrac{1}{3}x - \dfrac{2}{9}$    **68.** $4x^2 + \dfrac{18}{7}x - \dfrac{10}{49}$

**69.** $0.2x^3 - 0.58x^2 + 1.18x - 1.14$

**70.** $-0.3x^3 + 0.13x^2 + 0.81x - 0.6$

## Section 5.6

**17.** $6x^2 - x - \dfrac{6}{25}$    **18.** $5x^2 - \dfrac{13}{4}x + \dfrac{3}{8}$    **51.** $x^4 - \dfrac{1}{49}$

**52.** $y^4 - \dfrac{16}{25}$    **53.** $\dfrac{9}{16}c^6 - d^2$    **54.** $\dfrac{4}{81}a^6 - b^2$

**61.** $b^3 + \dfrac{3}{2}b^2 + \dfrac{3}{4}b + \dfrac{1}{8}$    **62.** $c^3 - \dfrac{3}{5}c^2 + \dfrac{3}{25}c - \dfrac{1}{125}$

**73.** $a^4 - \dfrac{9}{49}$    **74.** $b^2 - \dfrac{16}{9}$    **75.** $\dfrac{1}{4}c^6 - 25d^2$

**76.** $4a^2 - \dfrac{1}{25}b^6$

**87.** We cannot just square each term. We must apply the rule for squaring a binomial. $(3x - 8)^2 = 9x^2 - 48x + 64$

**88.** We cannot just cube each term. We must first square $(5y + 2)$ and then multiply this result by $(5y + 2)$.
$(5y + 2)^3 = (5y + 2)^2(5y + 2)$
$= (25y^2 + 20y + 4)(5y + 2)$
$= 125y^3 + 50y^2 + 100y^2 + 40y + 20y + 8$
$= 125y^3 + 150y^2 + 60y + 8$

## Section 5.7

**26.** $1 - \dfrac{4}{x} + \dfrac{2}{x^5}$    **27.** $-\dfrac{y^2}{2} + y - 1$    **28.** $-\dfrac{y^2}{2} + y - \dfrac{1}{2}$

**29.** $\dfrac{y}{4} - 2 - \dfrac{1}{2y^2}$    **30.** $\dfrac{y}{7} - \dfrac{9}{7} + \dfrac{2}{7y^3}$    **45.** $2x - 19 + \dfrac{140}{x + 7}$

**46.** $3x - 12 + \dfrac{5}{x + 10}$    **47.** $3y^2 + 2y - 5 + \dfrac{7}{2y - 3}$

**48.** $2y^2 - 5y + 6 - \dfrac{8}{4y + 1}$    **59.** $7x^2 - 15x + 30 - \dfrac{46}{x + 2}$

**60.** $3x^2 + 9x + 7 + \dfrac{13}{x - 3}$

**65.** $\dfrac{5x^2 + 2x - 3}{5x^2} = \dfrac{5x^2}{5x^2} + \dfrac{2x}{5x^2} - \dfrac{3}{5x^2} = 1 + \dfrac{2}{5x} - \dfrac{3}{5x^2}$

**66.** $\dfrac{2x^3 - 4x^2 - 1}{2x^3} = \dfrac{2x^3}{2x^3} - \dfrac{4x^2}{2x^3} - \dfrac{1}{2x^3} = 1 - \dfrac{2}{x} - \dfrac{1}{2x^3}$

# CHAPTER 5 GROUP ACTIVITY

**1.** $P = \dfrac{A \cdot \dfrac{r}{12}}{1 - \left(1 + \dfrac{r}{12}\right)^{-12t}} = \dfrac{\dfrac{r}{12} \cdot A}{1 - \left(1 + \dfrac{r}{12}\right)^{-12t}}$

$= \dfrac{\dfrac{r}{12}}{1 - \left(1 + \dfrac{r}{12}\right)^{-12t}} \cdot \dfrac{A}{1} = \dfrac{\dfrac{r}{12}}{1 - \left(1 + \dfrac{r}{12}\right)^{-12t}} \cdot A$

**3.** The coefficient of $A$ is
$$\dfrac{\dfrac{r}{12}}{1 - \left(1 + \dfrac{r}{12}\right)^{-12t}} = \dfrac{\dfrac{0.045}{12}}{1 - \left(1 + \dfrac{0.045}{12}\right)^{-12(30)}} = 0.00507$$

**6.** Answers vary, but we will choose $200,000. The total paid back over 30 yr is $360(1014) = \$365{,}040$. The total interest paid is $\$365{,}040 - 200{,}000 = \$165{,}040$.

# CHAPTER 5 REVIEW EXERCISES

## Section 5.6

**70.** $c^4 - \dfrac{4}{25}$    **71.** $\dfrac{4}{9}a^6 - 25b^2$

# CHAPTERS 1–5 CUMULATIVE REVIEW EXERCISES

**15. a.**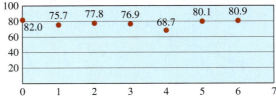

$(20, \infty)$
$\{x \mid x > 20\}$

**b.**
$-14.75$

$-14.75 < a < -7$
$\{a \mid -14.75 < a < -7\}$

**16. a.** $(0, 82.0), (1, 75.7), (2, 77.8), (3, 76.9), (4, 68.7), (5, 80.1), (6, 80.9)$

**b.** In 2003, 82.0% of flight operations for US Airways arrived on time. In 2009, 80.9% of flight operations for US Airways arrived on time.

**c.** The percentage of reported flight operations arriving on time for US Airways was the highest in 2003 and lowest in 2007.

**d.**
**Percent of Flight Operations Arriving on Time as Defined by DOT for US Airways (2003–2009)**

**17. a.**

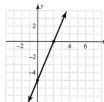

**b.**

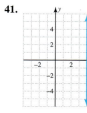

**c.**

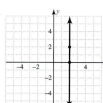

**41.**

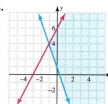

**42.**

**43.**

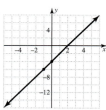

**56. d.** $-11x^2 + 3xy - 7y^2$

**18. a.** (0, 330.4); The airline consumed 330.4 million gal of fuel in 2005.
   **b.** The airline consumed 520.4 million gal of fuel in 2010.
   **c.** The airline consumed 634.4 million gal of fuel in 2013.
   **d.**

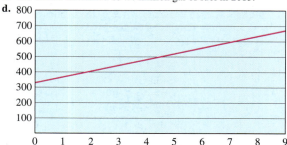

**20.** $m = -0.281$; The operating revenue is decreasing by \$0.281 billion per year. $y$-intercept is (0, 12.01); In 2006, the total operating revenue was \$12.01 billion.

**21. a.**

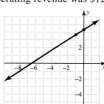

**b.**

**25. a.** Domain = $\{-4, -2, 0, 1, 3\}$; Range = $\{-4, 1, 3, 5\}$; a function
   **b.** Domain = $\{3\}$; Range = $(-\infty, \infty)$; not a function
   **c.** Domain = $(-\infty, \infty)$; Range = $(-\infty, \infty)$; a function

**27. a.** $f(x) = -\dfrac{4}{5}x + \dfrac{24}{5}$; $\left(-1, \dfrac{28}{5}\right), \left(0, \dfrac{24}{5}\right), \left(3, \dfrac{12}{5}\right)$
   **b.** $f(x) = 2|x - 1|$; (-1, 4), (0, 2), (3, 4)

**28. a.** $\left\{\left(\dfrac{8}{3}, \dfrac{1}{3}\right)\right\}$; consistent with independent equations
   **b.** $\{(-1, -4)\}$; consistent with independent equations

**29. b.** In 6 months, the total cost to rent a 42-in. LCD TV will be the same for both Rent-Me-Cheap and Rent to Go companies.
   **c.** George should choose the Rent-Me-Cheap company because it will be \$72 cheaper in 12 months.

**39.**

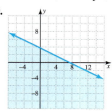

**40.**

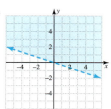

**59. a.** $15x^4y^3 - 5x^3y^2$    **b.** $25 - 4x^2$    **c.** $a^3 + 64b^3$
   **d.** $\dfrac{3}{16}x^2 + \dfrac{3}{4}x - 6$    **e.** $0.4x^3 - 1.6x^2 + 3.42x - 2.88$
   **f.** $6x^4 - 8x^3 - 29x^2 + 11x - 1$

## CHAPTER 6

### Section 6.1

**1.** The greatest common factor of a set of integers is the largest integer that is a factor of each integer.
**2.** Find the GCF of the integer coefficients and the GCF of the variable factors of the terms. The GCF of the terms is the product of the GCF of the integer coefficients and the GCF of the variable factors.
**5.** For polynomials with four or more terms, group pairs of terms by factoring out the GCF. Factor out the resulting binomial as the GCF.
**7.** The reverse operation of factoring is multiplication. For example, $(3x + y)(a + 2b) = 3ax + ay + 6bx + 2by$.
**8.** The inverse operation for the GCF is multiplication using the distributive property. For example, $2xy(3x - 5y) = 6x^2y - 10xy^2$.
**73.** Cannot be factored using grouping
**74.** Cannot be factored using grouping
**87.** Cannot be factored using grouping
**88.** Cannot be factored using grouping
**111.** Answers vary; $12x^2 - 6x = 6x(2x - 1)$
**112.** Answers vary; $40a^3 - 24a = 8a(5a^2 - 3)$
**113.** Answers vary; $4y^3 - 6y^2 = 2y^2(2y - 3)$
**114.** Answers vary; is $21b^6 + 14b^4 = 7b^3(3b^3 + 2b)$
**115.** Answers vary; $-10a^2b + 12ab^2 = -2ab(5a - 6b)$
**116.** Answers vary; $-20x^3y^2 - 4x^2y = -4x^2y(5xy + 1)$

### Section 6.2

**1.** Answers vary. For example, factor $x^2 + 7x + 10$. We need to find the factors of 10 whose sum is 7. Since 10 is positive and the sum is positive, we will consider only positive factors of 10: 1 and 10, 2 and 5. Since $2 + 5 = 7$, the expression is factored as $(x + 2)(x + 5)$.
**2.** Answers vary. For example, factor $x^2 - 4x - 12$. We need to find the factors of $-12$ whose sum is $-4$. Since the last term is negative, the factors are of opposite sign and give a negative sum. The relevant factors of $-12$ are 1 and $-12$, 2 and $-6$, 3 and $-4$. Since $2 + (-6) = -4$, the expression is factored as $(x + 2)(x - 6)$.
**3.** Answers vary. In Exercise 7, we need to find the factors of 6 whose sum is 5. Since the last term 6 is positive and the sum is positive, we will consider only positive factors of 6. In Exercise 8, we need to find the factors of 20 whose sum is 9. We will consider only positive factors of 20.
**4.** Answers vary. In Exercise 33, we need to find the factors of $-42$ whose sum is $-11$. The factors must be of opposite signs. We will consider factors of $-42$ that give a negative sum of $-11$. In Exercise 34, we need to find the factors of $-40$ whose sum is $-6$. The factors must be of opposite signs. We will consider factors of $-40$ that give a negative sum of $-6$.

**84.** The factors of 15 are 1 and 15, 3 and 5, −5 and −3, and −15 and −1. So, for the last term to be positive, both factors have to have the same sign. Since $b$ is the sum of the factors, the value of $b$ could be 16, 8, −8, or −16. So, an example is $x^2 + 8x + 15 = (x + 3)(x + 5)$.

**85.** The factors of 28 are 1 and 28, 2 and 14, 4 and 7, −7 and −4, −14 and −2, and −28 and −1. So, for the last term to be positive, both factors have to have the same sign. Since $b$ is the sum of the factors, the value of $b$ could be 29, 16, 11, −11, −16, or −29. So, an example is $x^2 − 16x + 28 = (x − 2)(x − 14)$.

**86.** The factors of −12 are 1 and −12, 2 and −6, 3 and −4, 4 and −3, 6 and −2, and 12 and −1. Since $b$ is the sum of the factors, the value of $b$ could be −11, −4, −1, 1, 4, or 11. An example is $x^2 − 4x − 12 = (x + 2)(x − 6)$.

**87.** The factors of −30 are 1 and −30, 2 and −15, 3 and −10, 5 and −6, 6 and −5, 10 and −3, 15 and −2, and 30 and −1. Since $b$ is the sum of the factors, the value of $b$ could be −29, −13, −7, −1, 1, 7, 13, or 29. An example is $x^2 + 13x − 30 = (x + 15)(x − 2)$.

**88.** The factors of 14 are 1 and 14, 2 and 7, −7 and −2, and −14 and −1. For the trinomial to be factorable with a last term of 14, the value of $b$ would have to be 15, 9, −9, or −15. So, as long as $b$ is different from these values, the trinomial is prime. An example is $x^2 + 2x + 14$.

**89.** The factors of −8 are 1 and −8, 2 and −4, 4 and −2, and 8 and −1. For the trinomial to be factorable with a last term of −8, the value of $b$ would have to be −7, −2, 2, or 7. So, as long as $b$ is different from these values, the trinomial is prime. An example is $x^2 + 5x − 8$.

### Section 6.3

**1.** Trial and error requires us to try every combination of factors of the first term, $a$, with factors of the last term, $c$, until the correct middle term, $b$, is obtained.

**2.** Grouping is a methodical process to find factors of the product of the leading coefficient, $a$, and the last term, $c$, that add to the middle coefficient, $b$.

**3.** If the last term in the trinomial is positive, then the signs of the last terms of the binomial factors are both positive or both negative. If the last term in the trinomial is negative, then the signs of the last terms of the binomial factors have opposite signs.

**5.** We write $6x^2 = (3x)(2x)$ and $−3 = (−1)(3)$. Therefore, $6x^2 + 7x − 3 = (3x − 1)(2x + 3)$. We check the factoring by multiplying.

**7.** In Exercise 17, we need to find every combination of factors of 4 with factors of 15 that add to −16. In Exercise 18, we need to find every combination of factors of 9 with factors of 4 that add to 15.

### Section 6.4

**1.** Factor out any GCF. Write the first term as $a^2$ and the second term as $b^2$. Factor the binomial as a product of conjugates, $(a + b)(a − b)$.

**2.** Write the first term as $a^3$ and the second term as $b^3$. The difference of two cubes is a product of a binomial and a trinomial. The sign of the binomial is the same as the sign of the given binomial, $a − b$. The trinomial is $a^2 + ab + b^2$.

**3.** Write the first term as $a^3$ and the second term as $b^3$. The sum of two cubes is a product of a binomial and a trinomial. The sign of the binomial is the same as the sign of the given binomial, $a + b$. The trinomial is $a^2 − ab + b^2$.

**37.** $\left(\frac{1}{3}a + \frac{2}{7}\right)\left(\frac{1}{3}a − \frac{2}{7}\right)$    **38.** $\left(\frac{1}{2}b + \frac{4}{5}\right)\left(\frac{1}{2}b − \frac{4}{5}\right)$

**51.** $\left(\frac{1}{3}s − 1\right)\left(\frac{1}{9}s^2 + \frac{1}{3}s + 1\right)$    **52.** $\left(\frac{1}{2}t − 5\right)\left(\frac{1}{4}t^2 + \frac{5}{2}t + 25\right)$

**63.** $\left(a + \frac{1}{3}\right)\left(a^2 − \frac{1}{3}a + \frac{1}{9}\right)$    **64.** $\left(b + \frac{2}{5}\right)\left(b^2 − \frac{2}{5}b + \frac{4}{25}\right)$

**71.** $\left(r^2 + \frac{9}{4}\right)\left(r + \frac{3}{2}\right)\left(r − \frac{3}{2}\right)$    **72.** $\left(s^2 + \frac{1}{25}\right)\left(s + \frac{1}{5}\right)\left(s − \frac{1}{5}\right)$

**97.** $\left(2x − \frac{1}{5}y\right)\left(4x^2 + \frac{2}{5}xy + \frac{1}{25}y^2\right)$

**98.** $\left(3x + \frac{1}{2}y\right)\left(9x^2 − \frac{3}{2}xy + \frac{1}{4}y^2\right)$

**99.** $\left(\frac{1}{4}x^2 + \frac{1}{9}\right)\left(\frac{1}{2}x − \frac{1}{3}\right)\left(\frac{1}{2}x + \frac{1}{3}\right)$

**100.** $\left(\frac{1}{25}y^2 + \frac{1}{4}\right)\left(\frac{1}{5}y − \frac{1}{2}\right)\left(\frac{1}{5}y + \frac{1}{2}\right)$

### Section 6.5

**2.** A quadratic equation is an equation that can be written in the form $ax^2 + bx + c = 0$, where $a$, $b$, and $c$ are real numbers and $a \neq 0$.

**3.** Write the quadratic equation in standard form. Factor the resulting polynomial. Use the zero products property to solve.

**4.** Write the equation in standard form. Factor the polynomial. Apply the zero products property to set each factor to zero. Solve the resulting equations.

**5.** In $(x − 5)(x − 1) = 12$, you need to write the equation in standard form by multiplying the binomials on the left and subtracting 12 from both sides. Now factor the left side and apply the zero products property to solve. In $(x − 5)(x − 1) = 0$, you can apply the zero products property to solve.

**6.** Applying the zero products property to $4x(x − 5)(x − 1) = 0$, you get $4x = 0$, $x − 5 = 0$, and $x − 1 = 0$. The solutions are $x = 0$, $x = 5$, and $x = 1$. Applying the zero products property to $4(x − 5)(x − 1) = 0$, you get $x − 5 = 0$ and $x − 1 = 0$. The solutions are $x = 5$ and $x = 1$.

**13.** $\left\{−2, \frac{1}{2}\right\}$    **14.** $\left\{−\frac{4}{3}, \frac{1}{4}\right\}$    **17.** $\left\{−12, \frac{3}{2}\right\}$

**18.** $\left\{−5, \frac{5}{4}\right\}$    **53.** $\left\{−\frac{3}{2}, −\frac{1}{7}, \frac{1}{7}, \frac{3}{2}\right\}$    **54.** $\left\{−\frac{5}{4}, −\frac{2}{3}, \frac{2}{3}, \frac{5}{4}\right\}$

**60.** $\left\{0, \frac{11}{2}\right\}$    **61.** $\left\{−\frac{10}{3}, 3\right\}$    **62.** $\left\{\frac{12}{5}, 6\right\}$

**63.** $\left\{−1, \frac{5}{4}\right\}$    **66.** $\left\{−4, \frac{3}{2}\right\}$    **71.** $\left\{−3, −\frac{1}{2}, \frac{1}{2}, 3\right\}$

**72.** $\left\{−1, −\frac{2}{3}, \frac{2}{3}, 1\right\}$    **81.** $\left\{−5, −\frac{4}{3}\right\}$    **82.** $\left\{−\frac{9}{2}, −\frac{1}{5}\right\}$

### Section 6.6

**1.** First, read the problem and identify the known and unknown. Assign a variable to the unknown and set up any formula. Second, translate the words into a mathematical equation.

**3.** If $x$ is an odd integer, then $x, x + 2, x + 4, \ldots$ represent consecutive odd integers. If $x$ is an even integer, then $x, x + 2, x + 4, \ldots$ represent consecutive even integers.

**4.** Revenue is an amount of monies generated by sales, i.e., product of price per item and the number of items sold. Profit is the difference of revenue and cost.

## CHAPTER 6  GROUP ACTIVITY

**1.** The blue graph represents consumer spending on VHS/UMD in billions of dollars. The red graph represents consumer spending on DVDs in billions of dollars. The green graph represents consumer spending on digital movies in billions of dollars. The graph shows that the spending on VHS/UMDs declined as spending on DVDs increased and eventually the spending on VHS/UMDs reached $0. The graph also shows that the spending in DVDs reached a maximum in 2006 and has been declining ever since. Lastly, the graph shows that the spending on digital movies is slowly but steadily rising.

**2.** The three points of intersection are approximately (0, 1), (3, 9), and (7, 0.7). The points mean that in 1999, the spending on DVDs and digital movies was approximately $1 billion. In 2002, the spending on VHS/UMDs and DVDs was approximately $9 billion. In 2007, the spending on VHS/UMDs and digital movies was approximately $0.7 billion.

**3. a.** $0.0305x^3 - 0.4393x^2 - 0.0712x + 11.5697 = 0$
  **b.** $-0.0411x^3 + 0.306x^2 + 2.5999x - 0.5456 = 0$
  **c.** $0.0716x^3 - 0.7453x^2 - 2.6711x + 12.1153 = 0$
**4.** The point of intersection of the red and green graphs is approximately (12.44, 3.15). This means that in 12 yr after 1999 or in 2011, the consumer spending on DVDs and digital movies will be approximately \$3.15 billion.

## CHAPTER 6  REVIEW EXERCISES

### Section 6.4

**47.** $(x^2y^2 + z^2)(x^4y^4 - x^2y^2z^2 + z^4)$
**49.** $(a - 10bc)(a^2 + 10abc + 100b^2c^2)$
**50.** $(7x - 2)(49x^2 + 14x + 4)$
**52.** $(a^5b^4 - 6c^2)(a^{10}b^8 + 6a^5b^4c^2 + 36c^4)$
**53.** $\left(\dfrac{1}{2}a + bc\right)\left(\dfrac{1}{4}a^2 - \dfrac{1}{2}abc + b^2c^2\right)$
**54.** $\left(x^2y^2 + \dfrac{1}{4}z^2\right)\left(xy - \dfrac{1}{2}z\right)\left(xy + \dfrac{1}{2}z\right)$

### Section 6.5

**60.** $\left\{2, \dfrac{7}{2}\right\}$    **61.** $\left\{-3, 0, \dfrac{1}{2}\right\}$

## CHAPTERS 1–6  CUMULATIVE REVIEW EXERCISES

**5. a.**
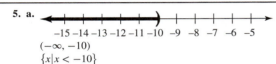
$(-\infty, -10)$
$\{x \mid x < -10\}$

**b.**
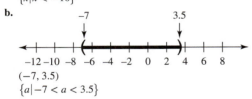
$(-7, 3.5)$
$\{a \mid -7 < a < 3.5\}$

**6. a.**
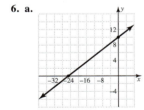
**b.**
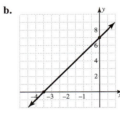

**7. a.** (0, 68.6), In 2005, about 68.6% of the U.S. households owned desktops and laptops.
**d.**
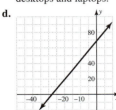

**8. a.** $y = -2x + 15, \ 2x + y = 15$
**10. a.** $(1, -3)$, system is consistent, equations are independent
  **b.** $(-1, -3)$, system is consistent, equations are independent

**13. a.**
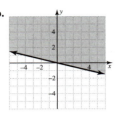
**b.**
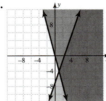

**14. a.**
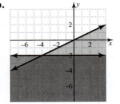
**b.**

**22. a.** $-5x^2 + 19x + 5$    **b.** $12x^3 + 12x^2 + 6x$
  **c.** $-7x^2 + 10xy - 22y^2$
**24. a.** $12x^3y^3 - 48x^2y$    **b.** $-12x^2 + 35x + 13$
**25. a.** $20x^2 + 7x - 3$    **b.** $125b^3 + 900b^2 + 2160b + 1728$
  **c.** $p^3 - 9p^2 + 27p - 27$
**27. a.** $x(x + 12)(x - 9)$    **b.** $5y(2y + 3)(y - 4)$
  **d.** $-7(x + 3)(x - 3)(x + 2)(x - 2)$
**29. a.** $(x - 9)(x - 6)$    **c.** $-6x(2x - 1)(x + 5)$
  **d.** $(a - 12b)(a - 4b)$
**31. a.** $(5p + 9)(p + 3)$    **b.** $p(5q + 9)(q - 3)$
  **c.** $(2x - 1)(15x + 7)$    **d.** $(4x + 7y)^2$
**33. b.** $2(5m + 4n)(5m - 4n)$    **c.** $2(3x - 4y)(9x^2 + 12xy + 16y^2)$
  **d.** $2(y + 5z)(y^2 - 5yz + 25z^2)$    **f.** $(9a^2 + 4)(3a + 2)(3a - 2)$
  **g.** $3(c^2 - 2)(c^4 + 2c^2 + 4)$    **h.** $(x^2y^2 + 4z^2)(xy + 2z)(xy - 2z)$
**34. a.** $\left\{-\dfrac{3}{4}, 2\right\}$    **b.** $\left\{-\dfrac{2}{5}, 1\right\}$    **c.** $\left\{-\dfrac{1}{2}\right\}$    **e.** $\left\{-\dfrac{3}{5}, 2\right\}$
  **f.** $\left\{\dfrac{5}{3}, 4\right\}$    **g.** $\{-3, -1, 1, 3\}$    **h.** $\{-2, 0, 5\}$
  **j.** $\left\{-\dfrac{5}{4}, -\dfrac{7}{6}\right\}$

## CHAPTER 7

### Section 7.1

**2.** To determine the values that make a rational expression undefined, set the denominator equal to zero and solve the resulting equation.
**3.** Examples are $\dfrac{x}{x + 1}$, $\dfrac{x^2 + 1}{2x}$, and $3x - 4$.
**4.** Examples are $\dfrac{1}{\sqrt{x}}$, $\dfrac{\sqrt[3]{x}}{x + 1}$, and $\dfrac{1}{\sqrt{x - 3}}$.
**5.** No, not every rational expression is a polynomial because the denominator must be a constant.
**6.** Yes, every polynomial is a rational expression because we can always use 1 as the denominator.
**7.** Since the expression is undefined at $x = 0$, $x$ must be part of the denominator. An example is $\dfrac{2x + 1}{x}$.
**8.** Since the expression is undefined for $x = 2$ and $x = -2$, then $x - 2$ and $x + 2$ must be part of the denominator. An example is $\dfrac{1}{x^2 - 4}$.
**9.** If the expression is defined for all real numbers, then the denominator does not equal to zero. An example is $\dfrac{1}{x^2 + 9}$.
**10.** Yes, it is in simplified form because it cannot be reduced further.

**11.** No, you can factor $x$ out from the numerator and reduce the fraction to $4x + 6$.

**12.** No, you can factor the numerator $x^2 - 9$ as $(x + 3)(x - 3)$ and reduce the fraction to $x + 3$.

**57.** $\dfrac{3x + a}{4y^2 + 2y + 1}$    **58.** $\dfrac{a^2 + 5a + 25}{2b + 3}$

**61.** $\dfrac{x - 3}{-2x - 4}$ or $\dfrac{-x + 3}{2x + 4}$    **62.** $\dfrac{2x - 3}{-4}$ or $\dfrac{-2x + 3}{4}$

**63.** $\dfrac{-3x + 4}{-x + 1}$ or $\dfrac{3x - 4}{x - 1}$    **64.** $\dfrac{-x + 8}{-2x + 1}$ or $\dfrac{x - 8}{2x - 1}$

**65.** $\dfrac{3x - 1}{-x + 4}$ or $\dfrac{-3x - 1}{x - 4}$    **66.** $\dfrac{4x - 8}{-x + 2}$ or $\dfrac{-4x - 8}{x - 2}$

**67.** $\dfrac{4x + 9}{-2x - 1}$ or $\dfrac{-4x + 9}{2x + 1}$    **68.** $\dfrac{2x - 15}{-3x + 12}$ or $\dfrac{-2x - 15}{3x - 12}$

**71.** No, $\dfrac{\sqrt{x + 1}}{x + 3}$ is irrational because $\sqrt{x + 1}$ is irrational.

**72.** No, $\dfrac{1}{\sqrt{x + 5}}$ is irrational because $\sqrt{x + 5}$ is irrational.

**89.** $x = 5, -2;\ \dfrac{x^2 - 2x + 4}{x - 5}$    **90.** $x = 4, -7;\ \dfrac{x^2 + 4x + 16}{x + 7}$

**92.** $x = 0, -5;\ \dfrac{-3x + 3}{x + 5}$    **93.** $\dfrac{9x - 18}{-x + 10}$ or $-\dfrac{9x - 18}{x - 10}$

**94.** $\dfrac{2x - 7}{-x - 1}$ or $-\dfrac{2x - 7}{x + 1}$    **95.** $\dfrac{-x^2 - 3x}{4x - 1}$ or $\dfrac{x^2 + 3x}{-4x + 1}$

**96.** $\dfrac{-x^2 + x}{-3x + 2}$ or $\dfrac{x^2 - x}{3x - 2}$

**99.** $\dfrac{81 - x^2}{x + 9} = -\dfrac{x^2 - 81}{x + 9} = -\dfrac{(x + 9)(x - 9)}{x + 9} = -(x - 9) = -x + 9$

**100.** $\dfrac{x^2 + 3x - 10}{x^2 + 9x + 20} = \dfrac{(x - 2)(x + 5)}{(x + 4)(x + 5)} = \dfrac{x - 2}{x + 4}$

## Section 7.2

**1.** Factor the numerator and denominator completely. Divide out the common factors. Multiply the remaining factors.

**3.** Change the division to multiplication of the reciprocal, factor all numerators and denominators completely, divide out the common factors, and multiply the remaining factors.

**4.** When converting units, multiply the expression to be converted by a form of 1. The form of 1 involves a ratio of two equivalent quantities. You divide out the units to be converted.

**11.** $\dfrac{2(x - 5)}{x + 1}$    **12.** $\dfrac{(x + 7)}{5(x - 1)}$    **15.** $-\dfrac{2}{3x^2(x - 5)}$

**16.** $-\dfrac{4x}{3(x + 6)^2}$    **20.** $\dfrac{2}{(x^2 - x + 1)(x + 2)}$    **22.** $\dfrac{2(x - 1)}{x - 2}$

**53.** $\dfrac{(x + 3)(x - 5)}{2x^2}$    **54.** $\dfrac{4}{x + 3}$

**78.** $\dfrac{x^2 - 9}{2x + 8} \div \dfrac{x + 3}{4x} = \dfrac{(x + 3)(x - 3)}{2(x + 4)} \div \dfrac{x + 3}{4x}$

$= \dfrac{(x + 3)(x - 3)}{2(x + 4)} \cdot \dfrac{4x}{x + 3} = \dfrac{(x - 3) \cdot 2x}{(x + 4)} = \dfrac{2x(x - 3)}{(x + 4)}$

**79.**

Since the $y$-values are the same, the quotient is correct.

**80.**

Since the $y$-values are the same, the quotient is correct

**81.**

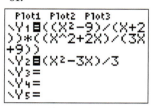

Since the $y$-values are the same, the product is correct.

**82.**

Since the $y$-values are the same, the product is correct.

## Section 7.3

**2.** We use the fundamental property of rational expressions to write the equivalent forms of rational expressions by multiplying a form of 1.

**3.** Two rational expressions are equivalent if one rational expression can be obtained from the other rational expression by multiplying a form of 1.

**4.** To find the equivalent form of a given rational expression, you multiply the numerator and denominator of a rational expression by the same nonzero value.

**5.** Factor each denominator completely. The LCD will include all factors in the first denominator and any additional ones from other denominators that are not in the first denominator.

**6.** $-\dfrac{1}{x - 1}$ is equivalent to $\dfrac{1}{-x + 1}$ which is not the same as $\dfrac{1}{-x - 1}$.

**104.** $\dfrac{4x + 3}{2x - 5} - \dfrac{6x + 1}{2x - 5} = \dfrac{4x + 3 - 6x - 1}{2x - 5} = \dfrac{-2x + 2}{2x - 5}$

## Section 7.4

1. To add or subtract rational expressions with different denominators, you need to find the least common denominator, convert each fraction to an equivalent fraction with the LCD as its denominator, add or subtract the numerators, and simplify, if possible.

2. The key step in subtracting rational expressions is to subtract the entire expression in the numerator of the fraction being subtracted.

11. $\dfrac{7x + 8}{(x - 1)(x + 2)}$

12. $\dfrac{3x - 4}{(x + 4)(x - 2)}$

13. $\dfrac{5x(x + 4)}{(x + 1)(x + 5)}$

14. $\dfrac{x^2 + 2x - 1}{(x - 6)(x + 1)}$

16. $\dfrac{-3x + 8}{(x + 1)(x - 1)}$

17. $\dfrac{x^2 + 5x + 12}{(x + 3)(x + 1)}$

18. $\dfrac{3x(x - 3)}{(x - 1)(x - 4)}$

19. $\dfrac{x(2x - 7)}{(x - 2)(x - 3)}$

20. $\dfrac{x(5x - 33)}{(x + 9)(x - 4)}$

21. $\dfrac{4(x^2 - x + 1)}{x(x - 2)}$

22. $\dfrac{3(x^2 - 2x + 3)}{x(x - 3)}$

36. $\dfrac{3x^2 - 2x + 3}{(x - 3)(x + 3)}$

37. $\dfrac{7x - 10}{(x - 10)(x + 2)}$

44. $\dfrac{7x^2 - 47x - 30}{(x + 4)(x - 10)}$

49. $\dfrac{150(5r - 11)}{(r - 10)(r + 3)}$

50. $\dfrac{30(7r + 180)}{(r + 40)(r + 20)}$

52. $\dfrac{2x}{4x + 1} - \dfrac{x}{4x - 1} = \dfrac{2x(4x - 1) - x(4x + 1)}{(4x + 1)(4x - 1)}$

$= \dfrac{8x^2 - 2x - 4x^2 - x}{(4x + 1)(4x - 1)}$

$= \dfrac{4x^2 - 3x}{(4x + 1)(4x - 1)}$

53.

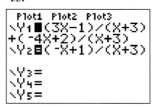

Since the y-values are the same, the addition has been performed correctly.

54.

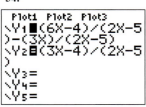

Since the y-values are the same, the subtraction has been performed correctly.

55.

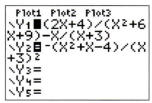

Since the y-values are the same, the subtraction has been performed correctly.

56.

 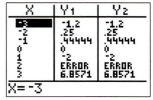

Since the y-values are the same, the addition has been performed correctly.

## Section 7.5

1. A complex fraction has fractions in its numerator and/or denominator.

2. You need to multiply the LCD, $abcd$, by the complex fraction. It is simplified to be $\dfrac{abcd\left(\dfrac{1}{a} + \dfrac{1}{b}\right)}{abcd\left(\dfrac{1}{c} + \dfrac{1}{d}\right)} = \dfrac{cd(b + a)}{ab(d + c)}$.

3. Convert the complex fraction to a division problem. Change the division to multiplication of the reciprocal of the denominator. For example, $\dfrac{\dfrac{6}{y^2}}{\dfrac{18}{y^3}} = \dfrac{6}{y^2} \cdot \dfrac{y^3}{18} = \dfrac{y}{3}$.

4. Multiply both the numerator and denominator of the complex fraction by the LCD. Apply the distributive property and simplify each product by dividing out common factors. For example,

$\dfrac{\dfrac{1}{2}}{\dfrac{2}{3} + \dfrac{5}{2}} = \dfrac{\left(\dfrac{1}{2}\right)6}{\left(\dfrac{2}{3} + \dfrac{5}{2}\right)6} = \dfrac{3}{4 + 15} = \dfrac{3}{19}$.

34. $\dfrac{x(x - 4)}{(x + 2)(x + 2)}$

63. $\dfrac{1}{12}$ of the pie

64. $\dfrac{3}{4}$ of a cup

68. $1\dfrac{2}{7}$ hr

70. $\dfrac{\dfrac{3x}{x + 1}}{\dfrac{2x}{1}} = \dfrac{\dfrac{3x}{x + 1}}{\dfrac{2x}{1}} = \dfrac{3x}{x + 1} \cdot \dfrac{1}{2x} = \dfrac{3}{2(x + 1)}$

71. $\dfrac{1 - \dfrac{3}{x}}{2 + \dfrac{4}{3x}} = \dfrac{\dfrac{x - 3}{x}}{\dfrac{6x + 4}{3x}} = \dfrac{x - 3}{x} \cdot \dfrac{3x}{6x + 4} = \dfrac{3(x - 3)}{6x + 4} = \dfrac{3(x - 3)}{2(3x + 2)}$

72. $\dfrac{1 - \dfrac{x}{2x + 1}}{\dfrac{2x + 1}{3x}} = \dfrac{\dfrac{2x + 1}{2x + 1} - \dfrac{x}{2x + 1}}{\dfrac{2x + 1}{3x}}$

$= \dfrac{\dfrac{2x + 1 - x}{2x + 1}}{\dfrac{2x + 1}{3x}} = \dfrac{\dfrac{x + 1}{2x + 1}}{\dfrac{2x + 1}{3x}}$

$= \dfrac{x + 1}{2x + 1} \cdot \dfrac{3x}{2x + 1} = \dfrac{3x(x + 1)}{(2x + 1)(2x + 1)}$

**73.** $\dfrac{\dfrac{2}{a}+\dfrac{1}{b}}{\dfrac{4}{a^2}-\dfrac{1}{b^2}}=\dfrac{\dfrac{2b+a}{ab}}{\dfrac{4b^2-a^2}{a^2b^2}}=\dfrac{2b+a}{ab}\div\dfrac{4b^2-a^2}{a^2b^2}$

$=\dfrac{2b+a}{ab}\cdot\dfrac{a^2b^2}{(2b-a)(2b+a)}$

$=\dfrac{2b+a}{ab}\cdot\dfrac{a^2b^2\,ab}{(2b-a)(2b+a)}$

$=\dfrac{ab}{2b-a}$

**74.** $\dfrac{a+b}{\dfrac{1}{a}+\dfrac{1}{b}}=\dfrac{ab(a+b)}{ab\left(\dfrac{1}{a}+\dfrac{1}{b}\right)}=\dfrac{ab(a+b)}{b+a}=\dfrac{ab(a+b)}{b+a}=ab$

## Section 7.6

1. A rational equation is an equation that contains a rational expression.
2. Determine the LCD of the denominators in the equation, multiply the LCD to both sides of the equation, solve the resulting equation, check for extraneous solutions, and write the solution set.
3. We must check the answers because if one of the solutions is a value that makes the LCD zero, then it is an **extraneous solution** and must be discarded from the solution set.
4. A rational expression is a ratio of two polynomials. A rational equation is an equation that contains a rational expression.
5. The LCD for both problems is the same, $(x-3)(x+5)$. To add fractions, you convert the fraction $\dfrac{4}{(x+5)}$ to an equivalent fraction with the LCD as the denominator. To solve the equation, you multiply both sides by the LCD and solve the resulting equation.
6. The LCD is $abc$. We multiply both sides of the equation by $abc$ and apply distributive property to simplify each product. The resulting equation is $bc+2ac=3ab$. We group the terms with $a$ on the right side by subtracting $2ac$ from both sides to obtain $bc=3ab-2ac$. Factor out $a$ on the right side and solve.

**57.** $\left\{-\dfrac{1}{2},3\right\}$   **58.** $\left\{0,\dfrac{-4}{7}\right\}$

**75.** $(x+1)(x+4)\dfrac{3}{x+1}=\dfrac{3}{x+4}(x+1)(x+4)$

$3(x+4)=3(x+1)$

$3x+12=3x+3$

$0=-9$

So there is no solution.

**76.** $(x-1)\dfrac{x^2}{x-1}=\dfrac{1}{x-1}(x-1)$

$x^2=1$

$x^2-1=0$

$(x-1)(x+1)=0$

$x-1=0 \qquad x+1=0$

$x=1 \qquad\quad x=-1$

The original equation is undefined for $x=1$. Therefore, the solution is just $x=-1$.

**77.** $\dfrac{2}{x+2}-\dfrac{5}{3x-1}=\dfrac{4x}{(3x-1)(x+2)}$

$(3x-1)(x+2)\dfrac{2}{x+2}-\dfrac{5}{3x-1}(3x-1)(x+2)$

$=\dfrac{4x}{(3x-1)(x+2)}(3x-1)(x+2)$

$2(3x-1)-5(x+2)=4x$

$6x-2-5x-10=4x$

$x-12=4x$

$-12=3x$

$x=-4$

So, the solution is $x=-4$.

**78.** $\dfrac{2}{x+1}-\dfrac{1}{2x-1}=\dfrac{4x}{2x^2+x-1}$

$(2x-1)(x+1)\dfrac{2}{x+1}-\dfrac{1}{2x-1}(2x-1)(x+1)$

$=\dfrac{4x}{(2x-1)(x+1)}(2x-1)(x+1)$

$2(2x-1)-(x+1)=4x$

$4x-2-x-1=4x$

$3x-3=4x$

$x=-3$

So, the solution is $x=-3$.

## Section 7.7

1. A ratio relates the measurement of two quantities.
2. A proportion is an equation with two ratios set equal to each other.
3. A ratio is an expression and a proportion is an equation.
4. We multiply both sides of the proportion by the LCD or cross multiply.
5. Two triangles that are similar have the same shape but different sizes. Their corresponding angles are equal and the lengths of their corresponding sides are proportional.
6. Identify the amount of time it takes for each person to complete the work and together. Write an expression to represent the part of the job completed in 1 hr by each person and the part of the job completed in 1 hr if they work together. Add the parts of the job completed in 1 hr by each person and set the sum equal to the part of the job completed in 1 hr if they work together.

**60.** $\dfrac{x}{7.8125}=\dfrac{30}{1}$

$x=234.375$ or $234.38$ Hong Kong dollars

**61.** $\dfrac{1}{5}+\dfrac{1}{9}=\dfrac{1}{x}$

$x=\dfrac{45}{14}$ hr

**62.** $\dfrac{6}{10}=\dfrac{a}{11}$

$a=\dfrac{33}{5}$

## CHAPTER 7  REVIEW EXERCISES

### Section 7.1

**10.** $\dfrac{3x+1}{-x+4}$ or $-\dfrac{3x+1}{x-4}$   **11.** $\dfrac{x^2-7x}{x^2-5}$ or $\dfrac{-x^2+7x}{-x^2+5}$

**12.** $\dfrac{2x^2-x}{3x^2-2}$ or $\dfrac{-2x^2+x}{-3x^2+2}$

### Section 7.4

**42.** $\dfrac{2x^2-4x-8}{(x-2)(x+2)}$   **43.** $\dfrac{12x^2+9x}{(5x+2)(2x+5)}$   **45.** $\dfrac{5x}{(x-12)(x+8)}$

**46.** $-\dfrac{x+6}{x+1}$   **47.** $\dfrac{3x^2+8x-4}{(x-7)(x+1)}$   **48.** $-\dfrac{x(x^2-3x-3)}{(2x+1)(x-3)}$

**50.** $\dfrac{x+27}{(x+3)(x-3)^2}$

### Section 7.5

**56.** $\dfrac{4}{3(3a-b)}$

### Section 7.6

**77.** $y_1=\dfrac{2A-hy_0-hy_2}{2h}$   **78.** $h=\dfrac{2A}{y_0+2y_1+y_2}$

**79.** $R_2=\dfrac{RR_1R_3}{R_1R_3-RR_3-RR_1}$

## CHAPTER 7 TEST

**9. a.** $-\dfrac{1}{3x+5}$    **b.** $\dfrac{2x-7}{x+4}$    **c.** $\dfrac{4a^2+6a+9}{2a+3}$

## CHAPTERS 1–7 CUMULATIVE REVIEW EXERCISES

**1. a.** $\dfrac{67}{60}$    **b.** $1\dfrac{5}{12}$    **c.** 54

**7. a.** $h=\dfrac{S-2\pi r^2}{2\pi r}$    **b.** $l=\dfrac{C-8w}{4}$    **c.** $y=-\dfrac{1}{5}x+2$

**d.** $x=70-\dfrac{1}{2}y$    **10. a.** $y=-\dfrac{3}{4}x-1;\ 3x+4y=-4$

**14. a.** 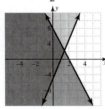    **b.**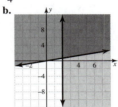

**16. a.** $\dfrac{16}{x^{12}}$    **b.** $\dfrac{1}{27x^6y^{21}}$    **c.** $\dfrac{1}{4c^4d^{10}}$

**19. a.** $3a^3b-18a^2b^2+15ab^3$
  **b.** $28x^2+23x-15$    **c.** $a^4+8a^3+24a^2+32a+16$
  **d.** $8b^3-12b^2+6b-1$
**21. b.** $5y(2y-3)(y+4)$    **d.** $-7(a^2+4)(a^2+9)$
**23. a.** $3y(y-4)^2$    **b.** $-6x(2x+1)(x-5)$    **c.** $(a-3b)(a-12b)$
  **d.** $2y(x+8)(x+9)$
**25. a.** $(5a+3)(7a+2)$    **b.** $5a(8b+9)(b-2)$
  **c.** $2(3x-5y)(2x+3y)$    **d.** $(3x+2y)^2$
**26. a.** $2(7a-3b)(7a+3b)$    **b.** $5(x+2y)(x^2-2xy+4y^2)$
  **c.** $(y-10z)(y^2+10yz+100z^2)$
**27. a.** $\{-4,10\}$    **b.** $\{-4,5\}$
**29. a.**

| $x$ | $y$ |
|---|---|
| $-2$ | Undefined |
| $0$ | $0$ |
| $3$ | $0$ |

**b.**

| $x$ | $y$ |
|---|---|
| $-1$ | $2$ |
| $0$ | $0$ |
| $1$ | Undefined |

**33. b.** $x=-6,1;\ \dfrac{x+1}{x+6}$    **c.** $x=-5,2;\ \dfrac{x^2+2x+4}{x+5}$

**34. a.** $-\dfrac{3x-6}{x+4}$ or $\dfrac{3x-6}{-x-4}$    **b.** $\dfrac{-1+x}{2x-3}$ or $\dfrac{1-x}{-2x+3}$

**39. b.** $\dfrac{9(x-2)}{(x-6)(x+3)}$    **c.** $\dfrac{x+6}{x-2}$    **d.** $\dfrac{-10x}{(x+5)(x-5)}$

  **f.** $\dfrac{(2x+3)(x+1)}{(x+10)(x-3)}$

**40. a.** $\dfrac{5}{4x^2}$    **b.** $\dfrac{4b-a}{2b+a}$    **c.** $\dfrac{2y+3}{4y}$    **d.** $-\dfrac{1}{6a}$

  **e.** $\dfrac{4(x+3y)}{3}$    **f.** $\dfrac{xy}{3x+4y}$

**44. a.** $R_1=\dfrac{RR_2R_3}{R_2R_3-RR_3-RR_2}$    **b.** $x_1=\dfrac{mx_2-y_2+y_1}{m}$

## CHAPTER 8

### Section 8.1

**2.** If $a>0$, $\sqrt{-a}$ is not a real number since $-a$ represents a negative number and the square root of a negative number is not a real number.
**3.** If the index is even, the radicand $x$ must be greater than or equal to zero for $\sqrt[n]{x}$ to be a real number.
**4.** If the index is odd, the radicand $x$ can be any real number for $\sqrt[n]{x}$ to be a real number.
**7.** To simplify the expression, we must find the expression that when cubed is $8x^{12}y^3$. Since $(2x^4y)^3=8x^{12}y^3$, $\sqrt[3]{8x^{12}y^3}=2x^4y$.
**8.** We must apply the quotient rule and find the fourth root of the numerator and denominator. Since $(3a^3)^4=81a^{12}$ and $(b^2)^4=b^8$,
$$\sqrt[4]{\dfrac{81a^{12}}{b^8}}=\dfrac{3a^3}{b^2}.$$

**48.** rational, $\dfrac{1}{5}$    **63.** $-\dfrac{4r^2}{s^6}$    **77.** $\dfrac{0.5u^2}{v}$    **85.** $-\dfrac{2a^4}{5}$

### Section 8.2

**3.** The product rule states that the square root of a product is the same as the product of the square root of the factors of the radicand.
**4.** The quotient rule states that the square root of a quotient is the same as the quotient of the square roots of the numerator and denominator.
**46.** $2r^2s\sqrt[4]{6r^2}$    **49.** $\dfrac{2x^2\sqrt[3]{20x}}{5}$

**95.** The distance formula has a square on each of the differences inside the radical.
$$d=\sqrt{(4-6)^2+[3-(-1)]^2}=\sqrt{(-2)^2+(4)^2}$$
$$=\sqrt{4+16}=\sqrt{20}=\sqrt{4\cdot5}=2\sqrt{5}$$
**96.** The order of operations requires us to simplify the expression inside the radical before we apply the square root.
$$\sqrt{(11-3)^2+(-16+1)^2}=\sqrt{(8)^2+(-15)^2}$$
$$=\sqrt{64+225}$$
$$=\sqrt{289}$$
$$=17$$

### Section 8.3

**3.** We must simplify each radical. Then if the radicals are like, we add or subtract their coefficients and keep the radical the same.

### Section 8.4

**1.** To multiply square root expressions, we multiply the coefficients of the square root expressions and multiply the radicands.
**3.** When conjugates are multiplied, the outer and inner terms add to zero and the first and last terms are squared.
**4.** To divide square root expressions, we divide the coefficients of the square root expressions and then divide the radicands.
**5.** It must be multiplied by $\dfrac{\sqrt{3}}{\sqrt{3}}$. This makes the denominator $\sqrt{9}$ which is 3.
**6.** To rationalize the denominator, we must multiply the numerator and denominator of the expression by the conjugate of $\sqrt{5}+\sqrt{3}$, which is $\sqrt{5}-\sqrt{3}$.

**22.** $5xy\sqrt[3]{2y^2}$    **24.** $4u^2\sqrt[3]{3v^2}$
**48.** $5\sqrt{3}+5\sqrt{10}$    **50.** $\sqrt{33}+\sqrt{21}$
**141.** The like radicals were combined incorrectly.
$$(\sqrt{2}+\sqrt{5})^2=(\sqrt{2}+\sqrt{5})(\sqrt{2}+\sqrt{5})$$
$$=\sqrt{4}+\sqrt{10}+\sqrt{10}+\sqrt{25}$$
$$=2+2\sqrt{10}+5$$
$$=7+2\sqrt{10}$$

### Section 8.5

**2.** The square root expression must be isolated on one side of the equation, then we square each side. We solve the resulting equation and check the solutions in the original equation.

**13.** $\left\{-\dfrac{2}{3}\right\}$

**70.** The equation was solved correctly but the answer was not checked.

*Check*: $\sqrt{(-1)^2 - 5(-1) + 3} \overset{?}{=} (-1) - 2$

$3 \neq -3$

Since the solution does not check, there is no solution.

### Section 8.6

**109.** $\dfrac{27}{a^{1/2}b^{11}}$    **110.** $\dfrac{64}{r^{11}s^{7/2}}$

**135.** $\sqrt[3]{x} \cdot \sqrt{x} = x^{1/3} \cdot x^{1/2}$
$= x^{1/3 + 1/2}$
$= x^{2/6 + 3/6}$
$= x^{5/6}$
$= \sqrt[6]{x^5}$

**136.** Sophia worked the problem correctly.
$\dfrac{\sqrt{x}}{\sqrt[4]{x}} = \dfrac{x^{1/2}}{x^{1/4}}$
$= x^{1/2 - 1/4}$
$= x^{2/4 - 1/4}$
$= x^{1/4}$
$= \sqrt[4]{x}$

## CHAPTER 8 REVIEW EXERCISES

### Section 8.2

**29.** $\dfrac{2x^4\sqrt[4]{7x}}{y^3}$

### Section 8.4

**63.** $\dfrac{8 - 2\sqrt{7}}{3}$

## CHAPTERS 1–8 CUMULATIVE REVIEW EXERCISES

**5. b.**

| $x$ | $y$ |
|---|---|
| $-1$ | $-\dfrac{1}{3}$ |
| $0$ | $-\dfrac{1}{4}$ |
| $1$ | $0$ |
| $2$ | Undefined |

**7. a.** $\{(0, 83.93), (1, 90.51), (2, 92.47), (3, 98.04), (4, 87.05)\}$
**b.** The average fare per revenue passenger flying the AirTran Airways was about \$83.93 in 2005 and \$87.05 in 2009.
**c.** The average fare per revenue passenger was the highest in 2008 and the lowest in 2005.
**d.**

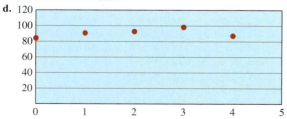

**9. a.** $\{(-3, 2)\}$, the system is consistent, the equations are independent

**b.** $\varnothing$, two parallel lines, the system is inconsistent

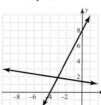

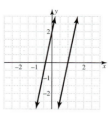

**10. a.**

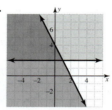

**b.**

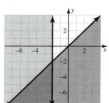

**12. a.** $\dfrac{32}{p^{10}}$   **b.** $\dfrac{x^6 y^{15}}{64}$   **c.** $-\dfrac{1}{27a^{15}d^6}$

**13. b.** $14x^3y - 7x^2y^2 + 21xy^3$   **c.** $16x^2 - 81y^2$

**14. a.** $2y(5x - 2y)^2$   **b.** $(a + 5)(a - 5)(a - 1)(a + 1)$
**c.** $(3x + 1)(12x - 5)$   **d.** $(3a + 2b)(9a^2 - 6ab + 4b^2)$
**f.** $(10p - q)(100p^2 + 10pq + q^2)$   **16. b.** $\{-2, 2, -3, 3\}$

**18. a.**

| $x$ | $y$ |
|---|---|
| $-1$ | $0$ |
| $0$ | $0$ |
| $2$ | Undefined |

**b.** $\dfrac{4y}{2y - 3}$ for $y = -1, 0, 1.5$

| $x$ | $y$ |
|---|---|
| $-1$ | $\dfrac{4}{5}$ |
| $0$ | $0$ |
| $1.5$ | Undefined |

**20. a.** $x = \dfrac{2}{3}, -4; \dfrac{1}{3x - 2}$   **b.** $x = -3, 0, 3; \dfrac{2}{x - 3}$

**21. a.** $-\dfrac{x + 5}{x - 7}$ or $\dfrac{-x - 5}{x - 7}$   **b.** $\dfrac{x - 2}{7x + 3}$ or $\dfrac{2 - x}{-7x - 3}$

**25. b.** $\dfrac{2x^2 - 5x - 4}{(x - 6)(x + 4)}$   **c.** $\dfrac{6x + 28}{(x - 3)(x + 7)}$   **d.** $\dfrac{-x + 13}{(x - 1)(x + 1)^2}$

**26. c.** $\dfrac{xy}{4y - 3x}$   **33. e.** $5a^2\sqrt[3]{2b} + 30a^3\sqrt[3]{2b}$   **f.** $10a\sqrt{ab}$

## CHAPTER 9

### Section 9.1

**1.** First isolate the squared term to one side of the equation and constant to the other side. Take the square root of both sides of the equation and solve the resulting equations.

**3.** For example, $(x - 3)^2 = 25$. We apply the square root property and get $x - 3 = 5$ or $x - 3 = -5$. Solving each equation, we obtain $x = 8$ or $x = -2$.

**29.** $\left\{\dfrac{3 - 2\sqrt{2}}{2}, \dfrac{3 + 2\sqrt{2}}{2}\right\}$   **30.** $\left\{\dfrac{-1 - 2\sqrt{5}}{5}, \dfrac{-1 + 2\sqrt{5}}{5}\right\}$

**53.** $\left\{\dfrac{-3 - 6\sqrt{2}}{2}, \dfrac{-3 + 6\sqrt{2}}{2}\right\}$   **54.** $\left\{\dfrac{1 - 2\sqrt{13}}{7}, \dfrac{1 + 2\sqrt{13}}{7}\right\}$

**57.** When applying the square root property, there are two solutions.
$a = \sqrt{16}$   or   $a = -\sqrt{16}$
$a = 4$     or   $a = -4$

**58.** When applying the square root property, there are two solutions.

$$x + 1 = \sqrt{9} \quad \text{or} \quad x + 1 = -\sqrt{9}$$
$$x + 1 = 3 \qquad\qquad x + 1 = -3$$
$$x = 2 \qquad\qquad\quad x = -4$$

**59.** The squared variable must be isolated before applying the square root property. So, subtract 1 from each side first.

$$4t^2 + 1 = 9 \Rightarrow 4t^2 = 8 \Rightarrow t^2 = 2 \Rightarrow t = \pm\sqrt{2}$$

**63.**

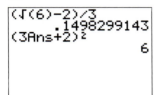

**64.**

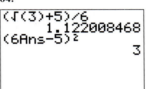

## Section 9.2

**2.** The method of completing the square can be applied to solve all quadratic equations while not all quadratic equations can be solved by the factoring method.

**3. a.** If the coefficient of the squared term is 1, then apply the square root property and solve for the unknown.

   **b.** If the coefficient of the squared term in not 1, then divide the coefficient into both sides of the equation. Apply the square root property and solve the resulting equations.

**4.** Since the square of the coefficient of $x$ is 144, then the coefficient must be twice the square root of 144. The binomial must be $x^2 + 24x$ or $x^2 - 24x$.

**11.** $a^2 + 3a + \dfrac{9}{4} = \left(a + \dfrac{3}{2}\right)^2$     **12.** $a^2 + 5a + \dfrac{25}{4} = \left(a + \dfrac{5}{2}\right)^2$

**14.** $x^2 - 11x + \dfrac{121}{4} = \left(x - \dfrac{11}{2}\right)^2$     **23.** $\left\{\dfrac{3 - \sqrt{17}}{2}, \dfrac{3 + \sqrt{17}}{2}\right\}$

**24.** $\left\{\dfrac{5 - \sqrt{57}}{2}, \dfrac{5 + \sqrt{57}}{2}\right\}$     **36.** $\left\{-1 - \dfrac{\sqrt{14}}{2}, -1 + \dfrac{\sqrt{14}}{2}\right\}$

**37.** $\left\{-1 - \dfrac{\sqrt{42}}{6}, -1 + \dfrac{\sqrt{42}}{6}\right\}$     **42.** $a^2 - 13a + \dfrac{169}{4} = \left(a - \dfrac{13}{2}\right)^2$

**45.** $a^2 + 15a + \dfrac{225}{4} = \left(a + \dfrac{15}{2}\right)^2$     **46.** $a^2 - 7a + \dfrac{49}{4} = \left(a - \dfrac{7}{2}\right)^2$

**47.** $x^2 + \dfrac{2}{3}x + \dfrac{1}{9} = \left(x + \dfrac{1}{3}\right)^2$     **48.** $y^2 - \dfrac{2}{5}x + \dfrac{1}{25} = \left(y - \dfrac{1}{5}\right)^2$

**55.** $\left\{\dfrac{3}{2} - \dfrac{\sqrt{205}}{10}, \dfrac{3}{2} + \dfrac{\sqrt{205}}{10}\right\}$     **56.** $\left\{\dfrac{2}{3} - \dfrac{\sqrt{13}}{3}, \dfrac{2}{3} + \dfrac{\sqrt{13}}{3}\right\}$

**59.** The error occurred when dividing the equation by 2. The middle coefficient should be $-1$ not $-2$.

$$2x^2 - 2x - 4 = 0$$
$$\dfrac{2x^2 - 2x - 4}{2} = \dfrac{0}{2}$$
$$x^2 - x - 2 = 0$$
$$x^2 - x = 2$$
$$x^2 - x + \dfrac{1}{4} = 2 + \dfrac{1}{4}$$
$$\left(x - \dfrac{1}{2}\right)^2 = \dfrac{9}{4} \Rightarrow x - \dfrac{1}{2} = \pm\dfrac{3}{2} \Rightarrow x = \dfrac{1}{2} \pm \dfrac{3}{2}$$
$$\Rightarrow x = 2 \quad \text{or} \quad -1$$

**60.** The number that completes the square is $\left(\dfrac{1}{2}\right)^2$ not $\dfrac{1}{2}$.

$$x^2 + x - 3 = 0$$
$$x^2 + x = 3$$
$$x^2 + 1x = 3$$
$$x^2 + 1x + \dfrac{1}{4} = 3 + \dfrac{1}{4}$$
$$\left(x + \dfrac{1}{2}\right)^2 = \dfrac{13}{4}$$
$$x + \dfrac{1}{2} = \pm\sqrt{\dfrac{13}{4}}$$
$$x = -\dfrac{1}{2} \pm \dfrac{\sqrt{13}}{2}$$

**61.** If the coefficient of $x$ is an odd number, say, $x^2 + ax$, then the pattern of the factor is $x^2 + ax = \left(x + \dfrac{a}{2}\right)^2$ and $\dfrac{a^2}{4}$ is added to both sides of the equation.

**62.** For this case, factoring method will be easier since there are factors of 24 whose sum is $-10$, namely $-4$ and $-6$.

$$x^2 - 10x + 24 = 0$$
$$(x - 4)(x - 6) = 0$$
$$x - 4 = 0 \quad \text{or} \quad x - 6 = 0$$
$$x = 4 \quad \text{or} \qquad x = 6$$

**63.**

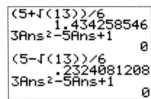

**64.**

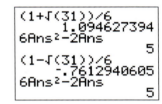

## Section 9.3

**1.** If $a$, $b$, and $c$ are real numbers, a quadratic equation written in the form $ax^2 + bx + c = 0$, $a \neq 0$, has solutions

$$x = \dfrac{-b \pm \sqrt{b^2 - 4ac}}{2a}.$$

**2.** If the leading coefficient $a$ is 1 and the coefficient of $x$ is an even number, then I prefer to solve by using completing the square. Otherwise, I prefer to solve by using the quadratic formula.

**3.** For example, solve the equation $2x^2 - 7x - 1 = 0$. Since the leading coefficient is 2 and the coefficient of $x$ is $-7$, it will be easier to use the quadratic formula.

$$x = \dfrac{-(-7) \pm \sqrt{(-7)^2 - 4(2)(-1)}}{2(2)} = \dfrac{7 \pm \sqrt{49 + 8}}{4} = \dfrac{7 \pm \sqrt{57}}{4}$$

**4.** For example, solve the equation $x^2 + 6x - 2 = 0$. Since the leading coefficient is 1 and the coefficient of $x$ is 6, it will be easier to complete the square.

$$x^2 + 6x = 2$$
$$x^2 + 6x + 9 = 2 + 9$$
$$(x + 3)^2 = 11$$
$$x + 3 = \pm\sqrt{11}$$
$$x = -3 \pm \sqrt{11}$$

**19.** $\left\{-1, \dfrac{9}{2}\right\}$     **20.** $\left\{-\dfrac{2}{3}, 2\right\}$     **21.** $\left\{-2, \dfrac{4}{3}\right\}$

**22.** $\left\{-\dfrac{3}{2}, 4\right\}$     **23.** $\left\{\dfrac{7 - \sqrt{29}}{2}, \dfrac{7 + \sqrt{29}}{2}\right\}$

**24.** $\{4 - \sqrt{14}, 4 + \sqrt{14}\}$    **25.** $\left\{\dfrac{1}{4}, 1\right\}$    **26.** $\left\{-\dfrac{2}{5}, 1\right\}$

**27.** $\left\{\dfrac{-3 - \sqrt{105}}{12}, \dfrac{-3 + \sqrt{105}}{12}\right\}$    **28.** $\left\{\dfrac{3 - \sqrt{37}}{7}, \dfrac{3 + \sqrt{37}}{7}\right\}$

**29.** $\left\{0, \dfrac{1}{10}\right\}$    **30.** $\left\{0, \dfrac{3}{8}\right\}$    **31.** $\left\{\dfrac{2 - 3\sqrt{6}}{5}, \dfrac{2 + 3\sqrt{6}}{5}\right\}$

**32.** $\left\{\dfrac{2 - \sqrt{13}}{2}, \dfrac{2 + \sqrt{13}}{2}\right\}$

**39.** $\left\{\dfrac{5 - \sqrt{15}}{3}, \dfrac{5 + \sqrt{15}}{3}\right\}$; $\{0.38, 2.96\}$

**40.** $\left\{\dfrac{4 - \sqrt{10}}{2}, \dfrac{4 + \sqrt{10}}{2}\right\}$; $\{0.42, 3.58\}$

**41.** $\left\{\dfrac{-1 - \sqrt{22}}{3}, \dfrac{-1 + \sqrt{22}}{3}\right\}$; $\{-1.90, 1.23\}$

**42.** $\left\{\dfrac{-5 - \sqrt{33}}{4}, \dfrac{-5 + \sqrt{33}}{4}\right\}$; $\{-2.69, 0.19\}$

**65.** 65; two nonrepeating real solutions; $\left\{\dfrac{-3 - \sqrt{65}}{4}, \dfrac{-3 + \sqrt{65}}{4}\right\}$; $\{-2.77, 1.27\}$

**66.** 45; two nonrepeating real solutions; $\left\{\dfrac{-1 - 3\sqrt{5}}{2}, \dfrac{-1 + 3\sqrt{5}}{2}\right\}$; $\{-3.85, 2.85\}$

**67.** 81; two nonrepeating real solutions; $\left\{-\dfrac{1}{2}, \dfrac{2}{5}\right\}$; $\{-0.5, 0.4\}$

**68.** 9; two nonrepeating real solutions; $\left\{0, \dfrac{1}{2}\right\}$; $\{0, 0.5\}$

**71.** 100; two nonrepeating solutions; $\{-2, 0\}$

**75.** 0; one repeating real solution; $\left\{\dfrac{2}{3}\right\}$

**77.** 76; two nonrepeating real solutions; $\left\{\dfrac{2 - \sqrt{19}}{3}, \dfrac{2 + \sqrt{19}}{3}\right\}$; $\{-0.79, 2.12\}$

**78.** 28; two nonrepeating real solutions; $\left\{\dfrac{5 - \sqrt{7}}{2}, \dfrac{5 + \sqrt{7}}{2}\right\}$; $\{1.18, 3.82\}$

**82.** The equation must be written in standard form first.
$$4x^2 + 16x - 13 = 0$$
$$x = \dfrac{-16 \pm \sqrt{(16)^2 - 4(4)(-13)}}{2(4)}$$
$$x = \dfrac{-16 \pm \sqrt{256 + 208}}{8}$$
$$x = \dfrac{-16 \pm \sqrt{464}}{8}$$
$$x = \dfrac{-16 \pm 4\sqrt{29}}{8}$$
$$x = \dfrac{-4 \pm \sqrt{29}}{2}$$

### Piece It Together  Sections 9.1–9.3

**3.** $\left\{\dfrac{-2 - 2\sqrt{7}}{3}, \dfrac{-2 + 2\sqrt{7}}{3}\right\}$    **4.** $\left\{\dfrac{1 - 2\sqrt{19}}{2}, \dfrac{1 + 2\sqrt{19}}{2}\right\}$

**10.** $\left\{-\dfrac{3}{2}, \dfrac{7}{3}\right\}$    **11.** $\left\{\dfrac{10 - \sqrt{2}}{3}, \dfrac{10 + \sqrt{2}}{3}\right\}$    **13.** $\left\{-3, \dfrac{5}{8}\right\}$

**15.** $\left\{\dfrac{2 - 5\sqrt{2}}{3}, \dfrac{2 + 5\sqrt{2}}{3}\right\}$    **16.** $\left\{\dfrac{-3 - 5\sqrt{3}}{4}, \dfrac{-3 + 5\sqrt{3}}{4}\right\}$

### Section 9.4

**1.** A complex number is a number of the form $a + bi$, where $a$ is the real part and $b$ is the imaginary part of the complex number.
**2.** Every imaginary number is a complex number because we can write 0 as the real part, that is, $bi = 0 + bi$.

**3.** No, not every complex number is an imaginary number because the imaginary part of the complex number can be 0.
**4.** Yes, every real number is a complex number because we can write 0 as the imaginary part, that is, $a = a + 0i$.
**5.** For example, to write $\sqrt{-169}$ as a complex number, rewrite it as $\sqrt{-1 \cdot 169}$. Now apply the product rule and simplify, $\sqrt{-1 \cdot 169} = \sqrt{-1}\sqrt{169} = i \cdot 13 = 0 + 13i$.
**6.** To add and subtract complex numbers, combine the real parts and imaginary parts.
**7.** To multiply complex numbers, use the distributive property or FOIL. Rewrite $i^2$ as $-1$ and combine like terms.
**8.** Multiply the product $(a + bi)(a - bi)$. We get $a^2 + abi - abi - b^2i^2 = a^2 + b^2$, which is always a real number.

**61.** $\left\{\dfrac{3 - 4i}{2}, \dfrac{3 + 4i}{2}\right\}$    **62.** $\left\{\dfrac{-1 - i}{3}, \dfrac{-1 + i}{3}\right\}$

**63.** $\left\{-\dfrac{3}{2} - \dfrac{\sqrt{3}}{2}i, -\dfrac{3}{2} + \dfrac{\sqrt{3}}{2}i\right\}$    **64.** $\left\{-\dfrac{1}{2} - \dfrac{\sqrt{7}}{2}i, -\dfrac{1}{2} + \dfrac{\sqrt{7}}{2}i\right\}$

**65.** $\left\{\dfrac{1 - \sqrt{17}i}{6}, \dfrac{1 + \sqrt{17}i}{6}\right\}$    **66.** $\left\{\dfrac{1 - \sqrt{31}i}{4}, \dfrac{1 + \sqrt{31}i}{4}\right\}$

**67.** $\left\{\dfrac{3 - \sqrt{11}i}{2}, \dfrac{3 + \sqrt{11}i}{2}\right\}$    **68.** $\left\{1 - \dfrac{i\sqrt{15}}{5}, 1 + \dfrac{i\sqrt{15}}{5}\right\}$

**83.** $\dfrac{49}{100} + \dfrac{1}{10}i$    **99.** $\left\{\dfrac{1 - \sqrt{3}i}{2}, \dfrac{1 + \sqrt{3}i}{2}\right\}$

**100.** $\left\{\dfrac{3 - \sqrt{6}i}{5}, \dfrac{3 + \sqrt{6}i}{5}\right\}$

### Section 9.5

**1.** To find solutions of the equation $y = ax^2 + bx + c$, make a table and assign values to $x$. Substitute each value of $x$ in the equation and solve for the corresponding $y$-value. Write the solutions in the form of $(x, y)$.
**2.** To find the solutions of the quadratic equation $ax^2 + bx + c = 0$, we can use factoring, the square root property, completing the square, and the quadratic formula.
**4.** The key points on the graph of a parabola are the highest or lowest point called the vertex, the $x$-intercepts, and the $y$-intercept.
**6.** The $x$-coordinate of the vertex is given as $-\dfrac{b}{2a}$ and the $y$-coordinate is found by substituting this value into the equation.
**7.** The graph of the equation never intersects the $x$-axis. This means that there are no real solutions of the equation, and the solutions will be complex numbers.
**8.** We need to find additional points by selecting any $x$-value and solving for the corresponding $y$-value.

**9.**

| $x$ | $y$ | $(x, y)$ |
|---|---|---|
| $-2$ | 10 | $(-2, 10)$ |
| $-1$ | 4 | $(-1, 4)$ |
| 0 | 0 | $(0, 0)$ |
| 1 | $-2$ | $(1, -2)$ |
| 2 | $-2$ | $(2, -2)$ |
| 3 | 0 | $(3, 0)$ |

**10.**

| $x$ | $y$ | $(x, y)$ |
|---|---|---|
| $-2$ | 6 | $(-2, 6)$ |
| $-1$ | 2 | $(-1, 2)$ |
| 0 | 0 | $(0, 0)$ |
| 1 | 0 | $(1, 0)$ |
| 2 | 2 | $(2, 2)$ |
| 3 | 6 | $(3, 6)$ |

**11.**

| $x$ | $y$ | $(x, y)$ |
|---|---|---|
| $-2$ | 9 | $(-2, 9)$ |
| $-1$ | 4 | $(-1, 4)$ |
| 0 | 1 | $(0, 1)$ |
| 1 | 0 | $(1, 0)$ |
| 2 | 1 | $(2, 1)$ |
| 3 | 4 | $(3, 4)$ |

**12.**

| $x$ | $y$ | $(x, y)$ |
|---|---|---|
| $-2$ | 0 | $(-2, 0)$ |
| $-1$ | $-4$ | $(-1, -4)$ |
| 0 | $-6$ | $(0, -6)$ |
| 1 | $-6$ | $(1, -6)$ |
| 2 | $-4$ | $(2, -4)$ |
| 3 | 0 | $(3, 0)$ |

**13.**

| x | y | (x, y) |
|---|---|---|
| −2 | −8 | (−2, −8) |
| −1 | −9 | (−1, −9) |
| 0 | −8 | (0, −8) |
| 1 | −5 | (1, −5) |
| 2 | 0 | (2, 0) |
| 3 | 7 | (3, 7) |

**14.**

| x | y | (x, y) |
|---|---|---|
| −2 | 5 | (−2, 5) |
| −1 | 0 | (−1, 0) |
| 0 | −3 | (0, −3) |
| 1 | −4 | (1, −4) |
| 2 | −3 | (2, −3) |
| 3 | 0 | (3, 0) |

**25.**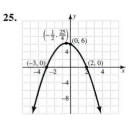

**26.**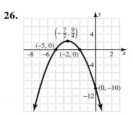

**15.**

| x | y | (x, y) |
|---|---|---|
| −2 | 25 | (−2, 25) |
| −1 | 16 | (−1, 16) |
| 0 | 9 | (0, 9) |
| 1 | 4 | (1, 4) |
| 2 | 1 | (2, 1) |
| 3 | 0 | (3, 0) |

**16.**

| x | y | (x, y) |
|---|---|---|
| −2 | 16 | (−2, 16) |
| −1 | 9 | (−1, 9) |
| 0 | 4 | (0, 4) |
| 1 | 1 | (1, 1) |
| 2 | 0 | (2, 0) |
| 3 | 1 | (3, 1) |

**27.**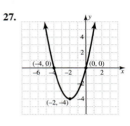

**28.**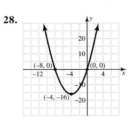

**17.**

| x | y | (x, y) |
|---|---|---|
| −2 | −11 | (−2, −11) |
| −1 | −4 | (−1, −4) |
| 0 | −1 | (0, −1) |
| 1 | −2 | (1, −2) |
| 2 | −7 | (2, −7) |
| 3 | −16 | (3, −16) |

**18.**

| x | y | (x, y) |
|---|---|---|
| −2 | −18 | (−2, −18) |
| −1 | −7 | (−1, −7) |
| 0 | −2 | (0, −2) |
| 1 | −3 | (1, −3) |
| 2 | −10 | (2, −10) |
| 3 | −23 | (3, −23) |

**29.**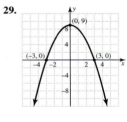

**30.**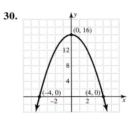

**19.**

| x | y | (x, y) |
|---|---|---|
| −2 | 4 | (−2, 4) |
| −1 | 7 | (−1, 7) |
| 0 | 8 | (0, 8) |
| 1 | 7 | (1, 7) |
| 2 | 4 | (2, 4) |
| 3 | −1 | (3, −1) |

**20.**

| x | y | (x, y) |
|---|---|---|
| −2 | 2 | (−2, 2) |
| −1 | 5 | (−1, 5) |
| 0 | 6 | (0, 6) |
| 1 | 5 | (1, 5) |
| 2 | 2 | (2, 2) |
| 3 | −3 | (3, −3) |

**31.**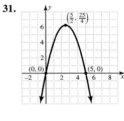

**32.**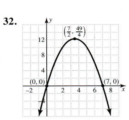

**21.**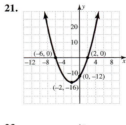

**22.**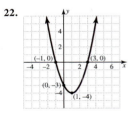

**33.**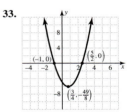

**34.**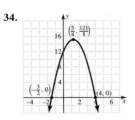

**23.**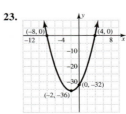

**24.**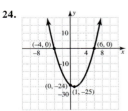

**35.**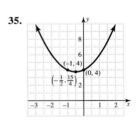

**36.**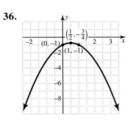

**47.**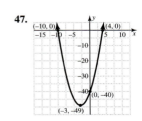

**48.**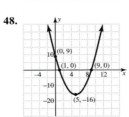

**49.** 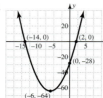    **50.**

**c.**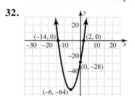

## CHAPTER 9 REVIEW EXERCISES

### Section 9.1

**5.** $\left\{ \dfrac{1 - 2\sqrt{6}}{2}, \dfrac{1 + 2\sqrt{6}}{2} \right\}$    **7.** $\left\{ -\dfrac{3}{5}, \dfrac{9}{5} \right\}$

### Section 9.2

**13.** $a^2 - 20a + 100 = (a - 10)^2$    **14.** $b^2 + 3.6b + 3.24 = (b + 1.8)^2$

**15.** $x^2 + 11x + \dfrac{121}{4} = \left( x + \dfrac{11}{2} \right)^2$    **16.** $y^2 - \dfrac{6}{5}y + \dfrac{9}{25} = \left( y - \dfrac{3}{5} \right)^2$

**19.** $\left\{ \dfrac{-3 - \sqrt{6}}{2}, \dfrac{-3 + \sqrt{6}}{2} \right\}$    **20.** $\left\{ \dfrac{2 - \sqrt{10}}{6}, \dfrac{2 + \sqrt{10}}{6} \right\}$

### Section 9.3

**29.** $\left\{ \dfrac{-1 - 3\sqrt{2}}{2}, \dfrac{-1 + 3\sqrt{2}}{2} \right\}$

### Section 9.4

**48.** $\dfrac{-4 + i\sqrt{17}}{6}$    **61.** $\left\{ \dfrac{1 - i\sqrt{3}}{2}, \dfrac{1 + i\sqrt{3}}{2} \right\}$

### Section 9.5

**69.** 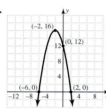    **70.**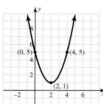

## CHAPTER 9 TEST

1. The four methods for solving a quadratic equation are factoring, the square root property, completing the square, and the quadratic formula. Completing the square and the quadratic formula can be used to solve any quadratic equation.

2. The discriminant determines the number and types of solutions of a quadratic equation. It can be found by evaluating $b^2 - 4ac$, where $ax^2 + bx + c = 0$, $a \neq 0$. If the discriminant is positive, the equation has two real solutions. If the discriminant is zero, the equation has one real solution. If the discriminant is negative, the equation has no real solutions (but two complex solutions).

**11. c.** $\left\{ \dfrac{-1 - i\sqrt{19}}{2}, \dfrac{-1 + i\sqrt{19}}{2} \right\}$ or $\left\{ -\dfrac{1}{2} - \dfrac{\sqrt{19}}{2}i, -\dfrac{1}{2} + \dfrac{\sqrt{19}}{2}i \right\}$

**18. a.** 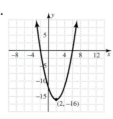    **b.**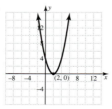

## CHAPTERS 1–9 CUMULATIVE REVIEW EXERCISES

**5. a.** (0, 87.646); The average fare per revenue passenger flying AirTran Airways was about $87.65 in 2005.
   **b.** The average fare, excluding transportation taxes, per revenue passenger flying AirTran Airways was $94.53 in 2010.

**8. a.** $\{(3, -2)\}$, consistent system with independent equations
   **b.** no solution, inconsistent system

**9. a.** $\begin{cases} y \le 21 - 0.4x \\ y \ge 11 \\ x \ge 0 \end{cases}$    **11. b.** $8x^4y^3 - 16x^3y^4 + 40x^2y^5$

   **c.** $4x^2 - 7xy - 15y^2$    **12. a.** $2xy(5x + 9y - 2y^2)$
   **b.** $(p - 1)(p + 1)(p - 4)(p + 4)$    **c.** $(4a - 3b)(4a - 3b)$

**13. a.** $\left\{ -7, \dfrac{5}{2} \right\}$    **14. d.** $\dfrac{2x^2 + 4x - 3}{(x - 3)(x + 6)}$    **e.** $\dfrac{6x - 16}{(x + 8)(x - 4)}$

   **f.** $\dfrac{5 - x}{2(x - 1)(x^2 + x + 1)}$    **17. a.** $c = \dfrac{3abd}{bd - 2ad - 4ab}$

   **b.** $y_2 = 2y_m - y_1$    **23. f.** $\left\{ \dfrac{5 - 6\sqrt{2}}{3}, \dfrac{5 + 6\sqrt{2}}{3} \right\}$

**27. a.** $\{5 - 2\sqrt{3}, 5 + 2\sqrt{3}\}$

**28. a.** two real solutions; $\{-4.24, 0.24\}$
   **b.** two real solutions; $\{-0.50, -0.33\}$    **30. b.** $8 - 20i$

**32.**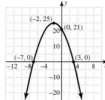

# Credits

## Photo Credits

### Design Icons

**Write about it:** © Blend Images/Getty RF; **You be the teacher:** © Rubberball/Getty RF; **Practice makes perfect, Think about it, Calculate it, Mix' em up:** © Getty RF; **Note, Piece it together:** © PhotoDisc/Getty RF; **Group Activity:** © Banana Stock/PictureQuest RF.

### Chapter S

**Opener:** © Purestock/Punchstock RF; **p. S2(top):** Library of Congress Prints and Photographs Division [LC-USW33-019093-C]; **p. S2(right):** © Getty RF; **p. S3:** © Ingram Publishing/Superstock RF; **p. S6(top):** © Janice Ortiz; **p. S6(right), S7, p. S8(top, bottom), p. S8(bottom), p. S9:** © Getty RF; **p. S10:** © Alamy RF; **p. S11(top):** © Todd Hendricks; **p. S11(right):** © Punchstock RF; **p. S11(bottom):** © Getty RF; **p. S12:** © Ingram Publishing RF; **p. S13:** © Corbis RF; **p. S14:** © Getty RF; **p. S15:** © Ingram Publishing RF; **p. S16, p. S18:** © Getty RF; **p. S19(top):** © Todd Hendricks; **p. S19(right):** © Veer RF; **p. S20:** © Corbis RF; **p. S23, p. S24:** © Getty RF; **p. S25:** © Getty RF; **p. S25(top):** Courtesy of Grace Shyu; **p. S25(bottom):** © Veer RF; **p. S25(right):** © Getty RF; **p. S26:** © Veer RF; **p. S27, p. S25, p. S28:** © Getty RF.

### Chapter 1

**Opener:** © Stockbyte/Getty RF; **p. 1(bottom):** © Brand X.Punchstock RF; **p. 2(top):** © Najlah Feanny/Corbis; **p. 2(bottom):** © Getty RF; **p. 13:** © The McGraw-Hill Companies, Inc./John Flournoy, photographer; **p. 15:** © Getty RF; **p. 27:** © Punchstock RF; **p. 34, p. 35:** © Getty RF; **p. 39:** © Corbis RF; **p. 40(top):** © Picturequest RF; **p. 40(bottom):** © RubberBall RF; **p. 41:** © Getty RF; **p. 46:** © Alamy RF; **p. 48:** © The McGraw-Hill Companies, Inc./Jill Braaten, photographer; **p. 49:** © Jupiter RF; **p. 50, p. 53:** © Alamy RF; **p. 56(top left):**© Corbis RF; **p. 56(bottom left):** © Getty RF; **p. 56(right):** © Punchstock RF; **p. 57:** © Corbis RF; **p. 64:** © The McGraw-Hill Companies, Inc./Mark Dierker, photographer; **p. 84:** © Getty RF; **p. 90:** © Getty RF.

### Chapter 2

**Opener:** © Corbis RF; **p. 91(bottom):** © Comstock RF; **p. 92:** Library of Congress, Prints and Photographs Division. (LC-USZ62-117122), **p. 96:** © Jupiter RF; **p. 99(left):** © Corbis RF; **p. 99(right), p. 101, p. 107(top):** © Getty RF; **p. 107(bottom):** © Comstock/Picturequest RF; **p. 110(top):** © Getty RF; **p. 110(bottom):** © The McGraw-Hill Companies, Inc./ Mark Dierker, photographer; **p. 112, p. 118:** © Brand X RF; **p. 136(top left):** © Getty RF; **p. 136(right):** © Brand X Pictures/PunchStock RF; **p. 136(bottom left):** © Getty RF; **p. 138, p. 139(bottom):** © Brand X/PunchStock RF; **139(top):** © Brand X Pictures/PunchStock RF; **p. 153:** © Alamy RF; **p. 154:** © The McGraw-Hill Companies, Inc./John Flournoy, photographer; **p. 157(top, middle):** © Getty RF; **p. 158:** © Comstock/Punchstock RF; **p. 162:** © Alamy RF; **p. 163(top, bottom):** © Getty RF; **p. 166:** © Corbis RF; **p. 174:** © Getty RF; **p. 175:** © Brand X/Punchstock RF; **p. 180(left):** © Getty RF; **p. 180(right):** © Copyright 1997 IMS Communications Ltd./Capstone Design. All Rights Reserved; **p. 181, p. 188:** © Corbis RF.

### Chapter 3

**Opener:** U.S. Department of Defense; **p. 193(bottom):** © Alamy RF; **p. 194(football):** © Getty RF; **p. 194(top):** © Eric Gay/AP Photo; **p. 200:** © Rubberball RF; **p. 206:** © Alamy RF; **p. 223:** © The McGraw-Hill Companies, Inc./John Flournoy, photographer; **p. 224(left):** © BananaStock/PunchStock RF; **p. 224(left):** © Creatas Images/PictureQuest RF; **p. 239:** © Getty RF; **p. 241:** © Corbis RF; **p. 247:** © Getty RF; **p. 249, 258:** © Corbis RF; **p. 264, p. 266(top, bottom), p. 273:** © Getty RF.

### Chapter 4

**Opener:** © Superstock RF; **p. 293(bottom):** © The McGraw-Hill Companies, Inc./Jill Braaten, photographer; **p. 294(top):** Photo courtesy of The Napoleon Hill Foundation; **p. 294(right):** © Rubberball RF; **p. 319, p. 327:** © The McGraw-Hill Companies, Inc./Jill Braaten, photographer; **p. 331:** © Corbis RF; **p. 332(top left):** © Brand X/Punchstock RF; **p. 332(bottom left), p. 333:** © Getty RF; **p. 334:** © Corbis RF; **p. 335:** © Stockbyte/Getty RF; **p. 336:** USDA Natural Resources Conservation Services/Photo by Jeff Vanuga; **p. 337:** © Corbis RF; **p. 340:** © Corbis RF; **p. 341, p. 343(bottom, right), p. 343(right), p. 344(bottom left):** © Getty RF; **p. 344(top left):** © Digital Vision/Punchstock RF; **p. 344(middle left):** © Getty RF; **p. 345:** © Digital Vision RF; **p. 346, 351:** © Punchstock RF; **p. 357, p. 359:** © Ingram Publishing RF; **p. 364:** © Getty RF; **369(top left):** © Getty RF; **p. 369:** © Corbis RF.

### Chapter 5

**Opener:** © Getty RF; **p. 375(bottom):** © Ingram Publishing RF; **p. 375(right):** © Getty RF; **p. 376(top):** Used with permission from Jim Rohn International ©2011; **p. 376(bottom), 381:** © Getty RF; **p. 385:** © Jupiter RF; **p. 392:** © Ingram Publishing RF; **p. 393:** © Getty RF; **p. 398(right):** © Punchstock RF; **p. 398(left), p. 401:** © Getty RF; **p. 402:** © Alamy RF; **p. 405, 411(top):** © Getty RF; **p. 411(bottom):** © Lars A. Niki; **p. 417, p. 420:** © Corbis RF; **p. 439:** © Goodshoot/PunchStock RF.

### Chapter 6

**Opener:** © Corbis RF; **p. 449(bottom):** © PictureQuest RF; **p. 450(top):** © Todd Hendricks; **p. 450(bottom), p. 454, p. 460, p. 463:** © Getty RF; **p. 467, p. 470:** © Lars A. Niki; **p. 484:** © Brand X Pictures/Superstock RF; **p. 495:** © Getty RF; **p. 497:** © Photolibrary RF; **p. 499(bottom):** © Getty RF; **p. 499(top):** © Lars A. Niki; **p. 501:** © Brand X Pictures/Superstock RF; **p. 504(left, right):** © Getty RF; **p. 505:** © Brand X RF; **p. 510:** © Getty RF.

### Chapter 7

**Opener:** © Getty RF; **p. 513(bottom):** © SuperStock RF; **p. 514(top):** Library of Congress Prints and Photographs Division[LC-USZ62-60242]; **p. 514(right):** © PictureQuest RF; **p. 525:** © The McGraw-Hill Companies, Inc./John Flournoy, photographer; **p. 528(top):** © Getty RF; **p. 528(bottom), p. 529:** © Corbis RF; **p. 531(top):** © IMS Communications Ltd./Capstone Design/FlatEarth Images; **p. 531(top left):** © Corbis RF; **p. 531(bottom):** © Getty RF; **p. 532:** © Alamy RF; **p. 549:** © Brand X RF; **p. 553, p. 558:** © Corbis RF; **p. 568, p. 570, p. 571, p. 573(top):** © Getty RF; **p. 573(bottom):** SuperStock RF.

### Chapter 8

**Opener:** © Ingram Publishing RF; **p. 587(bottom):**© SuperStock RF; **p. 588:** Photo courtesy Robbins Research International, Inc.; **p. 628:** © Punchstock/Stockbyte RF; **p. 639:** © Corbis RF; **p. 645:** © Getty RF; **p. 648, 649:** © SuperStock RF.

### Chapter 9

**Opener:** © Corbis RF; **p. 657(bottom):** © Getty RF; **p. 658:** © PictureQuest RF; **p. 661:** © Getty RF; **p. 662(top):** © PictureQuest RF; **p. 662(bottom):** © Corbis RF; **p. 664:** © Alamy RF; **p. 673:** © The McGraw-Hill Companies, Inc./Gerald Wofford, photographer; **p. 679:** © Getty RF; **p. 694:** © Passport Stck/AGE Fotostock RF; **p. 708:** © Corbis RF.

## Text Credits

### Chapter S
**Page S-2:** Reproduced with permission of Curtis Brown, London on behalf of the Estate of Sir Winston Churchill. Copyright © Winston S. Churchill.

### Chapter 1
**Page 2:** Stephen Covey. From The 7 habits of highly effective people. © 1989, 2004. Free Press, a division of Simon & Schuster. ISBN 0-7432-7245-5, page 161.

### Chapter 2
**Page 92:** Source: Harry S. Truman, United States President, The Harry S. Truman Library and Museum, Independence, MO.

### Chapter 3
**Page 194:** Source: From Quotable Edie Robinson, by Aaron S. Lee, page 30 ISBN: 978-1931249218. Used by permission of Taylor Trade Publishing, a member of RLPG.

### Chapter 4
**Page 294:** W. Clement Stone (Philanthropist, Author), The W. Clement & Jessie V. Stone Foundation, San Francisco, CA.

### Chapter 5
**Page 376:** Source: Article by Jim Rohan, America's Foremost Business Philosopher, reprinted with permisson from Jim Rohn International ©2010. As a world-renowned author and success expert, Jim Rohn touched millions of lives during his 46-year career as a motivational spearker and messager of positive life change. for more information on Jim and his popular personal achievement resources or to subscribe to the weekly Jim Rohn Newsletter, visit www.JimRohn.com.

### Chapter 6
**Page 450:** Source: Andrea Hendricks, 2011.

### Chapter 7
**Page 514:** ©1987-Current Year Hebrew University and Princeton University Press.

### Chapter 8
**Page 588:** Courtesy of Tony Robbins Productions, San Diego, CA. www.tonyrobbins.com

### Chapter 9
**Page 658:** Source: St. Augustine (Philosopher and theologian)

# Index

## A

absolute value, of real numbers, 8–9, 10
addition
  applications of, 45–46
  associative property of, 72, 75
  commutative property of, 72
  of complex numbers, 686–687, 691
  distributive property over, 73–74, 75
  of fractions
    with common denominators, 20–21, 24
    with unlike denominators, 21–23
  identity property of, 71
  inverse property of, 71–72
  of polynomials, 408–409, 412
  of radical expressions
    with like radicals, 609–610, 612
    not in simplified form, 610–611, 612
  of rational expressions
    with like denominators, 533, 539
    with unlike denominators, 543–544
  of real numbers
    applications of, 45–46
    with different signs, 43–45, 47
    order of operations for, 51–52
    with same signs, 42–43, 47
  translating expressions into symbols, 32–33
addition method. *See* elimination method
addition property of equality
  applications of, 106–107, 116–118
  definition of, 102
  in elimination method, 320–321
  linear equations in one variable solved by,
    102–104, 108, 114–118, 120
  multistep equations solved by, 104–106
addition property of inequality
  definition of, 169
  linear inequalities in one variable solved by,
    169–173
additive identity, 71
additive inverse. *See* opposites
algebraic equations
  applications of, 95–96
  definition of, 92
  solutions to, 93–94, 96
  translating phrases into, 94–95, 97
algebraic expressions
  coefficients in, 77
  definition of, 30
  evaluating, 30–31
    with signed numbers, 63–64
  identifying, 92–93, 96
  simplifying, 79–80
  terminology dealing with, 77
  terms in, 77
    combining, 78–79, 80
    like, 78
    unlike, 78
  translating into symbols, 32–33, 37
applications
  of addition property of equality, 106–107
  of algebraic equations, 95–96

of area, 495–497
of complex fractions, 553–554
of consecutive integers, 497–498, 502–503
of exponents, 381–382
of formulas, 151–165
of functions, 273
of initial value of $y$-intercept, 257–259
of linear equations in one variable, 106–107,
  116–118
  distance, 159–160
  mixtures, 156–159
  percent, 151–154
  simple interest, 154–156
of linear equations in two variables, 217–219
of linear inequalities in one variable, 173–175
of linear inequalities in two variables, 351–352
of models, 501–502
of motion, 573–574
of negative exponents, 392–394
of ordered pairs, 200–202
of polynomials, 411
  division of, 435
  multiplication of, 420–421
of profit, 499–500
of proportions, 570–571
of Pythagorean theorem, 500–501, 633–635
of quadratic equations, 495–505, 661–662
of quadratic formula, 679
of rational equations, 572–574
rational exponents, 645–646
of real numbers
  addition of, 45–46
  division of, 64–66
  multiplication of, 64–66
  subtraction of, 52–54
of revenue, 499–500
of scientific notation, 401–402
of slope, 241
of square root equations, 633–635
of systems of linear equations in two variables,
  302–303, 326–327
  distance, 337–339
  geometry, 339–341
  investment, 334–336, 341
  mixtures, 336–337, 342
  money, 331–334
of systems of linear inequalities in two
  variables, 359–360
of work, 572–573, 574
area applications, 495–497
area formula, 137–139, 140, 141
associative property of addition, 72, 75
associative property of multiplication, 72
average, 64
$ax^2 + bx + c$, factoring by trial and error, 467–476

## B

base
  in exponential expressions, 27
  negative
    raised to even power, 59
    raised to odd power, 59

binomials
  definition of, 405
  division of, by polynomials, 432–435
  factoring, 477–484
    difference of two cubes, 478–480, 482
    difference of two squares, 477–478, 481
    with powers greater than 3, 480–481
    sum of two cubes, 478–480
    sum of two squares, 478, 482
  higher powers of, 428
  multiplication of
    FOIL method for, 424
    by polynomials, 418–419, 421
    product of binomials, 424–425
    product of conjugates, 426–427, 429
    square of binomial, 425–426, 429
  perfect square trinomials from, 666–667
$b^{-m/n}$, evaluating, 641–642
$b^{m/n}$, evaluating, 640–641
$b^{1/n}$, evaluating, 639–640
boundary line, 348

## C

calculator. *See also* graphing calculator
  approximating square roots on, 4
Cartesian coordinate system. *See* rectangular
  coordinate system
circumference formula, 137–139
coefficient
  definition of, 77
  in scientific notation, 399
commitment, 193
common factors
  factoring trinomials with, 463, 464
  in simplifying fractions, 17
commutative property of addition, 72
commutative property of multiplication, 72
complementary angles
  definition of, 52
  equation for, 142, 146
completing the square
  with coefficient of 1 on $x^2$, 667–668, 670
  with coefficient other than 1 on $x^2$,
    669–670, 671
  definition of, 666–667
complex conjugate
  definition of, 688
  product of, 688
complex fractions
  definition of, 549
  simplest form of, 549
  simplification of
    applications of, 553–554
    by division, 549–551, 554
    by multiplication by the LCD, 551–553
complex numbers
  addition of, 686–687, 691
  conjugates of, 688
  definition of, 684
  division of, 688–690
  imaginary unit in, 684
  multiplication of, 687–688, 691

| Shape | Formulas for Area ($A$), and Perimeter ($P$), Circumference ($C$) |
|---|---|
| Triangle | $A = \dfrac{1}{2}bh = \dfrac{1}{2} \times \text{base} \times \text{height}$ <br> $P = a + b + c = \text{sum of sides}$ |
| Rectangle | $A = lw = \text{length} \times \text{width}$ <br> $P = 2l + 2w = \text{sum of twice the length and twice the width}$ |
| Trapezoid | $A = \dfrac{1}{2}(b_1 + b_2)h = \dfrac{1}{2} \times \text{sum of bases} \times \text{height}$ <br> $P = a + b_1 + c + b_2 = \text{sum of sides}$ |
| Parallelogram | $A = bh = \text{base} \times \text{height}$ <br> $P = 2a + 2b = \text{sum of twice the base and twice the side}$ |
| Circle | $A = \pi r^2 = \pi \times \text{square of radius}$ <br> $C = 2\pi r = 2 \times \pi \times \text{radius}$ <br> $C = \pi d = \pi \times \text{diameter}$ |

| Figure | Formulas for Volume ($V$) and Surface Area ($SA$) |
|---|---|
| Rectangular Prism | $V = lwh = \text{length} \times \text{width} \times \text{height}$ <br> $SA = 2lw + 2hw + 2lh$ <br> $\quad = 2(\text{length} \times \text{width}) + 2(\text{height} \times \text{width}) + 2(\text{length} \times \text{height})$ |
| General Prisms | $V = Bh = \text{area of base} \times \text{height}$ <br> $SA = \text{sum of the areas of the faces}$ |
| Right Circular Cylinder | $V = Bh = \text{area of base} \times \text{height}$ <br> $SA = 2B + Ch = (2 \times \text{area of base}) + (\text{circumference} \times \text{height})$ |
| Pyramid | $V = \dfrac{1}{3}Bh = \dfrac{1}{3} \times \text{area of base} \times \text{height}$ <br> $SA = B + \dfrac{1}{2}P\ell$ <br> $\quad = \text{area of base} + \left(\dfrac{1}{2} \times \text{perimeter of base} \times \text{slant height}\right)$ |
| Right Circular Cone | $V = \dfrac{1}{3}Bh = \dfrac{1}{3} \times \text{area of base} \times \text{height}$ <br> $SA = B + \dfrac{1}{2}C\ell$ <br> $\quad = \text{area of base} + \left(\dfrac{1}{2} \times \text{circumference} \times \text{slant height}\right)$ |
| Sphere | $V = \dfrac{4}{3}\pi r^3 = \dfrac{4}{3} \times \pi \times \text{cube of radius}$ <br> $SA = 4\pi r^2 = 4 \times \pi \times \text{square of radius}$ |